The chemistry of
phenols

Patai Series: The Chemistry of Functional Groups

A series of advanced treatises founded by Professor Saul Patai and under the general editorship of Professor Zvi Rappoport

The **Patai Series** publishes comprehensive reviews on all aspects of specific functional groups. Each volume contains outstanding surveys on theoretical and computational aspects, NMR, MS, other spectroscopical methods and analytical chemistry, structural aspects, thermochemistry, photochemistry, synthetic approaches and strategies, synthetic uses and applications in chemical and pharmaceutical industries, biological, biochemical and environmental aspects.
To date, over 100 volumes have been published in the series.

Recently Published Titles

The chemistry of the Cyclopropyl Group (2 volumes, 3 parts)
The chemistry of the Hydrazo Azo and Azoxy Groups (2 volumes, 3 parts)
The chemistry of Double-Bonded Functional Groups — (3 volumes, 6 parts)
The chemistry of Organophosphorus Compounds (4 volumes)
The chemistry of Halides, Pseudo-Halides and Azides (2 volumes, 4 parts)
The chemistry of the Amino, Nitro and Nitroso Groups (2 volumes, 4 parts)
The chemistry of Dienes and Polyenes (2 volumes)
The chemistry of Organic Derivatives of Gold and Silver
The chemistry of Organic Silicon Compounds (3 volumes, 6 parts)
The chemistry of Organic Germanium, Tin and Lead Compounds (2 volumes, 3 parts)
The chemistry of Phenols

Forthcoming Titles

The chemistry of Organolithium Compounds
The chemistry of Cyclobutanes
The chemistry of Peroxides (Volume 2)

The Patai Series Online

Starting in 2003 the **Patai Series** will be available in electronic format on Wiley InterScience. All new titles will be published online and a growing list of older titles will be added every year. It is the ultimate goal that all titles published in the **Patai Series** will be available in electronic format.
For more information see the **Patai Series Online** website:

www.interscience.wiley.com/bookfinder.html

The chemistry of
phenols
Part 1

Edited by

ZVI RAPPOPORT

The Hebrew University, Jerusalem

2003

An Interscience® Publication

Copyright © 2003   John Wiley & Sons Ltd, The Atrium, Southern Gate, Chichester,
                      West Sussex PO19 8SQ, England

                      Telephone (+44) 1243 779777

Email (for orders and customer service enquiries): cs-books@wiley.co.uk
Visit our Home Page on www.wileyeurope.com or www.wiley.com

All Rights Reserved. No part of this publication may be reproduced, stored in a retrieval system or transmitted in any form or by any means, electronic, mechanical, photocopying, recording, scanning or otherwise, except under the terms of the Copyright, Designs and Patents Act 1988 or under the terms of a licence issued by the Copyright Licensing Agency Ltd, 90 Tottenham Court Road, London W1T 4LP, UK, without the permission in writing of the Publisher. Requests to the Publisher should be addressed to the Permissions Department, John Wiley & Sons Ltd, The Atrium, Southern Gate, Chichester, West Sussex PO19 8SQ, England, or emailed to permreq@wiley.co.uk, or faxed to (+44) 1243 770620.

This publication is designed to provide accurate and authoritative information in regard to the subject matter covered. It is sold on the understanding that the Publisher is not engaged in rendering professional services. If professional advice or other expert assistance is required, the services of a competent professional should be sought.

Other Wiley Editorial Offices

John Wiley & Sons Inc., 111 River Street, Hoboken, NJ 07030, USA

Jossey-Bass, 989 Market Street, San Francisco, CA 94103-1741, USA

Wiley-VCH Verlag GmbH, Boschstr. 12, D-69469 Weinheim, Germany

John Wiley & Sons Australia Ltd, 33 Park Road, Milton, Queensland 4064, Australia

John Wiley & Sons (Asia) Pte Ltd, 2 Clementi Loop #02-01, Jin Xing Distripark, Singapore 129809

John Wiley & Sons Canada Ltd, 22 Worcester Road, Etobicoke, Ontario, Canada M9W 1L1

Wiley also publishes its books in a variety of electronic formats. Some content that appears in print may not be available in electronic books.

Library of Congress Cataloging-in-Publication Data

The chemistry of phenols / edited by Zvi Rappoport.
    p. cm.—(The chemistry of functional groups)
  Includes bibliographical references and indexes.
  ISBN 0-471-49737-1 (set: acid-free paper)
  1. Phenols.   I. Rappoport, Zvi. II. Series.

QD341.P5C524 2003
547′.632–dc21

2003045075

British Library Cataloguing in Publication Data

A catalogue record for this book is available from the British Library

ISBN 0-471-49737-1

Typeset in 9/10pt Times by Laserwords Private Limited, Chennai, India
Printed and bound in Great Britain by Biddles Ltd, Guildford, Surrey
This book is printed on acid-free paper responsibly manufactured from sustainable forestry in which at least two trees are planted for each one used for paper production.

Dedicated to

Gadi, Adina,

Sharon and **Michael**

Contributing authors

L. Ross C. Barclay	Department of Chemistry, Mount Allison University, Sackville, New Brunswick, Canada, E4L 1G8
M. Berthelot	Laboratoire de Spectrochimie, University of Nantes, 2, rue de la Houssiniere BP 92208, F-44322 Nantes Cedex 3, France
Volker Böhmer	Johannes Gutenberg-Universität, Fachbereich Chemie und Pharmazie, Abteilung Lehramt Chemie, Duesbergweg 10–14, D-55099 Mainz, Germany
Luis Castedo	Departamento de Química Orgánica y Unidad Asociada al C.S.I.C., Facultad de Química, Universidad de Santiago de Compostela, 15782 Santiago de Compostela, Spain
M. J. Caulfield	Polymer Science Group, Department of Chemical Engineering, The University of Melbourne, Victoria 3010, Australia
Victor Glezer	National Public Health Laboratory, Ministry of Health, 69 Ben Zvi Rd., Tel Aviv, Israel
J. Graton	Laboratoire de Spectrochimie, University of Nantes, 2, rue de la Houssiniere BP 92208, F-44322 Nantes Cedex 3, France
Concepción González	Departamento de Química Orgánica, Facultad de Ciencias, Universidad de Santiago de Compostela, 27002 Lugo, Spain
Poul Erik Hansen	Department of Life Sciences and Chemistry, Roskilde University, P.O. Box 260, DK-4000 Roskilde, Denmark
William M. Horspool	Department of Chemistry, The University of Dundee, Dundee DD1 4HN, Scotland, UK
Menahem Kaftory	Department of Chemistry, Technion—Israel Institute of Technology, Haifa 32000, Israel
Alla V. Koblik	ChemBridge Corporation, Malaya Pirogovskaya str., 1, 119435 Moscow, Russia
Eugene S. Kryachko	Department of Chemistry, University of Leuven, B-3001, Belgium, and Departement SBG, Limburgs Universitaire Centrum, B-3590 Diepenbeek, Belgium
Dietmar Kuck	Fakultät für Chemie, Universität Bielefeld, Universitätsstraße 25, D-33615 Bielefeld, Germany

C. Laurence	Laboratoire de Spectrochimie, University of Nantes, 2, rue de la Houssiniere BP 92208, F-44322 Nantes Cedex 3, France
Joel F. Liebman	Department of Chemistry and Biochemistry, University of Maryland, Baltimore County, 1000 Hilltop Circle, Baltimore, Maryland 21250, USA
Sergei M. Lukyanov	ChemBridge Corporation, Malaya Pirogovskaya str., 1, 119435 Moscow, Russia
P. Neta	National Institute of Standards and Technology, Gaithersburg, Maryland 20899, USA
Minh T. Nguyen	Department of Chemistry, University of Leuven, B-3001 Leuven, Belgium
Ehud Pines	Chemistry Department, Ben-Gurion University of the Negev, P.O.B. 653, Beer-Sheva 84105, Israel
G. K. Surya Prakash	Loker Hydrocarbon Research Institute and Department of Chemistry, University of Southern California, Los Angeles, California 90089-1661, USA
G. G. Qiao	Polymer Science Group, Department of Chemical Engineering, The University of Melbourne, Victoria 3010, Australia
V. Prakash Reddy	Department of Chemistry, University of Missouri-Rolla, Rolla, Missouri 65409, USA
Suzanne W. Slayden	Department of Chemistry, George Mason University, 4400 University Drive, Fairfax, Virginia 22030, USA
D. H. Solomon	Polymer Science Group, Department of Chemical Engineering, The University of Melbourne, Victoria 3010, Australia
Jens Spanget-Larsen	Department of Life Sciences and Chemistry, Roskilde University, P.O. Box 260, DK-4000 Roskilde, Denmark
S. Steenken	Max-Planck-Institut für Strahlenchemie, D-45413 Mülheim, Germany
Luc G. Vanquickenborne	Department of Chemistry, University of Leuven, B-3001 Leuven, Belgium
Melinda R. Vinqvist	Department of Chemistry, Mount Allison University, Sackville, New Brunswick, Canada, E4L 1G8
Masahiko Yamaguchi	Department of Organic Chemistry, Graduate School of Pharmaceutical Sciences, Tohoku University, Aoba, Sendai, 980-8578 Japan
Shosuke Yamamura	Department of Chemistry, Faculty of Science and Technology, Keio University, Hiyoshi, Yokohama 223-8522, Japan
Jacob Zabicky	Institutes for Applied Research, Ben-Gurion University of the Negev, P. O. Box 653, Beer-Sheva 84105, Israel

Foreword

This is the first volume in The 'Chemistry of Functional Groups' series which deals with an aromatic functional group. The combination of the hydroxyl group and the aromatic ring modifies the properties of both groups and creates a functional group which differs significantly in many of its properties and reactions from its two constituents. Phenols are important industrially, in agriculture, in medicine, in chemical synthesis, in polymer chemistry and in the study of physical organic aspects, e.g. hydrogen bonding. These and other topics are treated in the book.

The two parts of the present volume contain 20 chapters written by experts from 11 countries. They include an extensive treatment of the theoretical aspects, chapters on various spectroscopies of phenols such as NMR, IR and UV, on their mass spectra, on the structural chemistry and thermochemistry, on the photochemical and radiation chemistry of phenols and on their synthesis and synthetic uses and on reactions involving the aromatic ring such as electrophilic substitution or rearrangements. There are also chapters dealing with the properties of the hydroxyl group, such as hydrogen bonding or photoacidity, and with the derived phenoxy radicals which are related to the biologically important antioxidant behavior of phenols. There is a chapter dealing with polymers of phenol and a specific chapter on calixarenes — a unique family of monocyclic compounds including several phenol rings.

Three originally promised chapters on organometallic derivatives, on acidity and on the biochemistry of phenols were not delivered. Although the chapters on toxicity and on analytical chemistry deal with biochemistry related topics and the chapter on photoacidity is related to the ground state acidity of phenols, we hope that the missing chapters will appear in a future volume.

The literature coverage in the various chapters is mostly up to 2002.

I will be grateful to readers who draw my attention to any mistakes in the present volume.

Jerusalem
February 2003

ZVI RAPPOPORT

The Chemistry of Functional Groups
Preface to the series

The series 'The Chemistry of Functional Groups' was originally planned to cover in each volume all aspects of the chemistry of one of the important functional groups in organic chemistry. The emphasis is laid on the preparation, properties and reactions of the functional group treated and on the effects which it exerts both in the immediate vicinity of the group in question and in the whole molecule.

A voluntary restriction on the treatment of the various functional groups in these volumes is that material included in easily and generally available secondary or tertiary sources, such as Chemical Reviews, Quarterly Reviews, Organic Reactions, various 'Advances' and 'Progress' series and in textbooks (i.e. in books which are usually found in the chemical libraries of most universities and research institutes), should not, as a rule, be repeated in detail, unless it is necessary for the balanced treatment of the topic. Therefore each of the authors is asked not to give an encyclopaedic coverage of his subject, but to concentrate on the most important recent developments and mainly on material that has not been adequately covered by reviews or other secondary sources by the time of writing of the chapter, and to address himself to a reader who is assumed to be at a fairly advanced postgraduate level.

It is realized that no plan can be devised for a volume that would give a complete coverage of the field with no overlap between chapters, while at the same time preserving the readability of the text. The Editors set themselves the goal of attaining reasonable coverage with moderate overlap, with a minimum of cross-references between the chapters. In this manner, sufficient freedom is given to the authors to produce readable quasi-monographic chapters.

The general plan of each volume includes the following main sections:

(a) An introductory chapter deals with the general and theoretical aspects of the group.

(b) Chapters discuss the characterization and characteristics of the functional groups, i.e. qualitative and quantitative methods of determination including chemical and physical methods, MS, UV, IR, NMR, ESR and PES—as well as activating and directive effects exerted by the group, and its basicity, acidity and complex-forming ability.

(c) One or more chapters deal with the formation of the functional group in question, either from other groups already present in the molecule or by introducing the new group directly or indirectly. This is usually followed by a description of the synthetic uses of the group, including its reactions, transformations and rearrangements.

(d) Additional chapters deal with special topics such as electrochemistry, photochemistry, radiation chemistry, thermochemistry, syntheses and uses of isotopically labelled compounds, as well as with biochemistry, pharmacology and toxicology. Whenever applicable, unique chapters relevant only to single functional groups are also included (e.g. 'Polyethers', 'Tetraaminoethylenes' or 'Siloxanes').

This plan entails that the breadth, depth and thought-provoking nature of each chapter will differ with the views and inclinations of the authors and the presentation will necessarily be somewhat uneven. Moreover, a serious problem is caused by authors who deliver their manuscript late or not at all. In order to overcome this problem at least to some extent, some volumes may be published without giving consideration to the originally planned logical order of the chapters.

Since the beginning of the Series in 1964, two main developments have occurred. The first of these is the publication of supplementary volumes which contain material relating to several kindred functional groups (Supplements A, B, C, D, E, F and S). The second ramification is the publication of a series of 'Updates', which contain in each volume selected and related chapters, reprinted in the original form in which they were published, together with an extensive updating of the subjects, if possible, by the authors of the original chapters. A complete list of all above mentioned volumes published to date will be found on the page opposite the inner title page of this book. Unfortunately, the publication of the 'Updates' has been discontinued for economic reasons.

Advice or criticism regarding the plan and execution of this series will be welcomed by the Editors.

The publication of this series would never have been started, let alone continued, without the support of many persons in Israel and overseas, including colleagues, friends and family. The efficient and patient co-operation of staff-members of the publisher also rendered us invaluable aid. Our sincere thanks are due to all of them.

The Hebrew University	SAUL PATAI
Jerusalem, Israel	ZVI RAPPOPORT

Sadly, Saul Patai who founded 'The Chemistry of Functional Groups' series died in 1998, just after we started to work on the 100th volume of the series. As a long-term collaborator and co-editor of many volumes of the series, I undertook the editorship and I plan to continue editing the series along the same lines that served for the preceeding volumes. I hope that the continuing series will be a living memorial to its founder.

The Hebrew University   ZVI RAPPOPORT
Jerusalem, Israel
June 2002

Contents

1	General and theoretical aspects of phenols Minh Tho Nguyen, Eugene S. Kryachko and Luc G. Vanquickenborne	1
2	The structural chemistry of phenols Menahem Kaftory	199
3	Thermochemistry of phenols and related arenols Suzanne W. Slayden and Joel F. Liebman	223
4	Mass spectrometry and gas-phase ion chemistry of phenols Dietmar Kuck	259
5	NMR and IR spectroscopy of phenols Poul Erik Hansen and Jens Spanget-Larsen	333
6	Synthesis of phenols Concepción González and Luis Castedo	395
7	UV-visible spectra and photoacidity of phenols, naphthols and pyrenols Ehud Pines	491
8	Hydrogen-bonded complexes of phenols C. Laurence, M. Berthelot and J. Graton	529
9	Electrophilic reactions of phenols V. Prakash Reddy and G. K. Surya Prakash	605
10	Synthetic uses of phenols Masahiko Yamaguchi	661
11	Tautomeric equilibria and rearrangements involving phenols Sergei M. Lukyanov and Alla V. Koblik	713
12	Phenols as antioxidants L. Ross C. Barclay and Melinda R. Vinqvist	839
13	Analytical aspects of phenolic compounds Jacob Zabicky	909
14	Photochemistry of phenols	1015

William M. Horspool

15	Radiation chemistry of phenols P. Neta	1097
16	Transient phenoxyl radicals: Formation and properties in aqueous solutions S. Steenken and P. Neta	1107
17	Oxidation of phenols Shosuke Yamamura	1153
18	Environmental effects of substituted phenols Victor Glezer	1347
19	Calixarenes Volker Böhmer	1369
20	Polymers based on phenols D. H. Solomon, G. G. Qiao and M. J. Caulfield	1455
	Author index	1507
	Subject index	1629

List of abbreviations used

Ac	acetyl (MeCO)
acac	acetylacetone
Ad	adamantyl
AIBN	azoisobutyronitrile
Alk	alkyl
All	allyl
An	anisyl
Ar	aryl
Bn	benzyl
Bz	benzoyl (C_6H_5CO)
Bu	butyl (C_4H_9)
CD	circular dichroism
CI	chemical ionization
CIDNP	chemically induced dynamic nuclear polarization
CNDO	complete neglect of differential overlap
Cp	η^5-cyclopentadienyl
Cp*	η^5-pentamethylcyclopentadienyl
DABCO	1,4-diazabicyclo[2.2.2]octane
DBN	1,5-diazabicyclo[4.3.0]non-5-ene
DBU	1,8-diazabicyclo[5.4.0]undec-7-ene
DIBAH	diisobutylaluminium hydride
DME	1,2-dimethoxyethane
DMF	N,N-dimethylformamide
DMSO	dimethyl sulphoxide
ee	enantiomeric excess
EI	electron impact
ESCA	electron spectroscopy for chemical analysis
ESR	electron spin resonance
Et	ethyl
eV	electron volt

Fc	ferrocenyl
FD	field desorption
FI	field ionization
FT	Fourier transform
Fu	furyl(OC_4H_3)
GLC	gas liquid chromatography
Hex	hexyl(C_6H_{13})
c-Hex	cyclohexyl(c-C_6H_{11})
HMPA	hexamethylphosphortriamide
HOMO	highest occupied molecular orbital
HPLC	high performance liquid chromatography
i-	iso
ICR	ion cyclotron resonance
Ip	ionization potential
IR	infrared
LAH	lithium aluminium hydride
LCAO	linear combination of atomic orbitals
LDA	lithium diisopropylamide
LUMO	lowest unoccupied molecular orbital
M	metal
M	parent molecule
MCPBA	m-chloroperbenzoic acid
Me	methyl
MNDO	modified neglect of diatomic overlap
MS	mass spectrum
n	normal
Naph	naphthyl
NBS	N-bromosuccinimide
NCS	N-chlorosuccinimide
NMR	nuclear magnetic resonance
Pen	pentyl(C_5H_{11})
Ph	phenyl
Pip	piperidyl($C_5H_{10}N$)
ppm	parts per million
Pr	propyl (C_3H_7)
PTC	phase transfer catalysis or phase transfer conditions
Py, Pyr	pyridyl (C_5H_4N)

R	any radical
RT	room temperature
s-	secondary
SET	single electron transfer
SOMO	singly occupied molecular orbital
t-	tertiary
TCNE	tetracyanoethylene
TFA	trifluoroacetic acid
THF	tetrahydrofuran
Thi	thienyl(SC_4H_3)
TLC	thin layer chromatography
TMEDA	tetramethylethylene diamine
TMS	trimethylsilyl or tetramethylsilane
Tol	tolyl(MeC_6H_4)
Tos or Ts	tosyl(*p*-toluenesulphonyl)
Trityl	triphenylmethyl(Ph_3C)
Xyl	xylyl($Me_2C_6H_3$)

In addition, entries in the 'List of Radical Names' in *IUPAC Nomenclature of Organic Chemistry*, 1979 Edition, Pergamon Press, Oxford, 1979, p. 305–322, will also be used in their unabbreviated forms, both in the text and in formulae instead of explicitly drawn structures.

CHAPTER 1

General and theoretical aspects of phenols

MINH THO NGUYEN

Department of Chemistry, University of Leuven, B-3001 Leuven, Belgium
fax: 32-16-327992;   e-mail: minh.nguyen@chem.kuleuven.ac.be

EUGENE S. KRYACHKO*

Department of Chemistry, University of Leuven, B-3001, Belgium, and Departement SBG, Limburgs Universitaire Centrum, B-3590 Diepenbeek, Belgium
email: eugene.kryachko@luc.ac.be

and

LUC G. VANQUICKENBORNE

Department of Chemistry, University of Leuven, B-3001 Leuven, Belgium
email: luc.vanquickenborne@chem.kuleuven.ac.be

I. INTRODUCTION	3
A. Summary of Key Physico-chemical Properties of Phenol	4
B. The History of the Discovery of Phenol	6
C. Usage and Production	7
II. MOLECULAR STRUCTURE AND BONDING OF PHENOL	20
A. The Equilibrium Structure of Phenol in the Ground Electronic State . .	20
B. Molecular Bonding Patterns in the Phenol S_0	21
C. Atom-in-Molecule Analysis	31
D. Vibrational Modes	34
E. Three Interesting Structures Related to Phenol	38
III. STRUCTURES AND PROPERTIES OF SUBSTITUTED PHENOLS . . .	47
A. Intramolecular Hydrogen Bond in *ortho*-Halogenophenols	47
B. *meta*- and *para*-Halogenophenols	57

---

*On leave of absence from Bogoliubov Institute for Theoretical Physics, Kiev, 03143 Ukraine.

The Chemistry of Phenols.   Edited by Z. Rappoport
© 2003 John Wiley & Sons, Ltd   ISBN: 0-471-49737-1

C. The Bonding Trends in Monohalogenated Phenols in Terms
   of the Electronic Localization Function (*ELF*) 68
   1. Introduction to the *ELF* 68
   2. Topology of the *ELF* 68
   3. Vector gradient field $\nabla_r \eta(r)$ 70
   4. The bonding in benzene, phenol and phenyl halides 71
   5. Monohalogenated phenols: the bonding in terms of *ELF* 73
      a. The *ortho*-substituted phenols 73
      b. The *meta*-substituted phenols 76
      c. The *para*-substituted phenols 76
D. Some Representatives of Substituted Phenols 78
IV. ENERGETICS OF SOME FUNDAMENTAL PROCESSES 83
A. Protonation 83
   1. Protonation of phenol 83
   2. Proton affinities of halophenols 86
   3. Proton affinities of anisole and fluoroanisoles 88
   4. Two views on the protonation regioselectivity 89
   5. Interaction of phenol with Li^+, Na^+ and K^+ 92
B. Deprotonation 92
   1. Phenolate anion 93
   2. Gas-phase acidities 97
   3. Acidity in solution 101
   4. Correlation between intrinsic acidities and molecular properties . . . 101
   5. Alkali metal phenolates 103
C. Electronic Excitation 105
D. Ionization 110
   1. Molecular and electronic structure of phenol radical cation 111
   2. Relative energies of the $(C_6H_6O)^{\bullet +}$ radical cations 114
   3. The $(C_6H_6O)^{\bullet +}$ potential energy surface (PES) 116
   4. Mass spectrometric experiments 121
   5. Keto–enol interconversion 127
E. The O—H Bond Dissociation 129
   1. Phenoxyl radicals 129
      a. Electronic structure 129
      b. Geometry and vibrational frequencies 132
      c. Spin densities 135
      d. Decomposition of phenoxy radical 137
   2. Antioxidant activity of phenols 139
      a. The O—H bond dissociation energies 139
      b. Antioxidant activities 140
      c. Features of hydrogen atom abstraction from phenols 141
V. HYDROGEN BONDING ABILITIES OF PHENOLS 143
A. Introductory Survey 143
B. Phenol–(Water)$_n$, $1 \leq n \leq 4$ Complexes 147
   1. Introduction 147
   2. Interaction of phenol with water 149
   3. The most stable complexes of mono- and dihydrated phenol 149
   4. Lower-energy structures of $PhOH(H_2O)_3$ 156
   5. At the bottom of PES of $PhOH(H_2O)_4$ 160

C. Hydrogen Bonding between Phenol and Acetonitrile 170
  1. Introductory foreground 170
  2. Phenol–acetonitrile complex 171
  3. Phenol bonding with two acetonitrile molecules 174
  4. A rather concise discussion 177
D. Phenol–Benzonitrile Hydrogen-bonded Complex 177
E. A Very Short O—H · · · N Hydrogen Bond 178
VI. OPEN THEORETICAL PROBLEMS 178
VII. ACKNOWLEDGEMENTS 179
VIII. REFERENCES AND NOTES 179

Glossary of Acronyms

BDE	bond dissociation enthalpy	LIF	laser-induced fluorescence
BIPA	*trans*-butenylidene-isopropylamine	LUMO	lowest unoccupied MO
		MO	molecular orbital
N-BMA	benzylidenemethylamine	MP2	second-order Møller-Plesset perturbation theory
CCSD(T)	coupled cluster singles doubles (triples)	MW	microwave spectroscopy
DF	dispersed fluorescence spectroscopy	NBO	natural bond orbital
		PA	proton affinity
DFT	density functional method	PCA	1-pyrrolidinecarboxaldehyde
N,N-DMBA	dimethylbenzylamine	PES	potential energy surface
		Ph	phenyl C_6H_5
DPE	deprotonation energy	PhOH	phenol
DRS	double-resonance spectroscopy	R2PI	resonant two-photon ionization spectroscopy
ED	electron diffraction		
HF	Hartree-Fock method	SOMO	singly occupied MO
HOMO	highest occupied MO	TMA	trimethylamine
IR-UV	infrared-ultraviolet spectroscopy	ZPE-ZPVE	zero-point vibrational energy

I. INTRODUCTION

The chemistry of phenols has attracted continuing interest in the last two centuries. Compounds bearing this functional group have several applications indispensable in our daily life, as discussed in the following chapters of this book. Let us mention one example: phenols constitute, among others, an important class of antioxidants that inhibit the oxidative degradation of organic materials including a large number of biological aerobic organisms and commercial products. In human blood plasma, α-tocopherol, well-known as a component of vitamin E, is proved to be the most efficient phenol derivative to date to trap the damaging peroxy radicals (ROO•). Phenols owe their activity to their ability to scavenge radicals by hydrogen or electron transfer in much faster processes than radical attacks on an organic substrate.

In this chapter, we attempt to give an overview on the general and theoretical aspects of phenols, including a brief history of their discovery. However, in view of the very large wealth of related literature, the coverage is by no means complete. It is also not intended to be a comprehensive review of all the theoretical work in the area, and there are certainly many important studies of which we were unaware, for which we apologize.

We refer to the compilation *Quantum Chemistry Library Data Base* (QCLDB)[1] for an extended list of available theoretical papers.

The focus of this chapter is a presentation of representative physico-chemical and spectroscopic properties of phenols revealed by quantum chemical calculations, many of them carried out by us specifically for this chapter. In the discussion, the description of methodological details will be kept to a minimum. Unless otherwise noted, all reported computations were performed using the GAUSSIAN 98[2] and MOPAC-7[3] sets of programs. The natural bond orbital analysis[4] was conducted using the NBO (natural bond orbital) module[5] of the GAUSSIAN 98 software package.[2] For the vibrational analyses, the force constant matrices were initially obtained in terms of the cartesian coordinates and the non-redundant sets of internal coordinates were subsequently defined[6]. The calculation of potential energy distribution (PED) matrices of the vibrational frequencies[7] was carried out using the GAR2PED program[8].

A. Summary of Key Physico-chemical Properties of Phenol

Phenol shown in Chart 1 is the parent substance of a homologous series of compounds containing a *hydroxyl group* bound directly to the aromatic ring. Phenol, or PhOH in shorthand notation, belongs to the family of *alcohols* due to the presence of the OH group and it is in fact the simplest aromatic member of this family. The hydroxyl group of phenol determines its acidity whereas the benzene ring characterizes its basicity. Thus, it is formally the *enol* form of the *carbonyl group* (for a review, see ref. 9).

In this subsection we briefly outline the key physico-chemical properties of phenol. For its other properties consult with the NIST data located at URL http://webbook.nist.gov.

Phenol has a low melting point, it crystallizes in colourless prisms and has a characteristic, slightly pungent odor. In the molten state, it is a clear, colourless, mobile liquid. In the temperature range $T < 68.4\,°C$, its miscibility with water is limited; above this temperature it is completely miscible. The melting and solidification points of phenol are quite substantially lowered by water. A mixture of phenol and *ca* 10% water is called phenolum liquefactum, because it is actually a liquid at room temperature. Phenol is readily soluble in most organic solvents (aromatic hydrocarbons, alcohols, ketones, ethers, acids, halogenated hydrocarbons etc.) and somewhat less soluble in aliphatic hydrocarbons. Phenol forms azeotropic mixtures with water and other substances.

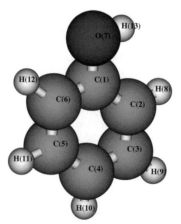

CHART 1. Chemical formulae of phenol: C_6H_5OH; early name: carbolic acid, hydroxybenzene; CAS registry number: 108-95-2

1. General and theoretical aspects of phenols

Other physical data of phenol follow below:

Molecular weight: 94.11 (molecular mass of C_6H_5OH is equal to 94.04186).
Weakly acidic: $pK_a(H_2O) = 9.94$ (although it varies in different sources from 9.89 to 9.95).
Freezing point: 40.91 °C.
Specific heats of combustion: $C_p = 3.06$ J mol^{-1} K^{-1}, $C_v = 3.07$ J mol^{-1} K^{-1}.
First ionization energy (IE_a): 8.47 eV (experimental), 8.49 ± 0.02 eV (evaluated).
Proton affinity (PA): 820 kJ mol^{-1} [10].
Gas phase basicity: 786.3 kJ mol^{-1} [10].
Gas-phase heat of formation $\Delta_f H_{298}$: -96.2 ± 8 kJ mol^{-1} (experimental); -93.3 kJ mol^{-1} (theoretical)[11].
Solvation free energy:
  Experimental: -27.7 kJ mol^{-1} [12], -27.6 kJ mol^{-1} [13].
  Theoretical: -17.3, -20.2, -16.4 kJ mol^{-1} (AMBER parameter[14]), -19.7, -23.8, -12.1 kJ mol^{-1} [1, 13–16].
Gas phase acidity: $\Delta_{acid} H_{298}$:
  Experimental: 1465.7 ± 10 kJ mol^{-1} [17, 18]; 1461.1 ± 9 kJ mol^{-1} [18, 19]; 1471 ± 13 kJ mol^{-1} [20].
  Theoretical: 1456.4 kJ mol^{-1} [20].
O—H bond dissociation energy $D_{298}(C_6H_5O-H)$:
  Experimental: 362 ± 8 kJ mol^{-1} [21]; 363.2 ± 9.2 kJ mol^{-1} [22]; 353 ± 4 kJ mol^{-1} [23]; 376 ± 13 kJ mol^{-1} [24]; 369.5 kJ mol^{-1} [25]; 377 ± 13 kJ mol^{-1} [26].
  Theoretical: 377.7 kJ mol^{-1} [20].

What else is worth noting, in view of the present review on the theoretical aspects of phenol, is that its electronic subsystem consists of 50 electrons and the ground state is a singlet closed-shell state designated as S_o.

Phenol can be considered as the enol of cyclohexadienone. While the tautomeric keto–enol equilibrium lies far to the ketone side in the case of aliphatic ketones, for phenol it is shifted almost completely to the enol side. The reason of such stabilization is the formation of the aromatic system. The resonance stabilization is very high due to the contribution of the *ortho*- and *para*-quinonoid resonance structures. In the formation of the phenolate anion, the contribution of quinonoid resonance structures can stabilize the negative charge.

In contrast to aliphatic alcohols, which are mostly less acidic than phenol, phenol forms salts with aqueous alkali hydroxide solutions. At room temperature, phenol can be liberated from the salts even with carbon dioxide. At temperatures near the boiling point of phenol, it can displace carboxylic acids, e.g. acetic acid, from their salts, and then phenolates are formed. The contribution of *ortho*- and *para*-quinonoid resonance structures allows electrophilic substitution reactions such as chlorination, sulphonation, nitration, nitrosation and mercuration. The introduction of two or three nitro groups into the benzene ring can only be achieved indirectly because of the sensitivity of phenol towards oxidation. Nitrosation in the *para* position can be carried out even at ice bath temperature. Phenol readily reacts with carbonyl compounds in the presence of acid or basic catalysts. Formaldehyde reacts with phenol to yield hydroxybenzyl alcohols, and synthetic resins on further reaction. Reaction of acetone with phenol yields bisphenol A [2,2-bis(4-hydroxyphenyl)propane].

The reaction in the presence of acid catalysts is used to remove impurities from synthetic phenol. Olefinic impurities or carbonyl compounds, e.g. mesityl oxide, can be polymerized into higher molecular weight compounds by catalytic quantities of sulphuric acid or acidic ion exchangers and can thus be separated easily from phenol, e.g. by its distillation.

Phenol readily couples with diazonium salts to yield coloured compounds. The latter can be used for the photometric detection of phenol as in the case of diazotized 4-nitroaniline. Salicylic acid (2-hydroxybenzoic acid) can be produced by the Kolbe–Schmitt reaction[26] (studied by the density functional method[27]) from sodium phenolate and carbon dioxide, whereas potassium phenolate gives the *para* compound. Alkylation and acylation of phenol can be carried out with aluminium chloride as catalyst; methyl groups can also be introduced by the Mannich reaction. Diaryl ethers can only be produced under extreme conditions.

With oxidizing agents, phenol readily forms a free radical which can dimerize to form diphenols or can be oxidized to form dihydroxybenzenes and quinones. Since phenol radicals are relatively stable, phenol is a suitable radical scavenger and can also be used as an oxidation inhibitor. Such a property can also be undesirable, e.g. the autoxidation of cumene can be inhibited by small quantities of phenol.

B. The History of the Discovery of Phenol

Phenol is a constituent of coal tar and was probably first (partly) isolated from coal tar in 1834 by Runge, who called it 'carbolic acid' (*Karbolsäure*) or 'coal oil acid' (Kohlenölsäure)[28–30].

Friedlieb Ferdinand Runge (born in Billwärder, near Hamburg, 8 February 1795—Oranienburg, died on 25 March 1867) began his career as a pharmacist and, after a long residence in Paris, became an associate professor in Breslau, Germany. Later, he served in the Prussian Marine in Berlin and Oranienburg. Runge published several scientific and technological papers and books (see References 31 and 32 and references therein). He rediscovered aniline in coal-tar oil and called it *kyanol*. He also discovered quinoline (*leukol*), pyrrole ($\pi \nu \rho \rho \sigma$), rosolic acid and three other bases.

Pure phenol was first prepared by Laurent in 1841. Auguste Laurent (La Folie, near Langres, Haute-Marne, 14 September 1808—Paris, 15 April 1853), the son of a wine-merchant, was assistant to Dumas at the Ecole Centrale (1831) and to Brongniart at the Sevres porcelain factory (1833–1835) in France. From 1835 until 1836, he lived in a garret in the Rue St. Andre, Paris, where he had a private laboratory. In December 1837 Laurent defended his Paris doctorate and in 1838 became professor at Bordeaux. Since 1845 he worked in a laboratory at the Ecole Normale in Paris. In his studies of the distillate from coal-tar and chlorine, Laurent isolated dichlorophenol (*acide chlorophénèsique*) $C^{24}H^8Cl^4O^2$ and trichlorophenol (*acide chlorophénisique*) $C^{24}H^6Cl^6O^2$, which both suggested the existence of phenol (phenhydrate)[33]. Laurent wrote: 'I give the name *phène* ($\varphi \alpha \tau \nu \omega$, I light) to the fundamental radical...'. He provided the table of 'general formulae of the derived radicals of phène' where phenol (*hydrate of phène*) was indicated by the incorrect formula $C^{24}H^{12} + H^4O^2$ (=C_6H_8O, in modern notation). In 1841, Laurent isolated and crystallized phenol for the first time. He called it 'hydrate de phényle' or 'acide phénique'[34]. His reported melting point (between 34 and 35 °C) and boiling point (between 187 and 188 °C) are rather close to the values known today. Apart from measuring these elementary physical properties, Laurent also gave some crystals to a number of persons with toothache to try it out as a possible pain killer. The effect on the pain was rather unclear, but the substance was 'very aggressive on the lips and the gums'. In the analysis of his experiments, Laurent applied the substitution hypothesis that was originally proposed by his former supervisor, Dumas. Apparently, however, Laurent went further than Dumas and assumed that the substitution reaction did not otherwise change the structural formula of the reactant and the product, whereas Dumas limited himself to the claim that the removal of one hydrogen atom was compensated by the addition of another group, leaving open the possibility of a complete rearrangement of the molecule[35].

1. General and theoretical aspects of phenols

The substitution hypothesis (especially in the form proposed by Laurent) was attacked rather strongly by Berzélius, who claimed that a simple replacement of the hydrogen atom by, for instance, the chlorine atom in an organic molecule should be utterly impossible 'due to the strong electronegative character' of chlorine[36, 37]. According to Berzélius, the very idea of Laurent contradicted the first principles of chemistry and 'seems to be a bad influence (une influence nuisible) in science' (see also Reference 32, p. 388). Instead, he reinterpreted all the results of Laurent by breaking up the reaction product into smaller (more familiar) molecules, satisfying the same global stoichiometry. It looks as if Berzélius was reluctant to accept the full richness of organic chemistry. He was unwilling to accept the existence of new molecules, if the atomic count (and a few other obvious properties) could be satisfied by known molecules. Dumas replied that Berzélius 'attributes to me an opinion precisely contrary to that which I have always maintained, viz., that chlorine in this case takes the place of the hydrogen.... The law of substitution is an empirical fact and nothing more; it expresses a relation between the hydrogen expelled and the chlorine retained. I am not responsible for the gross exaggeration with which Laurent has invested my theory; his analyses moreover do not merit any confidence'[38] (see also Reference 32, p. 388).

In 1843, Charles Frederic Gerhardt (Strasbourg, 21 August 1816—19 August 1856) also prepared phenol by heating salicylic acid with lime and gave it the name 'phénol'[39].

Since the 1840s, phenol became a subject of numerous studies. Victor Meyer studied desoxybenzoin, benzyl cyanide and phenyl-substituted methylene groups and showed that they have similar reactivities[31]. He subsequently published a paper on 'the negative nature of the phenyl group', where he noted how phenyl together with other 'negative groups' can make the hydrogen atoms in methylene groups more reactive. In 1867, Heinrich von Brunck defended his Ph.D. thesis in Tübingen under Adolph Friedrich Ludwig Strecker and Wilhelm Staedel on the theme 'About Derivatives of Phenol', where he particularly studied the isomers of nitrophenol[31].

The Raschig–Dow process of manufacturing phenol by cumene was discovered by Wurtz and Kekule in 1867, although the earlier synthesis was recorded by Hunt in 1849. Interestingly, Friedrich Raschig, working earlier as a chemist at BASF and known for his work on the synthesis of phenol and production of phenol formaldehyde adduct, later established his own company in Ludwigshafen.

It is also interesting to mention in this regard that in 1905, the BAAS subcommittee on 'dynamic isomerism' was established and included Armstrong (chairman), Lowry (secretary) and Lapworth. In the 1909 report, Lowry summarized that one of the types of isomerism involves the 'oscillatory transference' of the hydrogen atom from carbon to oxygen, as in ethyl acetoacetate (acetoacetic ester), or from oxygen to nitrogen, as in isatin, or from one oxygen atom to the other one, as in *para*-nitrosophenol[40, 41].

C. Usage and Production

Phenol is one of the most versatile and important industrial organic chemicals. Until World War II, phenol was essentially a natural coal-tar product. Eventually, synthetic methods replaced extraction from natural sources because its consumption had risen significantly. For instance, as a metabolic product, phenol is normally excreted in quantities of up to 40 mg L^{-1} in human urine. Currently, small amounts of phenol are obtained from coal tar. Higher quantities are formed in coking or low-temperature carbonization of wood, brown coal or hard coal and in oil cracking. The earlier methods of synthesis (via benzenesulphonic acid and chlorobenzene) have been replaced by modern processes, mainly by the Hock process starting from cumene, via the Raschig–Dow process and by sulphonation. Phenol is also formed during petroleum cracking. Phenol has achieved considerable importance as the starting material for numerous intermediates and final products.

Phenol occurs as a component or as an addition product in natural products and organisms. For example, it is a component of lignin, from which it can be liberated by hydrolysis. Lignin is a complex biopolymer that accounts for 20–30% of the dry weight of wood. It is formed by a free-radical polymerization of substituted phenylpropane units to give an amorphous polymer with a number of different functional groups including aryl ether linkages, phenols and benzyl alcohols[42]. Most pulp-processing methods involve oxidative degradation of lignin, since its presence is a limitation to the utilization of wood pulps for high end uses such as print and magazine grade paper. Such limitation is due to the photoinduced yellowing of lignin-rich, high-yield mechanical pulps and, as a result, the photooxidative yellowing has been extensively studied in the hope of understanding its mechanism and ultimately preventing its occurrence[42, 43]. Phenoxyl radicals are produced during the photooxidation of lignin and their subsequent oxidation ultimately leads to quinones, which are actually responsible for the yellow colour.

Phenol was first used as a disinfectant in 1865 by the British surgeon Joseph Lister at Glasgow University, Scotland, for sterilizing wounds, surgical dressings and instruments. He showed that if phenol was used in operating theatres to sterilize equipment and dressings, there was less infection of wounds and, moreover, the patients stood a much better chance of survival. By the time of his death, 47 years later, Lister's method of antiseptic surgery (Lister spray) was accepted worldwide. Its dilute solutions are useful antiseptics and, as a result of Lister's success, phenol became a popular household antiseptic. Phenol was put as an additive in a so-called carbolic soap. Despite its benefits at that time, this soap is now banned. In Sax's book *Dangerous Properties of Industrial Materials* (quoted in Reference 44), one finds frightening phrases like 'kidney damage', 'toxic fumes' and 'co-carcinogen'. Clearly, phenol is totally unsuitable for general use, but the benefits 130 years ago plainly outweighed the disadvantages. However, because of its protein-degenerating effect, it often had a severely corrosive effect on the skin and mucous membranes.

Phenol only has limited use in pharmaceuticals today because of its toxicity. Phenol occurs in normal metabolism and is harmless in small quantities according to present knowledge, but it is definitely toxic in high concentrations. It can be absorbed through the skin, by inhalation and by swallowing. The typical main absorption route is the skin, through which phenol is resorbed relatively quickly, simultaneously causing caustic burns on the area of skin affected. Besides the corrosive effect, phenol can also cause sensitization of the skin in some cases. Resorptive poisoning by larger quantities of phenol (which is possible even over small affected areas of skin) rapidly leads to paralysis of the central nervous system with collapse and a severe drop in body temperature. If the skin is wetted with phenol or phenolic solutions, decontamination of the skin must therefore be carried out immediately. After removal of contaminated clothing, polyglycols (e.g. lutrol) are particularly suitable for washing the skin. On skin contamination, local anesthesia sets in after an initial painful irritation of the area of skin affected. Hereby the danger exists that possible resorptive poisoning is underestimated. If phenol penetrates deep into the tissue, this can lead to phenol gangrene through damage to blood vessels. The effect of phenol on the central nervous system—sudden collapse and loss of consciousness—is the same for humans and animals. In animals, a state of cramp precedes these symptoms because of the effect phenol has on the motor activity controlled by the central nervous system. Caustic burns on the cornea heal with scarred defects. Possible results of inhalation of phenol vapour or mist are dyspnea, coughing, cyanosis and lung edema. Swallowing phenol can lead to caustic burns on the mouth and esophagus and stomach pains. Severe, though not fatal, phenol poisoning can damage inner organs, namely kidneys, liver, spleen, lungs and heart. In addition, neuropsychiatric disturbances have been described after survival of acute phenol poisoning. Most of the phenol absorbed by the body is excreted in urine as phenol and/or its metabolites. Only smaller quantities are excreted with faeces or exhaled.

The reactions are:

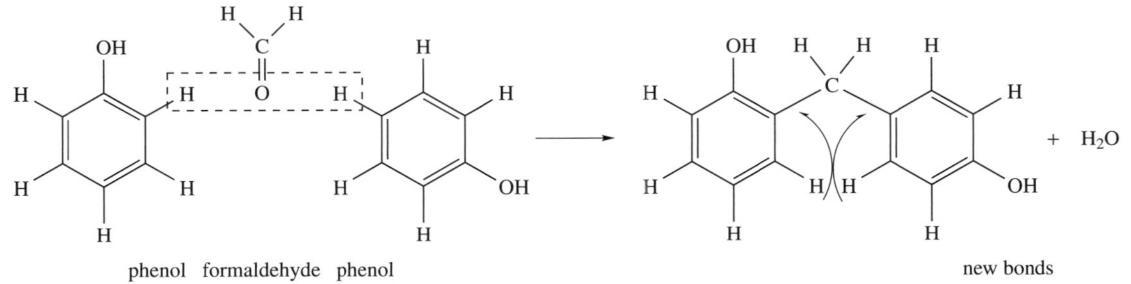

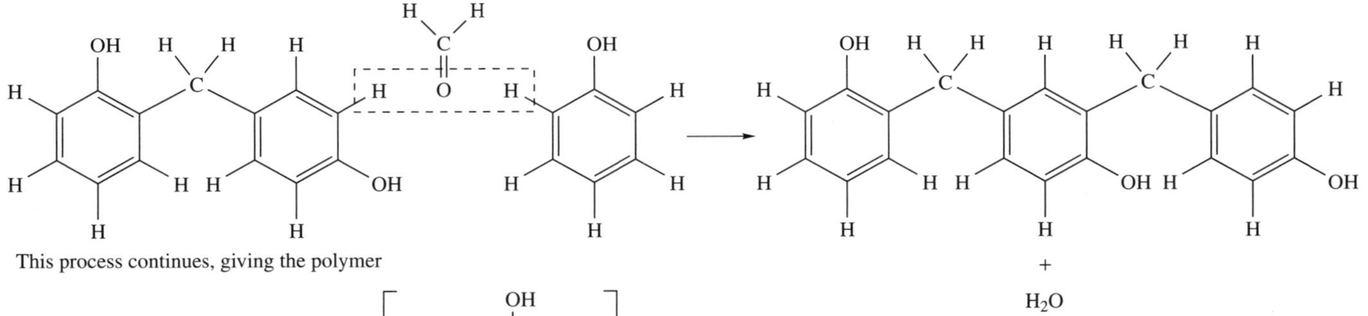

phenol   formaldehyde   phenol

+ H₂O

new bonds

This process continues, giving the polymer

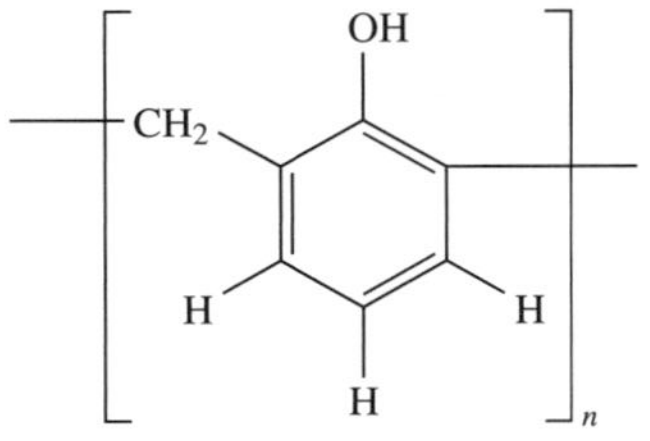

+ H₂O

CHART 2. Production of a phenolic resin

Phenol is a violent systemic poison. Less irritating and more efficient germicides (component of some plastics) replace phenol; nevertheless, it is widely used in the manufacture of phenolic resins (e.g. with formaldehyde—see Chart 2, with furfural etc.), epoxy resins, plastics, plasticizers, polycarbonates, antioxidants, lube oil additives, nylon, caprolactam, aniline insecticides, explosives, surface active agents, dyes and synthetic detergents, polyurethanes, wood preservatives, herbicides, fungicides (for wood preparation), gasoline additives, inhibitors, pesticides and as raw material for producing medical drugs like aspirin.

Acetylsalicylic acid was first synthesized by Bayer in 1897 and named Aspirin in 1899[45–47]. Nevertheless, its analgesic and antipyretic effects had been known long before. For example, in the 18th century, Stone discovered the medical effects of the salicin of willow bark and, since that time, salicylic acid was recognized as the active ingredient. Salicin is enzymatically hydrolysed to saligenin and glucose by β-glucosidase. Saligenin is then slowly oxidized to salicylic acid in the blood and in the liver. As is well known, the sodium salt of salicylic acid was used in the 19th century as a painkiller despite the fact that it causes stomach irritations. In his search for less-irritating derivatives of salicylic acid, the Bayer chemist Felix Hoffmann synthesized acetylsalicylic acid (Figure 1).

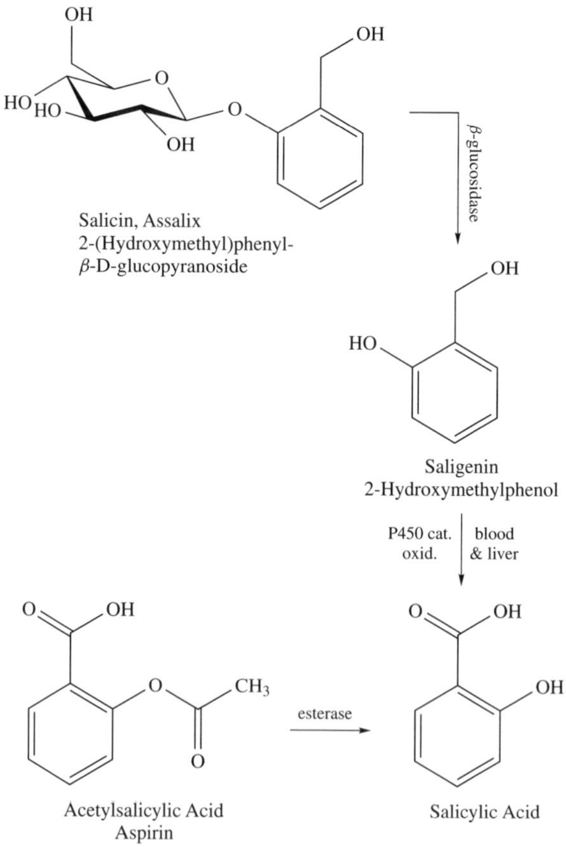

FIGURE 1. Salicin, saligenin, salicylic acid, and aspirin

The success of aspirin was terrific. In a 1994 article[48] in the *Medical Sciences Bulletin*, it was written that 'Americans consume about 80 billion aspirin tablets a year, and more than 50 nonprescription drugs contain aspirin as the principal active ingredient'. The Aspirin Foundation of America provides systematically scientific, regulatory, legislative and general educational information about aspirin to the medical community and the public[49]. In 1971, Vane[50] discovered that aspirin interferes with the biosynthesis of prostaglandins. In 1982 he was awarded the Nobel Prize in medicine in recognition of his work on the mechanism of the action of aspirin. In 1994, Garavito and coworkers[51, 52] elucidated the mechanism of aspirin interference with prostaglandin synthesis.

The crystal structure of aspirin was first determined by Wheatley[53] in 1964 and was refined later, in 1985, by Kim and coworkers[54]. Its crystal structure data can be obtained from the Cambridge Crystallographic Database[55]. The key features of the crystal structure of aspirin are shown in Figure 2. Quite recently, the potential energy surface of aspirin was studied using the B3LYP/6-31G(d) method and all its nine conformational isomers were located[56].

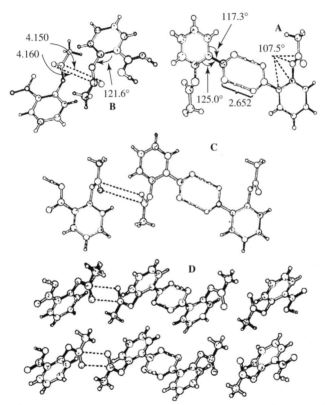

FIGURE 2. Hydrogen bonding patterns and dipole alignment in the crystal structure of aspirin. Two positions are shown for each of the hydrogen-bonded hydrogen atoms (**A**). Aspirin may also form another conformation of the dimer structure, a sort of inversion-symmetric dimer, with a perfect dipole–dipole alignment of the carbonyl groups of two ester functions (**B**). Actually, each aspirin is partly involved in a dimer of **A** and **B**. This is shown in **C**. **D** demonstrates the arrangement of the chains in the crystal. Adapted from Reference 56 with permission

Phenol is mainly used in the production of phenolic resins (plastics). These resins are important components of such items as appliance knobs, handles and housings, washing machine agitators and electrical devices. One example of its commercial usage is the phenol–formaldehyde polymer or phenol–formaldehyde resin called Bakelite (Formica, Micarta), first made in the USA in 1909. It took its name from its discoverer Leo Baekeland who developed it commercially between 1905 and 1910, and it was actually the first truly synthetic polymer. It is characterized by low cost, dimensional stability, high strength, stiffness and resistance to ageing; it is much safer than celluloid. It has insulating properties and could be moulded easily. Bakelite was the ideal plastic for electrical appliances, and in fact it was Bakelite which made possible the generation and distribution of electricity; it made electrical appliances safer for home utilization. It is also widely used in handles, table tops, cabinets and wall panels. The reaction between phenol and formaldehyde is a typical reaction of condensation polymerization, shown in Chart 2[57].

A phenol derivative, phenolphthalein is prepared by the reaction of phenol with phthalic anhydride in the presence of sulphuric acid and used as an indicator for acidity or alkalinity. Chlorinated phenol is much safer than phenol. Chlorine gas reacts with phenol to add one, two or three chlorine atoms and to form, respectively, chlorophenol, 2,4-dichlorophenol and 2,4,6-trichlorophenol[58]. The chlorination of phenol proceeds by electrophilic aromatic substitution. The latter two molecules are less soluble in water than phenol and appear to be a stronger antiseptic than phenol. Interestingly, in the first half of the past century, a bottle of antiseptic chlorophenols was a common attribute as a medicine in many homes. Its solution was used for bathing cuts, cleaning grazes, rinsing the mouth and gargling to cure sore throats. Nevertheless, it was revealed that its solution likely contains dioxins.

There are actually 31 different chloro- and polychlorophenols[57]. One of them, 2,4-dichlorophenoxyacetic acid (2,4-D), acts as a growth hormone. This makes it particularly effective as a weedkiller against broad-leaf weeds, even in a tiny drop. Surprisingly, it is actually a superb selective weedkiller for lawns and grain crops because it does not affect grass and cereals. Sometimes, 2,4-D is used to trick plants into flowering. This is widely used in Hawaii, where visitors are greeted with pineapple flowers during the whole year! It is safe for animals in low quantity, but 35 g of it is likely a fatal dose for an average person weighing about 70 kg. 2,4-D is quite inexpensive, effective, more selective than other weedkillers and much safer than the sodium arsenate and sodium chlorate which were popular weedkillers in the 1950s. In 1948, 2,4,5-trichlorophenoxyacetic acid (2,4,5-T) came into the market[44] and contained larger quantities of dioxin than 2,4-D[59]. It was used as a killer for tough weeds and was so successful in killing woody plants that it was deployed in the Vietnam War. From 1962 to 1969, at least 50,000 tonnes of a 50:50 mixture of 2,4-D and 2,4,5-T (called defoliant and widely known as Agent Orange) was sprayed from the air to destroy the dense foliage of trees covering the troops of the Vietnam National Front of Liberation. Agent Orange was contaminated with ca 2–4% of dioxins and for this reason it caused birth defects in new-born babies in Vietnam. It may also be linked to a form of acute myelogenous leukaemia, which represents 8% of childhood cancers among the children of Vietnam veterans, as the US Institute of Medicine (IOM) committee has recently reported[60].

Interestingly, phenols from peat smoke are included in the flavours of Scotch whisky to dry the malt[44].

Complex phenols are widespread in nature, although the simple ones are relatively uncommon. Phenol is particularly found in mammalian urine, pine needles and oil tobacco leaves. Abundant natural substances such as thymol (**1**) and carvacrol (**2**) are derivatives of phenols.

1. General and theoretical aspects of phenols

Natural phenols[57, 61, 62] arise in the three following manners[57]:

(i) Poly-β-ketones, for example (**3**), derived from the acid RCO_2H and three malonate units, are intermediates (enzyme-bound) in phenol biosynthesis. Cyclization can be envisaged as being similar to the aldol reaction (cf. **4**) or the Claisen condensation (cf. **5**) yielding phenolic acids like orsellinic acid (**6**), R = Me, or phenolic ketones, e.g. phloracetophenone (**7**), R = Me, respectively, after enolization of the carbonyl functions. Modification processes may ensue or intervene. The reduction of a carbonyl to secondary alcohol, away from the cyclization site, may thus afford a phenol with one less hydroxyl. However, such a mode of biogenesis[63–65] leads to phenols with *meta*-disposed hydroxyls. This character may be diagnostic of the origin.

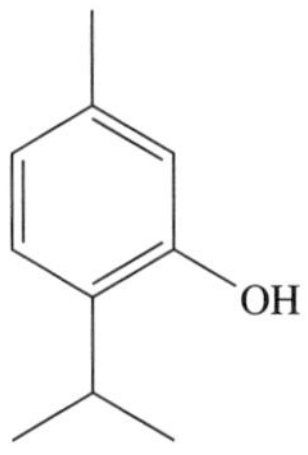

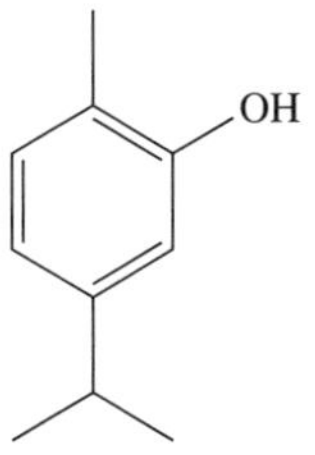

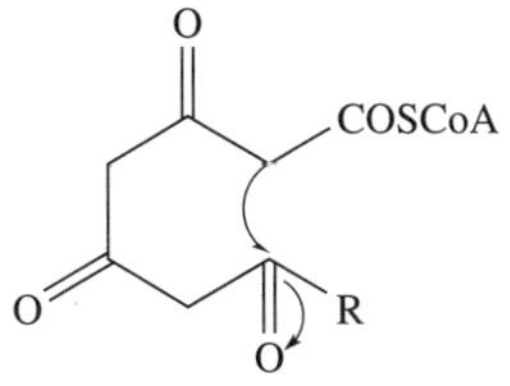

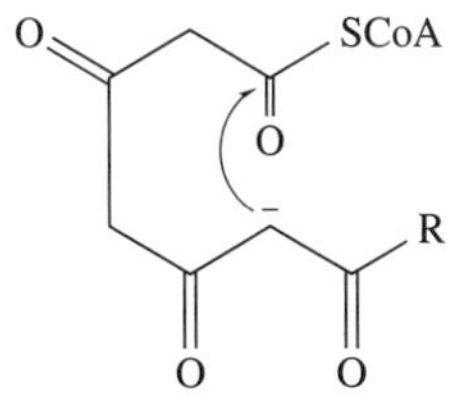

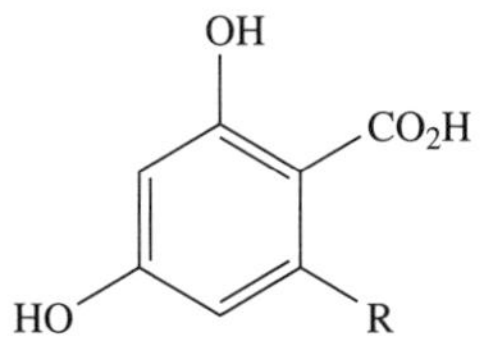

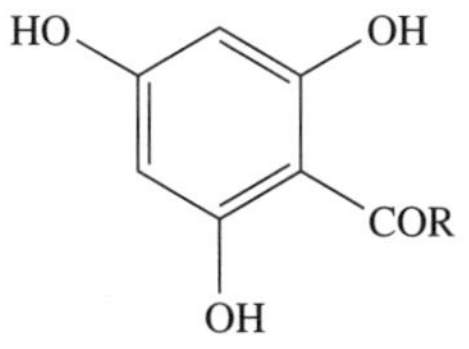

(ii) Aromatic rings may be hydroxylated in vivo by mono-oxygenases. Such reactions are often encountered in aromatics derived from the shikimate–prephenate pathway[66]. Phenylalanine (**8**) is thus *p*-hydroxylated to tyrosine (**9**) by phenylalanine mono-oxygenase using molecular oxygen. Cinnamic acid (**10a**) can be hydroxylated to *p*-hydroxycinnamic acid (**10b**), and on to di- and tri-hydroxy acids like, for instance, caffeic (**10c**) and gallic (**10d**) acids, with adjacent hydroxy functions. A useful list of micro-organisms and higher plant mono-oxygenases and phenolases is given elsewhere[67]. Hydroxylations such as

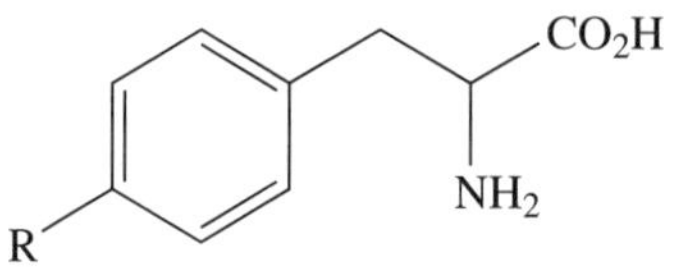

(8)

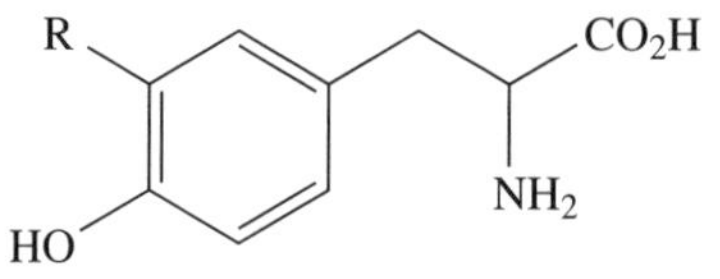

(9)

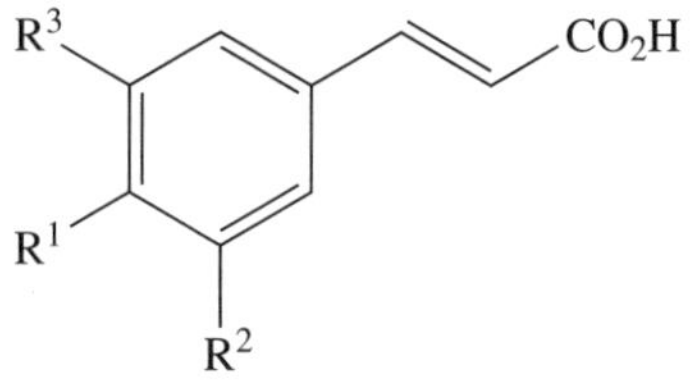

(10) (a) $R^1 = R^2 = R^3 = H$
(b) $R^1 = OH, R^2 = R^3 = H$
(c) $R^1 = R^2 = OH, R^3 = H$
(d) $R^1 = R^2 = R^3 = OH$

($8 \rightarrow 9$) may be accompanied by proton rearrangements as ($8, R = D$) $\rightarrow$ ($9, R = D$), the so-called 'National Institute of Health' ('NIH') shift, whose mechanism[68, 69] is displayed in Chart 3. Related 'NIH' shifts have been observed in vitro for various synthetic arene oxides and in oxidation of aromatics by permanganate and by chromyl compounds[70] such as CrO_2Cl_2 and $CrO_2(OAc)_2$.

(iii) Alicyclic rings with oxygen functions may be dehydrogenated to phenols. Compounds **1** and **2** are likely derived from monocyclic monoterpenes carrying a 3- or 2-oxygen function. Phenolic steroids like, for instance, estrone and equilenin can be derived in a similar way. This route to phenolic products is not yet well understood.

Phenol moieties are present in salvarsan (**11**) and neosalvarsan (**12**) synthesized by the German scientist Paul Ehrlich (1854–1915), considered as the father of chemotherapy for

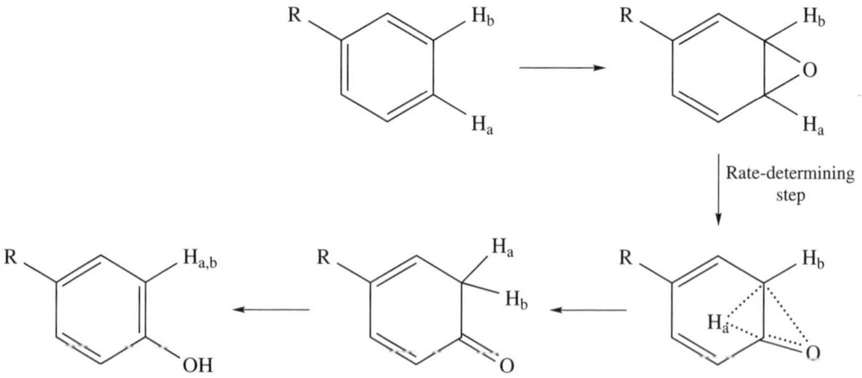

CHART 3. Mechanism of the so-called 'NIH'-shift

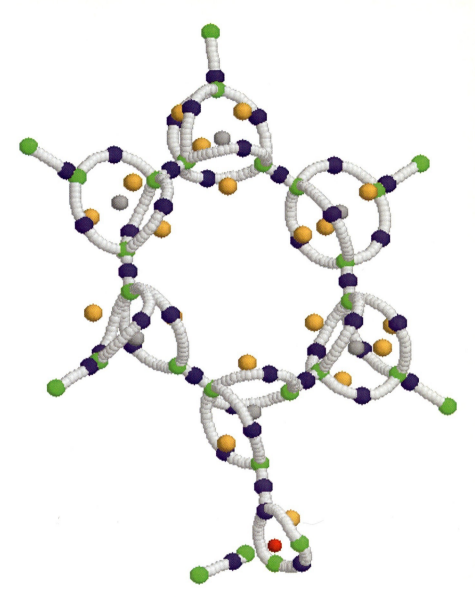

PLATE 1 (FIGURE 6). The VSCC graph for phenol. The oxygen atom is marked in red. The green spheres therein are the CPs $(3, -3)$ (maxima) in the phenolic $L(r)$ while the violet ones determine the $(3, -1)$ CPs. The yellow spheres correspond to the $(3, +1)$ CPs. The domain interaction lines (in light gray) link two $(3, -3)$ CPs via a $(3, -1)$ CP

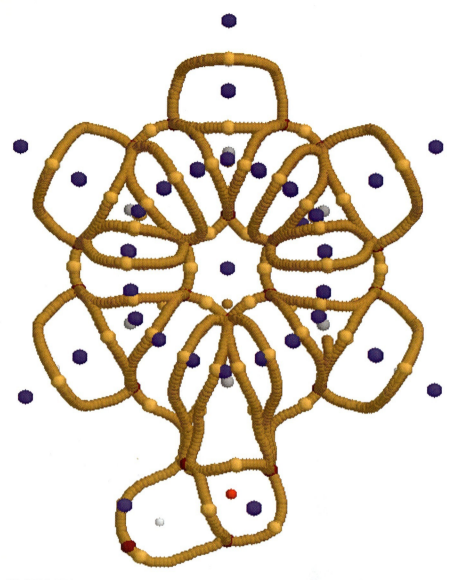

PLATE 2 (FIGURE 7). The VSCD graph for phenol. The oxygen atom is marked in red and the bonded hydrogen in white. The brown-red spheres therein are the CPs $(3,+3)$ (minima) in the phenolic $L(r)$ while the violet ones determine the $(3,-1)$ CPs. The yellow spheres correspond to the $(3,+1)$ CPs. The $(3, +1)$ CPs link the $(3, +3)$ CPs via a pair of gradient paths shown in light gray, each of which is repelled by a $(3, +3)$ CP

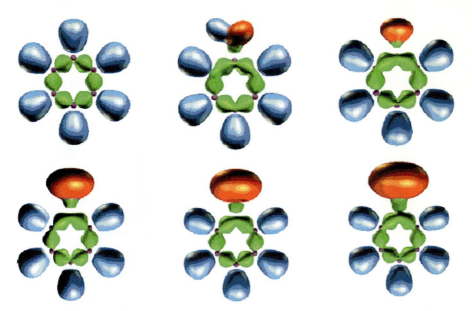

PLATE 3 (FIGURE 14). Localization domains of mono-X-substituted benzenes C_6H_5-X (from left to right top X= H, OH, F, bottom X= Cl, Br, I). The ELF value defining the boundary isosurface of 0.659 corresponds to the critical point of index 1 on the separatrix between adjacent V(C, C) basins of benzene. Color code: magenta = core, orange = monosynaptic, blue = protonated disynaptic, green = disynaptic

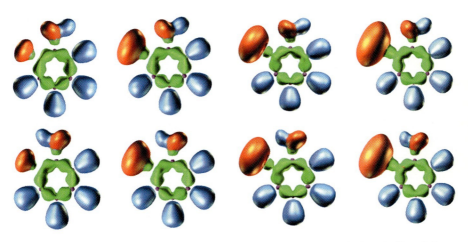

PLATE 4 (FIGURE 15). Localization domains of *ortho*-X-substituted phenols (from left to right X= F, Cl, Br, I; top – *trans* conformer, bottom – *cis* conformer). The ELF value defining the boundary isosurface of 0.659 corresponds to the critical point of index 1 on the separatrix between adjacent V(C, C) basins of benzene. Color code: magenta = core, orange = monosynaptic, blue = protonated disynaptic, green = disynaptic

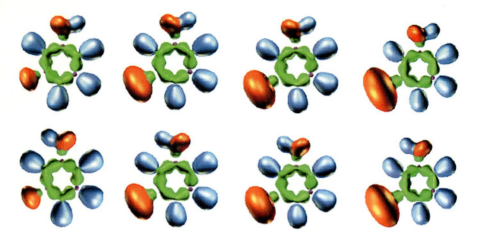

PLATE 5 (FIGURE 16). Localization domains of meta-X-substituted phenols (from left to right X= F, Cl, Br, I; top – *trans* conformer, bottom – *cis* conformer). The ELF value defining the boundary isosurface of 0.659 corresponds to the critical point of index 1 on the separatrix between adjacent V(C, C) basins of benzene. Color code: magenta = core, orange = monosynaptic, blue = protonated disynaptic, green = disynaptic

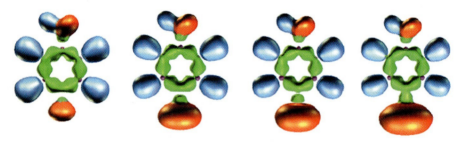

PLATE 6 (FIGURE 17). Localization domains of *para*-X-substituted phenols (from left to right X= F, Cl, Br, I). The ELF value defining the boundary isosurface of 0.659 corresponds to the critical point of index 1 on the separatrix between adjacent V(C, C) basins of benzene. Color code: magenta = core, orange = monosynaptic, blue = protonated disynaptic, green = disynaptic

use in syphilis treatments prior to the discovery of penicillin. He received a Nobel Prize in 1908 for his work.

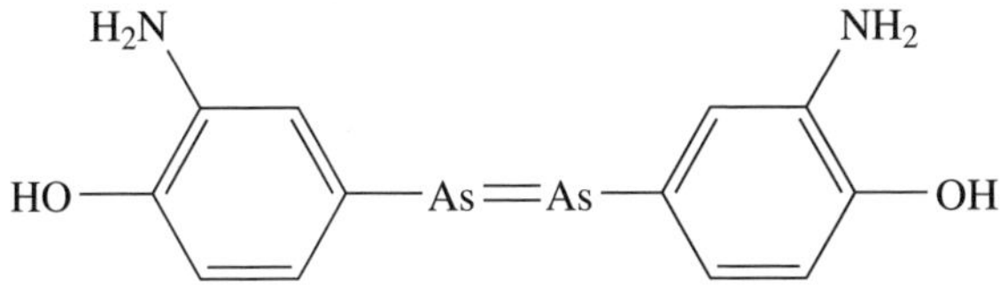

(11)

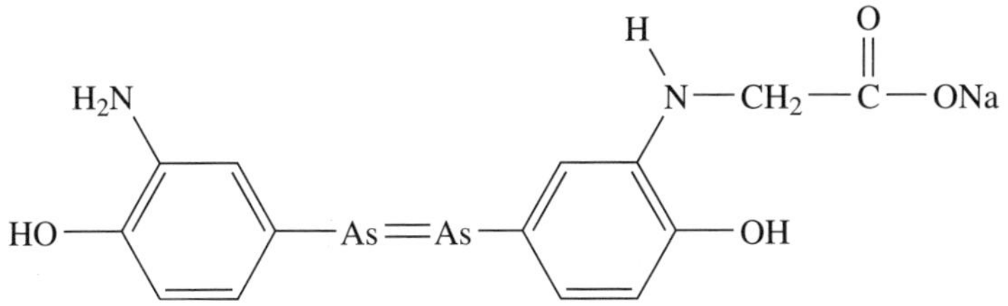

(12)

Phenol serves as a basic unit of larger molecules, e.g. tyrosine residues in proteins. The phenoxyl radical is treated as a model system for the tyrosyl radical whose formation via abstraction of the hydrogen atom from the hydroxyl group of tyrosine is a typical feature of oxidative stress in the physiological pH range[71,72].

Phenols are an extremely important class of antioxidants whose utilization in living organisms and synthetic organic materials reduces the rate of the oxidative degradation which all organic materials undergo by being exposed to air[73–77]. The antioxidant property can be related to the readily abstractable phenolic hydrogen as a consequence of the relatively low bond dissociation enthalpy of the phenolic O−H group [BDE(O−H)]. A large variety of *ortho-* and/or *para*-alkoxy-substituted phenols have been identified as natural antioxidants, such as α-tocopherol (13), which is known as the most effective lipid-soluble chain-breaking antioxidant in human blood plasma, and ubiquino-10 (14), both present in low-density lipid proteins. The mechanism of action of many phenolic antioxidants relies on their ability to transfer the phenolic H atom to a chain-carrying peroxyl radical at a rate much faster than that at which the chain-propagating step of lipid peroxidation proceeds[73–77]. Natural phenolic antioxidants can be also isolated from plants[78] such as sesamolinol (15), from sesame seeds and coniferyl alcohol (16), one of the three precursors for the biosynthesis of lignin. For example, Vitamin E (17) is a chain-breaking antioxidant that interferes with one or more of the propagation steps in autooxidation by atmospheric oxygen[79].

Phenolic compounds are also known to suppress the lipid peroxidation in living organisms. Furthermore, they are widely used as additives in food technology.

Regarding the production of phenol, small quantities of phenol are isolated from tars and coking plant water produced in the coking of hard coal and the low temperature carbonization of brown coal as well as from the wastewater from cracking plants. Most of the past and currently employed phenol syntheses are based on using benzene as a precursor which, however, is known as a volatile organic carcinogen. About 20% of the global benzene production is used for the manufacture of phenol[80]. By far the greatest proportion is obtained by oxidation of benzene or toluene. Although direct oxidation of

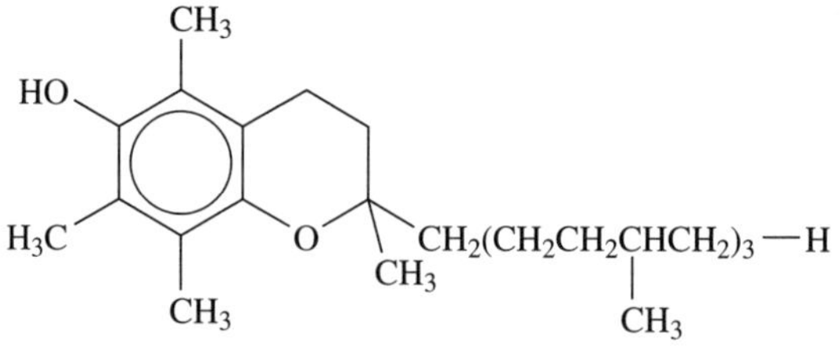
CH3
HO
H3C
O
CH2(CH2CH2CHCH2)3—H
CH3
CH3
CH3

(13)

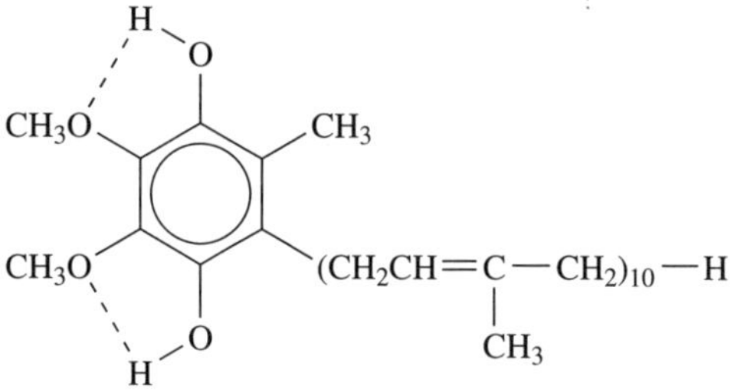
H
O
CH3O
CH3
CH3O
(CH2CH=C—CH2)10—H
CH3
O
H

(14)

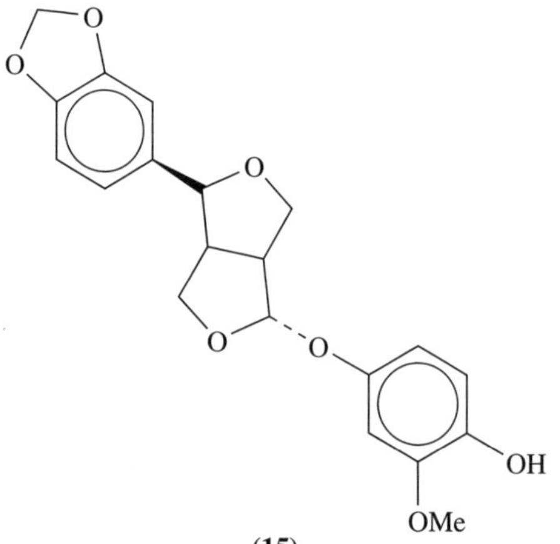
O
O
O
O
O
OH
OMe

(15)

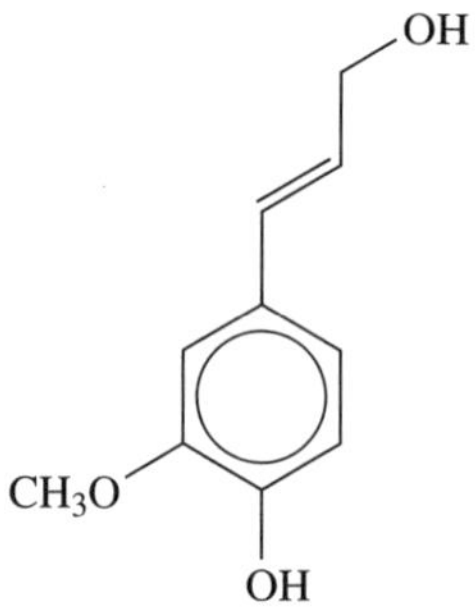
OH
CH3O
OH

(16)

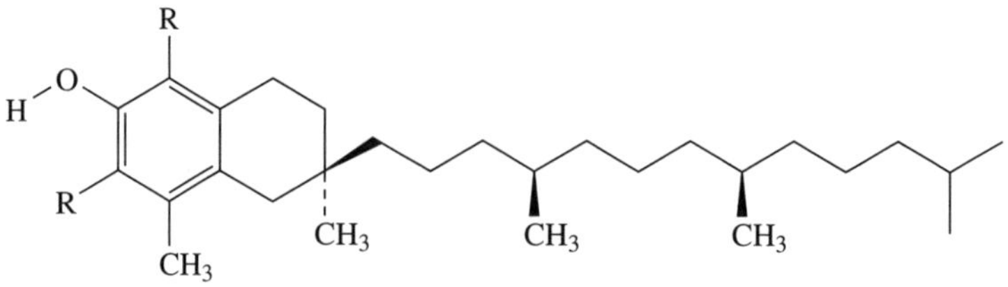
R
H
O
R
CH3
CH3
CH3
CH3

(17)

benzene is possible in principle, the phenol formed is immediately further oxidized. It is worth mentioning that a recent study[81] performed a thorough computational study of the potential energy surface for the oxidation reaction of benzene in the lowest-lying triplet state (equation 1)

$$C_6H_6 + O(^3P) \longrightarrow \text{Products} \quad (1)$$

followed by a kinetic analysis using the Rice–Ramsperger–Kassel–Marcus (RRKM) reaction theory[82] based on the electronic structure calculations employing the MP4/6-31G(d)//HF/6-31G(d) and B3LYP/cc-pVDZ computational levels. Below we outline the key results of this work.

Reaction 1 has a large number of energetically feasible product channels. In Figure 3, we display the theoretical triplet potential energy surface (PES) for reaction 1. The reaction initially proceeds via the addition of $O(^3P)$ to benzene, and this first step is exothermic by -37 kJ mol^{-1} and characterized by a barrier of approximately 21 kJ mol^{-1}. The chemically activated adduct reacts on the triplet PES in forming a number of products. The two lowest barriers lead to the formation of phenoxyl radical (-14 kJ mol^{-1}) and formylcyclopentadiene (-8 kJ mol^{-1}, both barriers taken relative to the reactant)[81]. The reaction route resulting in phenol is exothermic (-33 kJ mol^{-1}[81]). However, it has a rather high barrier of 100 kJ mol^{-1}. The calculated enthalpy of the reaction of the formation of

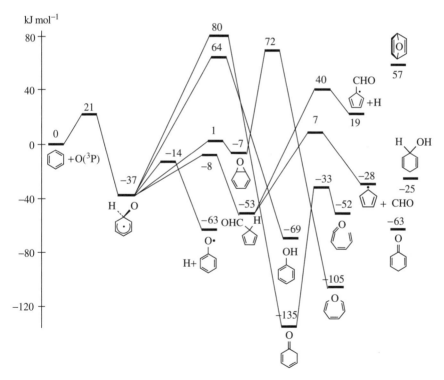

FIGURE 3. The potential energy profile of triplet products and transition structures in reaction 1. Adapted from Reference 81 with permission

phenol amounts to -433 kJ mol^{-1}, which agrees fairly well with the experimental value of -428 kJ mol^{-1}[81]. The theoretical singlet–triplet splitting of phenol (352 kJ mol^{-1}) is also very close to its experimental value of 341 kJ mol^{-1}. One may conclude that such high activation is likely sufficient to overcome the barrier in order to form phenoxy radical (372 kJ mol^{-1}), and therefore one might expect that the formation of the latter dominates on the singlet PES. This concurs with the flame data of Bittner and Howard[83] indicating that a direct reaction route to phenol is not possible.

It has been recently revealed that ZSM-5 zeolite exhibits an extremely high catalytic selectivity for the oxidation of benzene to phenol. The high reactivity of the zeolite should be ascribed to iron impurity arising in the intermediary steps in the zeolite synthesis[84, 85]. A surface oxygen (O) or α-oxygen, generated on Fe-ZSM-5 zeolite during N$_2$O decomposition[84, 85] (equation 2)

$$N_2O \xrightarrow{\text{zeolites}} (O) + N_2 \qquad (2)$$

takes part in the formation of phenol via equation 3

$$(O) + C_6H_6 \longrightarrow C_6H_5OH \qquad (3)$$

Reactions 2 and 3 have been thoroughly studied theoretically at the B3LYP computational level. In particular, a sound model of α-oxygen has been proposed[85, 86]. According to Reference 87, Solutia has recently developed a one-step technology producing phenol directly from benzene and N$_2$O. Due to the fact that such a process provides a very high yield and can use waste N$_2$O from the production of adipic acid, it is now considered to be a rather promising technology in the new millennium.

Therefore, alternative routes must be chosen for the production of phenol, e.g. via halogen compounds which are subsequently hydrolysed or via cumene hydroperoxide which is then cleaved catalytically. The following processes were developed as industrial syntheses for the production of phenol[88]:

1. Sulphonation of benzene and production of phenol by heating the benzenesulphonate in molten alkali hydroxide[89].
2. Chlorination of benzene and alkaline hydrolysis of the chlorobenzene.
3. Chlorination of benzene and catalytic saponification by Cu in the steam hydrolysis of the chlorobenzene[90, 91] (Raschig process, Raschig–Hooker, Gulf oxychlorination).
4. Alkylation of benzene with propene to isopropylbenzene (cumene), oxidation of cumene to the corresponding *tert*-hydroperoxide and cleavage to phenol and acetone (Hock process).
5. Toluene oxidation to benzoic acid and subsequent oxidizing decarboxylation to phenol (Dow process).
6. Dehydrogenation of cyclohexanol–cyclohexanone mixtures.

Among these processes, only the Hock process and the toluene oxidation are important industrially. The other processes were discarded for economic reasons. In the Hock process acetone is formed as a by-product. This has not, however, hindered the expansion of this process, because there is a market for acetone. New plants predominantly use the cumene process. More than 95% of the 4,691,000 m y^{-1} (m = metric tonnes) consumed is produced by the cumene peroxidation process. Phenol's consumption growth rate of 3% is primarily based on its use in engineering plastics such as polycarbonates, polyetherimide and poly(phenylene oxide), and epoxy resins for the electronic industry. The Mitsui Company is, for instance, the world's second largest producer of phenol. Japan's production output (in thousands of metric tonnes) is shown below[92].

Chemicals/Year	1996	1997	1998	1999	2000	Change 1999–2000
Phenol	768	833	851	888	916	3.2%
Phenolic resins	294	303	259	250	262	4.8%

The cumene process is based on the discovery of the oxidation of cumene with oxygen to cumene hydroperoxide and its acidic cleavage to phenol and acetone published in 1944[93]. This reaction was developed into an industrial process shortly after World War II by the Distillers Co. in the United Kingdom and the Hercules Powder Co. in the United States. The first plant was put into operation in 1952 by Shawinigan in Canada and had an initial capacity of 8000 t y^{-1} of phenol. Today, phenol is predominantly produced by this process in plants in the USA, Canada, France, Italy, Japan, Spain, Finland, Korea, India, Mexico, Brazil, Eastern Europe and Germany with an overall annual capacity of 5×10^6 tons[94, 95]. In addition to the economically favourable feedstock position (due to the progress in petrochemistry since the 1960s), the fact that virtually no corrosion problems occur and that all reaction stages work under moderate conditions with good yields was also decisive for the rapid development of the process. To produce cumene, benzene is alkylated with propene using phosphoric acid (UOP process) or aluminium chloride as catalyst.

The phenol-forming process via toluene oxidation developed originally by Dow (USA)[96–98] has been carried out in the USA, Canada and the Netherlands. Snia Viscosa (Italy) uses the toluene oxidation only for the production of benzoic acid as an intermediate in the production of caprolactam[99, 100]. The process proceeds in two stages. At the first stage, toluene is oxidized with atmospheric oxygen in the presence of a catalyst to benzoic acid in the liquid phase. At the second stage the benzoic acid is decarboxylated catalytically in the presence of atmospheric oxygen to produce phenol. This is a radical-chain reaction involving peroxy radicals. The activation energy of the exothermic oxidation of toluene to benzoic acid is 136 kJ mol^{-1}[99].

Most of the phenol produced is processed further to give phenol–formaldehyde resins. The quantities of phenol used in the production of caprolactam via cyclohexanol–cyclohexanone have decreased because phenol has been replaced by cyclohexane as the starting material for caprolactam. The production route starting from phenol is less hampered by safety problems than that starting from benzene, which proceeds via cyclohexane oxidation. Bisphenol A, which is obtained from phenol and acetone, has become increasingly important as the starting material for polycarbonates and epoxy resins. Aniline can be obtained from phenol by ammonolysis in the Halcon process. Adipic acid is obtained from phenol by oxidative cleavage of the aromatic ring. Alkylphenols, such as cresols, xylenols, 4-tert-butylphenol, octylphenols and nonylphenols, are produced by alkylation of phenol with methanol or the corresponding olefins. Salicylic acid is synthesized by addition of CO_2 to phenol (Kolbe synthesis). Chlorophenols are also obtained directly from phenol. All these products have considerable economic importance because they are used for the production of a wide range of consumer goods and process materials. Examples are preforms, thermosets, insulating foams, binders (e.g. for mineral wool and molding sand), adhesives, laminates, impregnating resins, raw materials for varnishes, emulsifiers and detergents, plasticizers, herbicides, insecticides, dyes, flavours and rubber chemicals.

It is worth noting the recent work on the benzene-free synthesis of phenol[101], which is actually a part of longstanding efforts[102] to elaborate the alternatives to benzene. This new alternative synthesis is based on the aromatization of shikimic acid which is now readily available by the elaboration of a microbe-catalysed synthesis from glucose in near-critical water, where phenol is the primary reaction product. An aqueous solution of shikimic acid is heated to and maintained at 350 °C for 30 min yielding 53% of phenol.

II. MOLECULAR STRUCTURE AND BONDING OF PHENOL

A. The Equilibrium Structure of Phenol in the Ground Electronic State

Until the mid-thirties of the 20th century electron diffraction or microwave studies of phenol had not yet been conducted and so, rather peculiarly, the equilibrium configuration of phenol remained uncertain although some indirect evidence suggested its ground electronic state S_0 to be certainly planar. The first X-ray structural data became available by 1938 for several phenolic compounds[103]. At that time, it was suggested that the C—O bond is about 1.36 Å, that is by ca 0.07 Å shorter than the C—O bond in aliphatic alcohols. This was accounted for by the decrease in the effective radius of the carbon atom due to the change of hybridization from sp^3 to sp^2, even though some degree of electron delocalization across the C—O bond could be assumed. Such increase in double-bond character favours a completely planar equilibrium configuration of phenol in its ground electronic state.

This character results from quinonoid resonance structures in addition to the more important Kekulé-type structures[104] and tends to cause the hydrogen atom to be placed in the molecular plane. This leads to two equivalent configurations with the hydrogen of the OH group being on one side of the other of the C—O bond[104]. It implies the existence of the activation barrier V_τ of the OH torsion motion around the C—O bond estimated in the mid-thirties as equal to 14 kJ mol^{-1}.

The molecular geometry of phenol was later determined experimentally by microwave spectroscopy[105–108] and electron diffraction[109] (ED). In 1960, MW experiments[105] of some phenol derivatives showed that their equilibrium configurations are planar (C_s symmetry). In 1966, two possible r_o-structures were determined by examining four new isotopic modifications of phenol[106], and three years later a partial r_s-structure was presented on the basis of the six monodeuteriated species[107]. The full r_s-structure of phenol was reported[108] in 1979 and is presented in Table 1[109]. Generally speaking, the structure of the phenyl ring in phenol deviates only slightly from the regular isolated phenyl ring. This is shown in Figure 4. All C—H distances are nearly equal, within the experimental uncertainties, although the *para*-distance seems to be shorter than the other ones. The CCC bond angles are slightly perturbed, viz. the bond angle $C_1C_3C_5$ is larger than 120° whereas the $C_2C_6C_4$ angle is smaller than 120°. The angle between the C_6O_7 bond and the C_1–C_4 axis was reported equal to 2.52°[108]. Our calculation performed at the B3LYP/6-31+G(d,p) computational method predicts it to be equal to 2.58°.

Since the first quantum mechanical calculation of phenol performed in 1967 using the CNDO/2 method[110], the phenol geometry was considered at a variety of computational levels[111–125] ranging from the HF to the MP2 method of molecular orbital theory and density functional theory (DFT) employed with several basis sets, mainly of the split valence type as, e.g. 6-31G(d,p) and 6-31+G(d,p). These computational results are summarized in Tables 1–3 and Figure 4. It seems noteworthy that the semi-empirical geometries listed in Table 1 are rather close to the experimental observations. Also, to complete the theoretical picture of the phenol molecule, its theoretical inertia moments calculated at the B3LYP/6-31+G(d,p) level are equal to 320.14639, 692.63671 and 1012.78307 a.u.

Table 3 summarizes the key properties of phenol[107–130]. Inspecting its rotational constants collected in Table 2, we may conclude that fair agreement between experiment and

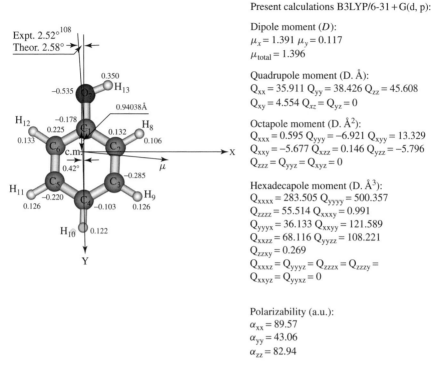

FIGURE 4. Key properties of the planar B3LYP/6-31+G(d,p) phenol molecule in the ground electronic state including the position of its centre of mass (c.m.), Mulliken charges and the direction of its total dipole moment

theory is provided by the MP2 and B3LYP methods (the mean absolute deviations are less than 0.2%) and the B3P86 method (<0.6%) whereas the HF and BLYP methods predict rather large values (*ca* 1.3% and *ca* 1.5%, respectively)[124]. The latter methods have well-known shortcomings, viz. the HF bond distances are too short while the BLYP distances are too large. Regarding in particular the length of the C–O bond, note that BLYP/6-31G(d) gives 1.384 Å although the corresponding MP2/6-31G(d) value of 1.396 Å is larger by 0.012 Å.

B. Molecular Bonding Patterns in the Phenol S_0

Let us start this subsection with somewhat simple arguments about the bonding in the phenol molecule. We may consider the two σ bonds of the oxygen atom as constituted of trigonal hybrids[131]. The third coplanar hybrid accommodates one sp^2 lone pair while the pure p orbital is also conjugated with the other p electrons of the phenyl ring.

Speaking at a higher theoretical level, the closed-shell electronic ground-state phenol molecule is described by the 25 occupied molecular orbitals whose 3D patterns are partly pictured in Figure 5. These 25 occupied MOs are partitioned into two classes, the first comprising the seven core orbitals (1s atomic orbitals on the carbon and oxygen atoms) and the second including 18 valence orbitals. The latter represent six σ C–C bonds (all

TABLE 1. Phenol geometry. Bond lengths in Å, bond angles in deg

	Experiment				Theory							
	MW[107]	MW[108]	ED[109]	MNDO[a]	MINDO/3[a]	AM1[a]	PM3[a]	HF/STO-3G[111]	HF/4-31G[111]	HF/6-31G[111]	HF/6-31G(d)[111]	HF/6-31G(d,p)[112]	HF/DZP[113]
Bond lengths													
C_1-C_2	1.398	1.3912	1.3969	1.420	1.419	1.402	1.401	1.397	1.381	1.385	1.385	1.410	1.389
C_2-C_3		1.3944	1.3969	1.405	1.406	1.394	1.390	1.386	1.385	1.389	1.387		1.392
C_3-C_4		1.3954	1.3969	1.405	1.404	1.394	1.390	1.390	1.381	1.385	1.382		1.387
C_4-C_5		1.3954		1.407	1.408	1.397	1.392	1.384	1.387	1.390	1.388		1.393
C_5-C_6		1.3922		1.403	1.403	1.391	1.388	1.382	1.389	1.383	1.381		1.386
C_1-C_6		1.3912		1.423	1.424	1.406	1.402	1.392	1.383	1.386	1.388		1.393
C_1-O_7	1.364	1.3745	1.3975	1.359	1.326	1.377	1.369	1.395	1.374	1.377	1.352	1.382	1.354
C_2-H_8		1.0856	1.081	1.090	1.105	1.099	1.096	1.082	1.073	1.074	1.077		
C_3-H_9	1.076	1.0835		1.091	1.106	1.100	1.095	1.083	1.072	1.073	1.075		
C_4-H_{10}	1.082	1.0802		1.90	1.104	1.099	1.096	1.082	1.071	1.072	1.074	1.093	
C_5-H_{11}		1.0836		1.091	1.107	1.100	1.096	1.083	1.072	1.072	1.075	1.092	
C_6-H_{12}		1.0813		1.090	1.104	1.099	1.096	1.082	1.069	1.070	1.074		
O_7-H_{13}	0.956	0.9574	0.953	0.948	0.951	0.968	0.949	0.989	0.950	0.949	0.947	0.977	0.944
Bond angles													
$C_1C_2C_3$		119.43	118.77	119.6	119.5	119.1	119.0	119.9	119.6	119.4	119.6		
$C_2C_3C_4$		120.48	120.57	120.6	121.0	120.4	120.4	120.5	120.5	120.4	120.5		
$C_3C_4C_5$		119.74	119.75	119.8	119.1	120.0	120.1	119.4	119.4	119.4	119.2		
$C_4C_5C_6$		120.79		120.7	121.4	120.6	120.6	120.8	120.7	120.6	120.7		
$C_1C_6C_5$		119.22		119.4	119.0	118.9	118.9	119.6	119.6	119.4	119.5		
$C_1C_2H_8$		120.01		121.2	121.3	120.4	120.9	120.4	120.2	120.3	120.0		
$C_2C_3H_9$		119.48		119.5	119.1	119.5	119.6	120.3	119.4	119.5	119.4		
$C_3C_4H_{10}$		120.25		120.1	120.5	120.1	120.0	120.4	120.3	120.3	120.4		
$C_6C_5H_{11}$		119.43		119.5	118.9	119.5	119.6						
$C_1C_6H_{12}$		119.23		120.8	121.7	119.5	120.4						
$C_1O_7H_{13}$	109.0	108.77	106.4	112.8	114.0	107.9	107.9	104.9	114.8	114.7	110.7	108.1	110.9

TABLE 1. (continued)

	Theory														
	CAS(8,7)/ cc-pVDZ[114]	CAS(8,9)/ cc-pVDZ[114]	CAS(8,8)/ cc-pVDZ[115]	B3LYP/ 6-31G(d)[116]	B3LYP/ 6-31G(d,p)[a,116]	BLYP/ 6-31G(d,p)[112]	B3LYP/ 6-311G(d,p)[117]	B3LYP/ 6-31+G(d,p)[a]	B3LYP/ DZP[113]	BLYP/ 6-311++G(d,p)[a]	BLYP/ 6-311++G(2df,2p)[a]	B3LYP/ cc-pVDZ[114]	MP2/ DZP[113]	MP2/ 6-31G(d,p)[111]	MP2/ 6-31G(d,p)[112]
	1.395	1.394	1.394	1.410	1.401	1.389	1.397	1.399	1.404	1.396	1.393	1.394	1.404	1.395	1.396
	1.400	1.400	1.400	1.403	1.401		1.390	1.398	1.402	1.394	1.391	1.400	1.404	1.395	
	1.395	1.394	1.394	1.409	1.400		1.395	1.397	1.400	1.393	1.390	1.394	1.403	1.393	
	1.400	1.401	1.400	1.405	1.403		1.392	1.397	1.404	1.396	1.392	1.401	1.406	1.396	
	1.394	1.393	1.394	1.406	1.397		1.393	1.395	1.400	1.391	1.388	1.393	1.400	1.392	
	1.399	1.399	1.399	1.410	1.401		1.396	1.399	1.405	1.396	1.392	1.399	1.404	1.396	
	1.356	1.355	1.355	1.384	1.395	1.351	1.367	1.372	1.373	1.370	1.367	1.355	1.378	1.374	1.372
	1.084	1.084	1.085	1.093	1.086	1.086		1.083	1.088		1.086	1.084	1.084	1.089	
	1.082	1.082	1.083	1.094	1.084	1.076	1.084	1.086		1.084	1.082	1.082		1.087	1.082
	1.081	1.081	1.083	1.093	1.083	1.075	1.083	1.085		1.083	1.081	1.081		1.086	1.081
	1.082	1.082	1.083	1.094	1.084		1.084	1.086		1.084	1.082	1.082		1.087	
	1.081	1.081	1.082	1.097	1.083		1.087	1.085		1.083	1.081	1.081		1.086	
	0.945	0.946	0.945	0.981	0.967	0.943	0.962	0.966	0.968	0.963	0.962	0.946	0.967	0.973	0.965
	119.9	119.9	120.0					119.7		119.7	119.8	119.9		119.7	
	120.4	120.4	120.4					120.5		120.5	120.5	120.4		120.5	
	119.3	119.3	119.3					119.3		119.3	119.3	119.3		119.4	
	120.6	120.6	120.6					120.8		120.8	120.8	120.6		120.6	
	119.8	119.8	119.8					119.5		119.6	119.6	119.8		119.6	
	120.1	120.1	120.0					120.1		120.0	120.0	120.0		120.1	
	119.4	119.4	119.4					119.3		119.3	119.3	120.1		119.2	
	120.3	120.4	120.4					120.3		120.3	120.3	119.4		120.3	
								119.3		119.3	119.3	120.4			
								119.0		119.0	119.1				
	110.2	110.2	110.3			110.9		109.9	108.9	109.7	109.9	110.2	108.3	108.4	108.5

[a] Present work.

TABLE 2. Rotational constants (in MHz) of phenol in its electronic ground state. The values in parentheses are the deviation from the experimental values in percent

	HF/6-31G(d,p)[124]	HF/6-311++G(d,p)[124]	CAS(8,7)/cc-pVDZ[126]	MP2/6-31G(d,p)[124]	BLYP/6-31G(d,p)[124]	B3LYP/6-31G(d,p)[124]	B3LYP/6-31+G(d,p)[a]	B3LYP/6-311++G(d,p)[a]	B3LYP/6-311++G(2df,2p)[a]	B3P86/6-31G(d,p)[124]	MW Expt.[108]	UV Expt.[118]	UV Expt.[127]
A	5750.0	5752.6	5659.3(0.16)	5650.6	5563.7	5650.4	5637.3	5667.2	5695.6	5679.9	5650.5154	5726.63	5650.515
B	2659.1	2660.0	2623.3(0.16)	2614.6	2573.7	2614.1	2607.3	2618.0	2629.6	2630.0	2619.2360	2660.0	2619.236
C	1818.3	1818.9	1792.4(0.14)	1787.5	1759.7	1787.3	1782.8	1790.8	1799.0	1797.3	1789.8520	1820.12	1782.855

[a]Present work.

TABLE 3. Dipole moment of phenol. Experimental data are partly reproduced from Reference 128[a]

Experiment			Theory	
μ(D)	Phase of solvent[b]	T(°C)	μ(D)	Method
1.40 ± 0.03	gas	20	1.73	CNDO/2[110]
1.41	gas	175	1.418	B3LYP/6-31+G(d,p)[129]
2.22	liq	20	1.16	HF/MiDi[130]
1.45	B	25	1.52	SM5.42R/HF/MiDi[130]
1.45	B	25		
1.45	B	30		
1.46	B	n.s.		
1.47 ± 0.02	B	20		
1.5	B	70		
1.53 ± 0.03	B	20		
1.53	B	26		
1.54	B	20		
1.54	B	25		
1.57	B	20		
1.59	B	(22)		
1.65	B	25		
1.72	B	25		
1.75	B	20		
1.80	D	20		
1.86	D	25		
1.92	D	20		
1.39	CCl_4	30		
1.40	CCl_4	20		
1.46 ± 0.03	CCl_4	10–60		
1.49	CCl_4	20		
1.50	CCl_4	20		
1.53	CCl_4	25		
1.55	CCl_4	27		
1.37	c-Hx	30		
1.32 ± 0.03	c-Hx	20		
1.33 ± 0.03	c-Hx	20		
1.39	c-Hx	30		
1.43	c-Hx	25		
1.37	Hp	20		
1.44	Hp	20		
1.86	Hp	20		
1.44 − 1.53	Tol	0–75		
1.46	Tol	30		
1.38	CS_2	20		
1.39	CS_2	30		
1.64	CS_2	25		
2.14	Ether	25		
2.14	Ether	20		
2.29	Ether	20		
1.45	ClB	20		
1.53[128]	B			

[a] See also Figure 4.
[b] B = benzene, D = dioxan, c-Hx = cyclohexane, Hp = n-heptane, Tol = toluene, ClB = chlorobenzene; n.s. = not specified.

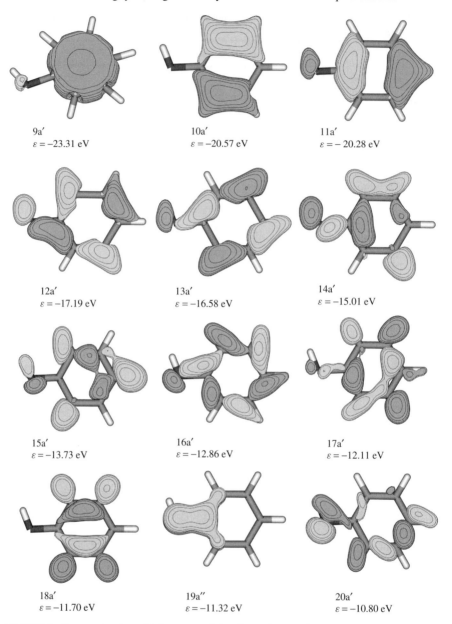

FIGURE 5. Some molecular orbital patterns of the electronic ground state of the phenol molecule. Due to the C_s symmetry of phenol, its MOs are characterized by the a' or a'' irreducible representations of this group; ε denotes the corresponding orbital energy in eV

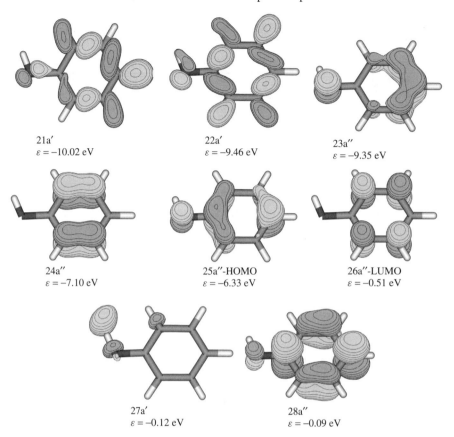

FIGURE 5. (*continued*)

of a' symmetry), five σ C−H bonds (also all of a' symmetry), the C−O σ bond (one a' orbital), the oxygen σ-type lone pair (one a' orbital), the oxygen p-type lone pair (19 a'' orbital) and, finally, the C−C π-bonds (three a'' orbitals, namely $23a''$, $24a''$ and the HOMO $25a''$). In addition, three unoccupied π molecular orbitals, the LUMO $26a''$, $27a'$ and $28a''$, are also shown in Figure 5.

In Table 4, we collect the natural atomic charges (nuclear charge minus the summed natural populations of the natural atomic occupancies, NAOs, on the atom) and the total core, valence and Rydberg populations on each atom. Table 4 presents a slightly larger positive charge on the hydroxyl hydrogen atom H_{13} relative to other atoms, arising due to the proximity of the electronegative oxygen atom. The other hydrogen atom H_8 next to the hydroxyl group is characterized by the lowest positive charge. This feature originates from electron donation from the ring to the corresponding C−H antibonding orbital taking place in order to decrease the electrostatic repulsion between the neighboring C−H and O−H bonds.

The HOMO and LUMO are of particular interest. As seen in Figure 5, the shape of the HOMO is generated by the out-of-phase overlap of the p_z AOs localized, on the one

TABLE 4. Natural atomic orbital (NAO) occupancies, natural population of the MOs, summary of natural population analysis and Mulliken atomic charges of the electronic ground state of phenol

NAOs N	Atom	N lm	Type(AO)	Occupancy	Energy (eV)
1	C	1 s	Cor(1s)	1.998	−9.95
2	C	1 s	Val(2s)	0.947	−0.16
3	C	1 s	Ryd(3s)	0.000	1.63
4	C	1 px	Val(2p)	1.170	−0.05
5	C	1 px	Ryd(3p)	0.004	1.11
6	C	1 py	Val(2p)	1.114	−0.04
7	C	1 py	Ryd(3p)	0.005	0.97
8	C	1 pz	Val(2p)	1.049	−0.10
9	C	1 pz	Ryd(3p)	0.000	0.80
10	C	2 s	Cor(1s)	1.998	−10.04
11	C	2 s	Val(2s)	0.832	−0.15
12	C	2 s	Ryd(3s)	0.000	1.58
13	C	2 px	Val(2p)	1.105	−0.06
14	C	2 px	Ryd(3p)	0.006	1.03
15	C	2 py	Val(2p)	0.762	−0.04
16	C	2 py	Ryd(3p)	0.008	1.01
17	C	2 pz	Val(2p)	0.991	−0.12
18	C	2 pz	Ryd(3p)	0.001	0.80
19	C	3 s	Cor(1s)	1.998	−9.96
20	C	3 s	Val(2s)	0.944	−0.17
21	C	3 s	Ryd(3s)	0.000	1.63
22	C	3 px	Val(2p)	1.165	−0.06
23	C	3 px	Ryd(3p)	0.004	1.12
24	C	3 py	Val(2p)	1.121	−0.06
25	C	3 py	Ryd(3p)	0.006	0.94
26	C	3 pz	Val(2p)	1.087	−0.11
27	C	3 pz	Ryd(3p)	0.000	0.79
28	C	4 s	Cor(1s)	1.998	−9.96
29	C	4 s	Val(2s)	0.944	−0.17
30	C	4 s	Ryd(3s)	0.000	1.62
31	C	4 px	Val(2p)	1.179	−0.05
32	C	4 px	Ryd(3p)	0.005	1.14
33	C	4 py	Val(2p)	1.100	−0.04
34	C	4 py	Ryd(3p)	0.004	0.94
35	C	4 pz	Val(2p)	0.989	−0.10
36	C	4 pz	Ryd(3p)	0.000	0.79
37	C	5 s	Cor(1s)	1.998	−9.95
38	C	5 s	Val(2s)	0.943	−0.16
39	C	5 s	Ryd(3s)	0.000	1.63
40	C	5 px	Val(2p)	1.070	−0.04
41	C	5 px	Ryd(3p)	0.005	0.84
42	C	5 py	Val(2p)	1.208	−0.05
43	C	5 py	Ryd(3p)	0.004	1.24
44	C	5 pz	Val(2p)	1.041	−0.10
45	C	5 pz	Ryd(3p)	0.000	0.80
46	C	6 s	Cor(1s)	1.998	−9.96
47	C	6 s	Val(2s)	0.946	−0.16
48	C	6 s	Ryd(3s)	0.000	1.63
49	C	6 px	Val(2p)	1.178	−0.05

TABLE 4. (continued)

NAOs N	Atom	N lm	Type(AO)	Occupancy	Energy (eV)
50	C	6 px	Ryd(3p)	0.005	1.15
51	C	6 py	Val(2p)	1.104	−0.04
52	C	6 py	Ryd(3p)	0.004	0.93
53	C	6 pz	Val(2p)	0.986	−0.10
54	C	6 pz	Ryd(3p)	0.000	0.79
55	O	7 s	Cor(1s)	1.999	−18.81
56	O	7 s	Val(2s)	1.664	−0.93
57	O	7 s	Ryd(3s)	0.000	3.22
58	O	7 px	Val(2p)	1.628	−0.30
59	O	7 px	Ryd(3p)	0.000	1.79
60	O	7 py	Val(2p)	1.467	−0.29
61	O	7 py	Ryd(3p)	0.001	1.62
62	O	7 pz	Val(2p)	1.850	−0.29
63	O	7 pz	Ryd(3p)	0.000	1.48
64	H	8 s	Val(1s)	0.765	0.06
65	H	8 s	Ryd(2s)	0.001	0.71
66	H	9 s	Val(1s)	0.757	0.09
67	H	9 s	Ryd(2s)	0.000	0.70
68	H	10 s	Val(1s)	0.757	0.09
69	H	10 s	Ryd(2s)	0.000	0.70
70	H	11 s	Val(1s)	0.756	0.09
71	H	11 s	Ryd(2s)	0.000	0.70
72	H	12 s	Val(1s)	0.745	0.09
73	H	12 s	Ryd(2s)	0.001	0.71
74	H	13 s	Val(1s)	0.546	0.05
75	H	13 s	Ryd(2s)	0.001	0.82

Natural population of the MOs

Core	13.990 (99.9319% of 14)
Valence	35.927 (99.7975% of 36)
Natural Minimal Basis	49.917 (99.8352% of 50)
Natural Rydberg Basis	0.082 (0.1648% of 50)

Summary of natural population analysis

Atom N	Charge	Core	Valence	Rydberg	Total
C 1	−0.252	1.999	4.234	0.018	6.252
C 2	0.315	1.998	3.662	0.022	5.684
C 3	−0.284	1.999	4.267	0.018	6.284
C 4	−0.183	1.999	4.165	0.018	6.183
C 5	−0.236	1.999	4.218	0.018	6.236
C 6	−0.182	1.999	4.165	0.017	6.182
O 7	−0.678	1.999	6.665	0.013	8.678
H 8	0.200	0.000	0.797	0.002	0.799
H 9	0.204	0.000	0.793	0.002	0.795
H 10	0.206	0.000	0.791	0.002	0.793
H 11	0.204	0.000	0.793	0.002	0.795
H 12	0.217	0.000	0.780	0.002	0.782
H 13	0.467	0.000	0.528	0.004	0.532
<Total>	0.000	13.994	35.863	0.142	50.000

(*continued overleaf*)

TABLE 4. (*continued*)

Mulliken charges on atoms	
1 C	−0.186
2 C	0.295
3 C	−0.223
4 C	−0.184
5 C	−0.195
6 C	−0.185
7 O	−0.607
8 H	0.173
9 H	0.187
10 H	0.182
11 H	0.187
12 H	0.199
13 H	0.357

hand, on the carbon atoms C_1, C_2 and C_6, and, on the other hand, on C_4 and the oxygen atom. The LUMO shape is quite different and composed of the out-of-phase overlap of the p_z AOs on the C_2, C_3, C_5 and C_6. Both HOMO and LUMO possess two nodal surfaces perpendicular to the phenolic ring. Both frontier orbitals have negative orbital energies: $\varepsilon_{HOMO} = -6.33$ eV and $\varepsilon_{LUMO} = -0.51$ eV. According to Koopmans' theorem[132, 133], the Koopmans ionization potential, which is simply the HOMO energy taken with the opposite sign, might be in general considered as a good approximation to the first vertical ionization energy. Therefore, in the case of phenol, ε_{HOMO} must be interpreted as the energy required to remove a π electron from phenol to form phenol radical cation PhOH$^{•+}$ (cf. **18** for one of its many possible resonance structures). As seen in Section 1, the experimental value of the adiabatic first ionization energy IE$_a$ of phenol is equal to 8.49 ± 0.2 eV and settled to 8.51 eV or 68639.4 cm^{-1}[134, 135] or 68628 cm^{-1}[136]. Interestingly, it is lower by nearly 71 kJ mol^{-1} than IE$_a$(benzene) = 74556.58 ± 0.05 cm^{-1}[137]. Summarizing, we may conclude that Koopmans' theorem is rather inadequate for phenol, even in predicting its vertical ionization energy (for a further discussion see Reference 131, p. 128).

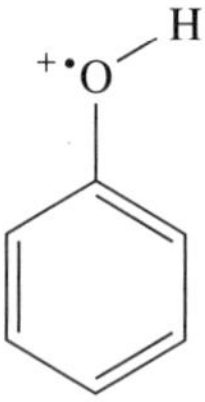

(18)

In order to theoretically determine the ionization energy of phenol, the same method/basis should be employed for both parent and cation. Table 5 summarizes the optimized geometries and the energies (including ZPVE) of phenol and phenol radical cation calculated using the B3LYP method in conjunction with 6-31G(d,p) and 6-311++G(d,p) basis sets. It is interesting to notice a rather drastic change in the geometry of phenol radical cation compared to the parent phenol molecule (Table 5), especially in the vicinity of the carbonyl group, whereas the difference between IE$_{vert}$ and IE$_{ad}$ is

TABLE 5. The B3LYP data of phenol and phenol radical cation[a,b]

Geometry	Phenol		Phenol radical cation	
	6-31G(d,p)	6-311++G(d,p)	6-31G(d,p)	6-311++G(d,p)
C_1-C_2	1.399	1.396	1.433	1.431(+0.035)
C_2-C_3	1.396	1.394	1.371	1.368(−0.026)
C_3-C_4	1.395	1.393	1.425	1.423(−0.030)
C_4-C_5	1.398	1.396	1.418	1.416(+0.020)
C_5-C_6	1.393	1.391	1.372	1.369(−0.021)
C_1-C_6	1.399	1.396	1.438	1.435(+0.039)
C_1-O_7	1.368	1.370	1.312	1.310(−0.060)
O_7-H_{13}	0.966	0.963	0.975	0.972(+0.009)
$C_1O_7H_{13}$	108.9	109.7	113.6	113.8(+4.1)
−Energy +307	0.478469	0.558732	0.183858	0.252608
−Energy$_{vert}$[c] + 307			0.176780	0.245650
ZPVE + 65	0.765	0.229	0.745	0.295
IE$_{ad}$			8.016	8.333
			7.03 HF/DZP[113]	
			8.70 MP2/DZP[113]	
			8.15 B3LYP/DZP[113]	
IE$_{vert}$			8.209	8.519

[a] The phenol radical cation have recently been studied theoretically[113, 138, 140]. See different properties in Reference 139.
[b] Bond lengths are given in Å, bond angle in degrees, energies in hartree, ZPVE in kJ mol^{-1} and ionization energy in eV. The atomic numbering is indicated in Chart 1. Deviations in the bond lengths of phenol radical cation from those of phenol are shown in parentheses.
[c] The energy$_{vert}$ of phenol radical cation is determined at the corresponding geometry of the parent phenol.

rather small. The potential energy surface of the ionized phenol will be discussed in a subsequent section.

C. Atom-in-Molecule Analysis

In this subsection, we briefly review the use of the function $L(\mathbf{r})$ of the electronic ground-state phenol which is defined as minus the Laplacian of its electron density, $\nabla_r^2 \rho(\mathbf{r})$, fully in the context of Bader's *'Atoms in Molecule'* (AIM) approach[141, 142] (the electronic localization function is discussed below). The topology of $L(\mathbf{r})$ can be almost faithfully mapped onto the electron pairs of the VSEPR model[143, 144]. The topology of the one-electron density $\rho(\mathbf{r})$ (see, e.g., Reference 145 and references therein for the definition) is fully understood within the AIM theory resulting in its partition which defines *'atoms'* inside a molecule or a molecular aggregate via the *gradient vector field* $\nabla_r \rho(\mathbf{r})$. Such a vector field is a collection of gradient paths simply viewed as curves in the three-dimensional (3D) space following the direction of steepest ascent in $\rho(\mathbf{r})$. Therefore, the meaning of a gradient path is absolutely clear: it starts and ends at those points where $\nabla_r \rho(\mathbf{r})$ vanishes. These points are called *critical points* (CPs). The CPs of $\rho(\mathbf{r})$ are special and useful points of the corresponding molecule.

The classification of the critical points is the following[142]. There are three types of CPs: maximum, minimum or saddle point. In 3D, one has two different types of saddle points.

CPs of the 3D function $\rho(r)$ can be classified in terms of the eigenvalues λ_i ($i = 1, 2$ and 3) of the Hessian of $\rho(r)$, which is defined as $\nabla^2 \rho(r)$ and is actually a 3×3 matrix evaluated at a given CP. Therefore, a given CP is classified by an (r,s) pattern, where r is the rank of this CP equal to the number of non-zero eigenvalues of the Hessian matrix and s is the signature equal to the sum of the signs of the eigenvalues. One example is worth discussing. One type of saddle point has two non-zero negative eigenvalues and one which is strictly positive, so its rank $r = 3$ and its signature $s = (-1) + (-1) + 1 = -1$, and therefore this CP is denoted as $(3, -1)$ CP. Such a CP is called a *bond critical point* because it indicates the existence of a bond between two nuclei of a given molecule. The bond critical points are linked to the adjacent nuclei via an *atomic interaction line*. This line in fact consists of a pair of gradient paths, each of which originates at the bond CP and terminates at a nucleus. The set of all atomic interaction lines occurring in a given molecule constitutes the *molecular graph*.

The AIM analysis of the electron density and the Laplacian of the electron density have been performed at the B3LYP/cc-pVDZ level using the MORPHY suite of codes[146]. The resulting AIM charges are given in Table 6. In Figures 6 and 7, we display the molecular graph $L(\mathbf{r})$ from different views of the one-electron density of the electronic ground-state phenol. Thus, the regions of local charge concentration correspond to the maxima in $L(\mathbf{r})$ and the regions of local charge depletion to minima in $L(\mathbf{r})$. Figure 6 shows the geometric positions of all the critical points in the valence shell charge concentration (VSCC) graph of phenol. The graph contains 87 CPs in total, 27 $(3, -3)$ CPs, 41 $(3, -1)$ CPs and 19 $(3, +1)$ CPs. The $(3, -3)$ CPs in $L(\mathbf{r})$ can be separated into three subsets: the two non-bonding maxima of oxygen; the bonding maxima between two carbons, oxygen and carbon, carbon and hydrogen and oxygen and hydrogen; the nuclear maxima, each virtually coincident with the hydrogen nucleus. The $(3, -1)$ CPs in general have a function which is analogous to a bond critical point, i.e. to link maxima. We trace the gradient paths in $L(\mathbf{r})$ starting from the $(3, -1)$ CPs. Usually, these would be expected to connect maxima and this is the case for the overwhelming majority of $(3, -1)$ CPs for phenol but, as may be seen occasionally in $\rho(\mathbf{r})^{142}$, we observe two $(3, -1)$ CPs connected in the vicinity of the oxygen atom. The presence of this unusual connectivity, generally only observed for 'conflict' structures, means that a planar graph cannot be drawn for the VSCC.

Figure 7 displays the geometric positions of all the CPs in the valence shell charge depletion (VSCD) graph of phenol. The graph contains 55 $(3, -1)$ CPs, 80 $(3, +1)$ CPs and 22 $(3, +3)$ CPs. The VSCD graph is considerably more complex than the VSCC one

TABLE 6. AIM charges of the ground-state phenol

	Charge	Dipole$_x$	Dipole$_y$	Dipole$_z$
C1	0.506	0.025	0.661	−0.000
C2	−0.014	0.086	0.014	0.000
C3	0.016	0.037	0.037	0.000
C4	0.000	−0.004	0.078	0.000
C5	0.014	−0.042	0.050	0.000
C6	0.004	−0.122	0.021	0.000
H8	−0.020	0.111	−0.074	0.000
H9	−0.002	0.114	0.065	0.000
H10	−0.005	−0.000	0.133	0.000
H11	−0.001	−0.115	0.064	0.000
H12	0.015	−0.117	−0.063	0.000
O7	−1.111	0.254	0.100	0.000
H13	0.600	0.158	−0.057	0.000

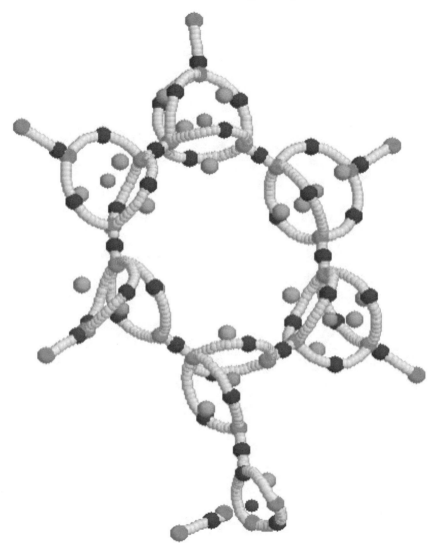

FIGURE 6 (PLATE 1). The VSCC graph for phenol. The oxygen atom is marked in red. The green spheres therein are the CPs $(3, -3)$ (maxima) in the phenolic $L(r)$ while the violet ones determine the $(3, -1)$ CPs. The yellow spheres correspond to the $(3, +1)$ CPs. The domain interaction lines (in light gray) link two $(3, -3)$ CPs via a $(3, -1)$ CP

and encompasses the whole molecule. In reality, of course, the separation of the VSCC and VSCD graphs is artificial; however, it allows for a much easier visual understanding of the significance of the two. The gradient paths belonging to the VSCC graph define the connectivities of the charge concentration maxima (*attractors*); the gradient paths belonging to the VSCD graph indicate the extensions of the *basins* of these attractors. Finally, the principal AIM properties of the atoms of phenol are collected in Table 7.

FIGURE 7 (PLATE 2). The VSCD graph for phenol. The oxygen atom is marked in red and the hydrogen in white. The brown spheres therein are the CPs $(3, +3)$ (minima) in the phenolic $L(\mathbf{r})$ while the purple ones determine the $(3, -1)$ CPs. The yellow spheres correspond to the $(3, +1)$ CPs. The $(3, +1)$ CPs link the $(3, +3)$ CPs via a pair of gradient paths shown in white, each of which is repelled by a $(3, +3)$ CP

D. Vibrational Modes

The phenol molecule has 13 atoms, and is therefore characterized by the 33 normal vibrational modes. Their overtone and combination bands are infrared active. The proper assignment of the fundamental vibrational modes of phenol in its electronic ground state

1. General and theoretical aspects of phenols

TABLE 7. The AIM properties of the ground-state phenol

Total volume	3727.90
Total molecular dipole moment	0.5124
Average L(Ω)[141,142]	−0.34E-03
Total K(Ω)[141,142]	304.76268
Total E(Ω)[141,142]	−307.49493
E(wave function)	−307.49478
Total charge	−49.999416285
Z + Q(total)[141,142]	0.000583715
Total dipole (in components)	−0.0001 0.3853 1.0577

has a long history that started in 1941 by assigning the observed Raman bands[147] of phenol confined to the region above 600 cm^{-1} followed by a study on the changes of its vibrational spectra under association[148]. The first examination of the phenol–OD infrared spectra was performed in 1954–1955[149,150]. In the electronic ground state S_o, the assignment of all fundamental vibrations of phenol was based on the earlier studies[151-153]. The lowest vibrational mode, a so-called mode 10b, had been assigned to 242 cm^{-1} in 1960[151] and to 241 cm^{-1} one year later[152] from the Raman spectra of molten phenol. In 1981, a slightly lower mode at 235 cm^{-1} was observed[154] by Raman spectroscopy in the gas phase. The frequency of the mode 10b in phenol and phenol-d1 were determined[155] at 225.2 and 211.5 cm^{-1}, respectively, and this led to the conclusion that the assignment of Reference 153 might be incorrect. Interestingly, during the last two decades, this mode and its correct value have not received much attention because the values predicted by a variety of *ab initio* methods appear to be lower than the experimental ones[151-155].

The vibrational modes of the ground-state phenol were examined by a number of spectroscopic techniques including UV-VIS[154,156-158], IR for the vapour[151,152,159,160], and the IR and Raman spectra in the solid and liquid phases[151,152,159,161,162], and microwave spectroscopy[105,107,163], see also References 164–166. They are collected in Table 8, where both nomenclatures by Wilson and coworkers[154] and Varsanýi[167] are used. Recently, the vibrational modes of phenol have become a benchmark for testing *ab initio* and density functional methods[111,124,168-170]. The Hartree–Fock calculations of the vibrational spectrum of phenol were first performed using the 6-31G(d,p) basis set[122]. An MP2 study with the same basis set was later carried out[121]. A combination[112] of three methods, viz. HF, MP2 and density functional BLYP, in conjunction with the 6-31G(d,p) basis was used to study the phenol spectrum and to make the complete and clear assignment of its vibrational modes (see Table 9).

In Figure 8 we display the normal displacements and in Table 10 we provide the corresponding vibrational assignments. Let us start from the end of Table 10 and Figure 8 where the stretching ν_{OH} mode is placed and its normal displacement is shown. It is a pure localized mode[111,112]. Furthermore, it is a well-known mode subject to numerous studies related to the hydrogen-bonding abilities of phenol[173]. Its second overtone in phenol and the phenol halogen derivatives has been studied experimentally[174].

The OH group of phenol participates in two additional modes, in-plane and out-of-plane bending vibrations. The latter is also called the torsional mode τ_{OH} observed near 300 cm^{-1} (see Table 8) in the IR spectra of phenol vapour and of dilute solutions of phenol in n-hexane[152]. In the associated molecules, it appears as a rather broad featureless band in the region of 600–740 cm^{-1} [149]. It results from the hydrogen-bonded association. The spectra of liquid and solid phenol–OD also exhibit a variety of broad bands near 500 cm^{-1}. The first overtone of the τ_{OH} was found at 583 cm^{-1} in the IR spectrum of phenol vapour[152]. This assignment of the torsional mode allows one to model the torsional motion of the OH group of phenol by assuming that it is described by the

TABLE 8. Experimental (infrared and Raman) and theoretical vibrational spectra of phenol

Nomenclature			Expt.							
Wilson and coworkers[154]	Vars ányi[167]	Sym	IR[153] Raman[154]		IR[171,172] Raman[167]	IR[111] Raman[111]	IR[169]	HF/6-31G(d,p)[112]		MP2/6-31G(d,p)[112]	
			ν	A	ν	ν	ν	ν	A	ν	A
11	10b	a"	244		241	242	225[a]	256.2	5	226.8	0.8
τ(OH)		a"	309	47	300	310		314.2	141	327.5	126
16a		a"	409	0.0	410	410	404[b]	461.2	0.3	403.3	0.9
18b	15	a"	403	5	408	420		440.6	11	404.9	10
16b		a'	503	26	500	503	504[b]	568.2	7	522.0	3
6a		a'	527	5	526	526	526	574.3	2	535.1	1
6b		a'	619		617	618	618	678.6	0.3	632.6	0.3
4		a'	686	50	688	687	686[b]	767.8	13	464.8	4
10b	11	a"	751	52	749	752		846.7	83	736.0	78
10a		a"	817	0.0	825	823		924.2	0.0	814.3	0.2
12	1	a"	823	20	810	810	820	892.0	20	837.1	18
17b		a"	881	12	881	881		996.0	13	849.7	0.7
17a		a"	973	1.0	958	956		1090.1	0.0	904.8	0.5
5		a"	995	5	978	973		1111.4	0.6	913.4	0.1
1	12	a'	1000		999	999	999	1085.2	3	1024.8	0.3
18a		a'	1025	8	1026	1026	1026	1122.8	4	1064.5	5
15	18b	a'	1072	10	1071	1070		1176.9	10	1117.0	12
9b		a'	1151	38	1145	1150		1197.0	33	1205.6	10
9a		a'	1169	70	1167	1176		1282.1	0.4	1218.1	0.5
β(COH)		a'	1177	80	1207	1197	1174	1291.0	85	1221.0	157
7a	13	a'	1261	62	1259	1261	1261	1404.4	114	1320.3	66
3		a'	1277	0	1313	1361		1488.6	36	1388.2	22
14		a'	1343	31	1354	1344	1349	1370.5	53	1478.5	12
19b		a'	1472	23	1465	1472		1635.2	29	1531.9	22
19a		a'	1501	54	1497	1501	1505	1671.2	76	1567.2	54
8b		a'	1610		1596	1604		1797.8	45	1681.2	27
8a		a'	1603	70	1604	1609		1810.6	66	1695.6	39
13		a'	3027		3030	3021		3326.4	16	3241.9	12
7b		a'	3049		3044	3046		3343.7	0.2	3261.4	0.1
2		a'	3063		3048	3052		3354.1	29	3269.6	16
20b		a'	3070		3076	3061		3370.9	28	3284.3	15
20a		a'	3087		3091	3074		3379.6	7	3290.6	5
ν(OH)		a'	3656	50	3623	3655		4197.2	84	3881.8	53

[a]Determined from the first and third overtone and the combination band with the mode 1a.
[b]Calculated from the first overtone of these normal modes.
[c]Present work (see page 37).

potential $V_\tau(1 - \cos 2\theta)/2$[152]. Here, θ is the torsional angle and V_τ is the corresponding barrier height. Within this model, the reduced moment of inertia can be chosen equal to 1.19×10^{-40} g cm^2.

The β_{COH} is the in-plane bending of the OH group placed at around 1175–1207 cm^{-1}. It is observed at 1176.5 cm^{-1} in the IR spectrum of phenol vapour[152–154]. This band is shifted to ca 910 cm^{-1} in dilute solution under deuteriation[153] and gives rise to a broad absorption ranging from 930 to 980 cm^{-1} in the spectrum of crystal. The first HF/6-31G(d,p) calculations[122] predicted it at 1197.3 cm^{-1} (the scaled value is 1081 cm^{-1} [175]).

Twenty-four vibrational modes of phenol are well assigned to the phenyl ring modes because they are not so sensitive to the nature of the substituent[176]. On the other hand, the six modes which involve a substantial motion of the phenyl and CO groups are rather sensitive to the isotopic substitution of OH by OD. These are the following modes[152]: 1260 (1253), 814 (808), 527 (523), 503 (503), 398 (380) and 242 (241) cm^{-1} for phenol and phenol–OD (in parentheses), respectively.

TABLE 8. (continued)

	Theor.						Expt.		
BLYP/6-31G(d,p)[112]		B3LYP/6-31G(d,p)[124]	B3LYP/6-311++G(d,p)[124]	B3LYP/6-311++G(2df,2p)^c		Expt-Phenol-d_1 IR[153]		Phenol-d_5 IR[153]	
ν	A			ν	A	ν	A	ν	A
225.4	0.2	234	227	229.5	1	232	2		
384.6	108	365	338	338.4	100	246	5	307	30
406.6	0.9	405	403	406.4	10				
394.0	9	421	414	421.7	1	382	6	386	3
499.2	6	518	508	514.9	14	503	15	431	20
520.7	1	536	537	537.5	2	522	3	513	2
616.1	0.3	633	633	634.3	0	617			
674.8	8	699	667	690.7	18	687	35	550	25
732.2	44	761	745	761.6	69	751	50	625	10
788.9	0.0	834	828	827.3	0				
805.0	17	822	816	830.5	21	805	8	754	2
846.2	3	884	834	894.0	6	881			
911.1	0.0	955	948	970.1	0				
939.5	0.1	981	969	986.5	0	997	4	960	
982.7	2	1013	1012	1017.6	3				
1017.4	3	1051	1043	1045.0	5	1025	2		
1069.8	10	1102	1094	1095.0	14				
1156.6	8	1183	1177	1177.8	36	1150	5		
1165.7	5	1197	1191	1191.9	90	1168	20		
1173.3	146	1200	1192	1193.4	32	917	44	1179	802
1254.0	64	1305	1275	1280.4	89	1257	90	1187	75
1329.7	7	1365	1349	1347.5	7			1021	15
1349.3	29	1378	1369	1375.2	23	1309	4	1300	12
1468.8	28	1514	1500	1505.3	23	1465		1372	40
1495.5	34	1547	1528	1533.2	53	1500	65	1405	40
1589.5	37	1654	1636	1636.0	48	1609		1578	
1602	32	1668	1646	1646.7	38	1603	60	1572	40
3079	19	3163	3152	3152.0	13	3024		2262	
3100.2	0.2	3183	3069	3170.1	0	3051		2283	
3107.9	27	3191	3178	3178.5	16	3060		2295	
3123.9	27	3207	3192	3192.1	15	3073		2302	
3131.4	7	3214	9198	3198.5	3	3087		2313	
3664.2	25	3827	3839	3835.2	62	2699		2700	35

Early work on the near-IR spectra of phenol has been focused on the study of the influence of the solvent or hydrogen-bond formation on the frequency of the first overtone of the ν_{OH} stretching vibration[177–179]. The frequency of the ν_{OH} vibration for the vibrational quantum numbers $v = 0$ to $v = 5$ has been reported, based on the photoacoustic spectroscopic measurements[180]. Recently, the near-IR spectrum between 4000 and 7000 cm^{-1} of phenol in solution has been investigated by conventional FT-IR spectroscopy[181]. Vibrational transitions in this range have also been detected by non-resonant two-photon ionization spectroscopy[182] and some of the transitions have been assigned to combinations involving mainly the ν_{OH} vibration and other fundamental modes of phenol. The interesting problem in this area is to resolve the origin of the cluster of peaks around 6000 cm^{-1} which were observed in solution and assigned to the first overtone of the ν_{CH} vibrations of phenol–OH because their fundamental vibrations are placed at 3000 cm^{-1} [181, 182] (Figure 9). The ν_{CH} absorptions of phenol–OH and phenol–OD and their first and second overtones are studied by a deconvolution procedure and the near-IR spectra are

TABLE 9. Theoretical assignments of the vibrational modes of phenol[112]. Potential energy distribution (PED) elements are given in parentheses, frequencies in cm^{-1}, IR intensities in km mol^{-1} [a]

Q1	11	a''	τ_3 ring(52) + τ_2 ring(18) + γCO(17) + τ_l ring(10)
Q2	OH torsion	a''	τ(O–H)(100)
Q3	18b	a'	δCO(81)
Q4	16a	a''	τ_2 ring (76) + τ_3 ring (24)
Q5	16b	a''	γCO(46) + τ_3 ring (30) + τ_l ring (13)
Q6	6a	a'	δ_2 ring def.(77) + ν(C–O) (12)
Q7	6b	a'	δ_3 ring def.(83)
Q8	4	a''	τ_l ring (90)
Q9	10b	a''	γC$_4$H(31) + γCO(23) + γC$_3$H(15) + γC$_2$H(12) + γC$_5$H(11)
Q10	10a	a''	γC$_2$H(53) + γC$_6$H(22) + γC$_5$H(17)
Q11	12	a'	ν(C–O)(25) + δ_1 ring def.(19) + ν(C$_1$–C$_2$)(17) + ν(C$_1$–C$_6$)(17) + δ_2 ring def.(14)
Q12	17b	a''	γC$_6$H(42) + γC$_4$H(26) + γC$_2$H(21) + γC$_3$H(17)
Q13	17a	a''	γC$_3$H(52) + γC$_5$H(22) + γC$_6$H(17) + γC$_2$H(12)
Q14	5	a''	γC$_5$H(44) + γC$_4$H(22) + τ_l ring(13) + γC$_6$H(12) + γC$_3$H(10)
Q15	1	a'	δ_1 ring def.(65) + ν(C$_1$–C$_6$)(10)
Q16	18a	a'	ν(C$_5$–C$_4$)(32) + ν(C$_4$–C$_3$)(26) + δCH(25)
Q17	15	a'	ν(C$_3$–C$_2$)(22) + ν(C$_6$–C$_5$)(19) + δC$_6$H(13) + ν(C$_4$–C$_3$)(11) + δC$_4$H(11) + δC$_2$H(10)
Q18	9b	a'	δC$_4$H(36) + δC$_5$H(23) + δC$_6$H(12) + δC$_3$H(11)
Q19	9a	a'	δC$_3$H(27) + δC$_2$H(26) + δC$_6$H(14) + δC$_5$H(10)
Q20	OH bend	a'	δOH(55) + ν(C$_1$–C$_6$)(13) + δC$_6$H(10)
Q21	7a	a'	ν(C–O)(52) + ν(C–C)(20)
Q22	3	a'	δC$_2$H(18) + δC$_6$H(18) + δC$_5$H(18) + δC$_3$H(14) + δC$_4$H(12)
Q23	14	a'	ν(C–C)(56) + δOH(21) + δC$_5$H(22)
Q24	19b	a'	δC$_4$H(25) + δC$_3$H(13) + ν(C$_6$–C$_5$)(13) + ν(C$_3$–C$_2$)(13) + δC$_6$H(10)
Q25	19a	a'	δC$_5$H(19) + δC$_2$H(16) + ν(C$_4$–C$_3$)(13) + δC$_3$H(12)
Q26	8b	a'	ν(C$_2$–C$_1$)(25) + ν(C$_5$–C$_4$)(22)
Q27	8a	a'	ν(C$_1$–C$_6$)(21) + ν(C$_6$–C$_5$)(17) + ν(C$_3$–C$_2$)(16) + ν(C$_4$–C$_3$)(14)
Q28	13	a'	ν(C$_2$–H)(90) + ν(C$_3$–H)(10)
Q29	7b	a'	ν(C$_5$–H)(52) + ν(C$_4$–H)(26) + ν(C$_3$–H)(13)
Q30	2	a'	ν(C$_3$–H)(58) + ν(C$_5$–H)(28)
Q31	20b	a'	ν(C$_4$–H)(50) + ν(C$_6$–H)(33) + ν(C$_3$–H)(17)
Q32	20a	a'	ν(C$_6$–H)(61) + ν(C$_4$–H)(19) + ν(C$_5$–H)(18)
Q33	OH stretch	a'	ν(O–H)(100)

[a] See footnote of Table 10.

reassigned[183]. At a concentration of 0.1 M, dimers of phenol and its higher associates might be present in solution. In the fundamental region, there appears a weak band at 3485 cm^{-1} in phenol–OH and at 2584 cm^{-1} in phenol–OD which originates from the dimer[111]. Weaker and broader bands around 3300 and 2500 cm^{-1} are assigned to higher associates of phenol. In the near-IR spectrum, a very weak absorption band at 6714 cm^{-1} refers to the dimer.

E. Three Interesting Structures Related to Phenol

Before ending the present section, we would like to briefly discuss the following three structures closely linked to the S_0-state phenol molecule.

It is well known that aliphatic carbonyl compounds with the hydrogens on C$_\alpha$ to the carbonyl group may undergo tautomeric transitions from the keto to the enol forms. The most stable tautomeric form of the S_0-state phenol molecule is in fact the enol form[184–186]. The reason why the enol form of phenol is favoured over the keto form is quite simple[131].

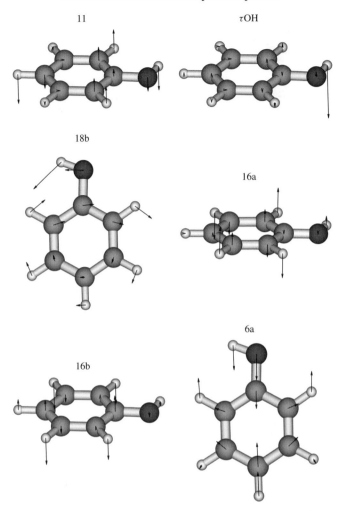

FIGURE 8. The normal displacements of the vibrational modes of phenol according to the Wilson's nomenclature. The B3LYP/6-31+G(d,p) method is employed. The assignments of the vibrational modes of phenol are presented in Table 10

On the one hand, due to the virtual absence of the electronic delocalization in the keto form, it has a larger intrinsic stability which can easily be accounted for in terms of the sum of the bond energies (*ca* 59 kJ mol^{-1}). On the other hand, the enol form is characterized by a larger resonance energy, by *ca* 126 kJ mol^{-1}, compared to that of the keto form. Therefore, the enol form is more stable by *ca* 67 kJ mol^{-1}. Such simple arguments are pretty well confirmed by the B3LYP/6-31+G(d,p) calculations performed in the present work (cf. also Reference 186) resulting in that the enol–keto tautomeric energy difference amounts to 69 kJ mol^{-1} after ZPVE. In Figure 10 we display the most stable keto form of phenol (cyclohexa-2,5-dienone) together with its most characteristic

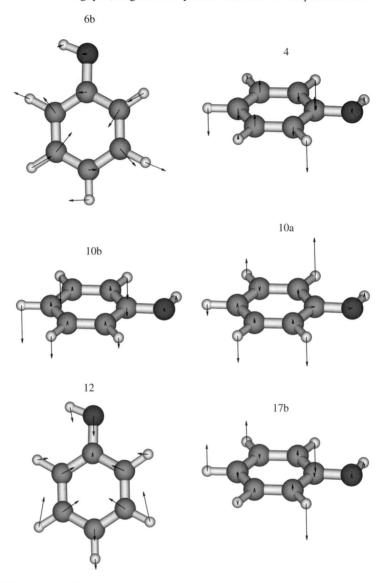

FIGURE 8. (*continued*)

vibrational modes. Interestingly, the keto form possesses a total dipole moment of 5.0 D and thus it is more polar than the favourable enol form. The standard heats of formation of both cyclohexa-2,4- and -2,5-dienones have recently been re-evaluated as −31 and −34 kJ mol^{-1}, respectively, in better agreement with theoretical estimates[187].

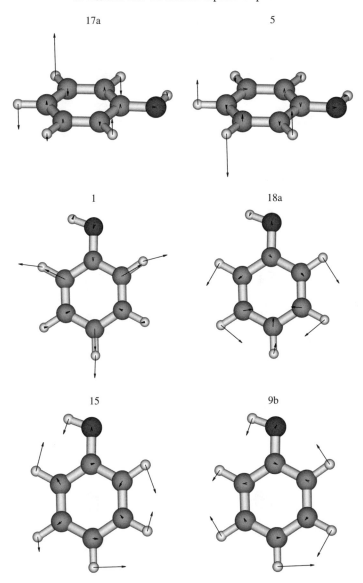

FIGURE 8. (*continued*)

In Figure 11 we display two other theoretical structures. The TS$_\tau$ structure is the transition state governing the torsional motion of the OH group of phenol between its equi-energetical structures shown in Chart 4. The energy difference between this structure and the S_0-state phenol molecule determines the torsional barrier V_τ as equal to 13 kJ mol^{-1}

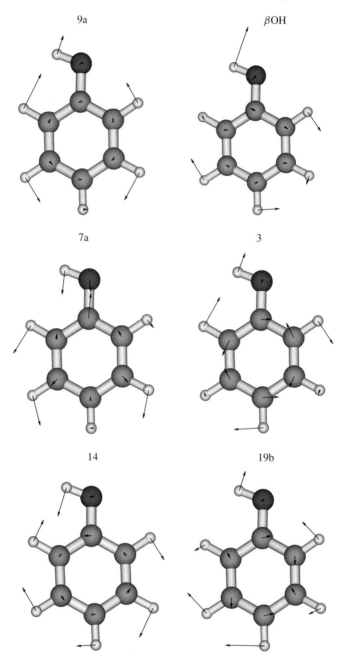

FIGURE 8. (*continued*)

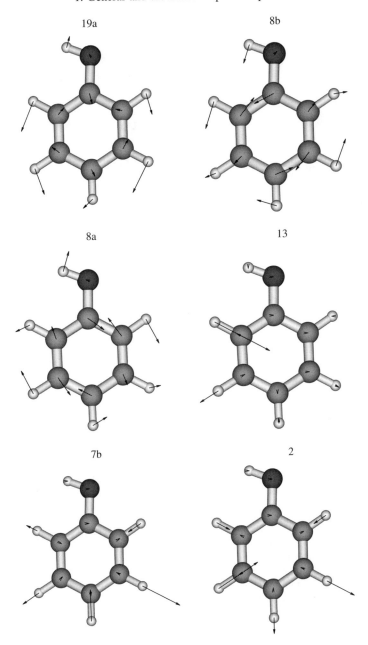

FIGURE 8. (*continued*)

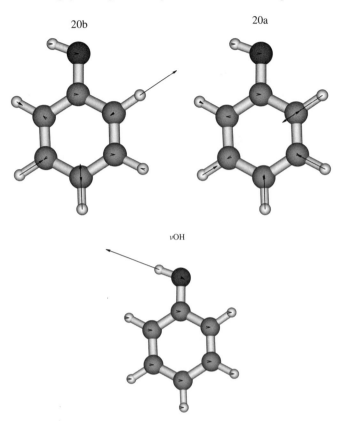

FIGURE 8. (*continued*)

after ZPVE at the B3LYP/6-311++G(d,p) computational level. The MP2/cc-pVTZ calculation recently performed yields 15 kJ mol^{-1} [120]. Note that the imaginary frequency characterizing this saddle point is predicted at 343 i cm^{-1}.

The second structure shown in Figure 11 is the saddle point of second order lying 113 kJ mol^{-1} above the phenol molecule at the B3LYP/6-31+G(d,p) level taking ZPVE into account. As a second-order saddle structure, it has two imaginary frequencies, 1222 i and 1150 i cm^{-1}. The former describes the in-plane hindered rotation of the OH group whereas in the latter its rotation is perpendicular to the phenyl ring. We suppose that both these structures are directly linked to the gas-phase bond dissociation enthalpy (BDE) of phenol defined (see, e.g., Reference 188 and references therein) as the enthalpy change for the reaction

$$C_6H_5O—H \longrightarrow C_6H_5O^{\bullet} + H^{\bullet} \qquad (4)$$

where the bond indicated by the horizontal line breaks, yielding the radicals as the products. The experimental and theoretical determination of the BDE of phenol and phenol

TABLE 10. Harmonic vibrational frequencies, IR intensities and assignments for phenol[a]

No.	Freq.	IR	Sym.	Assignment, PED(%)
1	227.8	1	A''	$\tau_2\text{rg}(65)$, $\tau_1\text{rg}(12)$
2	311.4	111	A''	$\tau\text{OH}(93)$ Expt: 310[b], 310[c]
3	405.1	11	A'	$\beta\text{CO}(77)$, $\beta_3\text{rg}(11)$
4	416.8	1	A''	$\tau_3\text{rg}(83)$
5	508.8	14	A''	$\tau_2\text{rg}(38)$, $\gamma\text{CO}(37)$, $\gamma\text{C}_4\text{H}(11)$
6	536.0	2	A'	$\beta_2\text{rg}(77)$
7	632.6	0	A'	$\beta_3\text{tg}(85)$
8	668.5	10	A''	$\tau_1\text{rg}(69)$, $\gamma\text{CO}(12)$
9	749.5	85	A''	$\gamma\text{C}_4\text{H}(27)$, $\gamma\text{C}_2\text{H}(16)$, $\gamma\text{CO}(15)$, $\tau_1\text{rg}(12)$, $\gamma\text{C}_3\text{H}(11)$
10	819.5	0	A''	$\gamma\text{C}_2\text{H}(44)$, $\gamma\text{C}_6\text{H}(25)$, $\gamma\text{C}_5\text{H}(20)$
11	827.4	23	A'	$\nu\text{CO}(26)$, $\beta_2\text{rg}(20)$, $\beta_1\text{rg}(16)$
12	878.1	5	A''	$\gamma\text{C}_6\text{H}(33)$, $\gamma\text{C}_2\text{H}(24)$, $\gamma\text{C}_4\text{H}(17)$
13	951.9	0	A''	$\gamma\text{C}_3\text{H}(53)$, $\gamma\text{C}_4\text{H}(18)$, $\gamma\text{C}_6\text{H}(10)$
14	971.8	0	A''	$\gamma\text{C}_5\text{H}(54)$, $\gamma\text{C}_4\text{H}(18)$, $\gamma\text{C}_6\text{H}(16)$
15	1012.6	2	A'	$\beta_1\text{rg}(65)$
16	1043.1	6	A'	$\nu\text{C}_4\text{C}_5(31)$, $\nu\text{C}_3\text{C}_4(24)$
17	1093.0	15	A'	$\nu\text{C}_2\text{C}_3(18)$, $\nu\text{C}_5\text{C}_6(15)$, $\beta\text{C}_6\text{H}(12)$, $\beta\text{C}_4\text{H}(11)$, $\nu\text{C}_3\text{C}_4(11)$, $\beta\text{C}_2\text{H}(10)$
18	1176.5	26	A'	$\beta\text{C}_4\text{H}(26)$, $\beta\text{C}_5\text{H}(16)$, $\beta\text{C}_6\text{H}(11)$, $\beta\text{C}_3\text{H}(10)$
19	1190.5	2	A'	$\beta\text{C}_2\text{H}(26)$, $\beta\text{C}_5\text{H}(17)$, $\beta\text{C}_3\text{H}(15)$, $\beta\text{C}_6\text{H}(11)$
20	1192.0	128	A'	$\beta\text{OH}(41)$, $\nu\text{C}_1\text{C}_6(13)$, $\beta\text{C}_3\text{H}(12)$, $\beta\text{C}_4\text{H}(10)$
21	1274.9	91	A'	$\nu\text{CO}(52)$, $\beta_1\text{rg}(12)$
22	1348.4	6	A'	$\nu\text{C}_2\text{C}_3(14)$, $\nu\text{C}_5\text{C}_6(14)$, $\nu\text{C}_3\text{C}_4(14)$, $\nu\text{C}_4\text{C}_5(13)$, $\nu\text{C}_1\text{C}_2(11)$, $\nu\text{C}_1\text{C}_6(10)$
23	1368.3	28	A'	$\beta\text{C}_5\text{H}(22)$, $\beta\text{OH}(18)$, $\beta\text{C}_3\text{H}(16)$, $\beta\text{C}_6\text{H}(13)$, $\beta\text{C}_2\text{H}(10)$
24	1499.4	23	A'	$\beta\text{C}_4\text{H}(26)$, $\beta\text{C}_3\text{H}(13)$, $\nu\text{C}_2\text{C}_3(13)$, $\nu\text{C}_5\text{C}_6(12)$, $\beta\text{C}_6\text{H}(10)$
25	1526.8	59	A'	$\beta\text{C}_5\text{H}(19)$, $\beta\text{C}_2\text{H}(16)$, $\nu\text{C}_3\text{C}_4(12)$, $\beta\text{C}_3\text{H}(12)$
26	1635.4	49	A'	$\nu\text{C}_1\text{C}_2(24)$, $\nu\text{C}_4\text{C}_5(21)$, $\nu\text{C}_5\text{C}_6(10)$
27	1645.9	39	A'	$\nu\text{C}_1\text{C}_6(23)$, $\nu\text{C}_3\text{C}_4(17)$, $\nu\text{C}_2\text{C}_3(13)$, $\nu\text{C}_5\text{C}_6(13)$
28	3149.0	14	A'	$\nu\text{C}_2\text{H}(88)$, $\nu\text{C}_3\text{H}(10)$
29	3167.4	0	A'	$\nu\text{C}_5\text{H}(51)$, $\nu\text{C}_4\text{H}(27)$, $\nu\text{C}_3\text{H}(11)$
30	3176.0	17	A'	$\nu\text{C}_3\text{H}(56)$, $\nu\text{C}_5\text{H}(26)$
31	3190.0	16	A'	$\nu\text{C}_4\text{H}(41)$, $\nu\text{C}_6\text{H}(41)$, $\nu\text{C}_3\text{H}(16)$
32	3196.6	4	A'	$\nu\text{C}_6\text{H}(47)$, $\nu\text{C}_4\text{H}(24)$, $\nu\text{C}_5\text{H}(21)$
33	3836.0	62	A'	$\nu\text{OH}(100)$

[a]Present calculations performed at B3LYP/6-311++G(d,p) computational level. Values taken from Reference 244 with permission. Frequencies in cm^{-1}, IR intensities in km mol^{-1}. Glossary of vibrational mode acronyms: ν, stretch; β, in-plane bend; γ, out-of-plane bend; τ, torsion; rg, ring; β_1, β_2 and β_3, ring deformations and τ_1, τ_2 and τ_3, ring torsions. PED elements $\geq 10\%$ only are included.
[b]The gas-phase IR experiment[171].
[c]The IR experiment in solution[172].

derivatives has been a matter of enormous interest[125, 140, 189–196]. The BDE of phenol is rather low and is estimated experimentally at 356.9 kJ mol^{-1} (NIST Standard Reference Database), 365.3 ± 6.3 kJ mol^{-1} [191], and 371.3 ± 2.3 kJ mol^{-1} [194] while the accurate theoretical estimations fell within 363.2 kJ mol^{-1} (DFT) and 364.4 kJ mol^{-1} [140]. Note finally that the BDE of phenol gives the reference value for all phenolic antioxidants[140, 197–201]. This property and the relevant reaction will be discussed in a subsequent section.

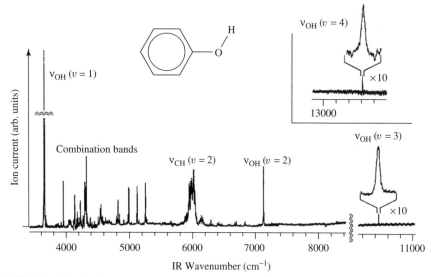

FIGURE 9. Vibrational spectrum of jet-cooled phenol measured by the non-resonant ionization-detected IR spectroscopy[182] fixing ν_{UV} to 34483 cm^{-1}. All peaks are attributed to the vibrational transitions of the phenol molecule in its ground electronic state S_0. The strongest peak at 3656 cm^{-1} is assigned to the fundamental of the ν_{OH} stretch. The cluster of peaks around 6000 cm^{-1} is assigned to the first overtone of the ν_{CH} modes. The sharp peaks at 7143, 10461 and 13612 cm^{-1} are assigned to the first ($2\nu_{OH}$), second ($3\nu_{OH}$) and third ($4\nu_{OH}$) overtones of the ν_{OH} stretch, respectively. Reproduced with permission from Reference 182

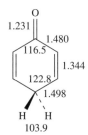

FIGURE 10. The keto tautomeric form of phenol viewed at the B3LYP/6-31+G(d,p) computational level. Bond lengths in Å, bond angles in degrees

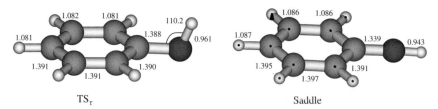

FIGURE 11. Calculated transition structure TS$_\tau$ (B3LYP/6-311++G(d,p)) and the second-order saddle structure (B3LYP/6-31+G(d,p)). Bond lengths are given in Å, bond angles in degrees

1. General and theoretical aspects of phenols

III. STRUCTURES AND PROPERTIES OF SUBSTITUTED PHENOLS

During the 160 years since the discovery of phenol, thousands of studies were conducted on halophenols, partly due to their significance in the theory of hydrogen bonding; indeed their hydrogen bonding abilities can be varied nearly continuously over a wide pK_a range from 10.2 to 0.4[202-211].

A. Intramolecular Hydrogen Bond in *ortho*-Halogenophenols

One of the most remarkable moments in the history of mono-halogen-substituted phenols occurred in 1936 when Pauling[104, 212] suggested the co-existence of two inequivalent rotational isomers (rotamers or conformers) of the *ortho*-Cl-substituted phenol in order to explain the experimental splitting of the first overtone of its ν_{OH} vibrational mode observed in the CCl$_4$ solution[213-217]. Instead of phenol whose first overtone $\nu_{OH}^{(1)}$ is sharply peaked at 7050 cm^{-1}, o-ClC$_6$H$_4$OH reveals a doublet at 7050 and 6910 cm^{-1} resulting in a band splitting $\Delta\nu_{OH}^{(1)} = 140$ cm^{-1} and having the former band placed at the same wavenumbers as in phenol. Almost two decades later, a splitting of the fundamental ν_{OH} mode by 83 cm^{-1} was observed in CCl$_4$ solvent[218]. What then lies behind Pauling's suggestion?

Let us consider Figure 12, which displays two conformers *cis* and *trans* of o-ClC$_6$H$_4$OH (computational details are given elsewhere[219, 220]). The former possesses the intramolecular hydrogen bond O–H $\cdots$ Cl whereas the latter does not. This makes (as long believed) the *cis* conformer energetically favoured, with a gain of energy $\Delta_{cis-trans}E_{ortho}^{Cl} = 12.5$ kJ mol^{-1}. Pauling's estimation of the corresponding free energy difference derived from the ratio of the areas of the peaks was 5.8 kJ mol^{-1}[104] in CCl$_4$ solution (a more precise value is 6.1 kJ mol^{-1}[170]; another value is 7.5 kJ mol^{-1}[221]). Our calculated energy difference agrees fairly well with the free energy difference of 14.2–16.3 kJ mol^{-1} in the vapour[222] bounded by 16.3 ± 3.0 kJ mol^{-1}[223] and 14.3 ± 0.6 kJ mol^{-1}[224]. However, there is yet another feature that distinguishes *cis* and *trans* conformers from each other: the *trans* form is more polar (3.0 vs 1.04 D). The directions of the total dipole moments of the *cis* and *trans* conformers are shown in Figure 12. Nevertheless, the gross difference between the *cis* and *trans* conformers consists, as mentioned, in the presence of the intramolecular hydrogen bond. Hence, $\Delta_{cis-trans}E_{ortho}^{Cl}$ can be interpreted as the energy of its formation. Indeed, it looks rather weak for *cis* o-ClC$_6$H$_4$OH.

Inspection of Table 11, which gathers the harmonic vibrational modes of both conformers with the corresponding potential energy distribution patterns, reveals that the *trans* ν_{OH} is calculated at 3835.4 cm^{-1}, which is almost identical to ν_{OH} of phenol in Table 10, while its *cis* partner is red-shifted (as expected according to the theory of hydrogen bonding[225, 226]) by $\Delta_{cis-trans}\nu_{OH}^{Cl} = 69$ cm^{-1}. This calculated value lies rather close to the experimental red shifts ranging from 58[227] to 60[228] and 63 cm^{-1}[222, 229], depending on the solvent. On the other hand, we note that our red shift is smaller, by 91 cm^{-1}, compared to that observed by Wulf and coworkers[217] for $\nu_{OH}^{(1)}$ that might be attributed to anharmonic effects[230]. After all, it is worth mentioning another indication of the rather weak intramolecular hydrogen bond in *cis* o-ClC$_6$H$_4$OH, namely the value of the corresponding hydrogen bridge stretching vibration ν_σ (mode 2 in Table 11) compared to mode 2 in its *trans* partner.

In this regard, the more than two decades following the appearance of Pauling's work[104] deserve to be recalled. Indeed, on the one hand, they were full of criticism[227] of the earlier experimental results[213-217] because it was believed that the higher frequency band appears 'more likely due to a trace of phenol impurity than to the presence of *trans* isomer'[218] and the new experiment demonstrated the ratio of the absorptions being much smaller and equal to $1/56 \approx 17.9 \times 10^{-3}$, which anyway is about three times larger than

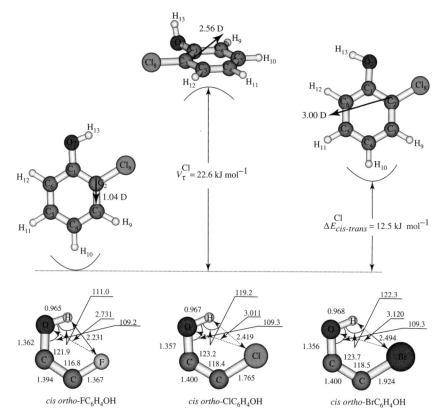

FIGURE 12. The portion of the potential energy surface of o-ClC$_6$H$_4$OH governing the *cis–trans* conversion is displayed at the top. Numbering of atoms follows Chart 1. Five-member sub-ring sections of the *cis ortho*-halogenophenols with the intramolecular hydrogen bond are shown at the bottom. Bond lengths are in Å, bond angles in degrees. Adapted from Reference 220 with permission

our theoretical magnitude. On the other hand, these years were also characterized by a further development of the Pauling model[231, 232] and its further experimental support[217] although, alas, the 'unsatisfactory state of affairs' in the area of the *cis–trans* doublet paradigm[217] remained at that time. Paradoxically, it still remains nowadays, even widening the gap between the experiments originating at the end of the 1950s and modern high-level theoretical studies[220]. This particularly concerns o-fluorophenol.

In 1958, it was verified experimentally[227] that the *cis–trans* doublet could not be detected for o-FC$_6$H$_4$OH: the *trans* ν_{OH} band was suggested to be too weak to show up in IR experiments and $\Delta_{cis-trans}\nu_{OH}^{F}$ to be too small (<20 cm^{-1}; it is estimated at 18 cm^{-1} [222]). Our prediction is $\Delta_{cis-trans}E_{ortho}^{F} = 11.4$ kJ mol^{-1}, which demonstrates that indeed the intramolecular O−H···F hydrogen bond in o-FC$_6$H$_4$OH is weaker (by 1.09 kJ mol^{-1}) than its analogue in o-ClC$_6$H$_4$OH. Furthermore, as follows from Table 12, the theoretical splitting $\Delta_{cis-trans}\nu_{OH}^{F} = 30$ cm^{-1} is larger than 20 cm^{-1}, as predicted by IR experiments. We also note that the dipole moment of *trans* o-FC$_6$H$_4$OH (2.95 D) exceeds that of the *cis* form (1.0 D) by almost a factor of three.

TABLE 11. Harmonic vibrational frequencies, IR intensities and assignments for *cis* and *trans ortho*-chlorophenols[a]

No.	Freq.	IR	Sym.	Assignment, PED(%)	No.	Freq.	IR	Sym.	Assignment, PED(%)
1	155.6	0	A''	$\tau_2 rg(49)$, $\tau_3 rg(19)$, $\gamma CCl(13)$, $\tau_1 rg(11)$	1	152.6	1	A''	$\tau_2 rg(50)$, $\tau_3 rg(19)$, $\tau_1 rg(13)$, $\gamma CCl(10)$
2	249.3	3	A'	$\beta CCl(75)$, $\beta CO(15)$	2	239.7	2	A'	$\beta CCl(72)$, $\beta CO(17)$
3	262.8	1	A''	$\gamma CCl(31)$, $\tau_2 rg(29)$, $\gamma C_6 H(11)$, $\gamma CO(10)$	3	260.3	11	A''	$\tau_2 rg(27)$, $\gamma CCl(27)$, $\tau_3 rg(13)$, $\gamma CO(10)$
4	375.8	3	A'	$\nu CCl(27)$, $\beta_3 rg(19)$, $\beta CO(17)$, $\beta_2 rg(16)$	4	317.7	96	A'	$\tau OH(88)$
5	407.2	108	A''	$\tau OH(90)$		Expt: 373[b], 361[c]			
	Expt: 407[b], 396[c], 396[d]				5	380.8	3	A'	$\nu CCl(26)$, $\beta_3 rg(20)$, $\beta CO(17)$, $\beta_2 rg(15)$, $\beta CCl(10)$
6	447.2	0	A''	$\tau_3 rg(60)$, $\gamma CCl(24)$	6	445.4	7	A''	$\tau_3 rg(58)$, $\gamma CCl(22)$
7	493.6	12	A'	$\beta CO(57)$, $\nu CCl(16)$, $\beta CCl(11)$	7	492.5	8	A'	$\beta CO(55)$, $\nu CCl(12)$, $\beta CCl(11)$
8	542.5	1	A''	$\tau_2 rg(39)$, $\gamma CO(29)$	8	548.5	1	A''	$\tau_2 rg(35)$, $\tau_1 rg(20)$, $\gamma CO(18)$
9	563.9	5	A'	$\beta_2 rg(61)$, $\beta_3 rg(10)$	9	566.0	4	A'	$\beta_2 rg(57)$, $\beta_3 rg(12)$
10	672.6	0	A''	$\tau_1 rg(63)$, $\gamma CO(17)$, $\gamma CCl(10)$	10	690.8	19	A'	$\beta_3 rg(54)$, $\nu CCl(25)$, $\nu ClC2(11)$
11	685.1	25	A'	$\beta_3 rg(56)$, $\nu CCl(24)$, $\nu ClC2(10)$	11	702.6	0	A''	$\tau_1 rg(58)$, $\gamma CO(19)$, $\gamma CCl(13)$
12	758.6	74	A'	$\gamma C_4 H(39)$, $\gamma C_5 H(24)$, $\gamma C_3 H(20)$, $\gamma C_6 H(12)$	12	751.6	82	A''	$\gamma C_4 H(32)$, $\gamma C_5 H(23)$, $\gamma C_6 H(22)$, $\gamma C_3 H(11)$
13	842.7	13	A'	$\beta_1 rg(27)$, $\nu CO(24)$, $\beta_2 rg(19)$, $\nu C_1 C_6(10)$	13	835.1	0	A''	$\gamma C_6 H(47)$, $\gamma CCL(22)$, $\gamma C_4 H(12)$, $\tau_1 rg(10)$
14	850.2	1	A''	$\gamma C_6 H(45)$, $\gamma C_3 H(28)$	14	841.4	21	A'	$\beta_1 rg(28)$, $\nu CO(24)$, $\beta_2 rg(19)$
15	942.2	3	A''	$\gamma C_3 H(41)$, $\gamma C_4 H(22)$, $\gamma C_6 H(20)$	15	929.4	2	A''	$\gamma C_5 H(38)$, $\gamma C_3 H(35)$, $\gamma C_6 H(17)$
16	974.2	0	A''	$\gamma C_5 H(49)$, $\gamma C_4 H(24)$, $\gamma C_6 H(13)$	16	962.6	0	A''	$\gamma C_4 H(42)$, $\gamma C_5 H(24)$, $\gamma C_3 H(22)$
17	1042.9	45	A'	$\beta_1 rg(34)$, $\nu C_4 C_5(21)$, $\nu C_5 C_6(13)$, $\nu CCl(11)$	17	1055.2	26	A'	$\nu C_4 C_5(29)$, $\beta_1 rg(19)$, $\nu C_5 C_6(15)$, $\nu C_3 C_4(11)$
18	1060.9	13	A'	$\beta_1 rg(29)$, $\nu C_4 C_5(16)$, $\beta C_3 H(10)$, $\nu CCl(10)$	18	1071.1	25	A'	$\beta_1 rg(40)$, $\nu CCl(13)$

(*continued overleaf*)

TABLE 11. (continued)

No.	Freq.	IR	Sym.	Assignment, PED(%)	No.	Freq.	IR	Sym.	Assignment, PED(%)
19	1140.7	5	A'	$\nu C_3 C_4(23)$, $\beta C_5 H(14)$, $\nu C_6 H(14)$, $\beta C_4 H(12)$	19	1139.9	75	A'	$\nu C_3 C_4(20)$, $\beta C_4 H(13)$, $\beta COH(12)$
20	1179.5	3	A'	$\beta C_4 H(34)$, $\beta C_5 H(28)$, $\nu C_6 H(10)$, $\beta C_3 H(10)$, $\nu C_4 C_5(10)$	20	1182.1	3	A'	$\beta C_5 H(33)$, $\beta C_4 H(31)$, $\beta C_3 H(10)$
21	1211.0	115	A'	$\beta COH(38)$, $\nu C_1 C_6(17)$, $\nu C_6 H(12)$, $\beta C_3 H(10)$	21	1189.6	85	A'	$\beta COH(40)$, $\beta C_6 H(15)$, $\beta C_3 H(11)$, $\nu C_1 C_6(10)$
22	1274.7	75	A'	$\nu CO(31)$, $\nu C_2 C_3(25)$, $\beta C_3 H(21)$	22	1285.4	23	A'	$\nu CO(33)$, $\nu C_2 C_3(24)$, $\beta C_3 H(15)$
23	1323.3	28	A'	$\nu C_5 C_6(17)$, $\nu C_1 C_2(17)$, $\nu C_6 H(15)$, $\nu CO(14)$	23	1316.5	77	A'	$\beta C_6 H(24)$, $\nu CO(15)$, $\nu C_1 C_2(12)$, $\nu C_5 C_6(11)$
24	1366.3	23	A'	$\beta COH(26)$, $\beta C_5 H(12)$, $\nu C_4 C_5(10)$, $\nu C_2 C_3(10)$, $\nu C_3 C_4(10)$	24	1353.0	45	A'	$\beta COH(17)$, $\nu C_3 C_4(14)$, $\nu C_1 C_6(13)$, $\beta C_3 H(11)$, $\nu C_4 C_5(10)$, $\nu C_5 C_6(11)$,
25	1489.1	4	A'	$\beta C_4 H(30)$, $\beta C_5 H(15)$, $\nu C_2 C_3(12)$	25	1477.9	48	A'	$\beta C_4 H(26)$, $\beta C_5 H(23)$, $\nu C_6 C_6(11)$, $\nu C_2 C_3(10)$
26	1510.0	136	A'	$\beta C_3 H(19)$, $\nu C_6 H(16)$, $\nu C_1 C_2(15)$,	26	1524.7	61	A'	$\beta C_3 H(18)$, $\beta C_6 H(18)$, $\nu C_1 C_2(11)$, $\nu C_4 C_5(11)$
27	1624.0	15	A'	$\nu C_4 C_5(22)$, $\nu C_1 C_2(18)$, $\nu C_1 C_6(15)$, $\nu C_3 C_4(12)$	27	1622.7	14	A'	$\nu C_3 C4(23)$, $\nu C_1 C_6(23)$
28	1634.5	38	A'	$\nu C_5 C_6(25)$, $\nu C_2 C_3(17)$, $\nu C_1 C_6(11)$, $\beta_2 rg(10)$	28	1635.7	27	A'	$\nu C_5 C_6(20)$, $\nu C_1 C_2(20)$, $\nu C_4 C_5(16)$, $\nu C_2 C_3(13)$
29	3174.5	2	A'	$\nu C_5 H(61)$, $\nu C_4 H(24)$, $\nu C_6 H(10)$	29	3151.3	11	A'	$\nu C_6 H(92)$
30	3188.2	6	A'	$\nu C_4 H(43)$, $\nu C_3 H(26)$, $\nu C_6 H(17)$ $\nu C_5 H(13)$	30	3178.0	4	A'	$\nu C_5 H(44)$, $\nu C_4 H(43)$
31	3196.5	5	A'	$\nu C_6 H(63)$, $\nu C_3 H(21)$, $\nu C_5 H(15)$	31	3190.5	8	A'	$\nu C_5 H(40)$, $\nu C_3 H(38)$ $\nu C_4 H(19)$
32	3203.5	4	A'	$\nu C_3 H(49)$, $\nu C_4 H(31)$, $\nu C_5 H(10)$	32	3202.2	5	A'	$\nu C_3 H(54)$, $\nu C_4 H(36)$
33	3766.7	93	A'	$\nu OH(100)$	33	3835.4	73	A'	$\nu OH(100)$

[a] See footnote a in Table 10.
[b] The gas-phase IR experiments[171].
[c] IR experiments in solution[172].
[d] IR experiments in solution[165].

TABLE 12. Harmonic vibrational frequencies, IR intensities and assignments for *cis* and *trans ortho*-fluorophenols[a]

No.	ω	A	Sym.	Assignment, PED(%)	No.	ω	A	Sym.	Assignment, PED(%)
1	190.4	0	A''	$\tau_2 rg(59)$, $\tau_1 rg(17)$, $\tau_3 rg(16)$	1	183.1	0	A''	$\tau_2 rg(59)$, $\tau_1 rg(20)$, $\tau_3 rg(15)$
2	290.9	1	A''	$\gamma CF(28)$, $\tau_3 rg(24)$, $\gamma CO(18)$, $\gamma C_6 H(11)$	2	274.2	60	A''	$\tau OH(43)$, $\tau_3 rg(20)$, $\gamma CF(15)$
3	296.9	8	A'	$\beta CO(42)$, $\beta CF(40)$, $\beta_3 rg(10)$	3	297.7	2	A'	$\beta CO(42)$, $\beta CF(41)$
4	396.3	126	A''	$\tau OH(90)$	4	308.2	51	A''	$\tau OH(41)$, $\tau_3 rg(16)$, $\gamma CO(14)$, $\gamma CF(11)$
5	443.3	1	A'	$\beta_2 rg(29)$, $\beta CF(22)$, $\beta CO(19)$, $\beta_3 rg(10)$	5	447.1	6	A'	$\beta_2 rg(29)$, $\beta CF(23)$, $\beta CO(20)$, $\beta_3 rg(10)$
6	454.7	0	A''	$\tau_3 rg(45)$, $\gamma CF(24)$, $\gamma CO(15)$	6	456.6	4	A''	$\tau_3 rg(44)$, $\gamma CF(22)$, $\gamma CO(18)$
7	555.8	4	A'	$\beta_2 rg(25)$, $\beta CO(23)$, $\beta CF(21)$	7	547.4	8	A'	$\beta_2 rg(28)$, $\beta CO(22)$, $\beta CF(20)$
8	557.8	1	A''	$\tau_2 rg(30)$, $\tau_1 rg(26)$, $\gamma CO(10)$	8	556.8	0	A''	$\tau_1 rg(36)$, $\tau_2 rg(25)$
9	584.4	5	A'	$\beta_3 rg(50)$, $\beta_2 rg(25)$	9	587.8	3	A'	$\beta_3 rg(53)$, $\beta_2 rg(22)$
10	683.6	0	A''	$\tau_1 rg(55)$, $\gamma CO(19)$, $\gamma CF(17)$	10	692.9	0	A''	$\tau_1 rg(50)$, $\gamma CO(20)$, $\gamma CF(19)$
11	757.8	84	A''	$\gamma C_4 H(36)$, $\gamma C_5 H(29)$, $\gamma C_3 H(18)$, $\gamma C_6 H(12)$	11	750.5	91	A''	$\gamma C_5 H(32)$, $\gamma C_4 H(28)$, $\gamma C_6 H(23)$, $\gamma C_3 H(10)$
12	772.7	42	A'	$\nu C_1 C_2(25)$, $\gamma CF(21)$, $\beta_3 rg(17)$, $\nu C_2 C_3(10)$	12	773.4	21	A'	$\nu C_1 C_2(27)$, $\nu CF(17)$, $\beta_3 rg(15)$, $\nu CO(11)$, $\nu C_2 C_3(11)$
13	847.3	0	A''	$\gamma C_6 H(39)$, $\gamma C_3 H(31)$, $\tau_1 rg(10)$	13	829.5	0	A''	$\gamma C_6 H(45)$, $\gamma C_3 H(24)$, $\tau_1 rg(10)$, $\gamma C_4 H(10)$
14	857.1	17	A'	$\beta_1 rg(44)$, $\nu CO(17)$, $\beta_2 rg(16)$, $\nu CF(11)$	14	860.0	27	A'	$\beta_1 rg(48)$, $\nu CO(15)$, $\beta_2 rg(15)$, $\nu CF(10)$
15	931.6	5	A''	$\gamma C_3 H(38)$, $\gamma C_4 H(26)$, $\gamma C_6 H(21)$	15	920.3	4	A''	$\gamma C_3 H(37)$, $\gamma C_5 H(31)$, $\gamma C_6 H(18)$
16	959.2	0	A''	$\gamma C_5 H(52)$, $\gamma C_4 H(22)$, $\gamma C_6 H(14)$	16	945.3	1	A''	$\gamma C_4 H(46)$, $\gamma C_5 H(27)$, $\gamma C_3 H(17)$
17	1043.6	16	A'	$\nu C_4 C_5(37)$, $\nu C_5 C_6(15)$, $\beta C_3 H(15)$, $\nu C_3 C_4(14)$, $\beta C_6 H(12)$	17	1052.9	5	A'	$\nu C_4 C_5(36)$, $\nu C_3 C_4(17)$, $\beta C_3 H(16)$, $\nu C_5 C_6(15)$, $\beta C_6 H(10)$

(continued overleaf)

TABLE 12. (continued)

No.	ω	A	Sym.	Assignment, PED(%)	No.	ω	A	Sym.	Assignment, PED(%)
18	1106.2	28	A'	$\beta_1$rg(27), νC_3C_4(10), νCF(10)	18	1111.2	64	A'	$\beta_1$rg(22), βCOH(12), νC_3C_4(11)
19	1175.4	2	A'	βC_5H(34), βC_4H(26), βC_6H(12), νC_4C_5(12)	19	1175.8	33	A'	βC_4H(31), βC_3H(21), βC_5H(17), βCOH(12)
20	1182.0	58	A'	βC_6H(21), νCF(17), βCOH(13), νC_1C_6(10)	20	1183.4	33	A'	βC_6H(28), βCOH(20), βC_5H(20), νC_5C_6(10)
21	1234.9	104	A'	βCOH(25), νCF(16), νCO(14), $\beta_1$rg(12)	21	1240.1	124	A'	νCF(40), $\beta_1$rg(19), βCOH(12)
22	1284.3	130	A'	βC_3H(27), νCO(26), νC_2C_3(12)	22	1298.9	61	A'	νCO(29), βC_3H(20), νC_2C_3(13)
23	1324.0	21	A'	βC_6H(17), νC_5C_6(17), νC_1C_2(15)	23	1317.4	78	A'	βC_6H(26), νC_1C_2(11), νCO(10), νC_5C_6(10)
24	1380.8	12	A'	βCOH(24), νC_4C_5(15), βC_5H(13), νC_2C_3(12)	24	1362.0	49	A'	βCOH(15), νC_1C_6(15), νC_4C_5(14), νC_3C_4(11), νC_2C_3(11), νC_5C_6(10)
25	1499.3	4	A'	βC_4H(26), βC_5H(20), νC_5C_6(13), νC_3C_4(11)	25	1488.2	29	A'	βC_5H(27), βC_4H(21), νC_3C_4(11), νC_1C_6(10)
26	1530.1	189	A'	βC_3H(16), βC_6H(15), νC_1C_2(13), νCO(11)	26	1545.1	131	A'	βC_6H(17), νC_4C_5(13), βC_3H(13), νC_2C_3(10)
27	1640.1	6	A'	νC_1C_6(25), νC_3C_4(18), νC_1C_2(10)	27	1636.0	19	A'	νC_1C_6(21), νC_3C_4(20), νC_2C_3(12)
28	1653.7	47	A'	νC_2C_3(21), νC_5C_6(18), νC_1C_2(17), νC_4C_5(11)	28	1655.9	28	A'	νC_1C_2(26), νC_4C_5(16), νC_5C_6(14), νC_2C_3(13)
29	3176.7	2	A'	νC_5H(59), νC_4H(25), νC_6H(12)	29	3152.6	10	A'	νC_6H(92)
30	3189.4	8	A'	νC_4H(40), νC_3H(25), νC_6H(25)	30	3179.6	4	A'	νC_4H(45), νC_5H(41)
31	3197.1	5	A'	νC_6H(53), νC_3H(28), νC_5H(18)	31	3191.4	9	A'	νC_3H(47), νC_5H(39), νC_4H(11)
32	3204.4	3	A'	νC_3H(43), νC_4H(34), νC_5H(13)	32	3202.4	4	A'	νC_3H(43), νC_4H(43), νC_5H(13)
33	3807.0	105	A'	νOH(100)	33	3837.1	75	A'	νOH(100)

[a]See footnote a in Table 10.

Regarding the transition state between the *cis* and *trans* isomers of o-FC$_6$H$_4$OH, we obtain that it has nearly the same slope as in the case of Cl, viz. 347 i cm^{-1}, although its barrier $V_\tau^F = 20.3$ kJ mol^{-1} is by 2.2 kJ mol^{-1} smaller than V_τ^{Cl}. Since $\Delta_{cis-trans}E_{ortho}^F < \Delta_{cis-trans}E_{ortho}^{Cl}$, we might expect that the equilibrium constant $k_{cis \rightleftharpoons trans}^F$ is larger than $k_{cis \rightleftharpoons trans}^{Cl}$, which is indeed found to be true: $k_{cis \rightleftharpoons trans}^F = 10.1 \times 10^{-3}$. On the contrary, no known IR experiment has ever revealed a *cis–trans* transition in o-FC$_6$H$_4$OH[223–229, 233–235]. The question is: Why?

The disparity between the older IR experiments and modern high-level theory becomes even sharper if we turn to the o-Br-substituted phenols whose harmonic vibrational modes are presented in Table 13. It is then easy to obtain $\Delta_{cis-trans}\nu_{OH}^{Br} = 94$ cm^{-1}, which agrees with the experimental values ranging from 74 to 93 cm^{-1}[218, 224, 229] (Tables 1 and 5 of Reference 222). On the other hand, the calculated $\Delta_{cis-trans}E_{ortho}^{Br} = 12.9$ kJ mol^{-1} (the experimental free energy difference in the vapour is 13.1 ± 14.6 kJ mol^{-1}[224]) implies that, first, the intramolecular hydrogen bond is slightly stronger with Br than with Cl, which surely contradicts the common order of the hydrogen bond acceptors[155, 171, 236, 237], and, second, the equilibrium constant $k_{cis \rightleftharpoons trans}^{Br} = 5.2 \times 10^{-3} < k_{cis \rightleftharpoons trans}^{Cl}$, although the experiments show the reverse trend[233, 234]. Altogether, this was dubbed as an 'anomalous' order in the strength of the intramolecular hydrogen bond[223, 224, 229, 231, 238–240]; the 'state of affairs' was summarized by Sandorfy and coauthors[229] in their 1963 work: 'Nothing emerges from our work, however, to explain this order. ... For a more thorough treatment we shall likely have to wait until the next stage in the development of quantum chemistry'. What modern calculations might tell us in this context is briefly outlined below:

(i) Under the assumption that $\Delta_{cis-trans}E_{ortho}^X$ (X = F, Cl, Br) defines the energy of formation of the intramolecular hydrogen bond in *cis ortho*-X-substituted phenols, the order of its strength in the gas phase (in kJ mol)$^{-1}$ appears to be that given in equation 5.

$$\text{Br} \overset{0.46}{\approx} \text{Cl} \overset{1.09}{>} \text{F}. \quad (5)$$

The numbers in equation 5 indicate the corresponding difference (in kJ mol^{-1}) in the energies of formation of the intramolecular hydrogen bond between the left-hand complex and its right-hand one. This order is confirmed to a certain extent by the order of red shifts $\Delta_{cis-trans}\nu_{OH}^X$ (in cm^{-1}) given in equation 6.

$$\text{Br} \overset{25}{>} \text{Cl} \overset{39}{>} \text{F}. \quad (6)$$

By comparing equations 5 and 6 it is seen that $\Delta_{cis-trans}\nu_{OH}^X$ is not proportional to $\Delta_{cis-trans}E_{ortho}^X$[224]. The order in equation 6 more likely resembles the van der Waals radii of the halogen atoms: Br(1.85Å) > Cl(1.75Å) > F(1.47Å) rather than their electronegativity trend (in Pauling units): F(3.98) > Cl(3.16) > Br(2.96), which is usually chosen to differentiate the strength of the conventional intermolecular hydrogen bonds[225, 226].

Both equations 5 and 6 unambiguously imply that in *cis ortho*-XC$_6$H$_4$OH, the strength of the O—H · · · X intramolecular hydrogen bond decreases as Br $\approx$ Cl > F (cf. Table 2 in Reference 236), which is completely opposite to that widely accepted for usual intermolecular hydrogen bonds[225, 226]. Such variance was in fact a matter of numerous investigations in the past[155, 236, 237]. Here, we could offer an explanation[239] relying on the geometrical criteria of the hydrogen bond[225, 226] that are simply expressed in terms of the elongation of the O—H bond length and the value of the ∠O—H · · · X bond angle: the larger they are the stronger the hydrogen bond[222, 240]. The fact that the strength of the intramolecular hydrogen bond in *cis ortho*-X-substituted phenols exactly follows the order of equations 5 and 6

TABLE 13. Harmonic vibrational frequencies, IR intensities and assignments for *cis* and *trans ortho*-bromophenols[a]

No.	Freq.	IR	Sym.	Assignment, PED(%)	No.	Freq.	IR	Sym.	Assignment, PED(%)
1	140.8	0	A''	$\tau_2\text{rg}(42), \gamma\text{CBr}(20), \tau_3\text{rg}(18), \tau_1\text{rg}(10)$	1	137.7	2	A''	$\tau_2\text{rg}(44), \tau_3\text{rg}(19), \gamma\text{CBr}(17), \tau_1\text{rg}(12)$
2	208.7	1	A'	$\beta\text{CBr}(83)$	2	197.8	2	A'	$\beta\text{CBr}(78), \beta\text{CO}(11)$
3	253.7	1	A''	$\tau_2\text{rg}(34), \gamma\text{CBr}(32), \gamma\text{C}_6\text{H}(11)$	3	250.7	13	A''	$\tau_2\text{rg}(33), \gamma\text{CBr}(27)$
4	292.8	1	A'	$\nu\text{CBr}(56), \beta_3\text{rg}(13), \beta_2\text{rg}(10)$	4	299.2	0	A'	$\nu\text{CBr}(53), \beta_3\text{rg}(14), \beta_2\text{rg}(10)$
5	417.6	84	A''	$\tau\text{OH}(54), \tau_3\text{rg}(29), \gamma\text{CBr}(11)$	5	318.5	94	A''	$\tau\text{OH}(87)$ Expt: $372^b 361^c$
	Expt: $404^b, 395^c, 395^d$					Expt: $372^b, 361^c$			
6	443.4	13	A''	$\tau_3\text{rg}(55), \gamma\text{CBr}(21), \tau\text{OH}(12)$	6	441.7	6	A''	$\tau_3\text{rg}(61), \gamma\text{CBr}(21)$
7	472.5	14	A'	$\beta\text{CO}(68), \nu\text{CBr}(10)$	7	469.3	9	A'	$\beta\text{CO}(68), \beta\text{CBr}(10)$
8	539.1	1	A''	$\tau_2\text{rg}(39), \gamma\text{CO}(30)$	8	545.1	1	A''	$\tau_2\text{rg}(35), \gamma\text{CO}(20), \tau_1\text{rg}(18)$
9	556.5	4	A'	$\beta_2\text{rg}(69)$	9	558.0	4	A'	$\beta_2\text{rg}(67)$
10	660.0	0	A''	$\tau_1\text{rg}(63), \gamma\text{CO}(18)$	10	668.7	15	A'	$\beta_3\text{rg}(67), \nu\text{CBr}(17)$
11	664.7	19	A'	$\beta_3\text{rg}(68), \nu\text{CBr}(17)$	11	685.1	1	A''	$\tau_1\text{rg}(59), \gamma\text{CO}(19), \gamma\text{CBr}(11)$
12	757.5	72	A''	$\gamma\text{C}_4\text{H}(40), \gamma\text{C}_5\text{H}(22), \gamma\text{C}_3\text{H}(21), \gamma\text{C}_6\text{H}(12)$	12	751.8	80	A''	$\gamma\text{C}_4\text{H}(32), \gamma\text{C}_5\text{H}(23), \gamma\text{C}_6\text{H}(22), \gamma\text{C}_3\text{H}(11)$
13	840.4	13	A'	$\nu\text{CO}(25), \beta_1\text{rg}(24), \beta_2\text{rg}(19), \nu\text{C}_1\text{C}_6(11), \nu\text{C}_1\text{C}_2(10)$	13	834.6	0	A''	$\gamma\text{C}_6\text{H}(49), \gamma\text{C}_3\text{H}(21), \gamma\text{C}_4\text{H}(14)$
14	848.7	1	A''	$\gamma\text{C}_6\text{H}(46), \gamma\text{C}_3\text{H}(28)$	14	838.7	20	A'	$\nu\text{CO}(25), \beta_1\text{rg}(25), \beta_2\text{rg}(19), \nu\text{C}_1\text{C}_6(11)$
15	941.3	2	A''	$\gamma\text{C}_3\text{H}(42), \gamma\text{C}_4\text{H}(24), \gamma\text{C}_6\text{H}(19)$	15	932.8	1	A''	$\gamma\text{C}_5\text{H}(38), \gamma\text{C}_3\text{H}(36), \gamma\text{C}_5\text{H}(16)$
16	974.6	0	A''	$\gamma\text{C}_5\text{H}(51), \gamma\text{C}_4\text{H}(22), \gamma\text{C}_6\text{H}(13)$	16	964.3	0	A''	$\gamma\text{C}_4\text{H}(41), \gamma\text{C}_5\text{H}(24), \gamma\text{C}_3\text{H}(23)$
17	1028.2	48	A'	$\beta_1\text{rg}(54), \nu\text{CBr}(13)$	17	1041.7	43	A'	$\beta_1\text{rg}(50), \nu\text{CBr}(14)$

18	1056.6	5	A'	$\nu C_4C_5(33)$, $\beta C_6H(13)$, $\beta C_5H(11)$, $\nu C_3C_4(11)$	18	1064.8	5	A	$\nu C_4C_5(29)$, $\nu C_3C_4(14)$, $\beta_1\text{rg}(13)$, $\beta C_3H(12)$
19	1135.5	2	A'	$\nu C_3C_4(24)$, $\beta C_5H(14)$, $\beta C_4H(13)$, $\beta C_6H(12)$, $\nu C_5C_6(10)$	19	1132.1	69	A'	$\nu C_3C_4(19)$, $\beta COH(15)$, $\beta C_4H(14)$
20	1179.9	4	A'	$\beta C_4H(33)$, $\beta C_5H(28)$, $\beta C_6H(12)$	20	1183.3	2	A'	$\beta C_5H(35)$, $\beta C_4H(28)$, $\beta C_6H(13)$
21	1213.7	117	A'	$\beta COH(35)$, $\nu C_1C_6(19)$, $\beta C_6H(12)$, $\beta C_3H(11)$	21	1189.2	74	A'	$\beta COH(40)$, $\beta C_5H(13)$, $\beta C_3H(12)$, $\nu C_1C_6(10)$
22	1273.9	63	A'	$\nu CO(31)$, $\nu C_2C_3(27)$, $\beta C_3H(20)$	22	1283.7	21	A'	$\nu CO(34)$, $\nu C_2C_3(24)$, $\beta C_3H(14)$
23	1323.7	27	A'	$\nu C_1C_2(17)$, $\nu C_5C_6(17)$, $\beta C_6H(14)$, $\nu CO(14)$	23	1316.8	73	A'	$\beta C_6H(24)$, $\nu CO(15)$, $\nu C_1C_2(12)$, $\nu C_5C_6(10)$
24	1366.4	26	A'	$\beta COH(29)$, $\beta C_5H(12)$, $\beta C_3H(11)$, $\nu C_3C_4(10)$	24	1350.5	49	A'	$\beta COH(18)$, $\nu C_3C_4(14)$, $\nu C_1C_6(13)$, $\beta C_3H(13)$, $\nu C_5C_6(11)$, $\nu C_4C_5(10)$
25	1486.4	3	A'	$\beta C_4H(30)$, $\beta C_5H(14)$, $\nu C_2C_3(12)$	25	1474.5	51	A'	$\beta C_4H(26)$, $\beta C_5H(23)$, $\nu C_1C_6(11)$, $\nu C_2C_3(10)$
26	1504.9	129	A'	$\beta C_3H(20)$, $\beta C_6H(16)$, $\nu C_1C_2(15)$, $\nu CO(11)$	26	1520.3	49	A'	$\beta C_3H(20)$, $\beta C_5H(18)$, $\nu C_1C_2(12)$, $\nu C_4C_5(11)$
27	1618.5	10	A'	$\nu C_4C_5(20)$, $\nu C_1C_6(19)$, $\nu C_3C_4(15)$, $\nu C_1C_2(14)$	27	1617.2	13	A'	$\nu C_1C_6(23)$, $\nu C_3C_4(23)$
28	1630.4	43	A'	$\nu C_5C_6(27)$, $\nu C_2C_3(17)$, $\beta_2\text{rg}(10)$	28	1632.9	27	A'	$\nu C_5C_6(21)$, $\nu C_1C_2(19)$, $\nu C_4C_5(17)$, $\nu C_2C_3(12)$
29	3174.5	2	A'	$\nu C_5H(57)$, $\nu C_4H(27)$, $\nu C_6H(12)$	29	3150.6	10	A'	$\nu C_6H(93)$
30	3187.6	7	A'	$\nu C_4H(41)$, $\nu C_6H(26)$, $\nu C_3H(22)$, $\nu C_5H(11)$	30	3177.9	4	A	$\nu C_4H(46)$, $\nu C_5H(41)$
31	3195.9	5	A'	$\nu C_6H(55)$, $\nu C_3H(23)$, $\nu C_5H(21)$	31	3190.8	8	A'	$\nu C_5H(43)$, $\nu C_3H(35)$, $\nu C_4H(19)$
32	3203.3	5	A'	$\nu C_3H(51)$, $\nu C_4H(31)$, $\nu C_5H(11)$	32	3202.3	5	A'	$\nu C_3H(57)$, $\nu C_4H(34)$
33	3739.7	97	A'	$\nu OH(100)$	33	3833.8	71	A'	$\nu OH(100)$

[a] Footnotes a–d are identical to those in Table 11.

is clearly seen in Figure 12: due to a larger van der Waals radius, the Br atom slightly better accommodates the intramolecular bond, even 'overcoming the innate lower H-bonding tendency to Br'[240] than Cl which, in turn, does better than F. Such a conclusion is also supported by the inequalities in equation 7.

$$\text{OH bond length (Å):} \quad \text{Br} \stackrel{0.001}{>} \text{Cl} \stackrel{0.002}{>} \text{F} \tag{7}$$

$$\angle \text{O}-\text{H} \cdots \text{X(deg):} \quad \text{Br} \stackrel{3.1}{>} \text{Cl} \stackrel{9.2}{>} \text{F}$$

(ii) The gas-phase theoretical equilibrium constants $k^X_{cis \rightleftharpoons trans}$ follow the order in equation 8,

$$\text{F} \stackrel{1.56}{>} \text{Cl} \stackrel{1.27}{>} \text{Br} \tag{8}$$

where the quantity above the inequality indicates the ratio of the equilibrium constants between the left-hand complex and the right-hand complex. Such order in the equilibrium constants is mirrored in the order of the calculated *cis–trans* barriers V_τ^X (equation 9):

$$\text{F} \stackrel{0.43}{<} \text{Cl} \stackrel{0.15}{<} \text{Br} \tag{9}$$

It would be expected that the *trans/cis* ratio follows the order of equation 5 for the hydrogen bond energies, but surprisingly the opposite is known. It has even been argued[227] that 'the fact that both the *trans/cis* ratio and the $\Delta \nu$ shift increase in the same order appears to argue against the applicability of Badger's rule[241] which stated that the progressive shift to lower frequencies is an indication of increasing strength of the hydrogen bond. If the rule is valid here ...'.

In order to resolve the longstanding controversy between experiment and theory, let us first suggest that the dipole moments of the *cis* and *trans* forms and their polarizability might play a key role, bearing in mind that all aforementioned experiments were conducted in a solvent although its role in theory was underestimated. This is clearly seen from the inequalities between the *trans/cis* ratio of the total dipole moments: $2.95_F > 2.87_{Cl} > 2.77_{Br}$. A similar ratio was also determined elsewhere[223, 238] (for a discussion see Reference 229). By analogy, we have the corresponding *trans/cis* ratio for the mean polarizability $\alpha = (\alpha_{xx} + \alpha_{yy} + \alpha_{zz})/3$ (in a.u.) in equation 10.

$$\frac{92.19}{91.77}_{Br} > \frac{84.63}{84.00}_{Cl} > \frac{71.60}{71.13}_{F} \tag{10}$$

The experimental data for the equilibrium constants in CCl$_4$ solution (equation 11)[233, 234],

$$\text{Br} \stackrel{1.47}{>} \text{Cl} \tag{11}$$

are in complete disagreement with the theoretical expectations based on equations 8 and 9. In order to explain this discrepancy, one must take into account a stabilizing effect of the solvent on the *trans* form[231], and we propose the following model[220].

The presence of the *ortho*-halogen atom in a phenol generates two distinct *cis* and *trans* conformers and changes the shape of the torsional transition barrier V_τ, making it partly asymmetric. Within the *cis* form, the halogen atom is capable of forming an intramolecular hydrogen bond, rather bent and quite weak. Its formation has a stabilizing effect on the *cis* (particularly in the gas phase) over the *trans* form. On the other hand, due to the larger polarity and larger polarizability of the *trans* oX-C$_6$H$_4$OH, the latter conformer might, in some rather polar solvents, be favoured over the *cis* form. We suggest that solvent

1. General and theoretical aspects of phenols

TABLE 14. AM1 and SM5.4/AM1 data on ortho-XC_6H_4OH (X = F, Cl and Br) and their cis–trans transition state (TS) including the heat of formation ΔH (kJ mol^{-1}), free solvation energy ΔG^{solv} (kJ mol^{-1}) and ν_{OH} stretching frequency (in cm^{-1})

		cis ortho-XC_6H_4OH		
		r_{OH} (Å)	∠O–H···X(deg)	$r_{H···X}$(Å)
F	gas phase	0.970	111.2	2.325
	solvent	0.979	110.5	2.335
Cl	gas phase	0.970	117.1	2.506
	solvent	0.975	116.3	2.524
Br	gas phase	0.971	119.8	2.617
	solvent	0.975	119.0	2.634

	Gas phase		CCl_4		
	$-\Delta H$	ν_{OH}	$-\Delta H$	$-\Delta G^{solv}$	ν_{OH}
cis o-FC_6H_4OH	280	3431	300	22	3316
trans o-FC_6H_4OH	273	3452	297	27	3309
cis-trans TS	266		289		
cis o-ClC_6H_4OH	120	3420	144	25	3350
trans o-ClC_6H_4OH	112	3451	142	32	3313
cis-trans TS	105		133		
cis o-BrC_6H_4OH	61	3407	96	28	3346
trans o-BrC_6H_4OH	61	3448	94	35	3311
cis-trans TS	54		85		

^{a}Compare with the free energy of hydration: AM1-SM2: -20 kJ mol^{-1}; PM3-SM3: -20 kJ mol^{-1} (Reference 243). Values are taken from Reference 220 with permission.

stabilizes the *trans* more strongly than the *cis* and hence decreases $\Delta_{cis-trans}E^X_{ortho}$, thus making it more accessible than in the gas phase.

In order to describe theoretically the *cis* and *trans* ortho-XC_6H_4OH in a solvent mimicking CCl_4, we invoke a rather simple but accurate computational model[242]. Its results are summarized in Table 14, which displays the following three key effects of the solvent. First, the solvent reduces the gas-phase $\Delta_{cis-trans}\nu^X_{OH}$ to 7, 37 and 35 cm^{-1} for F, Cl and Br, respectively. We think that this is a satisfactory explanation of why the *cis–trans* ν_{OH} doublet in o-FC_6H_4OH was not observed in CCl_4. Second, the solvent strongly stabilizes the *trans* form so that the *cis–trans* gap $\Delta_{cis-trans}E^X_{ortho}$ appears to be equal to 3.4, 2.8 and 2.0 kJ mol^{-1} for F, Cl and Br, respectively. This straightforwardly implies an increase in the equilibrium constants $k^X_{cis \rightleftarrows trans}$ in the series of F, Cl and Br equal to 0.25, 0.33 and 0.45, respectively with respect to that in the parent phenol. Third, the solvent reduces the *cis–trans* barrier V_τ to 11.8, 11.4 and 11.7 kJ mol^{-1} for F, Cl and Br, respectively. Altogether, we may conclude that even a rather simple modelling of solvent is able to resolve the aforementioned controversial 'state of affairs' in the *ortho*-X-substituted phenols.

B. *meta*- and *para*-Halogenophenols

The corresponding substituted phenols are displayed symbolically in Figure 13 and their characteristic vibrational modes, showing a rather strong dependence on the X substitution,

cis m-XC$_6$H$_4$OH   trans m-XC$_6$H$_4$OH

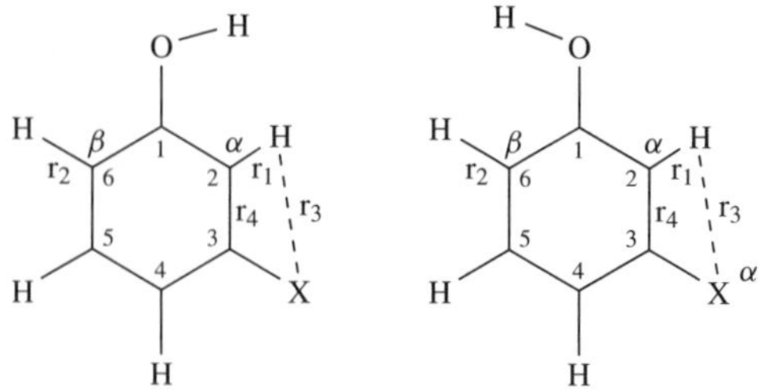

cis–trans differences of some geometrical parameters in m-XC$_6$H$_4$OH

	Δα (deg)	Δβ (deg)	Δr$_1$ (Å)	Δr$_2$ (Å)	Δr$_3$ (Å)	Δr$_4$ (Å)
X=F	1.2	−1.0	0.002	−0.003	−0.021	0.004
X=Cl	1.2	−1.1	0.002	−0.002	−0.021	0.004
X=Br	1.1	−1.1	0.002	−0.002	−0.021	0.004

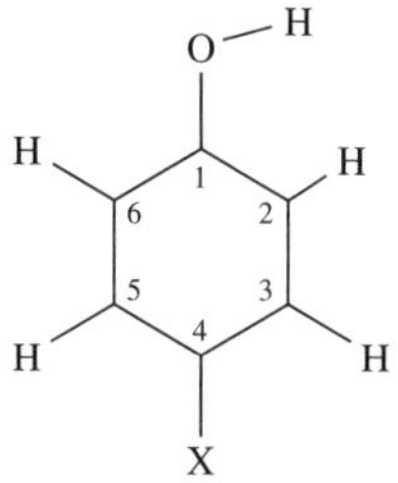

p-XC$_6$H$_4$OH

FIGURE 13. The minimum energy structures of *cis* and *trans meta*- and *para*-XC$_6$H$_4$OH (X = F, Cl and Br). Bond lengths are in Å, bond angles in degrees. Adapted from Reference 220 with permission

are presented in Tables 11–13 and 15–20[244]. Note that the spectra of p-ClC$_6$H$_4$OH and p-BrC$_6$H$_4$OH have been analyzed critically on the basis of DFT computations[170, 245]. It follows from these Tables that, first, a *para* substitution by fluorine downshifts the torsional vibrational mode τ_{OH} by 29 cm^{-1} in perfect agreement with the experimental red shift[172] of 30 cm^{-1}. In m-XC$_6$H$_4$OH, the mode τ_{OH} is placed higher than in the corresponding *para*-halophenols. This observation is partly supported by the NBO analysis, demonstrating a strong conjugative interaction of the p-type oxygen lone pair with the π-antibond of the ring, viz. $n_p \to \pi^*(C_1\text{-}C_2)$, a little increased in all *meta* structures, resulting in upshifting of the τ_{OH} in m-XC$_6$H$_4$OH with respect to p-XC$_6$H$_4$OH. This concurs with the earlier experimental findings[172].

Furthermore, one obtains certain subtle features in the spectra of m-XC$_6$H$_4$OH whose origin can only be explained by the co-existence of two very slightly inequivalent conformers of the *cis* and *trans* types. This is seen, for example, from the magnitude of $\Delta_{cis-trans}E^X_{meta}$ ranging from −0.8 kJ mol^{-1} for F to +0.08 kJ mol^{-1} for Cl, and finally to +0.04 kJ mol^{-1} for Br. If the difference is extremely small for Cl and Br, F is then an exception. Contrary to Cl and Br, we obtain that the *trans* conformer of m-FC$_6$H$_4$OH is a little more stable than its *cis* conformer. The *cis–trans* differences in the geometrical

TABLE 15. Harmonic vibrational frequencies, IR intensities and assignments for *cis* and *trans meta*-fluorophenols[a]

No.	Freq.	IR	Sym.	Assignment, PED(%)	No.	Freq.	IR	Sym.	Assignment, PED(%)
1	224.0	0	A''	$\tau_2 rg(23)$, $\tau_1 rg(22)$, $\tau_1 rg(18)$, $\tau_3 rg(14)$	1	223.2	3	A''	$\tau_1 rg(24)$, $\tau_2 rg(23)$, $\gamma C_2 H(22)$, $\tau_3 rg(10)$
2	239.9	2	A''	$\tau_2 rg(50)$, $\tau_3 rg(30)$	2	238.0	0	A''	$\tau_2 rg(49)$, $\tau_3 rg(33)$
3	314.0	111	A''	$\tau OH(90)$	3	320.7	109	A''	$\tau OH(90)$
	Expt: 318.5[b], 318[c], 317[d]								
4	328.7	9	A'	$\beta CF(40)$, $\beta CO(38)$	4	330.1	2	A'	$\beta CF(39)$, $\beta CO(39)$
5	460.1	12	A''	$\tau_3 rg(47)$, $\tau_2 rg(13)$, $\gamma CF(11)$, $\gamma CO(10)$	5	459.7	2	A''	$\tau_3 rg(47)$, $\tau_2 rg(12)$, $\gamma CF(10)$
6	480.3	5	A'	$\beta CO(43)$, $\beta CF(40)$	6	476.4	11	A'	$\beta CO(41)$, $\beta CF(39)$
7	520.5	8	A'	$\beta_3 rg(40)$, $\beta_2 rg(36)$, $\nu CF(8)$	7	520.7	6	A'	$\beta_3 rg(38)$, $\beta_2 rg(36)$
8	536.9	4	A'	$\beta_1 rg(44)$, $\beta_3 rg(25)$	8	538.2	4	A'	$\beta_2 rg(41)$, $\beta_3 rg(28)$
9	612.2	0	A''	$\gamma CF(35)$, $\gamma CO(33)$, $\tau_2 rg(23)$	9	614.5	1	A''	$\gamma CO(40)$, $\gamma CF(27)$, $\tau_2 rg(23)$
10	666.5	14	A'	$\tau_1 rg(67)$, $\gamma CO(12)$, $\gamma CF(11)$	10	657.6	11	A''	$\tau_1 rg(65)$, $\gamma CF(16)$
11	750.8	6	A'	$\beta_3 rg(25)$, $\beta_1 rg(18)$, $\nu CF(12)$, $\nu CO(10)$	11	748.2	5	A'	$\beta_3 rg(25)$, $\beta_1 rg(17)$, $\nu CF(12)$, $\nu CO(10)$
12	775.9	38	A''	$\gamma C_4 H(36)$, $\gamma C_6 H(25)$, $\gamma C_5 H(21)$	12	758.4	67	A''	$\gamma C_6 H(39)$, $\gamma C_4 H(22)$, $\gamma C_5 H(16)$, $\tau_1 rg(12)$
13	827.5	50	A''	$\gamma C_2 H(58)$, $\tau_1 rg(19)$	13	844.4	10	A''	$\gamma C_2 H(36)$, $\gamma C_6 H(25)$, $\gamma C_4 H(15)$
14	876.6	0	A''	$\gamma C_4 H(41)$, $\gamma C_6 H(40)$	14	862.7	22	A''	$\gamma C_2 H(38)$, $\gamma C_4 H(26)$, $\gamma CF(10)$, $\tau_1 rg(10)$
15	967.1	0	A''	$\gamma C_5 H(61)$, $\gamma C_6 H(18)$, $\gamma C_4 H(10)$	15	945.2	0	A''	$\gamma C_5 H(63)$, $\gamma C_4 H(17)$
16	970.3	54	A'	$\nu CF(17)$, $\nu CO(16)$, $\nu C_1 C_6(14)$, $\nu C_1 C_2(10)$	16	968.6	88	A'	$\nu CF(16)$, $\nu C_1 C_6(15)$, $\nu CO(15)$
17	1014.4	7	A'	$\beta_1 rg(60)$	17	1013.9	5	A'	$\beta_1 rg(61)$
18	1088.0	13	A'	$\beta C_4 H(27)$, $\nu C_4 C_5(24)$, $\beta C_6 H(18)$, $\nu C_5 C_6(16)$	18	1095.1	3	A'	$\beta C_4 H(32)$, $\nu C_4 C_5(24)$, $\nu C_5 C_6(18)$, $\beta C_6 H(13)$
19	1152.2	94	A'	$\beta C_2 H(29)$, $\nu CF(21)$, $\beta C_4 H(12)$, $\beta C_6 H(10)$, $\nu CO(10)$	19	1141.4	168	A'	$\beta C_2 H(38)$, $\nu CF(20)$, $\nu CO(15)$
20	1181.4	17	A'	$\beta C_6 H(22)$, $\beta COH(17)$, $\beta C_5 H(13)$, $\beta C_2 H(12)$	20	1179.2	14	A'	$\beta C_6 H(27)$, $\beta C_5 H(23)$, $\beta C_4 H(12)$, $\nu C_5 C_6(11)$, $\beta COH(11)$

(*continued overleaf*)

TABLE 15. (continued)

No.	Freq.	IR	Sym.	Assignment, PED(%)	No.	Freq.	IR	Sym.	Assignment, PED(%)
21	1192.6	175	A'	$\beta COH(38)$, $\beta C_5H(19)$, $\nu CO(12)$	21	1209.2	33	A'	$\beta COH(42)$, $\beta C_5H(16)$, $\nu CF(14)$, $\nu C_1C_2(11)$
22	1299.3	75	A'	$\nu CO(22)$, $\nu CF(18)$, $\beta_1 rg(13)$, $\beta C_4H(13)$, $\beta C_2H(10)$	22	1299.2	45	A'	$\nu CO(19)$, $\nu CF(18)$, $\beta C_4H(14)$, $\beta_1 rg(13)$
23	1334.5	7	A'	$\beta C_5H(18)$, $\beta C_6H(15)$, $\beta COH(14)$, $\nu CO(12)$, $\beta C_2H(11)$, $\beta C_4H(10)$	23	1322.7	82	A'	$\nu CO(21)$, $\beta C_6H(15)$, $\beta C_5H(13)$, $\beta C_2H(13)$, $\beta COH(11)$
24	1354.7	11	A'	$\nu C_2C_3(16)$, $\nu C_3C_4(16)$, $\nu C_5C_6(14)$, $\nu C_4C_5(14)$, $\nu C_1C_6(12)$, $\nu C_1C_2(11)$	24	1354.3	16	A'	$\nu C_5C_6(15)$, $\nu C_3C_4(15)$, $\nu C_4C_5(15)$, $\nu C_2C_3(15)$, $\nu C_1C_6(13)$, $\nu C_1C_2(10)$
25	1488.5	44	A'	$\beta C_6H(22)$, $\nu C_1C_2(14)$, $\nu C_2C_3(13)$, $\nu C_4C_5(10)$, $\beta C_4H(10)$	25	1507.4	18	A'	$\beta C_6H(19)$, $\nu C_2C_3(16)$, $\beta C_4H(13)$, $\nu C_1C_2(11)$
26	1529.0	81	A'	$\beta C_5H(22)$, $\nu C_3C_4(14)$, $\beta C_2H(14)$, $\nu C_1C_6(10)$	26	1515.9	125	A'	$\beta C_5H(26)$, $\beta C_2H(12)$, $\nu C_3C_4(12)$, $\nu C_1C_6(10)$
27	1637.3	147	A'	$\nu C_1C_2(23)$, $\nu C_4C_5(20)$	27	1642.1	72	A'	$\nu C_3C_4(23)$, $\nu C_1C_6(21)$, $\nu C_1C_2(13)$, $\nu C_4C_5(10)$
28	1657.9	98	A'	$\nu C_1C_6(19)$, $\nu C_2C_3(19)$, $\nu C_3C_4(18)$, $\nu C_5C_6(15)$	28	1652.2	164	A'	$\nu C_2C_3(23)$, $\nu C_5C_6(19)$, $\nu C_1C_2(13)$, $\beta_2 rg(10)$
29	3179.8	8	A'	$\nu C_5H(77)$	29	3160.4	10	A'	$\nu C_6H(89)$, $\nu C_5H(10)$
30	3181.5	1	A'	$\nu C_2H(91)$	30	3184.9	8	A'	$\nu C_5H(82)$, $\nu C_6H(10)$
31	3202.7	4	A'	$\nu C_6H(83)$, $\nu C_4H(10)$	31	3210.9	1	A'	$\nu C_4H(91)$
32	3211.1	1	A'	$\nu C_4H(84)$	32	3213.9	0	A'	$\nu C_2H(99)$
33	3836.4	70	A'	$\nu OH(100)$	33	3835.7	71	A'	$\nu OH(100)$

[a] Footnotes $a-d$ are identical to those in Table 11.

TABLE 16. Harmonic vibrational frequencies, IR intensities and assignments for *cis* and *trans* meta-chlorophenols[a]

No.	Freq.	IR	Sym.	Assignment, PED(%)	No.	Freq.	IR	Sym.	Assignment, PED(%)
1	180.4	0	A''	$\tau_3$rg(34), γCCl(30), γC_2H(13)	1	180.6	3	A''	$\tau_3$rg(33), γCCl(29), γC_2H(15)
2	228.8	3	A''	$\tau_2$rg(69)	2	229.2	1	A''	$\tau_2$rg(68), $\tau_1$rg(10)
3	247.6	2	A'	βCCl(69), βCO(16)	3	248.7	1	A'	βCCl(69), βCO(16)
4	311.9	110	A''	τOH(90) Expt: 312.5[b], 312[c], 313[d]	4	307.8	111	A''	τOH(93)
5	406.2	6	A'	νCCl(49), $\beta_3$rg(22)	5	407.3	5	A'	νCCl(50), $\beta_3$rg(23)
6	447.1	10	A'	βCO(59), βCCl(16), $\beta_3$rg(13)	6	443.8	9	A'	βCO(59), βCCl(16), $\beta_3$rg(12)
7	448.7	11	A''	$\tau_3$rg(57), γCCl(13)	7	449.7	2	A''	$\tau_3$rg(57), γCCl(12)
8	535.7	3	A'	$\beta_2$rg(75)	8	536.3	3	A'	$\beta_2$rg(73)
9	579.9	0	A''	γCO(34), $\tau_2$rg(27), γCCl(27)	9	580.1	1	A''	γCO(39), $\tau_2$rg(28), γCCl(24)
10	677.3	13	A''	$\tau_1$rg(69), γCO(13)	10	670.7	11	A''	$\tau_1$rg(68), γCO(11), γCCl(10), γC_5H(10)
11	697.1	7	A'	$\beta_3$rg(56), νCCl(20)	11	695.0	8	A'	$\beta_3$rg(56), νCCl(20)
12	780.9	45	A''	γC_4H(36), γCO(23), γC_5H(21), $\tau_1$rg(12)	12	764.3	61	A''	γC_6H(33), γC_4H(25), γC_5H(16), $\tau_1$rg(15)
13	835.3	28	A''	γC_2H(68), $\tau_1$rg(16)	13	859.3	4	A''	γC_6H(37), γC_4H(27), γC_2H(15)
14	886.1	0	A''	γC_6H(46), γC_4H(40)	14	866.9	17	A''	γC_2H(62), γC_4H(12), $\tau_1$rg(11)
15	893.7	82	A'	νCO(22), νCCl(22), $\beta_2$rg(15), $\beta_1$rg(11), νC_1C_6(10)	15	894.8	108	A'	νCO(23), νCCl(21), $\beta_2$rg(14), $\beta_1$rg(11), νC_1C_6(11)
16	975.6	0	A''	γC_5H(58), γC_6H(18), γC_4H(13)	16	954.1	0	A''	γC_5H(59), γC_4H(21)
17	1009.1	9	A'	$\beta_1$rg(64), νC_1C_6(11)	17	1009.5	6	A'	$\beta_1$rg(64), νC_1C_2(10)
18	1087.3	15	A'	νC_4C_5(29), βC_6H(16), νC_5C_6(10)	18	1087.8	30	A'	βC_2H(20), νC_3C_4(18), νC_4C_5(17), νCCl(13)
19	1107.6	38	A'	νC_4C_3(19), νC_2C_3(15), βC_4H(15), βC_2H(12)	19	1107.6	0	A'	βC_4H(23), νC_5C_6(18), βC_6H(12), νC_4C_5(12)
20	1180.4	43	A'	βC_6H(28), βC_5H(23), βC_4H(14)	20	1183.8	27	A'	βC_6H(23), βCOH(20), βC_5H(19), βC_4H(10)

(*continued overleaf*)

TABLE 16. (continued)

No.	Freq.	IR	Sym.	Assignment, PED(%)	No.	Freq.	IR	Sym.	Assignment, PED(%)
21	1191.1	89	A'	βCOH(46), $\beta C_5H(11)$, $\nu C_1C_2(10)$	21	1192.3	111	A'	βCOH(38), $\beta C_5H(17)$, $\nu C_1C_2(13)$, $\beta C_2H(11)$
22	1271.3	87	A'	νCO(42), $\beta C_2H(20)$, $\beta_1$rg(12)	22	1271.5	29	A'	νCO(40), $\beta_1$rg(12), $\beta C_2H(12)$
23	1333.6	3	A'	$\beta C_5H(21)$, $\beta C_4H(17)$, βCOH(16), $\beta C_2H(14)$	23	1322.3	81	A'	$\beta C_2H(17)$, βCOH(15), $\beta C_5H(15)$, νCO(13), $\beta C_4H(13)$, $\beta C_6H(10)$
24	1342.4	15	A'	$\nu C_3C_4(16)$, $\nu C_2C_3(16)$, $\nu C_5C_6(14)$, $\nu C_4C_5(14)$, $\nu C_1C_2(11)$, $\nu C_1C_6(10)$	24	1342.9	9	A'	$\nu C_2C_3(15)$, $\nu C_3C_4(15)$, $\nu C_4C_5(15)$, $\nu C_5C_6(14)$, $\nu C_1C_6(11)$, $\nu C_1C_2(10)$
25	1470.8	52	A'	$\beta C_6H(21)$, $\nu C_2C_3(15)$, $\beta C_4H(13)$, $\nu C_4C_5(12)$, $\nu C_1C_2(11)$	25	1487.1	24	A'	$\nu C_2C_3(21)$, $\beta C_4H(21)$, $\beta C_6H(14)$
26	1515.0	73	A'	$\beta C_5H(21)$, $\beta C_2H(17)$, $\nu C_3C_4(15)$, $\beta C_4H(11)$	26	1504.6	93	A'	$\beta C_5H(27)$, $\nu C_3C_4(12)$, $\beta C_2H(12)$
27	1623.7	117	A'	$\nu C_1C_2(24)$, $\nu C_4C_5(21)$	27	1625.5	62	A'	$\nu C_1C_6(27)$, $\nu C_3C_4(19)$
28	1640.5	81	A'	$\nu C_1C_6(22)$, $\nu C_5C_6(17)$, $\nu C_2C_3(15)$, $\nu C_3C_4(14)$	28	1637.6	133	A'	$\nu C_1C_2(20)$, $\nu C_5C_6(19)$, $\nu C_2C_3(15)$, $\nu C_4C_5(13)$
29	3177.4	9	A'	$\nu C_5H(49)$, $\nu C_2H(42)$	29	3159.2	10	A'	$\nu C_6H(87)$, $\nu C_5H(12)$
30	3178.5	1	A'	$\nu C_2H(57)$, $\nu C_5H(35)$	30	3182.9	9	A'	$\nu C_5H(81)$, $\nu C_6H(11)$
31	3201.6	3	A'	$\nu C_6H(84)$	31	3210.6	1	A'	$\nu C_4H(85)$
32	3210.6	2	A'	$\nu C_4H(86)$	32	3212.2	1	A'	$\nu C_2H(91)$
33	3833.7	68	A'	νOH(100)	33	3835.6	74	A'	νOH(97)

[a] Footnotes $a–d$ are identical to those in Table 11.

TABLE 17. Harmonic vibrational frequencies, IR intensities and assignments for *cis* and *trans meta*-bromophenols[a]

No.	Freq.	IR	Sym.	Assignment, PED(%)	No.	Freq.	IR	Sym.	Assignment, PED(%)
1	161.0	0	A''	γCBr(39), $\tau_3$rg(30), γC_2H(11)	1	160.9	4	A''	γCBr(38), $\tau_3$rg(29), γC_2H(13)
2	204.0	1	A'	βCBr(75), βCO(11)	2	204.7	1	A'	βCBr(76), βCO(11)
3	228.0	2	A''	$\tau_2$rg(68), $\tau_1$rg(10)	3	227.8	2	A''	$\tau_2$rg(68), $\tau_1$rg(10)
4	307.4	2	A'	νCBr(65), $\beta_3$rg(16)	4	308.1	2	A'	νCBr(65), $\beta_3$rg(16)
5	314.7	109	A''	τOH(89) Expt: 314[b], 312[c], 309[d]	5	310.4	110	A''	τOH(92)
6	438.3	13	A'	βCO(67), βCBr(13), $\beta_3$rg(10)	6	435.3	8	A'	βCO(66), βCBr(13)
7	443.3	12	A''	$\tau_3$rg(60), γCBr(15)	7	445.0	2	A''	$\tau_3$rg(61), γCBr(14)
8	535.6	2	A'	$\beta_2$rg(76)	8	535.9	3	A'	$\beta_2$rg(76)
9	564.5	0	A''	γCO(28), $\tau_2$rg(28), γCBr(26)	9	567.4	0	A''	γCO(31), $\tau_2$rg(28), γCBr(25)
10	668.4	11	A''	$\tau_1$rg(69), γCO(16)	10	664.7	11	A''	$\tau_1$rg(68), γCO(14)
11	679.5	7	A'	$\beta_3$rg(65), νCBr(17)	11	678.0	8	A'	$\beta_3$rg(66), νCBr(16)
12	779.0	48	A''	γC_4H(35), γC_6H(23), γC_5H(23), $\tau_1$rg(11)	12	762.5	61	A''	γC_6H(32), γC_4H(26), γC_5H(18), $\tau_1$rg(13)
13	832.5	22	A''	γC_2H(73), $\tau_1$rg(14)	13	863.1	0	A''	γC_6H(45), γC_4H(38)
14	874.1	76	A'	νCO(23), νCBr(17), $\beta_2$rg(17), $\beta_1$rg(15), νC_1C_6(10)	14	866.2	16	A''	γC_2H(74), $\tau_1$rg(12)
15	888.3	1	A''	γC_6H(46), γC_4H(40)	15	875.0	97	A'	νCO(23), νCBr(17), $\beta_2$rg(16), $\beta_1$rg(16), νC_1C_6(11)
16	974.5	0	A''	γC_5H(57), γC_6H(18), γC_4H(14)	16	953.9	0	A''	γC_5H(58), γC_4H(22)
17	1007.4	13	A'	$\beta_1$rg(64), νC_1C_6(11)	17	1007.9	9	A'	$\beta_1$rg(64), νC_1C_2(11), νC_1C_6(10)
18	1080.6	12	A'	νC_4C_5(27), νC_3C_4(13), βC_2H(11), βC_6H(11)	18	1079.4	18	A'	νC_3C_4(23), βC_2H(18), νC_4C_5(17)
19	1106.0	23	A'	νC_3C_4(16), βC_4H(15), νC_2C_3(14) νC_5C_6(12), βC_6H(12)	19	1108.0	1	A'	βC_4H(21), νC_5C_6(19), βC_6H(14), νC_4C_5(13)
20	1181.4	45	A'	βC_6H(27), βC_5H(23), βC_4H(15)	20	1184.5	32	A'	βC_6H(22), βCOH(20), βC_5H(18), βC_4H(10)

(*continued overleaf*)

TABLE 17. (continued)

No.	Freq.	IR	Sym.	Assignment, PED(%)	No.	Freq.	IR	Sym.	Assignment, PED(%)
21	1194.0	91	A'	βCOH(46), $\beta C_5H(10)$, $\nu C_1C_2(10)$, $\nu C_1C_6(10)$, $\beta C_2H(10)$	21	1192.4	116	A'	βCOH(34), $\beta C_5H(18)$, $\beta C_2H(13)$, $\nu C_1C_2(13)$
22	1269.1	86	A'	νCO(43), $\beta C_2H(21)$, $\beta_1$rg(12)	22	1269.2	26	A'	νCO(38), $\beta C_2H(12)$, $\beta_1$rg(12)
23	1334.0	2	A'	$\beta C_5H(20)$, $\beta C_4H(19)$, $\beta C_2H(16)$, βCOH(14), $\beta C_6H(10)$	23	1322.3	83	A'	$\beta C_2H(19)$, $\beta C_5H(15)$, $\beta C_4H(15)$, νCO(12), βCOH(12), $\beta C_6H(10)$
24	1339.6	18	A'	$\nu C_2C_3(17)$, $\nu C_3C_4(17)$, $\nu C_4C_5(14)$, $C_5C_6(14)$, $\nu C_1C_2(11)$, $\nu C_1C_6(10)$	24	1339.9	10	A'	$\nu C_2C_3(16)$, $\nu C_3C_4(15)$, $\nu C_4C_5(15)$, $\nu C_5C_6(14)$, $\nu C_1C_6(11)$, $\nu C_1C_2(10)$
25	1467.3	58	A'	$\beta C_6H(21)$, $\nu C_2C_3(15)$, $\beta C_4H(13)$, $\nu C_4C_5(12)$, $\nu C_1C_2(11)$	25	1483.1	21	A'	$\nu C_2C_3(21)$, $\beta C_4H(21)$, $\beta C_6H(14)$, βCOH(10), $\nu C_4C_5(10)$
26	1511.9	73	A'	$\beta C_5H(20)$, $\beta C_2H(18)$, $\nu C_3C_4(15)$, $\beta C_4H(12)$	26	1501.6	95	A'	$\beta C_5H(27)$, $\nu C_3C_4(13)$, $\beta C_2H(13)$
27	1618.5	123	A'	$\nu C_1C_2(24)$, $\nu C_4C_5(21)$	27	1621.9	54	A'	$\nu C_1C_6(28)$, $\nu C_3C_4(18)$, $\nu C_4C_5(10)$
28	1637.7	76	A'	$\nu C_1C_6(23)$, $\nu C_5C_6(17)$, $\nu C_2C_3(15)$, $\nu C_3C_4(13)$	28	1633.0	142	A'	$\nu C_5C_6(21)$, $\nu C_1C_2(17)$, $\nu C_2C_3(16)$, $\nu C_4C_5(11)$
29	3176.4	9	A'	$\nu C_5H(70)$, $\nu C_2H(18)$	29	3157.8	10	A'	$\nu C_6H(88)$, $\nu C_5H(11)$
30	3177.8	2	A'	$\nu C_2H(81)$, $\nu C_5H(15)$	30	3181.9	9	A'	$\nu C_5H(82)$, $\nu C_6H(11)$
31	3200.4	3	A'	$\nu C_6H(86)$	31	3210.5	1	A'	$\nu C_4H(85)$
32	3211.0	1	A'	$\nu C_4H(89)$	32	3212.4	1	A'	$\nu C_2H(91)$
33	3834.3	69	A'	νOH(100)	33	3835.3	75	A'	νOH(100)

[a] Footnotes $a-d$ are identical to those in Table 11.

TABLE 18. Harmonic vibrational frequencies, IR intensities and assignments for *para*-fluorophenol[a]

No.	Freq.	IR	Sym.	Assignment, PED(%)
1	155.6	0	A''	$\tau_2$rg(75)
2	282.3	113	A''	τOH(92) Expt: 280[b], 280[c], 283[d]
3	343.0	6	A'	βCF(44), βCO(40)
4	366.0	0	A''	$\tau_1$rg(36), γCF(29), γCO(14)
5	427.3	1	A''	$\tau_3$rg(83)
6	447.3	7	A'	βCO(35), βCF(32), $\beta_3$rg(27)
7	463.0	1	A'	$\beta_2$rg(76)
8	512.7	23	A''	$\tau_2$rg(33), γCF(29), γCO(29)
9	652.5	1	A'	$\beta_3$rg(74)
10	680.6	0	A''	$\tau_1$rg(64), γCO(16), γCF(14)
11	751.5	68	A'	$\beta_1$rg(36), vCF(24), vCO(21)
12	797.6	11	A''	$\gamma C_2H(49)$, $\gamma C_3H(27)$, $\gamma C_5H(10)$
13	836.1	63	A''	$\gamma C_6H(33)$, $\gamma C_5H(24)$, $\gamma C_2H(12)$, $\tau_2$rg(12), γCO(10)
14	861.1	0	A'	$\beta_2$rg(22), $vC_1C_6(13)$, $vC_1C_2(12)$, vCF(12), vCO(12)
15	907.1	3	A''	$\gamma C_3H(48)$, $\gamma C_2H(21)$, $\tau_1$rg(14)
16	949.4	0	A''	$\gamma C_5H(40)$, $\gamma C_6H(38)$
17	1024.7	1	A'	$\beta_1$rg(46), $vC_4C_5(10)$
18	1110.4	19	A'	$\beta C_5H(21)$, $\beta C_3H(17)$, $\beta C_6H(15)$, $vC_5C_6(13)$, $vC_2C_3(12)$, $\beta C_2H(10)$
19	1169.3	3	A'	$\beta C_3H(27)$, $\beta C_5H(20)$, $\beta C_2H(15)$, $\beta C_6H(14)$
20	1186.2	144	A'	βOH(52), $vC_1C_6(15)$
21	1229.0	174	A'	vCF(43), $\beta_1$rg(20)
22	1280.1	26	A'	vCO(49), $vC_5C_6(10)$, $vC_2C_3(10)$, vCF(7)
23	1317.3	3	A'	$\beta C_2H(18)$, $\beta C_6H(15)$, $vC_3C_4(14)$, $\beta C_5H(13)$, $vC_4C_5(12)$, $\beta C_3H(11)$
24	1355.6	26	A'	$vC_2C_3(14)$, βCOH(13), $vC_5C_6(12)$, $vC_1C_2(11)$, $vC_4C_5(11)$, $vC_3C_4(11)$
			A'	$vC_1C_6(10)$
25	1466.9	29	A'	$vC_5C_6(19)$, $vC_2C_3(16)$, $\beta C_6H(14)$, βOH(11)
26	1538.4	232	A'	$\beta C_2H(15)$, $vC_3C_4(12)$, $\beta C_5H(12)$, $\beta C_3H(11)$
27	1645.3	4	A'	$vC_4C_5(25)$, $vC_1C_2(23)$
28	1655.6	0	A'	$vC_1C_6(21)$, $vC_3C_4(20)$, $vC_5C_6(13)$, $vC_2C_3(13)$
29	3159.5	12	A'	$vC_2H(96)$
30	3191.6	3	A'	$vC_6H(57)$, $vC_5H(42)$
31	3201.3	2	A'	$vC_3H(95)$
32	3205.3	1	A'	$vC_5H(56)$, $vC_6H(42)$
33	3840.2	66	A'	vOH(100)

[a]Footnotes *a–d* are identical to those in Table 11.

TABLE 19. Some harmonic vibrational frequencies, IR intensities and assignments for *para*-chlorophenols[a]

Freq.	IR	Sym.	Assignment, PED(%)
300.1	104	A''	$\underline{\tau\text{OH}}$ (89)
836.0	2	A'	$\beta_2$rg(23), $\underline{v\text{CO}}$(21), $vC_1C_2(13)$, $vC_1C_6(13)$, $\beta_1$rg(11)
1279.1	107	A'	$\underline{v\text{CO}}$(53), $\beta_1$rg(10)
3836.0	73	A'	$\underline{v\text{OH}}$(100)

[a]See Footnote *a* in Table 10.

TABLE 20. Some harmonic vibrational frequencies, IR intensities and assignments for para-bromophenols[a]

Freq.	IR	Sym.	Assignment, PED(%)
303.1	51	A''	$\underline{\tau OH}(50)$, $\gamma CBr(22)$, $\tau_2 rg(11)$, $\tau_1 rg(10)$
312.0	63	A''	$\underline{\tau OH}(43)$, $\gamma CBr(21)$, $\tau_2 rg(11)$, $\gamma CO(10)$
831.7	1	A'	$\beta_2 rg(22)$, $\underline{\nu CO}(22)$, $\beta_1 rg(15)$, $\nu C_1 C_2(13)$, $\nu C_1 C_6(13)$
1279.1	125	A'	$\underline{\nu CO}(53)$, $\beta_1 rg(10)$
3835.3	76	A'	$\underline{\nu OH}(100)$

[a] See Footnote a in Table 10.

parameters of these conformers are demonstrated in Figure 13. This is also manifested in the vibrational spectra.

Let us deal first with the torsional mode τ_{OH}. In both cis m-ClC$_6$H$_4$OH and cis m-BrC$_6$H$_4$OH, it is predicted to be at higher wavenumbers compared to their trans partners while in m-fluorophenols it is placed higher, at 320.7 cm^{-1} ($\tau_{OH}^{expt} = 319$ cm$^{-1\,246}$), in the trans conformer than in the cis one, viz. 314.0 ($\tau_{OH}^{expt} = 311$ cm$^{-1\,246}$). Due to a small difference of about 7 cm^{-1}, it would be premature to offer a theoretical explanation of such 'misbehaviour' of τ_{OH} in m-XC$_6$H$_4$OH until it is fully proved or disproved experimentally, particularly in the related overtones where such a difference could be more pronounced. However, we suggest that such features are presumably related to the changes in the electrostatic repulsion between the O—H bond and its cis ortho C—H bond due to a different electron-withdrawing vs. electron-donating ability of the X atoms and a possible weak interaction between this ortho C—H bond and the halogen atom. The former repulsion might make the potential well more shallow for the planar orientation of the OH bond and thus cause a red shift of the τ_{OH}. Noteworthy is a rather strong dependence of τ_{OH} on the $C_1C_{2(6)}H$ angle of this C—H bond which partly determines the strength of this repulsive interaction. Thus, a positive departure of this angle from the phenolic one by 3° produces a blue shift of the τ_{OH} by about 5 cm^{-1}, while a negative deviation moves it downward by nearly the same value. Interestingly, the analogous Hartree–Fock calculations lead to approximately the same frequency alterations, thus indicating the dominant electrostatic origin of the cis–trans non-similarity.

The CO stretch internal coordinate in XC$_6$H$_4$OH is involved in several vibrational modes. Similarly to the parent phenol, ν_{CO} contributes dominantly to the two modes whose atomic displacements are inherent to modes 13 and 1 of benzene, according to Varsanýi nomenclature[167]. While the latter characterized by a lower frequency retains its radial skeleton character and describes a ring breathing, the former can be likely interpreted as the CO stretch due to a larger contribution of ν_{CO}. In the theoretical spectrum of the parent phenol (Table 10), it is centred at 1274.8 cm^{-1} (expt: 1259–1262 cm$^{-1\,153}$) and characterized by IR intensity of 91 km mol^{-1}. The X-substitution of phenol affects both its position and the IR intensity. Analysis of Tables 11–13 and 15–20 leads to the following conclusions: (a) all cis ortho-substituted phenols have this mode at lower frequencies and larger IR intensities compared to their trans partners; (b) in all cis meta-substituted structures, it is more IR active than in the corresponding trans-substituted ones, while its frequency in each pair of conformers is nearly the same. In ortho-substituted forms, it develops into a rather intense band placed at 1284.3 cm^{-1} (130 km mol^{-1}) and 1298.8 cm^{-1} (61 km mol^{-1}) in cis and trans o-FC$_6$H$_4$OH, respectively, 1274.8 cm^{-1} (75 km mol^{-1}) (expt: 1255 cm$^{-1\,167}$) and 1285.6 cm^{-1} (23 km mol^{-1}) in cis and trans o-ClC$_6$H$_4$OH, respectively, and 1273.9 cm^{-1} (63 km mol^{-1}) (expt: 1247 cm$^{-1\,167}$) and 1282.9 cm^{-1} (21 km mol^{-1}) in cis and trans o-BrC$_6$H$_4$OH, respectively.

The ring breathing vibrational mode predicted at 1010.5 cm^{-1} (expt: 993.1 cm$^{-1\,247-250}$) in the prototype benzene downshifts to 827.3 cm^{-1} (23 km mol^{-1}) (expt: 823 cm^{-1} (20 km mol$^{-1\,153}$)) upon substitution of one hydrogen atom by the OH group. In phenol and its halo-derivatives, this mode is mixed with the stretching vibrations of the light substituents, namely ν_{CO} and ν_{CF}. In halophenols, it is placed at higher wavenumbers compared to phenol, in particular at 861.1 cm^{-1} (expt: 854 cm^{-1}), 836.0 cm^{-1} (expt: 836 cm^{-1}) and 831.7 cm^{-1} (expt: 825 cm^{-1})[251] in the spectra of *para*-fluoro-, chloro- and bromophenols, respectively. This supports the earlier assignment of this vibrational mode in a series of *para*-substituted phenols[251] (cf. also Reference 245).

Further, if in all *para*-substituted phenols the CO stretching vibration is mainly localized on these two fundamental modes, in some *ortho*- and *meta*-phenols it appears coupled with the mode corresponding to the fundamental three of benzene whose displacements resemble a distortion towards a 'Catherine wheel' type of structure. Such vibration appears to be rather sensitive to the position (i.e. either *cis* or *trans*) of the X atom, being almost independent of its nature. In all *trans ortho*- and *meta*-substituted phenols, it is placed at slightly lower wavenumbers and characterized by a consistently larger IR intensity compared to the *cis* conformers. Consider the following example. For all *trans m*-XC$_6$H$_4$OH, it is centred at *ca* 1322 cm^{-1} (81–83 km mol^{-1}) and blue shifts to *ca* 1334 cm^{-1} (2–7 km mol^{-1}) for the *cis* conformer. In *trans ortho*-substituted forms, it is found at 1316–1317 cm^{-1} (73–78 km mol^{-1}), while in the *cis* forms it is at 1323–1324 cm^{-1} (21–28 km mol^{-1}).

We end this subsection with a surprise which is quite obsolete, since it is about twenty years old[252-254]. However, wise people always say that a forgotten surprise is often better than a new one. Anyway, we think that wrapping it within the present theoretical method is worth mentioning to complete our understanding of the stability of XC$_6$H$_4$OH. In equation 12 we present the relative energies (in kJ mol^{-1}) of all forms of the monohalogeno-substituted phenols.

$$\text{F:} \quad \textit{trans m} \overset{0.79}{>} \textit{cis m} \overset{3.64}{>} \textit{cis o} \overset{1.63}{>} \textit{para} \overset{9.75}{>} \textit{trans o}$$

$$\text{Cl:} \quad \textit{cis o} \overset{3.31}{>} \textit{cis m} \overset{0.08}{\approx} \textit{trans m} \overset{1.84}{>} \textit{para} \overset{7.24}{>} \textit{trans o}$$

$$\text{Br:} \quad \textit{cis o} \overset{1.16}{>} \textit{cis m} \overset{0.04}{\approx} \textit{trans m} \overset{1.05}{>} \textit{para} \overset{6.99}{>} \textit{trans o} \qquad (12)$$

Its analysis leads to the following conclusions. First, the intramolecular hydrogen bond in the *cis o*-Cl- and *cis o*-Br-phenols is rather strong and leads all *meta*- and *para*-chloro- and bromophenols to fall energetically between their *cis ortho*- and *trans ortho*-conformers. Such order of stability breaks down for FC$_6$H$_4$OH where the *trans meta*-conformer appears to be the most stable one and reluctant to be engaged in the intramolecular hydrogen bonding and is followed by the *cis meta*-conformer. Surprisingly, the *cis ortho*-conformer occupies only the third place in the rank of the most energetically stable ones being by 4.4 kJ mol^{-1} lower than the most stable conformer. The *para*-conformer falls between the *cis* and *trans ortho*-conformers. Interestingly, the earlier orders of stability of FC$_6$H$_4$OH obtained at rather lower (from the present point of view) computational levels are given in kJ mol^{-1} in equation 13:

$$\textit{cis o} \overset{0.17}{\approx} \textit{cis m} \overset{0.17}{\approx} \textit{trans m} \overset{0.75}{>} \textit{para} \overset{15.3}{>} \textit{trans o}^{252}$$

$$\textit{cis m} \overset{0.55}{>} \textit{trans m} \overset{4.98}{>} \textit{para} \overset{2.09}{>} \textit{cis o} \overset{7.07}{>} \textit{trans o}^{145}$$

$$\textit{cis m} \overset{1.30}{>} \textit{trans m} = \textit{cis o} \overset{4.73}{>} \textit{para} \overset{13.9}{>} \textit{trans o}^{146} \qquad (13)$$

In summary, although we have succeeded in explaining the order of the strength of the intramolecular hydrogen bond in ortho-XC_6H_4OH in the gas phase and in the model solvent mimicking CCl_4 and reconcile the longstanding conflict between experiment and theory on the basis of a generalized solvent-including Pauling model, we still feel that our explanation looks rather incomplete. Therefore, we attempt to build such a bridge in the next subsection using the concept of the electronic localization function.

C. The Bonding Trends in Monohalogenated Phenols in Terms of the Electronic Localization Function (ELF)

1. Introduction to the ELF

Nearly a decade ago, Becke and Edgecombe in their seminal paper[255] introduced the electron localization function (ELF) $\eta(\mathbf{r})$ of an arbitrary N-electron system (equation 14) as

$$\eta(\mathbf{r}) = (1 + [(t - t_W)/t_{TF}]^2)^{-1} \qquad (14)$$

where $t = \frac{1}{2}\sum_{i=1}^{N} |\nabla \psi_i|^2$ is the kinetic energy density of the studied system within the Hartree–Fock or Kohn–Sham approach and ψ_i ($i = 1, \ldots, N$) are the corresponding molecular orbitals. Here, $t_W[\rho(\mathbf{r})] = (\nabla \rho)^2/8\rho$ is the Weizsäcker kinetic energy density determined by the one-electron density $\rho(\mathbf{r}) = \sum_{i=1}^{N} |\psi_i(\mathbf{r})|^2$, and finally $t_{TF}[\rho(\mathbf{r})] = \alpha_{TF}[\rho(\mathbf{r})]^{5/3}$ is the Thomas–Fermi kinetic energy density with numerical coefficient $\alpha_{TF} = 3(6\pi^2)^{2/3}/5$ derived within the uniform electron gas approximation[145].

The ELF $\eta(\mathbf{r})$ has a rather simple normalized Lorentzian-type form and thus its domain lies in the interval $0 \leqslant \eta(\mathbf{r}) \leqslant 1$. The upper limit of $\eta(\mathbf{r}) = 1$ corresponds to the electron system whose kinetic energy density becomes identical to the Weizsäcker one. Bearing in mind that the latter was derived on the basis of the Pauli principle, $\eta(\mathbf{r}) = 1$ implies that all electrons are paired if $2/N$, and there is only one unpaired electron in the opposite case. Its value $\eta(\mathbf{r}) = \frac{1}{2}$ determining the FWHM ($\equiv$ full width at half maximum) describes a case when $t = t_W[\rho(\mathbf{r})] \pm t_{TF}[\rho(\mathbf{r})]$, where the lower sign is valid if $t_W[\rho(\mathbf{r})] \geqslant t_{TF}[\rho(\mathbf{r})]$.

2. Topology of the ELF

The purpose of the topological analysis of the electron localization function is to provide a sound mathematical model of the Lewis[256, 257], and VSEPR[143, 144, 258, 259] theories which removes the contradictions that the latter present with quantum mechanics. The ELF analysis therefore attempts to provide a mathematical bridge between chemical intuition and quantum mechanics. Since both Lewis and VSEPR phenomenological models describe the bonding within a molecule in the usual 3D space, the mathematical model should make a partition of this space into regions related to chemical properties. The theory of dynamical systems[260–262] then provides a very convenient mathematical framework to achieve the partition of the molecular space into such regions. The simplest dynamical systems are the gradient dynamical systems in which the vector field is the gradient field of a scalar function, say $V(\mathbf{r})$, called the potential function. The theory of atoms in molecules (AIM)[141] discussed above uses the gradient dynamical field of the charge density $\rho(\mathbf{r})$ to determine atomic basins. In order to provide evidence of electronic domains one has to choose another local function related to the pair-electron density. Unfortunately, the pair-electron functions depend on two space variables and therefore cannot be used directly as potential function.

The ELF defined in equation 14 is a local function which describes to what extent the Pauli repulsion is efficient at a given point of the molecular space. Originally, the ELF was derived from the Laplacian of the conditional probability $[\nabla_{\mathbf{r}_1}^2 P_{cond}(\mathbf{r}_1, \mathbf{r}_2)]_{\mathbf{r}_1=\mathbf{r}_2}$. An alternative interpretation was later proposed[263] in terms of the local excess kinetic energy density due to the Pauli exclusion principle. This interpretation not only gives a deeper physical meaning to the *ELF* function but also allows one to generalize the *ELF* to any wave function, in particular to the exact one. Therefore, the *ELF* provides a rigorous basis for the analysis of the wave function and of the bonding in molecules and crystals. In 1994, it was proposed to use the gradient field of *ELF* in order to perform a topological analysis of the molecular space[264] in the spirit of AIM theory. The attractors of *ELF* determine basins which are either core basin encompassing nuclei or valence basin when no nucleus except a proton lies within it. The valence basins are characterized by the number of core basins with which they share a common boundary; this number is called the valence basin synaptic order[265]. There are therefore asynaptic, monosynaptic, disynaptic and polysynaptic valence basins. Monosynaptic basins usually correspond to the lone pair regions whereas di- and polysynaptic basins characterize chemical bonds. An advantage of such representation is that it provides a clear criterion to identify multicentric bonds. In a way, this is a complementary view to the traditional valence representation: instead of counting bonds from a given centre which only accounts for two-body links, the count is performed from the 'piece of glue' which sticks the atoms one to another.

From a quantitative point of view a localization basin (core or valence) is characterized by its population, i.e. the integrated one-electron density $\rho(\mathbf{r})$ over the basin (equation 15)

$$\bar{N}(\Omega_i) = \int_{\Omega_i} d^3\mathbf{r}\rho(\mathbf{r}) \tag{15}$$

where Ω_i is the volume of the basin. It is worthwhile to calculate the variance of the basin population by equation 16,

$$\sigma^2(\bar{N};\Omega_i) = \int_{\Omega_i} d^3\mathbf{r}_1 \int_{\Omega_i} d^3\mathbf{r}_2 P(\mathbf{r}_1,\mathbf{r}_2) - [\bar{N}(\Omega_i)]^2 + N(\Omega_i)] \tag{16}$$

where $P(\mathbf{r}_1,\mathbf{r}_2)$ is the spinless pair-electron density[145]. It has been shown that the variance can readily be written as a sum of contributions arising from the other basins (covariance)[266] (equation 17)

$$\sigma^2(\bar{N};\Omega_i) = \sum_{j\neq i}\bar{N}(\Omega_i)\bar{N}(\Omega_j) - \int_{\Omega_i} d^3\mathbf{r}_1 \int_{\Omega_j} d^3\mathbf{r}_2 P(\mathbf{r}_1,\mathbf{r}_2) \tag{17}$$

In equation 17, $\bar{N}(\Omega_i)\bar{N}(\Omega_j)$ is the number of the electron pairs classically expected from the basin population whereas $\bar{N}(\Omega_i,\Omega_j)$ is the actual number of pairs obtained by integration of the pair-electron function over the basins Ω_i and Ω_j. The variance is then a measure of the quantum mechanical uncertainty of the basin population which can be interpreted as a consequence of the electron delocalization, whereas the pair covariance indicates how much the population fluctuations of two given basins are correlated. Within the AIM framework, the atomic localization and delocalization indices $\lambda(A)$ and $\delta(A, B)$ have been introduced[267] and defined by equations 18 and 19:

$$\lambda(A) = \bar{N}(\Omega_A) - \sigma^2(\bar{N};\Omega_A) \tag{18}$$

$$\delta(A, B) = 2\bar{N}(\Omega_A)\bar{N}(\Omega_B) - 2\int_{\Omega_A} d^3\mathbf{r}_1 \int_{\Omega_B} d^3\mathbf{r}_2 P(\mathbf{r}_1,\mathbf{r}_2) \tag{19}$$

The AIM delocalization indices are sometimes referred to as bond orders[268, 269]. The above notation[265] can be generalized to any partition in the direct space and therefore is adopted in the present work. Within the *ELF* approach, the core population variance and the core valence delocalization indices can be used to decide if a given core contributes to the synaptic order of an adjacent valence basin. For example, in the LiF molecule, the variances of the C(Li) and C(F) basins are 0.09 and 0.38, respectively, whereas $\delta(C(Li), V(F)) = 0.16$ and $\delta(C(F), V(F)) = 0.74$, where C stands for core and V for valence.

The concept of localization domain has been introduced[265] for graphical purposes and also in order to define a hierarchy of the localization basins which can be related to chemical properties. A localization domain is a volume limited by one or more closed isosurfaces $\eta(\mathbf{r}) = f$. A localization domain surrounds at least one attractor—in this case it is called *irreducible*. If it contains more than one attractor, it is *reducible*. Except for atoms and linear molecules, the irreducible domains are always filled volumes whereas the reducible ones can be either filled volumes, hollow volumes or donuts. Upon the increase in the value of $\eta(\mathbf{r})$ defining the boundary isosurface, a reducible domain splits into several domains, each containing less attractors than the parent one. The reduction of localization occurs at the turning points, which are critical points of index 1 located on the separatrix of two basins involved in the parent domain. Ordering these turning points (localization nodes) by increasing $\eta(\mathbf{r})$ enables one to build tree diagrams reflecting the hierarchy of the basins. A core basin is counted in the synaptic order of valence basins if there exists a value of the localization function which gives rise to a hollow volume localization domain (containing the considered valence basin attractors) with the core domain in its hole.

Before proceeding further with bridging the *ELF* with the key properties of monohalophenols, we pause briefly to analyse analytically the vector gradient field of *ELF*.

3. Vector gradient field $\nabla_\mathbf{r} \eta(\mathbf{r})$

Applying the gradient to $\eta(\mathbf{r})$ defined by equation 14, we derive equation 20,

$$\nabla \eta(\mathbf{r}) = \frac{2(t - t_W) t_{TF}}{[(t - t_W)^2 + t_{TF}^2]^2} [(t - t_W) \nabla t_{TF} - t_{TF} \nabla (t - t_W)] \tag{20}$$

where $\nabla_\mathbf{r} \equiv \nabla$ for short. Assuming molecular orbitals to be real valued, equation 20 is then easily transformed to equation 21,

$$\frac{\rho^{1/3}[(t - t_W)^2 + t_{TF}^2]^2}{2\alpha_{TF}(t - t_W) t_{TF}} \nabla \eta(\mathbf{r}) \tag{21}$$

$$= \sum_{i,j,k=1}^{N} \left[\frac{8}{3} \nabla \psi_i \psi_j \psi_k \nabla \psi_k (\psi_i \nabla \psi_j - \psi_j \nabla \psi_i) + \psi_i \psi_j^2 \nabla \psi_k (\psi_k \nabla^2 \psi_i - \psi_i \nabla^2 \psi_k) \right]$$

$$= \frac{8}{3} \nabla \rho \sum_{i<j}^{N} (\psi_i \nabla \psi_j - \psi_j \nabla \psi_i)^2 - \rho \sum_{i<j}^{N} (\psi_i \nabla \psi_j - \psi_j \nabla \psi_i)(\psi_i \nabla^2 \psi_j - \psi_j \nabla^2 \psi_i).$$

Therefore, we finally obtain equation 22[260, 261],

$$\nabla \eta(\mathbf{r}) = -\frac{\alpha_{TF}(t - t_W) t_{TF}}{[(t - t_W)^2 + t_{TF}^2]^2} \rho^{10/3} \nabla (J^2/\rho^8/3) \tag{22}$$

where J^2 is given by equation 23[270, 271],

$$J^2 = \frac{1}{4} \sum_{i<j}^{N} (\psi_i \nabla \psi_j - \psi_j \nabla \psi_i)^2 \tag{23}$$

Summarizing, the vector field $\nabla \eta(\mathbf{r})$ of the *ELF* vanishes at those $\mathbf{r} \in \mathbf{R}^3$ which obey the condition $t(\mathbf{r}) = t_W[\rho(\mathbf{r})]$ or equation 24,

$$J^2(\mathbf{r}) = C\rho^{8/3}(\mathbf{r}) \qquad (24)$$

where C is a constant in $\mathbf{R}^3$.

For one purpose let us rewrite equation 23 as equation 25,

$$J^2 = \sum_{i<j}^{N} |\mathbf{j}_{ij}|^2 \qquad (25)$$

where $\mathbf{j}_{ij} = (\psi_i \nabla \psi_j - \psi_j \nabla \psi_i)/2$ is the real time-independent electron transition current density between the ith and jth molecular orbitals. Hence, J^2 determines the square of the net charge transferred between all occupied molecular orbitals. Thus, the zero-flux surfaces of the *ELF* are defined by the condition that net charge or, in other words, the electron transition current density $Q_{tr}(\mathbf{r}) \equiv \sqrt{J^2(\mathbf{r})}$ associated with the transitions between all occupied molecular orbitals, is proportional to the electron density to the four-thirds power. This is the key difference in the vector gradient fields of $\rho(\mathbf{r})$ underlying the AIM theory and the *ELF*[272, 273].

4. The bonding in benzene, phenol and phenyl halides

In order to get some insight on how *ELF* works, we will analyse a number of parent molecules C_6H_5X (X = H, OH, F, Cl, Br and I). Their localization domains are displayed in Figure 14. Except for the substituent itself, all these molecules have 6 V(C, C), 5 V(C, H) and one V(C, X) basins. The differences are to be found in the hierarchy of the V(C, C) basins which is ruled by the nature of the substituent. In benzene, all the V(C, C) basins are equivalent and therefore the six critical points of index 1 between these basins have the same value, i.e. $\eta(\mathbf{r}_c) = 0.659$. In the phenyl halides where the molecular symmetry is lowered from D_{6h} to C_{2v}, the former critical points are then distributed in four sets according to the common carbon position: *ipso*, *ortho*, *meta* and *para*. In phenol with a C_s symmetry, the two *ortho* and the two *meta* positions are not totally equivalent. In all studied molecules, the $\eta(\mathbf{r}_c)$ values are enhanced in the *ipso*, *ortho* and *para* positions and decreased in the *meta* position. It has been remarked that the electrophilic substitution sites correspond to the carbon for which $\eta(\mathbf{r}_c)$ is enhanced[274]. Moreover, it is worthwhile to introduce *electrophilic substitution positional indices* defined by equation 26,

$$RI_c(S) = \eta(C_i; S) - \eta(C_i; H) \qquad (26)$$

where the subscript c denotes the position of the carbon labeled by i, i.e. *ortho*, *meta* or *para*. Interestingly, there exists a rather good correlation between the $RI_c(S)$ indices and the Hammett constants. Moreover, the positional indices are additive, enabling one to predict their values in a di-substituted molecule from the mono-substituted data.

The V(C_i, C_j) basin populations, their variance and the electrophilic substitution positional indices of the studied C_6H_5X molecules are listed in Table 21. The V(C, X) populations and their variance are close to their values in the CH_3X series. As expected the V(C, C) basin populations are intermediate between those inherent to a single and a double C—C bond and subject to a large fluctuation of the charge density. The classical meaning of the variance is the square of the standard deviation; though the standard deviation cannot be defined for a quantum system, the classical limit provides at least qualitative information about the delocalization. In the present case $\sigma \sim 1.16$, which is consistent with the resonance picture involving the Kekulé structures.

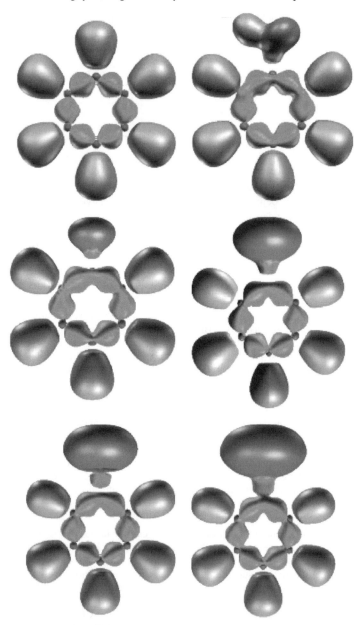

FIGURE 14 (PLATE 3). Localization domains of mono-X-substituted benzenes C_6H_5X (from left to right top X = H, OH, F, bottom X = Cl, Br, I). The *ELF* value defining the boundary isosurface, $\eta(\mathbf{r}) = 0.659$ corresponds to the critical point of index 1 on the separatrix between adjacent V(C, C) basins of benzene. Colour code: magenta = core, orange = monosynaptic, blue = protonated disynaptic, green = disynaptic. Adapted from Reference 220 with permission

TABLE 21. Basin populations $\bar{N}(V)$, variance of the basin populations $\sigma^2(V)$ and electrophilic substitution positional indices RI_c of the C_6H_5X molecules

	H	OH	F	Cl	Br	I
$\bar{N}(V(C_1, X))$	2.09	1.50	0.99	1.50	1.47	1.32
$\sigma^2(V(C_1, X))$	0.65	0.61	0.71	0.93	0.94	0.84
$\bar{N}(V(C_1, C_2))$	2.81	2.86	2.85	2.85	2.86	2.85
$\sigma^2(V(C_1, C_2))$	1.36	1.36	1.31	1.33	1.34	1.34
$\bar{N}(V(C_1, C_6))$	2.81	2.82	2.85	2.85	2.85	2.85
$\sigma^2(V(C_1, C_6))$	1.32	1.32	1.31	1.33	1.34	1.34
$\bar{N}(V(C_2, C_3))$	2.81	2.93	2.90	2.92	2.91	2.91
$\sigma^2(V(C_2, C_3))$	1.32	1.41	1.37	1.37	1.38	1.37
$\bar{N}(V(C_3, C_4))$	2.81	2.82	2.88	2.85	2.85	2.85
$\sigma^2(V(C_3, C_4))$	1.32	1.32	1.35	1.33	1.34	1.34
$\bar{N}(V(C_4, C_5))$	2.81	2.82	2.88	2.85	2.85	2.85
$\sigma^2(V(C_4, C_5))$	1.31	1.32	1.35	1.33	1.34	1.34
$\bar{N}(V(C_5, C_6))$	2.96	2.96	2.90	2.92	2.91	2.91
$\sigma^2(V(C_5, C_6))$	1.38	1.32	1.35	1.33	1.34	1.34
RI_1	0.0	0.032	0.077	0.059	0.053	0.049
RI_2	0.0	0.039	0.017	0.008	0.007	0.010
RI_3	0.0	−0.007	−0.005	−0.003	−0.003	0.003
RI_4	0.0	0.015	0.008	0.002	0.001	0.005
RI_5	0.0	−0.008	−0.005	−0.003	−0.003	0.003
RI_6	0.0	0.025	0.017	0.008	0.007	0.010

Values taken from Reference 220 with permission.

In phenol we reveal a noticeable increase in the $V(C_o, C_m)$ population with respect to benzene (0.11 e) whereas the populations of the other basins remain almost unchanged. Indeed, the net charge transfer towards the aromatic ring amounts to 0.20 e. The halogen atoms induce a larger net charge transfer: 0.34, 0.32, 0.32 and 0.30 for F, Cl, Br and I, respectively. However, this transfer is charged by all basins although the $V(C_o, C_m)$ populations are more enhanced than the $V(C_i, C_o)$ and $V(C_m, C_p)$ ones. The RI_s's are positive in the *ipso*, *ortho* and *para* positions and negative (except for I) in the *meta* ones. In the halogen series F–Br, the RI_c absolute values decrease with the electronegativity.

5. Monohalogenated phenols: the bonding in terms of ELF

The substitution of the CH group by the CX one (X = F, Cl, Br, I) in phenol is expected to be felt by the aromatic ring as a rather weak perturbation which would enhance the electron donation and modify the electrophilic substitutional indices according to the additive law[274]. As we have shown in Subsections III.A and III.B, in the *ortho* and *meta* substituted phenols the orientation of the OH bond in the molecular plane permits the existence of two conformers (Figures 12 and 13).

a. The ortho-substituted phenols. The localization domains of the *ortho*-substituted species are displayed in Figure 15: the *cis* conformers with the intramolecular hydrogen bond O–H···X are represented in the bottom row, the *trans* ones in the top row. Their basin populations and electrophilic substitution positional indices are given in Table 22.

Let us consider first the *trans* conformers in which the halogen substituent is not perturbed by an extra intramolecular interaction. In all molecules the $V(C_1, O)$ basin population is slightly increased with respect to phenol: the largest effect occurs for X = Cl, whereas for X = Br and I this effect is weaker than for the fluorinated species. The

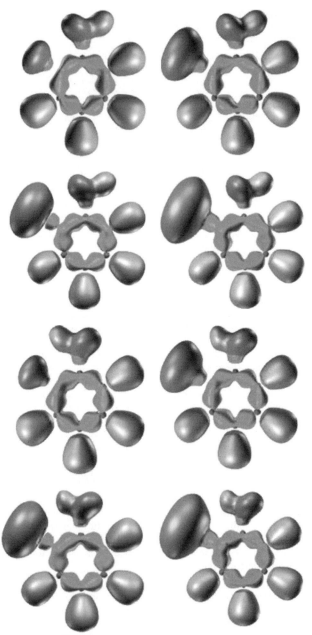

FIGURE 15 (PLATE 4). Localization domains of *ortho*-X-substituted phenols (from left to right X = F, Cl, Br, I; top—*trans* conformer, bottom—*cis* conformer). The *ELF* value defining the boundary isosurface, $\eta(\mathbf{r}) = 0.659$ corresponds to the critical point of index 1 on the separatrix between adjacent V(C, C) basins of benzene. Colour code: magenta = core, orange = monosynaptic, blue = protonated disynaptic, green = disynaptic. Adapted from Reference 220 with permission

TABLE 22. Basin populations $\bar{N}(V)$ and electrophilic substitution positional indices RI_c of ortho-substituted phenols

	trans conformation				cis conformation			
	F	Cl	Br	I	F	Cl	Br	I
				Populations				
$V(C_1, O)$	1.54	1.54	1.58	1.52	1.52	1.55	1.57	1.51
$V(C_2, X)$	1.05	1.49	1.45	1.39	1.0	1.42	1.33	1.29
$V(C_1, C_6)$	2.82	2.87	2.82	2.82	2.79	2.77	2.79	2.73
$V(C_1, C_2)$	2.92	2.78	2.80	2.74	2.97	2.85	2.80	2.82
$V(C_6, C_5)$	2.90	2.96	2.94	2.93	3.01	2.98	2.97	2.95
$V(C_5, C_4)$	2.97	2.86	2.86	2.89	2.90	2.75	2.76	2.80
$V(C_4, C_3)$	2.76	2.86	2.83	2.96	2.78	2.97	2.97	2.92
$V(C_3, C_2)$	2.98	3.05	2.83	2.96	2.99	3.04	3.0	3.04
Net transfer	0.43	0.46	0.16	0.26	0.52	0.44	0.37	0.34
				Positional indices				
RI_1	0.047	0.040	0.039	0.043	0.044	0.036	0.035	0.039
RI_2	0.100	0.082	0.076	0.072	0.113	0.099	0.094	0.092
RI_3	0.010	0.0	−0.001	0.002	0.009	0.00	−0.001	0.013
RI_4	0.010	0.012	0.013	0.018	0.011	0.013	0.014	0.009
RI_5	0.0	−0.006	−0.007	−0.003	0.0	−0.007	−0.008	−0.004
RI_6	0.035	0.037	0.037	0.042	0.020	0.022	0.023	0.028

Values taken from Reference 220 with permission.

$V(C_6, X)$ populations are close to their values in the corresponding halobenzene; however, there is a small electron transfer towards this basin for X = F, whereas the iodine atom undergoes an opposite effect. With respect to phenol, the regioselectivity of the electrophilic substitution is softened because as the OH and X = F, Cl, Br groups are both ortho–para directors, they contribute in opposite directions. As all the positional indices of C_6H_5I are positive, they are enhanced in the trans ortho-iodophenol. The additive rule works satisfactorily for all positions as the largest discrepancy between estimated and calculated value does not exceed 0.002.

In the cis conformer, the charge transfer towards the $V(C_1, O)$ basin is close to that calculated for the trans partner, as the population difference between the two conformers is of the order of the precision of the employed integration procedure. Within the OH group, the formation of the intramolecular hydrogen bond yields a small decrease of ca 0.005 e, whereas the V(O) basin population is increased by almost the same amount of electron density. The $V(C_6, X)$ populations are always significantly lower for the cis conformer than in the trans one; the difference increases from F to Br. This should be due to the formation of the intramolecular hydrogen bond which enhances the electron donation towards the V(X) basins. With respect to the basin population criterion, the $V(C_6, X)$ basin appears to be more perturbed than the $V(C_1, O)$ one, and we could therefore expect that the additivity of the reactivity indices no longer holds for the cis conformer because the halogen atom is perturbed in this case. Indeed, the maximum deviation between the estimated and calculated indices does not exceed 0.002 in the trans case while it is ten times larger for the cis conformer. The overall charge transfer towards the aromatic ring is always less than the sum of the substituent contributions arising from phenol and benzene halides, and it is larger for the cis conformer.

The strength of the intramolecular hydrogen bond can be estimated within the ELF analysis by the core valence bifurcation index ϑAHB^{275}. This index is defined as the difference between the values of *ELF* calculated at the index 1 critical point of the separatrix of the V(A, H) and V(B) basin and at the core valence boundary of the proton donor moiety. It is nicely correlated with the proton donor stretching frequency, namely negative values indicate a weak hydrogen bonding such as in the FH $\cdots$ N$_2$ complex whereas positive values indicate stronger hydrogen bonds such as in FH $\cdots$ NH$_3$. For the *cis ortho*-fluoro-, chloro- and bromo-phenols, we find the following values of the core valence bifurcation index: -0.06, -0.02 and -0.01, respectively. These values correspond to very weak or weak hydrogen bonds. On the other hand, they show that the hydrogen bond strength increases from F to Br, which is counterintuitive if one considers the halogen electronegativity. However, it completely explains the order reported in equation 5. This also indicates that the strength of the intramolecular hydrogen bond is driven by geometrical strains which hamper the formation of these bonds with the lightest halogens. A similar conclusion is drawn in Subsection III.A (see also Figure 12) although from a different point of view.

b. The meta-substituted phenols. Figure 16 displays the localization domains of the *trans* and *cis meta*-substituted phenols whereas quantitative information is provided by Table 23. In these derivatives the interaction of the two substituents is expected to be weaker than in the *ortho* case. The V(C$_1$, O) basin population is smaller than its value in phenol for all molecules except *cis* iodophenol. In the latter case the discrepancy could be due to the use of a large core pseudopotential on the iodine atom (in practice, the *ELF* analysis requires the explicit presence of core basins, at least determined by a small core pseudopotential). On the halogen side, the V(C$_4$, X) basin populations are also smaller (except for iodine) than in halobenzene. There is a net enhancement of the electron donation towards the ring which is evidenced by the calculated charge transfer which is larger than the value given by an additive assumption.

Except for iodine, the additivity of the electrophilic positional indices is nicely verified. With respect to phenol, the indices of the carbon in *ortho* and *para* positions are noticeably increased whereas that of carbon C$_3$ is more negative, because it corresponds to a *meta* position for both substituents.

c. The para-substituted phenols. In the *para*-substituted phenols presented in Figure 17, the two substituents act in the opposite directions. From Table 24 it becomes clear that the substitution of the hydrogen atom by a *para*-halogen induces a small increase in the V(C, O) basin population with respect to phenol as well as in the V(C, X) populations with respect to halobenzene. The additive estimate of the electrophilic substitution positional indices is verified (except in some cases for iodine). As expected, the orientational effects are smoothed.

The population of the V(C, H) basins are all close to 2.10 within the accuracy of the integration scheme, and therefore it is not possible to draw any conclusion about their behaviour.

The *ELF* population analysis enables one to show the following cooperative trends, which are in agreement with chemical intuition:

(i) In the *ortho*- and *para*-substituted species, the V(C, O) population is increased with respect to phenol.

(ii) In the *ortho*- and *para*-substituted species, the orientational effects are weakened except for *ipso* positions.

(iii) In the *meta*-substituted species, the V(C, O) and the orientational effects are enhanced.

(iv) The formation of the intramolecular hydrogen bond in the *ortho* species softens the additivity of the orientational effects.

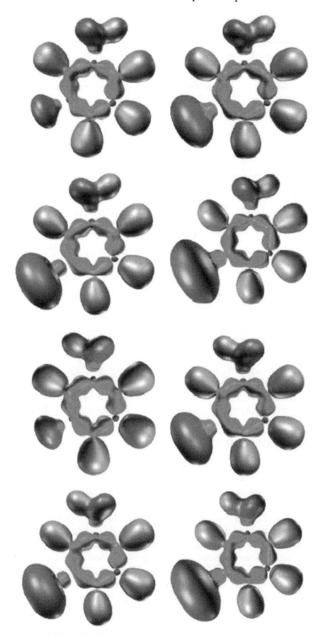

FIGURE 16 (PLATE 5). Localization domains of *meta*-X-substituted phenols (from left to right X = F, Cl, Br, I; top—*trans* conformer, bottom—*cis* conformer). The *ELF* value defining the boundary isosurface, $\eta(\mathbf{r}) = 0.659$ corresponds to the critical point of index 1 on the separatrix between adjacent V(C, C) basins of benzene. Colour code: magenta = core, orange = monosynaptic, blue = protonated disynaptic, green = disynaptic. Adapted from Reference 220 with permission

TABLE 23. Basin populations $\bar{N}(V)$ and electrophilic substitution positional indices RI_c of meta-substituted phenols

	trans conformation				cis conformation			
	F	Cl	Br	I	F	Cl	Br	I
				Populations				
$V(C_1, O)$	1.44	1.46	1.47	1.49	1.49	1.47	1.49	1.65
$V(C_3, X)$	0.97	1.45	1.45	1.40	1.0	1.45	1.45	1.39
$V(C_1, C_6)$	2.86	2.80	2.77	2.64	2.79	2.76	2.73	2.70
$V(C_1, C_2)$	2.72	2.84	2.84	3.05	3.03	3.03	3.02	2.85
$V(C_6, C_5)$	2.95	2.95	2.93	3.05	2.99	2.96	3.02	3.02
$V(C_4, C_3)$	2.85	2.80	2.74	2.77	2.92	2.91	2.89	2.69
$V(C_3, C_2)$	3.16	2.98	3.01	2.91	2.82	2.77	2.76	3.03
Net transfer	0.61	0.38	0.34	0.34	0.58	0.36	0.41	0.31
				Positional indices				
RI_1	0.027	0.028	0.029	0.035	0.027	0.028	0.029	0.035
RI_2	0.044	0.035	0.034	0.050	0.048	0.049	0.048	0.036
RI_3	0.069	0.051	0.044	0.041	0.070	0.052	0.045	0.040
RI_4	0.033	0.023	0.023	0.026	0.033	0.024	0.023	0.025
RI_5	−0.012	−0.010	−0.010	−0.005	−0.013	−0.011	−0.010	−0.004
RI_6	0.047	0.042	0.040	0.029	0.033	0.027	0.025	0.044

Values taken from Reference 220 with permission.

Finally, some of the unexpected results revealed for iodophenols warn against the use of large core pseudopotentials in the *ELF* analysis. It is noteworthy that the analysis of the topology of the *ELF* enables us to predict favoured protonation sites with the help of a 'least topological change principle'[276] which will be discussed in a following section.

D. Some Representatives of Substituted Phenols

We conclude this Section with a few words on nitrophenols and cyanophenols (CP or NCC_6H_4OH). For instance, the experimental K_a value for the proton separation of p-NCC_6H_4OH in both the ground and excited electronic states measured in solution[277] is higher than that of phenol by one order of magnitude. This implies that cyanophenols may form much stronger hydrogen bonds. And this fact has been particularly confirmed by an observation[278] of sharp vibronic bands in the R2PI spectrum with the electronic origin at ca 35410 cm^{-1} of the complex of p-NCC_6H_4OH with two water molecules. Cyanophenols are also rather convenient compounds for ultrafast experimental studies[279]. The p-NCC_6H_4OH in its ground state has been discussed theoretically[280] and its vibrational spectrum has been collected by Varsanýi[167]. Recently, *ab initio* calculations of p-cyanophenol have been performed in its ground and first excited states[280]. Strong evidence of the existence of a conical intersection in the excited state of p-cyanophenol following the proton dissociation coordinate has been shown[115]. The LIF and IR/UV double-resonance experiments have also been conducted on the hydrogen-bonded complexes between o-CP and one or two water molecules, combined with B3LYP/cc-pVTZ calculations[281, 282].

Figure 18 displays the optimized geometries of cyanophenols where it is seen particularly that the *cis ortho*-CP has a relatively weak intramolecular hydrogen bond. Similar

1. General and theoretical aspects of phenols 79

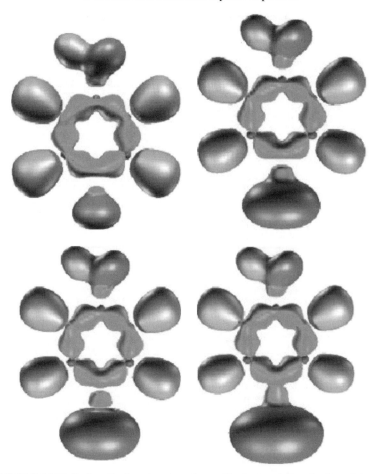

FIGURE 17 (PLATE 6). Localization domains of *para*-X-substituted phenols (from left to right X = F, Cl, Br, I). The *ELF* value defining the boundary isosurface, $\eta(\mathbf{r}) = 0.659$ corresponds to the critical point of index 1 on the separatrix between adjacent V(C, C) basins of benzene. Colour code: magenta = core, orange = monosynaptic, blue = protonated disynaptic, green = disynaptic. Adapted from Reference 220 with permission

to Subsection III.A, we can estimate its energy of formation as the energy difference between the *cis ortho*- and *trans ortho*-CPs which, at the present computational level, is 10.0 kJ mol^{-1} after including ZPVE corrections. It is worthwhile to deduce the order of stability of cyanophenols similar to that given in equation 12. We thus obtain equation 27, where the values are given in kJ mol^{-1}:

$$cis\ o \overset{2.8}{>} p \overset{4.4}{>} cis\ m \overset{0.75}{\approx} trans\ m \overset{2.0}{>} trans\ o \qquad (27)$$

It shows that, energetically, all cyanophenols fall into the interval of stability between the *cis ortho*- and *trans ortho*-CPs. Some characteristic vibrational modes are collected in Table 25 accompanied by their assignments based on the PEDs.

TABLE 24. Basin populations $\bar{N}(V)$ and electrophilic substitution positional indices RI_c of *para*-substituted phenols

	F	Cl	Br	I
	Populations			
$V(C_1, O)$	1.52	1.60	1.56	1.62
$V(C_4, X)$	1.0	1.53	1.47	1.41
$V(C_1, C_2)$	2.98	2.87	2.85	2.74
$V(C_1, C_6)$	2.68	2.66	2.68	2.80
$V(C_2, C_3)$	2.96	3.01	3.0	3.02
$V(C_3, C_4)$	3.0	3.0	2.99	2.69
$V(C_5, C_6)$	3.06	3.13	3.10	3.02
Net transfer	0.57	0.48	0.43	0.35
	Positional indices			
RI_1	0.040	0.033	0.032	0.036
RI_2	0.035	0.037	0.037	0.027
RI_3	0.010	0.001	0.0	0.002
RI_4	0.090	0.075	0.071	0.066
RI_5	0.008	0.0	−0.001	0.003
RI_6	0.020	0.022	0.022	0.042

Values taken from Reference 220 with permission.

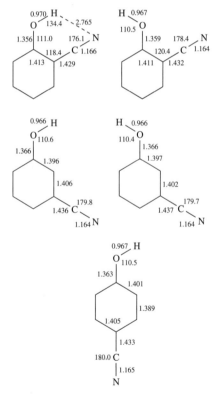

FIGURE 18. The B3LYP/6-31+G(d,p) geometries of cyanophenols in the ground electronic state. Bond lengths are in Å, bond angles in degrees

TABLE 25. Characteristic vibrational modes of cyanophenols, p-nitrophenol and pentachlorophenol. Frequencies are given in cm^{-1} and IR activities in km mol^{-1}

para-cyanophenol

	Freq.	IR	PED, %
ν_{OH}	3823	87	ν_{OH} (100)
τ_{OH}	366	118	τ_{OH} (95)
ν_{CO}	1303	130	ν_{CO} (55)

cis *ortho*-cyanophenol

	Freq.	IR	PED, %
ν_{OH}	3764	81	ν_{OH} (100)
τ_{OH}	438	112	τ_{OH} (91)
ν_{CO}	1281	49	ν_{CC} (29) ν_{CO} (25) β_{CH} (24)
	1341	11	β_{CH} (18) ν_{CO} (18) ν_{CC} (24)

trans *ortho*-cyanophenol

	Freq.	IR	PED, %
ν_{OH}	3826	81	ν_{OH} (100)
τ_{OH}	351	117	τ_{OH} (94)
ν_{CO}	1294	27	ν_{CO} (30) ν_{CC} (27) β_{CH} (16)

cis *meta*-cyanophenol

	Freq.	IR	PED, %
ν_{OH}	3826	70	ν_{OH} (100)
τ_{OH}	339	118	τ_{OH} (92)
ν_{CO}	1309	95	ν_{CO} (33) β_{CC} (14) β_{CH} (22)
	954	16	ν_{CO} (19) ν_{CC} (42) β_{CC} (10)

trans *meta*-cyanophenol

	Freq.	IR	PED, %
ν_{OH}	3828	75	ν_{OH} (100)
τ_{OH}	332	118	τ_{OH} (93)
ν_{CO}	1306	57	ν_{CO} (30) β_{CC} (14) β_{CH} (11) ν_{CC} (10)
	953	46	ν_{CO} (18) ν_{CC} (29)

para-nitrophenol

	Freq.	IR	PED, %
ν_{OH}	3821	96	ν_{OH} (100)
τ_{OH}	382	119	τ_{OH} (94)
ν_{CO}	1305	188	ν_{CO} (54)

Pentachlorophenol

	Freq.	IR	PED, %
ν_{OH}	3688	96	ν_{OH} (100)
τ_{OH}	429	108	τ_{OH} (93)
ν_{CO}	1454	149	ν_{CO} (27) ν_{CC} (40)

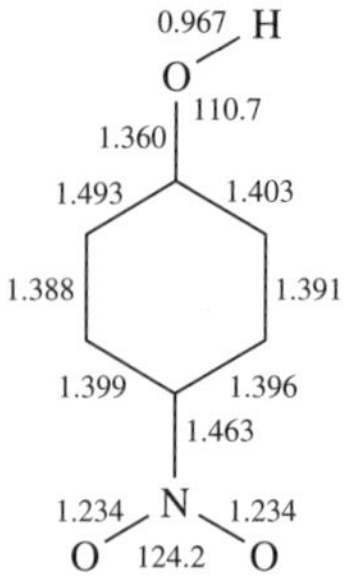

FIGURE 19. The B3LYP/6-31+G(d,p) geometry of *para*-nitrophenol in the ground electronic state. Bond length are in Å, bond angles in degrees

Figure 19 displays another representative of substituted phenols, namely *p*-nitrophenol, whose history of discovery was mentioned in Section I. A knowledge of its structure and IR spectrum is important for the study of inter- and intra-molecular interactions via a variety of spectroscopic methods.

To our knowledge, the first theoretical study of *p*-nitrophenol, at HF/3-21G computational level, was conducted in 1988[283]. The molecular structure of *o*-nitrophenol[284, 285] and its IR spectra in the gas phase, solution and solid[286] were reported. For *p*-nitrophenol, only the IR spectrum was available in the solid state[287]. Recently, a thorough study[288] of *p*- and *o*-nitrophenols using B3LYP/6-31G(d,p) calculations has been reported which consists, first, in obtaining their geometries and, second, in calculating the harmonic vibrational frequencies and making their assignments for *p*-nitrophenol. In Table 25, we collect the key harmonic vibrational modes of *p*-nitrophenol together with their PED analysis.

Finally, we briefly mention pentachlorophenol (PCP), which is the most complex substituted phenol whose structure is reported so far in the present review and which is widely used in studies on the hydrogen bonding abilities of phenols. Its optimized geometry is demonstrated in Figure 20 and Table 25 lists its characteristic vibrational modes (cf. Reference 289). Except for the vibrations involving the OH and OD bonds, agreement between experimental and calculated values exists for the fundamental wavenumbers between 3600 and 400 cm^{-1}. The infrared spectra between 3600 and 10000 cm^{-1} have also been studied and the overtones or combination bands were assigned by comparing the spectra of both isotopomers PCP-OH and PCP-OD. The anharmonicities of the OH

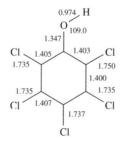

FIGURE 20. The B3LYP/6-31+G(d,p) geometry of pentachlorophenol in the ground electronic state. Bond lengths are in Å, bond angle in degrees

and OD stretching modes were determined and the binary or ternary combinations characterized by the highest coupling constants, and the highest intensities are those involving the OH and CO vibrations[289].

IV. ENERGETICS OF SOME FUNDAMENTAL PROCESSES

A. Protonation

Protonation is a simple but important chemical process. The primary protonated form is usually a pivotal intermediate that guides the subsequent steps of an entire chemical transformation. Biomolecules such as DNA and proteins can often exist in numerous protonated forms. In a molecular system having several basic sites, the protonation usually turns out to be regioselective yielding predominantly one protonated species. The attachment of proton to a molecule A is quantified by its proton affinity, PA(A)[290], which is defined as the negative standard enthalpy ($\Delta H°$) of the reaction $A + H^+ \rightarrow AH^+$. The PA is a measure of the basicity of the molecule which is one of the fundamental concepts in chemistry. In the most general sense, basicity is the ability of a substance to accept a positive charge. In the Lewis definition, the charge is transferred by gain or loss of an electron pair. In the Brønsted definition, the charge is transferred by gain or loss of a proton; therefore, the basicity is conventionally defined as the negative standard free energy ($\Delta G°$) of the protonation reaction. Although the PA of a functional group is definitely influenced by the presence of substituents, any given functional group is more or less characterized by a certain range of proton affinities and a simple comparison of their values could often allow the most favoured protonation site of a polyfunctional substrate to be determined.

Let us consider in some detail the protonation of the parent phenol, a series of monohalogenated phenols (XC_6H_4OH, X=H, F, Cl, Br, I), and for a further control, the fluoroanisoles, $FC_6H_4OCH_3$. The interaction of the alkali metal cations including Li^+, Na^+ and K^+ is also probed. In what follows, only the processes taking place in the gaseous phase are considered.

1. Protonation of phenol

Phenol contains both phenyl and hydroxyl functional groups. While the PA of the phenyl moiety could be estimated from that of benzene, the PA of water provides an estimate for that of the hydroxyl group. The experimental PA(H_2O)[291] is well established at 697 ± 4 kJ mol^{-1} whereas the PA of benzene[292] is experimentally evaluated as 753 kJ mol^{-1}. In other words, the PA(C_6H_6) exceeds the PA(H_2O) by as much as 56 kJ mol^{-1}. Such a difference suggests that the preferential protonation of phenol should occur on the ring moiety, even though it is not always true[293]. In reality, the experimental gas-phase PA(PhOH) of 816–818 kJ mol^{-1}, as determined by either pulsed ion cyclotron resonance equilibrium experiments[294] or high pressure mass spectrometry[295] (for a recent compilation, see Reference 296), turns out to be substantially larger than those mentioned above, implying that the OH group markedly affects the protonation of the phenyl moiety. In fact, it was demonstrated experimentally that the gas-phase phenol protonation occurs predominantly on the ring, and the oxygen PA is about 55–84 kJ mol^{-1} smaller than the carbon PA[297]. These findings were subsequently supported by *ab initio* MO calculations[298,299]. The O-protonated form was calculated to lie 81 kJ mol^{-1} higher in energy than its *para*-C-protonated isomer[299]. Recently, the existence of at least two protonated phenol isomers corresponding to proton attachment at oxygen and at the aromatic ring has been confirmed convincingly by using IR spectroscopy[300].

In contrast to these gas-phase findings, the oxygen protonation was found to be favoured in various solutions[297]. The influence of the solvent is known to be a crucial factor determining the strength of bases. In some cases, the relative basicity ordering is even reversed by external effects.

The presence of a hydroxyl group induces four different positions on the ring susceptible for an electrophilic attack, namely the *ipso*-C_1, *ortho*-C_2, *meta*-C_3 and *para*-C_4 carbons, relative to the hydroxyl position, and one of these carbon centres will show the largest attraction for the proton. For the sake of convenience, the term '*ortho*-protonation' stands hereafter for a protonation occurring at the carbon C_2 etc. All theoretical methods agreed with each other in predicting the *para*-position as the most favourable protonation site[298–300] followed by the *ortho* position with a rather small difference of ca 10 kJ mol^{-1}. The *meta*-protonated phenol is placed ca 60 kJ mol^{-1} above the *para*-counterpart, whereas the *ipso*-protonated species lies consistently much higher in energy. The difference between the PAs of both *meta*-C_3- and O-protonated forms is calculated to be small, approximately 15 kJ mol^{-1}[298–300].

At the B3LYP/6-311++G(d,p) + ZPE level of theory, the local PAs of phenol at different sites are evaluated in kJ mol^{-1} as follows: 820 for *para*-C_4, 809 for *ortho*-C_2, 757 for *meta*-C_3, 699 for *ipso*-C_1 and 743 for oxygen[299]. The coupled-cluster CCSD(T) approach in conjunction with the 6-311++G(d,p) basis set yields a PA(C_4) of 819 kJ mol^{-1}. When using an appropriate basis set, the calculated PAs thus compare reasonably well with the experimental value quoted above.

The potential energy surface (PES) of the protonated phenol species possesses seven local energy minima all displayed in Figure 21, which vividly illustrates the migration of the excess proton between the adjacent heavy atoms. This portion of the energy surface also includes four transition structures (TS) for 1,2-hydrogen migrations. Starting from the highest-energy *ipso*-protonated form, the excess H$^+$ almost freely migrates to the *ortho*-protonated form passing through a small barrier of 8 kJ mol^{-1} described by TS$_3$. The barriers for proton migration between the other adjacent carbon atoms are substantially larger, viz. 31 kJ mol^{-1} for the *meta*-to-*para* (TS$_1$) and 45 kJ mol^{-1} for the *meta*-to-*ortho* migration (TS$_2$). The activation barrier governing the *ipso*-to-oxygen migration amounts to 121 kJ mol^{-1} (TS$_4$). The corresponding transition frequencies of 773i, 869i, 960i and 1599i cm^{-1}, respectively, are assigned to the vibrational modes of the excess migrating proton. The large energy separation between the *para*-C_4 and O-protonations clearly demonstrated in Figure 21 constitutes a key difference from the protonation process in aniline ($C_6H_5NH_2$) where both the *para*-C_4- and N-protonated species have comparable energy content[301–303]. Nevertheless, a substantial energy barrier of 159 kJ mol^{-1} for H-shift has been found separating the O-protonated phenol from its nearest C-isomers. This result provides us with a rationalization for the recent experimental observations using IR spectroscopic techniques[300]. It appears that in this experiment, protonation initially occurs at several positions, but eventually only the O- and one C-protonated form were stabilized and spectroscopically detected. Due to the ease with which the proton scrambled around the ring, it is rather difficult to observe, for example, a *meta*-form even though it is thermodynamically more stable than the O-isomer. In contrast, the latter was able to resist unimolecular rearrangements, thanks to the more difficult oxygen-to-carbon proton migration (Figure 21), and thus it lived long enough to be detectable within the time frame of an IR experiment.

The regioselectivity of the gas-phase protonation of phenol can be understood in simple terms of its resonance structures. Drawing them, we may figure out that a positive π-charge of the protonated form is mainly localized in the *para*- and *ortho*-positions with respect to the protonation site. If the OH group is attached to one of these positions, the relevant molecule is then described by four resonance structures, resulting in the positive π-charge

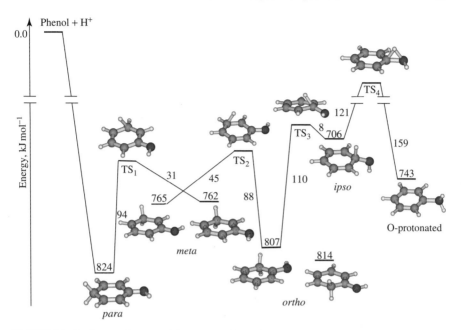

FIGURE 21. Portion of the potential energy surface of the protonated phenol showing the proton migration between the adjacent heavy atoms. Values given in kJ mol^{-1} were obtained from B3LYP/6-31+G(d,p)+ZPE computations[299]. Adapted from Reference 299 with permission

to be distributed over all atoms. Otherwise, only three structures are allowed. The presence of a positive charge in direct conjugation with the oxygen atom favours the electron density shift from the oxygen lone pairs to the ring and strengthens a stabilization of the arenium ion. Spectroscopically, it is manifested in a blue-shifting $\Delta\nu_{CO}$ of the fundamental mode 13 with the dominant contribution of the ν_{CO} stretching vibration which accompanies a shortening of the CO bond (Δr). In particular, the calculated Δr and $\Delta\nu_{CO}$ take the following values: 0.06 Å and 112 cm^{-1} in *para*-, 0.06 Å and 46 cm^{-1} in *ortho*- and 0.03 Å and 27 cm^{-1} in *meta*-protonated phenol. The other indicative frequency shifts showing the contribution of the resonance structures with the doubly-bonded oxygen atom, i.e., **19** and **20**, are associated with the OH stretching and torsional vibrations. The contribution of both structures **19** and **20** is expected to weaken the OH bond and shift the corresponding ν_{OH} mode to lower frequencies. Also, it likely determines the torsional barrier describing the rotation of the OH group around the single conjugated CO bond[304,305] and therefore increases the τ_{OH} frequency. The low-energy *para*- and *ortho*-protonated structures reveal the most pronounced and rather similar red shifts of the ν_{OH} mode by 89 cm^{-1} and 93 cm^{-1}, and also the blue shifts of the τ_{OH} mode by 281 cm^{-1} and 277 cm^{-1}, respectively. This implies participation of the lone pairs of oxygen in stabilizing the arenium ion that leads to increase in the PA of the phenyl moiety. The *meta*-protonation shifts the corresponding vibrations by only 30 cm^{-1} and 69 cm^{-1} compared to those in the neutral molecule. A similarity in frequency shifts of the ν_{OH} and τ_{OH} modes in both *para*- and *ortho*-protonated structures and also in their relative energies suggests that the regioselectivity of the protonation of phenol is primarily governed by resonance factors.

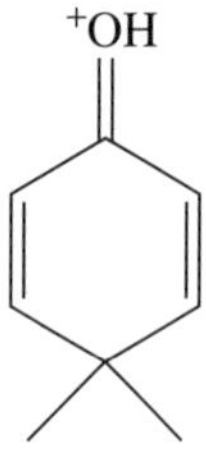

(19) 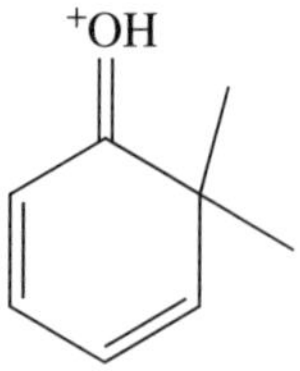   (20)

2. Proton affinities of halophenols

The calculated PAs for mono-fluorinated phenols listed in Table 26, obtained by using the B3LYP/6-31+G(d,p)+ZPE level, are found to be in reasonable agreement with the recent ion cyclotron resonance data[306], namely 797 kJ mol^{-1} (expt. 788 kJ mol^{-1}) for 2-fluorophenol, 813 kJ mol^{-1} (expt. 802 kJ mol^{-1}) for 3-fluorophenol and 787 (expt. 776 kJ mol^{-1}) for 4-fluorophenol, and thus approach the experimental PA with a quasi-systematic overestimation of ca 10–12 kJ mol^{-1}. To our knowledge, no experimental PAs of Cl-, Br- and I-substituted phenols have been available so far.

For the 2- and 3-halophenols, the *para*-position remains the most attractive protonation site, irrespective of the nature of the X-atom, followed by two *ortho*-positions, C_6 and C_2, respectively. All structures protonated at these sites lie within 20 kJ mol^{-1} above the corresponding global C_4 minima (Table 26). The other sites are less accessible for protonation. As envisaged by the classical resonance model, the lower-energy protonated structures always have the OH and X groups in the *para*- and *ortho*-positions relative to the protonation site. Among them, the structures where the OH group is attached in *para* and the X atom in *ortho* reach the global minimum on the PES of a given X-substituted protonated phenol, featuring the largest PAs in the whole series, viz. 813 kJ mol^{-1} in 3-fluorophenol,

TABLE 26. The B3LYP/6-31+G(d,p) proton affinities (kJ mol^{-1}) of halogenated phenols[a]

Protonation site Substitution	C_1	C_2	C_3	C_4	C_5	C_6	O
2-F	711	749	764	797	761	784	731
3-F	672	802	683	813	732	804	721
4-F	709	787	753	756	—	—	729
2-Cl	700	756	757	801	760	790	715
3-Cl	679	798	699	815	735	811	724
4-Cl	709	789	756	771	—	—	727
2-Br	702	763	760	806	763	795	736
3-Br	683	801	716	818	738	815	727
4-Br	713	792	759	784	—	—	728
2-I	710	791	767	813	769	803	743
3-I	691	807	—	823	746	820	730
4-I	719	791	765	816	—	—	731

[a] Atoms numbering is shown in Chart 1. In *meta*-fluorophenols, the OH bond is leaned away from the substituent, and in all other *meta*-substituted phenols, towards it, providing the most stable neutral structures. In case of *para*-X-phenols, two pairs of structures with the protonation sites on C_2–C_6 and C_3–C_5 atoms, respectively, are energetically close. Values taken from Reference 299.

815 kJ mol^{-1} in 3-chlorophenol, 818 kJ mol^{-1} in 3-bromophenol, and 823 kJ mol^{-1} in 3-iodophenol. Such behaviour can in part be accounted for by a better conjugation of the oxygen lone pairs with the ring compared to those of the X groups.

The 3-X-phenols (X=Cl, Br, I) protonated at the C$_4$ and C$_6$ positions are nearly isoenergetic; their PAs are equal to 815 and 811 kJ mol^{-1} in 3-chlorophenol, 818 and 815 kJ mol^{-1} in 3-bromophenol, and 823 and 820 kJ mol^{-1} in 3-iodophenol, whereas the C$_2$-protonated species lie slightly higher in energy due to a steric repulsion with the OH group.

In 4-halophenols (X =F, Cl, Br), the excess proton tends to reside in *ortho*-positions. On the other hand, in *para*-iodophenol, the protonated structure with both I and the excess H$^+$ residing in the *para*-site has the lowest energy.

As for a correlation between PAs and molecular properties, Table 27 lists the characteristic frequencies of the hydroxyl torsional τ_{OH} and stretching ν_{OH} vibrational modes in the neutral and protonated fluorophenols. The τ_{OH} vibration is directly related to distortions in the π-electronic system which was demonstrated experimentally for a wide variety of substituted phenols[307]. The π-electron donor substituents at the *para*-position lower the τ_{OH} frequency compared to unsubstituted phenol, while the π-electron acceptor substituents act in the opposite way. In neutral fluorophenols, the τ_{OH} mode is centred at 304 cm^{-1} for *para*-fluorophenol, 330 cm^{-1} for *meta*-fluorophenol and is blue-shifted to 411 cm^{-1} for *ortho*-fluorophenol due to the hydrogen bonding (see Table 27; 330 cm^{-1} in unsubstituted phenol). The τ_{OH} frequency is blue-shifted upon protonation depending on the protonation site. In *para*- and *ortho*-protonated phenols which are resonance-stabilized via the structures with the doubly-bonded oxygen atom of the types 19 and 20 exhibiting the highest PA, these shifts are very pronounced and yield values of 317 cm^{-1} in the C$_6$-protonated *para*-fluorophenol, 283 cm^{-1} in C$_2$-protonated, 264 cm^{-1} in C$_4$-protonated and 231 cm^{-1} in C$_6$-protonated *meta*-fluorophenols. In *meta*-protonated structures, the blue shift of the τ_{OH} becomes smaller, viz. 2 cm^{-1} in *para*-fluorophenol and 70 cm^{-1} in *meta*-fluorophenol. In *ortho*-fluorophenols, the effect of the protonation site on τ_{OH} is less evident due to its interplay with the effects of hydrogen bonding.

The ν_{OH} frequency behaves in a similar manner with respect to the protonation site, although shifts are in the opposite direction. By analogy with the torsional frequency, the maximal shifts are found in *para*- and *ortho*-protonated structures, viz. 98 cm^{-1} in C$_6$-protonated *para*-fluorophenol, 93 cm^{-1} in C$_2$-protonated, 84 cm^{-1} in C$_4$-protonated and 76 cm^{-1} in C$_6$-protonated *meta*-fluorophenols. In the hydrogen-bonded systems, both the hydrogen bonding and the distortions in the π-electronic system caused by protonation behave coherently in weakening of the OH bond and thus shifting the ν_{OH} to lower frequencies. The predicted red shifts of the ν_{OH} mode in these systems become even more pronounced: 116 cm^{-1} in C$_4$-protonated and 110 cm^{-1} in C$_6$-protonated 2-fluorophenols.

In the case of 3-X-phenols, the X-protonated structures are local minima, but they are consistently above the high-energy *ipso*-protonated phenols, except for 3-iodophenol in which an *ipso*-protonation is less favourable by 13 kJ mol^{-1} than an I-protonation. The calculated PAs for the X-protonated 3-halophenols are the following: 613 kJ mol^{-1} for 3-fluorophenol, 676 kJ mol^{-1} for 3-chlorophenol, 680 kJ mol^{-1} for 3-Br-phenol and 704 kJ mol^{-1} for 3-iodophenol, using the same level of theory.

It is well known that in halobenzenes, the *para*-position relative to the halogen is the more basic site and the *meta*-position the least basic[308]. The higher activity for the *para*-position in fluorobenzene results from the need to add a proton to a position that is not disfavoured by the σ-electron withdrawal by fluorine atom, due to its strong inductive effect. The effect is smaller for chlorine, bromine and iodine. Thus when there is competition between the hydroxy group and a halogen atom in directing the ring protonation,

TABLE 27. Frequencies (cm^{-1}) of the torsional and stretching vibrations in the protonated and unprotonated fluorophenols[299]

Structure[a]	τ_{OH}	ν_{OH}	Structure[a]	τ_{OH}	ν_{OH}	Structure[a]	τ_{OH}	ν_{OH}
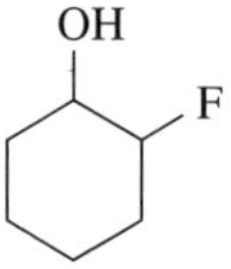 neutral fluorophenol	411	3802	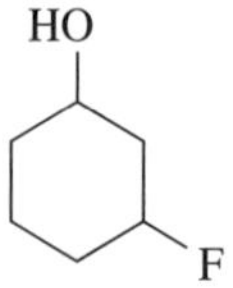 C2-protonated (ortho to F)	330	3831	 C4-protonated (para to F)	304	3833
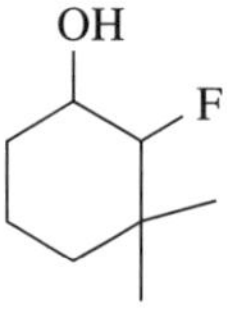 C3-protonated	399	3780	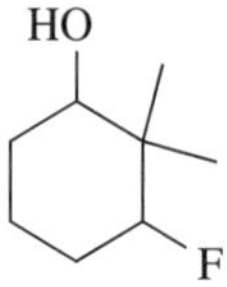 C3-protonated	613	3738	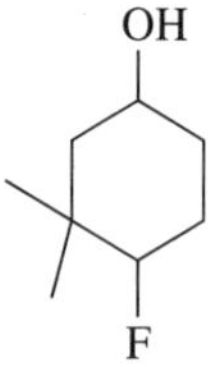 C3-protonated (alt)	306	3809
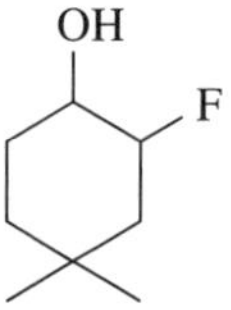 C4-protonated	684	3686	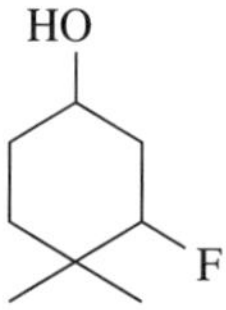 C4-protonated	594	3747	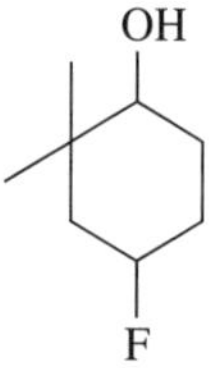 C2-protonated	621	3735
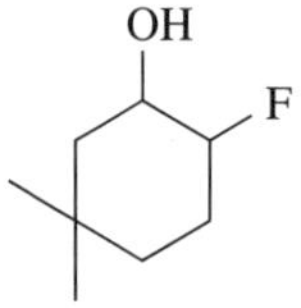 C5-protonated	370	3780	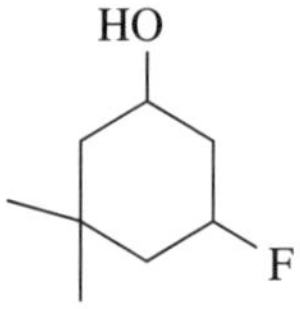 C5-protonated	400	3798			
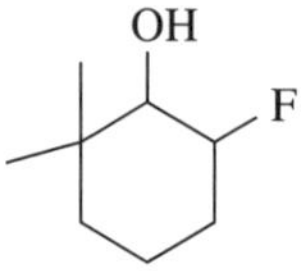 C6-protonated	655	3692	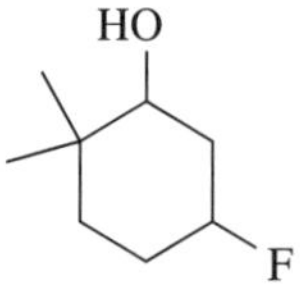 C6-protonated	561	3755			

[a]The first structure refers to the neutral fluorophenol. The other species are protonated fluorophenols at different positions.

as in the case of halophenols, the outcome turns out to be in favour of the hydroxy group which, as discussed above, consistently leads to a C_4-protonation (Table 26).

3. Proton affinities of anisole and fluoroanisoles

Anisoles are phenol derivatives in which the OH is replaced by the OCH$_3$ group. As expected, anisole reveals the same protonation pattern as phenol, although all of its local PAs appear to be larger, namely 845 kJ mol^{-1} in *para*-C_4-protonation, 836 kJ mol^{-1} in *ortho*-C_6-protonation and 780 kJ mol^{-1} in *meta*-C_3-protonation. Similarly, a correlation has been observed between the local PAs and the C−O bond shortening (Δr) and the blue-shifting ($\Delta \nu_{CO}$) of the fundamental mode with the dominant contribution of ν_{CO} vibration. The Δr and $\Delta \nu_{CO}$ changes take the following values: 0.03 Å and 28 cm^{-1} in C_3-protonated anisole, 0.07 Å and 103 cm^{-1} in C_4-protonated anisole, 0.03 Å and 29 cm^{-1} in C_5-protonated anisole and finally 0.07 Å and 81 cm^{-1} in C_6-protonated anisole[299].

The PAs of fluoroanisoles are equal to 820 kJ mol^{-1} (expt. 807[306]) for 2-fluoroanisole, 835 kJ mol^{-1} (expt. 826[306]) for 3-fluoroanisole and 809 kJ mol^{-1} (expt. 796[306]) for 4-fluoroanisole. An average overestimation of ca 12 kJ mol^{-1} by the B3LYP/6-31+G(d,p)+ZPE calculations can again be noted[299].

4. Two views on the protonation regioselectivity

It is now legitimate to pose the question as to whether there exists a clear-cut but simple theoretical approach to predicting the protonation regioselectivity solely on the basis of molecular properties of the neutral substrate. Theoretical chemists persist in their continuing endeavour to search for such a reactivity index. The relative gas-phase acidity and basicity data collected in the last several decades have been analysed and correlated with a variety of atomic and molecular parameters. Examples include the atomic charges, charge-induced dipole field or polarizabilities, electrostatic potentials surrounding a base, core ionization energies or 1s-orbital energies, electronegativities, hybridization, bond energies, electron affinities etc. The main idea is to design a way of partitioning the molecular charge distribution into atomic properties that show acceptable correlations with PA[309–315]. The most representative approach is the atom-in-molecule theory[309]. Use of the components of wave functions constructed by either multi-configurational[313] or spin-coupled[312] methods was also put forward in support of an interpretation in terms of resonance structures. However, all these approaches to identifying the protonation sites either were not quite successful[314–315] or could not be extended to a larger sample of compounds[315]. We will consider two of the most recent attempts including the use of the topological analysis based on the electron localization function (ELF)[272, 273, 316, 317], discussed in Section III.C, and the density functional theory-based reactivity descriptors, in both a global and a local sense[301, 302, 318–340].

As seen above, the topology of the ELF suggests that the most favoured protonation site can be found by using a 'least topological change principle'[275, 317] which states that:

(i) the protonation occurs in the most populated, accessible valence basin for which there is the least topological change of the electron localization function, and

(ii) in the protonated base, the V(B,H) population cannot be noticeably larger than 2.5 electrons.

In all cases, except for *ortho*-Cl and *ortho*-Br phenols, it is the V(O,H) basin which is favoured over the V(X) basin. In the two aforementioned molecules, the intramolecular hydrogen bond is strong enough to perturb the topology of the halogen valence shell having three basins, and the ELF predicts that the favoured protonation site is one of the most populated V(X) halogen basins. In other words, the ELF could correlate the relative basicities between heteroatoms but is apparently unable to account for the preference of the ring *para*-C_4 carbon in the protonation process[220].

We now turn to the reactivity indices defined within the framework of density functional theory (DFT). The validity and applicability of these indices have been discussed in several recent studies by different groups[301, 302, 318–340]. This is a different way of decomposing a molecular electronic distribution into global and/or local indices coupled with an account of the frontier molecular orbitals. Starting from the electronegativity equalization principle[318], the global descriptors such as 'group hardness' and 'group electronegativity' were defined[319] and correlated with PAs. Nevertheless, their scope of applicability was quite limited. More recently, the more local descriptors, including the Fukui function, local atomic softness or even orbital softness, have been employed in order to interpret the protonation sites[299, 301, 302]. The definitions[320, 321] and evaluations[322–325] of DFT-based reactivity indices are well established.

The condensed Fukui functions f_k of a kth atom in a molecule with N electrons are defined by equations 27a and 27b:

$$f_k^+ = [q_k(N+1) - q_k(N)] \quad \text{for nucleophilic attack} \qquad (27a)$$

$$f_k^- = [q_k(N) - q_k(N-1)] \quad \text{for electrophilic attack} \qquad (27b)$$

where q_k is the electronic population of atom k in the molecule under consideration. The local softness parameter can then be defined as $s_k^i = f_k^i \times S$ in which i stands for $+$ or $-$. Within the finite difference approximation[322], the global softness, S, can be approximated by

$$S = 1/(\text{IE} - \text{EA})$$

where IE and EA are the first vertical ionization energy and electron affinity of the molecule, respectively.

The local softness has been applied with much success in interpreting and predicting the regio-selectivities of different types of organic reactions including radical additions[326], nucleophilic additions[327–329], pericyclic [2 + 1][330–333], [2 + 2][334] and [3 + 2][335–341] additions, hydrogen shifts[342] and internal rotations.[343, 344]

In the parent phenol for which the local indices are summarized in Table 28, the values for the C_5 and C_6 atoms are also close to those for C_3 and C_2, respectively, and

TABLE 28. Calculated local softnesses of phenol

Property	B3LYP/cc-pVTZ
vert-IE (eV)	8.45
vert-EA (eV)	−1.66
S^a	2.69
s_k^+	
C_1	0.04
C_2	0.39
C_3	0.25
C_4	0.31
O	0.12
s_k^-	
C_1	0.08
C_2	0.36
C_3	−0.04
C_4	0.83
O	0.48
s_k^-/s_k^+ ratio	
C_1	1.94
C_2	0.92
C_3	−0.17
C_4	2.64
O	3.91

$^a S$ is the global softness. The Fukui functions f_k can be obtained using $s_k = f_k \cdot S$.

thus omitted for the sake of simplification. In the present case, the *local softness for electrophilic attack* s^- is to be used to probe the protonation mechanism, that is, the larger the local softness, the more basic the site. It is clear that the C_4 carbon atom bears the largest softness ($s^- = 0.83$), a value much larger than that of oxygen ($s^- = 0.48$). While the C_2 carbon has a significant softness ($s^- = 0.36$), the C_1 and C_3 atoms do not show much affinity for electrophiles. These observations are in accord with the proton affinities discussed above which unambiguously indicate the preferential protonation at the C_4 carbon of phenol, followed by that at C_2 carbon and oxygen. Table 28 lists the s_k values and the quantities s_k^-/s_k^+ and shows that *the latter ratio also does not hold true for phenol protonation*. In fact, the oxygen atom is characterized by the largest ratio followed by C_4. Among the ring carbon atoms, while C_4 has the largest ratio (which is correct), C_1 has a larger ratio than C_2 (which is not correct according to the calculated PAs).

The calculated local softnesses and Fukui functions of the fluoro- and chloro-substituted phenols (values of s_k^-) suggest the following protonation ordering (versus the real ordering found from calculated proton affinities).

(a) Fluorophenols: 2-F: $O > C_4 > C_6$ versus $C_4 > C_6 > C_3 > C_5 > C_2 > O$,
                   3-F: $C_4 > C_6 > O$ versus $C_4 > C_6 > C_2 > C_5 > O$,
         and 4-F: $O > C_4 > C_6 > C_2$ versus $C_2 > C_4 > C_3 > O$.

(b) Chlorophenols: 2-Cl: $C_4 > O > C_6$ versus $C_4 > C_6 > C_5 > C_3 \approx C_2 > O$,
                    3-Cl: $C_4 > C_6 > O$ versus $C_4 > C_6 > C_2 > C_5 > O$,
         and 4-Cl: $O > C_6 > C_2$ versus $C_2 > C_4 > C_3 > O$.

In comparison with the calculated PAs mentioned above, a few points are worth noting[299]:

(i) The local softnesses of atoms having different atomic numbers cannot be compared to each other (for example, a comparison of a carbon and an oxygen atom is not relevant). Similarly to the shortcomings of net atomic charges or electrostatic potentials[314, 315, 345, 346], this local descriptor is apparently unable to differentiate the relative basicities of heteroatoms. A comparable conclusion was drawn from an analysis of the *orbital local softnesses*[302, 342]. Such behaviour differs somewhat from that of the *ELF* discussed in Section III.C.

(ii) The local softness behaves more regularly among the ring carbon atoms. In fact, for both 2-X and 3-X phenols, the local softness points towards a *para* protonation in agreement with explicit computations of PAs. While for 4-Cl the local softness correctly predicts the preference of C_6 and C_2, the situation is more confusing in 4-F where the s^- values of all carbons are similar to each other, with a marginally larger value for C_4 followed by C_6 and C_2 (if oxygen is omitted).

(iii) The s_k^-/s_k^+ ratio is nowhere able to unravel the preferable protonation site.

(iv) There is no correlation between the absolute values of local softnesses with the PAs at the ring carbon centres.

These drawbacks of either *ELF* or DFT-based indices raise the question as to whether it is meaningful to use the local properties of reactants in distinguishing the protonation of atoms of different nature. Similar to the case of two different atoms, such as O and C, when a X-substituent strongly modifies the electronic environment of the carbon, a perturbative treatment could also no longer be applied to the C(H) and C(X) centres.

Although the local softness includes, by definition, both the differences of frontier orbitals of the neutral substrate and the differential electron densities between the neutral and ionized states, as expressed in the global softness and Fukui functions, the actual computations of these quantities suffer from some severe practical limitations[299].

In summary, neither the *ELF* nor the DFT-based reactivity indices are capable of accurately predicting the most preferably protonated sites of phenols as well as the order of the local PAs. Similar to the many well-known static indices, their performance is expected to be limited in other classes of compounds as well. Thus the discovery of a good protonation index remains a formidable challenge for theoretical chemists. The difficulty lies in the fact that any quantitative correlation between a molecular property and the PAs of a series of compounds is based on the assumption that the relaxation energy involved should practically be constant for the entire series. After all, the proton is strongly electrophilic and very hard, and its approach polarizes the whole medium due to its small size and basically modifies the molecular and electronic structure of the substrate. As the local indices are usually defined at unperturbed neutral substrates, it is obvious that they are not sensitive enough to predict the realistic situations characterized by drastic changes following the protonation process.

5. Interaction of phenol with Li^+, Na^+ and K^+

Properties of the complexes of alkali metal cations with various bases are important in understanding ion–molecule interactions, solvation effects, biomedical and physiological phenomena related to ion channels and relevant in medical treatments. Reliable experimental bond dissociation enthalpies, and thereby gas-phase alkali ion affinities, could now be obtained using various mass spectrometry techniques such as the Fourier-transform ion cyclotron resonance (FT-ICR), collision-induced dissociation and photodissociation methods. However, these methods do not provide direct information on the adduct structures.

The Li^+ cation exhibits a vacant p-orbital and its interaction with benzene occurs with the π-electrons giving rise to a symmetrical bridging complex in which the cation is placed on the C_6 axis, about 2.0–2.1 Å from the centre of the ring[347]. When approaching phenol, the cation could thus associate either with the ring or the oxygen lone pair. It has been argued that both the ion–dipole and polarizability interactions would strongly favour an alignment of the cation along the dipole axis of the compound[348]. Indeed, calculations point out that, in contrast to the protonation, the lithiation occurs preferentially at the position of the oxygen lone pair of phenol. The heavier alkali cations Na^+ and K^+ show a similar behaviour. The resulting complexes are nearly planar with a marginal torsion of the hydroxyl hydrogen atom. Some selected geometrical parameters are displayed in Figure 22. Significant lengthening of the C–O bond (up to 0.05 Å) is found upon complexation. The oxygen–cation distances are longer in the ring complexes. At the B3LYP/6-311++$G(d, p)$+ZPE level of theory, the alkali cation affinities of phenol amount to 149, 101 and 68 kJ mol^{-1} for Li^+, Na^+ and K^+, respectively. Thus, the heavier the cation, the smaller the binding energy and the weaker the ion–phenol complex becomes.

There is only a small charge transfer in the complexes in which the alkali metal retains from 0.75 to 0.95 electronic unit of its original positive charge. This supports the general view that ion–molecule bonding is due to a predominantly electrostatic interaction with a large contribution from the bond dipole.

B. Deprotonation

The Brønsted acidity of a molecule is its capacity to give up a proton. It can be expressed either by the equilibrium constant, the pK_a value, the change of standard free energy (ΔG_T°) or simply the energy of the deprotonation reaction: AH → A$^-$ + H$^+$. The acidities of phenols were measured experimentally[349–351], including a series of 38 *meta*-

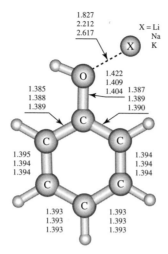

FIGURE 22. Selected B3LYP/6-311++G(d,p) geometrical parameters of the complexes between phenol and alkali metal cations. Bond lengths are in Å

and *para*-substituted phenols using the ion cyclotron resonance (ICR) equilibrium constant method[351]. Theoretical evaluations of acidity usually involve energy calculations of both the neutral substrates and conjugate anions.

1. Phenolate anion

Geometries and vibrational frequencies of phenolate anion (PhO$^-$) in the ground, triplet and excited states were analysed in details[115, 352–361]. Figure 23 displays selected optimized geometrical parameters of the free PhO$^-$ in both lowest-lying singlet and triplet electronic states. Although several crystal structures of phenolates have been reported[362–364], different degrees of aggregation and solvation prevent a direct comparison. The geometry of PhO$^-$ is quite close to that of the benzyl anion (PhCH$_2^-$). In both cases the

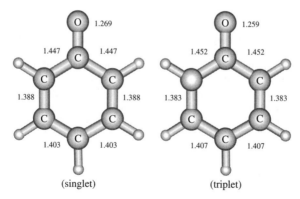

FIGURE 23. Selected (U)B3LYP/6-311++G(d,p) optimized bond lengths (Å) of the phenolate anion in both lowest-lying singlet and triplet states

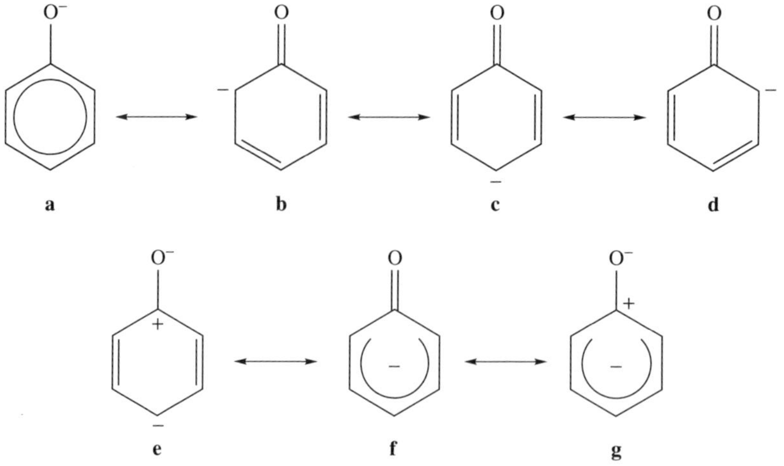

CHART 4. Resonance hybrids of the phenolate ion

p-π delocalization apparently causes a small bond alternation (up to 0.06—0.07 Å) in the anion ring. On this simple basis, PhO$^-$ has thus ca 60% of the aromatic character of PhOH[356]. The C—O distance of 1.27 Å of the anion lies between those of 1.37 Å in PhOH and 1.22 Å in *para*-benzoquinone, giving the CO bond of PhO$^-$ a partial double-bond character which could be understood in terms of simple resonance structures (Chart 4)[357, 358]. Considering the geometry, a quinoidal resonance form **c** with alternating double and single CC bonds may well be a depiction of PhO$^-$.

Since the charges on oxygen are −0.9 electron and on the *ipso*-carbon C_1 +0.5 electron, the dipolar forms are also expected to contribute significantly to the electronic structure of the anion. A certain similarity exists between the phenolate and enolate anions regarding the C—O distances. Quantum chemical calculations[115, 353, 358, 360] of vibrational frequencies for free PhO$^-$ in the ground state did show some discrepancies with experimental data[365, 366]. While IR frequencies determined using DFT methods compare reasonably with the FTIR results in the case of the modes v_4 and v_5, the frequency of the C—O stretching mode is overestimated in all calculations. In addition, large deviations were also found for most modes on isotopic ^{13}C and ^{18}O shifts, as well as on relative IR intensities. Using appropriate scaling factors on computed frequencies at different levels of theory led to the estimated values of 1594, 1495 and 1353 cm^{-1} for the modes v_4, v_5 and v_6, respectively. While the former two are close to the IR absorption peaks at 1585 (or 1592) and 1483 cm^{-1}, the latter deviates from the observed v_6 value of 1273 cm^{-1} by a larger amount of 80 cm^{-1}. Multi-reference CASSCF(10,10) calculations resulted equally in a CO bond distance of 1.285 Å and a v_6 frequency of 1450 cm^{-1}. Thus, the discrepancy between experiment and theory cannot be attributed to a failure of quantum chemical methods, but presumably results from the formation of a complex of PhO$^-$ with either solvent molecules or counterions, weakening the CO bond and inducing a down shift of the corresponding stretching mode. This point will be considered in a subsequent paragraph.

The delocalization of the negative charge from the oxygen to the ring affects the aromaticity of the latter. The magnetic properties of the ring carbons show in fact some marked changes upon deprotonation. Using the GIAO-HF/6-311+G(d,p) method, the

^{13}C NMR chemical shifts (δ in ppm) of phenol and phenolate anion are calculated as follows[352]:

C_1:156 (PhOH)/182 (PhO$^-$), C_2:111/115, C_3:131/132, C_4:118/91,

C_5:133/132 and C_6:115/115.

The C_1 (shielded) and C_4 (deshielded) atoms obviously experience the largest variations. The proton chemical shifts remain almost unchanged, varying by less than 2 ppm.

The nucleus-independent chemical shifts (NICS)[367], calculated as the negative of the absolute magnetic shieldings at ring centres, could be used as a probe for aromaticity. While the phenol in-plane NICS(0) value of -10.8 is greater than that of benzene (-9.7), the NICS value for PhO$^-$ is much smaller (-6.3), only about 58% of the phenol value. This reduction in aromaticity is apparently due to the predominance of the quinoidal structure having alternate CC distances c (Chart IV). It is worth noting that while the PhOH NICS(1) of -11.3 is only slightly larger than the corresponding NICS(0), the PhO$^-$ NICS(1) of -7.6 is larger than its NICS(0) counterpart. This indicates a larger concentration of π-electrons in the anion.

The decreasing aromaticity in the anion is also manifested in a smaller magnetic susceptibility exaltation (Λ)[368], which is defined as the difference between the bulk magnetic susceptibility (χ_M) of a compound and the susceptibility ($\chi_{M'}$) estimated from an increment system for the same structure without cyclic conjugation ($\Lambda = \chi_M - \chi_{M'}$ in units of ppm cgs). Thus, the value $\Lambda = -9.1$ for PhO$^-$ is equal to only 59% of the $\Lambda = -15.5$ for phenol. The computed values for the diamagnetic susceptibility anisotropy (χ_{anis}) follow the same trend, indicating that PhO$^-$ has actually about 60% of the aromaticity of PhOH[352].

It is perhaps interesting to examine here the NICS values for a series of halogenophenols. The influence of one halogen atom on the PhOH NICS is already noticeable: F increases it by 0.2 (-11.0 in *ortho*-F-phenol) whereas Cl reduces it by 1.3 (-9.6 in *ortho*-Cl-phenol) and Br reduces it further by 1.5 (-9.3 in *ortho*-Br-phenol). The effect of multiple X-substituents is appreciable in increasing the NICS to -13.0 in 2,4-di-F- and -14.6 in 2,4,6-tri-F-phenol. The 2,4-di-Cl and 2,4,6-tri-Br species have NICS values approaching that of PhOH. Although the halogen effect is quantitatively more important in phenolates, the trend of the variations is parallel to that in the neutral series, suggesting a significant effect of fluorine.

The electron affinity of the phenoxy radical has received considerable attention. Experimentally, a 2.36 eV upper limit was obtained in 1975[369]. Later, the UV photoelectron spectroscopy of PhO$^-$ was recorded[370] from which an adiabatic ionization energy $IE_a(PhO^-) = 2.253 \pm 0.006$ eV was determined. This low value implies that the valence excited states of phenolate are autoionizing. Evidence for an autoionizing state was found at about 3.5 eV in the photoelectron experiment[370] and at 3.65 eV (340 nm) in the photodetachment spectrum[369]. In other words, there is no evidence for singlet excited states of PhO$^-$ below the ionization threshold. The S_1 and S_2 states belong to the A_1 and B_1 irreducible representations of the C_{2v} symmetry group and can be labelled as ^1L_a and ^1L_b, respectively. Both S_1 and S_2 excited states of PhO$^-$ were calculated to have comparable vertical energies[115, 356, 361]. Recent large CASPT2 computations[357, 371, 372] suggested an adiabatic $S_1 \leftarrow S_0$ energy gap of about 3.69 eV[357]. The latter is further increased to 4.2 eV in aqueous medium, thus corresponding to a blue shift of 1817 cm^{-1}. Experimentally, the first two peaks in the phenolate UV absorption spectrum in aqueous solution are located at 4.32 and 5.30 eV[373]. Molecular dynamics simulations on excited states in solvents were also carried out[371]. A comparison of the oscillator strengths of both states seems to indicate that the $^1A_1$ state, which enjoys a much larger stabilization following geometry relaxation, actually corresponds to the lower-lying state (at least in aqueous solution) of the anion. There is thus a reversed ordering of excited singlet ^1L_a and ^1L_b states in going from phenol to its conjugate anion. The inversion of singlet states is further confirmed in cyanophenols, irrespective of the substitution position[115].

While the S_0 and S_2 ($^1B_1$) states are characterized by a similar charge distribution, they strongly differ from the S_1 ($^1A_1$). A large amount of negative charge (0.45 e) was estimated to be transferred from the oxygen to the ring centre upon the $S_1 \leftarrow S_0$ transition corresponding to a $\pi^* \leftarrow n$ character. This fact allows for the qualitative deprotonation behaviour of both diabatic states ^1L_a and ^1L_b to be understood in terms of electrostatic interactions when the O−H distance becomes sufficiently large. The approach of the positive charge to the anion does not modify the transition energy of ^1L_b due to the small difference in both ground and excited state dipole moments. In contrast, the ^1L_a transition energy changes, due to a significant charge transfer in the anion, reducing the negative charge on oxygen. At a certain O−H distance, both states eventually cross each other implying that, in a reduced symmetry, namely C_s rather than C_{2v} along the proton dissociation coordinate, a conical intersection in the excited states of phenol becomes possible. The centre of such a conical intersection, if it exists, should be located on the C−O axis at a distance of ca 2.6 Å from the oxygen atom. Although these features need to be confirmed by more accurate calculations than those reported[115], it seems that the presence of an avoided crossing along the physically relevant O−H direction, and a conical intersection along the C−O approach of the proton, is of importance per se, as well as, more generally, in the dynamics of the excited state proton transfer reaction from phenol to, for example, water.

The lowest-lying PhO$^-$ triplet state shows marginal deviations from planarity. Some important geometrical features of the T_1 state of the parent are also shown in Figure 23. It is of particular importance that the C−O distance remains almost unchanged with respect to the corresponding singlet state, and that the ring keeps the quinoidal shape (Figure 23). At the B3LYP/6-311++G(d,p)+ZPE level, the $T_1 \leftarrow S_0$ energy gaps are calculated to be around 2.4–2.5 eV for PhO$^-$ and the p-XC$_6$H$_4$O$^-$ anions. These values are slightly larger than the corresponding ionization energies. The triplet anion has not yet been experimentally observed. The T_1 state is readily formed with a dominant configuration arising from a single excitation from the ground state, and rapidly undergoes autodetachment.

The lower-lying singlet and triplet excited states of PhO$^-$ in the environment of photoactive yellow proteins (PYP) were recently simulated by placing point charges to represent the electrostatic field of the seven amino acids and explicit interaction of the anion with two water molecules to account for the hydrogen bonds[357]. The most interesting results are that while the hydrogen bonds were found to exert a minor influence for the lower-excited states of the embedded PhO$^-$, the electrostatic environment of the PYP protein is essential in providing the dominant stabilization, shifting the lowest singlet excited state below the first ionization energy of the system. This effect is also reinforced by a substantial increase of about 4 eV in the anion ionization energy, on passing from the free PhO$^-$ to the protein-bound anion, and then further increasing by up to 0.9 eV for the protein-bound anion−water complex. This feature is significant as it approaches more closely the spectral data for biological chromophores in their native environments.

In halophenolate anions, the *meta* isomers (Figure 24) turn out to be consistently the more stable ones followed by the *ortho* and *para* derivatives, irrespective of the nature of the substituents. The effect is more pronounced in fluoro-anions where the *meta* isomer is about 16 kJ mol^{-1} more stable than the *ortho* counterpart (Figure 25). This energy gap is reduced to 7 and 6 kJ mol^{-1} in chlorinated and brominated phenolate anions, respectively. The energy differences between the *ortho*- and *para*-anions are rather small (*ca* 2 kJ mol^{-1}). On the other hand, the phenolate anion (charge at oxygen) is calculated

1. General and theoretical aspects of phenols

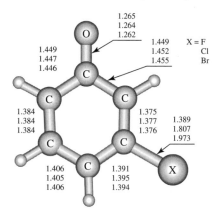

FIGURE 24. Selected B3LYP/6-311++G(d,p) optimized bond lengths (Å) of the metahalophenolate anions in their ground singlet state

to be remarkably more stable than the ring carbon anions by a large amount ranging from 150 to 200 kJ mol^{-1}.

Within the series of ring carbanions (Figure 25), the *ortho*-anions situated at the C_2 positions relative to the hydroxy group (except for *o*-FC$_6$H$_4$OH where the anion is on C_6) are found to be favoured regardless of halogen position. This is no doubt due to the strong interaction between the OH-hydrogen and the negatively charged carbon centre. This implies that the *ortho*-carbon is the most acidic atom within the ring, and this fact has also been verified even in the case of the electron-donating methyl group[374]. Bearing in mind that the *para*-carbon constitutes the most basic ring centre (cf. the preceding section), the difference can be understood by the fact that a ring deprotonation is foreshadowed by its σ-electron skeleton whereas a ring protonation is rather directed by its π-electron distribution. Overall, the deprotonation energies (DPE) of polysubstituted benzenes apparently follow a simple and transparent additivity of the independent substituent effects, implying these DPEs could be deduced using the pre-determined increments of monosubstituents[374].

Regarding the ionization energies of phenolate ions, or conversely the electron affinities of phenoxy radicals (XPhO$^\bullet$), calculated results of some simple substituted species are summarized in Table 29. Density functional theory, in particular when using the hybrid B3LYP functionals, could reproduce the EAs of aromatic radicals with an absolute error of 0.03 eV with respect to the experimental estimates[358, 359]. As substituents on the ring, the halogen atoms tend to increase this quantity by up to 0.4 eV, in the decreasing order: *meta* > *ortho* > *para* position, relative to the value for the parent radical. In contrast, OH and NH$_2$ groups on the *para*-C$_4$ position of the phenolate ion consistently reduce the ionization energy by 0.25 and 0.50 eV, respectively[359].

2. Gas-phase acidities

A convenient measure of the gas-phase acidity is the proton affinity (PA) of the anion, or conversely, the deprotonation energy (DPE) of the acid. For the parent phenol, the experimental value can be deduced from equation 28 for the PA of phenolate,

$$PA(PhO^-) = IE(H) + D(PhO\text{-}H) - EA(PhO^\bullet) \tag{28}$$

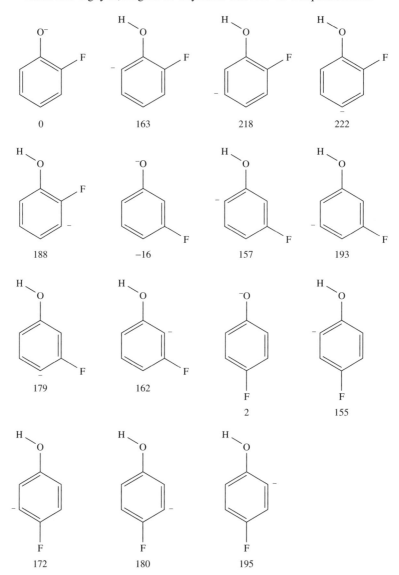

FIGURE 25. Relative energies (in kJ mol^{-1}) obtained from B3LYP/6-311++G(d,p)+ZPE calculations of different isomers of fluorophenolate ions

where IE(H) = 13.606 eV is the ionization energy of the hydrogen atom and EA(PhO•) = 2.253 eV is the electron affinity of the phenoxy radical[370]. The PA is thus dependent on D(PhO−H), being the PhO−H bond energy. Taking the most recent recommended value of D(PhO−H) = 3.838 eV[375], we obtain PA(PhO$^-$) = DPE(PhOH) = 15.191 eV, which is slightly larger than the value of 15.169 eV in an earlier compilation[376]. Indeed,

1. General and theoretical aspects of phenols

TABLE 29. Ionization energies of halophenolate anions

Phenolate anion	IE_a (eV)[a]
Phenolate	2.23 (2.25)
o-Fluorophenolate	2.40
m-Fluorophenolate	2.52
p-Fluorophenolate	2.27
o-Chlorophenolate	2.52
m-Chlorophenolate	2.61
p-Chlorophenolate	2.45
o-Bromophenolate	2.57
m-Bromophenolate	2.64
p-Bromophenolate	2.50

[a] Values were obtained from B3LYP/6-311++G(d,p)+ZPE. In parentheses is the experimental value taken from Reference 370.

TABLE 30. Deprotonation energy (DPE)[a] of phenol derived from various levels of the calculation method

Level of theory[b]	DPE (eV)
B3LYP/6-311++G(d,p)	14.98
B3LYP/6-311++G(3df,2p)	14.90
MP2/6-311++G(d,p)	15.05
MP2/6-311++G(3df,2p)	14.94
CCSD(T)/6-311++G(d,p)	15.17
CCSD(T)/6-311++G(3df,2p)	15.10
Experiment[c]	15.20

[a] Including zero-point energies (ZPE).
[b] Geometries were optimized at B3LYP/6-311++G(d,p) level.
[c] Experimental value, see text.

calculated values in Table 30 provide a support for this estimate with the DPE of phenol lying in the range of 15.1–15.2 eV. In the gas phase, phenol is thus by far more acidic than water (DPE = 16.95 eV) and methanol (16.50 eV), but slightly less acidic than formic acid (14.97 eV) and acetic acid (15.11 eV). Phenol also has a greater acidity than vinyl alcohol (DPE = 15.51 eV) thanks to a more extensive delocalization of the negative charge in the phenolate ion and a greater polarizability of the larger phenyl group.

Results derived from coupled-cluster calculations for halophenols are summarized in Table 31. It is remarkable that even the small variations due to substituents (as detected by experiments[351]) are correctly reproduced by the calculations. Accordingly, the *meta*-halophenols are consistently more acidic than the *para*-counterparts, in contrast to the pattern found for the cyano (CN) group, another strong electron-withdrawing one which tends to reduce the DPE to 14.56, 14.64 and 14.48 eV for *ortho*-, *meta*- and *para*-cyanophenols. The gas-phase acidity scale of cyanophenols is thus *para* > *ortho* > *meta*.

The effect of fluorine substitution on phenol acidities was examined in detail[351, 377, 378]. Through a charge analysis, the F-effect could classically be explained by invoking both resonance and induction effects. In the *meta* position, the halogen tends to stabilize preferentially the phenolate anion due to the resonance effects, resulting in a smaller

TABLE 31. Deprotonation energies of halophenols

Phenol	DPE	
	(calc, eV)[a]	(expt, eV)[b]
Phenol	15.10	15.20
o-Fluorophenol	14.98	
m-Fluorophenol	14.85	14.97
p-Fluorophenol	15.00	15.10
o-Chlorophenol	14.88	
m-Chlorophenol	14.80	14.89
p-Chlorophenol	14.84	14.94
o-Bromophenol	14.85	
m-Bromophenol	14.74	
p-Bromophenol	14.80	

[a] Calculated values from CCSD(T)/6-311++G(3df,2p)+ZPE based on B3LYP/6-311++G(d,p) geometries and frequencies.
[b] Based on the DPE(phenol) = 15.2 eV and relative acidities given in Reference 351.

DPE and a greater acidity. This pattern is confirmed by the energies of fluorophenolate anions shown in Figure 24 pointing towards a greater stability of the *meta*-derivatives.

The characteristics emphasized above for the halogens and the cyano group are actually relevant to other substituents as well. Indeed, it has been shown[377] that the effects of substituents on acidities are largely dominated by those occurring in the phenolate anion and only marginally by those in neutral phenol. Substituents which interact favourably in the *meta* position of phenol act unfavourably in the *para* position (the halogens), and vice versa (the cyano group). Both π and σ charge transfers are important in determining interaction energies. The σ acceptance by a substituent stabilizes OH and O$^-$ more effectively at the *para* position than at the *meta* position, due to a π-inductive mechanism. Stabilization by π acceptors and destabilization by π donors are the results of direct π delocalization, which is inherent of the *para* substituents (see also Reference 380). On the one hand, groups exhibiting a competition of both π-donating and σ-accepting effects, such as NH$_2$, OH and F, cause an increase in acidity at the *meta* position and a decrease at the *para* one (except for F). On the other hand, accepting groups such as CN, CHO, NO$_2$ and CF$_3$ provide an enhanced acidity following either *meta* or *para* substitution, with a preference for the *para* position[377]. There is also little evidence for direct steric strain in the series of *ortho*-phenols[379].

Overall, the acidities of the substituted phenols are largely determined by the stabilization of the corresponding phenolate ions, i.e. the energies of the phenolate HOMOs. There is a similarity between the substituent effect on the latter and the LUMOs of substituted benzenes; both can be understood by simple perturbative PMO treatment[381].

From a more quantitative point of view, it is more difficult to achieve accurate computations for DPEs than for PAs of neutral substrates, because molecular anions are involved in the former case. However, when using second-order perturbation theory (MP2), a coupled-cluster theory (CCSD(T)) or a density functional theory (DFT/B3LYP), in conjunction with a moderate atomic basis set including a set of diffuse and polarization functions, such as the 6-311++G(d,p) or cc-aug-pVDZ sets, the resulting DPE errors appear to be fairly systematic. To some extent, the accuracy rests on a partial but uniform cancellation of errors between the acid and its conjugate base. Therefore, use of appropriate linear regressions between experimental and calculated values allows the DPEs for new members of the series to be evaluated within the 'chemical accuracy' of ±0.1 eV or ±10 kJ mol^{-1}.

3. Acidity in solution

The situation is more complex for the acidities in condensed phases. The relevant quantities are rather estimated using a thermodynamic cycle[382] involving the experimental gas-phase PAs and solvation free energies for the neutral species along with the observed aqueous pK_a values. Using this approach, the experimental hydration free energy of the phenoxide ion[382] was estimated to be -301 kJ mol^{-1}, which is far larger than the corresponding value of -28 kJ mol^{-1} found for phenol[383, 384].

On the other hand, while the basic features of neutral solvation energies could, in general, be fairly well reproduced by continuum solvent models, similar treatments of the anions are less successful. Theoretical approaches to the solvation usually involve a combination of quantum and classical mechanical methods. The molecular responses in the presence of solvent are often handled classically. The most important ingredients in determining solvation energies are the charge distribution and dipole moment of the solute. Evaluation of the electron distribution and dipole moments of charged species is quite troublesome, as they are also quite sensitive to the polarity of the environment. As a consequence, the errors committed on predicted solvation energies for most of the anions are significantly larger than for the neutrals, and this makes quantitative prediction for pK_a values a more difficult task[371]. Similar to the treatment of electron correlation in polyatomic systems, modelling of the impact of the surrounding medium on different entities could hardly be carried out in a balanced way. A small error in the electrostatic terms for long-range interactions easily leads to a large variation in the relative scale. In addition, the difficulties associated with modelling also arise from the account for non-electrostatic interactions, that include among others the cavitation, dispersion and repulsion terms. In practice, a good fit between experimental and theoretical estimates for a category of acids could be established and the predicted values might be useful in establishing, in particular, the acidity order[385].

These general remarks could be applied to the phenol acidities in the aqueous phase that were studied using different combined theoretical methods for evaluating free energies of solvation[115, 374, 378, 385]. In fact, the relative acidities were reproduced with variable success. For example, while excellent agreement was obtained for the *ortho*-fluorophenol, a larger error of 12 kJ mol^{-1} was seen for the *para* isomer[378]. Similarly, experimental acidity trends of both ground and excited singlet states were found for phenol and cyanophenols, but the calculated differences between the ground and excited state pK_a values were only in qualitative agreement with experimental results, with errors up to 4 pK_a units[115].

Nevertheless, the analysis of the charge distribution and hydration behaviour revealed some interesting features. The effect of fluorine substitution on the charge density was found to be not greatly perturbed by the presence of an aqueous solution. The changes in the charge distribution upon substitution are found to be similar in both gaseous and aqueous phases. Thus the observed attenuation of the F-effect on phenol acidities in solution is likely to arise from a hydrophobic shift introduced by the substituent, which finally balances the effects on the hydration free energies of phenol and its conjugate anion.

The enhanced phenol acidity in excited states will be discussed in a subsequent section.

4. Correlation between intrinsic acidities and molecular properties

Understanding substituent effects on molecular properties in a quantitative way has long been a goal of physical organic chemistry and dates back to the 1930s with the introduction of the Hammett σ constants[386]. For phenol derivatives, a variety of correlations have in fact been established between their physical properties in different forms[349-351, 387-391]. The general-purpose Hammett constants yield a reasonable representation of the acidities. A decreasing value of DPE corresponds to an increasing acidity, and hence an increasing value of σ_p^-. We consider here in particular the correlations involving the intrinsic phenol acidities with quantum chemical reactivity descriptors.

The most obvious property related to acidity is the atomic charge on the acidic hydrogen of the neutrals[389–391] and on the deprotonated oxygen of the anions[390, 391]. Use of the atomic charges derived from either the simple Mulliken population analyses, $Q_M(H)$ and $Q_M(O^-)$, or the more advanced natural orbital population analyses, $Q_N(H)$ and $Q_N(O^-)$, leads to linear regression equations with acidities[391], expressed in terms of pK_a, of the type shown in equations 29a and 29b,

$$pK_a = -aQ(H) + b \quad (29a)$$

$$pK_a = -cQ(O^-) + d \quad (29b)$$

where Q is either Q_M or Q_N. A more positively charged hydrogen corresponds to a more acidic hydrogen and is arguably associated with lower pK_a values. In the same manner, delocalization of the negative charge of the phenolate oxygen tends to impart stability to the anion, favouring its formation and increasing the acidity. It is crucial to have a consistent atomic charge definition in order to describe the acid–base properties of the hydroxy group. Correlations between relative acidities and changes in the dipole moments were also attempted[387], but the regression was not very good.

For a given family of compounds, there exists a certain relationship between the proton affinity and the core ionization energy of the atom which is protonated[392–395]. The latter could be approximated by the 1s-orbital energy, $\varepsilon(1s)$, of the relevant atom of the conjugate anion, which is relatively stable with respect to the small variations in the basis sets. Calculations[389] demonstrated that for phenol derivatives, a linear relationship (equation 30) equally exists:

$$PA(X-C_6H_4O^-) = -A \cdot \varepsilon(O_{1s}) + B \quad (30)$$

in which $\varepsilon(O_{1s})$ are the oxygen core orbital energies.

The acidity is expressed above in terms of the PA of the anion. The basicity of the anion somehow describes the ease with which core electrons are removed. In fact, both quantities depend on two terms, namely the electrostatic potential at the site to which the proton is to be attached, and the ease with which the positive charge can be delocalized over the entire substrate by rearrangement of valence electrons.

The first term is known as the inductive effect and is determined by the charge distribution of the initial base. The electrostatic potential minima around the basic centres, V_{min}, needs to be considered. In view of the reasonable behaviour of atomic charges for a series of simple XC_6H_4OH, a good correlation was found between $V_{min}(oxygen)$ and σ_p^-, and thereby the acidities. This was verified with X = H, F, CH_3, NH_2, CN, CF_3, NO and NO_2[381]. Because the experimental σ_p^- were determined in aqueous solutions or in water/alcohol mixtures, their good correlation points out again that the substituent effects in phenolate ions in the gas phase and solution are linearly related. Good interpolations could therefore be made without using solvent models to evaluate unknown or uncertain σ_p^- values.

The second term, known as the relaxation or polarization, depends on the polarizability of the surrounding entity. An inductive effect which favours removal of an electron is expected to hinder the removal of a proton. It is thus logical that there is a negative correlation between the PA of the anion and the core-ionization energy. The higher the core-ionization energy, the lower the PA and the stronger the acid. This view points out the importance of considering both electrostatic and relaxation terms when evaluating the PAs.

In the same vein, the acidity could equally be related to the first ionization energy (IE), which can be estimated from the HOMO energy given by Hartree–Fock wavefunctions. Good linear relationships (equation 31) have been obtained between acidities and frontier orbital energies,

$$pK_a = C \cdot \varepsilon(\text{HOMO}) + D \tag{31}$$

No strong correlations could be found for either the $\varepsilon(\text{LUMO})$ or the absolute hardness (η), or the absolute electronegativity (χ) defined in the previous section. The poor result for the LUMO energies is probably due to their incorrect evaluation using the Hartree–Fock wavefunctions. Calculations[390] revealed that when the acidity of a *para*-substituted phenol decreases, its electronegativity (χ) decreases and its global hardness (η) increases. Conversely, an increasing basicity of the phenolate anion induces an increasing global hardness. This is in line with the original proposal[396] that basicity bears a direct relationship to the hardness of a base. Nevertheless, because the hardness is a global property, it cannot fully account for the changes in basicity/acidity, which is rather a site-specific problem. In fact, the changes in the hardness do not follow a regular pattern and the regression coefficients are lower than those involving other parameters[391].

The local descriptors for the oxygen centre, including the Fukui function (f_O^-) and local softness (s_O^-) whose definitions are given in the preceding section (equation 27), are expected to perform better for this purpose. Both indices tend to increase upon increasing basicity of the anion. Linear relationships were obtained for both indices with pK_a with higher correlation coefficients[391]. This supports the view that the basicity of phenolate ions depends on how the oxygen negative charge could be delocalized into the ring. If the charge cannot be delocalized, the base is getting destabilized and becomes more basic, and vice versa. As a consequence of an increasing oxygen charge, its nucleophilic Fukui function (f_O^-, always positive) and condensed softness (s_O^-) also increase, implying that the oxygen centre becomes more polarizable and softer, in the sense of the original softness definition[396].

A more direct measure of changes in acidity could be determined using the relative proton transfer between substituted phenolate ions and phenol[377, 391] (equation 32)

$$C_6H_5O^- + XC_6H_4OH \longrightarrow XC_6H_4O^- + C_6H_5OH \qquad \Delta E_{\text{prot}} \tag{32}$$

A positive value of ΔE_{prot} indicates that the substituted phenol is less acidic than phenol itself, and vice versa. As a correlation descriptor, ΔE_{prot} performs quite well, giving again a linear relationship with pK_a.

5. Alkali metal phenolates

The structures, energies and reactivities of polar organometallic species are often determined by the metal counterions. Solvation and aggregation also influence their stability and mechanism in condensed phases. The largely dominating electrostatic interactions of both ions outweigh the other modes of stabilization of the anions such as π-delocalization, hyperconjugation, polarization and inductive effects, and basically modify the behaviour of the ion pair relative to the free anion. Phenolate ions with different alkali metal gegenions also show varying reactivity, which has been attributed to the structural changes in the presence of the metal. A case in point is the Kolbe–Schmitt reaction[397] in which sodium phenolate is carboxylated by CO_2 mostly in the *ortho*-position whereas potassium phenolate yields predominantly a *para*-carboxylation product. Charge localization due to the metal ion tends to reduce the stabilization energies of phenolate ion. The metallation reactions (equation 33)

$$\text{PhO—H} + \text{MOH} \longrightarrow \text{PhO—M} + H_2O \tag{33}$$

are much less exothermic than those involving OH⁻/PhO⁻, amounting to −34, −51, −54, −59 and −52 kJ mol^{-1} for M = Li, Na, K, Rb and Cs, respectively (values obtained at B3LYP/6-311++G(d,p)+ZPE)[352] as compared with that of −173 kJ mol^{-1} for free PhO⁻. This emphasizes that the presence of the metal cation in a contact pair counteracts the stabilization of the free anion[398]. Similar behaviour was observed for the analogous enolate anions[352]. The metal ions in the ion pair retain a near unit positive charge (+0.97 to +0.99) pointing towards the pure ionic M−O bonds. Such electrostatic charge localization is no doubt responsible for a higher oxygen charge in the ion-paired species[399].

The geometrical parameters of the PhO−M species are displayed in Figure 26. The C−O−M moiety is actually linear. The C−C distance is shortest in phenol, longest in free phenolate anion and intermediate in metallated compounds. The bond angle around the *ipso*-carbon, $C_6C_1C_2$, is smallest in free phenolate ion [114°, charge $q(C_1) = 0.50$] and largest in phenol [120°, $q(C_1) = 0.38$]. The corresponding angles in PhO−M lie in between, ranging from 118° for M = Li [$q(C_1) = 0.43$] to 116° for Rb [$q(C_1) = 0.46$]. Both the angle and charge at the *ipso*-C_1 atom are somehow related to each other. Even at large M−O distances, the charge localizing effect remains effective because the electrostatic interaction energies decrease with the inverse of the distance, $d(M-O)^{-1}$. Even at a distance $d = 4$ Å, the negative charge on oxygen is already increasing (−0.97), suggesting that the counterion effect is significant in solvent-separated ion pairs.

The magnetic properties of C_1 and C_4 ring atoms are most affected by ion pairing. The calculated $\delta(^{13}C)$ chemical shifts in ppm [obtained from GIAO-HF/6-311+G(d,p) calculations] in PhO−M vary as follows:

$\delta(C_1)$: M = Li: 167 (expt: 168), Na: 171, K: 180, Rb: 180,

Cs: 179 and free anion: 182.

$\delta(C_4)$: M = Li: 111 (expt: 115), Na: 108, K: 106, Rb: 106,

Cs: 107 and free anion: 91.

Deshielding of the atom C_4 in the ion-paired structures is thus obvious. The chemical shifts of other carbon atoms remain almost unchanged upon deprotonation or ion pairing.

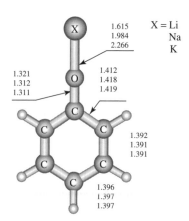

FIGURE 26. Selected B3LYP/6-311++G(d,p) optimized bond lengths (Å) of some alkali metal phenolates in their ground singlet state

The NICS(0) values of the alkali phenolates increase down the group from −9.9 in Li, −9.2 in Na, −8.8 in K, −8.0 in Rb, −7.5 in Cs and −6.3 for free phenolate anion. Thus the charge localization is still effective for cesium phenolate, which has a more aromatic character than the free anion. The other criteria yield a similar pattern[352]. The loss of aromaticity in the free phenolate anion, 60% of the neutral phenol, due to a p−π delocalization discussed above, could largely be restored by ion pair formation with alkali metal cations, thanks to a charge localization effect of the latter.

We now turn back to the CO stretching frequency (ν_6), where there is a discrepancy between observed and computed values (Section IV.B.1). Calculations[353] indicated that the C−O bond length is only slightly elongated by 0.012 Å upon complexation of PhO$^-$ with a water molecule. Such anion−molecule interactions induce only a weak downshift of at most 13 cm^{-1} for modes containing significant CO character. In contrast, as seen in Figure 26, the C−O distances are lengthened to a larger extent (up to 0.047 Å) following interaction with Li$^+$, Na$^+$ and K$^+$, and now have a more significant single-bond character (1.31−1.32 Å). The scaled frequencies of the mode ν_6 are calculated at 1310, 1306 and 1290 cm^{-1} in PhOLi, PhONa and PhOK, respectively [B3LYP/6-311++G(d,p) values]. Complexation with the heavier ion induces a larger downshift of the C−O stretching ν_6 mode, up to 53 cm^{-1} with the K$^+$ counterion. The latter values thus become closer to the experimental value[365, 366] of 1273 cm^{-1} than that derived from free PhO$^-$. More important perhaps is the fact that the ν_6 frequency is now associated with the most intense IR absorption in this region, in agreement with the FTIR data[366]. However, the theoretical overestimation of the C−O stretching mode frequencies remains significant, and some of the ^{13}C and d$_5$ isotope shifts are still large[353]. This suggests that an oligomer of the complex may actually be formed in solution and is responsible for the larger frequency downshift. Dimers and tetramers of lithium enolates[400] and lithium phenolates[401] have in fact been found experimentally.

Finally, it is noteworthy that, along with phenol, phenolate anion has been used as the simplest model to mimic the active site of the tyrosine protein residues. Its interaction with thiol (CH$_3$SH), a model of the cysteine side chain of glutathione, was studied using *ab initio* calculations[402] in order to examine the role of active site tyrosine in glutathione S-transferases. The location of the key proton of the enzyme−glutathione binary complex, O−H−S, was predicted to be near the phenolic oxygen, and this proton position could be manipulated by changing the acidity of the tyrosine. This could be accomplished either by introducing a substituent, such as a fluorine atom, on the phenol moiety, or by changing the protein environment. The hydrogen bond between phenolate anion and thiol is very strong (up to 80 kJ mol^{-1}) and the phenol OH group in the residue of the enzyme complexed by a water molecule in a mutant is related to the notion of substrate-assisted catalysis[403]. In conclusion, the use of PhOM species in order to initiate polymerization and/or to catalyse the chain growth in polycarbonates has been studied[354].

C. Electronic Excitation

Although the valence $\pi-\pi^*$ excitation spectra of benzene derivatives have been extensively studied over the past 65 years both experimentally and theoretically, much less is known about that of phenol, apart from its lowest excited state. In general, absorption and fluorescence spectroscopy of a benzene ring can be used to detect its presence in a larger compound and to probe its environment. While the relative constancy of the valence $\pi-\pi^*$ excitation spectrum allows a qualitative identification of spectral bands by a correspondence with those in free benzene, detailed quantitative differences could indicate the nature of substituents, ligands or medium. Key information on substituted benzene includes the excitation energies, transition moments and their direction, and electrostatic properties of the excited states. Although experimental transition dipole directions could be determined by aligning the molecule in a crystal or stretched film, their interpretation is not straightforward and needs the help of accurate calculations.

Thus, knowledge of the transition moment direction of a phenol band could help in interpreting the fluorescence spectrum of a tyrosine chromophore in a protein in terms of orientation and dynamics. The absorption spectrum of the first excited state of phenol was observed around 275 nm with a fluorescence peak around 298 nm in water. The tyrosine absorption was reported at 277 nm and the fluorescence near 303 nm. Fluorescent efficiency is about 0.21 for both molecules. The fluorescent shift of phenol between protic and aprotic solvents is small, compared to indole, a model for tryptophan-based protein, due to the larger gap between its first and second excited states, which results in negligible coupling[404].

A mono-substituted benzene has traditionally a number of singlet excited valence states, or pairs of states, of $\pi^* \leftarrow \pi$ type. The valence $\sigma^* \leftarrow \pi$ or $\pi^* \leftarrow \sigma$ excitations require much larger energies. Below the first ionization level, a number of Rydberg $\pi^* \leftarrow \pi$ and $\sigma^* \leftarrow \pi$ states could also be expected. Each open-shell singlet state also has a triplet companion situated at slightly lower energy. The corresponding vacuum UV singlet spectrum can be subdivided into three bands, the first denoted as ^1L_b centered at about 2600 Å, the second ^1L_a at ca 2050 Å and the third ^1B band at ca 1850 Å[405]. Note that the notations ^1L_b and ^1L_a mean that their dipole transition moment are approximately perpendicular and parallel, respectively, to the main axis.

The lower-lying singlet states of phenol exhibit a $^1A'$ symmetry. As mentioned above, the lowest ^1L_b band of phenol was well established experimentally to have an origin at 4.507 eV (275 nm or 36349 cm^{-1} with an oscillator strength $f = 0.02$)[118]. This first singlet excited state S_1 closely corresponds to the covalent $^1B_{2u}$ state of benzene and has a transition dipole in the x direction. The vertical ^1L_a absorption due to the second excited state S_2 was found at 5.82 eV[406], whereas the corresponding adiabatic value was estimated at 5.77 eV (with $f = 0.13$)[119] and is correlated to the more ionic $^1B_{1u}$ state of benzene. The identity of the third excited state of phenol inducing the appearance of its ^1B band was more problematic[119, 406], but it now appears that the observed band, centred at ca 6.66 eV, arises from the lower component of a splitting of the degenerate benzene $^1E_{1u}$ state and is associated with a fairly large transition moment ($f = 1.1$)[119]. A small and static splitting of this band is usually found in most mono-substituted benzenes with approximately equal intensities. As for the benzene $^1E_{2g}$ band, CASSCF/CASPT2 calculations[119, 407, 408] revealed a significantly larger splitting giving two components centred now at 7.14 and 7.72 eV. Although the E_{2g} states are formally characterized as covalent, they are in reality strongly mixed with a multitude of higher states.

The Rydberg states have not yet been detected experimentally, but CASPT2 calculations[119, 408] indicated the existence of at least six $\pi^* \leftarrow \pi$ Rydberg states that range from 6.3 to 7.6 eV and arise from the promotion of 3π and 4π electrons to 3p and 3d orbitals. There are also no less than twelve $\sigma^* \leftarrow \pi$ Rydberg states ranging from 5.8 to 7.8 eV.

The measured rotational constants of the first excited S_1 state[127] suggested rather moderate changes of the geometrical parameters upon electronic excitation. The $S_1 \leftarrow S_0$ excitation tends to enlarge the carbon ring and reduce the C–H and C–O bond lengths. The O–H bond length and the C–O–H bond angle are almost invariant upon excitation. The constants vary as follows: S_0/S_1 (in MHz): A; 5650/5314; B; 2619/2620; and C; 1782/1756. Multi-reference CASSCF computations reproduced these quantities reasonably well and suggest a planar structure[114, 115, 126, 139, 356, 372, 407, 408]. In particular, the CASSCF(8,7)[407] study provided the rotational constants of $A = 5338$, $B = 2572$ and $C = 1736$ MHz for phenol S_1. The changes in rotational constants could be understood as arising from a deformation of the molecule in the S_1 state along the in-plane mode

6a or mode 8a. CASSCF geometry optimizations[126, 372, 407, 408] showed a rather modest shortening of 0.006 Å of the CO distance and a somewhat more important lengthening of 0.03 Å of all CC bonds. Nevertheless, a comparison between the dispersed fluorescence spectrum of phenol and its Franck–Condon simulation[114] indicated that the CC bond length actually increases on average by 0.027 Å, whereas the CO bond distance decreases by 0.023 Å upon excitation. The most significant geometrical relaxation could also be deduced from the experimentally observed intensity pattern.

For the second excited S_2 state of phenol, a quite different geometry was found with larger variations of up to 0.11 Å for the CC bond lengths and the COH bond angle (opening by 10°), and a non-negligible shortening of 0.02 Å of the CO bond (relative to phenol S_0). This suggests a considerable charge delocalization from the oxygen into the ring.

The S_1 vibrational frequencies were also observed[153, 156, 169, 409] and analysed in detail by means of quantum chemical computations[114, 115, 126, 139, 356, 372, 407, 408]. Frequency shifts up to 100 cm^{-1} were detected for in-plane modes. While the σ(OH) mode decreases from 3656 to 3581 cm^{-1}, the CH-stretching modes 20a and 20b increase from 3087 and 3070 to 3186 and 3136 cm^{-1}, respectively, following excitation. Out-of-plane modes show much more scrambling in going from S_0 to S_1, and several original modes[409] needed to be re-assigned[114]. In particular, the Kekule mode 14 should have a larger wave number in the S_1 state (1572 cm^{-1}) than in the S_0 state (1343 cm^{-1}). This mode has CH-bending and CC-stretching character in the ground state but becomes a CC-stretching plus a small component of the OH-bending mode in the excited state. The relaxation of the OH-stretching vibrations in the S_1-S_0 transition could also be followed in examining the IR-UV double resonance spectra recorded after pumping to the OH stretching level[410]. These techniques provided us with valuable information on the intramolecular vibrational redistribution (IVR) of the corresponding vibrations.

The phenol dipole moment remains almost unchanged upon excitation to S_1 but shows a marked variation in S_2, in line with a more ionic character of the latter. The ratio of oscillator strengths for both S_2 and S_1 transitions amounts to 6.6 and, as evidenced by the $\langle z^2 \rangle$ values, both valence excited states have no relevant mixing with Rydberg states. Cyanophenols show a similar behaviour where the S_1 charge distribution is close to the ground state and the S_2 counterpart appears to have an appreciable charge transfer from the oxygen[115].

Solvent effects were found to have minimal influence on the excitation energies of phenol in aqueous solution using a quantum Monte Carlo simulation[372], which is in line with experimental observations on its absorption spectra[411]. Reaction field calculations of the excitation energy also showed a small shift in a solution continuum, in qualitative agreement with fluorescent studies of clusters of phenol with increasing number of water molecules[412a]. The largest fluorescent shift of 2100 cm^{-1} was observed in cyclohexane.

In substituted phenols, the excited S_1 states are again dominated by the LUMO ← HOMO and LUMO + 1 ← HOMO − 1 transitions and the corresponding excitation energies apparently differ from that of phenol by, at most, 0.6 eV. Results obtained using time-dependent density functional theory computations in conjunction with a systematic empirical correction are recorded in Table 32. CASSCF(8,7) calculations on both S_0 and S_1 of monochlorophenols[412b] also point to a similar trend. The frontier orbital energies are only weakly but uniformly stabilized by the halogens or the cyano group, or else they are destabilized by electron-donor groups such as methyl. While the fluorine atoms do not exert any significant effect, multiple substitutions by chlorine and bromine induce a significant decrease in the transition energies[412b]. The chlorine atom makes the C–O bond shorter and the methyl group makes a marginal modification; the cyano shows a detectable effect when introduced at the 2-*ortho* position.

TABLE 32. Lowest excitation energies of substituted phenols[a]

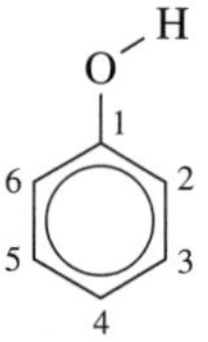

$S_1 \leftarrow S_0$ Transition energy (eV)

Substituent	Energy
Phenol	4.5 (4.5)
2-F	4.5
3-F	4.5
4-F	4.3
2,3-di-F	4.6
2,4-di-F	4.3
2,5-di-F	4.5
2,6-di-F	4.5
4,5-di-F	4.3
2,4,5-tri-F	4.3
2,4,6-tri-F	4.4
2,3,4,6-tetra-F	4.4
2,3,4,5,6-penta-F	4.5
2-Cl	4.4
3-Cl	4.4
4-Cl	4.2
2,3-di-Cl	4.3
3,4-di-Cl	4.1
4,6-di-Cl	4.1
3,4,5-tri-Cl	4.0
2,4,6-tri-Cl	4.0
2,3,4,6-tetra-Cl	3.9
2,3,4,5,6-penta-Cl	3.8
2-Br	4.3
3-Br	4.4
4-Br	4.2
2,3-di-Br	4.2
2,4-di-Br	4.0
2,5-di-Br	4.2
2,6-di-Br	4.2
2,4,6-tri-Br	3.9
2-CH$_3$	4.5
3-CH$_3$	4.5
4-CH$_3$	4.4
2-CN	4.2 (4.2)
3-CN	4.3
4-CN	(4.5)

[a]Estimated values using TD-DFT/B3LYP/6-311++G(d,p) calculations and a systematic correction based on a comparison of the calculated and experimental values for phenol. Experimental values are in parentheses.

The acidities of phenols were found to be greatly increased upon electronic excitation. Due to a change of about 40 kJ mol^{-1} in the free energy of deprotonation, phenol is intrinsically 7 pK_a units more acidic in the S_1 than in the S_0 state in the gas phase. Similarly, intermolecular proton transfer in solution from an S_1 excited phenol to, e.g., a solvent base is typically characterized by a pK_a value of some 6–7 units less than that of the corresponding ground state. In aqueous solution, the pK_a of phenol amounts to 10.0 in the ground state and 3.6 in the S_1 state.

It is natural to ask whether the enhanced acidity in the excited state arises from an electronic effect of the neutral acid or from the product anion. As seen in Section IV.B above, it has been shown in various ways[377, 413] that the changes in ground-state acidity resulting from several substitutions are due to the corresponding phenolate anions. The same argument could equally be applied to the difference between ground- and excited-state acidities. The pK_a modification could be understood by the fact that the gas-phase proton affinity of the phenolate anion, a measure of the phenol acidity, amounts to 15.2 eV in the ground state but decreases to 14.3 eV in the S_1 state[414]. This anion also has a large blue shift of the vertical excitation energy (1800 cm^{-1}) in solution. Monte Carlo simulations[372] demonstrated that the excited states of phenol and phenolate anion are better solvated than the ground states by ca -2 and -11 kJ mol^{-1} in water, respectively. The experimental value p$K_a = 3.6$ of phenol in the S_1 state in solution is likely to arise from a cancellation of the intrinsic energy difference (ca 50 kJ mol^{-1}) of the excitation energy of phenol and phenolate anion, and by the differential solvent spectral shift (ca 25 kJ mol^{-1}). The energetic outcome leads to a change of -5 in the pK_a value, which is roughly in accord with the experimental estimate[373].

It has been observed that the magnitude of the electrostatic potential (V_{min}) around the oxygen atom undergoes a much larger reduction for the anions than for the neutrals in going from S_0 to S_1[115]. In other words, from a purely electrostatic point of view, the increase in S_1 phenol acidity can better be understood by the fluctuations of the phenolate anions.

Relatively little is known about the phosphorescent phenol[415]. The experimental T_1-S_0 transition energy was found at 28500 cm^{-1}, confirming that the triplet state is, in general, lower in energy than its singlet counterpart[356]. The selected optimized geometrical parameters of the lowest triplet T_1 state of phenol is displayed in Figure 27. The molecule is no longer planar but shows a small ring deformation with stretched and compressed CC distances, and a marginal out-of-plane OH torsion.

The quenching mechanism of the first excited states of phenol and phenolate anion differ significantly from each other. The fluorescent neutral S_1 state lies substantially higher in energy than T_1 and could be inhibited from quenching by the energy gap (ca 8000 cm^{-1}) as well as the small one-electron spin–orbit coupling. At the anion-S_1 geometry, both

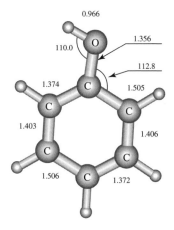

FIGURE 27. Selected UB3LYP/6-311++G(d,p) optimized parameters of the lowest triplet state of phenol. Bond lengths are in Å, bond angles in deg

singlet and triplet states of the anion are shown to be dominated by the same electronic configuration, thus allowing for a direct spin–orbit coupling[356]. As a consequence, the lifetime for fluorescence is short in the anion.

At the neutral-S_0 geometry, the spin–orbit coupling is expected to increase, but there was no evidence of a change in the fluorescence efficiency as a function of the excitation energy in the first singlet excited band[416]. Quenching in the singlet S_1 state to the T_1 triplet was reported[417]. The weak spin–orbit coupling is likely to account for an observation of the neutral triplet. In this case the corresponding anion triplet is not observed, due to the fact that its energy is larger than the electron affinity of the phenoxy radical and it is readily autodetached.

Finally, the phenol super-excited states, which are electronic states of neutral species with energy above the first ionization energy, were also identified at about 9 eV above the ground state[418, 419]. Some of these super-excited states could be mapped spectroscopically out on a picosecond and femtosecond time scale.

D. Ionization

Owing to their relatively low ionization energies (IE) of ca 8.0–8.5 eV, phenols are also good electron donor solutes. Recent experimental studies of phenols in non-protic solvents[420–423] showed that ionized solvent molecules react with phenol to yield not only phenol radical cations by electron transfer, but also phenoxy radicals by hydrogen transfer. An obvious question is whether, under these conditions, the latter radicals were formed from ionized phenols rather than by direct hydrogen abstraction, because proton transfer reactions could be facilitated upon ionization. This also raises a question about the influence of solvent properties, both by specific and non-specific interactions, on the mechanism and kinetics of deprotonation processes[424, 425].

Gas-phase properties of a molecule have, by definition, an intrinsic character and they could be modified by the environment. Although the formation and reactions of gaseous ionized phenol **21** (cf. Chart 5) and its cyclohexa-2,4-dienone isomer **22** have been studied in numerous ionization and mass spectrometric studies[182, 426–438], thermochemical parameters of these isomers[439–447] as well as information on other non-conventional isomers, such as the distonic ion **23**, were rather scarce. Conventional cations of analogous aromatic systems $(X-C_6H_5)^{\bullet+}$ and their distonic isomers generated by simple 1,2-hydrogen shifts within the ring were demonstrated to be observable gas-phase species[448–451]. In addition, the mechanism of the CO-loss upon phenol ionization has only recently been unraveled[452].

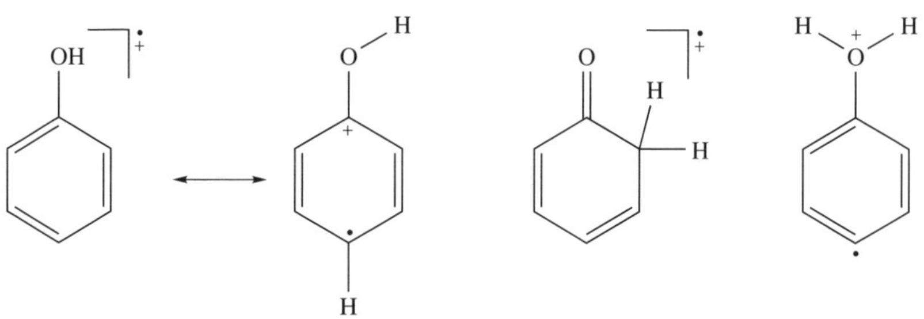

CHART 5. Two isomers of phenol radical cation

1. Molecular and electronic structure of phenol radical cation

The molecular structure, vibrational frequencies and spin densities of ionized phenol **21** in its ground and lower-lying excited electronic states have been investigated intensively using different MO and DFT methods[138, 168, 182, 425, 426, 453]. For the purpose of comparison, Figure 28 shows again a selection of (U)B3LYP/6-311++G(d,p) geometrical parameters of both neutral and ionized structures (c.f. Table 5). The lowest-energy electronic state of **21** exhibits a planar geometry and a $^2A''$ symmetry arising from removal of an electron from the π-system; therefore, its ground state can be qualified as a $^2\Pi$-state. Following such an ionization, the quasi-equal C—C bond (1.40 Å) framework in the neutral phenyl ring becomes longer (1.43 Å) and shorter (1.37 Å) bonds. The latter distance becomes now closer to that of a typical C=C double bond (1.35 Å). Although the absolute changes in the bond lengths vary with the methods employed, they consistently point out that the C—O bond is shortened in going from 1.37 Å in the neutral to 1.31 Å in the ionized phenol, but it remains longer than that of a typical C=O double bond (1.22 Å)[138]. Such distance changes can be understood from the shape of the HOMO of neutral phenol as displayed in Figure 5. Accordingly, the C—O bond is characterized by antibonding orbital 2p-lobes; therefore, electron removal is expected to shorten the C—O distance. The same argument could be applied to the changes in the ring C—C distances. In fact, electron removal from the bonding C_6—C_1—C_2 and C_3—C_4—C_5 components leads to bond stretching, whereas a decrease in the antibonding C_2—C_3 and C_4—C_5 components results in bond compression. Because the unpaired electron occupies a π-orbital and exerts a marginal effect on the σ framework, the C—H and OH distances are not significantly affected and the COH bond angle opens by only 4° upon ionization (cf. Figure 28). Although the changes in geometry are a clear-cut manifestation of the oxidation, it is not possible to correlate these alterations completely with all the accompanying intramolecular reorganization energies[454]. This reorganization is global rather than a local phenomenon.

To some extent, the geometry confers on the phenol ion a quinone-like distonic character as seen in **21a** (Chart 5) in which the charge and radical centres are located at two different sites. This picture is supported by the charge distribution according to the Mulliken population analysis suggesting that the *para*-C_4 carbon of the ring bears the largest

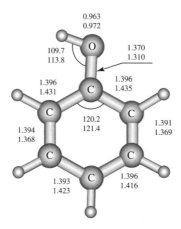

FIGURE 28. Selected (U)B3LYP/6-311++G(d,p) bond distances (Å) and angles (deg) of the neutral ($^1A'$, upper values) and ionized ($^2A''$, lower values) phenol. See also Table 5

part of the excess electron spin (ca 0.5 e). The positive charge is, as expected, delocalized over the entire ring skeleton but with a substantial part on the oxygen region[138, 168].

Bear in mind that the HOMO-1 is equally a phenyl orbital with the 2p(π)-lobes centred on the *ortho* and *meta* carbon atoms (Figure 5). As a consequence, ejection of an electron from this orbital is expected to yield a $^2\Pi$ excited state of phenol ion in which the C_2-C_3 and C_5-C_6 distances likely become longer than the corresponding values in neutral phenol whereas the C−O distance likely remains unchanged. Removal of an electron from the HOMO-2 again leads to a $^2\Pi$ excited state. The HOMO-3 of phenol is the first in-plane orbital (a') thus leading to a $^2A'$ excited state of phenol ion.

The recorded He(I) and He(II) photoelectron spectra of phenol[443−447] contain several peaks ranging from 8.56 to 22.67 eV. It appears that the reported value of 8.56 eV[443] is actually the first phenol vertical IE whereas that of 8.47 eV[444] corresponds to its first adiabatic IE_a. Geometry relaxation of the vertical ion results in a small stabilization. For comparison, note that the IE_a(phenol) is computed to be 8.37 and 8.42 eV using B3LYP and CCSD(T), respectively, in conjunction with the aug-cc-pVTZ basis set. This leads to a standard heat of formation of $\Delta H_f^\circ(\mathbf{21}) = 724 \pm 6$ kJ mol^{-1}[441].

The vertical lowest-lying excited state A^2A'' of phenol radical cation **21** lies only 0.72 eV above the ground X^2A'' state whereas the vertical B^2A'' state is identified at 2.96 eV from the photoelectron spectrum[443]. MCSCF/FOCI computations[455] yielded a value of 3.32 eV (373 nm) for this vertical transition. A recent photoinduced Rydberg ionization spectroscopic study[426] revealed a gap of 2.62 eV (21129 cm^{-1}), which is assigned to the B-state of **21**. Geometry relaxation apparently induces a larger stabilization in this B-state. Electronic spectra of the phenolcation–water complex also suggested a certain transition in this region[456].

Coupled-cluster CCSD(T)/6-311++G(d,p) electronic energy computations of the $^2A'$ state using the $^2A''$ ground-state geometry leads to an estimation of 3.6 eV for the vertical $C^2A' \leftarrow X^2A''$ transition, which compares reasonably well with the PE of 3.37 eV[443]. The lowest-lying quartet state of phenol ion was found to be a dissociative state giving a triplet phenyl cation plus OH radical that lie about 5.3 eV above the ground-state $^2A''$. Overall, the calculated results point towards the following energy ordering of electronic states of **21**: $X^2A''(0.0) < A^2A''(0.5) < B^2A''(2.6) < C^2A'$ (3.1), where values given in parentheses are energy gaps in eV.

A deprotonation of the phenol ion giving the phenoxy radical **21** → $C_6H_5O^\bullet(^2B_1)$ + H^+ is a barrier-free endothermic scission. Due to the small size of the proton, the stabilizing through-bond delocalization during the cleavage, if any, is likely to be small[441]. The process is characterized by a DPE of 857 kJ mol^{-1} (at 0 K) and 863 kJ mol^{-1} (at 298 K) derived from B3LYP computations. The latter value compares well with the experimental proton affinity of 860 kJ mol^{-1} previously determined for the phenoxy radical[442]. This is by far smaller than the corresponding value of neutral phenol, DPE(PhOH) = 1464 kJ mol^{-1} (15.16 eV), discussed above. Electron removal from a neutral system tends to facilitate effectively its deprotonation. For the sake of comparison, remember that the PAs (0 K) of phenol and anisole amount to PA(phenol) = 820 kJ mol^{-1} and PA(anisole) = 842 kJ mol^{-1} (cf. Section IV.A). From a technical point of view, the hybrid density functional B3LYP method appears to provide the most accurate DPE values[138].

The effect of substituents on the DPE and IE also depends on their nature and position. For a series of mono-halophenol ions, the DPEs (in kJ mol^{-1}) calculated using the B3LYP/6-311++G(d,p)+ZPE level are as follows:

F: *ortho*: 843;   *meta*: 840;   *para*: 849.
Cl: *ortho*: 852;   *meta*: 848;   *para*: 861.
Br: *ortho*: 858;   *meta*: 853;   *para*: 867.

Relative to the value of 857 kJ mol^{-1} for **21**, fluorine consistently tends to reduce the DPE up to 17 kJ mol^{-1}, whereas chlorine and bromine could either enhance or reduce it by *ca* 10 kJ mol^{-1}. The *meta*-C$_3$ position is peculiar in having the smallest DPE, irrespective of the nature of the halogen. This is due to the fact that the *meta*-X-phenol radical cation corresponds to the least stable isomer within each series, lying up to 20 kJ mol^{-1} above the most stable *para*-C$_4$ counterpart. In the *ortho* position, the *cis*-C$_2$ conformer is more stable than *trans*-C$_6$ and energetically close to the *para*-C$_4$ one. A direct consequence of the lower stability of the *meta*-X-phenol ions is the higher IE$_a$ of the corresponding neutral molecules whereas the *para*-X-phenols, on the contrary, exhibit the smallest IE$_a$. The IE$_a$s of the series of mono-halophenols are evaluated as follows (bearing in mind that the relevant value for the parent phenol is actually 8.48 eV):

F: *ortho*: 8.68; *meta*: 8.70; *para*: 8.48.
Cl: *ortho*: 8.61; *meta*: 8.63; *para*: 8.39.
Br: *ortho*: 8.55; *meta*: 8.57; *para*: 8.32.

The observed changes in both quantities could partly be rationalized in classical terms of electron-donating and electron-withdrawing effects[439, 440].

We now turn to the hyperfine coupling constants (hfcc) of **21** that were determined using EPR spectroscopy techniques[457]. It is believed that these properties could be used with enough accuracy to distinguish phenol radical cations from phenol radicals in tyrosine-derived species[138]. Isotropic hfcc values are a sensitive measure of the electronic spin distribution, as they are directly proportional to the spin density at the position of nucleus N, $\rho(r_N)$. According to the McConnell relation[458], the spin density at the H nucleus is well known to depend on the spin polarization of the σ(C—H) electrons by virtue of the unpaired carbon π-electron density. Therefore, it suggests the repartition of the excess electron among the ring carbon atoms. Measured hfcc values included 5.3, 0.8 and 10.7 Gauss for the protons at the C$_2$, C$_3$ and C$_4$, respectively. This agrees qualitatively with the spin distribution from simple resonance terms, where the highest spin density is on the *para*-C$_4$, followed by the *ortho* C$_2$ and C$_5$ carbons. The values for the hydroxyl proton, ^{13}C and ^{17}O hfcc values, as well as the sign of spin polarization at each proton were not reported[457].

Table 33 lists the hfcc values calculated at the UB3LYP/6-311++G(d,p) level for both phenol radical cation and phenoxy radical. A few points are noteworthy.

(i) There are significant differences between the hfcc values of both doublet species which are perfectly distinguishable on the basis of this spectroscopic parameter. Protonation of the symmetrical phenoxy radical induces some large shifts on the ^{13}C constants, in particular the *ipso*-carbon, and to a lesser extent the *ortho*-carbons. The odd-alternate pattern of spin densities is thus more pronounced in the radical cation than in the radical.

(ii) A large asymmetry is manifested in the hfcc values of ion **21**.

(iii) Calculated hfcc values for **21** agree qualitatively with the EPR results mentioned above. Thus the calculated $a(H_4) = -9.9$ G, $a(H_2) = -4.1$ G and $a(H_3) = 0.7$ G are close to the experimental magnitude of 10.7, 5.3 and 0.8 G, respectively.

(iv) Calculations reveal a substantial hfcc for the hydroxyl proton (-6.9 G).

The difference in structural and bonding properties of both neutral and ionized species also manifests itself strongly in their vibrational motions. Most of the 11 experimentally measured vibrational frequencies for **21** and 10 frequencies for its deuterated analogue **21**–*d*5 correspond to the CH bending, CC and CO stretching[425]. The highest frequency observed at 1669 cm^{-1} was assigned to a CC stretching mode (the Wilson 8a mode) and the lowest frequency of 169 cm^{-1} describes an out-of-plane ring torsion. No surprises were noted in the measured isotopic frequency shifts; all modes of **21** shift to lower frequencies

TABLE 33. Hyperfine coupling constants (G) of phenoxyl radical and phenol radical cation[a]

Atom	Phenoxyl radical Isotropic Fermi Contact Couplings	Phenol radical cation 21 Isotropic Fermi Contact Couplings
C-1	−12.5	−2.5
C-2	6.6	1.6
C-3	−8.8	−6.4
C-4	10.3	10.0
C-5	−8.8	−7.4
C-6	6.6	2.8
O	−7.3	−6.9
H(C-2)	−6.7	−4.1 (5.3)[b]
H(C-3)	2.6	0.7 (0.8)
H(C-4)	−8.8	−9.9 (10.7)
H(C-5)	2.6	1.5
H(C-6)	−6.8	−5.0
H(O)		−6.9

[a] Results obtained from UB3LYP/6-311++G(d,p).
[b] In parentheses are experimental values from Reference 424.

upon deuteriation and the largest observed frequency shift of -359 cm^{-1} appears for a CH bending motion. Calculations[138] have helped to reassign several observed bands[425]. Most importantly, the band observed at 1500 cm^{-1} is due to the CO stretching (rather than a CC stretching as originally assigned) and the band at 1395 cm^{-1} to a CH bending (rather than a CO stretching).

Although the atomic masses remain unchanged, the force constants, frequencies and normal modes are modified significantly upon electron loss. We note that the most important shifts arise from the C−O−H torsion mode (upshift of 256 cm^{-1}), the C−O−H bending (downshift of 57 cm^{-1}) and the CO stretching (upshift of 101 cm^{-1}). It is possible not only to identify these changes, but also to quantify them in terms of the percentage of a neutral mode present in that of the ion by making use of a vibrational projection analysis technique[168]. Figure 29 displays a qualitative graphic representation of the hydrogen displacements in the C−H stretching normal modes calculated for both neutral and ionized phenol. While the highest and lowest C−H stretching modes of **21** are clearly assignable to the respective modes of phenol, the middle three modes show a higher degree of changes and mixing.

2. Relative energies of the $(C_6H_6O)^{\bullet+}$ radical cations

There are obviously a large number of possible isomers of phenol ion. Let us consider only the isomers where the six-membered ring framework is preserved. Starting from **21**, one hydrogen atom could be displaced from either O or one C atom to another atom and this exercise results in the creation of the various isomeric groups presented in Figure 30: group 1 includes ions having a CH_2 group at the *para* (C_4) position, group 2 at the *meta* (C_3), group 3 at the *ortho* (C_2) and group 4 at the *ipso* (C_1) and oxygen positions.

Calculated energies relative to the phenol ion given in Figure 30 indicate that **21** represents the most stable form among the six-membered ring group of isomers. Keto-forms **22** and **24** are low-lying isomers which are situated 146 and 133 kJ mol^{-1}, respectively, above **21**. This energy ordering within the pair **21** and **22** (or **24**) is reminiscent of that encountered for simple keto−enol tautomers[459, 460]. For example, ionized vinyl alcohol

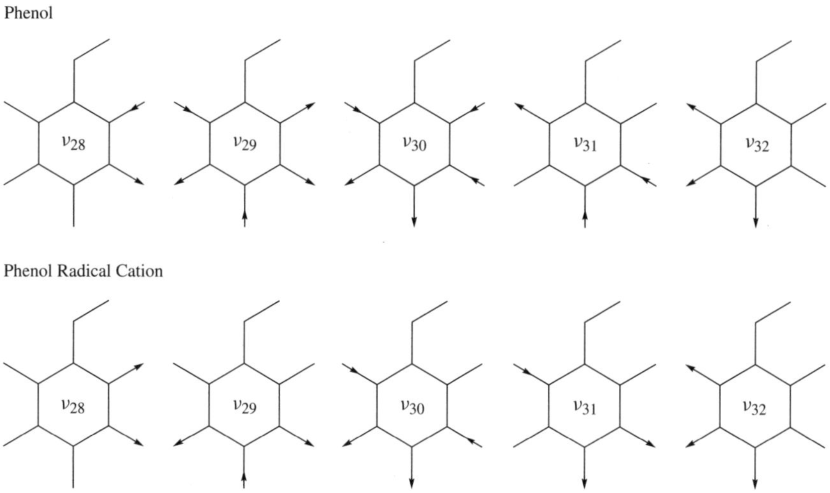

Phenol

[Figure: hexagonal ring diagrams labeled ν_{28}, ν_{29}, ν_{30}, ν_{31}, ν_{32} with arrows indicating hydrogen displacements]

Phenol Radical Cation

[Figure: hexagonal ring diagrams labeled ν_{28}, ν_{29}, ν_{30}, ν_{31}, ν_{32} with arrows indicating hydrogen displacements]

FIGURE 29. Qualitative graphic representation of the hydrogen displacements in the C−H stretching normal modes calculated for both neutral and ionized phenol

is significantly more stable (about 60 kJ mol^{-1}) than its keto ion counterpart[461, 462]. The difference in energy observed here between ionized phenol and its keto tautomers is, however, more pronounced; this point will be examined below. The distonic oxonium species **23** (Chart 5) belongs to the high-energy group of isomers; its energy, relative to **21**, equals 241 kJ mol^{-1}. The distonic species **25** (Figure 30) turns out to be the lowest lying isomer of group 2. This situation is opposite to the situation met in the ionized aniline system in which the ammonium distonic ion is found to be only 80 kJ mol^{-1} above ionized aniline[303]. The other *meta*- and *ortho*-distonic ions have similar energy and are separated from each other by high-energy barriers for 1,2-hydrogen shifts (Chart 6).

In order to evaluate the effect of ionization on the relative stabilities of phenol isomers, a selected set of neutral species is considered whose relative energies are displayed in Figure 31. It is remarkable that, in the neutral state, only three six-membered ring structures are in a *ca* 70 kJ mol^{-1} energy range, namely phenol **26** and its keto-forms **27** and **28**. The carbene, allene or biradical isomeric forms are strongly destabilized and lie more than 200 kJ mol^{-1} above **26**. In contrast, the five-membered ring containing a ketene or a ketone moiety are only 90 to 140 kJ mol^{-1} above phenol. As expected, phenol **26** is more stable than its tautomers **27** and **28**, and this is partly at the origin of the large difference in stability of the corresponding ionized species. In fact, in the phenol series, the aromaticity renders the enol tautomer more stable; this situation is opposite to that observed in the aliphatic series. For example, neutral acetaldehyde is *ca* 40 kJ mol^{-1} below its enol form, namely the vinyl alcohol[459, 461]. After removal of one electron, the enol structure becomes more stable than the keto form by 60 kJ mol^{-1} as recalled above[462]. This stability reversal is due to the large difference in IE$_a$ values between the two structures, namely 9.14 eV for vinyl alcohol and 10.23 eV for the acetaldehyde, in keeping with the fact that it consists of a $\pi_{C=C}$ ionization in the former case and an ionization of an oxygen lone pair in the latter. A comparable situation arises for the phenol (IE = 8.5 eV) and its keto tautomers **27** and **28** (IE = 10.8 eV). This difference, added to the difference in energy between the neutral molecules (in favour of the phenol molecule), explains the large energy gaps of **22** and **24** with respect to **21**.

3. The $(C_6H_6O)^{•+}$ potential energy surface (PES)

The essential features of the portion of the $(C_6H_6O)^{•+}$ PES starting from **21** were constructed and illustrated schematically in Figure 32. The shape of the most interesting intermediates are defined in Figure 33 and X/Y denotes a transition structure (TS) linking two equilibrium radical cation structures X and Y. The ion fragments $C_5H_6^{•+}$ resulting from elimination of CO, labelled as **31, 33, 38, 40** and **43** in Figure 32, are omitted for the sake of simplicity. Their actual shape can easily be deduced from the structures of

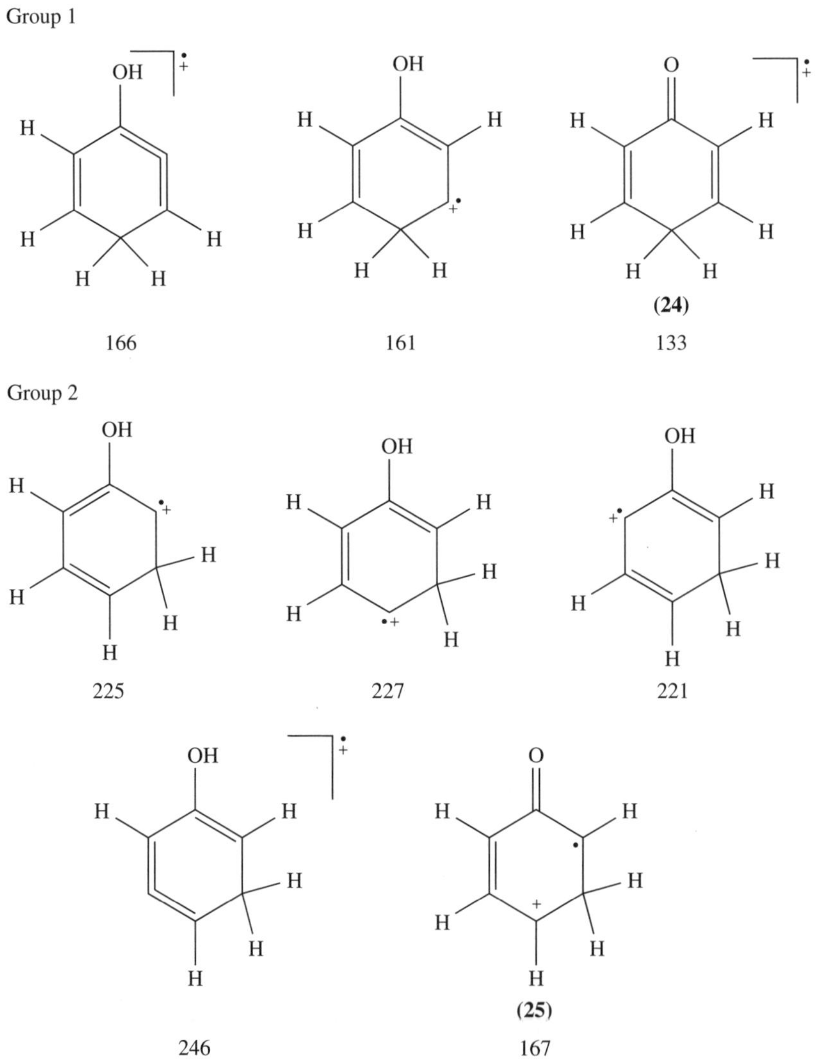

FIGURE 30. Relative energies of selected isomers of phenol radical cation containing a six-membered ring. Values given in kJ mol^{-1} relative to **21** were obtained from UB3LYP/6-311++G(d,p)+ZPE calculations. Adapted from Reference 452 with permission

1. General and theoretical aspects of phenols

Group 3

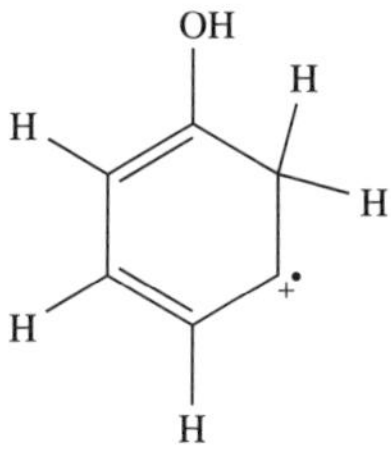

181 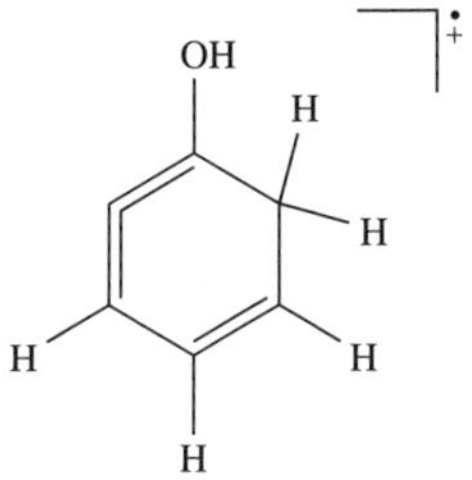   193 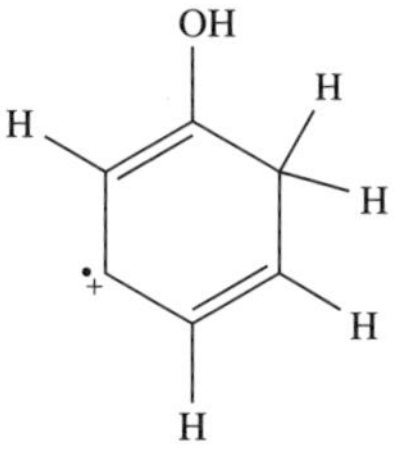   176

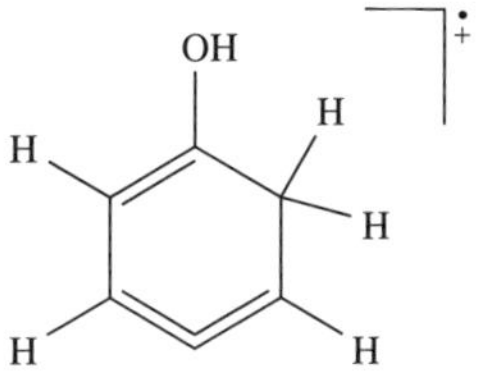

186 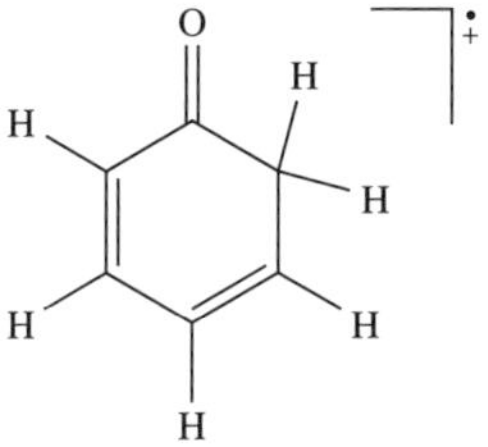   (22) 146

Group 4

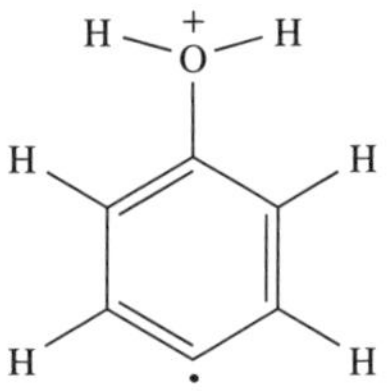

(23) 241   282

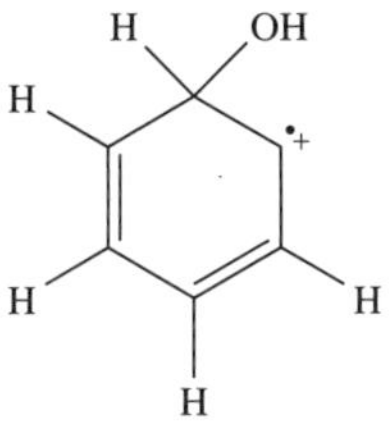

FIGURE 30. (continued)

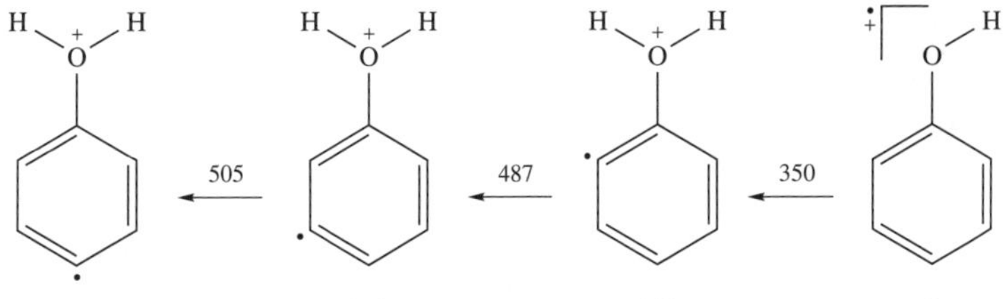

241 ← 505 — 240 ← 487 — 240 ← 350 — 0

CHART 6. B3LYP/6-311++G(d,p)+ZPE energies (in kJ mol^{-1}) of the oxonium distonic isomers and the transition structures connecting them relative to phenol ion **21**. Adapted from Reference 452 with permission

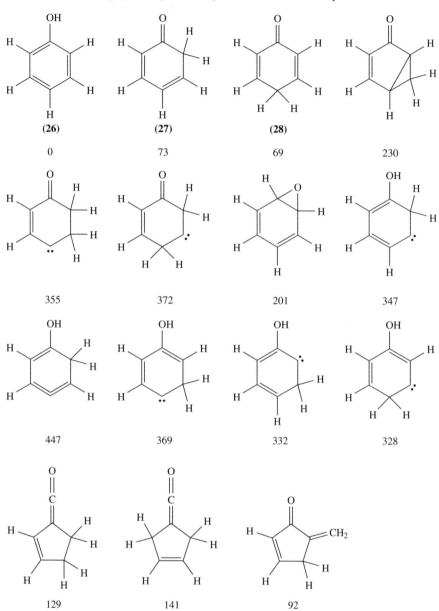

FIGURE 31. Relative energies of selected isomers of neutral phenol. Values given in kJ mol^{-1} were obtained from B3LYP/6-311++G(d,p)+ZPE calculations. Adapted from Reference 452 with permission

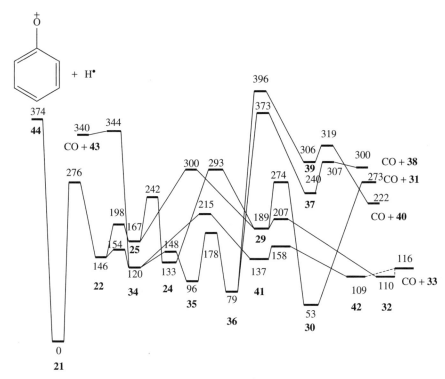

FIGURE 32. Schematic representation of the $(C_6H_6O)^{\bullet+}$ potential energy surface showing the rearrangements of phenol radical cation leading to a CO loss. Relative energies given in kJ mol^{-1} were obtained from B3LYP/6-311++G(d,p)+ZPE calculations. Adapted from Reference 452 with permission

the corresponding radical cations **30**, **32**, **37**, **39** and **11**, respectively. The phenoxy cation **44**, results from hydrogen loss from the phenol radical cation **21**.

The numerous reaction pathways found in Figure 32 invariably lead to an elimination of CO giving $(C_5H_6)^{\bullet+}$ ion fragments (m/z 66). The PES can be divided into two distinct parts: while the first part involves the three cyclohexanone ion isomers **22**, **24** and **25**, the second consists in the conversion of the cyclic keto-ions into either the various open-chain distonic forms **34** (or its conformers **35** and **36**), **37** and **39**, or the five-membered cyclic derivatives **29**, **30** and **41**. There are also some weak hydrogen bond complexes between CO and the CH bond of ionized cyclopentadienes such as **32** and **42**.

The first step thus corresponds to a 1,3-hydrogen shift via the transition structure (TS) **21/22** which is associated with a rather high energy barrier of 276 kJ mol^{-1} relative to the phenol radical cation **21**. The corresponding energy barrier for the neutral system amounts to 278 kJ mol^{-1}. Thus, there is practically no reduction in the barrier height following ionization, in contrast to the case of the propene ion[463]. It appears that, once formed, the keto ion **22** easily undergoes a ring opening via TS **22/34** yielding the open-chain distonic ketene radical cation **34**. The successive 1,2-hydrogen shifts within the ring can also give **25** and finally the most stable keto form **24**. From here, the six-membered cyclic framework could be converted into the five-membered ring **29** lying 189 kJ mol^{-1} above

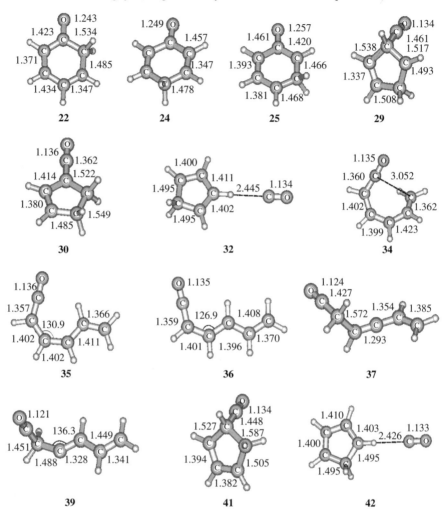

FIGURE 33. Selected B3LYP/6-311++G(d,p) geometric parameters of the $(C_6H_6O)^{\bullet+}$ equilibrium structures considered. Bond distances are given in Å and bond angles in deg. Adapted from Reference 452 with permission

21 by direct one-step rearrangements via the TSs **24/29** and **25/29**. From **29**, an almost spontaneous CO loss with an energy barrier of only 18 kJ mol^{-1} could thus occur, giving the complex **32**, which dissociates to the fragment products CO+ cyclopentadiene ion **33**, 116 kJ mol^{-1} less stable than phenol ion **21**.

Although the ketene ring **30** is found to be only 53 kJ mol^{-1} above **21** and by far more stable than acetyl ion **29**, it turns out that the CO loss from an indirect process, finally giving the five-membered ion **31**, namely **29** → **30** → (CO + **31**), constitutes a substantially more difficult route. It is apparent from Figure 32 that the cyclic isomer **29** could be a possible intermediate in the CO-eliminative process of phenol cation **21**. Nevertheless, the high energy content of both TSs **24/29** and **25/29** at 293 and 300 kJ mol^{-1}, respectively, above **21** but actually 20 kJ mol^{-1} above the TS **21/22** for the initial 1,3-hydrogen shift, makes the rearrangement through **29** less competitive than other routes. The latter is, however, more favoured than a hydrogen atom elimination characterized by a dissociation energy of 374 kJ mol^{-1} for a direct O−H bond cleavage (Figure 32).

The alternative route comprises the open-chain ketene ion **34** and its conformers **35** and **36** formed by ring opening of the ketone ion **22**. From here, the super-system could either rearrange to the open-chain acetyl cations **37** and **39** or undergo a cyclization, forming back the five-membered ring **41** which is significantly more stable than **37** and **39** (137, 240 and 306 kJ mol^{-1} above **21**, respectively). Figure 32 clearly points out that the CO loss via **41** is beyond any doubt the lowest energy route[452].

Figure 34 illustrates the lowest energy rearrangement path for the CO-loss process of ionized phenol. It involves, in a first step, the enol–keto conversion **21**–**22**. Starting from **22**, a ring opening leads to structure **34** which, in turn, by ring closure produces ion **41**. A direct and concerted isomerization **22** → **41** was not found[428]. The CO loss from **41** involves the slightly stabilized ion/neutral intermediate **42**. The rate-determining step of the overall process **21** → **22** → **34** → **41** → **42** → (CO + C$_5$H$_6^{\bullet+}$) is the 1,3-hydrogen shift **21/22**.

The processes suggested by calculations are in good agreement with experimental mass spectrometric studies[429–435] which demonstrated that the CO loss (m/z 66) corresponds to the least energy demanding fragmentation. Furthermore, it was found earlier that the kinetic energy released during the CO loss from the keto ion **22** was less than that involved during the dissociation of the phenol ion **21** itself[434]. This is clearly in keeping with the potential energy profile presented in Figure 34. The appearance energy of the [M−CO]$^{\bullet+}$ ions has been determined by time-resolved electron impact[435] and photoionization[436] experiments and by photoelectron photoion coincidence[444].

From a comparison of the data and after consideration of the kinetic shift, an energy threshold of 11.4 ± 0.1 eV at 298 K has been deduced. Considering an adiabatic ionization energy value of 8.47 ± 0.02 eV[444, 464] and a correction for the 298 K enthalpy of ca 0.1 eV for the phenol, the energy barrier separating **21** from its fragments is thus ca 3.0 ± 0.15 eV, i.e. 290 ± 15 kJ mol^{-1}. This value is in excellent agreement with the calculated 0 K energy barrier **21** → **22** of 276 kJ mol^{-1}.

It may be noted that the energy amount involved in the CO-loss process is by far smaller than that needed for a deprotonation of phenol cation as mentioned above, namely 857 kJ mol^{-1}. This suggests that the ease with which a deprotonation of phenol radical cations occurs in different solutions[419, 423, 424] was likely to arise from either a specific participation of the solvent molecules in the supermolecule or a strong continuum effect.

4. Mass spectrometric experiments

The state-of-the-art mass spectrometric experiments described below were designed to search for a possible production of (C$_6$H$_6$O)$^{\bullet+}$ isomers, such as dehydrophenyloxonium ions or cyclohexadienone ions. They were performed on a large-scale tandem mass spectrometer of E$_1$B$_1$E$_2$qcE$_3$B$_2$cE$_4$ geometry (E stands for electric sector, B for magnetic sector, q for a radio-frequency-only quadrupole collision cell and c for the 'conventional' collision cell)[465, 466]. The following three MS experiments have been carried out:

(a) First, both 4-bromophenol (**45**) and 4-bromoanisole (**46**) were protonated in the chemical ionization ion source. It was expected that collisional debromination of protonated 4-bromophenol (**47**) could be an interesting source of a distonic isomer of ionized phenol if protonation takes place at oxygen. Alternatively, phenol ions should be produced in the case of ring protonation. The same behaviour was expected for protonated 4-bromoanisole (**48**).

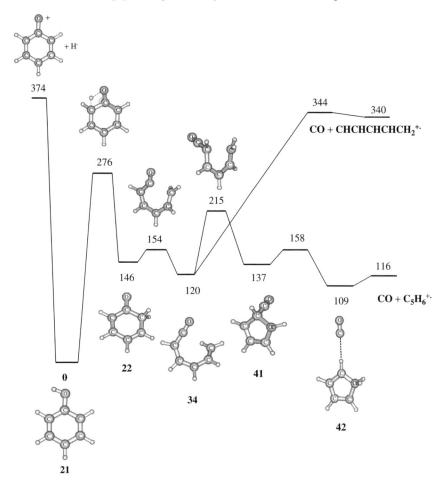

FIGURE 34. Schematic representation of the $(C_6H_6O)^{\bullet+}$ potential energy surface showing the lowest energy path for CO loss of phenol radical cation. Relative energies given in kJ mol^{-1} were obtained from B3LYP/6-311++G(d,p)+ZPE calculations. Adapted from Reference 452 with permission

The high-energy collisional activation (CA) spectra of the $C_6H_6O^{\bullet+}$ ions (m/z 94) or $C_7H_8O^{\bullet+}$ ions (m/z 108) were recorded and the resulting spectra depicted in Figure 35 were found identical to the corresponding spectra of ionized phenol or anisole, respectively.

This observation is in line with the preferential protonation at the ring, not at the oxygen atom, of phenol or anisole (cf. Section IV.A). Distonic dehydro-oxonium ions **50** are therefore not generated in these chemical ionization experiments, in line with the fact that they are more than 200 kJ mol^{-1} less stable than ions **49** (Chart 7). A major fragmentation of ions **50** should be a loss of HOH or ROH with the production of benzyne ions (m/z 76), but the relative intensity of this peak is not increased, thus confirming that ions **50** are not produced to a significant extent in the protonation–debromination sequence.

1. General and theoretical aspects of phenols

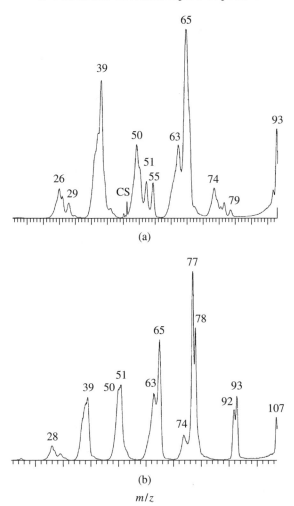

FIGURE 35. CA spectra of (a) [MH-Br]•+ radical cations (nitrogen collision gas) generated by low-energy collisional activation (argon collision gas) of mass-selected protonated 4-bromophenol **47**, and (b) protonated 4-bromoanisole **48**. CS refers to a charge stripping. Adapted from Reference 452 with permission

Such behaviour clearly contrasts with the case of 4-iodoaniline, where protonation in a chemical ionization source occurred not only on the ring but also on the nitrogen atom[450]. Nitrogen protonation was indicated by ion–molecule reactions with dimethyl disulphide consecutive to collisional dehalogenation (FT-ICR experiments)[467] or by an increase in the intensity of the peak at m/z 76 following high-energy collisional activation[450].

(b) Given the fact that a ring protonation was identified in the preceding experiment, unsubstituted anisole was also protonated under methane chemical ionization conditions with the expectation that if the methyl group could subsequently be expelled collisionally

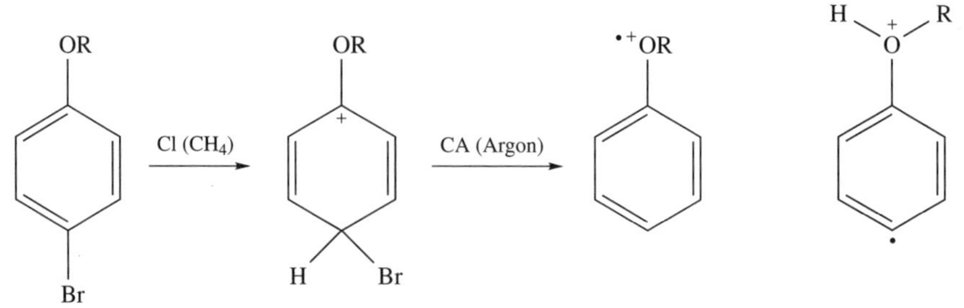

OR — Cl (CH₄) → OR(+) → CA (Argon) → •+OR → H–O(+)–R

(45) R = H
(46) R = CH₃

(47) R = H
(48) R = CH₃

(49)

(50)

CHART 7. Protonation and Debromination

within the quadrupole collision cell, a cyclohexadienone radical ion (*ortho* **22** and/or *para* **24**) should be produced. The protonation occurs on the ring as indicated by the experiments described above on 4-bromoanisole and a demethylation was indeed a prominent fragmentation of the protonated anisole (Figure 36a), but the CA spectrum of the re-accelerated m/z 94 ions (Figure 36b) was found identical to the CA spectrum of the phenol radical cations, not to that of cyclohexadienone ions.

A similar observation has also been made using another MS/MS/MS experiment, where the demethylation step was realized in the high kinetic energy regime. Demethylation of protonated anisole is evaluated to be less endothermic by about 146 kJ mol^{-1} if ionized phenol **21** was formed rather than ionized cyclohexadienone **22** (cf. Chart 8, where values given are estimated heats of formation).

Computations on the interconversion of protonated anisoles indicate that the demethylation of the latter invariably involves formation of its O-protonated form and ends up with the production of **21**. The O-protonation is about 57 kJ mol^{-1} less favoured than the ring *para*-C_4 protonation and the entire process is associated with an energy barrier of 232 kJ mol^{-1} relative to the most stable protonated form, a value comparable to that required for a direct C–O bond cleavage of O-protonated anisole.

(c) In the last experiment, 2-hydroxybenzaldehyde (salicylaldehyde) was submitted to electron ionization. Due to an *ortho* effect, carbon monoxide is, *inter alia*, expelled from the metastable molecular ions (MIKE spectrum, the concerned field-free region being the quadrupole cell, Figure 37a). The CA spectrum of the m/z 94 ions detected in the mass spectrum (Figure 37b) is depicted in Figure 37c. This spectrum indicates that these ions are actually *not* phenol ions. Moreover, when the m/z 94 ions are generated collisionally in the quadrupole, the CA spectrum is very significantly modified (Figure 37d) with the appearance of an intense signal at m/z 76, corresponding to a loss of water.

In summary, a debromination of protonated 4-bromophenol and 4-bromoanisole essentially produces phenol and anisole radical cations, respectively; no less conventional molecular ions were detected. Similarly, collisional demethylation of protonated anisole gives rise to ionized phenol. Only an electron ionization of salicylaldehyde appears to produce an *ortho*-oxonium distonic isomer of the phenol ion. Quantum chemical calculations suggest predominant stability of **21** lying at least 130 kJ mol^{-1} below the other six-membered isomers. Its preponderant fragmentation is a CO loss occurring via different intermediates, namely its keto six-membered ring, open-chain ketene and five-membered cyclopentadiene isomers. The rate-determining step corresponds to the enol–ketone interconversion of the phenol ion with a barrier height of 276 kJ mol^{-1} relative to phenol ion,

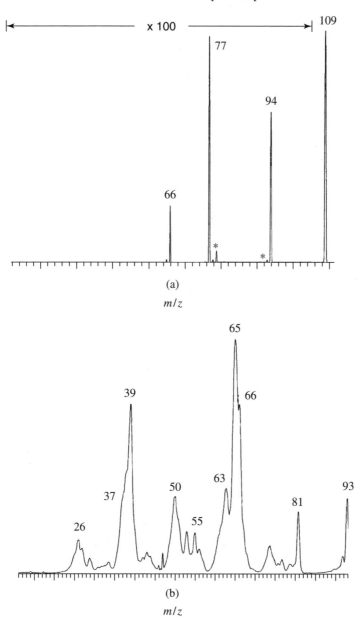

FIGURE 36. (a) MIKE spectrum of protonated anisole m/z 109 and (b) CA (nitrogen) spectrum of the m/z 94 ions. Adapted from Reference 452 with permission

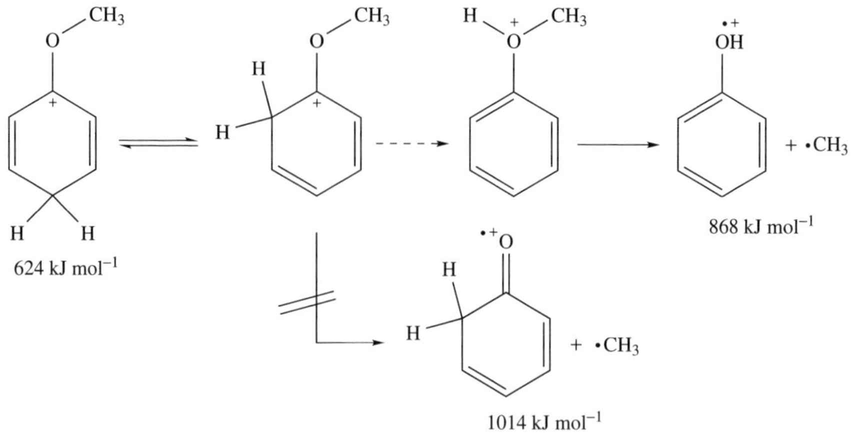

CHART 8. Protonation of anisole

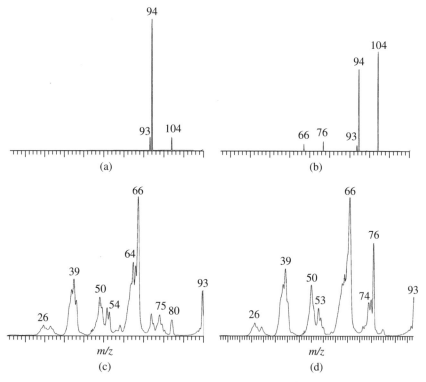

FIGURE 37. (a) MIKE and (b) CA spectra of the m/z 122 ions of ionized salicylaldehyde (peaks at m/z 121, ca 5 × more intense, not shown), and (c) and (d) CA (nitrogen) spectra of the m/z 94 ions produced in these conditions. Adapted from Reference 452 with permission

1. General and theoretical aspects of phenols   127

which is markedly smaller than that required for hydrogen atom loss or deprotonation. This suggests that the solvent plays an important role in assisting the deprotonation of phenol ions in non-polar media.

5. Keto–enol interconversion

As discussed in a previous section, thanks to the aromatic stabilization, the phenol–cyclohexadienone pair thus represents a specific case in which the enol form is actually more stable than its keto tautomers. Hydrogen transfer from oxygen to carbon indeed disrupts the phenyl ring and this disfavours the ketone form. However, the latter intervene as crucial intermediates during the phenol decomposition, in the oxidative metabolism of aromatic compounds (the 'NIH-shift'), in the reactions of arene oxides, the photo-Fries rearrangement, the Kolbe–Schmitt and the Reimer–Tiemann reactions[184, 185, 468]. Both cyclohexa-2,4-dien-1-one **27** and cyclohexa-2,5-dien-1-one **28** have been generated experimentally by flash photolysis of appropriate precursors in aqueous solution. Based on kinetic results, logarithms of the equilibrium constants for the enolization **27** → **26** and **28** → **26** were evaluated to be $pK_E(27, 25°C) = -12.73$ and $pK_E(28, 25°C) = -10.98$. Combination with the acidity constant of phenol **26** also defines the acidity of both ketones which are characterized as strong carbon acids with $pK_a(27) = -2.89$ and $pK_a(28) = -1.14$, all with errors of ±0.15. The common conjugate base is the phenolate anion discussed in a preceding section. Both ketone forms disappeared by proton transfer to the solvent with estimated lifetimes of $\tau(27) = 260$ μs and $\tau(28) = 13$ μs[468].

Let us remember that the energy difference between phenol **26** and both keto isomers **27** and **28** amount to 73 and 69 kJ mol^{-1}, respectively (Figure 31). The contribution of entropy is small, amounting to $\Delta S = -9$ and -1 J mol^{-1}K^{-1}, for both ketonization reactions, respectively, and this also leads to an estimate for the equilibrium constant of the enolization, pK_E, ranging from -12 to -13, of the same order of magnitude as the experimental results in aqueous solution[186, 187, 469]. It should be stressed that such similarity of values in both gaseous and condensed phases should not be considered as an 'agreement' and need to be treated with much caution, due to the fact that the solvent effect on the equilibrium has not been taken into account.

The results discussed above clearly demonstrate that the keto–enol energy difference is further enlarged upon ionization at the expense of the keto form (Figure 34), due to the higher IE$_a$ of the latter, namely 804 kJ mol^{-1} for **26** and 878 kJ mol^{-1} for **27**. Figure 38a shows a remarkable effect of the methyl substituent on the energy differences. Although the group placed either at the *meta* or *para* position does not induce large changes in the relative energies of the neutral species (a reduction of 3–5 kJ mol^{-1}), it strongly modifies those in the ionized state, in particular in the *para*-substituted system: the IE$_a$ of phenol is effectively reduced whereas the IE$_a$ of cyclohexadienone has increased. This results in a further destabilization of 18 kJ mol^{-1} of the ionized ketones.

The phenomenon is also manifested, albeit to a lesser extent, in the amino-substituted pairs as illustrated in Figure 38b. In this system, the IE$_a$s are substantially decreased due to the presence of the amino group, which confers an 'aniline' character to the ionized species.

It is also well known that the keto–enol equilibrium is modified fundamentally in aqueous solution due to the specific interaction of solvent molecules with the substrates through hydrogen bonds[470–472]. Calculated results summarized in Figure 39a indicate that the keto–enol equilibrium is markedly modified in the bimolecular neutral systems in which each tautomer interacts with one water molecule. In particular, the energy barrier for hydrogen transfer from oxygen to carbon is reduced appreciably, in going from

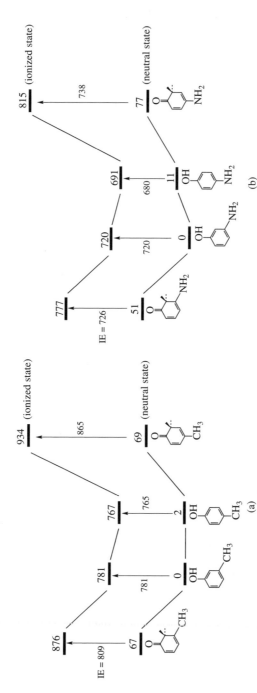

FIGURE 38. Relative and ionization energies of *meta*- and *para*-X-substituted phenol and cyclohexa-2,4-dienone: (a) X = methyl and (b) X = NH$_2$. Values given in kJ mol^{-1} were obtained from B3LYP/6-311++G(d,p)+ZPE computations

1. General and theoretical aspects of phenols 129

175 kJ mol^{-1} in the unimolecular system to 76 kJ mol^{-1} in the water-assisted hydrogen transfer. The displacement of the equilibrium in favour of the enol form is further accentuated in the ionized counterparts in which the ionized keto form virtually disappears. The relevant calculated results are illustrated in Figure 39b.

We also mention that the ionized phenol–water complex has been observed and examined in depth[113, 455, 473–476]. Complexes of phenol radical cation with ammonia[477] and molecular nitrogen[478] have also been produced. The existence of an intramolecular hydrogen bond in *ortho*-substituted phenol radical cations has also been demonstrated[479].

E. The O—H Bond Dissociation

1. Phenoxyl radicals

Owing to the relatively facile oxidation of phenols, phenoxyl radical (PhO•) and their substituted derivatives occur widely in nature and are involved in many biological and industrial processes as crucial intermediates[480]. The phenoxyl radical is a simple prototype of a substituted aromatic radical and a model for tyrosyl radicals [TyrO• = $p-(H_2N)(CO_2H)CHCH_2C_6H_4O^•$] in oxidized proteins. Tyrosyl radicals were found as stable cofactors in several metalloenzyme active sites including ribonucleotide reductase R2 protein[481], cytochrome c peroxidase, prostaglandin synthase[482], and the oxygen evolving complex of photosystem II[483]. Covalently modified analogues of TyrO• were detected in galactose oxidase[484] and amine oxidase[485]. While the biological function of these radicals is not always well established, they are believed to form covalent cross-links between DNA and proteins[486], to be involved in the catalytic cycles of a number of biosynthetic reactions and to serve as an electron transfer intermediate in photosynthesis[483].

Phenoxyl derivatives also play a primordial role in the antioxidant activities of the phenolic components of Vitamin E[76]. Because phenols are produced in the early stage of high temperature oxidation of benzenes, phenoxyl radicals are again postulated as key intermediates in the combustion of many aromatic compounds that are used as additives in lead-free fuels due to their high octane value[487]. In spite of their highly reactive nature which precluded direct structure determinations, a plethora of careful spectroscopic studies of phenoxyl radicals have been scattered throughout the literature. A considerable amount of information on the structure and properties of PhO• has thus been gained from numerous experimental electron paramagnetic resonance (EPR)[457–459, 488, 489], vibrational (IR, resonance Raman)[490–498] and absorption (UV, visible)[416, 499–506] spectroscopy studies.

a. Electronic structure. The unsubstituted PhO• radical exhibits a C_{2v} point group symmetry. The unpaired electron is expected to reside in a π-orbital which is anti-symmetric with respect to the two-fold axis and the reflection in the molecular plane. In this case, the notation of the corresponding irreducible representations depends on the choice of axes. Depending on whether the molecular plane is taken to be the first or the second plane of reflection, the ground state is denoted $^2B_2$ or $^2B_1$. In the literature both labels $^2B_1^{359, 455, 507-514}$ and $^2B_2^{507, 508}$ have been used equally. Although this is a simple symmetry notation problem, it might cause a certain confusion!

We adopt here an axis convention in which the ground state of the phenolate anion (PhO$^-$) is described by the following basic orbital configuration: $\ldots(13a_1)^2\ldots(8b_2)^2\ldots(3b_1)^2(1a_2)^2$. The reference configurations for the $^2A_2$, $^2B_1$ and $^2B_2$ electronic states of the neutral radical can hence be formed from this, making an electron hole in the $1a_2$, $3b_1$ and $8b_2$ orbitals, respectively. The shapes of the singly-occupied orbitals b_1, b_2 and a_2 are displayed in Figure 40. Numerous *ab initio* calculations[509, 510, 514] have indicated that, within this notation, the ground state π radical has $^2B_1$ symmetry. We are mainly concerned with the nature of the electronic states.

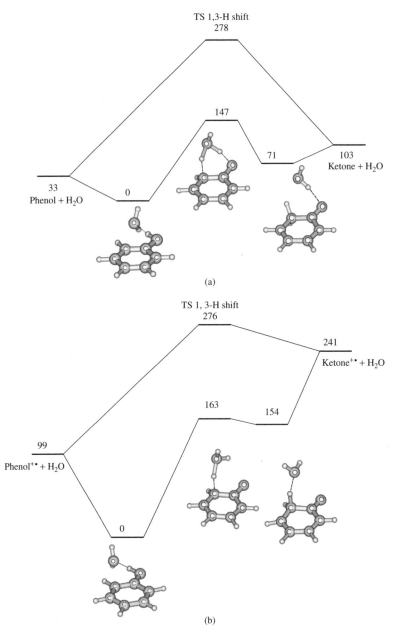

FIGURE 39. Schematic potential energy profiles showing the interconversion between phenol and cyclohexa-2,4-dienone in free and water-assisted systems: (a) in the neutral state and (b) in the ionized state. Values given in kJ mol^{-1} were obtained from B3LYP/6-31G(d,p)+ZPE computations

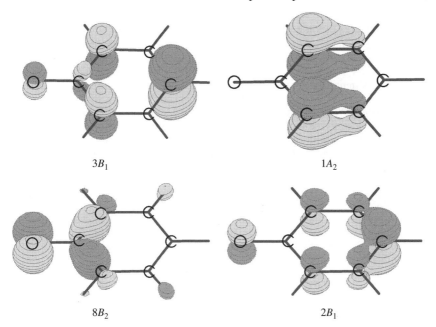

FIGURE 40. A representation of four different singly-occupied orbitals (SOMO) of the phenoxyl radical in the corresponding electronic states

Several electronic excitations have been identified experimentally. The early gas-phase absorption spectra[499, 500] showed bands with λ_{max} at 395 nm (3.1 eV) and 292 nm (4.2 eV). A subsequent experimental study in a nitrogen matrix observed the analogues of these bands and an additional higher energy band with λ_{max} at 240 nm (5.2 eV)[501]. A weak and broad band was detected in the 600 nm region with a peak centred at 611 nm (2.0 eV) and several other regularly spaced peaks whose spacings of about 500 cm^{-1} were presumably due to a vibrational progression[416, 502–506]. An ultraviolet photoelectron spectroscopy study[370] suggested, however, that the first excited state of phenoxyl radical appears rather at 1200 nm (1.06 eV). The identity of the 600 nm absorption band of PhO• and some of its derivatives was the subject of a subsequent study[507] which also used the calculated transition energies and oscillator strengths to help the assignments.

When comparing all the available observed absorption bands and the energies calculated using the multi-reference CASSCF methods with large active space[510, 515], the following assignments of the observed transitions can be proposed: (i) the band at 1200 nm is due to the $1^2B_2 \leftarrow X^2B_1$ transition, (ii) 611 nm to $1^2A_2 \leftarrow X^2B_1$, (iii) 395 nm to $2^2B_1 \leftarrow X^2B_1$, (iv) 292 nm to $2^2B_2 \leftarrow X^2B_1$ and finally (v) 240 nm to $2^2A_2 \leftarrow X^2B_1$.

A possible problem concerns the transition $^2B_2 \leftarrow {}^2B_1$, which is symmetry forbidden under C_{2v} symmetry and might cast doubt on the assignment of the 292 nm band. Experimentally, this band was observed to be weak and the relevant peak is almost completely obscured by the strong peak centred at 240 nm[503]. The CASSCF excitation energies were found to be overestimated by up to 0.5 eV, indicating the importance of dynamic electron correlation for a reasonable description of the excited states. Calculations on PhO• using small atomic basis sets turned out to give incorrect results.

b. Geometry and vibrational frequencies. There has been a persistent disagreement over the CO bond length of PhO• and its stretching frequency[353, 358–360, 370, 455, 508–519]. Indeed, values ranging from 1.22 Å to 1.38 Å were reported for the CO distance from a variety of wave functions. While both CASSCF(9,8)[510] and UMP2[515] treatments, in conjunction with various basis sets, resulted in a short distance of 1.22–1.23 Å, density functional methods yielded a consistently longer distance of 1.25–1.28 Å[358, 359] (cf. Figure 41). Despite a variance between CASSCF and DFT results which might be due to the choice of the active spaces in CAS computations, it seems reasonable to admit that the CO distance in the radical is close to the length typical of a double bond (1.23 Å in *p*-benzoquinone), which is also in line with the inference from the observed CO stretching frequency[498]. As in the phenolate anion, PhO• possesses a quinoidal structure with alternating long (1.45 and 1.40 Å) and shorter (1.37 Å) CC distances (Figure 41). The geometries of the neutral and the anion are in fact quite similar, with a noticeable difference being an increase in the $C_6C_1C_2$ angle of about 3° from the anion to the radical (cf. Figure 23, Section IV.B.1). The geometrical parameters remain almost unchanged upon halogenation, irrespective of the substitution position of the halogen (Figure 2). Even the *p*-amino[509, 512] or *p*-methoxy[517] phenoxyl radicals, having a strong π-donor group, also do not represent a special case; their structure is found to be similar to that of the parent radical with very small modifications of the parameters.

In the lower-lying excited states, the molecular frame remains planar (Figure 42). The $^2B_2$ state has a longer CO distance, stretched up to 0.13 Å, becoming close to that of a single bond. In going from the ground state to the $^2A_2$ state, the C_2C_3 distance also increases by 0.09 Å whereas the change of the CO remains small. This could be understood in examining the shape of the corresponding singly-occupied orbitals involved in the electronic transition[509]. In both excited states, the CCC bond angles deviate significantly

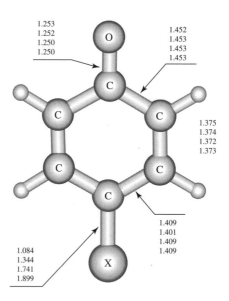

FIGURE 41. A comparison of the distances in (Å) for the phenoxyl radical and its *para*-halogenated derivatives. The entries are X = H (upper), F, Cl and Br (lower). Values were obtained from UB3LYP/6-311++G(d,p) optimizations

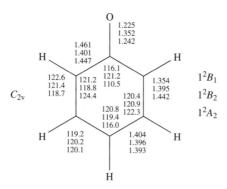

FIGURE 42. Comparison of the bond distances (Å) and angles (deg) of three lowest-lying electronic states of phenoxyl radical. Values obtained from UMP2/6-31G(d,p) optimizations

from the benzene value of 120°. Remarkably, all the CC distances of the $^2B_2$ state are close to 1.4 Å. All these changes suggest that the aromaticity of the benzene ring is probably preserved in $^2B_2$ but not in either $^2B_1$ or $^2A_2$ states.

The C_{2v} symmetry of PhO• leads to 21 in-plane modes (11 a_1 and 10 b_2) and 9 out-of-plane modes (3 a_2 and 6 b_1). The assignment of the associated frequencies was also the subject of considerable discussion among experimental[492–498] and theoretical[358, 359, 498, 509–518] chemists, in particular as regards the location of the CO stretching frequency. The resonance Raman spectra were observed[492–494] for the phenoxyl-h_5, phenoxyl-2,4,6-d_3 and phenoxyl-d_5 isotopomers. Thus, ten in-plane fundamental vibrations including eight totally symmetric a_1 modes and two non-symmetric b_2 modes were observed and now assigned. These fundamental vibrations are sketched in Figure 43 along with the experimental frequencies. High level calculation[358, 359, 498, 510, 514, 518] agreed on the identity and absolute values of most of these modes. There is now a large consensus that the band observed near 1505 cm^{-1}, characterized by the strongest intensity in the resonance Raman spectra, should be assigned to a primary CO stretching, whereas the band centred at 1398 cm^{-1}, which was assigned earlier to the CO stretch[31], corresponds rather to the CC stretch. The observed band near 1552 cm^{-1} is confirmed to arise from the C=C stretching vibration. These assignments were further supported by the downshifts upon deuteriation and the larger shift of the C=C stretch relative to the CO stretch (the CO stretch occurs at 1487 cm^{-1} in phenoxyl-2,4,6-d_3 and 1489 cm^{-1} in phenoxyl-d_5). In addition, it was found that the resonance Raman spectrum for near-resonance with the excited 2^2B_1 state is dominated by the CO stretch mode[510]. As mentioned above, the latter state is responsible for the absorption band centred at 400 nm. This finding was believed to lend further support for the assignment of the CO stretch band at 1505 cm^{-1}.

A correlation between the CO bond properties in the closed-shell molecules (single and double bonds) was proposed to estimate the bond lengths and stretching frequencies of open-shell phenoxyl radicals[512]. Nevertheless, while it is possible to estimate the CO force constants using the Badger-type relations, it is difficult to relate them to the experimental frequencies that do not represent the stretching of a single bond.

The CC and CO vibrations are also sensitive to the molecular environment by virtue of electrostatic and hydrogen bonding interactions. The frequencies of phenoxyl and tyrosyl radicals complexed by macrocyclic ligands[514] and generated *in vivo*[516] were measured by resonance Raman and FTIR techniques. Thus a selective enhancement of the vibrational CC and CO stretch modes of the phenoxyl chromophores in metal-coordinated radical

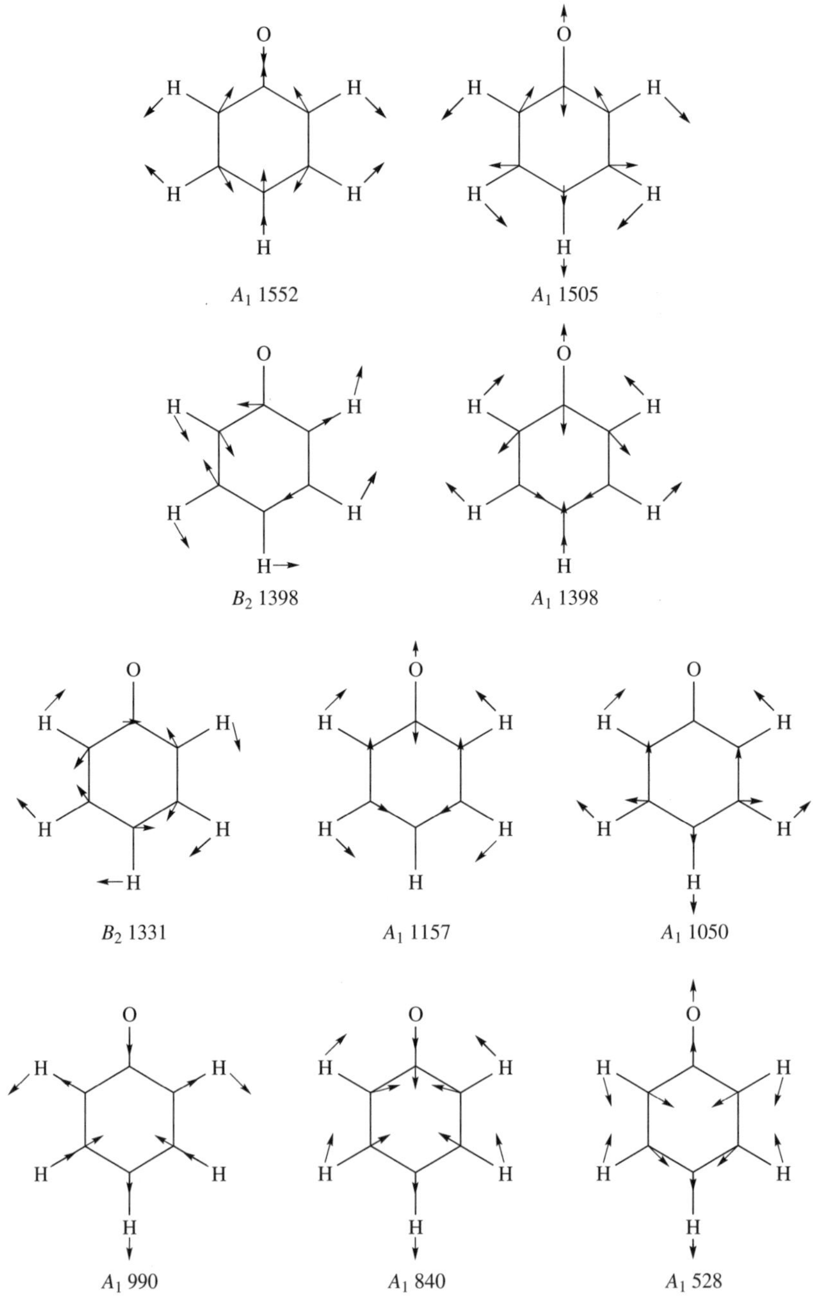

A_1 1552 A_1 1505

B_2 1398 A_1 1398

B_2 1331 A_1 1157 A_1 1050

A_1 990 A_1 840 A_1 528

FIGURE 43. In-plane vibrational modes of phenoxyl radical and the experimental frequencies (values in cm^{-1})

complexes was achieved upon excitation in resonance with the transition at 410 nm. The CO stretch mode is found at 1505 cm^{-1} in aqueous alkaline solution, but at 1518 cm^{-1} for neutral pH[514], which indicates a certain H-bonding interaction with water molecules. These CC and CO modes are of special interest in as much as they could be used as sensitive spectral indicators for the semi-quinoidal structural and electronic properties of the coordinated phenoxyl radicals. Accordingly, an upshift of these frequencies should reflect an increased double bond character of the bonds, which in turn is paralleled by a contraction of the bond distance and also by a decrease in the spin density at the oxygen atom. For example, the C=C frequency increases in the order: PhO$^{\bullet}$ (1562 cm^{-1}) < p-CH$_3$C$_6$H$_4$O$^{\bullet}$(1578 cm^{-1}) < p-CH$_3$OC$_6$H$_4$O$^{\bullet}$ (1595 cm^{-1}).

It is remarkable that the CO stretch frequencies calculated using DFT methods for free substituted phenoxyl radicals are invariably underestimated by 25–45 cm^{-1} with respect to the experimental values observed *in vivo* or in metal-coordinated complexes. This led to a proposition that the phenoxyl and related tyrosyl radicals exist as ion complexes *in vitro*[516]. Computations on model systems such as PhO–M$^+$ or PhO–(H$_2$O)$_2$ provide some support for this view. In spite of the fact that the CO distance is somewhat lengthened following complexation with an alkali metal cation (M = Li$^+$, Na$^+$, K$^+$; see geometrical parameters displayed in Figure 44), the resulting CO stretching frequency turns out to be enhanced by 60–70 cm^{-1} relative to the uncomplexed system, likely due to the underlying electrostatic interaction. Specific H-bonding interaction of the radical with water molecules also induces an enhanced CO stretching, but to a lesser extent, by about 30 cm^{-1}.

c. Spin densities. The EPR spectrum of PhO$^{\bullet}$ has been studied in considerable detail, and the different sets of experimental hyperfine splitting constants (hfcc values) obtained for hydrogen atoms[457–459, 488, 489] consistently offered the following picture: $a(ortho\text{-H}_2) = 6.6 - 6.9$ G, $a(meta\text{-H}_3) = 1.8-1.9$ G, and $a(para\text{-H}_4) = 10.1 - 10.2$ G.

In general, density functional methods in conjunction with the unrestricted formalism could satisfactorily reproduce the characteristics of the spin distribution and the

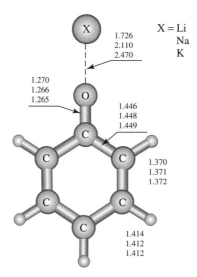

FIGURE 44. Comparison of the distances (Å) in phenoxyl radical—alkali cation complexes. The entries are X = Li$^+$ (upper), Na$^+$ and K$^+$ (lower), UB3LYP/6-311++G(d,p) values

experimental values within the errors of at most ±15%, depending on the basis set employed[511, 517, 518]. For example, the popular UB3LYP/6-311++G(d,p) method provides the following values, including the occurrence of negative spin densities on both *ortho* and *para* hydrogens: $a(H_2) = -6.8$ G, $a(H_3) = 2.6$ G and $a(H_4) = -8.8$ G. This constitutes a good performance bearing in mind that the spin densities at nuclei (Fermi contact terms) are known to be difficult to determine from molecular orbital wave functions (due to the cusp problem and spin contamination in UHF references). Calculations[518] showed that the corrections for vibrational averaging and polarization by the solvent are rather small. While a negligible correction (<0.1G) was estimated for the vibrational effect, a slight reduction of at most 0.6G is due to the effect of a bulk solvent.

The spin density on oxygen $a(O)$ is calculated to vary from -8 to -10 G. Nevertheless, the lack of a significant coupling at the oxygen site in radical–radical reactions is consistent with a dominant odd-alternate cyclic resonance structure in which the radical centre is displaced into the ring. The absolute hfcc values are only moderately changed upon the introduction of a halogen substituent into the benzene ring. The largest effects are found for a fluorine substitution at the *meta* position, which induces a decrease of 0.7 G on $a(H_2)$ and an increase of 0.4 G on $a(H_4)$. The methyl substituent also induces a marginal effect. As a consequence, spin densities of the phenoxyl side-chain in TyrO• radicals are very close to those of free PhO•. There is thus no evidence for a spin delocalization onto the tyrosyl peptide chain[513].

The general trend found earlier[519] for the aromatic hydrogen hfcc values is confirmed, namely $a(H_4) > a(H_2) > a(H_3)$. In view of the empirical McConnell relationship, the spin population on the adjacent carbon atoms could be taken to be proportional to the hfcc values of hydrogen atoms bound to them. Thus, the experimental hfcc values of phenoxyl radical show much larger spin density on the *para* and *ortho* carbons ($\rho_{para}/\rho_{ortho} = 1.5$) than on the *meta* carbon ($\rho_{para}/\rho_{meta} = 5.3$). While calculations are able to account for the ratio of *para* and *ortho* carbons, the trend for the *meta* carbon spin densities is not consistent with that suggested by the McConnell relationship.

As for a possible reason for this disagreement, we consider the spin densities in terms of different components[518]. In general, the spin densities can be decomposed into three contributions: (i) a delocalization, or direct term which is always positive, (ii) a spin polarization or indirect term, arising from the singly-excited configurations and (iii) a correlation term originating from the contribution of higher excitations[520, 521]. The spin polarization term arises from the fact that the unpaired electron interacts differently with the two electrons of a spin-paired bond; the exchange interaction is only operative for electrons with parallel spin. The shorter average distance between parallel spin than between antiparallel spins leads to a spin polarization illustrated by the map of spin densities in the molecular plane (Figure 45) whose sign is governed by some general rules[520]. Because the molecular plane is actually the nodal plane of the SOMO, the only contribution to spin density at nuclei should come from indirect spin polarization terms. The latter can again be decomposed into different first-order and second-order components. As the SOMO (b_1) is mainly localized on *ortho* and *para* carbon atoms leading to large π-spin populations on these atoms, large positive spin densities are thus induced at the corresponding nuclei and negative short-range hfcc values at *ortho* and *para* hydrogens. The positive spin population at an *ortho* carbon induces for its part a negative spin population at the *meta* carbon (first-order effect) and thereby a positive but weak (of second-order character) spin density at the *meta* hydrogen. The same mechanism is operative for the *para* carbon, yielding an additional contribution to the *meta* hydrogen. Overall, the *meta* hydrogens receive non-negligible positive spin densities resulting from cumulative second-order effects. If the oxygen atom was replaced by a more electronegative group, the hfcc values of *ortho* and *para* hydrogens would increase whereas the hfcc values of *meta* hydrogen would remain roughly unchanged due to cancellation of effects.

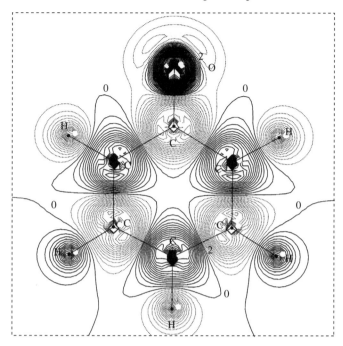

FIGURE 45. Isocontour spin density plot in the molecular plane of phenoxyl radical. Contour levels are spaced by 0.0005 a.u.

The McConnell relationship[520] basically converts spin population due to delocalization (direct term) into spin polarization (indirect term). It could strictly be applied to the first-order spin polarization effects and thus correctly account for the *ortho* and *para* carbon ratio spin densities of phenoxyl radical from hydrogen hfcc values. On the contrary, it could hardly be applied to more subtle second-order mechanism such as is the case of *meta* carbon and hydrogen atoms, and this is the probable reason for the disagreement revealed above. In the unrestricted spin formalism (UHF, UB3LYP), the spin polarization is directly included in the wave function together with delocalization. As a consequence of the unavoidable spin contamination by higher spin states, unrestricted methods tend to overestimate the spin polarization terms. That is the likely reason for a larger calculated value of the hfcc of the *meta* hydrogen compared with the observed values.

d. Decomposition of phenoxy radical. Under combustion conditions, this radical undergoes a thermal decomposition whose primary products are found to be cyclopentadiene radical (C_5H_5) and carbon monoxide[487]. Two mechanisms have been proposed[21, 522] to rationalize the decarbonylation. Results of kinetic measurements, thermochemical considerations[21] and quantum chemical computations of the potential energy surfaces[523,524] concur with each other and point towards the dominance of the molecular mechanism depicted in Figure 46. In brief, this involves the formation of the bicyclic intermediate **A** via the transition structure **TS-A**, followed by an α-CC bond cleavage via **TS-B** yielding the five-membered ring **B**. Finally, the elimination of the CO moiety from **B** through **TS-C**, producing the main products **C**, is an obvious

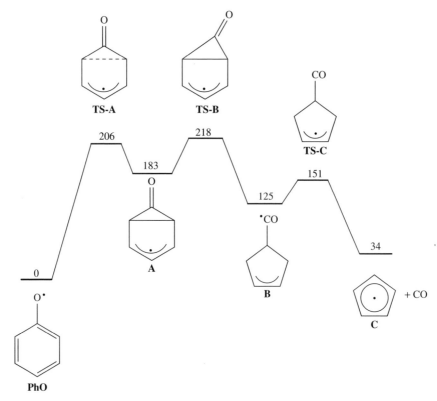

FIGURE 46. A schematic potential energy profile showing the CO elimination from phenoxyl radical. Relative energies, given in kJ mol^{-1}, were obtained from CASPT2/CASSCF(8,7)/6-311G(d,p)+ZPE computations. Adapted from Reference 524 with permission

step with low energy barrier. The rate-determining step corresponds to the formation of the five-membered cycle **B**. Using a modified G2M scheme based on coupled-cluster energies[524], the transformation is associated with an energy barrier of 218 kJ mol^{-1}, which is significantly larger than the experimental estimate of 184 kJ mol^{-1}[121]. Note that the first step PhO• → **A** is a symmetry-forbidden process, which could take place either via a non-symmetrical transition structure or through an avoided crossing mechanism. The energy barrier in both cases are close to each other (205 kJ mol^{-1}) and slightly smaller than for the rate-controlling step. The mechanism found for the PhO• decomposition is thus comparable to, but simpler than, the decarbonylation of the phenol radical cation discussed in a previous section. The key intermediate in both cases is in fact a high-energy five-membered cyclic species. Kinetic evaluations using the RRKM method in conjunction with the computed energetic and geometric parameters yielded rate constants close to the experimental values, especially for temperatures below 1200 K.

Let us also mention that interest in atmospheric chemistry and combustion chemistry of PhO• led to a number of theoretical studies of its reactions with simple radicals such as atomic oxygen[525], HOO• radical[526], NO and NO$_2$ radicals[527, 528] and molecular oxygen[528]. In all cases, computation of the potential energy surfaces has helped a great deal in the interpretation of reaction mechanisms and/or provided necessary parameters for appropriate kinetic analyses.

2. Antioxidant activity of phenols

a. The O−H bond dissociation energies. As discussed in previous sections, the adiabatic electron affinity and the proton affinity of phenoxyl radical were determined quite reliably and they amount to $EA_a(PhO^\bullet) = 2.25$ eV[370] and $PA(PhO^\bullet) = 860$ kJ mol^{-1}[442], respectively. The substituent effect on the PAs has been examined in a previous section. Note also that a *para*-methyl group induces an increase of 20 kJ mol^{-1} on the proton affinity of phenoxyl radicals[529]. Concerning the adiabatic ionization energy, a tentative value of 8.56 eV was suggested[529]. Nevertheless, our high-level coupled-cluster computations revealed that this value is likely somewhat too low and suggested a higher value of $IE_a(PhO^\bullet) = 8.8 \pm 0.2$ eV[530].

Combination of the phenol acidity $\Delta H_{acid}(PhOH) = 1458 \pm 8$ kJ mol^{-1} and the EA value given above yields the gas-phase bond dissociation energy of phenol $BDE(PhO-H) = 362 \pm 8$ kJ mol^{-1}[124,370]. Photoacoustic calorimetry studies in various solvents having different hydrogen-bond accepting properties provided values ranging from 360 to 369 kJ mol^{-1}[191,531]. A spectroscopic ESR equilibrium method for measuring differences in BDEs of substituted phenols yielding transient phenoxyl radicals led to a value of 369 kJ mol^{-1}[532]. The BDE is thus not very sensitive to the environmental properties.

Use of the above values together with the standard heats of formation $\Delta H_f^\circ(PhOH) = -96 \pm 1$ kJ mol^{-1} and $\Delta H_f^\circ(H) = 218$ kJ mol^{-1} leads to the heats of formation $\Delta H_f^\circ(PhO^\bullet) = 48 \pm 8$ kJ mol^{-1} for the neutral radical, and $\Delta H_f^\circ(PhO^+) = 897 \pm 8$ kJ mol^{-1} for the cation.

The quantity $BDE(PhO-H)$, which constitutes a measure of the O−H bond strength, is by far smaller than $BDE(HO-H) = 498$ kJ mol^{-1}, which is well established for water. Its magnitude is closer to that of $BDE(C-O)$ in phenyl ethers[533]. Electron donor groups such as CH_3 and CH_3O tend to cause destabilization in phenols, but stabilization in the corresponding phenoxyl radicals and the combined effect usually lead to a markedly reduced $BDE(O-H)$. An electron-withdrawing group has the opposite effect. Use of a multiple substitution of electron donor groups results in substantial O−H bond weakening due to the radical stabilizing effect. The BDE of α-tocopherol, the major and bioactive component of vitamin E, was measured by photoacoustic calorimetry to be 40 kJ mol^{-1} lower than that of phenol obtained by the same technique[191]. Similarly, the value for δ-tocopherol, which is the minor and least bioactive component, was measured to be 10 kJ mol^{-1} larger than that of the α-component. Thus, a small difference of 10 kJ mol^{-1} on the BDEs of the phenolic bond already makes a marked variation on the bioactivity[191]. Amino groups in the *ortho* position appear to induce a large O−H bond weakening of more than 59 kJ mol^{-1} and thus represent a peculiar group.

In general, the effect that a substituent exerts on the phenoxyl radical is by far more important than that on the corresponding phenol. An empirical equation[534] relating the differences in phenolic O−H strengths (in kJ mol^{-1}) to the sums of the σ-constants for all the ring substituents has been proposed (equation 34),

$$\Delta BDE(O-H) = 30\left[\sum(\sigma_{ortho}^+ + \sigma_{meta}^+ + \sigma_{para}^+)\right] - 2 \quad (34)$$

where the relationship $\sigma_{ortho}^+ = 0.66\sigma_{para}^+$ is presupposed. A simple group additivity scheme also allowed the BDE to be evaluated with high accuracy[116,192]. This quantitative consideration confirms the ease with which substituted phenols lose their phenolic hydrogens and points towards the main reason for their inherent antioxidant activities. The BDEs should thus be used as a reliable primary indicator in the search for novel antioxidants more active than vitamin E[116, 140, 192, 198, 201].

The radical stabilization energies (RSE) in a compound of the type ROH can be defined as

$$RSE(ROH) = BDE(O-H)_{ref} - BDE(O-H)_{ROH}$$

When taking the value $BDE(O-H)_{ref}$ of 440 kJ mol^{-1} in a saturated alcohol as reference, RSE(PhOH) is found to be 80 kJ mol^{-1}, which is in line with the view that in PhO$^{\bullet}$ is rather a resonance-stabilized radical in which the radical center is not fully centered on the oxygen atom. Regarding the substituent effects, a few general remarks can be noted: (i) In substituted radicals, the stability is influenced not only by the polar effect but also by the spin delocalization. While the polar contribution is related to the ability of the substituent to delocalize the lone pair on the phenolic oxygen, the spin delocalization is more characteristic of the radical stabilization. (ii) There are various approaches for estimating both effects using isodesmic reactions or charge distributions (electrostatic potentials, spin densities)[125, 193]. It has been found that the polar contribution is more important than the spin delocalization. (iii) For electron donor groups, both effects tend to stabilize the radical. (iv) In contrast, electron-withdrawing groups considerably destabilize the radical by virtue of the polar effect; although the spin delocalization tends to stabilize it, the destabilizing polar effect remains dominant. In this regard, the difference in reactivity between the isoelectronic phenoxyl (PhO$^{\bullet}$) and benzyl (PhCH$_2^{\bullet}$) radicals resides in the fact that oxygen is a strong π-acceptor whereas methylene is a poor electron-withdrawing group. As a result, the stability of the benzyl radicals is less sensitive to the polar effect of a substituent.

b. Antioxidant activities. The reaction of molecular oxygen with organic molecules under mild conditions is usually referred to as autooxidation. It can be represented by the following simplified reaction scheme (equations 35–39).

$$\text{initiation:} \quad \text{production of RO}^{\bullet} \tag{35}$$

$$\text{propagation:} \quad R^{\bullet} + O_2 \longrightarrow ROO^{\bullet} \tag{36}$$

$$ROO^{\bullet} + RH \longrightarrow ROOH + R^{\bullet} \tag{37}$$

$$\text{termination:} \quad ROO^{\bullet} + RO^{\bullet} \longrightarrow \text{products} \tag{38}$$

$$RO^{\bullet} + PhOH \longrightarrow ROOH + PhO^{\bullet} \tag{39}$$

While reaction 36 is very fast, having a rate constant of ca 10^9 M^{-1} s^{-1}, reaction 38 is much slower at 10^1 M^{-1} s^{-1}. All organic materials exposed to the air undergo oxidative degradation. Reduction of the rate of such degradation utilizing low concentrations of 'antioxidants' is important for all aerobic organisms and for many commercial products. In this respect, phenols turn out to represent a primordial family of antioxidants. Their activity arises from their ability to trap chain-carrying peroxyl radicals by donating the phenolic hydrogen atom (reaction 39), which is a much faster reaction than the attack of the peroxyl radicals on the organic substrate (reaction 37), thanks to the smaller BDE(PhO–H) values as discussed above.

The idea that autooxidation affects humans (and other mammals) was put forward in the mid-1950s by the so-called free-radical theory of ageing[535]. It was suggested that ageing is the result of endogenous oxygen radicals generated in cells during the normal course

1. General and theoretical aspects of phenols 141

of metabolism, disrupting the structure of biopolymers and resulting in cellular damage. This theory provided a mechanistic link between the metabolic rate and ageing. This link was noticed nearly a century before, when it was observed that animals with higher metabolic rates often have a shorter life span. A careful analysis further demonstrated that production of free radical species rather than metabolic rates provides the strongest correlation with longevity[536].

The relevant free radicals can be either produced endogenously as a consequence of metabolic activities or generated from different environmental sources such as ionizing radiation, ultraviolet light, chemotherapeutics, inflammatory cytokines or environmental toxins[537]. The balance of free-radical production and antioxidant defence determines the degree of oxidative stress. When the stress is severe, survival depends on the ability of the cell to resist the stress and to repair or replace the damaged molecules. If the oxidative stress and the ability to respond appropriately is important for ageing, then it follows that factors that increase resistance to stress should have anti-ageing benefits and lead to enhanced life span. After many years of research, it has been shown that mammalian maximum life span cannot be significantly increased with antioxidants, but the mean life span for mammals can be increased. In the light of these results, a 'disease-specific free-radical theory of ageing'[537] has been formulated, in which free radicals are involved in the etiology and development of many of the chronic diseases that contribute to shorten the (maximum) life span potential for a species. For humans, these chronic diseases particularly include atherosclerosis, emphysema and cancer.

At this point the antioxidants which are expected to protect key cell components from damage intervene by scavenging free radicals and are therefore to attenuate—in part—the diseases. Much progress has been achieved in our understanding of the role played by antioxidants in the maintenance of optimal health. It is now well established that vitamin E is the major lipid soluble, peroxyl radical-trapping chain-breaking antioxidant in human blood plasma[76, 538, 539] and in normal and cancerous tissues[540].

The naturally occurring vitamin E consists of four components, namely α, β, γ and δ tocopherols (TOH). These four molecules, which differ from each other by the number and position of methyl groups attached to the phenol ring, reveal a rather different antioxidant activity. The following results show that the ordering of antioxidant activity of the tocopherols *in vitro*[541] is $\alpha > \beta$, $\gamma > \delta$, which is almost the same order as their *in vivo* activities ($10^4 \cdot k_4$ values in $M^{-1} s^{-1}$ are 320 for α-TOH, 130 for β-TOH, 140 for γ-TOH and 44 for δ-TOH).

In other words, the α-TOH is the most active component of the vitamin E, responsible for its high antioxidant activity. The reason for this phenomenon could be found in the difference in BDE(O−H) values discussed above.

c. Features of hydrogen atom abstraction from phenols. In order to have a deeper appreciation of the remarkable aptitude of vitamin E as antioxidant[541], the details of the mechanism of reaction 39 will be examined.

Let us consider Figure 47, which vividly shows the reaction profile of the hydrogen atom abstraction (reaction 39) from structurally related model compounds—phenols with various numbers of methyl groups in the ring—by the simplest peroxyl radical •OOCH$_3$. In the case of the parent phenol **I**, the classical reaction barrier separating the reactant and product H-bonded complexes amounts to 28 kJ mol^{-1}, whereas the two minima of the corresponding H-atom double-well potential are nearly isoenergetic. The presence of methyl groups in the phenol ring stabilizes the phenoxyl radicals, lowers the barrier and makes the reaction certainly exothermic. In particular, substitution of two methyl groups in the *ortho* positions reduces the reaction barrier by about 8 kJ mol^{-1}, whereas a third CH$_3$ group in the *meta* position decreases it further by *ca* 3.0 kJ mol^{-1}. Invoking the Hammond

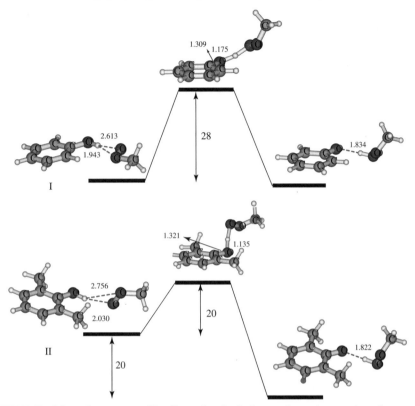

FIGURE 47. Schematic energy profiles illustrating the hydrogen abstraction reaction of a peroxy radical ˙CH$_3$OO with (I) phenol and (II) 2,6-dimethylphenol. Relative energies (given in kJ mol^{-1}) were obtained from UB3LYP/6-31+G(d,p)+ZPE calculations. Adapted from Reference 551 with permission

postulate, one can in fact relate the stability of the phenoxyl radical to the stability of the transition state structure. In the case of 2,6-dimethylphenol **II**, the corresponding phenoxyl radical is stabilized with respect to the parent molecule by 20 kJ mol^{-1} and the transition state structure is more reactant-like than the counterpart of the unsubstituted phenol and occurs at the phenolic C−O and O−H distances of 1.321 Å and 1.135 Å (cf. 1.309 Å and 1.175 Å in phenol **I**, respectively).

What make other structural factors of α-TOH such a good antioxidant? Extensive investigation on the effect of various substituents on the k values for reaction 39 in simple phenols[542] led to the conclusion that the 'best pattern' of substituents in the phenol ring required to facilitate this reaction is optimally the methoxy group residing in the *para* and four methyl groups in the remaining positions. Many years later, it was found that 4-methoxy-2,3,5,6-tetramethylphenol, which structurally approximates α-TOH, is actually a much less active compound than all tocopherols[76, 538]. A clue that helps us to rationalize this marked difference is provided by the X-ray structures of related molecules[76]. The oxygen's π-type lone pair of the methoxy group can stabilize the phenoxyl radical by resonance overlap with its singly-occupied molecular orbital and the degree of such interaction depends on the angle (θ) between this pair and aromatic π-orbitals.

Knowing the extremely important role of phenolic antioxidants in both biological and commercial systems, extensive experimental and theoretical studies have been conducted in the past on these species[543–552]. However, an unambiguous understanding of the physical mechanism of the reaction of phenols with free radicals was hindered by insufficient knowledge about the potential energy surface of reaction 39. By analyzing the geometries displayed in Figure 47, one can easily see that while the minimum-energy hydrogen-bonded complexes are characterized by a planar orientation of the phenol OH group, in transition state structures this bond is twisted out-of-plane. Such twisting occurs due to the fact that the TSs for such reactions are formed by the avoided crossing of two lower-lying electronic states of phenoxyl radical, which takes place at some angle τ between the OH bond and the aromatic ring plane, while the in-plane reaction pathway ($\tau = 0$) is characterized by the intersection of these surfaces[551]. In view of this fact, it is interesting to note that the barrier to internal rotation of the OH group (V_τ) which partly contributes to the activation barrier of reaction 39 is also influenced by a stereoelectronic effect of the lone pair of the *para*-alkoxy oxygens. When the latter is oriented perpendicular to the ring, the overlap is maximal and resonance structures with the doubly bound methoxyl oxygen prohibit a simultaneous conjugation of phenolic OH group with the ring, which results in decreasing V_τ. Correlations of V_τ and k values for reaction 39 with known or expected θ were established experimentally[552].

V. HYDROGEN BONDING ABILITIES OF PHENOLS

A. Introductory Survey

Molecular design requires detailed knowledge of hydrogen bond strengths, at least as much as knowledge of the polar atoms participating in such bonding. Phenol is rather specific in this respect because it involves the phenolic oxygen atom which is usually regarded as a major hydrogen acceptor due to its lone pairs. However, on comparison, for example, with furan, the hydrogen bond ability of phenol is determined by the degree of delocalization of the oxygen lone pair electrons into the π-system of the phenol ring.

On the other hand, phenols as proton donors actually occupy a very particular position among organic acids due to the well-known fact that by changing the substituents in the phenyl ring, we can readily regulate, almost continuously pK_a values from 10 to 0. For example, 4-$CH_3OC_6H_4OH$ is characterized by a pK_a equal to 10.21. Furthermore, we are also able to record readily the extent of proton transfer because it evokes a change in the electronic spectrum of phenol. The long-wavelength ^1L_b phenolic band is rather sensitive to the hydrogen bond formation. The stronger the hydrogen bond, the stronger the bathochromic shift and hyperchromic effects, and after the proton transfer, a further bathochromic shift and increase in intensity take place on increasing the charge separation. The largest bathochromic shifts of the ^1L_b bands are observed for free phenolic anions. The UV-VIS spectra of hydrogen-bonded complexes with phenols reflect not only the proton transfer process, but also a continuous displacement of the proton along the hydrogen bond bridge[553].

The literature on the hydrogen-bonded complexes of phenols with various proton acceptors and the corresponding proton transfer equilibria covers literally thousands of papers. First of all, it is worth mentioning the monograph by Davies[202], the reviews by Zeegers-Huyskens and Huyskens[210] and by Müller and coworkers[553]. Several groups[203–209, 554–573] made important contributions to elucidate the nature of the hydrogen bonding and proton transfer in complexes with phenols. Hydrogen-bonded complexes with phenol have been the subject of numerous studies at both experimental (e.g. molecular beam spectroscopy[164, 409, 412, 574–591]) and theoretical levels. Surveying briefly the achievements in this area, we would like to mention that the hydrogen-bonded complexes of phenol with proton-accepting molecules such as ethers and alcohols are known to shift the spectra to longer wavelengths from that of the parent phenol by 200–400 cm^{-1}, depending on the proton-accepting strength of the bases[33, 553, 574, 592–596]. Clusters of phenols with ammonia[473,577,597] and amines have been studied[25, 164, 473, 550, 568, 577, 588, 598–607]. Among these studies, it is worth mentioning a recent work[608] using BLYP/6-31G(d,p) calculations on complexes of ammonia with phenol, and its *p*-nitro, pentafluoro-, 2,6-difluoro-, 4-nitro- and 2-fluoro-4,6-dinitro derivatives. Under complexation with ammonia, these phenol derivatives show a growing acidity which, as expected, may lead to proton transfer in the gas phase, but which was observed in solution and the condensed phase. Alas, contrary to the growing acidity due to the pK_a change from 9.95 to 2, no proton transfer along the hydrogen bond O−H···N towards ammonia has been predicted. Interestingly, the interaction between the very strong proton sponge bases and phenols was studied in non-aqueous solutions using UV-VIS and IR spectroscopy[609]. The present survey continues in Table 34.

Mannich bases formed from formaldehyde, secondary amines and *ortho*-derivatives of phenol and Schiff bases derived from aromatic *ortho*-hydroxyaldehydes are treated as rather convenient model systems to study intramolecular proton transfer[25, 621, 647–676].

The hydrogen bonded clusters of phenol with water and methanol have been investigated rather thoroughly, both experimentally and theoretically, for several reasons. The key reason is that they can be considered as model systems for larger aggregates. We will discuss phenol–water clusters in Section V.B while the discussion of the phenol–methanol clusters will only be confined to listing the corresponding references[474, 574, 575, 596, 677–679] (note that the complex between PhOH and the NH$_2$ radical has recently been studied[680]). We will tell a more exciting story about hydrogen bonding between phenol and acetonitrile, and two brief stories about a very short O−H···N hydrogen bond recently determined in the 1:1 crystalline adduct of 2-methylpyridine and pentachlorophenol and about the hydrogen-bonded complex of phenol and benzonitrile. Before doing so, let us start with some interesting observations.

Phenol may also interact with some molecules directly via its aromatic ring due to a so-called π-bonding. For instance, spectroscopic measurements have revealed that phenols form π-bonded complexes in their ground electronic states with rare gas atoms (Rg) and methane[164, 576, 681–686]. On the other hand, phenols form only hydrogen bonds with ligands such as, CO and N$_2$ which have nonvanishing dipole and/or quadrupole moment[164, 478, 686–688]. As shown recently[689] in IR experiments and *ab initio* calculations, phenol cation may form two stable complexes with Ar: one is hydrogen bonded whereas the other is π-bonded. The former occupies the global minimum. A similar situation occurs with the phenol cation–N$_2$ complex.

If phenol forms hydrogen-bonded complexes with some molecules, it is natural to study proton transfer along these hydrogen bonds if the proton transfer PES has a double-well character. However, it has been stressed that an enhanced pK_a of the hydrogen-bonded complex upon electron transfer favours a concerted proton-coupled electron-transfer mechanism[690]. It implies that after electron transfer, a double-well proton potential is converted to a single minimum potential corresponding to proton transfer. For instance, recent *ab initio* studies of the radical cation complexes of phenol with water[476, 691] and molecular nitrogen[478] gave group distances which are substantially shorter compared to those in neutral complexes. This suggests[690] that the proton PES might have a vanishing or rather small barrier. Adding more water molecules to the phenol–water cation radical complexes leads to the stabilization of the proton-transferred forms[113]. Regarding hydrogen-bonded complexes of phenol with ammonia, only the proton-transferred structures were found to be stable[472, 597].

TABLE 34. References for some experimental and theoretical data on the hydrogen-bonding ability of phenol and its derivatives. The pK_a value of phenol and its derivatives is indicated in parentheses

Phenol	Hydrogen bond partner[a]	Method of study	Reference
Phenol (9.94)	N,N,9-Trimethyladenine	IR	610
	1,10-Phenanthroline derivatives	IR	611
	Pyridine, 3-I-pyridine	IR	611, 612[b]
	Conjugated imines	IR	613
	BIPA	IR	614
	PCA	IR	615
	N-Heterocyclic bases	IR	616
	Triethyl thiophosphate	IR	617
	Caffeine, 1,3-dimethyluracil	IR	618
	Pyridazine, pyrimidine, pyrazine	IR	619, 620
	N,N-DMBA, N-BMA	IR	621
	Dioxane, water, methanol, dimethyl ether, cyclohexene, benzene, tetrahydrofuran	Fluorescence excitation spectra	574
	$(HCOOH)_n$ $n = 1, 2$	R2PI, IR-UV, DF HF/6-31G(d,p)	622
	$(CH_3COOH)_n$ $n = 1-4$	R2PI, IR-UV, DF HF/6-31G(d,p)	623
	Phenoxides, TMA oxide	IR	624
	Ethanol	DF	33, 574
	Acetonitrile	IR	612
	TMA	IR, B3LYP/6-31G(d,p)	625
	Methanol	IR	626, 627
	Quinuclidine	IR, NMR	628
	N-Mono- and N,N'-dioxides	IR	629–631
	TMA N-oxide	IR	632, 633
	TMA acetate	IR	634, 635
		PhOH rotational coherence spectroscopy + B3LYP/6-31G(d)	636
4-F-Phenol			
	Bathocuproine	IR	611
	Triethyl thiophosphate	IR	617
	Pyridazine, pyrimidine, pyrazine	IR	619
4-Cl-Phenol			
	N-Heterocyclic bases	IR	616
	Triethyl thiophosphate	IR	617
	N,N-DMBA, N-BMA	IR	621
	n-Propylamine	IR	637[c]

(*continued overleaf*)

TABLE 34. (*continued*)

Phenol	Hydrogen bond partner[a]	Method of study	Reference
3-Br-Phenol (9.03)			
	$N,N,9$-trimethyladenine	IR	610
	Conjugated imines	IR	613
	BIPA	IR	614
	PCA	IR	615
	Caffeine, 1,3-dimethyluracil	IR	618
	Pyridazine, pyrimidine, pyrazine	IR	619
4-Br-Phenol (9.34)			
	$N,N,9$-trimethyladenine	IR	610
	1,10-Phenanthroline derivatives	IR	611
	Conjugated imines	IR	613
	BIPA	IR	614
	PCA	IR	615
	Triethyl thiophosphate	IR	617
	Caffeine, 1,3-dimethyluracil	IR	618
	N,N-DMBA, N-BMA	IR	621
3,4-di-Cl-Phenol (8.58)			
	$N,N,9$-trimethyladenine	IR	610
	Bathocuproine	IR	611
	Conjugated imines	IR	613
	BIPA	IR	614
	PCA	IR	615
	Triethyl thiophosphate	IR	617
	Caffeine, 1,3-dimethyluracil	IR	618
	pyridazine, pyrimidine, pyrazine	IR	619
3,5-di-Cl-Phenol (8.18)			
	$N,N,9$-Trimethyladenine	IR	610
	Bathocuproine	IR	611
	BIPA	IR	614
	PCA	IR	615
	Triethyl thiophosphate	IR	617
	Caffeine, 1,3-dimethyluracil	IR	618
3,4,5-tri-Cl-Phenol (7.75)			
	$N,N,9$-Trimethyladenine	IR	610
	Bathocuproine	IR	611
	BIPA	IR	614
	PCA	IR	615
	Triethyl thiophosphate	IR	617
	Caffeine, 1,3-dimethyluracil	IR	618
Pentachlorophenol			
	Pyridine betaine	X-ray, FTIR	638
	4-Methylpyridine	MNDO, PM3	639
	4-Acetylpyridine	AM1, PM3	640
	Formaldehyde	NMR	641
	Pyrimidine derivatives	IR	642

1. General and theoretical aspects of phenols 147

TABLE 34. (continued)

Phenol	Hydrogen bond partner^a	Method of study	Reference
2-NO$_2$–Phenol (7.17)			
	Methanol	IR, NMR	643–645
3-NO$_2$–Phenol (8.28)			
	Pyridazine, pyrimidine, pyrazine	IR	619
4-NO$_2$–Phenol (7.15)			
	BIPA	IR	614
	TMA	IR, PM3	646
2,4-di-NO$_2$–Phenol			
	Methanol	IR	644

^{a}BIPA = *trans*-butenylidene-isopropylamine; PCA = 1-pyrrolidinecarboxaldehyde; N,N-DMBA = dimethylbenzylamine; N-BMA = benzylidenemethylamine; TMA = trimethylamine.
^{b}For pyridine.
^{c}See Reference 573 for a recent review.

B. Phenol–(Water)$_n$, $1 \leqslant n \leqslant 4$ Complexes

1. Introduction

Knowledge of the potential energy surface of a molecular complex is always a key goal in the study of its vibrational pattern and dynamics. The PES of the interaction of water clusters with phenol is rather particular for several reasons. The prime reason is that phenol–water complexes are formed via hydrogen bonds and can thus be treated as prototypes for hydrogen-bonded aromatic systems and models of diverse important chemical and biological processes such as, e.g., solute–solvent interactions involving a participation of hydrogen bonds.

Hydrogen-bonded phenol–water complexes PhOH(H$_2$O)$_n$ ($\equiv$ PhOH-w_n) have been thoroughly studied experimentally[122, 164, 175, 412, 574, 578, 580, 585–587, 590, 596, 692–719] by standard spectroscopic methods, particularly by laser-induced fluorescence, resonance-enhanced multiphoton ionization, high-resolution UV spectroscopy, single vibronic level dispersed fluorescence and hole burning spectroscopy. The mass-selected multiphoton ionization studies[585–587, 693, 694] of these complexes with $n \leqslant 4$ suggested that the ground-state global minimum structure of PhOH(H$_2$O)$_2$ is realized when water molecules form a ring (defined hereafter as S_2)[164, 412, 574, 578, 585–587, 596, 692–702]. A comparison of the spectra of the PhOH(H$_2$O)$_{1-3}$ complexes led to the conclusion that these three complexes should not be treated as a sequence of additive derivatives and, moreover, that they might even have different geometries[585–587]. Two-colour photoionization and cluster ion dip spectroscopy of PhOH(H$_2$O)$_{n \leqslant 4}$ were carried out[590, 708] showing the existence of two isomers of PhOH(H$_2$O)$_4$. The Raman spectrum of PhOH(H$_2$O)$_1$ was also observed[164].

The infrared (IR) and Raman UV double-resonance spectroscopy of PhOH(H$_2$O)$_{n \leqslant 4}$ in the OH-stretching vibration region was also studied[580, 703–705]. These studies led to the conclusion that, on the one hand, the symmetric water ν_1 and phenolic OH-stretching (ν_{OH}) vibrations are downshifted considerably upon the formation of phenol–water complexes (compared with those inherent for bare water and phenol molecules). On the other hand, the antisymmetric ν_3 vibration of the water molecule is only weakly affected. This results in the appearance of a transparent 'window' region[704] in the IR spectrum of PhOH(H_2O)$_{n=2-4}$ which widens as n increases, having a width of ca 280 cm^{-1} for $n = 4$, and disappears in the spectrum of the PhOH(H_2O)$_5$ complex[203]. An explanation was proposed[704] for the origin of the 'window' region by the presence of the cyclic S_n arrangements of water molecules in these complexes with $n \leqslant 4$. Interestingly, these authors also observed a completely different IR pattern for PhOH(H_2O)$_4$ in the region of the OH-stretching vibrations where four bands fall into the 'window' region[704, 705]. It has been particularly suggested that such a pattern is attributed to the second isomer of PhOH(H_2O)$_4$[590] which might have a substantially different structure of water molecules compared to the cyclic structure[705]. A recent resonant two-photon ionization study[697] of PhOH(H_2O)$_{2-5}$ and PhOH-d-(D_2O)$_{2-5}$-d_1 complexes led to the conclusion that this second isomer of PhOH(H_2O)$_4$ might have a non-cyclic, more compact water arrangement that can only be expected for cage-, prism-, boat- and book-like structures of water clusters around PhOH (for the nomenclature of water cluster structures see, e.g., References 720–723 and references therein). This is somewhat similar to the book-like structure of water molecules in the global-minimum PhOH(H_2O)$_5$ complex, where one water molecule forms an anchor-type π H-bond with the aromatic ring[700, 702].

The first *ab initio* calculations of PhOH(H_2O)$_1$ were performed at the Hartree–Fock (HF) level[699, 701] and the second-order correlated Møller–Plessett (MP2) level[121] with the 6-31G(d,p) basis set within a frozen core ($\equiv$ fc) approximation[404, 724–726]. Density functional B3LYP calculation of PhOH(H_2O)$_n$ was recently carried out by different groups[473, 727]. The ground-state PhOH(H_2O)$_2$ complex was first optimized in 1994–1995[696, 710] (see also References 113 and 725–728). The structure and vibrations of PhOH(H_2O)$_3$ in the singlet ground and its first excited state, and the lowest triplet state were investigated by two groups[695, 711] at the HF/6-31G(d,p) computational level who reported that several local minima on the ground-state PES of PhOH(H_2O)$_3$ are situated above the global-minimum structure with the cyclic S_3 water arrangement by 33.5–58.5 kJ mol^{-1}.

Theoretical study of the PhOH(H_2O)$_n$ complexes calculated preliminarily at the HF/6-31G(d) computational level[729] suggested that the 'window' region originates from the spectra of the PhOH(H_2O)$_4$ isomer with the cyclic water structure S_4. Another, experimentally observed IR pattern of PhOH(H_2O)$_4$ does not fit the theoretical spectra of any complex found in the study and may probably be attributed to a mixture of certain complexes with more compact water arrangements. The proton-transferred PhOH(H_2O)$_4$ complex suggested earlier[704, 705] as a possible candidate for the second isomer was subsequently rejected[697, 701, 730]. This problem remains unsolved.

We performed a rather thorough search of the ground-state PES of the PhOH(H_2O)$_{n=3,4}$ complexes in the vicinity of the global minimum. We describe here the lower-energy minimum structures and offer a new, hopefully sound explanation of the origin of two different 'window' patterns in the IR spectra of the PhOH(H_2O)$_4$ complex[731]. Actually, the 'window' region measures the strength of hydrogen bonding: the larger the 'window', the stronger the bonding[732]. We also use a canonical indication of the strength of hydrogen bonding in terms of the stretching vibration ν_σ of the hydrogen-bond bridge[266] although the blue-shifted torsion vibration τ_{OH} of phenol can be applied for this purpose as well.

The present section is organized in the following manner. Computational methodology is outlined elsewhere[733, 734]. In Section V.B.3, we briefly report two lowest-energy structures of PhOH(H_2O)$_{n=1,2}$ and their theoretical spectra. Section V.B.4 demonstrates the existence of five lower-energy structures on the PES of PhOH(H_2O)$_3$ lying above the global minimum by less than 12.5 kJ mol^{-1}. On the one hand, this shows a rather rich landscape of the PES of PhOH(H_2O)$_3$ in comparison to the reported PES[711] and the three lower-energy structures found later[729] at the same computational level and located within 27.8 kJ mol^{-1} above the PES bottom. On the other hand, it also reveals a novel structure where one of the water molecules forms a so-called π hydrogen bond with the π-electrons of the phenol ring. Such a structure partly resembles the analogous structure named as Leg2 type and found for the benzene–water complex[735, 736]. Section V.B.5 considers ten lower-energy local minimum structures of the PhOH(H$_2$O)$_4$ complex compared with the five reported in Reference 729 and located in nearly the same interval of energies, 15.9 kJ mol^{-1}, above the global energy minimum. This section provides a novel interpretation[704, 705, 731] of the experiments on the existence of two different IR patterns in the IR spectra of this complex and confirms other observations[590].

2. Interaction of phenol with water

We know already that the chosen computational methods accurately describe the properties of phenol, particularly its vibrational spectrum. The frequencies of the OH stretching vibrations of phenol and water molecule are collected in Table 35. It is interesting to note that the HF/A frequency of 4118 cm^{-1} assigned to the ν_{OH} stretching vibration of bare phenol corresponds to its highest frequency. Therefore, it can be treated as the most accepting mode of phenol. Moreover, this frequency lies between the frequencies of the ν_1 (4070 cm^{-1}) and ν_3 (4188 cm^{-1}) OH-stretching vibrational modes of water molecules (equation 40),

$$\nu_1 \overset{48}{<} \nu_{OH} \overset{70}{<} \nu_3 \qquad (40)$$

Here, a value above the inequality sign indicates the corresponding frequency difference in cm^{-1} between its left- and right-hand side quantities. Notice that the first difference $\Delta \nu = \nu_{OH} - \nu_1$ is 48 cm^{-1}.

3. The most stable complexes of mono- and dihydrated phenol

Phenol is certainly more acidic than water and, for this reason, the energetically most favourable binding site of phenol is with its OH group acting as a hydrogen bond donor. Such a phenol donor–water acceptor structure, hereafter designated as PhOH-w_1-1 and shown in Figure 48, lies at the bottom of the PES of PhOH(H$_2$O)$_1$. Its binding energy of 30.8 kJ mol^{-1} calculated at the HF/A level rises to 39.9 kJ mol^{-1} when the MP2(sp)/A calculation is carried out (see Table 36). Note that the latter value agrees with the binding energy of 38.9 kJ mol^{-1} obtained at the MP2 level in conjunction with the D95* Dunning basis set[473]. Due to the donor function of the phenolic O–H group in PhOH-w_1-1, its bond length is slightly elongated by 0.006 Å compared to that in bare phenol. The oxygen atoms are calculated to be 2.901 Å apart from each other, which correlates rather well with the experimental separation of 2.93 ± 0.02 Å[697] or 2.88 Å[699], and also with the HF/6-31G(d,p) result of 2.90 Å[175]. The O–H$\cdots$O$_1$ hydrogen bond is practically linear: the corresponding angle $\angle$OHO$_1$ is 174.1° (the MP2/A value is 175.3°). The phenolic hydrogen donation to the water molecule only affects the geometries of the composing partners.

However, a major effect of the hydrogen bond in the PhOH-w_1-1 complex is anticipated to occur in its vibrational spectrum. It is primarily manifested by a significant red shift of ca 109 cm^{-1} as compared with ν_{OH} of bare phenol. Furthermore, the IR intensity of ν_{OH} gradually increases by a factor of 6.6. The HF/A red shift agrees rather satisfactorily with the experimental results[703, 705], showing a red shift of 133 cm^{-1}. Notice that the MP2/6-31G red shift amounts to 186 cm^{-1} [729] whereas its B3LYP/DZP value is larger and equal to 244 cm^{-1} [113]. The stretching vibrations of water are predicted to be much less affected. More specifically, its ν_1 and ν_3 frequencies are changed by only 1 and

TABLE 35. The OH-stretching frequencies (in cm^{-1}) of water and phenol, and phenol–water$_{1,2}$ complexes calculated via the HF/A and MP2/A (in parentheses) computational methods. Infrared intensity is in km mol^{-1}, Raman (R) activity in Å^4 amu^{-1}

	ν_1			ν_3			ν_{OH}		
	Frequency	IR	R	Frequency	IR	R	Frequency	IR	R
H$_2$O	4070.0 3658^a	18	76	4188.2 3756^a	58	39			
PhOH							4118.1 3657^b	81	79
							4197.2 (3881.8)^c	84 (53)	
PhOH-w_1-1	4068.6 (3764.1) 3650^b	22 (18)	69	4182.0 (3897.4) 3748^b	102 (81)	54	4008.9 (3597.8) 3524^b	537 (645)	144
PhOH-w_1-2	4057.2	94	89	4170.2	134	41	4114.3	94	73
PhOH-w_2-1	3973.2 (3560.7)	308 (419)	47	4147.1 (3846.9)	121 (99)	81	3916.6 (3420.9)	393 (501)	156
	4021.7 (3662.7)	237 (282)	58	4154.7 (3850.2)	137 (69)	40			

^{a}Experimental frequencies of water are taken from Reference 738.
^{b}Experimental frequencies for phenol and phenol–water clusters are taken from References 704 and 705. See also Table 10 for the phenol vibrational modes.
^{c}Calculated frequency at the HF/6-31G(d,p) and MP2/6-31G(d,p) (in parentheses) levels (cf. Table 10).

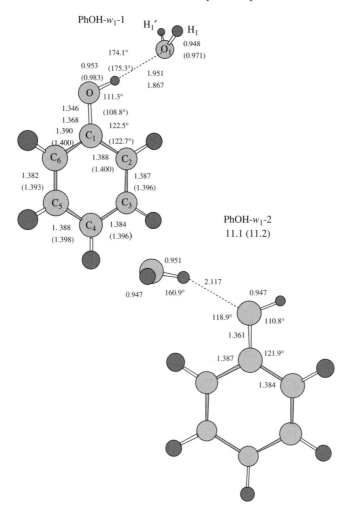

FIGURE 48. Two lower-energy structures of the phenol–water$_1$ complex. The HF/A bond lengths are in Å. The geometrical parameters of the global minimum structure are paired: the first value corresponds to the HF/A level while the MP2/A value is given in parentheses. The HF/A relative energy with respect to the global-minimum structure PhOH-w_1-1 is given in kJ mol^{-1}. Its MP2(sp)/A analogue is followed in parentheses. Numbering of the carbon atoms of phenol is as in Chart 1. Adapted from Reference 731 with permission

6 cm^{-1}, respectively. Besides, their IR intensities increase by 4 and 44 km mol^{-1} whereas the Raman activity decreases by 7 Å^4 amu^{-1} for v_1 and increases by 15 Å^4 amu^{-1} for v_3. Therefore, we may conclude that the hydrogen bond donation of phenol to the water molecule in the global minimum energy PhOH-w_1-1 structure has the following effects. It decreases the phenolic OH stretching vibration and breaks the order of the OH frequencies of the isolated phenol and water deduced in equation 40 in such a way that the phenolic

TABLE 36. Relative energies, ZPVEs, enthalpies (in kJ mol^{-1}) and entropies (in cal mol^{-1} K^{-1}) of PhOH(H$_2$O)$_n$ complexes. Relative energy of PhOH(H$_2$O)$_n$ is defined as -[E(PhOH(H$_2$O)$_n$)] [E(PhOH)-$n \times$ E(H$_2$O)]. The relative energies with respect to structure 1 of the phenol-water$_n$ complex are given[a,b,c]

PhOH(H$_2$O)$_n$	ΔEnergy$_{HF}$	ΔEnergy$_{MP2(sp)}$	ZPVE	ΔEnthalpy	ΔEntropy
PhOH-w_1-1	30.75	39.92	0.0	0.0	0.0
PhOH-w_1-2	19.66	28.70	-0.13	10.84	3.26
PhOH-w_2-1,2	43.64	61.92	0.0	0.0	0.0
PhOH-w_3-1,2	48.74	63.81	0.0	0.0	0.0
PhOH-w_3-3,4	46.57	61.25	-0.63	1.88	-6.02
PhOH-w_3-5	45.06	59.45	-0.46	3.14	-11.97
PhOH-w_3-6	37.70	51.38 (10.296[a])	-2.34	9.62	-12.18
PhOH-w_4-1	40.04	51.71	0.0	0.0	0.0
PhOH-w_4-2	33.81	52.17 (-4.56[a]; -4.92[b]; -0.29[c])	-2.22 (4.48[c])	7.24	-26.69
PhOH-w_4-3	33.52	53.56 (-3.72[a]; 1.88[c])	-2.18 (4.56[c])	5.36	-30.08
PhOH-w_4-4	33.68	51.25 (-2.30[a]; 2.80[b]; 3.89[c])	-0.08 (1.38[c])	6.52	-15.94
PhOH-w_4-5	33.26	50.38	0.33	6.82	-12.38
PhOH-w_4-6	36.11	49.87	-0.96	4.56	-8.91
PhOH-w_4-7	35.56	49.87	0.42	5.27	-7.28
PhOH-w_4-8	34.73	49.33	-1.17	5.77	-12.76
PhOH-w_4-9	37.66	48.12	1.05	1.97	10.96
PhOH-w_4-10	28.62	38.99	0.42	11.30	8.62
PhOH-w_4-11	21.55	36.02	1.46	17.99	4.60

[a]MP2/A.
[b]MP2/A$^+$.
[c]B3LYP/A. Values taken from Reference 731 with permission.

OH stretching vibration is characterized by a lower wavenumber than ν_1 (equation 41),

$$\text{Expt: } \nu_{OH} \overset{126}{<} \nu_1^a \overset{98}{<} \nu_3^a$$

$$\text{HF/A: } \nu_{OH} \overset{60}{<} \nu_1^a \overset{113}{<} \nu_3^a$$

$$\text{MP2/A: } \nu_{OH} \overset{166}{<} \nu_1^a \overset{133}{<} \nu_3^a \tag{41}$$

where the superscript a stands for an *acceptor* of hydrogen bonding, emphasizing the role of the water molecule. This merges into a 'window' region of *ca* 113–133 cm^{-1} width. The hydrogen bonding between phenol and the water molecule also gives rise to the hydrogen bond stretching ν_σ mounting at 158.5 (182.2) cm^{-1} (the experimental value ranges between 151 and 163 cm^{-1}; see in particular Table 2 in Reference 473). Interestingly, the torsional mode τ_{OH} of phenol is blue-shifted substantially to 719.3 (775.5) cm^{-1} (the B3LYP/D95* value[473] is 447 cm^{-1}).

The next lowest energy local minimum on the PES of PhOH(H$_2$O)$_1$ is occupied by the PhOH-w_1-2 structure shown in Figure 48. Here, phenol acts as an acceptor of the hydrogen bond and, compared to the hydrogen bond donor structure, it is less favourable, by 1.11 kJ mol^{-1} at the HF/A level[729]. The energy gap between PhOH-w_1-1 and PhOH-w_1-2 decreases slightly to 10.8 kJ mol^{-1} after ZPVE correction and increases to 11.2 kJ mol^{-1} when both structures are recalculated at the MP2(sp)/A level.

It is particularly unfavourable that the O—H$\cdots$O$_1$ bond length elongates by 0.12 Å in PhOH-w_1-2 compared to that in PhOH-w_1-1, and appears more bent by 13.2°. The hydrogen bond in this case also causes the elongation of the C—O bond by *ca* 0.1 Å compared to its value in bare phenol. In both mentioned structures, there is a very weak interaction between the oxygen atom of the water molecule and the *ortho* hydrogen atom of the phenol ring that is indicated by the corresponding distances of 2.875 Å and 2.727 Å for PhOH-w_1-1 and PhOH-w_1-2, respectively. The rotational constants and the total dipole moment of both reported PhOH-w_1 structures are gathered in Table 37. As seen there, the hydrogen-bond donor structure is more polar than the hydrogen-bond acceptor structure. There is still another feature which distinguishes the two studied structures of phenol with a water molecule from each other: if, in the global minimum energy structure, the oxygen atom of a water molecule resides in the phenol plane, in PhOH-w_1-2, on the contrary, it lies out-of-plane forming a dihedral angle of 95.0°. We explain this by the directionality of the lone pair of the phenolic oxygen. It implies that there are actually two isomers of PhOH-w_1-2: one where the oxygen atom of a water molecule is placed above the phenol ring and the other where it lies below it. Such a feature remains if more water molecules interact with phenol. We consider this as one of the reasons for the appearance of π hydrogen bonding after adding a sufficient number of water molecules to phenol: the cyclic arrangement of water molecules becomes exhausted and the energetic favour turns to 3D water patterns.

Compared with PhOH-w_1-1, the symmetric ν_1 and asymmetric ν_3 vibrations in PhOH-w_1-2 are red shifted by 13 and 18 cm^{-1} while the phenol ν_{OH} stretching vibration is downshifted by only 4 cm^{-1}. Therefore, the stretching IR pattern of PhOH-w_1-2 appears to be that given in equation 42

$$\nu_1^d \overset{57}{<} \nu_{OH} \overset{56}{<} \nu_3^d \tag{42}$$

Notice that the IR pattern inherent for isolated phenol and water molecules (equation 40) is nearly retained in the PhOH-w_1-2 structure. The H-bond vibrational mode $\nu_\sigma = 125.5$ cm^{-1} is lower than in PhOH-w_1-1, implying that the hydrogen bonding in the PhOH-w_1-1 structure is stronger.

TABLE 37. Theoretical rotational constants A, B and C (in GHz) and total dipole moment (in D) of PhOH(H$_2$O)$_n$ complexes calculated via the HF, MP2[a] and B3LYP[b] methods in conjunction with basis set A

PhOH(H$_2$O)$_n$	A	B	C	Dipole
PhOH-w_1-1	4.38507	1.08337	0.87222	3.92
	4.25523[a]	1.11400[a]	0.88657[a]	3.89[a]
PhOH-w_1-2	4.09796	1.11817	0.88142	3.56
PhOH-w_2-1	2.70968	0.73097	0.63654	1.15
	2.53870[a]	0.83238[a]	0.75134[a]	1.10[a]
PhOH-w_3-1	1.94209	0.50448	0.42647	1.16
	1.89925[a]	0.54336[a]	0.46239[a]	1.14[a]
PhOH-w_3-3	1.91922	0.51563	0.44259	1.13
PhOH-w_3-5	1.86586	0.52443	0.45787	1.52
PhOH-w_3-6	1.45663	0.69364	0.59343	1.98
	1.46994[a]	0.78406[a]	0.66417[a]	1.94[a]
PhOH-w_4-1	1.31037	0.37687	0.31183	0.96
	1.21338[a]	0.44044[a]	0.36928[a]	1.17[a]
	1.32264[b]	0.41360[b]	0.34133[b]	1.25[b]
PhOH-w_4-2	1.14775	0.53379	0.47190	2.58
	1.11526[a]	0.70260[a]	0.61283[a]	2.34[a]
	1.21478[b]	0.57219[b]	0.49779[b]	2.55[b]
PhOH-w_4-3	1.51720	0.43861	0.41398	3.55
	1.55673[a]	0.49177[a]	0.46151[a]	3.82[a]
	1.59183[b]	0.46947[b]	0.44238[b]	3.69[b]
PhOH-w_4-4	1.21216	0.51396	0.45024	2.48
	1.25108[a]	0.56229[a]	0.50145[a]	2.92[a]
	1.29591[b]	0.54344[b]	0.48837[b]	2.42[b]
PhOH-w_4-5	1.18647	0.52433	0.45296	1.73
PhOH-w_4-6	1.58475	0.34963	0.30846	3.23
PhOH-w_4-7	1.03920	0.52807	0.49064	1.11
PhOH-w_4-8	1.14855	0.47892	0.40965	2.35
PhOH-w_4-9	1.26070	0.37569	0.310086	0.92
PhOH-w_4-10	1.26992	0.35619	0.31544	2.37
PhOH-w_4-11	1.13115	0.49384	0.47181	1.56

Values taken from Reference 731 with permission.

Let us now proceed to the PES of PhOH(H$_2$O)$_2$ whose lower-energy portion is displayed in Figure 49. Two ring isomers, PhOH-w_2-1 and PhOH-w_2-2, reside at its global energy minimum. They are equivalent because PhOH-w_2-2 is obtained from PhOH-w_2-1 by applying the reflection relative to the phenol plane. In these structures, the OH group of phenol acts bifunctionally, both as the hydrogen-bond donor and acceptor. The three hydrogen bonds in PhOH-w_2-1 are rather bent, as indicated by the values of the corresponding O—H···O angles: 143.59°, 149.66° and 156.14° taken clockwise. The hydrogen bond formed between the phenol hydrogen-bond acceptor and the water molecule donor (w_{ad1}) is quite long and comprises 2.138 Å, although the corresponding oxygen–oxygen separation of 2.96 Å is reasonable and shorter than in PhOH-w_1-2. The other O—O distances are typical for such hydrogen bonds: $r(O-O_2) = 2.813$ Å and $r(O_1-O_2) = 2.848$ Å.

Five calculated OH-stretching vibrations of the PhOH-w_2-1 structure are presented in Table 35. By analogy with the PhOH-w_1-1 complex, the hydrogen-bonded phenolic v_{OH} vibration is red-shifted significantly by 202 cm^{-1} and its IR intensity is enhanced by a factor of 4.9 while its Raman activity only doubles. The other four vibrations are simply assigned to the v_1 and v_3 of water molecules w_{ad1} and w_{ad2}, although their collective nature (essential for larger water clusters) should be noted. One pair of them, v_1^{ad1} and

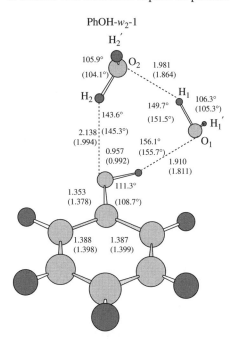

FIGURE 49. The lowest-energy structure of the phenol–water$_2$ complex. Bond lengths are in Å. The geometrical parameters are paired: the former value corresponds to the HF/A level while the MP2/A value is given in parentheses. Adapted from Reference 731 with permission

ν_3^{ad1}, at 3973.2 and 4147.1 cm^{-1}, corresponds to symmetric and asymmetric stretchings of the water molecule w_{ad1}, accepting the phenolic hydrogen bond and donating the hydrogen bond to water dimer. The other one, ν_1^{ad2} and ν_3^{ad2}, centred at 4021.7 and 4154.7 cm^{-1}, describes the symmetric and asymmetric OH-stretching vibrations of the water molecule w_{ad2}, donating the hydrogen bond to phenol and accepting the other one from w_{ad1}. Altogether, they are red-shifted and considerably enhanced compared with the similar vibrations in water and monohydrated phenol. Summarizing, the IR stretching region assumes the pattern shown in equation 43,

$$\text{Expt: } \nu_{OH} \stackrel{117}{<} \nu_1^{ad1} \stackrel{48}{<} \nu_1^{ad2} \stackrel{169}{<} \nu_3^{ad1} \stackrel{3}{<} \nu_3^{ad2}$$

$$\text{HF/A: } \nu_{OH} \stackrel{57}{<} \nu_1^{ad1} \stackrel{49}{<} \nu_1^{ad2} \stackrel{125}{<} \nu_3^{ad1} \stackrel{6}{<} \nu_3^{ad2}$$

$$\text{MP2(fc)/A: } \nu_{OH} \stackrel{140}{<} \nu_1^{ad1} \stackrel{102}{<} \nu_1^{ad2} \stackrel{184}{<} \nu_3^{ad1} \stackrel{3}{<} \nu_3^{ad2} \quad (43)$$

Here, we thus observe the MP2/A 'window' region of 184 cm^{-1} width. Compared to the value reported above for the phenol–water$_1$ complex and demonstrated in equation 41, it is extended by 51 cm^{-1}. It is clearly seen from Table 35 that its extension follows, first, from a further red shift by 177 cm^{-1} of the phenolic OH-stretching compared to PhOH-w_1-1 as a result of a stronger hydrogen-bonding donation of the OH group of phenol to water dimer. Despite the fact that the corresponding ν_σ frequency is less by 21 cm^{-1} than in PhOH-w_1-1, the hydrogen bonding is stronger since the phenolic O–H bond keeps elongating by 0.009 Å. Second, the 'window' extension also follows from a rather substantial red shift of 203 cm^{-1} in the water dimer, where the corresponding hydrogen-bridge stretching frequency reaches the value of 245.2 cm^{-1}. And finally, third, it stems from a strengthening of the hydrogen-bonding donation of water dimer to the lone pair electrons of the phenolic OH group as indicated particularly by the ν_σ frequency of 201.2 cm^{-1}, which exceeds the analogous one in PhOH-w_2-1 by a factor of 1.8. Note in conclusion that the ν_1 mode of the water molecule w_{ad2} (as donor of a hydrogen bond to phenol) borders the left-hand side edge of the 'window' region. This is a typical feature for the cyclic arrangements of water molecules bonded to phenol. We will observe it also for the PhOH(H$_2$O)$_3$ complex in the following subsection.

4. Lower-energy structures of PhOH(H$_2$O)$_3$

Adding a third water molecule to the PhOH(H$_2$O)$_2$ complex significantly enriches the PES landscape of PhOH(H$_2$O)$_3$. This is clearly seen in Figure 50, which displays six lower-energy structures of phenol bonded to three water molecules. The global minimum is occupied by two isoenergetic structures, PhOH-w_3-1 and PhOH-w_3-2, converting into each other via the plane containing the CO group, and perpendicular to the phenol ring. These structures possess a closed cyclic water pattern S_3 to which the phenolic OH group simultaneously donates and accepts hydrogen bonds. A similar water pattern is inherent for the other three structures PhOH-w_3-3, PhOH-w_3-4 (actually the isomer of PhOH-w_3-3) and PhOH-w_3-5 lying within ca 4.2 kJ mol^{-1} above the global minimum and reported in the present work for the first time. Their difference from the global minimum isomers originates from the flippings of the free OH groups of water molecules which can be classified by the u and d symbols[702]. In this regard it is worth mentioning that the structure reported as the most energetically close to the global minimum[729] is misplaced by 10.8 kJ mol^{-1}. By analogy with the existence of two isoenergetic global-minimum structures, there are actually three additional structures deduced from PhOH-w_3-3, PhOH-w_3-4 and PhOH-w_3-5 by applying the same reflection operation of bare phenol.

Analysis of the global minimum structures in Figures 48, 49 and 50 reveals a tendency towards systematic shortening of the phenol–water hydrogen bonds upon adding an extra water molecule. The length of the phenol donor–water acceptor hydrogen bond varies from 1.95 Å in PhOH-w_1 to 1.91 Å in PhOH-w_2 and, finally, to 1.83 Å in PhOH-w_3. This correlates fairly with recent experimental findings[636]. On the other hand, passing from PhOH-w_2 to PhOH-w_3, the water donor–phenol acceptor phenol–water hydrogen bond decreases by 0.18 Å.

Table 38 collects seven theoretical OH-stretching vibrations of the five relevant lower-energy PhOH-w_3 structures to discuss a 'window' region. Inspection of Table 38 shows that they are actually gathered in two rather well separated groups. Considering the PhOH-w_3-1 structure as an example, we find that the first group consists of four highly intense IR vibrations placed between 3835 and 3983 cm^{-1} and describing cooperative stretching vibrations of the intra-ring OH bonds. The first two are predominantly assigned to the coupled OH-stretching vibration of phenol and its nearest-neighbour OH bond O_1-H_1 (see Figure 50). The lower of these two, corresponding to the symmetric stretch of these OH bonds, is rather Raman active and red-shifted by 283 cm^{-1} with respect to the OH-stretching frequency of bare phenol. The other one is less red-shifted, by 223 cm^{-1}. The second group of vibrations consists of three vibrations lying between 4142 and 4148 cm^{-1}. The OH-stretching vibrations of three free OH groups of water molecules contribute predominantly to this group. They are shifted to lower wavenumbers relative to the ν_3 vibration of the water molecule by approximately 40 cm^{-1}. The separation between these

1. General and theoretical aspects of phenols

groups which determines a width of the 'window' region amounts to 307 cm^{-1} at the HF/A level and decreases to 267 cm^{-1} after performing the MP2/A calculation. In other words, the stretching IR pattern of the PhOH-w_3-1 structure are those in equation 44,

$$\text{MP2/A:} \quad \nu_{\text{OH}} \overset{145(109)}{<} \nu_1^{ad1} \overset{77(56)}{<} \nu_1^{ad2} \overset{83(50)}{<} \nu_1^{ad3} \overset{267(264)}{<} \nu_3^{ad1} \overset{5(4)}{<} \nu_3^{ad2} \overset{3(3)}{<} \nu_3^{ad3} \quad (44)$$

where the experimental spacings[705] are given in parentheses.

The sixth structure of the PhOH-w_3 complex reported in the present work for the first time and displayed in Figure 50 is rather peculiar in the following sense. As shown in Figure 50, one of its water molecules accepts the phenolic OH group. Another one, O$_3$H$_3'$H$_3''$, lies above the phenol ring. It forms a so-called π hydrogen bond with the

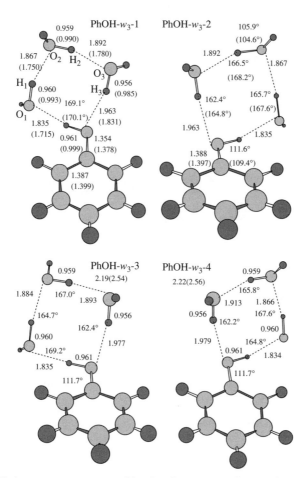

FIGURE 50. Six lower-energy structures of the phenol–water$_3$ complex. Bond lengths are in Å. The geometrical parameters are paired for some particular structures: the former value corresponds to the HF/A level while the MP2/A value is presented in parentheses. The HF/A [MP2(sp)/A] relative energy with respect to the global-minimum structure PhOH-w_3-1 is given in kJ mol^{-1}. Adapted from Reference 731 with permission

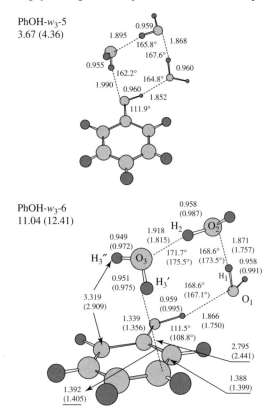

FIGURE 50. (*continued*)

π cloud of this ring, partly similar to the Leg2-type benzene–water structure discussed elsewhere[736]. The shortest MP2/A distance of 2.441 Å is predicted between the H'_3 and the carbon atom C_3 (see Figure 50). The other one, $r(H''_3-C_6) = 2.909$ Å, almost coincides with the sum of van der Waals radii of the corresponding atoms. Compared to a free water molecule, both O–H bond lengths undergo tiny elongations, about 0.003–0.006 Å, although, contrary to the other water molecules belonging to this structure as well as to all water molecules in the aforementioned structures, the water molecule participating in the π hydrogen bonding with phenol ring has its bond angle $\angle$HOH decreased by 1.3°. This is in turn manifested in the scissor vibrations of water molecules. If two of them, w_1 and w_2, are characterized by the scissor frequencies ν_2 centred at 1762 and 1788 cm^{-1}, which are red-shifted by 27 and 52 cm^{-1} compared to that in water monomer, the third water molecule w_3 possesses the scissor frequency at 1742 cm^{-1}, resulting in a blue shift of 7 cm^{-1}.

The novel PhOH-w_3-6 structure has the largest total dipole moment (1.98 D) among all reported lower-energy PhOH-w_3 structures[731]. It is also a more compact structure, as follows from a comparison of the rotational constants of all PhOH-w_3 structures. Energetically speaking, PhOH-w_3-6 is 11.0 kJ mol^{-1} (HF/A) and 10.3 kJ mol^{-1} (MP2/A) above the global minimum structure PhOH-w_3-1. These values are modified to 8.7 and

TABLE 38. The OH-stretch frequencies (in cm^{-1}) of phenol–water$_3$ complexes calculated at the HF/A and MP2/A (in parentheses) computational levels. Infrared intensity is in km mol^{-1}, Raman (R) activity in Å^4 amu^{-1}. Partial contributions are evaluated as the ratio of total displacements. The contribution of the first reported mode is referred to 100%

Frequency	IR	Raman	Assignment
PhOH-w_3-1,2			
3834.7 (3273.4)	537 (821)	213	ν_{OH}, $\nu_{O_1H_1}$ (33.8%)
3895.4 (3418.8)	602 (755)	50	$\nu_{O_1H_1}$, ν_{OH} (69.9%), $\nu_{O_2H_2}$ (56.3%)
3929.3 (3496.1)	571 (973)	43	$\nu_{O_2H_2}$, $\nu_{O_1H_1}$ (52.7%), $\nu_{O_3H_3}$ (10.1%)
3982.8 (3578.9)	418 (616)	114	$\nu_{O_3H_3}$
4142.2 (3845.4)	136 (85)	64	$\nu_{O_1H'_1}$, $\nu_{O_2H'_2}$ (11.9%)
4143.8 (3850.8)	81 (80)	65	$\nu_{O_2H'_2}$, $\nu_{O_1H'_1}$ (12.8%)
4148.3 (3853.3)	171 (107)	53	$\nu_{O_3H'_3}$
PhOH-w_3-3,4			
3840.2	531	219	ν_{OH}, $\nu_{O_1H_1}$ (24.5%), $\nu_{O_2H_2}$ (6.4%)
3901.4	629	49	$\nu_{O_1H_1}$, $\nu_{O_2H_2}$ (93.1%), ν_{OH} (66.5%)
3931.0	553	38	$\nu_{O_2H_2}$, $\nu_{O_1H_1}$ (82.7%)
3983.5	362	82	$\nu_{O_3H_3}$
4144.3	99	55	$\nu_{O_3H'_3}$
4145.2	110	47	$\nu_{O_1H'_1}$, $\nu_{O_2H'_2}$ (57.9%)
4146.2	129	88	$\nu_{O_2H'_2}$, $\nu_{O_1H'_1}$ (60.3%)
PhOH-w_3-5			
3854.9	382	215	ν_{OH}, $\nu_{O_1H_1}$ (67.2%), $\nu_{O_2H_2}$ (19.4%)
3903.9	739	30	ν_{OH}, $\nu_{O_1H_1}$ (63.4%), $\nu_{O_2H_2}$ (46.2%)
3932.0	533	48	$\nu_{O_2H_2}$, $\nu_{O_1H_1}$ (55.9%)
3988.9	333	66	$\nu_{O_3H_3}$
4141.5	115	66	$\nu_{O_2H'_2}$
4145.8	112	44	$\nu_{O_3H'_3}$
4152.5	107	52	$\nu_{O_1H'_1}$
PhOH-w_3-6			
3870.2 (3351.5)	459 (683)	188	ν_{OH}, $\nu_{O_1H_1}$ (74.1%)
3919.8 (3467.4)	773 (994)	33	ν_{OH}, $\nu_{O_1H_1}$ (80.3%), $\nu_{O_2H_2}$ (31.2%)
3950.9 (3549.4)	305 (462)	58	$\nu_{O_2H_2}$, $\nu_{O_1H_1}$ (25.7%)
4054.8 (3732.9)	91 (104)	56	$\nu_{O_3H'_3}$, $\nu_{O_3H''_3}$ (56.0%)
4142.2 (3841.9)	108 (72)	94	$\nu_{O_1H'_1}$
4147.9 (3858.4)	115 (76)	57	$\nu_{O_2H'_2}$
4156.6 (3852.0)	99 (73)	34	$\nu_{O_3H''_3}$, $\nu_{O_3H'_3}$ (55.3%)

Values taken from Reference 731 with permission.

7.7 kJ mol^{-1}, respectively, after taking the ZPVE corrections into account. Comparing the free energies of the lower-lying PhOH-w_3 structures determined by their enthalpies and entropies listed in Table 36, we conclude that at $T \geqslant 262.8$ K, PhOH-w_3-5 becomes energetically the most favourable structure. In terms of free energy, it also lies below the PhOH-w_3-3,4 structures when $T \geqslant 209.7$ K. The latter becomes more favourable than PhOH-w_3-1,2 at $T \geqslant 315.3$ K. At room temperature (298.15 K), the PhOH-w_3-6 structure is only 6.4 kJ mol^{-1} higher than PhOH-w_3-3,4.

Regarding the novel PhOH-w_3-6 structure, its seven OH-stretching vibrations are not separable into two distinct groups. It is also worth mentioning that, in contrast to the IR stretching pattern of PhOH-w_3-1 which spans over a region of 580 wavenumbers, the IR pattern of PhOH-w_3-6 is somewhat narrower, about 500 wavenumbers. Its most red-shifted vibration predicted at 3870 (3352) cm^{-1} is mainly attributed to the collective stretching vibration of the phenolic OH group and the OH group of the water molecule, which plays the role of hydrogen-bond acceptor of phenol (see Table 38). This feature looks drastically different from what we have already observed for the PhOH-w_3-1 complex, where the most red-shifted stretching vibration is essentially localized on the OH group of phenol. The second vibration of PhOH-w_3-6, placed at ca 3920 (3467) cm^{-1}, is characterized by the most intense IR absorption, equal to 773 (994) km mol^{-1}, among all reported PhOH-w_3 structures. Together with the third vibration at 3951 (3549) cm^{-1}, these vibrations describe the coupled stretchings of phenolic and water OH bonds. The fourth vibrational mode with the frequency of 4055 (3733) cm^{-1} is assigned to the symmetric π-OH stretching mode of the π hydrogen-bonded O_3H_3' and O_3H_3'' groups, whereas the corresponding π-OH asymmetric stretch amount to 4157 (3852) cm^{-1}. Their MP2/A red shifts are rather small and amount to, respectively, 41 and 63 cm^{-1} compared to a free water molecule. This is a typical feature of weak hydrogen bonds, such as we consider here as π bonds. The other vibrations of PhOH-w_3-6 found at 4142 (3842) and 4147 (3858) cm^{-1} describe, as usual, the stretching vibrations of free OH groups of water molecules. Altogether, these seven OH-stretching vibrations give rise to the IR pattern in equation 45,

$$\text{MP2/A: } \nu_{OH} \stackrel{116}{<} \nu_1^{ad1} \stackrel{82}{<} \nu_1^{ad2} \stackrel{184}{<} \nu_{sym}^{\pi} \stackrel{99}{<} \nu_3^{ad1} \stackrel{9}{<} \nu_{asym}^{\pi} \stackrel{6}{<} \nu_3^{ad2} \tag{45}$$

On inspecting equations 44 and 45, we note a narrowing of the 'window' region for the π hydrogen-bonded structure PhOH-w_3-6 compared to the conventional one with the S_3 arrangement of water molecules. This implies that some modes of the former structure might fall in the 'window' region of the latter. In the present case, these are two modes: one corresponds to ν_{sym}^{π} and the other to ν_3^{ad1}.

In concluding this subsection, it appears that all global minimum energy structures involve water molecule(s) arranged in a ring manner. Nevertheless, it seems that such a structure for PhOH(H_2O)$_3$ becomes somewhat exhausted in the sense that a more compact arrangement of water molecules emerges. We believe that the primary reason for this is that when $n \geqslant 3$, the hydrogen-bond acceptor ability of the phenolic OH group becomes competitive with the π hydrogen-bond acceptor ability of the phenol ring. This is seen more transparently in the next subsection for $n = 4$ which, in a certain sense, can be treated as a border between the global minimum energy structures where water molecules are arranged into a ring ($n \leqslant 3$) and those where water molecules form a 3D one with π hydrogen bonding ($n \geqslant 5$)[700, 702].

5. At the bottom of PES of PhOH(H_2O)$_4$

Analysis of the PES of the interaction of phenol with four water molecules reveals eleven lower-energy structures lying within an interval of less than 15.7 kJ mol^{-1} (MP2(sp)/A) above the global minimum. They are displayed in Figure 51. The landscape of the lower-energy portion of the PES of PhOH(H_2O)$_4$ is the following.

At the HF/A level, we find that the global minimum is occupied by the PhOH-w_4-1 structure with water molecules forming a ring S_4 via five typical hydrogen bonds. This is in fact a conventional structure already reported in the literature[729, 730]. It is characterized by a rather small total dipole moment of 0.96 D. Moving upward on this PES, we arrive at two energetically close structures, PhOH-w_4-3 and PhOH-w_4-2, which are placed above the global minimum one by 4.5 and 6.3 kJ mol^{-1}, respectively, after ZPVE correction. In PhOH-w_4-3, water molecules are arranged in a sort of cage-like pattern[720–723] having

1. General and theoretical aspects of phenols

six typical hydrogen bonds O—H···O and the additional O—H···π directed downward to the phenol ring[735–737]. In PhOH-w_4-2, water molecules form a S_4-like pattern with seven hydrogen bonds characterized by the following properties: first, the water molecule w_2 participates in three hydrogen bonds and, second, w_3 also takes part in π hydrogen bonding. One of the most interesting features of these structures is the appearance of double-donor water molecules, such as w_2 in PhOH-w_4-2 and w_3 in PhOH-w_4-3. Furthermore, the PhOH-w_4-3 structure has a rather peculiar pair of non-bonded oxygen atoms of water molecules, O$_1$ and O$_3$, separated from each other by 3.426 Å, a distance which is

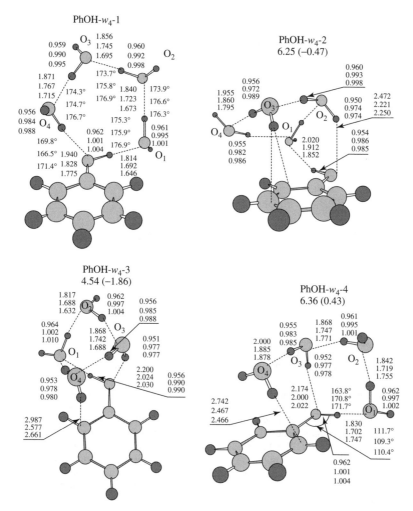

FIGURE 51. Eleven lower-energy structures of the phenol–water$_4$ complex. Bond lengths are in Å. The geometrical parameters are tripled for the lowest-energy structures in the following order (from the bottom to the top): the HF/A, MP2/A and B3LYP/A values. The HF/A relative energy with respect to the global-minimum structure PhOH-w_4-1 is given in kJ mol^{-1}. Its MP2(sp)/A analogue is followed in parentheses. Adapted from Reference 731 with permission

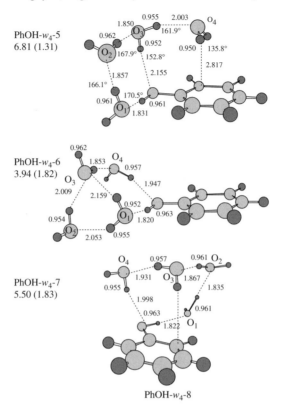

FIGURE 51. (*continued*)

smaller by about 0.2 Å than the first minimum of the radial oxygen–oxygen distribution function g_{oo} of liquid water widely used to define its first coordination shell[723].

The next, energetically less stable structures are PhOH-w_4-4 and PhOH-w_4-5. They are quite remarkably different from those studied above. Three water molecules are arranged in a cyclic structure whereas the fourth one forms two π hydrogen bonds of Leg1-type with the π-electrons of the phenol ring. This water molecule resides above the phenol ring with the distances $r(O_4-C_2) = 3.35$ Å and $r(O_4-C_3) = 3.32$ Å. The energy separations of PhOH-w_4-4 and PhOH-w_4-5 from the global minimum are 6.4 and 6.8 kJ mol^{-1}, respectively. The remainder of the lower-energy portion of the PES of the PhOH-w_4 complex is the following. The PhOH-w_4-6 structure has six hydrogen bonds and a total dipole moment of 3.23 D; it is 3.9 kJ mol^{-1} above the global minimum. Its water pattern also partly resembles a book. A similar structure is also inherent for PhOH-w_4-7 at 1.5 kJ mol^{-1} above PhOH-w_4-6. The next structure, PhOH-w_4-8, is quite particular in that its OH phenolic group functions only as a hydrogen bond donor, in contrast to all other reported PhOH-w_4 structures. The PhOH-w_4-9 structure is separated from the global minimum by 2.4 kJ mol^{-1}. Its four water molecules form a ring similar to the PhOH-w_4-1 structure and differs from the latter by flippings of free OH groups of water molecules. A similar water pattern is seen for PhOH-w_4-10 whereas PhOH-w_4-11 partly mimics the PhOH-w_4-3 structure.

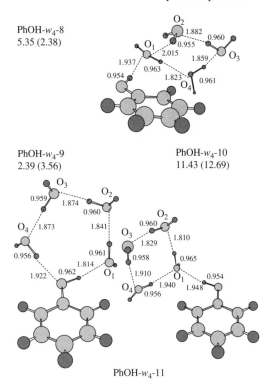

FIGURE 51. (*continued*)

Compared to HF/A, the MP2 and B3LYP/A PESs of PhOH(H$_2$O)$_4$ have somewhat different topologies, which is reflected in the geometries of the phenol–water$_4$ complexes. For example, the MP2/A level reverses the order between the PhOH-w_4-1–3 structures in such a way that PhOH-w_4-2 becomes the global minimum; PhOH-w_4-3 is only 0.8 kJ mol^{-1} higher, and PhOH-w_4-2 is 4.6 kJ mol^{-1} higher (neglecting ZPVE). As for the B3LYP/A geometries, we may note that, for instance, in PhOH-w_4-3 the oxygen atom O$_4$ is separated from the carbon atom C$_2$ of phenol by 3.410 Å whereas $r(\mathrm{H}_4'-\mathrm{C}_2) = 2.661$ Å. In PhOH-w_4-2, the distances $r(\mathrm{O}_4-\mathrm{C}_3) = 3.345$ Å and $r(\mathrm{H}_4'-\mathrm{C}_2) = 2.627$ Å. The latter is smaller by about 0.3 Å than the sum of van der Waals radii of the corresponding atoms. Summarizing and taking into account that the expected margin error of the computational methods employed in the present work is *ca* ±8 kJ mol^{-1}, we conclude that these four structures PhOH-w_4-1–4 are placed at the very bottom of the PES of PhOH(H$_2$O)$_4$ and are actually nearly isoenergetic.

In order to interpret the experimentally determined IR pattern of phenol interacting with four water molecules, we now consider theoretical OH-stretching modes of the PhOH-w_4-1–4 structures (Table 39). Contrary to the PhOH-w_1-1 and PhOH-w_2-1 structures studied above, the vibrational assignments are particular for each structure of the PhOH-w_4 complex. The most red-shifted OH-stretching vibration at 3772 (2970.1) cm^{-1} is predicted for the PhOH-w_4-2 structure. It is predominantly assigned to the hydrogen-stretching vibration of the O$_1$–H$_1 \cdots$ O$_2$ bond and is significantly enhanced by a factor

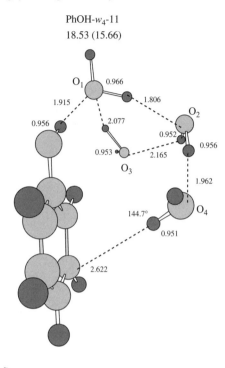

FIGURE 51. (*continued*)

of 24 in comparison with the IR intensity of the ν_1 vibration of the water molecule. The analogous OH-stretching vibration of PhOH-w_4-3 is placed at 3798 (3008.4) cm^{-1}. It is also predominantly assigned to the symmetric hydrogen-stretching vibration of the $O_1-H_1 \cdots O_2$ and $O_2-H_2 \cdots O_3$ bonds. The corresponding asymmetric vibrational mode is found at 3874 (3201.8) cm^{-1}. Its IR intensity exceeds that of the ν_3 vibrations of the water molecule by a factor of 12. Interestingly, the phenolic OH-stretching vibration contributes only to the fourth, 3988 (3476.8) cm^{-1}, and to the third, 3958 (3394.4) cm^{-1}, vibrations of PhOH-w_4-2 and PhOH-w_4-3, respectively. It is therefore red-shifted by *ca* 230 and 160 cm^{-1}, respectively, from that of bare phenol and their IR intensities are increased by *ca* 4-fold.

It follows from Table 38 that the quintessential feature of OH-stretching vibrations of the PhOH-w_4-3 and PhOH-w_4-2 is that they are not separable into groups of vibrations. For example, in the case of PhOH-w_4-3, the inter-vibrational separations take the following values: 75 (192), 84 (192), 26 (67), 41 (133), 71 (76), 41 (101), 6 (11) and 4 (7) cm^{-1}. We suggest that such vibrational non-separability occurs due to the cage-type arrangements of water molecules and the existence of π hydrogen bonding between one of the water molecules and the phenol ring. Such π hydrogen bonding results in that the corresponding π-OH stretching vibrations of this particular water molecule for the PhOH-w_4-3 structure at 4024.8 (3593.9) (symmetric) and 4146.9 (3771.1) (asymmetric) cm^{-1}. Compared with the ν_1 and ν_3 stretching vibrations of the water molecule, the former is red-shifted by 45 (180) cm^{-1} whereas the latter is red shifted by 41 (144) cm^{-1}.

TABLE 39. The OH-stretch frequencies (in cm^{-1}) of phenol–water$_4$ complexes calculated at the HF/A and B3LYP/A (in parentheses) computational levels. Infrared intensity is in km mol^{-1} and Raman (R) activity in Å^4 amu^{-1}. The partial contributions are evaluated as the ratio of the total displacements. The contribution of the first reported mode is referred to 100%

Frequency	IR	Raman	Assignment
PhOH-w$_4$-1			
3811.6 (3077.8)	724 (1260)	261	ν_{OH}, $\nu_{O_1H_1}$ (31.4%)
3869.0 (3217.1)	849 (1442)	68	$\nu_{O_2H_2}$, ν_{OH}(81.1%), $\nu_{O_1H_1}$(60.6%), $\nu_{O_3H_3}$(32.5%)
3898.5 (3287.7)	854 (1521)	39	$\nu_{O_1H_1}$, $\nu_{O_3H_3}$(78.0%), $\nu_{O_2H_2}$(21.8%)
3926.8 (3354.0)	330 (683)	73	$\nu_{O_1H_1}$, $\nu_{O_3H_3}$(87.6%), $\nu_{O_2H_2}$(16.1%)
3976.6 (3468.8)	314 (551)	77	$\nu_{O_4H_4}$
4140.7 (3796.0)	112 (42)	67	$\nu_{O_1H'_1}$
4143.4 (3797.5)	105 (46)	56	$\nu_{O_2H'_2}$, $\nu_{O_3H'_3}$(19.1%)
4143.8 (3798.7)	114 (40)	39	$\nu_{O_4H'_4}$
4144.8 (3800.3)	92 (44)	70	$\nu_{O_3H'_3}$, $\nu_{O_2H'_2}$(16.3%)
PhOH-w$_4$-2			
3771.8 (2970.1)	431 (772)	91	$\nu_{O_1H_1}$
3915.1 (3299.0)	309 (544)	52	$\nu_{O_2H_2}$
3961.7 (3429.8)	277 (315)	37	$\nu_{O_3H_3}$, $\nu_{O_4H_4}$(90.4%)
3987.7 (3476.8)	308 (987)	55	$\nu_{O_3H_3}$, $\nu_{O_4H_4}$(88.7%), ν_{OH}(22.7%)
4005.6 (3515.8)	416 (373)	110	ν_{OH}, $\nu_{O_4H_4}$(18.2%)
4109.6 (3721.9)	130 (80)	33	$\nu_{O_2H'_2}$
4132.7 (3763.5)	164 (115)	47	$\nu_{O_3H'_3}$
4133.6 (3794.8)	96 (51)	95	$\nu_{O_1H'_1}$
4151.9 (3802.8)	126 (51)	80	$\nu_{O_4H'_4}$
PhOH-w$_4$-3			
3798.4 (3008.4)	448 (77)	153	$\nu_{O_1H_1}$, $\nu_{O_2H_2}$(22.9%)
3873.6 (3201.8)	719 (1309)	36	$\nu_{O_2H_2}$, $\nu_{O_1H_1}$(23.8%)
3957.6 (3394.4)	361 (572)	73	ν_{OH}
3983.7 (3461.4)	244 (402)	54	$\nu_{O_3H'_3}$, $\nu_{O_3H''_3}$(16.8%)
4024.8 (3593.9)	163 (300)	54	$\nu_{O_4H'_4}$, $\nu_{O_4H''_4}$(30.5%)
4095.7 (3670.3)	218 (247)	56	$\nu_{O_3H'_3}$, $\nu_{O_3H''_3}$(17.9%)
4136.5 (3792.2)	111 (38)	77	$\nu_{O_1H'_1}$
4142.9 (3800.7)	112 (42)	63	$\nu_{O_2H'_2}$
4146.9 (3771.1)	122 (104)	42	$\nu_{O_4H''_4}$, $\nu_{O_4H'_4}$(28.4%)
PhOH-w$_4$-4			
3814.1 (3064.2)	468 (815)	214	ν_{OH}, $\nu_{O_1H_1}$(53.2%), $\nu_{O_2H_2}$(21.8%)
3866.3 (3198.2)	1022 (1708)	24	$\nu_{O_2H_2}$, ν_{OH}(77.3%), $\nu_{O_1H_1}$(28.1%)
3899.5 (3270.9)	379 (760)	53	$\nu_{O_1H_1}$, $\nu_{O_2H_2}$(63.4%), ν_{OH}(12.0%)
3991.0 (3516.8)	225 (342)	54	$\nu_{O_3H'_3}$, $\nu_{O_3H''_3}$(23.8%)
4058.1 (3691.8)	27 (156)	38	$\nu_{O_4H'_4}$, $\nu_{O_4H''_4}$(69.6%)
4095.0 (3648.2)	218 (180)	53	$\nu_{O_3H''_3}$, $\nu_{O_3H'_3}$(25.5%)
4139.4 (3795.8)	114 (50)	85	$\nu_{O_1H'_1}$
4145.3 (3796.9)	89 (29)	61	$\nu_{O_2H'_2}$
4163.4 (3804.6)	79 (47)	29	$\nu_{O_4H''_4}$, $\nu_{O_4H'_4}$(62.2%)
PhOH-w$_4$-5			
3822.5	288	234	ν_{OH}, $\nu_{O_1H_1}$(78.8%), $\nu_{O_2H_2}$(46.2%)

(*continued overleaf*)

TABLE 39. (continued)

Frequency	IR	Raman	Assignment
3865.8	1227	7	$v_{O_2H_2}$, v_{OH}(73.5%)
3902.7	346	54	$v_{O_1H_1}$, $v_{O_2H_2}$(26.1%), v_{OH}(17.1%)
3988.7	228	49	$v_{O_3H_3'}$, $v_{O_3H_3''}$(23.9%)
4060.3	19	28	$v_{O_4H_4}$, $v_{O_4H_4''}$(76.0%), $v_{O_3H_3''}$(10.3%)
4090.7	229	58	$v_{O_3H_3''}$, $v_{O_3H_3'}$(26.6%)
4144.8	79	68	$v_{O_2H_2'}$, $v_{O_1H_1'}$(56.8%)
4146.1	126	99	$v_{O_1H_1'}$, $v_{O_2H_2'}$(59.6%)
4163.5	69	25	$v_{O_4H_4'}$, $v_{O_4H_4}$(76.7%)

PhOH-w_4-6

Frequency	IR	Raman	Assignment
3815.7	534	211	v_{OH}
3866.0	906	59	$v_{O_2H_2}$
3972.1	514	110	$v_{O_3H_3}$
3992.4	145	108	$v_{O_4H_4}$, $v_{O_1H_1''}$(32.9%)
4007.3	190	27	$v_{O_1H_1''}$, $v_{O_1H_1'}$(48.3%), $v_{O_4H_4}$(34.5%)
4088.7	252	45	$v_{O_1H_1'}$, $v_{O_1H_1''}$(33.3%)
4140.6	112	69	$v_{O_2H_2'}$, $v_{O_3H_3'}$(12.4%)
4141.4	165	54	$v_{O_3H_3'}$, $v_{O_2H_2'}$(14.6%)
4154.8	128	60	$v_{O_4H_4'}$

PhOH-w_4-7

Frequency	IR	Raman	Assignment
3809.1	571	188	v_{OH}, $v_{O_1H_1}$(28.9%)
3869.4	894	46	$v_{O_2H_2}$, $v_{O_1H_1}$(68.2%), v_{OH}(55.6%)
3905.6	440	51	$v_{O_1H_1}$, $v_{O_2H_2}$(84.0%)
3954.8	227	50	$v_{O_3H_3}$, $v_{O_2H_2}$(11.5%), $v_{O_4H_4}$(11.4%)
3999.4	263	51	$v_{O_4H_4}$
4126.9	117	60	$v_{O_3H_3'}$
4143.1	111	95	$v_{O_1H_1'}$
4148.4	101	60	$v_{O_2H_2'}$
4155.5	126	42	$v_{O_4H_4'}$

PhOH-w_4-8

Frequency	IR	Raman	Assignment
3824.3	302	147	$v_{O_1H_1}$, $v_{O_2H_2}$(26.6%)
3884.9	765	40	$v_{O_3H_3}$, $v_{O_2H_2}$(83.4%), $v_{O_1H_1}$(61.5%)
3919.1	434	46	$v_{O_3H_3}$, $v_{O_2H_2}$(94.9%)
3977.9	334	63	v_{OH}, $v_{O_4H_4}$(26.9%)
4003.4	300	90	$v_{O_4H_4}$, v_{OH}(35.9%)
4131.9	202	39	$v_{O_4H_4'}$, $v_{O_1H_1'}$(14.9%)
4135.9	81	91	$v_{O_1H_1'}$, $v_{O_4H_4'}$(11.0%)
4142.9	130	47	$v_{O_2H_2'}$, $v_{O_3H_3}$(14.5%)
4144.4	84	83	$v_{O_3H_3'}$, $v_{O_2H_2}$(16.1%)

PhOH-w_4-9

Frequency	IR	Raman	Assignment
3817.5	732	243	v_{OH}, $v_{O_1H_1}$(33.8%)
3876.7	693	84	$v_{O_1H_1}$, v_{OH}(94.2%), $v_{O_2H_2}$(93.9%), $v_{O_3H_3}$(34.2%)
3903.3	941	38	$v_{O_3H_3}$, $v_{O_1H_1}$(81.5%), $v_{O_2H_2}$(22.7%)
3932.0	302	60	$v_{O_2H_2}$, $v_{O_3H_3}$(57.0%), $v_{O_1H_1}$(15.3%)

TABLE 39. (continued)

Frequency	IR	Raman	Assignment
3975.6	382	107	$\nu_{O_4H_4}$, $\nu_{O_3H_3}$ (11.6%)
4142.1	105	69	$\nu_{O_1H_1'}$
4146.2	116	67	$\nu_{O_4H_4'}$, $\nu_{O_3H_3'}$ (17.9%)
4146.6	128	27	$\nu_{O_3H_3'}$, $\nu_{O_2H_2'}$ (62.9%), $\nu_{O_4H_4'}$ (37.1%)
4147.5	118	74	$\nu_{O_2H_2'}$, $\nu_{O_3H_3'}$ (55.8%)

Values taken from Reference 731 with permission.

The PhOH-w_4-4 structure also has a rather peculiar and non-separable OH-stretching vibrational pattern. Its three most red-shifted vibrations are located at 3814 (3064.2), 3866 (3198.2) and 3900 (3270.9) cm^{-1}. Altogether, they describe the coupled OH-stretching vibrations of the trimeric water ring and the phenolic OH group. The second one is the most IR active among all OH-stretching vibrations of all reported PhOH-w_4 structures. Its IR intensity is 13 (18) times larger that of the OH-stretching vibration of bare phenol (ν_3 of the water molecule). The symmetric and asymmetric stretches of the water molecule connecting the water ring with the terminated water molecule placed above the phenol ring are found at 3991 (3516.8) and 4095 (3648.2) cm^{-1}. Between them, at 4058 (3691.8) cm^{-1}, there exists the symmetric π OH stretch whose asymmetric vibration has the highest frequency of 4163 (3804.6) cm^{-1}. These two vibrations are separated by the OH stretches at 4139 (3795.8) and 4145 (3796.9) cm^{-1}, assigned to free OH groups of water molecules.

As we would expect, the pattern of the OH-stretching vibrations of PhOH-w_4-1 is absolutely different from those of PhOH-w_4-2, PhOH-w_4-3 and PhOH-w_4-4 and resembles the typical S_4 pattern of the PhOH-w_1, PhOH-w_2 and PhOH-w_3-1–5 structures. It is clearly seen from Table 39 that the nine OH-stretching vibrations of the PhOH-w_4-1 structure are well separated into two groups in that way forming the 'window' region of width about 164 (327) cm^{-1}. Note that the B3LYP/A width agrees satisfactorily with the experimental one[705]. The former group spans the region between 3812 (3077.8; expt: ca 3135[705]) and 3977 (3468.8; expt: 3430[705]) cm^{-1} and consists of five rather IR and Raman active OH-stretching vibrations assigned to the coupled stretches of the water ring and phenolic OH groups. Its highest OH-stretching vibration is dominantly composed of the hydrogen stretch of the water molecule donating the hydrogen bond to phenol. The latter group is rather narrow with a width of only 4 (4) cm^{-1}. The OH-stretching vibrations of free water OH groups contribute to this group. Its lowest wavenumber stretch at 4141 (3796.0) cm^{-1} corresponds to the free OH group of the water molecule which accepts the phenolic hydrogen bond.

Summarizing the B3LYP/A IR patterns in the stretching region of the four most energetically stable structures PhOH-w_4-1, PhOH-w_4-2, PhOH-w_4-3 and PhOH-w_4-4, we illustrate them in equation 46.

PhOH-w_4-1: $\nu_{OH}(3077.8) \stackrel{139}{<} \nu_1^{ad2}(3217.0) \stackrel{81}{<} \nu_1^{ad13'}(3287.7) \stackrel{66}{<}$
$\nu_1^{ad13''}(3354.0) \stackrel{115}{<} \nu_1^d(3468.8) \stackrel{327}{<} \nu_3^{ad1}(3796.0) \stackrel{2}{<}$
$\nu_3^{ad2}(3797.5) \stackrel{1}{<} \nu_3^d(3798.7) \stackrel{2}{<} \nu_3^{ad3}(3800.3)$

PhOH-w_4-2: $\nu_1^{ad1}(2970.1) \stackrel{319}{<} \nu_1^{add2}(3299.0) \stackrel{131}{<} \nu_1^{add3}(3429.8) \stackrel{47}{<}$
$\nu_1^{add3,ad4}(3476.8) \stackrel{39}{<} \nu_{OH}(3515.8) \stackrel{106}{<} \nu_3^{add2}(3721.9) \stackrel{42}{<}$
$\nu_3^{\pi}(3763.5) \stackrel{31}{<} \nu_3^{ad1}(3794.8) \stackrel{8}{<} \nu_3^{ad4}(3802.8)$

PhOH-w_4-3: $\quad \nu_1^{ad1}(3008.4) \overset{193}{<} \nu_1^{ad2}(3201.8) \overset{192}{<} \nu_{OH}(3394.4) \overset{67}{<}$

$\nu_1^{add3}(3461.4) \overset{133}{<} \nu_{sym}^{\pi}(3593.9) \overset{76}{<} \nu_3^{add3}(3670.3) \overset{101}{<}$

$\nu_{asym}^{\pi}(3771.1) \overset{21}{<} \nu_3^{ad1}(3792.2) \overset{9}{<} \nu_3^{ad2}(3800.7)$

PhOH-w_4-4: $\quad \nu_{OH}(3064.2) \overset{134}{<} \nu_1^{ad2}(3198.2) \overset{73}{<} \nu_1^{ad1}(3270.9) \overset{246}{<}$

$\nu_1^{add3}(3516.8) \overset{131}{<} \nu_3^{add3}(3648.2) \overset{44}{<} \nu_{sym}^{\pi}(3691.8) \overset{104}{<}$

$\nu_3^{ad1}(3795.8) \overset{1}{<} \nu_3^{ad2}(3796.9) \overset{8}{<} \nu_{asym}^{\pi}(3804.6).$ (46)

The 'window' region of the PhOH-w_4-1 structure spreads from 3468.8 to 3796.0 cm^{-1} and covers an area of 327 cm^{-1} (the experimental value is 281 cm^{-1} [705]. It follows from equation 46 that, in this region, the isomer PhOH-w_4-4 has four OH-stretching modes placed at 3516.8, 3648.2, 3691.8 and 3795.8 cm^{-1}. PhOH-w_4-3 also has four OH-stretching modes there, i.e. 3593.9, 3670.3, 3771.1 and 3792.2 cm^{-1}, whereas PhOH-w_4-2 exhibits five modes: 3476.8, 3515.8, 3721.9, 3763.5 and 3794.8 cm^{-1}. In other words, the PhOH-w_4-3 and PhOH-w_4-4 have precisely that number of vibrational modes which was revealed experimentally[704, 705]. Due to theoretical and experimental uncertainties, the structure PhOH-w_4-2 might also be included into this class. Therefore, these three lower-energy structures of phenol with four water molecules characterized by the formation of the π hydrogen bond are likely referred to as the class of structures revealed in Reference 590. It is worth mentioning that the lowest stretching mode of the non-ring structure of PhOH(H_2O)$_4$ is calculated 73 cm^{-1} below the analogous mode in the ring S_4 structure PhOH-w_4-1[705]. It then follows from equation 46 that PhOH-w_4-2 and PhOH-w_4-3 have a similar feature, i.e. 108 and 69 cm^{-1}, respectively.

In order to obtain some insight into the formation of the π hydrogen bonding in the PhOH-w_4-2–PhOH-w_4-4 structures in terms of the molecular orbital (MO) or electron density patterns, we draw in Figure 52 the lowest unoccupied molecular orbital (LUMO), the highest one (HOMO) and HOMO-1 of the PhOH-w_4-4 structure. As seen vividly there, the π hydrogen bonding between the π cloud of the phenol ring and the water molecule w_4 reshapes the HOMO-1, HOMO and LUMO of bare phenol and slightly lowers the HOMO-1 orbital energy but, on the contrary, raises the orbital energies of the HOMO and LUMO by ca 0.2 eV. For example, we observe a small portion of the charge transfer from the HOMO-1 to the s orbital of the hydrogen atoms of this water molecule and to the lone pairs of the oxygen atom. This raises their population to 0.06 for H and to 0.15 for O and results in the appearance of a small hollow in the HOMO-1 of bare phenol precisely in the front of the water molecule w_4. A slightly smaller charge, ca 0.13, is transferred from the π-HOMO of phenol to the lone pair MO of the oxygen atom, whereas a substantial charge transfer from the LUMO to the s MO of the oxygen atom of the water molecule is predicted by the present B3LYP/A level.

What are the essential conclusions of the present subsection? As we have already mentioned above, the last decade was unprecedentedly successful, primarily from the experimental point of view, in studying the interaction between phenol and water molecules. In particular, it was discovered that phenol favours a *2D* ring type of arrangement of water molecules if there are less than *three* water molecules and, on the contrary, the *3D* ring type if these are *five* or more water molecules. It was therefore thought that *four* looks like the 'magic' number for the phenol–water$_n$ interaction, and this was really a sort of exclusive number thanks, first of all, to the experimental work by different groups[590, 704, 705] who revealed experimentally the existence of the phenol–water$_4$ isomer with a *3D* arrangement

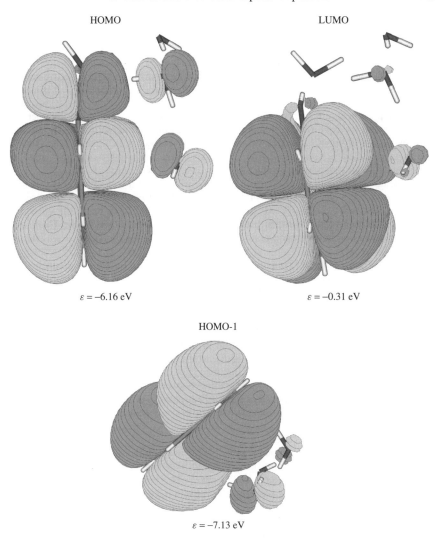

FIGURE 52. The shape (contour spacing of 0.008 e au^{-3}) of the highest occupied (HOMO), lowest unoccupied (LUMO) and HOMO-1 of the complex PhOH-w_4-4 using the B3LYP/A wave function; ε denotes the orbital energy

of water molecules. They showed that it was only one particular isomer which is capable of explaining the puzzling 'window' region in the IR stretching spectra. Logically, the 'magic' of the number *four* stems from the fact that this is just the borderline where the *2D* water pattern ($n \leqslant three$) meets the *3D* pattern ($n \geqslant five$). The analysis of the potential energy surface of the phenol–water$_4$ complex conducted above and its juxtaposition with the PESs of the phenol–water$_{1-3}$ complexes demonstrates vividly this point of view.

C. Hydrogen Bonding between Phenol and Acetonitrile

1. Introductory foreground

Acetonitrile (ACN) possesses some unique properties, such as a high dielectric constant (35.95) and the solubilization of many inorganic and organic materials[738, 739]. It is actually one of the few simple aprotic solvents miscible in water at any ratio. X-ray diffraction studies of pure acetonitrile revealed that ACN molecules do not strongly interact with themselves and are only weakly associated via dipole–dipole interaction[740]. The IR spectrum of pure acetonitrile includes two major bands placed at 2257 and 2295 cm^{-1}[740]. The former, called v_2, originates from the C≡N stretching mode while the latter is a combination band composed of the CCH bend v_3 and C−C stretch v_4 modes[741].

For the last forty years the acetonitrile molecule was, and still is, a 'work horse' in many laboratories worldwide, in experimental studies of the hydrogen bonding with nitriles. It is obvious that ACN possesses two sites for accepting a hydrogen bond: the one on the lone-pair electrons of the nitrogen atom (σ-bonding) and the other on the C≡N triple bond (π-bonding). The hydrogen bond formation in phenol–nitrile systems was initially examined by several authors[742-745] in inert solvents such as CCl$_4$ or C$_2$Cl$_4$[746-750] who all recorded that their IR spectra contain an additional band placed on the low-frequency side of the free phenol O−H stretching band v(OH) as the concentration of the nitrile increases. The Δv(OH) shift varies from 148.5 cm^{-1} at 0.119 M of ACN to 156.5 cm^{-1} when the ACN concentration reaches 0.687 M[742]. These authors then suggested that this new band results from the O−H stretching mode of a hydrogen-bonded complex involving the OH group of phenol and the nitrogen atom of the nitrile. It was at that time when the existence of a 1:1 complex between phenol and nitrile in inert solvents was postulated[743, 744]. The appearance of an unusual blue shift of the C≡N stretching vibration by about 12.5 cm^{-1} was noted[742] when the nitrogen atom of the nitrile group is complexed with the OH group of phenol, implying a σ-type hydrogen bonding between the nitrogen lone pair and the phenol OH. The increased frequency of the C≡N stretching vibration in the complex gave rise to a shoulder on the high-frequency side of the C≡N peak.

At nearly the same time, on the basis of the well-known Buckingham formula describing the frequency shift in a medium[751], it was deduced[752, 753] that if the fundamental stretching mode v(OH) of free phenol in the gas phase is fitted at 3655 cm^{-1}, it must be extrapolated in the phenol–acetonitrile complex to 3540 cm^{-1}, and therefore the red shift due to complexation becomes equal to 115 cm^{-1}. This value looks much smaller than expected[209] although, as we have already mentioned, a red shift of 148.5−156.5 cm^{-1} was found[742] and similar red shifts of 152[745] and 160[746] cm^{-1} were also detected. The origin of the frequency shift of the v(OH) mode of phenol was also noted[754] in the phenol–ACN complex from 3460 to 3409 cm^{-1} on increasing the concentration of acetonitrile from 0.19 to 100% in CCl$_4$ (interestingly, the change proceeds stepwise: between 0.19 and 0.39% no shift was detected, between 0.78 and 1.8% it is equal to −5 cm^{-1}, a further dilution to 4% results in −10 cm^{-1} etc.). It is not entirely clear and suggests a possible formation of 2:1 phenol–acetonitrile complexes due to the increased basicity of the oxygen atom of phenol. A similar trend was recently observed[755] for the pentachlorophenol–acetonitrile complex. Such a puzzling effect has not been so well appreciated by theoreticians despite the fact that it still annoys the experimentalists, although it is worth recollecting the mid-1980's theoretical work[756] (see also References 757−759) which suggested that the most favourable hydrogen bond formation with nitriles occurs via σ type hydrogen bonding. However, this is not the case with the hydrogen-bonded complexes of water with benzonitrile, where the π-bonding is slightly superior over the σ-type[729]—we could actually agree with some authors[743] that 'benzonitrile... is found to be anomalous'. Nevertheless, other authors concluded that this is just the case for hydrogen bonding with nitriles[746, 757–759], and also a quite recent B3LYP/6-31G(d,p) study[612] of the phenol–acetonitrile and phenol–pyridine complexes mainly focused on the anharmonicity contribution to their dipole moments.

Summarizing, what else we can tell the reader from a theoretical point of view? There are certainly some as yet unclear points related to routine use of quantum chemical programs for obtaining the optimized structure of the 1:1 complex between phenol and acetonitrile and somehow exploring the calculated frequencies to discuss, again routinely, agreement between experiment and theory. It seems as if what remains is the existence of the 2:1 complex and its structure and the puzzling dependence of the shift of the ν(OH) mode of phenol on the ACN concentration although, impressed by the rampant experimentalists arguments, this was likely a way to almost nowhere and does not deserve to be published at all. Nevertheless, we have performed a rather exhaustive search[760] of the PES of the phenol–acetonitrile interaction and its results and the consequent attempt to explain the experiments is presented below[761].

2. Phenol–acetonitrile complex

The PES of the interaction of the phenol and acetonitrile molecules consists of two lower-energy minimum structures[761] displayed in Figure 53. The first, named PhOH-ACN-1, is the conventional structure which has been explored by experimentalists for four decades. It occupies the global minimum on that PES and is characterized by a binding energy $E_{HB}^{(1)}$(PhOH–ACN) of 22.3 kJ mol^{-1} (see Table 40). It agrees fairly with the experimental value of 18.8 kJ mol^{-1} for the reported enthalpy of formation[746, 757]. The BSSE correction comprises only 0.7 kJ mol^{-1} and is hereafter neglected. The second minimum-energy structure, PhOH-ACN-2, is reported here for the first time and placed higher by 16.5 kJ mol^{-1}, and therefore has a binding energy $E_{HB}^{(2)}$(PhOH–ACN) of 5.8 kJ mol^{-1}.

If the conventional structure is formed due to the typical medium-strength hydrogen bond between the OH group of phenol and the lone pair of the nitrogen atom of acetonitrile, respecting all canonical though still somewhat loosely defined rules[173] which will be later thoroughly discussed, the structure PhOH–ACN-2 is quite peculiar in the sense that its formation is provided by two weaker bonds which could also be referred to with some caution as some sort of hydrogen bonds. One of them is a C—H···O hydrogen bond between the methyl group of acetonitrile and the oxygen atom of phenol, while the other seems to be much weaker and is formed between the π-electrons of the C≡N bond of acetonitrile and the CH group of phenol. The fact that this is affirmatively a π hydrogen bond is confirmed by the value of the bonding angle $\angle$C—H—N = 79.1°.

Let us first analyse by a routine procedure what are the substantial changes in the geometries of the precursors[762] and their characteristic vibrational modes which accompany the formation of the σ-type O—H···N hydrogen bond between phenol and acetonitrile. Obviously, this is primarily the elongation of the O—H bond length by 0.008 Å as manifested in a red shift of the ν(OH) stretching vibration by 158 cm^{-1} (in fair agreement with the experimental values[742, 745]) and a significant enhancement of its IR activity, viz. from 57 km mol^{-1} in phenol to 873 km mol^{-1} in PhOH–ACN-1 (Table 41). The formed hydrogen bond has a typical length of 1.997 Å and is rather linear with a bond angle $\angle$OHN of 171.6°. The hydrogen-bond stretching vibration ν_σ(O—H···N) appears at 111.7 cm^{-1}. It is also worth mentioning two lower-frequency modes centred at 59.0 and 69.5 cm^{-1}, referring to the hydrogen-bond bending motions and originating due to the molecular dipole rotation, by analogy with the band at 90 cm^{-1} in the phenol–pyridine complex[612].

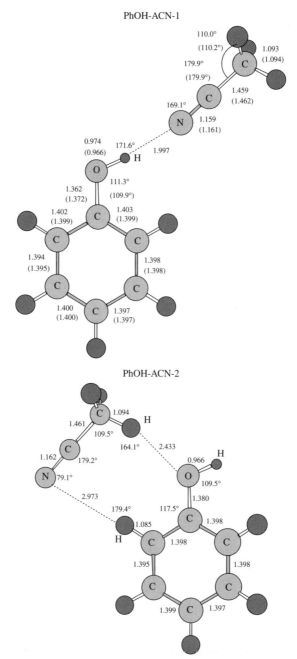

FIGURE 53. Complexes of phenol with acetonitrile. The bond lengths are in Å. Values in parentheses correspond to the optimized geometries of the free phenol and acetonitrile molecules. Adapted from Reference 761 with permission

TABLE 40. Some key features of 1:1 phenol–acetonitrile complexes

Feature	PhOH-ACN-1	PhOH-ACN-2
−Energy + 440, hartree	0.269694	0.262872
ZPVE + kJ mol^{-1}	396.83	395.46
E_{HB}, kJ mol^{-1}	22.3 (21.6)^a	5.77 (5.65)^a
Dipole moment, D	6.77	4.86
Frequencies, cm^{-1} and IR intensities, km mol^{-1}		
ν_σ(N···H–C)	—	35 (7)
ν_σ(O···H–C)	—	81 (13)
ν_σ(N···H–O)	112 (4) [117 (4)]^c	—
τ(OH) 330 (115)^b	645 (98) [596 (106)]^c	323 (108)
ν(C–O) 1284 (95)^b	1300 (94) [1397(68)]^c	1270 (96)
ν(C≡N) 2364 (12)^b	2378 (32)	2359 (13)
ν(C–H···O) 3137 (1)^b	—	3135 (3)
ν(C–H···N) 3213 (5)^b	—	3217 (3)
ν(OH) 3831(57)^b	3673 (873) [3679 (850)]^c	3826 (53)

^{a}BSSE corrected.
^{b}In the free phenol and acetonitrile molecules.
^{c}The theoretical B3LYP/6-31G(d,p) results^{731}.

TABLE 41. Some key features of the stable complexes of phenol with two acetonitrile molecules

Feature	PhOH-ACN$_2$-1	PhOH-ACN$_2$-2
−Energy + 573, hartree	0.045325	0.039648
ZPVE + kJ mol^{-1}	519.20	518.12
−Enthalpy + 572, hartree	0.832155	0.825477
Entropy, kJ mol^{-1}	559.6	645.5
E_{HB}, kJ mol^{-1}	44.60	30.75
Dipole moment, D	1.97	10.89
Quadrupole, D·Å	75.9 61.5 81.2	71.0 56.6 81.2
Polarizability, au	178.3 134.5 82.0	170.8 139.8 82.8
Frequencies, cm^{-1} and IR intensities, km mol^{-1}		
ν_σ(O···H–C)^a	96 (3)	91 (12)
ν_σ(N···H–O)^a	138 (12)	121 (5)
τ(OH)	681 (58)	660 (83)
ν(C–O)	1293 (90)	1289 (85)
ν(C≡N)	2357 (27) 2370 (47)	2358 (15) 2378 (36)
ν(C–H···O)	3048 (13) 3128 (37)	3047 (31) 3129 (32)
ν(OH)	3587 (958)	3658 (905)

^{a}Both modes are coupled to each other.

The out-of-plane bending mode mimicking the τ(OH) of phenol is characterized by a frequency at 645 cm^{-1}. Less substantial changes are predicted by the present *ab initio* method in the phenol geometrical patterns in the vicinity of the OH group. The COH angle increases slightly, by 2.2°. The elongation of the C=O bond by 0.009 Å makes it weaker and causes a blue shift of the tackled ν(C=O) stretching mode by 16 cm^{-1}. Interestingly, about the same elongation is predicted for a much lighter O–H bond. No significant changes occur in the phenol bonded counterpart, except perhaps the blue-shifted ν(CN) mode by 14 cm^{-1}, related to a shortening of the C≡N triple bond by 0.002 Å. The present value fairly matches the experimentally detected blue shift of 12.5 cm^{-1}[742].

As we mentioned earlier, two weak hydrogen bonds play a major role in the formation of the PhOH–ACN-2 structure. Figure 53 shows the bond lengths of 2.433 and 2.973 Å for the C–H···O and C–H···N bonds, respectively. Naturally, their stretching modes are characterized by lower frequencies, i.e. 81 and 35 cm^{-1}. If the C–H bond participating in the former bond is slightly lengthened by 0.0004 Å, the opposite is observed for the other one for which the C–H bond becomes shorter by 0.0002 Å. This involves the stretching mode placed at 3135 cm^{-1} (see Table 41). Participating in the π hydrogen bonding, the C≡N bond slightly elongates by 0.001 Å and its stretching mode ν(CN) is red-shifted by about 5 cm^{-1}.

3. Phenol bonding with two acetonitrile molecules

After discovering above the existence of two lower-energy structures of phenol and acetonitrile (there are certainly more structures via formation of C–H···N on the periphery of the OH group, although a π complex between the methyl group of acetonitrile and the phenol ring should be firmly ruled out), we shall explain the experimental results via modelling microscopically an increase in the acetonitrile concentration. Before doing so, it is worthwhile briefly discussing the acetonitrile dimer because it may be anticipated that combining the locations of acetonitrile molecules in the PhOH–ACN-1 and PhOH–ACN-2 structures leads to their partial dimerization whenever another acetonitrile molecule is added to either PhOH–ACN-1 or PhOH–ACN-2. The two possible structures of the acetonitrile dimer are a cyclic one whereas the other is built in a 'head-to-tail' manner[763–766]. The latter ACN dimer structure seems not to be very important (it plays a role beyond the second solvation shell) and nearly twice as weak as the cyclic dimer[763, 765, 766]. This is why we confine the present study to the cyclic dimer. Its optimized structure given in Figure 54 looks similar to that in Figure 2 of Reference 765 and in Figure 7 of Reference 763. Its binding energy is 17.1 kJ mol^{-1} and 14.1 kJ mol^{-1} after ZPVE corrections, and agrees satisfactorily with the MP2/cc-pVDZ and MP2/6-311+G(d) values[763, 765].

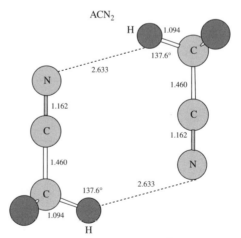

FIGURE 54. The acetonitrile dimer. Bond lengths are in Å. Adapted from Reference 761 with permission

The cyclic ACN dimer is formed thanks to two weak C—H···N hydrogen bonds characterized by N···H bond lengths of 2.633 Å and a bond angle of 137.6°. They are manifested spectroscopically by the appearance of two far-IR bands v_σ^{sym}(C—H···N) and v_σ^{asym}(C—H···N) at 87 and 89 cm^{-1}, respectively. Two CN stretching vibrations are also organized into the symmetric and asymmetric bands placed very close to each other, at 2358.1 and 2358.9 cm^{-1}. Consequently, we conclude that the formation of the cyclic ACN dimer leads to a red shift of v(C≡N) of the free acetonitrile molecule by 5–6 cm^{-1}.

Let us now consider the lower-energy stable structures, PhOH–ACN$_2$-1 and PhOH–ACN$_2$-2, of phenol with two acetonitrile molecules. Both are displayed in Figure 55 and, when supplied by the optimized geometrical parameters, PhOH–ACN$_2$-1 possesses a partially dimerized acetonitrile moiety (see Figure 54). The former appears to be the most stable at OH with a binding energy $E_{HB}^{(1)}$ (PhOH–ACN$_2$) = 44.6 kJ mol^{-1} compared to the latter whose binding energy is only 30.8 kJ mol^{-1}. Increasing the temperature reverses their order due to an entropy effect, because the entropy of PhOH–ACN$_2$-2 exceeds that of PhOH–ACN$_2$-1 by 85.8 J mol^{-1}. When $T > 204$ K, the temperature at which their enthalpy difference is precisely cancelled by their entropy difference, complex PhOH–ACN$_2$-2 becomes more favourable and, at room temperature, the free-energy difference between the former and latter complexes comprises

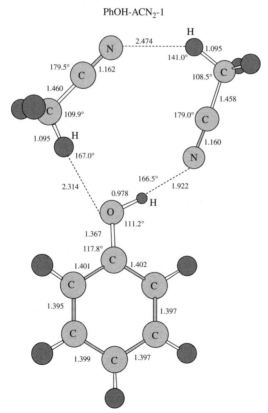

FIGURE 55. Complexes of phenol with two acetonitrile molecules. Bond lengths are in Å. Adapted from Reference 761 with permission

PhOH-ACN$_2$-2

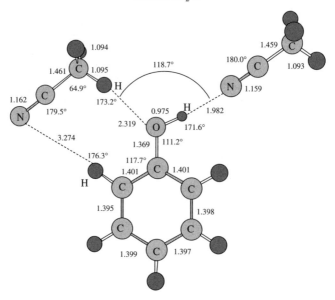

FIGURE 55. (*continued*)

8.1 kJ mol^{-1}. Another effect conferring a higher stability on the complex PhOH–ACN$_2$-2 mostly plays a role in polar solvents such as acetonitrile, since this complex has a huge total dipole moment of 10.89 D, 5.5-fold larger than for PhOH–ACN$_2$-1 (their polarizabilities and quadrupole moments are nearly the same, as shown in Table 41). After clearing up the role which the complex PhOH–ACN$_2$-2 might play in modelling an experimental setup with increasing concentration of the acetonitrile, let us consider whether it looks somewhat peculiar in comparison to the other complex of phenol with two acetonitrile molecules. Surprisingly, it has precisely what we are looking for. It follows from Table 41 that the ν(OH) stretch of phenol shifts further by 173 cm^{-1} towards lower wavenumbers compared with the free phenol and by -15 cm^{-1} compared to its frequency in PhOH–ACN-1. This is in line with a stepwise effect of dilution on the shift noted in the Introduction. What would also be interesting and deserves experimental verification is that the same stretch mode in PhOH–ACN$_2$-1 is red-shifting more strongly, by 244 cm^{-1} compared to that in PhOH and by 86 cm^{-1} compared to PhOH–ACN-1. Both red shifts could be ascribed to a somewhat stronger C–H$\cdots$O bond formed between the methyl group of acetonitrile and the lone-pair of the phenolic oxygen in PhOH–ACN$_2$-1 than in PhOH–ACN$_2$-2. This effect weakens more the O–H bond in PhOH–ACN$_2$-1 which participates in the other hydrogen bonding, and it is seen in Figure 55 that the O–H bond in PhOH–ACN$_2$-1 is longer by 0.003 Å than that in PhOH–AC$_2$-2. However, why has such a tremendous shift not yet been detected experimentally? We think that the reason is that the complex PhOH–ACN$_2$-1 is not favourable at room temperatures and in polar solvents, and therefore an increase in the acetonitrile concentration primarily leads to the formation of the complex PhOH–ACN$_2$-2. Our suggestion can readily be verified by determining the location of the ν(CN) bands in both complexes. As mentioned above, such mode shifts by 12 cm^{-1} to higher frequencies in the complex PhOH–ACN-1 is in perfect agreement with the experimental shift of 12.5 cm^{-1}^{553}. A similar shift of 13 cm^{-1} is predicted in the complex PhOH–ACN$_2$-2, where it appears at the lower-frequency wing with the red shift of 7 cm^{-1}, mimicking that found in the complex PhOH–ACN-2. On the contrary, in complex PhOH–ACN$_2$-1, the higher frequency band is placed by only 5 cm^{-1} aside that in the free acetonitrile molecule. Apparently, the other characteristic frequencies gathered in Table 41 might be of use to differentiate both complexes of phenol with two acetonitrile molecules.

4. A rather concise discussion

We have found the novel structure by which phenol complexes with the acetonitrile molecule. Such a structure has an absolutely different hydrogen bonding pattern, which certainly makes it less favourable on comparison with the conventional one attributed to the σ-type hydrogen bonding. A phenol–acetonitrile complex formation via π hydrogen bonding between the OH group of phenol and the C≡N bond should be ruled out affirmatively.

However, we have shown that the novel bond formation between phenol and acetonitrile plays a role on increasing the concentration of the acetonitrile. By postulating its existence under conditions in which phenol interacts with two acetonitrile molecules, we were able to explain the experimental data that have seemed to be rather unclear during the last four decades. Moreover, we have predicted the existence of another structure formed from phenol and two molecules of acetonitrile, which is characterized by a significant downshift by 244 cm^{-1} of the ν(OH) stretching mode of phenol, never observed experimentally in phenol–acetonitrile complexes. We have suggested that it is likely to exist in the gas phase and non-polar solvents at lower temperatures and showed its 'fingerprints' in order to facilitate its possible experimental detection.

D. Phenol–Benzonitrile Hydrogen-bonded Complex

The complex between phenol and benzonitrile is another, structurally speaking, rather complicated representative of the class of phenol–nitrile systems which are always associated by means of the π-electrons of the CN triple bond[732]. Note that the IR spectra of a variety of phenol–nitrile systems have been reported[767]. Experiments on the vibrational relaxation of benzonitrile in solutions were also studied by different groups[759, 768, 769].

In Figure 56, we display the B3LYP/6-31+G(d,p) structure of the phenol–benzonitrile associate. It undoubtedly shows that its formation is due to a σ-type bonding between the triple bond of benzonitrile and the OH group of phenol. The energy of formation of the bond is 22.8 kJ mol^{-1} after ZPVE corrections. Noteworthy are the vibrational features of

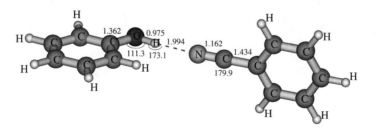

FIGURE 56. The complex of phenol with benzonitrile. Bond lengths are in Å, bond angles in deg

the studied complex. First, the ν_{CN} stretch undergoes a blue shift by 10 cm^{-1} whereas the ν_{OH} stretch of phenol is downshifted by 162 cm^{-1}. Second, the torsional mode τ_{OH} of phenol nearly doubles its frequency: 330 vs. 648 cm^{-1}.

E. A Very Short O−H · · · N Hydrogen Bond

Recently, neutron diffraction experiments[770] have demonstrated the existence of a very short O−H · · · N hydrogen bond in the crystalline adduct of 2-methylpyridine and pentachlorophenol which is discussed in Subsection 4.5: the O−H bond length is equal to 1.068(7) Å, the H · · · N bond length to 1.535(7) Å.

Figure 57 shows the complex of 2-methylpyridine and pentachlorophenol obtained at the B3LYP/6-31G(d) computational level. It is formed due to the O−H · · · N hydrogen bond whose O−H bond length is 1.004 Å, the H · · · N bond length is 1.795 Å and the ∠O−H · · · N bond angle is 153.0(8)°. We also note that these two molecules in the formed complex are twisted with respect to each other by an angle of 63.3°, which resembles the experimental structure shown in Figure 1 of Reference 770. It is clear that the discrepancy between the geometry of the O−H · · · N hydrogen bond in the studied complex and in the calculation is due to the difference between the gas phase and the crystal phase.

VI. OPEN THEORETICAL PROBLEMS

In spite of the great effort made in the last several decades, a large number of problems concerning the chemistry of phenols remain open wide for theoretical studies.

The significance of the reaction of phenol with hydrogen has a number of important facets. First, the selective hydrogenation of phenol yields cyclohexanone, which is a key raw material in the production of both caprolactam for nylon 6 and adipic acid for nylon 6[771]. Second, due to the fact that phenol is an environmental toxin[772] and phenolic waste has a variety of origins from industrial sources including oil refineries, petrochemical units, polymeric resin manufacturing and plastic units[773], catalytic hydrogenation of phenol is nowadays the best practicable environmental option[774].

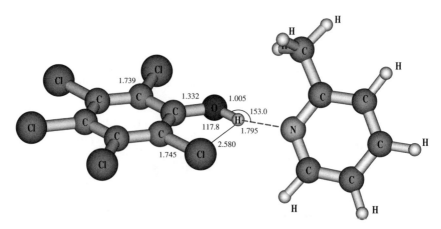

FIGURE 57. The complex of pentachlorophenol with 2-methylpyridine optimized at the B3LYP/6-31G(d) computational level. Bond lengths are in Å, bond angles in deg

1. General and theoretical aspects of phenols

The behaviour of the tyrosyl radicals involved in different processes and environments is not yet well understood[491, 546, 775]. Relatively little is known about the structure and selectivity of aryloxylium cations (Ar−O$^+$) that are produced in the phenolic oxidation reactions and implicated in biological processes such as isoflavone synthesis[776]. The thermochemistry[197] which is relevant to the antioxidant properties of phenols as well as the solvent effects on their reactivity[777-780] remain also a largely under-explored topic. Finally, the structure of phenol dimers and oligomers[781] or even of some specific phenols[782] also deserve more attention. We expect that these problems will be subjects for theoretical research in the coming years.

VII. ACKNOWLEDGEMENTS

The authors gratefully thank Therese Zeegers-Huyskens, Asit Chandra, Sergei Bureiko, Kiran Boggavarapu, Alexander Koll, Zdislaw Latajka, Noj Malcolm, Bernard Silvi, Lucjan Sobczyk, Raman Sumathy and Georg Zundel for warm and useful discussions and suggestions. We also thank Oksana Tishchenko, Le Thanh Hung, Alexei Arbuznikov, Nguyen Thanh Loc, Alk Dransfeld, Robert Flammang, Pascal Gerbaux, Pham-Tran Nguyen-Nguyen, Guy Bouchoux, Pham-Cam Nam and Nguyen Thi Minh Hue for their great help in the preparation of this chapter at all its stages.

We are indebted to the KU Leuven Research Council for financial support through the Program for Concerted Research Action (GOA). E. S. Kryachko also thanks the "Bijzonder Onderzoeksfunds" of the Limburgs Universitair Centrum.

VIII. REFERENCES AND NOTES

1. *Quantum Chemistry Library Data Base* (QCLDB), *Japan Association for International Chemical Information*, Tokyo, Japan, 2001.
2. M. J. Frisch, G. W. Trucks, H. B. Schlegel, G. E. Scuseria, M. Robb, J. R. Cheeseman, V. G. Zakrzewski, J. A. Montgomery, Jr., R. E. Stratmann, J. C. Burant, S. Dapprich, J. M. Millam, A. D. Daniels, K. N. Kudin, M. C. Strain, O. Farkas, J. Tomasi, V. Barone, M. Cossi, R. Cammi, B. Mennucci, C. Pomelli, C. Adamo, S. Clifford, J. Ochterski, G. A. Petersson, P. Y. Ayala, Q. Cui, K. Morokuma, D. K. Malick, A. D. Rabuck, K. Raghavachari, J. B. Foresman, J. Cioslowski, J. V. Ortiz, B. B. Stefanov, G. Liu, A. Liashenko, P. Piskorz, I. Komaromi, R. Gomperts, R. L. Martin, D. Fox, T. Keith, M. A. Al-Laham, C. Y. Peng, A. Nanayakkara, C. Gonzalez, M. Challacombe, P. M. W. Gill, B. Johnson, W. Chen, M. W. Wong, J. L. Andres, C. Gonzalez, M. Head-Gordon, E. S. Replogle and J. A. Pople, *GAUSSIAN 98* (Revision A.5), Gaussian Inc., Pittsburgh, PA, 1998.
3. J. J. P. Stewart, *MOPAC-7*, Quantum Chemistry Program Exchange, Bloomington, Ind., 1993.
4. (a) A. E. Reed, L. A. Curtiss and F. Weinhold, *Chem. Rev.*, **88**, 899 (1988).
   (b) F. Weinhold, in *Encyclopedia of Computational Chemistry* (Ed. P. v. R. Schleyer), Wiley, Chichester, 1998, pp. 1792–1811.
5. E. D. Glendening, J. K. Badenhoop, A. E. Reed, J. E. Carpenter and F. Weinhold, *NBO 4.0*, Theoretical Chemistry Institute, University of Wisconsin, Madison, 1996.
6. G. Fogarasi and P. Pulay, in *Vibrational Spectra and Structure*, Vol. 13 (Ed. J. R. Durig), Elsevier, New York, 1985.
7. S. Califano, *Vibrational States*, Wiley, New York, 1985.
8. J. M. L. Martin and K. Van Alsenoy, *GAR2PED*, University of Antwerp, Belgium, 1995.
9. S. Patai (Ed.), *The Chemistry of the Carbonyl Group*, Wiley, London, 1966.
10. E. P. Hunter and S. G. Lias, *J. Phys. Chem. Ref. Data*, **27**, 413 (1998).
11. S. A. Kafari and El-S. R. H. El-Gharkawy, *J. Phys. Chem. A*, **102**, 3202 (1998). See also S. A. Kafari, *J. Phys. Chem. A*, **102**, 10404 (1998).
12. D. Sitkoff, K. A. Sharp and B. Honig, *J. Phys. Chem.*, **98**, 1978 (1994).
13. F. J. Luque and M. Orozco, *J. Phys. Chem. B*, **101**, 5573 (1997).
14. B. Jayaram, D. Sprous and D. L. Beveridge, *J. Phys. Chem. B*, **102**, 9571 (1998).

15. J. Flórián and A. Warshel, *J. Phys. Chem. B*, **101**, 5583 (1997).
16. G. Schüürmann, *Quantum Struct.-Act. Relat.*, **15**, 121 (1996).
17. J. B. Cumming and P. Kebarle, *Can. J. Chem.*, **56**, 1 (1978).
18. J. E. Bartmess, *NIST Negative Ion Database, NIST Standard Reference Database 19B, Version 3.01*, National Institute of Standards and Technology, Washington, DC, 1991.
19. M. Fujio, R. T. McIver, Jr. and R. W. Taft, *J. Am. Chem. Soc.*, **103**, 4017 (1981).
20. V. F. DeTuri and K. M. Ervin, *Int. J. Mass Spectrom. Ion. Proc.*, **175**, 123 (1998).
21. A. J. Colussi, F. Zabel and S. W. Benson, *Int. J. Chem. Kinet.*, **9**, 161 (1977).
22. D. J. DeFrees, R. T. McIver, Jr. and W. J. Hehre, *J. Am. Chem. Soc.*, **102**, 3334 (1980).
23. K. S. Peters, *Pure Appl. Chem.*, **58**, 1263 (1986).
24. F. G. Bordwell, J.-P. Cheng and J. A. Harrelson, Jr., *J. Am. Chem. Soc.*, **110**, 1229 (1988)
25. J. A. Walker and W. Tsang, *J. Phys. Chem.*, **94**, 3324 (1990).
26. See, for example, M. Kunert, E. Dinjus, M. Nauck and J. Sieler, *Chem. Ber. Recl.*, **130**, 1461 (1997) and references therein
27. P. Ballone, B. Montanari and R. O. Jones, *J. Phys. Chem. A*, **104**, 2793 (2000).
28. F. F. Runge, *Ann. Phys. Chem.*, **31**, 65 (1834).
29. F. F. Runge, *Ann. Phys. Chem.*, **32**, 308 (1834).
30. C. F. v. Reichenbach, *Ann. Phys. Chem.*, **32**, 497 (1834).
31. G. Bugge, *Das Buch der Grossen Chemiker*, Verlag-Chemie, Weinheim, 1929.
32. J. R. Partington, *A History of Chemistry*, Vol. 4, The MacMillan Press, New York, 1972.
33. A. Laurent, *Ann. Chim. Phys.*, **63**, 27 (1836).
34. A. Laurent, *Ann. Chim.*, **iii**, 195 (1841).
35. A. Laurent, *Ann. Chim. Phys.*, **72**, 383 (1839).
36. Letter of Berzélius to Pelouze, *Ann. Chim. Phys.*, **67**, 303 (1838).
37. Letter of Berzélius to Pelouze, *Compt. Rend. Acad. Sci.*, **vi**, 629 (1838).
38. Dumas, *Compt. Rend. Acad. Sci.*, **vi**, 645 (1838).
39. Ch. Gerhardt, *Ann. Chim.*, **vii**, 215 (1843).
40. T. Lowry, *Dynamic Isomerism*, Report of the Committee, consisting of Professor H. E. Armstrong (Chairman), Dr. T. M. Lowry (Secretary), Professor Sydney Young, Dr. C. H. Desch, Dr. J. J. Dobbie, Dr. M. O. Forster and Dr. A. Lapworth, *BAAS Rep. Winnipeg* (1909), pp. 135–143 (1910). Cited in Reference 41, p. 176.
41. M. J. Nye, *From Chemical Philosophy to Theoretical Chemistry*, University of California Press, Berkeley, 1993.
42. C. Heitner and J. C. Scaiano (Eds.), *Photochemistry of Lignocellulosic Materials*, ACS Symposium Series, Vol. 531, American Chemical Society, Washington DC, 1993.
43. G. J. Leary, *J. Pulp Paper Sci.*, **20**, J154 (1994).
44. J. Emsley, *The Consumer's Good Chemical Guide. A Jargon-Free Guide to the Chemists of Everyday Life*, W. H. Freeman, Oxford, 1994.
45. J. R. Vane and R. M. Bottling (Eds.), *Aspirin and Other Salicylates*, Chapman and Hall, London, 1992.
46. See also Bayer's Aspirin web site at URL http://www.aspirin.de; see Hoffmann's laboratory record 10.10.1857 (source: Bayer AG).
47. *Aspirin, Molecule of the Month*, February 1996. Featured as the second molecule at MOTM site at URL http://www.bris.ac.uk/Depts/Chemistry/MOTM/aspirin/aspirin.htm.
48. America's 80 Billion-Aspirin Habit, *Medical Sciences Bulletin*, The Internet-Enhanced Journal of Pharmacology and Therapeutics, http://pharminfo.com/pubs/msb/aspirin.html.
49. Home page of the '*Aspirin Foundation of America*' at URL http://www.aspirin.org/.
50. J. Vane, *Nature-New Biol*, **231**, 232 (1971); *Nature*, **367**, 215 (1994); *Annu. Rev. Pharmacol.*, **38**, 97 (1998).
51. D. Picot, P. J. Loll and R. M. Garavito, *Nature*, **367**, 243 (1994); P. J. Loll, D. Picot and R. M. Garavito, *Nature Struct. Biol.*, **2**, 637 (1995).
52. P. J. Loll and R. M. Garavito, *Opin. Invest. Drugs*, **3**, 1171 (1994).
53. P. J. Wheatley, *J. Chem. Soc.*, 6036 (1964).
54. Y. Kim, K. Machida, T. Taga and K. Osaki, *Chem. Pharm. Bull.*, **33**, 2641 (1985).
55. *Cambridge Structural Database (CSD), Cambridge Crystallographic Data Center*, 12 Union Road, Cambridge, CB2 1EZ, UK, 1995. Its URL http://www.ccdc.cam.ac.uk/index.html.
56. R. Glasser, *J. Org. Chem.*, **66**, 771 (2001).
57. M. W. Miller, *The Pfizer Handbook of Microbial Metabolites*, McGraw-Hill, New York, 1961.

1. General and theoretical aspects of phenols 181

58. D. A. Whiting, in *Comprehensive Organic Chemistry. The Synthesis and Reactions of Organic Compounds. Vol. 1. Stereochemistry, Hydrocarbons, Halo Compounds, Oxygen Compounds* (Ed. J. F. Stoddart), Chap. 4.2, Pergamon Press, Oxford, 1979.
59. C. Hogue, *Chem. Eng. News*, May 1, 49 (2000).
60. (a) *Nature*, **410**, 1016 (2001).
60. (b) For a recent report on the use of Agent Orange in Vietnam, see: J. M. Stellman, S. D. Stellman, R. Christian, T. Weber and C. Tomasallo in *Nature*, **422**, 681 (2003).
61. W. Karrer, *Konstitution und Vorkommen der organischen Pflanzenstoffe*, Birkhäuser, Basel, 1958.
62. T. K. Devon and A. I. Scott, *Handbook of Naturally Occurring Compounds*, Academic Press, New York, 1975.
63. R. W. Rickards, in *Chemistry of Natural Phenolic Compounds* (Ed. W. D. Ollis), Pergamon Press, Oxford, 1961, pp. 1ff.
64. A. C. Neish, in *Biochemistry of Phenolic Compounds* (Ed. J. B. Harborne), Academic Press, London, 1964, pp. 295ff.
65. A. J. Birch, in *Progress in the Chemistry of Organic Natural Products* (Ed. L. Zechmeister), Vol. 14, Springer, Vienna, 1957. p. 186.
66. E. Halsam, *The Shikimate Pathway*, Butterworths, London, 1974.
67. S. K. Erikson, J. Schadelin, U. Schmeling, H. H. Schott, V. Ullrich and H. Staudinger, in *The Chemistry of the Hydroxyl Group, Part 2*, (Ed. S. Patai), Wiley-Interscience, New York, 1971. p. 776.
68. G. J. Kasperek and T. C. Bruice, *J. Am. Chem. Soc.*, **94**, 198 (1972).
69. G. J. Kasperek, T. C. Bruice, H. Yagi, N. Kausbisch and D. M. Jerina, *J. Am. Chem. Soc.*, **94**, 7876 (1972).
70. K. B. Sharpless and T. C. Flood, *J. Am. Chem. Soc.*, **93**, 2316 (1971).
71. B. A. Diner, D. A. Force, D. W. Randall and R. D. Britt, *Biochemistry*, **37**, 17931 (1998).
72. G. Xiao, A. L. Tsai, G. Palmer, W. C. Boyar, P. J. Marshall and R. J. Kulmacz, *Biochemistry*, **36**, 1836 (1997).
73. J. D. Spikes, H. R. Shen, P. Kopeckova and J. Kopecek, *Photochem. Photobiol.*, **70**, 130 (1999).
74. L. R. Mahoney, *Angew. Chem., Int. Ed. Engl.*, **8**, 547 (1969).
75. K. U. Ingold, *Spec. Publ. Chem. Soc.*, **24**, 285 (1970).
76. G. W. Burton and K. U. Ingold, *Acc. Chem. Res.*, **19**, 194 (1986). For a current review on the anaerobic degradation of phenolic compounds, see: B. Schink, B. Phillip and J. Müller, *Naturwissenschaften*, **87**, 12 (2000).
77. E. T. Denisov and I. V. Khudyakov, *Chem. Rev.*, **87**, 1313 (1987).
78. O. Potterat, *Curr. Org. Chem.*, **1**, 415 (1997).
79. J. R. Thomas, *J. Am. Chem. Soc.*, **85**, 2166 (1963).
80. H. H. Szmant, *Organic Building Blocks of the Chemical Industry*, Wiley, New York, 1989, p. 434.
81. D. Hodgson, H.-Y. Zhang, M. R. Nimlos and J. T. McKinnon, *J. Phys. Chem. A*, **105**, 4316 (2001).
82. R. A. Marcus, *J. Phys. Chem.*, **43**, 2658 (1965).
83. D. Bittner and J. B. Howard, in *Soot in Combustion Systems and Its Toxic Properties* (Eds. J. Lahaye and G. Prado), Plenum Press, New York, 1983. See also:
(a) K. H. Becker, I Barnes, L. Ruppert and P. Wiesen, in *Free Radicals in Biology and Environment* (Ed. F. Minisci), Chap. 27, Kluwer, Dordrecht, 1997, pp. 365–385.
(b) G. Ghigo and G. Tonachini, *J. Am. Chem. Soc.*, **120**, 6753 (1998).
84. G. I. Panov, V. I. Sobolev and A. S. Kharitonov, *J. Mol. Catal.*, **61**, 85 (1990).
85. A. M. Volodin, V. A. Bolshov and G. I. Panov, *J. Phys. Chem.*, **98**, 7548 (1994).
86. K. Yoshizawa, Y. Shiota, T. Yumura and T. Yamabe, *J. Phys. Chem. B*, **104**, 734 (2000)
87. *Chem. Eng. News*, April 6, 21 (1998). See also the recent experimental work on the Fe-silicalite catalyst for the N_2O oxidation of benzene to phenol: R. Leanza, I. Rossetti, I. Mazzola and L. Forni, *Appl. Catal. A—Gen.*, **205**, 93 (2001).
88. *Ullmann's Encyclopedia of Industrial Chemistry*, 6th edn., Wiley, New York, 2001.
89. R. K. Eikhman, M. M. Shemyakin and V. N. Vozhdaeva, *Anilinokrasochnaya Promyshl.*, **4**, 523 (1934).
90. V. Tishchenko and A. M. Churbakov, *J. Appl. Chem. (USSR)*, **7**, 764 (1934).

91. N. N. Vorozhtzov, Jr. and A. G. Oshuev, *Anilinokrasochnaya Promyshl.*, **3**, 245 (1933).
92. J.-F. Tremblay, *Chem. Eng. News*, December 11, 31 (2000). See also *Chem. Eng. News*, June 25, 78 (2001).
93. H. Hock and S. Lang, *Ber. Dtsch. Chem. Ges. B*, **77**, 257 (1944).
94. J. Wallace, in *Kirk-Othmer Encyclopedia of Chemical Technology*, Vol. 18, (Eds. J. I. Kroschwitz and M. Howe-Grant), 4th edn., Wiley, New York, 1992, pp. 592–602.
95. H.-G. Franck and J. W. Stadelhofer, *Industrial Aromatic Chemistry*, Springer, New York, 1987, pp. 146–183.
96. W. W. Kaeding, R. O. Lindblom and R. G. Temple, *Ind. Eng. Chem.*, **53**, 805 (1961).
97. W. W. Kaeding, *Hydrocarbon Process.*, **43**, 173 (1964).
98. W. W. Kaeding, R. O. Lindblom, R. G. Temple and H. I. Mahon, *Ind. Eng. Chem., Process Des. Dev.*, **4**, 97 (1965).
99. M. Taverna, *Riv. Combust.*, **22**, 203 (1968).
100. I. Donati, G. S. Sioli and M. Taverna, *Chim. Ind. (Milan)*, **50**, 997 (1968).
101. J. M. Gibson, P. S. Thomas, J. D. Thomas, J. L. Barker, S. S. Chandran, M. K. Harrup, K. M. Draths and J. W. Frost, *Angew. Chem., Int. Ed.*, **40**, 1945 (2001).
102. K. M. Draths and J. W. Frost, *J. Am. Chem. Soc.*, **116**, 399 (1994); *ACS Symp. Ser.*, **577**, 32 (1994); *J. Am. Chem. Soc.*, **117**, 2395 (1995); K. Li and J. W. Frost, *J. Am. Chem. Soc.*, **120**, 10545 (1998).
103. R. M. Badger and L. R. Zumwalt, *J. Chem. Phys.*, **6**, 711 (1938).
104. L. Pauling, *J. Am. Chem. Soc.*, **58**, 94 (1936).
105. T. Kojima, *J. Phys. Soc. Jpn.*, **15**, 284 (1960).
106. H. Forest and B. P. Dailey, *J. Chem. Phys.*, **45**, 1736 (1966).
107. T. Pedersen, N. W. Larsen and L. Nygaard, *J. Mol. Struct.*, **4**, 59 (1969).
108. N. W. Larsen, *J. Mol. Struct.*, **51**, 175 (1979).
109. G. Portalone, G. Schultz, A. Domenicano and I. Hargittai, *Chem. Phys. Lett.*, **197**, 482 (1992).
110. J. A. Pople and M. Gordon, *J. Am. Chem. Soc.*, **89**, 4253 (1967).
111. G. Keresztury, F. Billes, M. Kubinyi and T. Sundius, *J. Phys. Chem. A*, **102**, 1371 (1998).
112. D. Michalska, D. C. Bienko, A. J. Abkowicz-Bienko and Z. Latajka, *J. Phys. Chem.*, **100**, 17786 (1996).
113. S. Re and Y. Osamura, *J. Phys. Chem. A*, **102**, 3798 (1998).
114. S. Schumm, M. Gerhards, W. Roth, H. Gier and K. Kleinermanns, *Chem. Phys. Lett.*, **263**, 126 (1996).
115. G. Granucci, J. T. Hynes, P. Millié and T.-H. Tran-Thi, *J. Am. Chem. Soc.*, **122**, 12242 (2000).
116. J. S. Wright, D. J. Carpenter, D. J. McKay and K. U. Ingold, *J. Am. Chem. Soc.*, **119**, 4245 (1997).
117. B. J. C. C. Cabrol, R. G. B. Fonesca and J. A. Martinho Simões, *Chem. Phys. Lett.*, **258**, 436 (1996).
118. S. J. Martinez III, J. C. Alfano and D. H. Levy, *J. Mol. Spectrosc.*, **152**, 80 (1992).
119. J. Lorentzon, P.-A. Malmqvist, M. Fülscher and B. O. Roos, *Theor. Chim. Acta*, **91**, 91 (1995).
120. S. Tsuzuki, H. Houjou, Y. Nagawa and K. Hiratani, *J. Phys. Chem. A*, **104**, 1332 (2000).
121. D. Feller and M. W. Feyereisen, *J. Comput. Chem.*, **14**, 1027 (1993).
122. M. Schütz, T. Bürgi and S. Leutwyler, *J. Mol. Struct. (THEOCHEM)*, **276**, 117 (1992).
123. K. Kim and K. D. Jordan, *Chem. Phys. Lett.*, **218**, 261 (1994).
124. H. Lampert, W. Mikenda and A. Karpfen, *J. Phys. Chem. A*, **101**, 2254 (1997).
125. Y.-D. Wu and D. K. W. Lai, *J. Org. Chem.*, **61**, 7904 (1996).
126. S. Schumm, M. Gerhards and K. Kleinermanns, *J. Phys. Chem. A*, **104**, 10648 (2000).
127. G. Berden, W. L. Meerts, M. Schmitt and K. Kleinermanns, *J. Chem. Phys.*, **104**, 972 (1996).
128. A. L. McClellan, *Tables of Experimental Dipole Moments*, W. H. Freeman, San Francisco, 1963, p. 251.
129. A. Koll, H. Ratajczak and L. Sobczyk, *Ann. Soc. Chim. Polonorum*, **44**, 825 (1970).
130. C. J. Cramer and D. G. Truhlar, *Chem. Rev.*, **99**, 2161 (1999).
131. B. Pullman and A. Pullman, *Quantum Biochemistry*, Wiley-Interscience, New York, 1963.
132. T. A. Koopmans, *Physica*, **1**, 104 (1933).
133. R. S. Mulliken, *Phys. Rev.*, **74**, 736 (1948).
134. R. J. Lipert and S. D. Colson, *J. Chem. Phys.*, **92**, 3240 (1990).
135. K. Müller-Dethlefs and E. W. Schlag, *Annu. Rev. Phys. Chem.*, **42**, 109 (1991).

136. O. Dopfer, G. Lembach, T. G. Wright and K. Müller-Dethlefs, *J. Chem. Phys.*, **98**, 1933 (1993).
137. R. G. Neuhauser, K. Siglow and H. J. Neusser, *J. Chem. Phys.*, **106**, 896 (1997).
138. Y. Qin and R. A. Wheeler, *J. Phys. Chem.*, **100**, 10554 (1996).
139. http://webbook.nist.gov/cgi/cbook.cgi?ID=C108952&Units=SI&Mask=20.
140. J. S. Wright, E. R. Johnson and G. A. DiLabio, *J. Am. Chem. Soc.*, **123**, 1173 (2001).
141. R. F. W. Bader, *Atoms in Molecules. A Quantum Theory*, Clarendon, Oxford, 1990.
142. P. L. A. Popelier, *Atoms in Molecules. An Introduction*, Pearson Education, Harlow (1999).
143. R. J. Gillespie and E. A. Robinson, *Angew. Chem., Int. Ed. Engl.*, **35**, 495 (1996).
144. R. J. Gillespie and I. Hargittai, *The VSEPR Model of Molecular Geometry*, Allyn and Bacon, London, 1991.
145. E. S. Kryachko and E. V. Ludeña, *Energy Density Functional Theory of Many-Electron Systems*, Kluwer, Dordrecht, 1990.
146. N. Malcolm, Private communication, May–June, 2001. All calculations were performed using the program MORPHY 98, a topological analysis program written by P. L. A. Popelier with a contribution from R. G. A. Bone (UMIST, Manchester, U.K.).
147. K. W. F. Kohlrausch and H. Wittek, *Monatsh. Chem.*, **74**, 1 (1941).
148. M. M. Davies, *J. Chem. Phys.*, **16**, 274 (1948).
149. M. M. Davies and R. L. Jones, *J. Chem. Soc.*, 120 (1954).
150. R. Mecke and G. Rossmy, *Z. Elektrochem.*, **59**, 866 (1955).
151. J. C. Evans, *Spectrochim. Acta*, **16**, 1382 (1960).
152. J. H. S. Green, *J. Chem. Soc.*, 2236 (1961).
153. H. D. Bist, J. C. D. Brand and D. R. Williams, *J. Mol. Spectrosc.*, **24**, 402, 413 (1967).
154. H. W. Wilson, R. W. MacNamee and J. R. Durig, *J. Raman Spectrosc.*, **11**, 252 (1981).
155. N. W. Larsen and F. M. Nicolaisen, *J. Mol. Struct.*, **22**, 29 (1974).
156. H. D. Bist, J. C. D. Brand and D. R. Williams, *J. Mol. Spectrosc.*, **21**, 76 (1966).
157. J. Christoffersen, J. M. Hollas and G. H. Kirby, *Proc. Roy. Soc. London A*, **307**, 97 (1968).
158. E. V. Brown, *Opt. Spectrosc.*, **23**, 301 (1967).
159. V. N. Sarin, M. M. Rai, H. D. Bist and D. P. Khandelwal, *Chem. Phys. Lett.*, **6**, 473 (1970).
160. H. W. Wilson, *Anal. Chem.*, **46**, 962 (1974).
161. P. V. Huong, J. Lascombe and M. C. Josien,, *J. Chem. Phys.*, **58**, 694 (1961).
162. E. Mathier, D. Welti, A. Bauder and Hs. H. Gunthard, *J. Mol. Spectrosc.*, **37**, 63 (1967).
163. See also different rotational studies in references 164–166.
164. G. V. Hartland, B. F. Henson, V. A. Venturo and P. M. Felker, *J. Phys. Chem.*, **96**, 1164 (1992).
165. J. H. S. Green, D. J. Harrison and W. Kynaston, *Spectrochim. Acta*, **27A**, 2199 (1971).
166. B. Nelander, *J. Chem. Phys.*, **72**, 771 (1980).
167. G. Varsanýi, *Assignments for Vibrational Spectra of 700 Benzene Derivatives*, Wiley, New York, 1974.
168. A. K. Grafton and R. A. Wheeler, *J. Comput. Chem.*, **19**, 1663 (1998).
169. W. Roth, P. Imhof, M. Gerhards, S. Schumm and K. Kleinermanns, *Chem. Phys.*, **252**, 247 (2000).
170. D. Michalska, W. Zierkiewicz, D. C. Bienko, W. Wojciechowski and Th. Zeegers-Huyskens, *J. Phys. Chem. A*, **105**, 8734 (2001).
171. G. R. Carlson and W. G. Fateley, *J. Phys. Chem.*, **77**, 1157 (1973).
172. W. G. Fateley, G. L. Carlson and F. F. Bentley, *J. Phys. Chem.*, **79**, 199 (1975).
173. G. Pimentel and A. L. McClellan, *The Hydrogen Bond*, Freeman, San Francisco, 1960.
174. O. R. Wulf and E. J. Jones, *J. Chem. Phys.*, **8**, 745 (1940) and references therein.
175. M. Schütz, T. Bürgi, S. Leutwyler and T. Fisher, *J. Chem. Phys.*, **98**, 3763 (1993).
176. D. H. Whiffen, *J. Chem. Soc.*, 1350 (1956).
177. G. Durocher and C. Sandorfy, *J. Mol. Spectrosc.*, **15**, 22 (1965).
178. M. Asselin, G. Bèlanger and C. Sandorfy, *J. Mol. Spectrosc.*, **30**, 96 (1969).
179. M. Couzi and P. V. Huong, *Spectrochim. Acta*, **26A**, 49 (1970).
180. J. Davison, J. H. Gutow and R. N. Zare, *J. Phys. Chem.*, **94**, 4069 (1990).
181. M. Rospenk, N. Leroux and Th. Zeegers-Huyskens, *J. Mol. Spectrosc.*, **183**, 245 (1997).
182. S. I. Ishiuchi, H. Shitomi, K. Takazawa and M. Fujii, *Chem. Phys. Lett.*, **283**, 243 (1998).
183. M. Rospenk, B. Czarnik-Matusewicz and Th. Zeegers-Huyskens, *Spectrochim. Acta A*, **57**, 185 (2001).

184. M. Capponi, I. Gut and J. Wirz, *Angew. Chem., Int. Ed. Engl.*, **25**, 344 (1986).
185. O. S. Tee and N. R. Inyengar, *J. Am. Chem. Soc.*, **107**, 455 (1985).
186. T. A. Gadosy and R. A. McClelland, *J. Mol. Struct. (THEOCHEM)*, **369**, 1 (1996).
187. D. Santoro and R. Louw, *J. Chem. Soc., Perkin Trans. 2*, 645 (2001).
188. K. U. Ingold and J. S. Wright, *J. Chem. Educ.*, **77**, 1062 (2000).
189. L. R. Mahoney, F. C. Ferris and M. A. DaRooge, *J. Am. Chem. Soc.*, **91**, 3883 (1969).
190. M. Lucarini, G. F. Pedulli and M. Cipollone, *J. Org. Chem.*, **59**, 5063 (1994).
191. D. D. M. Wayner, E. Lusztyk, K. U. Ingold and P. Mulder, *J. Org. Chem.*, **61**, 6430 (1996).
192. G. A. DiLabio, D. A. Pratt, A. D. LoFaro and J. S. Wright, *J. Phys. Chem. A*, **103**, 1653 (1999).
193. T. Brinck, M. Haeberlein and M. Jonsson, *J. Am. Chem. Soc.*, **119**, 4239 (1997).
194. R. M. B. dos Santos and J. A. Martinho Simões, *J. Phys. Chem. Ref. Data*, **27**, 707 (1998).
195. M. I. De Heer, H.-G. Korth and P. Mulder, *J. Org. Chem.*, **64**, 6969 (1999).
196. F. Himo, L. A. Eriksson, M. R. A. Blomberg and P. E. M. Siegbahn, *Int. J. Quantum Chem.*, **76**, 714 (2000).
197. L. J. J. Laarhoven, P. Mulder and D. D. M. Wayner, *Acc. Chem. Res.*, **32**, 342 (1999).
198. D. A. Pratt, G. A. DiLabio, G. Brigati, G. F. Pedulli and L. Valgimigli, *J. Am. Chem. Soc.*, **123**, 4625 (2001).
199. T. N. Das, *J. Phys. Chem. A*, **105**, 5954 (2001).
200. D. A. Pratt, M. I. de Heer, P. Mulder and K. U. Ingold, *J. Am. Chem. Soc.*, **123**, 5518 (2001).
201. M. Lucarini, G. F. Pedulli, L. Valgimigli, R. Amoraati and F. Minisci, *J. Org. Chem.*, **66**, 5456 (2001).
202. M. M. Davies, *Acid-Base Behaviour in Aprotic Organic Solvents*, Natl. Bur. Standards Monograph 105, Washington D.C., 1968.
203. S. Nagakura and H. Baba, *J. Am. Chem. Soc.*, **74**, 5693 (1952).
204. S. Nagakura, *J. Am. Chem. Soc.*, **76**, 3070 (1954).
205. H. Baba and S. Suzuki, *J. Chem. Phys.*, **41**, 895 (1964).
206. H. Baba, A. Matsuyama and H. Kokubun, *Spectrochim. Acta A*, **25**, 1709 (1969).
207. R. Scott, D. De Palma and S. Vinogradov, *J. Phys. Chem.*, **72**, 3192 (1968).
208. R. Scott and S. Vinogradov, *J. Phys. Chem.*, **73**, 1890 (1969).
209. R. A. Hudson, R. Scott and S. Vinogradov, *J. Phys. Chem.*, **76**, 1989 (1972).
210. Th. Zeegers-Huyskens and P. Huyskens, in *Molecular Interactions Vol. 2*, (Eds. H. Ratajczak and W. J. Orville-Thomas), Wiley, New York, 1981.
211. L. Sobczyk, *Ber. Bunsenges. Phys. Chem.*, **102**, 377 (1998).
212. L. Pauling, *The Nature of the Chemical Bond*, Cornell University Press, Ithaca, 1939.
213. O. R. Wulf and U. Liddel, *J. Am. Chem. Soc.*, **57**, 1464 (1935).
214. O. R. Wulf, U. Liddel and S. B. Hendricks, *J. Am. Chem. Soc.*, **58**, 2287 (1936).
215. G. E. Hilbert, O. R. Wulf, S. B. Hendricks and U. Liddel, *Nature*, **135**, 147 (1935).
216. O. R. Wulf and E. J. Jones, *J. Chem. Phys.*, **8**, 745 (1940).
217. O. R. Wulf, E. J. Jones and L. S. Deming, *J. Chem. Phys.*, **8**, 753 (1940).
218. G. Rossmy, W. Lüttke and R. Mecke, *J. Chem. Phys.*, **21**, 1606 (1953).
219. All computations were performed at the density functional hybrid B3LYP potential in conjunction with the split-valence 6-311++G(d,p) basis set using a GAUSSIAN 98 suit of packages. The tight convergence criterion was employed in all geometrical optimizations. Harmonic vibrational frequencies were kept unscaled. ZPVEs and thermodynamic quantities were also calculated at $T = 298.15$ K. Throughout the present section, the energy comparison was made in terms of the total energy + ZPVE.
220. B. Silvi, E. S. Kryachko, O. Tishchenko, F. Fuster and M. T. Nguyen, *Mol. Phys.*, **100**, 1659 (2002).
221. M. M. Davies, *Trans. Faraday Soc.*, **36**, 333 (1940).
222. E. A. Robinson, H. D. Schreiber and J. N. Spencer, *Spectrochim. Acta A*, **28**, 397 (1972).
223. A. W. Baker and W. W. Kaeding, *J. Am. Chem. Soc.*, **81**, 5904 (1959).
224. T. Lin and E. Fishman, *Spectrochim. Acta A*, **23**, 491 (1967).
225. See various chapters in Reference 173.
226. P. Schuster, G. Zundel and C. Sandorfy (Eds.), *The Hydrogen Bond, Recent Developments in Theory and Experiments*, North-Holland, Amsterdam, 1976.
227. A. W. Baker, *J. Am. Chem. Soc.*, **80**, 3598 (1958).
228. M. St. C. Flett, *Spectrochim. Acta*, **10**, 21 (1957).

1. General and theoretical aspects of phenols 185

229. H. Bourassa-Bataille, P. Sauvageau and C. Sandorfy, *Can. J. Chem.*, **41**, 2240 (1963).
230. C. Sandorfy, in *The Hydrogen Bond. Recent Developments in Theory and Experiments* (Eds. P. Schuster, G. Zundel and C. Sandorfy, Chap. 13, North-Holland, Amsterdam, 1976.
231. L. R. Zumwalt and R. M. Badger, *J. Am. Chem. Soc.*, **62**, 305 (1940).
232. M. M. Davies, *Trans. Faraday Soc.*, **34**, 1427 (1938).
233. E. A. Allan and L. W. Reeves, *J. Phys. Chem.*, **66**, 613 (1962).
234. E. A. Allan and L. W. Reeves, *J. Phys. Chem.*, **67**, 591 (1963).
235. D. A. K. Jones and J. G. Watkinson, *Chem. Ind.*, 661 (1960).
236. R. A. Nyquist, *Spectrochim. Acta*, **19**, 1655 (1963).
237. L. Radom, W. J. Hehre, J. A. Pople, G. L. Carlson and W. G. Fateley, *J. Chem. Soc., Chem. Commun.*, 308 (1972).
238. J. H. Richards and S. Walker, *Trans. Faraday Soc.*, **57**, 412 (1961).
239. S. W. Dietrich, E. C. Jorgensen, P. A. Kollman and S. Rothenberg, *J. Am. Chem. Soc.*, **98**, 8310 (1976).
240. H. H. Jaffe, *J. Am. Chem. Soc.*, **79**, 2373 (1957).
241. R. M. Badger and S. H. Bauer, *J. Chem. Phys.*, **5**, 839 (1937).
242. G. D. Hawkins, D. J. Giesen, G. C. Lynch, C. C. Chambers, I. Rossi, J. W. Storer, J. Li, P. Winget, D. Rinaldi, D. A. Liotard, C. J. Cramer and D. G. Truhlar, *AMSOL, Version 6.6*, University of Minnesota, Minneapolis, 1997, based in part on *AMPAC, Version 2.1* by D. A. Liotard, E. F. Healy, J. M. Ruiz and M. J. S. Dewar.
243. J. J. Urban, R. L. v. Tersch and G. R. Famini, *J. Org. Chem.*, **59**, 5239 (1994).
244. O. Tishchenko, E. S. Kryachko and M. T. Nguyen, *Spectrochim. Acta A*, **58**, 1951 (2002).
245. W. Zierkiewicz, D. Michalska and Th. Zeegers-Huyskens, *J. Phys. Chem. A*, **104**, 11685 (2000).
246. F. A. Miller, *Molecular Spectroscopy*, Institute of Petroleum, London, 1969, p. 5.
247. A. B. Hollinger and H. L. Welsh, *Can. J. Phys.*, **56**, 974 (1978).
248. A. B. Hollinger and H. L. Welsh, *Can. J. Phys.*, **56**, 1513 (1978).
249. A. B. Hollinger, H. L. Welsh and K. S. Jammu, *Can. J. Phys.*, **57**, 767 (1979).
250. H. B. Jensen and S. Brodersen, *J. Raman Spectrosc.*, **8**, 103 (1979).
251. R. J. Jakobsen and E. J. Brewer, *Appl. Spectrosc.*, **16**, 32 (1962).
252. M. H. Palmer, W. Moyes, M. Speirs and J. N. A. Ridyard, *J. Mol. Struct.*, **52**, 293 (1979).
253. A. Pross and L. Radom, *Prog. Phys. Org. Chem.*, **13**, 1 (1981).
254. E. L. Mehler and J. Gerhards, *J. Am. Chem. Soc.*, **107**, 5856 (1985).
255. A. D. Becke and K. E. Edgecombe, *J. Chem. Phys.*, **92**, 5397 (1990).
256. G. N. Lewis, *J. Am. Chem. Soc.*, **38**, 762 (1916).
257. G. N. Lewis, *Valence and the Structure of Atoms and Molecules*, Dover, New York, 1966.
258. R. J. Gillespie and R. S. Nyholm, *Quart. Rev. Chem. Soc.*, **11**, 339 (1957).
259. R. J. Gillespie, *Molecular Geometry*, Van Nostrand Reinhold, London, 1972.
260. R. Thom, *Stabilité Structurelle et Morphogénèse*, Interéditions, Paris, 1972.
261. R. H. Abraham and C. D. Shaw, *Dynamics: The Geometry of Behavior*, Addison Wesley, New York, 1992.
262. R. H. Abraham and J. E. Marsden, *Foundations of Mechanics*, Addison Wesley, New York, 1994.
263. A. Savin, O. Jepsen, J. Flad, O. K. Andersen, H. Preuss and H. G. v. Schnering, *Angew. Chem., Int. Ed. Engl.*, **31**, 187 (1992).
264. B. Silvi and A. Savin, *Nature*, **371**, 683 (1994).
265. A. Savin, B. Silvi and F. Colonna, *Can. J. Chem.*, **74**, 1088 (1996).
266. S. Noury, F. Colonna, A. Savin and B. Silvi, *J. Mol. Struct. (THEOCHEM)*, **450**, 59 (1998).
267. X. Fradera, M. A. Austen and R. F. W. Bader, *J. Phys. Chem. A*, **103**, 304 (1998).
268. J. Cioslowski and S. T. Mixon, *J. Am. Chem. Soc.*, **113**, 4142 (1991).
269. J. G. Ángyán, M. Loos and I. Mayer, *J. Phys. Chem.*, **98**, 5244 (1994).
270. J. F. Dobson, *J. Chem. Phys.*, **94**, 4328 (1991).
271. J. F. Dobson, *J. Chem. Phys.*, **98**, 8870 (1993).
272. H. L. Schmider and A. L. Becke, *J. Chem. Phys.*, **109**, 8188 (1998).
273. P. L. A. Popelier, *Coord. Chem. Rev.*, **197**, 169 (2000).
274. F. Fuster, A. Savin and B. Silvi, *J. Phys. Chem. A*, **104**, 852 (2000).
275. F. Fuster and B. Silvi, *Theoret. Chem. Acc.*, **104**, 13 (2000).
276. F. Fuster and B. Silvi, *Chem. Phys.*, **252**, 279 (2000).

277. S. G. Schulman, W. R. Vincent and J. M. Underberg, *J. Phys. Chem.*, **85**, 4068 (1981).
278. S. Leutwyler, T. Bürgi, M. Schütz and A. Taylor, *Faraday Discuss.*, **97**, 285 (1994).
279. R. A. Slavinskaya, Kh. Kh. Muldagaliev and T. A. Kovaleva, *Izv. Akad. Nauk. Kaz. SSR, Ser. Khim. Engl. Transl.*, 23 (1990).
280. W. Roth, P. Imhof and K. Kleinermanns, *Phys. Chem. Chem. Phys.*, **3**, 1806 (2001).
281. P. Imhof and K. Kleinermanns, *J. Phys. Chem. A*, **105**, 8922 (2001).
282. M. Broquier, F. Lahmani, A. Zehnacker-Rentien, V. Brenner, P. Millé and A. Peremans, *J. Phys. Chem. A*, **105**, 6841 (2001).
283. P. Polizer and N. Sukumar, *J. Mol. Struct. (THEOCHEM)*, **179**, 439 (1988).
284. K. B. Borisenko, C. W. Boch and I. J. Hargittai, *J. Phys. Chem.*, **98**, 1442 (1994).
285. P. C. Chen and C. C. Huang, *J. Mol. Struct. (THEOCHEM)*, **282**, 287 (1993). See also P. C. Chen and S. C. Chen, *Int. J. Quantum Chem.*, **83**, 332 (2001).
286. A. Kovacs, V. Izvekow, G. Keresztury and G. Pongor, *Chem. Phys.*, **238**, 231 (1998).
287. Y. Kishore, S. N. Sharma and C. P. D. Dwivedi, *Indian J. Phys.*, **48**, 412 (1974).
288. A. J. Abkowicz-Bieńko, Z. Latajka, D. C. Bieńko and D. Michalska, *Chem. Phys.*, **250**, 123 (1999).
289. B. Czarnik-Matusewicz, A. K. Chandra, M. T. Nguyen and Th. Zeegers-Huyskens, *J. Mol. Spectrosc.*, **195**, 308 (1999).
290. R. P. Bell, *The Proton in Chemistry*, Cornell University Press, Ithaca, 1959, p. 20.
291. C. Y. Ng, D. J. Trevor, P. W. Tiedemann, S. T. Ceyer, P. L. Kroebusch, B. H. Mahan and Y. T. Lee, *J. Chem. Phys.*, **67**, 4235 (1977).
292. D. Kuck, *Angew. Chem., Int. Ed.*, **39**, 125 (2000).
293. E. S. Kryachko and M. T. Nguyen, *J. Phys. Chem. A*, **105**, 153 (2001).
294. R. T. McIver, Jr. and R. C. Dunbar, *Int. J. Mass Spectrom. Ion Phys.*, **7**, 471 (1971).
295. Y. Lau and P. Kebarle, *J. Am. Chem. Soc.*, **98**, 7452 (1976).
296. (a) E. P. Hunter and S. G. Lias, *J. Phys. Chem. Ref. Data*, **27**, 413 (1998).
296. (b) For the most recent value, see G. Bouchoux, D. Defaye, T. McMahon, A. Likhohyot, O. Mó and M. Yáñez, *Chem. Eur. J.*, **8**, 2899 (2002).
297. D. J. DeFrees, R. T. McIver and W. J. Hehre, *J. Am. Chem. Soc.*, **99**, 3853 (1977).
298. M. Eckert-Maksić, M. Klessinger and Z. B. Maksić, *Chem. Phys. Lett.*, **232**, 472 (1995). For semiempirical calculations see R. Voets, J.-P. Francois, J. M. L. Martin, J. Mullers, J. Yperman and L. C. Van Poucke, *J. Comput. Chem.*, **11**, 269 (1990).
299. O. Tishchenko, N. N. Pham-Tran, E. S. Kryachko and M. T. Nguyen, *J. Phys. Chem. A*, **105**, 8709 (2001).
300. N. Solca and O. Dopfer, *Chem. Phys. Lett.*, **342**, 191 (2001).
301. R. K. Roy, F. De Proft and P. Geerlings, *J. Phys. Chem. A*, **102**, 7035 (1998).
302. N. Russo, T. Toscano, A. Grand and T. Mineva, *J. Phys. Chem. A*, **104**, 4017 (2000).
303. H. T. Le, R. Flammang, M. Barbieux-Flammang, P. Gerbaux and M. T. Nguyen, *Int. J. Mass Spectrom.*, **217**, 45 (2002).
304. G. E. Campagnaro and J. L. Wood, *J. Mol. Struct.*, **6**, 117 (1970).
305. N. W. Larsen and F. M. Nicolaisen, *J. Mol. Struct.*, **22**, 29 (1974).
306. B. Bogdanov, D. van Duijn and S. Ingemann, *Proceedings of the XX Congress on Mass Spectrometry*, Madrid, Spain, July 2000.
307. W. G. Fateley, G. L. Carlson and F. F. Bentley, *J. Phys. Chem.*, **79**, 199 (1975).
308. K. B. Wiberg and P. R. Rablen, *J. Org. Chem.*, **63**, 3722 (1998).
309. R. F. W. Bader and C. Chang, *J. Phys. Chem.*, **93**, 2946, 5095 (1989) and references therein.
310. A. Bagno and G. Scorrano, *J. Phys. Chem.*, **100**, 1536 (1996) and references therein.
311. D. Kovacek, Z. B. Maksić and I. Novak, *J. Phys. Chem. A*, **101**, 1147, 7448 (1997).
312. G. Raos, J. Gerratt, P. B. Karadakov, D. L. Cooper and M. Raimondi, *J. Chem. Soc., Faraday Trans.*, **91**, 4011 (1995).
313. M. T. Nguyen, A. F. Hegarty, T. K. Ha and G. R. De Mare, *J. Chem. Soc., Perkin Trans. 2*, 147 (1986).
314. T. K. Ha and M. T. Nguyen, *J. Mol. Struct.*, **87**, 355 (1982).
315. M. T. Nguyen and A. F. Hegarty, *J. Chem. Soc. Perkin. Trans. 2*, **2037**, 2043 (1984).
316. See the ELF analysis on protonation in Section III.C and Reference 220.
317. E. S. Kryachko and M. T. Nguyen, unpublished results.
318. R. T. Sanderson, *Polar Covalence*, Academic Press, New York, 1983.
319. A. Baeten, F. De Proft and P. Geerlings, *Int. J. Quantum Chem.*, **60**, 931 (1996).

320. R. G. Parr and W. Yang, *J. Am. Chem. Soc.*, **106**, 4049 (1984).
321. H. Chermette, *J. Comput. Chem.*, **20**, 129 (1999).
322. W. Yang and W. J. Mortier, *J. Am. Chem. Soc.*, **108**, 5708 (1986).
323. T. Mineva, N. Russo and E. Silicia, *J. Am. Chem. Soc*, **120**, 9093 (1998).
324. P. Geerlings, F. De Proft and W. Langenaeker, *Adv. Quantum Chem.*, **33**, 301 (1999).
325. T. Mineva, N. Neshev, N. Russo, E. Silicia and M. Toscano, *Adv. Quantum Chem.*, **33**, 273 (1999).
326. A. K. Chandra and M. T. Nguyen, *J. Chem. Soc., Perkin Trans. 2*, 1415 (1997).
327. G. Raspoet, M. T. Nguyen, S. Kelly and A. F. Hegarty, *J. Org. Chem.*, **63**, 9669 (1998).
328. M. T. Nguyen, G. Raspoet and L. G. Vanquickenborne, *J. Phys. Org. Chem.*, **13**, 46 (2000).
329. M. T. Nguyen and G. Raspoet, *Can. J. Chem.*, **77**, 817 (1999).
330. A. K. Chandra, P. Geerlings and M. T. Nguyen, *J. Org. Chem.*, **62**, 6417 (1997).
331. L. T. Nguyen, N. T. Le, F. De Proft, A. K. Chandra, W. Langenaeker, M. T. Nguyen and P. Geerlings, *J. Am. Chem. Soc.*, **121**, 5992 (1999).
332. L. T. Nguyen, F. De Proft, M. T. Nguyen and P. Geerlings, *J. Chem. Soc., Perkin Trans. 2*, 898 (2001).
333. L. T. Nguyen, F. De Proft, M. T. Nguyen and P. Geerlings, *J. Org. Chem.*, **66**, 4316 (2001).
334. D. Sengupta, A. K. Chandra and M. T. Nguyen, *J. Org. Chem.*, **62**, 6404 (1997).
335. A. K. Chandra and M. T. Nguyen, *J. Comput. Chem.*, **19**, 195 (1998).
336. T. N. Le, L. T. Nguyen, A. K. Chandra, F. De Proft, M. T. Nguyen and P. Geerlings, *J. Chem. Soc., Perkin Trans. 2*, 1249 (1999).
337. A. K. Chandra and M. T. Nguyen, *J. Phys. Chem. A*, **102**, 6181 (1998).
338. A. K. Chandra, A. Michalak, M. T. Nguyen and R. Nalewajski, *J. Phys. Chem. A*, **102**, 10188 (1998).
339. A. K. Chandra, T. Uchimaru and M. T. Nguyen, *J. Chem. Soc., Perkin Trans. 2*, 2117 (1999).
340. M. T. Nguyen, A. K. Chandra, S. Sakai and K. Morokuma, *J. Org. Chem.*, **64**, 65 (1999).
341. L. T. Nguyen, F. De Proft, A. K. Chandra, T. Uchimaru, M. T. Nguyen and P. Geerlings, *J. Org. Chem.*, **66**, 6096 (2001).
342. E. Silicia, N. Russo and T. Mineva, *J. Phys. Chem. A*, **115**, 442 (2001).
343. A. Toro-Labbe, *J. Phys. Chem. A*, **103**, 4398 (1999).
344. T. Uchimaru, A. K. Chandra, S. I. Kawahara, K. Matsumura, S. Tsuzuki and M. Mikami, *J. Phys. Chem. A*, **105**, 1343 (2001).
345. M. T. Nguyen and A. F. Hegarty, *J. Chem. Soc. Perkin Trans. 2*, **1991**, 1999 (1985).
346. M. T. Nguyen and A. F. Hegarty, *J. Chem. Soc., Perkin Trans. 2*, 2005 (1985).
347. T. Fujii, H. Tokiwa, H. Ichikawa and H. Shinoda, *J. Mol. Struct. (THEOCHEM)*, **277**, 251 (1992).
348. P. Kollman and S. Rothenberg, *J. Am. Chem. Soc.*, **99**, 1333 (1977).
349. R. T. McIver and J. H. Silver, *J. Am. Chem. Soc.*, **95**, 8462 (1973).
350. T. B. McMahon and P. Kebarle, *J. Am. Chem. Soc.*, **99**, 2222 (1977).
351. M. Jujio, R. T. McIver and R. W. Taft, *J. Am. Chem. Soc.*, **103**, 4017 (1981).
352. T. Kremer and P. v. R. Schleyer, *Organometallics*, **16**, 737 (1997).
353. H. U. Suter and M. Nonella, *J. Phys. Chem. A*, **102**, 10128 (1998).
354. P. Ballone and R. O. Jones, *J. Phys. Chem. A*, **105**, 3008 (2001).
355. M. Schlosser, E. Marzi, F. Cottet, H. H. Buker and N. M. M. Nibbering, *Chem. Eur. J.*, **7**, 3511 (2001).
356. M. Krauss, O. J. Jensen and H. F. Hameka, *J. Phys. Chem.*, **98**, 9955 (1994).
357. Z. He, C. H. Martin, R. Birge and K. F. Freed, *J. Phys. Chem. A*, **104**, 2939 (2000).
358. Y. H. Mariam and L. Chantranupong, *J. Mol. Struct. (THEOCHEM)*, **454**, 237 (1998).
359. J. C. Rienstra-Kiracofe, D. E. Graham and H. F. Schaefer III, *Mol. Phys.*, **94**, 767 (1998).
360. O. Nwobi, J. Higgins, X. Zhou and R. Liu, *Chem. Phys. Lett.*, **272**, 155 (1997).
361. J. Danielsson, J. Ulicny and A. Laaksonen, *J. Am. Chem. Soc.*, **123**, 9817 (2001).
362. T. M. Krygowski, R. Anulewicz, B. Pniewska, P. Milart, C. W. Bock, M. Sawada, Y. Takai and T. Hanafusa, *J. Mol. Struct.*, **324**, 251 (1994).
363. M. Van Beylen, B. Roland, G. S. King and J. Aerts, *J. Chem. Res. (S)*, 388 (1985).
364. P. A. van den Schaff, M. P. Hogenheide, D. Grove, A. L. Spek and G. van Koten, *J. Chem. Soc., Chem. Commun.*, 1703 (1992).
365. A. Murkherjee, M. L. McGlashen and T. G. Spiro, *J. Phys. Chem.*, **99**, 4912 (1995).
366. C. Berthomieu and A. Boussac, *Biospectroscopy*, **1**, 187 (1995).

367. P. v. R. Schleyer, C. Maerker, A. Dransfeld, H. Jiao and N. R. J. van Eikema Hommes, *J. Am. Chem. Soc.*, **118**, 6317 (1996).
368. H. J. Dauben, J. D. Wilson and J. L. Laity, *J. Am. Chem. Soc.*, **90**, 811 (1968); *J. Am. Chem. Soc.*, **91**, 1991 (1969).
369. J. H. Richardson, L. M. Stephenson and J. I. Brauman, *J. Am. Chem. Soc.*, **97**, 2967 (1975).
370. R. F. Gunion, M. K. Gilles, M. L. Polak and W. C. Lineberger, *Int. J. Mass Spectrom. Ion Processes*, **117**, 601 (1992).
371. J. Gao, N. Li and M. Freindorf, *J. Am. Chem. Soc.*, **118**, 4912 (1996).
372. M. V. Vener and S. Iwata, *Chem. Phys. Lett.*, **292**, 87 (1998).
373. E. F. G. Herington and W. Kynaston, *Trans. Faraday Soc.*, **53**, 238 (1957).
374. Z. B. Maksic, D. Kovacek, M. Eckert-Maksic and I. Zrinski, *J. Org. Chem*, **61**, 6717 (1996).
375. R. M. Borges dos Santos and J. A. Martinho Simões, *J. Phys. Chem. Ref. Data*, **27**, 707 (1998).
376. S. G. Lias, J. E. Bartmess, J. F. Liebman, J. L. Holmes, R. D. Levin and W. G. Mallard, *J. Phys. Chem. Ref. Data*, **17**, Supplement 1 (1988).
377. A. Pross, L. Radom and R. W. Taft, *J. Org. Chem.*, **45**, 818 (1980).
378. J. J. Urban, R. L. von Tersch and G. R. Famini, *J. Org. Chem.*, **59**, 5239 (1994).
379. J. Niwa, *Bull. Chem. Soc. Jpn.*, **62**, 226 (1989).
380. T. Silvestro and R. D. Topsom, *J. Mol. Struct. (THEOCHEM)*, **206**, 309 (1990).
381. S. Chowdhury, H. Kishi, G. W. Dillow and P. Kebarle, *Can. J. Chem.*, **67**, 603 (1989).
382. R. G. Pearson, *J. Am. Chem. Soc.*, **108**, 6109 (1986).
383. J. Hine and P. K. Mookerjee, *J. Org. Chem.*, **40**, 287 (1975).
384. S. Cabani, P. Gianni, V. Mollica and L. Leprori, *J. Solution Chem.*, **10**, 563 (1981).
385. W. H. Richardson, C. Peng, D. Bashford, L. Noodleman and D. A. Case, *Int. J. Quantum. Chem.*, **61**, 207 (1997).
386. L. P. Hammett, *J. Am. Chem. Soc.*, **59**, 96 (1937); *Trans. Faraday Soc.*, **34**, 156 (1938).
387. J. Catalan and A. Macias, *J. Mol. Struct.*, **38**, 209 (1977).
388. W. J. Hehre, M. Taagepera, R. W. Taft and R. D. Topsom, *J. Am. Chem. Soc.*, **103**, 1344 (1981).
389. J. Catalan and A. Macias, *J. Chem. Soc., Perkin Trans. 2*, 1632 (1979).
390. F. Mendez, M. de L. Romero and P. Geerlings, *J. Org. Chem.*, **63**, 5774 (1998).
391. K. C. Gross and P. G. Seybold, *Int. J. Quantum Chem.*, **80**, 1107 (2000); **85**, 569 (2001).
392. R. L. Martin and D. A. Shirley, *J. Am. Chem. Soc.*, **96**, 5299 (1974).
393. D. W. Davis and J. W. Rabalais, *J. Am. Chem. Soc.*, **96**, 5305 (1974).
394. S. R. Thomas and T. D. Thomas, *J. Am. Chem. Soc.*, **100**, 5459 (1978).
395. G. Klopman, in *Chemical Reactivity and Reaction Paths* (Ed. G. Klopman), Wiley, New York, 1974, pp. 72–74.
396. R. G. Pearson, *Hard and Soft Acids and Bases*, Dowden, Hutchinson and Ross, Stroudsville, PA, 1973.
397. J. March, *Advanced Organic Chemistry*, Wiley, New York, 1992, pp. 546–547.
398. P. v. R. Schleyer, *Pure Appl. Chem.*, **59**, 1647 (1987).
399. C. Lambert, Y. D. Wu and P. v. R. Schleyer, *J. Chem. Soc., Chem. Commun.*, 255 (1993).
400. D. Seebach, *Angew. Chem., Int. Ed. Engl.*, **27**, 1624 (1988).
401. L. M. Jackman and B. D. Smith, *J. Am. Chem. Soc.*, **110**, 3829 (1988).
402. Y. J. Zheng and R. L. Ornstein, *J. Am. Chem. Soc.*, **119**, 1523 (1997).
403. M. Haeberlein and T. Brinck, *J. Phys. Chem.*, **100**, 10116 (1996).
404. J. B. A. Ross, W. R. Laws, K. W. Rousslang and H. R. Wyssbrod, in *Topics in Fluorescence Spectroscopy, Vol. 3, Biochemical Applications* (Ed. J. R. Lakowicz), Plenum Press, New York, 1992, pp. 1–62.
405. J. Petruska, *J. Chem. Phys.*, **34**, 1120 (1961).
406. K. Kimura and S. Nagakura, *Mol. Phys.*, **9**, 117 (1965).
407. W. H. Fang, *J. Chem. Phys.*, **112**, 1204 (2000).
408. A. L. Sobolewski and W. Domcke, *J. Phys. Chem. A*, **105**, 9275 (2001).
409. H. Abe, N. Mikami, M. Ito and Y. Udagawa, *Chem. Phys. Lett.*, **93**, 217 (1982).
410. T. Ebata, A. Iwasaki and N. Mikami, *J. Phys. Chem. A*, **104**, 7974 (2000).
411. I. Berlman, *Handbook of Fluorescence Spectra of Aromatic Molecules*, Academic Press, New York, 1969.

412. (a) K. Fuke and K. Kaya, *Chem. Phys. Lett.*, **94**, 97 (1983).
    (b) S. Hirokawa, T. Imasaka and T. Imasaka, *J. Phys. Chem. A*, **105**, 9252 (2002).
413. G. Kemister, A. Pross, L. Radom and R. W. Taft, *J. Org. Chem.*, **45**, 1056 (1980).
414. S. Martrenchard-Barra, C. Dedonder-Lardeux, C. Jouvet, D. Solgadi, M. Vervloet, G. Gregoire and I. Dimicoli, *Chem. Phys. Lett.*, **310**, 173 (1999).
415. J. E. LeClaire, R. Anand and P. M. Johnson, *J. Phys. Chem.*, **106**, 6785 (1997).
416. G. Kohler and N. Getoff, *J. Chem. Soc., Faraday Trans. 1*, **73**, 2101 (1976).
417. D. V. Bent and E. Hayon, *J. Am. Chem. Soc.*, **97**, 2599 (1975).
418. C. P. Schick and P. M. Weber, *J. Phys. Chem. A*, **105**, 3725 (2001).
419. C. P. Schick and P. M. Weber, *J. Phys. Chem. A*, **105**, 3735 (2001).
420. M. R. Ganapathi, R. Hermann, S. Naumov and O. Brede, *Phys. Chem. Chem. Phys.*, **2**, 4947 (2000); *Chem. Phys. Lett.*, **337**, 335 (2001).
421. F. G. Bordwell and J. P. Cheng, *J. Am. Chem. Soc.*, **113**, 1736 (1991).
422. O. Brede, H. Orthner, V. E. Zubarev and R. Hermann, *J. Phys. Chem.*, **100**, 7097 (1996).
423. H. Mohan, R. Hermann, S. Maunov, J. P. Mittal and O. Brede, *J. Phys. Chem. A*, **102**, 5754 (1998).
424. R. Hermann, S. Naumov, G. R. Mahalaxmi and O. Brede, *Chem. Phys. Lett.*, **324**, 265 (2000); *J. Mol. Struct. (THEOCHEM)*, **532**, 69 (2000).
425. T. A. Gadosy, D. Shukla and L. J. Johnston, *J. Phys. Chem. A*, **103**, 8834 (1999).
426. S. C. Anderson, L. Goodman, K. Krogh-Jespersen, A. G. Ozkabak and R. N. Zare, *J. Chem. Phys.*, **82**, 5329 (1985).
427. F. Borchers, K. Levsen, C. B. Theissling and N. M. M. Nibbering, *Org. Mass Spectrom.*, **12**, 746 (1977).
428. R. Caballol, J. M. Poblet and P. Sarasa, *J. Phys. Chem*, **89**, 5836 (1985).
429. D. H. Russel, M. L. Gross and N. M. M. Nibbering, *J. Am. Chem. Soc.*, **100**, 6133 (1978).
430. D. H. Russel, M. L. Gross, J. Van der Greef and N. M. M. Nibbering, *Org. Mass Spectrom.*, **14**, 474 (1979).
431. A. Y. Van Haverbeke, R. Flammang, C. De Meyer, K. G. Das and G. S. Reddy, *Org. Mass Spectrom.*, **12**, 631 (1977).
432. A. Maquestiau, R. Flammang, G. L. Glish, J. A. Laramee and R. G. Cooks, *Org. Mass Spectrom.*, **15**, 131 (1980).
433. A. Maquestiau, R. Flammang, P. Pauwels, P. Vallet and P. Meyrant, *Org. Mass Spectrom.*, **17**, 643 (1982).
434. F. Turecek, D. E. Drinkwater, A. Maquestiau and F. W. McLafferty, *Org. Mass Spectrom.*, **24**, 669 (1989).
435. C. Lifshitz and S. Gefen, *Org. Mass Spectrom.*, **19**, 197 (1984).
436. C. Lifshitz and Y. Malinovich, *Int. J. Mass Spectrom. Ion Processes*, **60**, 99 (1984).
437. E. L. Chronister and T. H. Morton, *J. Am. Chem. Soc.*, **112**, 9475 (1990).
438. V. Nguyen, J. S. Bennett and T. H. Morton, *J. Am. Chem. Soc.*, **119**, 8342 (1997).
439. E. L. Mehler and J. Gerhards, *J. Am. Chem. Soc.*, **107**, 5856 (1985).
440. G. A. DiLabio, D. A. Pratt and J. S. Wright, *J. Org. Chem.*, **65**, 2195 (2000).
441. D. M. Camaioni, *J. Am. Chem. Soc.*, **112**, 9475 (1990).
442. S. H. Hoke, S. S. Yang, R. G. Cooks, D. A. Hrovat and W. T. Borden, *J. Am. Chem. Soc.*, **116**, 4888 (1994).
443. M. H. Palmer, W. Moyes, M. Speirs and J. N. A. Ridyard, *J. Mol. Struct.*, **52**, 293 (1979).
444. M. L. Fraser-Monteiro, L. Fraser-Monteiro, J. de Wit and T. Baer, *J. Phys. Chem.*, **88**, 3622 (1984). For recent data see: O. Dopfer, K. Müller-Dethlefs, *J. Chem. Phys.*, **101**, 8508 (1994); O. Dopfer, M. Melf and K. Müller-Dethlefs, *Chem. Phys.*, **207**, 437 (1995); T. G. Wright, E. Cordes, O. Dopfer and K. Müller-Dethlefs, *J. Chem. Soc. Faraday Trans.*, **89**, 1609 (1993); E. Cordes, E. Dopfer and K. Müller-Dethlefs, *J. Phys. Chem.*, **97**, 7471 (1993); O. Dopfer, T. G. Wright, E. Cordes and K. Muller-Dethlefs, *J. Am. Chem. Soc.*, **116**, 5880 (1994); M. C. R. Cockett, M. Takahashi, K. Okuyama and K. Kimura, *Chem. Phys. Lett.*, **187**, 250 (1991).
445. T. P. Debies and J. W. Rabalais, *J. Electron Spectrosc. Relat. Phenom.*, **1**, 355 (1972).
446. J. P. Maier and D. W. Turner, *J. Chem. Soc., Faraday Trans. 2*, **69**, 521 (1973).
447. T. Kobayashi and S. Nakakura, *Bull. Chem. Soc. Jpn.*, **47**, 2563 (1974).
448. R. Flammang, M. Barbieux-Flammang, E. Gualano, P. Gerbaux, H. T. Le, M. T. Nguyen, F. Turecek and S. Vivekananda, *J. Phys. Chem. A*, **105**, 8579 (2001).

449. R. Flammang, M. Barbieux-Flammang, P. Gerbaux, H. T. Le, F. Turecek and M. T. Nguyen, *Int. J. Mass Spectrom.*, **217**, 65 (2002).
450. R. Flammang, J. Elguero, H. T. Le, P. Gerbaux and M. T. Nguyen, *Chem. Phys. Lett.*, **356**, 239 (2002).
451. D. J. Lavorato, J. K. Terlouw, G. A. McGibbon, T. K. Dargel, W. Koch and H. Schwarz, *Int. J. Mass Spectrom.*, **179**, 7 (1998).
452. H. T. Le, R. Flammang, P. Gerbaux, G. Bouchoux and M. T. Nguyen, *J. Phys. Chem. A*, **105**, 11582 (2001).
453. C. Trindle, *J. Phys. Chem. A*, **104**, 5298 (2000).
454. S. Jakabsen, K. V. Mikkelsen and S. U. Pedersen, *J. Phys. Chem.*, **100**, 7411 (1996).
455. M. Krauss, *J. Mol. Struct. (THEOCHEM)*, **307**, 47 (1994).
456. N. Mikami, S. Sato and M. Ishigaki, *Chem. Phys. Lett.*, **202**, 431 (1993).
457. W. T. Dixon and D. Murphy, *J. Chem. Soc., Faraday Trans. 2*, **72**, 1221 (1976).
458. W. Gordy, *Theory and Applications of Electron Spin Resonance*, Wiley, New York, 1980.
459. F. Turecek and C. J. Cramer, *J. Am. Chem. Soc.*, **117**, 12243 (1995).
460. B. J. Smith, M. T. Nguyen, W. J. Bouma and L. Radom, *J. Am. Chem. Soc.*, **113**, 6452 (1991).
461. W. Bertrand and G. Bouchoux, *Rapid Commun. Mass Spectrom.*, **12**, 1697 (1998).
462. Y. Apeloig, in *The Chemistry of Enols* (Ed. Z. Rappoport), Chap. 1, Wiley, Chichester, 1990.
463. M. T. Nguyen, L. Landuyt and L. G. Vanquickenborne, *Chem. Phys. Lett.*, **182**, 225 (1991).
464. NIST Webbook: http://webbook.nist.gov/chemistry.
465. R. Flammang, Y. Van Haverbeke, C. Braybrook and J. Brown, *Rapid Commun. Mass Spectrom.*, **9**, 795 (1995).
466. P. Gerbaux, Y. Van Haverbeke and R. Flammang, *J. Mass Spectrom.*, **32**, 1170 (1997).
467. L. J. Chyall and H. I. Kenttämaa, *J. Am. Chem. Soc.*, **116**, 3135 (1994); *J. Mass Spectrom.*, **30**, 81 (1995).
468. M. Capponi, I. Gut, B. Hellrung, G. Persy and J. Wirz, *Can. J. Chem.*, **77**, 605 (1999).
469. K. Mandix, A. Colding, K. Elming, L. Sunesen and I. Shim, *Int. J. Quantum. Chem.*, **46**, 159 (1993).
470. D. Delaere, G. Raspoet and M. T. Nguyen, *J. Phys. Chem. A*, **103**, 171 (1999).
471. M. T. Nguyen and G. Raspoet, *Can. J. Chem. A.*, **77**, 817 (1999).
472. G. Raspoet, M. T. Nguyen and L. G. Vanquickenborne, *J. Phys. Org. Chem.*, **13**, 46 (2000).
473. M. S. Sodupe, A. Oliva and J. Bertran, *J. Phys. Chem. A*, **101**, 9142 (1997).
474. A. Courty, M. Mons, I. Dimicoli, F. Piuzzi, V. Brenner and P. Millie, *J. Phys. Chem. A*, **102**, 4890 (1998).
475. R. J. Lipert and S. D. Colson, *J. Phys. Chem.*, **93**, 135 (1989); **92**, 188 (1988).
476. O. Dopfer, T. G. Wright, E. Cordes and K. Muller-Dethlefs, *J. Am. Chem. Soc.*, **116**, 5880 (1994).
477. R. J. Green, H. T. Kim, J. Qian and S. L. Anderson, *J. Phys. Chem.*, **113**, 4158 (2000).
478. S. R. Haines, W. D. Geppert, D. M. Chapman, M. J. Watkins, C. E. H. Dessent, M. C. R. Cockett and K. Muller-Dethlefs, *J. Chem. Phys.*, **109**, 9244 (1998).
479. I. Rozas, I. Alkorta and J. Elguero, *J. Phys. Chem. A*, **105**, 10462 (2001).
480. B. Halliwell and J. M. C. Gutteridge, *Free Radicals in Biology and Medicine*, 2nd edn., Oxford University Press, Oxford, 1989.
481. J. Stubbe, *Annu. Rev. Biochem.*, **58**, 257 (1989).
482. R. Karthien, R. Dietz, W. Nastainczyk and H. H. Ruf, *Eur. J. Biochem.*, **171**, 313 (1988).
483. B. Barry, M. K. El-Deeb, P. O Sandusky and G. T. Babcock, *J. Biol. Chem.*, **265**, 20 (1990).
484. M. M. Whitakker and J. M. Whitakker, *Biophys. J.*, **64**, 762 (1993).
485. S. M. Janes, D. Mu, D. Wemmes, A. J. Smith, S. Kaus, D. Maltby, A. L. Burlingame and J. P. Klinman, *Science*, **248**, 981 (1990).
486. M. G. Simic and T. Dizdaroglu, *Biochemistry*, **24**, 233 (1985).
487. C. Y. Lin and M. C. Lin, *J. Phys. Chem.*, **90**, 425 (1986).
488. T. J. Stone and W. A. Waters, *Proc. Chem. Soc. London*, 253 (1962); *J. Chem. Soc.*, 213 (1964).
489. W. T. Dixon, M. Moghimi and D. Murphy, *J. Chem. Soc., Faraday Trans. 2*, **70**, 1713 (1974).
490. S. Un, C. Gercz, E. Elleingand and M. Fontecave, *J. Am. Chem. Soc.*, **123**, 3048 (2001).
491. T. Maki, Y. Araki, Y. Ishida, O. Onomura and Y. Matsumaru, *J. Am. Chem. Soc.*, **123**, 3371 (2001).
492. S. M. Beck and L. E. Brus, *J. Chem. Phys.*, **76**, 4700 (1982).

1. General and theoretical aspects of phenols

493. C. R. Johnson, M. N. Ludwig and S. A. Asher, *J. Am. Chem. Soc.*, **108**, 905 (1986).
494. G. N. R. Tripathi and R. H. Schuler, *J. Phys. Chem.*, **92**, 5129 and 5133 (1988); *J. Chem. Phys.*, **81**, 113 (1984); *Chem. Phys. Lett.*, **98**, 594 (1983).
495. G. N. R. Tripathi, in *Time Resolved Spectroscopy* (Eds. R. J. H. Clark and R. E. Hester), Wiley, New York, 1989, p. 157.
496. A. Mukherjee, M. L. McGlashen and T. G. Spiro, *J. Phys. Chem.*, **99**, 4912 (1995).
497. L. D. Johnston, N. Mathivanan, F. Negri, W. Siebrand and F. Zerbetto, *Can. J. Chem.*, **71**, 1655 (1993).
498. J. Spanget-Larsen, M. Gil, A. Gorski, D. M. Blake, J. Waluk and J. G. Radziszewski, *J. Am. Chem. Soc.*, **123**, 11253 (2001).
499. G. Porter and F. G. Wright, *Trans. Faraday Soc.*, **51**, 1469 (1955).
500. E. J. Land, G. Porter and E. Strachan, *Trans. Faraday Soc.*, **57**, 1885 (1961).
501. J. L. Roebber, *J. Chem. Phys.*, **37**, 1974 (1962).
502. B. Ward, *Spectrochim. Acta, Part A*, **24**, 813 (1968).
503. R. H. Schuler and G. K. Buzzard, *Int. J. Radiat. Phys. Chem.*, **8**, 563 (1976).
504. D. Pullin and L. Andrews, *J. Mol. Struct.*, **95**, 181 (1982).
505. Y. Kajii, K. Obi, N. Nakashima and K. Yohihara, *J. Chem. Phys.*, **87**, 5059 (1984).
506. H. M. Chang, H. H. Jaffe and C. A. Masmandis, *J. Phys. Chem.*, **79**, 1118 (1975).
507. V. B. Luzhkov and A. S. Zyubin, *J. Mol. Struct. (THEOCHEM)*, **170**, 33 (1988).
508. H. Yu and J. D. Goddard, *J. Mol. Struct. (THEOCHEM)*, **233**, 129 (1991).
509. R. Liu and X. Zhou, *Chem. Phys. Lett.*, **207**, 185 (1993); *J. Phys. Chem.*, **97**, 9613 (1993).
510. D. Chipman, R. Liu, X. Zhou and P. Pulay, *J. Chem. Phys.*, **100**, 5023 (1994).
511. Y. Qin and R. A. Wheeler, *J. Chem. Phys.*, **102**, 1689 (1995).
512. G. N. R. Tripathi, *J. Phys. Chem. A*, **102**, 2388 (1998).
513. Y. Qin and R. A. Wheeler, *J. Am. Chem. Soc.*, **117**, 6083 (1995).
514. R. Schnepf, A. Sokolowski, J. Muller, V. Bachler, K. Wieghardt and P. Hildebrandt, *J. Am. Chem. Soc.*, **116**, 9577 (1994).
515. P. J. O'Malley, *Chem. Phys. Lett.*, **325**, 69 (2000).
516. C. Berthomieu, C. Boullais, J. M. Neumann and A. Boussac, *Biochim. Biophys. Acta*, **1365**, 112 (1998).
517. L. A. Eriksson, *Mol. Phys.*, **91**, 827 (1997).
518. C. Adamo, R. Subra, A. Di Matteo and V. Barone, *J. Chem. Phys.*, **109**, 10244 (1998).
519. H. Fischer, *Z. Naturforsch.*, **20**, 488 (1965).
520. H. M. McConnell and D. B. Chesnut, *J. Chem. Phys.*, **18**, 107 (1958).
521. T. A. Claxton, *Chem. Soc. Rev.*, 437 (1995).
522. A. M. Schmoltner, D. S. Anex and Y. T. Lee, *J. Phys. Chem.*, **96**, 1236 (1992).
523. S. Olivella, A. Sole and A. Garcia-Raso, *J. Phys. Chem.*, **99**, 10549 (1995).
524. R. Liu, K. Morokuma, A. M. Mebel and M. C. Lin, *J. Phys. Chem.*, **100**, 9314 (1996).
525. A. M. Mebel and M. C. Lin, *J. Am. Chem. Soc.*, **116**, 9577 (1994).
526. G. Ghio and G. Tonachini, *J. Am. Chem. Soc.*, **120**, 6753 (1998).
527. F. Berho, F. Caralp, M. T. Rayez, R. Lesclaux and E. Ratajczak, *J. Phys. Chem.*, **102**, 1 (1998).
528. J. Platz, O. J. Nielsen, T. J. Wallington, J. C. Ball, M. D. Hurley, A. M. Straccia, W. F. Schneider and J. Sehested, *J. Phys. Chem. A*, **102**, 7964 (1998).
529. F. P. Lossing and J. L. Holmes, *J. Am. Chem. Soc.*, **106**, 6917 (1984).
530. M. T. Nguyen, unpublished results (2001).
531. D. D. M. Wayner, E. Lusztyk, D. Page, K. U. Ingold, P. Mulder, L. J. J. Laarhoven and H. S. Aldrich, *J. Am. Chem. Soc.*, **117**, 8737 (1995).
532. M. Lucariri, P. Pedrielli, G. F. Pedulli, S. Cabiddu and C. Fattuoni, *J. Org. Chem.*, **61**, 9259 (1996).
533. W. van Scheppingen, E. Dorrestijn, I. Arends, P. Mulder and H. G. Korth, *J. Phys. Chem. A*, **101**, 5404 (1997).
534. M. Jonsson, J. Lind, T. E. Ericksen and G. Merenyi, *J. Chem. Soc., Perkin Trans. 2*, 1557 (1993).
535. D. J. Harman, *J. Geontol.*, **11**, 298 (1956); *Proc. Natl. Acad. Sci. U.S.A.*, **11**, 7124 (1981).
536. T. Finkel and N. J. Holbrook, *Nature*, **408**, 239 (2000) and references therein.
537. W. A. Pryor, in *Modern Biological Theories of Aging* (Eds. H. R. Warner, R. N. Butler and R. L. Sprott), Raven, New York, 1987, p. 89.

538. V. W. Bowry and K. U. Ingold, *Acc. Chem. Res.*, **32**, 27 (1999).
539. K. U. Ingold, A. C. Webb, D. Witter, G. W. Burton, T. A. Metcalf and D. P. R. Muller, *Arch. Biochem. Biophys.*, **259**, 224 (1987).
540. T. F. Slater, K. H. Cheeseman, C. Benedetto, M. Collins, S. Emery, S. P. Maddrix, J. T. Nodes, K. Proudfoot, G. W. Burton and K. U. Ingold, *Biochem. J.*, **265**, 51 (1990).
541. L. J. Machlin, *Vitamin E. A Comprehensive Treatise*, Marcel Dekker, New York, 1980.
542. J. A. Howard and K. U. Ingold, *Can. J. Chem.*, **40**, 1851 (1962).
543. J. A. Howard and K. U. Ingold, *Can. J. Chem.*, **41**, 1744 (1963).
544. J. A. Howard and K. U. Ingold, *Can. J. Chem.*, **41**, 2800 (1963).
545. J. A. Howard and K. U. Ingold, *Can. J. Chem.*, **42**, 1044 (1964).
546. K. Tanaka, S. Sakai, S. Tomiyama, T. Nishiyama and F. Yamada, *Bull. Chem. Soc. Jpn.*, **64**, 2677 (1991).
547. S. Tomiyama, S. Sakai, T. Nishiyama and F. Yamada, *Bull. Chem. Soc. Jpn.*, **66**, 299 (1993).
548. M. Perrakyla and T. A. Pakkanen, *J. Chem. Soc., Perkin Trans 2*, 1405 (1995).
549. M. J. Lundqvist and L. A. Eriksson, *J. Phys. Chem., B*, **104**, 848 (2000).
550. E. Mvula, M. N. Schuchmann and C. von Sonntag, *J. Chem. Soc., Perkin Trans. 2*, 264 (2001).
551. O. Tishchenko, E. Kryachko and M. T. Nguyen, *J. Mol. Struct. (THEOCHEM)*, **615**, 247 (2002).
552. G. Gilchrist, G. W. Burton and K. U. Ingold, *Chem. Phys.*, **95**, 473 (1985).
553. A. Müller, H. Ratajczak, W. Junge and E. Diemann (Eds.), *Electron and Proton Transfer in Chemistry and Biology*, Elsevier, Amsterdam, 1992.
554. G. S. Denisov and V. M. Schreiber, *Doklady Akad. Sci. USSR, (Engl. Transl.)*, **215**, 627 (1974).
555. G. S. Denisov, A. I. Kulbida and V. M. Schreiber, in *Molecular Spectroscopy*, Vol. 6, Leningrad State University, Leningrad, 1983, p. 124.
556. M. Rospenk, I. G. Rumynskaya and V. M. Schreiber, *J. Appl. Spectrosc. (USSR) (Engl. Transl.)*, **36**, 756 (1982).
557. G. S. Denisov, S. F. Bureiko, N. S. Golubev and K. Tokhadze, in *Molecular Interactions*, Vol. 2 (Eds. H. Ratajczak and W. J. Orville-Thomas), Wiley, New York, 1980, p. 107.
558. S. F. Bureiko, V. P. Oktyabrskii and K. Pihlaya, *Kinetika i Katalyz (USSR)*, **34**, 430 (1993).
559. S. F. Bureiko, N. S. Goluben, J. Mattinen and K. Pihlaya, *J. Mol. Liq.*, **45**, 139 (1990).
560. S. F. Bureiko, N. S. Golubev and K. Pihlaya, *J. Mol. Struct. (THEOCHEM)*, **480–481**, 297 (1999).
561. J. P. Dupont, J. D'Hondt and Th. Zeegers-Huyskens, *Bull. Soc. Chim. Belg.*, **80**, 369 (1971).
562. M. Rospenk and Th. Zeegers-Huyskens, *Spectrochim. Acta, Part A*, **42**, 499 (1986).
563. M. Rospenk and Th. Zeegers-Huyskens, *J. Phys. Chem.*, **91**, 3974 (1987).
564. M. Szafran, *J. Mol. Struct. (THEOCHEM)*, **381**, 9 (1996).
565. Z. Dega-Szafran, A. Kania, M. Grunwald-Wyspianska, M. Szafran and E. Tykarska, *J. Mol. Struct. (THEOCHEM)*, **381**, 107 (1996).
566. Z. Dega-Szafran and M. Szafran, *J. Chem. Soc., Perkin Trans. 2*, 897 (1987).
567. H. Ratajczak and L. Sobczyk, *J. Chem. Phys.*, **50**, 556 (1969).
568. Z. Malarski, M. Rospenk, E. Grech and L. Sobczyk, *J. Phys. Chem.*, **86**, 401 (1982).
569. H. Romanowski and L. Sobczyk, *J. Phys. Chem.*, **79**, 2535 (1975).
570. L. Sobczyk, in *Radio and Microwave Spectroscopy* (Eds. N. Piślewski and A. Mickiewicz), University Press, Poznan, 1985, p. 55.
571. M. Ilczyszyn and H. Ratajczak, *J. Chem. Soc., Faraday Trans.*, **91**, 1611, 3859 (1995).
572. M. Ilczyszyn, H. Ratajczak and K. Skowronek, *Magn. Reson. Chem.*, **26**, 445 (1988).
573. G. Zundel, *Adv. Chem. Phys.*, **111**, 1 (2000).
574. H. Abe, N. Mikami and M. Ito, *J. Phys. Chem.*, **86**, 1768 (1982).
575. H. Abe, N. Mikami, M. Ito and Y. Udagawa, *J. Phys. Chem.*, **86**, 2567 (1982).
576. N. Gonohe, H. Abe, N. Mikami and M. Ito, *J. Phys. Chem.*, **89**, 3642 (1985).
577. A. Schiefke, C. Deusen, Ch. Jacoby, M. Gerhards, M. Schmitt and K. Kleinermanns, *J. Chem. Phys.*, **102**, 9197 (1995).
578. A. Oikawa, H. Abe, N. Mikami and M. Ito, *J. Phys. Chem.*, **87**, 5083 (1983).
579. M. Ito, *J. Mol. Struct.*, **177**, 173 (1988).
580. T. Ebata, M. Furukawa, T. Suzuki and M. Ito, *J. Opt. Soc. Am. B*, **7**, 1890 (1990).
581. K. Fuke and K. Kaya, *Chem. Phys. Lett.*, **91**, 311 (1982).

582. K. Fuke, H. Yoshiuchi, K. Kaya, Y. Achiba, K. Sato and K. Kimura, *Chem. Phys. Lett.*, **108**, 179 (1984).
583. A. Sur and P. M. Johnson, *J. Phys. Chem.*, **84**, 1206 (1986).
584. J. L. Knee, L. R. Khundkar and A. H. Zewail, *J. Chem. Phys.*, **87**, 115 (1987).
585. R. J. Lipert, G. Bermudez and S. D. Colson, *J. Phys. Chem.*, **92**, 3801 (1988).
586. L. J. Lipert and S. D. Colson, *J. Chem. Phys.*, **89**, 4579 (1989).
587. R. J. Lipert and S. D. Colson, *Chem. Phys. Lett.*, **161**, 303 (1989).
588. D. Solgadi, C. Jouvet and A. Tramer, *J. Phys. Chem.*, **92**, 3313 (1988).
589. J. Steadman and J. A. Syage, *J. Chem. Phys.*, **92**, 4630 (1990).
590. R. J. Stanley and A. W. Castleman, *J. Chem. Phys.*, **94**, 7744 (1991).
591. G. Reiser, O. Dopfer, R. Lindner, G. Henri, K. Müller-Dethlefs, E. W. Schlag and S. D. Colson, *Chem. Phys. Lett.*, **181**, 1 (1991).
592. D. L. Gerrard and W. F. Maddams, *Spectrochim. Acta, Part A*, **34**, 1205, 1213 (1978).
593. G. Nemethy and A. Ray, *J. Phys. Chem.*, **77**, 64 (1973).
594. H. Baba and S. Suzuki, *J. Chem. Phys.*, **35**, 1118 (1961).
595. M. Ito, *J. Mol. Spectrosc.*, **4**, 106 (1960).
596. M. Gerhards, K. Beckmann and K. Kleinermanns, *Z. Phys. D*, **29**, 223 (1994).
597. M. Yi and S. Scheiner, *Chem. Phys. Lett.*, **262**, 567 (1996).
598. A. Iwasaki, A. Fujii, T. Watanabe, T. Ebata and N. Mikami, *J. Phys. Chem.*, **100**, 16053 (1996).
599. N. Mikami, A. Okabe and I. Suzuki, *J. Phys. Chem.*, **92**, 1858 (1988).
600. C. Jouvet, C. Lardeux-Dedonder, M. Richard-Viard, D. Solgadi and A. Tramer, *J. Phys. Chem.*, **94**, 5041 (1990).
601. C. Crepin and A. Tramer, *Chem. Phys. Lett.*, **156**, 281 (1991).
602. M. Ilczyszyn, H. Ratajczak and J. A. Ladd, *Chem. Phys. Lett.*, **153**, 385 (1988).
603. M. Ilczyszyn, H. Ratajczak and J. A. Ladd, *J. Mol. Struct.*, **198**, 499 (1989).
604. M. Wierzejewska and H. Ratajczak, *J. Mol. Struct. (THEOCHEM)*, **416**, 121 (1997).
605. I. Majerz and L. Sobczyk, *J. Chim. Phys.*, **90**, 1657 (1993).
606. I. Majerz, Z. Malarski and L. Sobczyk, *Chem. Phys. Lett.*, **274**, 361 (1997).
607. G. Albrecht and G. Zundel, *J. Chem. Soc., Faraday Trans. 1*, **80**, 553 (1984).
608. A. J. Abkowicz-Bieńko and Z. Latajka, *J. Phys. Chem. A*, **104**, 1004 (2000).
609. B. Brzezinski, E. Grech, Z. Malarski, M. Rospenk, G. Schroeder and L. Sobczyk, *J. Chem. Res. (S)*, 151 (1997).
610. J. De Taeye, J. Parmentier and Th. Zeegers-Huyskens, *J. Phys. Chem.*, **92**, 4555 (1988).
611. G. G. Siegel and Th. Zeegers-Huyskens, *Spectrochim. Acta, Part A*, **45**, 1297 (1989).
612. S. M. Melikova, D. N. Shchepkin and A. Koll, *J. Mol. Struct. (THEOCHEM)*, **448**, 239 (1998).
613. P. Mighels, N. Leroux and Th. Zeegers-Huyskens, *Vibrational Spectrosc.*, **2**, 81 (1991).
614. P. Mighels and Th. Zeegers-Huyskens, *J. Mol. Struct. (THEOCHEM)*, **247**, 173 (1991).
615. K. De Wael, J. Parmentier and Th. Zeegers-Huyskens, *J. Mol. Liq.*, **51**, 67 (1992).
616. M. Goethals, B. Czarnik-Matusewicz and Th. Zeegers-Huyskens, *J. Heterocycl. Chem.*, **36**, 49 (1999).
617. D. Reyntjens-Van Damme and Th. Zeegers-Huyskens, *J. Phys. Chem.*, **84**, 282 (1980).
618. J. De Taeye and Th. Zeegers-Huyskens, *J. Pharm. Sci.*, **74**, 660 (1985).
619. O. Kasende and Th. Zeegers-Huyskens, *J. Phys. Chem.*, **88**, 2132 (1984).
620. J. Fritsch and G. Zundel, *Spectrosc. Lett.*, **17**, 41 (1984).
621. A. Filarowski and A. Koll, *Vib. Spectrosc.*, **17**, 123 (1998).
622. P. Imhof, W. Roth, C. Janzen, D. Spangenberg and K. Kleinermanns, *Chem. Phys.*, **242**, 141 (1999).
623. P. Imhof, W. Roth, C. Janzen, D. Spangenberg and K. Kleinermanns, *Chem. Phys.*, **242**, 153 (1999).
624. D. Hadži, A. Novak and J. E. Gordon, *J. Phys. Chem.*, **67**, 1118 (1963).
625. S. M. Melikova, A. Ju. Inzebejkin, D. N. Shchepkin and A. Koll, *J. Mol. Struct. (THEOCHEM)*, **552**, 273 (2000).
626. S. F. Bureiko, G. S. Denisov and R. M. Martsinkovsky, *React. Kinet. Catal. Lett.*, **2**, 343 (1975).
627. S. F. Bureiko, N. S. Golubev and I. Ya. Lange, *React. Kinet. Catal. Lett.*, **16**, 32 (1981).
628. A. Rabold, B. Brzezinski, R. Langner and G. Zundel, *Acta Chim. Slov.*, **44**, 237 (1997).

629. T. Keil, B. Brzezinski and G. Zundel, *J. Phys. Chem.*, **96**, 4421 (1992).
630. B. Brzezinski, G. Schroeder, G. Zundel and T. Keil, *J. Chem. Soc., Perkin Trans. 2*, 819 (1992).
631. B. Brzezinski, H. Maciejewska and G. Zundel, *J. Phys. Chem.*, **94**, 6983 (1990).
632. B. Brzezinski, B. Brycki, G. Zundel and T. Keil, *J. Phys. Chem.*, **95**, 8598 (1991); see also R. J. Alvarez and E. S. Kryachko, *J. Mol. Struct. (THEOCHEM)*, **433**, 263 (1998).
633. B. Brycki, B. Brzezinski, G. Zundel and T. Keil, *Magn. Reson. Chem.*, **30**, 507 (1992).
634. H. Merz, U. Tangermann and G. Zundel, *J. Phys. Chem.*, **90**, 6535 (1986).
635. H. Merz and G. Zundel, *Chem. Phys. Lett.*, **95**, 529 (1983).
636. A. Weichert, C. Riehn and B. Brutschy, *J. Phys. Chem. A*, **105**, 5679 (2001).
637. G. Zundel and A. Nagyrei, *J. Phys. Chem.*, **82**, 685 (1978).
638. G. Buczak, Z. Dega-Szafran, A. Katrusiak and M. Szafran, *J. Mol. Struct. (THEOCHEM)*, **436**, 143 (1997).
639. A. Koll and I. Majerz, *Bull. Soc. Chim. Belg.*, **103**, 629 (1994).
640. I. Majerz and A. Koll, *Polish J. Chem.*, **68**, 2109 (1994).
641. S. F. Bureiko and N. S. Golubev, *Zh. Strukt. Khim. (USSR) (Engl. Transl.)*, **28**, 171 (1987).
642. B. Brzezinski and G. Zundel, *J. Mol. Struct. (THEOCHEM)*, **380**, 195 (1996).
643. S. F. Bureiko, N. S. Golubev, G. S. Denisov and I. Ya. Lange, *Izv. Akad. Sci. Latv. SSR (Engl. Transl.)*, **3**, 369 (1980).
644. S. F. Bureiko and V. P. Oktyabrskii, *React. Kinet. Catal. Lett.*, **31**, 245 (1986).
645. S. F. Bureiko, N. S. Golubev and I. Ya. Lange, *Kinetika i Katalyz (USSR) (Engl. Transl.)*, **1**, 209 (1982).
646. A. Rabold and G. Zundel, *J. Phys. Chem.*, **99**, 12158 (1995).
647. E. Haslinger and P. Wolschann, *Monatsh. Chem.*, **11**, 563 (1980).
648. V. M. Schreiber, A. Koll and L. Sobczyk, *Bull. Acad. Polon. Sci., Ser. Sci. Chim.*, **24**, 651 (1978).
649. K. Rutkowski, S. M. Melikova and A. Koll, *Vib. Spectrosc.*, **7**, 265 (1994).
650. A. Koll, M. Rospenk and L. Sobczyk, *J. Chem. Soc., Faraday Trans. 1*, **77**, 2309 (1981).
651. A. Fedorowicz, J. Mavri, P. Bala and A. Koll, *Chem. Phys. Lett.*, **289**, 457 (1998).
652. A. Koll, M. Rospenk, E. Jagodzinska and T. Dziembowska, *J. Mol. Struct. (THEOCHEM)*, **552**, 193 (2000).
653. M. Rospenk, L. Sobczyk, A. Rabold and G. Zundel, *Spectrochim. Acta, Part A*, **55**, 85 (1999).
654. A. Filarowski, T. Glowiaka and A. Koll, *J. Mol. Struct. (THEOCHEM)*, **484**, 75 (1999).
655. M. Przeslawska, A. Koll and M. Witanowski, *J. Phys. Org. Chem.*, **12**, 486 (1999).
656. A. Koll and P. Wolschann, *Monatsh. Chem.*, **127**, 475 (1996).
657. K. Rutkowski and A. Koll, *J. Mol. Struct. (THEOCHEM)*, **322**, 195 (1994).
658. A. Szemik-Hojniak and A. Koll, *J. Photochem. Photobiol. A: Chem.*, **72**, 123 (1993).
659. A. Filarowski, A. Koll and T. Glowiak, *J. Chem. Cryst.*, **27**, 707 (1997).
660. A. Filarowski, A. Koll, T. Glowiak, E. Majewski and T. Dziembowska, *Ber. Bunsenges. Phys. Chem.*, **102**, 393 (1998).
661. A. Mandal, A. Koll, A. Filarowski, D. Majumder and S. Mukherjee, *Spectrochim. Acta, Part A*, **55**, 2861 (1999).
662. L. Sobczyk, *Appl. Magn. Reson.*, **18**, 47 (2000).
663. L. Sobczyk, *J. Mol. Struct.*, **177**, 111 (1988).
664. H. Ptasiewicz-Bak, R. Tellgren, I. Olovsson and A. Koll, *Z. Kristallogr.*, **212**, 126 (1997).
665. A. Filarowski, A. Koll and T. Glowiak, *Monatsh. Chem.*, **130**, 1097 (1999).
666. A. Mandal, A. Koll, A. Filarowski, D. Majumder and S. Mukherjee, *Spectrochim. Acta, Part A*, **55**, 2868 (1999).
667. A. Koll and P. Wolschann, *Monatsh. Chem.*, **130**, 983 (1999).
668. D. Guha, A. Mandal, A. Koll, A. Filarowski and S. Mukherjee, *Spectrochim. Acta, Part A*, **56**, 2669 (2000).
669. A. Fedorowicz and A. Koll, *J. Mol. Liq.*, **87**, 1 (2000).
670. M. Rospenk, A. Koll and L. Sobczyk, *Chem. Phys. Lett.*, **261**, 283 (1996).
671. B. Brzezinski, P. Radziejewski, A. Rabold and G. Zundel, *J. Mol. Struct. (THEOCHEM)*, **355**, 185 (1995).
672. B. Brzezinski, J. Olejnok, G. Zundel and R. Krämer, *J. Mol. Struct. (THEOCHEM)*, **212**, 247 (1989).
673. B. Brzezinski, J. Olejnik and G. Zundel, *J. Mol. Struct. (THEOCHEM)*, **238**, 89 (1990).

674. H. Schmideder, O. Kasende, H. Merz, P. P. Rastogi and G. Zundel, *J. Mol. Struct. (THEOCHEM)*, **161**, 87 (1987).
675. B. Brzezinski, H. Maciejewska and G. Zundel, *J. Phys. Chem.*, **96**, 6564 (1992).
676. B. Brzezinski, H. Urjasz, F. Bartl and G. Zundel, *J. Mol. Struct. (THEOCHEM)*, **435**, 59 (1997).
677. T. G. Wright, E. Gordes, O. Dopfer and K. Müller-Dethlefs, *J. Chem. Soc., Faraday Trans.*, **89**, 1601 (1993).
678. M. Schmitt, H. Müller, U. Henrichs, M. Gerhards, W. Perl, C. Deusen and K. Kleinermanns, *J. Chem. Phys.*, **103**, 584 (1995).
679. J. Küpper, A. Westphal and M. Schmitt, *Chem. Phys.*, **263**, 41 (2001).
680. O. Tishchenko, E. S. Kryachko and V. I. Staninets, *Theor. Expt. Khim. (Engl. Transl.)*, **35**, 331 (1999).
681. E. J. Bieske, M. W. Rainbird, I. M. Atkinson and A. E. W. Knight, *J. Chem. Phys.*, **91**, 752 (1989).
682. M. Schmidt, M. Mons and J. Le Calve, *Z. Phys. D*, **17**, 153 (1990).
683. M. Mons, J. Le Calve, F. Piuzzi and I. Dimicoli, *J. Chem. Phys.*, **92**, 2155 (1990).
684. X. Zhang and J. L. Knee, *Faraday Discuss.*, **97**, 299 (1994).
685. C. E. H. Dessent, S. R. Haines and K. Müller-Dethlefs, *Chem. Phys. Lett.*, **315**, 103 (1999).
686. A. Fujii, M. Miyazaki, T. Ebata and N. Mikami, *J. Chem. Phys.*, **110**, 11125 (1999).
687. S. R. Haines, C. E. H. Dessent and K. Müller-Dethlefs, *J. Chem. Phys.*, **111**, 1947 (1999).
688. D. M. Chapman, K. Müller-Dethlefs and J. B. Peel, *J. Chem. Phys.*, **111**, 1955 (1999).
689. N. Solcá and O. Dopfer, *Chem. Phys. Lett.*, **325**, 354 (2000).
690. R. I. Cukier, *J. Phys. Chem. A*, **103**, 5989 (1999).
691. P. Hobza, R. Burel, V. Špirko, O. Dopfer, K. Müller-Dethlefs and E. W. Schlag, *J. Chem. Phys.*, **101**, 990 (1994).
692. A. Goto, M. Fujii, N. Mikami and M. Ito, *J. Phys. Chem.*, **90**, 2370 (1986).
693. R. J. Lipert and S. D. Colson, *J. Phys. Chem.*, **93**, 3894 (1389) (1989).
694. R. J. Lipert and S. D. Colson, *J. Phys. Chem.*, **94**, 2358 (1990).
695. M. Schmitt, H. Mueller and K. Kleinermanns, *Chem. Phys. Lett.*, **218**, 246 (1994).
696. M. Gerhards and K. Kleinermanns, *J. Chem. Phys.*, **103**, 7392 (1995).
697. Ch. Jacoby, W. Roth, M. Schmitt, Ch. Janzen, D. Spangenberg and K. Kleinermanns, *J. Phys. Chem. A*, **102**, 4471 (1998).
698. G. Berden, W. L. Meerts, M. Schmitt and K. Kleinermanns, *J. Chem. Phys.*, **104**, 972 (1996).
699. M. Gerhards, M. Schmitt, K. Kleinermanns and K. Stahl, *J. Chem. Phys.*, **104**, 967 (1996).
700. K. Kleinermanns, M. Gerhards and M. Schmitt, *Ber. Bunsenges. Phys. Chem.*, **101**, 1785 (1997).
701. M. Schmitt, Ch. Jacoby and K. Kleinermanns, *J. Chem. Phys.*, **108**, 4486 (1998).
702. W. Roth, M. Schmitt, Ch. Jacoby, D. Spangenberg, Ch. Janzen and K. Kleinermanns, *Chem. Phys.*, **239**, 1 (1998).
703. S. Tanabe, T. Ebata, M. Fujii and N. Mikami, *Chem. Phys. Lett.*, **215**, 347 (1993).
704. N. Mikami, *Bull. Chem. Soc. Jpn.*, **68**, 683 (1995).
705. T. Watanabe, T. Ebata, S. Tanabe and N. Mikami, *J. Chem. Phys.*, **105**, 408 (1996).
706. A. Fujii, T. Sawamura, S. Tanabe, T. Ebata and N. Mikami, *Chem. Phys. Lett.*, **225**, 104 (1994).
707. T. Ebata, A. Fujii and N. Mikami, *Int. Rev. Phys. Chem.*, **17**, 331 (1998).
708. R. J. Stanley and A. W. Castleman, Jr., *J. Chem. Phys.*, **98**, 796 (1993).
709. M. Schütz, T. Bürgi, S. Leutwyler and H. B. Bürgi, *J. Chem. Phys.*, **99**, 5228 (1993).
710. S. Leutwyler, T. Bürgi, M. Schütz and A. Taylor, *Faraday Disc. Chem. Soc.*, **96**, 456 (1994).
711. T. Bürgi, M. Schütz and S. Leutwyler, *J. Chem. Phys.*, **103**, 6350 (1995).
712. O. Dopfer and K. Müller-Dethlefs, *J. Chem. Phys.*, **101**, 8508 (1994).
713. R. M. Helm and H. J. Neusser, *Chem. Phys.*, **239**, 33 (1998).
714. H.-D. Barth, K. Buchhold, S. Djafari, B. Reimann, U. Lommatzsch and B. Brutschy, *Chem. Phys.*, **239**, 49 (1998).
715. R. M. Helm, H.-P. Vogel and N. J. Neusser, *J. Chem. Phys.*, **108**, 4496 (1998).
716. H. J. Neusser and K. Siglow, *Chem. Rev.*, **100**, 3921 (2000).
717. C. E. H. Dessent and K. Müller-Dethlefs, *Chem. Rev.*, **100**, 3999 (2000).

718. E. W. Schlag, in *Adv. Chem. Phys., Volume 101: Chemical Reactions and Their Control on the Femtosecond Time Scale, XXth Solvay Conference on Chemistry* (Eds. P. Gaspard, I. Burghardt, I. Prigogine and S. A. Rice), Wiley, New York, 1997, pp. 607–623.
719. J. A. Syage, *J. Phys. Chem.*, **99**, 5772 (1995).
720. K. Kim, K. D. Jordan and T. S. Zwier, *J. Am. Chem. Soc.*, **116**, 11568 (1994).
721. K. Liti, M. G. Brown, C. Carter, R. J. Saykally, J. K. Gregory and D. C. Clary, *Nature* **381**, 501 (1996).
722. E. S. Kryachko, *Int. J. Quantum Chem.*, **70**, 831 (1998).
723. E. S. Kryachko, *Chem. Phys. Lett.*, **314**, 353 (1999) and references therein.
724. D. M. Benoit, A. X. Chavagnac and D. C. Clary, *Chem. Phys. Lett.*, **283**, 269 (1998).
725. D. M. Benoit and D. C. Clary, *J. Phys. Chem. A*, **104**, 5590 (2000).
726. D. C. Clary, D. M. Benoit and T. Van Mourik, *Acc. Chem. Res.*, **33**, 441 (2000).
727. R. C. Guedes, B. J. Costa Cabral, J. A. Martinho Simões and H. P. Diogo, *J. Phys. Chem. A*, **104**, 6062 (2000).
728. W.-H. Fang and R.-Z. Liu, *J. Chem. Phys.*, **113**, 5253 (2000).
729. H. Watanabe and S. Iwata, *J. Chem. Phys.*, **105**, 420 (1996).
730. H. Watanabe and S. Iwata, *Int. J. Quantum Chem., Quantum Chem. Symp.*, **30**, 395 (1996).
731. E. S. Kryachko and H. Nakatsuji, *J. Phys. Chem. A*, **106**, 73 (2002).
732. E. S. Kryachko and M. T. Nguyen, *J. Chem. Phys.*, **115**, 833 (2001).
733. By analogy with the earlier studies (Reference 729), the PES search of the phenol–(water)$_n$ complexes was initially performed by using a split-valence double-zeta 6-31G(d) basis set via a GAUSSIAN 98 suit of packages.
734. We distinguish five computational levels of theory/basis sets used for geometry optimizations although a 6-31G(d) basis set denoted throughout the present work as A plays a key role. The ground level corresponds to the common HF/A one, which is also employed for calculating harmonic frequencies, ZPVE, and thermodynamic properties. Empirical scaling factor of 0.8907 employed in Reference 729 was not used in the present work. Single-point (sp) energy calculations of the lower- energy PhOH(H$_2$O)$_n$ complexes were then performed at the MP2(sp)/A level in order to investigate the effect of correlation on their energy differences. The most stable PhOH(H$_2$O)$_{n=1-4}$ structures as the key structures in the present study were further refined at the MP2(fc)/A (fc is hereafter omitted) and, besides, the four lowest-energy PhOH(H$_2$O)$_4$ structures were also reoptimized at the MP2/6-31+G(d) ($\equiv$MP2/A$^+$) and B3LYP/A levels. The latter one was also used to recalculate their harmonic frequencies.
735. Z. Bačič and R. E. Miller, *J. Phys. Chem.*, **100**, 12945 (1996).
736. K. S. Kim, J. L. Lee, H. S. Choi, J. Kim and J. H. Jang, *Chem. Phys. Lett.*, **265**, 497 (1997).
737. K. S. Kim, P. Tarakeshwar and J. Y. Lee, *Chem. Rev.*, **100**, 4145 (2000).
738. J. M. Flaud, C. Camy-Payret and J. P. Maillard, *Mol. Phys.*, **32**, 499 (1976).
739. J. A. Riddick, W. B. Bungh and T. K. Sakano, *Organic Solvents*, 4th edn., Wiley, New York, 1986.
740. T. Takamuku, M. Tabata, M. Yamaguchi, J. Nishimoto, M. Kumamoto, H. Wakita and T. Yamaguchi, *J. Phys. Chem. B*, **102**, 8880 (1998) and references therein.
741. D. J. Jamroz, J. Stangret and J. Lingdren, *J. Am. Chem. Soc.*, **115**, 6165 (1993).
742. S. C. White and H. W. Thompson, *Proc. Roy. Soc. London A*, **291**, 460 (1966).
743. M. S. Sousa Lopes and H. W. Thompson, *Spectrochim. Acta, Part A*, **24**, 1367 (1968).
744. S. S. Mitra, *J. Chem. Phys.*, **36**, 3286 (1962).
745. A. Allerhand and P. v. R. Schleyer, *J. Am. Chem. Soc.*, **85**, 371 (1963).
746. T. Gramstadt and J. Sandström, *Spectrochim. Acta, Part A*, **25**, 31 (1969).
747. J. M. Campbell, Y. S. Park and H. F. Shurvell, *Can. J. Spectrosc.*, **36**, 6 (1991).
748. J. C. F. Ng, Y. S. Park and H. F. Shurvell, *J. Raman Spectrosc.*, **23**, 229 (1992).
749. J. C. F. Ng, Y. S. Park and H. F. Shurvell, *Spectrochim. Acta, Part A*, **48**, 1137 (1992).
750. S. M. Quadri and H. F. Shurvell, *Spectrochim. Acta, Part A*, **51**, 1355 (1995).
751. D. A. Buckingham, *Proc. Roy. Soc. London A*, **248**, 169 (1958).
752. M. Horak, J. Polakova, M. Jakoubkova, J. Moravec and J. Pliva, *Collect. Czech. Chem. Commun.*, **31**, 622 (1966).
753. M. Horak and J. Moravec, *Collect. Czech. Chem. Commun.*, **36**, 2757 (1971).

754. J. Yarwood, in *Spectroscopy and Structure of Molecular Complexes* (Ed. J. Yarwood), Plenum, London, 1973, pp. 182ff.
755. B. Czarnik-Matusewicz and Th. Zeegers-Huyskens, *J. Phys. Org. Chem.*, **13**, 237 (2000).
756. H. Figeys, P. Geerlings, D. Berckmans and C. Van Alsenoy, *J. Chem. Soc., Faraday Trans. 2*, **77**, 721 (1981).
757. T. Gramstadt and K. Tjessem, *J. Mol. Struct.*, **41**, 231 (1977).
758. M.-I. Baraton, *J. Mol. Struct.*, **10**, 231 (1971).
759. H. Abramczyk, W. Reimschüssel, H. Barańska and A. Labudzińska, *Chem. Phys.*, **94**, 435 (1985).
760. It was not actually our intention to explore the PES using a large basis set and more sophisticated computational level, so we have confined our PES search to the use of a rather simple density functional hybrid B3LYP computational level in conjunction with a split-valence double-zeta 6-31+G(d,p) basis set with the help of a GAUSSIAN 98 suit of packages[2]. The chosen computational level, which by no means could not be considered as rather inaccurate, was further employed for calculating harmonic frequencies and, therefore, for identifying the stationary points on the studied PES and also obtaining zero-point vibrational energy (ZPVE) in order to deduce the binding energy of the hydrogen-bonded complex AB as E_{HB} (AB) $= -([E(AB) - ZPVE(AB)] - ([E(A) - ZPVE(A)] + [E(B) - ZPVE(B)]))$ expressed throughout the present work in kJ mol^{-1}. The effect of the basis set superposition error (BSSE) was only tested for the phenol- acetonitrile complexes using the standard counterpoise procedure.
761. E. S. Kryachko and M. T. Nguyen, *J. Phys. Chem. A*, **106**, 4267 (2002). For a recent study of the $\pi-H$ bonding between phenol and HCN, HOCN, HF and HCl see E. S. Kryachko and M. T. Nguyen, *Polish J. Chem.*, **76**, 1233 (2002).
762. The B3LYP/6-31+G(d,p) optimized geometry of the acetonitrile molecule shown parenthetically in Figure 6 is fairly consistent with the microwave data [J. Demaison, A. Dubrelle, D. Boucher, J. Burie and V. Typke, *J. Mol. Spectrosc.*, **76**, 1 (1979) and C. C. Costain, *J. Chem. Phys.*, **29**, 864 (1958)]: $r(C-N) = 1.157$ Å, $r(C-C) = 1.462$ (1.458) Å, $r(C-H) = 1.095$ (1.102) Å, and $\angle C-C-H = 109.8°$ (109.5°) and appears to be more accurate than that obtained at the B3LYP/DZVP2 level [D. H. Barich, T. Xu, W. Song, Z. Wang, F. Deng and J. F. Haw, *J. Phys. Chem. B*, **102**, 7163 (1998)]. See also J. R. Reimers, J. Zeng and N. S. Hush, *J. Phys. Chem. A*, **100**, 1498 (1996) and compare with the values in Table 1 in Reference 763. It becomes evident there that the MP2/6-311++G(d,p) level (Reference 763) overestimates the C−N and C−C bond lengths by *ca* 0.005–0.017 Å and underestimates the C−H one by 0.010 Å.
763. E. M. Cabaleiro-Lago and M. Ríos, *J. Phys. Chem. A*, **101**, 8327 (1997) and references therein.
764. U. P. Agarwal, R. S. Green and J. Yarwood, *Chem. Phys.*, **74**, 35 (1983).
765. J. R. Reimers and L. E. Hall, *J. Am. Chem. Soc.*, **121**, 3730 (1999).
766. According to the MP2/cc-pVDZ calculations (Reference 765), the cyclic dimer has the binding energy of 18.7 kJ mol^{-1} whereas the 'head-to-tail' one is 8.7 kJ mol^{-1}. The HF/STO-3G computational level substantially underestimates the former value by a factor of 2.6 as quoted by A. Wakisaka, Y. Shimizu, N. Nishi, K. Tokumaru and H. Sakuragi, *J. Chem. Soc., Faraday Trans.*, **88**, 1129 (1992) based on the incorrect Reference 209.
767. J. Jawed, *Bull. Chem. Soc. Jpn.*, **49**, 659 , 1155 (1976).
768. K. Tanabe, *Chem. Phys.*, **63**, 135 (1981).
769. H. Abramczyk and W. Reimschüssel, *Chem. Phys.*, **100**, 243 (1985).
770. T. Steiner, C. C. Wilson and I. Majerz, *Chem. Commun.*, 1231 (2000).
771. I. Dodgson, K. Griffen, G. Barberis, F. Pignatoro and G. Tauszik, *Chem. Ind.*, 830 (1989).
772. E. Y. Bezuglaya, A. B. Shchutskaya and I. V. Smirnova, *Atmos. Environ.*, **27**, 773 (1998).
773. S. K. Ong and A. R. Bowers, *J. Environ. Eng.*, **116**, 1013 (1990).
774. E.-J. Shin and M. A. Keane, *Ind. Eng. Chem. Res.*, **39**, 883 (2000).
775. M. Engstrom, F. Himo and H. Angren, *Chem. Phys. Lett.*, **319**, 191 (2000).
776. A. F. Hegarty and J. P. Keogh, *J. Chem. Soc. Perkin Trans. 2*, 758 (2001) and references therein.

777. B. J. C. Cabral, R. G. B. Fonseca and J. A. Martinho Simøes, *Chem. Phys. Lett.*, **258**, 436 (1996).
778. J. Florian and A. Warshel, *J. Phys. Chem. B*, **101**, 5583 (1997).
779. W. Siebrand, M. Z. Zgierski and Z. K. Smedarchina, *Chem. Phys. Lett.*, **279**, 377 (1997).
780. C. Chipot, *J. Phys. Chem. B*, **105**, 5987 (2001).
781. A. Weichert, C. Riehn and B. Brutschy, *J. Phys. Chem. A*, **105**, 5679 (2001) and references therein.
782. Z. Bikadi, G. Keresztury, S. Holly, O. Egyed, I. Mater and M. Simonyi, *J. Phys. Chem. A*, **105**, 3471 (2001) and references therein.

CHAPTER 2

The structural chemistry of phenols

MENAHEM KAFTORY

Department of Chemistry, Technion — Israel Institute of Technology, Haifa 32000, Israel
Fax: (972) 48293761; e-mail: kaftory@tx.technion.ac.il

I. INTRODUCTION	199
II. STRUCTURAL CHEMISTRY OF MONO- AND POLYHYDROXYBENZENES	200
III. STRUCTURAL CHEMISTRY OF SUBSTITUTED PHENOLS (2–13) . . .	204
A. Substituted 1,2-Dihydroxybenzene (3)	204
B. Substituted 1,3-Dihydroxybenzene (4)	207
C. Substituted 1,4-Dihydroxybenzene (5)	209
D. Substituted 1,3,5-Trihydroxybenzene (6)	211
E. Substituted 1,2,3-Trihydroxybenzene (7)	211
F. Substituted 1,2,4-Trihydroxybenzene (8)	212
IV. STERIC AND ELECTRONIC EFFECTS ON THE STRUCTURAL CHEMISTRY OF PHENOLS	212
A. *ortho*-Substituted Phenols (39)	213
B. *meta*-Substituted Phenols (40)	213
C. *para*-Substituted Phenols (41)	213
V. SPECIAL SUBSTITUTED PHENOLS	216
A. Nitrophenols	216
B. Fluoro, Chloro and Bromo Phenols	219
VI. REFERENCES	220

I. INTRODUCTION

Phenols are organic compounds that contain a hydroxyl group (−OH) bound directly to carbon atom in a benzene ring. The structural moiety of phenols in the context of the present chapter is given by structure **1**.

The Chemistry of Phenols. Edited by Z. Rappoport
© 2003 John Wiley & Sons, Ltd ISBN: 0-471-49737-1

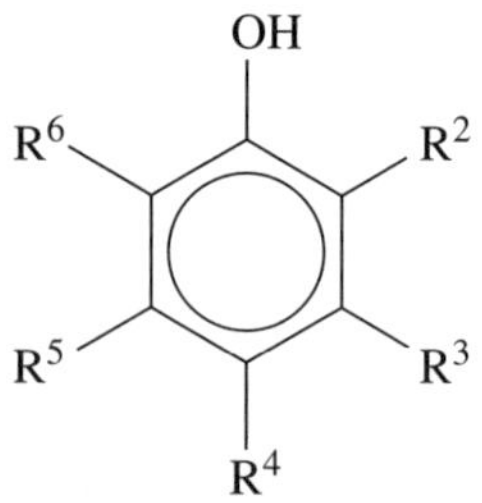

                            OH
                            |
            R⁶ ╲   ╱ R²
                 ╲   ╱
                  ╲ ╱
                  ╱ ╲
                 ╱   ╲
            R⁵ ╱   ╲ R³
                            |
                            R⁴

where R^2–R^6 are H, C, N, O, S, F, Cl, Br

(1)

93,460 publications can be found in one of the databases for scientific references under the word 'phenols'; when adding the words 'structural chemistry' the number of publications drops dramatically to 732. None of these publications summarizes or discusses the molecular geometry and intermolecular geometry of solid phenols. However, in a chapter entitled 'Solid state chemistry of phenols and possible industrial applications'[1] the geometries of phenols are described. In the present chapter we summarize the molecular structure of phenols mostly with regard to the geometry at the hydroxyl group. The best source of structural data is the Cambridge Crystallographic Structural Data Centre[2]. The analysis was conducted using geometrical data from crystal structures that were refined to $R < 0.075$, omitting organometallic compounds. Statistical analysis was executed in most cases for the relevant geometric parameters. The statistical analysis was performed with the Origin Program[3]; an average value of a geometric parameter and its standard deviation (s.d.) were calculated according to equations 1 and 2:

$$d(\text{mean}) = (1/N)\Sigma d_i \quad (1)$$

where d(mean) is the calculated average, N is the number of data points, the sum is taken over all data points and d_i is the experimental value;

$$\text{s.d.} = \{[1/(N-1)]\Sigma[d_i - d(\text{mean})]^2\}^{1/2} \quad (2)$$

where s.d. is the standard deviation.

II. STRUCTURAL CHEMISTRY OF MONO- AND POLYHYDROXYBENZENES

Before we discuss the structural chemistry of phenols it is important to describe the geometry of mono- and polyhydroxyphenols compounds (**2–13**) as observed in their crystal structures. Careful examination of the crystal structures of **2–6**[4–7] shows that hydrogen bonding is not only an important factor in controlling the packing arrangement of phenols but also affects the molecular geometry. Although the crystal structure of 1,2,3-trihydroxybenzene (**7**) is known both in its pure solid state and in its complex with two molecules of 8-hydroxyquinoline[8], no geometrical details have been published. The crystal structures of 1,2,4-trihydroxybenzene (**8**) and of tetra-, penta- and hexa-hydroxybenzenes (**9–13**) are unknown. In five compounds (**2–6**), hydrogen bonding is the dominant factor in determining the molecular packing in the crystals. Hydrogen bonds also affect the molecular geometry, especially the HO−C bond lengths and the bond angles involving this bond. Therefore, we start our discussion with a description of the hydrogen bonding in these compounds.

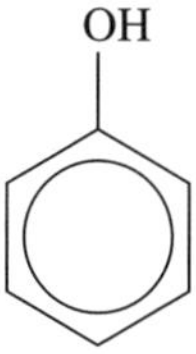

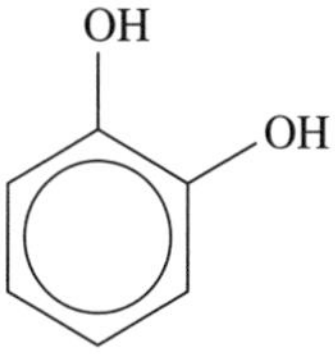

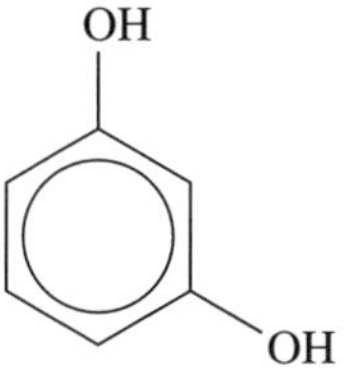

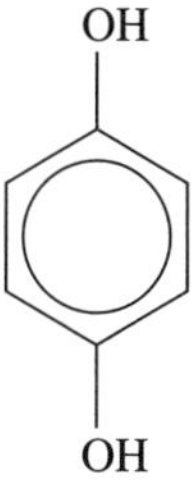

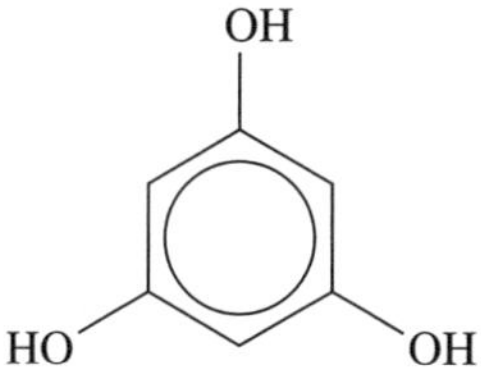

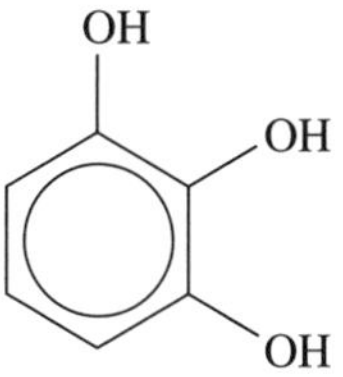

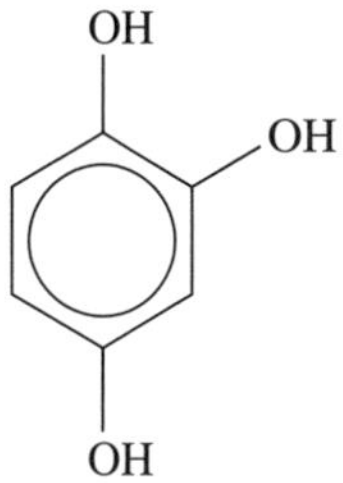

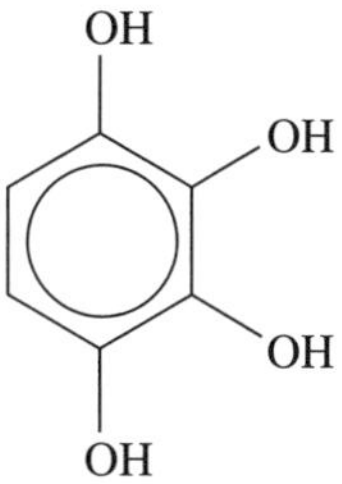

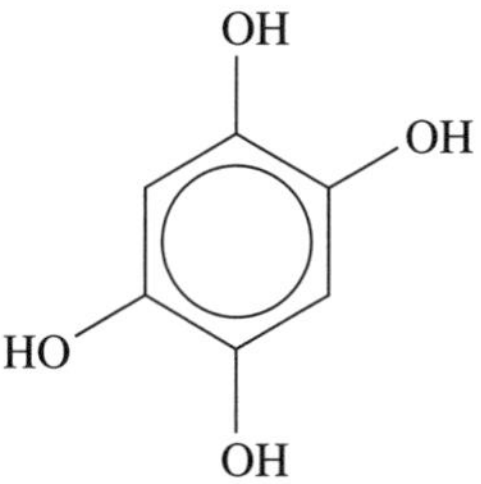

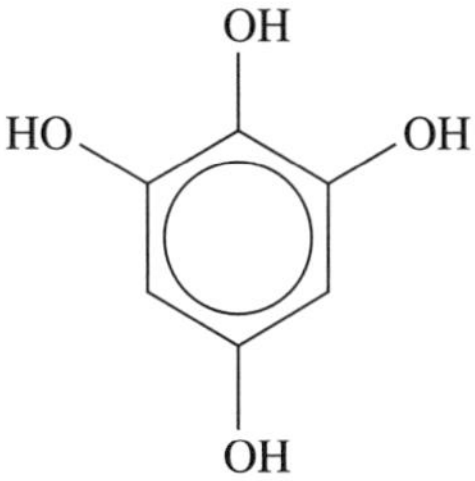

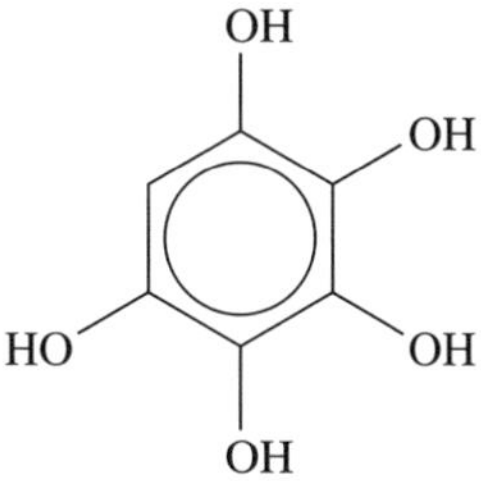

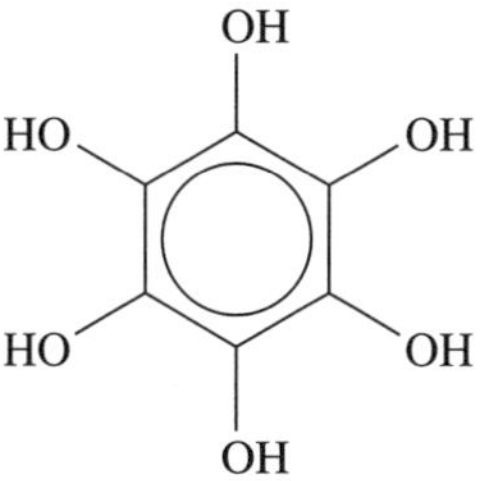

Figure 1 shows the hydrogen bonding in the five compounds. The hydrogen bond geometry is given in Table 1. With the exception of **3**, each hydroxyl oxygen atom plays the roles of both an acceptor and a donor for hydrogens. The hydrogen bonding schemes of **2** and **4** are very similar. Three crystallographically independent molecules of **2** form an infinite one-dimensional hydrogen-bond pattern (see Table 1 for the geometry of the hydrogen bonding); **4** forms a two-dimensional hydrogen-bonding pattern by using the two hydroxyl groups. The crystal structure of **6** shows that each molecule is hydrogen bonded to six neighbors. Molecules of **5** form an infinite arrangement of hexagons made up of six molecules.

The bond distances (Å) and bond angles in compounds **2–6** are shown in Figure 2 and their average values are given in Table 2. It is clearly seen that the averages of all bond lengths within the aromatic ring are practically equal and that the average C—OH bond is 1.371(2) Å. The outer-ring bond angles a1 and a2, on the other hand, are very sensitive

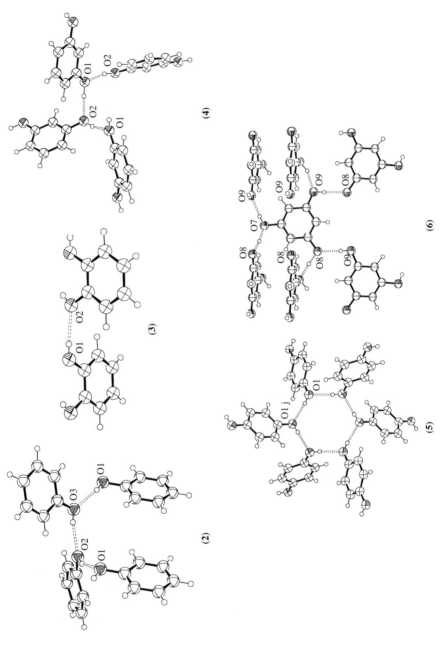

FIGURE 1. Hydrogen bonding in the crystal structures of **2–6** (O1j in (**5**) appears as O2 in Table 1.)

TABLE 1. Hydrogen bond geometry (Å, deg) in **2–6**[a]

Compound	D–H A	D–H	H ⋯ A	D ⋯ A	D–H ⋯ A
2	O1–H1 O3	0.82	1.89	2.655	156.0
	O3–H3 O2	0.74	1.74	2.693	164.7
	O2–H2 O1	0.91	1.81	2.664	157.2
3	O1–H1 O2	0.81	1.99	2.796	169.7
4	O1–H1 O2	0.98	1.73	2.714	175.9
	O2–H2 O1	0.98	1.76	2.718	165.6
5	O1–H1 O2	0.78	1.91	2.678	167.4
6	O7–H7 O9	0.97	1.83	2.763	158.4
	O8–H8 O7	1.27	1.49	2.750	170.8
	O9–H8 O8	0.85	1.92	2.730	160.1

[a]D and A are the donor and acceptor for hydrogen, respectively.

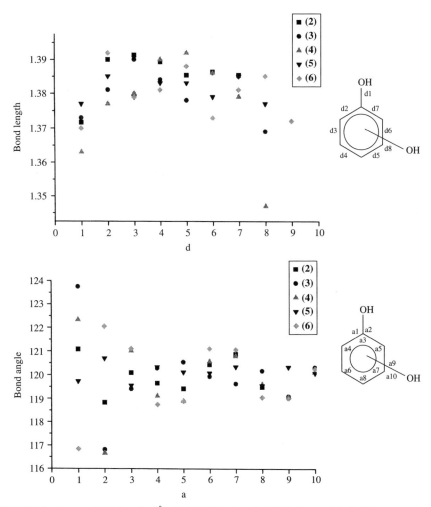

FIGURE 2. Average bond lengths (Å) (top) and bond angles (deg) (bottom) in **2–6**

TABLE 2. Average bond distances (d in Å) and bond angles (a in deg) and their standard deviations (s.d.) and standard errors (s.e.)

	d1	d2	d3	d4	d5	d6	d7	d8		
mean	1.371	1.385	1.384	1.385	1.385	1.382	1.383	1.370		
s.d.	0.005	0.006	0.006	0.004	0.005	0.006	0.003	0.164		
s.e.	0.002	0.003	0.003	0.002	0.002	0.003	0.001	0.008		
	a1	a2	a3	a4	a5	a6	a7	a8	a9	a10
mean	120.7	119.0	120.2	119.6	119.6	120.4	120.5	119.6	119.4	120.2
s.d.	2.6	2.4	0.8	0.7	0.7	0.5	0.6	0.4	0.6	0.1
s.e.	1.2	1.2	0.4	0.3	0.3	0.2	0.3	0.2	0.3	0.1

to the position of the hydrogen atom relative to the ring. The bond angle C−C−O (a1 or a2) *syn* to the C−O−H bond angle is in all five compounds larger than the bond angle *anti* to the C−O−H bond angle. Therefore, there is a significant scattering of a1 and a2 as seen in Figure 2 (bottom), and expressed by the large standard deviation in the mean values shown in Table 2.

The O−H bond is practically co-planar with the aromatic ring. The range of the absolute values of the rotation angle (expressed by H−O−C−C torsion angle) is 0.2−12.9° with the exception of 1,3,5-trihydroxybenzene (**6**), where a larger torsion angle was found (38.7°).

III. STRUCTURAL CHEMISTRY OF SUBSTITUTED PHENOLS (2–13)

It is interesting to compare the structures of the parent compounds **2–13** with their substituted analogues. The geometry data were obtained for the analogues where the substituents are H, C, O, N, F, Cl or Br. The structural chemistry of the most interesting systems is given below.

A. Substituted 1,2-Dihydroxybenzene (3)

The mean value of the C−OH bond length (d1) (see notation in Figure 2) calculated from 144 experimental values is 1.365 Å (s.d. = 0.014, s.e. = 0.001). The mean value of the HOC−COH bond length (d3) is 1.396 Å (s.d. = 0.015, s.e. = 0.001). The mean value of d2 and d5, which are chemically symmetry-related bonds, is 1.381 Å (s.d. = 0.016, s.e. = 0.0009). The mean bond length of d6 is 1.398 Å (s.d. = 0.020, s.e. = 0.002). While the histogram of the above bond lengths shows clearly a single maximum, the histogram of bond lengths d4 and d7 shows a double maximum (see Figure 3).

It turned out that d4 and d7 are longer in 31 compounds, all consisting of 1,2-dihydroxynaphthalene skeleton such as **14**[9] and **15**[10].

The mean value of d4 and d7 bonds in the 1,2-dihydroxybenzenes is 1.395 Å (s.d. = 0.014, s.e. = 0.001) while the mean value of d4 and d7 in the naphthalene analogue is 1.440 Å (s.d. = 0.020, s.e. = 0.003). The most interesting bond angles are the outer-ring angles involved with the hydroxyl group. The four bond angles a1, a2, a9 and a10 (O1−C1−C2, O1−C1−C6, O2−C6−C1 and O2−C6−C5, respectively, as shown in **16**) are strongly dependent on the local conformation of the O−H bond relative to the ring plane. In most of the 1,2-dihydroxybenzenes, the O−H bond is coplanar with the ring as expressed by the conformations shown in **16** and **17**. In the conformation presented by **16**

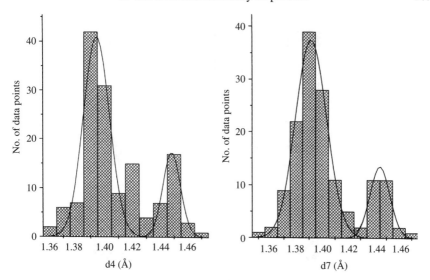

FIGURE 3. Histogram of d4 and d7 bond lengths (Å) in substituted **3**

the expected torsion angles are 0° and 180° for H1−O1−C1−C6 and H2−O2−C6−C1, respectively. The conformation presented by **17** is characterized by a single torsion angle of 180°. In the conformation of **18**, on the other hand, one of the torsion angles is 0° and the other is 90°. Figure 4 shows the conformation map of 1,2-dihydroxybenzenes

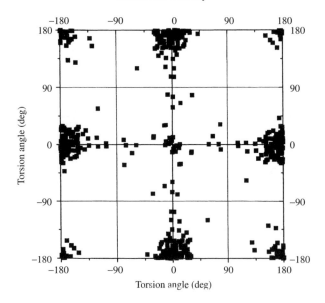

FIGURE 4. Conformation map in 1,2-dihydroxybenzenes

expressed by the two torsion angles mentioned above. There are 145 data points, which were expanded to 590 data points by the use of symmetry considerations.

As shown in Figure 4, most of the compounds adopt the conformation shown schematically by **16** (0° and 180°). There are only few compounds that adopt the conformation shown in **17** (180° and 180°) and **18** (0° and 90°). The conformation presented by **16** is dominant, due to the ability to form intramolecular hydrogen bonds. All three conformations are observed in the crystal structure of 10,15-dihydro-2,3,7,8,12,13-hexahydroxy-5H-tribenzo(a,d,g)cyclononane dipropanolate clathrate (**19**)[11].

The effect of the conformation is best seen when comparing bond angles a1, a2, a9 and a10 of compounds adopting the conformations presented by **16** and **17** (see Table 3 calculated from 52 and 18 data points, respectively). The bond angles at C1 (a1 and a2) or at C6 (a9 and a10) are larger at the side of the hydrogen due to steric congestion with

(**19**)

TABLE 3. Average bond distances (d in Å) and bond angles (a in deg) and their standard deviations in two different conformations of 1,2-dihydroxybenzenes

Conformation	d1(C1−O1)	d2(C6−O2)	a1	a2	a9	a10	No. of data points
syn-anti (16)	1.370(10)	1.359(15)	119.6(1.3)	119.8(2.2)	116.1(1.6)	123.3(1.3)	52
anti-anti (17)	1.365(13)	1.364(10)	123.1(9)	116.9(8)	117.1(1.4)	123.1(8)	18

a neighboring hydrogen atom. Therefore, it is expected that a1 and a10 (see Table 3) in compounds adopting the *anti-anti* (with respect to bond C1−C6) conformation of **17** will be larger than a2 and a9. In compounds adopting the *syn-anti* (with respect to bond C1−C6) conformation of **16**, a10 is indeed larger than a9; however, a1 is practically equal to a2. This finding is attributed to the intramolecular hydrogen bond formed between the two hydroxyl groups. It is also important to notice the difference between the two C−OH bond lengths in compounds having the *syn-anti* conformation. This bond is longer whenever the oxygen atom plays the role of acceptor for hydrogen [1.370(10) Å compared with 1.359(15) Å].

B. Substituted 1,3-Dihydroxybenzene (4)

The majority of substituted 1,3-dihydroxybenzenes adopt either *syn-anti* (with respect to atom C6, with torsion angles of 0° and 180° at the C1−O and C5−O bonds, respectively) or *anti-anti* (with respect to atom C6, with torsion angles of 180° and 180° at the C1−O and C5−O bonds, respectively) conformation as shown by the conformation map in Figure 5. The *syn-syn* conformation (with respect to atom C6, with torsion angles of 0° and 0° at C1−O and C5−O, respectively) was observed for 23 compounds.

As in substituted 1,2-dihydroxybenzenes, the bond angles involved with the OH group are larger at the side of the hydrogen atom, therefore a2 and a9 (see notation in Figure 2) are larger than their counterparts a1 and a10 in compounds having the *syn-anti* conformation. In the compounds adopting the *anti-anti* conformation, a1 and a10 are larger than a2 and a9 (Table 4).

2,6-Dihydroxybenzoic acid crystallizes in two polymorphic forms[12,13], monoclinic and orthorhombic. The molecule in the monoclinic form adopts the *syn-anti* conformation (**20**) while it adopts the *syn-syn* conformation in the orthorhombic form (**21**).

In the crystal structure of 2,2′,4,4′-tetrahydroxybenzophenone[14] there are two crystallographically independent molecules in the asymmetric unit, each adopting a different conformation as shown in **22** and **23**. Intermolecular hydrogen bonds determine the conformations of the two compounds.

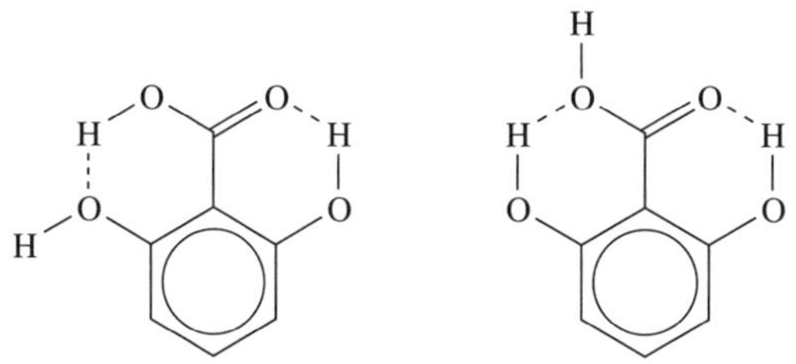

(20)   (21)

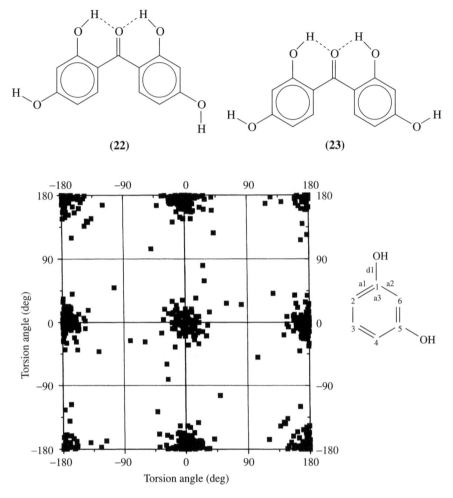

FIGURE 5. Conformation map in 1,3-dihydroxybenzenes

TABLE 4. Average bond distances (d in Å) and bond angles (a in deg) and their standard deviations in three different conformations of 1,2-dihydroxybenzenes

Conformation	d1	d2	a1	a2	a9	a10	No. of data points
syn-anti	1.362(14)	1.360(13)	117.2(1.1)	121.7(1.4)	118.7(2.5)	121.8(1.4)	87
anti-anti	1.356(14)	1.354(16)	121.6(1.6)	117.0(1.1)	117.7(1.5)	121.4(1.8)	50
syn-syn	1.367(17)	1.361(15)	117.7(1.7)	121.6(1.8)	117.5(1.0)	121.5(1.5)	23

C. Substituted 1,4-Dihydroxybenzene (5)

Statistical analysis of the bond lengths in substituted 1,4-dihydroxybenzenes shows that it has C_{2v} symmetry. The histograms are given in Figure 6, using the notation given in Figure 2. The mean value of the C–OH bond distance is 1.365 Å (s.d. = 0.018, s.e. = 0.001, $N = 296$). The mean bond length of d4 and d5 is 1.392 Å (s.d. = 0.019, s.e. = 0.001, $N = 296$), and the mean bond length of d2, d3, d6 and d7 is 1.392 Å (s.d. = 0.013, s.e. = 0.001, $N = 592$).

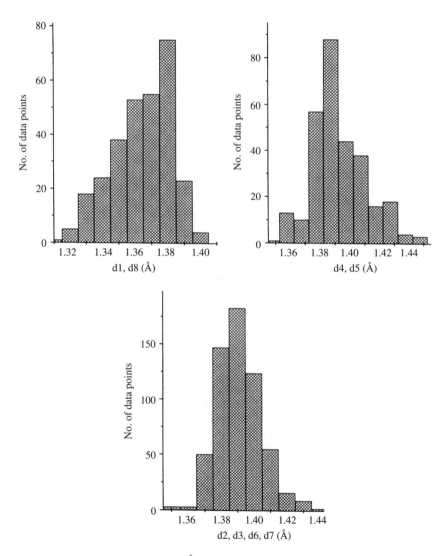

FIGURE 6. Histogram of bond lengths (Å) in substituted **5**

Five different conformations (**24–28**) might be expected to be observed in substituted 1,4-dihydroxybenzenes. The rotation of the O—H bond relative to the ring plane is expressed by the torsion angles shown in Figure 7.

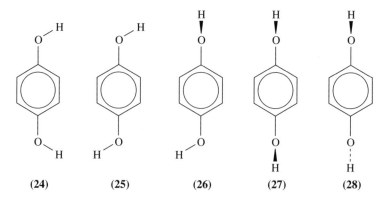

It is clearly shown that, as in the previously mentioned substituted dihydroxybenzenes, most of the compounds adopt the two conformations **24** and **25** (expressed by torsion angles of 0° and 180°). There are, however, compounds that adopt conformation **28**. In most cases the conformation is determined by the substituents. For example, in the crystal structure of tris(hydroquinone) methyl isocyanide clathrate[15] the hydroquinone adopts the conformation of **25** with the expected opening of the bond angle at the side of the hydrogen atom (a2 in **16**), as a result of the steric repulsion by the neighboring hydrogen atom and a closing of the other bond angle (a1 in **16**) (123.4° and 116.6°, respectively).

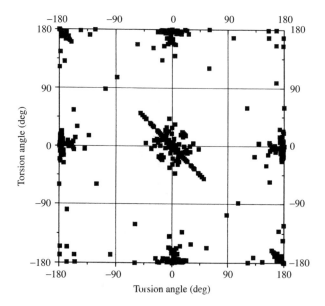

FIGURE 7. Conformation map in 1,4-dihydroxybenzenes

However, although the same conformation was found also in the structure of chloranilic acid **(29)** with pyrazine[16], the difference between the bond angles is reversed, namely a1 (122.3°) is larger than a2 (117.7°) as a result of the attractive hydrogen bonding with the carbonyl oxygen and the repulsion between the hydroxyl oxygen and the electronegative chlorine atom.

The parent compound adopts a different conformation when another intramolecular hydrogen bonding is available, such as in **30**[17], and yet another conformation when intermolecular hydrogen bonding are available, such as in the crystal structure of 2,5-dibromohydroquinone[18] **(31)**.

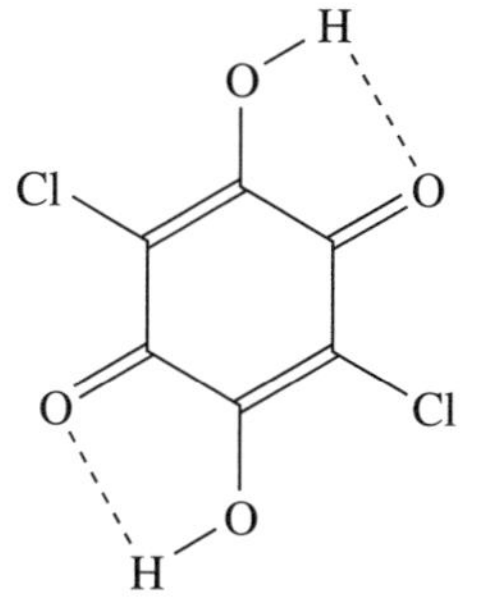

(29) 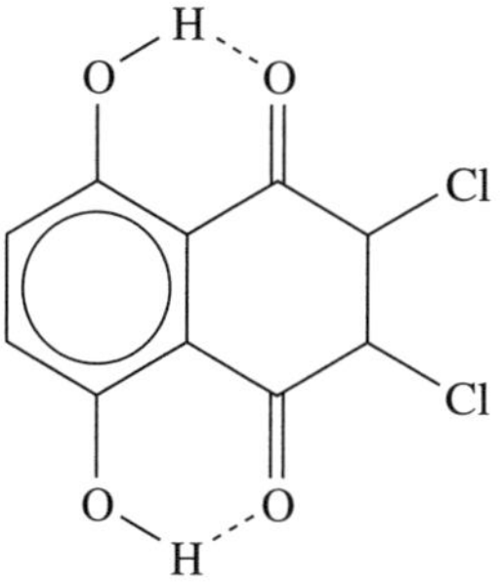 **(30)** 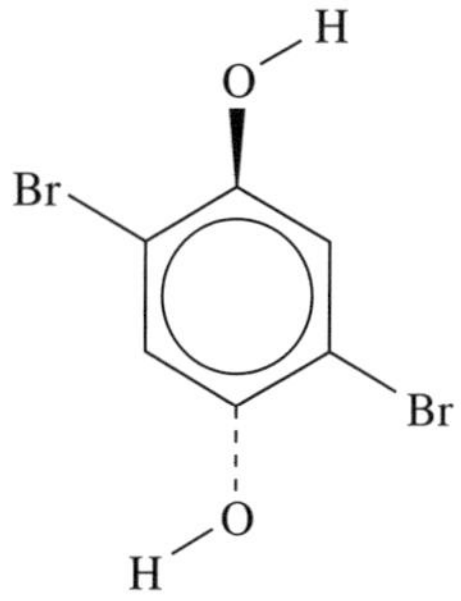 **(31)**

D. Substituted 1,3,5-Trihydroxybenzene (6)

The crystal structures of only seven compounds of substituted 1,3,5-trihydroxybenzene (including the nonsubstituted parent compound) are known. The mean value of the C−OH bond length is 1.358(20) Å. Five of these compounds adopt the conformation represented by macrocarpal[19] **(32)** and by 2,4,6-trinitro-1,3,5-benzenetriol **(33)**[20]. In the complex between 1,3,5-trihydroxybenzene and 4-methylpyridine[21] the conformation is different, as shown in **34**.

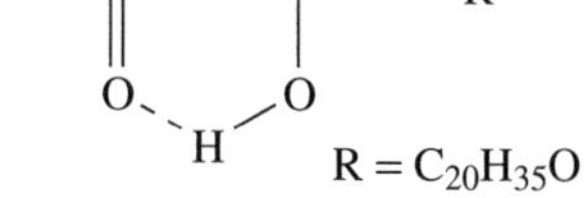

$R = C_{20}H_{35}O$

(32) **(33)** 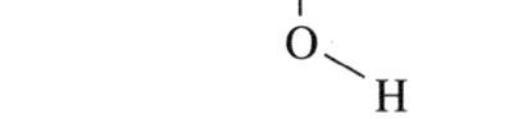 **(34)**

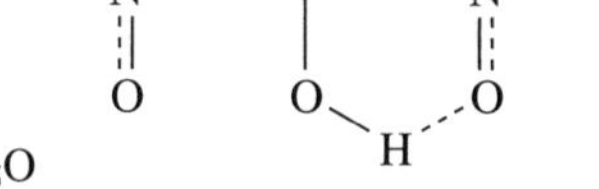

E. Substituted 1,2,3-Trihydroxybenzene (7)

The crystal structures of seven substituted 1,2,3-trihydroxybenzenes are known. The average C−OH bond length is 1.367(11) Å. The two different conformations observed among this class of compounds are represented by 3,4,5-trihydroxybenzoic acid (**35**) monohydrate[22], and by 3,4,5-trihydroxybenzohydroxamic acid (**36**) monohydrate[23] and 2,3,4-trihydroxyacetophenone[24] (**37**). There is severe steric congestion in 1,2,3-trihydroxybenzenes caused by the neighboring hydroxyl groups. One of the hydroxyl groups is rotated from the ring plane to minimize this steric hindrance. Therefore, the central C—OH bond is rotated by 24.5° in **35** and by 29.0° in **37**.

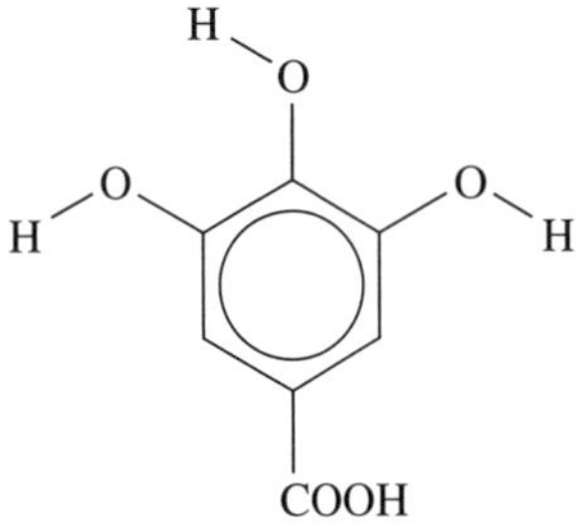

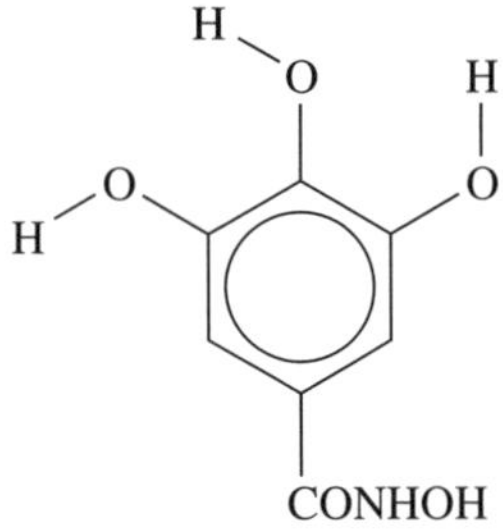

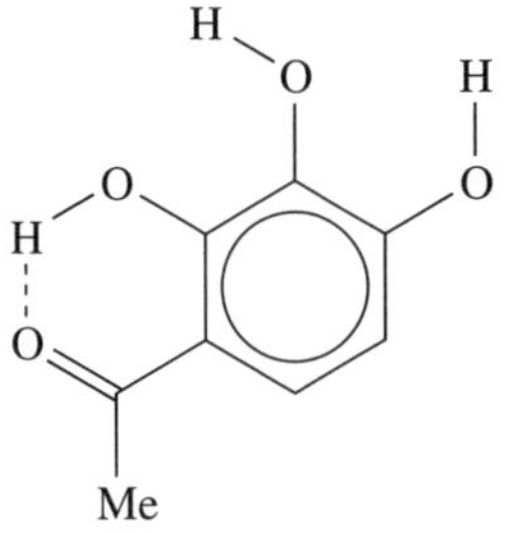

F. Substituted 1,2,4-Trihydroxybenzene (8)

The crystal structures of only three substituted 1,2,4-trihydroxybenzenes are known. All three have the same conformation, determined by intramolecular hydrogen bonding such as in (2′S,4aS)-4,4a-dihydro-5,6,8-trihydroxy-7-(2′-hydroxypropyl)-1,2,4a-trimethylphenanthrene-3,9-dione[25] (**38**).

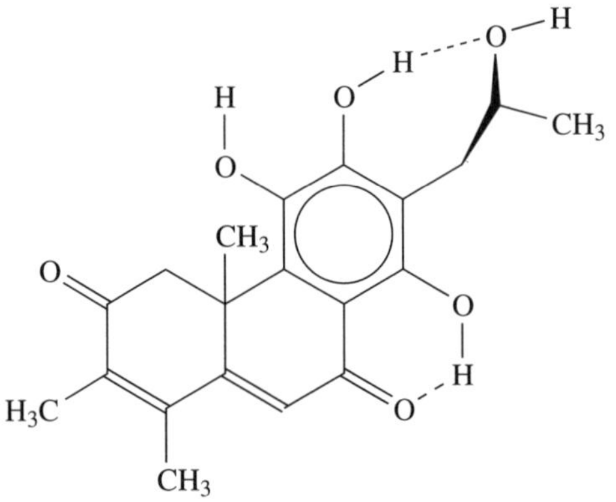

IV. STERIC AND ELECTRONIC EFFECTS ON THE STRUCTURAL CHEMISTRY OF PHENOLS

The structural chemistry, namely bond lengths and bond angles, are subject to electronic and congestion effects. In this paragraph we compare the structural parameters in compounds of type **39–41** where R^2, R^3 and R^4 are N, O or C atoms.

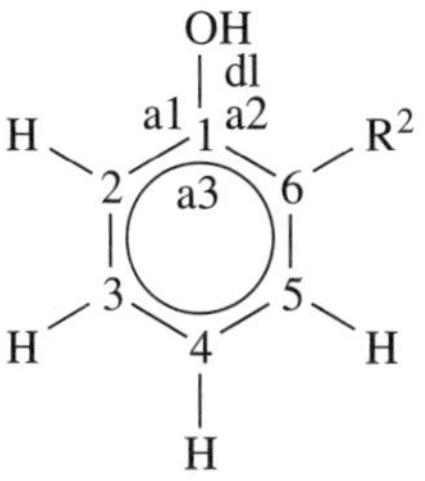

2. The structural chemistry of phenols

(39) 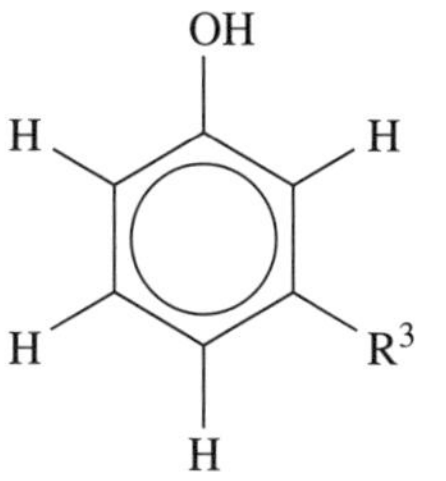 (40) 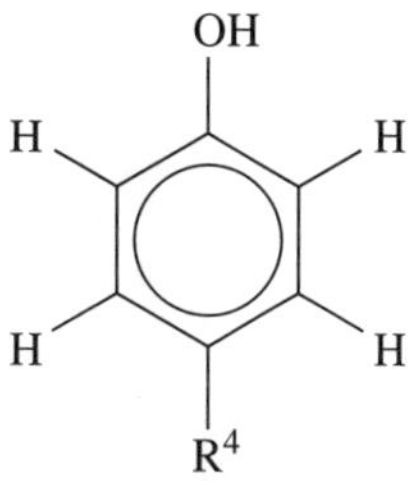 (41)

A. *ortho*-Substituted Phenols (39)

The effect on the geometry of substituted phenol is most pronounced upon substitution at the next-neighboring carbon to the hydroxyl group (i.e. in the *ortho* position) such as in **39**. The bond angles a1, a2 and a3 are highly dependent on the orientation of the O—H (expressed by the torsion angle H—O—C1—C6). When $R^2 = N$ there are 8 compounds with H—O—C1—C6 torsion angle of 0° (or close to 0°) (*cis* conformation) and 19 with torsion angle close to 180° (*trans* conformation). The average d1 is not significantly longer (1.362 Å, s.d. 0.008) in the former than in the latter (1.358 Å, s.d. 0.011). The position of the hydrogen atom with respect to the nitrogen atom has a major effect on the bond angles a1 and a2. Therefore, the average bond angle a1 is smaller than the average of the bond angle a2 [118.4(1.1)° and 121.8(1.7)°, respectively] when H—O—C1—C6 is close to 0°, but the average bond angle a1 is larger than the average bond angle a2 when this torsion angle is close to 180° [123.5(0.7)° and 117.1(1.0)°, respectively]. Very similar geometry was found in compounds where $R^2 = O$. The average bond length of d1 is practically equal and is not affected by the position of the hydrogen atom [1.369(6) Å]. The average bond angle a1 [118.9(5)° for 7 data points] is smaller than the average of the bond angle a2 [121.4(1.0)° for 6 data points] when the conformation is *cis*, and the average of a1 is larger than the average of a2 [123.9(6)° and 116.5(8)°, respectively] when the conformation is *trans*.

There are 176 reference codes in the Cambridge Crystallographic Structural Database of phenols of type **39** where $R^2 = C$. The histograms of the C—O bond length (d1) and bond angles a1 and a2 are shown in Figures 8a and 8b, respectively. The average of d1 bond length is 1.355(9) Å for 135 data points when the conformation is *cis*, and 1.362(9) Å for 65 data points when the conformation is *trans*. The average bond angles a1 and a2 are 117.9(1.2)° and 121.9(1.0)°, respectively, for the *cis* conformation and 121.9(9)° and 117.7(9)°, respectively, for the *trans* conformation.

B. *meta*-Substituted Phenols (40)

The small number of known crystal structures of phenols of type **40** does not provide meaningful statistical averaging of structural parameters and their dependence on the substituent R^3 and on the conformation with regard to the O—H bond. The average bond length d1 is 1.367(9) Å for 29 observations.

C. *para*-Substituted Phenols (41)

There are over 200 crystal structures of substituted phenols of type **41** in the CCSD. In 23 of them, $R^4 = N$. The C1—OH bond length (d1) is somewhat shorter [1.356(14) Å] than

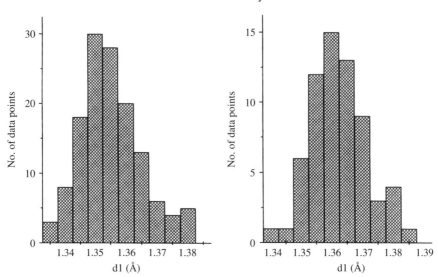

FIGURE 8a. Histogram of d1 in compounds of type **39** where $R^2 = C$ and the conformation is *cis* (left) and *trans* (right)

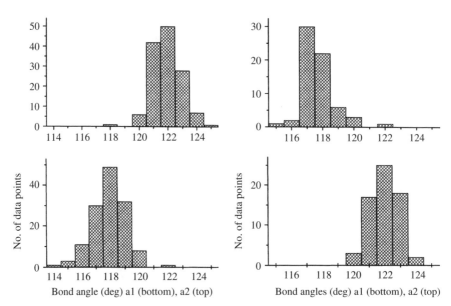

FIGURE 8b. Histogram of bond angles a1 (top) and a2 (bottom) in compounds of type **39** where $R^2 = C$ and the conformation is *cis* (left) and *trans* (right)

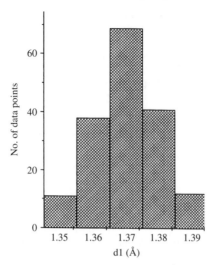

FIGURE 9. Histogram of d1 in compounds of type **41** where $R^4 = C$

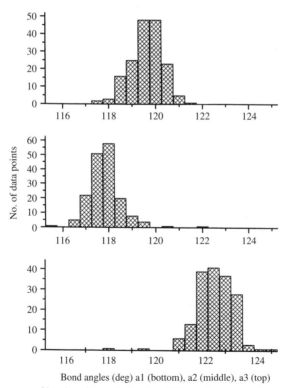

FIGURE 10. Histogram of bond angles a1 (top), a2 (middle) and a3 (bottom) in compounds of type **41** when $R^4 = C$

for the compounds with different substituents such as O [1.376(9) Å for 31 observations] and C [1.371(10) Å for 171 observations]; see also Figure 9. The average bond angles a1, a2 and a3 are very similar and independent of the substituent R^4. The averages of a1 are 122.5(6)°, 122.4(7)° and 122.5(8)°, the averages of a2 are 117.8(6)°, 117.9(5)° and 117.8(7)°, and the averages of a3 are 119.7(4)°, 119.6(5)° and 119.6(7)° for phenols of type **41** with $R^4 =$ N, O and C, respectively. Histograms of the bond length and bond angles when $R^4 =$ C are shown in Figures 9 and 10.

V. SPECIAL SUBSTITUTED PHENOLS

The effect of special substituents, such as nitro groups and halogens, on the geometry of phenols deserves special attention.

A. Nitrophenols

The presence of an acceptor for protons (the nitro group) and a donor for protons (the OH group) on the same molecule may affect the structure of the molecule as well as the molecular arrangement in the solid state. It can adopt either an intramolecular hydrogen bond as shown for o-nitrophenol[26] (see Figure 11, top left) or intermolecular hydrogen bonds as shown for m-nitrophenol[27] (see Figure 11, top right) and in the two polymorphs of p-nitrophenol[28]. The geometrical parameters of the hydrogen bonding in m-nitrophenol are: the OH···O distance is 2.181 Å, the O···O distance is 2.935 Å, the O−H···O angle is 178.5°. There are small but significant differences in the relative geometry of molecules connected by intermolecular hydrogen bonds in the two polymorphs of p-nitrophenol. In the β-phase the two molecules are coplanar (see Figure 11, bottom left), the OH···O distance is 1.908 Å, the O···O distance is 2.831 Å and the O−H···O angle is 160.6°. In the α-phase the two molecules are inclined to each other (see Figure 11, bottom right), the hydrogen bond is much weaker, and the geometrical parameters are: OH···O distance is 2.461 Å, O···O distance is 3.196 Å, O−H···O angle is 133.2°.

2,4,6-Trinitrophenol (picric acid) and its substituents are good examples to demonstrate the effect of intramolecular hydrogen bonding on the molecular structures of the compounds. The molecular structures of five compounds possessing different substituents: 2,4,6-trinitrophenol (picric acid)[29] (Figure 12a), 3,5-dimethylpicric acid[30] (Figure 12b), 3,5-dichloropicric acid[31] (Figure 12c), 2,4,6-trinitro-1,3,5-benzenetriol[20] (Figure 12d) and 3,5-diaminopicric acid[32] (Figure 12e), are shown in Figure 12. An intramolecular hydrogen bond between the hydroxyl and one of the o-nitro groups exists in all five compounds, therefore the H−O bond and the hydrogen-bonded nitro group are coplanar with the aryl ring. The second o-nitro group that is not involved in the hydrogen bonding is rotated with respect to the ring plane. The rotation angles are 50.8, 52.8, 73.6, 60.8 and 52.5° for the five compounds, respectively.

While the nitro group in the p-position is coplanar with the ring in picric acid, it is rotated whenever the neighboring carbon atom is substituted by a bulky group, such as methyl in 3,5-dimethylpicric acid and chlorine in 3,5-dichloropicric acid (83.3° and 83.7°). In 2,4,6-trinitro-1,3,5-benzenetriol, all the donors are involved with intramolecular hydrogen bonding. In 3,5-diaminopicric acid, on the other hand, one of the NH groups is not hydrogen bonded to the neighboring nitro group that is rotated out of the ring plane by 52.5°.

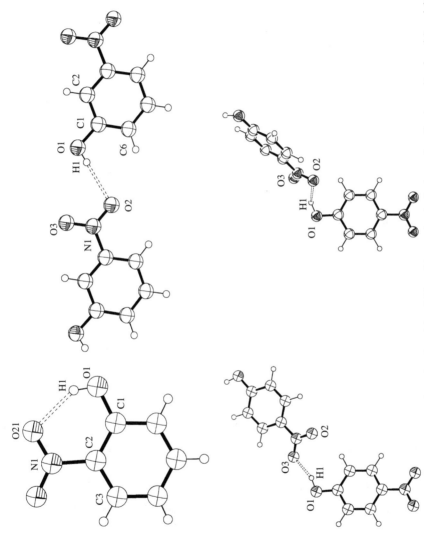

FIGURE 11. Intramolecular hydrogen bonding in *o*-nitrophenol (top left), and intermolecular hydrogen bonding in *m*-nitrophenol (top right), *p*-nitrophenol (bottom left) and *p*-nitrophenol (bottom right)

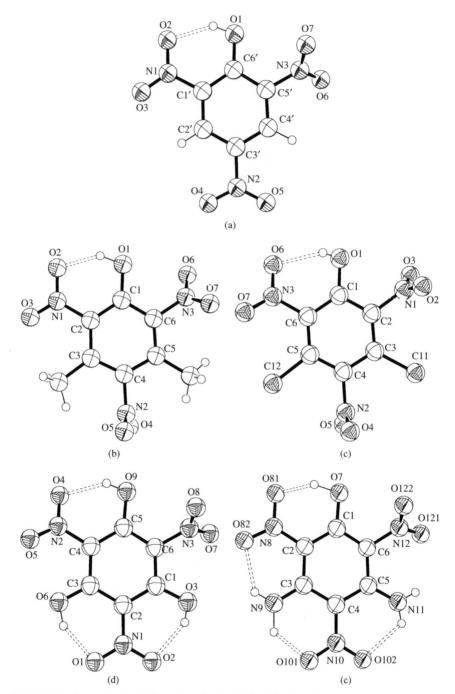

FIGURE 12. Intramolecular hydrogen bonding in 2,4,6-trinitrophenols

2. The structural chemistry of phenols

TABLE 5. Comparison of the average bond length (Å) and bond angle (deg) at the hydroxyl group in nitro-substituted phenols (the notation is given in **42**)

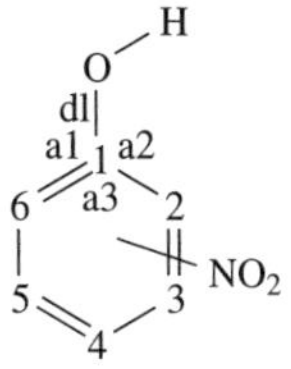

	2-Nitro	3-Nitro	4-Nitro	2,4,6-Trinitro
d1	1.343(8)	1.352(13)	1.346(9)	1.323(10)
a1	118.7(2.9)	116.9(6)	117.0(1.1)	118.3(1.3)
a2	123.8(2.8)	123.4(8)	123.0(1.9)	125.1(1.1)
a3	117.5(1.4)	119.7(7)	120.0(1.5)	116.6(1.3)
No. of data points	24	8	25	20

Comparison of the C—OH bond length and the bond angles at C1 in nitro-substituted phenols is given in Table 5. The presence of a nitro group as a substituent causes a dramatic decrease in the C—OH bond length. In all the compounds discussed in previous paragraphs, the range of the C—OH bond lengths was 1.356–1.371 Å, while this bond decreases to 1.323 for 2,4,6-trinitrophenols. It also seems that the bond angles at the hydroxyl group (a1, a2 and a3) are affected by the positions of the nitro groups. The most significant effect is observed for 2,4,6-trinitrophenols, where a3 is the smallest angle [116.6(1.3)°] and a2 is the largest [125.1(1.2)°].

B. Fluoro, Chloro and Bromo Phenols

Shortening of the C—OH bond length is also observed in halogen-substituted phenols. An average bond length of 1.343(6) Å was obtained from seven complexes, such as bis(pentafluorophenol) dioxane[33]. The average of the inner bond angle (a3) is 117.6(1.6)°. However, the crystal structure of five of these compounds has been solved with data collected at liquid nitrogen temperature, which might be the reason for the shortening of the C—OH bond. In 3,5-difluorophenol[34] (**43**) and in 2,3,5,6-tetrafluorohydroquinone[35] (**44**) these bond lengths are 1.375 and 1.362 Å, respectively.

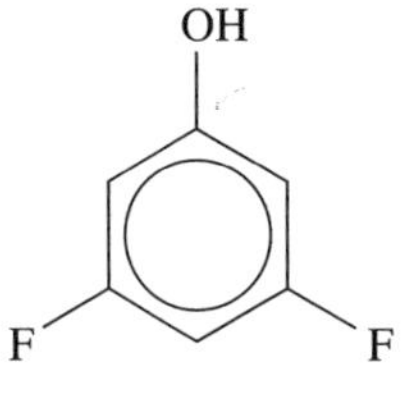

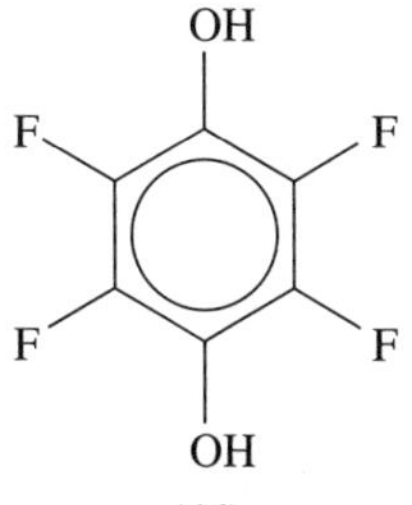

The crystal structures of many o- and p-chlorophenols, but only of a few m-chlorophenols, are known. Representative examples are 1,5-dichloro-2,6-dihydroxynaphthalene[36] (**45**), a complex between 3,5-dichlorophenol (**46**) and 2,6-dimethylphenol[37], and a complex of p-chlorophenol (**47**) with 1,4-phenylenediamine[38]. The C—OH bond lengths in **45** and **47** are normal (1.364 and 1.361 Å, respectively). The same bond in **46** is significantly longer (1.387 Å) for unknown reasons.

Comparison of the average geometrical parameters in o- and p-chloro and bromophenols is given in Table 6. The average bond angles in o-chlorophenols and o-bromophenols as well as the average bond angles in p-chlorophenols and p-bromophenols are the same.

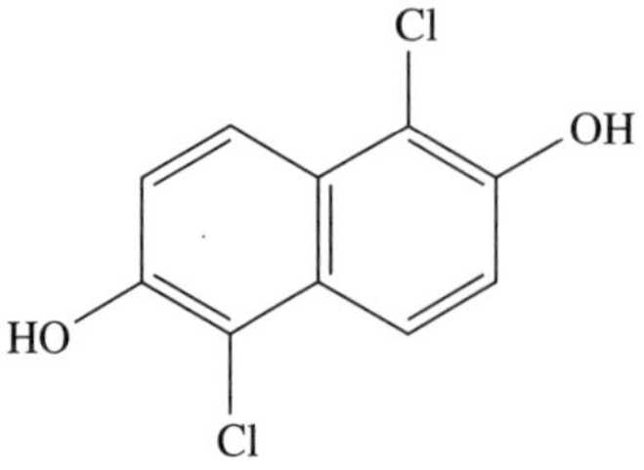

TABLE 6. Comparison of the average bond length (Å) and bond angle (deg) at the hydroxyl group in o- and p-chloro and bromophenols (the notation is given in **48**)

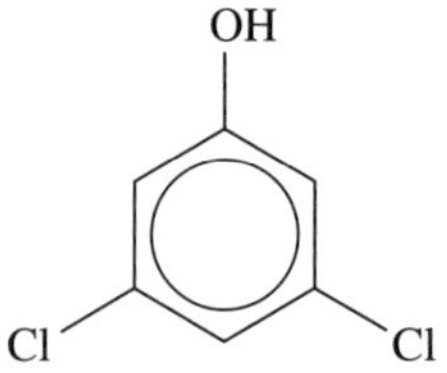

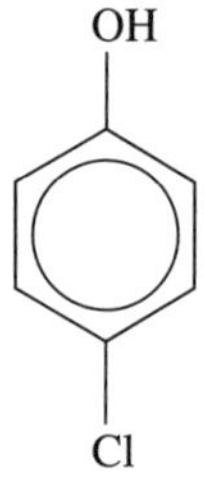

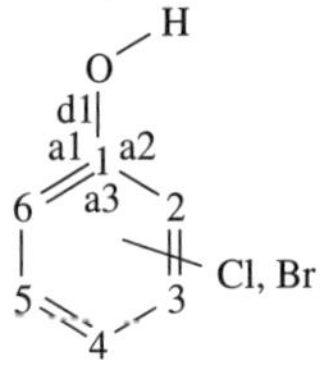

	o-Chloro	p-Chloro	o-Bromo	p-Bromo
d1	1.349(14)	1.358(19)	1.356(14)	1.362(20)
a1	118.7(1.1)	117.6(1.8)	118.5(1.2)	117.8(2.0)
a2	123.3(1.4)	121.9(1.8)	123.7(1.3)	121.5(1.1)
a3	118.0(1.1)	120.5(1.7)	117.8(1.3)	120.7(1.9)
No. of data points	32	74	29	23

As expected, substituents at the o-position will affect the bond angles. Therefore, a2 in both o-chlorophenols and o-bromophenols is larger than in the p-substituted phenols. The other bond angles are adjusted accordingly.

VI. REFERENCES

1. R. Perrin, R. Lamartine, M. Perrin and A. Thozet, in *Organic Solid State Chemistry* (Ed. G. R. Desiraju), Elsevier, Amsterdam, 1987, p. 271.
2. CCSD, Cambridge Crystallographic Structural Data Centre, Cambridge, U.K.
3. Origin, software for technical graphics and data analysis, Microcal Software Inc.
4. V. E. Zavodnik, V. K. Bel'skii and P. M. Zorkii, *Zh. Strukt. Khim.*, **28**, 175 (1987); *Chem. Abstr.*, **108**, 29805a (1987).
5. H. Wunderlich and D. Mootz, *Acta Crystallogr., Sect. B*, **27**, 1684 (1971).
6. G. E. Bacon and R. J. Jude, *Z. Kristallogr.*, **138**, 19 (1973).
7. S. V. Lindeman, V. E. Shklover and Yu. T. Struchkov, *Cryst. Struct. Commun.*, **10**, 1173 (1981).
8. P. Becker, H. Brusset and H. Gillier-Pandraud, *C. R. Acad. Sci. Paris, Ser. C*, **274**, 1043 (1972).
9. J. A. Herbert and M. R. Truter, *J. Chem. Soc., Perkin Trans. 2*, 1253 (1980).
10. S. A. Talipov, B. T. Ibragimov, G. B. Nazarov, T. F. Aripov and A. S. Sadikov, *Khim. Prir. Soedin*, 835 (1985); *Chem. Abstr.*, **102**, 42891w (1984).
11. J. A. Hyatt, E. N. Duesler, D. Y. Curtin and I. C. Paul, *J. Org. Chem.*, **45**, 5074 (1980).
12. L. R. MacGillivray and M. J. Zaworotko, *J. Chem. Cryst.*, **24**, 703 (1994).
13. M. Gdaniec, M. Gilski, and G. S. Denisov, *Acta Crystallogr., Sect. C*, **50**, 1622 (1994).
14. E. O. Schlemper, *Acta Crystallogr., Sect. B*, **38**, 554 (1982).
15. T. L. Chan and T. C. Mak, *J. Chem. Soc., Perkin Trans. 2*, 777 (1983).
16. H. Ishida and S. Kashino, *Acta Crystallogr., Sect. C*, **55**, 1923 (1999).
17. P. Rubio, S. Garcia-Blanco and J. G. Rodriguez, *Acta Crystallogr., Sect. C*, **41**, 1797 (1985).
18. S. Hoger, K. Bonrad and V. Enkelmann, *Z. Naturforsch., Teil B*, **53**, 960 (1998).
19. M. Nishizawa, M. Emura, Y. Kan, H. Yamada, K. Ogawa and N. Hamanaka, *Tetrahedron Lett.*, **33**, 2983 (1992).

20. J. J. Wolff, F. Gredel, H. Imgartinger and D. Dreier, *Acta Crystallogr., Sect. C*, **52**, 3225 (1996).
21. K. Biradha and M. J. Zaworotko, *J. Am. Chem. Soc.*, **120**, 6431 (1998).
22. Ren-Wang Jiang, Dong-Sheng Ming, P. P. H. But and T. C. W. Mak, *Acta Crystallogr., Sect. C*, **56**, 594 (2000).
23. B. B. Nielsen, *Acta Crystallogr., Sect. C*, **49**, 810 (1993).
24. E. O. Schlemper, *Acta Crystallogr., Sect. C*, **42**, 755 (1986).
25. A. C. Alder, P. Ruedi, J. H. Bieri and C. H. Eugster, *Helv. Chim. Acta*, **69**, 1395 (1986).
26. F. Iwasaki and Y. Kawano, *Acta Crystallogr., Sect. B*, **34**, 1286 (1978).
27. F. Pandarese, L. Ungaretti and A. Coda, *Acta Crystallogr., Sect. B*, **31**, 2671 (1975).
28. G. U. Kulkarni, P. Kumaradhas and C. N. R. Rao, *Chem. Mater.*, **10**, (1998).
29. F. H. Herbstein and M. Kaftory, *Acta Crystallogr., Sect. B*, **32**, 387 (1976).
30. M. K. Chantooni Junior and D. Britton, *J. Chem. Cryst.*, **28**, 329 (1998).
31. M. K. Chantooni Junior and D. Britton, *J. Chem. Cryst.*, **27**, 237 (1997).
32. S. K. Bhattacharjee and H. L. Ammon, *Acta Crystallogr., Sect. B*, **37**, 2082 (1981).
33. T. Gramstad, S. Husebye and K. Maartman-Moe, *Acta Chem. Scand. Ser. B*, **39**, 767 (1985).
34. M. Shibakami and A. Sekiya, *Acta Crystallogr., Sect. C*, **50**, 1152 (1994).
35. V. R. Thalladi, H. C. Weiss, R. Boese, A. Nangia and G. R. Desiraju, *Acta Crystallogr., Sect. B*, **55**, 1005 (1999).
36. K. Nakasuji, K. Sugiura, K. Kitagawa, J. Toyoda, H. Okamoto, K. Okaniwa, T. Mitani, H. Yamamoto, I. Murata, A. Kawamoto and J. Tanaka, *J. Am. Chem. Soc.*, **113**, 1862 (1991).
37. C. Bavoux and A. Thozet, *Cryst. Struct. Commun.*, **9**, 1115 (1980).
38. J. H. Loehlin, K. J. Franz, L. Gist and R. H. Moore, *Acta Crystallogr., Sect. B*, **54**, 695 (1998).

CHAPTER 3

Thermochemistry of phenols and related arenols

SUZANNE W. SLAYDEN

Department of Chemistry, George Mason University, 4400 University Drive, Fairfax, Virginia 22030, USA
Phone: (+1)703-993-1071; Fax: (+1)703-993-1055; e-mail: sslayden@gmu.edu

and

JOEL F. LIEBMAN

Department of Chemistry and Biochemistry, University of Maryland, Baltimore County, 1000 Hilltop Circle, Baltimore, Maryland 21250, USA
Phone: (+1)520-455-2549; Fax: (+1)410-455-2608; e-mail: jliebman@umbc.edu

I. INTRODUCTION: SCOPE AND DEFINITIONS	224
A. Thermochemistry	224
B. Definition of Phenols and Arenols: Comparisons with Related Compounds	225
II. ARENOLS	227
A. Unsubstituted Arenols	227
1. The OH/H increment exchange energies: δ(OH/H) and $\delta^\&$(OH/H)	228
2. Comparison of phenol with alkanols	228
3. Naphthols and anthrols	229
B. Carbon-bonded Substituents	229
1. Monoalkylated phenols: methyl and *tert*-butyl substituents	229
2. The amino acid tyrosine and its derivatives	231
3. Carboxylic acids and their derivatives	232
4. Acylphenols and their derivatives	232
5. Cyanophenols	233
C. Nitrogen-bonded Substituents	233
1. Aminophenols	234
2. Nitrosophenols	235

The Chemistry of Phenols. Edited by Z. Rappoport
© 2003 John Wiley & Sons, Ltd   ISBN: 0-471-49737-1

	3. Nitrophenols	235
D.	Oxygen-bonded Substituents	236
	1. Hydroxy derivatives	236
	2. Alkoxy derivatives	236
E.	Sulfur-bonded Substituents	237
F.	Halogen Substituents	237
	1. Monohalophenols	237
	2. Dihalophenols	239
	3. Polyhalophenols	240
III. ARENEDIOLS		241
A.	Unsubstituted Benzenediols	242
B.	Alkylated Benzenediols	242
C.	Otherwise Substituted Benzenediols	243
D.	Naphthalenediols and Other Arenediols	246
IV. ARENETRIOLS		247
V. ARENOLQUINONES		248
VI. TAUTOMERIC ARENOLS		250
A.	Obstacles and Opportunities	250
B.	Unsubstituted Arenols	250
C.	Nitrosophenols and Nitrosonaphthols (Quinone Oximes)	251
D.	Arylazo Derivatives of Phenol and the Naphthols	252
E.	Ambiguous Arenepolyols	252
VII. REFERENCES AND NOTES		254

I. INTRODUCTION: SCOPE AND DEFINITIONS

A. Thermochemistry

As has been the approach for most of the authors' other reviews on organic thermochemistry[1], the current chapter is primarily devoted to the relatively restricted property, the 'molar standard enthalpy of formation', $\Delta_f H_m °$, often called the 'heat of formation', ΔH_f or $\Delta H_f °$. This chapter foregoes discussion of other thermochemical properties such as Gibbs energy, entropy, heat capacity and excess enthalpy. We also avoid discussion of bond dissociation energies (e.g. of the phenolic O—H bond) and gas phase clustering energies (e.g. with halide or metal ions). Likewise, we ignore questions of acid strength (in either solution or gas phase) or of any intermolecular complexation energies except for hydrogen bonding in the pure condensed phase. The temperature and pressure are assumed to be 25 °C ('298 K') and 1 atmosphere or one bar (101,325 or 100,000 Pa) respectively. The energy units are kJ mol^{-1} (where 4.184 kJ ≡ 1 kcal).

Unreferenced enthalpies of formation are taken from the now 'classic' thermochemical archive by Pedley and his coworkers[2]. These thermochemical numbers are usually for comparatively simple and well understood species where we benefit from the data evaluation performed by these authors, rather than using the raw, but much more complete, set of data found in a recent, evolving, on-line data base[3]. Where there are more recently published values in the literature, we include those as well.

Again following our earlier chapters as precedent, we continue to emphasize gas phase species in the discussions. Condensed phases in general are complicated, and phenols the more so because these solids may be intra- or intermolecularly hydrogen bonded, and the resulting thermochemical results are often idiosyncratic. For example, under the thermochemical idealized conditions, 3-methylphenol (*m*-cresol) is a liquid while its isomers,

2- and 4-methylphenol (*o*- and *p*-cresols), are solids. No answer is apparent as to why the phases are not the same other than to note that had the standard temperature been 5° or 35 °C instead of 25 °C (i.e. closer to the water/ice divide and 'normal' human body temperature, respectively), all three isomers would be solids or liquids, respectively.

Enthalpies of vaporization (ΔH_{vap}) and of sublimation (ΔH_{sub}) are necessary to interrelate gas phase data with those for the liquid or solid state that characterizes most organic compounds as they are customarily synthesized, reacted, purified and thermochemically investigated. These are defined by equations 1 and 2,

$$\Delta_{vap}H \equiv \Delta_f H_m°(g) - \Delta_f H_m°(lq) \tag{1}$$

$$\Delta_{sub}H \equiv \Delta_f H_m°(g) - \Delta_f H_m°(s) \tag{2}$$

where g, lq and s refer to gas, liquid and solid, respectively[4]. While we accept the values of these quantities at any temperature, we endeavor to choose those that correspond to the above idealized conditions. Experimentally measured enthalpies of vaporization and/or sublimation of phenols are affected by the diminished vapor pressure by Raoult's law. More importantly, the enthalpy of formation of most gas phase species is found by summing the enthalpy of formation of the liquid or solid phase compound with the appropriate phase change enthalpy. It is very rare that enthalpies of formation of gas phase species are obtained by measuring the enthalpy of combustion of the gas.

It is occasionally necessary to use data for a species as liquid when the compound is 'normally' a solid, or as a solid when it is 'normally' a liquid. These two phases are numerically interrelated by the enthalpy of fusion[5] as defined by equation 3.

$$\Delta_{fus}H \equiv \Delta_f H_m°(lq) - \Delta_f H_m°(s) \tag{3}$$

This last quantity is quite temperature independent and so values most conveniently and most often measured at the melting point are used without correction.

Finally, phenols have a tendency to autooxidize and so form quinones and thereby condense to form ill-defined polymers. The 'label on the bottle' and the stoichiometry and structure do not completely correspond. Thus, the measured enthalpy of combustion and the derived enthalpy of formation are for an impure sample.

B. Definition of Phenols and Arenols: Comparisons with Related Compounds

In this chapter an arenol is taken to be any carbocyclic aromatic species in which one or more C—H units have been replaced by C—OH. The aromatic species is most generally a benzene ring, in which case the compound is a phenol. Phenols dominate the discussion because benzene derivatives of any type are more prevalent than derivatives of any other type of aromatic species. Only occasionally are there thermochemical data for derivatives of naphthalene and still rarer are derivatives of other benzenoid hydrocarbons. We discuss the parent and substituted phenols, naphthols, anthrols, arenepolyols and tautomerically ambiguous species. The substituent groups encompass carbon-bonded (e.g. alkyl, carboxy, carbonyl), nitrogen-bonded (e.g. amino, nitro, nitroso, azo), oxygen-bonded, sulfur-bonded and the halogens. Although our earlier review[1] published in 1993 lists over 100 phenols and arenols, the focus there was on alcohols. We deemed it desirable in this chapter to analyze and compare the data with an intent to provide insights and interrelations along with enthalpies.

Arenols, and phenols in particular, are not best understood as ordinary alcohols, any more than carboxylic acids are understood as either alcohols or ketones. As such, the change in enthalpy of formation on oxygenating benzene to phenol is not the same as, for example, oxygenating butane to n- or sec-butyl alcohol. The hydroxyl group affects the enthalpies of formation differently when attached to saturated vs. unsaturated carbon. When attached to saturated carbon, the oxygen is σ-electron withdrawing; when attached to unsaturated carbon the oxygen is simultaneously σ-electron withdrawing and π-electron donating. The three classical zwitterionic/dipolar resonance structures for phenol portray the π-donation and provide a 'textbook' rationalization for the preferred o- and p-substitution of phenol by electrophilic reagents. Indeed, it is this *ortho, para* proclivity that no doubt accounts for so many of the isomer 'choices' in the thermochemical literature. Calorimetrists are rarely synthetic chemists.

We would like to compare phenols with the corresponding isoelectronic methyl, amino and fluoro aromatic derivatives, as well as with the corresponding valence isoelectronic aromatic thiols and chloro derivatives in order to probe the steric and electronic properties of substituents. However, although many substituted phenols have been thermochemically investigated, such ancillary comparisons are almost never possible because of the absence of thermochemical data for most of the desired nonphenolic compounds. Indeed, it is only for phenol itself that all of these comparisons can be made. As such, we generally limit comparisons of the phenol with the corresponding deoxygenated species and to isomers formed by relocating the $-$OH group and/or whatever other substituents there are already on the aromatic ring. That is, we discuss the enthalpy of the formal reactions 4 and 5.

$$\text{Ar}-\text{H} \longrightarrow \text{Ar}-\text{OH} \quad (4)$$

$$\text{Ar}'-\text{OH} \longrightarrow \text{Ar}-\text{OH} \quad (5)$$

The experimental enthalpies of formation for the phenol and arenol compounds appear in tables within the section in which they are discussed. Because we make extensive use of their deoxygenated counterparts, as in equation 4, these species appear in Table 1 below in the order in which they are introduced in the text.

TABLE 1. Enthalpies of formation of ancillary deoxygenated compounds related to arenols (kJ mol^{-1})

Compound	Solid	Liquid	Gas	Reference[a]
Benzene	39.1[b]	49.0 ± 0.6	82.6 ± 0.7	—
Naphthalene	77.9 ± 1.2	—	150.3 ± 1.5	—
Anthracene	129.2 ± 1.8	—	230.9 ± 2.2	—
Toluene	5.8[b]	12.4 ± 0.6	50.4 ± 0.6	—
tert-Butylbenzene	−79.1[b]	−70.7 ± 1.2	−22.6 ± 1.2	—
tert-Butyltoluene	—	—	—	
m-	—	—	−54 ± 2	6
p-	—	—	−57 ± 2	6
L-Phenylalanine	−466.9 ± 0.9	—	—	—
3-Benzyl-2,5-piperazinedione (cycloglycylphenylalanyl)	−345.4 ± 1.7	—	—	—
Benzoic acid	−385.2 ± 0.5	—	−294.1 ± 2.2	—
Phenyl benzoate	−241.6 ± 2.1	—	−142.6 ± 2.2	—
Benzamide	−202.1 ± 0.6	—	—	7
Benzanilide (*N*-Phenylbenzamide)	−93	—	—	8

TABLE 1. (continued)

Compound	Solid	Liquid	Gas	Reference[a]
Benzaldehyde	−97.2[b]	−87.0 ± 2.1	−36.7 ± 2.9	—
Acetophenone	−158[c]	−142.5 ± 1.0	−86.7 ± 1.6	—
Benzaldoxime	25	—	—	9
Benzalaniline N-oxide (N-(phenylmethylene)benzenamine-N-oxide)	148.0 ± 2.0	—	263.0 ± 2.1	10
Benzonitrile	152.2 ± 1.3[b]	163.2 ± 1.3	215.7 ± 2.1	—
Aniline	20.8[b]	31.3 ± 1.0	87.1 ± 1.0	—
Nitrobenzene	−0.5[b]	12.5 ± 0.5	67.5 ± 0.6	—
m-Dinitrobenzene	−27.4 ± 0.5	−6.9 ± 0.7	53.8 ± 1.8	—
1,3,5-Trinitrobenzene	−37.2 ± 0.5	—	—	—
Fluorobenzene	—	−150.6 ± 1.4	−116.0 ± 1.4	—
Chlorobenzene	1.4[b]	11.0 ± 1.3	52.0 ± 1.3	—
Bromobenzene	—	60.9 ± 4.1	105.4 ± 4.1	—
Iodobenzene	—	117.2 ± 4.2	164.9 ± 5.9	—
Dichlorobenzene				
o-	−30.4[b]	−17.5 ± 1.3	30.2 ± 2.1	—
m-	−33.3[b]	−20.7 ± 1.3	25.7 ± 2.1	—
p-	−42.3 ± 1.3	—	22.5 ± 1.5	—
Pentafluorobenzene	−852.7 ± 1.6	−841.8 ± 1.6	−806.5 ± 1.7	—
Pentachlorobenzene	−127 ± 9[d]	—	−40.0 ± 8.7	11, 12
Isopropylbenzene	—	—	4.0 ± 1.0	—
p-Cymene	—	—	−28[e]	—
1,3-Di-tert-butylbenzene	—	—	−125.6[e]	13
Anisole	—	−114.8 ± 1.8	−67.9 ± 0.9	—
Phenanthrene	113.0 ± 2.1[f]	—	204.7 ± 2.9[f]	7
Naphthoquinone				
1,2-	−163.5	—	—	14
1,4-	−188.5 ± 1.7	—	−97.9 ± 1.9	15
	−186.9	—	—	14
9,10-Anthraquinone	−188.5 ± 2.8	—	−75.7 ± 2.9	15
Anthracene	129.2 ± 1.8	—	230.9 ± 2.2	—
(E-) Azobenzene	310.2 ± 3.4[g]	—	—	—

[a] Data are from Reference 2 unless otherwise stated.
[b] The enthalpy of formation of the solid was obtained from the enthalpy of formation of the liquid and the enthalpy of fusion from Reference 5.
[c] The solid phase enthalpy of formation was derived using the parameters suggested in Reference 1.
[d] The solid phase enthalpy of formation was derived from the enthalpy of formation of the gas and the sublimation enthalpy of 87.1 ± 0.4 kJ mol^{-1} from Reference 12.
[e] The gaseous enthalpy of formation was derived from the liquid phase enthalpy of formation and an estimated enthalpy of vaporization from Reference 4.
[f] The solid phase enthalpy of formation is the mean of the two most recent values, 109.8 ± 1.6 kJ mol^{-1} from Reference 7 and 116.2 ± 1.3 kJ mol^{-1} as found in Reference 2. The enthalpy of sublimation is the mean of the values given in Reference 2, 91.7 ± 2.0 kJ mol^{-1}.
[g] This compound is mislabeled as the (Z-) stereoisomer in Reference 2.

II. ARENOLS

A. Unsubstituted Arenols

The enthalpies of formation of the unsubstituted arenols, phenol, 1- and 2-naphthol and 9-anthrol appear in Table 2.

TABLE 2. Enthalpies of formation of unsubstituted arenols (kJ mol^{-1})

Compound	Solid	Liquid	Gas	Reference
Phenol	-165.1 ± 0.8	-153.6	-96.4 ± 0.9	2
Naphthol				
1-	-121.0 ± 1.0	—	-29.9 ± 1.1	16
	-122.0 ± 1.5	—	-30.8 ± 1.6	17
2-	-124.2 ± 1.0	—	-30.0 ± 1.1	16
	-124.1 ± 1.6	—	-29.9 ± 1.7	17
9-Anthrol	—	—	45	a

[a] See discussion in text.

1. The OH/H increment exchange energies: δ(OH/H) and δ$^&$(OH/H)

The enthalpy of formal reaction 4 is the difference between the enthalpies of formation of the two substances, $\delta \Delta H$, where * denotes the chosen phase of interest (s, lq or g):

$$\delta \Delta H(*; \text{Ar}) \equiv \Delta_f H_m°(*, \text{ArOH}) - \Delta_f H_m°(*, \text{ArH}) \qquad (6)$$

The difference quantity for benzene and phenol, $\delta \Delta H$(g; Ph), is -179.0 ± 1.2 kJ mol^{-1}. This value figures prominently in this review and so we rewrite $\delta \Delta H$(g; Ph) as the more streamlined and simple δ(OH/H) to reflect its seminal importance in the current context. The corresponding liquid and solid phase differences, $\delta \Delta H$(lq; Ph) and $\delta \Delta H$(s; Ph), are the nearly identical -202.6 and -204.2 kJ mol^{-1}. It is quite fortuitous as well as fortunate that the enthalpies of fusion of benzene and phenol are so close. The consensus value of -203.4 for the condensed phase difference is denoted by δ$^&$(OH/H), where the '&' was chosen to convey it is for liquids & solids. From these data alone, an error bar of ±0.8 may seem appropriate. However, for the general use of this quantity, given the vagaries of condensed phases (the idiosyncrasies of crystal packing and the difficulties of describing hydrogen bonded liquids), it seems unequivocal that a larger uncertainty should be appended but we have an inadequate sense of how big. These two quantities, also known as the OH/H increment exchange energies, are used throughout the current study as simple additive constants.

The enthalpy of reaction 7 is mathematically equivalent to generating any deviations for other aromatic nuclei from the previously calculated -179.0 and -203.4 kJ mol^{-1} derived for benzene itself.

$$\text{ArH} + \text{PhOH} \longrightarrow \text{ArOH} + \text{PhH} \qquad (7)$$

The difference between experimental results and the simplistic estimates, that is, the deviations from δ(OH/H) and δ$^&$(OH/H), will generally be rationalized or reconciled by acknowledging steric and/or electronic interactions.

2. Comparison of phenol with alkanols

If phenols were alcohols, would they be like methanol? Or would they be more like primary, secondary or tertiary alcohols? Said differently, is there a simple alkyl group that most resembles phenyl? The gas phase OH/H increment exchange energies for R = methyl, ethyl, isopropyl or *tert*-butyl, derived analogously to equation 6, are respectively -126.0 ± 0.5, -151.4 ± 0.6, -168.1 ± 0.7 and -178.3 ± 1.1 kJ mol^{-1}. Numerically, the answer appears to be *tert*-butyl alcohol. Structurally, *tert*-butyl alcohol resembles phenol only in that the substituted carbon has its remaining bonds to other carbons. Is this a coincidence?

As employed in Reference 1, consider the formal exchange reaction 8 and the enthalpy of reaction 9.

$$\text{RMe} \longrightarrow \text{ROH} \qquad (8)$$

$$\delta\Delta H(\text{g; R}) = \Delta_f H_m°(\text{g, ROH}) - \Delta_f H_m°(\text{g, RMe}) \qquad (9)$$

For R = phenyl, the difference is -146.8 ± 1.1 kJ mol^{-1}. For R = methyl, ethyl, isopropyl and *tert*-butyl, the differences are -117.7 ± 0.6, -130.5 ± 0.6, -138.6 ± 0.9 and -144.3 ± 1.1 kJ mol^{-1}. Again, *tert*-butyl and phenyl correspond. Is this significant? Perhaps it is. Is it useful? The OH/H increment exchange energy is clearly so because it compares phenols and the related deoxygenated arene. In principle, the OH/Me increment exchange energy also should be useful because this probes the unique interactions of OH with other substituents by comparing, where possible, substituted phenols with correspondingly substituted, and also isoelectronic and isostructural, toluenes. This interrelation, however, is rarely employed in the current chapter because of the paucity of data for the requisite methylated species.

3. Naphthols and anthrols

The enthalpies of formation for both isomeric naphthols are nearly identical from either of two modern sources[16,17]. From the archival values for the enthalpy of formation of the parent naphthalene and the difference enthalpies, δ(OH/H) and $\delta^{\&}$(OH/H), we would have predicted values for the solid and gaseous forms of either naphthol of -125.6 and -28.7 kJ mol^{-1}, respectively, in wonderful agreement.

Of the three isomeric anthrols, thermochemical data are available only for 9-anthrol. While more discussion will appear in Section VI (*vide infra*), from the average of the literature values for the enthalpy of formation of gaseous 9-anthrone[18,19] of 22.2 ± 2.6 kJ mol^{-1} and a recommended 9-anthrol/9-anthrone enthalpy of formation difference[20] of 23 ± 8 kJ mol^{-1}, the enthalpy of formation of gaseous 9-anthrol is deduced to be *ca* 45 kJ mol^{-1}. Using the OH/H exchange increment value of 179 kJ mol^{-1} for gaseous phenols and hence arenols, together with the enthalpy of formation of anthracene, the estimated value would be 51 kJ mol^{-1}, in very good agreement with our derived value.

B. Carbon-bonded Substituents

The enthalpies of formation for phenols with carbon-bonded substituents appear in Table 3.

1. Monoalkylated phenols: methyl and tert-butyl substituents

There are enthalpy of formation data for numerous alkylated phenols—some 40 are found in Reference 23 alone. Rather than either archiving all of them or discussing all of them, we limit our attention to a subset of species, those with the smallest and those with almost the largest substituent groups, the methyl and *tert*-butylated phenols[32].

Of the three isomeric methylphenols (cresols), the *m*-isomer is the most stable, and in the gas phase, by considerably more than for the isoelectronic isomeric xylenes for which the enthalpy of formation difference spans less than 2 kJ mol^{-1}. The variation in these three cresol values is at least partly due to the larger partial negative charge on the ring

TABLE 3. Enthalpies of formation of phenols with carbon-bonded substituents (kJ mol^{-1})

Compound	Solid	Liquid	Gas	Reference[a]
Methylphenol (cresol)				
o-	−204.6 ± 1.0	−188.8[b]	−128.6 ± 1.3	—
m-	−205.4[c]	−194.0 ± 0.7	−132.3 ± 1.3	—
p-	−199.3 ± 0.8	−188.6[b]	−125.4 ± 1.6	—
tert-Butylphenol				
o-	—	−252.6 ± 2.4	−184.7 ± 2.6	21
	—	−254.8 ± 1.6	−191.6 ± 1.6	22
m-	−286.9 ± 2.0	—	−198.0 ± 2.1	21
	−286.5 ± 1.4	−272.0	−200.5 ± 1.5	22
p-	−270	—	—	23
	−310.5 ± 1.2	—	—	24
	−276.7 ± 2.2	—	−187.3 ± 3.3	21
	−289.7 ± 1.5	−270.6	−203.8 ± 1.6	22
tert-Butylmethylphenol				
2,4-	−306.9 ± 1.6	−289.8	−224.0 ± 1.0	22
2,5-	−304.9	−293.0 ± 1.7	−225.9 ± 1.7	22
4,2-	—	−305.7 ± 1.5	−233.6 ± 1.6	22
L-Tyrosine	685.1 ± 1.6	—	—	—
3-(4-Hydroxybenzyl)-2,5-piperazinedione (cycloglycyltyrosyl)	512.3 ± 0.6	—	—	—
Hydroxybenzoic acid				
o-	−589.7 ± 1.1	—	—	25
	−592.1 ± 1.3	—	−494.6 ± 1.8[d]	26
m-	−590.6 ± 1.0	—	—	25
	−594.1 ± 1.1	—	−467.3 ± 1.7[d]	26
p-	−594.5 ± 1.0	—	—	25
	−606.6 ± 2.1	—	−486.5 ± 2.4[d]	26
Phenyl salicylate	−436.6 ± 4.6	—	−344.5 ± 6.2	—
2-Hydroxybenzamide	−402.7 ± 2.2	—	—	27
2-Hydroxybenzanilide	−308.2 ± 3.0	—	—	27
Hydroxybenzaldehyde	—	—	—	—
o-	—	−283.2	—	28
p-	−297.1	—	−198.9	28, 29
Hydroxyacetophenone				
o-	−357.6 ± 3.8	—	—	—
m-	−370.6 ± 4.2	—	—	—
p-	−364.3 ± 4.2	—	—	—
2-Hydroxybenzaldoxime	−183.7 ± 0.8	—	—	—
2-Hydroxybenzalaniline N-oxide	−62.6 ± 2.0	—	53.9 ± 2.4	—
Cyanophenol				
o-	−56.5 ± 1.8	—	32.8 ± 2.1	30
m-	−56.5 ± 2.0	—	37.8 ± 2.2	30
p-	−59.1 ± 1.2	—	35.1 ± 2.5	30
	−60.9	—	—	31

[a]Data are from Reference 2 unless otherwise stated.
[b]The enthalpy of formation of the liquid was obtained from the enthalpy of formation of the solid and the enthalpy of fusion from Reference 5.
[c]The enthalpy of formation of the solid was obtained from the enthalpy of formation of the liquid and the enthalpy of fusion from Reference 5.
[d]The gas phase enthalpy of formation was derived from the average of the solid phase enthalpies of formation and the enthalpy of sublimation from Reference 26.

3. Thermochemistry of phenols and related arenols 231

carbon bonded to the σ-electron-donating methyl group when methyl is *ortho* or *para* to the π-electron-donating hydroxyl group. Applying the OH/H increment exchange energies, δ(OH/H) and $\delta^{\&}$(OH/H), to toluene we would predict gaseous, liquid and solid enthalpies of formation for any methylphenol of (g) -128.6, (l) -191.0 and (s) -197.6 kJ mol^{-1}. The gas and liquid phase predictions are identical to the corresponding isomer-averaged enthalpies of the three cresols. Both the *o*- and *m*-cresol solid enthalpies of formation are *ca* 7 kJ mol^{-1} more exothermic than the prediction. While this may indicate stabilization by intermolecular hydrogen bonding, it is unclear why the *para* isomer's enthalpy would not also benefit by such an interaction. From the isomer-averaged gas phase enthalpy of formation of xylene, 18.1 ± 0.9 kJ mol^{-1}, and the OH/CH$_3$ exchange increment from equation 9 of 146.8 ± 1.1 kJ mol^{-1}, the enthalpy of formation of any cresol is predicted to be -128.8 kJ mol^{-1}, a result identical to that above. Moreover, the OH frequencies of the cresols in the infrared are very close to that of free phenol[33] (3657 cm^{-1}), suggesting that the methyl substitution results in a negligible perturbation of the force field of the hydroxyl group.

The solid phase enthalpies of formation of the *m*- and *p*-*tert*-butylphenols from the various sources range from nearly identical to discordant. Reference 22 suggests that the hygroscopic nature of the compounds accounts for the discrepancies. From the average enthalpies of formation of each isomer in the gaseous phase, the stability order is *p*- $\geqslant$ *m*- > *o*-. The *o*-*tert*-butylphenol is less stable than its isomers, presumably because of steric interference between the two substituent groups. In the liquid and solid phases, the *para* and *meta* isomers are again of comparable stability[34], but the enthalpy difference between them and the *ortho* isomer in the liquid phase has approximately doubled, which is suggestive of hindrance of intermolecular hydrogen bonding as well. That the *ortho*-substituted phenol is liquid under standard conditions while its *meta* and *para* counterparts are solid is also corroborative of weakened hydrogen bonding. From the archival values for the enthalpies of formation of *tert*-butylbenzene and the increment exchange energies, the enthalpy of formation of any of the gaseous *tert*-butylphenols is -202 kJ mol^{-1}, for any liquid phase species it is -274 kJ mol^{-1} and for any solid phase species it is -283 kJ mol^{-1}. These predicted values are nearly the same as for the *meta* and *para* isomers in the respective phases while the *ortho* isomer is relatively destabilized from prediction.

Very nearly the same 10 kJ mol^{-1} destabilization for gas phase *ortho*- vs. *para*-*tert*-butylation of phenol is seen in the enthalpies of formation of variously substituted *tert*-butylmethylphenols shown in Table 3, and indeed the difference between the enthalpies of formation of these species and their demethylated counterparts, *ca* 30 kJ mol^{-1}, reflects the 33 kJ mol^{-1} difference between the enthalpies of formation of gaseous toluene and benzene. Said differently, the δ(OH/H) increment satisfactorily reproduces the enthalpy of formation of these *tert*-butylmethylphenols when acknowledgment is made for the *ca* 10 kJ mol^{-1} destabilization or strain associated with *tert*-Bu and OH *ortho* to each other[35].

2. The amino acid tyrosine and its derivatives

The amino acid tyrosine is related to the amino acid phenylalanine in the same way as phenol is related to benzene. The enthalpy difference between the amino acids is -219 ± 1.8 kJ mol^{-1}, somewhat larger than the $\delta^{\&}$(OH/H) increment of -203 kJ mol^{-1}. However, in the solid phase, tyrosine may be stabilized by additional hydrogen bonding sites unavailable to phenylalanine.

The enthalpy of formation difference for the solid phenolic cyclic dipeptides, 3-(4-hydroxybenzyl)-2,5-piperazinedione (cycloglycyltyrosyl) and its deoxygenated analog, 3-benzyl-2,5-piperazinedione (cycloglycylphenylalanyl) is -166.9 ± 1.8 kJ mol^{-1}, much smaller than either the above difference for the monopeptides or the difference for simple arenols. Either the numerical values and/or our understanding of the tyrosine/phenylalanine difference is suspect.

3. Carboxylic acids and their derivatives

The three isomeric hydroxybenzoic acids represent a well-defined set of phenols: the *o*-isomer, long recognized as salicylic acid, is one of the oldest and best known organic compounds. We might not expect the solid phase OH/H exchange increment of -203.4 kJ mol^{-1} to be of much value here because of additional hydrogen bonding sites available in the solid phase compared to those in benzoic acid itself. Nonetheless, from the enthalpy of formation of benzoic acid, the predicted value for any hydroxybenzoic acid of -588.6 kJ mol^{-1} shows that the *ortho* and *meta* isomers are very slightly stabilized. The *para* isomer average value is *ca* 12 kJ mol^{-1} more negative than predicted, which could reflect stabilization in the solid phase from the ordered cyclic hydrogen-bonded dimers which are linked together through hydrogen-bonded phenolic groups[36].

The stability order of the isomers in the gas phase is clearly *o*- > *p*- > *m*-. Using δ(OH/H), the predicted enthalpy of formation for any gaseous hydroxybenzoic acid is -473.1 kJ mol^{-1}, close to the experimental value for the *m*-isomer which has no stabilizing resonance structures. The large 22 kJ mol^{-1} difference for the *o*-isomer could be ascribed to stabilization by intramolecular hydrogen bonding of the type [HO—C=O $\cdots$ HO] or [O=COH $\cdots$ OH]. However, the difference between the predicted and experimental values for the *para* isomer is about the same in the gas as in the solid phase. We don't expect any intermolecular hydrogen bonding in the gas phase. It is tempting to suggest a dipolar resonance structure for the *p*-isomer not found in the *m*-, analogous to *p*- vs. *m*-nitroaniline, and so provide a mechanism for considerable stabilization for only one isomer. However, for the gaseous *m*- and *p*-substituted anilines[37], the difference between the enthalpies of formation is 7.3 ± 2.5 kJ mol^{-1}, very similar to the difference between *m*- and *p*-methoxybenzoic acids[38] of 5.8 ± 1.5 kJ mol^{-1}.

An ester and its deoxygenated analog for which there are enthalpies of formation are phenyl salicylate and phenyl benzoate. From their enthalpies of formation and the OH/H increment exchange energies, the predicted enthalpies of formation of phenyl salicylate are (s) -445.0 and (g) -321.6 kJ mol^{-1}. The difference between the predicted and experimental values for the solid is very slightly greater than for *o*-cresol but less than for salicylic acid. There is undoubtedly much less opportunity for intermolecular hydrogen bonding for the salicylate. The gas phase enthalpy difference shows a stabilization of *ca* 23 kJ mol^{-1} for the salicylate which is comparable to that for salicylic acid, *ca* 22 kJ mol^{-1}, and so we can postulate intramolecular hydrogen bonding in the ester as well. The hydrogen bonding in the ester would necessarily be a [C=O $\cdots$ HO] interaction.

There is a recently determined value for the enthalpy of formation of solid 2-hydroxybenzamide which is identical to the -405.5 kJ mol^{-1} derived from the enthalpy of formation of the parent benzamide and $\delta^{\&}$(OH/H). The recently reported value for solid 2-hydroxybenzanilide is also in satisfactory accord with the -296 kJ mol^{-1} derived from the ancient value[8] of -93 kJ mol^{-1} for the parent benzanilide. It is unfortunate that there are no gas phase measurements to test our supposition about intramolecular hydrogen bonding.

4. Acylphenols and their derivatives

The simplest members of the acylphenols are the isomeric hydroxybenzaldehydes (formylphenols). Using the OH/H exchange increments and the enthalpy of formation and of fusion for liquid benzaldehyde, the predicted enthalpies of formation for any hydroxybenzaldehyde would be (lq) −291.3 and (s) −300.6 kJ mol^{-1}, respectively. A slight destabilization is indicated in the liquid phase for the o-isomer. There is a negligible difference between the measured and predicted enthalpies for the solid p-isomer. It is unfortunate there are no thermochemical data for the m-isomer which is known to form infinite hydrogen-bonded chains in the solid state[39]. Solid 2,4-dihydroxybenzaldehyde exhibits intramolecular hydrogen bonding[40]. From δ(OH/H) and the gas phase enthalpy of formation of benzaldehyde, the predicted enthalpy of formation of the gas phase hydroxy derivative is −215.7 kJ mol^{-1}. It is not clear how to account for the 17 kJ mol^{-1} estimated destabilization in a compound where the *para* substituents should produce favorable resonance contributions.

Archival enthalpies of formation are available for all three acetylphenols as solids. The derived enthalpy of formation is −361 ± 8 kJ mol^{-1}, in good agreement with experiment. It is very surprising that there is no measured enthalpy of fusion for acetophenone and so we estimated this quantity using the parameters suggested in Reference 1. Because of the large uncertainty, it is impossible to state which of the acetylphenols are stabilized or destabilized relative to the model compound. However, the relative instability of the o-isomer is most likely due to steric hindrance in the solid phase. Akin to the situation with the hydroxybenzaldehydes, it is surprising the *para* compound is not more stable compared to its isomers.

Among the classical derivatives of aldehydes and ketones are oximes. We are fortunate to be able to compare the archival enthalpy of formation of solid 2-hydroxybenzaldoxime with the ancient measurement of the solid parent benzaldoxime. The difference is 208 kJ mol^{-1}, comfortably close to $\delta^{\&}$(OH/H) derived for benzene and phenol.

Other carbonyl derivatives are imines and their N-oxide derivatives, the so-called nitrones. One relevant example involves benzaniline N-oxide and its 2-hydroxy derivative. From the appropriate OH/H exchange increments, we would have predicted enthalpies of formation of the phenol nitrone of (s) −55.4 and (g) 84.0 kJ mol^{-1}, respectively. We lack understanding as to why the measured and predicted values for the gas are so disparate except to note that the benzaniline N-oxide and its 2-hydroxy derivative have very nearly identical enthalpies of sublimation, 116.5 ± 1.4 and 115.0 ± 0.8 kJ mol^{-1}, as do benzoic acid and salicylic acid.

5. Cyanophenols

The enthalpies of combustion and of sublimation of the three isomeric solid cyanophenols have been measured very recently, together with a theoretical study[30]. The stability order of the isomers in the gas phase is the same as for the related hydroxybenzoic acids although the enthalpy of formation differences between them are much smaller. Using the OH/H exchange increments for condensed and gas phases, the predicted enthalpies of formation for any cyanophenol are (s) −51.2 and (g) 36.7 kJ mol^{-1}. That the experimental enthalpies for the solids are all more exothermic by ca 5–8 kJ mol^{-1} may suggest intermolecular hydrogen bonding. At least for o-cyanophenol, [O−H···NC] hydrogen bonds connect the individual molecules into infinite chains[41]. From the small 4 kJ mol^{-1} discrepancy for the gaseous *ortho* isomer, intramolecular hydrogen bonding would not seem to be indicated. However, the theoretical estimate from Reference 30 for such an interaction is ca 11.5 kJ mol^{-1} which agrees with the IR spectroscopic experimental value of 7.2 kJ mol^{-1} [42].

C. Nitrogen-bonded Substituents

The enthalpies of formation of phenols with nitrogen-bonded substituents appear in Table 4.

TABLE 4. Enthalpies of formation of phenols with nitrogen-bonded substituents (kJ mol^{-1})

Compound	Solid	Liquid	Gas	Reference[a]
Aminophenol				
o-	−191.0 ± 0.9	—	−87.1 ± 1.3	43
	−201.3 ± 1.5	—	−104.4 ± 1.7	44
m-	−194.1 ± 1.9	—	−89.4 ± 1.6	43
	−200.2 ± 1.2	—	−98.6 ± 1.6	44
p-	−190.6 ± 0.9	—	−81.5 ± 1.7	43
	−194.1 ± 0.9	—	−90.5 ± 1.2	44
Nitrophenol				
o-	−204.6 ± 1.4	—	−132.3 ± 1.4	33
	−202.4 ± 1.0	—	−128.8 ± 1.6	45
m-	−200.5 ± 1.0	—	−109.3 ± 1.1	45
	−205.7 ± 1.7	—	−105.5 ± 1.8	46
p-	−207.1 ± 1.1	—	−114.7 ± 1.2	33
	−212.4 ± 1.0	—	−117.7 ± 2.0	45
Dinitrophenol				
2,4-	−232.7 ± 3.1	—	−128.1 ± 5.2	—
	−235.5	−211.3[b]	−130.9 ± 4.2	47, 48
2,6-	−209.9 ± 2.7	—	−97.8 ± 5.0	—
	−209.6 ± 3.3	−190.0 ± 3.3[b]	−97.5 ± 5.3	47, 48
2,4,6-Trinitrophenol (picric acid)	−214.3 ± 1.4	—	—	—

[a]Data are from Reference 2 unless otherwise stated.
[b]The enthalpy of formation of the liquid was obtained from the enthalpy of formation of the solid and the enthalpy of fusion from Reference 5.

1. Aminophenols

Before presenting the experimentally measured enthalpies of formation, we first ask what intuition suggests. We expect competing resonance-derived destabilization when the π-electron-donating hydroxy and amino groups are *ortho* or *para* to each other. The *o*-isomer has the possibility of weakly stabilizing intramolecular [O−H···N] hydrogen bonding which could mitigate the destabilization, although IR spectroscopic analysis indicated its absence[49]. The *m*-isomer has neither of these means for stabilization or destabilization.

In 1986, Pilcher and his coworkers measured the enthalpies of combustion and of sublimation of all three isomers[43]. The *ortho* and *para* isomers are less stable and, most probably, the *meta* is more stable. For the gas phase, both *ortho* and *para* amino substitution is destabilizing relative to *meta*. However, there are two other sets of measurements. The first consists of a value from early in the last century[50] for the *p*-isomer, −168 kJ mol^{-1}. This value is so discordant from the others, as well as so ancient, that it is easily disqualified. However, such early values are the only ones available for some phenols and other interesting and important compounds. Late in the last century, another thermochemical study[44] also reported the enthalpies of formation for all three isomers from measured enthalpies of combustion and of sublimation. From this source it is much more decisive that in the gas phase the *o*-isomer is the most stable and the *p*- is the least. The individual enthalpies of formation and of sublimation from the two contemporary sources for the solid phenols differ by 4–17 kJ mol^{-1} with no apparent explanation for the large isomeric disparities. If there were no interaction between the amino and hydroxy substituents, the exchange reaction 10 would be nearly thermoneutral, and the gas phase enthalpy of formation of the three aminophenols would be −92 kJ mol^{-1}, a value close to but still discrepant to both sets of contemporary results.

$$PhNH_2 + PhOH \longrightarrow C_6H_6 + NH_2C_6H_4OH \qquad (10)$$

Part of the above discrepancies may arise from problems with sample purity. Aminophenols autooxidize even more readily than most other classes of phenols. They readily form quinones and quinonimines and then these combine, polymerize, dehydrate and otherwise contaminate samples and confound chemists. Or, at least, that is how we understand the over 40 kJ mol^{-1} difference between the enthalpies of formation of solid phase 3,3′-diamino-4,4′-dihydroxydiphenylmethane and its isomer in which the locations for the amino and hydroxy groups are exchanged[51].

2. Nitrosophenols

The *o*- and *p*-nitrosophenols enjoy the possibility of resonance stabilization by π-electron donation from the phenolic hydroxyl group to the nitroso group, and the *o*-isomer could also be stabilized by an intramolecular hydrogen bond. These species are also tautomeric with benzoquinone oximes. All of this could confound interpretation of enthalpy of formation values if only they were available—there are seemingly no measured enthalpy of formation values for *o*-nitrosophenol. The value for *p*-nitrosophenol will be discussed later in Section VI because of tautomeric ambiguity. The *m*-species lacks the stabilizing conjugate NO/OH interaction, and so the monomer–dimer equilibrium as found in other nitroso compounds becomes problematic—should the measurement of enthalpy of combustion be available.

3. Nitrophenols

Unlike the aminophenols, the *o*- and *p*-nitrophenols should reflect the expected strong resonance stabilization by π-electron donation from the phenolic hydroxyl group to the strongly π-electron-withdrawing nitro group with additional stabilization in the *o*-isomer from intramolecular hydrogen bonding. All three nitrophenols have been thermochemically investigated with two contemporary calorimetric measurements for each of the isomers. The order of gas phase stability is decidedly *o* > *p* > *m*. From the archival enthalpy of formation of gaseous nitrobenzene and δ(OH/H), a gas phase enthalpy of formation of any nitrophenol of -111.5 kJ mol^{-1} can be derived. The gas phase enthalpy of formation of the *m*-isomer shows this species to be a little destabilized and the *p*-isomer likewise stabilized compared to the predicted value. If the *o*-isomer is stabilized by dipolar resonance by about the same amount, then the *ca* 14 kJ mol^{-1} stabilization for the *ortho* isomer suggests intramolecular hydrogen bonding. The same conclusion is reached by taking the difference between the enthalpies of formation of the *o*- and *p*-isomers. This value is much smaller than a theoretical hydrogen-bond strength of *ca* 53 kJ mol^{-1} in *o*-nitrophenol found as the difference between the energies of the *cis* and *trans* O—H conformers[52]. The enthalpy difference between the *meta* and *para* isomers is very close to the corresponding difference for the nitroanilines mentioned earlier.

That all three isomers have very nearly the same value for the solid phase enthalpy of formation indicates that intermolecular hydrogen bonding in the *o*-isomer is approximately the same strength as for the other two isomers. The predicted enthalpy of formation of any solid nitrophenol is -203 kJ mol^{-1}, identical to the values observed for the *o*- and *m*-isomers. This, of course, does not imply that *o*- and *m*-nitrophenols lack hydrogen bonding in the condensed phase but rather the hydrogen bonding in the various nitrophenols is not particularly different from that found in the parent phenol. The *p*-isomer is stabilized by *ca* 7 kJ mol^{-1}, only slightly more than in the gas phase.

Of the six isomeric dinitrophenols, there are thermochemical data for only two, the 2,4- and 2,6-species. Both are related to the same deoxygenated parent, m-dinitrobenzene, and so, in the absence of hydrogen bonding or steric effects, the two dinitrophenols should have the same enthalpy of formation. From the enthalpies of formation of m-dinitrobenzene, $\delta^\&(OH/H)$ and $\delta(OH/H)$, the predicted enthalpy values for any dinitrophenol are (s) -227.8 and (g) -125.2 kJ mol^{-1}. The large discrepancies for the 2,6-isomer would seem to be due to steric interference by one or both nitro groups with the hydroxyl. However, all of the mononitrophenols as well as 2,4- and 2,6-dinitrophenol have been found to be planar by *ab initio* and density functional theory[53] with substantial intramolecular hydrogen bonding, consistent with experimental data. The stabilization of the 2,4-isomer is only *ca* 4 kJ mol^{-1} in the gas phase, very different from the large stabilization observed for *o*-nitrophenol which it should resemble. Comparing related compounds, the difference between the solid phase enthalpies of formation of 2,4- and 2,6-dinitroaniline[54] is 15 kJ mol^{-1}, of 2,4- and 2,6-dinitrotoluene[55] is 22 kJ mol^{-1}, and of 2,4- and 2,6-dinitrophenol is 25 kJ mol^{-1}.

We now turn to the trinitrophenols, of which only one of the six isomers has been thermochemically characterized. This is the 2,4,6-species, most commonly known as picric acid. Again, there is significant destabilization: from the enthalpy of formation of solid 1,3,5-trinitrobenzene and $\delta^\&(OH/H)$, the predicted enthalpy of formation is -240.6 kJ mol^{-1}. The calculated destabilization is nearly 27 kJ mol^{-1}. From the point of view of steric hindrance at C2−C1−C6, this compound should not be any worse than 2,6-dinitrophenol. However, it is calculated to be a nonplanar compound with intramolecular hydrogen bonding[53].

D. Oxygen-bonded Substituents

The enthalpies of formation of phenols with oxygen-bonded substituents appear in Table 5.

1. Hydroxy derivatives

The three monohydroxy derivatives of phenol are all well-known compounds, the *o*-, *m*- and *p*-species with the long-established, trivial names catechol, resorcinol and hydroquinone. These compounds are all benzenediols and, as such, they and their substituted derivatives will be discussed later in this text.

2. Alkoxy derivatives

For reasons to be discussed later, we are doubtful of the enthalpy of formation measurement[56] of solid 2,6-dimethoxyphenol, -518.4 kJ mol^{-1}. Another species which

TABLE 5. Enthalpies of formation of phenols with oxygen-bonded substituents (kJ mol^{-1})

Compound	Solid	Liquid	Gas	Reference
2,6-Dimethoxyphenol	−518.4	—	—	56
2-Methoxy-4-methylphenol	—	−291.9	−362.8 ± 2.2	57
4-Allyl-2-methoxyphenol	−659.2	—	—	58
4-(1-Propenyl)-2-methoxyphenol	−696.8	—	—	58
Morphine hydrate	−711	—	—	59

3. Thermochemistry of phenols and related arenols

has been studied[57] is 2-methoxy-4-methylphenol. An immediate question is the extent of the interaction, if any, between the two oxygens. One probe is the thermicity of the gas phase substituent exchange reaction

$$\text{PhOH} \longrightarrow \text{PhOMe} \qquad (11)$$

which is endothermic by 28.5 kJ mol^{-1}. As will be discussed later in the benzenediol section, this formal increment is somewhat more positive when both hydroxyl groups in 1,2-dihydroxybenzene are converted to 1,2-dimethoxybenzene. It does not seem credible, therefore, that the formal increment converting 4-methyl-1,2-dihydroxybenzene ($\Delta H_f =$ -298.4 ± 1.6 kJ mol^{-1})[60] to 2-methoxy-4-methylphenol is only $+6.5$ kJ mol^{-1}. We view the literature enthalpy of formation of 2-methoxy-4-methylphenol as suspect.

Two other alkoxyphenols are the isomeric 4-allyl- and 4-(1-propenyl)-2-methoxyphenols with the ancient[58] enthalpies of combustion of 5384.4 and 5346.8 kJ mol^{-1}. We are automatically troubled by these values. The 38 kJ mol^{-1} derived difference between the enthalpies of formation of the two isomers is rather much larger than the ca 22 kJ mol^{-1} derived[61] for their oxygen-defunctionalized counterparts allyl and 1-propenylbenzene.

Another species that qualifies as an alkoxy derivatized phenol is morphine. Because of the multifunctional complexity and solid phase of the compound, as well as the dates of the literature citations[59] (1899, 1900) the result is essentially without use in our thermochemical context.

E. Sulfur-bonded Substituents

Neither sulfur-substituted phenols nor benzenethiols have been much studied by the thermochemist. The only thermochemical data for a sulfur-derivatized phenol that is known to the authors is 'sulfosalicylic acid' (2-hydroxy-5-sulfobenzoic acid dihydrate) and some of its salts[62]. The difference between the solid phase enthalpies of formation of sulfosalicylic acid dihydrate (-1982 ± 3 kJ mol^{-1}) and salicylic acid is ca -1392 kJ mol^{-1}. Correcting for two molecules of water (-286 kJ mol^{-1}, assumed uncomplexed liquid) changes the value to -820 kJ mol^{-1}, while assuming Handrick's universal hydrate correction[63] of ca 19 kJ mol^{-1} per water suggests a value of ca -780 kJ mol^{-1} for the free acid. This last value has been suggested as problematic[64].

F. Halogen Substituents

The enthalpies of formation of halogenated phenols appear in Table 6.

1. Monohalophenols

The four halogens F, Cl, Br, I form an interesting and well-ordered set of substituents. Along with hydrogen, they change monotonically in many key properties: in size H < F < Cl < Br < I; in polarizability H < F < Cl < Br < I; in electronegativity, H ≈ I < Br < Cl ≪ F; in hydrogen bonding ability H < I < Br < Cl < F. How do their enthalpies of formation depend on the halogen and its position on the ring relative to OH?

We begin with fluorophenols. Disappointingly, the data are old[65], and because the calorimeter was not a rotating bomb and the products were not analyzed, the results are not particularly to be trusted[70]. In any case, the enthalpies of formation are only for the condensed phase. That the enthalpy of fusion is always endothermic means the enthalpy of formation of a liquid must be less negative than the corresponding solid.

TABLE 6. Enthalpies of formation of halogenated phenols (kJ mol^{-1})

Compound	Solid	Liquid	Gas	Reference[a]
Fluorophenol				
o-	−302	—	—	65
m-	—	−340	—	65
p-	−334	—	—	65
Chlorophenol				
m-	−206.5 ± 8.4	−189.3 ± 8.4	−153.3 ± 8.7	—
p-	−197.7 ± 8.4	−181.3 ± 8.4	−145.5 ± 8.7	—
Iodophenol				
o-	−95.8 ± 4.2	—	—	—
m-	−94.5 ± 4.2	—	—	—
p-	−95.4 ± 4.2	—	—	—
5-Iodosalicylic acid	−512.5	—	—	66
Dichlorophenol				
2,3-	−223.3 ± 1.1	—	−151.6 ± 2.5	67
2,4-	−226.4 ± 1.5	—	−156.3 ± 1.9	67
2,5-	−232.0 ± 1.2	—	−158.4 ± 2.4	67
2,6-	−222.1 ± 1.1	—	−146.3 ± 1.5	67
3,4-	−231.6 ± 1.1	—	−150.3 ± 2.5	67
3,5-	−231.0 ± 1.0	—	−150.3 ± 2.3	67
2,4-Dibromo-6-methylphenol	−159 ± 6	—	—	68
3,5-Diiodosalicylic acid	−397.1	—	—	66
Pentafluorophenol	−1024.1 ± 2.1	−1007.7 ± 2.1	−956.8 ± 2.7	—
Pentachlorophenol	−292.5 ± 3.0	—	−225.1 ± 3.6	—
2,4,6-Tribromophenol	−100 ± 5	—	−0.9 ± 2.5	68, 69
2,4,6-Tribromo-3-methylphenol	−131 ± 5	—	—	68

[a]Data are from Reference 2 unless otherwise stated.

And so the enthalpy of formation of liquid p-fluorophenol is less negative than the solid phase value of −334 kJ mol^{-1}, and the enthalpy of formation of solid m-fluorophenol is more negative than −340 kJ mol^{-1}, the value for its liquid phase. Equivalently, the m-isomer is more stable than the p- in both phases. This result is consistent with the thermochemistry of amino and methyl phenols which are also species containing π- or σ-electron donating substituents. It is surprising that the o-isomer is seemingly so much less stable than the m-isomer. Ab initio computations[71] indicate weak intramolecular hydrogen bonding in the *ortho* isomer which supported observations from IR[72] and gas electron diffraction[73] measurements. The experimentally-determined energy difference between the intramolecular hydrogen-bonded *syn* conformer and the *anti* conformer was 6.8 ± 0.3 kJ mol^{-1} by the former method and 2 kJ mol^{-1} by the latter. From $\delta^\&$(OH/H) and the archival (and trusted) enthalpy of formation of liquid fluorobenzene, we derive an enthalpy of formation of *ca* −354 kJ mol^{-1} for any of the three isomers. If the thermochemical measurements are reasonably accurate, it seems they are all less stable than predicted.

There are apparently no data for the o-isomer of chlorophenol. Disregarding the error bars, it appears the m-isomer is more stable than the p- by *ca* 9 kJ mol^{-1}, the stability order predicted for a π-donating substituent on a phenolic ring. Including the error bars allows for the possibility that the relative stability of the two isomers is reversed. Again using the OH/H exchange increments and the enthalpies of formation of chlorobenzene, any chlorophenol would have an enthalpy of formation of (s) −202.0, (lq) −192.4 and (g) −127.0 kJ mol^{-1}. In the solid phase, the apparent stabilization of the *meta* isomer and the apparent destabilization of the *para* isomer are very small and within the experimental uncertainties. In the liquid phase, the *para* isomer is also seemingly stabilized. Even considering the experimental uncertainty, the gaseous chlorophenols are apparently much more stable than predicted. While the *meta* isomer has more favorable resonance structures and its stabilization is understandable, it is not clear why the *para* isomer, with its less favorable resonance structures, should be so apparently stabilized. Given the importance of chlorinated aromatics, we eagerly await new and more precise measurements for both isomers, as well as for the *o*-isomer.

There are no thermochemical data for the bromophenols. For the iodophenols, there are enthalpy data only for the solid phases. Our estimation procedure, using the enthalpy of formation of iodobenzene and the OH/H exchange increment, predicts -86 kJ mol^{-1}. How the iodophenols could be stabilized by *ca* 10 kJ mol^{-1} is not clear, except that the experimental uncertainty is somewhat large and we know very little about the solid phase.

2. Dihalophenols

There are no reported enthalpies of formation of any of the isomeric difluorophenols. In contrast, the enthalpies of formation of all six of the dichlorophenols are available[67]. Assuming the general applicability of the exchange energies, from the archival enthalpies of formation of the three dichlorobenzenes we would predict values of -234 and -149 kJ mol^{-1} for the solid and gaseous states of both the 2,3- and 3,4-dichlorophenol; -237 and -153 kJ mol^{-1} for the solid and gaseous states of the 2,4-, 2,6- and 3,5-dichlorophenol and -246 and -157 kJ mol^{-1} for solid and gaseous 2,5-dichlorophenol. The enthalpies of formation of the gaseous phenols are predicted somewhat more reliably than those of the solids which are all $2-15$ kJ mol^{-1} less stable than predicted. We are neither surprised nor disappointed—the intricacies of solids usually are problematic and we have no handle on the vagaries of intermolecular hydrogen bonding. The exception to reliable gas phase prediction is for 2,6-dichlorophenol which presumably suffers from adverse steric effects of the hydroxyl group buttressed between two chlorine atoms, an effect not present in the similarly substituted 2,3-dichlorophenol.

The only enthalpy of formation data[68] for any dibromophenol is that of solid phase 2,4-dibromo-6-methylphenol. In the absence of an experimental enthalpy of formation of *m*-dibromobenzene, we must assess the reliability of the phenol derivative in another way. The reaction in equation 12 for estimating the enthalpy of formation of the deoxygenated 3,5-dibromotoluene might be approximately thermoneutral for all phases, assuming no steric or electronic interactions between substituents:

$$2PhBr + PhMe \longrightarrow 1,3,5-C_6H_3MeBr_2 + 2C_6H_6 \qquad (12)$$

From archival enthalpies of formation and of fusion, the estimated enthalpy of formation of solid 3,5-dibromotoluene is 28 ± 9 kJ mol^{-1}. This value, combined with $\delta^\&$(OH/H), gives a predicted enthalpy of formation of the corresponding phenol of -175 kJ mol^{-1}. A bromine atom and methyl group crowd the intervening OH, which could account for at least some of the *ca* 16 kJ mol^{-1} difference between the predicted and experimental values, and we don't expect 2,4-dibromo-6-methylphenol to participate in intermolecular hydrogen bonding. The remainder of the difference is accounted for by the error bars. Altogether, the value is plausible.

Diiodophenols are represented only by one very old study[66] of solid 3,5-diiodosalicylic acid. Lacking an enthalpy of formation for the deoxygenated parent to make a prediction, we calculate the enthalpy of the reaction involving this diiodo species and the corresponding monoiodo[66] and parent acid from equation 13.

$$2[\text{iodosalicylic acid}] \longrightarrow \text{diiodosalicylic acid} + \text{salicylic acid} \quad (13)$$

This reaction is found to be ca 36 kJ mol^{-1} endothermic. Intuition suggests that the iodine in iodosalicylic acid is p- to the OH and in the diiodo compound they are o- and p-. While we acknowledge that (a) the thermochemistry of organoiodine compounds is often problematic, (b) there is considerable crowding by the adjacent carboxyl, hydroxyl and iodo groups in the diiodo species and (c) predictions of the enthalpy of formation of solids remain precarious, nonetheless, we recommend the remeasurement of the enthalpy of combustion of the iodosalicylic acids, and for that matter, of iodophenols in general.

3. Polyhalophenols

The polyhalogenated phenols are species with three or more halogen atoms. The first such species is pentafluorophenol. The enthalpies of formation predicted from the related pentafluorobenzene are (s) −1056.1, (lq) −1045.2 and (g) −985.5 kJ mol^{-1}. These values are some 30–40 kJ mol^{-1} more negative than the measured values, the largest destabilization observed so far. Before questioning the reliability of the data, consider the thermochemical differences for the gas phase reaction 14 where X = CH$_3$, OH, F, Cl, Br and I:

$$C_6H_5X + C_6HF_5 \longrightarrow C_6H_6 + C_6F_5X \quad (14)$$

The endothermic enthalpies of reaction indicate that the C$_6$F$_5$X species are destabilized from prediction by 4, 30, 50, 20, 72 and 175 kJ mol^{-1}, respectively.

Based on the above experience with pentafluorophenol, we would expect some destabilization for pentachlorophenol. However, the calculated enthalpies of formation for this species using the appropriate OH/H increment exchange energies are (s) −330 and (g) −219 kJ mol^{-1}. We are surprised that the values from 'the literature' and our estimate for the gas are so close and those for the solid are so disparate, respectively.

We close this discussion with two tribrominated phenols, the 2,4,6-tribromo derivatives of phenol and 3-methylphenol. We might expect equation 15 for estimating the enthalpy of formation of the deoxygenated 1,3,5-tribromobenzene to be approximately thermoneutral for all phases, assuming no interactions among substituents.

$$3\text{PhBr} \longrightarrow 1,3,5\text{-}C_6H_3Br_3 + 2C_6H_6 \quad (15)$$

From the archival enthalpies of formation and of fusion for benzene and bromobenzene, the estimated enthalpies of formation of 1,3,5-tribromobenzene are (s) 72 kJ mol^{-1} and (g) 151 kJ mol^{-1}. From these values and the appropriate OH/H exchange increments, we would predict enthalpies of formation for 2,4,6-tribromophenol of −131 kJ mol^{-1} for the solid and −28 kJ mol^{-1} for the gas phase species. The predicted results are both ca 30 kJ mol^{-1} more exothermic than the experimental ones. Since we don't expect the solid tribromophenol to participate in intermolecular hydrogen bonding in the same way as solid phenol does, on that basis the estimated values are seemingly too negative.

Comparing 2,4,6-tribromophenol with its 3-methylated derivative, methylation decreases the solid phase enthalpy of formation by some 31 kJ mol^{-1}. This can be compared to the decrease of 40 kJ mol^{-1} for the parent solid phenol when it is methylated to 3-methylphenol (m-cresol). Given the uncertainties in many of the measured quantities as well as derived values, and lack of quantitation of buttressing effects, we consider these last differences to be consistent.

III. ARENEDIOLS

The enthalpies of formation for a variety of arenediols and arenetriols (triols to be discussed in Section IV) appear in Table 7.

TABLE 7. Enthalpies of formation of arenediols and arenetriols (kJ mol^{-1})

Compound	Solid	Gas	Reference[a]
Benzenediol			
o-	−353.1 ± 1.1	−271.6 ± 2.0	33
	−354.1 ± 1.1	−267.5 ± 1.9	60
	−362.3 ± 1.1	−274.8 ± 1.2	74
m-	−370.7 ± 1.1	−284.7 ± 1.2	74
	−368.0 ± 0.5	−275	75
p-	−371.1 ± 1.3	−277.0 ± 1.4	74
1,2-Benzenediol			
3-methyl	−392.5 ± 1.1	−299.3 ± 1.6	60
4-methyl	−393.3 ± 1.2	−298.4 ± 1.6	60
3-isopropyl	−447.8 ± 1.6	−350.0 ± 2.3	60
3-isopropyl,6-methyl	−475.7 ± 1.6	−379.1 ± 1.8	60
4-*tert*-butyl	−474.0 ± 1.6	−375.7 ± 2.1	60
3,5-(*tert*-butyl)$_2$	−570.6 ± 2.6	−470.5 ± 2.7	60
4-nitro	−411.1 ± 1.1	−290.0 ± 1.8	76
3-methoxy	−510.2 ± 1.2	−418.5 ± 1.4	76
1,3-Benzenediol			
2,4-dinitro	−422.8 ± 2.7	—	77
	−415.6 ± 2.5	—	78
4,6-dinitro	−443.4 ± 2.7	—	77
	−439.5 ± 2.5	—	78
4-acetyl	−573.5 ± 3.8	—	—
1,4-Benzenediol			
2-chloro	−383.0 ± 8.4	−314.0 ± 11.8	—
2,3-dichloro	−416.0 ± 8.4	—	—
2,5-dichloro	−427.3 ± 8.4	—	—
2,6-dichloro	−423.4 ± 8.4	−331.5 ± 11.8	—
2,3,5-trichloro	−440.7 ± 8.4	−339.4 ± 11.8	—
2,3,5,6-tetrachloro	−453.6 ± 8.4	—	—
Naphthalenediol			
1,2-	−309.8 ± 1.6	−200.5 ± 2.8	17
1,3-	−327.2 ± 1.4	−211.2 ± 1.9	17
1,4-	−317.4 ± 1.5	−197.0 ± 1.8	17
	−339.4 ± 7	—	79
2,3-	−302.4 ± 1.7	−192.8 ± 2.0	17
	−316.4 ± 1.5	−207.0 ± 1.6	80
2,7-	−326.1 ± 1.7	—	80
Phenanthrene-9,10-diol	−243.5	—	14
Benzenetriol			
1,2,3-	−551.1 ± 0.9	−434.2 ± 1.1	76
1,2,4-	−563.8 ± 1.1	−444.0 ± 1.6	76
1,3,5-	−584.6 ± 1.1	−452.9 ± 1.5	76
5-Carboxy-1,2,3-benzenetriol	−1013 ± 5.0	—	81

[a] Data are from Reference 2 unless otherwise stated.

A. Unsubstituted Benzenediols

The unsubstituted 1,2-, 1,3- and 1,4-benzenediols are historically and trivially known as catechol, resorcinol and hydroquinone. The individual experimental enthalpy of formation values for both the solid catechol and the gaseous resorcinol are rather disparate. Assuming no interaction between the two hydroxyl groups, we would have anticipated an enthalpy of formation for any benzenediol of ca -368 kJ mol^{-1} for the solid and ca -275 kJ mol^{-1} for the gas. The solid *meta* and *para* diols exhibit no significant deviation from prediction while for the solid *o*-diol enthalpies indicate ca 6–15 kJ mol^{-1} destabilization. Presumably, the solid *meta*- and *para*-diols engage in hydrogen bonding of the same type and strength as phenol itself, although we might have expected some stabilization due to a more ordered array with additional hydrogen-bonding sites. The destabilization of solid catechol may be a combination of less stable resonance structures and sterically hindered intermolecular hydrogen bonding.

The experimental values for the gas phase diols are roughly comparable to the predicted enthalpy of formation. We expect some stabilization due to the more stable resonance structures for *meta* hydroxyl groups compared to *ortho* and *para*, but only one of the measured enthalpy of formation values for the *meta* isomer is more exothermic than either the predicted or the *para* values. From comparison with *m*-cresol and *m*-chlorophenol, examples of other compounds with electron-donating substituents *meta* to the hydroxyl group, we would expect stabilization comparable to theirs, ca 4–26 kJ mol^{-1}. The seeming absence of hydrogen-bond-derived stabilization for the gaseous *ortho*-diol[82] is somewhat surprising, unless there is a compensating destabilization that is ascribed to the less favorable resonance structures of *ortho* hydroxyl substituents. We would then expect destabilization in the gaseous *para*-isomer also. The situation here resembles that of the cresols where the *o*- and *p*-isomer enthalpies of formation did not deviate significantly from the prediction, while the *m*-isomer, with its more favorable resonance structures, is most stable. For reasons to be described in more detail at the end of this section, the best enthalpy of formation values for the *o*- and *m*-benzenediols are probably the average values.

In order to better understand the hydroxyl interactions in the diols, we can compare them with their methylated counterparts, the dimethoxybenzenes. Consider the exchange reaction 16 for the *o*-, *m*- and *p*-substituted compounds:

$$C_6H_4(OH)_2 + 2PhOMe \longrightarrow C_6H_4(OMe)_2 + 2PhOH \quad (16)$$

Using the dimethoxybenzene enthalpies of formation from Reference 83, the enthalpies of reaction 16 are (*o*-) 11.9, (*m*-) 1.1 and (*p*-) 8.5 kJ mol^{-1} in the gas phase and (*o*-) 11.2, (*m*-) 11.3 and (*p*-) 0.7 kJ mol^{-1} in the solid phase. The normalized enthalpy of methylation is one-half of the overall reaction enthalpy. *Ab initio* geometry optimizations indicate that the *m*- and *p*-dimethoxybenzenes are planar and the *o*-dimethoxybenzene is nonplanar[83]. For the *meta* isomer, the exchange of hydroxy for methoxy in the gas phase is essentially thermoneutral which indicates, because there are no steric effects or hydrogen bonding, that the electronic effects of hydroxy and methoxy substituent groups on the aromatic ring are comparable. The gas phase endothermicities for the *ortho* and *para* isomers are not very large and may reflect the increased electron donation by methoxy compared to hydroxy. At least for the *ortho* isomer, that the methylation reaction also introduces substituent steric effects is reflected in its greater endothermicity.

B. Alkylated Benzenediols

The differences between the predicted and experimental enthalpies of formation for the various alkylated benzenediols are in the range of ca 4–13 kJ mol^{-1} destabilization for the real compound. The derived destabilization for catechol itself is ca 4 kJ mol^{-1} and so we might expect substitution at C-3 to cause steric strain, which is manifested in a larger destabilization, and substitution at C-4 to have minimal effect. However, the results are not straightforward: a methyl group at either C-3 or C-4 increases the destabilization to ca 8–9 kJ mol^{-1} while the larger isopropyl substituent at C-3 has hardly any effect. The effect of a large *tert*-butyl group at C-4 is less than that of the methyl group in the same position. The electronic effects of the various alkyl groups are not expected to differ very much, except that they are *o-p-* or *m-* to a hydroxyl group. However, the calculated effects are small and experimental error bars accumulate in these calculations. Overall, the enthalpies of formation of any alkylated 1,2-benzenediol can be derived satisfactorily from its totally deoxygenated parent hydrocarbon.

C. Otherwise Substituted Benzenediols

From the parent nitrobenzene and the OH/H increment exchange energy, the predicted enthalpy of formation for 4-nitro-1,2-benzenediol is −290.5 kJ mol^{-1} which is identical to the experimental value. By comparison with the *m-* and *p*-nitrophenols from a previous section, this is not a surprising result. *p*-Nitrophenol is 8.8 ± 5 kJ mol^{-1} more stable than its *m*-isomer. Taking into account the ca 4 kJ mol^{-1} destabilization due to placing the two hydroxyls *ortho* to each other and the large error bar for the isomer stability difference, the favorable resonance effect of the *para* substituents is nearly nullified. For the solid phase, the predicted enthalpy of formation is −407.2 kJ mol^{-1} which differs from the experimental value by only 3.9 kJ mol^{-1}. The experimental *para/meta* isomer enthalpy of formation difference for the solid nitrophenols was ca 7 kJ mol^{-1}. Again, because of the experimental uncertainty and/or destabilization caused by the *ortho* hydroxyls, the effects seemingly cancel.

There are enthalpy of formation data for the 2,4- and 4,6-dinitro-1,3-benzenediols. From the archival enthalpy of formation of solid *m*-dinitrobenzene and $\delta^{\&}$(OH/H), the enthalpy of formation of either dinitrobenzenediol isomer is predicted to be −434.2 kJ mol^{-1}. Compared to this estimated value, the 2,4-isomer is ca 11 kJ mol^{-1} destabilized and the 4,6-isomer is ca 9 kJ mol^{-1} stabilized. Each of these compounds has a pair of *meta* hydroxy groups, a pair of *meta* nitro groups and two pairs of *ortho* hydroxy/nitro groups. The 2,4-isomer has an additional *ortho* hydroxy/nitro interaction.

The solid *o*-nitrophenol was neither stabilized nor destabilized relative to prediction; neither is ca 9 kJ mol^{-1} a very large stabilization for two possible pairs of hydrogen-bonding hydroxy/nitro substituents in 4,6-dinitro-1,3-benzenediol. Accordingly, the reaction depicted in equation 17 is only 3.6 kJ mol^{-1} endothermic.

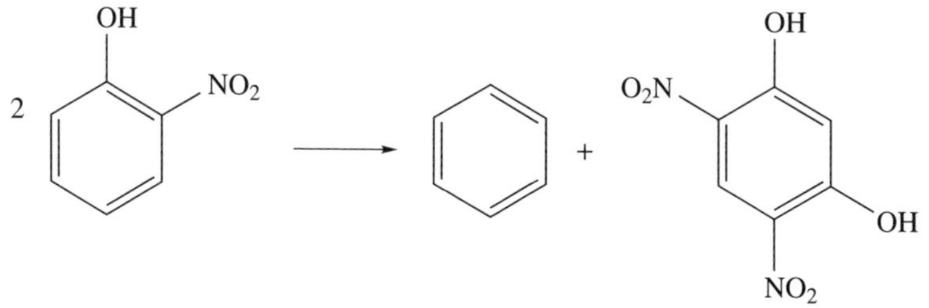

(17)

2,4-Dinitrophenol was only 6 kJ mol^{-1} more stable than predicted and the related reaction of equation 18

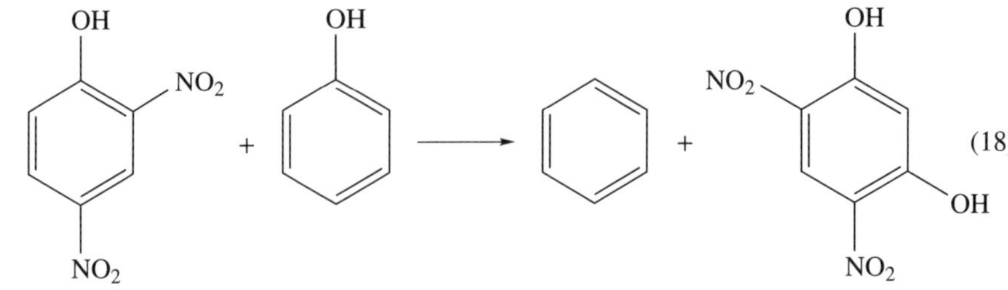

$$\text{2,4-dinitrophenol} + \text{phenol} \longrightarrow \text{benzene} + \text{2,4-dinitro-1,3-benzenediol} \quad (18)$$

is only 4.2 kJ mol^{-1} exothermic. The reactions corresponding to equations 17 and 18 to produce 2,4-dinitro-1,3-benzenediol have enthalpies of +24.2 and +16.4 kJ mol^{-1}, respectively. The related reaction 19

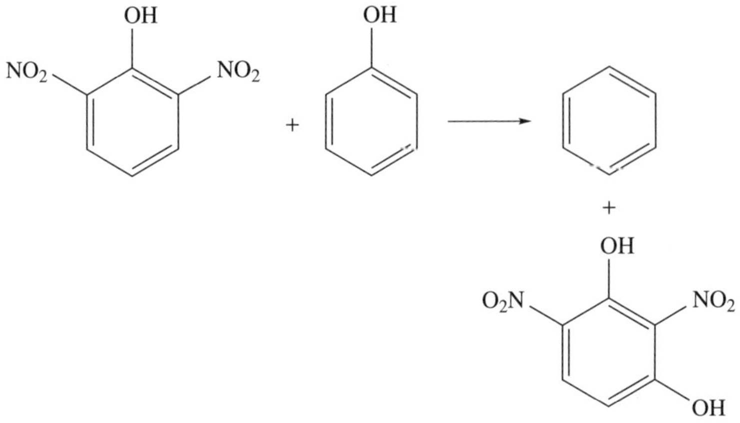

$$\text{2,6-dinitrophenol} + \text{phenol} \longrightarrow \text{benzene} + \text{4,6-dinitro-1,3-benzenediol} \quad (19)$$

is 8.1 ± 4.3 kJ mol^{-1} exothermic. The enthalpy of this reaction, about the same as that for reaction 18, demonstrates that most of the destabilization is due to the presumed steric effect of the two nitro groups flanking the hydroxy group. Recall that 2,6-dinitrophenol was destabilized from prediction by ca 18 kJ mol^{-1}. Introduction of the second hydroxy group appears to be slightly stabilizing. It is unfortunate that the gas phase enthalpy of formation is not available so that we can compare the thermochemical data with theoretical[84,85] and experimental[86,87] gas phase studies of intramolecular hydrogen bonding in 2-nitroresorcinol and 4,6-dinitroresorcinol.

The enthalpy of formation of solid 4-acetyl-1,3-benzenediol may be estimated from the enthalpy of formation of solid acetophenone and twice the OH/H exchange increment to be −565 kJ mol^{-1}. The destabilization of less than 10 kJ mol^{-1} may be due to uncertainty in the estimation of the enthalpy of formation of acetophenone and experimental uncertainty in the measurement of the diol. We would have expected resonance stabilization by the favorably situated acetyl and hydroxyl groups. The experimental enthalpy of formation is consistent with those of the acetylphenols, discussed in an earlier section. Their predicted values, estimated now by deoxygenating the diol, are both −370 kJ mol^{-1}, close to the −361 kJ mol^{-1} found as the average of the 2- and 4-hydroxy species.

Finally, we consider the numerous chlorinated benzene-1,4-diols. Just as with benzene-1,4-diol itself, introducing a second hydroxy group *para* to the first in *m*-chlorophenol should not appreciably affect the stability. Since there are no *o*-chlorophenol data to compare, we are unsure of the effect of introducing the second hydroxy group *ortho* to the chlorine. Qualitatively, we expect the predicted and experimental enthalpies of formation of 2-chloro-1,4-benzenediol to be comparable. The predicted gas phase enthalpy of formation for this compound is −306.0 kJ mol^{-1} calculated from chlorobenzene which is 8 kJ mol^{-1} less exothermic than the experimental value. However, the experimental uncertainty is larger than the difference. In contrast, the estimated solid phase enthalpy of formation, −405.4 kJ mol^{-1}, is 22.4 kJ mol^{-1} more exothermic than the measured enthalpy, indicating an extremely large destabilization for the real compound. Recall that the difference between the predicted (from chlorobenzene) and experimental enthalpies of formation of *m*-chlorophenol was the very large stabilization of −26.3 kJ mol^{-1}.

Among the dichloro derivatives, the 2,5-dichloro isomer is seemingly the most stable, avoiding the substituent crowding which is present in its isomers. However, the experimental uncertainties are quite large and so the actual isomer stability order is not known. Predicting the enthalpies of formation from dichlorobenzene and the OH/H increment exchange energies gives a gas phase value of −306.0 kJ mol^{-1} and solid phase value of −405.4 kJ mol^{-1}. The gas phase predicted and experimental enthalpies are indistinguishable for the 2,6-isomer, the only one for which a measured value is available. All of the solid enthalpies of formation are much more endothermic than predicted, *ca* 22–33 kJ mol^{-1}. In addition to the two *ortho* chloro/hydroxyl interactions, there are two *meta* chloro/hydroxyl and one each of dihydroxyl and dichloro interactions between the substituents on the ring in 2,5-dichlorobenzene-1,4-diol. Can we state which of these interactions are important to the predicted instability of this compound in the solid phase, despite the lack of solid phase enthalpy of formation of *o*-chlorophenol? Reaction 20, which redistributes the hydroxyl and chloro substituents, is 52.1 kJ mol^{-1} endothermic in the gas phase but thermoneutral (−0.4 kJ mol^{-1}) in the solid phase:

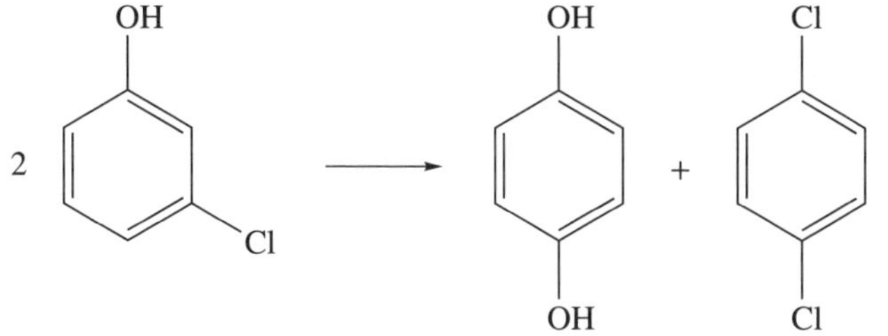

$$2 \text{ } m\text{-chlorophenol} \longrightarrow \text{1,4-benzenediol} + \text{1,4-dichlorobenzene} \quad (20)$$

Accordingly, in the solid phase, the enthalpies of reactions 21 and 22 are the same and equal *ca* 26 kJ mol^{-1}.

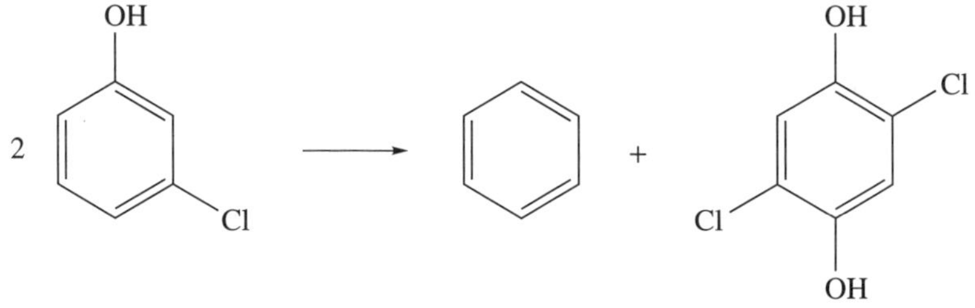

$$2 \text{ } m\text{-chlorophenol} \longrightarrow \text{benzene} + \text{2,5-dichlorobenzene-1,4-diol} \quad (21)$$

$$\text{HO-C}_6\text{H}_4\text{-OH} + \text{Cl-C}_6\text{H}_4\text{-Cl} \longrightarrow \text{C}_6\text{H}_6 + \text{(OH)}_2\text{C}_6\text{H}_2\text{Cl}_2 \quad (22)$$

The derived instability seemingly comes from the *ortho* relationship of two OH/Cl pairs of substituents. The *ca* 11 kJ mol^{-1} instability of 2,3-dichlorobenzene-1,4-diol, relative to the 2,5-isomer, is due to the additional crowding of substituents on the aromatic ring and whatever difference there may be between the electronic effects of *ortho* vs. *para* chlorines. For comparison, the stability difference between solid *o*- and *p*-dichlorobenzene is *ca* 9 kJ mol^{-1}. The stability of the 2,6-isomer is intermediate between its other two isomers. Relative to the 2,5-isomer, it is less stable by *ca* 4 kJ mol^{-1}, the same as the difference between *m*- and *p*-dichlorobenzene. Evidently, but surprisingly, additional steric effects are unimportant here.

There are no trichloro- or tetrachlorobenzenes to compare with 2,3,5-trichloro- or 2,3,5,6-tetrachlorobenzene-1,4-diol. Neither are there any completely satisfactory isodesmic reactions for which there are the necessary data to disentangle the myriad steric and electronic effects in these highly substituted aromatic rings.

3-Methoxycatechol, also considered as a derivative of benzenetriol, will be discussed in a later section.

D. Naphthalenediols and Other Arenediols

Although there are ten isomeric naphthalenediols, there are enthalpy of formation data for only five of them. The enthalpy of formation data for the 1,4-isomer from two sources are disparate, as are the data from the two sources for the 2,3-isomer. The 1,3-naphthalenediol is more stable than either the 1,2- or the 1,4-diols for the same reason that the *m*-benzenediol, resorcinol, is more stable than its isomers: more stable resonance structures for 1,3-dihydroxy substitution on an aromatic ring. From the appropriate OH/H increment exchange energies and the enthalpy of formation of naphthalene, we would have predicted a value of -329 kJ mol^{-1} and -208 kJ mol^{-1} for any solid and gaseous naphthalenediol, respectively. Only for 1,3- and 2,7-naphthalenediol is the expectation confirmed: the others with their less stable *ortho*- and *para*-type substitution are less negative.

Both the 1,2- and the 2,3-isomers contain adjacent hydroxyl groups, analogous to the *o*-benzenediol, catechol. However, the 1,2-diol might experience a small destabilization relative to the 2,3-isomer because of a *peri* substituent interaction in the former. For comparison, the enthalpies of formation of 1-naphthol are (s) -121.0 ± 1.0 and (g) -29.9 ± 1.0 and of 2-naphthol are (s) -124.2 ± 1.0 and (g) -30.0 ± 1.1 kJ mol^{-1}. The destabilization seemingly exists, at least in the solid phase. Of the two sets of data for the 2,3-isomer, one corresponds to greater stability and the other to lesser stability relative to the 1,2-isomer. One method of testing the data for both the naphthalenediols and the benzenediols, at least for consistency if not accuracy, is to compare the enthalpies of the two exchange reactions in equations 23 and 24,

$$\text{C}_6\text{H}_6 + \text{C}_{10}\text{H}_7\text{OH} \longrightarrow \text{C}_6\text{H}_5\text{OH} + \text{C}_{10}\text{H}_8 \quad (23)$$

$$C_6H_6 + C_{10}H_6(OH)_2 \longrightarrow C_6H_4(OH)_2 + C_{10}H_8 \qquad (24)$$

in their various isomeric combinations. The enthalpy of reaction 23 for 1-naphthol is (s) −5.3 and (g) 1.2 kJ mol^{-1}, and for 2-naphthol is (s) −2.1 and (g) 1.3 kJ mol^{-1}. That is, in the gas phase, the reactions are essentially thermoneutral and only slightly less so for 1-naphthol in the solid phase. For the similarly disubstituted 2,3-naphthalenediol and catechol in reaction 24, the enthalpy of reaction should also be thermoneutral. After calculating all 6 combinations for which there are data, the enthalpies of reaction which most closely fit the criterion are (s) 2.1 and (g) 3.1 kJ mol^{-1} using the enthalpy of formation data for the naphthalenediol from Reference 80 and for the benzenediol from Reference 33. Using the averages of the catechol enthalpies gives an almost identical enthalpy of reaction for the gas phase (3.4 kJ mol^{-1}) and a slightly negative enthalpy of reaction (−1.3 kJ mol^{-1}) for the solid phase. The enthalpies of reaction 24 for 1,2-naphthalenediol and the catechol average are −7.9 (s) and −3.1 (g) kJ mol^{-1}, slightly more negative than predicted. These results are consistent with our expectation that 2,3-naphthalenediol is more stable than 1,2-naphthalenediol. The enthalpy of reaction of reaction 24, calculated for 1,3-naphthalenediol and resorcinol, is (s) −3.4 and (g) −1.0 kJ mol^{-1}, again using the averages of the enthalpies of formation of the 1,3-benzenediol.

The sole other arenediol we know of, phenanthrene-9,10-diol, has an enthalpy of formation derived from the ancient calorimetric results in Reference 88. Acknowledging that there has been some dispute about the enthalpy of formation of the parent hydrocarbon, we adopt the value of 113.0 ± 2.1 kJ mol^{-1} for phenanthrene[89]. The estimated enthalpy of formation of the phenanthrenediol, based on twice the OH/H increment exchange energy, is −293.8 kJ mol^{-1}, ca 50 kJ mol^{-1} more negative than the actual measurement. The apparent destabilization of the real diol is considerably greater than that for the related vicinally dihydroxylated naphthalenes or for catechol.

IV. ARENETRIOLS

The only thermochemically characterized arenetriols known to the authors are the benzenetriols listed in Table 7: 1,2,3-(pyrogallol), 1,2,4- and 1,3,5-(phloroglucinol) and the 5-carboxy derivative of pyrogallol (gallic acid). For the three parent triols, an enthalpy of formation of −571 kJ mol^{-1} would have been expected for the solids and −454 kJ mol^{-1} in the gas phase by combining the appropriate OH/H increment exchange energy and the enthalpy of formation of benzene. Good agreement is found for the gas phase for the 1,3,5-isomer in which there are no unfavorable resonance structures or interhydroxylic interactions. The decreased stability for triols with o-hydroxyl groups (as also observed in the parent diols) is shown by the ca 10 kJ mol^{-1} successive increase in enthalpies of formation of gaseous 1,2,4- and 1,2,3-benzenetriol.

In order to better understand the hydroxyl interactions in the triols, they can be compared with their methylated counterparts, the trimethoxybenzenes (equation 25).

$$C_6H_3(OH)_3 + 3\,PhOMe \longrightarrow C_6H_3(OMe)_3 + 3\,PhOH \qquad (25)$$

The reaction should be thermoneutral if there is no net difference, steric or electronic, upon replacing the phenolic hydrogen with a methyl group. From the enthalpies of formation of gas and solid phase trimethoxybenzenes[83], the enthalpies of reaction are 9.7 kJ mol^{-1} for the 1,2,3- and −4.0 kJ mol^{-1} for the 1,3,5-benzenetriol in the gas phase. The solid phase reaction enthalpies are 1.9 kJ mol^{-1} for the 1,2,3- and 0.4 kJ mol^{-1} for the 1,3,5-isomer. The enthalpies of reaction per methyl replacement are one-third these values. All of these, except for the gas phase value for 1,2,3-trimethoxybenzene, are indistinguishable from thermoneutrality once the experimental uncertainties are considered. From *ab initio* geometry optimizations, the 1,3,5-trimethoxybenzene is shown to be a planar molecule, while its 1,2,3-isomer is nonplanar[83]. The slight endothermicity for the latter's reaction suggests a planar triol converted to a nonplanar triether.

3-Methoxycatechol, after exchanging two hydrogens of methoxybenzene (anisole) for hydroxyls, would be expected to have enthalpies of formation of (s) -534.5 and (g) -425.9 kJ mol^{-1}. The estimated destabilization in the gas phase of *ca* 7 kJ mol^{-1} is almost twice that for catechol itself, as might be expected for two *ortho* interactions in the tri-oxygenated derivative. The *ca* 24 kJ mol^{-1} calculated destabilization in the solid phase is also about twice that for catechol. Although the doubled destabilization of 3-methoxycatechol is not unreasonable, consider equation 26, now written for replacement of only one hydrogen with a methyl group:

$$C_6H_3(OH)_3 + 1\ PhOMe \longrightarrow C_6H_3(OH)_2OMe + 1\ PhOH \qquad (26)$$

The enthalpies of reaction are (s) 3.5 kJ mol^{-1} and (g) -12.8 kJ mol^{-1}. Although the solid phase reaction enthalpy is reasonable, the gas phase reaction enthalpy seems too exothermic, i.e. the gaseous enthalpy of formation of this compound, determined from its enthalpy of sublimation, is at least 13 kJ mol^{-1} too negative.

A similar assessment can be made for 2,6-dimethoxyphenol and its methyl exchange reaction (equation 27):

$$C_6H_3(OH)_3 + 2\ PhOMe \longrightarrow C_6H_3(OH)(OMe)_2 + 2\ PhOH \qquad (27)$$

There are data only for the solid phase, but the enthalpy of reaction, -42.1 kJ mol^{-1}, shows that the measurement[56] of -518.4 kJ mol^{-1} for 2,6-dimethoxyphenol must be inaccurate.

With regard to gallic acid, the expected enthalpy of formation value is -995 kJ mol^{-1}. The difference between its measured and estimated enthalpy of formation is similar to the difference for *p*-hydroxybenzoic acid.

V. ARENOLQUINONES

The enthalpies of formation of arenolquinones appear in Table 8. In reviewing the data there and in Table 1, note the identical solid phase enthalpies of formation of 1,4-naphthoquinone and 9,10-anthraquinone and the identical solid phase enthalpies of formation of 5,8-dihydroxy-1,4-naphthoquinone and 1,4-dihydroxy-9,10-anthraquinone. These

TABLE 8. Enthalpies of formation of arenolquinones (kJ mol^{-1})

Compound	Solid	Gas	Reference
5,8-Dihydroxy-1,4-naphthoquinone	-595.8 ± 2.1	-499.1 ± 3.2	15
9,10-Anthraquinone			
2-hydroxy	-453.1	—	14
1,2-dihydroxy (alizarin)	-590.3	—	14
1,4-dihydroxy	-595.1 ± 2.1	-471.0 ± 2.3	15
1,2,4-trihydroxy	-786.1	—	14
1,2,3,5,6,7-hexahydroxy	-1426.3	—	14

3. Thermochemistry of phenols and related arenols

values have been recalculated from the original enthalpy of combustion data from the sources cited. We are confident the data are accurate as reported in the most recent reference cited in Tables 1 and 8, because the enthalpy values of combustion data for these compounds are virtually identical to the results reported for the identical compounds in Reference 14 from 1925.

The simplest compounds which are both quinones and arenols are the hydroxynaphthoquinones. However, the only one of the many possible isomers which has been thermochemically characterized is 5,8-dihydroxy-1,4-naphthoquinone. OH/H increment exchange energies calculated for 1,4-dihydroxynaphthalene, rather than the OH/H increment exchange energies derived from phenol, are used to assess the relative stability of the dihydroxyquinone so that any hydroxyl substituent interactions on the aromatic ring parents cancel. Reaction 28 is mathematically equivalent to generating an increment exchange energy for two p-hydroxyl groups substituted on naphthalene as from equation 6, and then adding the increment to the enthalpy of formation of the naphthoquinone parent.

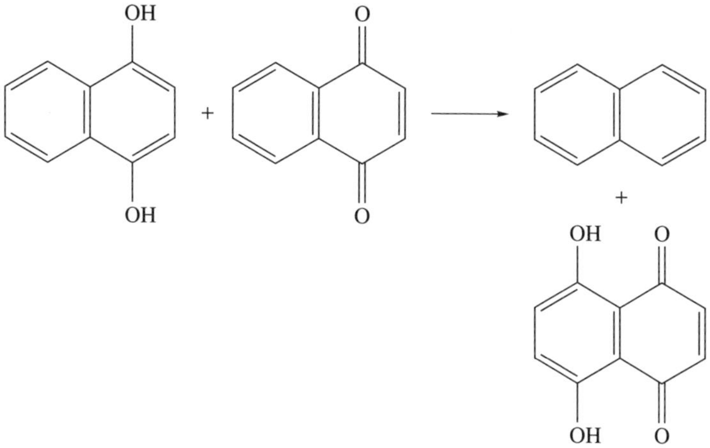

$$\text{(28)}$$

In the solid phase, the enthalpy of reaction 28 is a modest -12 kJ mol^{-1}, but in the gaseous phase it is the extremely large -53.9 kJ mol^{-1}, indicating significant stabilization. There may be dipolar resonance contributing structures and strong intramolecular hydrogen bonding between each pair of OH/C=O substituents on the *peri* positions which stabilize the p-hydroxylated p-quinone.

There are various hydroxy-9,10-anthraquinones for which thermochemical data exist, mainly in the solid phase. The 1,2- and 1,4-dihydroxy isomers have nearly identical enthalpies of formation. The difference between their enthalpies of formation, and the fact that the 1,4-isomer is more stable, is consistent with the enthalpy differences and relative stabilities of the similarly substituted 1,2- and 1,4-naphthalenediols. Although there are no hydroxy anthracenes with which to compare any of these compounds, we can estimate their enthalpies of formation by adding the OH/H increment exchange energy from a correspondingly substituted naphthalene to the enthalpy of formation of anthracene. The estimated enthalpies of formation for the mono- and di-substituted anthracenes are: 2-hydroxy (s), -72.9; 1,2-dihydroxy (s), -258.5; 1,4-dihydroxy (s), -266.1; and 1,4-dihydroxy (g), -116.4 kJ mol^{-1}. The 1,2,4-trihydroxyanthracene solid enthalpy of formation (-458 kJ mol^{-1}) is estimated from a 1,2,4-(OH/H) increment generated as the average of two increments obtained either by summing the *para*-OH/H increment and one-half the *ortho*-OH/H increment or by summing the *ortho*-OH/H increment and one-half the *para*-OH/H increment. The 1,2,3,5,6,7-hexahydroxyanthracene solid enthalpy of formation (-1034 kJ mol^{-1}) is derived from 2(1.5) *ortho*-OH/H increments. The enthalpies of reaction 28, recast for the anthracenes instead of the naphthalene, are now discussed in order of increasing 'perplexity'.

The 1,4-dihydroxy-9,10-anthraquinone is the only compound for which there are both solid and gaseous enthalpies of formation and for which the solid enthalpy of formation has been independently measured twice and the results found to be indistinguishable. The enthalpies of the recast reaction 28 are -11.3 (s) and -48 kJ mol^{-1} (g) which are essentially identical to the correspondingly substituted naphthalenes. The enthalpy of the recast reaction 28 for solid 1,2-dihydroxy-9,10-anthraquinone is -14.1 kJ mol^{-1}, a result which is consistent with the two previously derived. The stabilization in the solid phase exhibited by these three compounds evidently is not solely dependent on two [OH $\cdots$ C=O] intramolecular hydrogen bonds, since the 1,2-dihydroxy derivative has only one such interaction. The enthalpy of reaction 28 for solid 1,2,4-trihydroxy-9,10-anthraquinone is -10.4 kJ mol^{-1}, again consistent with the others in the solid phase. The solid enthalpies of reaction 28 for 2-hydroxy-9,10-anthraquinone and 1,2,3,5,6,7-hexahydroxy-9,10-anthraquinone are -62.5 and -74.6 kJ mol^{-1}, respectively. While these values are compatible with each other, they resemble the gas phase reaction enthalpies, not the solid phase ones. This excessive calculated stabilization in the solid phase is inexplicable, regardless of the method of generating increment exchange energies. The enthalpies of formation of the two quinones are 40–60 kJ mol^{-1} more negative than we would expect.

VI. TAUTOMERIC ARENOLS

A. Obstacles and Opportunities

Clarifying the subsection title, we say 'obstacles' because any significant presence of tautomers complicates the interpretation of the measured values. We say 'opportunities' because two substances may be understood for the experimentally measured price of one. The enthalpies of formation of the tautomeric arenols appear in Table 9.

B. Unsubstituted Arenols

The archetypal arenol, phenol, has two cyclohexadienone tautomers. Although interesting in their own right, we ignore the latter two species and the difference between their enthalpies of formation and that of the more stable and isolable phenol. We likewise ignore the various tautomers of 1- and 2-naphthol because only these arenols, and not their keto isomers, are isolable. The anthrols and their corresponding anthrone tautomers are of interest, however, because for the 9-isomer, both tautomers are isolable[88]. From this earliest study, it has been known that 9-anthrone is the more stable tautomer and perhaps because of its greater stability, it alone has had its enthalpy of formation determined calorimetrically[18,19]. Through the decades, solvent effects on the enthalpy of formation difference between 9-anthrone and 9-anthrol have been measured. Derived as a limiting result from the solvated species, it has been suggested[20] that the 9-anthrone tautomer is favored by 23 ± 8 kJ mol^{-1}. We thus obtain the enthalpy of formation of gaseous 9-anthrol as *ca* 45 kJ mol^{-1} as discussed in Section II. From the discussion of the isomeric naphthols, their gas phase enthalpies of formation are nearly equal and the hydroxy exchange reaction is almost thermoneutral. Thus the OH/H increment exchange reaction

TABLE 9. Enthalpies of formation of tautomeric arenols (kJ mol^{-1})

Compound	Solid	Gas	Reference[a]
9-Anthrone	−79.9 ± 2.1	22.3 ± 2.7[b]	18, 19
4-Nitrosophenol (p-benzoquinone oxime)	−90.6	—	14
2-Nitroso-1-naphthol	−61.8 ± 4.5	−5.4 ± 6.2	—
(1,2-naphthoquinone-2-monoxime)	−51.0		14
1-Nitroso-2-naphthol	−50.5 ± 2.2	36.1 ± 4.7	—
(1,2-naphthoquinone-1-monoxime)	−47.1		14
4-Nitroso-1-naphthol	−107.8 ± 2.5	−20.3 ± 4.9	—
(1,4-naphthoquinone monoxime)	−67.4		14
p-Phenylazophenol	+163	—	90
1-Phenylazo-2-naphthol	+246	—	90
4-Phenylazo-1-naphthol	+225	—	90
1-[(2,4-Dimethylphenyl)azo]-2-naphthol	+207	—	90
	−125 ± 20	—	81
2,4-Dinitrosobenzene-1,3-diol (5-cyclohexene-1,2,3,4-tetrone-1,3-dioxime)	−235	—	91

[a]Data are from Reference 2 unless otherwise stated.
[b]The enthalpy of formation is the average of the values found in the references cited, 23.4 ± 2.2 and 21.1 ± 1.5 kJ mol^{-1}, respectively.

between anthracene and either benzene (equation 29) or naphthalene (equation 30) should be nearly thermoneutral as well:

$$C_{14}H_{10} + PhOH \longrightarrow 9\text{-}C_{14}H_9OH + C_6H_6 \quad (29)$$

$$C_{14}H_{10} + 2\text{-}NpOH \longrightarrow 9\text{-}C_{14}H_9OH + C_{10}H_8 \quad (30)$$

From these, we conclude that the gas phase enthalpy of formation of 9-anthrol should be 52 kJ mol^{-1} and 51 kJ mol^{-1}, respectively, in good agreement with the estimation above, and far less negative than that measured for the tautomeric 9-anthrone.

C. Nitrosophenols and Nitrosonaphthols (Quinone Oximes)

The o- and p-nitrosophenols are tautomeric with o- and p-benzoquinone oxime, respectively. Some of the nitrosonaphthols are tautomers of naphthoquinone oximes. A recent publication summarizes the current knowledge of tautomeric equilibria in solution, the composition of the solid phase and the results of theoretical studies[92]. While tautomeric composition in solution is very much dependent on compound structure and solvent polarity, various nitrosophenols, 2-nitrosonaphthol, 1-nitroso-2-naphthol and 4-nitrosonaphthol exist exclusively as quinone oximes in the solid state.

Of the nitrosophenols/benzoquinone oximes, only one compound has been thermochemically studied and in only one phase, solid 4-nitrosophenol/p-benzoquinone oxime[14]. We welcome a new thermochemical investigation of this species, the 2-nitrosophenol, as well as of the m-isomer for which no oxime 'contamination' or ambiguity is possible because of the absence of stable m-benzoquinones and related derivatives.

The archival enthalpies of combustion for three nitrosonaphthol (naphthoquinone oxime) isomers are from a rather contemporary paper[93], while the earlier ones (reported over 40 years before[14]) are not referenced in our archival source. In the solid phase, the differences between the reported solid phase enthalpies of formation range from 5–40 kJ mol^{-1}. Two of the three results are roughly consistent between the two references and the third is considerably dissonant. For comparison, the enthalpy of formation difference between the solid o- and p-naphthoquinone parent isomers is ca 24 kJ mol^{-1}[94]. The stability order is 1,4- > 2,1- > 1,2-naphthoquinone oxime. At least for the p- vs. the o-isomers, the substituents are more accessible for intermolecular hydrogen bonding.

The solid phase enthalpies of formation are quite similar for the 2,1- and 1,2-compounds, but they have very different enthalpies in the gaseous phase due to the nearly 30 kJ mol^{-1} difference in their enthalpies of sublimation. The enthalpies of sublimation for 1-nitroso-2-naphthol, 4-nitroso-1-naphthol and 1,4-naphthoquinone are nearly the same. Recent ab $initio$ calculations[92] show that the phenolic form is favored. The energy increase is in the order 2-nitrosonaphthol < 1-nitrosonaphthol < 4-nitrosonaphthol with a corresponding increase in the energy difference between the nitrosophenol and quinone oxime tautomers. However, there was no significant calculated difference between the 2-nitrosonaphthol/quinone oxime tautomers. Whatever the tautomeric form of these species, we would have expected the experimental measurements to show the gaseous 2,1- and 1,2-compounds as more stable than their 1,4-isomer because of the presence of an intramolecular hydrogen bond found solely in the first two. The measurement of the enthalpies of formation of some tautomerically frozen nitrosoarenols and their quinone oxime counterparts (e.g. O-methyl ethers) would be most welcome[95].

D. Arylazo Derivatives of Phenol and the Naphthols

We now turn to arylazophenols and naphthols which may alternatively be described as benzo- and naphthoquinone phenylhydrazones. The structural ambiguities and resultant thermochemical problems which plagued us for nitrosoarenols return here but in a different way. We start with one of the archetypal species, p-phenylazophenol[90]. We have no isomer with which to compare the result, although based on earlier results in this study by the same author we are suspicious. The predicted enthalpy of formation using the condensed phase OH/H increment exchange enthalpy is ca 107 kJ mol^{-1}, very different from the published value. From the same source we find the enthalpy of formation of 1-phenylazo-2-naphthol and 4-phenylazo-1-naphthol. These values are some 70 ($\pm$10) kJ mol^{-1} higher than that of the azophenol while the difference between the unsubstituted naphthols and phenol is ca 42 kJ mol^{-1} and between the benzo- and naphthoquinone, indistinguishable. Again from the same source we find a value for the enthalpy of formation of 1-[(2,4-dimethylphenyl)azo]-2-naphthol which is 39 kJ mol^{-1} less than for the demethylated species. This difference is plausible in that the difference between the enthalpies of formation of solid benzene and the related dimethyl species, m-xylene, is 25 kJ mol^{-1}. However, our comfort is marred because we know of another, highly disparate, calorimetric measurement for this same azonaphthol[81].

With regard to the question of tautomers, the 1-phenylazo-2-naphthol/1,2-naphthoquinone phenylhydrazone equilibrium has been studied for a variety of substituted phenyl groups[96]. At least in CDCl$_3$ solution, the difference in stability is small (the parent compound favors the hydrazone by but 4.0 ± 0.3 kJ mol^{-1}).

E. Ambiguous Arenepolyols

We close this section with a brief mention of 2,4-dinitrosobenzene-1,3-diol or 5-cyclohexene-1,2,3,4-tetrone-1,3-dioxime or yet some other tautomer studied as a solid almost 100 years ago[91]. Is the cited enthalpy of formation plausible? Equation 31 is exothermic by 15 kJ mol^{-1}.

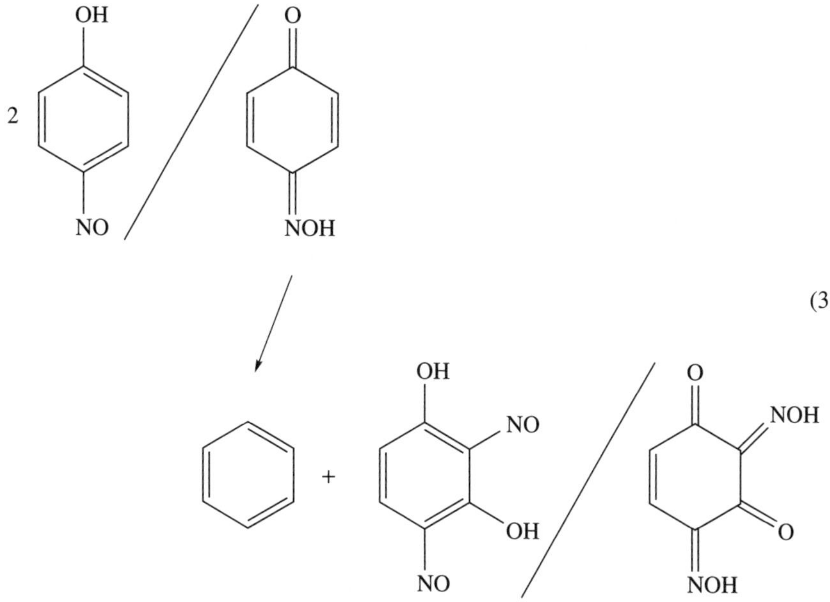

$$2 \; \text{HO-C}_6\text{H}_4\text{-NO} \; / \; \text{O=C}_6\text{H}_4\text{=NOH} \; \longrightarrow \; \text{C}_6\text{H}_6 \; + \; \text{(NO)(OH)}_2\text{C}_6\text{H}_2\text{(NO)(OH)} \; / \; \text{O=C}_6\text{H}_2(\text{=NOH})_2\text{=O}$$

(31)

Admitting considerable ambiguity as to the nature of the phenol and the diol and to an understanding of solid phase reactions, and doubting that the substituents are independent of each other, the result is not unreasonable. For comparison, the related 'nitro' reaction 32 is endothermic by 36 kJ mol^{-1}.

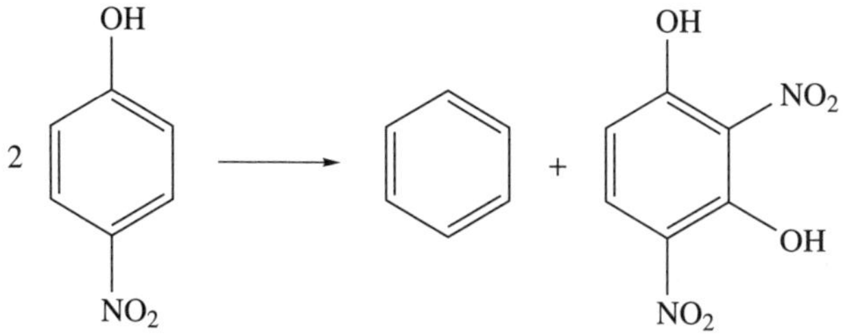

$$2 \; \text{HO-C}_6\text{H}_4\text{-NO}_2 \; \longrightarrow \; \text{C}_6\text{H}_6 \; + \; (\text{NO}_2)(\text{OH})_2\text{C}_6\text{H}_2(\text{NO}_2)(\text{OH})$$

(32)

We do not know how endothermic or exothermic the deoxygenated dinitroso reaction 33 is because the enthalpies of formation of nitrosobenzene (as solid monomer) and 1,3-dinitrosobenzene are not available.

$$2\text{PhNO} \longrightarrow 1,3\text{-C}_6\text{H}_4(\text{NO})_2 + \text{C}_6\text{H}_6$$

(33)

However, the corresponding dinitro reaction is endothermic by 13 kJ mol^{-1}. And yes, this is our phenol answer for this chapter as well as section therein.

VII. REFERENCES AND NOTES

1. See, for example, S. W. Slayden and J. F. Liebman, in *The Chemistry of Functional Groups Supplement E2: Chemistry of Hydroxyl, Ether and Peroxide Groups* (Ed. S. Patai), Wiley, Chichester, 1993.
2. J. B. Pedley, R. D. Naylor and S. P. Kirby, *Thermochemical Data of Organic Compounds* (2nd edn.), Chapman & Hall, New York, 1986.
3. H. Y. Afeefy, J. F. Liebman and S. E. Stein, 'Neutral Thermochemical Data', in *NIST Chemistry WebBook*, NIST Standard Reference Database Number 69 (Eds. W. G. Mallard and P. J. Linstrom), National Institute of Standards and Technology, Gaithersburg, MD 20899, July 2001 (http://webbook.nist.gov).
4. In the few cases where enthalpies of vaporization are needed, but not available from experiment, we use the estimation approach given in J. S. Chickos, D. G. Hesse and J. F. Liebman, *J. Org. Chem.*, **54**, 5250 (1989).
5. Where needed, the enthalpy of fusion will be obtained from J. S. Chickos, W. E. Acree, Jr. and J. F. Liebman, *J. Phys. Chem. Ref. Data*, **28**, 1535 (1999). Strictly, use of these fusion enthalpies are estimates. The quantities should be corrected to 298 K from the melting point. However, the error is generally small because changes in heat capacities of solids and liquids as functions of temperature are generally small.
6. For comparison, the difference between the enthalpies of formation of the *o*- and *m*- (or *p*-)-*tert*-butyltoluene is *ca* 22 kJ mol^{-1}; cf. E. J. Prosen, unpublished results, cited in H. C. Brown and L. Domash, *J. Am. Chem. Soc.*, **78**, 5384 (1956).
7. W. V. Steele, R. D. Chirico, A. Nguyen, I. A. Hossenlopp and N. K. Smith, *Am. Inst. Chem. Eng. Symp. Ser. (AIChE Symp. Ser.)*, **86**, 138 (1990).
8. G. Stohmann and R. Schmidt, *Ber. Verhändl. K. Sächs, Gss. Wiss. Math. Phys. Kl.*, **47**, 375 (1895); *J. Prakt. Chem.*, **52**, 59 (1895).
9. P. Landrieu, *Compt. Rend. Acad. Sci. Paris*, **140**, 867 (1905).
10. J. J. Kirchner, W. E. Acree, Jr., G. Pilcher and L. Shaofeng, *J. Chem. Thermodyn.*, **18**, 793 (1986).
11. V. A. Platonov, Y. N. Simulin and M. M. Rozenberg, *Russ. J. Phys. Chem. (Engl. Transl.)*, **59**, 814 (1985); *Chem. Abstr.*, **103**, 77002e (1985).
12. R. Sabbah and X. W. An, *Thermochim. Acta*, **179**, 81 (1991).
13. T. N. Nesterova, S. R. Verevkin, S. Ya. Karaseva, A. M. Rozhnov and V. F. Tsvetkov, *Russ. J. Phys. Chem. (Engl. Transl.)*, **58**, 297 (1984); *Chem. Abstr.*, **100**, 10124m (1984).
14. W. Swientoslawski and H. Starczedska, *J. Chim. Phys.*, **22**, 399 (1925).
15. M. A. V. Ribeiro da Silva, M. D. M. C. Ribeiro da Silva, J. A. S. Teixeira, J. M. Bruce, P. M. Guyan and G. Pilcher, *J. Chem. Thermodyn.*, **21**, 265 (1989).
16. M. Colomina, M. V. Roux and C. Turrión, *J. Chem. Thermodyn.*, **6**, 571 (1974).
17. M. A. V. Ribeiro da Silva, M. D. M. C. Ribeiro da Silva and G. Pilcher, *J. Chem. Thermodyn.*, **20**, 969 (1988).
18. L. El Watik and R. Sabbah, *Bull. Soc. Chim. Fr.*, **128**, 344 (1991).
19. S. P. Verevkin, *Thermochim. Acta*, **310**, 229 (1998).
20. See the summary study, B. Freiermuth, B. Hellrung, S. Peterli, M.-F. Schultz, D. Wintgens and J. Wirz, *Helv. Chim. Acta*, **84**, 3796 (2001); J. Wirz, personal communication.
21. M. A. V. Ribeiro da Silva, M. A. R. Matos, V. M. F. Morais and M. S. Miranda, *J. Org. Chem.*, **64**, 8816 (1999).
22. S. P. Verevkin, *J. Chem. Thermodyn.*, **31**, 559 (1999).
23. G. Berthelon, M. Giray, R. Perrin and M. F. Vincent-Falquet-Berny, *Bull. Soc. Chim. Fr.*, 3180 (1971).
24. M. M. Ammar, N. El Sayed, S. E. Morsi and A. El Azmirly, *Egypt. J. Phys.*, **8**, 111 (1977).
25. M. Colomina, P. Jiménez, M. V. Roux and C. Turrión, *J. Calorim. Anal. Therm.*, **11**, 1 (1980); *An. Quim.*, **77**, 114 (1981).
26. R. Sabbah and T. H. D. Le, *Can. J. Chem.*, **71**, 1378 (1993).
27. A. K. Ryskalieva, G. V. Abramova, R. S. Erkasov and N. N. Nurakhmetov, *Russ. J. Phys. Chem. (Engl. Transl.)*, **66**, 421 (1992); *Chem. Abstr.*, **117**, 119512t (1972).
28. M. Delépine and P. Rivals, *Compt. Rend.*, **129**, 520 (1899).
29. G. H. Parsons, C. H. Rochester and C. E. C. Wood, *J. Chem. Soc., B*, 533 (1971).

30. V. M. F. Morais, M. S. Miranda and M. A. R. Matos, unpublished results, submitted for publication.
31. D. Zavoianu, I. Ciocazanu, S. Moga-Gheorghe and C. Bornaz, *Rev. Chim. (Bucharest)*, **39**, 487 (1988).
32. M. A. V. Ribeiro da Silva, M. A. R. Matos, M. S. Miranda, M. H. F. A. Sousa, R. M. Borges dos Santos, M. M. Bizarro and J. A. Martinho Simões, *Struct. Chem.*, **12**, 171 (2001).
33. A. Finch, P. J. Gardner and D. Wu, *Thermochim. Acta*, **66**, 333 (1983).
34. Reaction calorimetry provides useful insights here even if direct enthalpy of formation measurements are absent. Liquid phase isomerization of the *o*- to *p-tert*-butylphenol has been shown to be exothermic by 16.9 ± 1.6 kJ mol^{-1} [T. N. Nesterova, S. P. Verevkin, T. N. Malova and V. A. Pil'shchikov, *Zh. Prikl. Khim.*, **58**, 827 (1985); *Chem. Abstr.*, **103**, 159918x (1985)] while the *p*- to *m*-isomerization in both the liquid and gas phase is exothermic by *ca* 1 kJ mol^{-1} [cf. V. A. Pil'shchikov, T. N. Nesterova and A. M. Rozhnov, *J. Appl. Chem. USSR*, **54**, 1765 (1981); *Chem. Abstr.*, **96**, 68487b (1982) and S. P. Verevkin, *Termodin. Organ. Soedin.*, **67** (1982); *Chem. Abstr.*, **99**, 157694k (1983), respectively].
35. The only disquieting note is that the enthalpies of formation (cf. Reference 6) of *m*- and *p-tert*-butyltoluene are -54 ± 2 and -57 ± 2 kJ mol^{-1}, which suggests the *para*-isomer is more stable, unlike the case for most other dialkylated benzenes.
36. E. A. Heath, P. Singh and Y. Ebisuzaki, *Acta Crystallogr.*, **48C**, 1960 (1992). The acetyl derivative of salicylic acid, aspirin, has a solid state structure somewhat similar to *p*-hydroxybenzoic acid in that a pair of molecules forms hydrogen-bonded dimers which are linked by an antiparallel dipole–dipole alignment of the ester carbonyl functional groups. See P. J. Wheatley, *J. Chem. Soc.*, 6039 (1964); Y. Kim, K. Machida, T. Taga and K. Osaki, *Chem. Pharm. Bull.*, **33**, 2541 (1985); R. Glaser, *J. Org. Chem.*, **66**, 771 (2001).
37. K. Nishiyama, M. Sakiyama and S. Seki, *Bull. Chem. Soc. Jpn.*, **56**, 3171 (1983).
38. M. Colomina, P. Jiménez, M. V. Roux and C. Turrión, *J. Chem. Thermodyn.*, **10**, 661 (1978).
39. J. A. Paixo, A. M. Beja, M. R. Silva, L. A. de Veiga and A. C. Serra, *Acta Crystallogr., Sect. C*, **56**, 1348 (2000).
40. Z. Y. Hy, M. J. Hardie, P. Burckel, A. A. Pinkerton and P. W. Erhardt, *J. Chem. Crystallogr.*, **29**, 185 (1999).
41. C. Beswick, M. Kubicki and P. W. Codding, *Acta Crystallogr., Sect. C*, **52**, 3171 (1996).
42. G. L. Carlson and W. G. Fateley, *J. Phys. Chem.*, **77**, 1157 (1973).
43. L. Nuñez, L. Barral, L. S. Gavelanes and G. Pilcher, *J. Chem. Thermodyn.*, **18**, 575 (1986).
44. R. Sabbah and M. Gouali, *Can. J. Chem.*, **74**, 500 (1996).
45. R. Sabbah and M. Gouali, *Aust. J. Chem.*, **47**, 1651 (1994).
46. M. A. V. Ribeiro da Silva, A. M. M. V. Reis, M. J. S. Monte, M. M. S. F. Bartolo and J. A. R. G. O. Rodrigues, *J. Chem. Thermodyn.*, **24**, 653 (1992).
47. A. Finch and A. E. Smith, *Thermochim. Acta*, **69**, 375 (1983). No error bars were assigned for 2,4-dinitrophenol.
48. The enthalpies of sublimation are from H. Hoyer and W. Peperle, *Z. Electrochem.*, **62**, 61 (1958).
49. P. J. Krueger, *Tetrahedron*, **26**, 4753 (1970).
50. P. Lemoult, *Compt. Rend.*, **143**, 772 (1906).
51. See, for example, N. V. Karyakin, G. P. Kamelova, V. N. Yurchenko, E. G. Kiparisova, V. V. Korshak, G. M. Tseitlin, V. N. Kulagin and I. B. Rabinovich, *Tr., Khim. Khim. Tekhnol*, p. 71 (1973); *Chem. Abstr.*, **80**, 121472r (1974) and K. V. Kir'yanov and N. V. Karyakin, *Fiz.-Khim. Osnovy Sinteza i Pererab. Polimerov, Gor'kii*, 109 (1987); *Chem. Abstr.*, **109**, 55268t (1988).
52. A. J. Abkowicz-Bienko, Z. Latajka, D. C. Bienko and D. Michalska, *Chem. Phys.*, **250**, 123 (1999).
53. P. C. Chen, S. C. Tzeng, *J. Mol. Struct. (Theochem)*, **467**, 243 (1999).
54. K. Macharacek, A. I. Zacharov and L. A. Aleshina, *Chem. Prumsyl*, **12**, 23 (1962); *Chem. Abstr.*, **75**, 1628g (1962).
55. The 2,4-isomer from C. Lenchitz, R. W. Velicky, G. Silvestro and L. P. Schlosberg, *J. Chem. Thermodyn.*, **3**, 689 (1971) and the 2,6-isomer from A. Barakat and A. Finch, *Thermochim. Acta*, **73**, 205 (1984).
56. J. J. Lindberg, T. P. Jauhiainen and A. Savolainen, *Pap. Puu*, **54**, 91 (1972); *Chem. Abstr.*, **77**, 5881n (1972).

57. D. A. Ponomarev, T. P. Oleinikova and T. N. Masalitinova, *Izv. Vyssh. Uchebn. Zaved, Khim. Khim. Tekhnol.*, **30**, 115 (1987); *Chem. Abstr.*, **107**, 122235a (1987).
58. F. Stohmann and H. Langbein, *J. Prakt. Chem.*, **46**, 530 (1893).
59. E. Leroy, *Compt. Rend.*, **128**, 107 (1899); *Ann. Chim. Phys.*, **21**, 87 (1900).
60. M. D. M. C. Ribeiro da Silva and M. A. V. Ribeiro da Silva, *J. Chem. Thermodyn.*, **16**, 1149 (1984).
61. We find -23.3 ± 0.5 kJ mol^{-1} from direct measurement of isomerization enthalpy, from E. Taskinen and N. Lindholm, *J. Phys. Org. Chem.*, **7**, 256 (1994), and 20.1 ± 1.9 kJ mol^{-1} from the difference between enthalpies of hydrogenation of allyl benzene, -126.0 ± 0.8 kJ mol^{-1}, from D. W. Rogers and F. J. McLafferty, *Tetrahedron*, **27**, 3765 (1971) and (*E*)-1-propenylbenzene, -105.9 ± 1.7 kJ mol^{-1}, from J.-L. M. Abboud, P. Jiménez, M. V. Roux, C. Turrión, C. Lopez-Mardomingo, A. Podosenin, D. W. Rogers and J. F. Liebman. *J. Phys. Org. Chem.*, **8**, 15 (1995).
62. K. V. Rajagopalan, R. Kalyanaraman and M. Sundaresan, *J. Indian Inst. Soc.*, **70**, 409 (1990).
63. G. R. Handrick, *Ind. Eng. Chem.*, **48**, 1366 (1956) and personal communication from the author.
64. This was discussed earlier in a 'Patai' chapter on general sulfonic acid thermochemistry, J. F. Liebman, in *The Chemistry of the Sulphonic Acids, Esters and their Derivatives* (Eds. S. Patai and Z. Rappoport), Wiley, Chichester, 1991, p. 283.
65. F. Swarts, *J. Chim. Phys.*, **17**, 3 (1919).
66. Using the original data from M. Berthelot, *Ann. Chim. Phys.*, **21**, 296 (1900), cautiously acknowledging for organoiodine compounds the suggested enthalpies of formation of the two iodinated salicylic acids from D. R. Stull, E. F. Westrum, Jr and G. C. Sinke, *The Chemical Thermodynamics of Organic Compounds*, Wiley, New York, 1969 and making a structural assignment assuming electrophilic iodination of the parent salicylic acid, we derived the enthalpies of formation and structural assignments presented in the text.
67. M. A. V. Ribeiro da Silva, M. L. C. C. H. Ferrao and F. Jiye, *J. Chem. Thermodyn.*, **26**, 839 (1994).
68. P. H. Allott and A. Finch, *Thermochim. Acta*, **146**, 371 (1989).
69. P. H. Allott and A. Finch, *J. Chem. Thermodyn.*, **19**, 771 (1987).
70. See the discussion of the difficulties of enthalpy of formation measurements of fluorinated species in A. J. Head and W. D. Good, in *Combustion Calorimetry* (Eds. S. Sunner and M. Mansson), Pergamon, Oxford, 1979. Complications include corrosion of the calorimeter, inadequate mixing of the solution formed from combustion (a uniform final aqueous solution of HF is required) and the formation of any perfluorocarbon during combustion.
71. A. Kovács, I. Macsári and I. Hargittai, *J. Phys. Chem. A*, **103**, 3110 (1999).
72. G. L. Carlson, W. G. Fateley, A. S. Manocha and F. F. Bentley, *J. Phys. Chem.*, **76**, 1553 (1972).
73. E. Vajda and I. Hargittai, *J. Phys. Chem.*, **97**, 70 (1993).
74. R. Sabbah and E. N. L. E. Buluku, *Can. J. Chem.*, **69**, 481 (1991).
75. P. D. Desai, R. C. Wilhoit and B. J. Zwolinski, *J. Chem. Eng. Data*, **13**, 334 (1968).
76. M. D. M. C. Ribeiro da Silva, M. A. V. Ribeiro da Silva and G. Pilcher, *J. Chem. Thermodyn.*, **18**, 295 (1986).
77. A. Finch and J. Payne, *Thermochim. Acta*, **189**, 109 (1991).
78. L. Medard and M. Thomas, *Mem. Poudres*, **36**, 97 (1954).
79. A. E. Alegria, C. Munoz and M. S. Rodriguez, *J. Phys. Chem.*, **94**, 930 (1990).
80. M. Colomina, M. V. Roux and C. Turrión, *J. Chem. Thermodyn.*, **8**, 869 (1976); M. Colomina, P. Jiménez, M. V. Roux and C. Turrión, *An. Quim.*, **75**, 620 (1979).
81. G. A. Lobanov and L. P. Karmanova, *Izv. Vyssh. Uchebn. Zaved., Khim. Khim. Tekhnol.*, **14**, 865 (1971); *Chem. Abstr.*, **76**, 26438s (1972).
82. From an *ab initio* and DFT investigation, 1,2-benzenediol is stabilized by *ca* 16 kJ mol^{-1} from intramolecular hydrogen bonding. See G. Chung, O. Kwon and Y. Kwon, *J. Phys. Chem. A*, **101**, 9415 (1997).
83. M. A. R. Matos, M. S. Miranda and V. M. F. Morais, *J. Phys. Chem. A*, **104**, 9260 (2000). The gas phase dimethoxybenzene enthalpy of formation values are (kJ mol^{-1}): -202.4 ± 3.4 (*o*-), -221.8 ± 2.4 (*m*-), -211.5 ± 3.0 (*p*-).
84. A. Kovács, G. Keresztury and V. Izvekov, *Chem. Phys.*, **253**, 193 (2000).
85. G. Chung, O. Kwo and Y. Kwon, *J. Phys. Chem. A*, **101**, 4628 (1997).
86. K. B. Borisenko and I. Hargittai, *J. Phys. Chem.*, **97**, 4080 (1993).

87. K. B. Borisenko, K. Zauer and I. Hargittai, *J. Phys. Chem.*, **99**, 13808 (1995).
88. K. H. Meyer, *Liebigs Ann. Chem.*, **379**, 37 (1911).
89. S. W. Slayden and J. F. Liebman, *Chem. Rev.*, **101**, 541 (2001).
90. P. Lemoult, *Ann. Chim. Phys.*, **14**, 289 (1908).
91. W. Swientoslawski, *Z. Phys. Chem.*, **72**, 49 (1910).
92. A. Kržan, D. R. Crist and V. Horák, *J. Mol. Struct. (Theochem)*, **528**, 237 (2000).
93. J. V. Hamilton and T. F. Fagley, *J. Chem. Eng. Data*, **13**, 523 (1968).
94. This number is derived from the measurements found in Reference 14. The value for the 1,4-isomer is within 2 kJ mol^{-1} from that found in contemporary Reference 15.
95. The interaction of nitroso groups with aromatic ethers (and by inference arenols) is also of considerable interest to chemists interested in nitroso groups; cf. B. G. Gowenlock, M. J. Maidment, K. G. Orrell, I. Prokěs and J. R. Roberts, *J. Chem. Soc., Perkin Trans. 2*, 1904 (2001).
96. R. Hässner, H. Mustroph and R. J. Borsdorf, *Prakt. Chem.*, **327**, 555 (1985).

CHAPTER 4

Mass spectrometry and gas-phase ion chemistry of phenols

DIETMAR KUCK

Fakultät für Chemie, Universität Bielefeld, Universitätsstraße 25, D-33615 Bielefeld, Germany
e-mail: dietmar.kuck@uni-bielefeld.de

I. INTRODUCTION	260
II. GASEOUS RADICAL CATIONS DERIVED FROM PHENOLS: THERMOCHEMISTRY OF SOME TYPICAL SPECIES AND REACTIONS	261
III. UNIMOLECULAR FRAGMENTATION REACTIONS OF PHENOL RADICAL CATIONS	263
A. Loss of CO from Simple Phenol Radical Cations	263
B. Fragmentation of Alkylphenols	265
C. Fragmentation of Di- and Trihydroxylated Benzenes and Alkylbenzenes	270
D. The Phenoxy Cation ('Phenoxenium Ion') and the Hydroxyphenyl Cations	273
IV. THE EFFECT OF THE HYDROXYL GROUP ON THE REVERSIBLE INTRAMOLECULAR HYDROGEN TRANSFER IN IONIZED PHENOLS	277
V. ORTHO EFFECTS IN SUBSTITUTED PHENOLS	280
VI. SECONDARY FRAGMENTATION REACTIONS OF PHENOL RADICAL CATIONS	289
VII. MISCELLANEOUS FRAGMENTATIONS OF PHENOL RADICAL CATIONS	290
VIII. CHEMICAL IONIZATION MASS SPECTROMETRY OF PHENOLS	291
IX. IONIZED PHENOL AND CYCLOHEXA-1,3-DIEN-5-ONE (*ortho*-ISOPHENOL) GENERATED BY FRAGMENTATION OF PRECURSOR IONS	298
A. The Radical Cations of *ortho*-Isophenol	298
B. Phenol Ions Generated within Transient Ion/Neutral Complexes during Mass Spectrometric Fragmentation of Alkyl Aryl Ethers	300
C. Allylphenols and Allyl Phenyl Ethers	301

The Chemistry of Phenols. Edited by Z. Rappoport
© 2003 John Wiley & Sons, Ltd ISBN: 0-471-49737-1

	D. Lignin Model Compounds: Eugenol, Dehydrodieugenol and Related Compounds	302
X.	ION/MOLECULE REACTIONS OF PHENOLS IN THE GAS PHASE	307
XI.	GASEOUS PHENOL ANIONS	310
	A. Gas-phase Acidities of Phenols	310
	B. Phenolic Anion/Molecule Adducts [ArO-H X$^-$] and [ArO$^-$ H-X]	312
	C. Negative Chemical Ionization Mass Spectrometry of Phenols	314
	D. Ion/Molecule Complexes Formed during the Unimolecular Fragmentation of Phenolic Anions	316
XII.	MISCELLANEOUS ANALYTICAL EXAMPLES	319
XIII.	MASS SPECTROMETRY OF CALIXARENES	320
XIV.	PHENOLIC COMPOUNDS AS MALDI MATRICES	323
XV.	ACKNOWLEDGEMENTS	325
XVI.	REFERENCES	325

I. INTRODUCTION

Phenols are electron-rich, polar, and acidic aromatic compounds. Therefore, the chemical behaviour of phenolic ions in the gas phase, and the mass spectrometric information resulting from it, are characterized by the relatively facile formation of stable but nevertheless reactive radical cations, protonated and cationized molecules, and more complex ionic adducts. For the same reasons, the gas-phase chemistry of phenolate anions and other negative ions derived from phenols is multifold and negative ion mass spectrometry is more extended for phenols than for many other classes of organic compounds. Phenols are important compounds in the chemistry of natural compounds but also in applied chemistry, including lignins and polyesters. Phenolic compounds, including chlorinated derivatives, represent omnipresent environmental pollutants. Design and synthesis of phenol-based compounds for supramolecular chemistry, such as the calixarenes, has recently inspired researchers to investigate the chemistry of gaseous ionic complexes and clusters of various types.

Remarkably, however, textbooks on mass spectrometry hardly comprise the multifaceted aspects of the gas-phase ion chemistry of phenols and of their consequences for the characteristics of the mass spectra of phenolic compounds[1,2]. A great many insights into the fundamentals and developments for analytical applications have been collected during the past four decades or so, and research is actively continued in this field. As a special circumstance, a number of phenolic compounds play a crucial—and yet not completely understood—role as energy-transferring media and protonating reagents in an extremely important, 'modern' ionization method of mass spectrometry, viz. matrix-assisted laser desorption/ionization (MALDI)[3-5].

This review article is intended to cover the above-mentioned topics by presenting and discussing selected examples of each of them. Although—or maybe because—mass spectrometry is mainly considered an analytical tool, emphasis is put on the gas-phase ion chemistry of phenolic species occurring in the mass spectrometer and in the dilute gas phase[6]. The archetypical fragmentation behaviour of phenol derivatives under electron ionization (EI) will be explained with respect to its chemical origins as will be the reactions of protonated and deprotonated phenols under the conditions of chemical ionization (CI) and related techniques. Bimolecular reactions of positively and of negatively charged phenolic ions will also be treated in some detail. Selected examples for analytical applications will be illustrated and the discussion on the role of phenolic matrices in MALDI mass spectrometry will be briefly highlighted in the final section of this review.

II. GASEOUS RADICAL CATIONS DERIVED FROM PHENOLS: THERMOCHEMISTRY OF SOME TYPICAL SPECIES AND REACTIONS

Ionization of phenol (IE = 8.5 eV) and simple alkylphenols by removal of a single electron leads to the corresponding radical cations and requires energies in the range of 8.5–7.8 eV (i.e. 195–180 kcal mol^{-1} ≈ 820–750 kJ mol^{-1})[7,9]. The presence of the hydroxyl group at the benzene or a simple alkylbenzene ring decreases its ionization energy by $\Delta IE = -(0.75-0.5)$ eV and the second OH group in the dihydroxybenzenes still contributes another ΔIE of $-(0.5-0.3)$ eV. In the presence of an electron-withdrawing ring substituent, such as NO$_2$, the effect of the OH group is equally strong as in benzene itself ($\Delta IE \approx -0.75$ eV). Thus, in the absence of other electron-rich or electropositive substituents or structural units, the molecular ions of phenolic compounds will bear the positive charge preferentially at the oxygenated arene nucleus.

The ease of addition of a hydrogen atom by, say, an intramolecular hydrogen rearrangement from an H$^{\bullet}$ donor group to a phenolic radical cation depends on the position of the acceptor site on the ring, since a protonated phenol results (Scheme 1). Thus, the local hydrogen atom affinities (*HA*) of the ring positions of phenolic radical cations influence their reactivity. The local *HA* values can be calculated from the thermochemistry of the corresponding ions, e.g. **1**$^{\bullet+}$ and [1 + H]$^+$. The thermochemical relations between the (gaseous) neutral molecule, **1**, its molecular cation **1**$^{\bullet+}$ and its most important tautomer, *o*-**2**$^{\bullet+}$, as well as of the various protonated conjugate [1 + H]$^+$ ions, are displayed in Scheme 1. A ladder of heats of formation (ΔH_f) is also included. From these data, it is evident that, for example, addition of H$^{\bullet}$ to the radical cation **1**$^{\bullet+}$ at its *para* position is exothermic by $-HA_{(p)} = -78$ kcal mol^{-1} but only by $-HA_{(p)} = -65$ kcal mol^{-1} at the hydroxyl group.

Protonation of phenols is governed by the relative energy-rich highest occupied molecular orbitals. The electronic structure of phenols gives rise to increased gas-phase basicities (*GB*) and proton affinities (*PA*) as compared to benzene. The experimentally determined gas-phase basicity of the parent compound is $GB(1) = 188$ kcal mol^{-1} = 786 kJ mol^{-1}, that is, by $\Delta GB = +14.5$ kcal mol^{-1} = +61 kJ mol^{-1} higher than that of the hydrocarbon (Table 1)[7-9]. The first experimentally determined value for the proton affinity of phenol was found to be $PA(1) = 195.0$ kcal mol^{-1} [10]. However, it is important to note that protonation of phenol in the gas phase occurs with a strong preference at the ring positions *para* to the hydroxy functionality, rather than on the oxygen atom[10,11]. Early *ab initio* calculations already suggested a good correlation of the stabilizing or destabilizing effect of electron-releasing and electron-withdrawing substituents on a protonated benzene ring (benzenium ion) with σ^+ values[11]. An OH group *para* to the protonation site was calculated to render the ion more stable by $\Delta PA = 16.0$ kcal mol^{-1} than the parent benzenium ion, whereas a *meta*-OH group was suggested to *destabilize* the ion by $\Delta PA = -5.3$ kcal mol^{-1}. However, the calculated gas-phase stabilization by the *p*- and *m*-OH groups was found to fall somewhat short of the value predicted on the basis of a linear free-energy correlation for protonation in solution, pointing to the additional stabilization gained by hydrogen bonding of the OH proton(s) to solvent molecules in the condensed phase[11,12].

Phenols are lucid examples for aromatic compounds displaying several protonation sites with individual 'local' proton affinities and gas-phase basicities. To date, an impressively large set of local *PA*'s has been determined for simple aromatic compounds by combined experimental and computational[10,13] and, more recently, purely computational techniques[14,15], and selected examples for simple phenol derivatives are collected in Table 1. For example, $PA(1) = 195.5$ kcal mol^{-1} = 817 kJ mol^{-1} reflects the negative enthalpy change associated with the addition of a proton to the *para* position. The other ring sites of phenol display significantly lower proton affinities, e.g. the *meta*

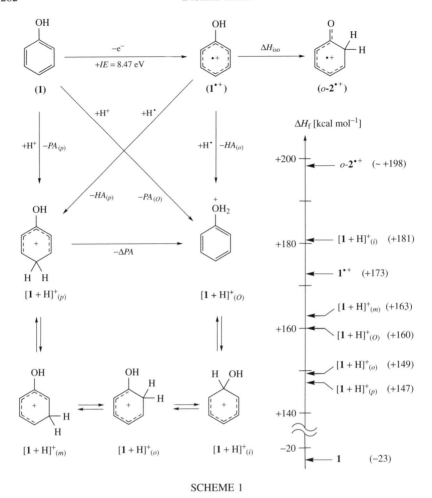

SCHEME 1

TABLE 1. Proton affinities (*PA*), gas-phase basicities (*GB*) and local proton affinities of phenol, toluene and *para*-cresol (in kcal mol^{-1})[a]

Compound Position of H$^+$	Phenol		Toluene PA (ΔPA) vs. benzene	*para*-Cresol PA ($\Sigma \Delta PA$) vs. benzene
	PA (GB)	$\Delta PA (\Delta GB)$ vs. benzene		
(experimental)	195.3 (187.9)	+16.0 (+14.5)	187.4 (+8.1)	(unknown)
para	195.5	+16.1	187.3 (+8.0)	195.1 (+15.8)
meta	179.9	+0.6	182.9 (+3.6)	185.4 (+7.5)
ortho	193.0	+13.7	186.2 (+6.9)	195.6 (+17.3)
ipso	162.2	−17.1	179.9 (+0.6)	—
OH	182.7	+3.4	—	—

[a]For references, see text.

positions and the OH group are by $\Delta PA(\mathbf{1})_{meta} = -15.6$ kcal mol^{-1} and by $\Delta PA(\mathbf{1})_O = -12.8$ kcal mol^{-1}, respectively, less strong H$^+$ binding sites, as calculated by *ab initio* methods[14,15]. Notably, early ICR mass spectrometric experiments using the 'bracketing' technique had already shown that the hydroxyl group is by 13–20 kcal mol^{-1} less strong a base than the ring[13].

A remarkable feature of the local proton affinities scales of simple arenes is the additivity of the substituent effects on the local *PA* values, as revealed for the first time by experiment for the methylbenzenes[16]. Computational approaches have confirmed the additivity rules. Thus, the 'overall' *PA* values and also the local *PA* values of the isomeric cresols and dihydroxybenzenes have been predicted by *ab initio* methods. The case of *para*-cresol is included in Table 1. Protonation *ortho* to the hydroxyl group (and *meta* to methyl) is calculated to be most favourable and protonation *para* to the hydroxy group (and *ipso* to methyl) is almost as favourable, in accordance with the additivity of the incremental contributions of an OH and a CH$_3$ substituent to *PA*(**1**). Protonation of *para*-cresol *meta* to the hydroxy and *ortho* to the methyl group is less favourable by *ca* 10 kcal mol^{-1}, again in agreement with incremental additivity (Table 1). Remarkably, and again in line with its negligible effect in *ipso*-protonated toluene, the methyl substituent of *para*-cresol does not affect the proton affinity of the methyl-substituted site: Protonation *para* to the OH group is still very favourable. Within the same scheme, the most basic sites of *ortho*-cresol and *meta*-cresol are predicted to be C-4 in each case: PA (*o*-cresol) $\approx$ 198 kcal mol^{-1} and PA (*m*-cresol) $\approx$ 201 kcal mol^{-1}. Hence, although the proton affinities (and gas-phase basicities) of the cresols and related simple phenols are not known experimentally to date, the available data allow us to estimate these important thermodynamic properties for these and many other phenol derivatives.

It may be also mentioned here that, in contrast to protonated benzene and the protonated alkylbenzenes[6,17], the 1,2-shift of protons (H$^+$ ring walk) in protonated phenols is rather energy demanding. This is due to the large differences between the thermochemical stabilities of the tautomeric forms of ions [**1** + H]$^+$. Thus, both intramolecular and intermolecular protonation of phenolic rings mostly occurs with high regioselectivity. As a consequence, the unimolecular fragmentation of alkylphenols is subject to pronounced substituent effects, and the bimolecular H/D exchange with deuteriated acids in the gas phase may be used to determine the number of basic ring positions and thus the position of the substituents at the ring (see below).

III. UNIMOLECULAR FRAGMENTATION REACTIONS OF PHENOL RADICAL CATIONS

A. Loss of CO from Simple Phenol Radical Cations

The hydroxyl group in the radical cations of phenols strongly facilitates the formation of transient intermediates and fragment ions whose structures correspond to ionized or protonated cyclohexadienones, quinomethanes or quinones. This is already evident for the most characteristic fragmentation path of the parent phenol radical cation **1**$^{\bullet+}$, viz. the expulsion of carbon monoxide, producing C$_5$H$_6$$^{\bullet+}$ ions with *m/z* 66 (Scheme 2). Very early, this reaction was found to release significant amount of the ions' internal energy as kinetic energy ('kinetic energy release', T_{kin}), as indicated by a broadened, flat-topped signal for the dissociation of the metastable ions[18,19]. Although the cyclic form **3** is generally assumed to be the product, it has been found by charge stripping (CS) mass spectrometry that C$_5$H$_6$$^{\bullet+}$ ions formed from **1** within the ion source, i.e. from short-lived, high-energy ions **1**$^{\bullet+}$, consist of a mixture of ions containing mainly *acyclic* isomers[20]. By contrast, long-lived, metastable ions **1**$^{\bullet+}$ generate exclusively the cyclic isomer, ionized cyclopentadiene **3**$^{\bullet+}$. Subsequent loss of H$^{\bullet}$ giving C$_5$H$_5$$^+$ ions **4** is a common secondary fragmentation.

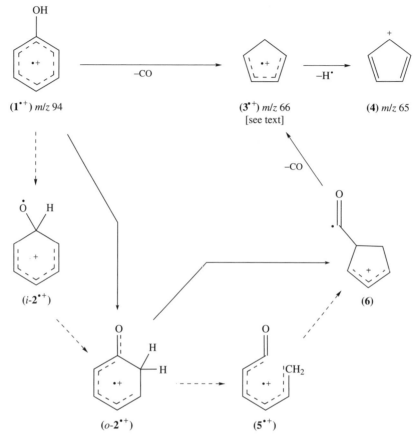

SCHEME 2

The observation of the [M − 28]•+ ions and the accompanying fragments [M − 29]+ is typical for **1**•+ and ionized phenols, naphthols etc., which bear additional functional groups attached directly at the ring, provided that less energy-demanding channels are absent.

Loss of CO from the parent ions **1**•+ is the least energy-demanding fragmentation path but it requires as much as 3.1 eV to occur fast enough to contribute to the normal EI mass spectrum, i.e. to ion formation within the ion source, and still ca 2.4 eV to occur after acceleration, i.e. in the metastable ions[21]. (The difference reflects a large part of the so-called kinetic shift of the fragmentation[2,22].) Thermochemically, however, the overall process is endothermic by only 1.3 eV (29.6 kcal mol^{-1})[7]; hence the formal 1,3-H shift in ions **1**•+, generating ionized cyclohexadienone o-**2**•+, is considered the energy- (and rate-) determining step. Ions o-**2**•+ lie in an energy minimum which has been estimated to be by $\Delta H_{iso} \approx 25$ kcal mol^{-1} above that of ions **1**•+ (Scheme 1)[21]. Ring contraction to the distonic ions **6** and/or the eventual expulsion from those involve another relatively high activation barrier. This is reflected in part by the release of kinetic energy during the expulsion of CO (ca 120 meV ≈ 2.8 kcal mol^{-1})[21]. In fact, the reverse reaction, i.e. the addition of CO to the cyclopentadiene radical cation **3**•+, should be associated with a considerable activation barrier. In fact, the reverse activation energy of the CO loss from ions $\mathbf{1^{\bullet+}}$ has been experimentally determined to be as high as 1.64 eV (37.8 kcal mol^{-1})[23,24]. The formation of the *ortho*-isophenol ions *o*-$\mathbf{2^{\bullet+}}$ in competition to that of ions $\mathbf{1^{\bullet+}}$ will be discussed in Section IX.

More details of the multistep mechanism leading to loss of CO deserve notice. For example, it appears questionable whether the ring contraction to ions **6** occurs in a concerted manner or via the intermediate ring-opened form $\mathbf{5^{\bullet+}}$. In contrast, a stepwise mechanism via the *ipso* tautomer of $\mathbf{1^{\bullet+}}$, viz. ions *i*-$\mathbf{2^{\bullet+}}$, involving two sequential 1,2-H shifts appears unlikely in view of the remarkably low thermochemical stability of the protonated *ipso*-tautomer $[1 + H]_i^+$ (Table 1). The gas-phase ion chemistry of ions *o*-$\mathbf{2^{\bullet+}}$, representing ionized *ortho*-isophenol in analogy to *ortho*-isotoluene[25], the corresponding *ortho*-tautomer of ionized toluene in the $C_7H_8^{\bullet+}$ series, has been explored in much detail (Section IX.A).

B. Fragmentation of Alkylphenols

If an aliphatic or alicyclic group is attached to the phenol ring, another typical, and in fact extremely frequent fragmentation channel is opened, viz. the benzylic cleavage (Scheme 3)[26,27]. This is particularly facile when the aliphatic group is positioned *ortho* or *para* to the phenolic hydroxyl group, allowing for the formation of thermodynamically stable *para*- and *ortho*-hydroxybenzyl cations *p*-**9** and *o*-**9**. In the simplest case, loss of H$^{\bullet}$ occurs from the molecular ions of *para*-cresol (*p*-$\mathbf{7^{\bullet+}}$) and *ortho*-cresol (*o*-$\mathbf{7^{\bullet+}}$) with particular ease giving ions *p*-**9** and *o*-**9**, respectively, which represent the $[M + H]^+$ ions of *para*- and *ortho*-quinomethane. Correspondingly, the radical cations of higher alkylphenols and the related functionalized hydroxybenzyl derivatives[8] lose the alkyl radical R$^{\bullet}$ or the functional group (e.g. a carboxyl radical) generating the stable hydroxylbenzyl ions **9**. Among these, the *meta*-isomer *m*-**9** is significantly less stable and the tendency to generate the ring-expanded hydroxytropylium ion **10** during the fragmentation is increased.

Phenols containing α-branched side chains undergo the benzylic cleavage reaction with particular ease, since secondary benzylic ions *o*-HOC$_6$H$_4$CH$^+$R and *p*-HOC$_6$H$_4$CH$^+$R, representing β-protonated hydroxystyrenes, are even more stable than primary ones. This allows reliable structural assignments, as shown for the mixture of six isomeric 2- and 4-(*sec*-octyl)phenols generated by octylation of phenol[28]. The same holds for *tert*-alkylphenols, which generate HOC$_6$H$_4$C$^+$RR$'$ ions. However, note that ionized higher *meta*-alkylphenols frequently undergo another, quite characteristic fragmentation reaction involving unimolecular hydrogen migration (see below).

For the reasons outlined above, the mass spectrometric fragmentation of cycloalkyl-substituted phenols, such as 2- and 4-hydroxyphenylcyclohexane *o*-**11** and *p*-**11**[29], is also governed by the favourable benzylic cleavage. However, this initial rupture of a benzylic C–C bond in ions *p*-$\mathbf{11^{\bullet+}}$ does not give rise to the direct formation of fragments. Rather, the isomeric distonic ion *p*-**12** thus formed suffers subsequent isomerization, such as 1,5-H$^{\bullet}$ transfer processes generating the conventional radical-cations *p*-$\mathbf{13b^{\bullet+}}$ and *p*-$\mathbf{13a^{\bullet+}}$ which, eventually, dissociate by benzylic and vinylogous benzylic C–C bond cleavages to give *p*-**9** and **14**, respectively (Scheme 4). The corresponding peaks at *m/z* 107 and *m/z* 133 reflect the major part of the in-source fragmentation of *p*-**11** under EI conditions. The formation of ionized hydroxystyrene (C$_8$H$_8$O$^{\bullet+}$, *m/z* 120) by loss of 56 Th represents another characteristic path starting from ions *p*-**12**. Owing to similar electronic factors, the same fragmentation pathways operate in the radical ions of the *ortho*-isomer *o*-$\mathbf{11^{\bullet+}}$ and its mass spectrum is similar to that of *p*-$\mathbf{11^{\bullet+}}$ (Scheme 5).

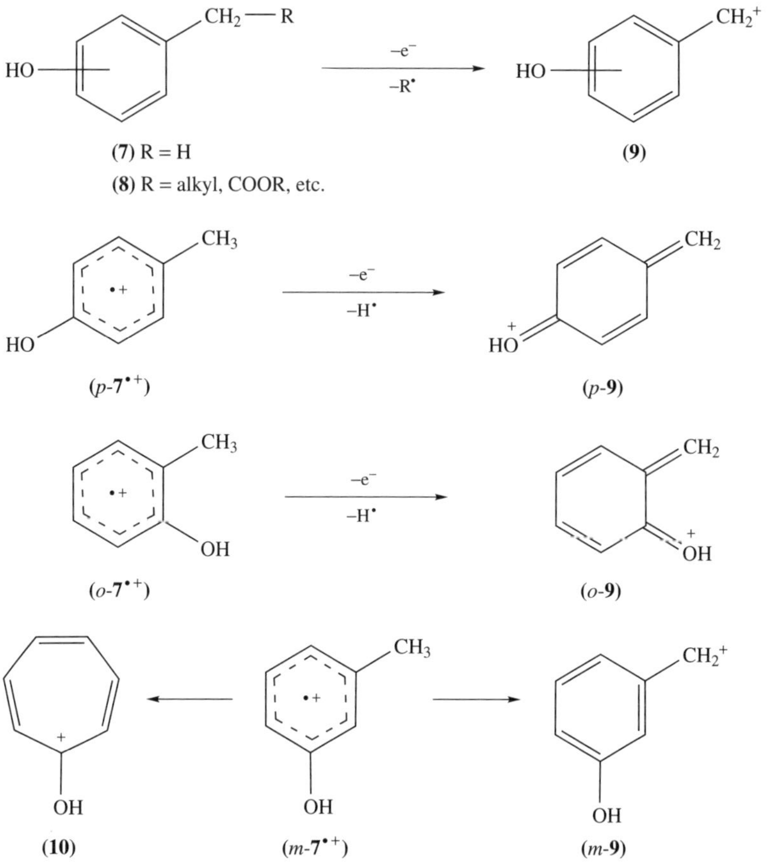

SCHEME 3

The facile, albeit hidden, benzylic cleavage is more important than generally recognized and potentially relevant in the fragmentation of many benzoannelated alicyclic compounds bearing phenolic hydroxyl groups. Estrogenic steroids, which contain a phenolic A ring, are prone to undergo this type of isomerization prior to fragmentation. Thus, the major primary fragmentation of ionized estradiol **15**$^{•+}$, that is, dismantling of the D ring by loss of $C_3H_7O^•$, may be triggered by initial benzylic cleavage giving the distonic ion **16**, which opens a multistep isomerization path via **17** to **18**, rather than by remote cleavage occurring in non-aromatic steroids (Scheme 6). Admittedly, it may be difficult to differentiate between the two valence-isomeric fragmentation ions **19** and **20**.

Alkylphenols containing the alkyl group in the *meta* position to the hydroxy functionality exhibit a highly characteristic fragmentation behaviour under EI conditions, which allows us to distinguish them easily from their *para-* and *ortho-*isomers. The reaction represents a special case of the McLafferty reaction, giving rise to the elimination of an alkene (or analogous unsaturated neutrals) through rearrangement of a hydrogen atom from the γ position relative to the aromatic nucleus (Scheme 7). In the case of the radical ions of 3-alkylphenols, such as *m*-**21**$^{•+}$ bearing at least one γ-H atom, the radical cations of

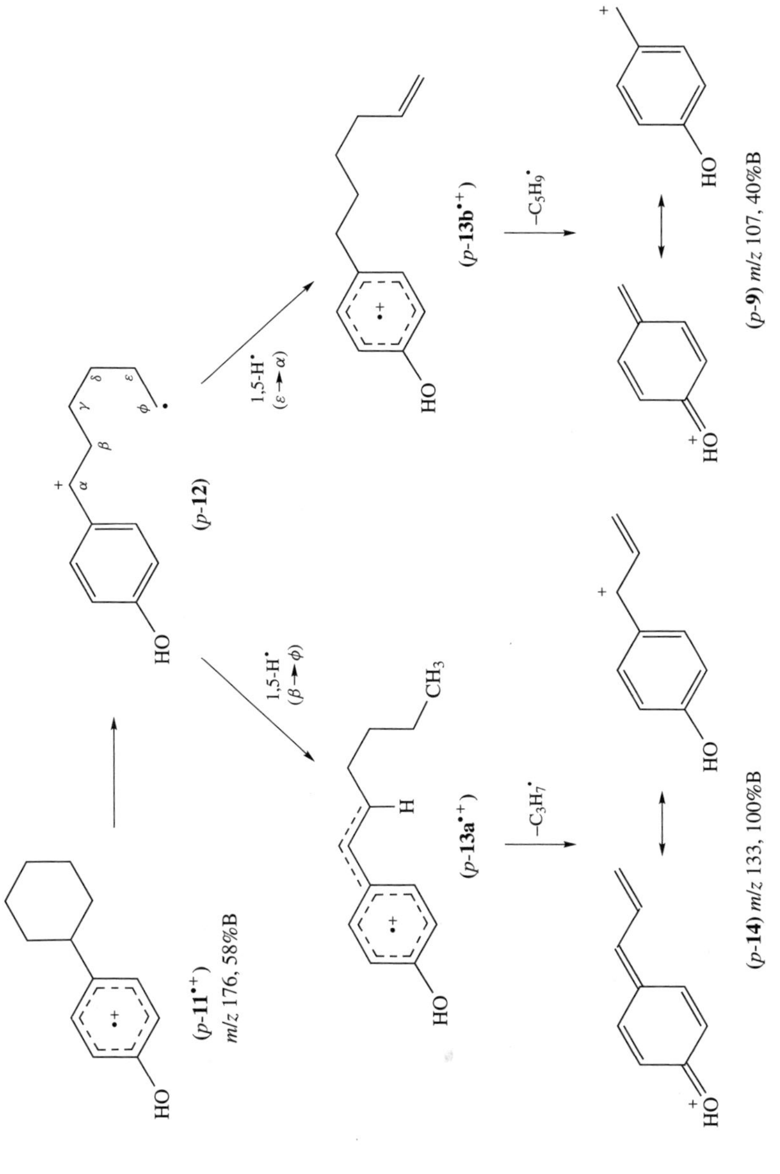

SCHEME 4

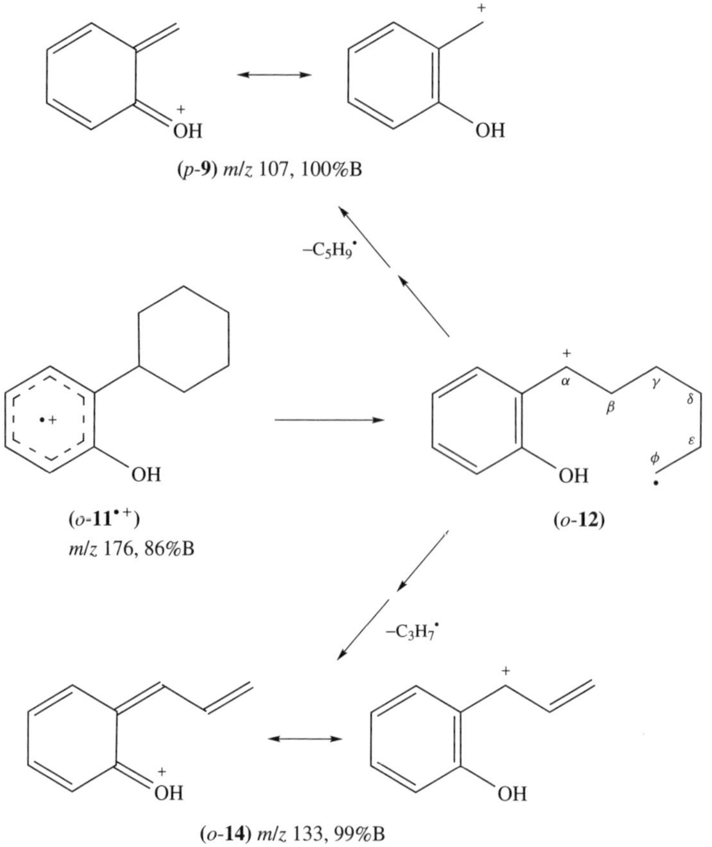

(*p*-9) *m/z* 107, 100%B

(*o*-11•+)
m/z 176, 86%B

–C₅H₉•

(*o*-12)

–C₃H₇•

(*o*-14) *m/z* 133, 99%B

SCHEME 5

3-hydroxy- and 1-hydroxy-substituted 5-methylene-1,3-cyclohexadiene, *m*-**22a**•+ and *m*-**22b**•+ ($C_7H_8O^{•+}$, *m/z* 108), are formed with high relative abundances. The corresponding *para*-isotoluene isomer, viz. ionized 1-hydroxy-3-methylene-1,4-cyclohexadiene, cannot be formed due to steric restrictions by the alkyl chain. The McLafferty reaction is much more dominant in the standard EI mass spectra than the more energy-demanding benzylic cleavage leading to ions *m*-**9** ($C_7H_7O^+$ *m/z* 107). As a consequence, the spectra of metastable (less excited, long-lived) alkylphenol ions often exhibit exclusively the signals due to this rearrangement reaction.

A concrete example for the competition between the rearrangement reaction and the simple benzylic cleavage is shown in Scheme 8. In the EI mass spectrum of 3-(*n*-butyl)phenol *m*-**23**, the McLafferty reaction of *m*-**23**•+ gives rise to the base peak at *m/z* 108 ($C_7H_8O^{•+}$), whereas the intensity of the *m/z* 107 peak ($C_7H_7O^+$) is only 55% of that of the base peak[27,29a,30]. (Such relative intensity data are denoted as '%B'.) The [*m/z* 108] : [*m/z* 107] ratio increases with increasing length of the chain, e.g. to 100 : 49 for 3-(*n*-pentadecyl)phenol[31]. This chain-length dependence is typical for alkylarenes in general but also subject to the ion source conditions[25,32,33]. By contrast to the *meta*-isomer, both the EI

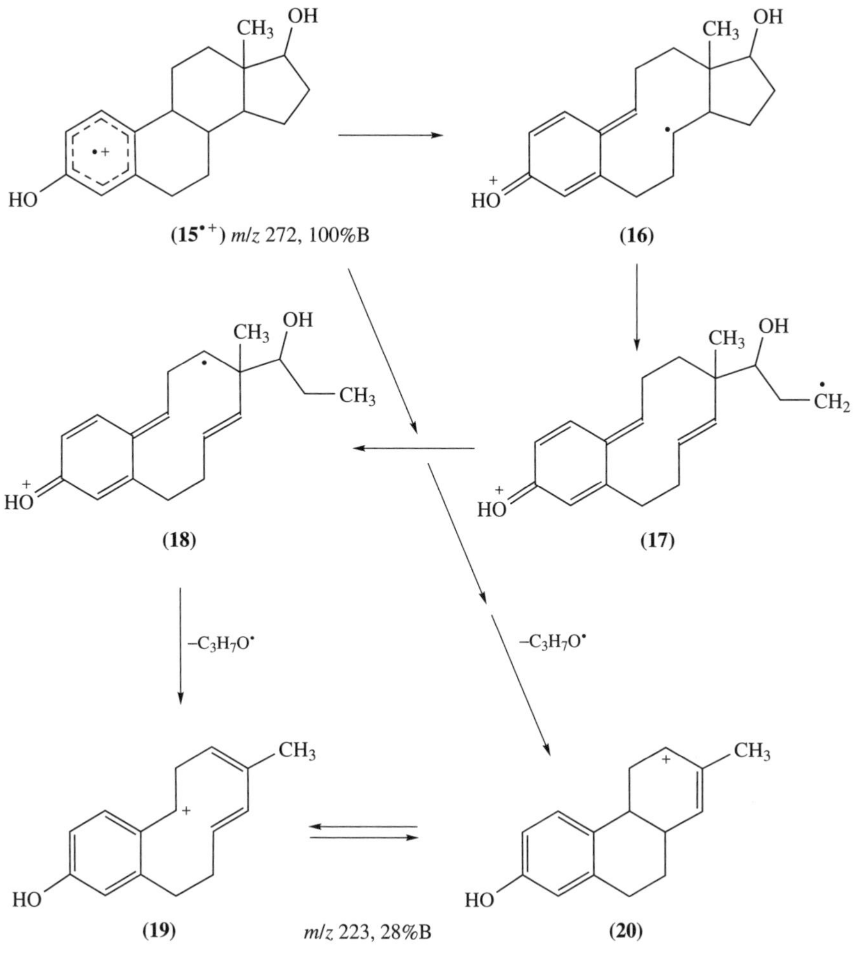

4. Mass spectrometry and gas-phase ion chemistry of phenols

SCHEME 6

mass spectra of 4-(n-butyl)phenol p-**23** and 2-(n-butyl)phenol o-**23** are dominated by the base peak at m/z 107 and the peaks at m/z 108 are only in the range corresponding to the naturally occurring ^{13}C contribution of the $C_7H_7O^+$ ions[27,29a]. The same drastic difference was found with more complex alkylphenols, such as the 2-(hydroxybenzyl)indanes (see below). A linear free-energy relationship connecting the log ratio $[C_7H_7X^{\bullet+}]/[C_7H_6X^+]$ and the Hammett parameter σ_x was unraveled from the mass spectra of a series *ortho*-, *meta*- and *para-n*-alkylphenols (X = OH) and (*n*-alkyl)anisoles (X = OCH_3) with three different chain lengths[34].

The McLafferty reaction in alkylphenols occurs stepwise: The migration of a γ-H$^{\bullet}$ to one of the *ortho* positions generates distonic ions **24** en route to the fragmentation products (Scheme 8). It has been shown that the relative stability of these reactive intermediates governs the competition between the McLafferty reaction and the benzylic cleavage in the EI source and that its influence is more important than the relative stability of the

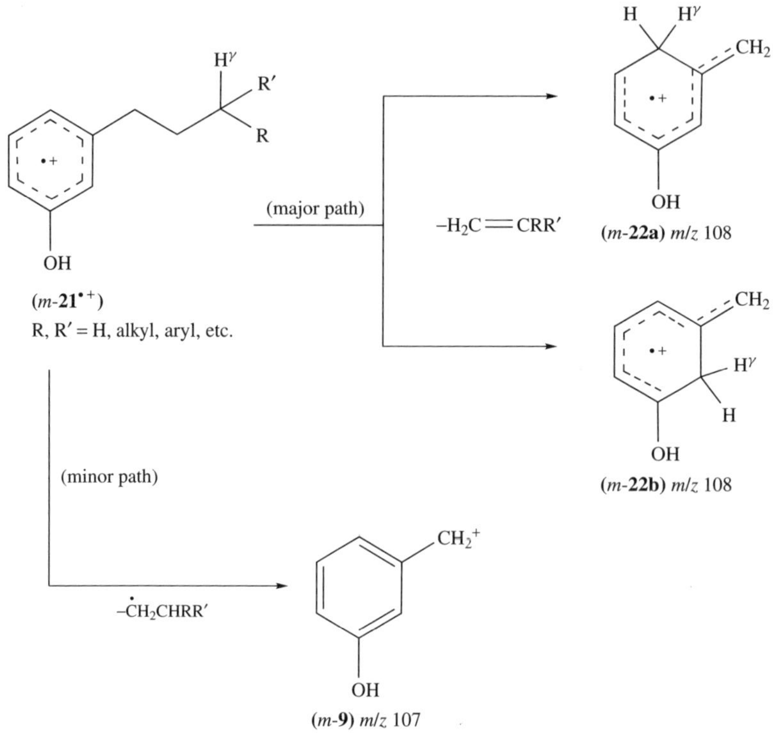

SCHEME 7

final fragmentation products **22**[25,35,36]. The 1,5-H• transfer is reversible (see below) and energetically much more favourable in the case of the *meta*-alkylphenols, such as *m*-**23**•+, than for the other isomers, in analogy to the related alkylanisoles[25,35]. This is a consequence of particular stability of the σ-complex units formed, within the distonic ions, by addition of the H• atom *para* or *ortho* relative to the hydroxyl group, such as in the conversion *m*-**23**•+ → *m*-**24**. Since mass spectrometric fragmentation occurs under kinetic rather than under thermodynamic control, the more facile formation of the distonic ions from the ionized *meta*-alkylphenols leads to an enhanced relative rate of the McLafferty reaction.

The close relations between odd- and even-electron cations is shown in Scheme 9. Viewed in a retrosynthetic manner, the protonated cresols [**7** + H]+, representing parent species for the ionic part of the distonic ions **24**, can be generated by addition of an H atom to the radical cations **7**•+ as well as by addition of a proton to the neutral arenes **7**. In the same way, the radical cations **22a**, representing the ionic fragments of the McLafferty reaction, can be formed both by addition of H• to the benzylic cations **9** and by addition of H+ to the benzylic radicals **25**.

C. Fragmentation of Di- and Trihydroxylated Benzenes and Alkylbenzenes

The two characteristic reaction channels of ionized monophenols dominate the EI-induced fragmentation of dihydroxyalkylbenzenes as well. Some examples are collected

4. Mass spectrometry and gas-phase ion chemistry of phenols

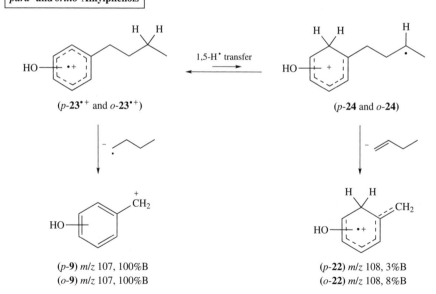

SCHEME 8

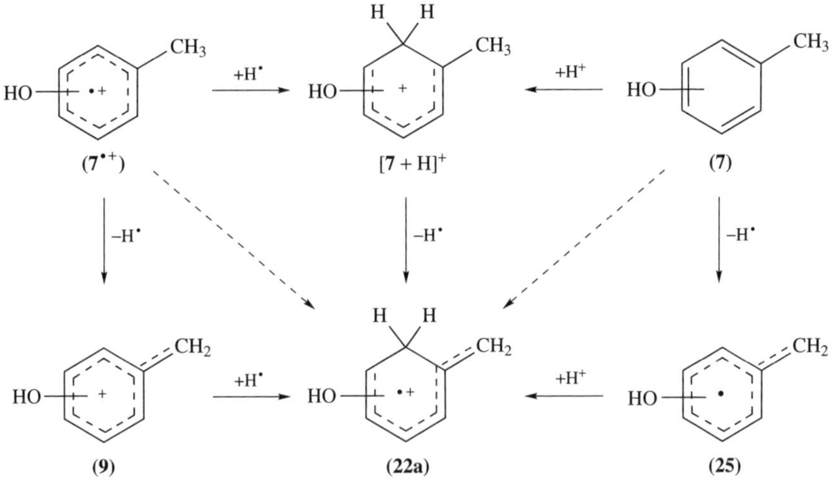

272   Dietmar Kuck

SCHEME 9

in Scheme 10. As can be expected from the above discussion on the role of the σ-complex intermediates, the two *meta*-hydroxyl groups in ionized 5-*n*-heptylresorcinol, *m,m*-**26**$^{•+}$, render the dominance of the McLafferty reaction even more pronounced, and an *ortho/para* combination in ionized 6-(*n*-hexyl)resorcinol, *o,p*-**27**$^{•+}$, suppresses the rearrangement reaction completely in favour of the simple benzylic cleavage[29a]. Similar to the EI mass spectrum of *m,m*-**26**, ions $C_7H_8O_2^{•+}$ (*m/z* 124) give also rise to the base peaks in 5-(*n*-pentadecyl)- and 5-(*n*-pentadec-10-ene-1-yl)resorcinols, with the ratios $[C_7H_8O_2^{•+}]/[C_7H_7O_2^+] \approx 100 : 17^{29}$. (The abundances ratios given here and in the following are corrected for the natural occurrence of ^{13}C).

What happens if both a *meta*- and a *para*- (or *ortho*-) hydroxyl group are present at the same aromatic nucleus, such as in the catechols and the hydroquinones? The two reaction pathways are followed in competition, albeit with a slight preference for the benzylic cleavage. For example, the radical cations of the two isomeric long-chain catechols *m,p*-**28** and *o,m*-**28** give nearly the same EI mass spectra with a ratio $[C_7H_8O_2^{•+}]/[C_7H_7O_2^+] \approx 45 : 100$ (Scheme 10)[37]. Thus, in spite of the particular length of the pentadecyl chain, which enhances the relative rate of the McLafferty reaction, the simple benzylic cleavage dominates in these cases. The rearrangement is even more attenuated in unsaturated analogues: The mass spectra of the isomeric 7(*Z*),10(*Z*)-heptadecadiene-1-ylcatechols *m,p*-**29** and *o,m*-**29** exhibit even smaller peaks at *m/z* 124, with the intensity ratio $[C_7H_8O_2^{•+}]/[C_7H_7O_2^+] \approx 17 : 100$ (Scheme 11). However, a competing rearrangement takes place here quite significantly, namely the elimination of a $C_{15}H_{28}$ neutral, leaving very probably, the radical ions of the dihydroxystyrenes *m,p*-**30**$^{•+}$ and *o,m*-**30**$^{•+}$ ($C_8H_8O_2^{•+}$, *m/z* 136) as the ionic fragments.

Interestingly, the presence of an ω-phenyl ring at the aliphatic chain affects the ratio $[C_7H_8O_2^{•+}]/[C_7H_7O_2^+]$ quite differently for electronically similar isomers. In the case of the radical ions derived from 4-(ω-phenylalkyl)catechols *m,p*-**31** and *m,p*-**32**, the EI mass spectra exhibit intensity ratios of 31 : 100 and 40 : 100, respectively. However, in the spectra of the isomeric 3-(ω-phenylalkyl)catechols *o,m*-**31** and *o,m*-**32**, ratios of 84 : 92 (as % B, i.e., 91 : 100) and 97 : 100, respectively, were reported (Scheme 12)[37]. It appears likely that the attractive interactions known to operate between the two aromatic rings of

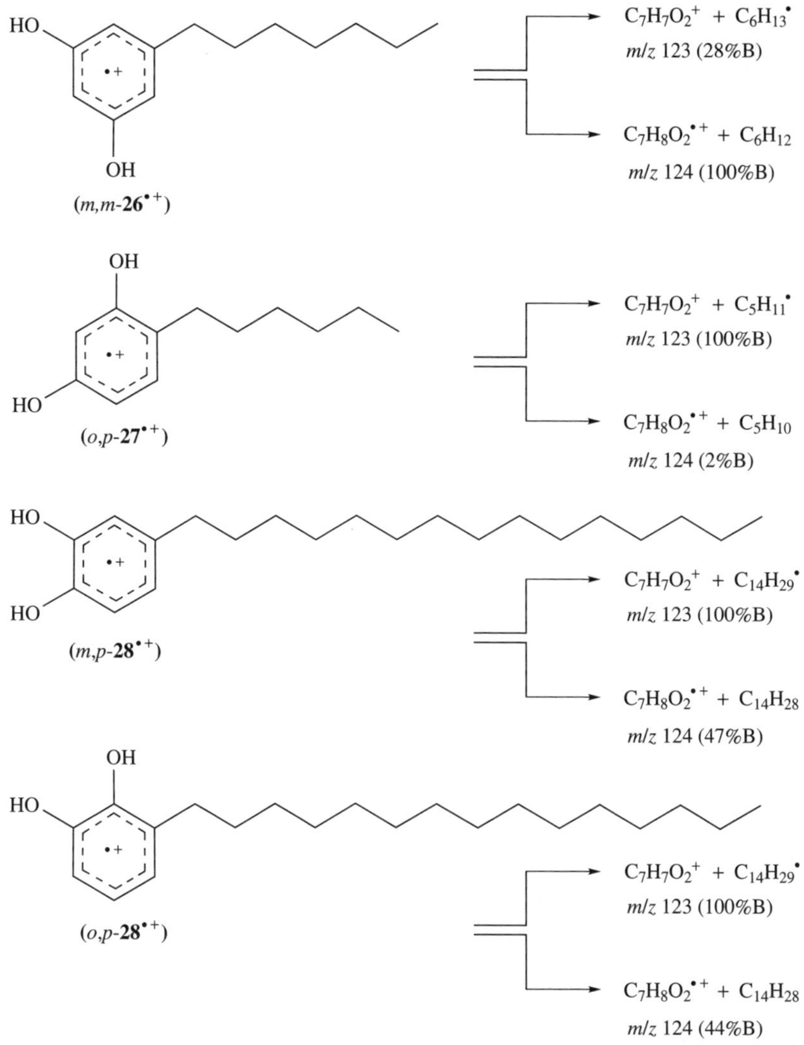

4. Mass spectrometry and gas-phase ion chemistry of phenols

SCHEME 10

(Structures shown:)
- $(m,m\text{-}26^{\bullet+})$: 3,5-dihydroxyphenyl with heptyl chain
  → $C_7H_7O_2^+ + C_6H_13^\bullet$, m/z 123 (28%B)
  → $C_7H_8O_2^{\bullet+} + C_6H_{12}$, m/z 124 (100%B)

- $(o,p\text{-}27^{\bullet+})$: 2,4-dihydroxyphenyl with pentyl chain
  → $C_7H_7O_2^+ + C_5H_{11}^\bullet$, m/z 123 (100%B)
  → $C_7H_8O_2^{\bullet+} + C_5H_{10}$, m/z 124 (2%B)

- $(m,p\text{-}28^{\bullet+})$: 3,4-dihydroxyphenyl with long alkyl chain
  → $C_7H_7O_2^+ + C_{14}H_{29}^\bullet$, m/z 123 (100%B)
  → $C_7H_8O_2^{\bullet+} + C_{14}H_{28}$, m/z 124 (47%B)

- $(o,p\text{-}28^{\bullet+})$: 2,4-dihydroxyphenyl with long alkyl chain
  → $C_7H_7O_2^+ + C_{14}H_{29}^\bullet$, m/z 123 (100%B)
  → $C_7H_8O_2^{\bullet+} + C_{14}H_{28}$, m/z 124 (44%B)

ionized α,ω-diarylalkanes[25,38] are enhanced by the presence of a hydroxyl group in the 3-phenylalkylcatechol ions.

D. The Phenoxy Cation ('Phenoxenium Ion') and the Hydroxyphenyl Cations

As already indicated in Scheme 2, fragmentation of ionized phenol and its simple derivatives is quite complex, certainly more complex than that of higher alkylphenols under EI-MS conditions. Therefore, the gas-phase ion chemistry of simple phenolic cations is of major importance for a sound understanding of mass spectrometry of phenols. A

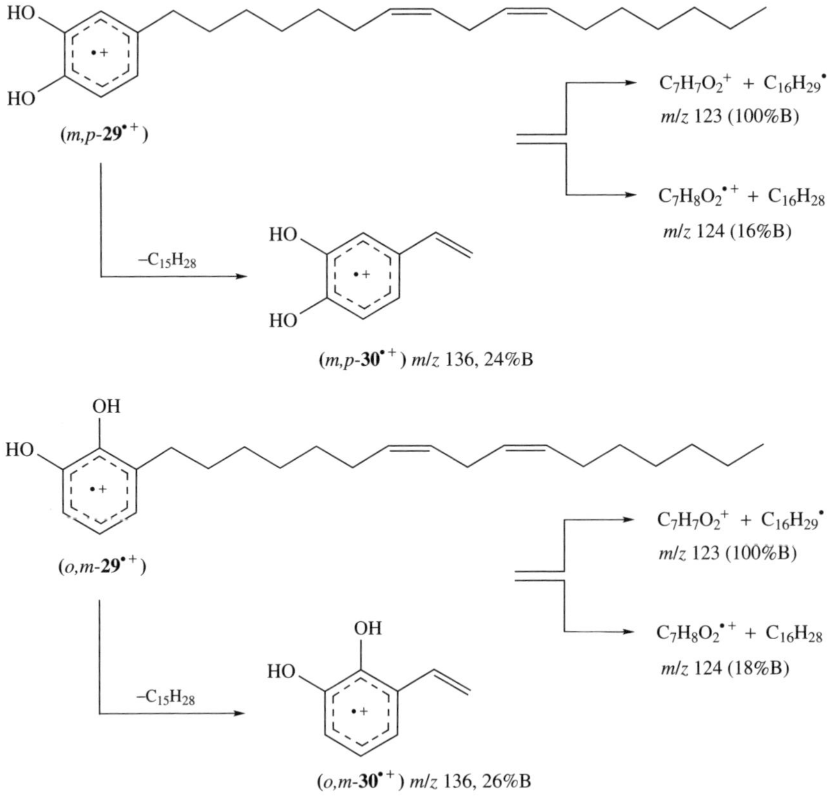

SCHEME 11

group of simple and ubiquitous phenol-derived, even-electron cations are the $C_6H_5O^+$ ions (m/z 93), the properties of which have interested gas-phase ion chemists during four decades because of their fundamental importance[39,40]. Clearly, the stability and reactivity of the phenoxy ('phenoxenium') cation (**34**, Scheme 13) should be quite distinct from that of the isomeric hydroxyphenyl cations, $[C_6H_4OH]^+$ (**38**), but difficult to predict by intuition.

Most of the previous papers on the properties of gaseous $C_6H_5O^+$ ions dealt with their heats of formation, relative stabilities and unimolecular and collision-induced fragmentation characteristics. An extended study on the trapping of the three isomeric hydroxyphenyl cations o-, m- and p-**38** in both the liquid and the gas phase has appeared recently[41]. In the liquid phase, o-**38** was found to isomerize rapidly to the relatively stable phenoxenium ion **34** within the hydrogen-bonded complex with methanol. In the gas phase, both methanol and chloromethane react as quenching reagents, affording the respective methoxyphenols and chlorophenols. 1,2-Hydride shifts in the hydroxyphenyl cations were invoked to explain deviations from the isomer distribution expected on statistical grounds. Ab initio calculations suggested the thermochemical stability order to be o-**38** < p-**38** < m-**38** and ≪ **34**, the latter isomer being by $\Delta E = -20.3$ kcal mol^{-1} more stable than m-**38**. The calculations also indicated the particularly high gas-phase acidity of isomer m-**38** in spite of its relatively high thermochemical stability. Two further recent papers have

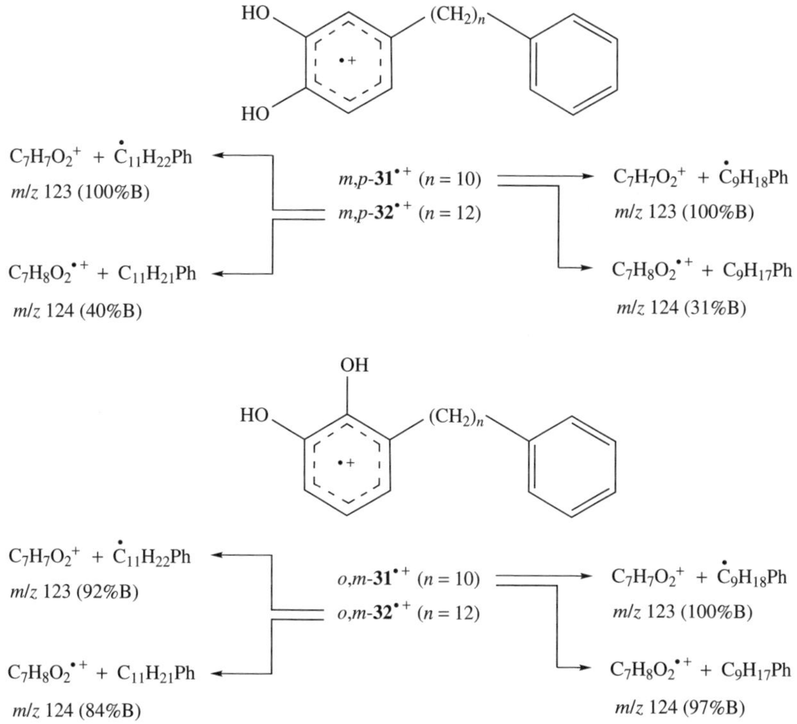

4. Mass spectrometry and gas-phase ion chemistry of phenols

[Scheme showing dihydroxyphenyl compound with (CH$_2$)$_n$-Ph substituent]

$C_7H_7O_2^+$ + $\dot{C}_{11}H_{22}Ph$ ← $m,p\text{-}31^{\bullet+}$ ($n = 10$) ⇌ $C_7H_7O_2^+$ + $\dot{C}_9H_{18}Ph$
m/z 123 (100%B) $m,p\text{-}32^{\bullet+}$ ($n = 12$) m/z 123 (100%B)

$C_7H_8O_2^{\bullet+}$ + $C_{11}H_{21}Ph$ ←   → $C_7H_8O_2^{\bullet+}$ + $C_9H_{17}Ph$
m/z 124 (40%B) m/z 124 (31%B)

[Scheme showing hydroxyphenyl compound with OH and (CH$_2$)$_n$-Ph substituents]

$C_7H_7O_2^+$ + $\dot{C}_{11}H_{22}Ph$ ← $o,m\text{-}31^{\bullet+}$ ($n = 10$) ⇌ $C_7H_7O_2^+$ + $\dot{C}_9H_{18}Ph$
m/z 123 (92%B) $o,m\text{-}32^{\bullet+}$ ($n = 12$) m/z 123 (100%B)

$C_7H_8O_2^{\bullet+}$ + $C_{11}H_{21}Ph$ ←   → $C_7H_8O_2^{\bullet+}$ + $C_9H_{17}Ph$
m/z 124 (84%B) m/z 124 (97%B)

SCHEME 12

also contributed substantially to the knowledge on gaseous $C_6H_5O^+$ ions[42,43], involving extended experimental work and semi-empirical computation. In agreement with the above-mentioned *ab initio* results, but again in contrast to previous work on the heats of formation of $C_6H_5O^+$ ions[44], the phenoxy cation **34** was calculated to be more stable by ca 13 kcal mol^{-1} than the *meta-* and *para-*hydroxyphenyl cations *m-***38** and *p-***38** and the *ortho-*isomer was estimated to be by some 5–10 kcal mol^{-1} less stable than the other two isomers, a result which had not been predicted by early *ab initio* calculations[45]. Thus, it appears that conjugation of the electron-deficient oxygen atom in **34** relieves much of the unfavourable situation, generating much double bond character. In contrast, the hydroxy group in ions **38** cannot contribute significantly to stabilization.

The four isomeric $C_6H_5O^+$ ions were generated by electron-impact induced methyl loss from anisole **33** and bromine loss from the isomeric bromophenols **37** (Scheme 13). Experimental determination of the heats of formation of ions **34** by appearance energy measurements gave a value of 207 kcal mol^{-1} but only estimations of the upper limits for ions **38** in the range of 221–233 kcal mol^{-1} due to the interference of large kinetic shifts[42]. The unimolecular fragmentation of the isomers, being governed by CO loss in each case, revealed significant differences except for *m-***38** and *p-***38**, whose unimolecular ('metastable') fragmentation), collision-induced dissociation (CID) and neutralization/re-ionization (NR) mass spectra were also found to be identical. However, the *ortho-*isomer *o-***38** and ion **34** could be readily distinguished by these methods. Also, charge-stripping (CS) mass spectrometry showed different behaviour of some isomers.

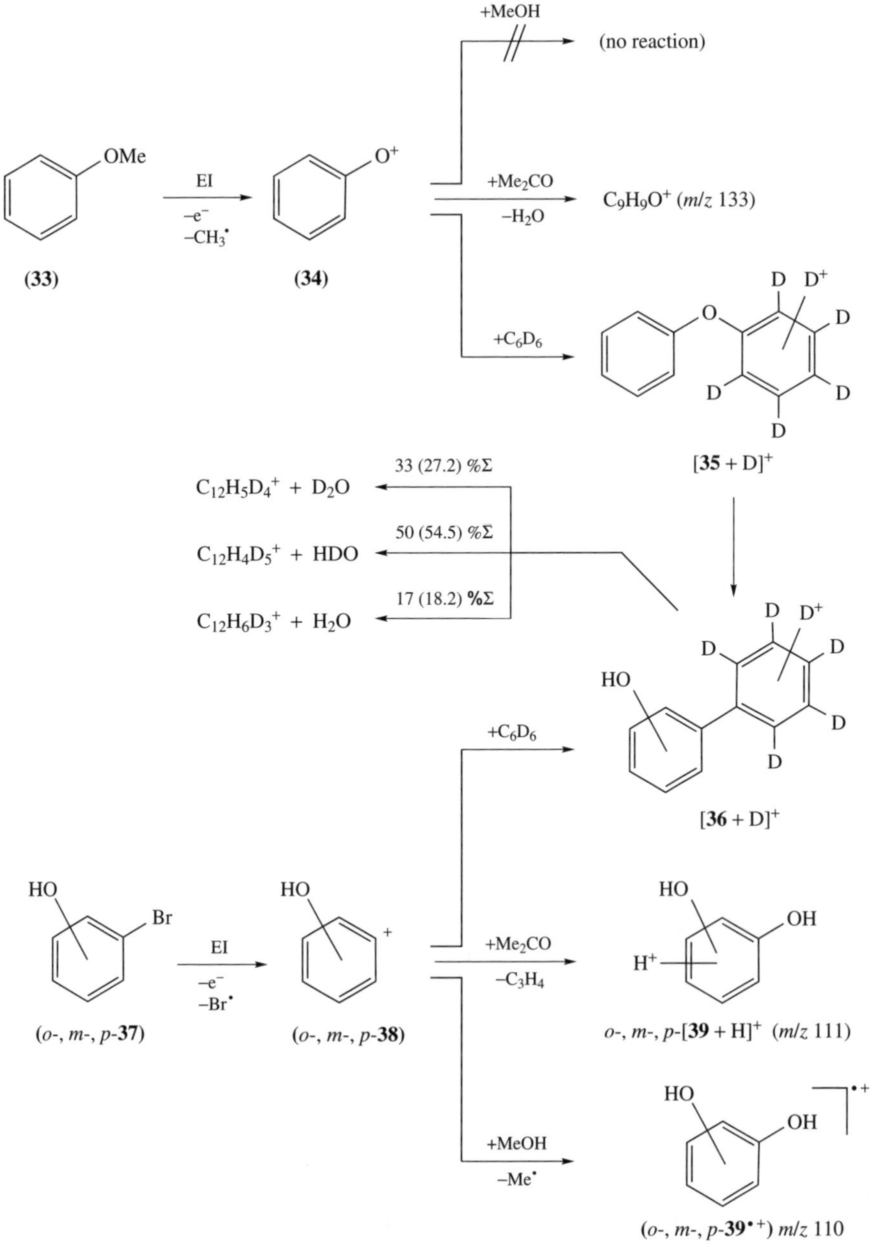
+MeOH
(no reaction)
OMe
EI
−e⁻
−CH3•
O+
+Me2CO
−H2O
C9H9O+ (m/z 133)
(33)
(34)
+C6D6
[35 + D]+
33 (27.2) %Σ
C12H5D4+ + D2O
50 (54.5) %Σ
C12H4D5+ + HDO
17 (18.2) %Σ
C12H6D3+ + H2O
HO
[36 + D]+
Br
−Br•
(o-, m-, p-37)
(o-, m-, p-38)
−C3H4
OH
o-, m-, p-[39 + H]+ (m/z 111)
−Me•
(o-, m-, p-39•+) m/z 110

SCHEME 13

From the previous and the recent results it appears obvious that CO loss from the 'phenolic' $C_6H_5O^+$ ions takes place via the *ortho*-isomer *o*-**38**, which undergoes further rearrangement to the phenoxy cation **34**, all these steps involving rather energy-demanding hydride shifts. The $C_6H_5O^+$ species from which CO is eventually expelled has been suggested to be (non-conjugated) cyclopentadienyl-5-carbonyl ion, *cyclo*-$C_5H_5CO^+$, whose heat of formation was calculated to be similar to that of ion **34**. The thermochemical minimum of the $C_6H_5O^+$ hypersurface was assigned to the conjugated cyclopentadienyl-1-carbonyl ion which was estimated to be far more stable than all the other isomers[42].

Partial distinction of the isomeric phenoxy and hydroxyphenyl cations was also achieved by reacting these ions with a variety of neutral reagents in the rf-only zone of a triple quadrupole mass spectrometer[43]. The bimolecular reactivity of the hydroxyphenyl cations **38** turned out to be similar but clearly distinct from that of the phenoxy cation **34** (Scheme 13). The reaction of ion **34** with methanol is unproductive but the hydroxyphenyl ions **38** reacted by formal transfer of a hydroxyl group, producing the dihydroxybenzene radical cations **39**$^{\bullet+}$. Acetone was found to form adducts which also react distinctly. In both cases, covalent $C_9H_9O^+$ species were formed but the intermediate formed from ion **34** expels water, whereas those formed from ions **38** eliminate allene (or propyne). Obviously, deep-seated rearrangement occur in these species, giving rise to fission of the acetone molecule by the highly reactive $C_6H_5O^+$ ions into its constituents, H_2O and C_3H_4. In contrast to these positive probe reactions, addition of benzene and hexadeuteriobenzene to ions **34** and **38** give rise to the same products. The reaction with C_6D_6 is particularly interesting in that the primary adducts, deuterated [ring-D_5] labelled diphenyl ether $[35 + D]^+$ and hydroxybiphenyls $[36 + D]^+$ both expel the three possible water isotopomers in a ratio close to that calculated for the complete scrambling of five hydrogen and six deuterium atoms. A review on related scrambling phenomena has appeared very recently[46]. Comparisons with the behaviour of the authentic deuteriated precursors suggests that ions $[36 + D]^+$ are the species from which water is eventually expelled.

IV. THE EFFECT OF THE HYDROXYL GROUP ON THE REVERSIBLE INTRAMOLECULAR HYDROGEN TRANSFER IN IONIZED PHENOLS

The proof for the reversibility of the 1,5-H$^{\bullet}$ transfer in *n*-alkylphenols originates from site-selective deuterium labelling experiments[46]. It has been shown that ionized *n*-alkylbenzenes, in general, suffer H/D exchange between the γ-position of the chain and (exclusively) the *ortho* positions of the ring, which can reach the 'statistical' distribution prior to fragmentation under favourable conditions. Whereas complete scrambling is generally not achieved in the ions fragmenting in the ion source, long-lived metastable ions have a chance to reach the statistical distribution of the H and D atoms over the sites involved. In such a situation, all ions must necessarily have undergone a certain minimum of 1,5-H$^{\bullet}$ transfer cycles, each of which involving a $\gamma \rightarrow$ *ortho* transfer and a reverse *ortho* $\rightarrow \gamma$ migration. As mentioned above, the stability of the distonic ion generated by the 1,5-H$^{\bullet}$ transfer relative to that of the conventional molecular ion plays a decisive role in the overall fragmentation and this is also reflected by the relative rate of the intermolecular H/D exchange in suitably labelled isotopomers.

The radical cations of *n*-alkylphenols are special because of the strongly different local hydrogen atom (or proton) affinities of the ring positions (Section II). A particularly telling case is presented here[46–48]: The molecular ions *p*-**40**$^{\bullet+}$ and *m*-**40**$^{\bullet+}$ generated from the corresponding isomeric 2-(hydroxybenzyl)indanes give drastically different EI mass spectra (Scheme 14). The peak at *m/z* 107 caused by the benzylic cleavage giving ions *p*-**9** dominates in the spectrum of the *para*-isomer and that at *m/z* 108 caused by the McLafferty

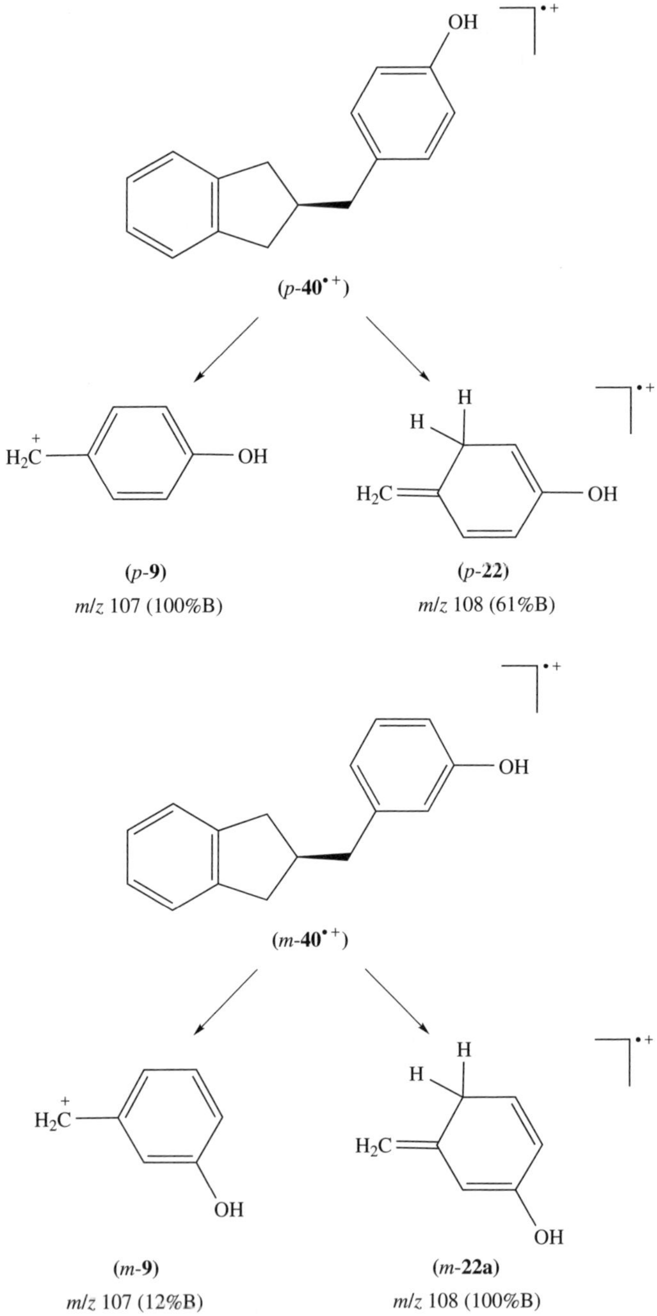

(p-**40**•+)

(p-**9**)
m/z 107 (100%B)

(p-**22**)
m/z 108 (61%B)

(m-**40**•+)

(m-**9**)
m/z 107 (12%B)

(m-**22a**)
m/z 108 (100%B)

SCHEME 14

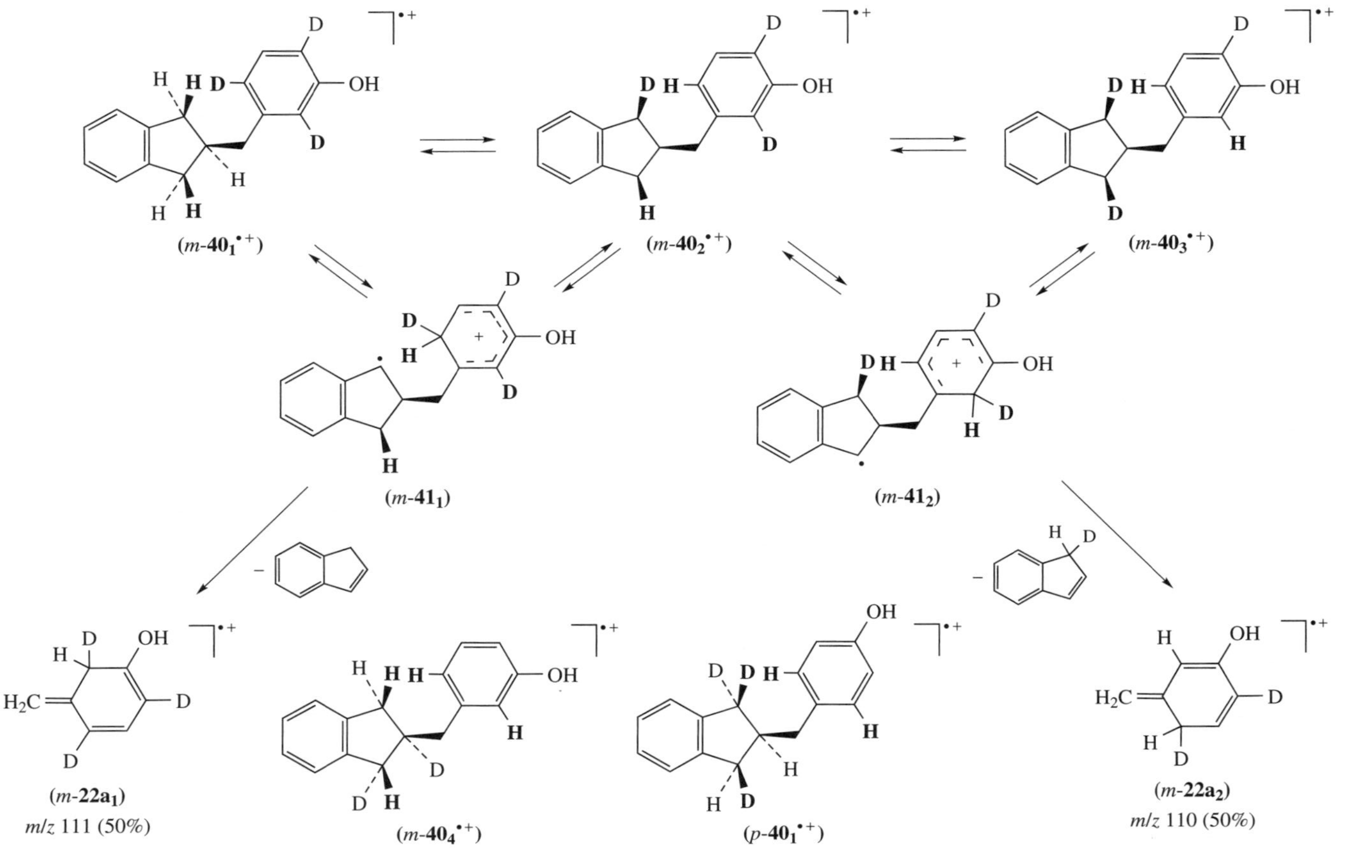

SCHEME 15

reaction giving ions m-**22a** governs the spectrum of the *meta*-isomer. With decreasing internal energy of the ions, the ratio $[C_7H_8O^{\bullet +}]/[C_7H_7O^+]$ strongly increases, since the less energy-demanding but slow rearrangement reaction gains importance. Metastable ions p-**40**$^{\bullet +}$ and m-**40**$^{\bullet +}$ dissociate exclusively by McLafferty reaction.

Extensive synthetical deuterium labelling of the neutral precursors of ions m-**40**$^{\bullet +}$ and p-**40**$^{\bullet +}$ was performed by using, in part, the different basicities of the phenolic ring positions in solution[48]. The results confirmed that only the two (benzylic) *cis*-H atoms at C-1 and C-3 of the indane ring and the two *ortho*-H atoms of the phenol ring are 'mobilized' during the McLafferty reaction. Thus, only four hydrogen atoms are involved in the intramolecular exchange process preceding the fragmentation. In the case of the trideuteriated radical cations m-**40**$_1^{\bullet +}$, for example, this leads to the formation of the isotopomers m-**40**$_2^{\bullet +}$ and m-**40**$_3^{\bullet +}$ via the corresponding distonic ions m-**41**$_1$ and m-**41**$_2$ (Scheme 15). The overall fragmentation of metastable ions m-**40**$_1^{\bullet +}$ leads to the fragment ions m-**22a**$_1$ (*m/z* 111) and m-**22a**$_2$ (*m/z* 110) in the ratio of 53:47, i.e. close to unity, indicating complete equilibration of the four H and D atoms. However, in the short-lived ions fragmenting already in the ion source, the ratio [*m/z* 111] / [*m/z* 110] is higher, e.g. 85:15 at 70 eV and 62:38 at 12 eV ionization energy. The parallel experiments with the corresponding labelled *para*-isomers, e.g. p-**40**$_1^{\bullet +}$, reveal that the exchange between the *cis*- and *ortho*-H atoms is much slower and proceeds only slightly with increasing lifetime of the ions. For example, metastable ions p-**40**$_1^{\bullet +}$ produce ions $C_7H_7DO^{\bullet +}$ (*m/z* 109) and $C_7H_6D_2O^{\bullet +}$ (*m/z* 110) in a ratio of 85:15 only.

V. ORTHO EFFECTS IN SUBSTITUTED PHENOLS

In many cases, the EI mass spectra of *ortho*-substituted phenols differ from those of the *meta*- and *para*-isomers by a dominant peak corresponding to the elimination of water or other stable molecules that incorporate elements of the phenolic hydroxy functionality. The reactive neighbouring group interaction in ionized 1,2-disubstituted arenes ('ortho effect') is often analytically valuable and has been studied mechanistically in much detail[49,50]. However, the effect is not always as pronounced as stated in textbooks and the structural assignment is not unambiguous if comparison with the spectra of the isomers is not possible. This holds in particular if the phenolic OH group in the molecular ion **42**$^{\bullet +}$ represents the hydrogen acceptor site (Scheme 16, path a) and has to be eliminated as water. In this case, in contrast to the alternative case (Scheme 16, path 6), in which 1,5-H transfer leads to fragile distonic ions **44**$^{\bullet +}$ (see below), dissociation of the intermediate distonic ion **43**$^{\bullet +}$ formed by 1,4-H transfer generates an incipient, energetically unfavourable phenyl cation, which may undergo subsequent isomerization. In addition, isomerization of the molecular ion **42**$^{\bullet +}$ by ring expansion to ionized hydroxycycloheptatrienes may obscure structure-specific fragmentation and thus attenuate the ortho effect.

For example, ionized *ortho*-cresol o-**7**$^{\bullet +}$ expels water by formal 1,4-H transfer from the methyl group, a reaction which should be largely suppressed in the *meta*- and *para*-cresol ions (Scheme 17). However, the [M − 18]$^{\bullet +}$ peak in the EI mass spectrum of o-**7** is only marginally larger (27%B) than the corresponding signals in the spectra of m-**7** (11%B) and p-**7** (8%B). The dominant fragmentation path is loss of H$^{\bullet}$ in all three cases (90, 80 and 100%B, respectively). It can be argued that the OH$_2$ group in the distonic ion intermediate **45** is too poor a leaving group to act as a sink for the dissociating H atom, in spite of the subsequent formation of ionized cyclopropabenzene **46** (*m/z* 90). Interestingly, the ortho effect registered for loss of H$_2$O from the isomeric cresol ions parallels the trend of the isomeric ions to expel CO: The relative intensity of the [M − 28]$^{\bullet +}$ peak in the EI mass spectrum of o-**7** is 21%B but only 6–8%B in the spectra of the other isomers. Note that this special 'ortho effect' is not initiated by transfer of an α-H atom from the

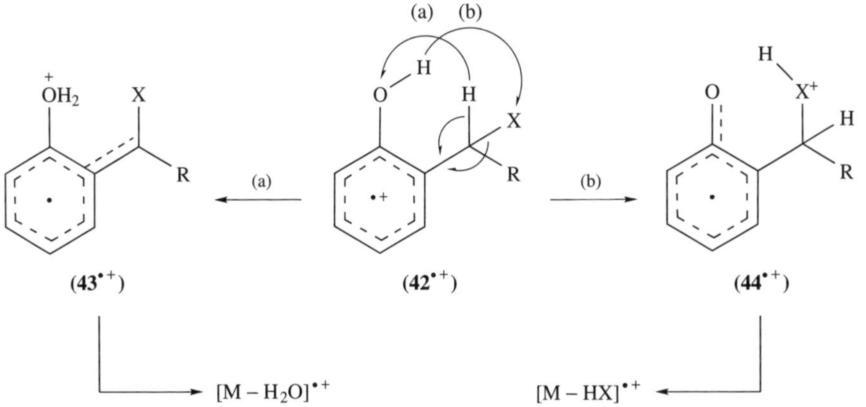

SCHEME 16

ortho-methyl substituent; rather, it is in line with the mechanism outlined in Scheme 2, in that the ring fission should be facilitated by the presence of an *ortho*-methyl group.

The cresols and other lower alkylphenols have been studied by multiphoton ionization mass spectrometry (MPI) with the aim of distinguishing positional isomers. Only slight differences were found in some cases, mainly concerning the low-mass region. Remarkably, the MPI mass spectrum of *ortho*-cresol was again distinct because water loss, i.e. a primary fragmentation reaction, from ions o-**7**$^{\bullet+}$ was found to be significantly more frequent than with the other isomers[51].

The ionized dihydroxybenzenes **39**$^{\bullet+}$ behave similarly (Scheme 17). Note that, in this series, one of the phenolic hydroxyl groups acts as a hydrogen acceptor and the other as an H donor. The EI mass spectrum of catechol (o-**39**) exhibits a significant ortho effect. While the intensity of the [M − H$_2$O]$^{\bullet+}$ peak in the EI spectrum of o-**39** is no greater than ca 15%B, the spectra of resorcinol (m-**39**) and hydroquinone (p-**39**) both show negligibly small [M − H$_2$O]$^{\bullet+}$ peaks ($\leqslant$2%B). It is likely that water loss from the intermediate **47** generates again bicyclic [M − H$_2$O]$^{\bullet+}$ ions, i.e. ionized benzoxirene **48**. And, notably, the CO losses does not parallel the ortho effect of the water elimination in this series, as it is the most pronounced ion the case of m-**39**.

Much more impressive, and analytically more reliable, ortho effects can be encountered in molecular radical cations where the phenolic hydroxyl group acts as an H donor instead of an H acceptor (Scheme 16, path b). In these cases, 1,4-shift of the phenolic H atom (or proton) in the molecular ions **42**$^{\bullet+}$ to a benzylic, sp^3-hybridized atom or group (X) within the *ortho* substituent generates a good leaving group (XH) in the distonic ion **44**$^{\bullet+}$. Numerous examples have been found for this situation, including the radical cations and [M + H]$^+$ ions of the dihydroxybenzoic acids (see below). It is also noted here that the fragmentation of *peri*-oriented groups falls into this category.

The fragmentation of the radical cations of the isomeric (hydroxymethyl)phenols **49** provides good examples for the ortho effect (Scheme 18). Loss of water via ion **50** generates a 30%B peak at m/z 106, due to ions **51**, in the EI mass spectrum of the *ortho*-isomer o-**49**$^{\bullet+}$ and the secondary fragmentation of ions **51**, viz. expulsion of carbon monoxide, gives rise to the base peak at m/z 78. Thus, the overall fragmentation of ions o-**49**$^{\bullet+}$ is induced by the initial 1,5-H transfer from the phenolic hydroxyl group. Successive losses of H$^{\bullet}$ and CO represent very minor pathways only. However, in order to assess the significance of *ortho*-specific fragmentation reactions, the behaviour of the isomeric ions

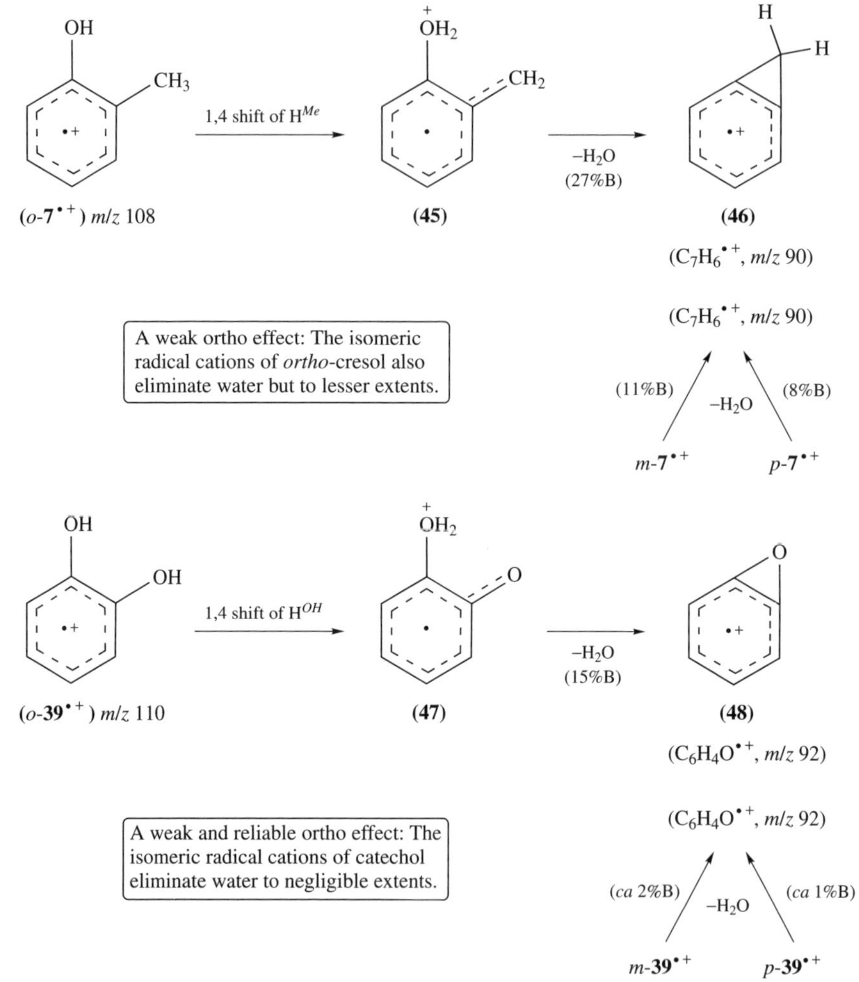

SCHEME 17

Top reaction:

OH, CH₃ (o-**7**·⁺) m/z 108 → [1,4 shift of H^Me] → OH₂⁺, CH₂ (**45**) → [−H₂O, (27%B)] → (**46**) (C₇H₆·⁺, m/z 90)

A weak ortho effect: The isomeric radical cations of ortho-cresol also eliminate water but to lesser extents.

m-**7**·⁺ (11%B) → [−H₂O] → (C₇H₆·⁺, m/z 90)
p-**7**·⁺ (8%B) → [−H₂O] → (C₇H₆·⁺, m/z 90)

Bottom reaction:

OH, OH (o-**39**·⁺) m/z 110 → [1,4 shift of H^OH] → OH₂⁺, =O (**47**) → [−H₂O, (15%B)] → (**48**) (C₆H₄O·⁺, m/z 92)

A weak and reliable ortho effect: The isomeric radical cations of catechol eliminate water to negligible extents.

m-**39**·⁺ (ca 2%B) → [−H₂O] → (C₆H₄O·⁺, m/z 92)
p-**39**·⁺ (ca 1%B) → [−H₂O] → (C₆H₄O·⁺, m/z 92)

---

have to be checked also. This is strikingly evident from the EI mass spectra of m-**49** and, in particular, of p-**49** (Scheme 18). In fact, the variety of fragmentation channels is much broader in both cases, rendering the mass spectrum of o-**49**·⁺ rather 'ortho-specific'. The moderate loss of water from the meta-isomer m-**49**·⁺ (20%B) indicates skeletal rearrangements. Protonated phenol, $C_6H_7O^+$ (m/z 95, 90%B), is the second dominant fragment ion, whereas ions $C_6H_5^+$ (m/z 77) give rise to the base peak. Completely unexpected, however, is the base peak in the EI mass spectrum of the para-isomer p-**49**, which corresponds to water loss! Although the competing fragmentations are again quite pronounced, this example demonstrates that interpretation of the EI mass spectra can be misleading. A reasonable explanation for the water loss from ions p-**49**·⁺ lies again in the favourable formation of a particularly stable hydroxybenzenium ion by 1,2-H· shift (p-**49**·⁺ → **52**)

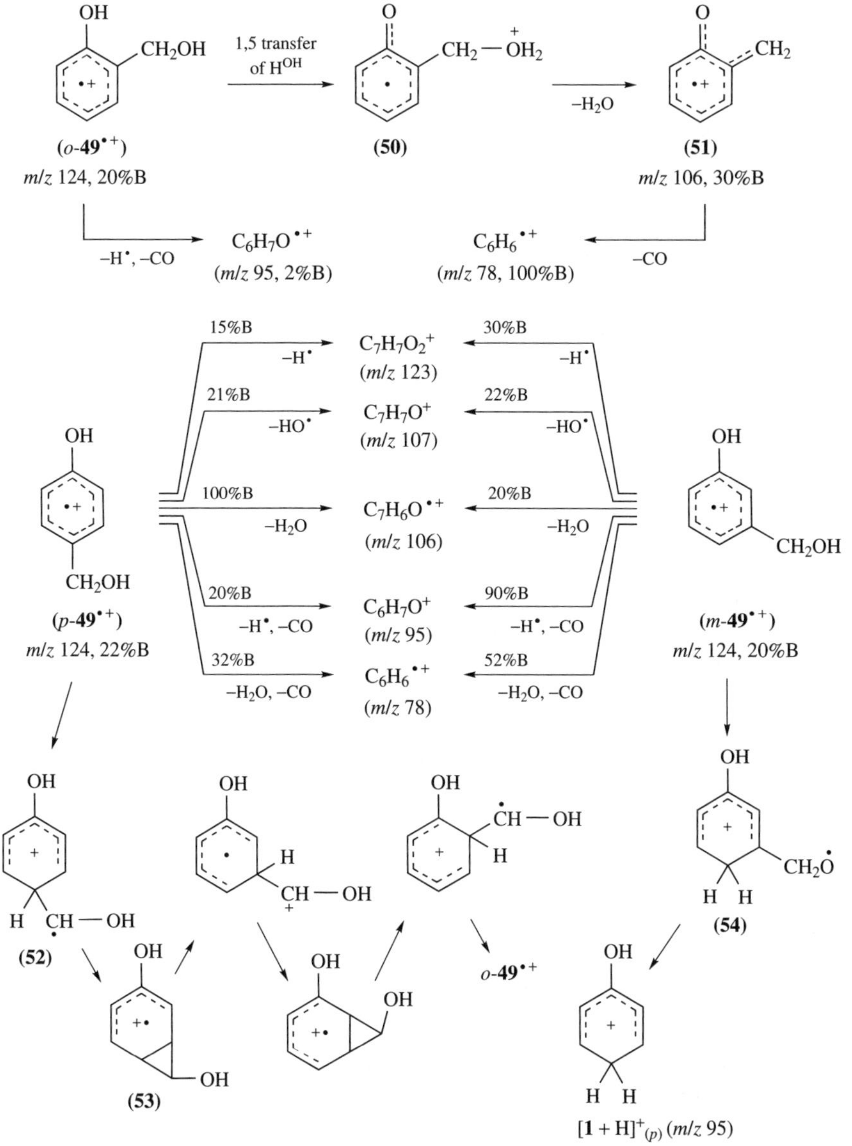

SCHEME 18

which, in a sequence of ring-walk isomerizations involving bicyclic isomers, e.g. **53**, may rearrange to the *ortho*-isomer *o*-**49**$^{•+}$ (Scheme 18). The role of such rearrangements has recently been determined in methoxymethyl-substituted naphthalenes[52]. The formation of ions $C_6H_7O^+$ (*m/z* 95) from the *meta*-isomer *m*-**49**$^{•+}$ may also be initiated by hydrogen rearrangement to the transient isomers **54** (Scheme 18).

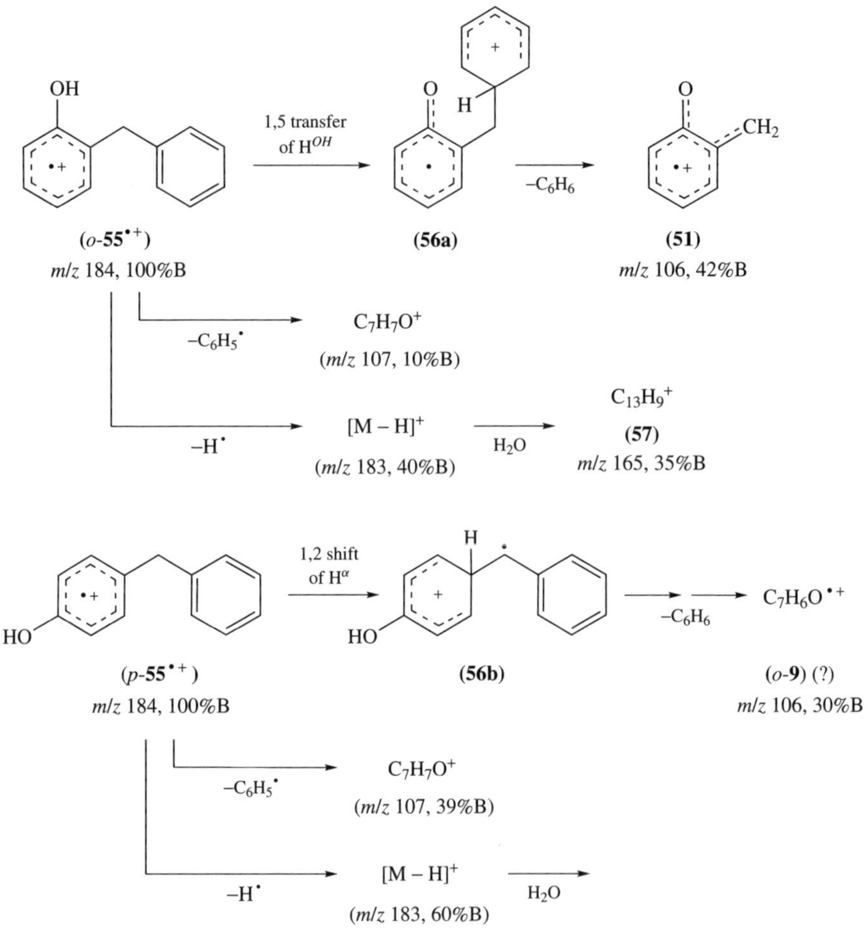

SCHEME 19

The phenolic H atom can also be transferred to another aromatic ring, which is then expelled as neutral benzene. Loss of benzene constitutes a characteristic fragmentation channel for many arylaliphatic radical cations and a major one for protonated alkylbenzenes[17,25,46,53,54]. A simple example, which again demonstrates an only moderately strong ortho effect, is shown in Scheme 19. The EI mass spectrum of *ortho*-benzylphenol *o*-**55** exhibits loss of benzene as a medium-size peak at m/z 106 (42%B), due to ions **51**, presumably via the distonic ion **56a**. The competing sequential losses of H$^\bullet$ and H$_2$O leading to fluorenyl cations (C$_{13}$H$_9^+$, m/z 165), give rise to similarly high peaks. Notably, the mass spectrum of the *para*-isomer *p*-**55** exhibits the same peaks but the relative abundance of the [M − C$_6$H$_6$]$^{\bullet+}$ ions is somewhat decreased. It is known that ionized diphenylmethanes undergo cyclization processes and extensive subsequent hydrogen scrambling prior to fragmentation[25,55]. In the case of the electron-rich hydroxy derivatives **55**, formation of protonated phenol intermediates, such as the distonic ion **56b**, appears again to be likely as an initial step. The only quantitatively different fragmentation of the isomers o-**55**$^{•+}$ and p-**55**$^{•+}$ points to the interplay of complex isomerization.

Particularly strong ortho effects are found in the EI mass spectra of salicylic acids and more highly hydroxylated benzoic acids (Schemes 20 and 21) and their derivatives, such as the benzamides. Throughout, the *ortho*-specificity of the fragmentation of the molecular radical cations of these compounds is much higher than in the cases discussed above.

1,5-H transfer in ionized salicylic acid o-**58**$^{•+}$ from the phenolic OH group to the carboxyl functionality can take place in two ways (Scheme 20). In contrast to migration to the carbonyl group, generating the stable distonic ion **59**, transfer of the hydroxyl group generates a highly fragile distonic ion, viz. **60**, which readily loses water to produce ion **61** (m/z 120) giving rise to the base peak in the spectrum. Subsequent loss of CO produces another significant fraction (75%B) of the total ion current. The otherwise ubiquitous fragmentation of carboxylic acids, viz. the successive losses of OH$^•$ and CO, is almost completely suppressed by the ortho interaction. In contrast, the EI mass spectra of the *meta*- and *para*-isomers of salicylic acid reflect a complementary fragmentation behaviour of the molecular ions, which react very similarly to each other. Thus, losses of H$_2$O and CO from the *meta* isomer m-**58**$^{•+}$ are almost negligible but the [M − OH]$^+$ ions give rise to the dominating fragment ion peaks at m/z 121 (77 and 100%B, respectively). Subsequent expulsion of CO still occurs to ca 25% in both cases.

2,6-Dihydroxybenzoic acid **63** and 3,4,5-trihydroxybenzoic acid **67** behave accordingly, with one remarkable exception (Scheme 21). Whereas the fragmentation of ionized gallic acid **67**$^{•+}$ is again dominated by the successive losses of OH$^•$ and CO, generating ions **68** and **69**, that of ionized γ-resorcylic acid **63**$^{•+}$ is strongly governed by loss of water and then CO, a sequence involving the distonic ion **64** and the *ortho*-quinoid ion **65**. In addition, however, ions **63**$^{•+}$ suffer decarboxylation, giving ions m-**39** (17%B). Obviously, the increased proton affinity of the *meta*-dihydroxy-substituted aromatic nucleus gives rise to a relatively facile transfer of the carboxylic proton to the ring, generating transient ions **66**.

There are many further examples for the specific effects of *ortho*-hydroxy substituents on the EI-induced fragmentation of phenolic compounds. Conceptually, many of them can be traced to either the increased acidity of the O−H bond in the radical cations or, more often, to the increased proton affinity or hydrogen atom affinity of the ring at positions *ortho* (and *para*) to the OH group. A last example concerns the strikingly distinct fragmentation behaviour of *ortho*-hydroxycinnamic acid o-**70** as compared to its *meta*- and *para*-isomers, m-**70** and p-**70** (Scheme 22). In this case, not only the specific behaviour of the *ortho*-isomer deserves notice but also that of the other isomers[56]. The EI mass spectrum of o-**70** exhibits a relatively small molecular ion peak but strong signals corresponding to the ions generated by loss of water (m/z 146) and subsequent single and double loss of CO (m/z 118 and m/z 90). Without any doubt, H$_2$O elimination yields ionized coumarin **71**$^{•+}$, possibly, or rather necessarily involving an oxygen atom from the carboxylic group (see below), and CO expulsion from this ion can be safely assumed to give ionized benzofuran **72**$^{•+}$. Loss of H$^•$ and OH$^•$, being typical reactions of the parent ionized cinnamic acid, do not occur. By contrast, these two processes give rise to characteristic peaks at m/z 163 and m/z 147 in the EI mass spectra of m-**70** and p-**70**, while the molecular ions generate the base peaks in both cases. Loss of H$^•$ takes place after the cyclization of the molecular ions to their isomers m-**74** and p-**74**, which represent the radical cations of electron-rich 1,3,5,7-octadiene derivatives[57]. The ionic products of the H$^•$ loss from ions m-**70**$^{•+}$ and p-**70**$^{•+}$ are the dihydroxy benzopyrylium ions m-**75** and p-**75**, respectively. Loss of OH$^•$ from m-**70**$^{•+}$ and p-**70**$^{•+}$ may occur, at least in part, in a straightforward manner, i.e. from the carboxyl group with concomitant cyclization to the [M − OH]$^+$ ions m-**73** and p-**73**, respectively, again followed by single and two-fold expulsion of CO. In fact,

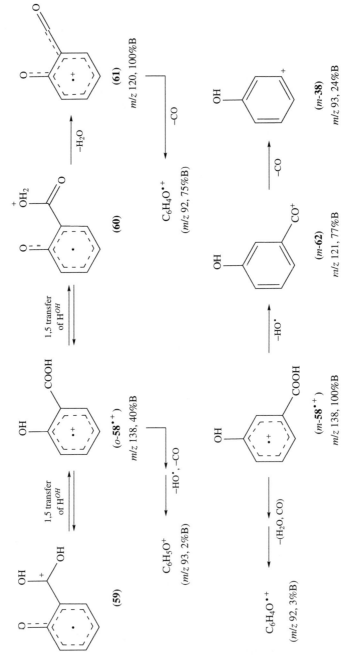

SCHEME 20

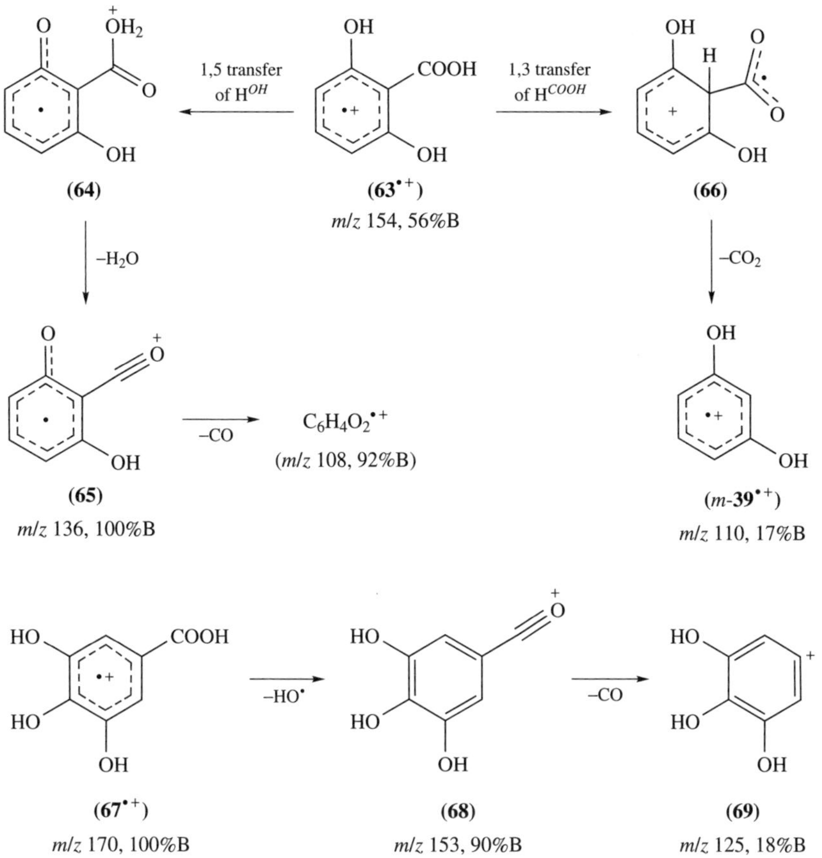

SCHEME 21

ionized cinnamic acid and many of its derivatives do lose a hydroxyl radical (Scheme 22). However, arguments have been invoked, in analogy to the EI-induced isomerization and fragmentation of other cinnamic acids and of benzylideneacetones, which point to the loss of the arene substituent, viz. the *phenolic* hydroxyl group in the case of m-**70**$^{\bullet+}$ and p-**70**$^{\bullet+}$. This requires a series of 1,2-H shifts, or even a sufficiently fast hydrogen ring walk, by which ions m-**74** and p-**74** are converted into their respective tautomers, m- and p-**74a** and m- and p-**74b**. It should be noted that not only the 1,2-H shift in the radical cationic π-systems of ions **74** is unusual (cf. its non-occurrence in the distonic ions **41**, Scheme 15) but also the loss of a phenolic OH$^{\bullet}$ radical, which leads to the benzopyrylium ions m-**76** and p-**76**. Different from even-electron species, where *ipso* protonation of phenol nuclei was found to be extremely unfavourable as compared to *para* and *ortho* protonation (cf. Scheme 1), both the formation and homolytic dissociation of odd-electron quinoid species bearing an sp^3-hybridized carbinol centre appear to be energetically reasonable.

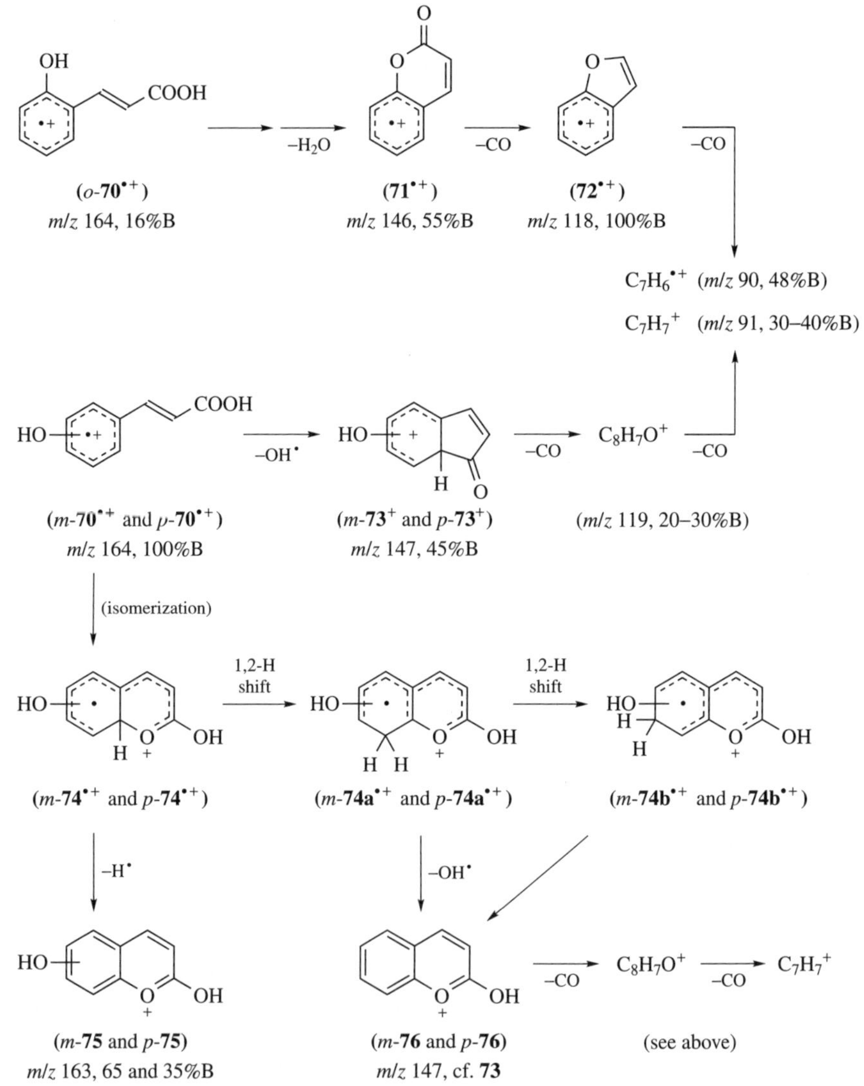

SCHEME 22

Finally, ortho effects involving the radical cations of various nitro-substituted alkylphenols are mentioned here. In these cases, two sequential ortho effects have been observed. For example, the EI mass spectra of 2-ethyl-4,6-dinitrophenol and its 2-cyclohexyl analogue exhibit pronounced peaks for the formation of $[M - H_2O]^{•+}$ and $[M - H_2O - OH]^+$ ions and the spectrum of 2-isopropyl-4,6-dinitrophenol even indicates that the secondary fragmentation step is faster than the primary one, because an ion abundance ratio $[M - H_2O - OH]^+/[M - H_2O]^{•+} = 25$ was found[58].

VI. SECONDARY FRAGMENTATION REACTIONS OF PHENOL RADICAL CATIONS

As mentioned earlier (Section III.A), the radical cations of simple phenols undergo the characteristic expulsion of carbon monoxide. The shift of a hydrogen atom is necessary to allow this elimination reaction to occur (Scheme 2). However, most of the CO losses observed in the mass spectra of the hydroxycinnamic acids (Scheme 22) have other origins and are mechanistically different from the behaviour of phenolic radical cations. In fact, many decarbonylation processes observed in the EI mass spectra of phenols derivatives take place from *even-electron* (closed-shell) primary fragment ions, which themselves are formed by loss of a radical from the open-shell precursor ion. Thus, while CO loss as a primary fragmentation of the molecular ions is often suppressed by less energy-demanding fragmentation channels, CO elimination as a secondary fragmentation is quite frequent—albeit analytically less obvious. Only a few examples will be given in the following, in part with respect to the notable prototype character of the reacting ions involved.

The EI mass spectra of the monomethyl ethers of catechol and hydroquinone, o-**77** and p-**77**, are very similar and exhibit two major fragment ion peaks at m/z 109 and m/z 81, indicating the sequential loss of $CH_3^{\bullet}$ and CO, respectively (Scheme 23). The most logical structures of the $[M - CH_3]^+$ ions are protonated *ortho*- and *para*-benzoquinone, $[o\text{-}\mathbf{78} + H]^+$ and $[p\text{-}\mathbf{78} + H]^+$, and the $C_5H_5O^+$ ions formed as secondary fragments by subsequent expulsion of CO should have the energetically favourable pyranylium structure **80**, rather than that of a protonated cyclopentadienone. Much in contrast to o-**77** and p-**77**, the EI mass spectrum of resorcinol monomethyl ether m-**77** exhibits almost no $[M - CH_3]^+$ signal but again a significant $[M - CH_3 - CO]^+$ peak. Major competing fragmentation channels (not shown in Scheme 23) are the loss of 29 Th (probably $H^{\bullet}$ and CO) and 30 Th (possibly CH_2O) via ring protonation. The apparent suppression of the methyl loss from ions m-**77**$^{\bullet+}$ is attributed to the energetically unfavourable structure of ion **79** which, in contrast to their isomers $[o\text{-}\mathbf{78} + H]^+$ and $[p\text{-}\mathbf{78} + H]^+$, represents an electronically destabilized phenoxy cation (or O-protonated benzene-1,3-dioxyl). As a consequence, the vanishingly low relative abundance of the $[M - CH_3]^+$ ions from **79** is attributed to its fast decomposition by CO loss to give ions **80**.

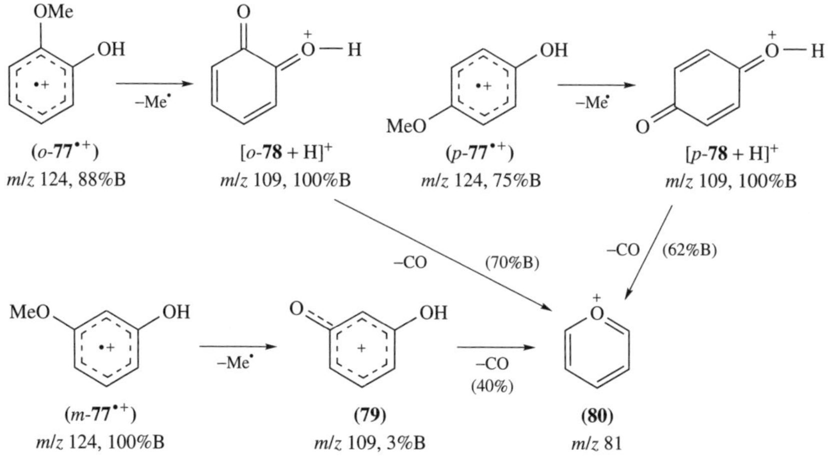

(o-**77**$^{\bullet+}$)
m/z 124, 88%B

$[o\text{-}\mathbf{78} + H]^+$
m/z 109, 100%B

(p-**77**$^{\bullet+}$)
m/z 124, 75%B

$[p\text{-}\mathbf{78} + H]^+$
m/z 109, 100%B

(m-**77**$^{\bullet+}$)
m/z 124, 100%B

(**79**)
m/z 109, 3%B

(**80**)
m/z 81

SCHEME 23

VII. MISCELLANEOUS FRAGMENTATIONS OF PHENOL RADICAL CATIONS

The $[M - NO]^+$ ions of *para*-nitrophenol were found to show kinetic energy release (T_{50}, evaluated from the strong peak broadening at half peak height) for the expulsion of CO from the metastable ions ($T_{50} = 0.52$ eV) as do the $[M - CH_3]^+$ ions of hydroquinone monomethyl ether *p*-**77** ($T_{50} = 0.50$ eV)[59,60]. Thus, the structure of the primary fragment ions is likely to be that of protonated *para*-benzoquinone $[p$-**78** $+ H]^+$ in both cases. The role of electron-withdrawing and electron-releasing substituent, including the hydroxyl group, in *para*-substituted nitrobenzenes on the kinetic energy release during NO• loss was studied in more detail[61]. The EI mass spectra of several *ortho*-nitrosophenols have been studied with respect to the tautomerism in the molecular radical cations prior to fragmentation[62].

The EI-induced fragmentation of the α,α,α-trifluorocresols **81** has been studied in detail and in comparison to the cresols **7** (Scheme 24). A pronounced ortho effect was observed

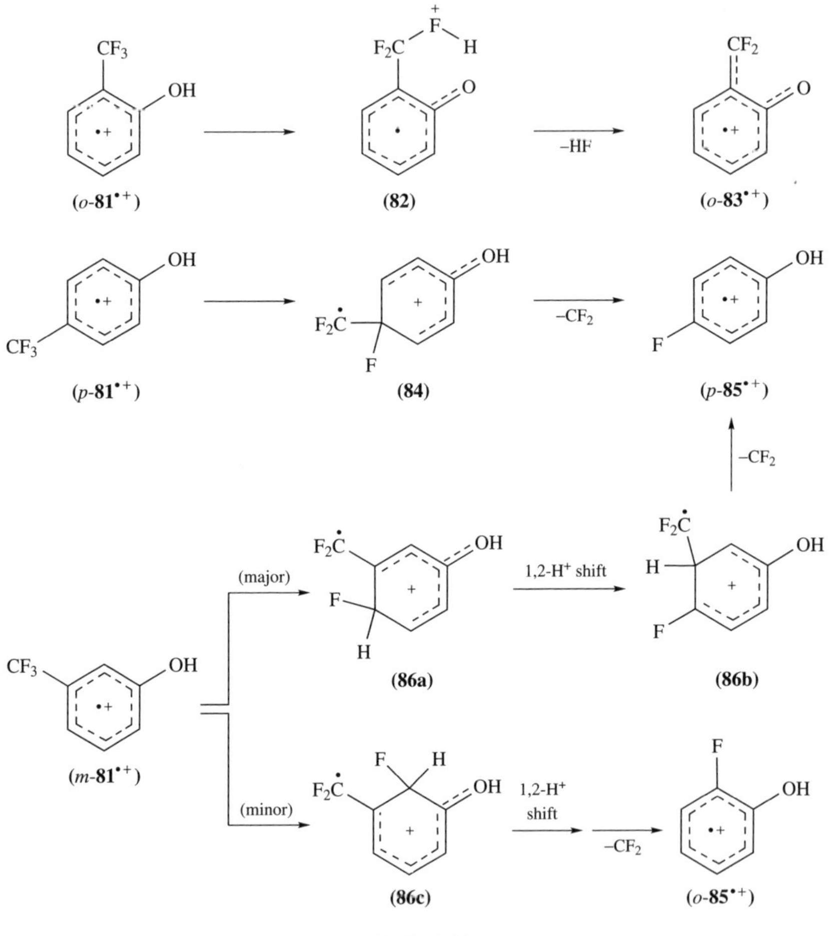

SCHEME 24

for the radical cations of the *ortho*-isomer *o*-**81**$^{\bullet+}$, which gives rise to the elimination of HF, presumably via **82**, along with some F$^{\bullet}$ loss in the standard ion-source EI mass spectra and exclusive elimination of HF from the metastable ions[63]. Investigation of the [*O*-D] isotopomer of *o*-**81**$^{\bullet+}$ revealed that the phenolic proton is transferred exclusively to the trifluoromethyl group prior to the primary fragmentation process. Subsequent fragmentation of the [M − HF]$^+$ ions consists of loss of CO. The *meta*-isomer *m*-**81**$^{\bullet+}$ was found to undergo loss of F$^{\bullet}$ rather than elimination of HF. In a related paper, the mass spectrometric behaviour of *meta*- and *para*-(α,α,α-trifluoro)cresol *m*-**81**$^{\bullet+}$ and *p*-**81**$^{\bullet+}$ was studied in view of the remarkable elimination of difluorocarbene, CF$_2$, which gives rise to intense peaks in the EI mass spectra of these two isomers, whereas the process does not occur in the spectrum of the *ortho*-isomer *o*-**81**$^{\bullet+}$ [64]. The pronounced directing effect of the hydroxyl group on migrating protons, which has been discussed above in several respects, is mirrored here for migrating fluorine atoms. The trifluoromethyl substituent in ionized *para*-(α,α,α-trifluoro)cresol *p*-**81**$^{\bullet+}$ disintegrates by leaving one of the fluorine atoms at the original *ipso* position (i.e. at the *para* position with respect to the hydroxy substituent), generating the radical cations of *para*-fluorophenol *p*-**85**$^{\bullet+}$ via **84**. Thus, the transition state is stabilized by the electron-donating OH group, similar to the stabilization of ions [**1** + H]$^+_{(p)}$. The identity of the [M − CF$_2$]$^{\bullet+}$ ion was demonstrated by energy-dependent CID mass spectrometry. Similarly, ionized *meta*-(α,α,α-trifluoro)cresol *m*-**81**$^{\bullet+}$ was found to react by 1,3-F$^{\bullet}$ shift, producing mainly ionized *para*-fluorophenol *p*-**85**$^{\bullet+}$ via ions **86a** and **86b** and minor amounts of ionized *ortho*-fluorophenol *o*-**85**$^{\bullet+}$ via ion **86c**. The intermediates **86a** and **86c** represent special cases of ionized *para* and *ortho*-isotoluene, respectively (cf. **107**$^{\bullet+}$, Scheme 29). In these cases, the hydroxyl group directs the migrating fluorine atom either to the *para* or to the *ortho* position with respect to its own[64].

VIII. CHEMICAL IONIZATION MASS SPECTROMETRY OF PHENOLS

Chemical ionization (CI) mass spectrometry of phenol and phenol derivatives has been studied using a number of reagent gases. In most cases, positive ion CI mass spectrometry was found to be governed by the different response of isomeric phenols toward proton addition and/or electrophilic attack by reactant ions of the CI plasma. In addition, intermolecular H$^+$/D$^+$ exchange was found to be a useful probe for structure elucidation, depending on the relative acidity of the proton-transferring reactant ions. The phenolic OH group undergoes fast proton exchange; however, those ring positions which have sufficiently high local proton affinities can also be subject to H$^+$/D$^+$ exchange. In several cases, this allows us to identify isomeric arenes which are indistinguishable by other mass spectrometric methods, such as EI. Whereas this effect was demonstrated for the first time by using ion cyclotron resonance (ICR) mass spectrometry of a number of substituted benzenes excluding phenols[65], systematic studies using water chemical ionization, CI(H$_2$O) and CI(D$_2$O), were performed with a variety of arenes[66,67], including phenols[68]. The results of the latter work were discussed in detail in view of the site of protonation and pointed to the preferred coordination of the water molecule to protonated phenolic OH group. Significant but only partial exchange was found to occur.

It appears that the tendency of the arenes to undergo intermolecular exchange of the ring hydrogens depends on the structure and relative stability of the cluster ions, such as {[**1** + H]$^+_{(p)}$ + HX} and {[**1** + H]$^+_{(o)}$ + HX} (cf. Scheme 1), with X = OH, OMe, NH$_2$ etc. Preferred coordination at polar groups, in particular the phenolic OH group, may suppress the H$^+$/D$^+$ exchange between otherwise reactive ring positions. On the other hand, the polar group may facilitate the formation of stable cluster ions. This point was discussed for CI(NH$_3$) of arenes including phenol which, interestingly, was found to be reluctant to formation of $[1 + NH_4]^+$ ions[69].

Fragmentation of protonated or cationized phenols occurs easily under CI-MS conditions if a good leaving group is formed in the $[M + H]^+$ or $[M + HX]^+$ ion. This is not the case when the phenolic OH group itself is protonated but this substituent can strongly influence the fragility of other groups attached to the aromatic nucleus, in accordance with the thermodynamic stability of phenolic ions in the gas phase. This can be used favourably for analytical purposes. For example, *meta*- and *para*-hydroxybenzyl alcohol show drastically different CI(NH$_3$) mass spectra, with the abundance ratios $[M + NH_4 - H_2O]^+/[M + NH_4]^+ = 14.5$ for the *para*-isomer but close to zero for the *meta*-isomer. Clearly, the phenolic OH group facilitates the heterolytic cleavage of the benzylic C−O bond in the former case but not in the latter[69].

Similarly drastical differences have been observed in the CI(MeOH) mass spectra of various oxygen-containing aromatic compounds, including phenols, naphthols and indanols (Scheme 25)[70]. Whereas *benzylic* alcohols of the same elemental composition, e.g. 1-indanol **87**, undergo facile loss of water from the $[M + H]^+$ ions, giving rise to intense $[M + H - H_2O]^+$ peaks, the corresponding phenolic isomers, e.g. 4- and 5-indanol **89**, gave characteristically strong $[M + H]^+$ signals. When perdeuteriated methanol was used as the reagent gas, the CI(CD$_3$OD) mass spectra exhibited clean mass shifts of 3 Th for the quasi-molecular ions in the case of the phenolic isomers. By contrast, the rather fragile quasi-molecular ions formed by deuteriation of the benzylic alcohols, e.g. $[87 + D]^+$, readily generate abundant indanyl fragment ions with the same *m/z* values as observed in the CI(MeOH) spectra, e.g. **88**. Clearly, the phenolic isomers are subject not only to deuteriation, giving ions $[89 + D]^+$, but also to a subsequent single H$^+$/D$^+$ exchange of the phenolic proton, yielding ions $[89_1 + D]^+$ and giving rise to the diagnostic $[M + 3]^+$ peaks at *m/z* 137. (These ions may eventually exist in both the *O*- and *ring*-deuteronated forms.) The analytical usefulness of the combined CI(CH$_3$OH)

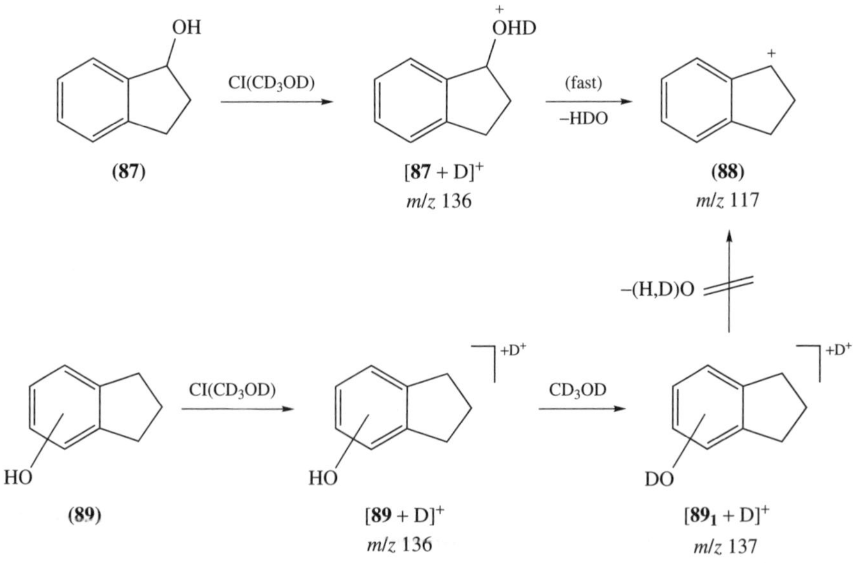

SCHEME 25

and CI(CD$_3$OD) mass spectrometry was demonstrated by GC/MS characterization of a complex, neutral-polar subfraction of coal-derived liquids[70].

In a related extensive study, application of CI(NH$_3$), CI(ND$_3$) and even CI(^{15}NH$_3$) mass spectrometry to the analysis of phenylpropanoids and substituted phenylalkyl ethers containing phenolic OH groups was demonstrated with the aim to model pyrolysis mass spectrometric experiments of lignin[71]. Similar to the indanols discussed above, hydroxycinnamyl alcohols and α-hydroxy-substituted phenylalkyl ethers containing *para*-hydroxybenzylic alcohol units showed intense peaks for the [M + H − H$_2$O]$^+$ and also for the [M + NH$_4$ − H$_2$O]$^+$ fragments. In contrast to the earlier report mentioned above[69], the CI(NH$_3$) spectra of these more complex phenol derivatives exhibited also cluster ions [M + NH$_4$]$^+$ which, in accordance with general expectation, gained relative abundance with increasing pressure of the CI reagent gas. It is evident that the [M + NH$_4$ − H$_2$O]$^+$ fragment ions are isobaric with the molecular radical cations M$^{\bullet +}$ generated by residual EI and/or charge transfer processes; however, use of CI(ND$_3$) helps a lot to remove any ambiguities in this respect.

Various other reagent gases have also been used in CI mass spectrometry of phenols. These include chloromethanes, CH$_{1+x}$Cl$_{3-x}$ ($x = 0 - 2$), tetramethylsilane, nitric oxide and acrylonitrile. Methylene chloride was used in negative ion chemical ionization (NCI) and found to produce abundant cluster ions [M + Cl]$^-$ and [2 M + Cl]$^-$ with phenol (Section XI.B). The formation of [M − Cl]$^-$ ions was not observed[72]. In the positive ion mode, CI(CH$_2$Cl$_2$) was studied with phenol **1** and its [*ring*-D$_5$] and the [D$_6$] isotopomer[73]. Besides the signals for M$^{\bullet +}$ and [2 M]$^{\bullet +}$ ions, which are probably due to charge transfer processes, the mass spectra of these compounds are dominated by the peaks for [M + 13]$^+$ and [2 M + 13]$^+$ ions. These ions are formed by substitution of a hydrogen by a methylene group, thus corresponding to the net attachment of a methine (CH) group to the ring. Similarly to other carbenium ions, e.g. benzyl cations[25], CH$_2$Cl$^+$ ions attack electron-rich arenes like phenol quite readily, generating the σ-complexes, such as [**90** + H]$^+$, which contains mobile protons[17,25,46,74]. Subsequent elimination of HCl leaves the corresponding hydroxybenzyl cations, probably mainly *p*-**9** and *o*-**9** (Scheme 26). A competing, minor fragmentation path allows the [M + CHCl$_2$]$^+$ adduct ions to expel dihydrogen, presumably generating the related α-chlorohydroxybenzyl cations (not shown in Scheme 26). Loss of H$_2$ is an ubiquitous reaction channel of protonated methylbenzenes[75]. The CI(CH$_2$Cl$_2$) mass spectra of the two labelled analogues of **1** are similar and show exclusive elimination of DCl and HD, as expected for an electrophilic substitution of phenols[73]. Notably, the CI(CH$_2$Cl$_2$) mass spectrum of **1** does not reflect the structure of the stable ions, i.e. of those which do not dissociate before detection. This follows from collision-induced dissociation (CID) measurements performed with the [M + CH$_2$Cl]$^+$ ions[76]. The CID mass spectrum indicates only little formation of the hydroxybenzyl cations **9** (*m/z* 107) but a strong signal for phenyl cations **93** (*m/z* 77). Thus, it has been argued that the stable adduct ions may have adopted a structure different from that formed by electrophilic attack at the ring. Scheme 26 offers a possible explanation for the predominant formation of C$_6$H$_5$$^+$ ions **93** as structure-specific fragmentation path of the putative adduct ions [**91** + H]$^+$ via the intermediate ions **92**. Notably, CID fragmentation of the [M + CHCl$_2$]$^+$ ions generated by CI(CH$_2$Cl$_2$) of **1** and also the [M + CCl$_3$]$^+$ ions generated by CI(CHCl$_3$) of **1** provide positive evidence for the occurrence of electrophilic attack of the corresponding, more highly chlorinated reactant ions[76].

Among the more rare CI reagent gases, acrylonitrile was studied and found to produce particularly abundant adduct ions [M + C$_3$H$_3$N]$^+$ with many aliphatic alcohols, along with the corresponding [M + C$_3$H$_3$N − H$_2$O]$^+$ and [M + H − H$_2$O]$^+$ fragment ions[77]. The CI(CH$_2$=CHCN) mass spectra of phenol were found to be special in that the [M + H]$^+$ ions gave rise to the base peak, the [M + C$_3$H$_3$N]$^+$ peak being only moderately intense.

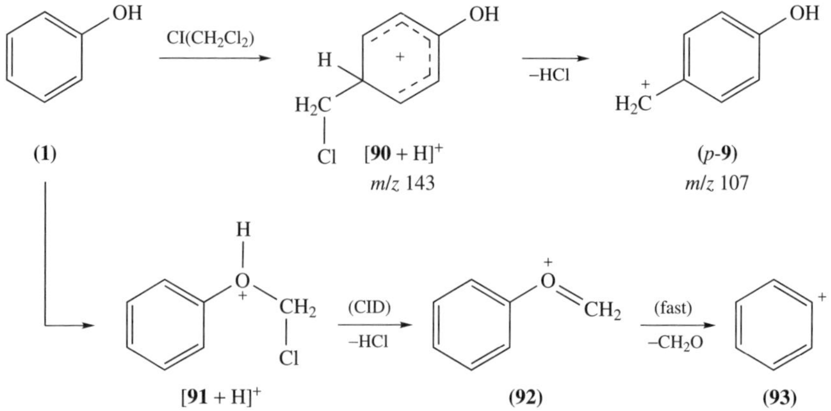

294   Dietmar Kuck

SCHEME 26

Use of tetramethylsilane (TMS) as a reagent gas in CI mass spectrometry enables the gas-phase trimethylsilylation of aromatic compounds, including phenol[78]. The Me_3Si^+ ions generated in the CI plasma give rise to abundant adduct ions $[M + 73]^+$. However, charge transfer processes lead also to the formation of large amounts of molecular radical cations $1^{•+}$, whereas the protonated phenol $[1 + H]^+$ ion is formed only in minor relative abundance. Comparison of the CID mass spectra of the adduct ions $[1 + Me_3Si]^+$ (m/z 167) with those of protonated trimethylsilyl phenyl ether $[Me_3SiOC_6H_5 + H]^+$ and protonated *para*-(trimethylsilyl)phenol $[p-Me_3SiC_6H_4OH + H]^+$, generated by CI(MeOH) of the corresponding neutral precursors, suggests that the stable adduct ions $[1 + Me_3Si^+]$ obtained by CI(TMS) are formed exclusively by attack at the phenolic OH group[78]. It has been shown that the efficient addition of Me_3Si^+ ions to various organic molecules can be used to detect compounds of low volatility under so-called direct chemical ionization (DCI) conditions[79]. For example, the $DCI(TMS/N_2)$ and/or $DCI(TMS/i - C_4H_{10})$ mass spectra of estradiol **15** and estrone were reported. The recognition of the $[M + Me_3Si^+]$ adduct ions and their fragment ions was shown to be facilitated by using mixtures of $[D_0]$-TMS and $[D_{12}]$-TMS as additives to the reagent gas, thus giving rise to adduct ion peaks and $[M + (CX_3)_3Si - H_2O]^+ (X = H$ or $D)$ fragment ion peaks as 'twin signals' being 9 Th apart. In contrast to the use of pure TMS as CI reagent gas, abundant $[M + H]^+$ and $[M + H - H_2O]^+$ ions were formed along with the silylated derivatives under $DCI(TMS/N_2)$[79]. Mechanistic aspects of the formation of the adduct ions under related CI(TMS/He) conditions, at least with aliphatic alcohols, and of the origin of the protons used to generate the $[M + H]^+$ quasi-molecular ions in the CI plasma have been discussed[80,81].

Several interesting papers have dealt with the use of nitric oxide as the reagent gas in chemical ionization mass spectrometry. Phenol derivatives are prone to show a strong response to electrophilic attack by NO^+ ions, yielding abundant $[M + NO]^+$ peaks, but charge transfer with electron-rich phenol derivatives giving rise to $M^{•+}$ ions is also frequent. Thus, the CI(NO) mass spectrum of the parent compound **1** exhibits the $[M + NO]^+$ and $M^{•+}$ peaks in ratios of *ca* $1:2.5$[82]. A linear correlation was found between this ratio and the σ_P^+ parameter comprising six orders of magnitude. Moreover, CI(NO) mass spectrometry was found to be highly diagnostic with respect to the substituent pattern of arenes. For example, the three cresols give structure-specific spectra owing to the individual relative abundances of [M + NO]$^+$, M$^{•+}$ and also [M − H]$^+$ ions. Again, in accordance with the relative stabilities of the hydroxybenzyl cations **9** (cf. Scheme 3), hydride abstraction by NO$^+$ is most pronounced in the CI(NO) mass spectrum of *para*-cresol *p*-**7** and least pronounced in that of the *meta*-isomer. The spectra of aminophenols exhibit only small differences since the charge transfer process dominates strongly here; however, the [M + NO]$^+$ peak is most intense, albeit only 0.4%B, with the *ortho*-isomer, probably owing to an ortho effect (see below). The CI(NO) mass spectra of the nitrophenols are surprising because of the significant occurrence of [M − H]$^+$ ions for the *ortho*- and *para*-isomers and the strong predominance of the [M + NO]$^+$ ions giving rise to the base peaks in all three cases[82]. In view of the protonation of nitrobenzenes (see below), it appears reasonable to assume that the NO$^+$ ion is attached to the nitro group of nitrophenols, rather than to the hydroxyl group or to the ring.

A detailed study on the protonation site and the fragmentation of nitrobenzene derivatives in CI(CH$_4$) mass spectrometry included phenol and the three nitrophenols[83]. The pronounced ortho effect observed for the *ortho*-isomer **94**, that is, strongly dominating water loss from the [**94** + H]$^+$ ions (Scheme 27), and the far suppressed reactivity of other substituents, which would normally accept the proton from the highly acidic CI plasma ions, indicate that the nitrophenols, as well as other nitrobenzene derivatives, are preferably protonated at an oxygen atom of the nitro functionality, at least in the reactive form [**94** + H]$^+_{(NO_2)}$. Note that the proton affinity of nitrobenzene, PA(C$_6$H$_5$NO$_2$) = 193.4 kcal mol^{-1}, is only slightly (by 2–3 kcal mol^{-1}) lower than PA(**1**). According to the additivity rule

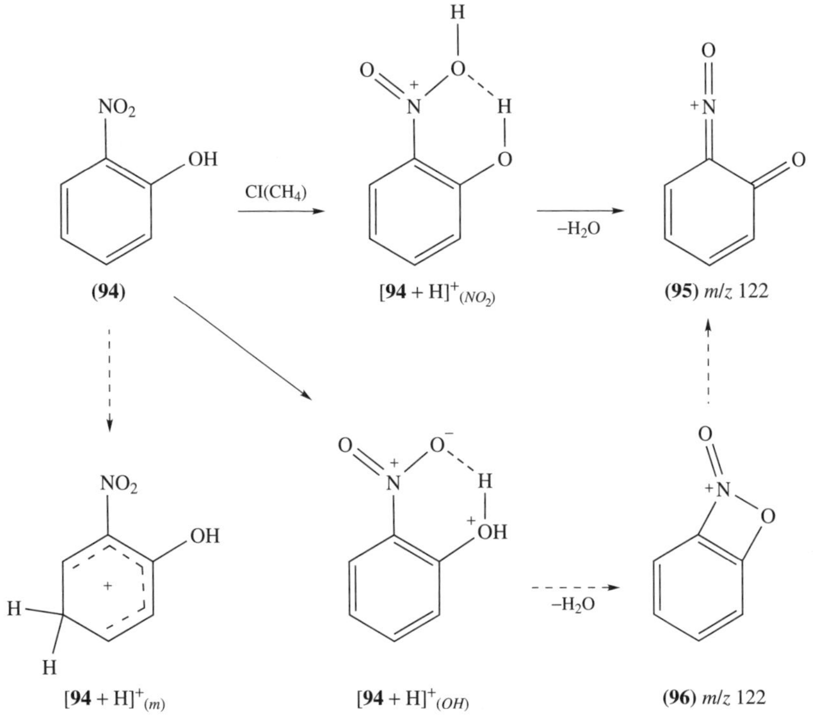

SCHEME 27

of the local PA increments[6,14-16], the C-4 position of *ortho*-nitrophenol can be estimated to have PA($94_{(C4)}$) ≈ 178.3 kcal mol^{-1} only [cf. PA(benzene) = 180.0 kcal mol^{-1}], being the most basic ring site (cf. ion [94 + H]$_m^+$). The characteristic [M + H − H$_2$O]$^+$ peak at *m/z* 122 in the CI(CH$_4$) mass spectrum of **94** is as intense as the [M + H]$^+$ peak, whereas it is negligibly small in the spectra of the other isomers and in that of phenol itself. Again, a quinoid structure (2-nitrosophenoxenium ion **95**) is ascribed to the [M + H − H$_2$O]$^+$ ions and its formation has been attributed to the proton transfer from the hydroxyl group to the protonated nitro group in [94 + H]$^+_{(NO_2)}$. However, it has been shown that *ortho*-nitroanisole also exhibits a pronounced ortho effect under the same CI conditions, and the [M + H − MeOH]$^+$ ions produced therein were shown to be structurally identical[83]. Therefore, the alternative path of water elimination, i.e. via intermediate ions [94 + H]$^+_{(OH)}$ and **96**, which has been suggested analogously for methanol loss from the methyl ether, may be followed in the case of *ortho*-nitrophenol as well.

Related ortho effects were studied by collision-induced dissociation (CID) of the [M + H]$^+$ ions and the [M + CH]$^+$ and [M + CH$_3$]$^+$ adduct ions generated by CI(Me$_2$O) or CI(oxirane) of the isomeric methoxyphenols **77**, hydroxybenzaldehydes **97** and hydroxyacetophenones **100**[84]. The spectra of the protonated *meta*- and *para*-isomers were found to be qualitatively indistinguishable. As the most remarkable result, which was corroborated by some deuterium labelling experiments, the products of methine transfer, [M + 13]$^+$ ions (see above), were found to provide different CI/CID spectra for the *ortho*-isomers. Thus, [*o*-77 + CH]$^+$ ions react by sequential loss of CH$_3^{\bullet}$ and CO, whereas the respective ions generated from *m*-77 and *p*-77 expel predominantly CO and CH$_2$O in competing pathways. In contrast to the [M + H]$^+$ ions of *o*-97, adduct ions [*o*-97 + CH]$^+$ generated from *ortho*-hydroxybenzaldehyde *o*-97 behave in a specific manner: They expel CO but do not undergo successive elimination of H$_2$ and CO, as do the respective isomers. Again in contrast, protonated *ortho*-hydroxyacetophenone [*o*-100 + H]$^+$ exhibits a specific sequence of water and CO losses. CI(Me$_2$O) mass spectrometry performed in a quadrupole ion trap mass spectrometer (ITMS) revealed competitive formation of the [M + 13]$^+$ and [M + 15]$^+$ adduct ions of the isomeric hydroxybenzaldehydes **97** and hydroxyacetophenones **100**, which pertained also to vanillin *p*-105 and *ortho*-vanillin *o*-105 (Scheme 28)[85]. Thus, while all compounds formed abundant [M + H]$^+$ ions, only *m*-97, *p*-97, *m*-100 and *p*-100 showed [M + CH$_3$]$^+$ but no [M + CH]$^+$ peaks, whereas *o*-97 and *o*-100 both exhibited the opposite behaviour. It appears reasonable to assume that the CH$_3$OCH$_2^+$ reactant ion generated in the CI(Me$_2$O) plasma transfers a CH$_3^+$ ion to the carbonyl oxygen atom generating the [M + 15]$^+$ ions *m*- and *p*-104, respectively. In contrast, the methine group transfer giving rise to ions [M + 13]$^+$, viz. *o*-99 and *o*-102, occurs by electrophilic aromatic substitution (cf. Scheme 26) if a sufficiently nucleophilic ring position is available to enable formation of the intermediate ions *o*-98 and *o*-101.

The examples discussed above refer to adduct ion formation, where covalent bonds are formed in the CI plasma. However, with the increasing importance of alternative ionization techniques, such as thermospray (TSI) and, in particular, electrospray ionization (ESI), a wealth of non-covalent ion/molecule adduct ions can be generated and studied nowadays. One recent example[86] concerns the formation of ion/solvent adducts, [M + So]$^+$, with M including 3-aminophenol, 3-(methylamino)phenol and 3-(dimethylamino)phenol, and several hydroxypyrimidines, among other aromatic molecules. The relative abundances of ions [M + H]$^+$, [M + So + H]$^+$ and [M + 2 So + H]$^+$ were studied as a function of the temperature and the pH, with the solvents being mixtures of methanol/water and acetonitrile/water which may contain ammonium acetate as an additive[86]. Quite in contrast to this empirical study on proton-bound ion/molecule complexes, the non-covalent, open-shell adduct ions [**1**$^{\bullet+}$ + NH$_3$] were investigated with respect to their intrinsic reactivity[87]. These adduct ions were generated from phenol and ammonia by laser ionization of a

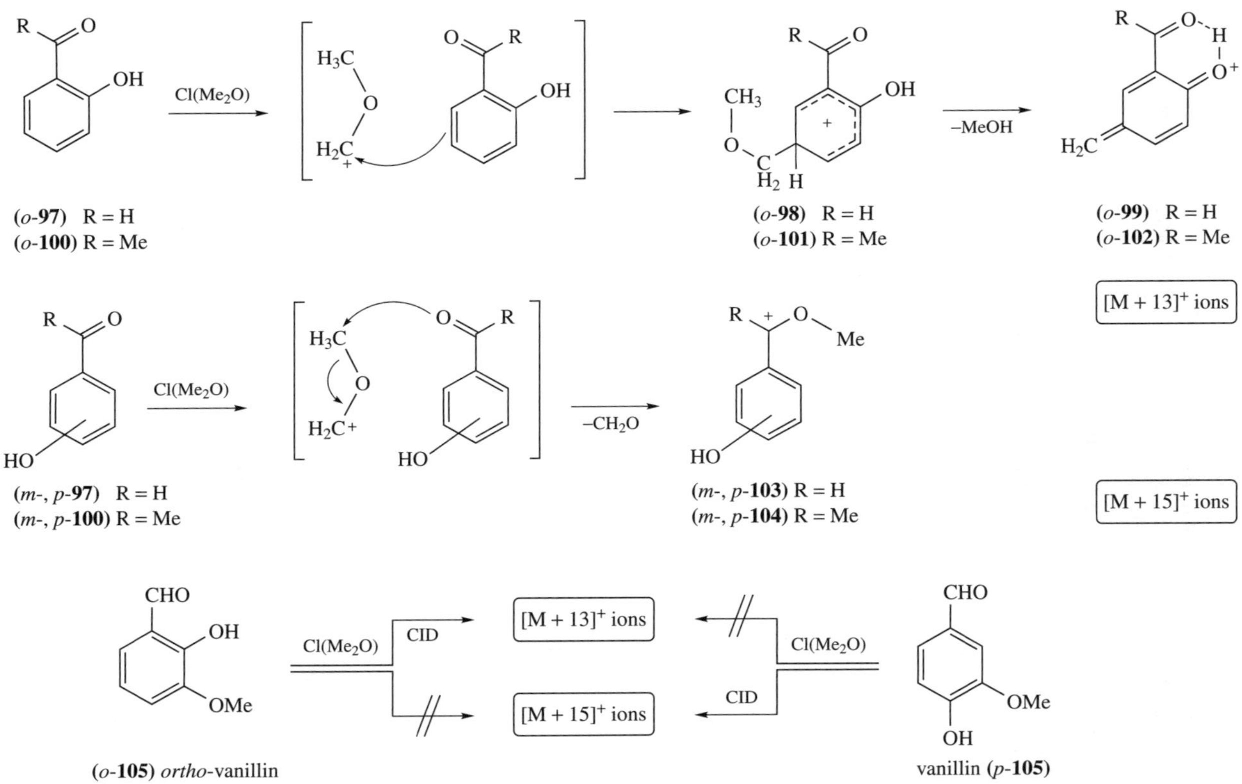

SCHEME 28

mixture of the neutral components and studied by photoelectron spectroscopy (PES) to estimate the height of the isomerization barrier to ions $[C_6H_5O^{\bullet} + NH_4^+]^{88}$.

IX. IONIZED PHENOL AND CYCLOHEXA-1,3-DIEN-5-ONE (ortho-ISOPHENOL) GENERATED BY FRAGMENTATION OF PRECURSOR IONS

A. The Radical Cations of ortho-Isophenol

The role of ionized cyclohexa-1,3-dien-5-one (*ortho*-isophenol), o-$2^{\bullet+}$, as a crucial intermediate in the expulsion of CO from ionized phenol, $1^{\bullet+}$, has been discussed above (Section III). The formation and properties of the radical cations of the '*ortho*-tautomers' of simple arenes such toluene, phenol and aniline (Scheme 29) has been investigated in much detail. Briefly, ionized *ortho*-isotoluene $107^{\bullet+}$ was found to be a stable species exhibiting fragmentation characteristics which are distinct from those of ionized toluene $106^{\bullet+\,25}$. These ions can be generated by McLafferty reaction of ionized *n*-alkylbenzenes and related α,ω-diphenylalkanes (Scheme 29, X = CH$_2$, R = alkyl, aryl), in the course of which a γ-H atom is transferred from the aliphatic chain to one of the *ortho* positions of the benzene ring, in analogy to the elimination of olefins from the corresponding ionized *n*-alkylphenols discussed above (cf. Scheme 7). When the benzylic (α-) methylene group is replaced by an oxygen atom, the molecular radical cations of the respective

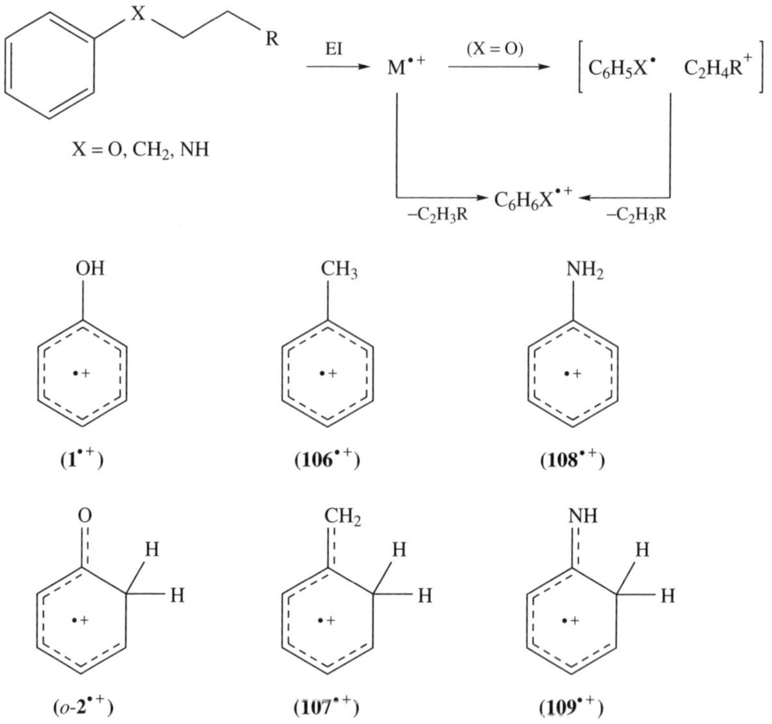

SCHEME 29

4. Mass spectrometry and gas-phase ion chemistry of phenols   299

n-alkyl phenyl ethers apparently undergo the same olefin elimination; however, all evidence has documented the formation of the phenol radical cation, $1^{\bullet+}$, rather than of the isophenol radical cation, $o\text{-}2^{\bullet+}$. Similarly, the EI-induced fragmentation of phenyl esters, such as phenyl acetate, gives rise to ions $1^{\bullet+}$ by H$^{\bullet}$ rearrangement followed by loss of the corresponding ketene[89-95]. Thus, the migrating hydrogen atom is accepted by the heteroatom rather than by one of the carbon atoms of the aromatic ring. A similar behaviour was found for the EI-induced fragmentation of aniline derivatives, such as N-alkylanilines and aliphatic anilides, which generates ionized aniline $108^{\bullet+}$, rather than ionized $ortho$-isoaniline $109^{\bullet+}$[95]. Again, the heteroatom acts as the preferential H$^{\bullet}$ acceptor site.

The characteristic features of the keto–enol tautomers $1^{\bullet+}$ and $o\text{-}2^{\bullet+}$ have been studied by a variety of mass spectrometric methods. Whereas the discovery of the formation of $C_6H_6O^{\bullet+}$ ions (m/z 94) from ionized alkyl phenyl ethers dates back to 1959[96], a wealth of papers have been published since, showing that the 'aromatic' tautomer $1^{\bullet+}$—actually being a 5π electron system only—is generated as a stable species likewise by EI of phenol 1 and of phenyl ether precursors, such as phenetole 110^{97}. These ions exhibit the same unimolecular and collision-induced fragmentation characteristics and also the same bimolecular reactivity, as studied by ion cyclotron resonance mass spectrometry (ICR-MS). The formation of the tautomeric ions $o\text{-}2^{\bullet+}$ was achieved by starting from a neutral precursor which is prone to undergo a facile (formal) retro-Diels–Alder reaction, viz. bicyclo[2.2.2]oct-2-en-5,7-dione 111. In fact, $C_6H_6O^{\bullet+}$ ions (m/z 94) generated by elimination of ketene from ions $111^{\bullet+}$ (Scheme 30) were found to behave in a manner distinct from ions $1^{\bullet+}$ in many ways. For example, the collision-induced dissociation (CID) mass spectra of ions $[111 - C_2H_2O]^{\bullet+}$ are clearly different from those of the ions $C_6H_6O^{\bullet+}$ generated from the aromatic precursors[98,99]. However, as pointed out in a critical discussion of ion/molecule reactions as a probe for ion structures[100], identical bimolecular reactivity of the presumably tautomeric forms of $C_6H_6O^{\bullet+}$ was also encountered[101]. In fact,

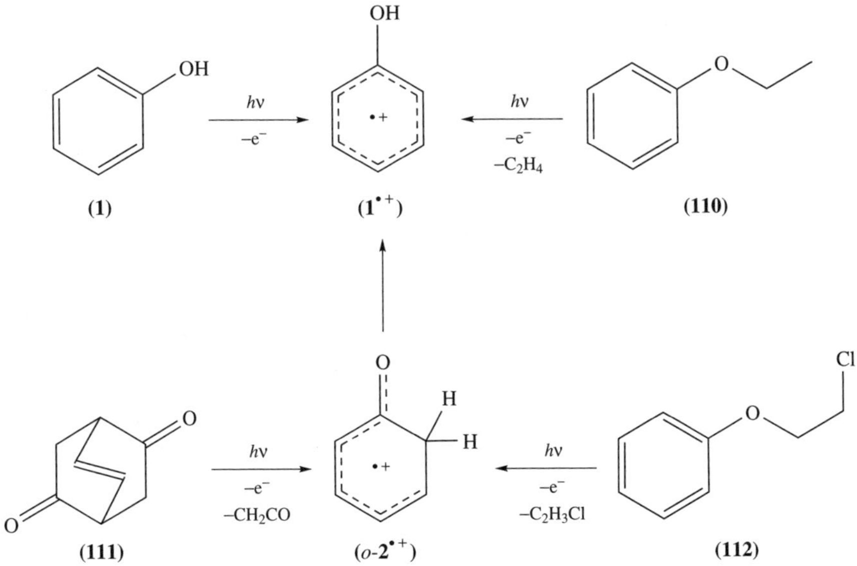

SCHEME 30

$C_6H_6O^{\bullet+}$ ions of both tautomeric forms could lose their structural identity by catalysed 1,3-H shift within an ion/molecule complex, as they can interconvert prior to CO loss[21]. Even photodissociation spectroscopy making use of the CO expulsion as the probe reaction revealed that, in contrast to **1** and phenetole **110**, from which almost pure **1**$^{\bullet+}$ ions can be produced (Scheme 30), the $C_6H_6O^{\bullet+}$ ions generated from **111** consist in fact of a mixture of the tautomers o-**2**$^{\bullet+}$ and **1**$^{\bullet+}$ [102]. According to these results, a similar mixture is formed even from β-chlorophenetole **112** as the neutral precursor.

B. Phenol Ions Generated within Transient Ion/Neutral Complexes during Mass Spectrometric Fragmentation of Alkyl Aryl Ethers

Extensive studies of the formation of ions **1**$^{\bullet+}$ and the radical cations of related hydroxyarenes have been performed with respect to the formation of ion/molecule complexes in the course of unimolecular fragmentation of organic ions in a mass spectrometer. Transient ion/molecule and ion/radical complexes ('ion/neutral' complexes) have both analytical and fundamental importance[102-108]. Owing to the fact that the ionic and the neutral fragment formed by the primary dissociation can move relatively freely with respect to each other, they behave like a (formally equivalent) aggregate generated by bimolecular encounter, and secondary processes may occur between the constituents of the complex, e.g. proton transfer, hydrogen atom abstraction and hydride transfer[109-115]. If these processes take place between groupings which, in the original molecular structure, were far apart from each other, the intermediary of reactive ion/neutral complexes during mass spectrometric fragmentation may open unexpected reaction paths and give rise to unusual (if not 'irritating') peaks[111,115-117] in the mass spectrum.

One of the most studied reactions occurring via ion/neutral complexes is the elimination of alkenes from the radical cations of alkyl phenyl ethers[118-125]. As shown in Scheme 29, a primary ion/radical complex $[C_6H_5O^{\bullet}\ C_2H_4R^+]$ is formed by cleavage of the $O-C^\alpha$ bond. It has been argued[119] that the alkyl cation bound to the phenoxy radical by mainly ion/dipole and ion/induced dipole interactions transfers a hydrogen atom, rather than a proton, in a nearly thermoneutral reaction to the oxygen atom, thus generating a second ion/neutral complex, $[C_6H_5OH\ C_2H_3R^{\bullet+}]$, from which the olefin is eventually released after charge transfer, giving rise to ions **1**$^{\bullet+}$ (m/z 94). Competitively, the latter ion/molecule complex may undergo another intra-complex reaction, this time a proton transfer, generating another ion/molecule complex, $\{[C_6H_5OH + H]^+\ C_2H_2R^{\bullet}\}$, which gives rise to protonated phenol ions $[1 + H]^+$ (m/z 95) with the concomitant elimination of an allylic neutral fragment. In general, these products of double hydrogen transfer have only very low relative abundance. However, if the relative proton and hydrogen atom affinities allow, as in the case of ionized alkyl pyridyl ethers, the double hydrogen transfer reaction may become the dominating channel. A most remarkable example in this respect is ionized 4-pyridyl cyclooctyl ether[126], which undergoes mainly or exclusively double hydrogen transfer, giving rise to protonated 4-pyridone (m/z 96) and $C_8H_{13}^{\bullet}$, presumably being the cycloocten-3-yl radical. Labelling experiments indicated symmetrization of the cyclooctyl ion associated to the pyridyl-4-oxy radical in the primary ion/neutral complex, in accordance with the non-classical structure of the cyclo-$C_8H_{15}^+$ ion. The mechanistic details of the reactivity of ionized aryl alkyl ethers and the ion/neutral complexes containing phenolic constituents have been described in great detail[118-125]. Interestingly, there is no evidence for intra-complex protonation of the relatively basic ortho and para positions of the phenoxy radical or phenol.

Related fragmentation and isomerization behaviour was unraveled for the unimolecular fragmentation of protonated alkyl phenyl ethers, $[C_6H_5OC_2H_4R + H]^+$ [127-129]. These closed-shell, even-electron analogues of ionized alkyl phenyl ethers also form ion/molecule complexes, in this case {$[1+H]^+$ C_2H_3R} as the primary and, after intra-complex proton transfer, [1 $C_2H_4R^+$] as the secondary complex. Again, detailed studies have been carried out with regard to the origin of the rearranged hydrogens and the isomerization of the alkyl cations within the latter complex. However, as the most remarkable result with respect to the gas-phase ion chemistry of phenols, proton exchange with ring hydrons and, thus, protonation of the electron-rich phenol ring, was excluded by experimental evidence[128]. Similar to the behaviour of the ion/neutral complexes generated during the fragmentation of the open-shell, odd-electron analogues, protonation or H^+ transfer within the complexes is restricted to the oxygen atom. However, although never proven in this case, thermodynamic reasons suggest that the actual fragment ions, $C_6H_7O^+$ (m/z 95), formed by alkene elimination from protonated alkyl phenyl ethers via ion/molecule complexes [1 $C_2H_4R^+$], should be a *ring*-protonated phenol, e.g. $[1+H]^+_{(p)}$.

C. Allylphenols and Allyl Phenyl Ethers

Whereas the fragmentation of ionized and protonated alkyl phenyl ethers generate phenolic ions, such as $1^{\bullet+}$ and $[1+H]^+$, together with neutral alkenes, ionized and protonated allyl phenyl ethers and related unsaturated analogues do not fragment via *reactive* ion/molecule complexes. The EI-induced fragmentation of a number of allyl phenyl ethers and the isomeric *ortho*-allylphenols and *ortho*-propenylphenols have been studied[130]. These ions undergo several competing fragmentation reactions, including the loss of the allylic side chain as C_3H_4, generating ions $C_6H_6O^{\bullet+}$ (m/z 94), ionized phenol $1^{\bullet+}$, and the loss of $C_2H_3^\bullet$, generating ions $C_7H_7O^+$ (m/z 107), probably *o*-9^+. It has been suggested from these findings that a part of the allyl aryl ether ions undergo Claisen rearrangement prior to fragmentation[130]. A later photoionization study suggested the occurrence of multistep skeletal and hydrogen rearrangement processes prior to fragmentation of ionized allyl phenyl ethers, mainly initiated by the particularly facile electrophilic attack of the ω-CH_2 group at an *ortho* position of the electron-rich aromatic ring[131]. Claisen rearrangement was also reported to precede the fragmentation of ionized allenyl phenyl ether and phenyl propargyl ether[132,133]. Again, the EI mass spectra of these compounds exhibit intense m/z 94 peaks indicating the formation of ionized phenol $1^{\bullet+}$. Similarly, $C_6H_6O^{\bullet+}$ ions give rise to the base peak in the EI mass spectrum of allenylmethyl phenyl ether [(buta-2,3-dien-1-yl) phenyl ether], which has been traced to the elimination of butatriene. Interestingly, ionized 2-(buta-1,3-dien-2-yl)phenol reacts quite differently in that the m/z 94 peak is only 20–25%[134].

Phenolic radical cations can also be generated in the absence of mobile hydrogens in the initial step of the fragmentation. Different from aliphatic anilides (see above), Claisen-type rearrangement represents also a major fragmentation route of ionized aroylanilides, $ArCONHAr'^{\bullet+}$, generating ionized phenols, $Ar'OH^{\bullet+}$ along with neutral arylisocyanides, ArNC. For example, the EI mass spectrum of N-(4-methoxyphenyl)benzamide (i.e., $Ar' = p$-anisyl) exhibits a significant peak at m/z 124, indicating the formation of ions 4-MeOC$_6$H$_4$OH$^{\bullet+}$, the structure of which has been proven by CID mass spectrometry[135]. The relative rate of this fragmentation channel is strongly affected by the electronic nature of the aryl nuclei[136].

The fragmentation of protonated allyl phenyl ethers, and their phenolic isomers, such as *ortho*-allylphenol **113**, is much simpler than that of their open-shell congeners generated by EI mass spectrometry. Under Cl(CH$_4$) conditions, both the protonated phenol, $[113+H]^+$, and the protonated parent ether, $[115+H]^+$, behave very similarly (Scheme 31)[137]. The by far major fragmentation path of these closed-shell ions is the elimination of ethene, generating ions *o*-**9** (m/z 107, see below). Ionized or protonated phenol (m/z 94 and 95) are formed in negligible amounts only. Nevertheless, Claisen rearrangement induced by

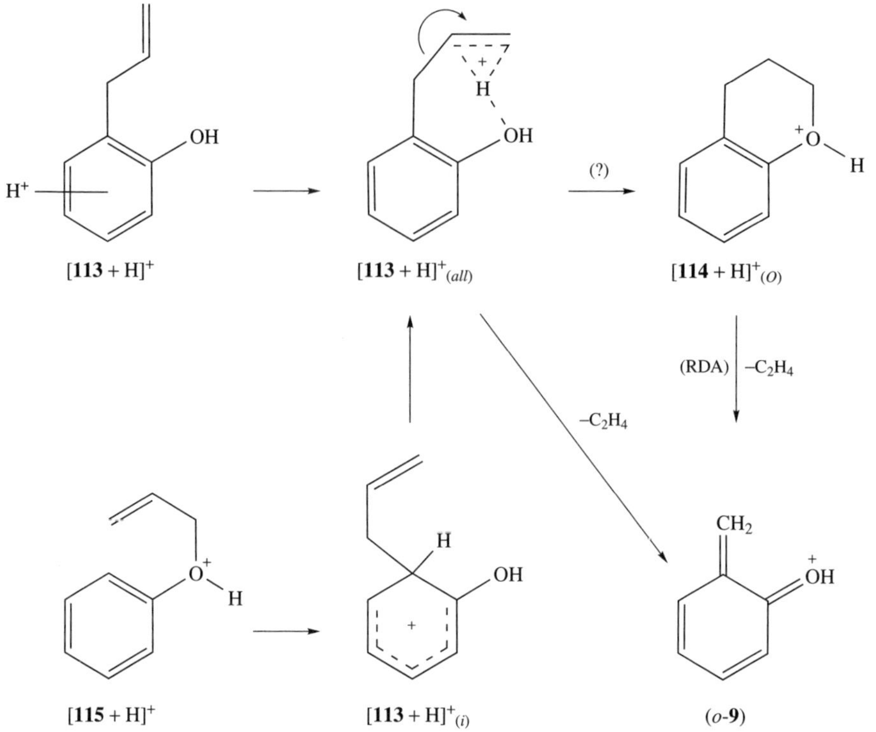

SCHEME 31

O-protonation in $[115 + H]^+$ is the dominant reaction path, giving rise to an intramolecular CH-group transfer (Scheme 31, cf. $[M + 13]^+$ ions discussed in Section VIII). Ethene loss has been suggested to occur via the olefin-protonated tautomer $[113 + H]^+_{(all)}$ [137]; however, a cycloreversion of the O-protonated dihydrobenzopyrane $[114 + H]^+_{(O)}$ could also be envisaged. The structure of the $[M + H - C_2H_4]^+$ ions from both **113** and **115** has been identified by collision-induced dissociation and by ion/molecule reactions as *ortho*-hydroxybenzyl ions, o-**9**[137,138]. This study is a lucid example for the need of the rigorous application of the instrumental tools of fundamental mass spectrometry[139]: Similar to an experience of the author of this review in a mass spectrometric study of a completely different class of oxygen-containing compounds, viz. 1,3-indanediones[140], the loss of 28 Th from protonated *ortho*-allylphenol $[113 + H]^+$ could be attributed, at first glance, to the loss of carbon monoxide, instead of ethene. In fact, ions $[113 + H]^+$ and also ions $[115 + H]^+$ were found to eliminate C_2H_4 and only minor amounts of CO. In contrast, protonated phenyl propargyl ether expels CO, rather than C_2H_4, after Claisen rearrangement[138].

D. Lignin Model Compounds: Eugenol, Dehydrodieugenol and Related Compounds

The complex polycyclic molecular frameworks of lignin comprise phenol and alkyl phenyl ether derivatives as major structural units. Chemical degradation and more or less undirected decomposition of lignin releases such relatively simple aromatic compounds, which can be identified by GC/MS or pyrolysis/gas chromatography/mass spectrometry (Py/GC/MS) and related methodologies. Recent examples concern the identification of various phenylpropanoid compounds of the guaiacyl and syringyl series in wood smoke[141] and in samples from wood casks used for wine ageing[142]. Using photoionization (PI) for improving the reproducibility, Curie-point Py/GC/PI-MS analysis of beech milled wood lignin led to the identification of more than forty phenols as pyrolysis products[143]. A method for the quantification of lignins in paper mill waste water by Curie-point Py/GC/MS was presented recently[144]. Furthermore, a Curie-point carbon-isotope-ratio (Py/GC/MS-C-IRMS) study on the turnover rate of specific organic compounds in plant soil was published recently, including lignins which were traced by the detected phenols[145].

For several decades, monomeric and dimeric building blocks of lignin have been studied with respect to their mass spectrometric fragmentation. The fragmentation of the molecular radical cations was found to be rather complicated but some of the major reaction channels reflect the fundamental gas-phase ion chemistry of phenols.

5-Propylguaiacol **116**, eugenol **117** and isoeugenol **118** are amongst the simplest pyrolysis degradation products of lignins and their fragmentation under EI-MS conditions is straightforward (Scheme 32)[146]. Owing to the presence of the *para*-hydroxy group, the saturated side chain in ions **116**$^{•+}$ cleaves preferentially by loss of the ethyl group giving rise to the base peak at m/z 137. Accordingly, the elimination of ethene by McLafferty reaction is largely suppressed (only 7%B after correction for the contribution of $^{13}C^{12}C_7H_9O_2^+$ ions to the peak at m/z 138), in spite of the presence of a *meta*-methoxy substituent (cf. Section III.C). The high hydrogen atom affinity of the guaiacol nucleus in ions **116**$^{•+}$ is reflected by the elimination of C_3H_6 giving ionized guaiacol (m/z 124, 9%B), which necessarily involves a hydrogen rearrangement to the ring position *para* to the hydroxyl group. EI mass spectra of the unsaturated analogues, **117** and **118**, are much distinct from that of **116**, but rather similar among each other. Benzylic cleavage of ionized eugenol **117**$^{•+}$ by loss of a vinyl radical is energetically much less favourable than the corresponding loss of the ethyl radical from ions **116**$^{•+}$ and cleavage of the methoxy group by loss of $CH_3^•$ can compete, as is evident from the peak at m/z 149 (35%B). Subsequent expulsion of CO from the [**117** – CH_3]$^+$ ions leads to ions $C_8H_9O^+$ (m/z 121, 15%B). Cleavage of the propenyl group in ionized isoeugenol **118**$^{•+}$ is even more difficult, and the relative abundance of ions [**118** – C_2H_3]$^+$ is further reduced. It is very likely that the allyl and propenyl side chains in ions **117**$^{•+}$ and **118**$^{•+}$ undergo not only partial interconversion but also cyclization with the adjacent phenol nucleus (see below). Cyclization of alkenylbenzene ions to indane-type isomers is a common isomerization channel[25].

The fragmentation of the two 'dimeric' derivatives, dehydrodiisoeugenol **119** and dehydrodiconiferyl alcohol **120**, under EI conditions is again governed by the characteristic reactivity of the phenolic moieties. However, the spectra are very different in that ions **119**$^{•+}$ are much more stable than ions **120**$^{•+}$ (Scheme 32). Again, hydrogen rearrangement to the pending guaiacyl group initiates the formation of an intact guaiacol molecule which, different from ions **116**$^{•+}$, is eliminated as a neutral fragment to give ions at m/z 202 (8%B). Elimination of neutral arenes is a major reaction channel of many ionized arylindanes and related aryl-substituted benzocycloalkanes[25,54]. The more highly hydroxylated congener **120**$^{•+}$ dissociates much more readily than ions **119**$^{•+}$ to generate the abundant (probably) benzylic fragment ions at m/z 137, similar to the fragmentation of ions **116**$^{•+}$ but necessarily involving an additional hydrogen rearrangement. It is reasonable to assume benzylic cleavage of the benzofuran unit of **120**$^{•+}$ as a first step of the isomerization cascade preceding the eventual fragmentation.

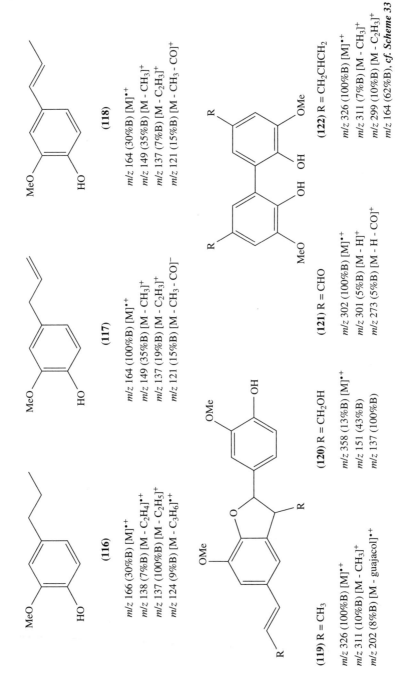

SCHEME 32

A particularly interesting case was found for the biaryl-type dehydrodieugenol **122** (Schemes 32 and 33). Whereas the EI mass spectrum of the related dehydrodivanillin **121** is dominated by the molecular ion peak, as expected, showing the typical fragmentation of ionized benzaldehydes, viz. loss of H$^{\bullet}$ and CO, the mass spectrum of **122** exhibits an intense peak at m/z 164 as a unique feature[146]. It has been suggested that this signal is due to the formation of ionized eugenol, again corresponding to the pronounced tendency of ionized phenols to initiate hydrogen rearrangement. Notably, deuterium labelling of the phenolic hydroxyl groups confirmed that the hydrogen atom being transferred between the two aromatic moieties originates from an allyl side chain. This finding again reflects the ability of ionized phenols to isomerize to distonic ions containing a protonated phenol ring.

A more detailed interpretation in view of the multifaceted reactivity of phenol radical cations is depicted in Scheme 33. It is well conceivable that ions **122**$^{\bullet+}$ fragment by formation of ionized eugenol **117**$^{\bullet+}$ (or ionized isoeugenol **118**$^{\bullet+}$), with the first step of this path being a 1,2-H shift from a benzylic methylene group to the basic *ipso* position of the same guaiacol ring. Subsequent ring walk to the biaryl junction with concomitant shift of a phenolic hydrogen of the same guaiacol ring could then effect the cleavage of the biaryl bond. In this case, a quinomethane-type neutral would be expelled as the neutral fragment during the generation of ions **117**$^{\bullet+}$.

An alternative and much more likely, albeit also quite complicated, isomerization path is also depicted in Scheme 33. While tautomerization of the eugenyl to the isoeugenyl moieties may occur independently, cyclization of one of them (or even both) should also take place. The distonic ion **123**$^{\bullet+}$ thus formed contains a protonated guaiacol ring bearing mobile protons and isomer **124**$^{\bullet+}$ should be readily accessible by proton ring walk. At this stage, at latest, the C–H bonds of the formerly remote methylene group become sufficiently acidic to transfer a proton to the other guaiacol ring, generating the next distonic ion **125**$^{\bullet+}$. Another proton ring walk opens an access to the highly fragile isomer **126**$^{\bullet+}$, from which ionized eugenol and/or isoeugenol are formed together with an energetically favourable hydroxymethoxyindene as the neutral fragment. This example of a (yet hypothetical) EI-induced fragmentation mechanism demonstrates the ability of hydroxy- (as well as methoxy-) substituted arenes to undergo complex isomerization reactions prior to fragmentation, owing to the fact that phenol (and anisole) rings are easily attacked by electrophiles, leading to unimolecular cyclization and protonation.

Mass spectrometric analysis of lignin building blocks, oligomers and polymers represents a challenge for future research efforts. It is important to note, in this context, that methods developed during the past two decades offer promising prospects for the generation and analysis of ions from highly polar and high-mass compounds. Progress has been made in many instances, even for the investigation of entire lignin polymers, by using matrix-assisted laser desorption/ionization (MALDI) mass spectrometry[147]. Electrospray mass ionization (ESI) mass spectrometry has been applied, and will be developed further, for the analysis of phenolic derivatives of lignin building blocks and adducts. This comprises both positive ESI mass spectrometry of various catechin/histidine adducts[148] and dopamine derivatives[149] and also negative ion ESI mass spectrometry of oligophenols of this sort[150]. For example, the collision-induced fragmentation of electrospray-generated oligophenolate ions (ESI-MS/MS) was shown to be highly structure-specific. Thus, the three dihydroxyphenols *o*-**39**, *m*-**39** and *p*-**39** give even qualitatively distinct CID mass spectra[150], which points to the fact that the reactivity of gaseous phenolate anions depends strongly on the electronic influence of the ring substituents. Thus, catechols from green tea were identified recently by using negative ion LC/ESI-MS/MS techniques[151].

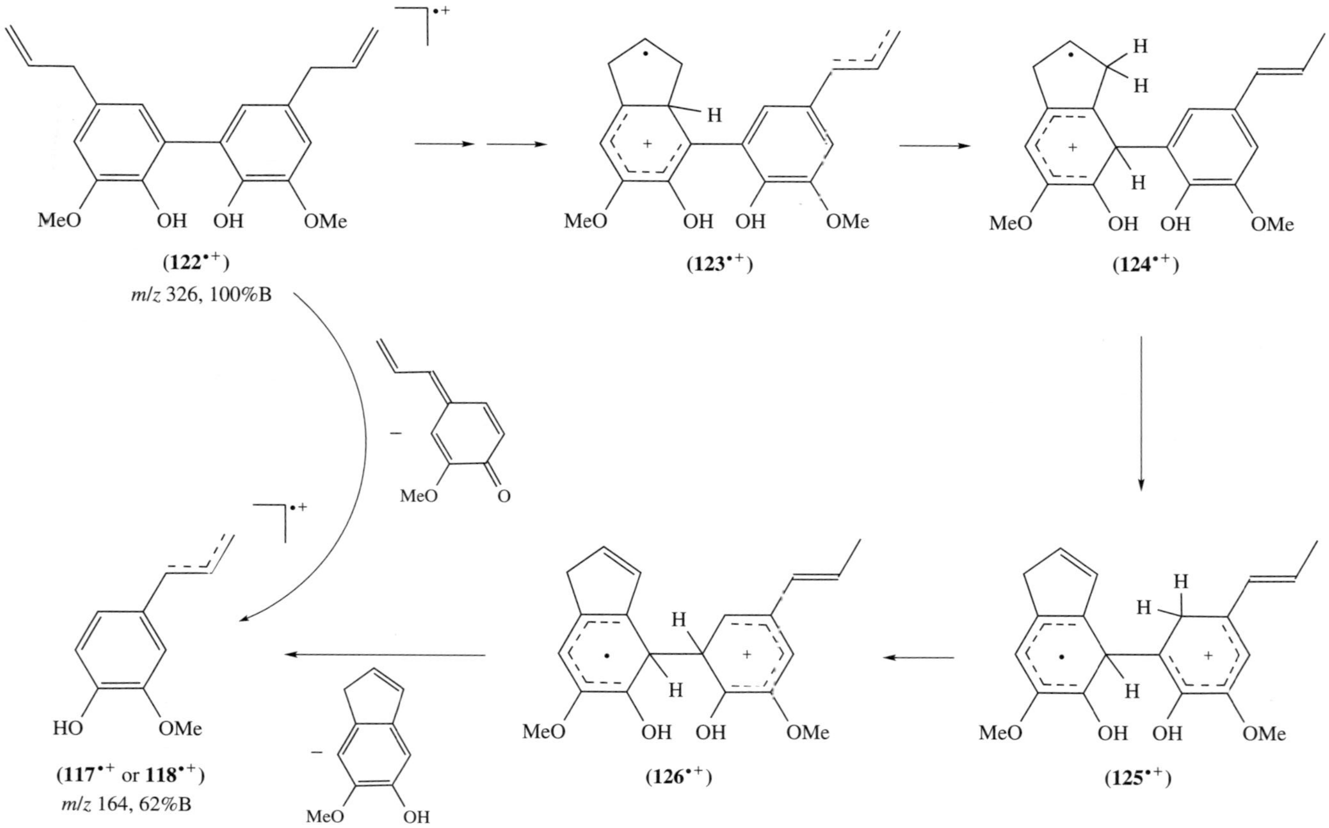

SCHEME 33

X. ION/MOLECULE REACTIONS OF PHENOLS IN THE GAS PHASE

Various bimolecular reactions of phenolic species occur in the CI plasma and in the condensed environment present in the ion sources operating under fast-atom bombardment (FAB), electrospray ionization (ESI) and matrix-assisted laser desorption/ionization (MALDI) conditions. As shown above, bimolecular ion/molecule reactions can be studied in the reactive complexes generated during unimolecular fragmentation of many even- and odd-electron phenol derivatives. In this section, a number of bimolecular reactions of phenols with cationic electrophiles in the gas phase are discussed. Similar to many processes occurring in positive ion mass spectrometry of phenols, the relevance of the attack of gaseous cations on electron-rich ambiphiles, such as phenols and anisoles, to our understanding of the fundamentals of electrophilic arene substitution is obvious. Notably, significantly less studies have been published concerning the gas-phase attack of Lewis acids on phenol than studies concerning the electronically related anisole. For example, early two-stage ion beam mass spectrometric work investigating the reactions of acetyl and nitronium cations on various arenes lacks phenol among the aromatic precursors but includes its methyl ether[152].

An early ion cyclotron resonance (ICR) study demonstrated the steric hindrance of bulky alkyl groups at the *ortho* positions of phenols[153]. While 3,5-di(*tert*-butyl)phenol underwent addition of an acetyl cation, 2,6-di(*tert*-butyl)phenol did not. In a subsequent work[154], it was demonstrated that encounter of acetyl cations with phenol in the highly diluted gas phase of an ICR mass spectrometer is not productive but that CH_3CO^+-transferring ions, such as O-acetylated acetone and O-acetylated butane-2,3-dione, give an acetylation product $[C_6H_6O + CH_3CO]^+$ in relatively high rates. It is clear that the exothermicity of the reaction requires a third body, e.g. acetone or butane-2,3-dione in this case, to take over a fraction of the energy released on acetylation. In fact, acetylation of phenol, the cresols and the xylenols under high-pressure conditions (380–760 Torr) by radiolysis of CH_3F/CO mixtures gave arenium ions which were sufficiently long-lived to undergo deprotonation, yielding the neutral acetylation products[155]. Competition between 'n-attack' and 'π-attack' at oxygen and the ring was found to be highly pressure-dependent but in all regimes to strongly favour O-acetylation. Under relatively low pressures, i.e. under increased thermodynamic control, *ortho*-acetylation gained importance over *para*-acetylation, whereas *meta*-attack proved to be of minor importance as expected for electrophilic substitution of electron-rich arenes[155].

Different from acetylation, benzoylation of phenol under gas-phase radiolysis conditions was found to occur exclusively at the functional group[156]. Owing to the experimental set-up, by which the $C_6H_5CO^+$ ions were generated from primarily formed $C_6H_5^+$ ions and excess CO, the relative rates of the competing attacks of both electrophiles could be assessed. While this competition turned out to be the same for phenol and anisole ($k_{phenyl}/k_{benzoyl} = 0.12$ and 0.13, respectively), anisole was found to undergo considerable ring benzoylation, in contrast to phenol (and aniline, too). In any case, it has become clear from these studies that the benzoyl cation is much softer an electrophile than the acetyl ion, as reflected by its pronounced regioselectivity[156].

More recent work focused on the use of the benzoyl cation as a chemoselective reagent for the detection of various aliphatic and alkylaromatic alcohols in ion-trap mass spectrometry (IT-MS)[157]. Different from purely aliphatic alcohols, benzyl alcohol and benzhydrol, phenol was found to be completely non-productive. This finding may again be viewed as reflecting the mildness of the $C_6H_5CO^+$ electrophile; however, it has to be traced to the good leaving group ability of the protonated phenoxy group in the primarily formed adduct, $C_6H_5(OH^+)(COC_6H_5)$, which suffers multiple collisional excitation with the bath gas within the reaction time (100 ms at 10^{-3} Torr He). Accordingly, even milder benzoyl cations, such as $4\text{-}CH_3C_6H_4CO^+$ and $4\text{-}t\text{-}C_4H_9C_6H_4CO^+$ ions, did not react either[158].

However, the pentafluorobenzoyl cation, $C_6F_5CO^+$, was found to convert phenol into the corresponding primary adduct, $C_6H_5(OH^+)$-(COC_6F_5), whose relatively stronger 'intra-ester' C—O bond apparently withstands the collisional excitation in the ion trap. The utility of this method in the selective detection of various (notably, mostly non-phenolic) hydroxy-functionalized compounds by GC/IT-MS has been demonstrated[158].

The electrophilic attack of $C_3H_5^+$ ions generated from EI-induced fragmentation of several C_3H_5Br precursors on phenol under low and high pressure conditions was studied with respect to the use of the neutral arene, in turn, as a probe for the structure of the reactant cation[159]. It was suggested that either pure allyl cations, pure 2-propenyl cations or mixtures of both isomers were generated. In fact, the $C_9H_{11}O^+$ (m/z 133) adduct ions formed on electrophilic attack on phenol were found to exhibit distinct CID spectra, depending in part on the pressure regime. Comparison with the CID spectra of $C_9H_{11}O^+$ model ions obtained by protonation of *ortho*-allylphenol and allyl phenyl ether indicated that allyl cations react with phenol preferentially by attack on the ring rather than on oxygen. It is obvious that allylation and propenylation of phenol produce $C_9H_{11}O^+$ ions which are structurally distinct from those generated by ion/molecule reaction of phenoxy and hydroxyphenyl cations with acetone (Section III.D). The differences in comparison with the unimolecular fragmentation of protonated allyl phenyl ether under CI-MS conditions are also remarkable.

The ion/molecule reactions between neutral phenol and various small reactant ions were also studied both in a conventional CI-MS source and in an ion-trap (IT) mass spectrometer[160]. Ethene, ethylene oxide and dimethyl ether were used and produced the products of formal methyne transfer, viz. $[M + 13]^+$ ions, along with the $[M + H]^+$ ions. Vanillin and 2- and 4-hydroxyacetophenones were found to behave similarly. Phenol formed also $[M + 27]^+$ ions with C_2H_4, but not with oxirane, as the reagent gas. CID spectra of the $[M + 13]^+$ ions were found to be indistinguishable, i.e. independent of the reagent gas. Notably, $[M + 41]^+$ ions were formed neither with phenol nor with one of its derivatives, whereas anisole did react by addition of $C_3H_5^+$. It is reasonable to assume that the $[M + C_3H_5]^+$ ions formed with phenol are more labile than those formed with anisole because the phenol adduct can easily expel C_2H_4 (cf. Scheme 31), generating $C_7H_7O^+$, i.e. $[M + 13]^+$ ions. Using dimethyl ether as the reagent gas, phenol was found to produce also $[M + 45]^+$ and $[M + 47]^+$ ions which, on the basis of their CID spectra, were interpreted as the covalent adducts of phenol and $H_2C=O^+-CH_3$ and proton-bound 'heterodimers' $[C_6H_5OH\ H^+\ OMe_2]$[160]. In a subsequent work, the competitive reactions of the reactant ions formed from dimethyl ether in an IT mass spectrometer with various phenols were evaluated with respect to steric and substituents effects[161].

As discussed in the first section of this chapter, the site of protonation of phenol is fundamental to its reactivity under many mass spectrometric conditions. Global and local alkyl cation affinities of aromatic compounds follow similar trends as do proton affinities, including the additivity rule, as shown in a recent *ab initio* computational work including phenols[162]. CID mass spectrometry was used to determine not only the site of protonation but also of methylation and ethylation under CI-MS conditions[163]. Whereas exclusive ring protonation was confirmed in agreement with the large local PA differences, alkylation of phenol was found to take place preferentially at the ring, but to occur also at the hydroxyl group. Aniline and thiophenol exhibited distinct behaviour. In a later work, charge-stripping (CS) mass spectrometry was used to deduce the sites of protonation and alkylation of phenol, aniline and thiophenol[164]. The results obtained by CID and CS were fully consistent and, in addition, the formation of doubly charged species was found to be favoured when the precursor phenolic ions were generated by *ring*- rather than by *O*-alkylation.

As may be expected, gas-phase methylation of the dihydroxybenzenes **39** in the plasma of a CI(CH_3F) or CI(CH_3Cl) source occurs also preferentially on the ring[165]. In most cases, methyl cation transfer from the reactant ions, $(CH_3)_2F^+$ and $(CH_3)_2Cl^+$, to the arenes gives ca 3:1 mixtures of the corresponding dihydroxytoluenium ions and the hydroxymethoxybenzenium ions $[77 + H]^+$, as reflected by the CID mass spectra of the methylation products and the protonated dihydroxytoluenes and hydroxyanisoles. For example, collision-induced dissociation of the protonated monomethyl ethers $[77 + H]^+$ yields abundant ions $[77 + H - CH_3]^{\bullet+}$ ions (m/z 110), whereas the protonated dihydroxytoluenes produce, in addition, ions $[M + H - H_2O]^+$ (m/z 107). In line with the synergetic effect of its mutually *meta*-oriented hydroxyl substituents, resorcinol (*m*-**39**) was found to undergo almost exclusively methylation at the ring, with the less reactive reactant ion, $(CH_3)_2Cl^+$, being most selective. Halomethylation of the dihydroxybenzenes by electrophilic attack of CH_2F^+ and CH_2Cl^+ was found to occur in competition with alkylation, again followed by elimination of the respective hydrogen halide, generating abundant $[M + 13]^+$ ions by net transfer of a CH^+ unit to the arenes[165]. Chloromethylation followed by loss of HCl was studied subsequently in detail with phenol and some cresols and dimethylphenols in an ion trap (IT) mass spectrometer[166].

Gas-phase methylation of phenol, benzene and anisole with dimethylhalonium ions $(CH_3)_2X^+$ (X = F, Cl, Br) was also performed under γ-radiolysis in the pressure range of 100–760 Torr in the presence of ammonia used as the quenching base[167,168]. With the most chemoselective electrophile, $(CH_3)_2Br^+$, phenol was found to react up to 40 times faster than benzene. The competition of *O*- and ring-methylation was found to be biased under kinetic control in favour of the former process. At low pressures and in the absence of NH_3, *ortho*-attack was found to dominate over *para*-attack. The formation of an intermediate chelate complex $[M + (CH_3)_2X]^+$ involving non-covalent bonding between a methyl group and the hydroxyl substituent, on one hand, and the second methyl group and the π-electron system of the arene ring, on the other, was suggested[168]. Predominant *O*-alkylation of phenol and anisole was also found previously when the radiolysis was performed with neopentane, giving rise to the transfer of a t-$C_4H_9^+$ ion preferentially to the hydroxyl group[169]. As expected, *tert*-butylation of the ring took place with high regioselectivity in favour of the *ortho* and *para* positions but without preference of the *ortho*-attack in the case of phenol[170]. In contrast to the *tert*-butylation of phenol, isopropylation under radiolytic conditions occurs with relatively low selectivity and in favour of the ring-substituted products in all pressure regimes (22–320 Torr). Further, *ortho*-alkylation was found to be dominant both at high and low pressures. The results were interpreted in terms of kinetically controlled *O*-attack in competition with dealkylation and skeletal isomerization of the protonated isopropyl phenyl ether to the protonated isopropylphenols[171].

In another series of investigations, the products of the ion/molecule reactions occurring in the plasma of a GC ion-trap mass spectrometer (GC/IT-MS) were studied with the particular aim to distinguish the reactivity of phenol, benzyl alcohol and the phenylethanols. In particular, ethylation and allylation of these substrates were studied under $CI(CH_4)$ conditions in the ion trap and, not unexpectedly, phenol turned out to be distinct from the arylaliphatic alcohols in that it gave abundant $[M + H]^+$ but no $[M + H - H_2O]^+$ ions[172]. A previous work dealt with the ion/molecule reactions of CF_3^+ ions with the same arenes, but using an ion-beam apparatus instead of an ion-trap mass spectrometer[173]. Under these conditions, phenol was found to undergo dissociative addition reactions involving attack at both the hydroxyl functionality and the aromatic ring. Thus, *O*-attack led to $C_6H_5O^+(H)CF_3$ ions, which then eliminate mainly CF_3OH and some CF_2O, giving phenyl cations and protonated fluorobenzene. Loss of HF is another major fragmentation channel of the adduct ions and, whereas *O*- and *ring*-attack were calculated to be similarly exothermic, this fragmentation appeared to be much more thermochemically favourable from the intermediates formed by electrophilic attack on the ring. Deuterium labelling experiments, which would help to determine the origin of the hydrogen lost in the HF fragment, and thus confirm the course of the CF_3^+ attack, have not been performed. Charge transfer from phenol to the electrophile was another channel observed[173]. Anisole was found to undergo similar reactions with CF_3^+ ions as does phenol, with electrophilic attack on the ring being the dominating reaction channel[174].

XI. GASEOUS PHENOL ANIONS

Different from many other classes of organic compounds, phenols can be particularly easily converted to gaseous anions under various mass spectrometric conditions. In addition to the classical technique for generating phenolate ions, i.e. chemical ionization using NH_3, CH_4/O_2 mixtures, CF_4, NF_3 and other reagent gases, deprotonation of phenols occurs in fast atom bombardment (FAB), or liquid secondary ion mass spectrometry (L-SIMS), matrix-assisted laser desorption/ionization (MALDI) and electrospray ionization (ESI) mass spectrometry. Therefore, the number of studies, in both fundamental and applied mass spectrometry, has increased with the advent of new and alternative ionization methodologies. Moreover, electron-capture (EC) mass spectrometry, generating radical anions in appropriate cases, represents a classical but still important technique for the mass spectrometric identification of phenols. Several reviews on negative ion chemical ionization (NICI or NCI) mass spectrometry and the gas-phase chemistry of anions have appeared[175-180]. In this section, some fundamental aspects of the gas-phase chemistry of phenolate ions will be presented together with selected examples for the application of negative ion mass spectrometry to analytical problems. Some additional examples will be mentioned in the last section of this chapter.

A. Gas-phase Acidities of Phenols

Similar to the intrinsic, gas-phase thermodynamic properties of phenolic cations discussed in the first section, the gas-phase properties of phenolic anions, in particular the heats of formation of phenolate ions, have been compiled and can be easily accessed nowadays[8] and new data are being determined frequently by using various mass spectrometric techniques. Although not driven as far as for the gas-phase chemistry of phenolic cations and radical cations, the intrinsic reactivity of phenolic anions and radical anions has also been traced to the 'local parameters', such as to the acidity of the ArO−H functionality or to the charge localization of proton acceptor sites. One such example concerns the loss of OH• from the radical anions of *ortho*-nitrophenol and related nitrobenzenes[181] — a case of ortho effects in anionic species derived from simple phenols. Also, the formation and reactivity of intermediate ion/neutral complexes generated during the fragmentation of the $[M-H]^-$ ions of fatty acid esters of hydroxybenzyl alcohols and even estradiols is governed by such local thermodynamic properties (see below).

The *absolute* gas-phase acidity of molecular species is defined as $\Delta H_{acid}^0(M) = \Delta H_f([M-H]^-) + \Delta H_f(H^+) - \Delta H_f(M)$[182]. In the case of the phenols, it can be calculated from the (homolytic) bond dissociation energies of the phenolic O−H bond, $BDE(ArO-H)$, the ionization energy of the hydrogen atom, $IE(H^\bullet)$, and the electron affinity of the phenoxy radical, $EA(ArO^\bullet)$ (Scheme 34)[182,183]. The *relative* gas-phase acidities $\Delta G_{acid}^0(M) = \Delta H_{acid}^0(M) - T\Delta S_{acid}^0(M) = -RT \ln K$ are accessible from equilibrium and kinetic measurements of ion/molecule reactions in the gas phase and fall short of the $\Delta H_{acid}^0(M)$ values by ca 7.0 kcal mol^{-1} (29 kJ mol^{-1}) in the case of simple phenols. The gas-phase acidity scale of phenols (as any class of molecular compounds) spreads over a much wider range than the acidities measured in solution[184,185]. From a recent empirical-theoretical treatment of the origins of the acidities of various compounds containing OH groups, it follows that the high intrinsic acidity

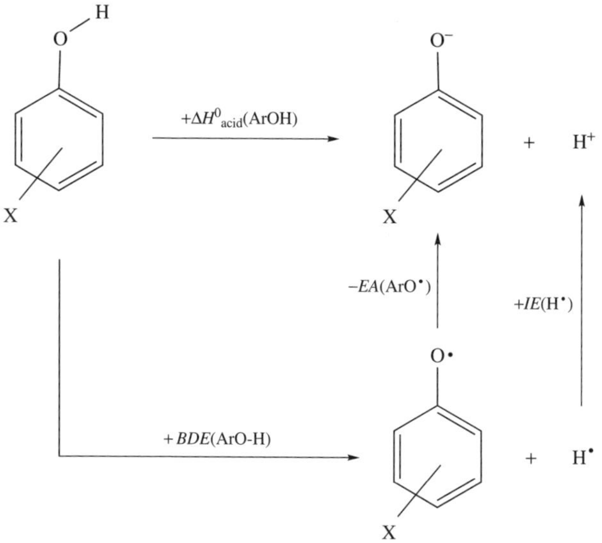

$\Delta H^0_{acid}(ArOH) = BDE(ArO\text{-}H) - EA(ArO^{\bullet}) + IE(H^{\bullet})$

SCHEME 34

of phenol, as compared to that of cyclohexanol $[\Delta G^0_{acid}(\mathbf{1}) - \Delta G^0_{acid}(c\text{-}C_6H_{11}OH) \approx -24 \text{ kcal mol}^{-1}]$, is mainly due to the π-electron delocalization in the phenolate anion $(ca - 15 \text{ kcal mol}^{-1})$ and also to field/inductive effects $(ca - 7 \text{ kcal mol}^{-1})$, but not to enhanced polarizability[186]. A theoretical study using semiempirical methods to determine the acidities of various monosubstituted phenols, $\Delta H^0_{acid}(M)$—taken there as proton affinities of the corresponding phenolate ions, $PA([M - H]^-)$—demonstrated good agreement with experimental data, particularly when the AM1 method was used[187]. An in-depth *ab initio* investigation on the structure and aromaticity of the parent phenolate ion $[\mathbf{1} - H]^-$ was shown to reproduce the experimental gas-phase data very well and also suggested a considerable degree of quinoid character of the highly delocalized π-electron system. In addition, the effect of the counterions on charge localization has been discussed[188].

Most of the dissociation energies of the phenolic O−H bond are in the range of $90 \pm 5 \text{ kcal mol}^{-1}$ [189,190]. Electron-withdrawing groups increase the $DBE(ArO\text{-}H)$ values and moderately electron-releasing groups decrease them within this range. Only strongly electron-releasing substituents, such as amino groups, weaken the ArO−H bond, e.g. $BDE(p\text{-}H_2NC_6H_4O\text{-}H) \approx 76 \text{ kcal mol}^{-1}$ [189]. The electron affinity of the phenoxy radical has the by far greatest effect on the gas-phase acidity of the phenols. The electron affinities of the parent radical and its simple alkyl derivatives are in a narrow range, e.g. $EA(C_6H_5O^{\bullet}) = 2.21 \text{ eV} \triangleq 51.0 \text{ kcal mol}^{-1}$ and $EA(p\text{-}CH_3C_6H_4O^{\bullet}) = 2.16 \text{ eV} \triangleq 49.8 \text{ kcal mol}^{-1}$, but nitro-substituted congeners have strongly increased electron affinities, e.g. $EA(m\text{-}O_2NC_6H_4O^{\bullet}) = 2.85 \text{ eV} \triangleq 65.7 \text{ kcal mol}^{-1}$. In contrast, amino groups are electronically indifferent, e.g. $EA(m\text{-}H_2NC_6H_4O^{\bullet}) = 2.15 \text{ eV} \triangleq 49.6 \text{ kcal mol}^{-1}$ [17].

The gas-phase acidity of phenol is $\Delta H^0_{acid}(\mathbf{1}) = 349.2 \text{ kcal mol}^{-1}$[7], *ca* 42 kcal mol^{-1} 'higher', i.e. stronger, than that of the aliphatic alcohols and of water $[\Delta H^0_{acid}(H_2O) =$

390.8 kcal mol^{-1}] and very close to those of acetic acid [ΔH^0_{acid}(CH$_3$COOH) = 348.7 kcal mol^{-1}] and α,α,α-trifluoroacetone [ΔH^0_{acid}(CF$_3$COCH$_3$) = 350.4 kcal mol^{-1}]^{7,191}. The gas-phase acidities of many simple phenol derivatives reflect the role of the *BDE* and, in particular, the *EA* values and were found to be quite different. For example, the three cresols all have the same gas-phase acidities as phenol within experimental error [ΔH^0_{acid}(**7**) = 349.4 – 350.4($\pm$3)kcal mol^{-1}]. Higher alkyl substituents decrease the gas-phase basicity only marginally, as does an amino group [e.g. ΔH^0_{acid}(*m*-H$_2$NC$_6$H$_4$OH) = 350.6 kcal mol^{-1}]. However, strongly electron-withdrawing substituents significantly increase the gas-phase acidities of phenols, thus lowering the ΔH^0_{acid}(M) values. For example, *para*-trifluoromethyl-, *para*-cyano- and *para*-nitrophenol (**125**) have ΔH^0_{acid}(*p*-F$_3$CC$_6$H$_4$OH) = 337.0 kcal mol^{-1}, ΔH^0_{acid}(*p*-NCC$_6$H$_4$OH) = 332.2 kcal mol^{-1} and ΔH^0_{acid}(*p*-O$_2$NC$_6$H$_4$OH) = 327.9 kcal mol^{-17}. An early ICR mass spectrometric study had revealed a good linear free-energy relationship between the gas-phase and aqueous-phase acidities of substituted phenols and demonstrated that the intrinsic effect of the substituents in the gaseous phenolate ions is greatly attenuated in the solvent medium^{192}.

The gas-phase acidities of extremely strong neutral Brønsted acids have been determined recently by equilibrium measurements in an FT-ICR mass spectrometer, including several phenols^{193}. On this extended scale, which fits very well to that comprising the numerous previous data^7, *para*-nitrophenol **128** represents only a moderately strong Brønsted acid (Scheme 35). 3,5-Bis(trifluoromethyl)phenol **127** is similarly acidic and, notably, has a very high electron affinity, *EA*(**127**) = 3.05 eV $\hat{=}$ 70.3 kcal mol^{-17}. 2-Chloro-4-nitrophenol **129** is more acidic than **128** by almost 5 kcal mol^{-1}, but a single trifluorosulfonyl substituent *para* to the hydroxyl group in **130** exerts at least the same strong acidification. Beyond 4-trifluorosulfonylphenol **130**, the benzologue of triflic acid, three considerably more acidic phenols, **131–133**, have been identified, including picric acid **132**. It is remarkable that 2,4-dinitrophenol **130** is far more acidic than its singly substituted congener **128** ($\Delta\Delta G^0_{acid}$ = 12.3 kcal mol^{-1}) and that another *ortho*-nitro substituent in picric acid **132** pushes the acidity further by only half of this difference ($\Delta\Delta G^0_{acid}$ = 5.8 kcal mol^{-1}). The bond dissociation energy and electron affinity of the latter compound were estimated to be *BDE*(**132**) = 88.3 kcal mol^{-1} and *EA*(**132**) < 88.3 kcal mol^{-1} (3.8 eV)^{194}. The record gas-phase acidity is held by 2,4,6-tris(trifluoromethyl)phenol, for which ΔG^0_{acid}(**133**) = 291.8 kcal mol^{-1} has been determined by experiment. Thus, the absolute gas-phase acidity should be ΔH^0_{acid}(**133**) = 298.8 kcal mol^{-1} and, assuming a similar *BDE* value as in **128**, the electron affinity can be estimated to exceed that of **128** by 31 kcal mol^{-1} and thus to be *EA*(**133**) $\approx$ 4.4 eV $\hat{=}$ 101 kcal mol^{-1}!

B. Phenolic Anion/Molecule Adducts [ArO-H X$^-$] and [ArO$^-$ H-X]

A topic related to that of the gas-phase acidities of phenols is the quest for quantitative data on the thermodynamic stability of hydrogen-bonded complexes, or 'clusters', [ArO-H X$^-$] and [ArO$^-$ H-X] between phenols and various anions derived from other Brønsted acids. The thermodynamics of cluster formation of the phenolate ion [**1** – H]$^-$ with water, ethanol and acetic acid have been determined by using a pulsed electron-beam mass spectrometer and their stability ΔH^0_D was found to increase with the gas-phase acidity of the Brønsted acid. For example, association of [**1** – H]$^-$ with H$_2$O is much weaker, ΔH^0 = $-$15.4 kcal mol^{-1}, than that of [**1** – H]$^-$ with CH$_3$COOH, ΔH^0_D = $-$27.4 kcal mol$^{-1\,195}$. The association enthalpy of the complex of phenol and fluoride ion, [**1** F$^-$], has been measured to be ΔH^0 = $-$41.3 kcal mol^{-1}, much stronger than

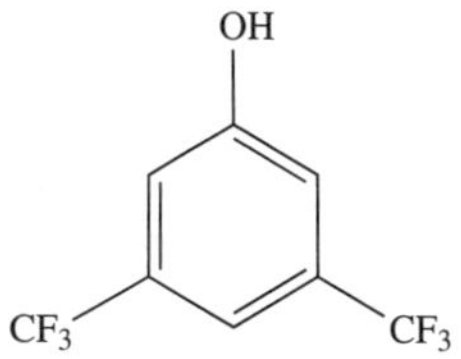

(127)
$\Delta G^0_{acid} = 322.9$ kcal mol^{-1}
$\Delta H^0_{acid} = 329.8$ kcal mol^{-1}

(128)
$\Delta G^0_{acid} = 320.9$ kcal mol^{-1}
$\Delta H^0_{acid} = 327.9$ kcal mol^{-1}

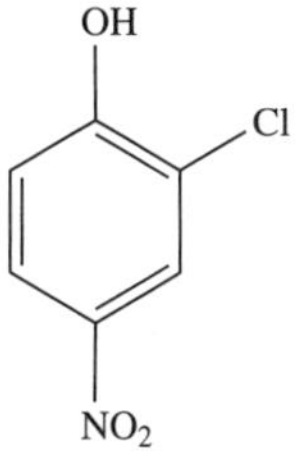

(129)
$\Delta G^0_{acid} = 316.1$ kcal mol^{-1}
$\Delta H^0_{acid} = 323.3$ kcal mol^{-1}

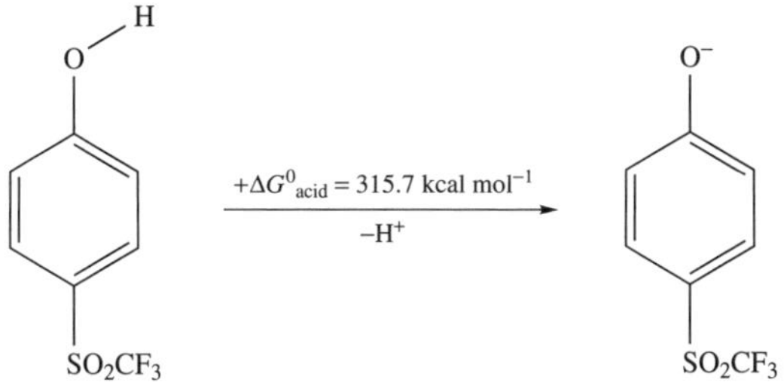

(130) $\xrightarrow[-H^+]{+\Delta G^0_{acid} = 315.7 \text{ kcal mol}^{-1}}$ [130 - H]$^-$

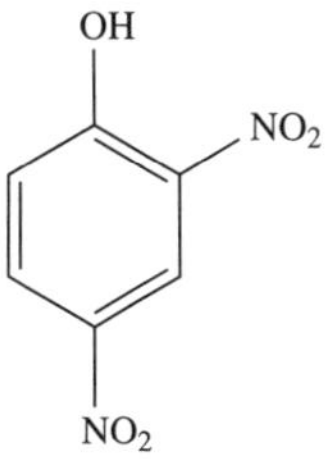

(131)
$\Delta G^0_{acid} = 308.6$ kcal mol^{-1}

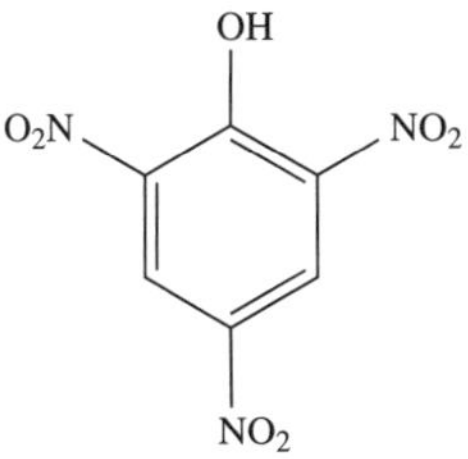

(132)
$\Delta G^0_{acid} = 302.8$ kcal mol^{-1}

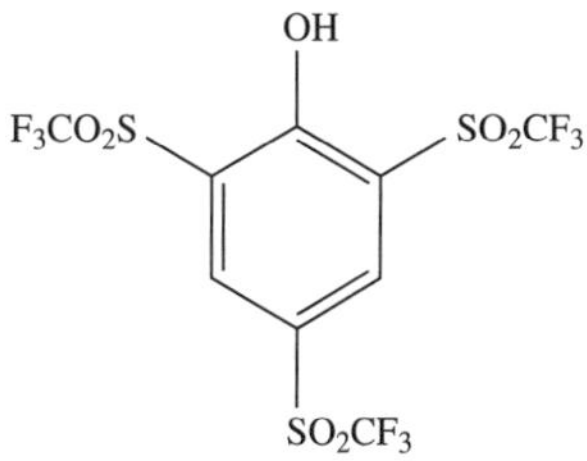

(133)
$\Delta G^0_{acid} = 291.8$ kcal mol^{-1}

SCHEME 35

that of the complex [**1** Cl$^-$], which is only $\Delta H^0 = -26$ kcal mol^{-1} [196,197]. Several *para*-substituted phenols were included in this study. The stabilities of the gaseous complexes formed from various substituted benzenes and Br$^-$ ions in pulsed-electron high-pressure equilibrium measurements were determined and, different from the other singly substituted benzene derivatives, phenol and also aniline were found to form much more stable complexes than expected from the correlation of the $\Delta\Delta G^0$ values and the dipole moments[198]. This indicates that the bonding between phenols and halide anions is governed by the hydrogen bond, in contrast to arenes which do not bear a highly acidic OH functionality. This is confirmed by an extended work focusing on the effects of the arene substituents on the stability of the complexes of, in total, twenty-six phenols with F$^-$, Br$^-$ and I$^-$ ions[199,200]. Within this large group of phenols, the $\Delta\Delta G^0$ values measured furnished a very good correlation with the Taft σ_R and σ_F parameters. Again, the more acidic the phenol, the stronger the stabilization of the complexes [ArO-H X$^-$]. For example, the adduct generated from *para*-nitrophenol and Br$^-$, [**128** X$^-$], is by $\Delta\Delta G^0 \approx +7.7$ kcal mol^{-1} more stable than that of the phenol, [**1** X$^-$], similar to that of *para*-cyanophenol and Br$^-$ ions ($\Delta\Delta G^0 \approx +7.1$ kcal mol^{-1}). By contrast, the complex of *para*-aminophenol and Br$^-$ is only slightly less stable than the parent complex ($\Delta\Delta G^0 \approx -1.3$ kcal mol^{-1}). Similar correlations were unraveled for the series of *meta*-substituted phenols, exhibiting a slightly compressed scale, and for the other halide ions[199]. The double minimum potential and the kinetics of crossing the intrinsic barrier towards proton transfer have been discussed in great detail, including the exothermic protonation of alkoxide ions by phenol, which occurs with high efficiency[201,202].

C. Negative Chemical Ionization Mass Spectrometry of Phenols

The facile attachment of halide ions to polar organic compounds, and in particular of compounds containing hydrogen donor functionalities, can be utilized to generate quasi-molecular [M + X]$^-$ ions under negative ion chemical ionization (NCI) conditions. Similar to the use of halogen-containing reagent gases in positive ion CI mass spectrometry (Section VIII), gases such as dichloromethane can serve as a source of chlorine-containing reactant ions in the CI plasma. Electron bombardment of CH$_2$Cl$_2$ under relatively high pressure (*ca* 1 Torr) generates Cl$^-$ ions, which are attached to the reagent molecules to give CH$_2$Cl$_3^-$ ions, which in turn may dissociate to HCl$_2^-$ ions and monochlorocarbene[203]. Owing to the relatively strong bonding interaction between the constituents, the NCI mass spectra of phenol, hydroquinone and other polar analytes exhibit the signals for the adduct ions [ArO-H Cl$^-$] as the base peaks, along with peaks due to the anion-bound dimers, [ArO-H Cl$^-$ H-OAr], in varying intensities[203]. Similarly, the NCI(CBr$_2$Cl$_2$) mass spectra of phenol, *para*-nitrophenol and *meta*-chlorophenol were found to exhibit the base peaks due to the [ArO-H Br$^-$] ions, along with intense peaks for the dimeric adducts[204]. The NCI mass spectra of phenol and several phenol derivatives generated with 1 : 4 mixtures of iodomethane/methane as the reagent gas were governed by the peak for the I$^-$ reactant ions. However, they also exhibited intense signals for the simple [ArO-H I$^-$] adduct ions, whereas only weak signals were found for the dimeric aggregates[204].

The attachment of halide ions to phenolic compounds affords abundant adduct ions which are valuable for selective detection and molecular mass determination of analyte compounds. For example, Cl$^-$ attachment has proven suitable for the GC/MS recognition of various phenols and naphthols in the acidic fractions of coal-derived liquids[205,206]. However, structure-specific fragmentation by mass spectrometry is hardly accessible with these quasi-molecular anions. To obtain analytically useful fragmentation in NCI mass spectrometry of phenolic compounds, in particular, deprotonation has to be performed in the NCI plasma, e.g. by NH$_2^-$ ions generated in the CI(NH$_3$) plasma[207]. The [ArOH − H]$^-$ ions formed in this manner can then be subjected to collision-induced dissociation (CID), giving rise to characteristic negatively charged fragment ions, or to charge-stripping (CS), yielding positively charged fragment ions.

In a fundamental study, the CID mass spectra of the parent phenolate anion was investigated by using exhaustive deuterium labelling[208]. Mechanisms for the formation of the C$_6$H$_3^-$, C$_5$H$_5^-$ and C$_2$HO$^-$ fragment ions were suggested. Formation of the C$_6$H$_3^-$ ions (*m/z* 75) is induced by isomerization of the conventional [**1** − H]$^-$ ion to the *ortho*-hydroxyphenyl anion, from which water is eliminated via an ion/molecule complex [*c*-C$_6$H$_4$ OH$^-$] containing 1,2-benzyne. Loss of CO from ions [**1** − H]$^-$ is the most important fragmentation, generating C$_5$H$_5^-$ ions (*m/z* 65). A previous ^{13}C-labelling study[209] on β-phenoxyethoxide ions, which yield [**1** − H]$^-$ ions by elimination of oxirane, had shown that the *ipso* carbon atom is lost with the neutral fragment. Expulsion of CO was proposed to occur via the bicyclo[3.1.0]hex-2-en-6-one-4-yl anion. The C_2HO^- (m/z 41) fragment ion from [1 − H]$^-$ was suggested to be generated via a Dewar benzene-type isomer of the phenolate anion, from which two molecules of acetylene are expelled sequentially[208].

In a related study, the CID mass spectra of several isomeric $C_7H_7O^-$ ions, including the [M − H]$^-$ ion of benzyl alcohol, deprotonated bicyclo[2.2.1]hept-2-en-5-one and the three cresolate ions, were studied[210]. Whereas the former anion was found to expel CH_2O through the major fragmentation channel, and the bicyclic isomer suffers a retro-Diels–Alder reaction to yield abundant C_2HO^-, viz. ethynolate ions (m/z 41, see above), the isomeric cresolate ions underwent a manifold of less characteristic collision-induced fragmentations, e.g. loss of $CH_3^•$, CH_4, CO, CHO$^•$ and CH_2O. Notably, however, the three CID spectra of the cresolate ions were found to be distinct from each other in the relative weights of the individual fragmentation processes, and a similar behaviour was observed for the CS mass spectra of these $C_7H_7O^-$ isomers[210].

Besides deprotonation and anion addition in the CI plasma, which generate stable [M − H]$^-$ and [M + X]$^-$ ions, respectively, the use of electron capture into low-lying π^* orbitals represents a principal approach to generate radical anions. However, if these M$^{•-}$ ions are relatively small, they are too short-lived and decay within $\tau \leqslant 10^{-13}$ s by ejection of an electron, thus suppressing any structure-specific fragmentation reaction. Larger radical anions, however, in which the excitation energy can be distributed over many internal degrees of freedom, may be sufficiently long-lived ($\tau \geqslant 10^{-6}$ s) to enable their mass spectrometric detection[176]. Electron transmission spectroscopy was used to determine the electron affinities of benzene and several of its derivatives containing electron-withdrawing substituents[211]. The electron affinity of phenol is strongly negative, $EA(1) = -1.01$ eV, only slightly less negative than that of benzene, $EA(C_6H_6) = -1.15$ eV, whereas chlorobenzene and bromobenzene are better electron acceptors [$EA(C_6H_5Cl) = -0.75$ eV and $EA(C_6H_5Br) = -0.70$ eV]. In fact, chlorophenols can be detected by electron-capture negative-ion (EC-NI) mass spectrometry[212], but favourably in the presence of a moderating gas[213] and/or after suitable esterification or etherification with an acid or alcohol, respectively, the electron affinity of which is positive, such as pentafluorobenzoic acid or pentafluorobenzyl alcohol[214].

When argon or other inert gases are used to decelerate the electrons, the efficiency of the overall electron capture process is much enhanced[176,178,213]. Under these conditions, chlorine-substituted phenols were found to undergo significant condensation reactions involving the molecular radical anions M$^{•-}$, or the corresponding [M − H]$^-$ ions, and the neutral precursor molecules. This reaction furnishes condensation products reminiscent of those formed from polychlorodibenzodioxins (PCDDs) under energetic neutral reaction conditions[215,216].

A recent study on the condensation of the monochloro- and dichlorophenols under NCI condition using argon as the moderating gas ('argon-enhanced NI-MS') revealed characteristically different reactivities of the isomers[217]. Whereas all of the monochlorophenols formed abundant [M + Cl]$^-$ ions, giving rise to the base peaks, and minor signals for the dimeric [2 M − H]$^-$ and [2 M + Cl]$^-$ adducts, only the *ortho*-isomer generated significant signals at m/z 220 and 222 for the condensation products, [2 M − H − Cl]$^{•-}$, obviously being the M$^{•-}$ ion of 2-chlorophenyl 2′-hydroxyphenyl ether. Evidently, an intermolecular nucleophilic attack of a phenolate ion occurs, followed by loss of Cl$^•$ and assisted by the adjacent *ortho*-hydroxy group. Similarly, the dichlorophenols gave abundant peaks for the products of intermolecular condensation, probably again [2 M − H − Cl]$^{•-}$ ions—rather than the products of HCl loss from [2 M]$^{•-}$ ions, as postulated[217]. In addition, the EC-NCI mass spectra of the 2,3- and the 2,5-isomer specifically gave significant peak clusters at m/z 252, 254 and 256, indicating a subsequent intramolecular cyclocondensation in the radical anion [2 M − H − Cl]$^{•-}$. From these ions, the radical anions [2 M − H − Cl − HCl]·⁻ are generated, to which the structure of the corresponding dichlorodibenzodioxins has been asigned. The occurrence of the Smiles rearrangement has been invoked to account for a putative formation of positional isomers of the initially formed dichlorodibenzodioxins[180,209,217].

D. Ion/Molecule Complexes Formed during the Unimolecular Fragmentation of Phenolic Anions

Returning to the gas-phase chemistry of phenolic steroids allows us to consider further examples for the formation of reactive ion/neutral complexes during the fragmentation of gaseous organic ions. Similar to the behaviour of positively charged ions of bifunctional steroids (and many other classes of organic compounds), negatively charged ions bearing a rigid steroid skeleton and two functional groups (or more) can be converted into ion/neutral complexes, which subsequently undergo intra-complex proton transfer or other processes that could never occur in the intact framework of the original molecular structure.

When 17β-estradiol C(17) fatty acid esters such as **134** are deprotonated under NCI(NH₃) conditions, two tautomeric [M − H]⁻ ions are generated, one being the 3-phenolate form [**134** − H]⁻_{(OH)} and the other the ester enolate [**134** − H]⁻_{(CH)} (Scheme 36)[218,219]. Deuterium labelling of the starting steroid allows one to distinguish the two forms and determine their fragmentation by CID mass spectrometry. The enolate ion [**134** − H]⁻_{(CH)} fragments by heterolysis of the ester bond to generate the complex **135**, which initially contains the C(17)-alcoholate and a neutral ketene. However, owing to its high basicity, the alcoholate group abstracts the acyl proton from the ketene to give complex **136**, and the corresponding propynolate ion may be released from the latter. Still, in a further intra-complex step, the acidic 3-OH group transfers its proton to the propynolate generating a third complex, **137**, and the phenolate fragment ion is liberated from that complex. The relative yields of the two tautomeric forms of the [M − H]⁻ ions were found to depend strongly on the ionization conditions. Under FAB conditions, deprotonation appears to occur mostly at the phenolic OH group, whereas NCI(NH₃) conditions give rise to deprotonation in the ester group.

Further investigation of the complex fragmentation behaviour of the steroid [M − H]⁻ ions led to the study of simpler model systems, viz. fatty acid esters containing a *para*-hydroxybenzyl (cf. **138**, Scheme 37) or β-(*para*-hydroxy)phenethyl moiety[220]. Again, the formation of intermediate anion/molecule complexes was demonstrated by the course of collision-induced dissociation of various deuterium-labelled [M − H]⁻_{(CH)} ions, where deprotonation had occurred at the α-position of the acyl methylene group, cf. [**138** − H]⁻_{(CH)}. The tautomeric ions [**138** − H]⁻_{(OH)} generated by deprotonation of the phenolic OH group may be assumed to form the anion/molecule complex **139** which, however, is non-reactive with respect to further tautomerization. Rather, this complex loses the entire carboxylate residue as the fragment ion (*m/z* 255), leaving the phenolic unit as a quinoid neutral fragment (Scheme 37)[220]. In a further work, 3,4-dihydroxybenzyl carboxylates derived from stearic acid (cf. **140**), dihydrocinnamic acid and phenylacetic acid were studied under NCI(NH₃)-MS/MS conditions. In these cases, deprotonation was found to take place exclusively at the phenolic sites, owing to the increased acidity of hydroxyl groups in a catechol nucleus, in contrast to simple phenols. Heterolysis of the benzylic C−O bond, e.g. in ions [**140** − H]⁻_{(OH)}, gives rise to *reactive* anion/molecule complexes, such as **141**. Here, the carboxylate ion is able to react with the second OH group in the quinoid neutral formed from the original catechol unit, giving complex **142**. Owing to its high acidity, this 2-hydroxyquinomethane component transfers a proton to the carboxylate to release, eventually, the neutral acid and produce the stable $C_7H_5O^-$ anion with *m/z* 121 (Scheme 37)[221].

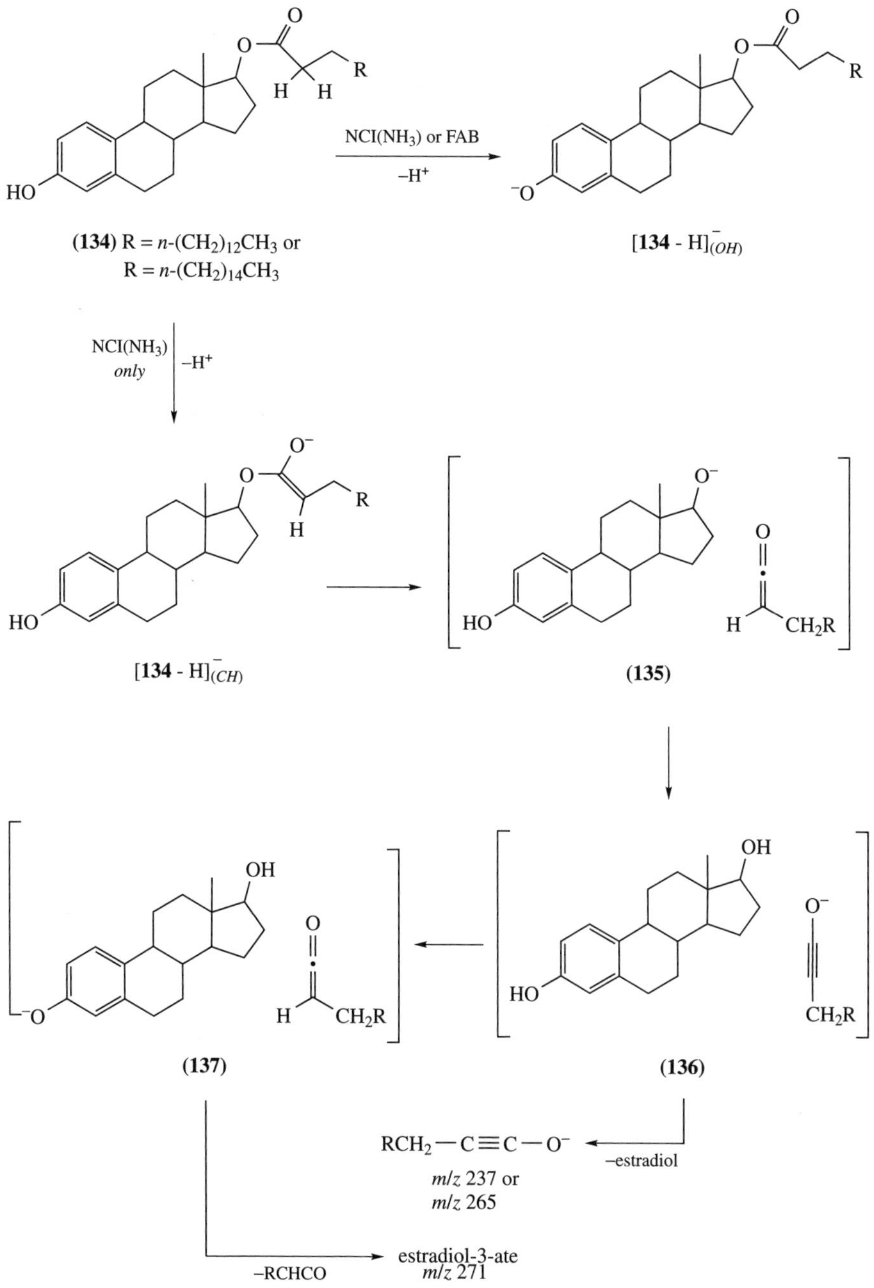

4. Mass spectrometry and gas-phase ion chemistry of phenols

(134) R = n-(CH$_2$)$_{12}$CH$_3$ or
R = n-(CH$_2$)$_{14}$CH$_3$

[134 - H]$^-_{(OH)}$

[134 - H]$^-_{(CH)}$

(135)

(136)

(137)

RCH$_2$—C≡C—O$^-$
m/z 237 or
m/z 265

−estradiol

−RCHCO

estradiol-3-ate
m/z 271

SCHEME 36

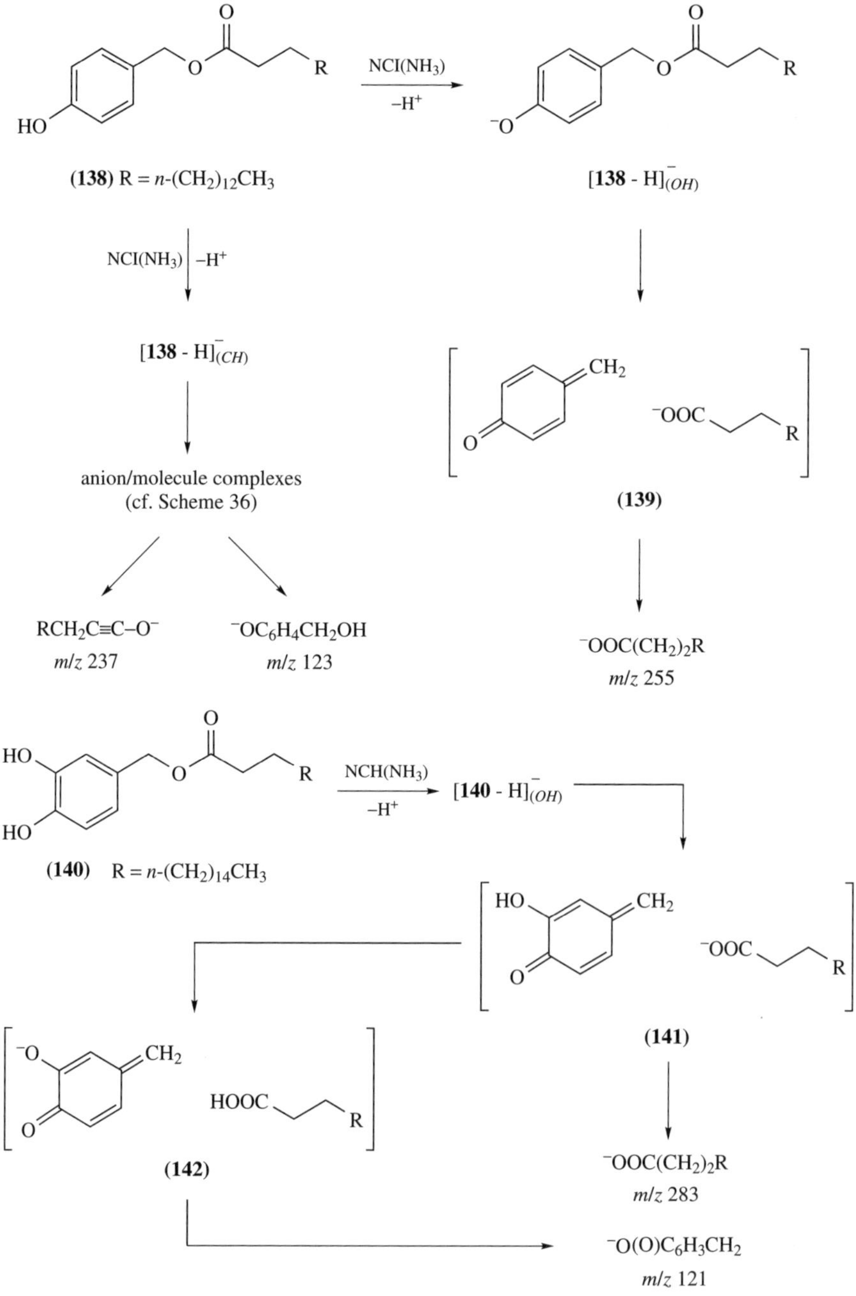
(138) R = n-(CH2)12CH3
NCI(NH3)
−H+
[138 - H]−(OH)
[138 - H]−(CH)
anion/molecule complexes
(cf. Scheme 36)
RCH2C≡C–O−
m/z 237
−OC6H4CH2OH
m/z 123
(139)
−OOC(CH2)2R
m/z 255
(140) R = n-(CH2)14CH3
NCH(NH3)
[140 - H]−(OH)
(141)
(142)
−OOC(CH2)2R
m/z 283
−O(O)C6H3CH2
m/z 121

SCHEME 37

XII. MISCELLANEOUS ANALYTICAL EXAMPLES

As mentioned above in the context of the analysis of lignin degradation products, gas chromatography/mass spectrometry and related methods have been developed as extremely powerful tools for the identification of phenolic compounds. Use of high-pressure liquid chromatography in combination with mass spectrometry adds to the analytical arsenal with respect to the detection of polar, non-volatile compounds but, in particular, the advent of modern ionization techniques, such as ESI and MALDI mass spectrometry, have continued to broaden the analytically governable field of organic chemistry. The latter methods diminish the need of derivatization of polar phenolics to increase the volatility of the analyte. In this section, a more or less arbitrary selection of examples for the application of mass spectrometric techniques in analytical chemistry is added to the cases already discussed above in the context of gas-phase ion chemistry.

Mixtures of alkylphenols are frequently obtained by electrophilic substitution of phenol and phenol derivatives, and GC/MS analysis of these mixtures can be highly useful. The products of octylation of phenols[28] and of *tert*-butylation of cresols[222] were analysed by GC/MS. An example for the successful (and essential) use of derivatization for the GC/MS identification of trace phenolic compounds in waste water was published recently[223]. In that case, on-column benzylation of the sample constituents was performed by using 3,5-bis(trifluoromethyl)benzyldimethylphenyl ammonium fluoride (BTBDMAF), followed by negative ion CI-MS. Another recent study demonstrated the detection of more than fifty phenol derivatives, including six phenolic pesticides, by conventional (positive ion) EI-MS after conversion to their *tert*-butyldimethylsilyl ethers using *N*-(*tert*-butyldimethylsilyl)-*N*-methyl trifluoroacetamide (MTBSTFA)[224]. A method for the quantitative determination of phenolic compounds in cigarette smoke condensates without using derivatization was developed recently[225]. Further, the discrimination of isomeric mono-, di- and trichlorophenols in aqueous samples by GC/MS using both positive ion (EI) and negative ion (NCI) mass spectrometry was studied recently[226]. The tetrachlorophenols and pentachlorophenol were also included. A systematic comparison was reported of the sensitivities of GC/MS analysis of various phenols using both EI and positive and negative ion CI methods[227]. The trace detection of halophenols in the presence of the related haloanisoles was investigated by use of deuteriodiazomethane derivatization prior to GC/MS analysis[228]. Three spray techniques, viz. thermospray (TSP), atmospheric pressure chemical ionization (APCI) and ion spray (ISP), were compared when applied to the identification of phenolic compounds by LC/MS analysis run in the negative ion mode[229]. As a further extension of the analytical manifold, the use of gas chromatography combined with both Fourier transform infrared spectroscopy and mass spectrometry (GC/FT-IR/MS) was demonstrated with fifty different phenolic compounds[230]. In this case, with another 'orthogonal' instrumental methodology being added to mass spectrometry, simple positive-ion EI-MS turned out to be sufficient to manage this analytical challenge.

LC/ESI-MS has been used to determine *trans*-resveratrol (*trans*-3,5,4'-trihydroxystilbene) in wines[231]. The identification of several hydroxylated polycyclic aromatic hydrocarbons was elaborated using LC/APCI mass spectrometry run in both the positive and negative ion modes[232]. Although not carrying far in the case of the phenols, in-source fragmentation was applied to increase the analytical specificity.

The combination of chromatography/mass spectrometry with MS/MS methods can in fact markedly enhance the analytical performance of the identification of phenols. This was demonstrated in the case of hydroxyaromatic components in coal-derived liquids[233]. The analytical performance can be further improved by using chemical derivatization, as also shown in an MS/MS study of some methylphenols and methylnaphthols[234]. In the course of GC/MS/MS analytical studies on nonylphenol in biological tissues, derivatization proved to be favourable in an indirect way: The EI mass spectrum of nonylphenol shows a moderately intense molecular ion peak at m/z 220, whereas the spectrum of the related acetate lacks a molecular ion signal. Loss of ketene from the ionized phenyl acetate, known to generate the corresponding phenol radical cations, having m/z 220, again (cf. Section IX.A), is surprisingly fast here but nevertheless leads to a markedly increased relative intensity of the m/z 220 peak. This increase was exploited to improve the performance of the MS/MS analysis of nonylphenol in tissue samples[235].

Finally, the importance of GC/MS techniques for the analysis of hydroxyaromatic compounds generated during microsomal hydroxylation of benzene derivatives is mentioned here. Using various partially ring-deuteriated substituted benzenes, including biphenyl, evidence for direct aromatic hydroxylation was gained from the careful mass spectrometric tracing of the fate of the label in the various silyl-derivatized hydroxylation products[236].

XIII. MASS SPECTROMETRY OF CALIXARENES

Calixarenes are polycyclic organic compounds with pronounced convex–concave, albeit flexible molecular shape, a property which renders them highly interesting molecular hosts[237–240]. Owing to the synthetic access to the [1.1.1.1]metacyclophane skeleton of calixarenes by oligocyclocondensation of several molecules of a phenol with the same number of aldehyde molecules, calix[n]arenes and resorc[n]arenes contain four or more (n) phenolic subunits within the macrocyclic framework (Scheme 38). The four hydroxyl groups in the 5,11,17,23-tetrakis(*tert*-butyl)calix[4]arene **143** can develop cooperative reactivity because of their mutual proximity at the 'lower rim' of the macrocyclic framework. The phenolic hydroxyl groups, or some of them, may be converted by etherification or esterification. Starting with resorcinols or pyrogallols instead of phenols, a variant of the same aufbau principle affords the related resorc[n]arenes and pyrogallo[n]arenes, such as the 2,8,14,20-tetra (n-alkyl)resorc[4]arenes **144** and **145** and the 2,8,14,20-tetra-(n-alkyl)pyrogallo[4]arenes **146** and **147**, respectively, which bear $2n$ or even $3n$ phenolic hydroxyl groups at the 'upper rim' of the skeleton. The strong capability of calixarene-type compounds to form various host–guest complexes and multiple adducts in the condensed phase renders them interesting objects for the investigation of the intrinsic properties of such aggregates in the gas phase. In this section, a brief introduction is given to some aspects which combine mass spectrometry and the gas-phase ion chemistry of calixarenes.

The unimolecular fragmentation of calixarene-derived ions will not be treated here, especially as studies on this topic are much restricted due to the fact that classical EI and CI techniques cannot be applied to these involatile and often quite polar polyphenols. Rather, mass spectrometric analysis is limited to the detection of positively or negatively charged quasi-molecular ions, such as $[M + H]^+$ and $[M - H]^-$, or molecular adduct ions, such as $[M + NR_4]^+$ and $[M + \text{metal}]^+$. In general, these ions can be readily generated by using matrix-assisted laser desorption (MALDI) and/or electrospray ionization (ESI) mass spectrometry.

In this context, it is noteworthy that calixarenes have been found to be suitable for mass calibration in ESI mass spectrometry, owing to their ability to form cluster ions in both the positive and negative ion mode[241]. In particular, a calix[4]arene derived from 4-n-octylpyrogallol, *rccc*-2,8,14,20-tetra (n-octyl)-5,11,17,23-tetrahydroxyresorc[4]arene **146** ($C_{60}H_{88}O_{12}$, MW 1001), containing twelve phenolic hydroxyl groups, was shown recently to generate cluster cations of the series $[xM + Na]^+ (x = 1-5)$ and $[x'M + 2 Na]^{2+} (x' = 7, 9, 11)$ and also cluster anions of the series $[yM - H]^- (y = 1-5)$ and $[y'M - 2 Na]^- (y' = 1-5)$. Different from other, conventional calibrants, compound **146** allows one to extend the 'mass' scale of the ESI mass spectrometer to mass-to-charge ratios as high as $m/z = 6000$ with notably abundant cluster ions, whose relative abundances do not drop off significantly

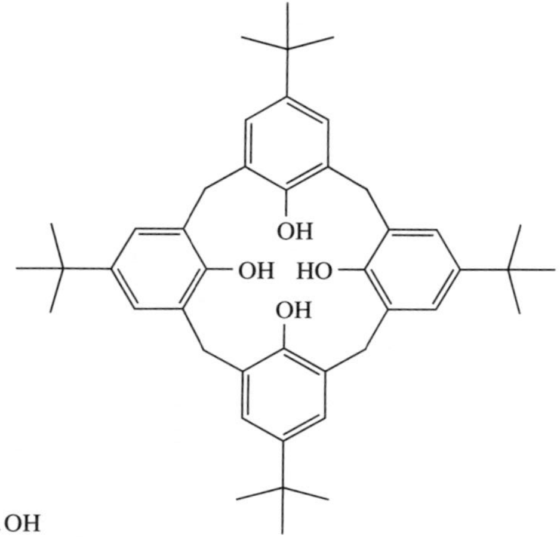

(143)

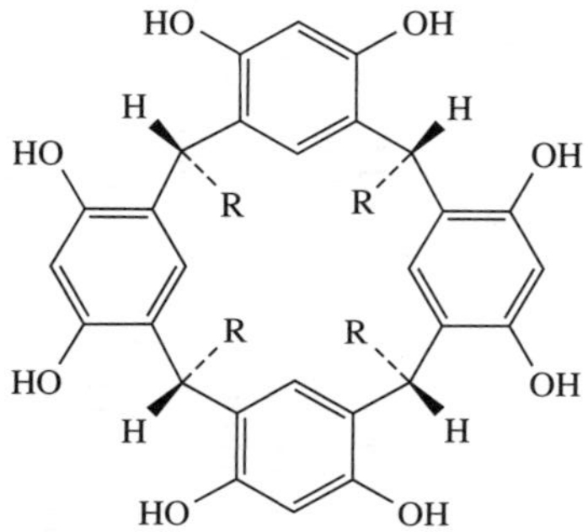

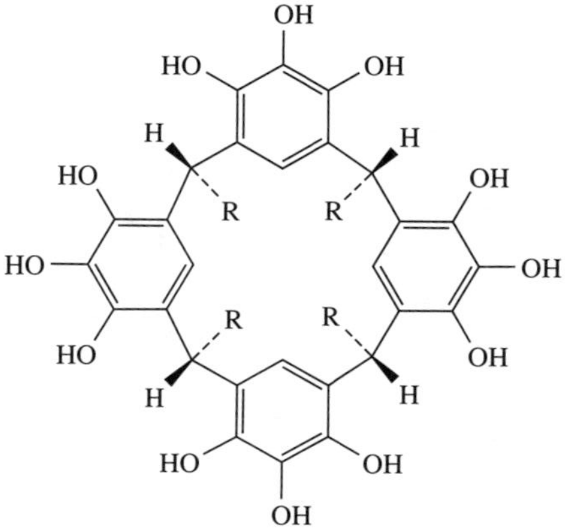

(146) R = *n*-octyl
(147) R = *n*-undecyl

(144) R = *n*-octyl
(145) R = *n*-undecyl

SCHEME 38

with increasing mass. As an additional advantage, the cluster ions of **146** were found to form without addition of modifiers, such as caesium salts[241].

Turning from application to fundamentals of calixarene gas-phase ion chemistry, the ability of these phenolic compounds to form stable adducts with alkali metal ions has also been investigated in detail recently. It is known that, different from proton attachment to aromatic molecules by σ-bonding, alkali cations coordinate with aromatic rings preferentially by π-cation interaction[242,243]. The gas-phase binding energies of Na^+, NH_4^+ and NMe_4^+ ions have been calculated to be close to the binding energies of these ions to benzene[242], suggesting that these cations are bound to the π-electron system rather than to the heteroatom. The experimentally determined lithium cation affinity of benzene is particularly high, $LCA(C_6H_6) = 38.3$ kcal mol^{-1} [242]. A recent combined experimental and theoretical study reports the theoretical binding enthalpy of Li^+ to the π-electron system of phenol to be $LCA(\mathbf{1})_{(6\pi)} = 39.2$ kcal mol^{-1} [244]. Coordination of the cation to the HO-C bond was calculated to be enthalpically less [$LCA(\mathbf{1})_{(OH)} = 36.8$ kcal mol^{-1}] but entropically more favourable. Equilibrium measurements in the FT-ICR mass spectrometer afforded the lithium cation basicity of phenol, $LCB(\mathbf{1}) = 28.1$ kcal mol^{-1}, close to the calculated values (29.2 and 28.0 kcal mol^{-1}, respectively) for the two coordination modes[244]. Thus, the pronounced tendency of calixarenes to form ionic aggregates in the gas phase may be tentatively attributed to the presence of several π-electron systems, being held in a cone-type orientation and thus being able to develop a high negative electrostatic potential[245,246] in the cavity of the [1.1.1.1]metacyclophane framework. Alternatively, the polar phenolic groups may act as the sites of attachment.

Only few investigations have been published on the gas-phase ion chemistry of host–guest complexes of calixarenes. With the advent of ESI mass spectrometry, especially when combined with ion-trap and FT-ICR mass spectrometry, this field has started to be developed. Binding selectivities of alkali metal ions to calixarene-based crown ethers and open-chain ethers have been studied[247–249], the inclusion of neutral guests into the protonated resorcarene-based cavitand hosts by gas-phase ion–molecule reactions with amines have been studied[250] and the formation of capsules from various calixarene tetraether derivatives and alkylammonium ions as ionic guests (notably enabling their detection by mass spectrometry) have been described recently[251].

In another recent study[252], ESI tandem mass spectrometry was used to generate cationized resorc[4]arenes and pyrogallo[4]arenes bearing eight or twelve hydroxyl groups, respectively, and n-octyl and n-undecyl residues at the benzylic positions. Competition experiments performed with calix[4]arene **145** for the series of alkali metal cations showed a strong preference for Cs^+ in both the cationized monomers and cationized dimers. The corresponding pyrogallo[4]arene **147** exhibited the same behaviour for the monomeric adduct but the homodimers of **147** were found to be most stable with Li^+ and Na^+. Tetramethylammonium ions such as ionic complexation partners of calixarene **146** gave much more abundant monomeric and dimeric adducts than higher tetraalkylammonium ions. These results point to the particularly favourable fit of the larger (but not too large) ions, such as Cs^+ and Me_4N^+, into the cavity of the monomeric and dimeric adducts. Collisional activation experiments (ESI-MS/MS) with various heterodimers, such as [**144** Li **146**]$^+$ and [**144** K **146**]$^+$, revealed the preferred bonding of the smaller alkali metal cations to the pyrogallo[4]arenes, as compared to the simple resorc[4]arenes, whereas K^+ and the larger metal ions showed no preference. Therefore, two different binding mechanisms were put forward: The larger alkali metal cations are insensitive to the number of hydroxyl groups at the outer rim of the calix[4]arenes because they are bound preferentially by the cavity of the macrocycles; by contrast, the smaller alkali metal cations are coordinated outside that cavity, in the vicinity of the polar hydroxyl functionalities[252].

XIV. PHENOLIC COMPOUNDS AS MALDI MATRICES

Mass spectrometry and gas-phase ion chemistry of phenols concerns this class of compounds and, in particular, the various types of gaseous ions formed from them, as objects of fundamental interest and analytical significance. However, in the special case of phenols, a mass spectrometry *'with'* phenols has been developed. As mentioned in the Introduction, one of the modern methodologies for the formation of ions from polar and/or high-molecular mass, and thus non-volatile, organic and bioorganic compounds, relies on the use of various phenolic compounds as matrices for ion generation. Matrix-assisted laser ionization/desorption (MALDI)[3,4,253–256] has become one of the major essential ionization methods in mass spectrometry and has widened the fields of application of analytical mass spectrometry enormously. In particular, the detection and identification of biopolymer samples (peptides and proteins, oligosaccharides and oligonucleotides) has gained extreme progress through the advent and application of MALDI mass spectrometry. The samples are co-crystallized with aromatic matrix compounds, which are able to absorb the energy of laser pulses and transfer parts of it to the analyte molecules, often with concomitant protonation. As already mentioned, the mechanism of the MALDI process is not well understood nowadays - in spite of its enormous analytical importance - and the success of an analysis by MALDI mass spectrometry depends strongly on the selection of the matrix compounds, possibly of some additives, and of its actual preparation. Thus, different matrix compounds have proven useful for different classes of analyte compounds and, despite the fact that certain classes of compounds are measurable with a high degree of confidence (e.g. peptides), quite some experience is required to choose the appropriate MALDI conditions in a given analytical case.

Remarkably, phenol derivatives are amongst the most useful matrix compounds. A list of the most frequently used organic matrices has been compiled in a recent review[4]. The formulae of the phenolic matrix compounds among these, **148**–**156**, are reproduced in Scheme 39. 2,5-Dihydroxybenzoic acid **148** (2,5-DHB, mostly addressed simply as 'DHB'), *trans*-3,5-dimethoxy-4-hydroxycinnamic acid **149** (sinapinic acid, SA), *trans*-3-methoxy-4-hydroxycinnamic acid **150** (ferulic acid, FA) and *trans*-4-hydroxy-α-cyanocinnamic acid **151** (4HCCA) have proven to be most useful. All the isomers of 'DHB' (see below) turned out to be much less efficient for the production of ions. Beyond the hydroxybenzoic acids and the hydroxycinnamic acids, notably being vinylogues of the former, phenol derivatives which lack the carboxyl group are good MALDI matrices, too. For example, 2,4,6-trihydroxyacetophenone **152** and 1,8,9-trihydroxyanthracene **153** (or its tautomer, 1,8-dihydroxyanthrone **154**, 'dithranol') are used frequently and nitrogen-containing phenols, such as 3-hydroxypicolinic acid **155** (3HPA) and 2-carboxy-4'-hydroxyazobenzene **156** [2-(4-hydroxyphenylazo)benzoic acid, HABA] have to be mentioned. The list of useful MALDI matrices containing phenolic hydroxyl groups will certainly increase further in the near future.

Various molecular and quasi-molecular ions can be formed under MALDI conditions. The formation of protonated analyte (A) molecules, $[A + H]^+$, is generally most important at least for samples containing slightly basic centres, such as the peptides and proteins, MALDI mass spectrometry of which is known to be most facile and reproducible. Therefore, proton transfer from the electronically excited, neutral or ionized, or protonated matrix species is considered to be crucial in the overall MALDI process[257–263]. Notably, proton transfer can occur already in the condensed phase, followed by desorption of the preformed ions[264–267]. However, the generation of the $[A + H]^+$ ions is believed to take place preferably in the so-called 'plume', that is, in the energized, short-lived and relatively dense vapour phase generated above the solid matrix upon excitation by the laser pulse[4]. The actual proton donor species (be it one or several) in a given case is still a matter of

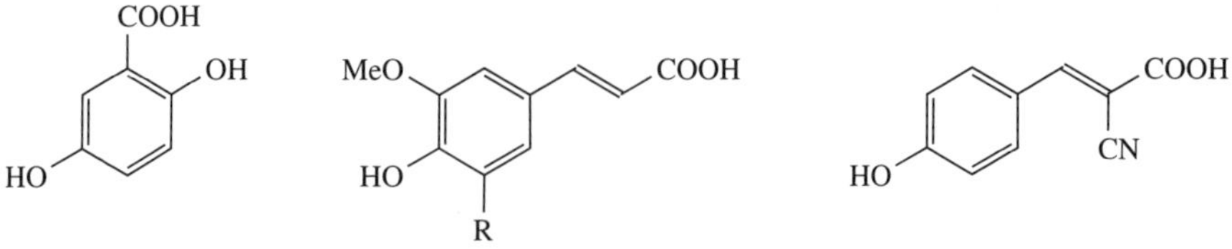

(**148**) 'DHB' (**149**) R = OMe (**150**) R = H (**151**)

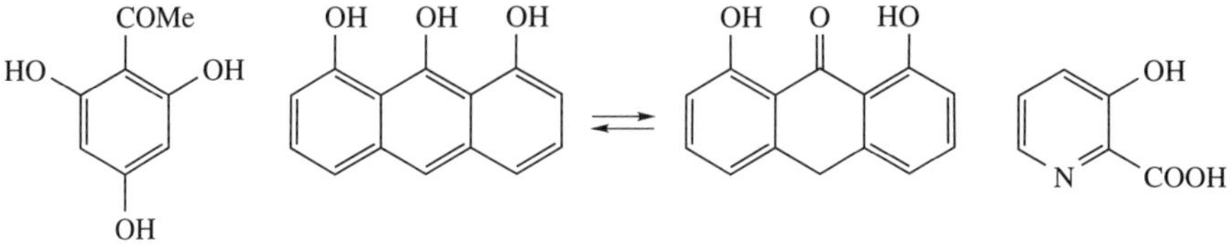

(**152**) (**153**) (**154**) (**155**)

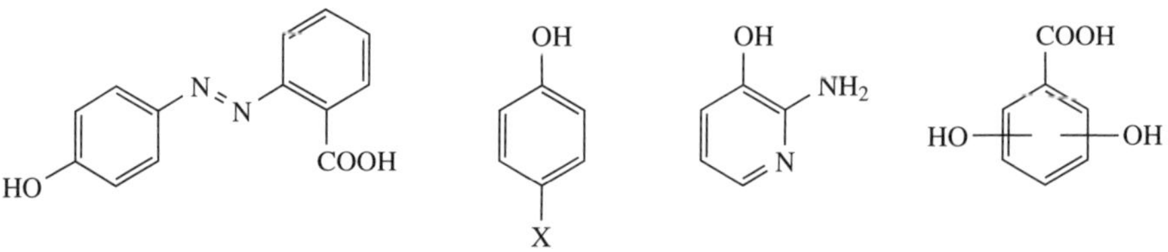

(**156**) (**157**) X = COOH (**158**) X = NH_2 (**159**)

(**160**) 2,3-DHB
(**161**) 2,4-DHB
(**162**) 2,6-DHB
(**163**) 3,4-DHB
(**164**) 3,5-DHB

SCHEME 39

research. Proton transfer from the neutral matrix molecule, from their radical cations or from the protonated forms, as well as from dimeric species have been considered.

It appears clear that a better understanding of the proton transfer processes in MALDI is crucial to the further development of the methodology[268,269]. Therefore, the determination of gas-phase acidities of neutral and cationic organic species is actively continued, and adds many details to the knowledge collected over several decades[270]. The intrinsic acidities of substituted phenols and benzoic acids in the gas phase have been studied in detail in the 1970s[271,272]. It was noted early that the phenolic OH group of (neutral) *para*-hydroxybenzoic acid is more acidic than the carboxyl group[271]. Also, the anomalous 'ortho acidities' of *ortho*-substituted benzoic acids were traced to the special interaction of the substituents in these isomers. These findings rely on the structural motifs of many phenolic matrix substances, such as 'DHB' **148** and 2,4,6-trihydroxyacetophenone **152**. In fact, the phenolic OH groups of several phenol-derived matrices were recently shown by labelling experiments to be the proton donor sites rather than the carboxyl functionalities[273,274]. In a recent FT-ICR work, it was demonstrated that, amongst a group of important MALDI matrices including *para*-hydroxybenzoic acid **157**, *para*-aminophenol **158**, 'DHB' **148**, 2-amino-3-hydroxypyridine **159** and 2,4,6-trihydroxyacetophenone **152**, the latter compound was found to be the most acidic one[275].

Obviously, the electronic effects of carbonyl groups oriented *ortho* or *para* to the phenolic hydroxyl groups, and the spatial influence of the *ortho* orientation, in particular, are attractive structural motifs to trace the origins of the MALDI processes. However, the situation has remained obscure, as also demonstrated by the recent systematic experimental determination of all the six isomeric dihydroxybenzoic acids, comprising the 2,5-isomer **148** and its isomers **160–164**, by FT-ICR mass spectrometry[5]. The gas-phase basicities, and the corresponding proton affinities, of the neutral molecules were found to span only a small range, from $GB(3,5\text{-DHB}) = 194.6$ kcal mol^{-1} for the least basic isomer to $GB(2,4\text{-DHB}) = 198.6$ kcal mol^{-1} for the most basic one, including that of the empirically most 'successful' isomer, $GB(2,5\text{-DHB}) = 196.5$ kcal mol^{-1}. This value and $PA(2,5\text{-DHB}) = 204.3$ kcal mol^{-1} are in excellent agreement with the previously published data of this particular isomer[276–278]. Furthermore, the gas-phase acidities of the corresponding radical cations of the six isomers were also determined[5]. In this series, the range is larger, spanning from $\Delta G_{\text{acid}}(3,4\text{-DHB}) = 194.8$ kcal mol^{-1} for the most acidic isomer to $\Delta G_{\text{acid}}(2,5\text{-DHB}) = 205.1$ kcal mol^{-1} for the least acidic one. Notably, the best-proven matrix, 2,5-dihydroxybenzoic acid, stands out as the least acidic isomer. Therefore, the results of this complete series of isomeric matrix compounds indicate that the ground-state proton transfer from the matrix radical cations to the analyte molecules may play a crucial role in the ionization process of MALDI, whereas proton transfer from the protonated matrix molecules can be excluded[5]. It appears most probable that dimeric and/or oligomeric species of the diverse matrices, be it in the ground state or electronically excited states, represent the key intermediates for ion formation in the MALDI process. The dissociative proton transfer in gaseous cluster ions of various phenol-derived carboxylic acids has been studied recently[279].

XV. ACKNOWLEDGEMENTS

I would like to thank Dr. Michael Mormann, Universität Münster, Münster/Germany, for his valuable help in screening the literature on mass spectrometry and gas-phase ion chemistry of phenols. Also, I would like to thank Prof. Jean-François Gal, Université de Nice - Sophia Antipolis, Nice/France, for providing a manuscript (Reference 244) prior to publication, and Prof. Minh Tho Nguyen, Katholike Universiteit Leuwen, Leuwen/Belgium, for providing a manuscript of his work on phenols. I am grateful to Prof. Susumu Tajima, Gunma College, Takasaki/Japan, for kindly providing additional material. Finally, I express my sincere appreciation to Prof. Zvi Rappoport, The Hebrew University of Jerusalem, for his encouragement, steadiness and unrestricted confidence in this project.

XVI. REFERENCES

1. H. Budzikiewicz, C. Djerassi and D. H. Williams, *Mass Spectrometry of Organic Compounds*, Holden-Day, San Francisco, 1967.
2. F. W. McLafferty and F. Tureček, *Interpretation of Mass Spectra*, 4th edn., University Science Books, Mill Valley, CA, 1993.
3. M. Karas, D. Bachmann, U. Bahr and F. Hillenkamp, *Int. J. Mass Spectrom. Ion Processes*, **78**, 53 (1987).
4. R. Zenobi and R. Knochenmuss, *Mass Spectrom. Rev.*, **17**, 337 (1999).
5. M. Mormann, S. Bashir, P. J. Derrick and D. Kuck, *J. Am. Soc. Mass Spectrom.*, **11**, 544 (2000).
6. For a recent essay on relations between solution and gas-phase ion chemistry, see: D. Kuck, *Angew. Chem.*, **112**, 129 (2000); *Angew. Chem., Int. Ed.*, **39**, 125 (2000).
7. S. G. Lias, J. E. Bartmess, J. F. Liebman, J. L. Holmes, R. D. Levin and W. G. Mallard, *J. Phys. Chem. Ref. Data*, **17**, Suppl. No. 1 (1988).

8. E. P. L. Hunter and S. G. Lias, *J. Phys. Chem. Ref. Data*, **27**, 413 (1998).
9. W. G. Mallard and P. J. Linstrom (Eds.), NIST Chemistry Webbook, NIST Standard Reference Database No. 69 - July 2001 Release, National Institute of Standards and Technology, Gaithersburg, MD 20899 (http://webbook.nist.gov).
10. Y. Lau and P. Kebarle, *J. Am. Chem. Soc.*, **98**, 7452 (1976).
11. J. M. McKelvey, S. Alexandratos, A. Streitwieser, Jr., J. L. M. Abboud and W. J. Hehre, *J. Am. Chem. Soc.*, **98**, 244 (1976).
12. H. H. Jaffé, *Chem. Rev.*, **53**, 191 (1953).
13. D. J. DeFrees, R. T. McIver, Jr. and W. J. Hehre, *J. Am. Chem. Soc.*, **99**, 3853 (1977).
14. Z. B. Maksić and M. Eckert-Maksić, in *Theoretical and Computational Chemistry*, Vol. 5 (Ed. C. Párkányi), Elsevier, Amsterdam, 1998, pp. 203–231.
15. M. Eckert-Maksić, A. Knežević and Z. B. Maksić, *J. Phys. Org. Chem.*, **11**, 663 (1998).
16. J. L. Devlin III, J. F. Wolf, R. W. Taft and W. J. Hehre, *J. Am. Chem. Soc.*, **98**, 1990 (1976).
17. D. Kuck, *Mass Spectrom. Rev.*, **9**, 583 (1990).
18. K. R. Jennings, *Int. J. Mass Spectrom.*, **1**, 227 (1968).
19. D. H. Williams and R. D. Bowen, *Org. Mass Spectrom.*, **11**, 223 (1976).
20. R. Flammang, P. Meyrant, A. Maquestiau, E. E. Kingston and J. H. Beynon, *Org. Mass Spectrom.*, **20**, 253 (1985).
21. D. H. Russell, M. L. Gross and N. M. M. Nibbering, *J. Am. Chem. Soc.*, **100**, 6133 (1978).
22. R. G. Cooks, J. H. Beynon, R. M. Caprioli and G. R. Lester, *Metastable Ions*, Chap. 4, Elsevier, Amsterdam, 1973, pp. 89–158.
23. Y. Malinovich and C. Lifshitz, *J. Phys. Chem.*, **90**, 4311 (1986).
24. M. L. Fraser-Monteiro, L. Fraser-Monteiro, J. de Wit and T. Baer, *J. Phys. Chem.*, **88**, 3622 (1984).
25. D. Kuck, *Mass Spectrom. Rev.*, **9**, 187 (1990).
26. T. Aczel and H. E. Lumpkin, *Anal. Chem.*, **32**, 1819 (1960).
27. P. Buryan, V. Kubelka, J. Mitera and J. Macák, *Collect. Czech. Chem. Commun.*, **44**, 2798 (1979).
28. M. R. Darby and C. B. Thomas, *Eur. Mass Spectrom.*, **1**, 399 (1995).
29. (a) Mass spectra taken from Reference 7.
(b) See also: J. Tateiwa, E. Hayama, T. Nishimura and S. Uemura, *J. Chem. Soc., Perkin Trans. 1*, 1923 (1997).
30. R. Nicoletti and D. A. Lightner, *J. Am. Chem. Soc.*, **90**, 2997 (1968).
31. J. L. Occolowitz, *Anal. Chem.*, **36**, 2177 (1964).
32. H. M. Grubb and S. Meyerson, in *Mass Spectrometry of Organic Ions* (Ed. F. W. McLafferty), Chap. 10, Academic Press, New York, 1963, pp. 453–527.
33. D. Kuck and H. F. Grützmacher, *Z. Naturforsch. B*, **34**, 1750 (1979).
34. M. I. Gorfinkel, L. Yu. Ivanosvskaia and V. A. Kopytug, *Org. Mass Spectrom.*, **2**, 273 (1969).
35. D. Kuck and H. F. Grützmacher, *Org. Mass Spectrom.*, **13**, 81 (1978).
36. D. Kuck and H. F. Grützmacher, *Org. Mass Spectrom.*, **13**, 90 (1978).
37. Y. Du, R. Oshima, Y. Yamauchi, J. Kumanotani and T. Miyakoshi, *Phytochemistry*, **25**, 2211 (1986).
38. M. Meot-Ner (Mautner), *Acc. Chem. Res.*, **17**, 186 (1984).
39. J. M. S. Tait, T. W. Shannon and A. G. Harrison, *J. Am. Chem. Soc.*, **84**, 4 (1962).
40. I. P. Fisher, T. F. Palmer and F. P. Lossing, *J. Am. Chem. Soc.*, **86**, 2741 (1964).
41. A. Filippi, G. Lilla, G. Occhiucci, C. Sparapani, O. Ursini and M. Speranza, *J. Org. Chem.*, **60**, 1250 (1995).
42. D. V. Zagorevskii, J. M. Régimbal and J. L. Holmes, *Int. J. Mass Spectrom. Ion Processes*, **160**, 211 (1997).
43. D. V. Zagorevskii, J. L. Holmes and J. A. Stone, *Eur. Mass Spectrom.*, **2**, 341 (1996).
44. H. Konishi, S. Kitagawa, H. Nakata, K. Sakurai, A. Tatematsu and C. Fujikawa, *Nippon Kagaku Kaishi*, 1650 (1986); *Chem. Abstr.*, **107**, 6653q (1987).
45. J. D. Dill, P. v. R. Schleyer and J. A. Pople, *J. Am. Chem. Soc.*, **99**, 1 (1977).
46. D. Kuck, *Int. J. Mass Spectrom.*, **213**, 101 (2002).
47. D. Kuck and H. F. Grützmacher, *Adv. Mass Spectrom.*, **8**, 867 (1980).
48. D. Kuck, unpublished results.
49. J. T. Bursey, M. M. Bursey and D. G. I. Kingston, *Chem. Rev.*, **73**, 191 (1973).

50. H. Schwarz, *Top. Curr. Chem.*, **73**, 231 (1978).
51. J. Y. Zhang, D. S. Nagra, A. P. L. Wang and L. Li, *Int. J. Mass Spectrom. Ion Processes*, **110**, 103 (1991).
52. N. Harting, G. Thielking and H. F. Grützmacher, *Int. J. Mass Spectrom.*, **167/168**, 335 (1997).
53. D. Kuck and U. Fastabend, *Int. J. Mass Spectrom. Ion Processes*, **179/180**, 147 (1998).
54. D. Kuck, in preparation.
55. U. Neuert, *Doctoral Thesis*, Universität Hamburg, 1975.
56. B. Schaldach and H. F. Grützmacher, *Org. Mass Spectrom.*, **15**, 175 (1980).
57. For a related review, see: D. Kuck and M. Mormann, in *The Chemistry of Functional Groups: The Chemistry of Dienes and Polyenes*, Vol.2 (Ed. Z. Rappoport), Chap. 1, Wiley, New York, 2000, pp. 1–57.
58. J. S. Riley, T. Baer and G. D. Marbury, *J. Am. Soc. Mass Spectrom.*, **2**, 69 (1991).
59. B. Davis and D. H. Williams, *J. Chem. Soc., Chem. Commun.*, 412 (1970).
60. H. Florêncio, P. C. Vijfhuizen, W. Heerma and G. Dijkstra, *Org. Mass Spectrom.*, **14**, 198 (1979).
61. J. H. Beynon, M. Bertrand and R. G. Cooks, *J. Am. Chem. Soc.*, **95**, 1739 (1973).
62. J. Charalambous, R. R. Fysh, C. G. Herbert, M. H. Johri and W. M. Shutie, *Org. Mass Spectrom.*, **15**, 221 (1980).
63. T. Yanagisawa, S. Tajima, M. Iizuka, S. Tobita, M. Mitani, H. Sawada and T. Matsumoto, *Int. J. Mass Spectrom.*, **125**, 55 (1993).
64. S. Tajima, S. Takahashi and O. Sekiguchi, *Rapid Commun. Mass Spectrom.*, **13**, 1458 (1999).
65. B. S. Freiser, R. L. Woodin and J. L. Beauchamp, *J. Am. Chem. Soc.*, **97**, 6893 (1975).
66. D. F. Hunt, C. N. McEwen and R. A. Upham, *Anal. Chem.*, **44**, 1292 (1972).
67. W. J. Richter, W. Blum and J. G. Liehr, quoted in Reference 68.
68. D. P. Martinsen and S. E. Buttrill Jr., *Org. Mass Spectrom.*, **11**, 762 (1976).
69. T. Keough and A. J. DeStefano, *Org. Mass Spectrom.*, **16**, 527 (1981).
70. M. V. Buchanan, *Anal. Chem.*, **56**, 546 (1984).
71. E. R. E. van der Hage, T. L. Weeding and J. J. Boon, *J. Mass Spectrom.*, **30**, 541 (1995).
72. H. P. Tannenbaum, J. D. Roberts and R. C. Dougherty, *Anal. Chem.*, **47**, 49 (1975).
73. D. C. Lane and M. McGuire, *Org. Mass Spectrom.*, **18**, 495 (1983).
74. D. Kuck, A. Petersen and U. Fastabend, *Int. J. Mass Spectrom. Ion Processes*, **179/180**, 129 (1998).
75. M. Mormann and D. Kuck, *Int. J. Mass Spectrom.*, **219**, 497 (2002).
76. D. J. Burinsky and J. E. Campana, *Org. Mass Spectrom.*, **23**, 613 (1988).
77. R. Srinivas, M. Vairamani, G. K. Viswanadha Rao and U. A. Mirza, *Org. Mass Spectrom.*, **24**, 435 (1989).
78. R. Srinivas, A. Rama Devi and G. K. Viswanadha Rao, *Rapid Commun. Mass Spectrom.*, **10**, 12 (1996).
79. R. N. Stillwell, D. I. Carroll, J. G. Nowlin and E. C. Horning, *Anal. Chem.*, **55**, 1313 (1983).
80. R. Orlando, F. Strobel, D. P. Ridge and B. Munson, *Org. Mass Spectrom.*, **22**, 597 (1987).
81. R. Orlando, D. P. Ridge and B. Munson, *Org. Mass Spectrom.*, **23**, 527 (1988).
82. S. Daishima, Y. Iida and F. Kanda, *Org. Mass Spectrom.*, **26**, 486 (1991).
83. Y. P. Tu, K. Lu and S. Y. Liu, *Rap. Commun. Mass Spectrom.*, **9**, 609 (1995).
84. T. Donovan and J. Brodbelt, *Org. Mass Spectrom.*, **27**, 9 (1992).
85. J. Brodbelt, J. Liou and T. Donovan, *Anal. Chem.*, **63**, 1205 (1991).
86. M. Honing, D. Barceló, B. L. M. van Baar and U. A. T. Brinkman, *J. Mass Spectrom.*, **31**, 527 (1996).
87. J. Steadman and J. A. Syage, *J. Am. Chem. Soc.*, **113**, 6786 (1991).
88. J. A. Syage and J. Steadman, *J. Phys. Chem.*, **96**, 9606 (1992).
89. R. H. Shapiro and K. B. Tomer, *Org. Mass Spectrom.*, **2**, 579 (1969).
90. S. A. Benezra and M. M. Bursey, *J. Chem. Soc., B*, 1515 (1971).
91. A. A. Gamble, J. R. Gilbert and J. G. Tillett, *Org. Mass Spectrom.*, **5**, 1093 (1971).
92. H. Nakata and A. Tatematsu, *Org. Mass Spectrom.*, **5**, 1343 (1971).
93. K. B. Tomer and C. Djerassi, *Tetrahedron*, **29**, 3491 (1973).
94. S. A. Benezra and M. M. Bursey, *J. Chem. Soc., Perkin Trans. 2*, 1537 (1972).
95. G. Bouchoux, *Int. J. Mass Spectrom. Ion Phys.*, **26**, 379 (1978).
96. F. W. McLafferty, *Anal. Chem.*, **31**, 82 (1959).
97. For pertinent quotations, see Reference 21.

98. A. Maquestiau, Y. V. Haverbeke, R. Flammang, C. D. Meyer, C. G. Das and G. S. Reddy, *Org. Mass Spectrom.*, **12**, 631 (1977).
99. F. Borchers, K. Levsen, C. B. Theissling and N. M. M. Nibbering, *Org. Mass Spectrom.*, **12**, 746 (1977).
100. K. Levsen, *Fundamental Aspects of Organic Mass Spectrometry*, Verlag Chemie, Weinheim, 1978, p. 251.
101. P. N. T. van Velzen, W. J. van der Hart, J. van der Greef and N. M. M. Nibbering, *J. Am. Chem. Soc.*, **104**, 1208 (1982).
102. P. Longevialle, *Mass Spectrom. Rev.*, **11**, 157 (1992).
103. T. H. Morton, *Tetrahedron*, **38**, 3195 (1982).
104. T. H. Morton, *Org. Mass Spectrom.*, **27**, 353 (1992).
105. D. J. McAdoo, *Mass Spectrom. Rev.*, **7**, 363 (1988).
106. R. D. Bowen, *Acc. Chem. Res.*, **24**, 264 (1991).
107. D. J. McAdoo and T. H. Morton, *Acc. Chem. Res.*, **26**, 295 (1993).
108. N. Heinrich and H. Schwarz, in *Ion and Cluster Ion Spectroscopy and Structure* (Ed. J. P. Meier), Elsevier, Amsterdam, 1989, pp. 329–372.
109. H. W. Leung and A. G. Harrison, *Org. Mass Spectrom.*, **12**, 582 (1977).
110. H. E. Audier, C. Monteiro and D. Robin, *New J. Chem.*, **13**, 621 (1989).
111. D. Kuck and U. Filges, *Org. Mass Spectrom.*, **23**, 643 (1988).
112. D. Kuck and C. Matthias, *J. Am. Chem. Soc.*, **114**, 1901 (1992).
113. C. Matthias, S. Anlauf, K. Weniger and D. Kuck, *Int. J. Mass Spectrom.*, **199**, 155 (2000).
114. C. Matthias and D. Kuck, *Int. J. Mass Spectrom.*, **217**, 131 (2002).
115. H. E. Audier, F. Dahhani, A. Milliet and D. Kuck, *J. Chem. Soc., Chem. Commun.*, 429 (1997).
116. D. Kuck and A. Mehdizadeh, *Org. Mass Spectrom.*, **27**, 443 (1992).
117. D. Kuck, in *Proceedings of the 8th ISMAS Symposium*, Vol. 1 (Ed. S. K. Aggarwal), Indian Society for Mass Spectrometry, Mumbai, 1999, pp. 245–260.
118. A. M. H. Yeo and C. Djerassi, *J. Am. Chem. Soc.*, **94**, 482 (1982).
119. T. H. Morton, *Org. Mass Spectrom.*, **26**, 18 (1991).
120. D. Harnish and J. L. Holmes, *J. Am. Chem. Soc.*, **113**, 9729 (1991).
121. T. A. Shaler and T. H. Morton, *J. Am. Chem. Soc.*, **113**, 6771 (1991).
122. G. Sozzi, H. E. Audier, P. Mourgues and A. Milliet, *Org. Mass Spectrom.*, **22**, 746 (1987).
123. J. P. Morizur, M. H. Taphanel, P. S. Mayer and T. H. Morton, *J. Org. Chem.*, **65**, 381 (2000).
124. G. Bouchoux, *Org. Mass Spectrom.*, **13**, 184 (1978).
125. O. Sekiguchi, T. Ayuzawa, M. Fujishige, T. Yanagisawa and S. Tajima, *Rapid Commun. Mass Spectrom.*, **9**, 75 (1995).
126. H. W. Biermann, W. P. Freeman and T. H. Morton, *J. Am. Chem. Soc.*, **104**, 2307 (1982).
127. F. M. Benoit and A. G. Harrison, *Org. Mass Spectrom.*, **11**, 599 (1976).
128. R. W. Kondrat and T. H. Morton, *J. Org. Chem.*, **56**, 952 (1991).
129. R. W. Kondrat and T. H. Morton, *Org. Mass Spectrom.*, **26**, 410 (1991).
130. V. I. Denisenko, R. G. Mirzoyan, P. B. Terent'ev, E. S. Agavelyan, G. G. Khudoyan and R. O. Matevosyan, *Zh. Org. Khim.*, **20**, 327 (1984); *Chem. Abstr.*, **101**, 6439a (1984).
131. V. V. Takhistov, D. A. Ponomareva, A. D. Misharev, V. M. Orlov and K. Pihlaja, *Zh. Obshch. Khim.*, **64**, 110 (1994); *Chem. Abstr.*, **122**, 159955z (1995).
132. D. V. Ramana and M. S. Sudha, *Org. Mass Spectrom.*, **27**, 1121 (1992).
133. D. V. Ramana and M. S. Sudha, *J. Chem. Soc., Perkin Trans. 2*, 675 (1993).
134. D. V. Ramana, K. K. Balasubramanian, M. S. Sudha and T. Balasubramanian, *J. Am. Soc. Mass Spectrom.*, **6**, 195 (1995).
135. L. Ceraulo, P. De Maria, M. Ferrugia, S. Foti, R. Saletti and D. Spinelli, *J. Mass Spectrom.*, **30**, 257 (1995).
136. L. Ceraulo, P. De Maria, M. Ferrugia, S. Foti, R. Saletti and D. Spinelli, *Eur. Mass Spectrom.*, **5**, 89 (1999).
137. E. E. Kingston, J. H. Beynon, J. G. Liehr, P. Meyrant, R. Flammang and A. Maquestiau, *Org. Mass Spectrom.*, **20**, 351 (1985).
138. H. van der Wel, N. M. M. Nibbering, E. E. Kingston and J. H. Beynon, *Org. Mass Spectrom.*, **20**, 537 (1985).
139. J. L. Holmes, *Org. Mass Spectrom.*, **20**, 169 (1985).
140. D. Kuck, *Org. Mass Spectrom.*, **29**, 113 (1994).

4. Mass spectrometry and gas-phase ion chemistry of phenols 329

141. J. Kjällstrand, O. Ramnäs and G. Petersson, *J. Chromotogr. A*, **824**, 205 (1998).
142. G. C. Galetti, A. Carnacini, P. Bocchini and A. Antonelli, *Rapid Commun. Mass Spectrom.*, **9**, 1331 (1995).
143. W. Genuit, J. J. Boon and O. Faix, *Anal. Chem.*, **59**, 508 (1987).
144. C. Schiegl, B. Helmreich and P. A. Wilderer, *Vom Wasser*, **88**, 137 (1997).
145. G. Gleixner, R. Bol and J. Balesdent, *Rapid Commun. Mass Spectrom.*, **13**, 1278 (1999).
146. V. Kováčik and J. Škamla, *Chem. Ber.*, **102**, 3623 (1969).
147. J. O. Metzger, C. Bicke, O. Faix, W. Tuszynski, R. Angermann, M. Karas and K. Strupat, *Angew, Chem.*, **104**, 777 (1992); *Angew. Chem., Int. Ed. Engl.*, **31**, 762 (1992).
148. J. L. Kerwin, F. Turecek, R. Xu, K. J. Kramer, T. L. Hopkins, C. L. Gatlin and J. R. Yates III, *Anal. Biochem.*, **268**, 229 (1999).
149. J. L. Kerwin, *Rapid Commun. Mass Spectrom.*, **11**, 557 (1997).
150. J. L. Kerwin, *J. Mass Spectrom.*, **31**, 1429 (1996).
151. P. Miketova, K. H. Schram, J. Whitney, M. Li, R. Huang, E. Kerns, S. Valcic, B. N. Timmermann, R. Rourick and S. Klohr, *J. Mass Spectrom.*, **35**, 860 (2000).
152. T. P. J. Izod and J. M. Tedder, *Proc. R. Soc. London, Ser. A*, **337**, 333 (1974).
153. S. A. Benezra and M. M. Bursey, *J. Am. Chem. Soc.*, **94**, 1024 (1972).
154. D. A. Chatfield and M. M. Bursey, *Int. J. Mass Spectrom. Ion Phys.*, **18**, 239 (1975).
155. M. Speranza and C. Sparapani, *Radiochim. Acta*, **28**, 87 (1981).
156. G. Occhiucci, F. Cacace and M. Speranza, *J. Am. Chem. Soc.*, **108**, 872 (1986).
157. C. S. Creaser and B. L. Williamson, *J. Chem. Soc., Chem. Commun.*, 1677 (1994).
158. C. S. Creaser and B. L. Williamson, *J. Chem. Soc., Perkin Trans. 2*, 427 (1996).
159. J. O. Lay, Jr. and M. L. Gross, *J. Am. Chem. Soc.*, **105**, 3445 (1983).
160. T. Donovan, C. C. Liou and J. Brodbelt, *J. Am. Soc. Mass Spectrom.*, **3**, 39 (1992).
161. G. F. Bauerle, Jr. and J. Brodbelt, *J. Am. Soc. Mass Spectrom.*, **6**, 627 (1995).
162. Z. B. Maksić, M. Eckert-Maksić and A. Knežević, *J. Phys. Chem.*, **102**, 2981 (1998).
163. K. V. Wood, D. J. Burinsky, D. Cameron and R. G. Cooks, *J. Org. Chem.*, **48**, 5236 (1983).
164. S. J. Pachuta, I. Isern-Flecha and R. G. Cooks, *Org. Mass Spectrom.*, **21**, 1 (1986).
165. I. Isern-Flecha, R. G. Cooks and K. V. Wood, *Int. J. Mass Spectrom. Ion Processes*, **62**, 73 (1984).
166. J. S. Brodbelt and R. G. Cooks, *Anal. Chim. Acta*, **206**, 239 (1988).
167. M. Speranza, N. Pepe and R. Cipollini, *J. Chem. Soc., Perkin Trans. 2*, 1179 (1979).
168. N. Pepe and M. Speranza, *J. Chem. Soc., Perkin Trans. 2*, 1430 (1981).
169. M. Attinà, F. Cacace, G. Ciranni and P. Giacomello, *J. Chem. Soc., Chem. Commun.*, 466 (1976).
170. M. Attinà, F. Cacace, G. Ciranni and P. Giacomello, *J. Am. Chem. Soc.*, **99**, 5022 (1977).
171. M. Attinà, F. Cacace, G. Ciranni and P. Giacomello, *J. Chem. Soc., Perkin Trans. 2*, 891 (1979).
172. M. Tsuji, T. Arikawa and Y. Nishimura, *Bull. Soc. Chem. Jpn.*, **73**, 131 (2000).
173. M. Tsuji, M. Aizawa and Y. Nishimura, *Bull. Soc. Chem. Jpn.*, **69**, 147 (1996).
174. M. Tsuji, M. Aizawa and Y. Nishimura, *J. Mass Spectrom. Jpn.*, **43**, 109 (1995).
175. M. von Ardenne, K. Steinfelder and R. Tümmler, *Elektronenanlagerungsmassenspektrometrie organischer Verbindungen*, Springer, Berlin, 1971.
176. H. Budzikiewicz, *Angew. Chem.*, **93**, 635 (1981); *Angew. Chem., Int. Ed. Engl.*, **20**, 624 (1981).
177. J. H. Bowie, *Mass Spectrom. Rev.*, **3**, 161 (1984).
178. H. Budzikiewicz, *Mass Spectrom. Rev.*, **5**, 345 (1986).
179. E. A. Stemmler and R. A. Hites, *Biomed. Environ. Mass Spectrom.*, **17**, 311 (1988).
180. J. H. Bowie, *Mass Spectrom. Rev.*, **9**, 349 (1990).
181. S. A. McLuckey and G. L. Glish, *Org. Mass Spectrom.*, **22**, 224 (1987).
182. J. E. Bartmess and R. T. McIver, Jr., in *Gas Phase Ion Chemistry*, Vol. 2 (Ed. M. T. Bowers), Chap. 11, Academic Press, New York, 1979, pp. 87–121.
183. J. I. Brauman and L. K. Blair, *J. Am. Chem. Soc.*, **92**, 5986 (1970).
184. J. E. Bartmess, J. A. Scott and R. T. McIver, Jr., *J. Am. Chem. Soc.*, **101**, 6056 (1979).
185. J. E. Bartmess, J. A. Scott and R. T. McIver, Jr., *J. Am. Chem. Soc.*, **101**, 6046 (1979).
186. R. W. Taft, I. A. Koppel, R. D. Topsom and F. Anvia, *J. Am. Chem. Soc.*, **112**, 2047 (1990).
187. R. Voets, J. P. François, J. M. L. Martin, J. Mullens, J. Yperman and L. C. Van Poucke, *J. Comput. Chem.*, **11**, 269 (1990).
188. T. Kremer and P. v. R. Schleyer, *Organometallics*, **16**, 737 (1997).

189. D. F. DeFrees, R. T. McIver, Jr. and W. J. Hehre, *J. Am. Chem. Soc.*, **102**, 3334 (1980).
190. For more recent work, see: F. G. Bordwell and W. Z. Liu, *J. Am. Chem. Soc.*, **118**, 10819 (1996).
191. J. I. Brauman and L. K. Blair, *J. Am. Chem. Soc.*, **92**, 5986 (1970).
192. R. T. McIver, Jr. and J. H. Silvers, *J. Am. Chem. Soc.*, **95**, 8462 (1973).
193. I. A. Koppel, R. W. Taft, F. Anvia, S. Z. Zhu, L. Q. Hu, K. S. Sung, D. D. DesMarteau, L. M. Yagupolskii, Yu. L. Yagupolskii, N. V. Ignat'ev, N. V. Kondratenko, A. Yu. Volkonskii, V. M. Vlasov, R. Notario and P. C. Maria, *J. Am. Chem. Soc.*, **116**, 3047 (1994).
194. I. Dzidic, D. I. Carroll, R. N. Stillwell and E. C. Horning, *J. Am. Chem. Soc.*, **96**, 5258 (1974).
195. M. Meot-Ner (Mautner) and L. W. Sieck, *J. Am. Chem. Soc.*, **108**, 7525 (1986).
196. T. Zeegers-Huyskens, *Chem. Phys. Lett.*, **129**, 172 (1986).
197. R. Yamdagni and P. Kebarle, *J. Am. Chem. Soc.*, **93**, 7139 (1971).
198. G. J. C. Paul and P. Kebarle, *J. Am. Chem. Soc.*, **113**, 1148 (1991).
199. G. J. C. Paul and P. Kebarle, *Can. J. Chem.*, **68**, 2070 (1990).
200. J. B. Cumming, A. French and P. Kebarle, *J. Am. Chem. Soc.*, **99**, 6999 (1977).
201. J. A. Dodd, S. Baer, C. R. Moylan and J. I. Brauman, *J. Am. Chem. Soc.*, **113**, 5942 (1991).
202. W. E. Farneth and J. I. Brauman, *J. Am. Chem. Soc.*, **98**, 7891 (1976).
203. H. P. Tannenbaum, J. D. Roberts and R. C. Dougherty, *Anal. Chem.*, **47**, 49 (1975).
204. G. W. Caldwell, J. A. Masucci and M. G. Ikonomou, *Org. Mass Spectrom.*, **24**, 8 (1989).
205. G. B. Anderson, R. B. Johns, Q. N. Porter and M. G. Strachan, *Org. Mass Spectrom.*, **19**, 584 (1984).
206. M. G. Strachan, G. B. Anderson, Q. N. Porter and R. B. Johns, *Org. Mass Spectrom.*, **22**, 670 (1987).
207. G. B. Anderson, R. G. Gillis, R. B. Johns, Q. N. Porter and M. G. Strachan, *Org. Mass Spectrom.*, **19**, 199 (1984).
208. R. W. Binkley, M. J. S. Tevesz and W. Winnik, *J. Org. Chem.*, **57**, 5507 (1992).
209. P. C. H. Eichinger, J. H. Bowie and R. N. Hayes, *J. Am. Chem. Soc.*, **111**, 4224 (1989).
210. P. C. H. Eichinger, J. H. Bowie and R. N. Hayes, *Aust. J. Chem.*, **42**, 865 (1989).
211. K. D. Jordan, J. A. Michejda and P. D. Burrow, *J. Am. Chem. Soc.*, **98**, 7189 (1976).
212. S. Erhardt-Zabik, J. T. Watson and M. J. Zabik, *Biomed. Environ. Mass Spectrom.*, **19**, 101 (1990).
213. I. K. Gregor and M. Guilhaus, *Int. J. Mass Spectrom. Ion Processes*, **56**, 167 (1984).
214. P. Vouros and T. M. Trainor, *Adv. Mass Spectrom.*, **10**, 1405 (1986).
215. H. Beck, A. Dross, K. Eckart, W. Mathar and R. Wittkowski, *Chemosphere*, **19**, 167 (1989).
216. H. R. Buser and C. Rappe, *Anal. Chem.*, **56**, 442 (1984).
217. M. J. Incorvia Mattina and L. J. Searls, *Org. Mass Spectrom.*, **27**, 105 (1992).
218. F. Fournier, J. C. Tabet, L. Debrauwer, D. Rao, A. Paris and G. Bories, *Rapid Commun. Mass Spectrom.*, **5**, 44 (1991).
219. L. Debrauwer, A. Paris, D. Rao, F. Fournier and J. C. Tabet, *Org. Mass Spectrom.*, **27**, 709 (1992).
220. F. Fournier, B. Remaud, T. Blasco and J. C. Tabet, *J. Am. Soc. Mass Spectrom.*, **4**, 343 (1993).
221. F. Fournier, M. C. Perlat and J. C. Tabet, *Rap. Commun. Mass Spectrom.*, **9**, 13 (1995).
222. C. E. Döring, D. Estel, W. Pehle, M. Gaikowski and K. Seiffarth, *J. Chromatogr.*, **348**, 430 (1985).
223. J. Cheung and R. J. Wells, *J. Chromatogr. A*, **771**, 203 (1997).
224. T. Heberer and H. J. Stan, *Anal. Chim. Acta*, **341**, 21 (1997).
225. T. J. Clark and J. E. Bunch, *J. Chromatogr. Sci.*, **34**, 272 (1996).
226. M. A. Crespín, S. Cárdenas, M. Gallego and M. Valcárcel, *J. Chromatogr. A*, **830**, 165 (1999).
227. M. A. Crespín, S. Cárdenas, M. Gallego and M. Valcárcel, *Rap. Commun. Mass Spectrom.*, **12**, 198 (1998).
228. J. P. G. Wilkins, C. P. Yorke and M. R. Winkler, *Rapid Commun. Mass Spectrom.*, **11**, 206 (1997).
229. D. Puig, D. Barcelo, I. Silgoner and M. Grasserbauer, *J. Mass Spectrom.*, **31**, 1297 (1996).
230. D. T. Williams, Q. Tran, P. Fellin and K. A. Brice, *J. Chromatogr.*, **549**, 297 (1991).

231. K. Gamoh and K. Nakashima, *Rapid Commun. Mass Spectrom.*, **13**, 1112 (1999).
232. T. Letzel, U. Pöschl, E. Rosenberg, M. Grasserbauer and R. Niessner, *Rapid Commun. Mass Spectrom.*, **13**, 2456 (1999).
233. M. Möder, D. Zimmer, J. Stach and R. Herzschuh, *Fuel*, **68**, 1422 (1989).
234. J. Stach, D. Zimmer, M. Möder and R. Herzschuh, *Org. Mass Spectrom.*, **24**, 946 (1989).
235. T. R. Croley and B. C. Lynn, Jr., *Rapid Commun. Mass Spectrom.*, **12**, 171 (1998).
236. R. P. Hanzlik, K. Hogberg and C. M. Judson, *Biochemistry*, **23**, 3048 (1984).
237. C. D. Gutsche, *Calixarenes*, The Royal Society of Chemistry, Cambridge, 1989.
238. V. Böhmer, *Angew. Chem.*, **107**, 785 (1995); *Angew. Chem., Int. Ed. Engl.*, **34**, 713 (1995).
239. J. Vicens and V. Böhmer (Eds.), *Calixarenes: A Versatile Class of Macrocyclic Compounds*, Kluwer Academic Publishers, Dordrecht, 1991.
240. A. Ikeda and S. Shinkai, *Chem. Rev.*, **97**, 1713 (1997).
241. M. C. Letzel, C. Agena and J. Mattay, *Eur. Mass Spectrom.*, **7**, 35 (2001).
242. J. C. Ma and D. A. Dougherty, *Chem. Rev.*, **97**, 1303 (1997).
243. K. Murayama and K. Aoki, *J. Chem. Soc., Chem. Commun.*, 119 (1997).
244. J. F. Gal, P. C. Maria, M. Decouzon, O. Mó and M. Yáñez, *Int. J. Mass Spectrom.*, **219**, 445 (2002).
245. M. Kamieth, F. G. Klärner and F. Diederich, *Angew. Chem.*, **110**, 3497 (1998); *Angew. Chem., Int. Ed.*, **37**, 3303 (1998).
246. F. G. Klärner, U. Burkert, M. Kamieth, R. Boese and J. Benet-Buchholz, *Chem. Eur. J.*, **5**, 1700 (1999).
247. M. T. Blanda, D. B. Farmer, J. S. Brodbelt and B. J. Goolsby, *J. Am. Chem. Soc.*, **122**, 1486 (2000).
248. B. J. Goolsby, J. S. Brodbelt, E. Adou and M. Blanda, *Int. J. Mass Spectrom.*, **193**, 197 (1999).
249. F. Allain, H. Virelezier, C. Moulin, C. K. Jankowski, J. F. Dozol and J. C. Tabet, *Spectroscopy*, **14**, 127 (2000).
250. J. M. J. Nuutinen, A. Irico, M. Vincenti, E. Dalcanale, J. M. H. Pakarinen and P. Vainiotalo, *J. Am. Chem. Soc.*, **122**, 10090 (2000).
251. C. A. Schalley, R. K. Castellano, M. S. Brody, D. M. Rudkevich, G. Siuzdak and J. Rebek, Jr., *J. Am. Chem. Soc.*, **121**, 4568 (1999).
252. M. C. Letzel, C. Agena and J. Mattay, *J. Mass Spectrom.*, **37**, 63 (2002).
253. M. Karas, D. Bachmann and F. Hillenkamp, *Anal. Chem.*, **57**, 2935 (1985).
254. M. Karas and F. Hillenkamp, *Anal. Chem.*, **60**, 2299 (1988).
255. F. Hillenkamp, M. Karas, R. C. Beavis and B. T. Chait, *Anal. Chem.*, **63**, A1193 (1991).
256. K. Tanaka, H. Waki, Y. Ido, S. Akita, Y. Yoshida and T. Yoshida, *Rapid Commun. Mass Spectrom.*, **2**, 151 (1988).
257. C. M. Land and G. R. Kinsel, *J. Am. Soc. Mass Spectrom.*, **12**, 726 (2001).
258. Y. Huang and D. H. Russell, *Int. J. Mass Spectrom. Ion Processes*, **175**, 187 (1998).
259. P. J. Calba, J. F. Muller and M. Inouye, *Rapid Commun. Mass Spectrom.*, **12**, 1727 (1998).
260. K. J. Wu, T. A. Shaler and C. H. Becker, *Anal. Chem.*, **66**, 1637 (1994).
261. M. P. Chiarelli, A. G. Sharkey, Jr. and D. M. Hercules, *Anal. Chem.*, **65**, 307 (1993).
262. V. Bökelmann, B. Spengler and R. Kaufmann, *Eur. Mass Spectrom.*, **1**, 81 (1995).
263. V. Karbach and R. Knochenmuss, *Rapid Commun. Mass Spectrom.*, **12**, 968 (1998).
264. M. Karas, M. Glückmann and J. Schäfer, *J. Mass Spectrom.*, **35**, 1 (2000).
265. R. Knochenmuss, E. Lehmann and R. Zenobi, *Eur. Mass Spectrom.*, **4**, 421 (1998).
266. E. Lehmann, R. Zenobi and S. Vetter, *J. Am. Soc. Mass Spectrom.*, **10**, 27 (1999).
267. E. Lehmann, R. Knochenmuss and R. Zenobi, *Rapid Commun. Mass Spectrom.*, **11**, 1483 (1997).
268. H. Ehring, M. Karas and F. Hillenkamp, *Org. Mass Spectrom.*, **27**, 472 (1992).
269. R. Knochenmuss, V. Karbach, U. Wiesli, K. Breuker and R. Zenobi, *Rapid Commun. Mass Spectrom.*, **12**, 529 (1998).
270. J. F. Gal, P. C. Maria and E. D. Raczyńska, *J. Mass Spectrom.*, **36**, 699 (2001).
271. T. B. McMahon and P. Kebarle, *J. Am. Chem. Soc.*, **99**, 2222 (1977).

272. T. B. McMahon and P. Kebarle, *J. Am. Chem. Soc.*, **98**, 3399 (1976).
273. A. Meffert and J. Grotemeyer, *Eur. J. Mass Spectrom.*, **1**, 594 (1995).
274. A. Meffert and J. Grotemeyer, *Ber. Bunsenges. Phys. Chem.*, **102**, 459 (1999).
275. K. Breuker, R. Knochenmuss and R. Zenobi, *Int. J. Mass Spectrom.*, **184**, 25 (1999).
276. R. J. J. M. Steenvorden, K. Breuker and R. Zenobi, *Eur. Mass Spectrom.*, **3**, 339 (1997).
277. T. J. D. Jørgensen, G. Bojesen and H. Rahbek-Nielsen, *Eur. Mass Spectrom.*, **4**, 39 (1998).
278. R. D. Burton, C. H. Watson, J. R. Eyler, G. L. Lang, D. H. Powell and M. Y. Avery, *Rapid Commun. Mass Spectrom.*, **11**, 443 (1997).
279. A. Meffert and J. Grotemeyer, *Int. J. Mass Spectrom.*, **210/211**, 521 (2001).

CHAPTER 5

NMR and IR spectroscopy of phenols

POUL ERIK HANSEN and JENS SPANGET-LARSEN

Department of Life Sciences and Chemistry, Roskilde University, P.O. Box 260, DK-4000 Roskilde, Denmark
Fax: +45 46743011; e-mail: poulerik@ruc.dk; spanget@ruc.dk

I. INTRODUCTION	335
II. NMR	335
A. Introduction	335
B. OH Exchange	336
C. δOH	336
D. ^{13}C Chemical Shifts	338
1. Reference values of chemical shifts	338
2. Chemical shift patterns	339
3. Anisotropy	339
E. ^{17}O Chemical Shifts	339
1. Substituent effects	339
2. Hydrogen bonding	340
3. Solvent effects	341
F. Deuterium Isotope Effects on Chemical Shifts	342
1. Experimental conditions	342
2. $^n\Delta C(D)$	342
a. $^n\Delta C(OD)$	342
b. $^2\Delta C(D)$	345
3. $^n\Delta OH(OD)$	345
4. Solvent isotope effects	346
5. $^1\Delta^{13}C(^{18}O)$ isotope effects	346
6. $^1\Delta O(OD)$	346
7. Primary isotope effects	347
G. Coupling Constants	349
1. $J(X,OH)$ coupling constants	349
a. $J(^{13}C,OH)$	349

The Chemistry of Phenols. Edited by Z. Rappoport
© 2003 John Wiley & Sons, Ltd ISBN: 0-471-49737-1

    b. $^nJ(C,OD)$ 349
    c. $^5J(OH,H)$ 349
    d. $^1J(H, ^{17}O)$ 349
   2. $^nJ(^{13}C,^{13}C)$ 350
   3. $^nJ(^{13}C,^{1}H)$ 350
 H. Hydrogen Bonding 350
   1. Hydrogen bond strength 350
   2. Ranking of substituents as hydrogen bond partners 350
   3. Multiple hydrogen bonding to the same acceptor 351
   4. Conformational equilibria 352
   5. Bifurcated hydrogen bonds to hydrogen 352
   6. Strong hydrogen bonds 352
   7. Rotation 353
 I. Steric Effects 354
 J. Proton Transfer 355
 K. Tautomeric Equilibria 356
   1. General introduction 356
   2. *o*-Hydroxyazo compounds 358
   3. Mannich bases 358
   4. Schiff bases 359
   5. Nitrosophenols 360
   6. Equilibrium isotope effects 361
 L. Complexes 363
   1. Proton transfer 363
   2. Weak complexes 364
 M. Theoretical Calculations 366
 N. Solid State NMR 366
 O. Fractionation Factors 367
III. IR 367
 A. Introduction 367
 B. The Characteristic Vibrations of the Phenolic OH Group 369
   1. OH stretching, ν(OH) 369
    a. Free OH groups 369
    b. Hydrogen bonded OH groups 369
   2. OH out-of-plane bending, γ(OH) 371
 C. The IR Spectrum of Phenol 372
   1. Low-temperature Argon matrix spectrum of phenol 372
   2. IR polarization spectra of phenol aligned in a liquid crystal 372
 D. Hydrogen Bonded Complexes 375
 E. Phenols with Intramolecular Hydrogen Bonds 377
   1. 2-Hydroxybenzoyl compounds 377
   2. Hydroxyquinones 382
   3. 2-Nitrophenols 384
   4. 2-Nitrosophenols 385
   5. Schiff bases 385
   6. *ortho*-Mannich bases 385
IV. ACKNOWLEDGEMENTS 386
V. REFERENCES 386

I. INTRODUCTION

Phenols are major constituents of many biological and naturally occurring compounds. The structural element C=C−O−H formally existing in phenols clearly indicates some of the key features of phenols, their acidity and related to that their ability to form complexes, the ability to take part in hydrogen bonding (intra- and intermolecular) and tautomerism. These subjects will be some of the key features in this review of IR and NMR spectroscopy of phenols. Hydrogen bonds make phenols interesting partners in self-association or self-organizing systems. As already indicated, phenols are almost ubiquitous in the plant and animal kingdom and therefore in transformation products such as tar, coal, oil, humic substances etc. Phenols are also often components of polymers. In order to limit the review no attention will be paid to quantitative analysis, nor will specific groups of compounds like the above-mentioned be treated specifically. The emphasis will be placed on general features covering phenols whether these are benzene derivatives, polycyclic or heteroaromatic hydroxy compounds with none or one or more additional functional groups. The phenol part must be a major constituent of the compounds, but the borderline is diffuse. Likewise are polyhydroxy heteroaromatics existing primarily in the keto forms not treated.

The presence of an aromatic moiety clearly has very important consequences for the NMR and IR spectra and the structural element mentioned above also illustrates the vibrational coupling between the hydroxy group and the aromatic ring vibrational modes. The NMR part will cover analysis of chemical shifts both in solution and in the solid. The nuclei immediately coming to mind are ^{1}H, ^{13}C and ^{17}O as these are vital parts of the phenol moiety, but others such as ^{15}N and ^{19}F can also be present. Furthermore, as isotope effects on chemical shifts depend on vibrations they combine NMR and IR theory. *Ab initio* calculations of NMR properties such as chemical shifts and isotope effects can be very useful in studying some of these systems. These types of calculations are likewise invaluable in interpreting vibrational spectra.

A few symbols are common to both NMR and IR literature; one of these is δ, meaning in NMR chemical shift and in IR an in-plane bending vibration.

II. NMR

A. Introduction

NMR is clearly a very versatile technique in the study of phenols and in particular the important charge distribution in them. An example is the titration of phenols leading to phenolate ions, which is accompanied by distinct chemical shift changes (Table 1). Differences between the chemical shifts of C-1 and C-4 of phenols upon titration is a useful parameter when estimating the extent of deprotonation of phenols. Examples are given for 2,3-dichloro-, 2,5-dichloro- and 2,3,5-trichlorophenol[1]. This sensitivity also means that pK_a values of phenols can be determined. This is of particular interest in compounds containing several phenol groups e.g. proteins having more than one tyrosine. The advantage of the NMR technique is the ability to determine the individual pK_a values simultaneously. Phenols of known pK_a values have also been used in co-titration studies[2]. The OH group plays a central role in NMR studies of phenols. The O^{1}H chemical shift is a key parameter. The orientational dependence of the OH group will be treated as well as its reorientation (kinetics of rotation). As the present chapter covers both NMR and IR spectroscopy, an obvious inclusion is the effects of isotope substitution on chemical shifts as these are of vibrational origin.

TABLE 1. NMR parameters for phenol and phenolate ion

No. of C	$\delta^{13}C$	$\delta^{13}C^a$	δ^1H^b	$\delta^{17}O$
1	155.6	168.3		77.3
2	116.1	120.5	6.70	
3	130.5	130.6	7.14	
4	120.8	115.1	6.81	

[a] Chemical shift for the phenolate anion.
[b] The OH chemical shift may vary both with the solvent and the concentration.

B. OH Exchange

The intermolecular exchange of the OH proton is of vital importance for the appearance and interpretation of NMR spectra of phenols. Intermolecular exchange determines the position (chemical shift) of the OH resonance (see Section II.C). Splitting due to the OH proton (or deuterium) is only seen if the exchange is slow on the corresponding NMR time scale. Coupling constants to OH protons are quite often not observed because of too fast exchange (see Section II.G.1) or may be removed by heating[3]. Isotope effects due to deuteriation at the OH position may likewise not be directly observable (see Section II.F.1). In order to slow down the exchange, dry solvents must be used. Hydrogen bonding solvents such as DMSO are also useful. Finally, the temperature can in some cases, solvent permitting, be lowered.

However, NMR measurements give quite often estimates of the exchange parameters. Electron-withdrawing groups at the *p*-position seem to increase the ease of exchange judging from the difficulty of observing deuterium isotope effects at the ^{13}C NMR spectra of 5-nitrosalicylaldehyde[4]. On the other hand, large alkyl substituents at the *ortho*-position to the OH group seem to slow down the exchange. This is most likely related to the exchange mechanism in which the OH group has to swing away from its hydrogen bond partner before exchange can take place[5,6].

C. δOH

The OH chemical shifts have been studied intensely. The shifts are clearly solvent and concentration dependent[7] and must be extrapolated to infinite dilution before comparisons can be made. They depend on intramolecular hydrogen bonding. δ(OH) of **1** is 12.26 ppm, that of **2** 14.74 (OH-2) and 14.26 ppm (OH-4), that of **3** 16.24 (OH-2), 14.5 (OH-4) and 10.4 (OH-6)[39,40]. They have been used to estimate hydrogen bond strength (see Section II.H.1). In ultimate intramolecularly hydrogen bonded non-tautomeric cases like **4** an OH chemical shift value as high as 17.09 ppm is found[39]. These values can clearly be used to estimate hydrogen bond strength (see Section II.H.1). As ring current effects may contribute, these also must be taken into account and subtracted if the values are to be used to estimate hydrogen bond strength. In intermolecular hydrogen bonded complex the shift can be even higher (see Section II.L.1).

The phenolic protons move to higher frequency upon cooling. A linear relationship between temperature and OH chemical shift is found. The temperature coefficient is close to −4 ppb per degree in chloroform, cyclohexane and acetone. Slightly numerically smaller values are found in acetonitrile and methylene chloride. A much more negative value is found in benzene. The variations are ascribed to the influence of resonance forms and variation in conformational changes resulting from different types of solute–solvent interactions[8]. A large set of data for *o*-hydroxyacyl aromatics show values between 2

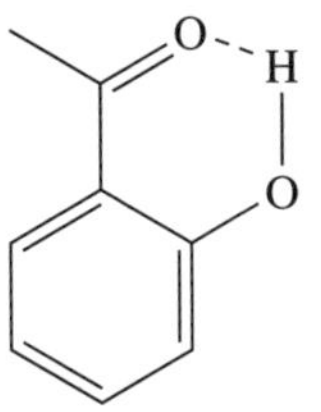

(1) 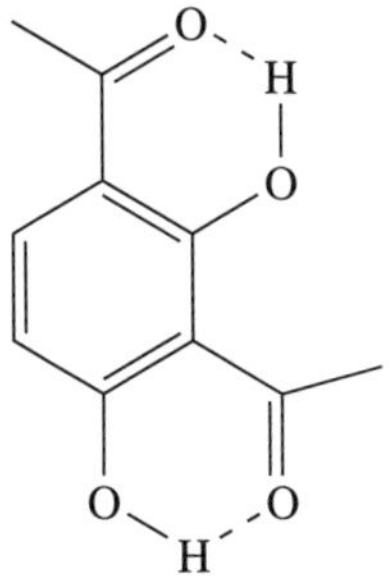 (2) 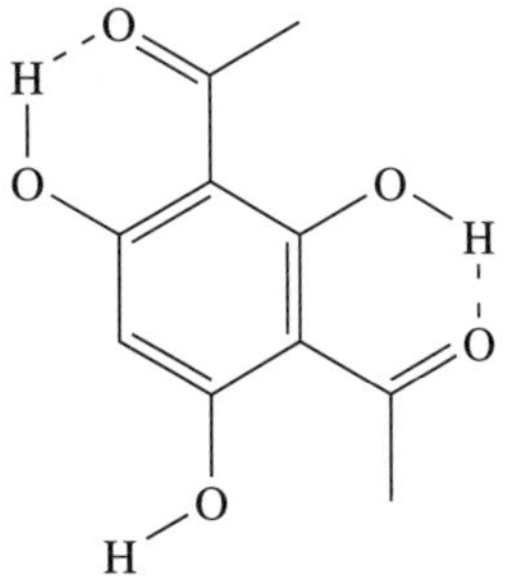 (3)

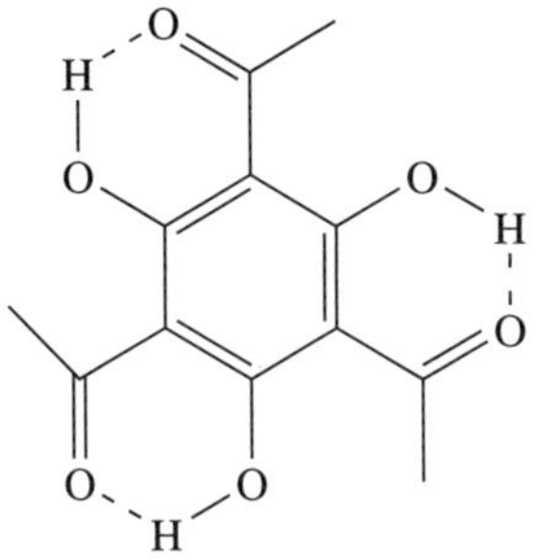

(4)

and 10 ppb. The largest temperature coefficients are seen for compounds with the lowest OH chemical shifts again pointing to a relation with hydrogen bond strength. The difference between hydrogen bonded and free OH protons with temperature is shown for the 2,6-dihydroxy derivatives **5**, R^2 = Me. It is found that the free OH proton has a larger temperature coefficient than the hydrogen bonded one at low temperature at which the two forms can be observed individually. For **6**, R^2 = H the average temperature coefficient is smaller than for **5**, R^2 = H[5], probably due to the hydrogen bond of the OH-6 group to the OR^1 moiety.

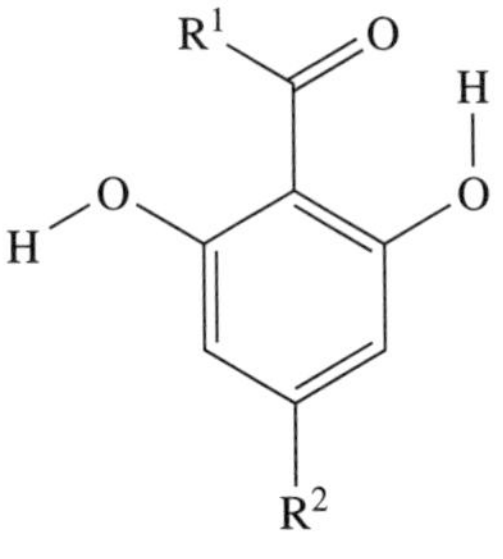

(5) 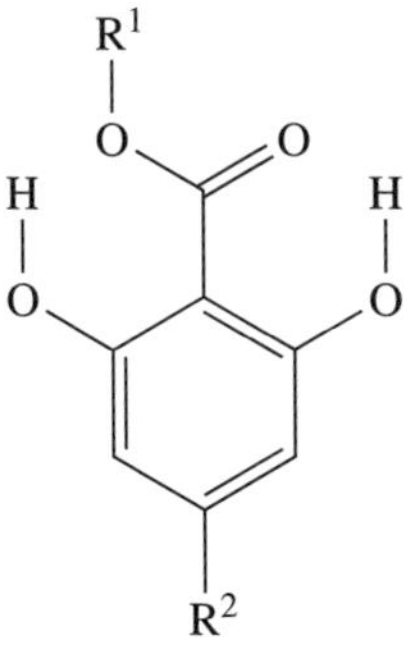 (6)

Bertolasi and coworkers[9] have related the OH chemical shifts of *o*-hydroxyacyl aromatics to the oxygen–oxygen distance. A plot of δOH vs. the oxygen–oxygen distance ($R_{O...O}$) in Å shows a reasonable linear relationship (equation 1).

$$\delta\text{OH} = -34.1\ (\pm 2.6)\ R_{O...O} + 100.3\ (\pm 6.4),\ r = -0.88 \tag{1}$$

The authors themselves point to the unusual correspondence considering the different conditions (solution and solid state) and that account must be taken of the fact that compounds with intramolecular hydrogen bonds in solution are intermolecular in the solid state. Furthermore, a number of the compounds are tautomeric and the predominant form in solution and in the solid state is not necessarily the same. A rather poor fit to equation 1 was seen for daunomycin[10].

D. ^{13}C Chemical Shifts

1. Reference values of chemical shifts

^{13}C chemical shifts for simple phenols are given by Kalinowski, Berger and Braun[11]. ^{1}H, ^{13}C and ^{17}O chemical shifts are likewise listed in Table 1. As mentioned in the section on anisotropy, these values are only valid as long as 'free' rotation occurs. Chemical shifts for special groups of compounds are given for prenylphenols[12], anthraquinones[13], acetophenones[14a], benzophenones[14b] and hydroxy derivatives of naphthalene[15]. Although phenols are generally well soluble in water, hydrophobic substituents may change this pattern. For hydrophobic compounds like, e.g., 2,6-di-*tert*-butyl-4-methylphenol (also known as butylated hydroxytoluene), a high frequency shift of all resonances is observed in water solutions compared to organic solvents due to emulsion formation[16].

Hydroxy-substituted polycyclic aromatics (PAH) with hydroxy substituents are not radically different from those of the corresponding benzenes except in their larger ability to form keto forms, to form complexes and their ability to delocalize charge. The latter is well documented in the long-range substituent effects, e.g. on ^{13}C chemical shifts of PAHs such as pyrene if the hydroxy group is in a well conjugated position. For hydroxypyrene[17] we observed for the 1-position (**7**) (well conjugated) and the 2-position (**8**) (poorly conjugated) the substituent effects shown in **7** and **8**.

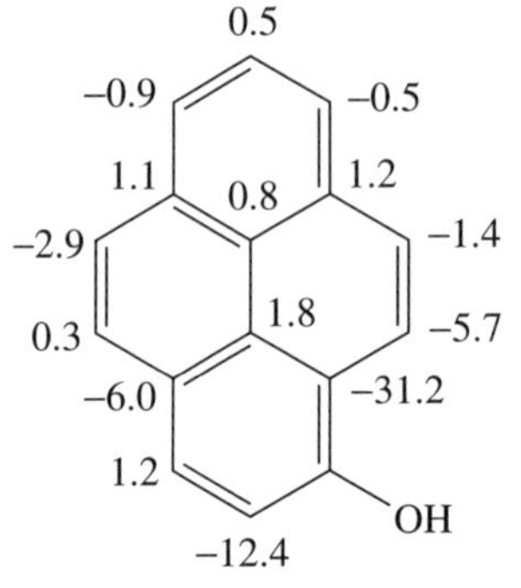

(**7**) 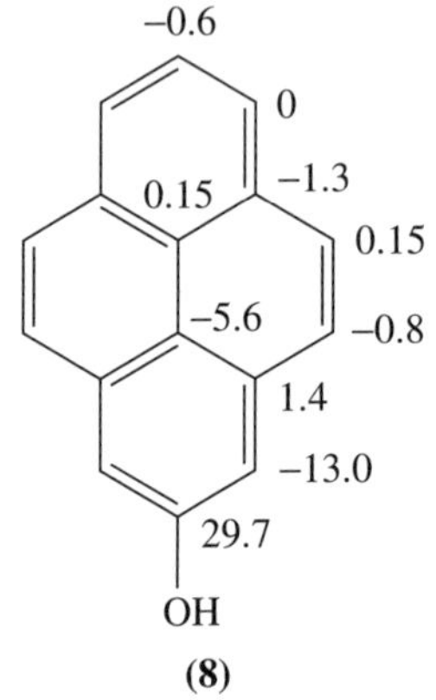   (**8**)

2. Chemical shift patterns

The OH group exerts a strong influence on both the ^{1}H and ^{13}C chemical shifts of the phenol ring. This will lead to very distinct chemical shifts (in ppm) for multiply substituted rings as often observed in biological material, as seen in **9–11**. Such patterns could help to identify commonly occurring patterns in non-homogeneous materials such as lignins or fulvic or humic acids[18,19].

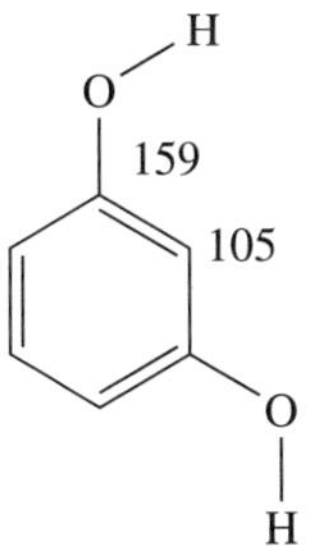

Values reported in chemical shift tables assume 'free' rotation of the OH group. The anisotropy of this can be judged either from hydrogen bonded cases, from solid state NMR spectra in which the two *ortho* protons or carbons have become non-equivalent, or from theoretical calculations of chemical shifts (see Section II.M).

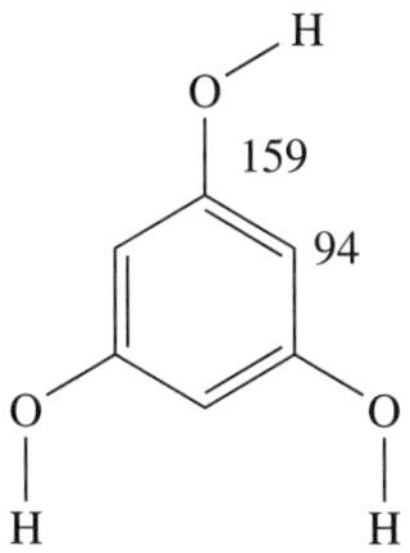

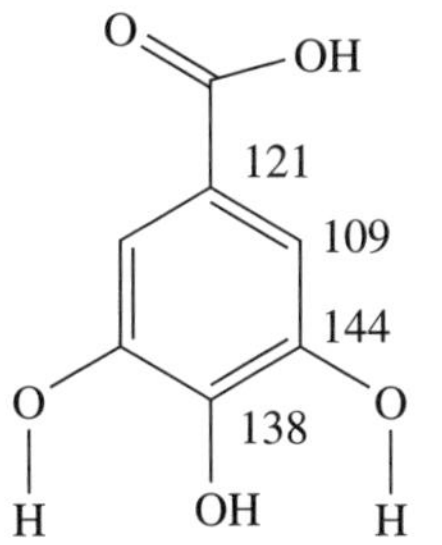

It is not possible, at least for hydrogen bonded cases, to correlate ^{13}C chemical shifts of phenols with other parameters[4].

3. Anisotropy

Of importance in understanding chemical shift patterns of phenols is, of course, also the effect of taking part in hydrogen bonding as, e.g., in salicylaldehyde, *o*-hydroxyacetophenones etc. Firstly, the anisotropy caused by the OH group but also the anisotropy effects of the other substituent (aldehyde, ketone etc.) lead to extensive non-additivity if using the standard values mentioned above.

The anisotropy of the OH group was obvious from measurements of splittings caused by isotopic perturbation (SIP) values in 2,6-dihydroxyacetophenones[20]. The anisotropy due to the OH group has been calculated in phenol. Depending slightly on the method and the basis set, the difference in chemical shifts between C-2 and C-6 is 3.5 to 5 ppm[21].

E. ^{17}O Chemical Shifts

The present review will concentrate on ^{17}OH chemical shifts. These have not been studied so intensely because measurement of ^{17}O resonances are best done on enriched compounds and the preparation of ^{17}O enriched phenols is not simple. For a review see Boykin's book[22].

1. Substituent effects

The effects of substituents are clearly demonstrated in the shifts for **12–15**[22]. The shift decreases with electron-attracting substituents and increases with electron-donating ones.

The effect of a nitro group is clearly diminished when the nitro group is twisted out of the ring plane, as seen by a comparison of 2-nitrophenol and 2-nitro-3-methylphenol[23]. The compound 3-hydroxy-9-fluorenone shows an OH chemical shift of 92.5 ppm[24]. This again can be related to the effects of the C=O group, possibly in conjunction with a formal biphenyl moiety.

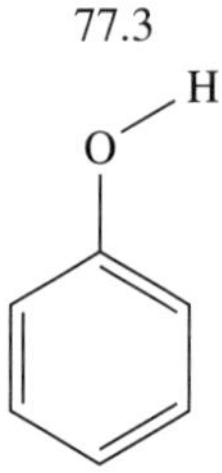

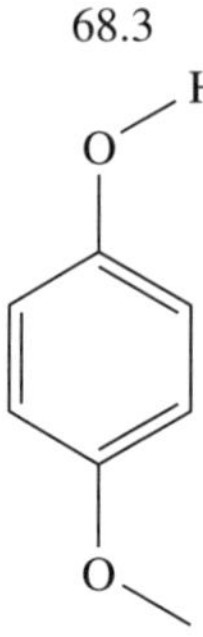

For 2-hydroxy-1-naphthaldehyde[25,26], 2-acetyl-1-naphthol and 1-acetyl-2-naphthol[26], the ^{17}OH chemical shift is larger than 92 ppm. This is distinctly larger than for the corresponding benzene derivatives. This can either be ascribed to stronger hydrogen bonding (see Section E.2) or to the more effective delocalization of the lone-pairs by the aromatic system.

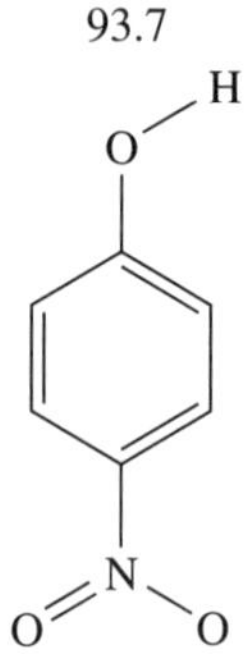

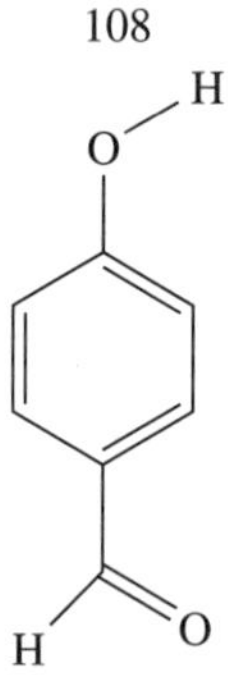

2. Hydrogen bonding

Related to substituent effects is the effect of intramolecular hydrogen bonding. The effect of hydrogen bonding has been discussed extensively for intramolecularly hydrogen bonded systems, but with emphasis on the hydrogen bond acceptor oxygen as found, e.g., in carbonyl groups[27,28].

For phenolic oxygens the picture is as seen for **15** (see above) and **16**. This shows a low frequency shift caused by hydrogen bonding. A similar picture is seen for intramolecularly hydrogen bonded nitro compounds[23]. However, these data consist of both intra- and intermolecular hydrogen bonding effects as well as proximity effects.

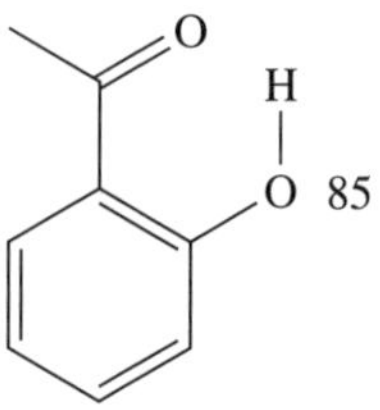

A more elaborate plot of the shifts of 5-substituted salicylaldehydes[29] vs. the corresponding *para*-substituted phenols showed likewise that the aldehyde group at the *ortho*-position caused only a moderate (5 ppm) high frequency shift, again indicating that hydrogen bonding is opposing the normal substituent effect (see above)[29].

Boykin[30] compared data of 1,4- and 1,2-dihydroxybenzenes and 1-hydroxy-4-methoxy- and 1-hydroxy-2-methoxybenzene (**13, 17–19**) and found that formation of a hydrogen bond to the singly bonded oxygen of both OH and OCH$_3$ causes shielding of the ^{17}O resonance. This effect is considerably smaller than for the C=O group, but in the latter the hydrogen bond is stronger. In the case of 1,2-dihydroxybenzene the effect of the OH being a hydrogen bond donor has not been taken into account. As seen above this effect is probably small, but it exists.

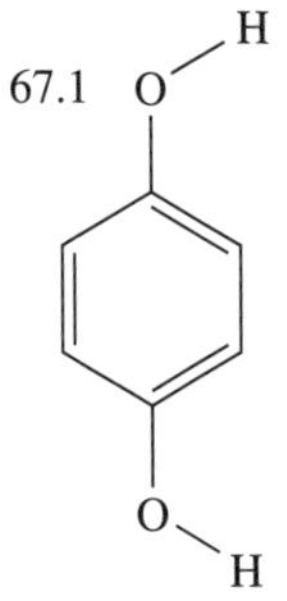

(**17**) 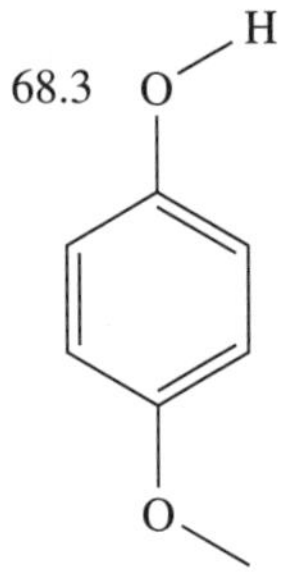   (**13**) 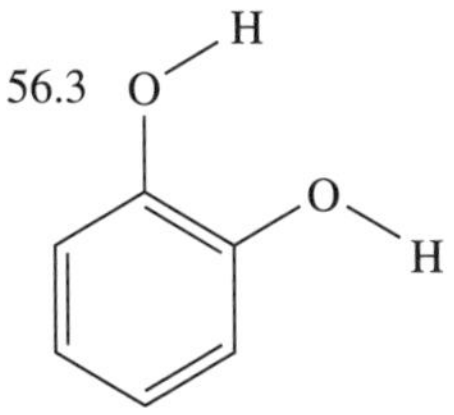   (**18**) 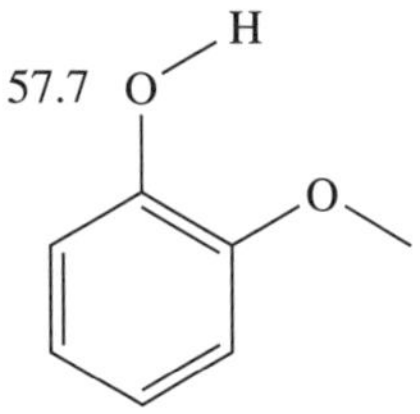   (**19**)

^{17}O chemical shifts of phenols hydrogen bonded to heteroaromatic nitrogens in systems like *o*-hydroxypyridines or similar compounds with one or more nitrogens or hydroxy groups show ^{17}OH chemical shifts that are very similar (94–97 ppm), with the exception of a *para*-substituted methoxy derivative (90 ppm)[26], but this can be ascribed to a simple substituent effect (see above).

Cerioni and coworkers[31] investigated calixarenes. For calix[6]arene, a shift very similar to that of 2,6-dimethylphenol was found, thereby showing that the steric hindrance was similar. For the corresponding calix[4]arene a dramatically higher value was observed. This was ascribed to stronger hydrogen bonding in the calix[4]arene.

3. Solvent effects

Many ^{17}O chemical shifts have been measured at high temperature and in a low viscosity solvent like acetonitrile. However, solvents play a small role as seen for 5-hydroxy-1,4-naphthoquinone: 84.5 ppm in toluene, 83 in acetonitrile and 84.1 in CDCl$_3$. A difference of 1 ppm between toluene and acetonitrile was also observed for 2′-hydroxypropiophenone, 2′-hydroxybenzophenone and 2′-hydroxyacetophenone[25]. For the conformationally flexible 2,2′-dihydroxybenzophenone δ^{17}O is 85 ppm in acetonitrile, 86 in toluene and 84.4 ppm in CDCl$_3$. In acetonitrile only one hydrogen bond exists (see Section II.H.4)[25]. Pyridine as solvent has a strong effect at the ^{17}O chemical shift of calix[6]arene, but not for calix[4]arene, whereas the effect at phenol and 2,6-dimethylphenol is similar and slightly smaller than that for calix[6]arene. This either points towards a stronger intramolecular hydrogen bond in the calix[4]arene or to a more shielding environment around the phenolic groups. The reason for these apparent different trends in different types of compounds is at present unclear[31].

F. Deuterium Isotope Effects on Chemical Shifts

These effects have recently been reviewed by Dziembowska and Rozwadowski[32,33] and, for the more specific cases of intramolecularly hydrogen bonded cases, by Perrin and Nielson[34] and by Bolvig and Hansen[35]. Therefore, only a brief summary will be given including more recent developments.

1. Experimental conditions

Deuterium isotope effects on chemical shifts of phenols of which the OH proton has been exchanged by deuterium can be measured in two different ways. If the OH(D) proton is exchanging slowly (see Section II.B) two different resonances are observed, one due to the protio and one due to the deuterio species (see Figure 1). The relative intensities will depend on the H : D ratio, perhaps not in a quantitative way due to fractionation (see Section II.O). If exchange is fast on the NMR time scale only one resonance for the X-nuclei (e.g. ^{13}C) is observed, the position of which depends on the H : D ratio. In order to determine the isotope effects properly, a series of experiments must be conducted varying the H : D ratios of the exchanging species, typically 1 : 5, 1 : 2, 1 : 1, 2 : 1 and pure solvent[36]. The exchanging species is typically H_2O : D_2O but could equally well be deuteriated alcohols, ROD.

2. $^n\Delta C(D)$

a. $^n\Delta C(OD)$. Deuterium isotope effects have been measured in simple phenols dissolved in DMSO-d_6. Relatively few non-intramolecularly hydrogen bonded phenols have been measured. The typical two-bond isotope effects are 0.1 to 0.15 ppm[37]. The difference between inter- and intramolecular hydrogen bonding can be seen in **20** (**B** and **A**).

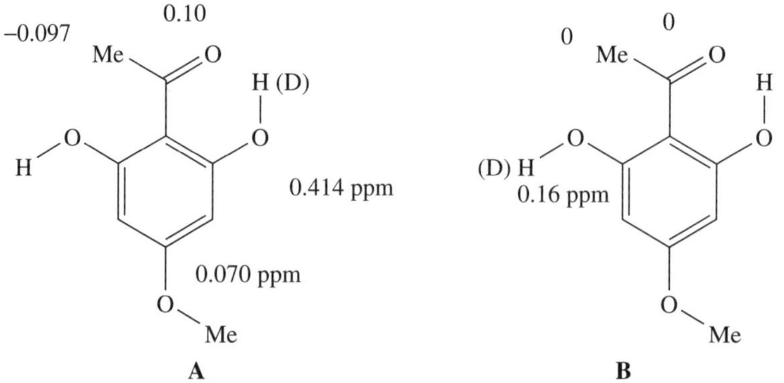

(20)

The intramolecularly hydrogen bonded phenols can be divided into two groups, the resonance-assisted hydrogen bonded (RAHB)[9] ones and those which are not. The RAHB case is the normal case in phenols (Figure 2). The resonance assistance depends on the double bond order of the double bond between the donor and the acceptor of the hydrogen bond. This is clearly seen in a plot of two-bond isotope effects vs. bond order[4] and is demonstrated in 4-hydroxy-6-methyl-3-carboxyethylpyridine and the corresponding 5-carboxyethyl derivative[38] (**21** and **22**). Two-bond isotope effects are shown to be a good

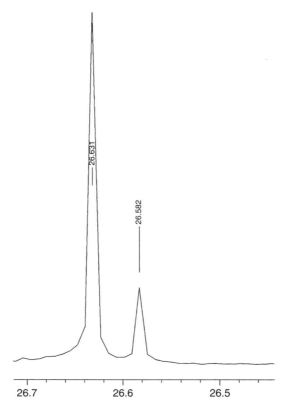

FIGURE 1. Part of a ^{13}C NMR spectrum of a methyl resonance showing splitting due to deuteriation. ^{13}C resonance of the deuteriated species appears at the low frequency

measure of hydrogen bond strength. This is related to the finding that the isotope effect depends strongly on the O···O and O−H distances[39,40].

In cases like **22**, the two-bond isotope effects can be really small as the double bond order is very low due to double bond fixation.

The two-bond isotope effects can be related to OH chemical shifts[4,41], and to other parameters such as five-bond isotope effects, $^5\Delta^{17}$OD isotope effects[42].

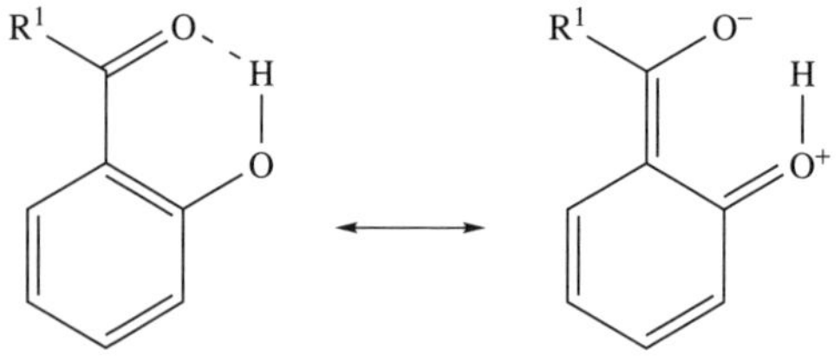

FIGURE 2. Resonance-assisted hydrogen bonding (RAHB)

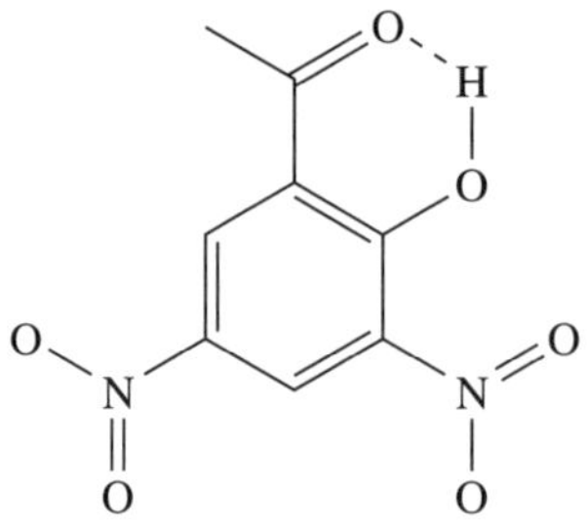

(23)

For isotope effects over three bonds $^3\Delta$C-2(OD)$_{trans}$ > $^3\Delta$C-2(OD)$_{cis}$[4,43]. Isotope effects over four bonds may be of different types: $^4\Delta$C=O(OD) or $^4\Delta$C(OD). In the former case the four-bond classification is formal as the effect is most likely transmitted via the hydrogen bond[44]. This means that for systems with multiple OH groups the hydrogen bond partner can be identified. In the latter case the $^4\Delta$C-4(OD) or $^4\Delta$C-6(OD) values are normally not large, except in cases in which the hydrogen bond is very strong[39,40]. This statement is generally true as most of the isotope effects increase in numerical size as the hydrogen bond strength increases. The exception is $^4\Delta$C = O(OD), as this is transmitted via the hydrogen bond. (An example is **23** in which $^4\Delta$C = O(OD) is 0.30 ppm[44].)

In a comparison of hydrogen bonded systems, salicylaldehydes, *o*-hydroxyacetophenones and *o*-hydroxyesters, a parallel phenomenon in the isotope effects is observed at different carbons and the effect can be described by **24**, showing how the isotope effects reflect the transmission pathway through the aromatic system.

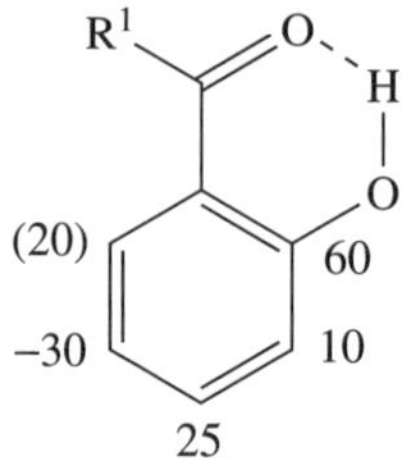

(24)

Another general finding is that, for phenols and RAHB systems, the OH group forms a stronger hydrogen bond than the OD group[4,20,45].

A very interesting case is that of gossypol (**25**). O'Brien and Stipanovich[46] reported very early unusual negative deuterium isotope effects on ^{13}C chemical shifts. These have been reinvestigated and found to be related to electric field effects[47]. Recently, the isotope effects were studied in α-(2-hydroxyaryl)-N-phenylnitrones[48].

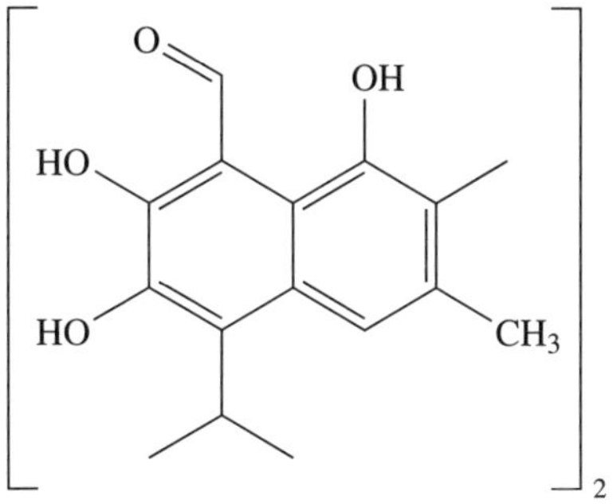

(**25**)

b. $^2\Delta C(D)$. Deuterium isotope effects of C-deuteriated phenols show that $^2\Delta C(D)$ isotope effects are roughly related to ^{13}C chemical shifts and to substituent effects on chemical shifts (SCS). Substitution always leads to a decrease of the isotope effect compared to phenol itself. Substitution at the *ipso* position gives the largest effects in parallel to the SCS. Steric interactions may play a role in cases having substituents like *t*-butyl[49].

3. $^n\Delta OH(OD)$

Long-range deuterium isotope effects at other OH protons are seen in a number of systems. They are normally small as deuterium isotope effects on ^{1}H chemical shifts in general. With regard to magnitude this follows the normal scheme, that the stronger the hydrogen bond the larger the isotope effect. The $^6\Delta OH(OD)$ of compounds such as those shown in **2–4**, in which the OH(D) group is part of a strong hydrogen bond, seem to be on the larger side provided the OH group is also hydrogen bonded[39,40]. For **2**, a value of 0.022 ppm is found. For **3**, we have 0.044 ppm for OH-3 and 0 for OH-6, whereas for **4** with the strongest hydrogen bond it is 0.056 ppm. For 2,6-dihydroxyacyl compounds (**26**) an effect is seen at low temperature when the acyl group is an ester (**6**), but not when it is an acetyl group (**5**, $R^1 =$ Me). For the ester, the OH-6 group is hydrogen bonded to the OR group whereas for the acetyl derivative, the OH-6 points towards C-5. The difference in geometry or transmission via two hydrogen bonds could explain the difference[5]. Very small isotope effects are found in 1,4-dihydroxy-9,10-anthraquinones. Ten values are also reported for perylenequinones and 1,4-dihydroxy-5,8-naphthoquinones. Both of these systems are equilibrium ones and the large values seen in the former over formally eleven bonds[10] could possibly be of equilibrium type (see Section II.K.6). A relatively large effect is seen for the 3-OH resonance of 6-methyl-1,3,8-trihydroxyanthraquinone (emodin)–23 ppb[50]. For the similar hypericin anion (**27**) the sign of the isotope effect is positive (19 ppb)[50]. In this case the position of the OH proton is strongly delocalized (see Section III.E.2).

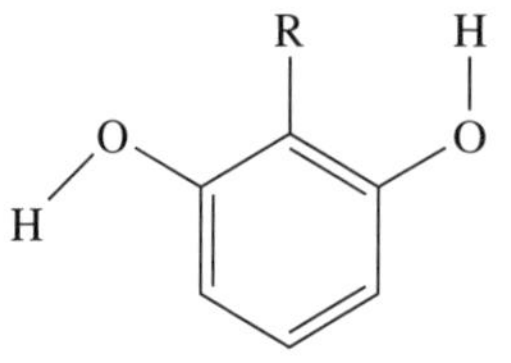

1,8-dihydroxyanthraquinones and 2,2′-dihydroxybenzophenones deuteriated at one OH position lead to high frequency shifts at the other position[47,51]. This has been termed a relay effect[51]. The suggested mechanism is that deuteriation leads to a weakening of the hydrogen bond in which it is involved, leading to a slightly stronger hydrogen bond for the other bond and therefore to a high frequency shift of the OH resonance. On the other hand, this also proves that the two OH groups are hydrogen bonded simultaneously (see also Section III.E.1)[47,51].

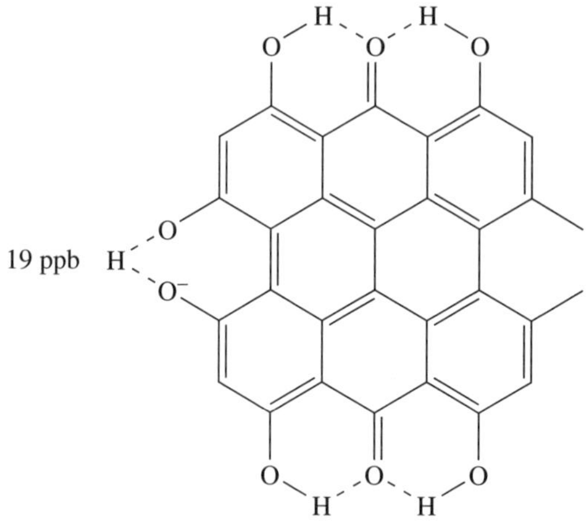

A negative effect is also observed for anthralin (dithranol) (**28**)[35]. For equilibrium systems, the effects can be larger and of both signs as seen for the enol form of o-hydroxydibenzoylmethane (**29**)[36] and for the tautomeric naphthalene (**30**)[47]. Isotope effects at the chelate proton are −0.047 ppm and 0.0126 ppm and at OH−2′ = −0.0916 ppm. In the former case it is due to deuteriation at OH−2′ and at CH_2, respectively. For **30** the isotope effect at OH−1 is 0.0295 ppm and that at OH−8 is 0.161 ppm.

4. Solvent isotope effects

Solvent isotope effects ($H_2O : D_2O$) on ^{19}F chemical shifts are much larger in o- and p-fluorophenolates than in the corresponding phenols and much larger than that in the m-fluorophenol, thereby relating the strength of the solvation of the fluorine to its electronegativity[52].

5. $^1\Delta^{13}C(^{18}O)$ isotope effects

These effects for acetyl groups have been correlated with ^{13}C chemical shifts of the carbonyl carbon[51]. A similar correlation was not found for single bonded C−O groups including phenols. In the single bonded case much smaller isotope effects are found (10−ca 30 ppb)[53].

6. $^1\Delta O(OD)$

For deuteriated phenols, isotope effects on ^{17}O chemical shifts are 2.3 and 1.7 ppm to lower frequency for salicylaldehyde and methyl salicylate, respectively. Interestingly, the

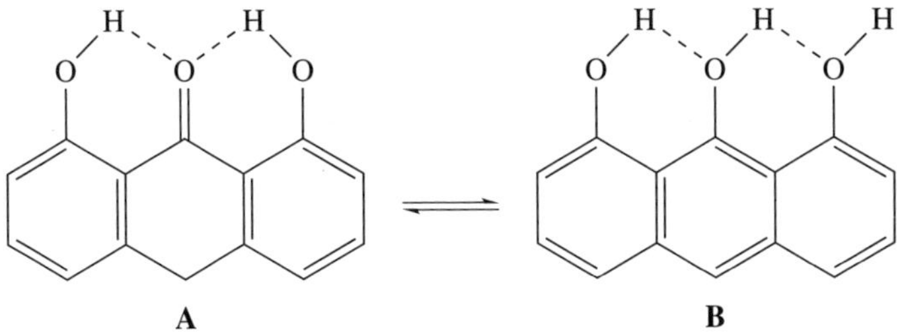

(28)

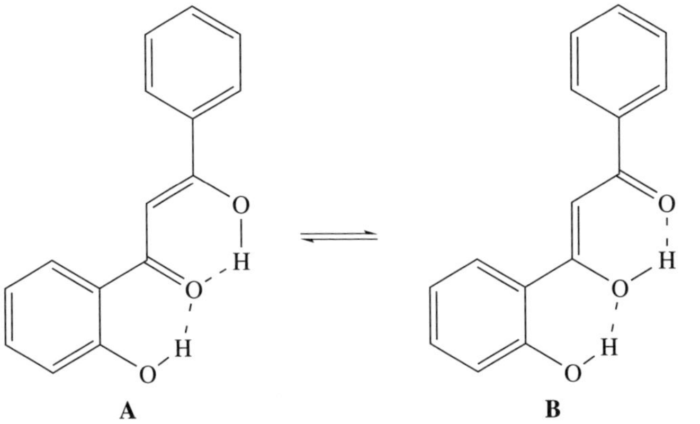

(29)

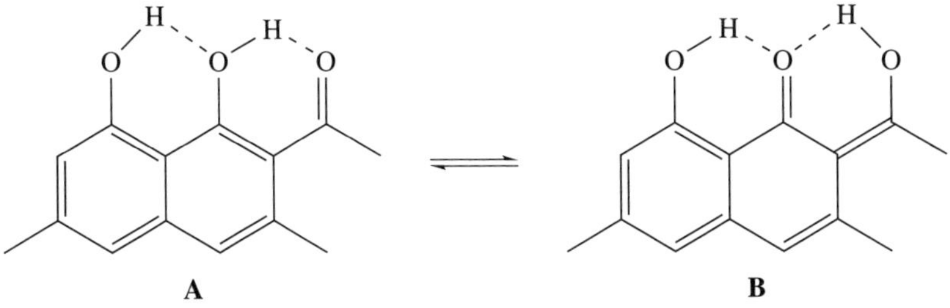

(30)

signs are opposite to those observed at the carbonyl oxygen[54]. For the carbonyl oxygen the size increases with the strength of the hydrogen bond[35,42]. Because of the large chemical shift difference between the ^{17}OH and the C=O chemical shifts, large equilibrium isotope effects are found[42].

7. Primary isotope effects

The primary deuterium, $^P\Delta(^1H, {}^2H)$, and tritium isotope effects, $^P\Delta(^1H, {}^3H)$, are proportional in general[55]. The primary isotope effects are proportional to the hydrogen bond strength and may be correlated with OH chemical shifts. A plot of $^P\Delta(^1H, ^3H)$ and the two-bond isotope effects, $^2\Delta COD$, revealed a very good correlation. Compounds like **2–4** as well as compounds like **31** fall out of this correlation. In the former cases steric compression is present. The primary tritium isotope effect is apparently more responsive due to a strongly asymmetric potential well (see Figure 3). For compounds with weak hydrogen bonds like salicylaldehyde or methyl 6-fluorosalicylaldehyde, $^P\Delta H(D)$ is less than 0.1 ppm. In strongly hydrogen bonded systems like **4** it can reach 0.44 ppm[55].

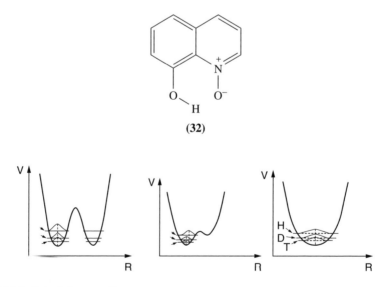

(31)

Primary deuterium isotope effects have also been measured in 8-hydroxyquinoline N-oxides (**32**)[56]. It was found that a plot of $^P\Delta(^1H, ^2H)$ vs. δOH had a different slope than observed for tautomeric hydroxyquinones[10] and β-diketones[57].

(32)

FIGURE 3. Potential energy diagrams

Primary isotope effects have been used to describe the shape of the potential well. Large positive values point towards a symmetric two-potential well, whereas negative values indicate a single potential well[57,58]. An example of the former is the rubazoic acid derivatives[59]. An example of the former is a small negative value observed for methanol[58]. In all cases equilibrium isotope effects (see Section II.K.6) should be ruled out as contributors before making such potential surface type assignments[55].

G. Coupling Constants

1. J(X,OH) coupling constants

These couplings X being ^{13}C or ^{1}H have been studied for ^{13}C in some detail in intramolecularly hydrogen bonded compounds[46,60–65].

a. $J(^{13}C,OH)$. These couplings have traditionally been used for assignment purposes[64]. The two-bond coupling constant is found to correlate only weakly with $^2\Delta COD$ and therefore with the hydrogen bond strength.

For the three-bond coupling constants, $^3J(C,OH)_{trans} > {}^3J(C,OH)_{cis}$ and a plot of $^3J(C,OH)_{cis}$ vs. δOH shows a good correlation for o-hydroxybenzoyl derivatives. The corresponding correlation line for olefinic derivatives is parallel. Data for naphthalene derivatives fall mainly in between[60]. Bond order is clearly an important parameter.

Couplings involving the OH proton can be transmitted via the carbon skeleton or, for hydrogen-bonded cases, via the hydrogen bond. The latter may be the case for $^4J(C=O,OH)$. A plot of $^4J(C=O,OH)$ vs. δOH shows a good correlation except for 2-hydroxy-1-acenaphthone. This was ascribed to transmission via the hydrogen bond as those compounds (**1–4**) have long OH bond and short O···O distances, leading to substantial orbital overlap. For the sterically hindered compounds the coupling is small, due to poor orbital overlap[60]. Interestingly, esters show very small $^4J(C=O,OH)$ couplings[60].

A similar situation is found in Schiff bases. Kurkovskaya found a coupling $^5J(^{15}N,OH)$ of 1.65[66].

b. $^nJ(C,OD)$. These couplings are proportional to $^nJ(C,OH)$ (factor of 1/6.51) and are usually too small to be observed directly. However, they will often be visible as a broadening of the C-2 and C-3 resonances of the deuterated species, thereby providing an assignment tool.

c. $^5J(OH,H)$. Hydrogen–hydrogen couplings involving the phenolic proton are small, but depend on the geometry of the coupling path. Five-bond couplings that have a W pathway are observable, whereas the corresponding coupling having a *cis* coupling pathway, e.g. OH, H-3 of phenol are zero. Based on this criterion the conformational preference of phenols has been investigated[67–71]. The same principle has been transferred to hydroxy derivatives of naphthols. For 1-naphthol, the OH group is pointing towards C-2 approximately 90% of the time[72]. For sterically hindered compounds like 2-*t*-butylphenol, the method may break down due to distortion of the COH geometry[73]. In a slightly more complex system, 2-hydroxybenzyl alcohols, three different rotamers are found with those involving hydrogen bonding dominating in non-polar solvents[74]. In D_2O as solvent, a complex is suggested in which the D_2O molecule forms a bridge (see Section II.H.5).

d. $^1J(H, {}^{17}O)$. The one-bond coupling to ^{17}O is obtained in a few cases. One example is the 8-hydroxyquinoline N-oxide[75] (**32**). They are also observed in a number of o-hydroxyacyl aromatics. The magnitude is 80 ± 25 Hz[29,30,76]. The $^1J(H, {}^{17}O)$ couplings depend on concentration, temperature and solvent[30], but a structural dependence has not yet been found, probably because of the difficulty of measuring these couplings accurately.

2. $^nJ(^{13}C,^{13}C)$

Hydroxy substitution has a major effect on $^1J(^{13}C,^{13}C)$ if the OH group is at one of the participating carbons. A considerable increase is observed[77,78]. For $^2J(C,C,OH)$ intra-ring couplings are slightly diminished numerically[79]. However, as the signs are not always determined, it is difficult to draw too extensive conclusions, but generally a decrease in the numerical magnitude occurs irrespective of the position of the electronegative substituent. Three-bond couplings can be of different types. Those within the same ring decrease slightly upon substitution at the coupling carbon, whereas substituents attached to carbons of the coupling pathway markedly decrease all three types of three-bond couplings.

3. $^nJ(^{13}C,^1H)$

The one-bond $^1J(^{13}C,^1H)$ coupling constant in mono-substituted benzenes does not correlate with the electronegativity, but can be correlated to a combination of σ_I and σ_P or to other Hammett parameters[62]. The lack of strict correlations can also be seen from 2-hydroxynaphthalene in which $^1J(^{13}C\text{-}1,^1H\text{-}1)$ is decreased slightly compared to naphthalene, whereas $^1J(^{13}C\text{-}3,^1H\text{-}3)$ is increased slightly[80]. For 1-hydroxynaphthalene, an increase is seen for the C-8,H-8 coupling constant of the peri bond[80]. For a general introduction see elsewhere[62].

H. Hydrogen Bonding

1. Hydrogen bond strength

Several attempts have been made to relate the OH chemical shift with the strength of the hydrogen bond. The original attempt is a correlation of corrected OH chemical shifts with OH torsional frequencies. The latter can be related to hydrogen bond energies[81]. More recently, Reuben suggested a logarithmic relation between hydrogen bond energy and $^2\Delta(^{13}COD)$[82].

A correlation is found between magnetic anisotropy corrected C-1 chemical shifts and the OH stretching frequency for complexes between phenols and π and n bases such as benzene, pyridines and picolines[83].

Mikenda and coworkers[84–86] have investigated o-hydroxyacyl and thioacyl derivatives and found very good agreement between δOH and the ν(OH) vibrational frequency (Section III.B.1). No distinct differences were seen between the thio derivatives and the corresponding oxygen ones, except that at a comparable ν(OH) frequency the δOH values of the oxygen compounds are typically slightly larger than those of the thio compounds, but the difference can possibly be ascribed to anisotropy effects[84].

2. Ranking of substituents as hydrogen bond partners

Two-bond isotope effects on ^{13}C chemical shifts are a good measure of hydrogen bond strength[39]. A simple example is seen for **33** and **34**. The phenol provides a common scaffold so that o-hydroxy-substituted phenols form a suitable way of ranking acceptor

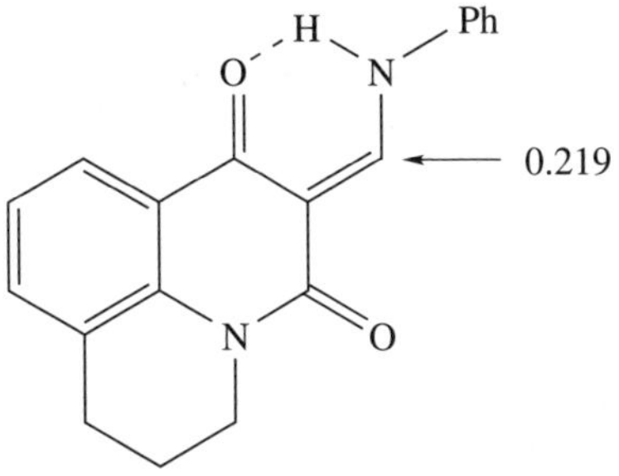

(33) 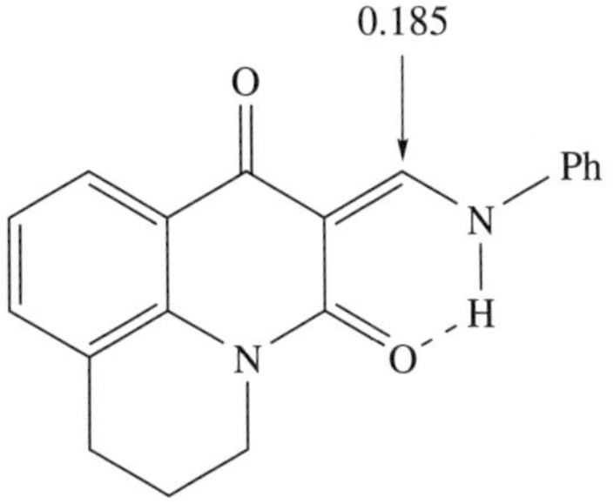   (34)

substituents of RAHB systems. The two-bond isotope effects decrease in the following order: RC=O > HC=O > C=O(OR).

3. Multiple hydrogen bonding to the same acceptor

Phenols quite often take part in multiple hydrogen bonding exemplified by **28, 29, 30, 35** and **36**, a system akin to a large number of dyes and indicators. The hydrogen bonding can be described by ^{17}O chemical shifts (see Section II.E.2), by $^{1}\Delta^{13}C\,(^{18}O)$ (Section II.F.5) or by $^{n}\Delta C(OD)$ isotope effects (Section II.F.2).

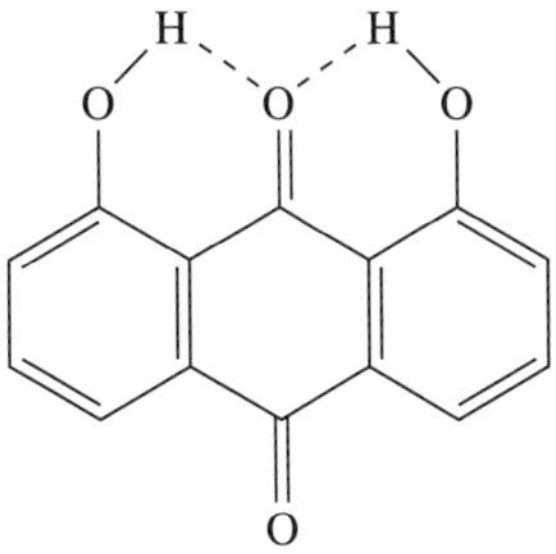

(35) 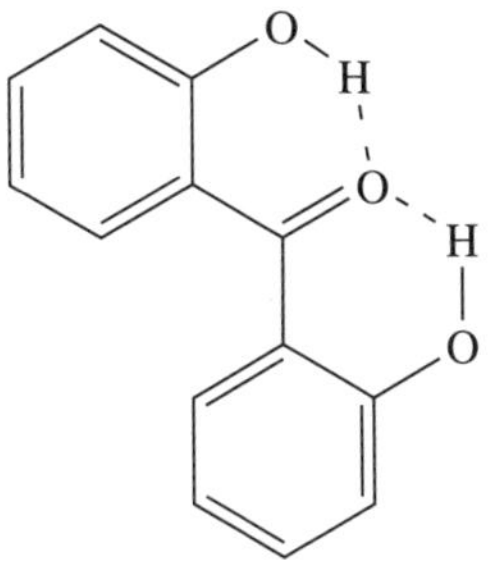   (36)

A different situation is seen in **37**, with R^1 and R^2 being either H or OH[87].

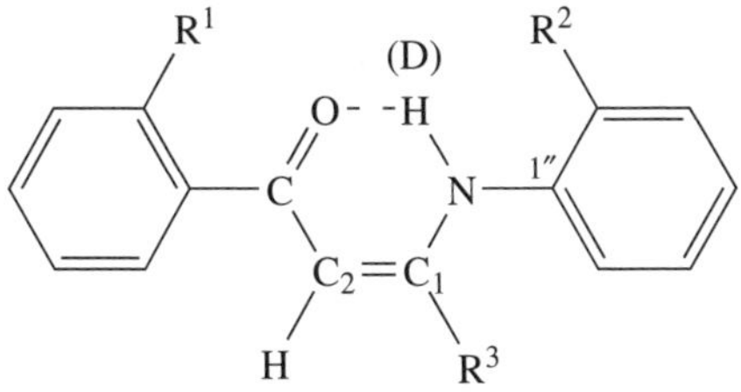

(37)

4. Conformational equilibria

Jaccard and Lauterwein concluded for 2,2′-dihydroxybenzophenone (**36**) that both OH groups are hydrogen bonded simultaneously[25]. A similar conclusion was reached based on $^1\Delta^{13}C(^{18}O)$ isotope effects (see Section II.F.5). Baumstark and Boykin found in acetonitrile, based on ^{17}O chemical shifts, that only one hydrogen bond existed[88].

5. Bifurcated hydrogen bonds to hydrogen

Bureiko and coworkers[89,90] have studied hydrogen bonding of intramolecularly bonded phenols (e.g. 2,6-dinitro- and 2,4,6-trinitrophenols) at low temperature in freons. At 90 K the rotation of the OH group is slow on the NMR time scale. Addition of proton acceptors causes a bifurcated bond, as evidenced by a shift to lower frequency of the OH resonance. An example is dioxane (**38**).

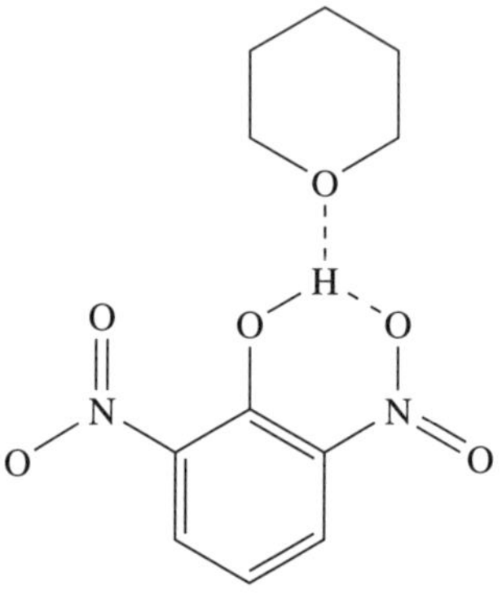

(**38**)

Grech and coworkers[91] studied 8-hydroxy-*N*,*N*-dimethyl-1-naphthylamine and suggested that the high frequency shift in dioxane compared to cyclohexane was due to formation of a bifurcated bond. This is somewhat unexpected. See also how multiple hydrogen bonding may affect equilibria (Section II.K).

6. Strong hydrogen bonds

A strong hydrogen bond of sodium 4,5-dihydroxynaphthalene-2,7-disulphonate (**39**) has been observed by NMR at a low temperature ($\delta(OH)$ 17.72 ppm)[92]. This value has been related to the O···O distance in relation to other measurements, in order to use OH chemical shifts to obtain oxygen–oxygen distances. One possible drawback of using values from the mono-ionized 1,8-dihydroxynaphthalene directly is that the chemical shift of the OH proton is likely to have a sizeable ring current contribution. Phenols are also involved in catalytic triads, e.g. in ketosteroid isomerase[93,94], leading to OH chemical shifts of 18.2 ppm.

Very favourable hydrogen bonding may occur in substituted 8-hydroxyquinoline *N*-oxides (substituted **32**) judging from the δOH value (in the 5,7-dinitro-8-quinolinol *N*-oxide a value of 20.38 ppm is found) as well as the deuterium isotope effects on the ^{13}C chemical shifts. Complicated substituent effects are found, because substituents such as bromine may interact with both the OH and the N−O group. No tautomerism was observed

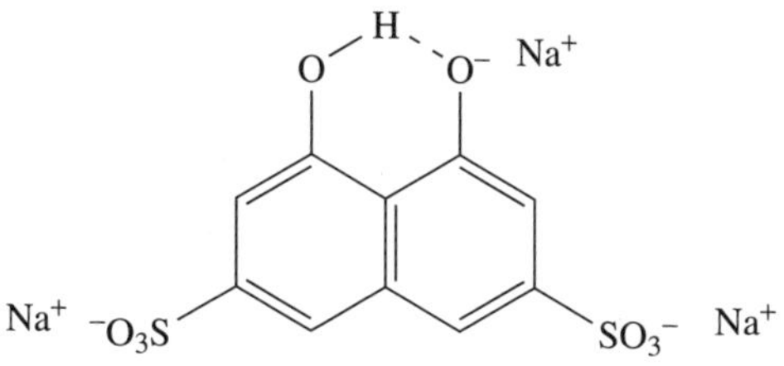

(39)

in these systems judged from deuterium isotope effects[56]. Brzezinski and Zundel[95] reached a different conclusion based on solvent effects. Solvent effects have been studied over a wide range of solvents[96]. δOH shifts to higher frequency with increasing solvent polarity. The correlation with the Onsager parameter $\varepsilon - 1/(2\varepsilon + 1)$ is poor, suggesting that specific interactions take place. A multiple regression analysis using E_T and DN parameters[97] gave a good correlation[96].

Geometry is clearly of great importance for the strength of hydrogen bonds. In N-oxides of Schiff bases much weaker hydrogen bonds are seen[98], as we are now dealing with a seven-membered hydrogen bond ring system.

In an N-substituted dihomoazacalix[4]arene, strong hydrogen bonding is found at low temperature between the OH and the N resulting in an OH resonance ultimately at 17.1 ppm[99].

7. Rotation

For phenols, rotation around the C—O bond is clearly assumed. This rotation may be slowed down by intramolecular hydrogen bonding. For compounds such as **40**, activation parameters for the rotation can be determined by dynamic NMR. A classic study is that of Koelle and Forsén[100] of aldehydes (**5**, $R^1 = H$). The activation energy E_a was determined as 37.9 kJ mol^{-1}, $\Delta H^{\ddagger} = 35.6$ kJ mol^{-1} and $\Delta S^{\ddagger} = 43.9$ J K^{-1} mol^{-1}.

In substituted **40** for $R^1 = CH_2NR_2$ (Mannich bases), E_a was 32.6–43.5 kJ mol^{-1} depending on the substituent at the *para* position. The rotational barrier increases with strengthening of the hydrogen bond. For $R^1 = CH=NR$ (Schiff bases) the activation energy was found as 48 kJ mol^{-1} [101].

In a similar, though different situation with 2,4-diaryl-6-(2-hydroxy-4-methoxyphenyl)-1,3,5-triazine, $\Delta H^{\ddagger} = 50$ kJ mol^{-1}; $\Delta S^{\ddagger}$ was found to be close to zero. The $\Delta H^{\ddagger}$ value is almost three times as large as that found for the intermolecular complex between phenol and pyrimidine in CCl$_4$[102].

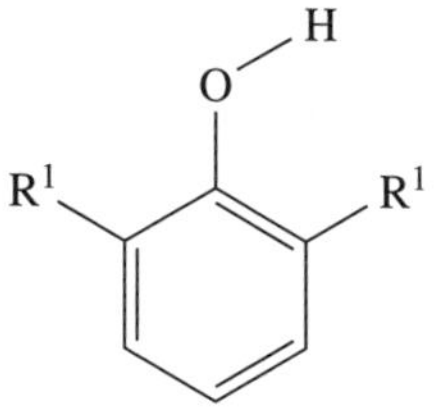

(40)

For the compounds **40**, $R^1 = N=N-Ph$, $\Delta H^\ddagger = 47.3$ kJ mol^{-1} and $\Delta S^\ddagger = -24$ J K^{-1} mol^{-1} in toluene-d$_8$. In CD$_2$Cl$_2$, $\Delta S^\ddagger$ decreased to -45 J K^{-1} mol^{-1}. Substitution at the *p*-position to the OH increased $\Delta H^\ddagger$. A similar value was found in the 3,5-di-*t*-butyl derivative. However, in this case the change of solvent to CD$_2$Cl$_2$ had a much smaller effect on $\Delta S^{\ddagger\,103}$, probably reflecting the hindered access to the OH groups.

Studies with $R^1 = $ acetyl[5,100] and methoxycarbonyl[5] in different solvents have been undertaken. Rather large $\Delta S^\ddagger$ values, *ca* -30 to -81 J K^{-1} mol^{-1}, are found[5]. For the esters, an additional hydrogen bond to the OR group is suggested[4] to account for the larger $\Delta S^\ddagger$ of the esters, indicating that two hydrogen bonds must be broken in the transition state. For the acetyl derivatives the non-intramolecularly hydrogen bonded OH-6 group points preferably towards C-5[5]. As seen above, the entropy plays a major role in some systems. For **41**, the OH group prefers to form a hydrogen bond to the nitro group (as in **41B**), although not exclusively, whereas at lower temperature the equilibrium is shifted fully towards hydrogen bonding to the acetyl group (**41A**). Entropy was suggested to play a role, as no similar effect was observed in derivatives with electronegative substituents at the 6-position[47].

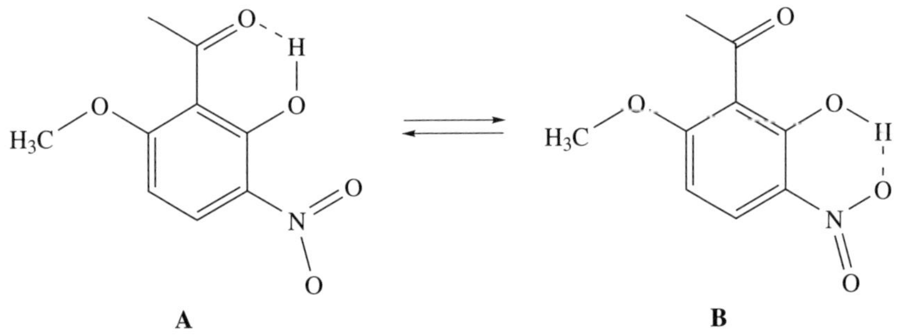

A   **B**

(**41**)

Rotation may clearly have a strong effect on spectra of compounds such as **2** and in similar natural products in terms of broadening of resonances at ambient temperature.

I. Steric Effects

Steric effects play an important role for phenols. The OH exchange may be influenced (see Section II.B). Hydrogen bonding in *o*-hydroxy derivatives where the *ortho* group is an aldehyde etc. (see above) is dominated by resonance-assisted hydrogen bonding. For this to be effective the six-membered ring involving the hydrogen bond must be planar. Non-planarity of the acceptor groups could be the case in systems in which the acceptor group is subject to steric interaction.

This problem has been addressed using isotope effects. Two different cases are found due to the interaction present. Two typical examples are seen in **42** and **43**. In **43** steric twist is observed and in **42** steric compression is found. Spectroscopically these effects are characterized by the pairs of ^{1}H chemical shifts and deuterium isotope effects on chemical shifts: δOH, $^1\Delta$X(OD), $^4\Delta$X=O(OD), $^5\Delta$CH$_3$(OD) and $^6\Delta$OH(CD$_3$). The latter is especially useful, but requires that deuterium is incorporated into methyl groups of, e.g., acetyl groups[51,40]. In **43** the number of intervening bonds are six (H–C8–C8a–C1–C(O)–C–H) whereas for the steric compression cases (**42**) the number of bonds is five (O–C2–C3–C(O)–C–H).

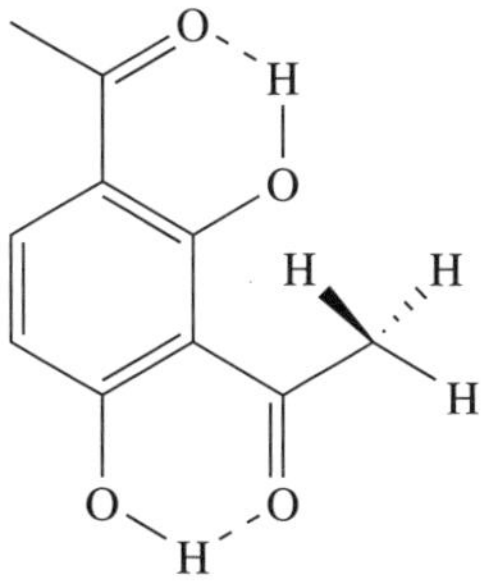

(42) 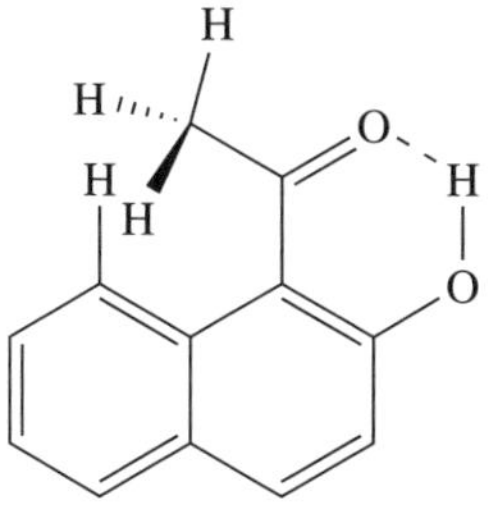   (43)

The twist is seen in a large number of polycyclic aromatic compounds[51]. For these compounds and for 2-hydroxyacenaphthophenone (43) one could wonder what happens to the OH group as the acetyl group is twisted out of the ring plane. In this case both X-ray and *ab initio* calculations indicate that the C-1–C=O bond is pushed out of the ring plane so that the acetyl C=O bond points back towards the OH group, which is in the aromatic ring plane[40]. Steric effects have been pointed out in Schiff bases[104]. NMR studies of deuterium isotope effects on chemical shift found a difference in the position and intensity of the $^2\Delta(^{13}\text{CXD})$ maximum as a function of the mole fraction[105].

Steric effects could also play a role in achieving planarity (conjugation) of non-hydrogen bonded phenolic hydroxy groups. This will be the case in, e.g., 2,6-di-*tert*-butylphenol leading to a low frequency resonance position of the OH proton. Large substituents, like *t*-butyl, next to the OH group of intramolecularly hydrogen bonded compounds have only little effect on the hydrogen bond strength as judged from $^2\Delta(^{13}\text{COD})$[47]. This is slightly peculiar as this seems to be the case in the more complicated systems like those of **1–4**[39,40].

Steric effects play a role in the ability of the OH group to exchange (see Section II.B).

J. Proton Transfer

The phenolic proton with its acidic properties is a good partner in proton transfer reactions leading to, e.g., tautomeric equilibria. Of interest in such situations is the barrier to interconversion which is related to the rate of interchange. The barrier height can be determined by means of NMR spectroscopy and the rate can be found from line shape analysis in suitable cases.

The characterization of the two species taking part in the equilibrium is of utmost importance. Infrared spectroscopy being such a 'fast' technique is obviously preferred, but is in a number of cases unsuitable due to strong coupling, leading to very broad resonances (see Section III). NMR data for a model situation are given in Table 1.

One of the most used ways to gauge the extent of proton transfer is to plot appropriate chemical shifts (O^1H or ^{13}C) vs. the difference in pK_a values of the donor and the acceptor, or simply the pK_a value of the phenol itself if the acceptor is the same for a series[106,107]. Another parameter used is $\Delta_{14} = (\delta\text{C-1} - \delta\text{C-1}_{\text{phenol}}) - (\delta\text{C-4} - \delta\text{C-4}_{\text{phenol}})$ (phenol refers to the unsubstituted compound). The normal type of plot is seen in Figure 4a, but also a plot of the type of Figure 4b may be found. This is ascribed to a homoconjugate system (NO···H$^+$···ON), e.g. in a system like 2,6-bis(diethylaminomethyl)phenol di-*N*-oxide[106] (see Section II.L.1).

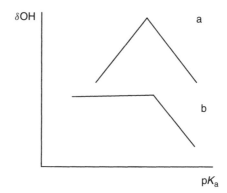

FIGURE 4. Plot of the OH chemical shift vs. pK_a values for a phenol

K. Tautomeric Equilibria

1. General introduction

Tautomerism involving phenols is most often seen for Schiff bases, Mannich bases, or *o*-hydroxyazo aromatics, but is also discussed for *o*-hydroxynitroso compounds. This is relatively seldom for phenols not having a nitrogen-containing substituent. An exception is **30** and other cases mentioned later.

The tautomerism has been described using a deuterium isotope on ^{13}C and ^{17}O chemical shifts as well as primary tritium isotope effects (0.90 ppm). The role of the hydroxy group at C-8 in the naphthalene system **30** could be important as this contributes strongly to hydrogen bonding in tautomer **B**. The methyl group at C-3 should lead to twist in tautomer **A** (see **30A**) thereby probably making the two tautomers more energetically similar[42,47]. ^{17}O chemical shifts have been used extensively to study tautomeric equilibria involving enolic groups like β-diketones. Also, for selected compounds it is relevant to use ^{17}OH chemical shifts of phenols. For compounds like **44** the tautomerism is clearly shown by the observation of only one ^{17}O chemical shift at 282.6 ppm[108]. The chemical shift corresponds to an average between the chemical shifts of a C=O and a C−O oxygen. The

(**44**)

5. NMR and IR spectroscopy of phenols

observation of two shifts in the ^{17}OH chemical shift range shows that ditranol (anthralin) is at form A, despite the fact that it is often depicted as form B (**28**).

Lios and Duddeck[109] studied substituted 1-(2-hydroxyphenyl)-3-naphthyl-1,3-propanediones. The ^{17}OH resonance falls in the range 93–102 ppm. The higher value was found in a derivative with a methoxy group *meta* to the OH group in question. The position of the tautomeric equilibrium could also influence the chemical shifts.

The ^{17}OH chemical shifts parallel the strength of the hydrogen bond as seen previously. Hydrogen bonding of enols has been investigated in great detail and is of course related to the present study. A case involving both is *o*-hydroxydibenzoylmethane (**29**) in which the extra hydrogen bond perturbs the enolic equilibrium.

A similar, though not identical, case is that of usnic acid (**45**) in which the equilibrium is markedly changed upon acetylation of the OH group at position 9^4.

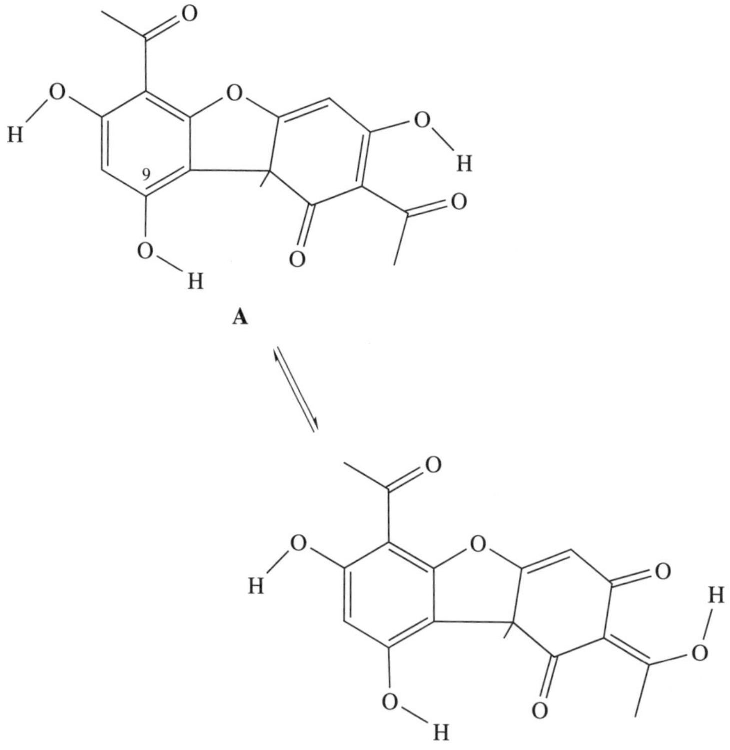

(**45**)

Another case that has been debated is that of 9-hydroxyphenalen-1-one[110]. Based on the large deuterium isotope effects of both signs, one of the present authors has suggested that these are of equilibrium type (see Section II.5).

Benzaurins and fuchsones are a new type of tautomeric species showing intermolecular exchange (**46**)[111].

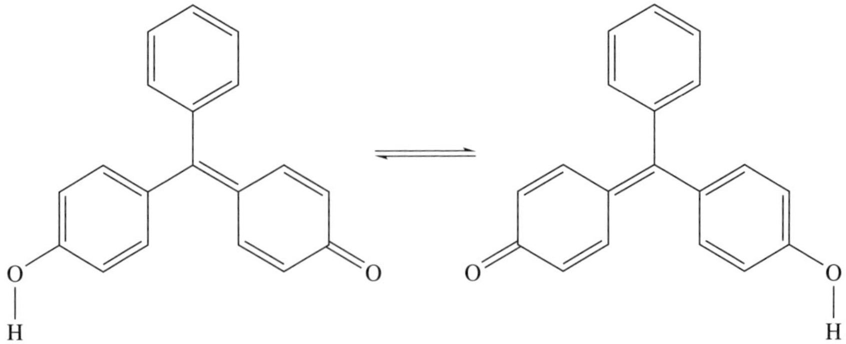

(46)

2. o-Hydroxyazo compounds

These compounds (47) are very widespread as both water-soluble and more hydrophobic dyes. The former group often have a SO_3H group as hydrophore. An example is FD&C Yellow no. 6, which is shown to exist primarily as a hydrazone below pH 12 and as an azo form as shown above[112].

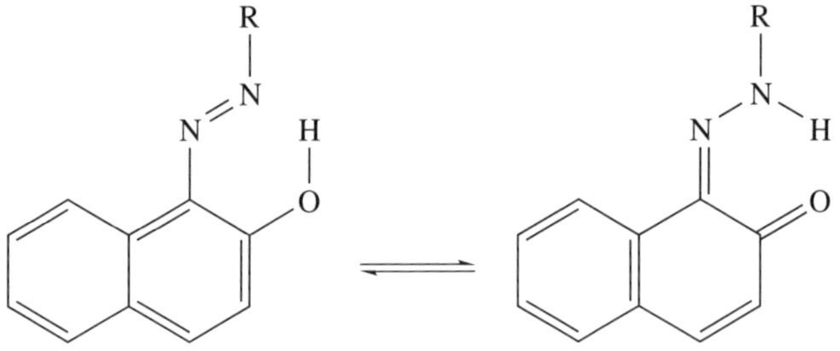

(47)

The tautomeric equilibrium of these has been described by several methods, i.e. $^{13}C-^{13}C$ couplings[113], $^1J(N,H)$ coupling constants[114] and deuterium isotope effects on ^{13}C[115,116] and ^{15}N chemical shifts as these are very different for the azo and hydrazo forms[116]. Isotope effects on ^{19}F chemical shifts are very sensitive due to the large chemical shift range (and, more importantly, the large difference in chemical shifts of the two tautomeric forms)[117].

3. Mannich bases

Mannich bases have been studied intensely by both IR and NMR techniques. These have been reviewed very recently[118-120] and will very briefly be touched upon. For Mannich bases the proton transfer leads to a moiety with separated charges. This may also be the case for Schiff bases (see below). Charge separation is clearly important in understanding the factors influencing proton transfer and the way the equilibrium responds to temperature and solvent.

4. Schiff bases

The equilibrium of Schiff bases (**48**) has been studied in detail because of their interesting properties both in the solid state (Section II.N) and in biological reactions[121]. This can be done as just described for o-hydroxy azo compounds (^1J(N,H) coupling constants[122,123] and deuterium isotope effects on ^{13}C and ^{15}N chemical shifts)[122–131]. Based on ^1J(N,H) it could be concluded that the Schiff bases form a conventional tautomeric equilibrium that can be described by two species[132].

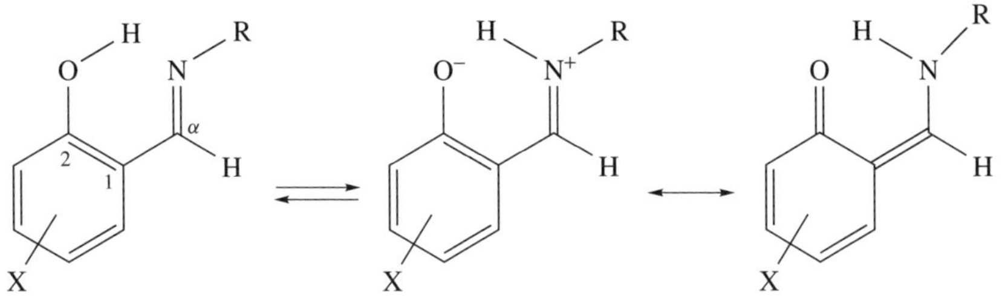

A   **C**   **B**
(48)

Of interest is also the interconversion barrier. These have been determined in N,N'-bis(salicylidene)phenylene diamine (**49**) as values of only 10 and 25 kJ mol^{-1} in the solid state (Section II.N); 10 kJ mol^{-1} refers to the first proton transfer and 25 kJ mol^{-1} to the second (see also Section II.N). For **50** the values are only 2 and 10 kJ mol^{-1}.

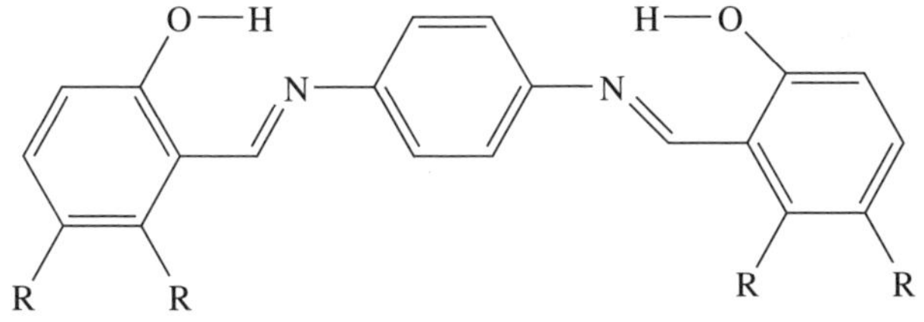

(**49**) R = R = H (DSP)
(**50**) R = R = C$_4$H$_4$ (DNP)

Zhuo[133] investigated ^{17}O chemical shifts of o-hydroxy Schiff bases. These systems are in some instances tautomeric. As described previously ^{17}O chemical shifts are very good indicators of tautomerism (see Section II.K.1). Provided that good reference values for the two tautomeric states exist, the equilibrium constant can be determined. Zhuo used the values for simple Schiff bases as models for the phenolic form (**48**). For the form **48B** a value from a simple enamine was chosen. This, however, is not a very appropriate choice, as it does not at all take into account the charged resonance form (**48C**). The equilibrium constant determined for N-(2-hydroxy-1-naphthalenylmethylene) amine is quite different from that derived by ^1J(N,H) coupling constants[132].

For Schiff bases a difficult question remains. To what extent has the proton transferred form B a formal charge separation (**48C**) or not (cf **48B**)? This problem is in principle approachable by NMR, but not easily solved. Using ^{13}C chemical shifts of C-1 the B and C forms are not sufficiently different. Dudek and Dudek[132] approached the problem using

^1J(N,H) coupling, but found that no conclusions could be drawn, partly owing to lack of proper model data.

A different approach is to use ^{17}O chemical shifts. These are very sensitive to differences in chemical surroundings. The ^{17}O chemical shifts of hydrogen bonded phenolates and quinones can be estimated. In addition, using a set of compounds with different equilibrium constants and extrapolating to a mole fraction of one, the ^{17}O chemical shift of the proton transferred form can be estimated. Using the above estimated ^{17}O chemical shifts it can be estimated that the *o*-quinonoid form **48B** contributed *ca* 65% to the proton transferred form[134].

5. Nitrosophenols

Tautomeric equilibria have been studied in nitrosophenols. An early study of 4-nitrosophenol showed an intermolecular tautomerism catalysed by slight traces of water. In dry dioxane both the *N*-oxide and the oxime form could be observed[135].

In acenaphthenequinonemonoxime in DMSO four different species could be observed: primarily the *cis* and *trans* forms of the oxime but also the nitroso isomers[136].

The 1-nitroso-2-naphthol and the 2-nitroso-1-naphthol have been studied by ^{1}H[137] and ^{13}C NMR[138]. In an early study Vainiotalo and Vepsäläinen[139] suggested that 1-nitroso-2-hydroxynaphthalene exist at the *trans* form **51C**. For the latter both the oxime and the nitroso (**51A**) form were suggested based on ^{1}H and ^{13}C NMR in CDCl$_3$. However, in a recent study Ivanova and Enchev[140] assigned the two different sets of resonances to two different rotamers of the oxime form, i.e. **51B** and **51C**. They also measured solid

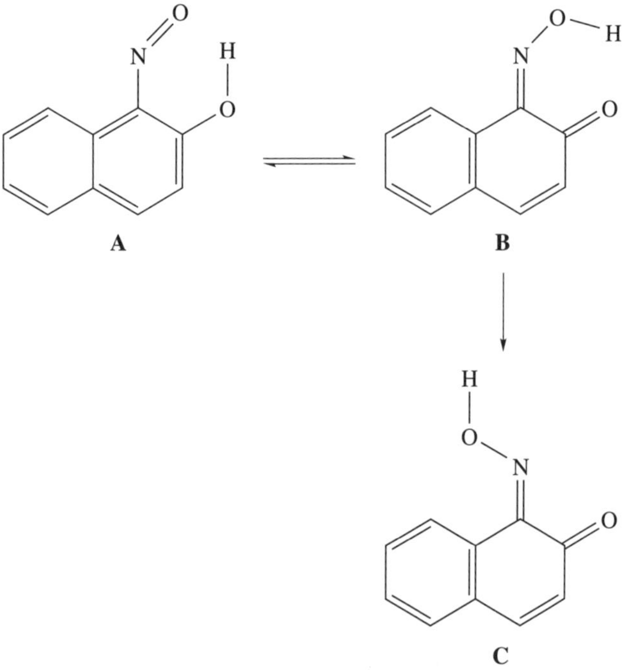

(51)

5. NMR and IR spectroscopy of phenols

state NMR spectra and found both compounds to exist in the oxime form in the solid. For the 2-nitroso compound they assigned this to the *anti* form. This is different from the 1-nitroso compound, which exists in the *syn* form in the solid state according to X-ray studies[141].

Theoretical calculations showed an energy difference of *ca* 17 kJ mol^{-1} and a barrier to interconversion of *ca* 37 kJ mol^{-1} for the 1-nitroso derivative[140]. From deuterium isotope effects on ^{13}C chemical shifts it was concluded that the 1-nitroso-2-naphthol was tautomeric[47].

6. Equilibrium isotope effects

Deuteriation at the XH position of tautomeric equilibria (XH being the transferred proton) leads to a shift (change) in the equilibrium. This has been demonstrated for, e.g., Schiff bases[122] and *o*-hydroxyazo[115] compounds. The change depends on the differences in zero point energies of the two tautomeric species. The observed equilibrium isotope effects (an intrinsic component is normally also present) depend besides the change in the equilibrium, also upon the chemical shift differences between the interconverting nuclei. Consequently, equilibrium isotope effects can be of both signs and be observed far from the centre of deuteriation. Observation of equilibrium isotope effects is thus a good way of establishing the presence of an equilibrium in cases of doubt. Furthermore, the presence of equilibrium isotope effects indicate that a two-potential well is at play. A typical feature observed for such systems is that the equilibrium isotope effect goes through either a maximum or a minimum as the mole fraction is increased/decreased from $x = 0.5$[122,142].

For Mannich bases, isotope effects on ^{13}C chemical shifts have also been observed. In this case the authors have chosen to ascribe this to a shift of the XH position as a single well potential is suggested[143,144].

Rubazoic acids (**52**) may occur in a Q = CH or a Q = N form. Deuterium isotope effects on ^{13}C chemical shifts are given in Table 2. For Q = CH the compounds are not tautomeric, but for Q = N they are in polar solvents[145] as can also be seen from the isotope effects. For the *N*-forms, the compounds can be divided into two groups according to symmetry. The symmetrical ones show only few isotope effects and those at C-5 and C-5' are of equal magnitude, pointing either to a symmetrical structure with the OH equally shared between the two oxygens or a tautomeric equilibrium. The OH chemical shifts are for all investigated Q = N compounds close to 17 ppm[145]. For those compounds having different substituents at *N*-1 and *N*-1*, the isotope effects are dramatically different (C-1 = 0.6 ppm, C-5' = −0.5 ppm). A large difference is found for C-4 = 0.65 ppm and C-4' = −0.6 ppm. However, the average value for C-5 and C-5' is equal to 0.25 ppm and

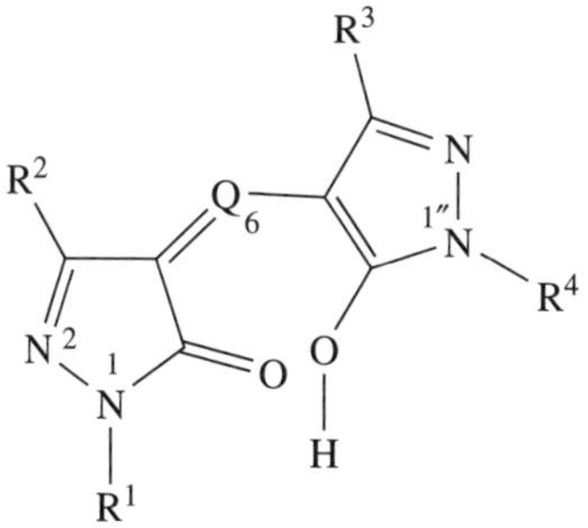

(**52**)

TABLE 2. Deuterium isotope effects on ^{13}C chemical shifts of rubazoic acids

Q	R^1	R^2	R^3	R^4	C-3	C-4	C-5	C-3'	C-4'	C-5'	Me-N	Ph_o-N	Ph_p-N	Ph_i-C	C-6
CH	Me	Me	Me	Me	0.03	0	0.20	a	a	a	0	0	0	0	0.22
CH	Me	Ph	Ph	Me			b								b
CH	Ph	Me	Me	Ph	0.06	0	0.22	a	a	a	0	0	0	0	0.24
N	Me	Me	Me	Me	b	0	0.25	a	a	a	0	0	0	0	—
N	Me	Ph	Ph	Me	0		0.27	a	a	a					—
N	Ph	Me	Me	Ph	0.05	0	0.27	0	0	0.27	0	0	0	0	—
N	Ph	Me	Ph	Ph	0.1	0.70	0.62	0	−0.65	−0.10	0.045	0.09	0.08	0	—
N	Me	Me	Me	Ph	0	0.71	0.58	0	−0.64	−0.09	0	0.08	0.05	0	—
N	Me	Me	Ph	Ph	−0.06	0.68	0.63	0.11	−0.70	−0.1	−0.08	0.06	0.06	0	—

[a]See the symmetrical carbon.
[b]Broad signal.

for C-4 and C-4' it is *ca* 0 ppm, similar to the values found for the symmetrical compounds. The isotope effects of vastly different signs for carbons related by symmetry, and the fact that the average is similar to those of the symmetrical compounds (in which equilibrium isotope effects cannot contribute)[142] shows that these compounds are tautomeric. The relatively large intrinsic two-bond isotope effects ($^2\Delta$C-5(OD))$_{int}$ (the value is about twice as large as measured, as it corresponds to an average) found in the symmetrical compounds corresponds well with the primary isotope effect (see Section II.F.7) and with the short O··· O distance (2.42–2.45 Å). In the solid state, this system shows almost total delocalization of the HO—C=C—X=C—C=O electrons[146].

L. Complexes

1. Proton transfer

Complexes involving phenols can clearly be of many kinds. Much of the effort in this review will be concentrated on complexes with bases leading possibly to proton transfer. Intramolecular proton transfer has been treated too for a number of types of compounds (see Section II.J). Bases in the complexes are typically pyridines, aromatic and aliphatic amines, amine N-oxides and phosphine oxides. This is one of the rather difficult areas to review due to the fact that it is not always clear whether a single or a double potential well type is at play. Complexes have also been studied in a few cases in the solid state (see Section II.N).

It has turned out that temperature and the ratio of acid : base molecules is rather important for the outcome. The following situations are studied: (phenol : base) 2 : 1, 1 : 1, 1 : 2, 1 : 5 and 1 : 10. Historically, an excess of base was used. Ilczyszyn and coworkers[1] assumed a 1 : 1 complex in a mixture of phenol with a five-fold excess of triethylamine and suggested that the situation could be described by a simple tautomeric equilibrium between a molecular complex (to the left) and an ion-pair (to the right) (equation 2)

$$\text{PhOH} \cdots \text{NR}_3 \rightleftharpoons \text{PhO}^- \cdots \text{HN}^+\text{R}_3 \qquad (2)$$

and by using variable-temperature measurements they could determine the difference Δ_{14} (see Section II.J). For the non-charged complex, Δ_{14} is roughly proportional to the pK_a of the phenol. The values of Δ_{14} for the ion-pairs are similar to the values of phenolate ions. A $\Delta H°$ of the order of -4.7 kJ mol^{-1} and a $\Delta S°$ of -29 J K^{-1} mol^{-1} could be determined for the process. The small $\Delta H°$ suggests an almost symmetrical double well potential. The large $\Delta S°$ confirms the suggestion that solvation helps to stabilize the ion-pair (see later). The approach is too simple, as the authors themselves have shown later (equation 3). From line-shape analysis the reaction rates could also be determined[147]. The tautomeric equilibrium depends on interaction with surrounding molecules. The proton transfer process has been analysed in further detail[148]. It was found that an extra molecule of amine plays a role and that the proton transfer proceeds through an intermediate with bifurcated hydrogen bonds (see Section II.H.5). The transition state corresponds to a homoconjugated situation. It is also shown that the amine molecules exchange, as judged from the CH$_2$ resonances of the triethylamine. For a 1 : 2 (phenol : base) complex the equilibrium of equation 3 can be written as follows:

$$\text{A—H} \cdots \text{B} + \text{B} \rightleftharpoons [\text{A} \cdots \text{H} \overset{\text{B}}{\underset{\text{B}}{\cdots}}] \rightleftharpoons \text{A}^- \cdots \text{H—B}^+ + \text{B} \qquad (3)$$

The rate constants depend very much on the ΔpK_a. In the so-called inversion region (ΔpK_a ca 2–3) one observes the highest δOH (Figure 4a) and the slowest exchange rates. For such complexes, the equilibrium may be frozen out on the NMR time scale and the δXH (X = N or O) of the two bases observed. Examples of frozen out equilibria are for the complexes 2,4-dichlorophenol-triethylamine[147]; 2,5-dichlorophenol-N-methylpiperidine[149]; 2,3,5,6-tetrachlorophenol-N,N-dimethylaniline[150,151].

The formation of (PhOH$\cdots$NR$_3$)$\cdots$NR$_3$ aggregates helps to explain why in phenols with pK_a of ca 7.7 the OH$\cdots$N exchange is slow on the NMR time scale, allowing both species to be observed. The (PhOH$\cdots$NR$_3$)$\cdots$NR$_3$ complex is probably not found at room temperature[148]. 1 : 1 Complexes have not been investigated in so much detail. They do not give rise to separate signals. The XH chemical shifts can be plotted vs. pK_a values as shown in Figure 4a[148].

The 2 : 1 situation is somewhat different. A 2 : 1 phenol–amine complex can be observed at a δOH of 14.8 ppm (OH$\cdots$N). For the 1 : 1 complex at an equilibrium constant for proton transfer K_{PT} close to 1, the δOH is 13.6 ppm. This is in very good agreement with an equilibrium between the non-molecular complex at ca 12 ppm and an ion-pair (taken as 14.8 ppm as for the 2 : 1 complex)[152]. A plot of δOH vs. the pK_a value of the phenol at 153 K in C$_2$H$_5$Br showed a characteristic shape (Figure 4a)[148].

This is explained by the authors by assuming that both δOH and δNH$^+$ increase as the XII distance increases. At low temperature and a reduced amount of amine two OH resonances of 2,4-dichlorophenol may be observed, one at ca 12 ppm and the other at ca 15 ppm. The former is ascribed to the molecular complex, the latter to the ion-pair form[147]. For the 2 : 1 complex, a δOH for the OH$\cdots$O situation could also be measured at a value approximately 1 ppm lower than for the OH$\cdots$N complex.

Plotting ΔpK_a values for complexes between phenols and pyridines and lutidines gave two different plots depending on temperature: a normal one of type as shown in Figure 4a at 230 K and one at 128 K having a much higher δOH value (as high as 18 ppm)[153]. The latter was ascribed to formation of a homoconjugated ion. However, this behaviour was only found in a very narrow ΔpK_a range of -2 to 1.5 and was not observed in other studies of complexes of phenols with tertiary amines[152]. Because of solubility problems at low temperature these complexes could not be studied further, but an evaluation was conducted with thiophenol[153]. The complexes observed at 230 K could be described by Δ_{14} and the degree of proton transfer in this tautomeric equilibrium could be determined.

For the pyridine complexes, effects due to complexation observed at the pyridine molecule (such as ^{15}N chemical shifts) can also be used[153,154]. Low temperature measurements have clearly been very useful in elucidating these reactions. An approach using ^{15}N and ^{1}H chemical shifts as well as deuterium isotope effects on ^{15}N chemical shifts and primary proton isotope effects (see Section II.F.7) at very low temperature in freons showed in the ^{15}N spectrum three different species: AHB, AHAHB and AHAHAHB. For the 1 : 1 complex an asymmetric single well potential is assumed[155], different from the approach taken above. Furthermore, a linear correlation was found between the ^{15}N chemical shift and the one-bond ^1J(N,H) coupling constant. This type of reaction has also been studied using fractionation factors (See Section II.O).

N-Dodecyl-N,N-dimethylamine oxide yields with phenols a typical sigmoidal curve when chemical shifts are plotted vs. ΔpK_a[156].

2. Weak complexes

When dealing with complexes in which no proton transfer has occurred, this could be due to self-association[156] or association in general. Albrecht and Zundel[157] have determined the degree of association (as) for pentachlorophenol with different pyridines in

CCl$_4$ solution. Log K_{as} increased with ΔpK_a. Aggregation has been studied in a Schiff base of diazafluorenone with a long linear N-alkyl chain[158]. The interaction could also be with typical solvents like alcohols, acetone or dioxane (see Section II.H.5). A study of thymol, carvacrol, eugenol and vanillin with a number of alcohols and ketones showed for the former two compounds a high frequency shift of the *ipso* and *ortho* carbon resonances and a small low frequency shift of the other carbons, indicative of the phenol hydrogen bonding to the alcohol or the ketone. In case of alcohols, hydrogen bonding to the phenolic oxygen is ruled out. For the eugenol, the effects are small probably due to intramolecular hydrogen bonding[159].

In order to test the effect of phenolic compounds on aromatic flavours, NOE experiments have been conducted and it was found that gallic acid forms a stronger complex than naringin (**53**) with aromatic flavours such as 2-methylpyrazine, vanillin and ethyl benzoate. The former two compounds form the strongest complexes[160].

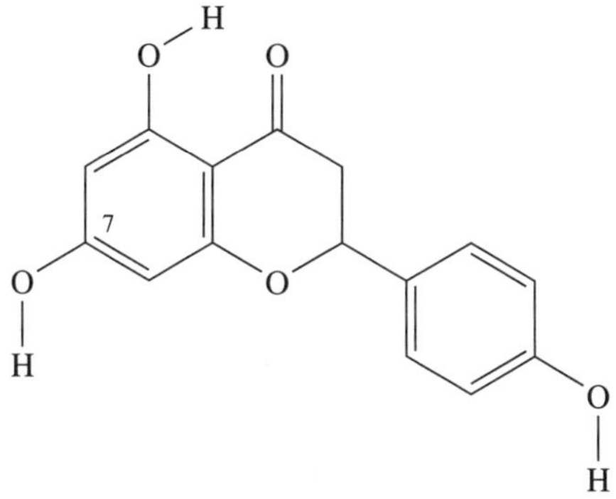

(**53**)

Complexes with β-cyclodextrins are well studied[52,161,162]. *m*-Fluorophenol showed that the fluorine is inside the cavity, but also that it formed a hydrogen bond with OH groups of the cyclodextrin judging from the isotope effects measured (see Section II.F.4)[52]. For Naringin-7-O-β-neohesperidoside, a structure is suggested in which the 4-keto and 5-OH group form hydrogen bonds to the secondary hydroxy groups at the rim of the wider end of the β-cyclodextrin cavity[161]. A study of hydroxyphenyl alkyl ketones with β-cyclodextrin showed a 1 : 1 complex of mixed complexation modes with the aryl or alkyl groups inside the cavity[162].

The tetra-anion of macrocycles made from resorcinol allows likewise host–guest complexes with positively charged organic compounds, but also with neutral molecules like diethyl ether[163].

Gels may be formed by mixing sodium bis(2-ethylhexyl) sulphosuccinate with phenols in non-polar solvents. Doping these gels with other phenols is claimed to yield information about the importance of hydrogen bonding[164]. Based on other methods the more acidic phenols are leading to the most stable gels. The OH chemical shifts are diminished at higher temperature. This is interpreted as a decrease of the hydrogen bonding. The temperature coefficients are largest for the more acidic phenols measured in the 20–30 °C range (-8 ppb K^{-1}). For the dopands like 4-cresol the temperature coefficients are much smaller. A large temperature coefficient is, however, supposed to indicate weak hydrogen bonding (see Section II.C). Furthermore, for doped gels separate OH resonances are observed for the various phenols. The question is whether NMR at all supports hydrogen bonding.

The complex between phenols and the stable radical 2,2,6,6-tetramethyl-1-piperidinyloxy radical (TEMPO) was studied by ^{13}C NMR. Having constant phenyl concentration the concentration of TEMPO was varied and a linear change of the carbon chemical shifts was observed. The *ipso* carbon was shifted to lower frequency, whereas all others were shifted to higher frequency. CH carbons showed larger shifts than the quaternary ones. For 2,4,6-trinitrophenol unusually large shifts were observed, suggesting a π-stacking. For the 2,5-dinitro and 2,6-di-*t*-butyl derivatives no hydrogen bonding to the TEMPO radical is seen[165].

M. Theoretical Calculations

Theoretical calculations have now reached a level that allows one to calculate both vibrational frequencies and NMR chemical shifts to a good accuracy. Such calculations offer great help in assigning NMR chemical shifts and providing reliable structures. Structural information is also available from X-ray and neutron diffraction studies. The neutron studies and *ab initio* method have the advantage of giving the OH positions, a parameter very important for understanding hydrogen bonding of phenols.

Overviews of theoretical calculations of chemical shifts using salicylaldehydes are given[21,166]. In these papers a large number of methods and basis sets are tested.

A very good correlation between calculated and experimental ^{1}H and ^{13}C chemical shifts are found for the series **1–4**[39]. Recently, this range has been extended[40]. In this context the change in chemical shifts is calculated as a function of the O—H bond length. The variation is found to be rather similar in the series. Deuterium isotope effects on ^{13}C chemical shifts are also calculated and it is shown that these originate very strongly from the change in the O—H bond length upon deuteriation[39].

^{17}O chemical shifts were calculated in phenol, anisole, 4-methoxyphenol and 2-methoxyphenol. Reasonable agreement is obtained with experimental results. In the case of 2-methoxyphenol the ^{17}OH chemical shift is 12 ppm different for the *cis* (hydrogen bonded) form and the *trans* conformation with the latter being at a higher frequency. This appears to be in very good agreement with experimental findings[167].

^{17}O chemical shifts were calculated (DFT BPW91, 6-31G(d) basis set; GIAO approach) for the C=O groups of *o*-hydroxyaromatics. A good correlation was found except for 1-propionyl-2-naphthol, which is sterically hindered[51].

N. Solid State NMR

Conformational effects and effects due to intermolecular interactions can often be measured in the solid state.

For strongly hydrogen bonded systems like compounds **1–4**, the rings are stacked and are only moderately taking part in strong intermolecular hydrogen bonding[40]. 1,3-Diacetyl-2,4,6-trihydroxybenzene (**2**) showed two sets of resonances. This is ascribed to the fact that of the two molecules in the asymmetric unit, one is forming a hydrogen bond to a water molecule. For **4**, the CO resonances are seen in a 2 : 1 ratio, indicating that the molecule in the solid has no C_3 axis.

One of the interesting questions is whether the proton transfer found in solution is also present in the solid state. A second, always relevant problem is to distinguish between centrosymmetric and tautomeric cases for symmetrical compounds. A classic example is naphthazarin (**44**).

The solid state of the Schiff bases is of great interest because of their photochromic and thermochromic properties. A few studies of Schiff bases in the solid state exist. Salman and coworkers[168] found for aniline Schiff bases of 2-hydroxy-1-naphthaldehyde that at equilibrium in the solid state about 85% are the ketoamine form judged from the C-α chemical shift. N-(2'-Hydroxybenzylidene)-2-hydroxyaniline was likewise found to show tautomerism in the solid state, whereas the corresponding 4-nitro derivative did not[169]. Residual dipolar couplings were studied in phenylazo-2-naphthols[170] (**47**).

A very extensive study of N,N'-di-(2-hydroxynaphthylmethylene)-p-phenylenediamine (**49**)[171] exploits both spin–lattice relaxation times of protons and ^{15}N CP-MAS spectroscopy at low temperature. Very low barriers are observed for the tautomeric processes: 8 kJ mol^{-1} for NH,NH → NH,OH (converting one of the NH forms to an OH form) and 2 kJ mol^{-1} for OH,OH → NH,OH. Furthermore, the effect of one hydrogen bond propagates to the other one[171].

In a study of complexes between triphenylphosphine oxide (TPPO) and substituted phenols, a good correlation between the pK_a of the phenols and the degree of hydrogen transfer was found in solution but not in the solid. This was ascribed to TPPO being too weak a base so that crystallographic influences obscured the acid–base effects[172]. Using a highly basic phosphine oxide like tris(2,4,6-trimethoxyphenyl)phosphine oxide gave better results, as determined by ^{13}C and ^{31}P CPMAS solid state NMR[173]. The authors find effects on Δ_{14} that are parallel to the solution data despite the crystal packing effects. However, several results are at least not quantitatively consistent. The ^{1}H NMR data in solution suggest a 50% proton transfer at a pK_a value of the phenol of ca 5.5. However, the ^{31}P results show that hardly any proton transfer takes place down to a pK_a of 3.8. Likewise, the ^{13}C results (Δ_{14}) indicates a value of 17.1 ppm for 2,4-dinitrophenol with a pK_a of 3.96. The 17.1 ppm is very close to that of picric acid, which is supposed to show full proton transfer.

The extent of proton transfer was also studied in complexes between genistein and piperazine. This was done by comparing solid and solution state ^{13}C spectra[174].

Studies of novolac-type resins (phenolic polyethylene oxide blends) show by ^{13}C NMR that a blend of 30 : 70 composition leads to a ca 2 ppm high frequency shift compared to a pure phenolic resin. This is ascribed by the authors to increased hydrogen bonding[175].

O. Fractionation Factors

Fractionation factors (the ratio between XD and XH in a H/D mixed solvent) can be determined by ^{13}C NMR[176]. For phenol, a value of 1.13 was found at 32 °C. This is slightly dependent on ionic strength[177]. For complexes between phenol and diamines, the fractionation factor is smallest for 1,2-propanediamine with a pK_a difference between donor and acceptor of −0.45. The fractionation factor increases as this difference becomes numerically larger.

For t-butylphenol and a series of other acids, fractionation factors were determined at low temperature in freons. A quasi-linear relationship between OH chemical shifts and fractionation factors was observed with different slopes for OH and NH bonds[178].

Tyrosine can be part of low barrier hydrogen bonds in enzymatic reactions. This is suggested for ketosteroid isomerases[179]. A fractionation factor of the COOH proton of Asp-99 (0.34) supports this[93,180]. The phenol proton having a hydrogen bond to the steroid shows a fractionation factor of 0.97. The fractionation factors can be related to the O···O distance[93].

III. IR

A. Introduction

Vibrational spectroscopy is a particularly useful tool in the study of phenols. Due to the polarity of the phenolic hydroxyl group, this structural element is associated with strong and characteristic IR absorption bands, and the appearance of these bands generally contains significant information on intra- and intermolecular interactions[181]. The most important of these interactions involve hydrogen bonding, and historically, IR spectroscopy has been the most important spectroscopic method in the study of hydrogen bonds[182,183]. IR spectroscopy, in combination with Raman spectroscopy, has thus found widespread chemical, analytical and technical application in the study of a variety of phenols[184-186]. These applications are facilitated by the presence of extensive collections of IR data in the literature, such as those by Varsányi[187], Nyquist[188] and by Pouchert[189]. These collections contain IR data, spectra and detailed assignments for a very large number of phenols; the volumes by Varsányi contain data for more than 100 phenols.

Among general methodological advances in the last couple of decades, we shall mention two, one experimental and one theoretical. The first is the application of IR polarization spectroscopy on partially aligned molecular samples[190,191]. The second is the development of new quantum theoretical procedures based on density functional theory (DFT)[192-194].

Traditional IR spectroscopy allows determination of transition energies (wavenumbers) and intensities, but it does not provide information on directional properties such as transition moment directions[190,191]. However, experimental determination of transition moment directions is of great significance, for example in the study of molecular symmetry aspects and in the assignment of observed transitions. Information on the polarization directions of vibrational transitions can be obtained by linear dichroism (LD) IR spectroscopy on oriented molecular samples. Molecular crystals are obvious examples of oriented molecular systems, but adequate crystalline samples for LD spectroscopy are frequently difficult to obtain, and the observed spectra are influenced by crystal effects. A much simpler procedure of obtaining oriented molecular samples is the use of anisotropic solvents, in particular stretched polymers and liquid crystals[190,191,195,196]. This technique is generally associated with significant baseline absorption from the anisotropic medium, and efficient application in the field of IR spectroscopy generally requires modern Fourier transform (FT) instrumentation with a high signal-to-noise ratio[197]. In the following sections we shall illustrate the results of IR polarization spectroscopy for phenol oriented in a nematic liquid crystal[198], and for 1,8-dihydroxy-9(10H)-anthracenone (anthralin, dithranol, **28**) partially aligned in a stretched polyethylene matrix[199].

The most important development in applied quantum chemistry in recent years is probably the successful implementation of computational procedures based on DFT[192-194] in several standard software packages, e.g. GAUSSIAN[200]. The DFT procedures offer the advantage of an adequate representation of electron correlation effects in the theoretical model at a moderate computational cost. A proper consideration of electronic correlation effects is crucial in the prediction of molecular vibrations, particularly in the description of effects associated with hydrogen bonding[201]. A variety of computational DFT procedures are available, but extensive surveys have shown that the functionals B3LYP and B3PW91 are particularly suitable for prediction of vibrational transitions[202-204]. It is notable that the performance of these procedures is not only much superior to that of traditional Hartree–Fock (HF) molecular orbital theory, but the DFT predictions are in better agreement with experiment[202-204] than those of post-HF MP2 perturbation theory[193,194] that requires much longer computation time. The availability of powerful and computationally feasible DFT procedures has inspired a number of recent re-investigations of the vibrational structure of phenolic model compounds, as indicated in the ensuing survey.

In the following sections, some recent work in this field is reviewed. In a number of cases, references are given to recent publications with discussions of earlier work. The main focus is on IR investigations of key phenols that serve as reference compounds, particularly in relation to the study of hydrogen bonding effects. IR spectroscopy of biological systems is considered to fall outside the scope of this survey. For an example of the application of IR spectroscopy in the study of biological systems, see the recent work by Berthomieu and collaborators[205–207] on Photosystem II of green plants.

B. The Characteristic Vibrations of the Phenolic OH Group

The IR spectra of phenols are characterized by a number of bands associated with the hydroxyl group, involving the stretching and bending motions of the O−H and C−O moieties. C−O stretching, ν(CO), and in-plane O−H bending, δ(OH), tend to couple strongly with aromatic CC and CH movements, giving rise to patterns of IR bands mainly in the 1500–1000 cm^{-1} region (see Section III.C.2). In contrast, the vibrational modes ν(OH) and γ(OH), corresponding to O−H stretching and O−H out-of-plane bending (or torsion), tend to be strongly localized in the OH moiety. They usually give rise to normal modes with effective masses close to 1 amu, indicating that the vibrational motion is essentially limited to the OH proton; these bands are therefore characterized by large isotope shifts in the corresponding OD isotopomers. The ν(OH) and γ(OH) vibrations generally give rise to strong IR transitions (but weak Raman bands) and are of great diagnostic value, particularly in the study of hydrogen bonding effects. We give a brief description of these vibrational modes below. For a comprehensive account, see the volume by Lin–Vien and coworkers[208].

1. OH stretching, ν(OH)

a. Free OH groups. The O−H stretching vibration of phenols with no substituents *ortho* to the hydroxyl group gives rise to a sharp band between 3700 and 3600 cm^{-1} in the gas phase (the corresponding O−D stretching band is observed between 2700 and 2600 cm^{-1}). The presence of *ortho* substituents frequently complicates the situation. In particular, the presence of a hydrogen bond acceptor group in this position leads to intramolecular hydrogen bonding effects (see below). Even alkyl groups may cause complication. For example, gaseous 2-*tert*-butylphenol (**54**) exhibits two O−H stretching bands at 3670 and 3642 cm^{-1}, indicating the presence of *cis* and *trans* −OH rotamers[181,188]. In a recent investigation of 2,6-diisopropylphenol (**55**) in CCl_4 solution, Bikádi and coworkers[209] concluded that five conformers, corresponding to isopropyl rotamers, contribute to the pattern of IR absorption in the O−H stretching region.

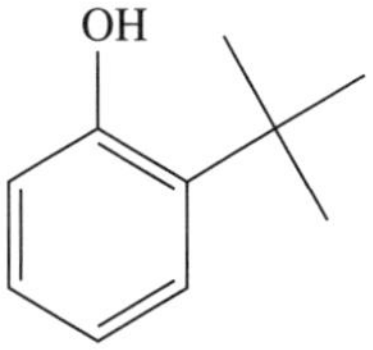

(54)

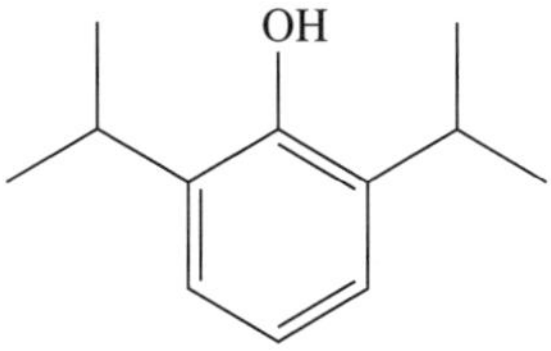

(55)

b. Hydrogen bonded OH groups. Participation of the OH proton in hydrogen bonding leads to a marked red shift of the O−H stretching band. It is observed that the stronger the hydrogen bonding, the larger the shift towards lower wavenumbers. At the same time, a broadening of the band is usually observed. IR spectroscopy is thus a very sensitive technique in the study of hydrogen bonding effects, and the wavelength shift, $\Delta\nu$, and half-height width, $\nu_{1/2}$, of the ν(OH) band are among the most important spectroscopic parameters in the characterization of these phenomena.

Intermolecular hydrogen bonding is frequently associated with an increase in the integrated IR intensity. As an example, Figure 5 shows the O−H stretching region for phenol in CCl$_4$ solutions with different concentration[198] (note that curve A in Figure 5 is shown on a five times expanded ordinate scale). The IR absorption indicates the coexistence of free and different associated forms. The sharp peak observed at 3611 cm^{-1} is due to free, non-complexed hydroxyl groups, while the broad band between 3600 and 3100 cm^{-1} is due to hydroxyl groups involved in hydrogen bonded dimer or polymer formation. Increasing the phenol concentration increases the relative concentration of self-associated forms, resulting in a rapid increase of the broad, continuous band belonging to hydrogen bonded OH groups.

Intramolecular hydrogen bonding is expected for those phenols that contain accessible hydrogen bond acceptor groups within the molecule. The formation of an intramolecular hydrogen bond usually results in the closing of a 5- or 6-membered pseudo-ring structure[188,210,211]. Weak effects are observed for phenols 2-substituted by halogen atoms (see, e.g., the recent investigation[212] of 2,6-difluorophenol, **56**), or by methoxy, thiomethoxy, amino, cyano, vinyl or allyl groups. For example, the ν(OH) band observed for 2-allylphenol[213,214] (**57**) is split into two peaks at 3656 and 3592 cm^{-1}; the red-shifted component is ascribed to the presence of a rotamer with hydrogen bonding between the hydroxyl and the π-bond of the adjacent allyl group. A similar splitting (3645 and 3508 cm^{-1}) is observed for 2-(hydroxymethyl)phenol[214] (**58**) but this time the red-shifted component is by far the more intense, indicating the predominance of the hydrogen bonded form. Much stronger interaction is observed for 2-(alkylaminomethyl)phenols (*ortho*-Mannich bases, see Section III.E.6), leading to complicated ν(OH) profiles in the 3500−2000 cm^{-1} range[215]. Strong, so-called 'resonance-enhanced' intramolecular

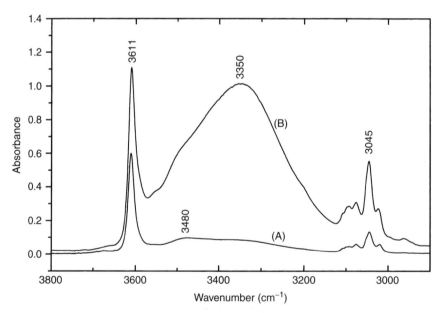

FIGURE 5. The OH stretching region of the IR absorption spectrum of phenol in CCl$_4$ solution[198]: (A) 1% solution (5 × ordinate expansion); (B) 5% solution. Reprinted with permission from Reference 198. Copyright (1998) American Chemical Society

(56)   (57)   (58)

hydrogen bonding is present for phenols with NO_2, R—C=O or R—C=N—R' groups in the 2-position[210] (see Figure 2). This kind of interaction is frequently referred to as 'chelation'. The chelated OH···X stretching vibration usually gives rise to a broad absorption band in the 3200–2500 cm^{-1} region. A large variation in intensity and shape for this absorption has been observed. The stronger the chelated hydrogen bond, the lower the recorded intensity, a situation that is opposite to that observed for intermolecular hydrogen bonding. The lowering of the IR intensity has been explained by the bending of the intramolecular OH···X linkage[216]. Sometimes, this absorption may be overlooked because of broad and weak features[32,208,210]. In Section III.E we consider the IR spectra of some compounds with chelated hydrogen bonding.

Participation of the OH group in hydrogen bonding increases the asymmetry of the O—H stretching potential, thereby increasing the importance of anharmonic effects. Strong interaction may lead to broad potentials of highly asymmetrical shape, or low-barrier double-minima potentials, possibly associated with proton transfer and tunnelling effects[217]. Non-rigid systems with easily polarizable hydrogen bonds (mobile protons) are frequently characterized by anomalous, broad or 'continuous' absorption bands[218]. In addition to affecting the fundamental of the OH stretching mode, the anharmonic effects tend to increase the intensity of overtone and combination bands observed in the near-IR (NIR) region. Hydrogen bonded phenols are generally characterized by rich NIR spectra[219-221]. Theoretical modelling of the molecular and vibrational structure of hydrogen bonded systems and the associated optical properties is an area of current research[204,216,222-233].

2. OH out-of-plane bending, $\gamma(OH)$

For phenols with free, uncomplexed hydroxyl groups, this torsional mode usually gives rise to an absorption band in the far-IR region. In the gas phase spectrum of phenol[188,198], the transition is observed as a strong band around 310 cm^{-1}. When the hydroxyl proton participates in hydrogen bonding, the force constant for the out-of-plane torsional motion is increased, resulting in a shift towards larger wavenumbers. This band thus moves in the opposite direction to that of the $\nu(OH)$ band: Stronger hydrogen bonding increases the $\gamma(OH)$ wavenumber and decreases the $\nu(OH)$ wavenumber. In strongly chelated phenols like *ortho*-hydroxybenzoyls (**24**), the $\gamma(OH)$ transition is observed in the 700–800 cm^{-1} region[208,234,235], and a band observed at 984 cm^{-1} in the spectrum of the salicylate anion can possibly be assigned to this transition (Section III.E.1). In these compounds, the $\gamma(OH)$ vibration becomes near-degenerate with other out-of-plane vibrations like $\gamma(CH)$. This may lead to mixing with these modes and the $\gamma(OH)$ intensity is frequently distributed over a number of vibrational transitions in this region. This detracts from its diagnostic value, but bands with large $\gamma(OH)$ character can frequently be recognized by their broader shape.

C. The IR Spectrum of Phenol

The vibrational structure of phenol and its main isotopomers has recently been the subject of several investigations[198,220,221,234–237]. A critical review of previous assignments of the fundamental transitions can be found in the treatise by Keresztury and coworkers[198].

1. Low-temperature Argon matrix spectrum of phenol

Figure 6 shows the IR spectrum of phenol isolated in an Argon matrix at 20 K[237]. This is a very inert medium and the observed wavenumbers are similar to those observed in the gas phase (see Table 3). The largest deviations concern the strong lines observed at 1343 and 1176 cm^{-1} which are red-shifted by about 20 cm^{-1} relative to the gas phase spectrum. The low-temperature matrix spectrum has the advantage of very sharp lines, in contrast to the spectrum of gaseous phenol, which is influenced by rotational line broadening. The splitting of some of the lines in the matrix spectrum is due to the occupation of different sites. The ν(OH) line, for example, is split into two major components at 3639 and 3634 cm^{-1}. The observed wavenumbers are well reproduced by the results of a B3LYP/cc-pVTZ calculation[200]. The theoretical wavenumbers listed in Table 3 have been scaled by a common scale factor, $\alpha = 0.9776$. This factor was determined by a regression analysis based on 14 strong peaks in the matrix spectrum between 700 and 1700 cm^{-1}, yielding a standard deviation of 3.1 cm^{-1}.

2. IR polarization spectra of phenol aligned in a liquid crystal

Keresztury and coworkers[198] have recently measured the IR LD spectra of phenol aligned in a uniaxially oriented liquid crystal nematic phase, thereby providing new

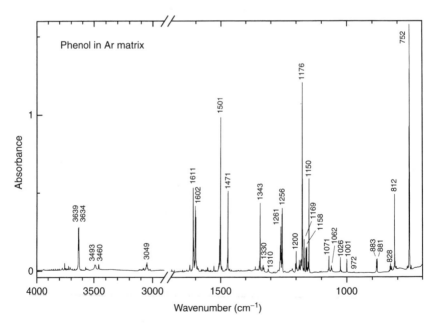

FIGURE 6. IR absorption spectrum of phenol isolated in an Argon matrix at 20 K[237]

5. NMR and IR spectroscopy of phenols

TABLE 3. Observed and calculated fundamental vibrational transitions for phenol

	Gas phase[198]		Ar matrix[237]		Nematic phase[198]			B3LYP/cc-pVTZ[200]			
	$\tilde{v}^a$	I^b	$\tilde{v}^a$	OD^c	$\tilde{v}^a$	OD^d	$|\phi|^e$	$\tilde{v}^{a,f}$	I^g	ϕ^e	Assignment[h]
1	3655	104.22	3634.1	0.28	3403	0.21	(0)	3732.3	55.8	(0)	v(OH)
2	3074	0.35						3124.8	5.8	0.9	v(CH)
3	3061	1.75						3118.6	18.6	73.5	v(CH)
4	3052	4.29						3104.7	17.7	−32.8	v(CH)
5	3046	11.83	3049.2	0.06	3043	0.05	50.5	3096.4	0.2	−34.6	v(CH)
6	3021	1.93						3078.1	14.4	10.5	v(CH)
7	1609	36.09	1610.7	0.53	1607	0.22	35.7	1612.9	35.1	20.2	v(CC), δ(CH)
8	1604	62.90	1602.0	0.42	1595	0.27	57.7	1601.9	45.7	56.3	v(CC), δ(CH)
9	1501	69.12	1501.0	0.98	1502	0.23	52.8	1500.7	52.2	42.8	v(CC), δ(CH)
10	1472	42.92	1471.1	0.51	1472	0.22	35.2	1473.7	26.1	24.3	δ(CH), v(CC)
11	1361	0.96	1342.6	0.43	1358	0.08	43.7	1346.1	26.6	40.0	δ(CH), δ(OH)
12	1344	17.52	1330.0	0.04				1319.0	7.7	1.9	v(CC), δ(CH)
13	1261	51.33	1255.7	0.40	1268	0.14	47.4	1257.3	79.6	35.0	v(CO), v(CC)
14	1197	106.26	1176.1	1.21	1219	0.21	47.4	1168.4	131.8	40.4	δ(OH), v(CC)
15	1176	10.34	1169.0	0.20	1165	0.08	46.0	1167.1	2.1	66.6	δ(CH), v(CC)
16	1150	13.87	1150.0	0.59	1151	0.05	63.0	1152.3	26.1	55.7	v(CC), δ(CH)
17	1070	12.18	1071.0	0.09	1069	0.05	29.0	1071.9	12.9	−6.0	v(CC), δ(CH)
18	1026	4.73	1025.9	0.08	1024	0.03	43.5	1023.0	3.7	38.9	v(CC), δ(CH)
19	999	3.68	1000.8	0.07	999	0.04	47.8	996.1	3.3	43.8	δ(CC), v(CC)
20	973	0.26	972.0	0.00				971.6	0.2	z	γ(CH), γ(CC)
21	956	0.18						950.6	0.1	z	γ(CH), γ(CC)
22	881	8.94	881.1	0.08	883	0.04	z	880.0	6.7	z	γ(CH), γ(CC)
23								810.1	0.0	z	γ(CH)
24	810	17.78	812.4	0.49	814	0.08	54.1	812.7	19.9	37.2	v(CC), δ(CC)
25	752	74.81	752.2	1.58	754	0.25	z	752.5	55.1	z	γ(CH), γ(CO)
26	687	42.92			692	0.24	z	691.0	25.5	z	γ(CC), γ(CH)
27	618	0.61						620.5	0.3	−77.0	δ(CC), v(CC)
28	526	1.93						525.8	1.8	50.1	δ(CC), v(CC)
29	503	20.5			509	0.07	z	507.2	12	z	γ(CC), γ(CO)
30	420	0						414.6	0.4	z	γ(CC), γ(CH)
31	410	8.23			415	0.03	35.7	397.8	9.5	36.9	δ(CO), v(CC)
32	310				620	0.06	z	346.4	99.9	z	γ(OH)
33	242							226.4	0.9	z	γ(CC), γ(CO)

[a] Wavenumber in cm^{-1}.
[b] Intensity in arbitrary units.
[c] Optical density.
[d] Isotropic optical density, $(E_\parallel + 2E_\perp)/3$.
[e] In-plane moment angles ϕ (deg) relative to the moment direction of transition no. 1, v(OH). The experimental sign of ϕ is unknown[198]. z indicates out-of-plane polarization.
[f] Scaling factor 0.9776.
[g] Intensity in km mol^{-1}.
[h] Approximate mode description: v = stretching, δ = in-plane bending, γ = out-of-plane bending.

experimental information on the molecular and vibrational structure of phenol. Figure 7 shows the absorption curves A and B recorded with the electric vector of the linearly polarized IR radiation parallel ($E_\parallel$ = curve A) and perpendicular ($E_\perp$ = curve B) to the director of the liquid crystalline sample. The smallest dichroic ratio $d = E_\parallel/E_\perp = 0.325$ is observed for the five peaks at 509, 620, 692, 754 and 883 cm^{-1}, corresponding to a common orientation factor[190,196,197,238] $K = d/(2 + d) = 0.14$. These peaks were assigned to out-of-plane polarized transitions in the C_s symmetric molecule. The remaining peaks with larger dichroic ratios d ranging from 0.79 to 3.71, corresponding to K values from 0.28 to 0.65, were assigned to transitions with different in-plane transition moment directions. The largest dichroic ratio $d = 3.71$ is observed for the broad OH stretching band with maximum around 3403 cm^{-1}. It was therefore assumed[198] that the transition moment of the ν(OH) fundamental is oriented preferentially along the director of the liquid crystal, probably due to hydrogen bonding with the terminal nitrile groups of the rod-like molecules forming the liquid crystalline phase. The effective molecular orientation axis[190] was thus taken to coincide with the ν(OH) transition moment direction, and by using the formulas of Thulstrup and Michl[190], the absolute values $|\phi|$ of the in-plane moment angles relative to that of ν(OH) could be derived. The results for the observed fundamentals are included in Table 3.

Keresztury and coworkers[198] compared the derived moment angles for phenol with those predicted by a B3P86/6–311G** DFT calculation. Corresponding results obtained with B3LYP/cc-pVTZ[200] are included in Table 3. The theoretical angles ϕ listed in Table 3 are relative to the predicted direction of the ν(OH) transition moment (the sign of ϕ is defined as in Reference 198). This direction forms a considerable angle with the O–H bond axis (30°), very roughly corresponding to an axis through C$_1$ and the OH proton. The calculated moment angles are in fair agreement with the experimental estimates. The analysis by Keresztury and associates[198] is based on a number of assumptions, and the derived numerical values are associated with experimental error limits that are difficult to estimate. In addition, the experimental results relate to phenol engaged in hydrogen bonding, whereas the calculated data refer to an isolated phenol molecule.

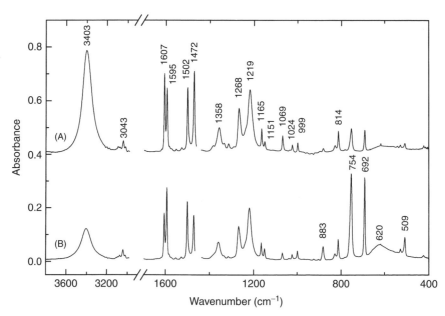

FIGURE 7. Linear dichroic (LD) absorption spectra of phenol partially aligned in a uniaxially oriented nematic liquid crystal[198]. The curves indicate absorption measured with the electric vector of the linearly polarized radiation parallel (A) and perpendicular (B) to the director of the liquid crystalline sample. Reprinted with permission from Reference 198. Copyright (1998) American Chemical Society

5. NMR and IR spectroscopy of phenols 375

The results of the LD investigation of phenol are significant for a number of reasons. The grouping of the observed orientation factors allows conclusions concerning the molecular symmetry. The observation of five individual peaks with precisely the same (small) K value supports the assumption that phenol is a planar molecule. In particular, observation of the same K for $\gamma(\text{CH})$ and $\gamma(\text{OH})$ transitions demonstrates that the hydrogen bonding OH group stays co-planar with the benzene ring. On the other hand, the variation of K values observed for in-plane polarized peaks demonstrates a significant symmetry lowering relative to a C_{2v} symmetrical model, a situation that for example complicates an unambiguous correlation with the modes of benzene (as attempted, e.g., by Varsányi[187] with reference to Wilson's notation[239]). The clear experimental distinction between in-plane and out-of-plane polarized transitions enabled Keresztury and associates[198] to suggest a reassignment of one transition: The weak peak observed at 829 cm^{-1} was previously assigned by most investigators to the out-of-plane polarized $\gamma(\text{CH})$ fundamental ν_{23}, but the observed $K = 0.36$ shows that the peak is in-plane polarized. The peak may be assigned[198] to $2\nu_{31}$, an overtone of the fundamental observed near 415 cm^{-1}. The overtone may gain intensity by Fermi coupling with the medium intense transition at 814 cm^{-1} (also with $K = 0.36$), which can be assigned to ν_{24}. The fundamental ν_{23} is predicted to be extremely weak and is not clearly observed.

Figure 7 illustrates the influence of weak intermolecular hydrogen bonding on the IR spectrum of phenol. The $\nu(\text{OH})$ transition is observed as a broad, nicely Gaussian-shaped band with maximum at 3403 cm^{-1}, red-shifted by 230–250 cm^{-1} relative to the transition in Argon matrix or in the gas phase. A similar broadening, but a shift in the opposite direction, is observed for $\gamma(\text{OH})$: In the liquid crystal this transition is observed at 620 cm^{-1}, a blue-shift of more than 300 cm^{-1} relative to the position in the gas phase spectrum. The remaining bands show much smaller shifts, but a significant broadening and a relatively large blue shift are observed for the transition at 1219 cm^{-1}. This transition is shifted by 43 cm^{-1} relative to the Argon matrix spectrum where it is found at 1176 cm^{-1}. It can be assigned to the fundamental ν_{14} which has substantial $\delta(\text{OH})$ character[198,234–236]. The peaks at 1268 and 1358 cm^{-1} are slightly broadened and are blue-shifted by 12–15 cm^{-1} relative to the Argon matrix spectrum. They are assigned to ν_{13} and ν_{11} which involve $\nu(\text{CO})$ and $\delta(\text{OH})$ contributions[198,234–236].

D. Hydrogen Bonded Complexes

IR spectroscopic investigation of intermolecular interactions with phenols has a long history[240]. Phenols are frequently used as convenient model proton donors in the study of intermolecularly hydrogen bonded systems. Differently substituted phenols are characterized by a range of different acidities, and complexes can be studied with a wide variety of proton acceptors. The most commonly adopted acceptors are O and N bases. A special case is carbon monoxide that in complexes with phenols apparently forms ArOH···CO contacts (rather than ArOH···OC)[241]. Here we mention some recent investigations with typical proton acceptors like water, alcohols and amines.

The vibrational structure of phenol or phenolate hydrates has been investigated for example by Leutwyler[242,243], Müller-Dethlefs[244], Ebata[245], Carabatos-Nédelec[246], Gerhards[247] and their coworkers. As an illustrative example of an IR spectrum of a crystalline hydrate we show in Figure 8 the spectrum of ellagic acid dihydrate (EA·2H$_2$O)[248,249]. Ellagic acid (**59**) is a plant phenol that is widely distributed in Nature. It has an extremely high melting point (>360 °C), an indication of strong intermolecular forces in the solid state. In the dihydrate crystal, the ellagic acid molecules are stacked, and the crystal water molecules act as hydrogen bond bridges in three directions[250]. The OH stretching region of the IR spectrum is characterized by a sharp peak close to 3600 cm^{-1}

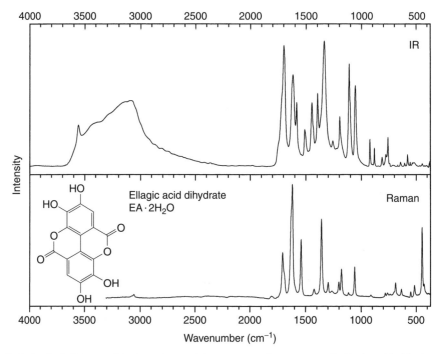

FIGURE 8. Solid state IR absorption and Raman scattering spectra of ellagic acid dihydrate (**59**)[248,249]

which can be assigned to free OH groups, followed by a strong, continuous absorption band with a maximum at 3100 cm^{-1} and a long tail down to around 2300 cm^{-1}. The considerable absorption intensity and anomalous band shape can probably be explained by coupling of the easily polarizable hydrogen bonds with low-frequency lattice phonons,

5. NMR and IR spectroscopy of phenols 377

in combination with strong anharmonic effects (multi-Fermi resonance)[227,233]. The absence of the OH stretching band in the Raman spectrum is characteristic. Raman spectroscopy (Figure 8) offers a window to the weak CH stretching bands that are buried below the OH continuum in the IR absorption spectrum. In the region below 2300 cm^{-1} the spectrum of ellagic acid seems 'normal'. Because of the centro-symmetric molecular structure (C_{2h} point group), the IR and Raman spectra are complementary: those transitions that are IR active are forbidden in Raman, and vice versa.

Like water and alcohols, phenols are prone to self-association, as indicated in Section III.B.1.b. An interesting example is the self-association of phenolic calixarene-like building blocks, which was investigated by IR spectroscopy by Lutz and coworkers[251], and most recently by Painter and associates[252]. The structure of the binary phenol–methanol cluster was investigated recently by Schmitt and coworkers[253].

Several investigations have considered intermolecular phenol complexes with ammonia and amines[216,230,254–264], and with aza-aromates, nitriles and Schiff bases[219,220,222,265–267]. The IR spectra of complexes with strong trialkylamine bases usually show continuous absorption bands characteristic of hydrogen bonded bridges with broad, asymmetric single or double minimum potentials. The interaction with very strong bases (proton sponges) leads to proton transfer effects; optical UV-VIS and IR spectroscopy are excellent tools in the study of these reactions[259,260]. For examples of complicated spectra see the recent publications by Wojciechowski, Brzezinski and their coworkers[263,264] on complexes between phenols and triazabicyclodecene bases; the observed broad and continuous IR profiles are interpreted in terms of strong, multiple hydrogen bonding, proton transfer and double minimum potentials with vibrational tunnelling splitting.

E. Phenols with Intramolecular Hydrogen Bonds

1. 2-Hydroxybenzoyl compounds

In these compounds, the phenolic OH group is situated next to a position with a carbonyl substituent, O=C–R. As in other conjugated β-hydroxycarbonyl compounds, these molecules are characterized by the formation of a stable, intramolecular hydrogen bond, OH $\cdots$ O=C–R, closing a six-membered chelate ring (see Figure 2).

The IR spectrum of salicylaldehyde, the simplest member of the series (R = H), has been the subject of several recent investigations[234,235,268,269]. The ν(OH) fundamental gives rise to a complicated band between 3500 and 3100 cm^{-1}. According to the analysis by Koll and coworkers[269], the band profile is influenced by Fermi coupling with overtones and combinations of δ(OH) bending vibrations, and by other anharmonic effects. Bands observed in the 760–700 cm^{-1} region have been assigned to γ(OH) vibrations; normal mode calculations predict significant coupling between near-degenerate γ(CH) and γ(OH) vibrations, giving rise to two or more modes with partial γ(OH) character. The ν(C=O) stretching band is observed around 1670 cm^{-1}, indicating a red shift of ca 40 cm^{-1} relative to the corresponding band in the spectrum of benzaldehyde.

Systematic investigation of the IR spectra of different 2-hydroxybenzoyls has been undertaken, in particular by Mikenda and coworkers[268] and by Palomar and coworkers[235]. The first group of investigators considered a series of 14 different 2-substituted phenols, with the following carbonyl substituents O=C–R : R = Cl, OH, SH, OCH$_3$, SCH$_3$, H, CH$_3$, C$_6$H$_5$, N(CH(CH$_3$)CH$_2$)$_2$CH$_2$, N(CH$_2$CH$_2$)$_2$CH$_2$, N(CH$_3$)$_2$, NHCH$_3$, NH$_2$ and NHNH$_2$. The second group investigated salicylaldehyde, 2-hydroxyacetophenone, methyl salicylate and salicylamide (R = H, CH$_3$, OCH$_3$ and NH$_2$). Observed ν(OH) and γ(OH) wavenumbers for these four compounds are listed in Table 4. Both groups[235,268] supported their investigations by comparison with data for pertinent reference compounds, and by

TABLE 4. Observed wavenumbers (cm^{-1}) for ν(OH) and γ(OH) vibrations in phenol and a number of 2-hydroxybenzoyl compounds with intramolecular hydrogen bonding[a]

Compound	ν(OH)	γ(OH)
Phenol	3655	322
Methyl salicylate	3258	714[b]
Salicylaldehyde	3190	714[b]
2-Hydroxyacetophenone	3100	787[b]
Salicylamide	3070	807[b]
Sodium salicylate	2910–1900	984[c]

[a] The data for sodium salicylate refer to the solid state spectrum and are taken from the work by Philip and coworkers[271]. The remaining data refer to gas phase spectra for ν(OH) and to CCl$_4$ or CS$_2$ solution spectra for γ(OH) and are taken from the compilation by Palomar and coworkers[235].
[b] According to Palomar and coworkers[235], additional modes with partial γ(OH) character are observed.
[c] This assignment differs from the one suggested by Philip and coworkers[271]; see Section III.E.1.

correlation with the results of DFT calculations. Both groups derived empirical relationships between spectral data and hydrogen bond parameters, particularly the energy E_{IMHB} of the intramolecular hydrogen bond. It is found that the observed ν(OH) and γ(OH) shifts closely parallel the calculated or otherwise estimated hydrogen bond strengths. We refer to these publications[235,268] for further discussion of the IR spectroscopic properties of these key compounds, and for references to earlier work in the field.

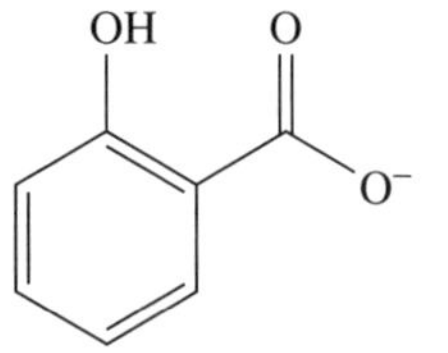

(60)

Very strong intramolecular hydrogen bonding is predicted[270] for the salicylate anion (R = O$^-$) (60). Philip and associates[271] recently published an IR and Raman spectroscopic investigation of sodium salicylate. Not surprisingly, a very complicated ν(OH) stretching band is observed with a broad IR profile between 3000 and 2300 cm^{-1}, and also broad features around 1900 cm^{-1}. These bands are absent in the Raman spectrum. A sharp IR peak at 537 cm^{-1} (KBr pellet) is assigned to γ(OH) by Philip and associates[271]. However, one would expect a blue shift of this band relative to the spectra of neutral 2-hydroxybenzoyl compounds where it is observed in the 700–800 cm^{-1} region (see Table 4). B3LYP/6-31G* calculations predict a blue shift of the γ(OH) vibration of no less than 280 cm^{-1} when passing from salicylaldehyde to the salicylate anion. We suggest that the broad, intense band observed[271] at 984 cm^{-1} in the spectrum of sodium salicylate may be assigned to γ(OH). This band does not seem to have a counterpart in the Raman spectrum.

Simperler and Mikenda[272] investigated a series of 2,6-disubstituted phenols containing two different carbonyl substituents (61). Five different substituents were considered:

COOH, COOMe, CHO, COMe, CONH$_2$, resulting in a series of ten phenols. These compounds are able to form two competitive kinds of intramolecular hydrogen bonds. According to the analysis by Simperler and Mikenda[272], the conformation of the most stable isomer is determined by the energetically most favourable non-bonded O···R−C interaction and not by the more favourable one of the two possible O−H···O=C hydrogen bond interactions.

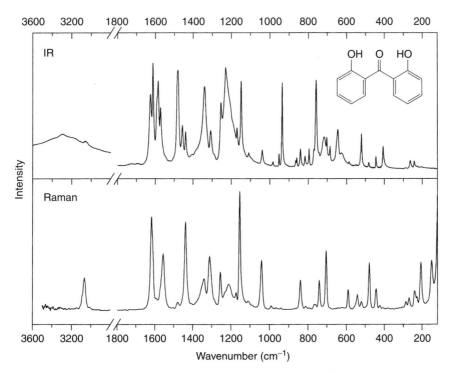

(61)   **(62)**

A different example is provided by 2,2'-dihydroxybenzophenone (**36**), where two equivalent hydroxyl groups simultaneously form hydrogen bonds with the same carbonyl group, resulting in a bifurcated arrangement. The solid state IR and Raman spectra of this compound are shown in Figure 9[273]. The two ν(OH) vibrations give rise to a broad IR band with maximum at 3300 cm^{-1}, overlapping a weaker CH stretching band at 3050 cm^{-1}.

FIGURE 9. Solid state IR absorption and Raman scattering spectra of 2,2'-dihydroxybenzophenone (**36**)[273]

The ν(OH) IR band of 2-hydroxybenzophenone is observed at a lower wavenumber (3080 cm^{-1}, liquid solution)[268], perhaps an indication that the single hydrogen bond in this compound is stronger. According to the IR LD analysis by Andersen and coworkers[274], a relatively broad IR band at 714 cm^{-1} can be assigned to the antisymmetric combination of the two γ(OH) vibrations (b symmetry in the C_2 point group of this compound). Neither ν(OH) nor γ(OH) transitions seem to have counterparts in the Raman spectrum. Two strong transitions at 1626 and 1584 cm^{-1}, polarized along the C_2 symmetry axis, could be assigned to modes involving coupling of ν(C=O) with δ(OH) and other motions. For comparison, the reported[268] ν(C=O) wavenumbers for benzophenone and 2-hydroxybenzophenone are 1660 and 1632 cm^{-1}. References to work on other hydroxybenzophenones can be found in the volume by Martin[15].

Anthralin (dithranol), an efficient drug in the treatment of psoriasis and other skin diseases, is closely related to 2,2'-dihydroxybenzophenone. The compound was for many years believed to be 1,8,9-anthracenetriol (**28B**), but on the basis of IR and other spectroscopic data Avdovich and Neville[275] could in 1980 show that the compound is 1,8-dihydroxy-9(10H)-anthracenone (**28A**). Solid state IR and Raman spectra of anthralin[273] are shown in Figure 10. The broad ν(OH) band is centred around 3000 cm^{-1}, completely blocking the ν(CH) bands. However, they are nicely resolved in the Raman spectrum, giving rise to an aromatic ν(CH) band with maximum at 3053 cm^{-1} and two peaks at 2910 and 2882 cm^{-1} that can be assigned to the two CH stretches of the methylene unit of anthralin. The analysis of the IR spectrum was supported by LD spectroscopy on a sample of anthralin partially aligned in a stretched polyethylene matrix. The observed LD absorbance curves[276] are shown in Figure 11. In this case, the interpretation of the LD

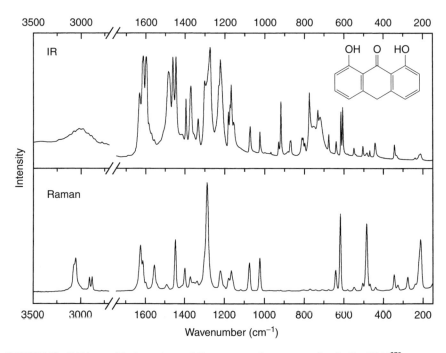

FIGURE 10. Solid state IR absorption and Raman scattering spectra of anthralin (**28A**)[273]

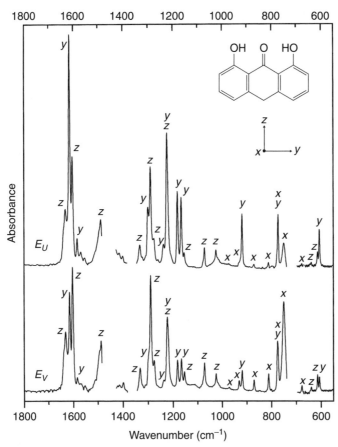

FIGURE 11. Linear dichroism (LD) absorbance curves for anthralin (**28A**) partially aligned in uniaxially stretched polyethylene[276]. E_U and E_V denote absorbance curves measured with the electric vector of the linearly polarized light parallel (U) and perpendicular (V) to the stretching direction. The regions 1480–1430, 1380–1350 and 740–700 cm^{-1} were blocked by strong polyethylene absorption

data is greatly simplified by the C_{2v} symmetry of the anthralin molecule, which limits the molecular transition moment directions to three mutually perpendicular directions defined by the symmetry axes x, y and z. The resulting assignment of moment directions[199] is indicated in Figure 11. It is evident that the results offer a unique insight into the vibrational structure. The strong x-polarized transition close to 750 cm^{-1} can be assigned to a γ(OH) vibration of b_1 symmetry. This and other transitions in the 1000–700 cm^{-1} region are very weak or absent in the Raman spectrum. It would be tempting to assign the strong IR transition at 1614 cm^{-1} to a C=O stretching vibration[275], but this transition is y-polarized and thus cannot be assigned to ν(C=O). However, the neighbouring peaks at 1632 and 1602 cm^{-1} are z-polarized and can be assigned to totally symmetric vibrations with significant ν(C=O) character[199].

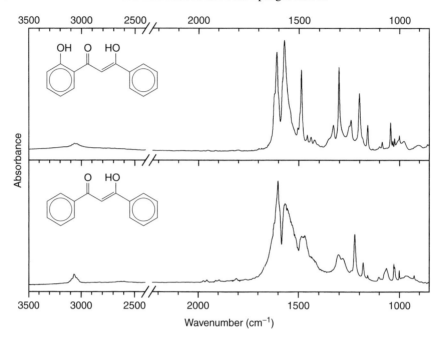

FIGURE 12. IR absorption spectra of (2-hydroxybenzoyl)benzoylmethane (**29A**, top) and dibenzoylmethane enol (**62**, bottom) in CCl$_4$ solution[277]

The prevailing enol form of (2-hydroxybenzoyl)benzoylmethane (**29A**) contains a bifurcated hydrogen bonding system similar to that of 2,2′-dihydroxybenzophenone (**36**) and anthralin (**28A**). The IR spectrum in CCl$_4$ solution (Figure 12, top)[277] show similarities with the spectra of those compounds, particularly in the region around 1600 cm^{-1} where the IR LD analysis[199,277] reveals the presence of four similar transitions in all three compounds. Replacement of the phenolic hydroxyl group by a hydrogen atom produces dibenzoylmethane enol (**62**). Somewhat surprisingly, the IR spectrum of the latter compound is more complex than that of the former (Figure 12, bottom). In particular, the 1800–1400 cm^{-1} region of the spectrum of **62** has comparatively broad and poorly resolved structures, with a curious tail towards higher wavenumbers. Similar spectra are recorded in other solvents and in the solid state[277]. Probably the IR spectrum of **62** is influenced by profound anharmonic effects associated with the symmetrical double-minimum OH stretching potential[32,210] for this compound.

2. Hydroxyquinones

Very recently, Rostkowska and coworkers[211] investigated the IR spectra of a series of compounds that form intramolecular hydrogen bonds closing five-membered rings, including 2-hydroxynaphthoquinone (**63**) and 2,5-dihydroxy-1,4-benzoquinone (**64**). The observed ν(OH) band maxima show a characteristic dependence on geometrical constraints, ranging from 3552 cm^{-1} in 3,4-dihydroxy-3-cyclobutene-1,2-dione (**65**) to 3120 cm^{-1} in tropolone (**66**) (Argon matrix), reflecting the increasing strength of the hydrogen bonding. At the same time, the γ(OH) band is shifted from 463 to 746 cm^{-1}.

5. NMR and IR spectroscopy of phenols   383

In the case of tropolone, with a seven-membered ring, a complicated OH stretching region is observed, and the positions of ν(OH) and γ(OH) are in the range typical for molecules with intramolecular hydrogen bonds forming six-membered rings, such as the 2-hydroxybenzoyls considered above, and in hydroxyquinones like naphthazarin (**44**)[276,278,279], quinizarin (**67**)[280–282] and chrysazin (**35**)[238,283].

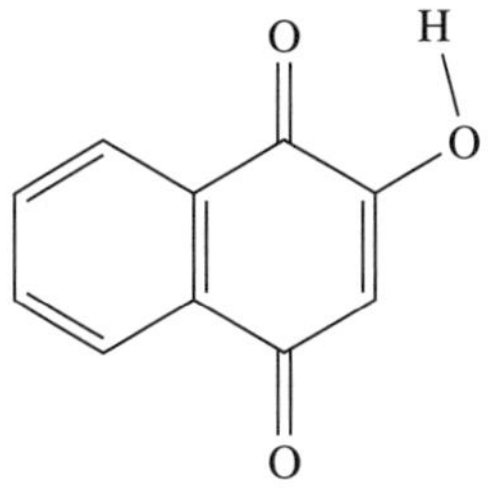

(**63**) 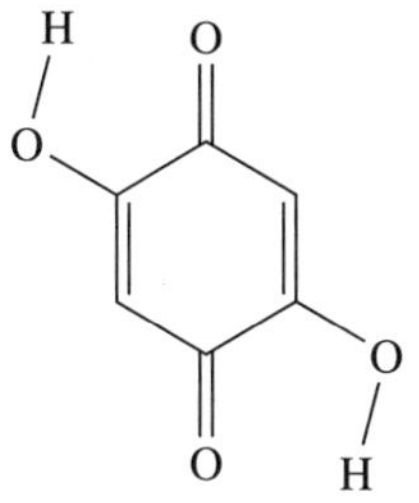   (**64**) 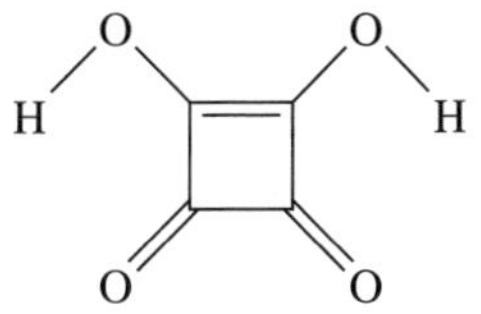   (**65**)

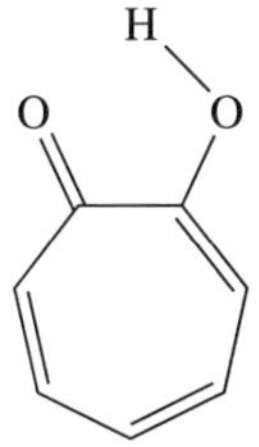

(**66**)

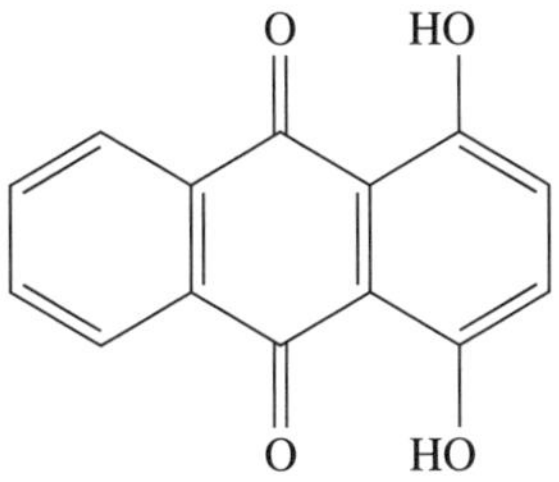

(**67**) 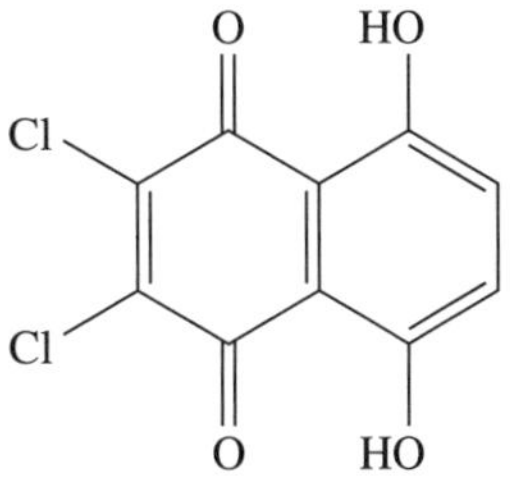   (**68**)

The molecular structure of naphthazarin (**44**) has been a subject of considerable interest, particularly because of the rapid intramolecular proton transfer effects observed for this species[284,285]. Andersen[276] recently investigated naphthazarin and its 2,3-dichloro derivative (**68**) by means of IR LD spectroscopy on samples aligned in stretched polyethylene. Unfortunately, no useful LD was observed for naphthazarin (partly because of low solubility), but the dichloro derivative was readily dissolved and aligned in stretched polyethylene[286]. The observed wavenumbers, IR intensities and polarization directions were well reproduced by the results of B3LYP/6–31G* calculations. Two strong, differently in-plane polarized bands at 1230 and 1204 cm^{-1} were assigned to transitions with significant δ(OH) character, and an out-of-plane polarized band at 775 cm^{-1} could be assigned to γ(OH).

Chrysazin (**35**) contains an intramolecular hydrogen bonding system similar to those of 2,2'-dihydroxybenzophenone (**36**) and anthralin (**28A**), and similar ν(OH) and γ(OH) transitions are observed[238,283]. A complicated spectrum is observed in the 1700–1550 cm^{-1} region with at least six overlapping transitions. Two transitions close to 1680 and 1627 cm^{-1} can be assigned to the ν(C=O) modes of the 'free' carbonyl group and the one involved in bifurcated hydrogen bonding, respectively. IR LD spectroscopy of crystalline chrysazin[283] and on a sample aligned in stretched polyethylene[238] revealed that the two transitions are polarized along the in-plane short axis (the symmetry axis) of the molecule, consistent with the assignment of ν(C=O) bands. For a few other bands in the IR spectrum of chrysazin, the results of the two investigations disagreed; e.g. two strong transitions close to 1200 cm^{-1} were assigned to in-plane short-axis polarized transitions in the crystal investigation[283], but the LD spectra measured in stretched polyethylene showed that these transitions are long-axis polarized[238].

A related species is hypericin, a polycyclic plant pigment that has attracted interest as a potent antiviral and antitumor agent. Deprotonated hypericin (**27**) forms an exceptionally short, linear hydrogen bond in the sterically constrained bay region. The vibrational structure of hypericin has been investigated by several investigators[287–290]. According to the DFT theoretical study by Uličný and Laaksonen[290], the short hydrogen bond is of covalent rather than ionic nature, and is characterized by a symmetric potential without any proton transfer barrier. The normal mode analysis predicted a wavenumber of 1800–1700 cm^{-1} for the OH stretching vibration of the covalent hydrogen bond, compared with wavenumbers close to 2600 cm^{-1} for the OH stretching modes in the peri area.

3. 2-Nitrophenols

The IR spectra of phenols with nitro substituents in the 2-positions show intramolecular hydrogen bonding effects that are similar to those observed for the corresponding carbonyl compounds. Abkowicz, Bienko and coworkers recently investigated 2- and 4-nitrophenol[291] and 2-fluoro-4,6-dinitrophenol[292]. Kovács and associates[293,294] performed a detailed IR and Raman spectroscopic investigation of 2-nitrophenol (**69**) including a critical discussion of previous investigations. In this work, the ν(OH) band maximum was observed at 3242 cm^{-1} and the γ(OH) band at 671 cm^{-1} (CCl$_4$ solution), corresponding to a red shift of 400 cm^{-1} and a blue shift of 380 cm^{-1}, respectively, relative to the spectrum of phenol. A band at 95 cm^{-1} in the solid state spectrum was assigned to the torsional vibration of the nitro group, indicating a blue shift of *ca* 50 cm^{-1} compared with the spectrum of nitrobenzene.

(**69**)

(**70**)

Kovács and collaborators[294] also investigated 2-nitroresorcinol (**70**), where the nitro group is involved in hydrogen bonding with two hydroxyl groups. The results of IR LD spectroscopy in an anisotropic liquid crystalline solvent indicated that the compound is planar and belongs to the C_{2v} symmetry point group. $\nu(OH)$ transitions were observed at 3252 and 3230 cm^{-1}, and a strong broad band at 663 cm^{-1} was assigned to $\gamma(OH)$ (CCl$_4$ solution). Four strong IR transitions were observed between 1700 and 1500 cm^{-1}, but only three fundamentals were predicted in this region by B3LYP/6–31G* calculations. Kovács and collaborators[294] assigned a peak at 1598 cm^{-1} to a combination band, gaining intensity by Fermi resonance with the nearby transitions at 1581 and 1553 cm^{-1}. This assignment was supported by the IR LD results, indicating that the transitions at 1598, 1581 and 1553 cm^{-1} all belong to symmetry species b_2 in the C_{2v} point group. In addition, the investigation was supported by Raman spectroscopy. The Raman activities predicted with B3LYP/6–31G* were in relatively poor agreement with the observed spectrum; a larger basis set with inclusion of diffuse functions seems to be required for the prediction of Raman activities.

4. 2-Nitrosophenols

In the case of 2-nitrosophenols, the main issue is the question of whether the species exist as an nitrosophenol (**71**) or as the quinone–monooxime tautomer (**72**). Both forms are stabilized by intramolecular hydrogen bonding. A similar tautomerism, but without intramolecular hydrogen bonding, is relevant for 4-nitrosophenols. According to IR spectroscopic data[295], 2-nitrosophenol is present in the quinonoid form while 4-nitrosophenol exists in equilibrium between both tautomeric forms. 1-Nitroso-2-naphthol exists in quinonoid form only, while the existence of both forms has been suggested for 2-nitroso-1-naphthol. For recent reviews, see the publications by Ivanova and Enchev[296] and Kržan and coworkers[297].

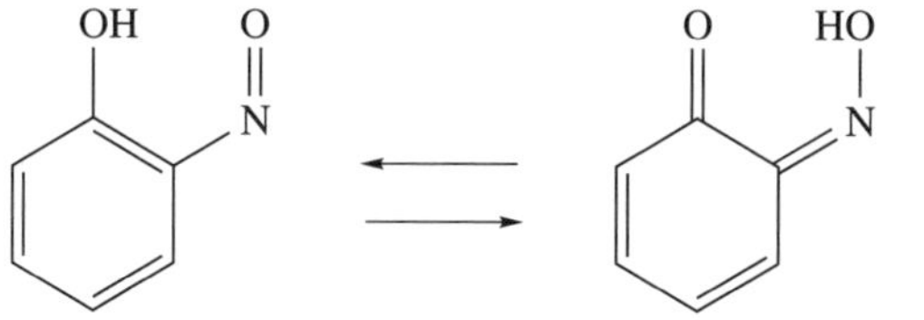

(**71**)   (**72**)

5. Schiff bases

Schiff bases (**50**) derived from aromatic 2-hydroxyaldehydes are characterized by strong, chelated intramolecular hydrogen bonding and by intriguing conformational and proton transfer phenomena. They have attracted recent interest as analytical agents[298] and as building blocks in the designing of novel molecular devices[299]. Their IR spectra have investigated by Cimerman and coworkers[298,300] and by Filarowski and Koll[216]. For additional perspectives see, for example, recent theoretical investigations[301,302].

6. ortho-Mannich bases

In these compounds, a strong, thermodynamically stable intramolecular hydrogen bond is formed between a phenolic OH group and an N,N-dialkylaminomethyl substituent in the 2-position (**73**). *ortho*-Mannich bases are excellent models for the investigation of

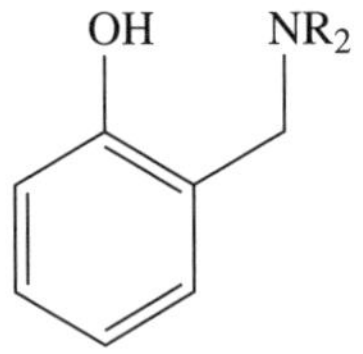

(73)

intramolecular hydrogen bonding and proton transfer phenomena, and their very complicated IR spectra continue to attract the interest of experimentalists and theoreticians[303-307]. For a recent review, see the account by Koll and Wolschann[215].

IV. ACKNOWLEDGEMENTS

The authors are grateful to J. George Radziszewski and Kristine B. Andersen for communication of spectra prior to publication, to Gábor Keresztury for providing the phenol spectra in Figures 5 and 7 and to Hans Fritz for providing data on rubazoic acids.

V. REFERENCES

1. M. Ilczyszyn, H. Ratajczak and K. Skowronek, *Magn. Reson. Chem.*, **26**, 445 (1988).
2. P. E. Hansen, H. Thiessen and R. Brodersen, *Acta Chem. Scand.*, **B33**, 281 (1979).
3. E. G. Sundhom, *Acta Chem. Scand.*, **B32**, 177 (1978).
4. P. E. Hansen, *Magn. Reson. Chem.*, **31**, 23 (1993).
5. P. E. Hansen, M. Christoffersen and S. Bolvig, *Magn. Reson. Chem.*, **31**, 893 (1993).
6. W. G. Antypas, L. M. Siukolan and D. A. Kleir, *J. Org. Chem.*, **46**, 1172 (1981).
7. J. C. Davis, Jr. and K. K. Deb, *Adv. Magn. Reson.*, **4**, 201 (1970).
8. N. Wachter-Jurszak and C. A. Detmer, *Org. Lett.*, **1**, 795 (1999).
9. V. Bertolasi, P. Gilli, V. Ferretti and G. Gilli, *J. Chem. Soc., Perkin Trans. 2*, 945 (1997).
10. S. Mazzini, L. Merlini, R. Mondelli, G. Nasini, E. Ragg and L. Scaglioni, *J. Chem. Soc., Perkin Trans. 2*, 2013 (1997).
11. H. O. Kalinowski, S. Berger and S. Braun, ^{13}C *NMR Spektroskopie*, Thieme Verlag, Stuttgart, 1984.
12. T. Fukai and T. Normura, *Heterocycles*, **42**, 911 (1996).
13. K. Danielsen, D. W. Aksnes and G. W. Frances, *Magn. Reson. Chem.*, **30**, 359 (1992).
14. (a) R. Martin, *Handbook of Hydroxyacetophenones*, Kluwer, Dordrecht, 1997.
    (b) R. Martin, *Handbook of Hydroxybenzophenones*, Kluwer, Dordrecht, 2000.
15. T. M. Alam, M. Rosay, L. Deck and R. Royer, *Magn. Reson. Chem.*, **32**, 61 (1994).
16. P. E. Hansen, U. Skibsted, J. Nissen, C. D. Rae and P. W. Kuchel, *Eur. Biophys. J.*, **30**, 69 (2001).
17. P. E. Hansen, O. K. Poulsen and A. Berg, *Org. Magn. Reson.*, **7**, 475 (1975).
18. M. Thomsen, P. Lassen, S. Dobel, P. E. Hansen, L. Carlsen and B. B. Mogensen, *Chemosphere*, in press.
19. I. R. Perminova, N. Yu. Grechischeva and V. S. Petrosyan, *Environ. Sci. Technol.*, **33**, 3781 (1999).
20. P. E. Hansen, *Acta Chem. Scand.*, **B42**, 423 (1988).
21. H. Lampert, W. Mikenda, A. Karpfen and H. Kählig, *J. Phys. Chem.*, **101**, 9610 (1997).
22. D. W. Boykin, ^{17}O *NMR Spectroscopy in Organic Chemistry*, Chap. 5, CRC Press, Boca Raton, 1991, pp. 95–115.
23. D. W. Boykin, *J. Mol. Struct.*, **295**, 39 (1993).
24. D. W. Boykin, *Magn. Reson. Chem.*, **29**, 152 (1991).
25. G. Jaccard and J. Lauterwein, *Helv. Chim. Acta*, **69**, 1469 (1986).

5. NMR and IR spectroscopy of phenols

26. V. V. Lapachev, I. Ya. Mainagashev, S. A. Stekhova, M. A. Fedotov, V. P. Krivopalow and V. P. Mamaev, *J. Chem. Soc., Chem. Commun.*, 494, (1985).
27. L. M. Gorodetsky, Z. Luz and Y. Mazur, *J. Am. Chem. Soc.*, **89**, 1183 (1967).
28. L. Kozerski, R. Kawecki, P. Krajewski, B. Kwiecien, D. W. Boykin, S. Bolvig and P. E. Hansen, *Magn. Reson. Chem.*, **36**, 921 (1998).
29. D. W. Boykin, S. Chandrasekan and A. L. Baumstark, *Magn. Reson. Chem.*, **31**, 489 (1993).
30. D. W. Boykin, in *Studies in Natural Product Chemistry* (Ed. Atta-ur-Rahman), Vol. 17, Elsevier, Amsterdam, 1995, pp. 549–600.
31. G. Cerioni, S. E. Biali and Z. Rappoport, *Tetrahedron Lett.*, **37**, 5797 (1996).
32. T. Dziembowska, *Polish J. Chem.*, **68**, 1455 (1994).
33. T. Dziembowska and Z. Rozwadowski, *Curr. Org. Chem.*, **5**, 289 (2001).
34. C. L. Perrin and J. B. Nielson, *Annu. Rev. Phys. Chem.*, **48**, 511 (1997).
35. S. Bolvig and P. E. Hansen, *Curr. Org. Chem.*, **3**, 211 (1999).
36. P. E. Hansen, *J. Mol. Struct.*, **321**, 79 (1994).
37. R. A. Newmark and J. R. Hill, *Org. Magn. Reson.*, **13**, 40 (1980).
38. P. E. Hansen, S. Bolvig and T. Kappe, *J. Chem. Soc., Perkin Trans. 2*, 1901 (1995).
39. J. Abildgaard, S. Bolvig and P. E. Hansen, *J. Am. Chem. Soc.*, **120**, 9063 (1998).
40. P. E. Hansen, S. Bolvig and K. Wozniak, *Chem. Phys. Phys. Chem.*, submitted.
41. E. Steinwender and W. Mikenda, *Monatsh. Chem.*, **125**, 695 (1994).
42. S. Bolvig, P. E. Hansen, D. Wemmer and P. Williams, *J. Mol. Struct.*, **509**, 171 (1999).
43. P. E. Hansen, *Magn. Reson. Chem.*, **24**, 903 (1986).
44. M. West-Nielsen, K. Wozniak, P. Dominiak and P. E. Hansen, *J. Org. Chem.*, to be submitted.
45. P. E. Hansen, in *Interactions of Water in Ionic and Nonionic Hydrates* (Ed. H. Kleenberg), Springer, Berlin, 1987, pp. 287–289.
46. D. H. O'Brien and R. D. Stipanovitch, *J. Org. Chem.*, **43**, 1105 (1978).
47. P. E. Hansen and S. Bolvig, *Magn. Reson. Chem.*, **35**, 520 (1997).
48. T. Dziembowska, Z. Rozwadowski, E. Majewski and K. Amborziak, *Magn. Reson. Chem.*, **39**, 484 (2001).
49. Y. Nakashima, A. Yoshino and K. Takahashi, *Bull. Chem. Soc. Jpn.*, **62**, 1401 (1989).
50. C. Etlzhofer, H. Falk, E. Mayr and S. Schwarzinger, *Monatsh. Chem.*, **127**, 1229 (1996).
51. P. E. Hansen, S. N. Ibsen, T. Kristensen and S. Bolvig, *Magn. Reson. Chem.*, **32**, 399 (1994).
52. P. E. Hansen, H. D. Dettman and B. D. Sykes, *J. Magn. Reson.*, **62**, 487 (1985).
53. J. R. Everett, *Org. Magn. Reson.*, **19**, 861 (1982).
54. E. Liepins, M. V. Petrova, E. Gudrinize, J. Paulins and S. L. Kutnetsov, *Magn. Reson. Chem.*, **27**, 907 (1989).
55. S. Bolvig, P. E. Hansen, H. Morimoto, D. Wemmer and P. Williams, *Magn. Reson. Chem.*, **38**, 525 (2000).
56. T. Dziembowska, Z. Rozwadowski and P. E. Hansen, *J. Mol. Struct.*, **436–437**, 189 (1997).
57. G. Gunnarson, H. Wennerstöm, W. Egan and S. Forsén, *Chem. Phys. Lett.*, **38**, 96 (1976).
58. L. J. Altman, D. Laungani, G. Gunnarson, H. Wennerström and S. Forsén, *J. Am. Chem. Soc.*, **100**, 8265 (1978).
59. A. C. Olivieri, D. Sanz, R. M. Claramunt and J. Elguero, *J. Chem. Soc., Perkin Trans. 2*, 1299 (1989).
60. E. V. Borsiov, W. Zhang, S. Bolvig and P. E. Hansen, *Magn. Reson. Chem.*, **36**, S104 (1998).
61. J. L. Marshall, in *Methods in Stereochemical Analysis* (Ed. A. P. Marchand), Vol. 2, Verlag Chemie, Deerfield Beach, Fl., 1983, pp. 1–241.
62. P. E. Hansen, *Prog. Nucl. Magn. Reson. Spectrosc.*, **14**, 175 (1981).
63. P. Äyras and C.-J. Widén, *Org. Magn. Reson.*, **11**, 551 (1978).
64. F. W. Wehrli, *J. Chem. Soc., Chem. Commun.*, 663 (1975).
65. C.-J. Chang, *J. Org. Chem.*, **41**, 1883 (1976).
66. L. N. Kurkovskaya, *Zh. Strukt. Khim.*, **19**, 946 (1977).
67. T. Schaefer and J. B. Rowbotham, *Can. J. Chem.*, **52**, 3037 (1974).
68. J. B. Rowbotham, M. Smith and T. Schaefer, *Can. J. Chem.*, **53**, 986 (1975).
69. J. B. Rowbotham and T. Schaefer, *Can. J. Chem.*, **54**, 2228 (1976).
70. T. Schaefer, J. B. Rowbotham and K. Chum, *Can. J. Chem.*, **54**, 3666 (1976).
71. T. Schaefer and W. J. E. Parr, *Can. J. Chem.*, **55**, 3732 (1977).
72. S. R. Salman, *Org. Magn. Reson.*, **22**, 385 (1984).
73. T. Schaefer, R. Sebastian and S. R. Salman, *Can. J. Chem.*, **62**, 113 (1984).

74. R. Laatikainen and K. Tupparinen, *J. Chem. Soc., Perkin Trans. 2*, 121 (1987).
75. D. W. Boykin, P. Balakrishnan and A. L. Baumstark, *J. Heterocycl. Chem.*, **22**, 981 (1985).
76. D. W. Boykin and A. Kumar, *J. Heterocycl. Chem.*, **29**, 1 (1992).
77. P. E. Hansen, O. K. Poulsen and A. Berg, *Org. Magn. Reson.*, **7**, 475 (1975).
78. V. Wray, L. Ernst, T. Lund and H. J. Jakobsen, *J. Magn. Reson.*, **40**, 55 (1980).
79. L. B. Krivdin and E. W. Della, *Prog. NMR Spectrosc.*, **23**, 301 (1991).
80. J. Seita, T. Drakenberg and J. Sandtröm, *Org. Magn. Reson.*, **11**, 239 (1978).
81. T. Schaefer, *J. Phys. Chem.*, **79**, 1888 (1975).
82. J. Reuben, *J. Am. Chem. Soc.*, **108**, 1735 (1986).
83. M.-C. M. Descoings, G. Goethals, J.-P. Seguin and J.-P. Doucet, *J. Mol. Struct.*, **177**, 407 (1988).
84. E. Steinwender and W. Mikenda, *Monatsh. Chem.*, **125**, 695 (1994).
85. E. Steinwender, F. Pertlik and W. Mikenda, *Monatsh. Chem.*, **124**, 867 (1993).
86. W. Mikenda, E. Steinwender and K. Mereiter, *Monatsh. Chem.*, **126**, 495 (1995).
87. D. K. Zheglova, D. G. Genov, S. Bolvig and P. E. Hansen, *Acta Chem. Scand.*, **51**, 1016 (1997).
88. A. L. Baumstark and D. W. Boykin, *New J. Chem.*, **16**, 357 (1992).
89. S. F. Bureiko, N. S. Golubev, J. Mattinen and K. Pihlaja, *J. Mol. Liq.*, **45**, 139 (1990).
90. S. F. Bureiko, N. S. Golubev and K. Pihlaja, *J. Mol. Struct.*, **480–481**, 297 (1999).
91. E. Grech, J. Nowicka-Sceibe, Z. Olejnik, T. Lis, Z. Pawelka, Z. Malarski and L. Sobczyk, *J. Chem. Soc., Perkin Trans. 2*, 343 (1996).
92. T. K. Harris, Q. Zhao and A. S. Mildvan, *J. Mol. Struct.*, **552**, 97 (2000).
93. T. K. Harris and A. S. Mildvan, *Proteins*, **35**, 275 (1999).
94. Q. Zhao, C. Abeygunwardana and A. S. Mildvan, *Biochemistry*, **36**, 14614 (1997).
95. B. Brzezinski and G. Zundel, *J. Magn. Reson.*, **48**, 361 (1982).
96. T. Dziembowska, Z. Malarski and B. Szczodrowska, *J. Solution Chem.*, **25**, 179 (1996).
97. M. T. Krygowski and W. R. Fawcett, *J. Am. Chem. Soc.*, **97**, 2143 (1975).
98. T. Dziembowska, E. Majewski, Z. Rozwadowski and B. Brzezinski, *J. Mol. Struct.*, **403**, 183 (1997).
99. H. Takemura, T. Shinmyozu, H. Miura, I. U. Khan and T. Inazu, *J. Incl. Phen. Mol. Rec. Chem.*, **19**, 193 (1994).
100. U. Koelle and S. Forsén, *Acta Chem. Scand.*, **A28**, 531 (1974).
101. T. Dziembowska, Z. Rozwadowsi, J. Sitkowski, L. Stefaniak and B. Brzezinski, *J. Mol. Struct.*, **407**, 131 (1997).
102. P. Fisher and A. Fettig, *Magn. Reson. Chem.*, **35**, 839 (1997).
103. F. Hibbert and R. J. Sellens, *J. Chem. Soc., Perkin Trans. 2*, 529 (1988).
104. A. Filarowski, A. Koll and T. Glowiak, *Monatsh. Chem.*, **130**, 1097 (1999).
105. T. Dziembowska, Z. Rozwadowski, A. Filarowski and P. E. Hansen, *Magn. Reson. Chem.*, **39**, S67 (2001).
106. B. Brzezinski, H. Maciejewska-Urjasz and G. Zundel, *J. Mol. Struct.*, **319**, 177 (1994).
107. B. Brzezinski, B. Bryzki, H. Maciejewska-Urjasz and G. Zundel, *Magn. Reson. Chem.*, **31**, 642 (1993).
108. S. Chandrasekan, W. D. Wilson and D. W. Boykin, *Org. Magn. Reson.*, **22**, 757 (1984).
109. J. L. Lios and H. Duddeck, *Magn. Reson. Chem.*, **38**, 512 (2000).
110. C. Engdahl, A. Gogoll and U. Edlund, *Magn. Reson. Chem.*, **29**, 54 (1991).
111. P. E. Hansen, A. S. Peregudov, D. N. Kravtsov, A. I. Krylova, G. M. Babakhina and L. S. Golovchenko, *J. Chem. Soc., Perkin Trans. 2*, submitted.
112. E. P. Mazzola, S. A. Turujman, S. J. Bell, J. N. Barrows, J. E. Bailey, Jr., E. G. Mouchahoir, S. Prescott, M. D. Archer, M. Oliver, W. F. Reynolds and K. W. Nielsen, *Tetrahedron*, **52**, 5691 (1996).
113. P. E. Hansen and A. Lycka, *Magn. Reson. Chem.*, **24**, 772 (1986).
114. A. Lycka and D. Snobl, *Collect. Czech. Chem. Commun.*, **46**, 892 (1981).
115. A. Lycka and P. E. Hansen, *Magn. Reson. Chem.*, **22**, 569 (1984).
116. J. Abildgaard, P. E. Hansen, J. Josephsen and A. Lycka, *Inorg. Chem.*, **33**, 5271 (1994).
117. P. E. Hansen, S. Bolvig, A. Buvari-Barcza and A. Lycka, *Acta Chem. Scand.*, **51**, 881 (1997).
118. L. Sobczyk, *Appl. Magn. Reson.*, **18**, 47 (2000).
119. A. Koll and P. Wollschann, *Monatsch. Chem.*, **127**, 475 (1996).
120. A. Koll and P. Wollschann, *Monatsch. Chem.*, **130**, 983 (1999).

5. NMR and IR spectroscopy of phenols 389

121. C.-J. Chang, T.-L. Shieh and H. G. Floss, *J. Med. Chem.*, **20**, 176 (1977).
122. P. E. Hansen, J. Sitkowski, L. Stefaniak, Z. Rozwadowski and T. Dziembowska, *Ber. Bunsenges. Phys. Chem.*, **102**, 410 (1998).
123. J. Sitkowski, L. Stefaniak, T. Dziembowska, E. Grech, E. Jagodzinska and G. A. Webb, *J. Mol. Struct.*, **381**, 177 (1996).
124. Z. Rozwadowski, E. Majewski, T. Dziembowska and P. E. Hansen, *J. Chem. Soc., Perkin Trans. 2*, 2809 (1999).
125. S. R. Salman, R. D. Farrant and J. C. Lindon, *Spectrosc. Lett.*, **24**, 1071 (1991).
126. Z. Rozwadowski and T. Dziembowska, *Magn. Reson. Chem.*, **37**, 274 (1999).
127. S. R. Salman, J. M. A. Al-Rawi and G. Q. Benham, *Org. Magn. Reson.*, **22**, 535 (1984).
128. K. Kishore, D. N. Sathyanarayana and V. A. Bhanu, *Magn. Reson. Chem.*, **25**, 471 (1987).
129. S. H. H. Alarcon, A. C. Olivieri and M. Gonzalez-Sierra, *J. Chem. Soc., Perkin Trans. 2*, 1067 (1994).
130. A. G. J. Ligtenbarg, R. Hage, A. Meetsma and B. L. Feringa, *J. Chem. Soc., Perkin Trans. 2*, 807 (1999).
131. D. K. Zheng, Y. Ksi, J. K. Meng, J. Zhou and L. Hunad, *J. Chem. Soc., Chem. Commun.*, 168 (1985).
132. G. Duddek and E. P. Duddek, *J. Chem. Soc. B*, 1356 (1971).
133. J.-C. Zhuo, *Magn. Reson. Chem.*, **37**, 259 (1999).
134. P. E. Hansen, Z. Rozwadowksi and T. Dziembowska, personal communication.
135. K. K. Norris and S. Sternhell, *Aust. J. Chem.*, **19**, 841 (1966).
136. V. Enchev, G. Ivanova, U. Ugrinov and G. D. Neykov, *J. Mol. Struct.*, **508**, 149 (1999).
137. T. Shono, Y. Hayashi and K. Shinra, *Bull. Chem. Soc. Jpn.*, **44**, 3179 (1971).
138. C. F. G. C. Geraldes and M. I. F. Silva, *Opt. Pura Apl.*, **21**, 71 (1988).
139. A. Vainiotalo and J. Vepsäläinen, *Magn. Reson. Chem.*, **24**, 758 (1986).
140. G. Ivanova and V. Enchev, *Chem. Phys.*, **264**, 235 (2001).
141. H. Saarinen and J. Korvenranta, *Finn. Chem. Lett.*, 233 (1978).
142. S. Bolvig and P. E. Hansen, *Magn. Reson. Chem.*, **34**, 467 (1996).
143. M. Rospenk, A. Koll and L. Sobczyk, *J. Mol. Liq.*, **76**, 63 (1995).
144. M. Rospenk, A. Koll and L. Sobczyk, *Chem. Phys. Lett.*, **261**, 283 (1996).
145. S. Bratan-Mayer, F. Strohbusch and W. Hänsel, *Z. Naturforsch.*, **31b**, 1106 (1976).
146. P. Gilli, V. Ferretti, V. Bertolasi and G. Gilli, in *Advances in Molecular Structure Research* (Ed. M. Hargittai), Vol. 2, JAI Press, Greenwich, CT, 1996, p. 67.
147. M. Ilczyszyn, H. Ratajczak and J. A. Ladd, *Chem. Phys. Lett.*, **153**, 385 (1988).
148. M. Ilczyszyn and H. Ratajczak, *J. Chem. Soc., Faraday Trans.*, **91**, 3859 (1995).
149. M. Ilczyszyn, *Bull. Pol. Acad. Sci. Chem.*, **48**, 91 (2000).
150. M. Ilczyszyn and H. Ratajczak, *J. Mol. Struct.*, **198**, 499 (1989).
151. M. Ilczyszyn, *J. Chem. Soc., Faraday Trans.*, **90**, 1411 (1994).
152. M. Ilczyszyn and H. Ratajczak, *J. Mol. Liq.*, **67**, 125 (1995).
153. M. Ilczyszyn and H. Ratajczak, *J. Chem. Soc., Faraday Trans.*, **91**, 1611 (1995).
154. G. Goethals, M. C. Moreau-Descoings, C. Sarazin, J. P. Sequin and J. P. Doucet, *Spectrosc. Lett.*, **22**, 973 (1989).
155. S. N. Smirnow, N. S. Golubev, G. S. Denisov, H. Benedict, P. Schah-Mohammedi and H.-H. Limbach, *J. Am. Chem. Soc.*, **118**, 4094 (1996).
156. B. Brycki and M. Szafran, *J. Mol. Liq.*, **59**, 83 (1994).
157. G. Albrecht and G. Zundel, *J. Chem. Soc., Faraday Trans.*, **80**, 553 (1984).
158. Z. Tai, H. Sun, X. Qian and J. Zou, *Spectrosc. Lett.*, **26**, 1805 (1993).
159. V. Castola, V. Mazzoni, M. Coricchiato, A. Bighelli and J. Casanova, *Can. J. Anal. Sci. Spectrosc.*, **42**, 90 (1997).
160. D. M. Jung, J. S. de Rop and S. E. Ebeler, *J. Agric. Food Chem.*, **48**, 407 (2000).
161. S. Divakar, *J. Inc. Phen. Mol. Recog. Chem.*, **15**, 305 (1993).
162. A. V. Veglia and R. H. de Rossi, *Can. J. Chem.*, **78**, 233 (2000).
163. U. Schneider and H.-J. Schneider, *Chem. Ber.*, **127**, 2455 (1994).
164. M. Tata, V. T. John, Y. Y. Waguepack and G. L. McPherson, *J. Am. Chem. Soc.*, **116**, 9464 (1994).
165. I. G. Shenderovich, Z. Kecki, I. Wawer and G. S. Denisov, *Spectrosc. Lett.*, **30**, 1515 (1997).
166. J. Abildgaard and P. E. Hansen, *Wiad. Chem.*, **54**, 845 (2000).

167. A. M. Orendt, R. R. Biekofsky, A. B. Pomio, R. H. Contreras and J. C. Facelli, *J. Phys. Chem.*, **95**, 6179 (1991).
168. S. R. Salman, J. C. Lindon, R. D. Farrant and T. A. Carpenter, *Magn. Reson. Chem.*, **31**, 991 (1993).
169. D. Maciejewska, D. Pawlak and V. Koleva, *J. Phys. Org. Chem.*, **12**, 875 (1999).
170. R. K. Harris, P. Johnsen, K. J. Packer and C. D. Campbell, *Magn. Reson. Chem.*, **24**, 977 (1986).
171. S. Takeda, T. Inabe, C. Benedict, U. Langer and H. H. Limbach, *Ber. Bunsenges. Phys. Chem.*, **102**, 1358 (1998).
172. C. M. Lagier, U. Scheler, G. McGeorge, M. G. Sierra, A. C. Olivieri and R. K. Harris, *J. Chem. Soc., Perkin Trans. 2*, 1325 (1996).
173. C. M. Lagier, A. Olivieri and R. K. Harris, *J. Chem. Soc., Perkin Trans. 2*, 1791 (1998).
174. W. Kolodziejski, A. P. Mazurek and T. Kaspzycka-Guttman, *Chem. Phys. Lett.*, **328**, 263 (2000).
175. P. P. Chu and H.-D. Wu, *Polymers*, **41**, 101 (2000).
176. R. M. Jarret and M. Saunders, *J. Am. Chem. Soc.*, **107**, 2648 (1985).
177. S. H. Hibdon, C. A. Coleman, J. Wang and C. J. Murray, *Inorg. Chem.*, **20**, 334 (1992).
178. S. N. Smirnow, M. Benedict, N. S. Golubev, G. S. Denisov, M. M. Kreevoy, R. L. Schowen and H.-H. Limbach, *Can. J. Chem.*, **77**, 943 (1999).
179. J. A. Gerlt and P. G. Gassman, *J. Am. Chem. Soc.*, **115**, 11552 (1993).
180. Q. Zhao, C. Abeygunawardana, P. Taylor and A. S. Mildvan, *Proc. Natl. Acad. Sci., USA*, **93**, 8220 (1996).
181. B. T. G. Lutz, M. H. Langoor and J. H. van der Maas, *Vibr. Spectrosc.*, **18**, 111 (1998).
182. M. D. Joesten and L. J. Schaad, *Hydrogen Bonding*, Marcel Dekker, New York, 1974.
183. S. N. Vinogradov and R. H. Linnell, *Hydrogen Bonding*, Van Nostrand Reinhold, New York, 1971.
184. W. T. Smith, Jr. and J. M. Patterson, in *The Chemistry of Hydroxyl, Ether and Peroxide Groups*, Suppl. E2 (Ed. S. Patai), Chap. 5, Wiley, Chichester, 1993.
185. F. K. Kawahara, in *Water Analysis*, Vol. 3 (Eds. R. A. Minear and L. H. Keith), Chap. 7, Academic Press, Orlando, 1984.
186. J. M. Chalmers and P. R. Griffiths (Eds.), *Handbook of Vibrational Spectroscopy*, Wiley, Chichester, 2001.
187. G. Varsányi, *Assignments for Vibrational Spectra of Seven Hundred Benzene Derivatives*, Vols. 1&2, Wiley, New York, 1974.
188. R. A. Nyquist, *The Interpretation of Vapor-Phase Infrared Spectra, Group Frequency Data*, The Sadtler Research Laboratories, 1985.
189. C. J. Pouchert, *The Aldrich Library of FT-IR Spectra*, Vol. 1, Aldrich Chemical Company, 1985.
190. E. W. Thulstrup and J. Michl, *Spectroscopy with Polarized Light. Solute Alignment by Photoselection, in Liquid Crystals, Polymers, and Membranes*, VCH Publishers, New York, 1995.
191. A. Rodger and B. Nordén, *Circular Dichroism and Linear Dichroism*, Oxford University Press, 1997.
192. W. Koch and M. C. Holthausen, *A Chemist's Guide to Density Functional Theory*, Wiley-VCH, Weinheim, Germany, 2000.
193. F. Jensen, *Introduction to Computational Chemistry*, John Wiley & Sons Ltd, Chichester, 1999.
194. J. B. Foresman and M. J. Frisch, *Exploring Chemistry with Electronic Structure Methods*, Gaussian Inc., Pittsburgh, PA, 1996.
195. M. Belhakem and B. Jordanov, *J. Mol. Struct.*, **218**, 309 (1990).
196. E. W. Thulstrup, J. Spanget-Larsen and J. Waluk, in *Encyclopedia of Spectroscopy and Spectrometry* (Eds. J. C. Lindon, G. E. Tranter and J. L. Holmes), Academic Press, New York, 2000, pp. 1169–1178.
197. J. G. Radziszewski and J. Michl, *J. Am. Chem. Soc.*, **108**, 3289 (1986).
198. G. Keresztury, F. Billes, M. Kubinyi and T. Sundius, *J. Phys. Chem. A*, **102**, 1371 (1998).
199. K. B. Andersen, M. Langgård and J. Spanget-Larsen, *J. Mol. Struct.*, **475**, 131 (1999).
200. M. J. Frisch, G. W. Trucks, H. B. Schlegel, G. E. Scuseria, M. A. Robb, J. R. Cheeseman, V. G. Zakrzewski, J. A. Montgomery, Jr., R. E. Stratmann, J. C. Burant, S. Dapprich, J. M. Millam, A. D. Daniels, K. N. Kudin, M. C. Strain, O. Farkas, J. Tomasi, V. Barone,

M. Cossi, R. Cammi, B. Mennucci, C. Pomelli, C. Adamo, S. Clifford, J. Ochterski, G. A. Petersson, P. Y. Ayala, Q. Cui, K. Morokuma, D. K. Malick, A. D. Rabuck, K. Raghavachari, J. B. Foresman, J. Cioslowski, J. V. Ortiz, A. G. Baboul, B. B. Stefanov, G. Liu, A. Liashenko, P. Piskorz, I. Komaromi, R. Gomperts, R. L. Martin, D. J. Fox, T. Keith, M. A. Al-Laham, C. Y. Peng, A. Nanayakkara, C. Gonzalez, M. Challacombe, P. M. W. Gill, B. Johnson, W. Chen, M. W. Wong, J. L. Andres, C. Gonzalez, M. Head-Gordon, E. S. Replogle and J. A. Pople, GAUSSIAN98, Revision A.7, Gaussian, Inc., Pittsburgh, PA, 1998.
201. M. J. Frisch, A. C. Scheiner, H. F. Schaefer III and J. S. Binkley, *J. Chem. Phys.*, **82**, 4194 (1985).
202. A. P. Scott and L. Radom, *J. Phys. Chem.*, **100**, 16502 (1996).
203. M. W. Wong, *Chem. Phys. Lett.*, **256**, 391 (1996).
204. J. Spanget-Larsen, *Chem. Phys.*, **240**, 51 (1999).
205. R. Hienerwadel, A. Boussac, J. Breton and C. Berthomieu, *Biochemistry*, **35**, 15447 (1996).
206. R. Hienerwadel, A. Boussac, J. Breton, B. A. Diner and C. Berthomieu, *Biochemistry*, **36**, 14712 (1997).
207. C. Berthomieu, R. Hienerwadel, A. Boussac, J. Breton and B. A. Diner, *Biochemistry*, **37**, 10548 (1998).
208. D. Lin-Vien, N. B. Colthup, W. G. Fateley and J. G. Graselli, *The Handbook of Infrared and Raman Characteristic Frequencies of Organic Molecules*, Academic Press, London, 1991.
209. Z. Bikádi, G. Keresztury, S. Holly, O. Egyed, I. Mayer and M. Simonyi, *J. Phys. Chem. A*, **105**, 3471 (2001).
210. F. Hibbert and J. Emsley, *Adv. Phys. Org. Chem.*, **26**, 255 (1990).
211. H. Rostkowska, M. J. Nowak, L. Lapinski and L. Adamowicz, *Phys. Chem. Chem. Phys.*, **3**, 3012 (2001).
212. I. Macsári, V. Izvekov and A. Kovács, *Chem. Phys. Lett.*, **269**, 393 (1997).
213. M. L. Rogers and R. L. White, *Appl. Spectrosc.*, **41**, 1052 (1987).
214. M. Mahmud, D. Cote and P. Leclerc, *Anal. Sci.*, **8**, 381 (1992).
215. A. Koll and P. Wolschann, in *Hydrogen Bond Research* (Eds. P. Schuster and W. Mikenda), Springer, Wien, 1999, p. 39.
216. A. Filarowski and A. Koll, *Vibr. Spectrosc.*, **17**, 123 (1998).
217. V. I. Minkin, B. Ya. Simkin and R. M. Minyaev, *Quantum Chemistry of Organic Compounds*, Springer, Berlin, 1990, pp. 217–237.
218. G. Zundel, in *The Hydrogen Bond*, Vol. 2 (Eds. P. Schuster, G. Zundel and C. Sandorfy), North-Holland, Amsterdam, 1976, p. 683.
219. M. Rospenk and Th. Zeegers-Huyskens, *J. Phys. Chem. A*, **101**, 8428 (1997).
220. M. Rospenk, N. Leroux and Th. Zeegers-Huyskens, *J. Mol. Spectrosc.*, **183**, 245 (1997).
221. M. Rospenk, B. Czarnik-Matusewicz and Th. Zeegers-Huyskens, *Spectrochim. Acta A*, **57**, 185 (2001).
222. Z. Smedarchina, W. Siebrand and M. Zgierski, *J. Chem. Phys.*, **103**, 5326 (1995).
223. V. A. Benderskii, E. V. Vetoshkin, S. Yu. Grebenshchikov, L. Von Laue and H. P. Trommsdorff, *Chem. Phys.*, **219**, 119 (1997).
224. S. Melikova, D. Shchepkin and A. Koll, *J. Mol. Struct.*, **448**, 239 (1998).
225. P. Blaise, O. Henri-Rousseau and A. Grandjean, *Chem. Phys.*, **244**, 405 (1999).
226. O. Henri-Rousseau and P. Blaise, *Chem. Phys.*, **250**, 249 (1999).
227. P. Schuster and W. Mikenda (Eds.), *Hydrogen Bond Research*, Springer, Wien, 1999.
228. V. A. Benderskii, E. V. Vetoshkin, I. S. Irgibaeva and H. P. Trommsdorff, *Chem. Phys.*, **262**, 393 (2000).
229. H. Ratajczak and A. M. Yaremko, *J. Mol. Struct. (THEOCHEM)*, **500**, 413 (2000).
230. S. M. Melikova, A. Ju. Inzebejkin, D. N. Shchepkin and A. Koll, *J. Mol. Struct.*, **552**, 273 (2000).
231. G. Gilli and P. Gilli, *J. Mol. Struct.*, **552**, 1 (2000).
232. W. A. P. Luck, *J. Mol. Struct.*, **552**, 87 (2000).
233. A. Haid and G. Zundel, *J. Mol. Struct. (THEOCHEM)*, **500**, 421 (2000).
234. H. Lampert, W. Mikenda and A. Karpfen, *J. Phys. Chem. A*, **101**, 2254 (1997).
235. J. Palomar, J. L. G. De Paz and J. Catalán, *Chem. Phys.*, **246**, 167 (1999).
236. D. Michalska, D. C. Bienko, A. J. Abkowicz-Bienko and Z. Latajka, *J. Phys. Chem. A*, **100**, 17786 (1996).
237. J. G. Radziszewski, personal communication, 2001.

238. F. Madsen, I. Terpager, K. Olskær and J. Spanget-Larsen, *Chem. Phys.*, **165**, 351 (1992).
239. E. B. Wilson, *Phys. Rev.*, **45**, 706 (1934).
240. W. Lüttke and R. Mecke, *Z. Elektrochemie*, **53**, 241 (1949).
241. A. Pierretti, F. Ramondo, L. Bencivenni and M. Spoliti, *J. Mol. Struct.*, **560**, 315 (2001).
242. S. Leutwyler, T. Bürgi, M. Schütz and A. Taylor, *Faraday Discuss.*, **97**, 285 (1994).
243. T. Bürgi, M. Schütz and S. Leutwyler, *J. Chem. Phys.*, **103**, 6350 (1995).
244. K. Müller-Dethlefs, *J. Electron Spectrosc. Relat. Phenom.*, **75**, 35 (1995).
245. T. Ebata, A. Fujii and N. Mikami, *Int. Rev. Phys. Chem.*, **17**, 331 (1998).
246. M. Ben Salah, P. Becker and A. Carabatos-Nédelec, *Vibr. Spectrosc.*, **26**, 23 (2001).
247. M. Gerhards, A. Jansen, C. Unterberg and K. Kleinermanns, *Chem. Phys. Lett.*, **344**, 113 (2001).
248. P. W. Thulstrup, M.Sc. Thesis, Roskilde University, Denmark, 1999.
249. Th. Thormann, M.Sc. Thesis, Roskilde University, Denmark, 1999.
250. M. Rossi, J. Erlebacher, J. D. E. Zacharias, H. L. Carrell and B. Iannucci, *Carcinogenesis*, **12**, 2227 (1991).
251. B. T. G. Lutz, G. Astarloa, J. H. van der Maas, R. G. Janssen, W. Verboom and D. N. Reinhoudt, *Vibr. Spectrosc.*, **10**, 29 (1995).
252. P. Opaprakasit, A. Scaroni and P. Painter, *J. Mol. Struct.*, **570**, 25 (2001).
253. J. Küpper, A. Westphal and M. Schmitt, *Chem. Phys.*, **263**, 41 (2001).
254. W. Siebrand and M. Z. Zgierski, *Chem. Phys. Lett.*, **334**, 127 (2001).
255. Y. Matsumoto, T. Ebata and N. Mikami, *J. Mol. Struct.*, **552**, 257 (2000).
256. J. P. Hawranek and A. S. Muszyński, *J. Mol. Struct.*, **552**, 205 (2000).
257. V. Schreiber, A. Kulbida, M. Rospenk, L. Sobczyk, A. Rabold and G. Zundel, *J. Chem. Soc., Faraday Trans.*, **92**, 2555 (1996).
258. V. M. Schreiber, M. Rospenk, A. I. Kulbida and L. Sobczyk, *Spectrochim. Acta A*, **53**, 2067 (1997).
259. B. Brzezinski, E. Grech, Z. Malarski, M. Rospenk, G. Schroeder and L. Sobczyk, *J. Chem. Res. (S)*, 151 (1997); (*M*), 1021 (1997).
260. P. L. Huyskens, T. Zeegers-Huyskens and Z. Pawelka, *J. Solution Chem.*, **28**, 915 (1999).
261. Z. Jin, D. Xu, Y. Pan, Y. Xu and M. Y.-N. Chiang, *J. Mol. Struct.*, **559**, 1 (2001).
262. T. Kuc, Z. Pawelka and L. Sobzyk, *Phys. Chem. Chem. Phys.*, **2**, 211 (2000).
263. S. W. Ng, P. Naumov, S. Chantrapromma, S. S. S. Raj, H. K. Fun, A. R. Ibrahim, G. Wojciechowski and B. Brzezinski, *J. Mol. Struct.*, **562**, 185 (2001).
264. T. Glowiak, I. Majerz, L. Sobzyk, B. Brzezinski, G. Wojciechowski and E. Grech, *J. Mol. Struct.*, **526**, 177 (2000).
265. P. Migchels and T. Zeegers-Huyskens, *Spectrosc. Lett.*, **25**, 733 (1992).
266. P. Migchels and T. Zeegers-Huyskens, *Vibr. Spectrosc.*, **2**, 89 (1991).
267. B. Brzezinski, J. Olejnik and G. Zundel, *J. Mol. Struct.*, **238**, 89 (1990).
268. H. Lampert, W. Mikenda and A. Karpfen, *J. Phys. Chem.*, **100**, 7418 (1996).
269. V. Schreiber, S. Melikova, K. Rutkowski, D. E. Shchepkin, A. Shurukhina and A. Koll, *J. Mol. Struct.*, **381**, 141 (1996).
270. C. Chen and S.-F. Shyu, *J. Mol. Struct. (THEOCHEM)*, **536**, 25 (2001).
271. D. Philip, A. John, C. Y. Panicker and H. T. Varghese, *Spectrochim. Acta A*, **57**, 1561 (2001).
272. A. Simperler and W. Mikenda, in *Hydrogen Bond Research* (Eds. P. Schuster and W. Mikenda), Springer, Wien, 1999, p. 59.
273. K. B. Andersen, LEO Pharmaceutical Products, DK-2710 Ballerup, Denmark; personal communication, 2001.
274. K. B. Andersen, M. Langgård and J. Spanget-Larsen, *J. Mol. Struct.*, **509**, 153 (1999).
275. H. W. Avdovich and G. A. Neville, *Can. J. Spectrosc.*, **25**, 110 (1980).
276. K. B. Andersen, *Polarization Spectroscopic Studies of β-Hydroxy Carbonyl Compounds with Intramolecular H-bonding*, PhD Thesis, Roskilde University, Denmark, 1999.
277. M. W. Jensen, M.Sc. Thesis, Roskilde University, Denmark, 1997.
278. C. J. H. Schutte, S. O. Paul and D. Röhm, *J. Mol. Struct.*, **232**, 179 (1991).
279. F. Ramondo and L. Bencivenni, *Struct. Chem.*, **5**, 211 (1994).
280. A. N. Anoshin, E. A. Gastilovich, M. V. Gorelik, T. S. Kopteva and D. N. Shigorin, *Russ. J. Phys. Chem.*, **56**, 1665 (1982).
281. G. Smulevich, L. Angeloni, S. Giovannardi and M. Marzocchi, *Chem. Phys.*, **65**, 313 (1982).

282. D. H. Christensen, J. Spanget-Larsen and E. W. Thulstrup, *Spectrochim. Acta A*, **45**, 431 (1989).
283. G. Smulevich and M. P. Marzocchi, *Chem. Phys.*, **94**, 99 (1985).
284. K. Wolf, A. Simperler and W. Mikenda, in *Hydrogen Bond Research* (Eds. P. Schuster and W. Mikenda), Springer, Wien, 1999, p. 87.
285. Y. H. Mariam and R. N. Musin, *J. Mol. Struct. (THEOCHEM)*, **549**, 123 (2001).
286. K. B. Andersen, *Acta Chem. Scand.*, **53**, 222 (1999).
287. S. M. Arabei, J. P. Galaup and P. Jardon, *Chem. Phys. Lett.*, **232**, 127 (1995).
288. S. M. Arabei, J. P. Galaup and P. Jardon, *Chem. Phys. Lett.*, **270**, 31 (1997).
289. M. Mylrajan, P. Hildebrandt and Y. Mazur, *J. Mol. Struct.*, **407**, 5 (1997).
290. J. Uličný and A. Laaksonen, *Chem. Phys. Lett.*, **319**, 396 (2000).
291. A. J. Abkowics, Z. Latajka, D. C. Bienko and D. Michalska, *Chem. Phys.*, **250**, 123 (1999).
292. A. J. Abkowics-Bienko, D. C. Bienko and Z. Latajka, *J. Mol. Struct.*, **552**, 165 (2000).
293. A. Kovács, V. Izvekov, G. Keresztury and G. Pongor, *Chem. Phys.*, **238**, 231 (1998).
294. A. Kovács, G. Keresztury and V. Izvekov, *Chem. Phys.*, **253**, 193 (2000).
295. D. Hadži, *J. Chem. Soc.*, 2725 (1956).
296. G. Ivanova and V. Enchev, *Chem. Phys.*, **264**, 235 (2001).
297. A. Kržan, D. R. Crist and V. Horák, *J. Mol. Struct. (THEOCHEM)*, **528**, 237 (2000).
298. Z. Cimerman, N. Galic and B. Bosner, *Anal. Chim. Acta*, **343**, 145 (1997).
299. E. Hadjoudis, *Mol. Eng.*, **5**, 301 (1995).
300. Z. Cimerman, R. Kiralj and N. Galic, *J. Mol. Struct.*, **323**, 7 (1994).
301. V. Guallar, V. S. Batista and W. H. Miller, *J. Chem. Phys.*, **113**, 9510 (2000).
302. J. L. Pérez-Lustres, M. Bräuer, M. Mosquera and T. Clark, *Phys. Chem. Chem. Phys.*, **3**, 3569 (2001).
303. A. Filarowski and A. Koll, *Vibr. Spectrosc.*, **12**, 15 (1996).
304. M. Rospenk, L. Sobczyk, A. Rabold and G. Zundel, *Spectrochim. Acta A*, **55**, 855 (1999).
305. S. M. Melikova, A. Koll, A. Karpfen and P. Wolschann, *J. Mol. Struct.*, **523**, 223 (2000).
306. A. Koll, S. M. Melikova, A. Karpfen and P. Wolschann, *J. Mol. Struct.*, **559**, 127 (2001).
307. G. Wojciechowski, Z. Rozwadowski, T. Dziembowska and B. Brzezinski, *J. Mol. Struct.*, **559**, 379 (2001).

CHAPTER 6

Synthesis of phenols

CONCEPCIÓN GONZÁLEZ

Departamento de Química Orgánica, Facultad de Ciencias, Universidad de Santiago de Compostela, 27002 Lugo, Spain
Fax: +34 982 22 49 04; e-mail: cgb1@lugo.usc.es

and

LUIS CASTEDO

Departamento de Química Orgánica y Unidad Asociada al C.S.I.C., Facultad de Química, Universidad de Santiago de Compostela, 15782 Santiago de Compostela, Spain
Fax: +34 981 59 50 12; e-mail: qocasted@usc.es

I. BY DISPLACEMENT OF OTHER FUNCTIONAL GROUPS	396
A. From Aryl Halides	396
B. From Sulphonic Acids	397
C. From Nitrogen Derivatives	399
1. Hydrolysis	399
2. Bucherer reaction	401
3. Diazotation reaction	402
II. BY OXIDATION	404
A. Hydroxylation	404
1. With hydrogen peroxide	404
2. With peroxides	405
3. With peroxyacids	408
a. Inorganic peroxyacids	408
b. Organic peroxyacids	410
4. Electrochemical hydroxylation	410
5. Biotransformations	411
6. Miscellaneous methods	413
B. Oxidation of Organometallic Derivatives	414
C. Oxidation of Nitrogen Derivatives	419
D. Oxidation of Carbonyl Groups	419

The Chemistry of Phenols. Edited by Z. Rappoport
© 2003 John Wiley & Sons, Ltd   ISBN: 0-471-49737-1

III. BY CONDENSATION 425
   A. Cyclization 425
   B. Claisen and Aldolic Condensations 426
      1. Intramolecular reaction 426
      2. Intermolecular reaction 433
         a. Synthesis of monophenols 433
         b. Synthesis of resorcinols 436
   C. Radical Cyclizations 438
IV. BY CYCLOADDITION 439
   A. Cycloaromatization 439
   B. Diels–Alder Reaction 440
   C. Benzannulation 450
V. BY REARRANGEMENT 459
   A. Alkyl and Benzyl Aryl Ethers 459
   B. Allyl Aryl Ethers. Aromatic Claisen Rearrangement 460
   C. Diaryl Ethers. Smiles Rearrangement 466
   D. Dienones. Dienone–Phenol Rearrangement 470
   E. Phenolic Esters. Fries Rearrangement 472
VI. ACKNOWLEDGEMENT 478
VII. REFERENCES 478

I. BY DISPLACEMENT OF OTHER FUNCTIONAL GROUPS

A. From Aryl Halides

The alkaline fusion is an important industrial method for the production of phenol (equation 1). For instance, bromobenzene in dilute sodium hydroxide gives an 89% yield of phenol at 236 °C in 2.5 h. Similarly, chlorobenzene affords a 97% yield of phenol at 370 °C in 30 min. Diphenyl ether and o- and p-hydroxybiphenyls, as well as other bicyclic compounds are some of the by-products of this type of reaction.

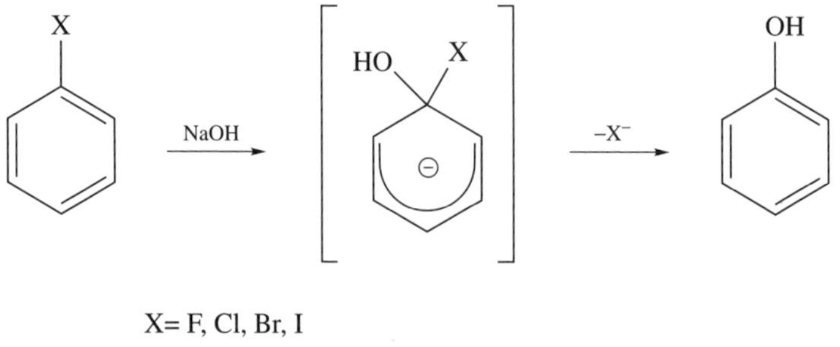

$$\text{PhX} \xrightarrow{\text{NaOH}} \left[\text{HO-C}_6\text{H}_5\text{X}^- \right] \xrightarrow{-X^-} \text{PhOH} \quad (1)$$

X= F, Cl, Br, I

Cuprous oxide accelerates the substitution reaction, and, for instance, chlorobenzene under these conditions affords a 92% yield of phenol in 1 h at 316 °C. Copper and barium chlorides catalyse also the steam hydrolysis of chlorobenzene over silica gel[1].

This type of reaction requires extremely high reaction temperatures, normally above 200 °C, and consequently the transformation is limited by the stability of the starting material.

6. Synthesis of phenols

Reactions of chlorotoluenes with aqueous alkalies give cresols, but the positions taken by the hydroxy groups are sometimes not the same as those vacated by the chlorine atoms[2].

In the case of polyhalogenated systems, a partial substitution to afford halogenated phenols can be achieved. For instance, treatment of 1,2,4-trichlorobenzene with sodium hydroxide at 130 °C gives 2,5-dichlorophenol in 93% yield[3].

Generally, the reaction rates of aryl halides follow the order: iodides > bromides > chlorides > fluorides. This fact can be used for the selective substitution in polyhalogenated systems. For instance, 2-bromo-4-chlorotoluene gives 76% of 5-chloro-2-methylphenol by treatment with sodium hydroxide at 200 °C. Nevertheless, polyhalogenated systems which contain fluorides have a variable behaviour depending on the reaction temperature. At lower temperatures preferential hydrolysis of the fluoride takes place and at >200 °C the usual reactivity order iodides > bromides > chlorides > fluorides is observed. For instance, 1,2-dibromo-3,4,5,6-tetrafluorobenzene affords 2,3-dibromo-4,5,6-trifluorophenol in 87% yield by treatment with potassium hydroxide at 85 °C. Under the same conditions, 1,4-dibromo-2,3,5,6-tetrafluorobenzene produces a 78% yield of 2,5-dibromo-3,4,6-trifluorophenol. However, 4-fluorobromobenzene with NaOH at 200 °C gives 4-fluorophenol in 70–79% yield[4].

As in other nucleophilic substitutions[5], electron-withdrawing groups (NO_2, CN, CO_2H, SO_3H) in *ortho* and *para* positions increase the reactivity of the aryl halide to hydrolysis. For instance, chlorobenzene is best hydrolysed above 300 °C, whereas 1-chloro-2,4-dinitrobenzene gives a 95% yield of 2,4-dinitrophenol at 100 °C[6].

Substituted 1,2-dichlorobenzenes **1** with an electron-withdrawing substituent in the 4-position react with sodium nitrite to afford 2-nitrophenols **5** in good yields (75–85%) (equation 2)[7]. The formation of nitrophenols **5** proceeds presumably according to equation 2. The electron-withdrawing substituent in the 4-position promotes a nucleophilic substitution of the 1-chlorine atom in compounds **1**. The second chlorine atom in compounds **2** is now easily replaced due to the activating effect of the *o*-nitro group, leading to compounds **3**. These compounds are unstable and rapidly react with the nitrite nucleophile resulting in the formation of an unstable nitrite ester **4**. Finally, compounds **4** are converted into the 2-nitrophenols **5** with dilute acid.

Sodium trimethylsilanolate has been reported as a convenient synthon for a hydroxy group in the *ipso* substitutions of fluoride in aromatic compounds[8]. The $S_N Ar$ displacement of the fluoride by the nucleophilic trimethylsilanolate leads to the silyl ether, which is immediately desilylated by the liberated fluoride ion yielding the sodium aryloxide salts. Acidification of these salts affords the hydroxylated product. For instance, 1,4-difluoroanthracene gives 1-hydroxy-4-fluoroanthracene in 90% yield by treatment with sodium trimethylsilanolate.

B. From Sulphonic Acids

Aryl sulphonic acids can be converted to phenols by alkali fusion through their salts. This method has been used for the industrial production of phenol. In spite of the extreme conditions, the reaction gives fairly good yields, except when the substrate contains other groups that are attacked at the fusion temperatures by the alkali. Milder conditions can be used when the substrate contains electron-withdrawing groups, but the presence of electron-donating groups hinders the reaction. The reaction mechanism (equation 3) has been proved to be a nucleophilic aromatic substitution by isotopic studies using benzenesulphonate specifically labelled with ^{14}C at its C-1 position and by the use of $K^{18}OH$[9].

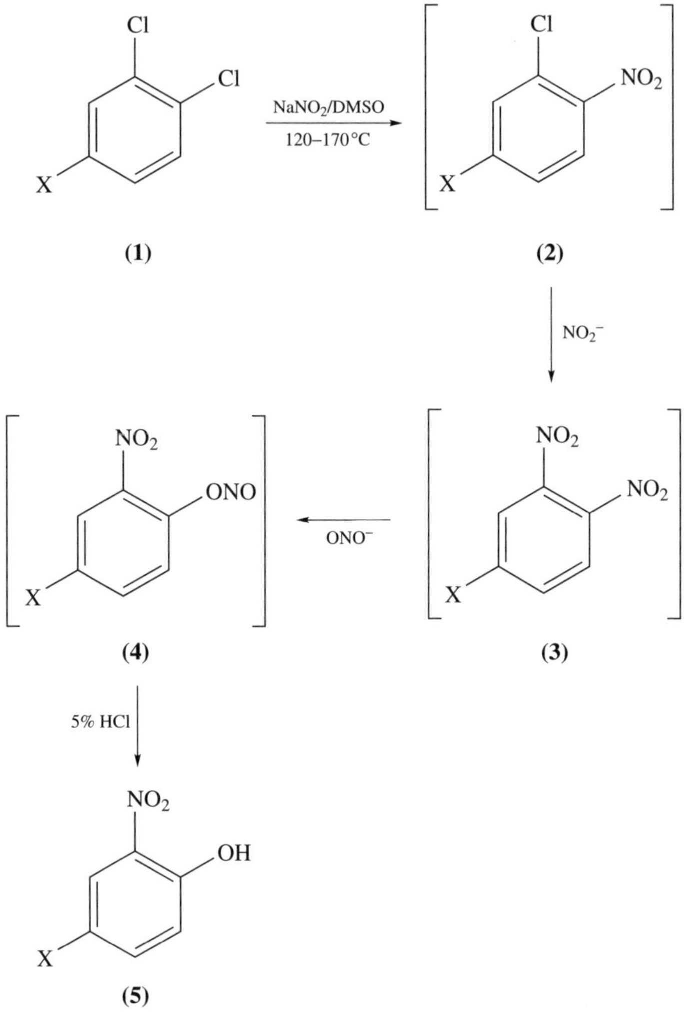

$$X = NO_2, PhCO, 3,4\text{-}Cl_2C_6H_3SO_2, PhSO_2, 4\text{-}MeC_6H_4SO_2, CF_3, MeCO$$

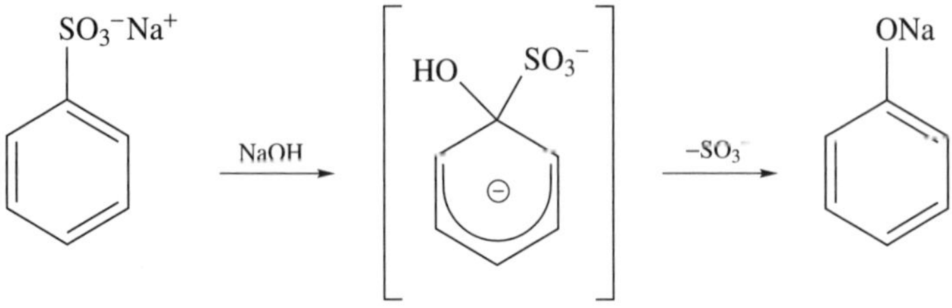

6. Synthesis of phenols

Some examples of this type of reaction are shown in equations 4–8.

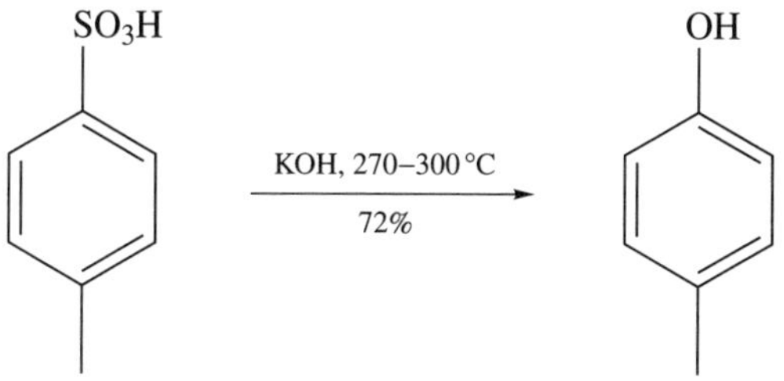

$$\text{4-MeC}_6\text{H}_4\text{SO}_3\text{H} \xrightarrow[72\%]{\text{KOH, 270–300 °C}} \text{4-MeC}_6\text{H}_4\text{OH} \quad (4)^{10}$$

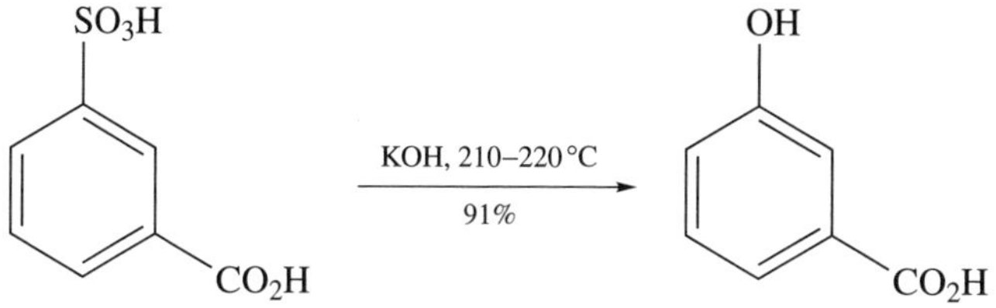

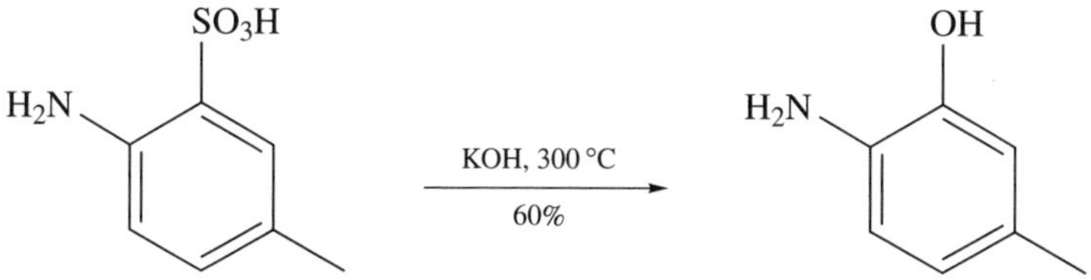

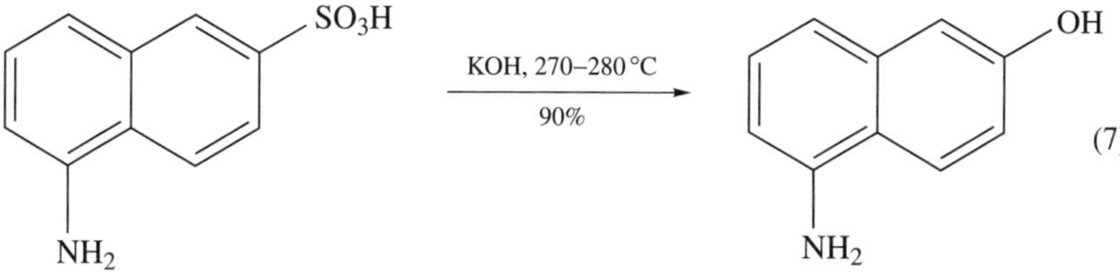

$$\text{3-HO}_2\text{C-C}_6\text{H}_4\text{SO}_3\text{H} \xrightarrow[91\%]{\text{KOH, 210–220 °C}} \text{3-HO}_2\text{C-C}_6\text{H}_4\text{OH} \quad (5)^{11}$$

$$\text{2-H}_2\text{N-5-Me-C}_6\text{H}_3\text{SO}_3\text{H} \xrightarrow[60\%]{\text{KOH, 300 °C}} \text{2-H}_2\text{N-5-Me-C}_6\text{H}_3\text{OH} \quad (6)^{12}$$

$$\text{5-H}_2\text{N-naphthalene-2-SO}_3\text{H} \xrightarrow[90\%]{\text{KOH, 270–280 °C}} \text{5-H}_2\text{N-naphthalen-2-ol} \quad (7)^{13}$$

$$\text{7-H}_2\text{N-1,3-(SO}_3\text{H)}_2\text{-naphthalene} \xrightarrow[84\%]{\text{NaOH, 220 °C}} \text{7-H}_2\text{N-1-OH-3-SO}_3\text{H-naphthalene} \quad (8)^{14}$$

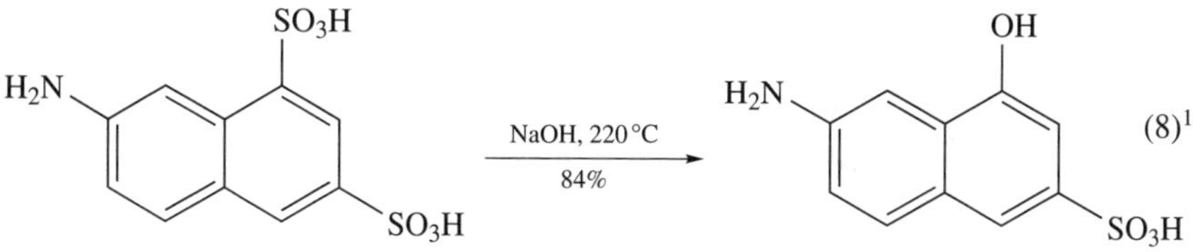

C. From Nitrogen Derivatives

1. Hydrolysis

Arylamines can undergo hydrolysis in acid or basic media to afford phenol and ammonia. Acid hydrolysis can be achieved under treatment with $ZnCl_2$, HCl, BF_3, H_2SO_4 or H_3PO_4 at very high temperatures (equations 9–11). Arylamines with *ortho* or *para* electron-withdrawing groups can also undergo hydrolysis in basic media by treatment with alkali (equations 12–14).

2. Bucherer reaction

The amino group of naphthylamines can be replaced by a hydroxy group by treatment with aqueous bisulphite[21]. The scope of the reaction is very limited. With very few exceptions, the amino group (NH_2 or NHR) must be on naphthalene or phenanthrene rings. The reaction is reversible and both the forward and reverse reactions are called the Bucherer reaction.

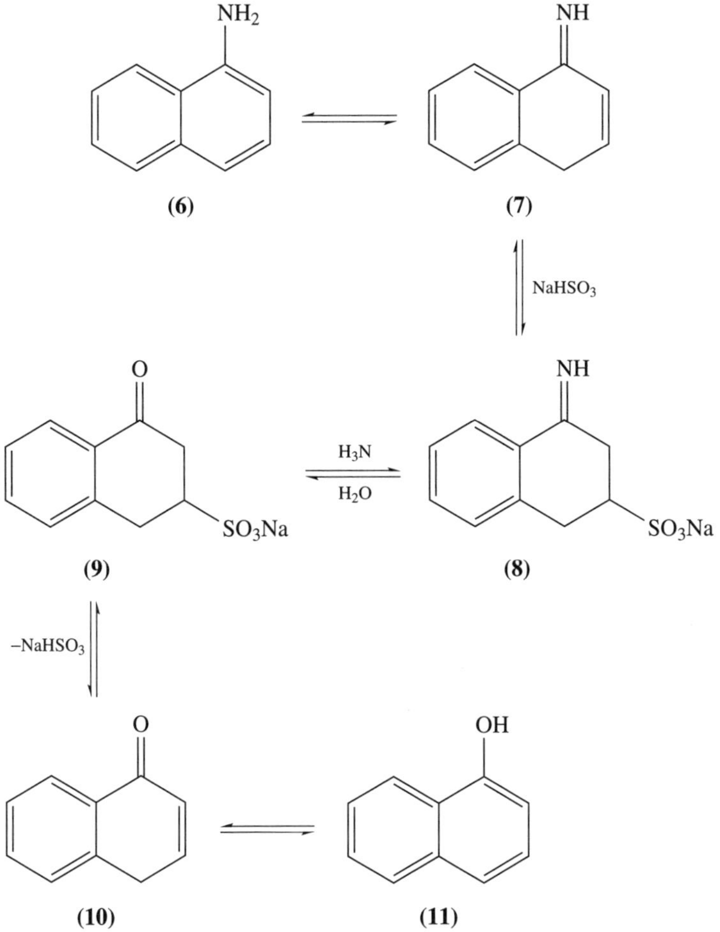

The mechanism seems to involve tetralone imine sulphonate **8**, which is formed by addition of $NaHSO_3$ to the C=C double bond of the tautomeric imine form **7** of the naphthylamine **6** (equation 15). Imine **8** undergoes hydrolysis to afford ketone **9** which, by elimination of $NaHSO_3$, yields ketone **10**, the tautomeric form of the final naphthol **11**. Equations 16 and 17 show examples of this type of transformation. *Para* electron-withdrawing substituents to the amino group accelerate the reaction. Only one substituent can be exchanged in diamino naphthalenes (or dihydroxynaphthalenes in the reverse reaction) (equations 18 and 19). If the two functional groups are attached to different rings, the replacement of the second one would require the dearomatization of the benzene ring in the tetralone sulphonate. If both substituents are attached to the same ring, a second addition of $NaHSO_3$ is no longer possible.

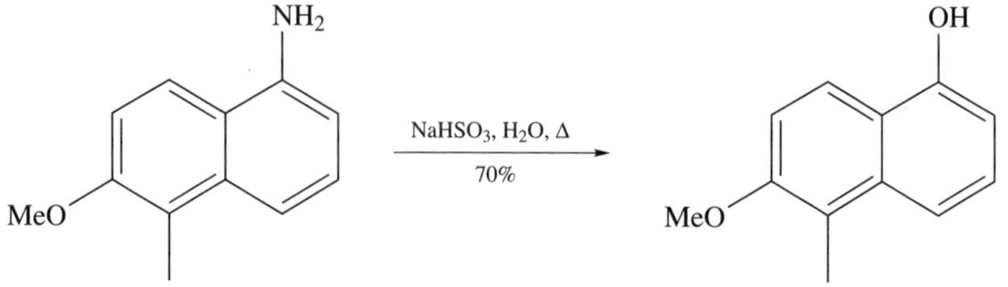

$$\text{(structure)} \xrightarrow[70\%]{NaHSO_3, H_2O, \Delta} \text{(structure)} \quad (16)^{22}$$

$$\text{(structure)} \xrightarrow[90\%]{NaHSO_3, H_2O, \Delta} \text{(structure)} \quad (17)^{23}$$

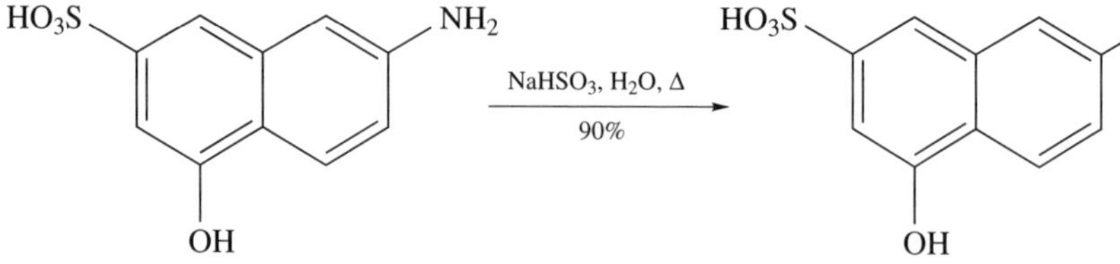

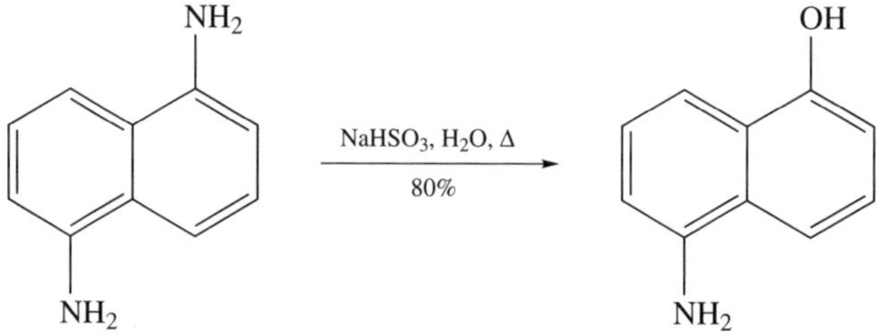

$$\text{(structure)} \xrightarrow[80\%]{NaHSO_3, H_2O, \Delta} \text{(structure)} \quad (18)^{23}$$

$$\text{(structure)} \xrightarrow[80\%]{NaHSO_3, H_2O, \Delta} \text{(structure)} \quad (19)^{24}$$

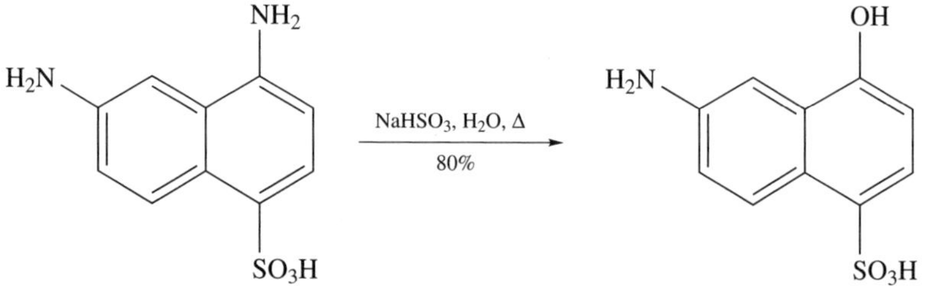

3. Diazotation reaction

Diazonium compounds can be converted to phenols by hydrolysis, under conditions where formation of the aryl cation takes place (equation 20)[25]. This reaction is usually accomplished synthetically by heating an aqueous solution of the diazonium salt[26]. Some examples of this type of reaction are given in equations 21–26.

$$ArNH_2 \xrightarrow{HNO_2} ArN_2^+ X^- \xrightarrow{N_2} Ar^+ X^- \xrightarrow{H_2O, H^+} ArOH + HX \quad (20)$$

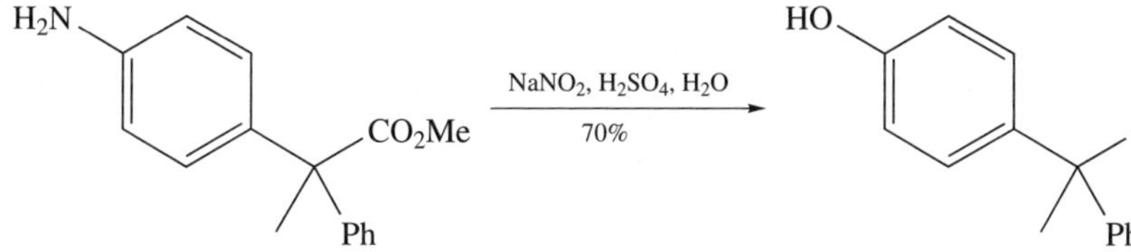

6. Synthesis of phenols

Reaction (21)[27]: 4-aminophenyl derivative with CO₂Me and Ph substituents → 4-hydroxyphenyl analog; NaNO₂, H₂SO₄, H₂O; 70%

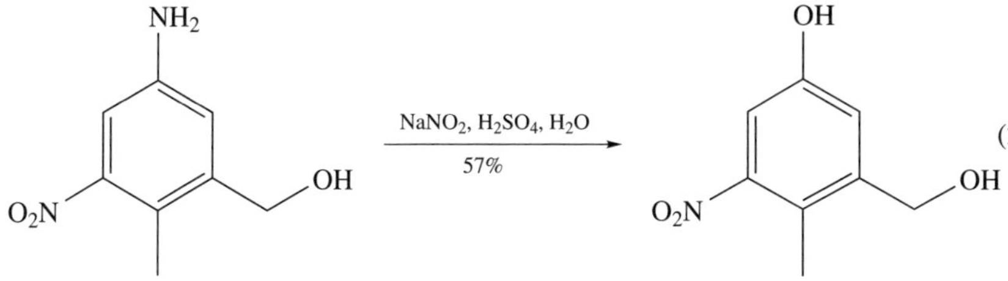

Reaction (22)[28]: aminonitromethylbenzyl alcohol → hydroxynitromethylbenzyl alcohol; NaNO₂, H₂SO₄, H₂O; 57%

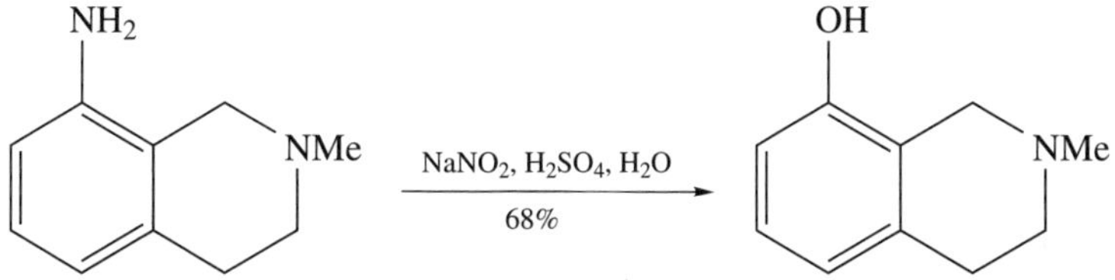

Reaction (23)[29]: 8-amino-N-methyl-tetrahydroisoquinoline → 8-hydroxy-N-methyl-tetrahydroisoquinoline; NaNO₂, H₂SO₄, H₂O; 68%

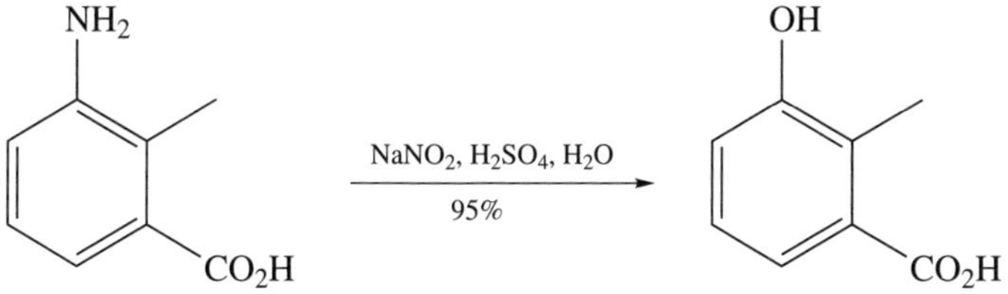

Reaction (24)[30]: 3-amino-2-methylbenzoic acid → 3-hydroxy-2-methylbenzoic acid; NaNO₂, H₂SO₄, H₂O; 95%

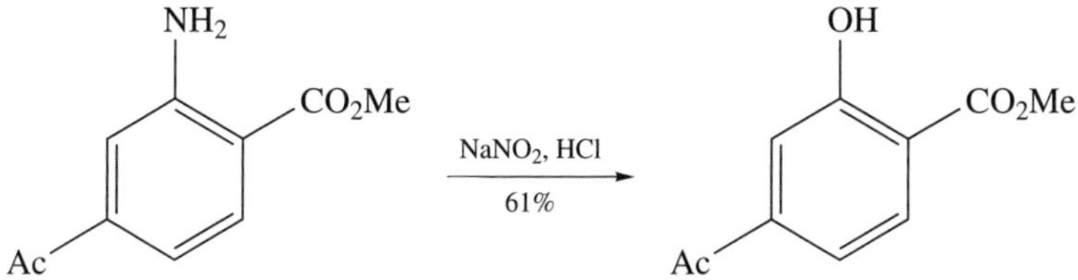

Reaction (25)[31]: methyl 2-amino-4-acetylbenzoate → methyl 2-hydroxy-4-acetylbenzoate; NaNO₂, HCl; 61%

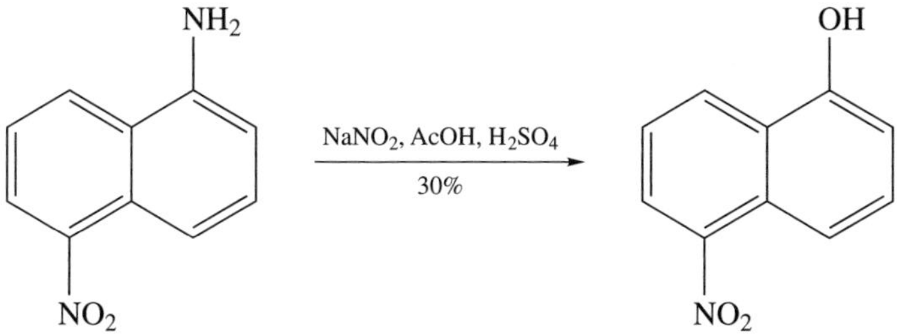

Reaction (26)[32]: 1-amino-5-nitronaphthalene → 1-hydroxy-5-nitronaphthalene; NaNO₂, AcOH, H₂SO₄; 30%

An alternative redox mechanism leads to the formation of phenols under rather mild conditions[33]. This reaction is initiated by Cu_2O, which effects reductive formation of an aryl radical. In the presence of Cu^{II} salts, the radical is oxidized to the phenyl cation by a reaction presumably taking place in the copper coordination sphere. The reaction is very rapid and gives good yields of phenols over a range of structural types. Equations 27–29 show some examples of this type of transformation[33].

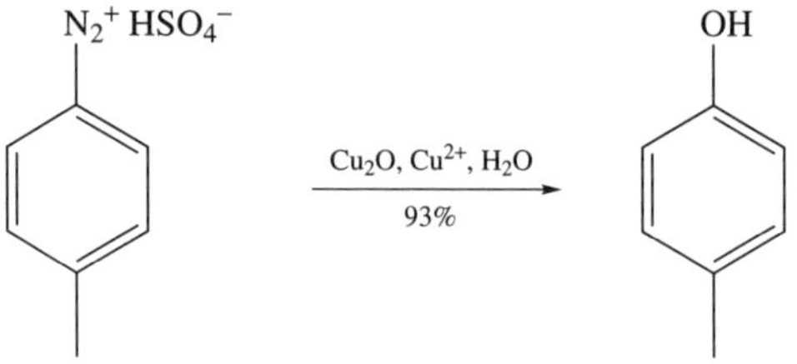

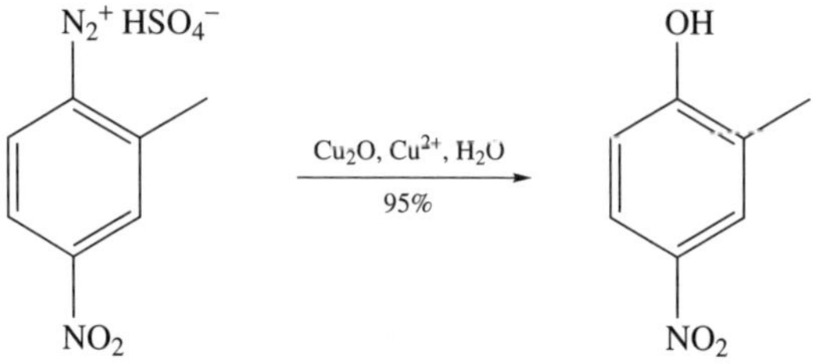

II. BY OXIDATION

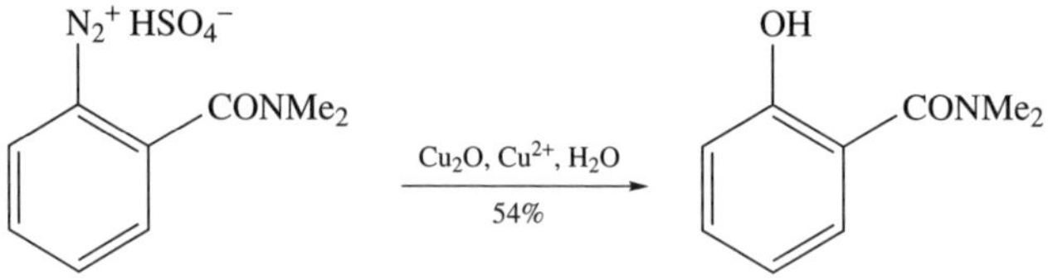

A. Hydroxylation

1. With hydrogen peroxide

The direct production of phenols from aromatic hydrocarbons (electrophilic aromatic hydroxylation) would presumably need a hydroxy cation HO^+, analogous to NO_2^+ or R^+. However, since the hydroxy group is strongly activating towards electrophilic substitution, further oxidations usually occur so that the yields are generally low[34]. Support for the proposal that HO^+ should be present in acidified solutions of hydrogen peroxide[35] was first provided by the hydroxylation of mesitylenes with hydrogen peroxide in acetic and sulphuric acids[36]. A possible mechanism of this reaction involves the displacement of water from protonated hydrogen peroxide by the reactive aromatic compound (equation 30).

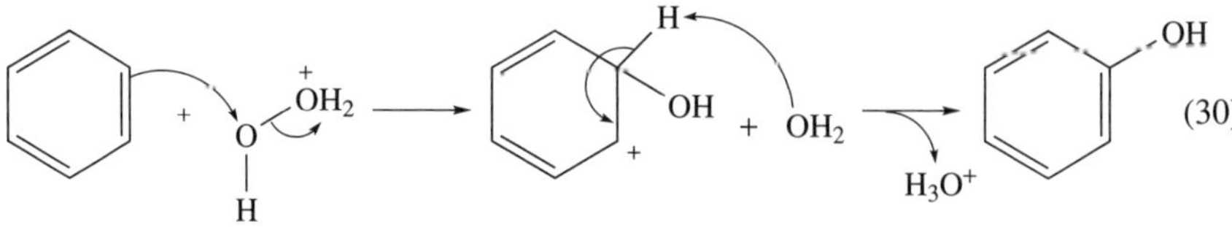

Other acids such as HF[37], HSO_3F-SbF_5/SO_2ClF, HF/BF_3[38] and HF/SbF_5[39] have been used with the advantage that the phenolic products are protonated and so do not undergo further electrophilic attack.

A variation of the above method uses a Lewis acid in place of the protic acid. Here the advantage is that the acid coordinates to the oxygen of the product, thus retarding further degradation. $AlCl_3$[40] is an effective catalyst to afford mainly *ortho* and *para* substitution.

Direct hydroxylation can be accomplished by free radical reagents, such as a mixture of hydrogen peroxide with a transition metal catalyst and a redox buffer [e.g. Fe^{2+} + H_2O_2 (Fenton's reagent[41]), Fe^{3+} + H_2O_2 + catechol (Hamilton's reagent)[42]]. The yields are usually poor, in the 5–20% range, and there are significant amounts of coupling products. A modification of the method, developed as a model for the biogenic oxidation of tyramine, has been introduced by Udenfriend and coworkers who used the system of O_2 + Fe^{2+} + ascorbic acid in the presence of EDTA (Udenfriend's reagent[43]). This method gives useful yields of *ortho* and *para* phenolic derivatives from phenylacetamide[44]. An update version of this oxidation uses anodic oxidation in the presence of Udenfriend's reagent, which converts tyramine to a mixture of hydroxytyramines and dihydroxytyramine (DOPA)[45].

2. With peroxides

Vicarious nucleophilic substitution (VNS) of hydrogen allows the direct introduction of substituents onto electrophilic aromatic rings (equation 31)[46]. A variety of carbo- and heterocyclic nitroarenes, as well as some electrophilic heterocycles lacking a nitro group, undergo this process with carbanions that contain a leaving group X at the carbanionic centre. The reaction proceeds according to the addition–elimination mechanism shown in equation 31[47]. Anions of alkyl hydroperoxides (ROOH)[48] can be considered to be nucleophiles that bear a leaving group (RO) at the anionic centre, like α-halocarbanions and anions of sulphenamides, etc. They can therefore undergo the VNS reaction with nitroarenes to produce nitrophenols (equation 32).

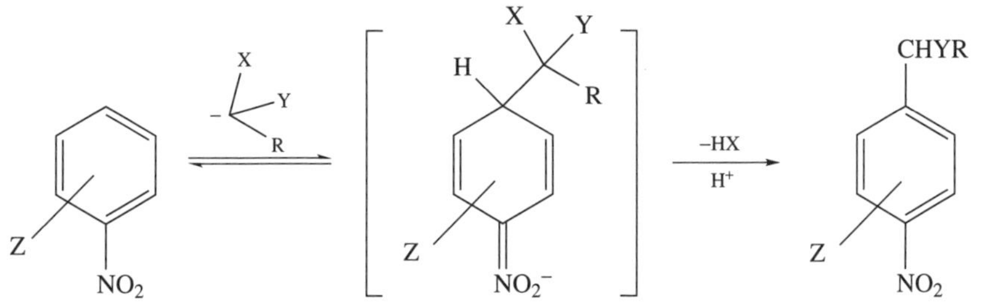

(31)

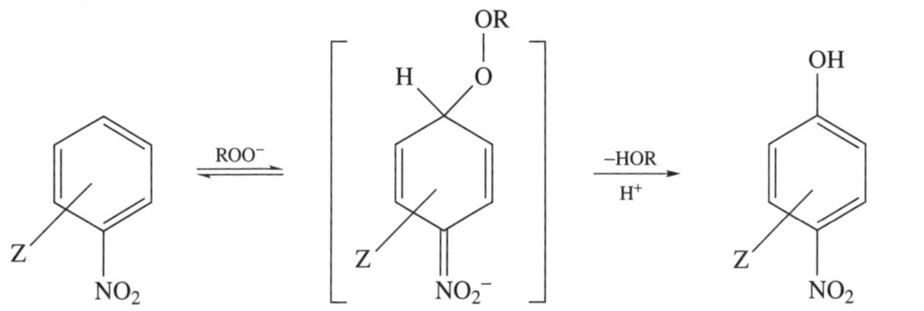

(32)

The reaction usually proceeds in high yields, and it is often possible to control the orientation of the hydroxylation. For instance, nitrobenzene derivatives **12** substituted at the *meta* position with electron-withdrawing groups, such as halogens, CF_3, SO_2Me, COPh, CN, NO_2, etc., easily underwent a regioselective VNS hydroxylation, giving the corresponding *p*-nitrophenols **13** (equation 33)[49].

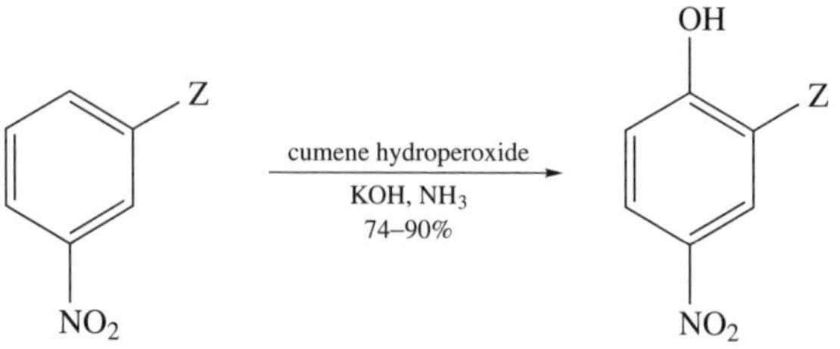

$$Z = F, Cl, Br, CF_3, SO_2Me, CN, NO_2, COPh, CO_2Me$$

A bicyclic aromatic ring system provides additional stabilization of the anionic σ-adducts; hence, nitronaphthalene derivatives show good reactivity in the VNS hydroxylation. 1-Nitronaphthalenes give 2- and 4-hydroxy derivatives in high yields. The orientation of the hydroxylation depends on the kind of base. For instance, treatment of 1-nitronaphthalene (**15**) with *t*-butyl hydroperoxide and potassium *t*-butoxide affords 1-nitro-2-naphthol (**16**) whereas using sodium hydroxide as base gives 4-nitro-1-naphtol (**14**) (equation 34)[49b].

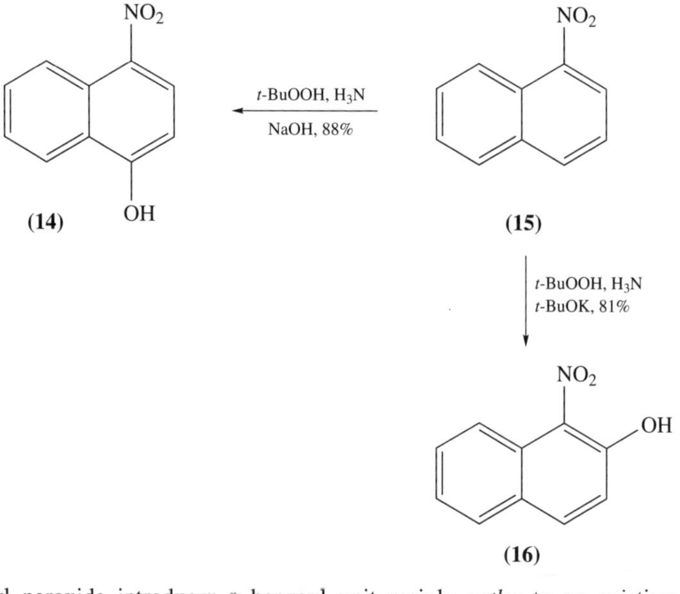

Benzoyl peroxide introduces a benzoyl unit mainly *ortho* to an existing hydroxyl (equation 35), but *para* products can be formed by [3,3] migration of the acyloxy group around the ring periphery of the dienone intermediate[50]. The benzoate esters can easily be hydrolysed to the corresponding phenols.

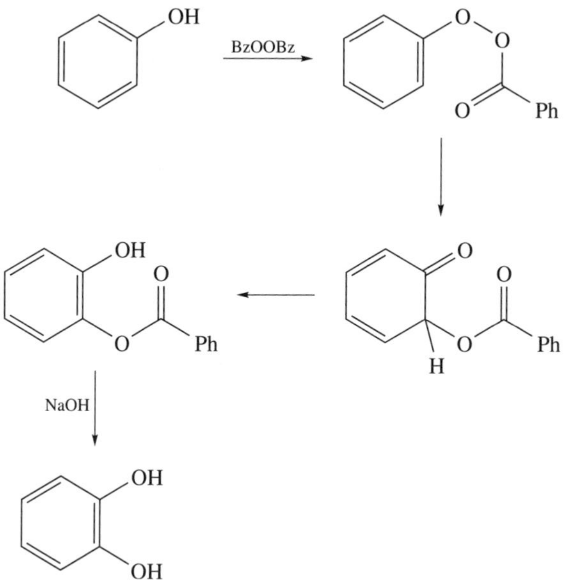

Phenols can also be formed by aromatic hydroxylation with the hydroxy radical generated from α-azo hydroperoxides in anhydrous organic media (equation 36)[51]. Photo- and thermal decomposition of α-azo hydroperoxides give hydroxy radicals, which can react with an aromatic ring generating a phenolic compound.

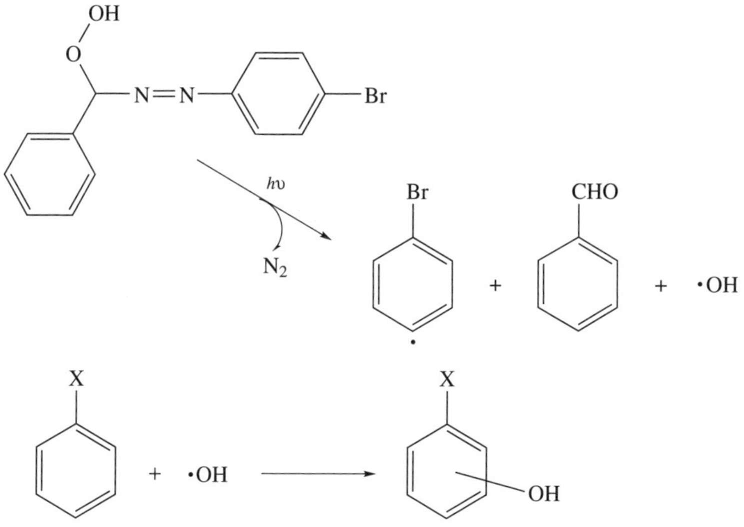

3. With peroxyacids

Inorganic and organic peroxyacids can be used as a source of hydroxy cations HO⁺ for the oxidation of aromatic rings.

a. Inorganic peroxyacids. Unstable inorganic peroxyacids can be generated *in situ* by oxidation of OsO_4, MoO_3, V_2O_5 or CrO_3 with hydrogen peroxide[52]. Peroxymonophosphoric acid also provides hydroxylation, taking place at a much faster rate than with perbenzoic acid[53]. A good example of the utilization of inorganic peracids is the Elbs reaction[54], which involves the oxidation of phenols with a persulphate in alkaline media. This reaction introduces a second hydroxy group into a phenol in the *para* position unless this is occupied (equation 37). In that case, an *ortho* hydroxylation occurs, but yields are, however, very poor. Potassium persulphate in alkaline medium is usually employed[55]. The initial oxidation product is the derivative **18**, which is then hydrolysed to hydroquinone (**19**).

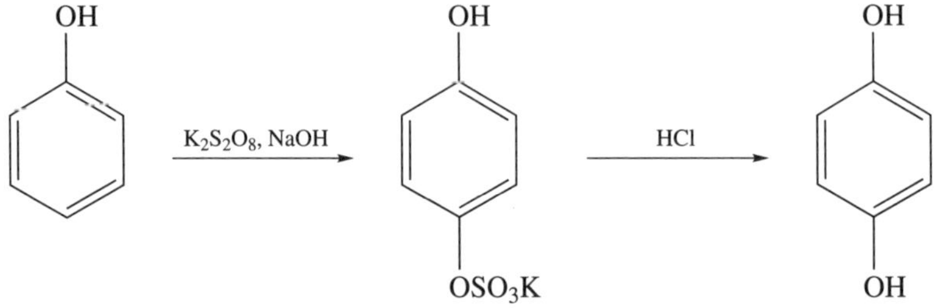

$$\text{(17)} \xrightarrow{K_2S_2O_8,\ NaOH} \text{(18)} \xrightarrow{HCl} \text{(19)} \quad (37)$$

The presence of electron-withdrawing groups on the aromatic ring improves the yield, but electron-donating groups can also be tolerated. Equations 38–43 show some examples of these peroxyacid oxidations.

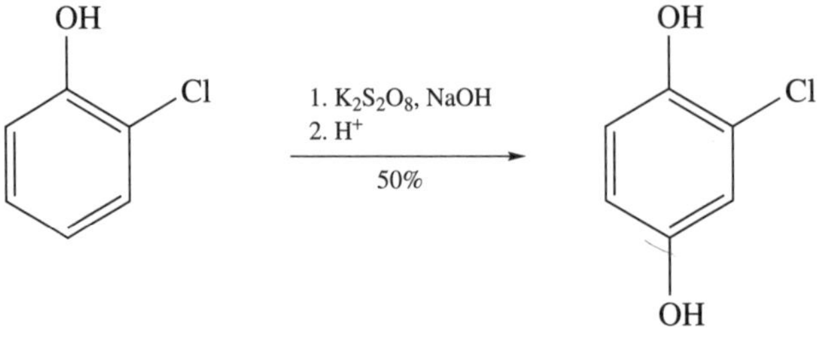

$$(38)^{56}$$

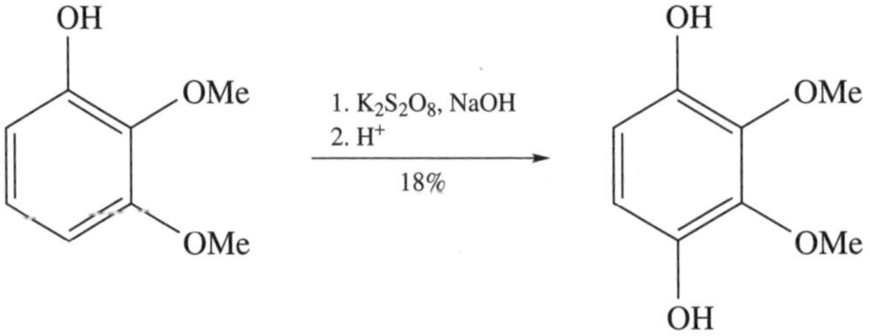

$$(39)^{57}$$

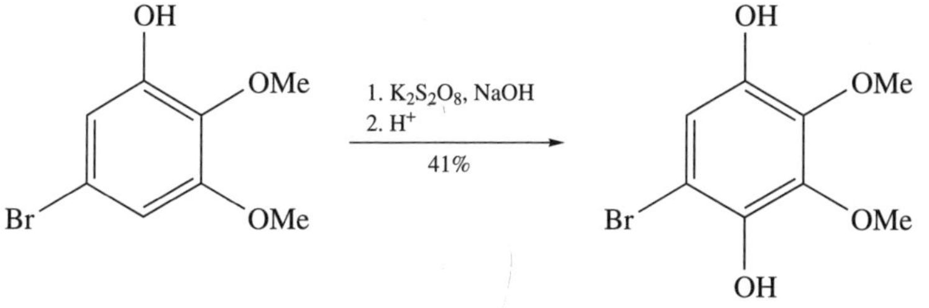

6. Synthesis of phenols

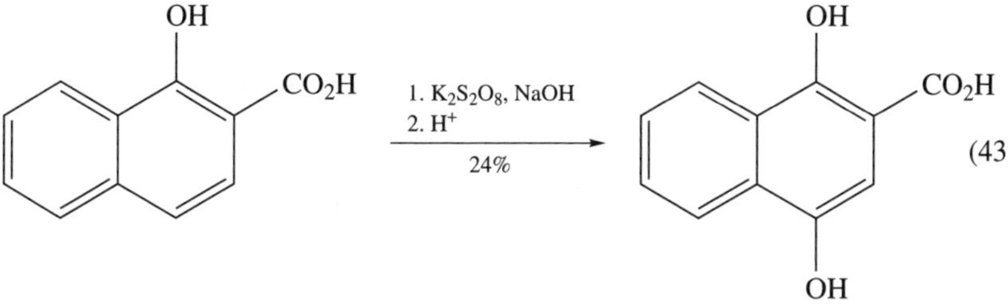

[Scheme showing four reactions with K$_2$S$_2$O$_8$, NaOH / H$^+$ as reagents, giving products (40)[58], 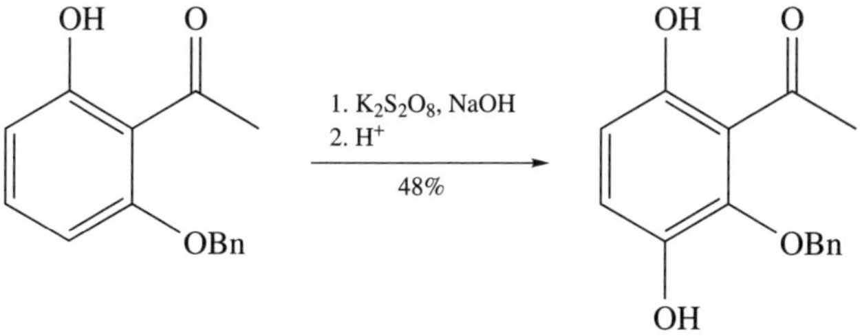 (41)[59], 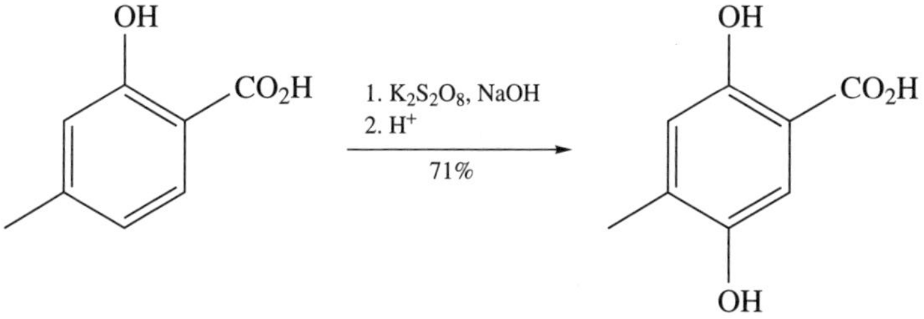 (42)[60], and (43)[61] in 41%, 48%, 71%, and 24% yields respectively.]

Analogously to the oxidation of phenols by the Elbs reaction, aromatic amines react with persulphate to give *o*-aminoaryl sulphates which can then be hydrolyzed to afford phenols. This reaction is known as the Boyland–Sims oxidation[62]. In this case, the substitution takes place exclusively *ortho* to the amino group just as in the phenol oxidation using the radical generator benzoyl peroxide. This is in contrast to the Elbs oxidation of phenols, which occurs predominantly in the *para* position[62a,63]. For instance, under these conditions, *N,N*-dimethylaniline, 4-methylaniline and 2-naphthylamine have been converted to the corresponding phenols *N,N*-dimethyl-2-hydroxyaniline, 2-hydroxy-4-methylaniline and 1-hydroxynaphthylamine in 40%, 28% and 45% yield, respectively[63b]. *Para* substitution takes place only if the *ortho* positions are occupied by substituents other than hydrogen.

b. Organic peroxyacids.

Oxidation with organic peroxyacids, such as peroxyacetic or trifluoroperoxyacetic acids, gives reasonable yields of phenol. Trifluoroperoxyacetic acid, usually prepared *in situ* from hydrogen peroxide and trifluoroacetic acid, is the most effective peroxyacid in aromatic oxidations[64]. Usually, the oxidation takes place preferentially in the *para* position. The yields of these reactions can be greatly increased by the addition of Lewis acids, such as BF_3. For instance, under these conditions mesitol can be obtained from mesitylene in 88% yield[65]. However, during hydroxylation of polymethylbenzenes with trifluoroperoxyacetic acid and BF_3 methyl groups can migrate, and this has been attributed to an *ipso* hydroxylation followed by a 1,2-shift of methyl[65b]. For instance, 1,2,3,4-tetramethylbenzene (**20**) on treatment with trifluoroperoxyacetic acid/BF_3 gives not only the expected 2,3,4,5-tetramethylphenol (**21**) but also small amounts of an isomeric phenol **22** and cyclohexadienone **23** (equation 44). Products **22** and **23** are obtained by assuming electrophilic attack on an already substituted position followed by Wagner–Meerwein methyl migration.

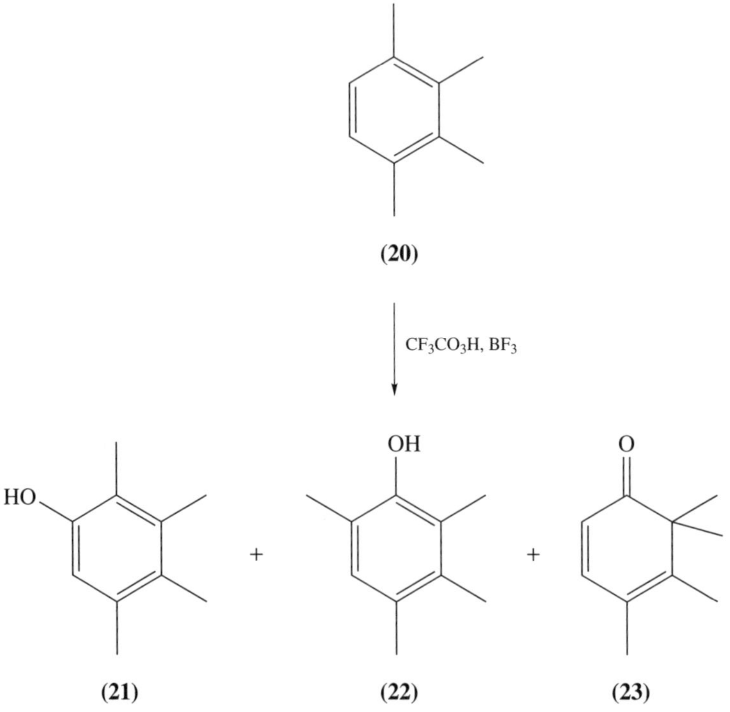

4. Electrochemical hydroxylation

Electrochemical oxidations proceed by a radical mechanism. Normally, the hydroxylated products are converted to quinones which undergo a further degradation process. Selective monohydroxylation of some aromatic compounds, such as chloro- and trifluoromethylbenzenes, has been achieved in trifluoroacetic acid containing sodium trifluoroacetate and trifluoroacetic anhydride[66]. Although under similar conditions benzene gives only 12–25% yield of phenol[67], Nishiguchi and coworkers[68] reported an improved version of the procedure. The Nishiguchi method involves the selective monohydroxylation of benzene and substituted benzenes **24** through anodic oxidation in a solvent mixture of trifluoroacetic acid and dichloromethane resulting in the corresponding phenols **25** in good yields, mainly substituted in the *ortho* and *para* positions (equation 45).

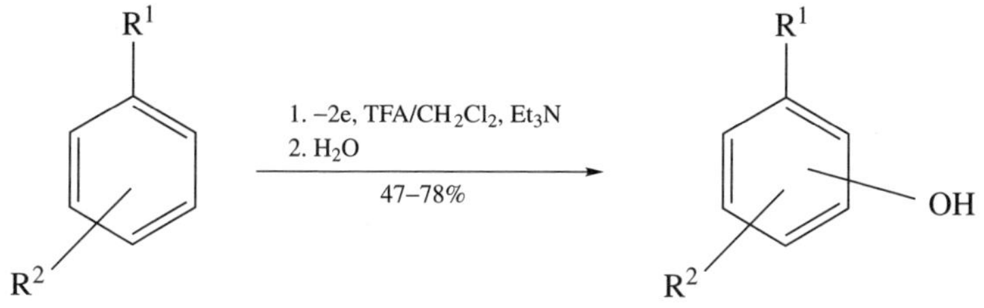

$$\underset{(24)}{R^2 \text{—}\!\!\!\!\bigcirc\!\!\!\!\text{—} R^1} \xrightarrow[47-78\%]{\text{1. } -2e, \text{ TFA/CH}_2\text{Cl}_2, \text{ Et}_3\text{N} \quad \text{2. H}_2\text{O}} \underset{(25)}{R^2\text{—}\!\!\!\!\bigcirc\!\!\!\!\text{—}(R^1)(\text{OH})} \quad (45)$$

$R^1, R^2 = H, Cl, Br, F, CF_3, Ac, CO_2Et, CHO, CN, NO_2$

5. Biotransformations

The selective hydroxylation of aromatic compounds is a difficult task in preparative organic chemistry. The problem is particularly severe when the compounds to be hydroxylated (or their products) are optically active and/or unstable, since in these instances the reaction should be conducted rapidly and under mild conditions in order to prevent racemization and decomposition. The selective hydroxylation of substituted phenols in the *ortho* and *para* positions can be achieved by using monooxygenases. In contrast, *meta*-hydroxylation is rarely observed[69]. For instance, phenolic compounds **26** can be oxidized selectively by *polyphenol oxidase*, one of the few available isolated oxygenating enzymes, to give *o*-hydroxylated products **27** (catechols) in high yields (equation 46)[70]. Usually, only *p*-substituted phenols can be oxidized since *m*- and *o*-substituted phenols are unreactive. The reactivity of the *p*-substituted phenols decreases as the nature of the group R is changed from electron-donating to electron-withdrawing substituents. The synthetic utility of this reaction has been demonstrated by the oxidation of amino acids and alcohols containing a *p*-hydroxyphenyl moiety. In this way, L-DOPA, D-3,4-dihydroxyphenylglycine and L-epinephrine have been synthesized from their *p*-monohydroxy precursors in good yield[70a].

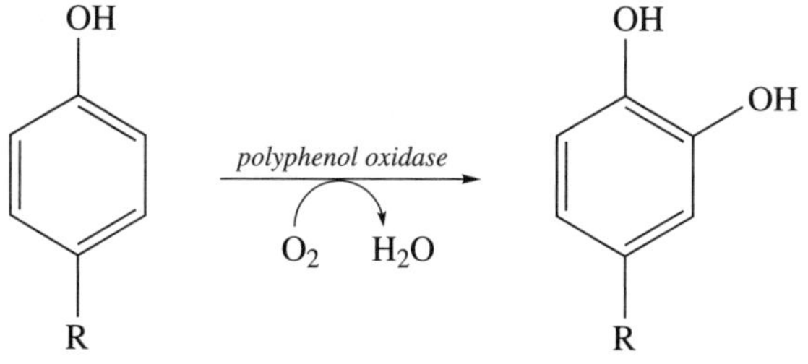

$$\underset{(26)}{\text{HO—}\!\!\!\!\bigcirc\!\!\!\!\text{—R}} \xrightarrow[\text{O}_2 \quad \text{H}_2\text{O}]{\text{polyphenol oxidase}} \underset{(27)}{\text{HO—}\!\!\!\!\bigcirc\!\!\!\!\text{(OH)—R}} \quad (46)$$

$R = H, Me, MeO, CH_2CH_2CO_2H, CH_2OH, CH_2CH_2OH, CH_2NHCOPh$

Mechanistically, it has been proposed that the reaction proceeds predominantly via epoxidation of the aromatic species **28**, which leads to unstable arene-oxides **29–31** (equation 47)[71]. Rearrangement of the arene-oxides **29–31** involving the migration of a hydride anion (NIH-shift) forms the phenolic product **32** or **33**[72]. Alternative flavin-dependent oxidases have been proposed to involve a hydroperoxide intermediate[73].

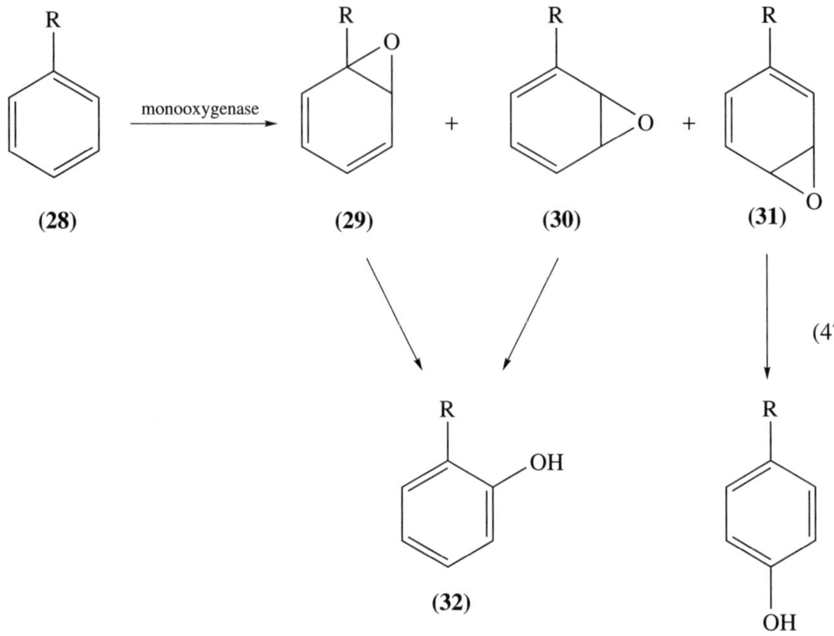

Regioselective hydroxylation of aromatic compounds can also be achieved by using whole cells[74]. For instance, 6-hydroxynicotinic acid (**35**) is produced industrially from nicotinic acid (**34**) by a *Pseudomonas* or *Bacillus sp* (equation 48)[75]. Racemic prenalterol (**37**) has been obtained by regioselective *p*-hydroxylation of (±)-1-isopropylamino-3-phenoxypropan-2-ol (**36**) using *Cunninghamella echinulata* (equation 49)[76].

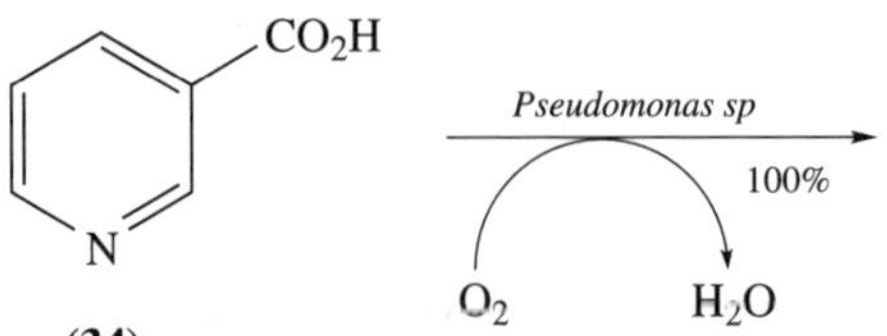

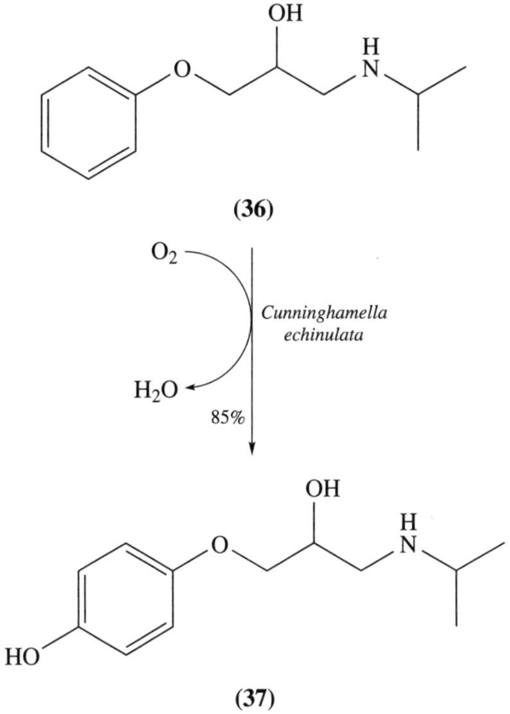

(49)

6. Miscellaneous methods

The oxidation of benzene to phenol can also be achieved using nitrous oxide as an oxidant in the presence of a catalytic system such as vanadium, molybdenum or tungsten oxides at 550 °C, and after addition of 30% of water to afford phenol in 10% yield[77]. More effective catalytic systems have been investigated and zeolites show promise to be good catalysts for the oxidation of benzene to phenol with nitrous oxide[78]. The use of zeolite catalysts has led to a reduction in the reaction temperature to 300–400 °C, to the exclusion of water addition to the reaction mixture and to an increase in the yields up to 25–30%[79]. Recently, direct oxidation of benzene to phenol by nitrous oxide has been commercialized[80].

Aromatic hydrocarbons can be oxidized to the corresponding phenols by transition metal peroxo complexes and, in particular, vanadium(V) peroxo complexes[81], which act either as electrophilic oxygen transfer reagents[82] or as radical oxidants[81,83], depending on the nature of the ligands coordinated to the metal and on the experimental conditions. Vanadium picolinato peroxo complex (VO(O$_2$)PIC(H$_2$O)$_2$) (**39**) (PIC = picolinic acid anion) has been reported to be particularly effective in the hydroxylation of benzene and substituted benzenes (equation 50)[81,84]. Accordingly, **39** smoothly oxidizes substituted benzenes **38** to the corresponding monophenols **40** in acetonitrile at room temperature.

The reaction proceeds also under catalytic conditions by using hydrogen peroxide as co-oxidant[85].

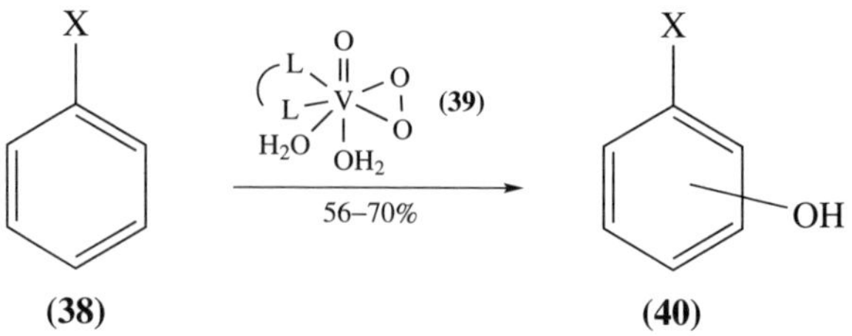

$$\underset{(38)}{\text{Ar-X}} \xrightarrow[56-70\%]{\text{L}_2\text{V(O)(O}_2\text{)(H}_2\text{O)(OH}_2\text{)} \ (39)} \underset{(40)}{\text{X-C}_6\text{H}_4\text{-OH}} \quad (50)$$

X = Me, F, Cl, Br, NO$_2$
L = PIC

Other type of complexes have also been used for the oxidation of hydrocarbons. For instance, Fujiwara and coworkers[86] employ a coordinated complex of palladium with *o*-phenanthroline as an efficient catalyst for the direct conversion of benzene into phenol. Moro-oka and coworkers[87] use an oxo-binuclear iron complex, whereas Machida and Kimura[88] work with macrocyclic polyamines. Sasaki and coworkers[89] employ Pd–Cu composite catalysts, which are prepared by impregnating the respective metal salts on silica gel.

Direct hydroxylation of aromatic rings with oxygen and hydrogen reported so far have been conducted by simultaneously mixing the aromatic compound, oxygen and hydrogen in the liquid phase in the presence of a multicomponent catalyst and additives[90]. However, these hydroxylations, besides the possibility of an explosion, give very low yields (below 1%). Mizukami and coworkers[91] have developed a more efficient and safe method, involving the direct hydroxylation in the gas phase with oxygen, activated by dissociated hydrogen obtained from a palladium membrane. Hydrogen atoms react with oxygen, producing species such as HOO• and HO• which cause hydroxylation.

B. Oxidation of Organometallic Derivatives

Autooxidation of an aryl Grignard or aryl lithium reagent gives a mixture of products which includes the phenol in variable yield[92]. Nevertheless, the controlled oxidation of aromatic lithium and magnesium derivatives with oxygen[93] or with hydroperoxides[94] produces the corresponding *o*-substituted phenols in yields that vary with the direct metalating group (DMG) of the ring (equation 51).

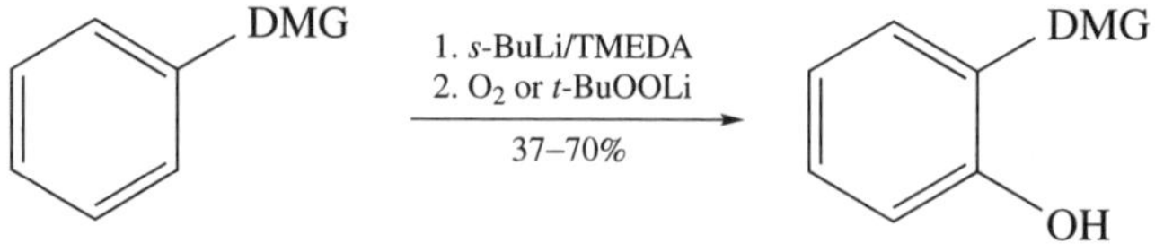

$$\text{Ar-DMG} \xrightarrow[37-70\%]{\text{1. }s\text{-BuLi/TMEDA} \quad \text{2. O}_2 \text{ or } t\text{-BuOOLi}} \text{(2-HO-Ar)-DMG} \quad (51)$$

DMG = CON(Pr-*i*)$_2$, OCONEt$_2$, OCON(Pr-*i*)$_2$, OMe

More efficiently, Grignard reagents[95] or lithium compounds[96] react with boronic esters to give borinic esters which can be oxidized with hydrogen peroxide or *t*-butyl hydroperoxide to give phenols in good yields (equation 52)[97]. The mechanism has been formulated as

6. Synthesis of phenols

involving an aryl rearrangement from boron to oxygen. The oxidation can also be achieved with oxygen[98], hydrogen peroxide/sodium perborate[99], hydrogen peroxide/sodium carbonate[100], ozone[101] or trimethylamine oxide, either anhydrous[102] or as dihydrate[103]. This method has been applied, for instance, for the preparation of phenol, α-naphthol or p-cresol, which have been obtained from the corresponding halides in 78, 75 and 60% yield, respectively[104]. Other examples of this type of oxidation are shown in equations 53–56.

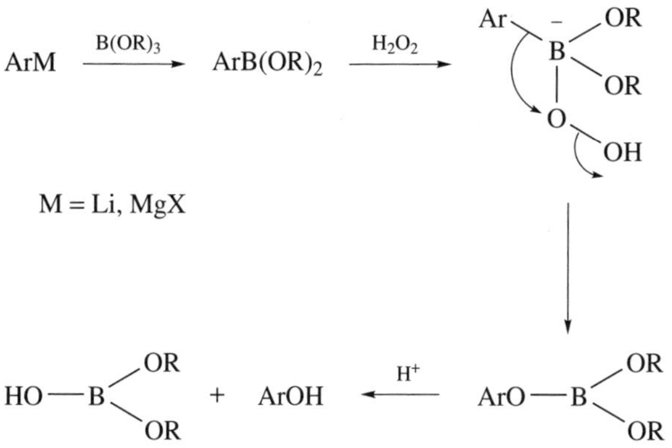

$$ArM \xrightarrow{B(OR)_3} ArB(OR)_2 \xrightarrow{H_2O_2} \cdots$$

M = Li, MgX

(52)

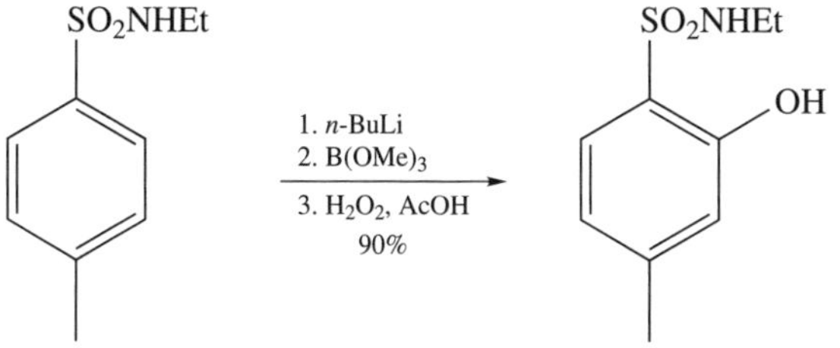

(53)[97d]

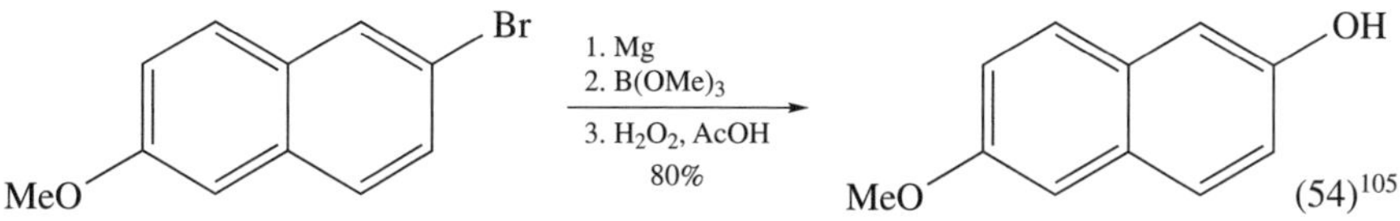

(54)[105]

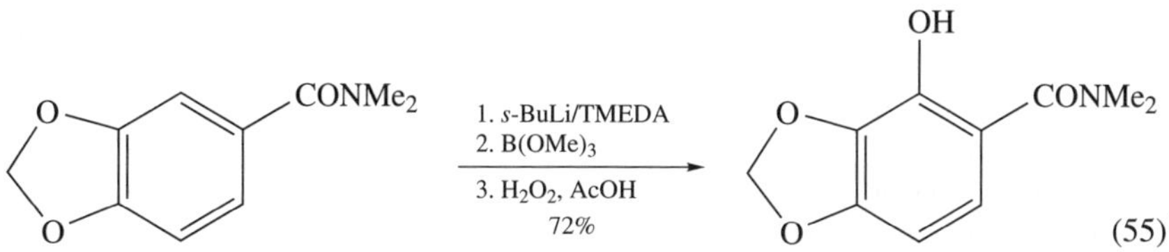

(55)[106]

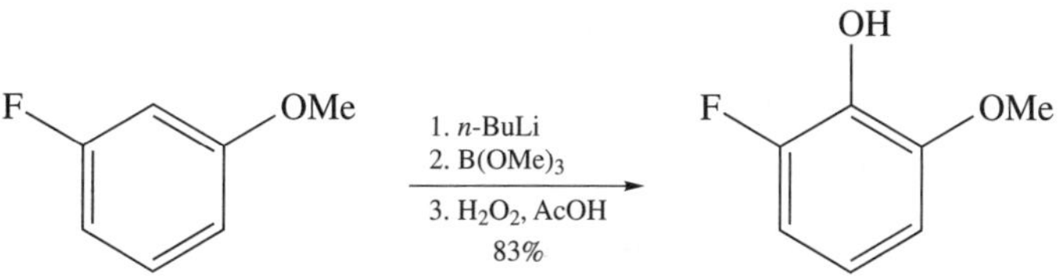

(56)[107]

Thallium(III)[108], particularly as its trifluoroacetate salt[109], has been successfully used for the synthesis of phenols. This method can be carried out in a single step and is subject to isomer orientation control[110]. The aromatic compound to be hydroxylated is first thallated with thallium trifluoroacetate (TTFA)[111] and, by treatment with lead tetraacetate followed by triphenylphosphine and then dilute NaOH, it is converted to the corresponding phenol (equation 57). Table 1 shows some examples of these transformations[108].

$$ArH \xrightarrow{TTFA} ArTl(OOCCF_3)_2 \xrightarrow[\text{2. PPh}_3]{\text{1. Pb(OAc)}_4} ArOOCCF_3 \xrightarrow{\text{dilute NaOH}} ArOH$$

(57)

TABLE 1. Formation of phenols according to equation 57

Substrate	Product	Yield (%)
Benzene	phenol	39
Toluene	p-cresol	62
o-Xylene	3,4-xylenol	78
m-Xylene	2,4-xylenol	70
p-Xylene	2,5-xylenol	68
Anisole	4-hydroxyanisole	41
Chlorobenzene	4-chlorophenol	56

An interesting alternative which combines both boron and thallium chemistry has been developed. The arylthallium compound is treated with diborane to provide the arylboronic acid which, by oxidation under standard conditions, yields the phenolic compound in good yield (equations 58 and 59)[112].

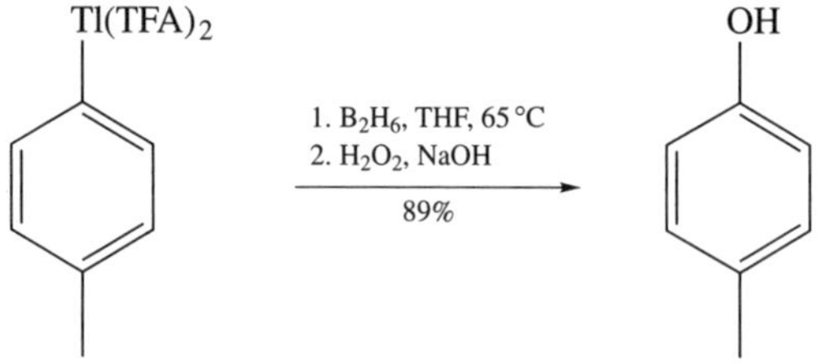

(58)

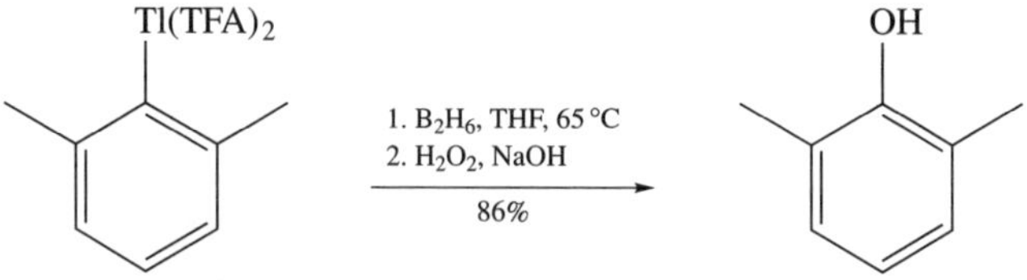

(59)

A convenient synthetic method for the conversion of aryl bromides to phenols is the reaction of the corresponding organometallic reagents with molybdenum peroxide–pyridine–hexamethylphosphoramide (MoO$_5$-Py-HMPA $\equiv$ MoOPH)[113]. This method provides a mild one-pot reaction sequence for the synthesis of phenols under basic conditions. Phenols are obtained in good to excellent yields with several prototype compounds. Other strongly basic carbanions have been hydroxylated with MoOPH, including aryllithium derivatives[114]. Table 2 shows some examples of this type of reaction[113,115].

Other oxidizing reagents such as MoOPH which produce direct hydroxylation of organometallic reagents are the 2-sulphonyloxaziridines **41** (equation 60)[116]. Both MoOPH

TABLE 2. Phenols obtained by oxidation of aryllithium derivatives by MoOPH

Entry	Substrate	Product	Yield (%)
1	Bromobenzene	phenol	89
2	1-Bromo-4-methoxybenzene	4-methoxyphenol	67
3	1-Bromo-4-ethylbenzene	4-ethylphenol	70
4	1-Bromonaphthalene	1-naphthol	85

and **41** have oxygens as part of a three-membered ring at their active site. These reagents have been suggested to transfer oxygen to neutral substrates by a similar $S_N 2$ reaction mechanism[117]. The organometallic reagent (Ar^2M) attacks the oxaziridine **41** to afford intermediate **42**, which collapses to N-benzylidenesulphonimine **43** and the phenol **44**. Oxidation of aryl lithium and Grignard reagents by **41** gives good to excellent yields of phenols, accompanied by the sulphonamide addition product **45**. Table 3 shows some examples of this type of oxidation[116].

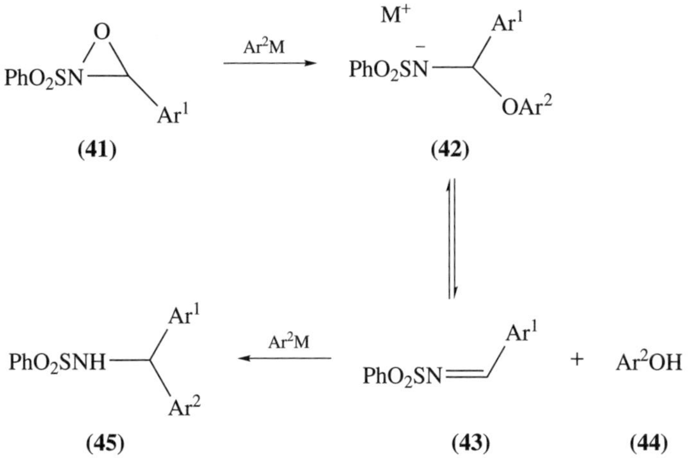

Ar^1 = p-tolyl, phenyl, 2-Cl-5-nitrophenyl

To avoid the formation of the addition product **45**, an oxaziridine that affords a sulphonimine resistant to addition by the organometallic reagent can be used. In this regard, oxidation of PhMgBr or PhLi with (+)-(camphorsulphonyl)oxaziridine (**46**)[118]

TABLE 3. Formation of phenols according to equation 60

Ar^1	Ar^2M	Ar^2OH	Yield (%)
p-Tolyl	PhMgBr	PhOH	90
p-Tolyl	PhLi	PhOH	62
p-Tolyl	p-MeOC$_6$H$_4$MgBr	p-MeOC$_6$H$_4$OH	29
Phenyl	o-MeOC$_6$H$_4$Li	o-MeOC$_6$H$_4$OH	70
Phenyl	PhNa	PhOH	56
2-Cl-5-nitrophenyl	PhMgBr	PhOH	49

gave phenol in 96% and 41% yield, respectively (equation 61).

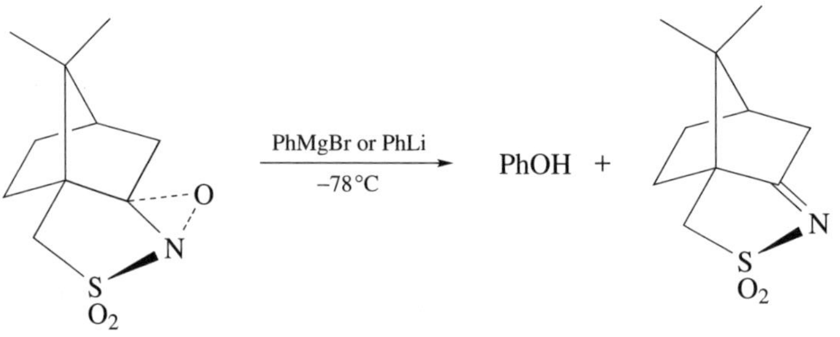

$$\text{(46)} \xrightarrow[-78\,^\circ\text{C}]{\text{PhMgBr or PhLi}} \text{PhOH} + \text{product} \quad (61)$$

Bis(trimethylsilyl)peroxide (TMSO)$_2$ can be considered as a source of TMSO$^+$ and consequently of HO$^+$. Reaction of (TMSO)$_2$ with aromatic lithium compounds **48** generated from the corresponding halides **47** gives the trimethylsilyloxy derivatives **49**, which under desilylation afford the corresponding phenols **50** in good yields (equation 62)[119].

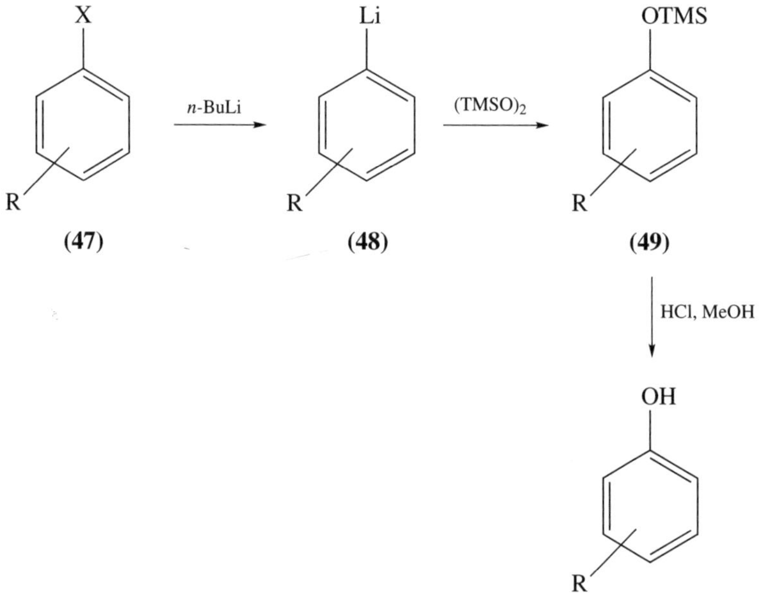

$$\text{(47)} \xrightarrow{n\text{-BuLi}} \text{(48)} \xrightarrow{(\text{TMSO})_2} \text{(49)} \xrightarrow{\text{HCl, MeOH}} \text{(50)} \quad (62)$$

R = H, OMe, Me, NMe$_2$

Perfluoroethyl-substituted stannanes **51** can be oxidized directly to the corresponding phenols **52** in excellent yield under mild conditions using potassium superoxide, sodium perborate, oxone or hydrogen peroxide/KHCO$_3$ (equation 63). The latter conditions give the best results[120].

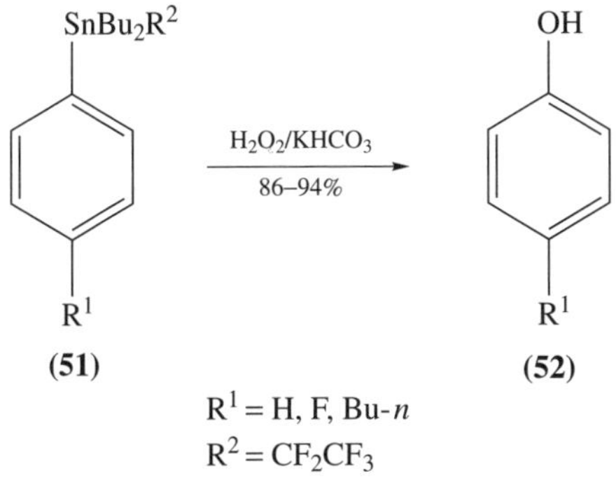

6. Synthesis of phenols

$$\text{SnBu}_2\text{R}^2\text{-C}_6\text{H}_4\text{-R}^1 \xrightarrow[86-94\%]{H_2O_2/KHCO_3} \text{HO-C}_6\text{H}_4\text{-R}^1 \quad (63)$$

(51) → (52)

$R^1 = H, F, Bu-n$
$R^2 = CF_2CF_3$

C. Oxidation of Nitrogen Derivatives

N-Arylhydroxylamines **53** readily rearrange in aqueous acid solution (HCl, HBr, H$_2$SO$_4$, HClO$_4$, etc.) to p-aminophenols **58** (equation 64)[121]. This reaction, known as the Bamberger rearrangement[122], occurs by an S_N1-type mechanism. Protonation of the hydroxy group to **54**, followed by dehydration, affords an intermediate nitrenium ion **55** ↔ **56**. This conjugated cation is trapped by water in the *para* position to give the intermediate **57**, the tautomer of the final p-aminophenol (**58**). Among the evidence[123] for this mechanism are the facts that other products are obtained when the reaction is run in the presence of competing nucleophiles, e.g. p-ethoxyaniline when ethanol is present, and that when the *para* position is blocked, compounds similar to **57** are isolated. In the case of 2,6-dimethylphenylhydroxylamine, the corresponding intermediate nitrenium ion **55** has been trapped, and its lifetime in solution was measured[124].

Nitrobenzenes also undergo the Bamberger rearrangement, being the most convenient and economical method for the synthesis of p-aminophenols, particularly on an industrial scale[125]. The process is normally carried out by catalytic hydrogenation under highly acidic conditions, where N-phenylhydroxylamine has been shown to be the intermediate (equation 65).

The conversion of azoxy compounds, on acid treatment, to p-hydroxy azo compounds (or sometimes the o-hydroxy isomers[127]) is called the Wallach rearrangement[128]. When both *para* positions are occupied, the o-hydroxy product may be obtained, but *ipso* substitution at one of the *para* positions is usually obtained. The mechanism of this reaction is not clear[129]. Equations 66–68 are examples of these transformations.

Nevertheless, azoxy compounds can be transformed into o-hydroxy azo derivatives by photolysis, the reaction being known as the photo-Wallach rearrangement[132]. Irradiation of these compounds leads to migration of the oxygen to the aromatic ring far from the original N-O function. For instance, (phenyl)4-methoxyphenyldiazene-1-oxide (**59**) under photolysis affords 2-hydroxy-4-methoxyphenylazobenzene (**60**) in 79% yield (equation 69)[133].

An intramolecular pathway shown in equation 70 has been postulated[134].

In strongly acid solution, irradiation of 3-substituted 2,1-benzisoxazols **61** gives 2-amino-5-hydroxyacylbenzenes **62** in good yield (equation 71)[135].

D. Oxidation of Carbonyl Groups

The conversion of benzaldehydes to phenols using alkaline hydrogen peroxide is generally known as the Dakin's oxidation[136,137]. However, this reaction is limited in general to *ortho-* and *para*-hydroxy or alkoxy benzaldehydes because in other cases the corresponding benzoic acid is formed instead[137]. The reaction consists in the oxidation of an aromatic aldehyde **63** via rearrangement of the hydroperoxide **64** to the formyl ester **65**, which is finally hydrolysed to yield the corresponding phenol **66** (equation 72). For instance, veratraldehyde[138], piperonal[139], isovanillin[140], 5-bromovanillin[141] and *p*-hydroxybenzaldehyde under Dakin's oxidation produce 3,4-dimethoxyphenol, 3,4-methylenedioxyphenol, 2,4-dihydroxyanisole, 3-bromo-2,5-dihydroxyanisole and hydroquinone in 45%, 67%, 49%, 92% and 78% yield, respectively. Hydrogen peroxide in the presence of acid can be used for the oxidation of benzaldehydes without an activating group at the *ortho* and *para* position[142]. This method represents an alternative to the Dakin's oxidation described above.

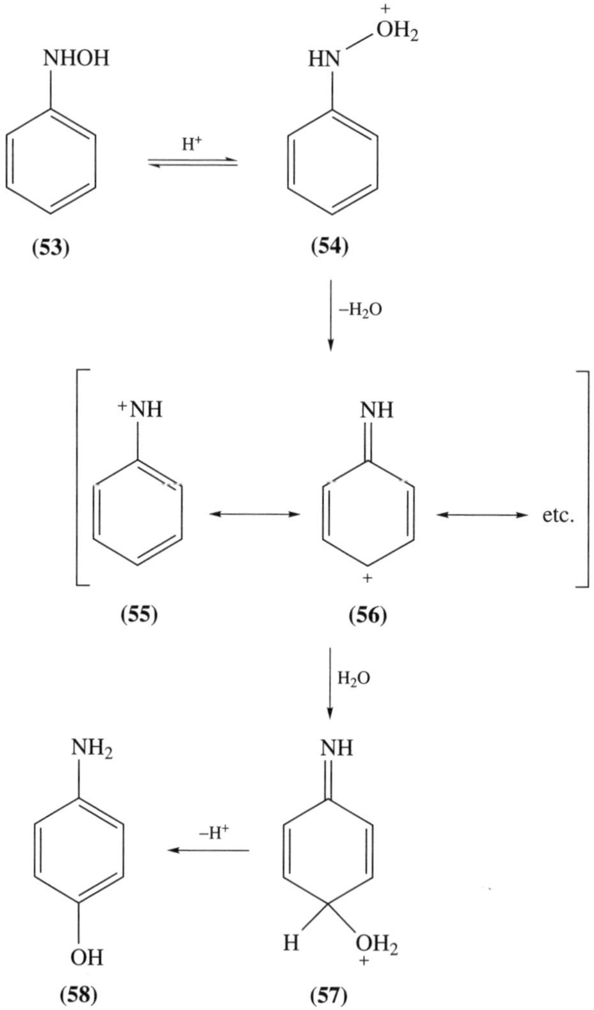

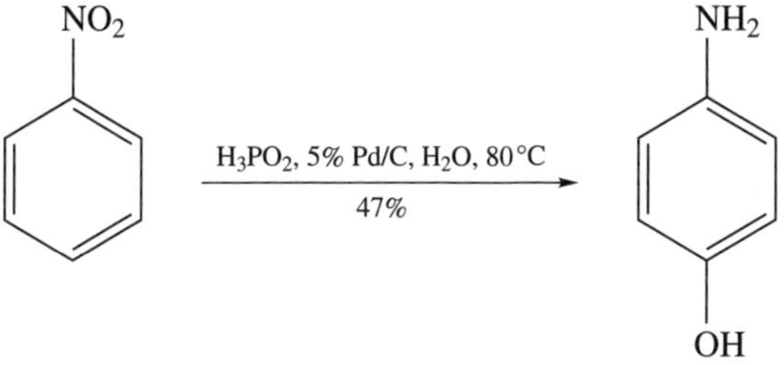

6. Synthesis of phenols

(65)[126]

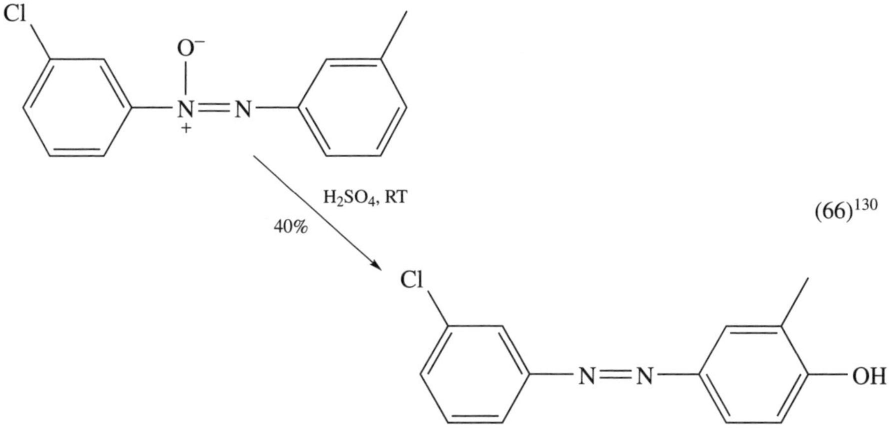

(66)[130]

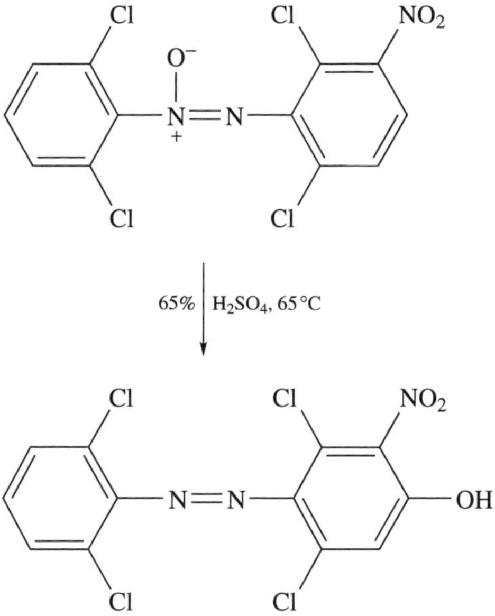

(67)[130]

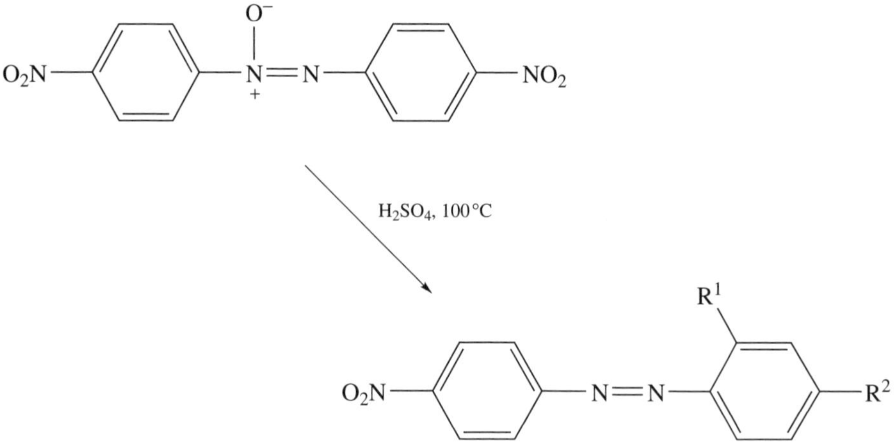

$R^1 = OH, R^2 = NO_2$ (54%)
$R^1 = H, R^2 = OH$ (11%) (68)[131]

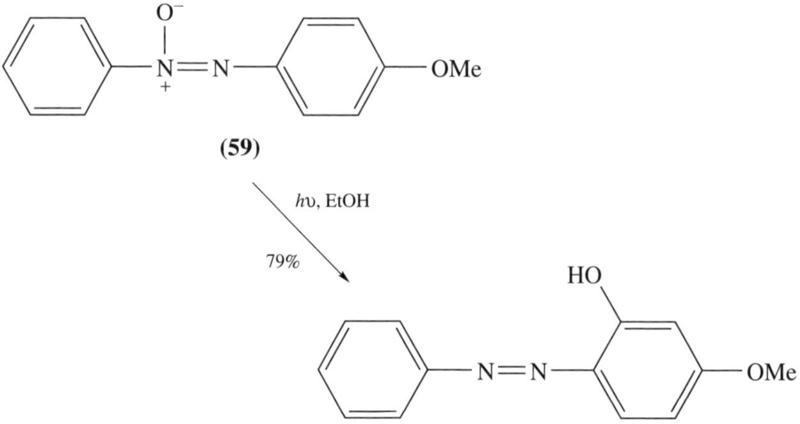

(59)

(69)

(60)

The solid-state oxidation of hydroxylated benzaldehydes has been reported with urea-hydrogen peroxide (UHP) adduct, which appears to be a superior alternative in terms of shorter reaction time, cleaner product formation and easier manipulation[143]. For instance, under these conditions p-hydroxybenzaldehyde has been transformed into hydroquinone in 75 min at 85 °C and with 82% yield[143].

Other reagents have been employed to oxidize aromatic aldehydes to arylformates; these include peroxyacetic acid[144], peroxybenzoic acid[145], m-chloroperoxybenzoic acid (MCPBA)[146] and organoperoxyseleninic acid[147]. Sodium perborate and sodium carbonate[148] have also been shown to be versatile activating reagents of hydrogen peroxide for similar transformations[149].

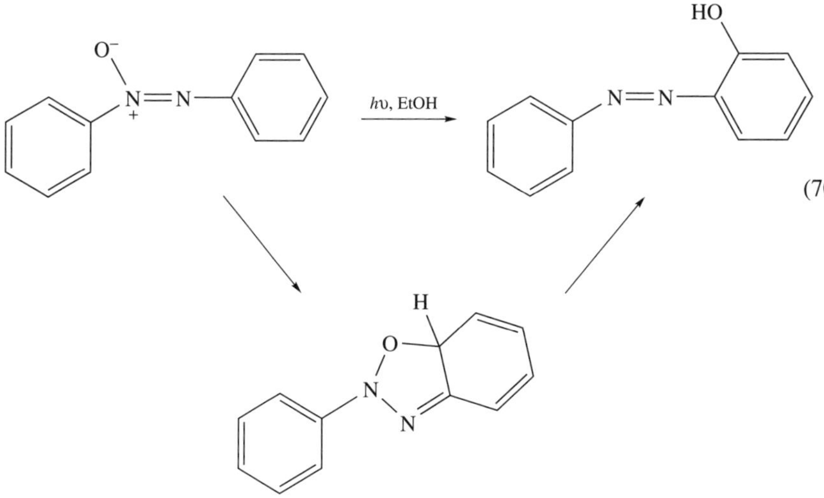

6. Synthesis of phenols 423

(70)

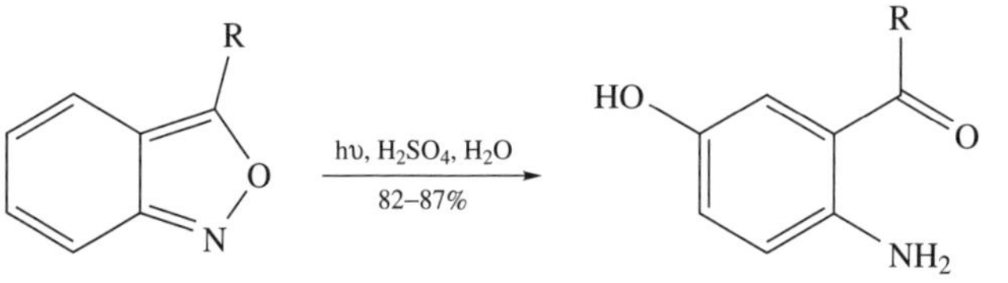

(71)

(61) → (62)

R = H, alkyl, aryl

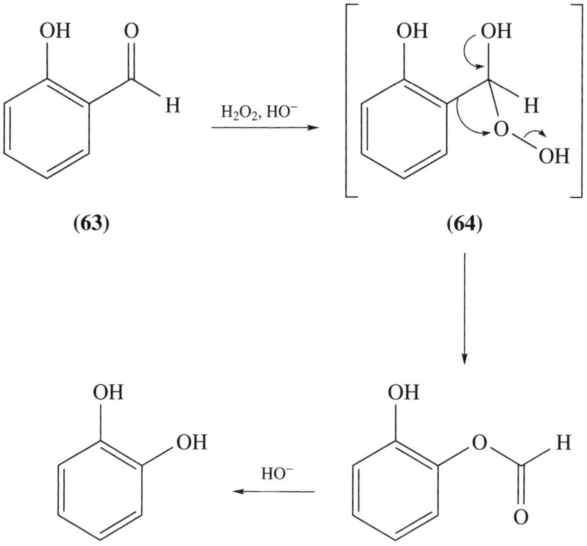

(72)

(63) → (64) → (65) → (66)

The Baeyer–Villiger oxidation of aromatic ketones by peroxyacids is a widely applicable method for the synthesis of phenols[150]. This oxidation can be carried out by organic peroxyacids such as peroxyacetic[151], trifluoroperoxyacetic[152], 4-nitro- and 3,5-dinitroperoxybenzoic acids[153]. However, m-chloroperoxybenzoic acid[154] is most frequently used. Hydrogen peroxide is sometimes used, but it works only in the presence of strong acids[155].

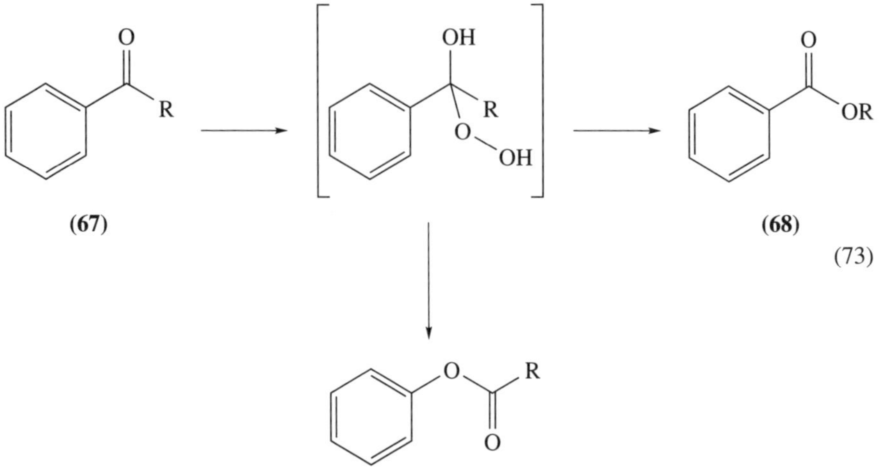

(67) → [intermediate] → (68)

(73)

↓

(69)

Alkyl aryl ketones **67** under treatment with peroxyacids undergo a Baeyer–Villiger reaction by a similar mechanism to the Dakin reaction (equation 73). In this case, migration of the alkyl or the phenyl group would occur to give the corresponding benzoate ester **68** or phenoxyester **69**, respectively. The relative ratio of esters **68** and **69** depends on the type of alkyl group. Usually, the reactivity increases in the order: t-alkyl > s-alkyl > primary alkyl > methyl. Migration of tertiary alkyl groups predominates against phenyl group and consequently almost no formation of phenoxyester is observed. For instance, acetophenone gives 90% of phenyl acetate by treatment with MCPBA. Nevertheless, t-butylacetophenone under the same reaction conditions produces 77% of t-butylbenzoate[156]. Each one of the acetophenones **70** with varied electron-withdrawing or attracting groups in the *meta* or *para* position to the acetyl function yields up to 80% of a single ester **71** (equation 74)[157].

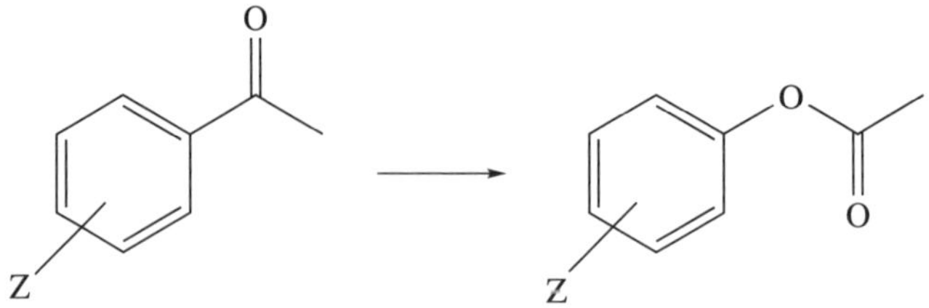

(70) → (71)   (74)

$Z = H, NO_2, CO_2H, CO_2Me, OMe, CF_3, Me$

In the case of asymmetric diaryl ketones, migration of aryl groups with electron-donating substituents occurs preferentially[158]. For instance, *p*-methoxybenzophenone affords 96% of (4-methoxy)phenyl benzoate by oxidation with trifluoroperacetic acid, whereas *p*-nitrobenzophenone under the same reaction conditions gives 95% of phenyl 4-nitrobenzoate.

III. BY CONDENSATION

A. Cyclization

The reaction involving cyclization between an acylium ion derived from an unsaturated carboxylic acid and an ethylenic double bond was first studied by Banerjee and coworkers[159] (equation 75). For example, PPA, P_2O_5 or $POCl_3$ in benzene or anhydrous HF are the reagents for this reaction.

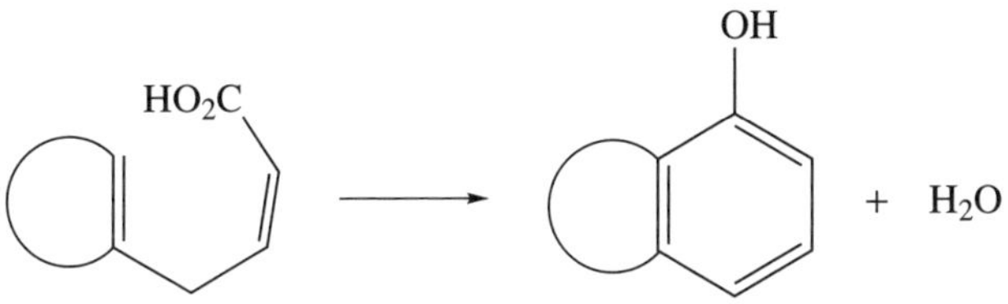

HO2C
OH
+ H2O

(75)

Some examples of this type of cyclization are given in equations 76–81.

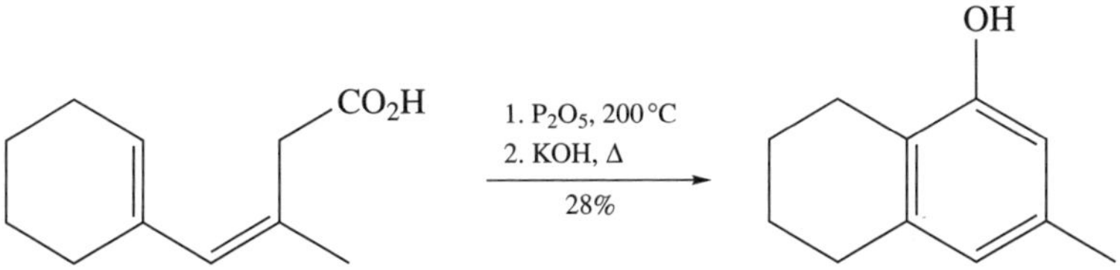

CO2H
1. P2O5, 200 °C
2. KOH, Δ
28%
OH

(76)[160]

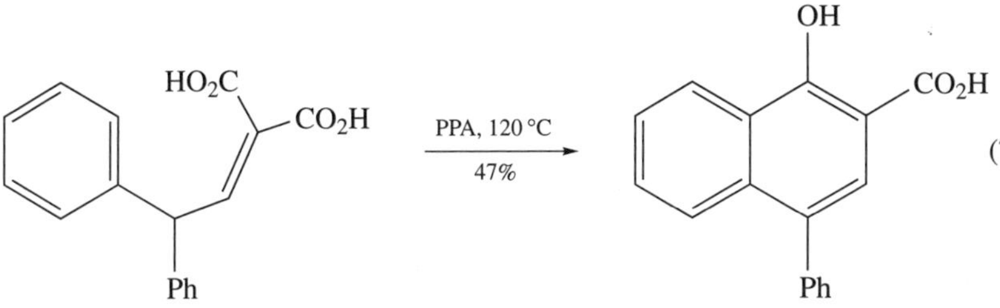

HO2C
CO2H
Ph
PPA, 120 °C
47%
OH
CO2H
Ph

(77)[161]

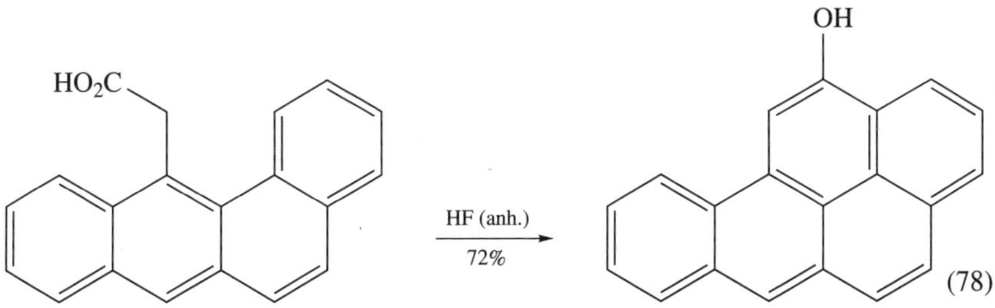

HO2C
HF (anh.)
72%
OH

(78)[162]

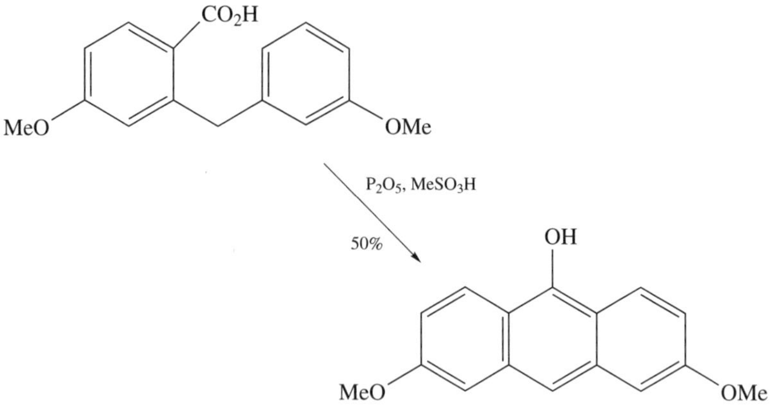

(79)[163]

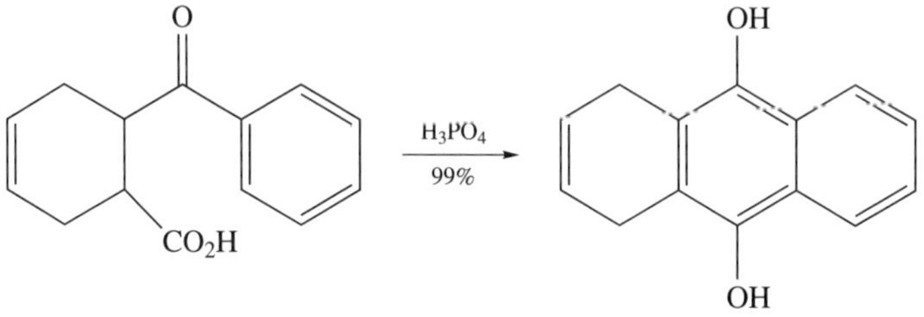

(80)[164]

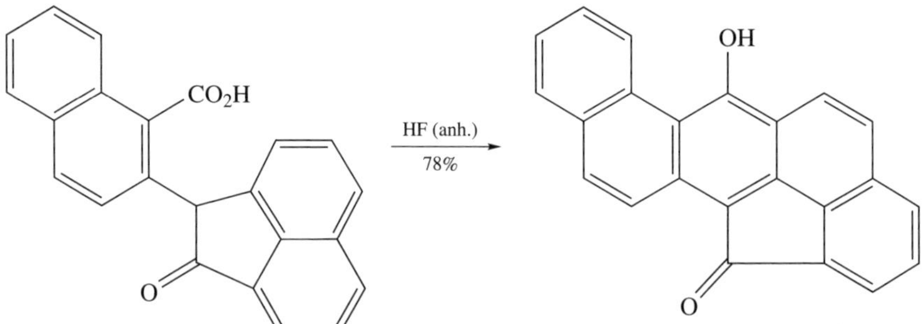

(81)[165]

B. Claisen and Aldolic Condensations

1. Intramolecular reaction

A large proportion of naturally occurring phenolic compounds (polyketides) may be derived by intramolecular condensation of a linear β-polyketo acid derivative **74** (polyacetate hypothesis) (equation 82)[166]. Structural analysis and tracer studies[166b] indicate that the activated forms of acetic, propionic and cinnamic acid act normally as chain-initiating units **72** whereas malonyl coenzyme A (**73**) is presumably the chain-building unit. Cyclization of 3,5,7-triketoacids (**75**) has been suggested to give aromatic compounds in two ways (equation 83). The first route involves an aldol condensation to form β-resorcylic acids (**76**) and the second one corresponds to an internal Claisen condensation to give

6. Synthesis of phenols

acylphloroglucinols (**77**)[166b]. The two models of cyclization can also occur in the same biological system.

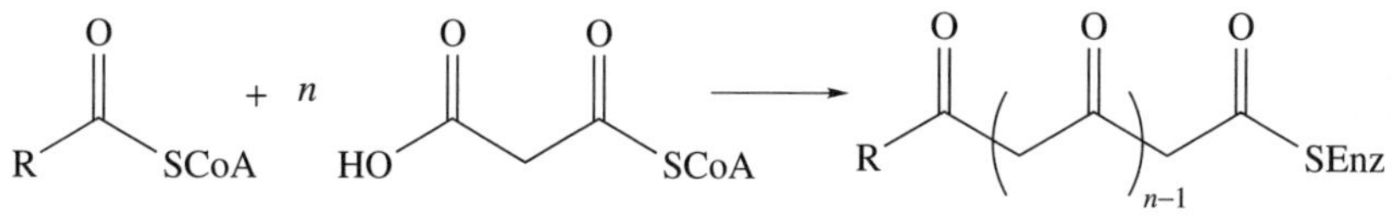

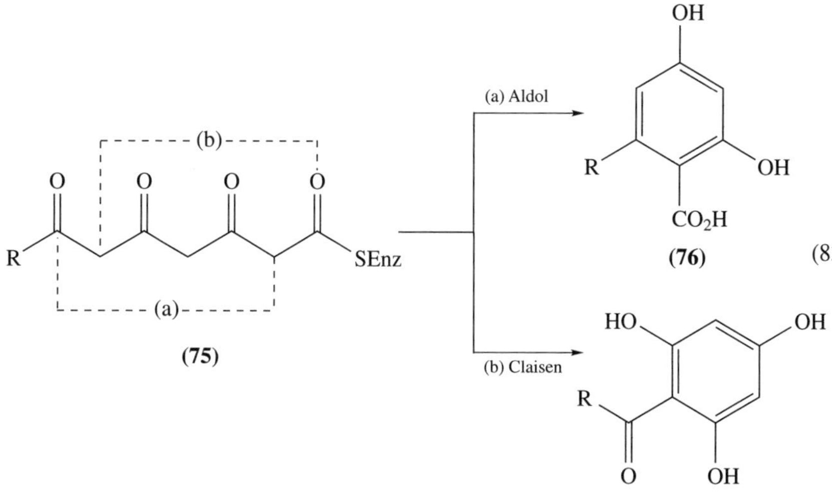

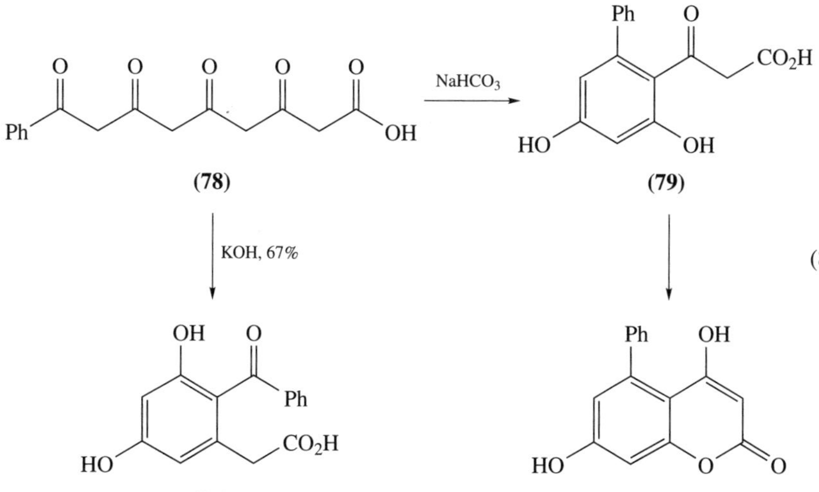

With increasing chain lengths the number of possible cyclization products rises rapidly. A tetraketoacid can undergo three aldol condensations, a Claisen and additional heterocyclic ring closures. Some of the initial cyclization products can undergo further cyclization reactions. For instance, ketoacid **78** under treatment with aqueous $NaHCO_3$ produces mainly the unstable resorcinol **79**, which cyclizes further to give the coumarin **80** (equation 84). With aqueous KOH, the resorcinol **79** became a minor product whereas the isomer **81** is the major product in the reaction[167].

Some examples of these intramolecular cyclizations are shown in equations 85–90.

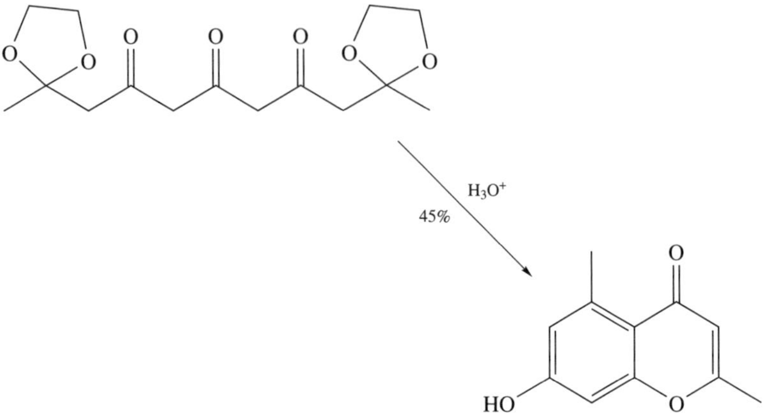

(85)[168]

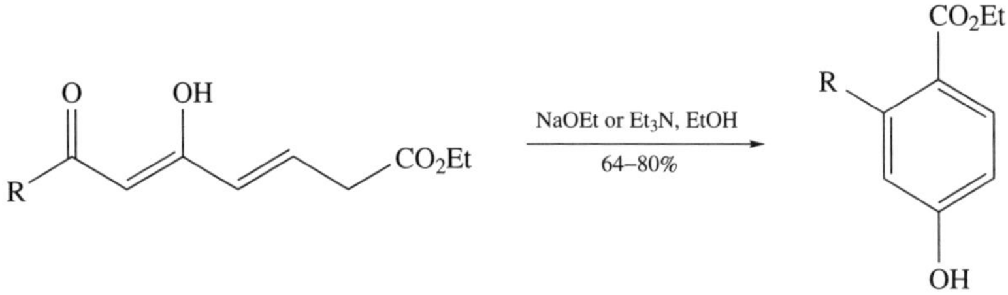

(86)[169]

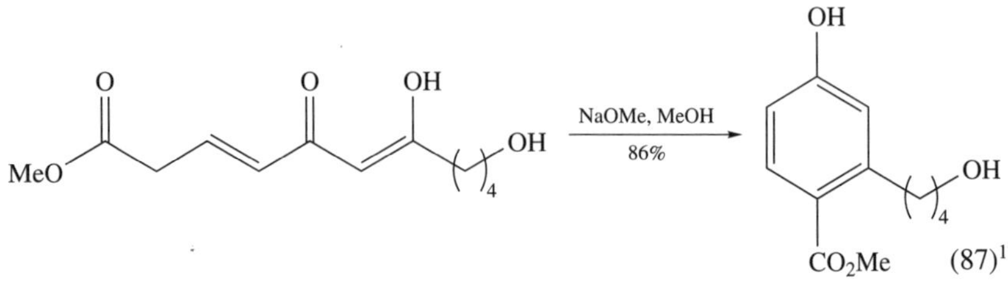

(87)[170]

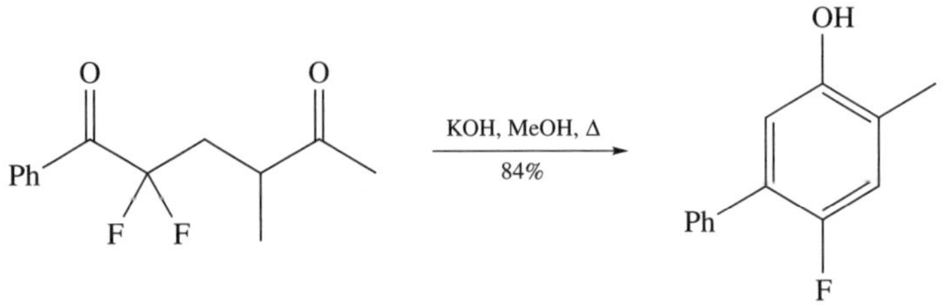

(88)[171]

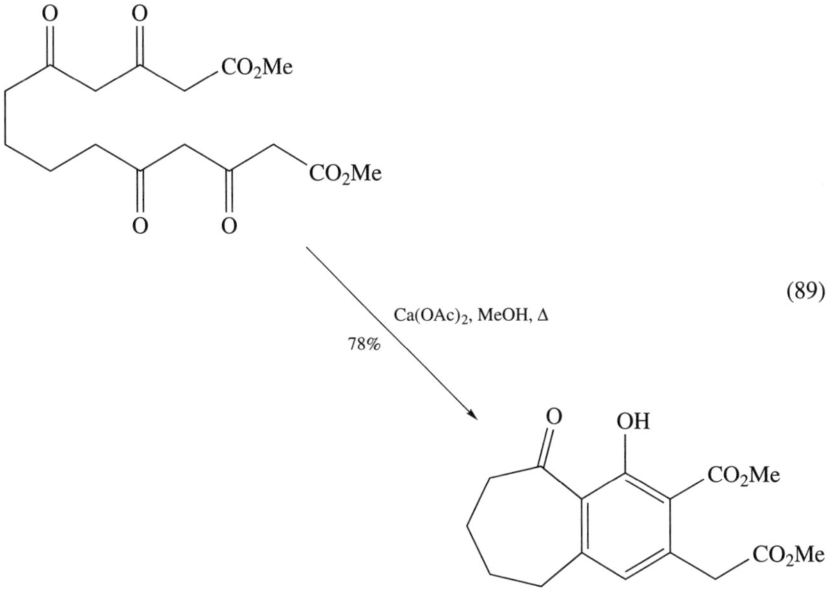

Although tetraketones would be the simplest starting materials for the synthesis of resorcinols, for instance, 2,4,6,8-nonatetraone (**82**) for resorcinol **83** (equation 91), protected forms of tetraketones are normally used to avoid different possible cyclizations. Protected forms of **82** include the 2-acetal **84**[174], the 2,8-bisacetal **85**[175], the 2,8-bisenamine **86**[176], the acetylenic ketone **87**[177], the 2,8-bis(hemithioacetal) **88**[178] and the pyrones **89–91** (Chart 1)[179].

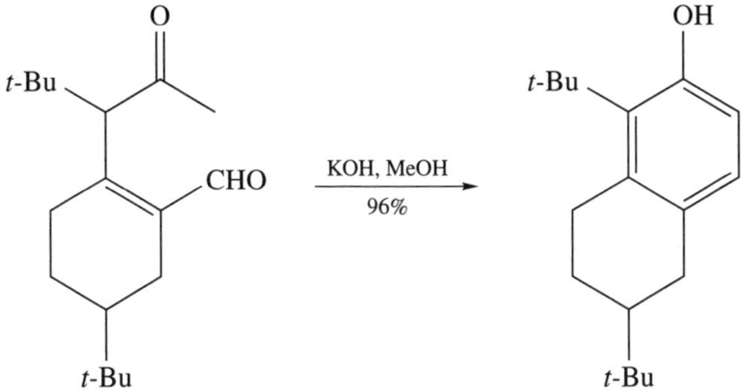

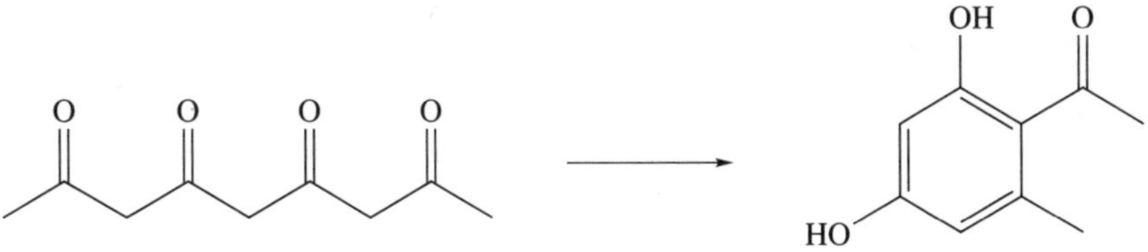

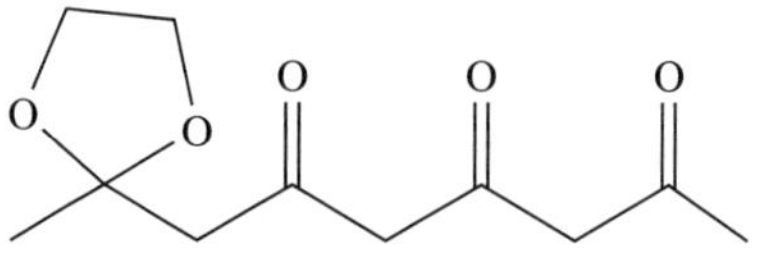

CHART 1

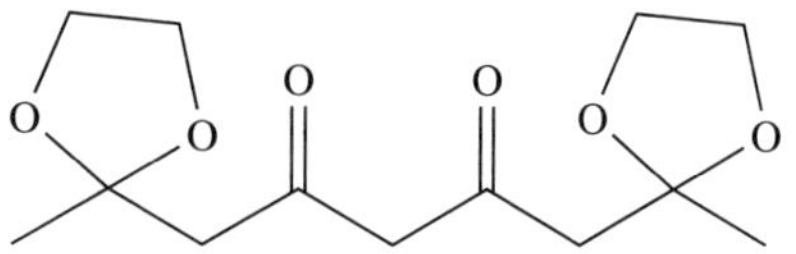

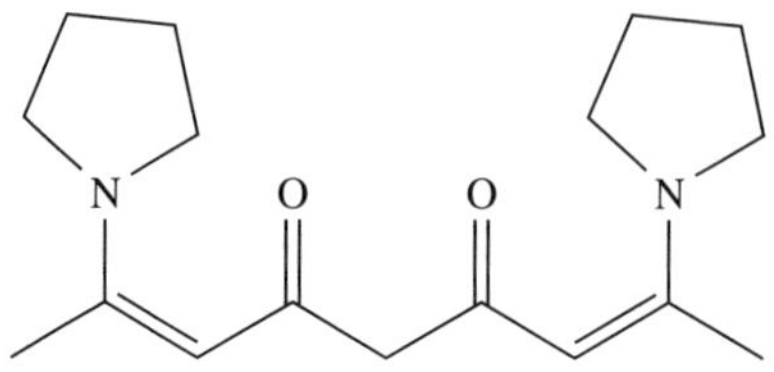

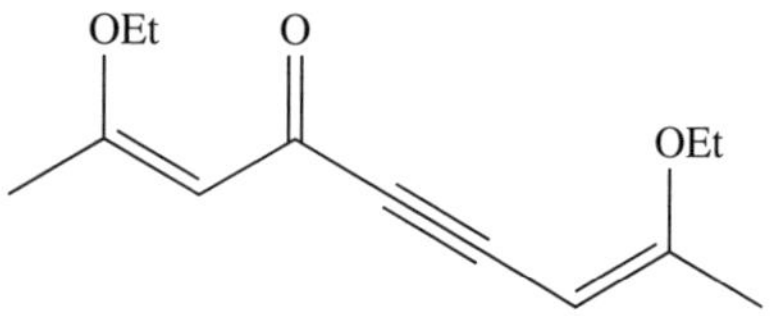

α-Pyrones undergo also intramolecular Claisen condensation. Control over the various phenolic compounds obtained could be achieved by choosing the appropriate reaction conditions. For instance, methanolic potassium hydroxide converted pyrone **93** into resorcylic ester **92** whereas treatment with methanolic magnesium methoxide afforded the phloroglucinol **94** (equation 92)[180]. In the same way, resorcylic acids **97** have been formed from pyrones **95** when potassium hydroxide has been used as the base, whereas phloroglucinol derivatives **98** have been produced when magnesium methoxide was employed (equation 93)[179,181]. These reactions are considered to involve ring opening to the triketo dicarboxylic acids or esters **96**, followed by cyclization.

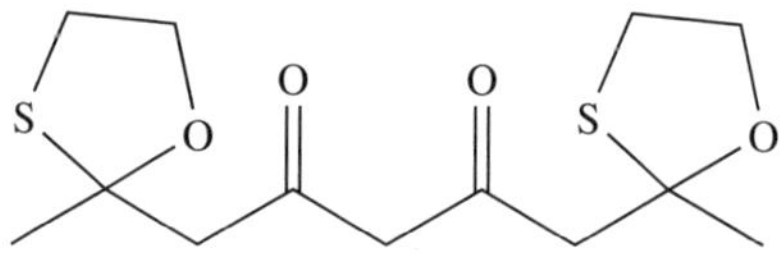

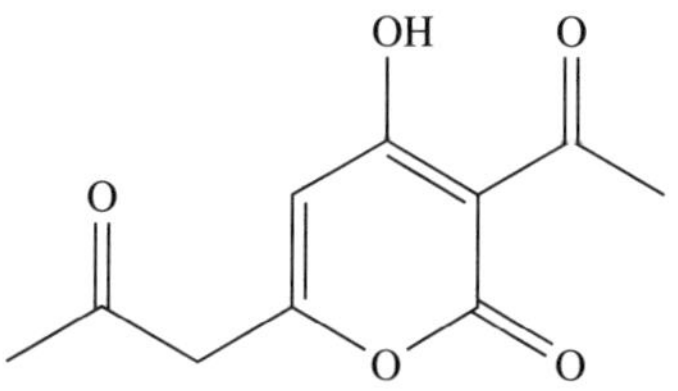

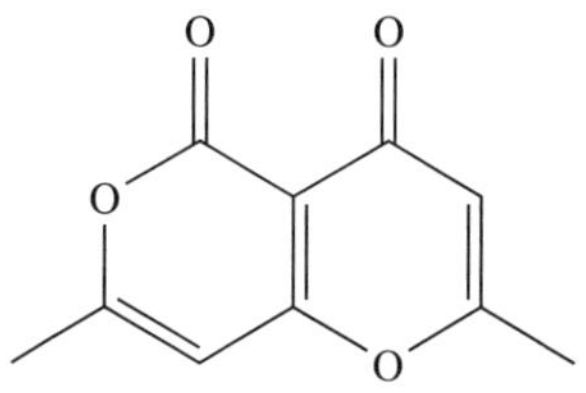

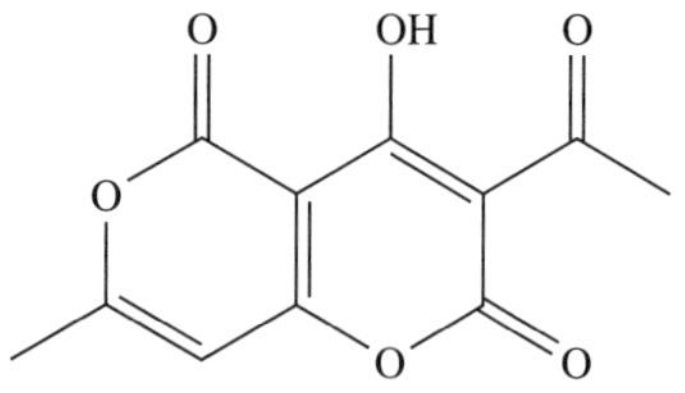

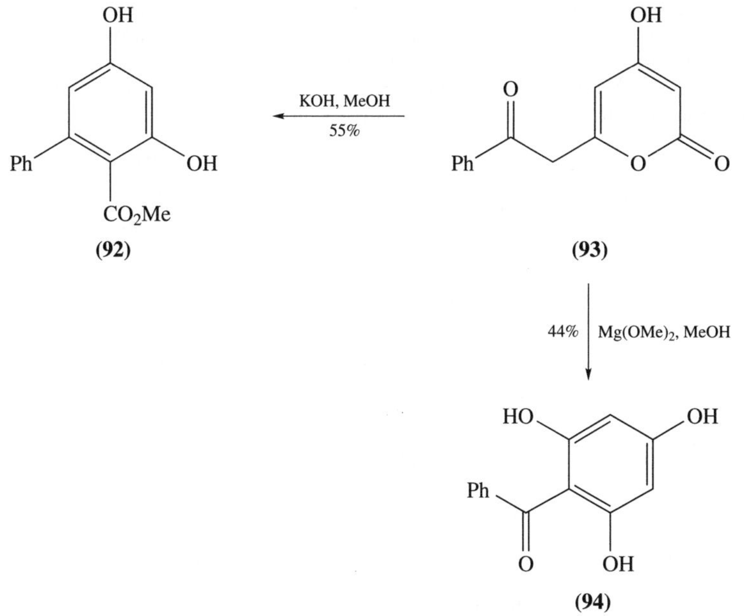

6. Synthesis of phenols

(92) ← KOH, MeOH / 55% — **(93)**

(93) → 44% Mg(OMe)$_2$, MeOH → **(94)** → (92)

(95) → **(96)**

From **(96)**:
- C(2)–C(7), KOH, H$_2$O → **(97)**
- C(1)–C(6), Mg(OMe)$_2$, MeOH → **(98)** (93)

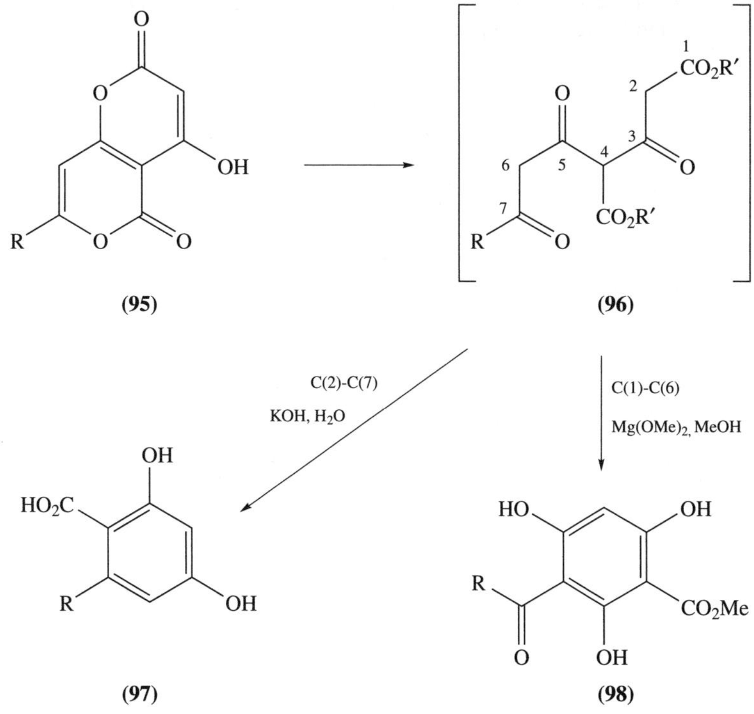

R, R′ = alkyl

Furans with suitable substituents in the 2-position can be transformed in acid conditions into phenols. The reaction proceeds through cyclic acetals of 1,4-dicarbonyl compounds, which then undergo an intramolecular condensation[182]. Equations 94 and 95 show some examples where different catechols have been prepared by refluxing several types of tetrahydrofuran dimethyl acetals with dilute hydrochloric acid.

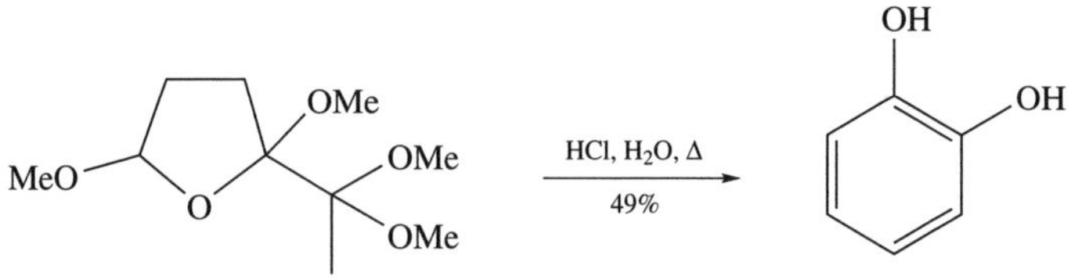

(94)[182c]

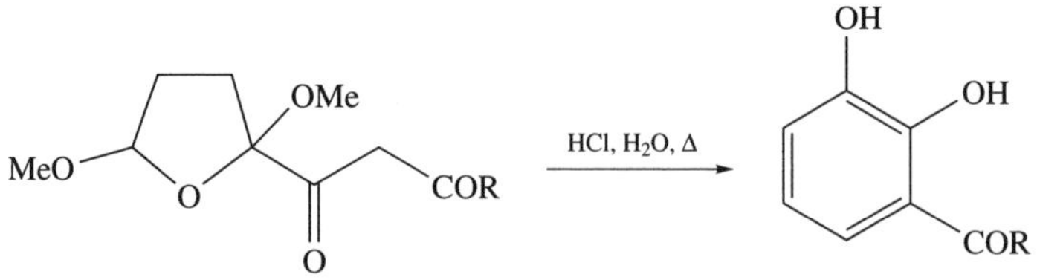

R = alkyl (75–91%), CO_2Me (89%), COMe (88%), COBu-t (81%)   (95)[183]

Under treatment with warm alkali, pyrylium salts with α-alkyl groups **99** undergo hydrolysis and subsequent aldol condensation of the acyclic intermediate **100** to give phenols **101** in moderate yield (equation 96)[184].

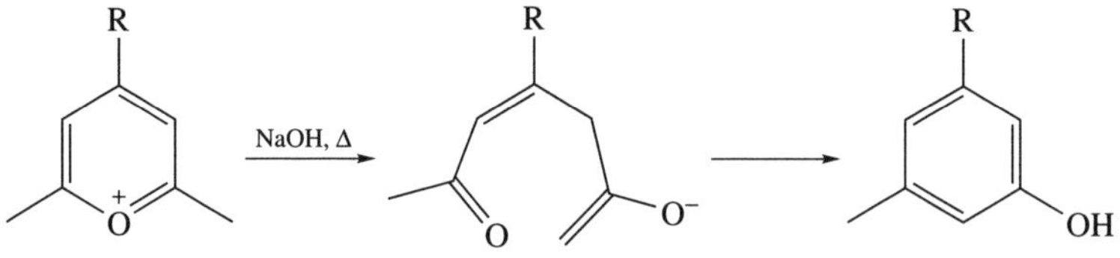

(96)

Some examples of this type of cyclization are given in equations 97–99.

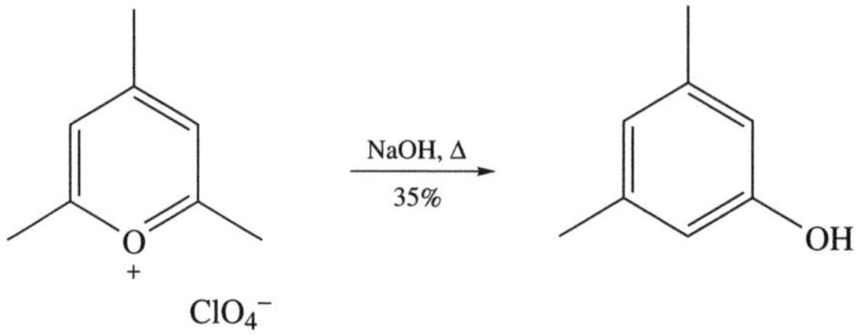

(97)[185]

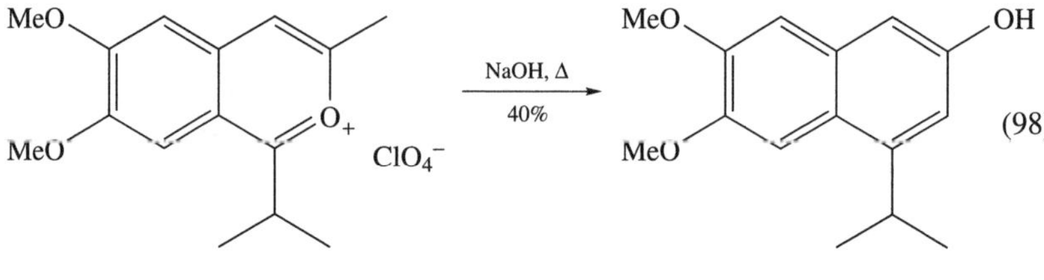

(98)[186]

(99)[187]

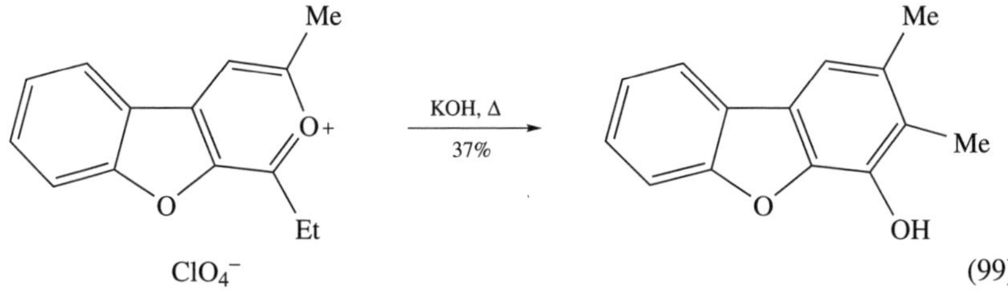

2. Intermolecular reaction

a. Synthesis of monophenols. Intermolecular condensation between ketones and 1,3-dicarbonyl compounds such as 1,3-oxoaldehydes or 1,3-diketones produces monophenols in good yields (equation 100)[188].

(100)

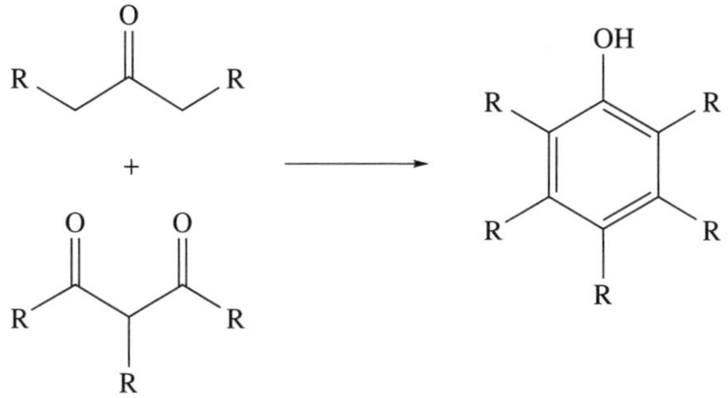

For instance, condensation of a variety of 1,3-dicarbonyl compounds **102** with diethyl *β*-ketoglutarate (**103**) afforded the phenols **104** by treatment with sodium in ethanol (equation 101)[189]. Naphthalene **106** has been obtained by self condensation of 2,4,6-heptatrione (**105**) (equation 102)[190].

(102) **(103)** **(104)**

(101)

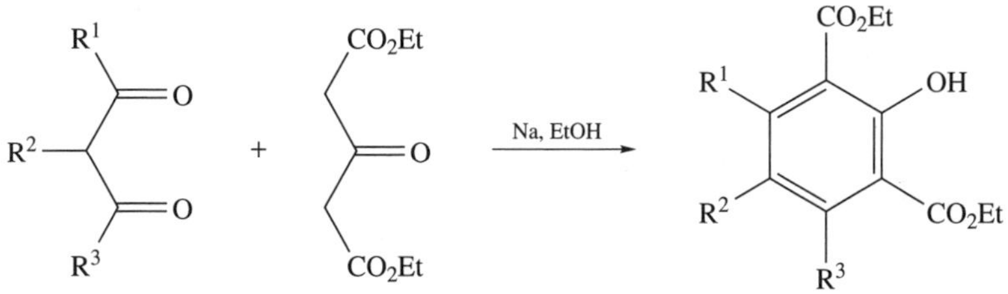

(a) $R^1 = R^3 = H$, $R^2 = COPh$; 77%
(b) $R^1 = R^3 = H$, $R^2 = CO_2Et$; 50%
(c) $R^1 = p\text{-}ClC_6H_4$, $R^2 = R^3 = H$; 53%
(d) $R^1 = H$, $R^2R^3 = (CH_2)_5$; 61%
(e) $R^1 = Me$, $R^2 = H$, $R^3 = Ph$; 47%
(f) $R^1 = R^3 = Me$, $R^2 = H$; 92%

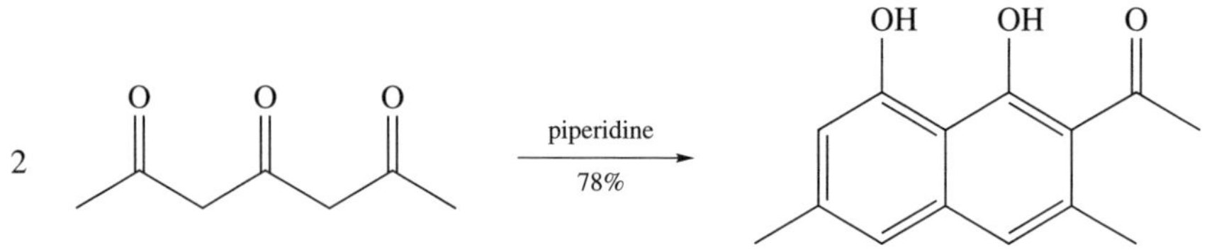

[Scheme showing: 2 × (105) triketone →(piperidine, 78%) (106) dihydroxy-acetyl-dimethylnaphthalene]   (102)

Acylketene dithioacetal **107** and the corresponding β-methylthio-α,β-enone **108** undergo self-condensation and aromatization in the presence of sodium hydride and methyl benzoates in refluxing xylene to give 2,6-bis(methylthio)-4-hydroxyacetophenone (**109**) and 4-hydroxyacetophenone (**110**), respectively, in good yields (equation 103)[191]. The possible pathway for the formation of **109** and **110** could involve base-catalysed condensation of either **107** or **108** with methyl benzoates followed by successive inter- and intramolecular Michael additions and elimination of SMe. No reaction is observed in the absence of methyl benzoates.

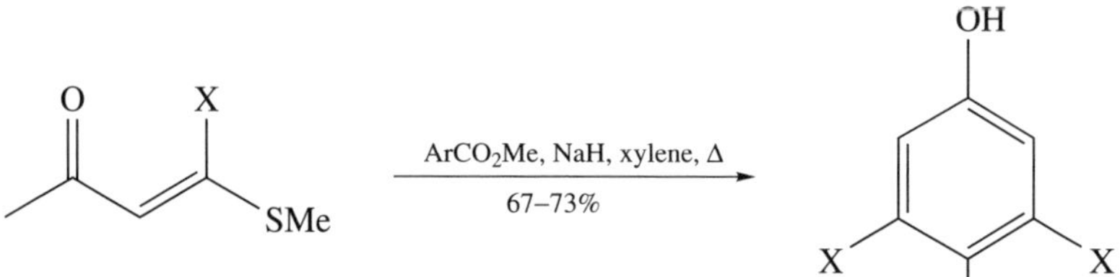

[Scheme: (107) X = SMe / (108) X = H →(ArCO₂Me, NaH, xylene, Δ, 67–73%) (109) X = SMe / (110) X = H]

Ar = Ph, 4-ClC$_6$H$_4$, 4-MeOC$_6$H$_4$, 4-MeC$_6$H$_4$, 3-MeOC$_6$H$_4$ (103)

Tandem Michael addition/aldol condensation of 1-(2-oxopropyl)pyridinium chloride (**112**) or 1-(3-ethoxycarbonyl-2-oxopropyl)pyridinium bromide (**113**) with chalcones **111** forms diketones **114** or **115**, respectively, which under condensation afford cyclohexanones that aromatize by the elimination of pyridinium chloride or bromide, respectively, to give 3,5-disubstituted phenols **116** and 4,6-disubstituted ethyl 2-hydroxybenzoates **117**, respectively (equation 104)[192]. This approach has been extended to solid-phase synthesis in order to prepare a phenol library (equation 105)[193].

An alternative tandem Michael addition/aldol condensation for the synthesis of 3,5-diaryl-substituted phenols **121** employs, instead of 1-(2-oxopropyl)pyridinium chloride (**112**), 1-(benzotriazol-1-yl)propan-2-one (**119**) in the presence of excess of NaOH in refluxing ethanol (equation 106)[194]. Under these conditions, several types of 3,5-diaryl-substituted phenols **121** have been obtained in 52–94% yield. The reaction proceeds by Michael addition of the enolate of **119** to the α,β-unsaturated ketone **118** to afford intermediate **120**, which then undergoes an intramolecular aldol condensation with elimination of benzotriazole.

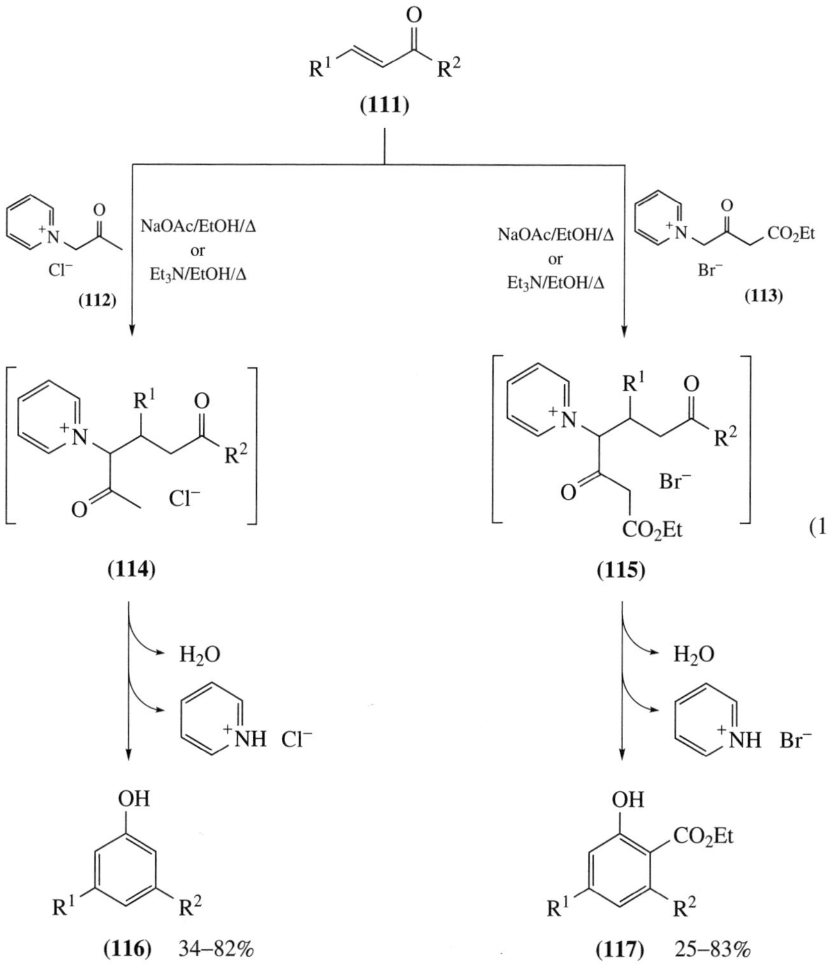

6. Synthesis of phenols

$R^1 = Ph, 4\text{-}ClC_6H_4, 4\text{-}MeOC_6H_4, 4\text{-pyridyl}, 4\text{-}O_2NC_6H_4, 4\text{-}Me_2NC_6H_4$
$R^2 = Me, 2\text{-furyl}, Ph, 4\text{-}MeOC_6H_4, 4\text{-}ClC_6H_4, 4\text{-}O_2NC_6H_4$

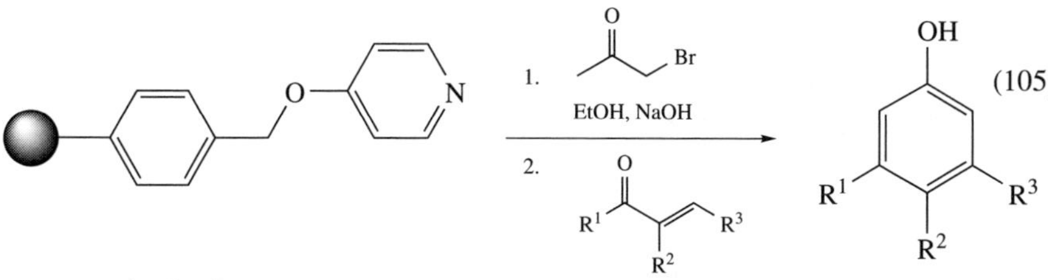

(104)

(105)

$R^1, R^2, R^3 = $ alkyl, aryl   53–85%

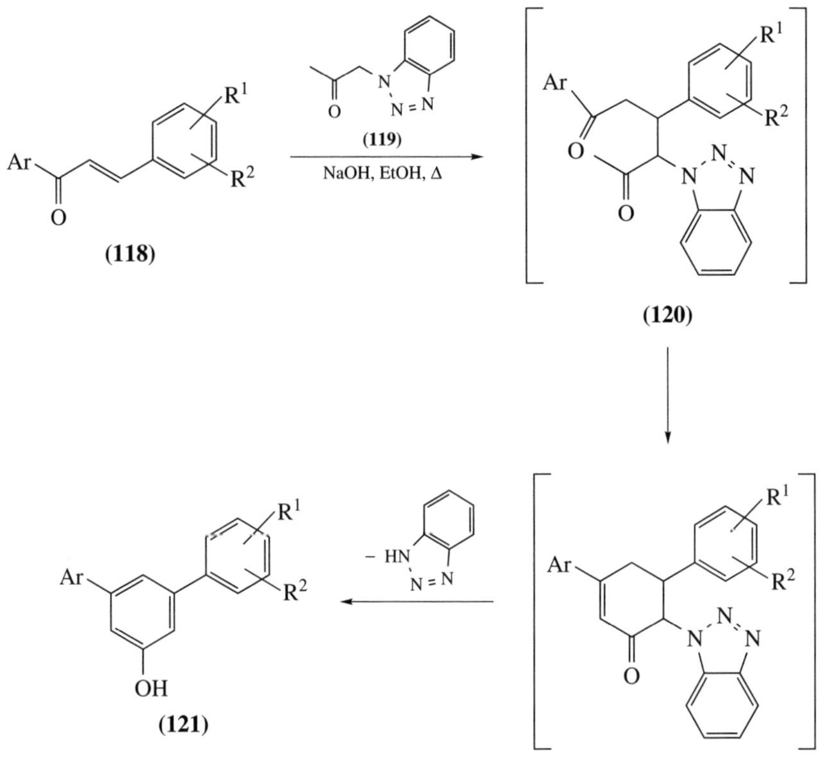

b. *Synthesis of resorcinols.* The conversion of dimethyl acetonedicarboxylate (DMAD) to resorcinols proceeds through the initial formation of a metal chelate compound[195]. The reaction proceeds readily with catalytic amounts of many metals (Na, Co(OAc)$_2$, MgCl$_2$·6H$_2$O, Pb(Ac)$_4$·3H$_2$O, CaCl$_2$, etc.) present either as the preformed metal chelate of DMAD or as a simple organic or inorganic metal compound. The yields of resorcinols varied considerably with the catalyst. For instance, in the presence of sodium metal, diethyl β-ketoglutarate (**103**) underwent a self-condensation to afford resorcinol **122** in 53% yield (equation 107)[196].

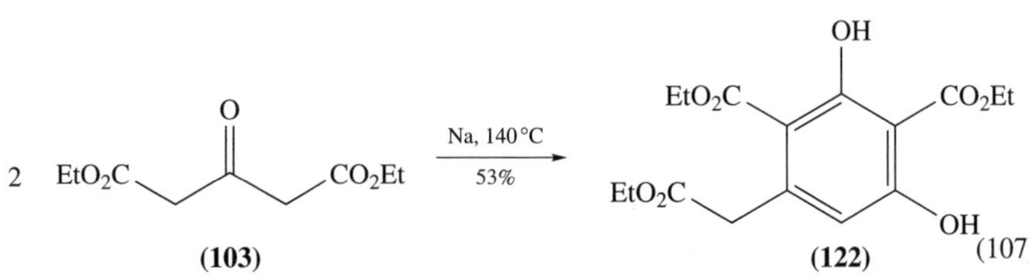

In the case of α,β-unsaturated esters, Michael addition and Claisen condensation are liable to proceed simultaneously. Equation 108 shows one of these cases where ethyl phenylpropiolate (**123**) reacted with dibenzyl ketone (**124**) by a combination of both types of reactions, leading to the formation of 2,4,5-triphenylresorcinol (**125**)[197].

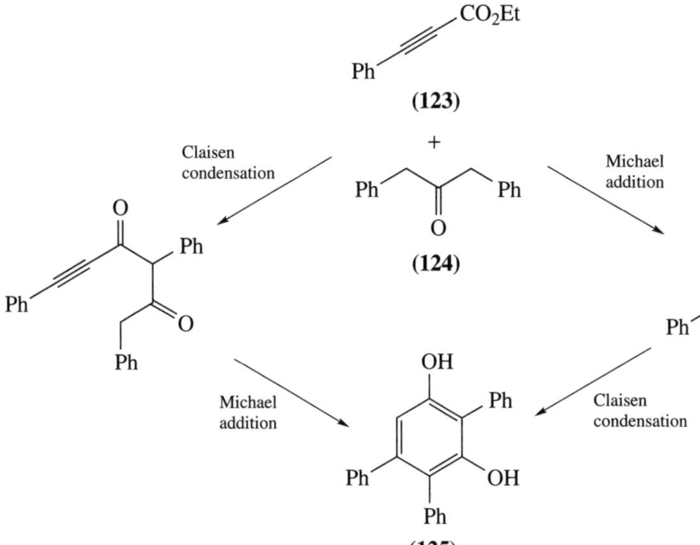

6. Synthesis of phenols

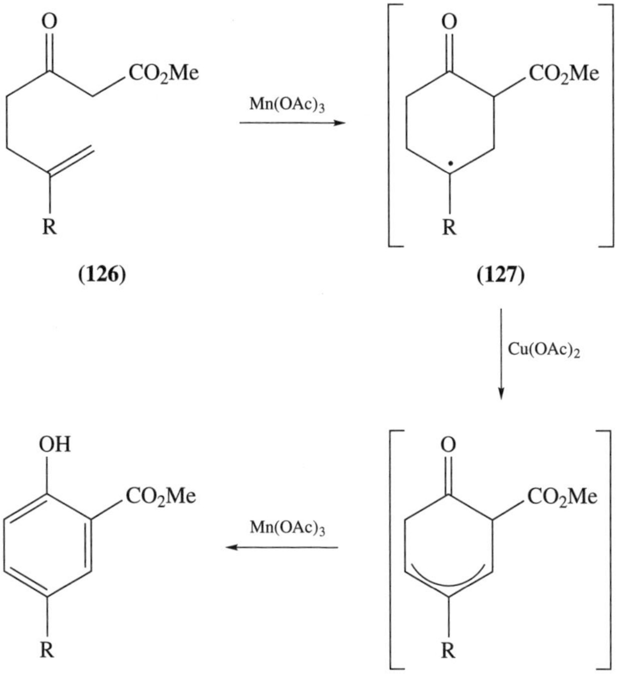

(108)

(109)

R = H, alkyl

C. Radical Cyclizations

Oxidative radical cyclization is an alternative method for the preparation of phenols from ω-unsaturated-β-dicarbonyl compounds. Usually, manganese(III) acetate is used as an efficacious oxidant of enolizable carbonyl compounds. For instance, β-ketoesters **126** with 4 equivalents of Mn(OAc)$_3$ and one equivalent of Cu(OAc)$_2$ afforded salicylate derivatives **129** in good yield (equation 109)[198]. It has been suggested that in the first stage of this reaction, the β-ketoester **126** forms a manganese enolate which then reacts with the double bond to give the cyclic radical **127** as a reactive intermediate. Then the radical **127** reacts with Cu(OAc)$_2$ to give a mixture of double-bond isomers **128**, which are then oxidized to salicylate **129** by a second equivalent of Mn(OAc)$_3$. Table 4 shows some salicylates and o-acetylphenol synthesized using this type of radical cyclization.

TABLE 4. Salicylates and o-acetylphenol obtained by radical cyclizations

Entry	Starting material	Reaction product	Yield (%)	Reference
1	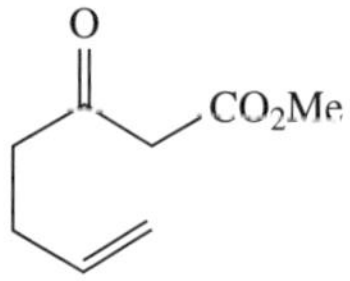 (O=C–CH$_2$–CO$_2$Me, with pent-4-enyl chain)	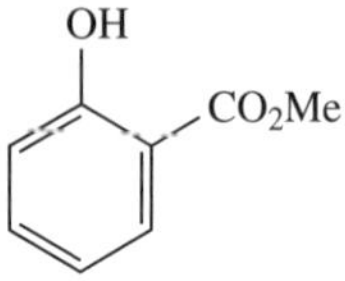 (2-hydroxyphenyl)–CO$_2$Me	78	199
2	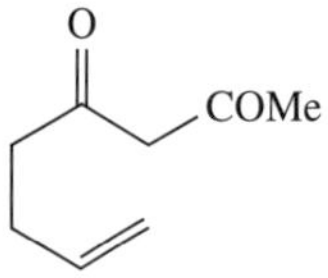 (O=C–CH$_2$–COMe, with pent-4-enyl chain)	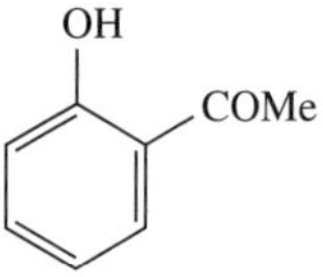 (2-hydroxyphenyl)–COMe	96	199
3	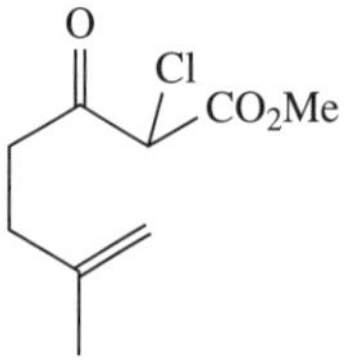 (O=C(Cl)–CH$_2$–CO$_2$Me, with methylallyl chain)	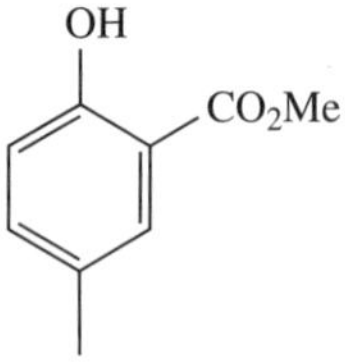 (2-hydroxy-4-methylphenyl)–CO$_2$Me	70	198
4	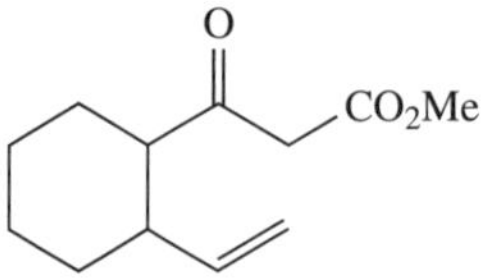 (cyclohexyl-vinyl β-ketoester)	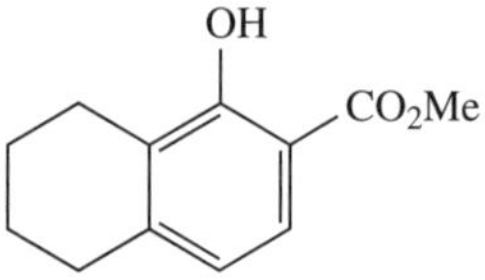 (5,6,7,8-tetrahydronaphthalen-2-ol)–CO$_2$Me	46	198
5	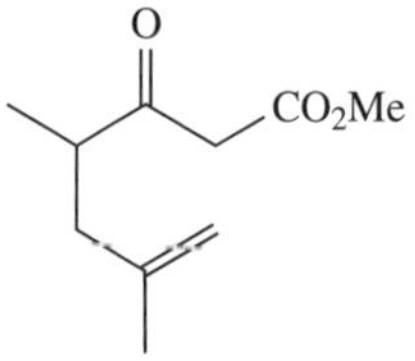 (branched β-ketoester with allyl group)	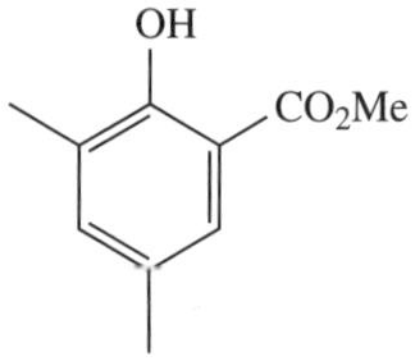 (substituted 2-hydroxyphenyl)–CO$_2$Me	91	198

6. Synthesis of phenols 439

IV. BY CYCLOADDITION

A. Cycloaromatization

The classical methods of constructing six-membered rings are the Diels–Alder reaction or the Robinson annulation, which consist of the union of two fragments, one with two carbon atoms and the other with four carbons. A conceptually different method from the above involves the condensation of two three-carbon units, one with two nucleophilic sites and the other containing two electrophilic sites. Furthermore, the regiochemistry of the reaction is controlled by the differential reactivities of these sites.

The 1,3-bis(trimethylsililoxy)butadienes **130–132**, as the equivalent of methyl acetoacetate dianion, constitute the three-carbon fragments with two nucleophilic sites (equation 110). Condensation of **130-132** with various equivalents of β-dicarbonyl compounds and titanium(IV) chloride gives substituted methyl salicylates. The differential reactivity of the electrophiles which increases in the order: conjugated position of enone > ketone > monothioacetal, acetal and of **130–132** (4-position > 2-position) ensures complete regioselectivity in this combination of two three-carbon units to form phenols such as **133** and **134**[200,201].

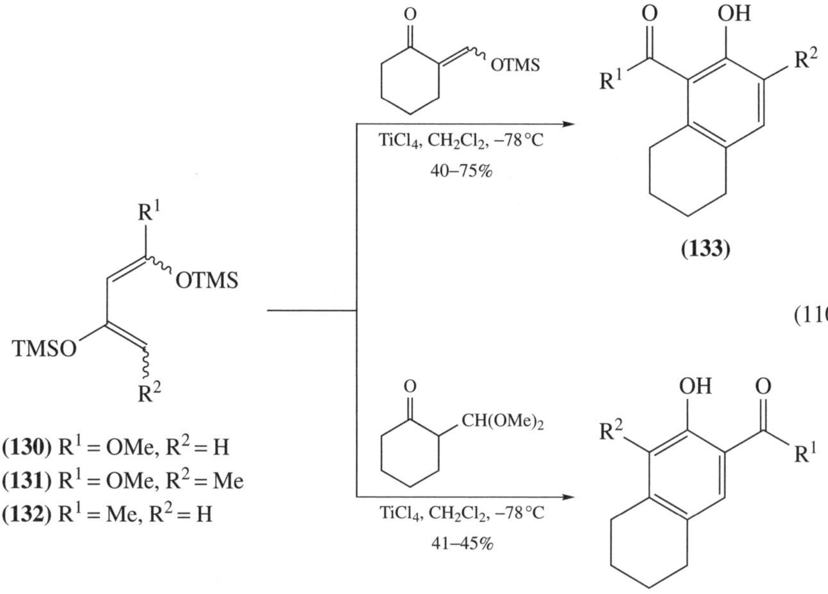

(130) $R^1 = OMe, R^2 = H$
(131) $R^1 = OMe, R^2 = Me$
(132) $R^1 = Me, R^2 = H$

(110)

(133)

(134)

The diene **130** undergoes an interesting reaction with the orthoesters **135** or the anhydrides **136** and titanium(IV) chloride: the 4-position is first acylated to give an intermediate **137** or **138**, which condenses with another molecule of **130** to produce 3-hydroxyhomophthalates **139** (equation 111)[202].

A synthesis of (−)-Δ^1-tetrahydrocannabinol has been achieved using the cycloaromatization reaction of the 1,3-bis(trimethylsililoxy)butadiene (**130**) with the β-dicarbonyl equivalent **140** to generate methyl olivetolate **141** with complete regioselectivity (equation 112)[203].

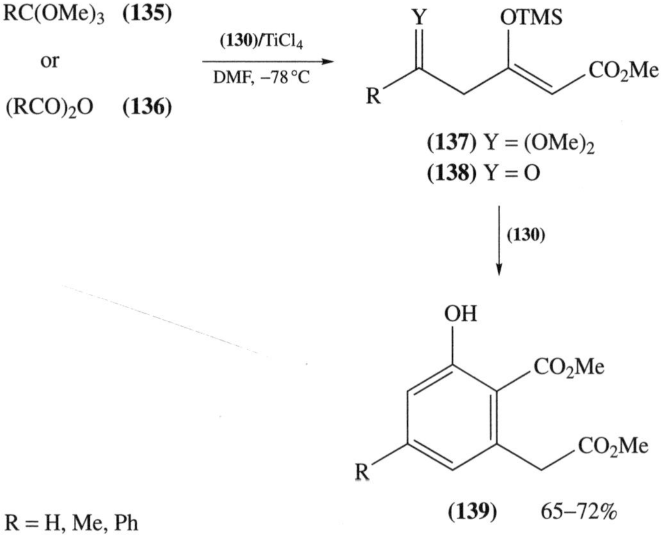
RC(OMe)3 (135)
or
(RCO)2O (136)
(130)/TiCl4
DMF, −78 °C
Y
OTMS
CO2Me
R
(137) Y = (OMe)2
(138) Y = O
(130)
OH
CO2Me
CO2Me
R
(139) 65–72%
R = H, Me, Ph

(111)

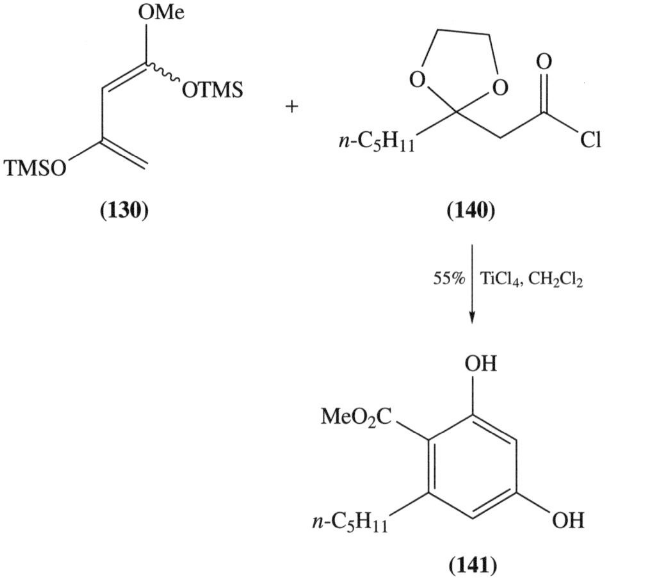
OMe
OTMS
TMSO
(130)
+
O
O
O
n-C5H11
Cl
(140)
55%
TiCl4, CH2Cl2
OH
MeO2C
n-C5H11
OH
(141)

(112)

B. Diels–Alder Reaction

Diels–Alder reactions have been used for the regioselective synthesis of phenols which are difficult to make by direct substitution. Aromatization of the initial Diels–Alder adducts can be effected by straightforward dehydrogenation, by elimination of suitably

6. Synthesis of phenols

(142) $R^1 = H$; $R^2 = H$
(143) $R^1 = Me$; $R^2 = H$
(144) $R^1 = H$; $R^2 = Me$
(145) $R^1 = R^2 = Me$

(146) $R^3 = OMe$
(147) $R^3 = OTMS$

CHART 2

placed substituents or by a retro-Diels–Alder step with loss of a small molecule such as carbon dioxide or nitrogen.

Trimethylsilyloxy dienes **142–147**[204] (Chart 2) have been used in the ring synthesis of substituted phenols[205]. For instance, phenol **149**, the aromatic unit of milbemycin $\beta 3$, has been obtained by reaction of the diene **143** with the alkyne **148** (equation 113). Ring aromatization with concomitant oxidation of the side chain was effected by treatment of the Diels–Alder adduct with Jones' reagent[206].

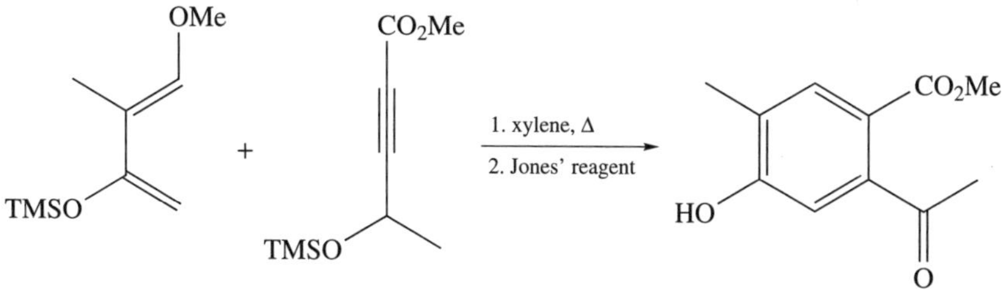

(**143**) (**148**) (**149**)   (113)

1,1-Dimethoxy-3-trimethylsilyloxybutadiene (**146**)[205,207] reacts even more rapidly than Danishefsky's diene (**147**), and with equally high regioselectivity[205,207]. For instance, dimethyl acetal **146** reacts with methyl propiolate to afford β-resorcylic ester **150** (equation 114)[208]. α-Resorcylic ester **152** has been obtained in a variation using methyl trans-β-nitroacrylate as dienophile. Here the orientation of the cycloaddition is controlled by the nitro group and elimination of nitrous acid from the adduct **151** leads exclusively to **152**[209]. β-Resorcylic ester **154**, a key intermediate in a synthesis of the plant growth inhibitor lasiodiplodin, has been obtained in 35% yield by reaction of butadiene **146** with the acetylene derivative **153** (equation 115)[210].

The cycloaddition reactions of allenes with trimethylsilyloxybutadienes produce phenols in good yield by regioselective cyclization and subsequent aromatization by acid-catalysed enolization (equation 116)[211], or by fluoride-induced cleavage of the trimethylsilyl groups and elimination of ethanol (equation 117)[212].

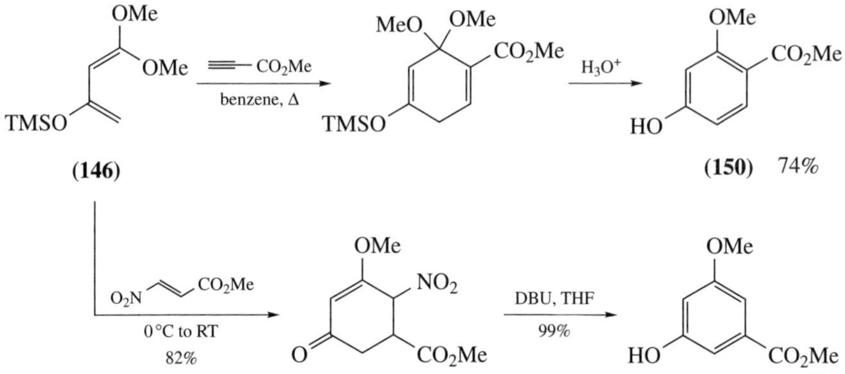

(114)

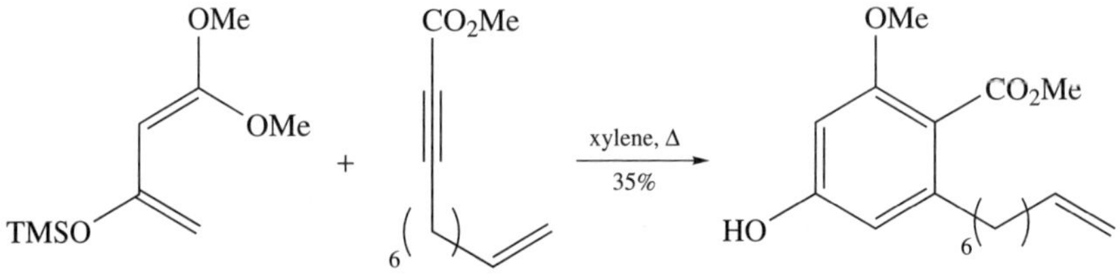

(115)

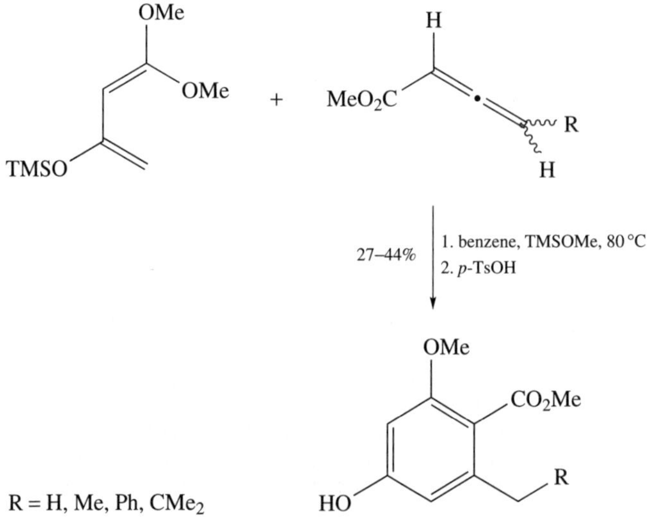

(116)

R = H, Me, Ph, CMe$_2$

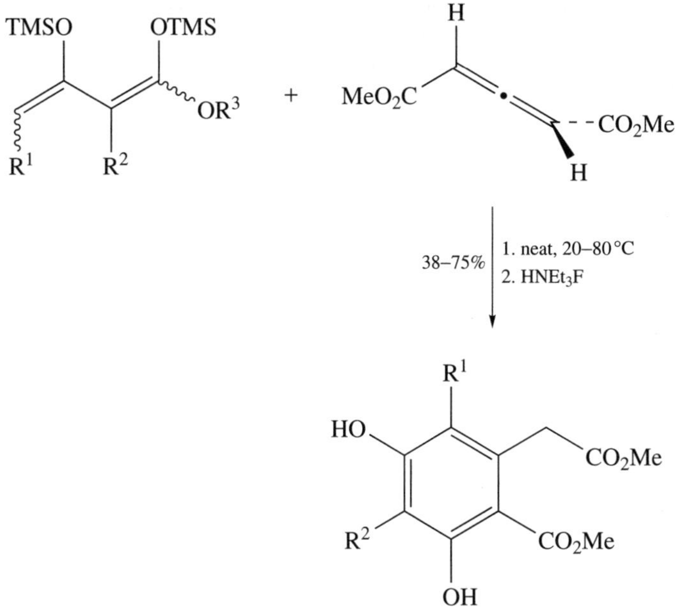

6. Synthesis of phenols

$$R^1 = H, Me, Et, Bu, All; R^2 = H, Me; R^3 = Me, Et$$

(117)

In contrast to diene **146**, the analogous compounds **147**[201a,b] or the pyrone **156**[213] are poor Diels–Alder dienes affording phenol **155** only in moderate yield (equation 118).

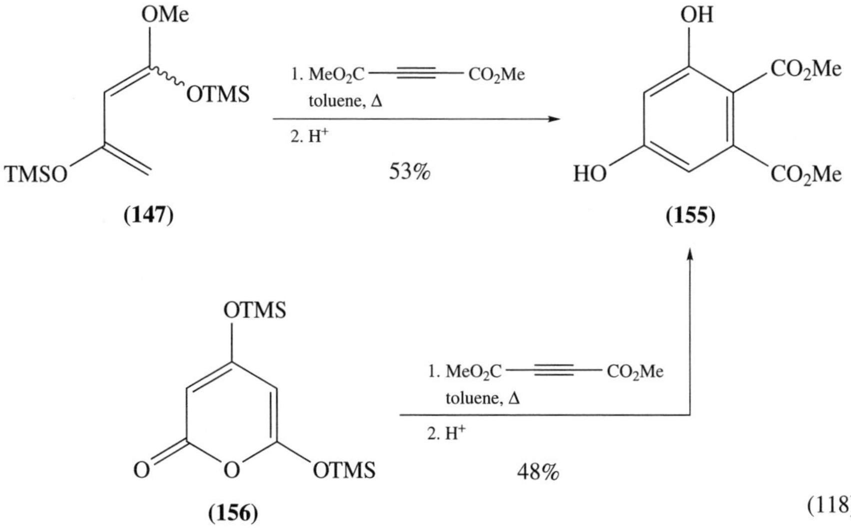

(118)

2,2-Dialkyl-2,3-dihydro-4H-pyran-4-ones **157** have also been shown to be good precursors for the *in situ* preparation of electron-rich dienes, affording highly substituted phenols **159** by reaction with electron-poor acetylenes **158** (equation 119)[214]. This reaction proceeds with a high degree of regioselectivity and under very mild conditions.

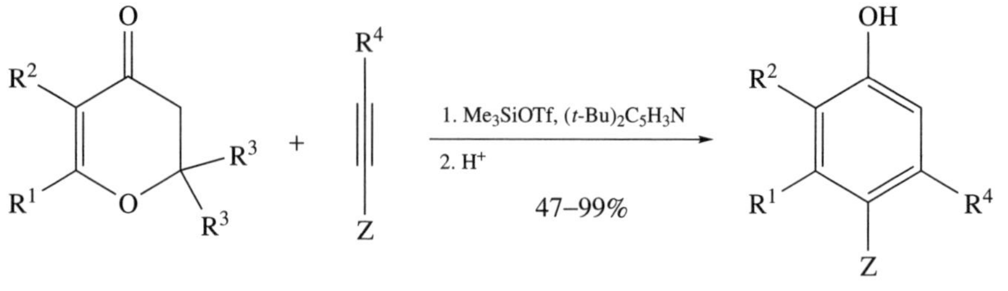

R^1 = Ph, t-Bu; R^2 = H, CO_2Et, Br; R^3 = Me
R^4 = H, Me, CO_2Me, CHO; Z = COPh, CO_2Me, CO_2Et (119)

Highly oxygenated butadienes have proven very useful for synthesizing anthraquinone natural products (e.g. aloesaponarins) and anthracyclinone antibiotics. Anthraquinones have been obtained by cycloaddition of 1,1-dioxygenated butadienes to appropriate chlorobenzoquinones and chloronaphthoquinones. The chloro substituents in the quinone dienophiles facilitate the reaction and control the regiochemistry of the addition. The best results have been obtained with vinyl ketene acetals, such as **161** which readily undergo cycloaddition reactions with quinones at room temperature (equation 120). Aromatization of the initial adduct is effected by pyrolysis, with evolution of hydrogen chloride, or better, by percolation through silica gel. These reactions could apparently give different products, depending on which of the acetal oxygen functions is eliminated during aromatization. In practice, the methoxy substituent is found to be eliminated preferentially, giving a phenol as the main product. The chrysophanol (**162**) has been obtained in one step from 3-chlorojuglone (**160**) and the acetal **161** (equation 120)[215] and the isomeric 2-chlorojuglone (**163**) gave ziganein (**164**), illustrating the well-established regioselectivity of these reactions (equation 121). Many naturally occurring naphthoquinones and anthraquinones have been synthesized by this convenient procedure[215c,216].

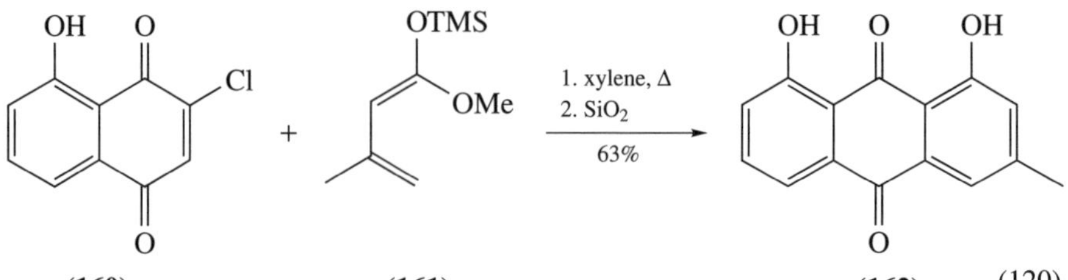

In the same way, exocyclic dienes **165**[217], **166**[218], **167**[219] and **168**[216e] (Chart 3) react readily with naphthoquinones to give products with the anthracyclinone skeleton.

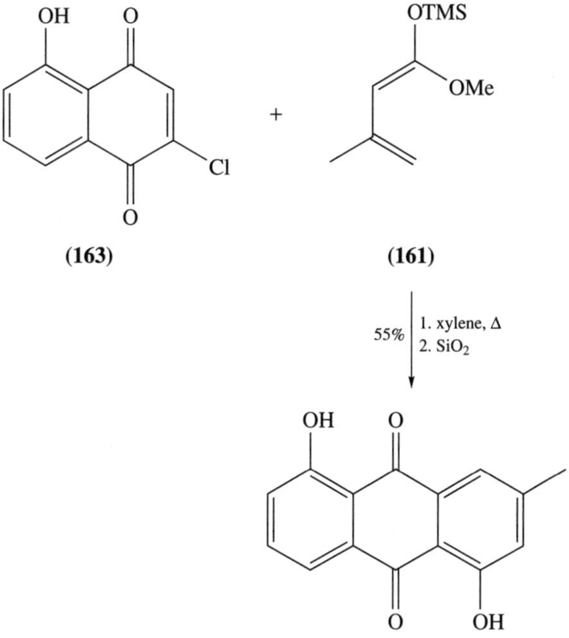

6. Synthesis of phenols

(163) + (161) →[1. xylene, Δ; 2. SiO₂][55%] (164)   (121)

(163): 5-hydroxy-3-chloro-1,4-naphthoquinone
(161): 1-OTMS-1-OMe-3-methyl-1,3-butadiene
(164): 1,8-dihydroxy-3-methyl-9,10-anthraquinone

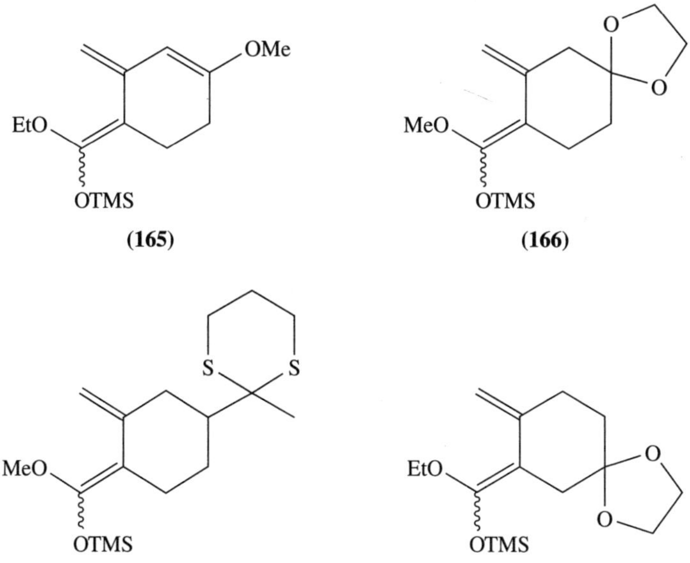

(165), **(166)**, **(167)**, **(168)**

CHART 3

(122)

6. Synthesis of phenols

Anthraquinones have also been obtained by reaction of *o*-quinodimethanes with substituted benzoquinones. Again, halogen substituents in the quinones control the regiochemistry of the cycloaddition. For instance, the unsymmetrical *o*-quinodimethane **169** and 2-bromo-6-methylbenzoquinone (**170**) gave the adducts **171** and **172** in a 92:8 ratio, respectively (equation 122), whereas the 3-bromo-6-methylquinone afforded the same products but in a 2:98 ratio. These cycloadducts have been converted in several steps into the anthraquinones islandicin (**173**) and digitopurpone (**174**)[220].

Homophthalic anhydrides undergo a strong-base induced [4 + 2] intra-[221] or intermolecular[222] cycloaddition reaction with dienophiles to afford various types of polycyclic *peri*-hydroxy aromatics in a single step (equation 123). This elegant strategy has been employed in the synthesis of many biologically important compounds such as fredericamycin A[222b], galtamycinone[223] and dynemycin A[224].

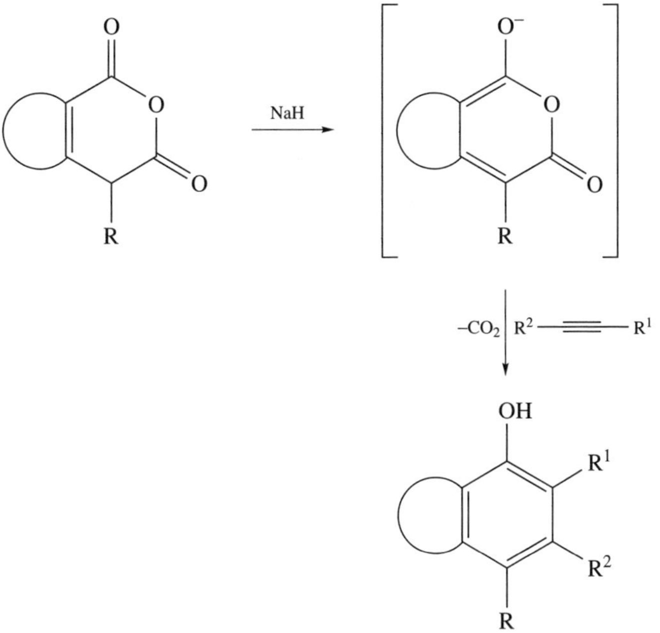

(123)

Equations 124–127 show some examples of this type of reaction.

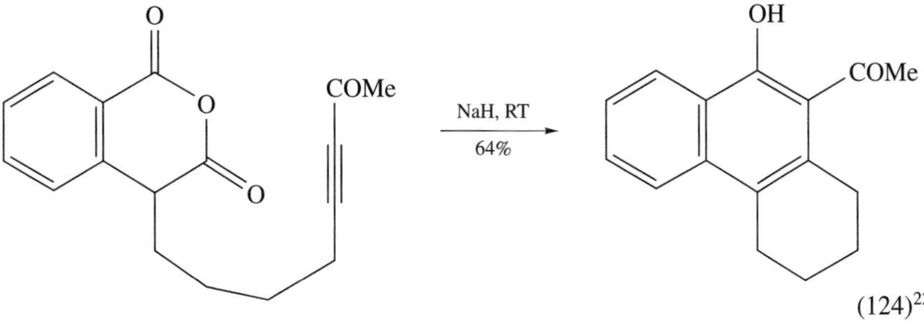

(124)[221a]

(125)[222b,225,226]

(126)[226,227]

(127)[228]

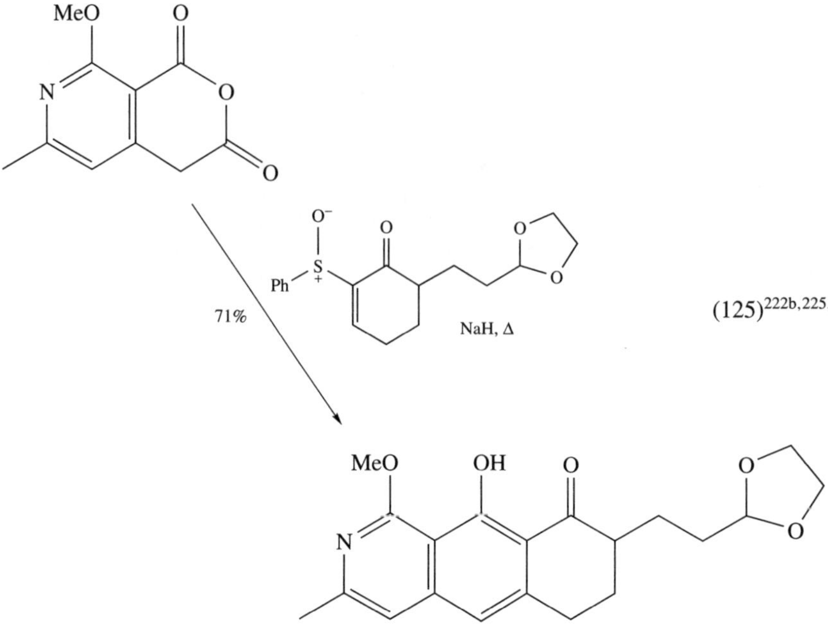

Furans and their substituted derivatives undergo Diels–Alder reactions and the resultant 7-oxabicyclo[2.2.1]heptanes can be further transformed to substituted phenols by the cleavage of the oxygen bridge, which is a crucial step in the transformation. Lewis acid catalysts[229], Brønsted acids[230], metals[231] or high pressure[232] catalyse the cycloaddition. The incorporation of an electron-donating group onto the 2-position of furans enhances the reactivity of the heteroaromatic ring system[233]. The major drawbacks of these protocols include lower regiochemical predictability and the intolerance of many functional groups in the ring-opening process.

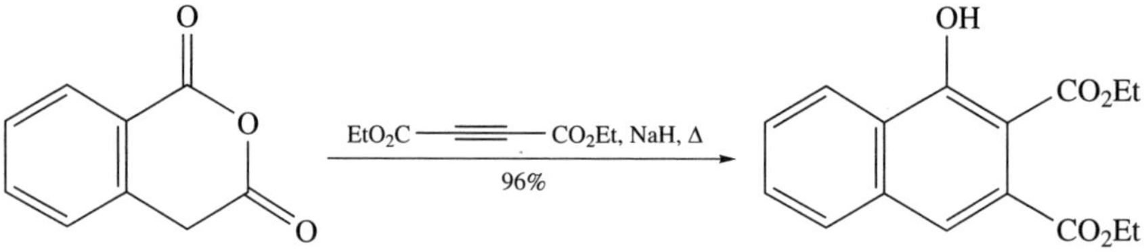

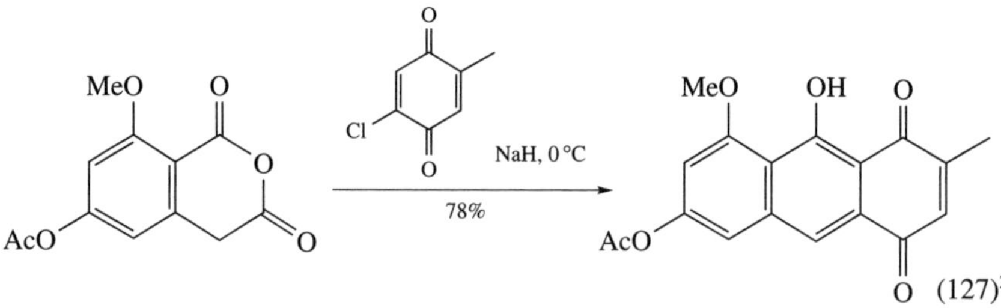

6. Synthesis of phenols

For instance, 2,5-bis(trimethylsililoxy)furans **175**, which are synthetic equivalents of the diketene **177**, are reactive Diels–Alder dienes undergoing cycloaddition reaction with dienophiles to give, after hydrolytic workup, *p*-hydroquinones **176** in high yield (equation 128)[234].

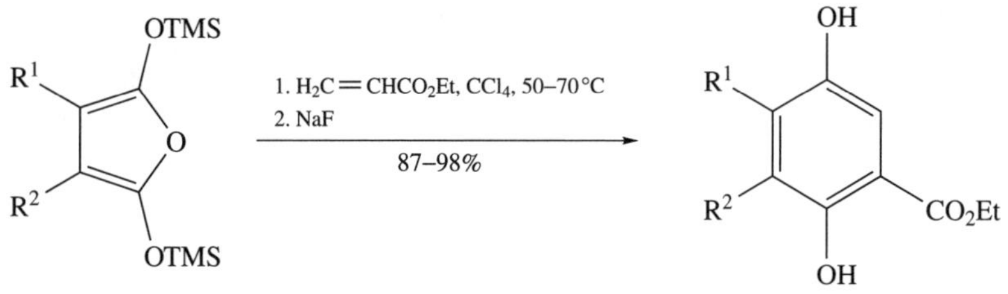

(a) $R^1 = R^2 = H$
(b) $R^1 = Me, R^2 = H$
(c) $R^1 = Ph, R^2 = H$
(d) $R^1R^2 = (CH_2)_4$
(e) $R^1R^2 = CH_2CH=CHCH_2$

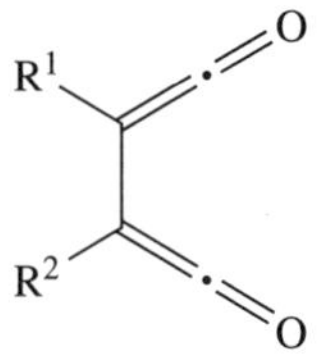

(128)

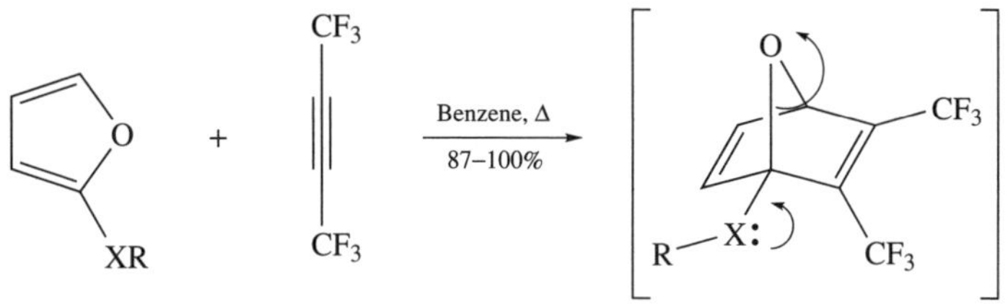

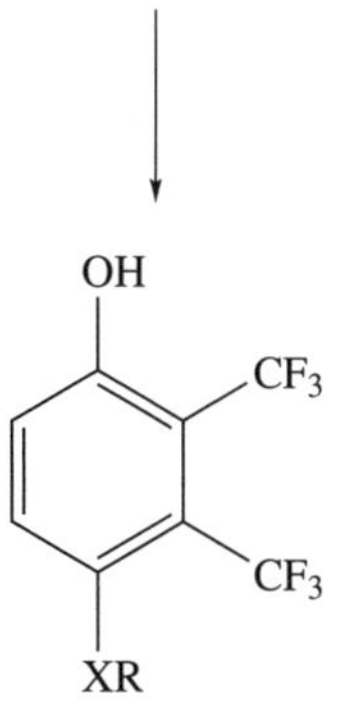

(129)

XR = NHBoc, OMe, OTMS, OCO$_2$Me, Sn(Bn-*n*)$_3$, O$_2$CBu-*t*

Zhu and coworkers[235] have reported a regioselective rearrangement of the Diels–Alder cycloadduct **180**, derived from furan **178** and acetylene **179**, to form the 1,4-difunctionalized 2,3-bis(trifluoromethyl)benzene system **181** in one chemical operation (equation 129).

Recently, Hashmi and coworkers[236] reported a selective Diels–Alder synthesis of phenolic compounds catalysed by Au(III) (equation 130). The mechanism has proven to include an intramolecular migration of the oxygen atom of the furan ring[237]. Several other transition metals with d^8 configuration (Pd^{II}, Pt^{II}, Rh^I, Ir^I) allow this conversion, but Au^{III} is shown to be the most active catalyst giving the cleanest conversion.

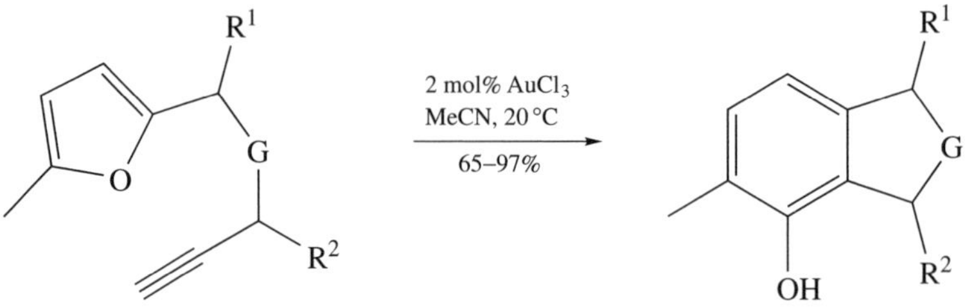

R^1, R^2 = H, Me
G = O, CH_2, NTs, N(Ts)CH_2, C(CO_2Me)$_2$

(130)

C. Benzannulation

One of the most powerful strategies for the construction of polysubstituted phenols is the reaction of dienylketenes[238], generated *in situ*, with heterosubstituted alkynes by a cascade of pericyclic reactions, affording the aromatic ring in one step and with predictable regioselectivity. There are two methods for the generation of such dienylketenes. One is based on the irradiation of cyclobutenones **182**[239], which triggers a four-electron electrocyclic ring opening (equation 131). The second method consists in a photochemical Wolff rearrrangement[240] of α,β-unsaturated α'-diazoketones **188**[241]. Equation 131 outlines the mechanistic course of this benzannulation reaction. The generated vinylketenes **185** react with an electron-rich acetylene **183** (X = OR, SR, NR_2) in a regioselective [2 + 2] cycloaddition to form **186**. Further irradiation (or warming) induces a second four-electron electrocyclic ring-opening reaction to generate the dienylketene **187**, which undergoes a rapid 6π electrocyclization, affording the desired substituted phenol **184** by tautomerization.

Table 5 shows some examples of benzannulation reactions with various cyclobutenones (entries 1–4), α,β-unsaturated α'-diazoketones (entries 5–7) and stable vinylketenes (entry 8).

The use of a metal carbene complex in benzannulations has become one of the most valuable synthetic applications of these organometallic reagents[245]. Because of its applicability to a broad spectrum of substituents, its regioselectivity and its mild experimental conditions, benzannulation has been employed as a key step in the synthesis of a series of natural compounds[246].

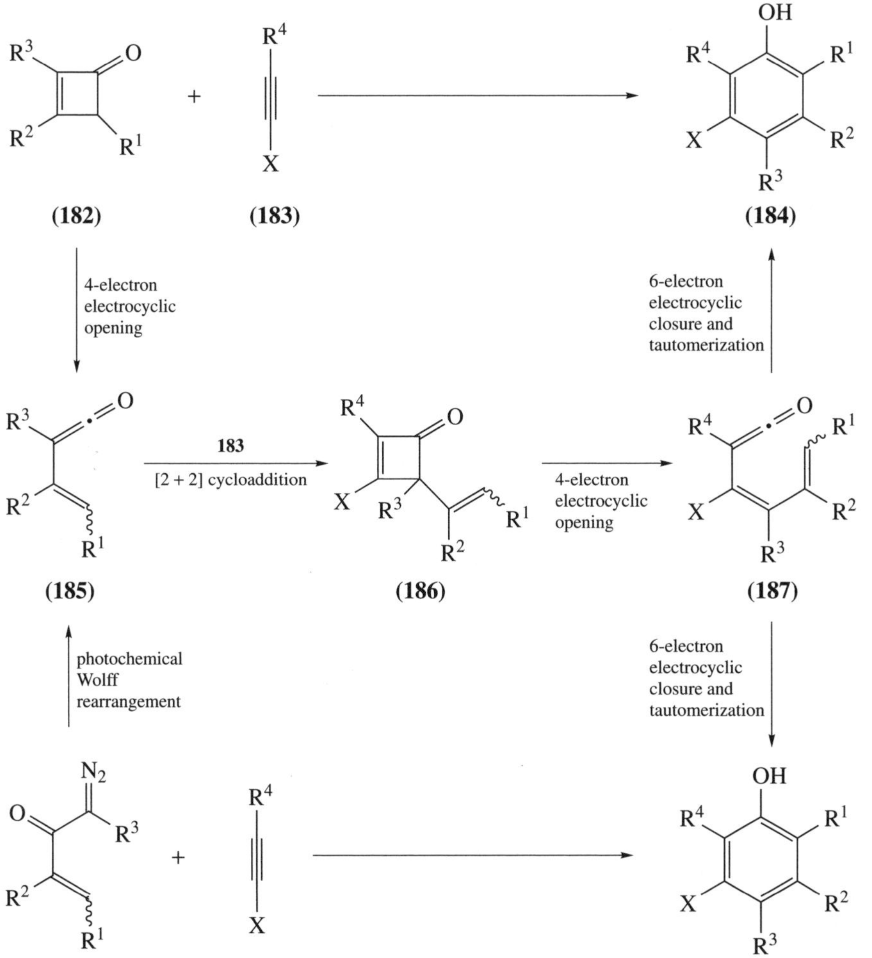

6. Synthesis of phenols

(131)

Several transition metal complexes (Co[247], Mo[248], W[249], Fe[250], etc.) have been used in benzannulation reactions, but vinyl- or aryl(alkoxy)carbene chromium complexes **189**, reported by Dötz, are the most generally employed (equation 132)[251]. The chromium tricarbonyl coordinated dienylketenes **190** generated *in situ* have been converted to the chromium complexes of polysubstituted phenols **191** in high yield. The reaction is a transition-metal-induced benzannulation, which corresponds formally to a [3 + 2 + 1] cycloaddition.

Carbocycles, heterocycles and polycyclic arenes can serve as carbene ligands for the synthesis of complexes with benzannulated arenes (equation 133)[251c,252].

TABLE 5. Phenols obtained by benzannulation reactions

Entry	Precursor	Alkyne	Reaction product	Yield (%)	Reference
1	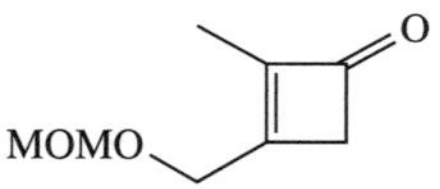	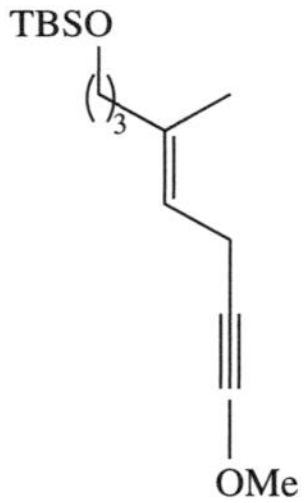	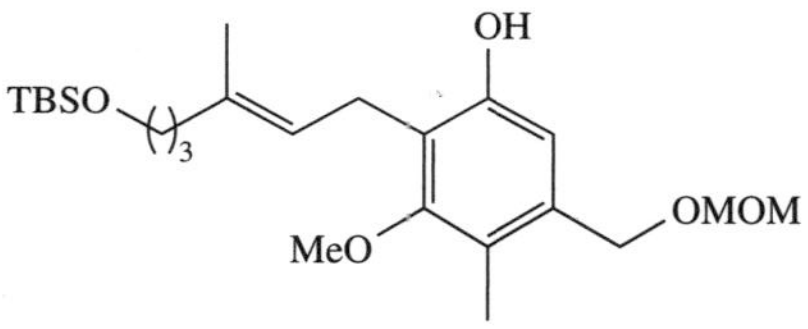	73	239c
2	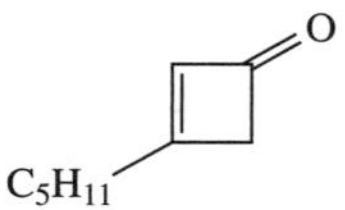	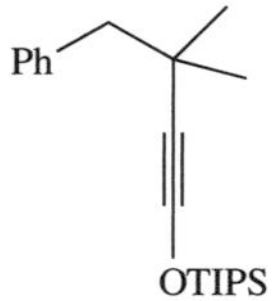	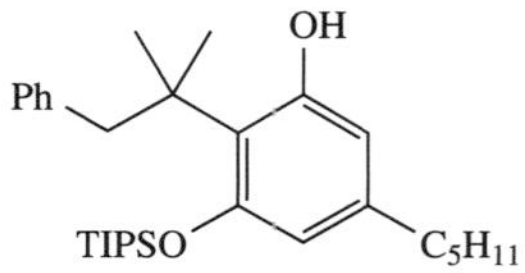	86	242
3	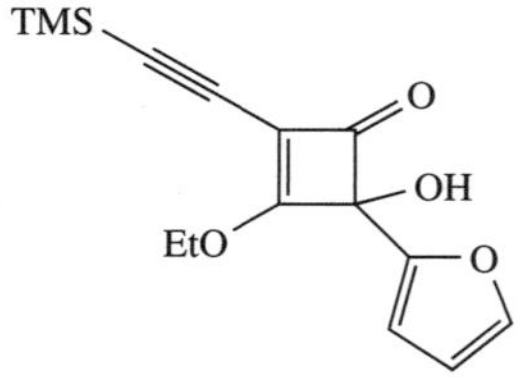		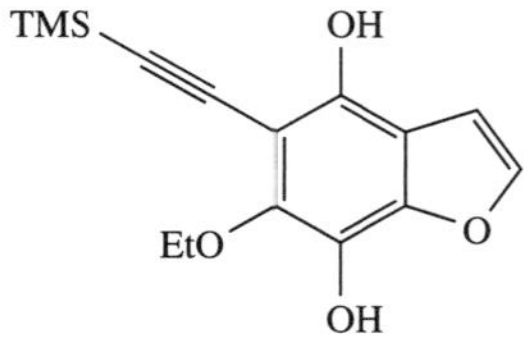	90	239b
4	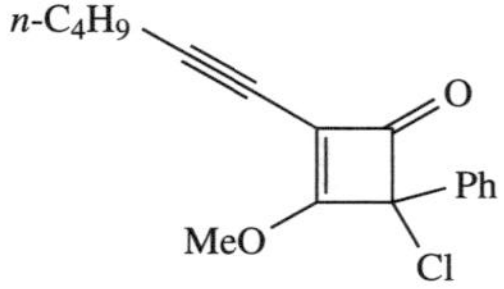		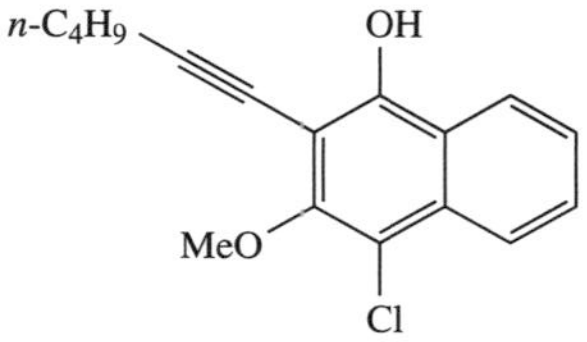	78	243

5	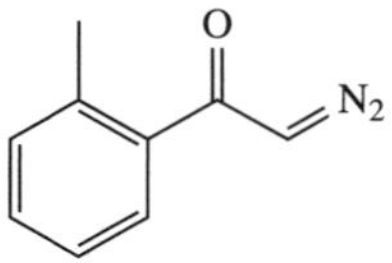 (2-methylphenyl)-C(O)-CH=N₂	 MeC≡C-OMe	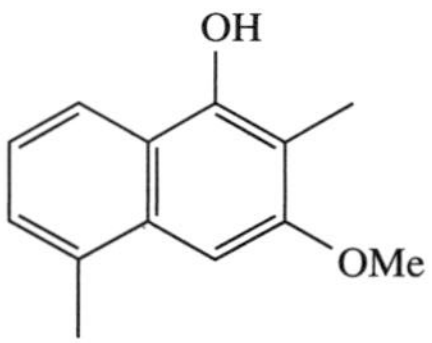 2-methyl-3-OMe-1-naphthol (with OH, Me, OMe substituents)	57	241c
6	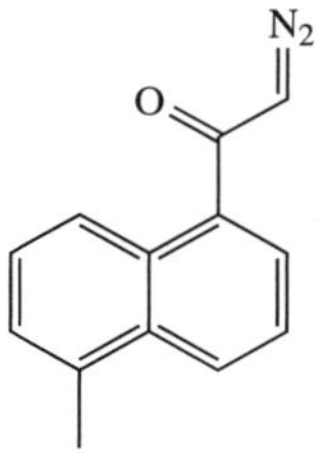 (5-methylnaphthyl)-C(O)-CH=N₂	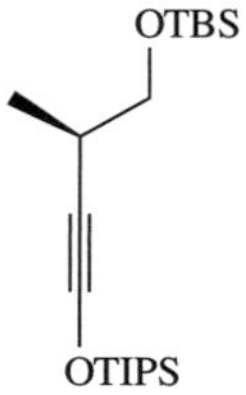 TBSOCH₂-CH(Me)-C≡C-OTIPS	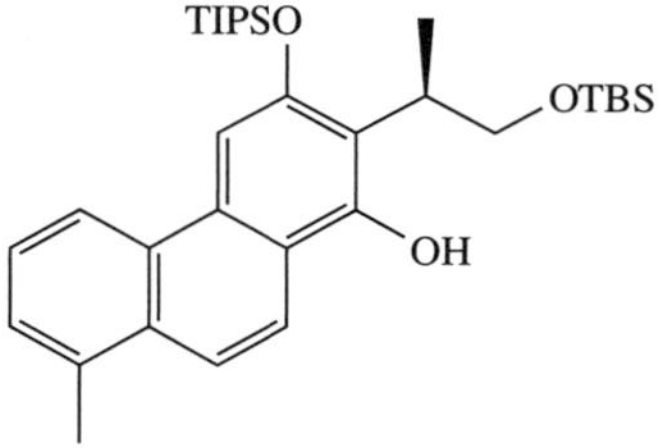 phenanthrene with TIPSO, OH, CH(Me)CH₂OTBS, Me substituents	70	241e
7	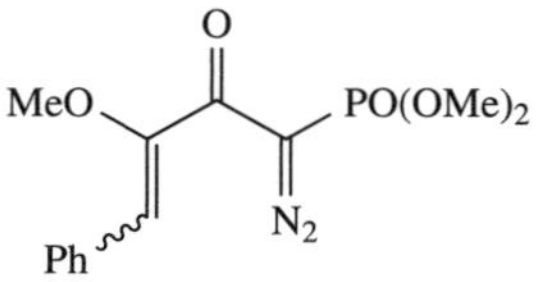 MeO-C(=CHPh)-C(O)-C(N₂)-PO(OMe)₂	—	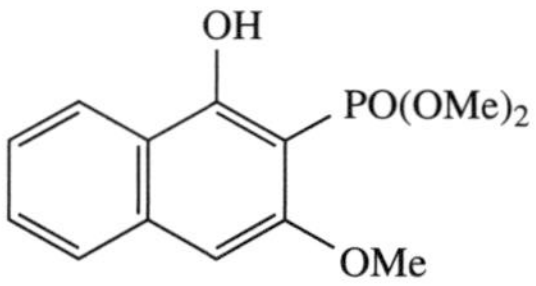 2-PO(OMe)₂-3-OMe-1-naphthol (OH)	75	244
8	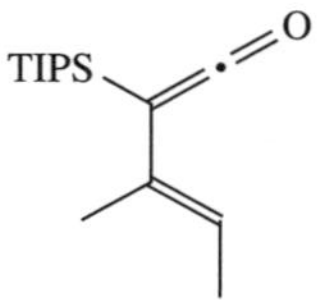 TIPS-C(=C=O)=C(Me)-CH=CHMe ketene	 MeO₂C-C≡C-CO₂Me	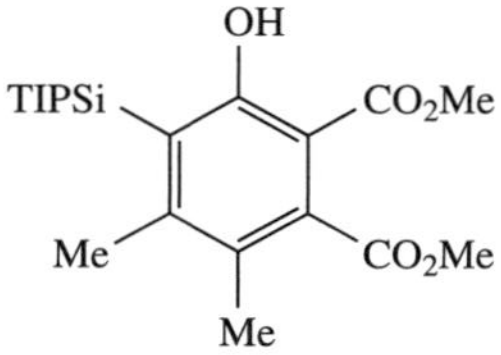 benzene with OH, TIPSi, 2×Me, 2×CO₂Me	95	239f

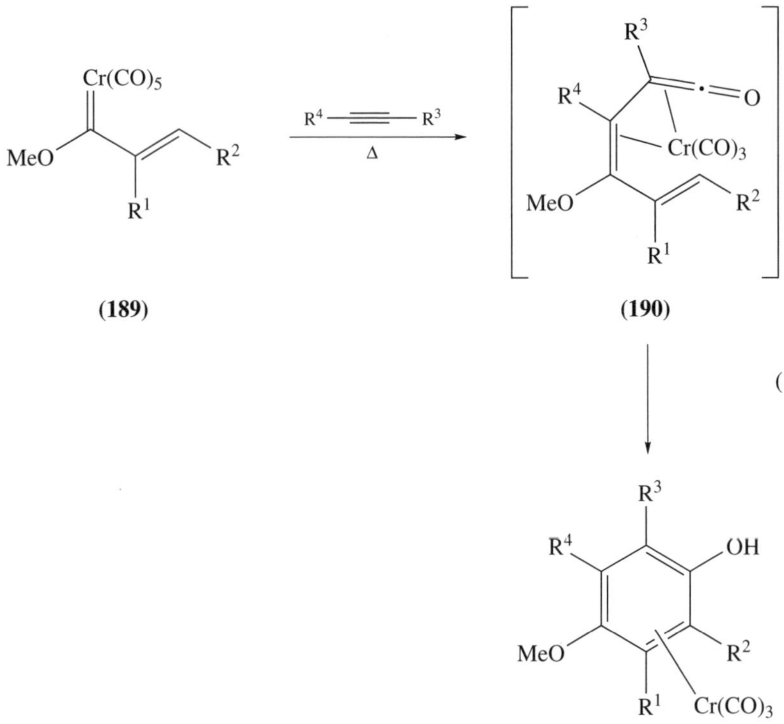

$$\text{(189)} \xrightarrow{R^4 \equiv\!\!\!\equiv\!\!\!\equiv R^3, \Delta} \text{(190)}$$

(132) ↓

(191)

Generally, arene(alkoxy)carbene chromium complexes react with aryl-, alkyl-, terminal or internal alkynes in ethers or acetonitrile to yield 4-alkoxy-1-naphthols, with the more hindered substituent *ortho* to the hydroxyl group[251,253]. Upon treatment with alkynes, aryl(dialkylamino)carbene chromium complexes do not yield aminonaphthols, but they form indene derivatives[254]. Vinyl(dialkylamino)carbene complexes, however, react with alkynes to yield aminophenols as the main products[249,255]. The solvent is one of the many factors that affects this type of reaction, for which the most important is the polarity and/or coordinating ability of the solvent. The Dötz benzannulation reaction yields either arene chromium tricarbonyl complexes or the decomplexed phenols, depending on the work-up conditions. Oxidative work-up yields either decomplexed phenols or the corresponding quinones.

Remarkable improvements have been reported experimentally regarding the optimization of the reaction yield, such as variations in the reaction temperature and solvent, and the introduction of special techniques (e.g. dry stage adsorption conditions[256], ultrasonication[257] and photoirradiation employing a Xenon lamp[258]).

Examples of the Dötz benzannulation reaction are given in Table 6.

The rate-determining step has been demonstrated to be the dissociation of a CO ligand from the carbene complex **192** and the newly formed coordination site of complex **193** is being occupied either by a solvent molecule (e.g. THF) or saturated intramolecularly by the vinyl group (for vinylcarbene complexes) (equation 134)[265]. When the η^2-alkyne complex **194** has been formed, insertion of the alkyne into the Cr—C double bond takes place to yield an η^3-vinyl-carbene complex **195**. Depending on the carbene substituent X, two different reaction pathways must be considered. Amino carbene complexes, which usually require higher temperatures to react with alkynes, tend to cyclize without incorporation of carbon monoxide to yield aminoindene complexes **196**. Alkoxy carbene complexes are generally more reactive and undergo fast CO-insertion to yield η^4-vinylketene complexes **197**. The latter intermediates can cyclize to cyclohexadienone complexes, which finally tautomerize to naphthols **198**.

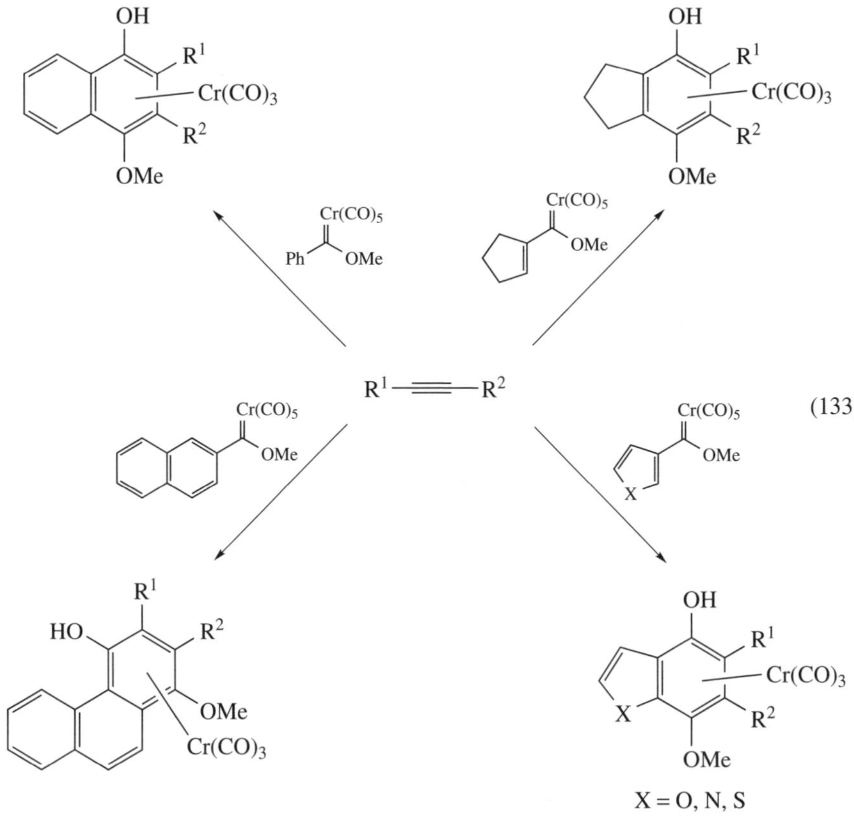

(133)

Recently, it has been suggested that the first step of the Dötz benzannulation reaction may not necessarily be the dissociation of one carbonyl ligand[266]. Alternatively, the [2 + 2] cycloaddition of the alkyne to the unsaturated chromium carbene complex **199** has been proposed to afford a cyclic complex **200**, which undergoes a four-electron electrocyclic opening to yield a 1-chroma-1,3,5-hexatriene **201** (equation 135)[267]. Dissociation of a carbonyl ligand gives chromium complex **202**, which then, as in the CO-dissociation mechanism, undergoes a CO insertion to yield ketene **203**, generating phenol **204** by electrocyclic ring closure and subsequent tautomerization.

TABLE 6. Phenols obtained by the benzannulation reaction

Entry	Carbene	Alkyne	Reaction Product	Yield (%)	Reference
1	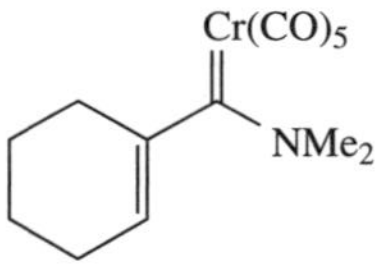 (CO)$_5$Cr=C(NMe$_2$)(cyclohexenyl)	1-pentyne	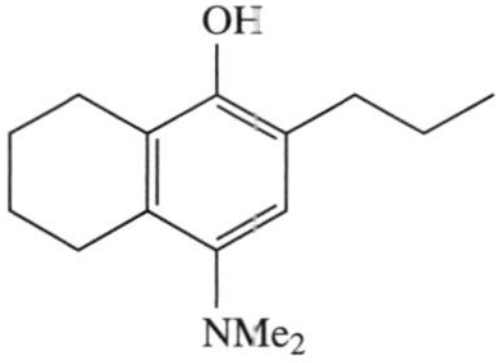 tetrahydronaphthalene with OH, propyl, NMe$_2$	66	255
2	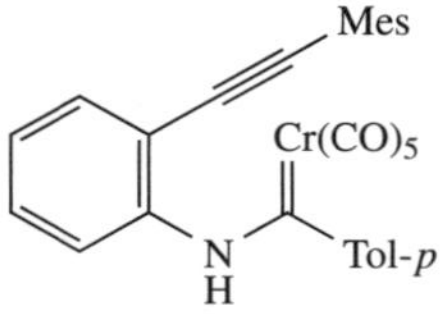 (CO)$_5$Cr=C(Tol-p)(2-alkynyl-NH-mesityl-anilide); Mes = 2,4,6-Me$_3$C$_6$H$_2$	—	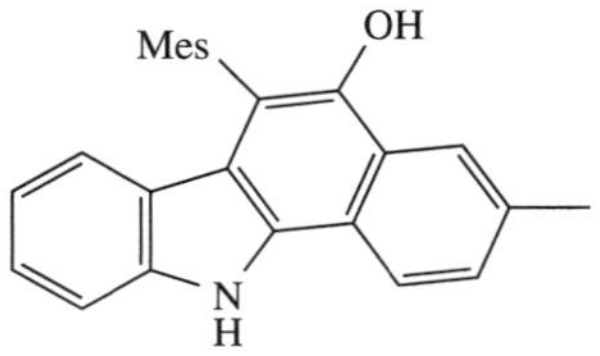 benzo-carbazole with OH, Mes	63	259
3	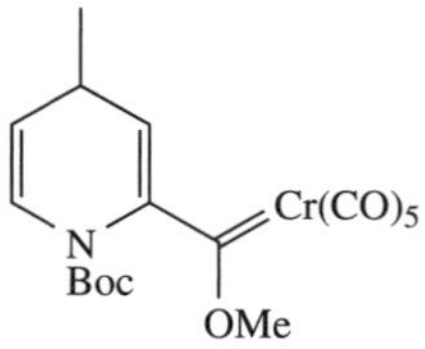 (CO)$_5$Cr=C(OMe)(N-Boc-dihydropyridinyl)	1-pentyne	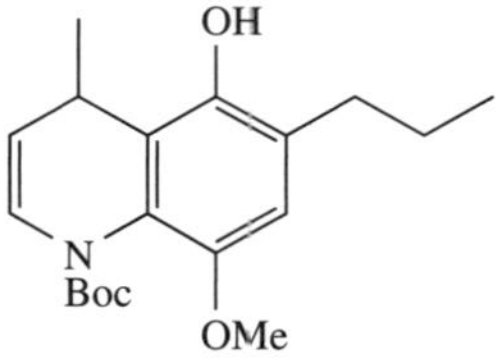 dihydroquinoline with OH, OMe, propyl, N-Boc	60	260
4	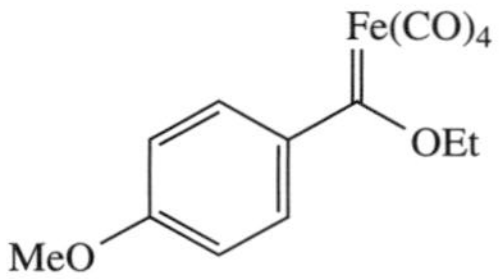 (CO)$_4$Fe=C(OEt)(4-MeO-C$_6$H$_4$)	 MeO$_2$C–C≡C–CO$_2$Me	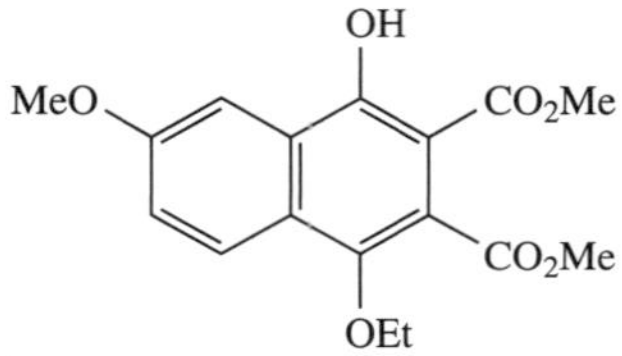 naphthalene with OH, OMe, OEt, two CO$_2$Me	93	261

5			90	262
6			61	263
7			81	264

[a] R = Z-(CH$_2$CH$_2$CH=CMe)$_2$Me.

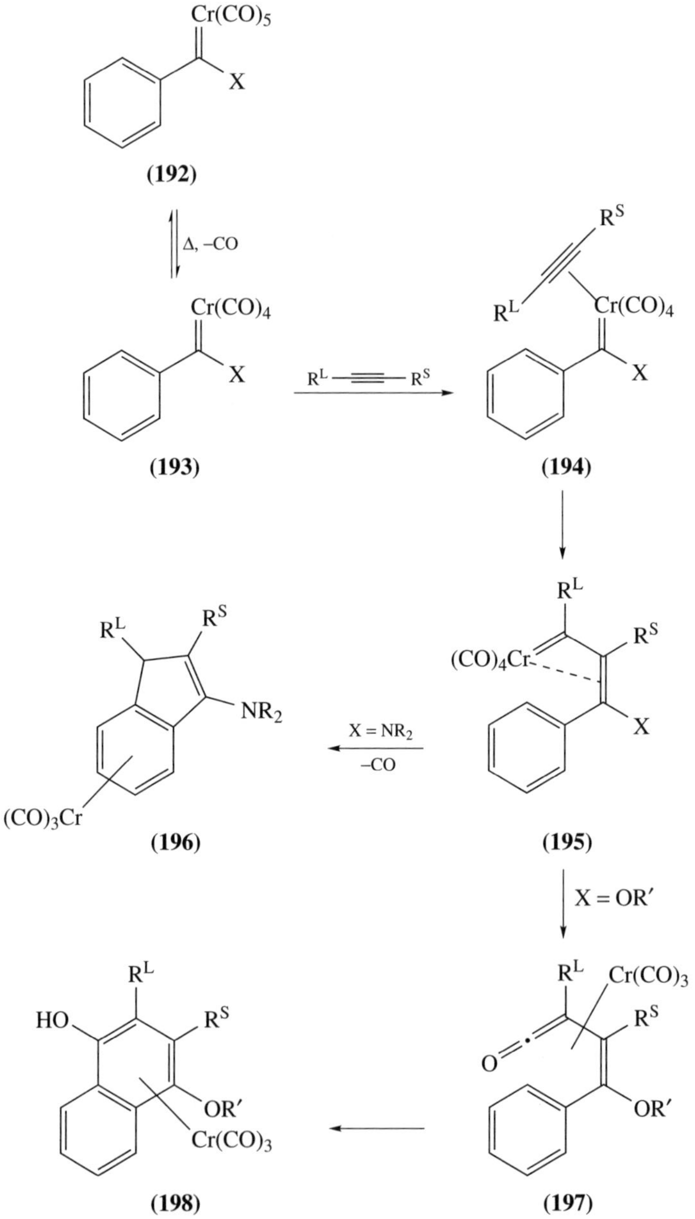

(134)

R^L = Large substituent
R^S = Small substituent

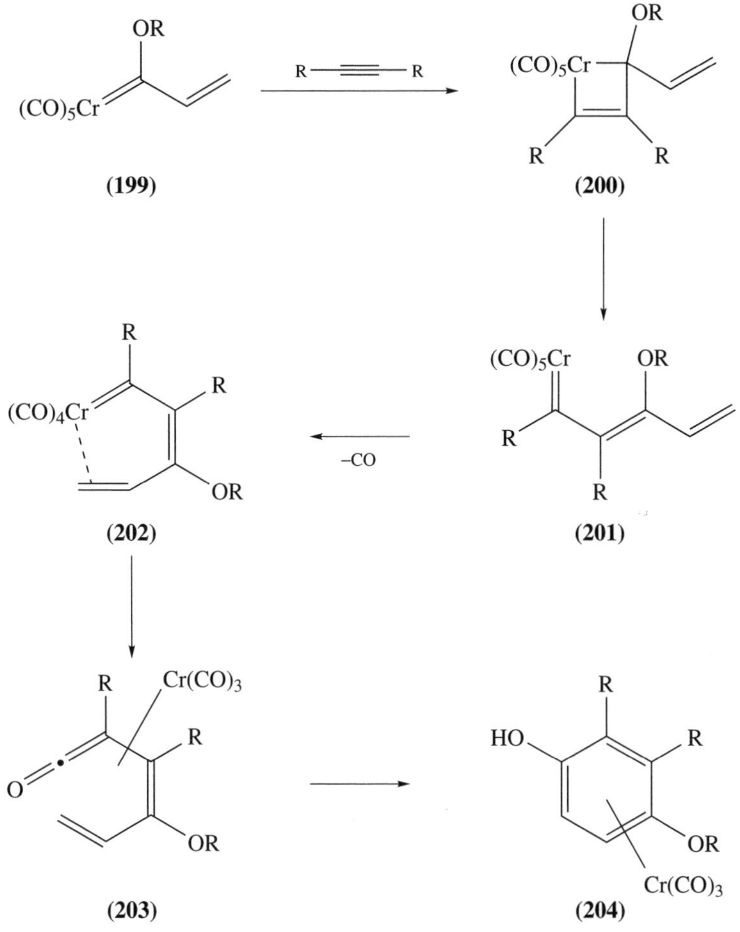

V. BY REARRANGEMENT

A. Alkyl and Benzyl Aryl Ethers

Alkyl and benzyl aryl ethers undergo acid-catalysed rearrangement to afford phenols. For instance, benzyl phenyl ether under treatment with AlBr$_3$ in dichloromethane yields exclusively 2-benzylphenol with simultaneous production of phenol[268]. The ratio of phenol and 2-benzylphenol is hardly affected by the solvent. Other type of catalysts have also been used successfully in this type of rearrangement. For instance, trifluoroacetic acid converts 4-(2′-methyl-but-2′-yl)phenyl benzyl ether (**205**) to the corresponding phenol **206** (equation 136)[269] and over montmorillonite clays, benzyl phenyl ether (**207**), is converted to 2-benzylphenol (**208**) (equation 137)[270].

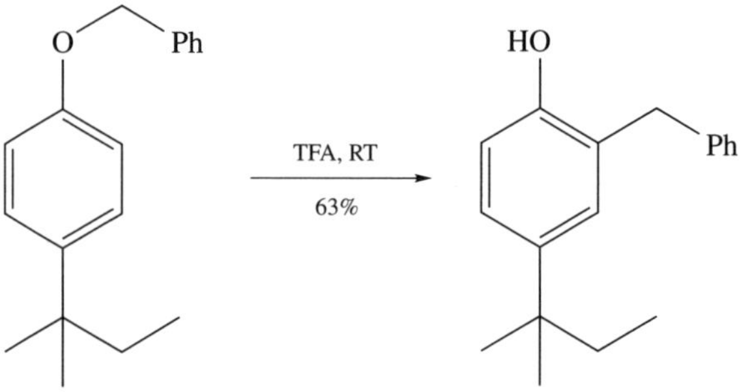

(136)

(205) → (206) TFA, RT, 63%

(137)

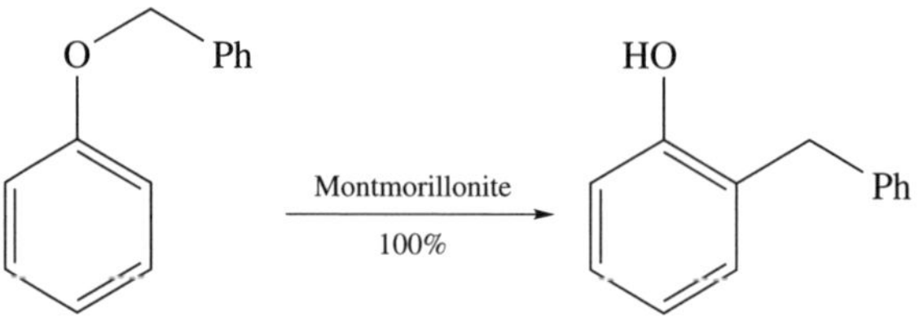

(207) → (208) Montmorillonite, 100%

B. Allyl Aryl Ethers. Aromatic Claisen Rearrangement

Claisen rearrangements of allyl phenyl ethers to *ortho*-allylphenols (aromatic Claisen rearrangement) were thoroughly studied before the analogous rearrangements of allyl vinyl ethers. The initial [3,3] step in the Claisen rearrangement of an allyl aryl ether **209** gives an *ortho*-cyclohexadienone **210**, which usually enolizes rapidly to the stable product, an *ortho*-allylphenol **211** (*ortho* Claisen rearrangement) (equation 138)[271]. If the rearrangement is to an *ortho* position bearing a substituent, a second [3,3] step followed by enolization leads to the *para*-allylphenol **212** (*para* Claisen rearrangement). The *ortho* Claisen rearrangements predominate in the majority of the cases, but the *para* process can compete even when both *ortho* positions are free.

Some examples of this type of transformation are indicated in equations 139–144.

Remarkable improvements have been achieved in the optimization of the rate and yield of these thermal reactions (typically 150–220 °C), such as the use of microwave irradiation or catalysts. For instance, allyl phenyl ether at 220 °C gives an 85% yield of 2-allylphenol in 6 h[278], but the reaction time drops to 6 min by using microwave ovens and the yields also increase up to 92%[279]. On the other hand, Lewis acids, such as BCl_3[280], $BF_3 \cdot Et_2O$[281], Et_2AlCl[282], $TiCl_4$[283] and $(i\text{-PrO})_2TiCl_2$[283] have been successfully used to catalyse this rearrangement reaction under mild conditions. Other catalysts such as Ag^I[284] and Pt^0[285] complexes or zeolites[286] have also been employed.

Few approaches for the development of enantioselective aromatic Claisen rearrangements have been reported. For instance, Trost and Toste[287] proved that europium complexes, $Eu(fod)_3$, induce the diastereoselective Claisen rearrangements for the synthesis of asymmetric phenols. For instance, the cyclic ethers **213** have been transformed into phenols **214** in high yields and excellent ee (equation 145). Rearrangement of acyclic system **215** proved to be a good yielding reaction under these conditions, producing phenol **216** with 91% ee and in 83% yield (equation 146). Taguchi and coworkers[288] used a catechol

6. Synthesis of phenols

monoallyl ether derivative which can form a σ-bond with a chiral boron reagent. For instance, under these conditions, phenol **217** has been converted into catechol **218** with 93% ee and in 97% yield (equation 147).

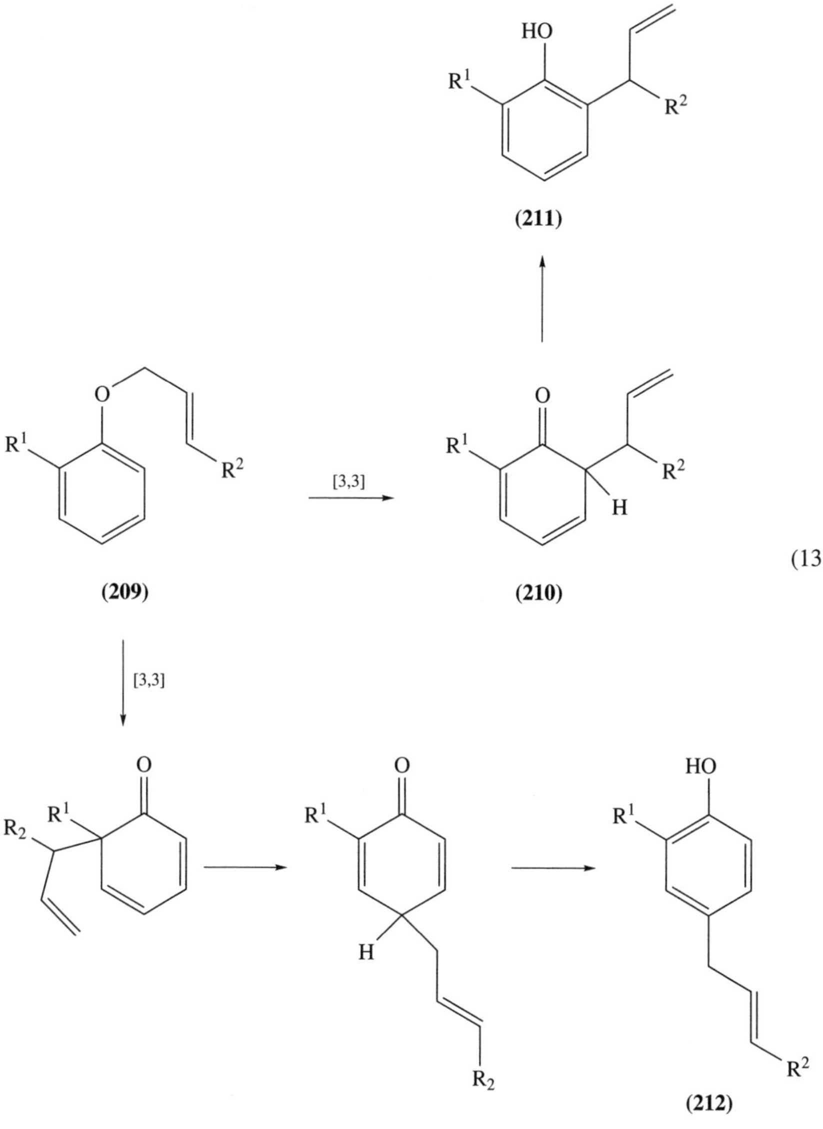

(138)

Recently, Wipf and Ribe[289] reported a novel tandem process in which water accelerates both a sigmatropic Claisen rearrangement catalysed by Erker's catalyst[290] and a subsequent carbometallation reaction with trimethylaluminium providing optically active phenols. Examples of this tandem process are shown in equations 148–150.

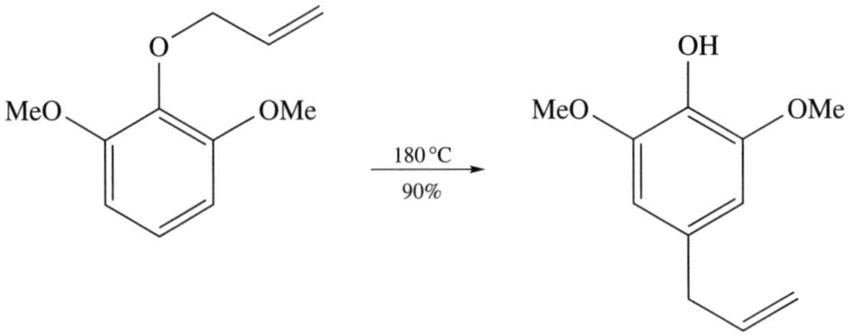

(139)[272]

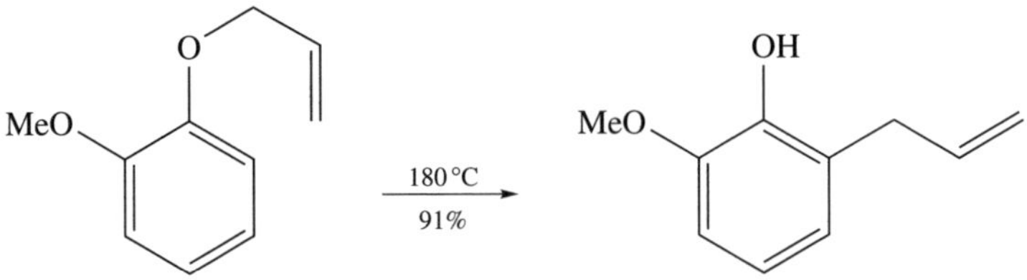

(140)[273]

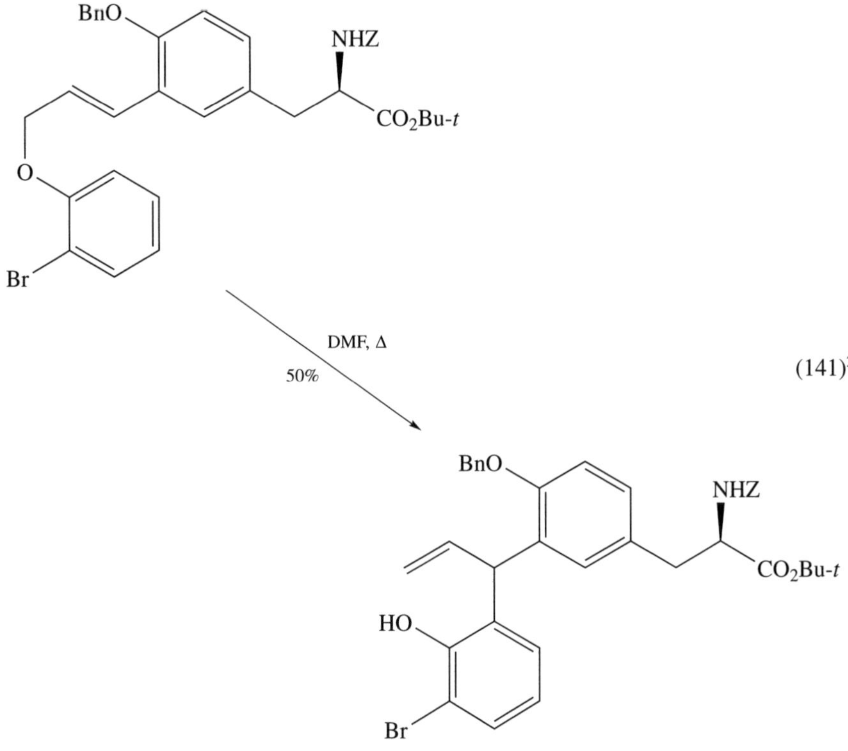

(141)[274]

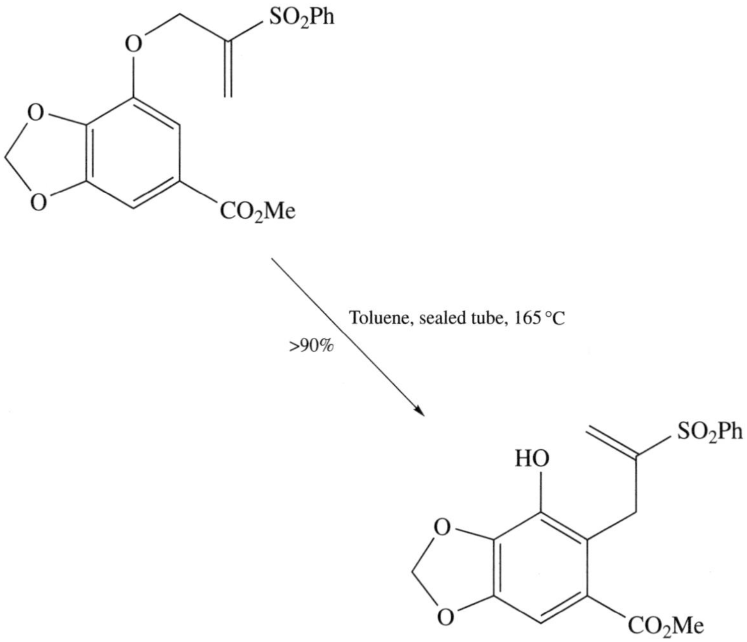

6. Synthesis of phenols

(142)[275] Toluene, sealed tube, 165 °C, >90%

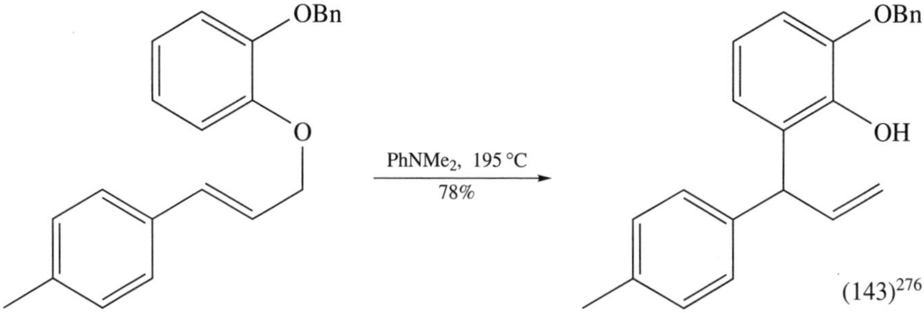

(143)[276] PhNMe₂, 195 °C, 78%

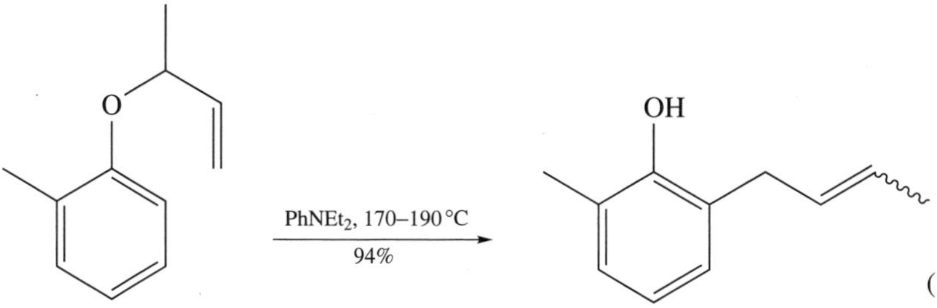

(144)[277] PhNEt₂, 170–190 °C, 94%

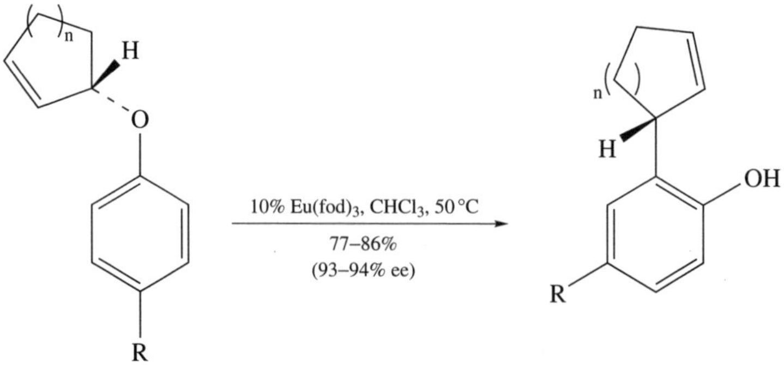

(213) → (214)   10% Eu(fod)₃, CHCl₃, 50 °C; 77–86% (93–94% ee); R = OMe, F; n = 1–3   (145)

(215) → (216)   10% Eu(fod)₃, CHCl₃, 50 °C; 83% (91% ee)   (146)

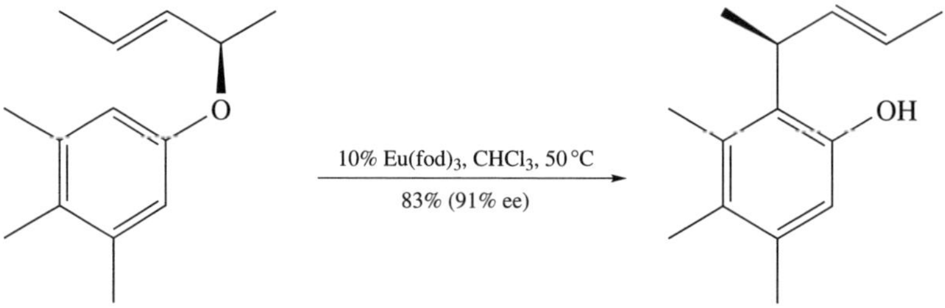

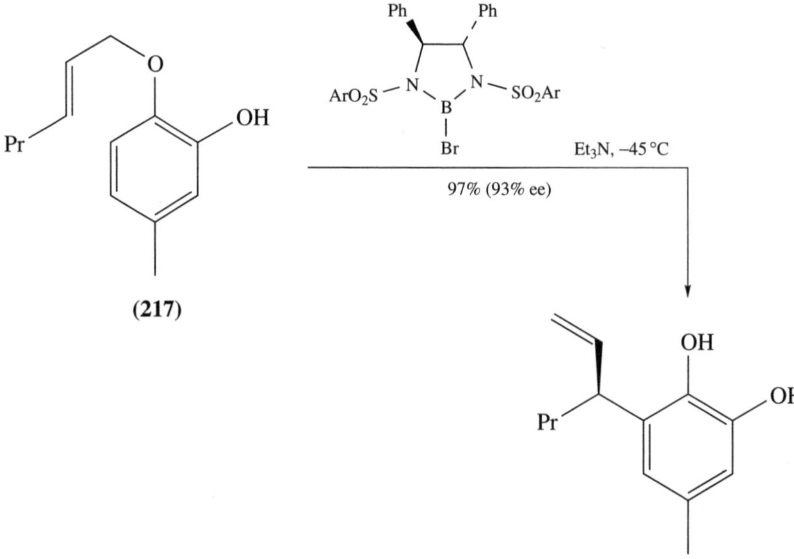

(217) → (218)   97% (93% ee); Et₃N, −45 °C   (147)

Ar = 4-methylphenyl (S,S); 3,5-bis (trifluoromethyl) phenyl (S,S)

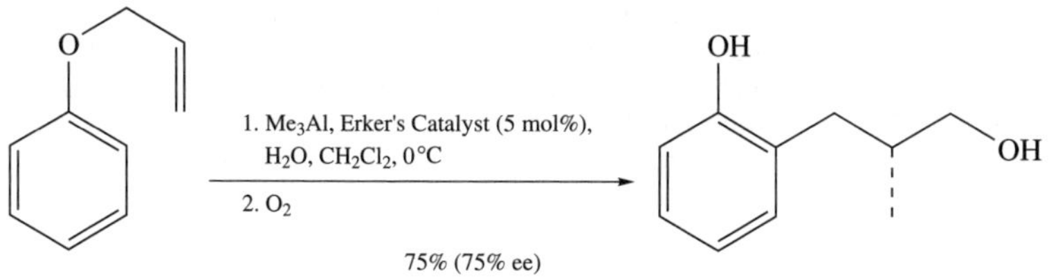

6. Synthesis of phenols

(148) Allyl phenyl ether → 1. Me₃Al, Erker's Catalyst (5 mol%), H₂O, CH₂Cl₂, 0 °C; 2. O₂ → 2-(3-hydroxy-2-substituted-propyl)phenol, 75% (75% ee)

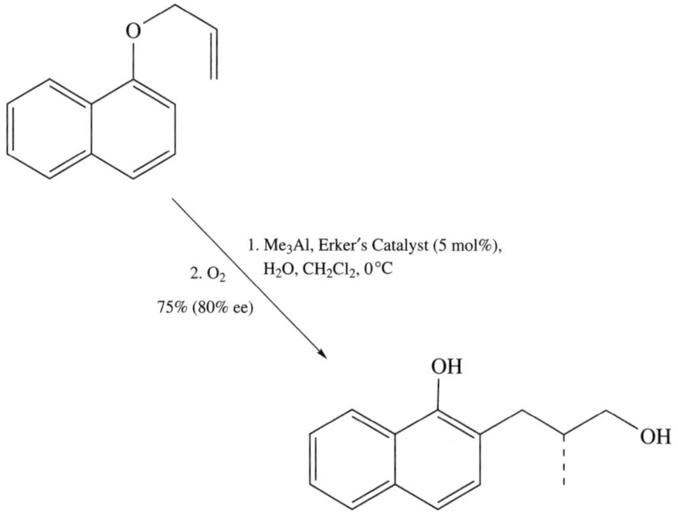

(149) 1-(allyloxy)naphthalene → 1. Me₃Al, Erker's Catalyst (5 mol%), H₂O, CH₂Cl₂, 0 °C; 2. O₂ → 75% (80% ee)

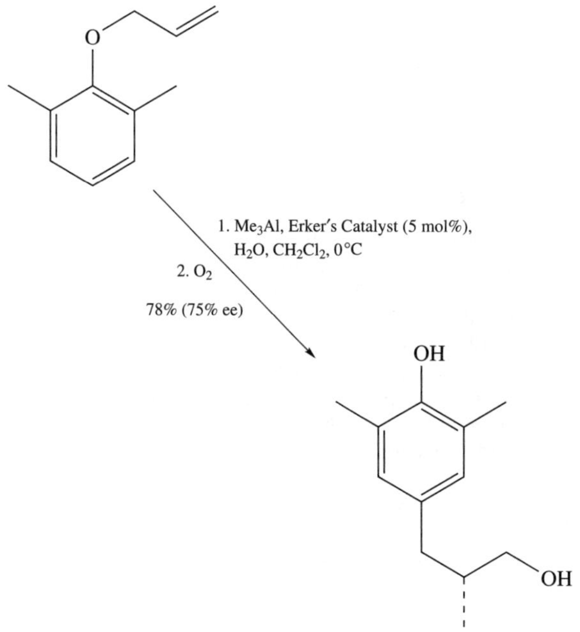

(150) 2,6-dimethyl allyl aryl ether → 1. Me₃Al, Erker's Catalyst (5 mol%), H₂O, CH₂Cl₂, 0 °C; 2. O₂ → 78% (75% ee)

In the same way, Brønsted acid catalysts such as trifluoroacetic acid substantially accelerate the Claisen rearrangement of allylphenyl ether. However, the initially formed allylphenols generally react further under the acidic reaction conditions. For instance, crotyl *p*-tolyl ether (**219**) in trifluoroacetic acid affords benzofuran **220** as the main reaction product derived from cyclization of the Claisen rearrangement product **221** (equation 151)[291].

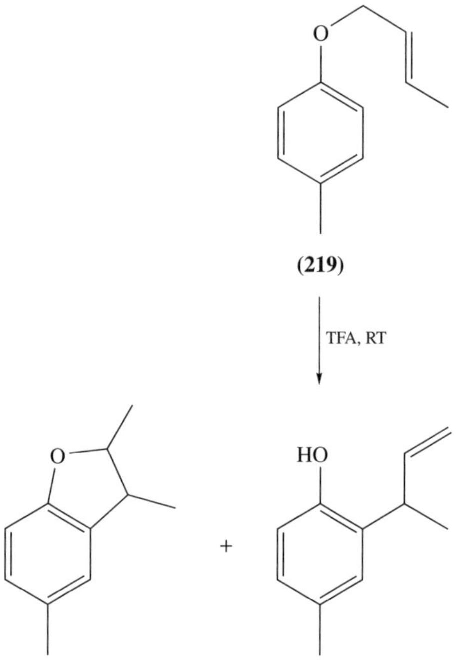

(**219**)

TFA, RT (151)

(**220**), 69%   +   (**221**), 8%   +   8%

C. Diaryl Ethers. Smiles Rearrangement

The ether linkage of aryl ethers is considered one of the more stable chemical bonds. In fact, the extreme stability of phenyl ethers has made them important heat-exchange fluids and high-temperature lubricants. However, at high temperatures (>400 °C), 2,6-dimethylphenyl phenyl ether (**222a**) undergoes an exothermic decomposition with the formation of 2-benzyl-6-methylphenol (**223a**) in 70% yield (equation 152)[292]. Similarly, ethers **222b–e** undergo the same type of transformation yielding the corresponding phenols **223b–e** in moderate yield. The mechanism appears to be a radical process initiated by abstraction of a hydrogen atom from a methyl group. The generated benzyl radical undergoes a rearrangement reaction to afford a phenoxy radical, which abstracts a hydrogen atom from another molecule of the starting ether to continue the process.

Under treatment with phenyl sodium, diphenyl ether (**224**) affords 57% of 2-hydroxybiphenyl (**225**) (equation 153)[293]. Equation 154 outlines the mechanistic course of this reaction. The first step is the abstraction of an *ortho*-hydrogen to the oxygen to afford

226, which generates benzyne **227** by elimination of sodium phenoxide[294]. Benzyne (**227**) then reacts with intermediate **226** to give aryl sodium salt **228**, which gives **229** by transmetallation. Finally, intermediate **229** affords 2-biphenyloxy sodium (**230**) and regenerates benzyne (**227**) to continue the process. Other examples of this rearrangement are given in equations 155–157[293].

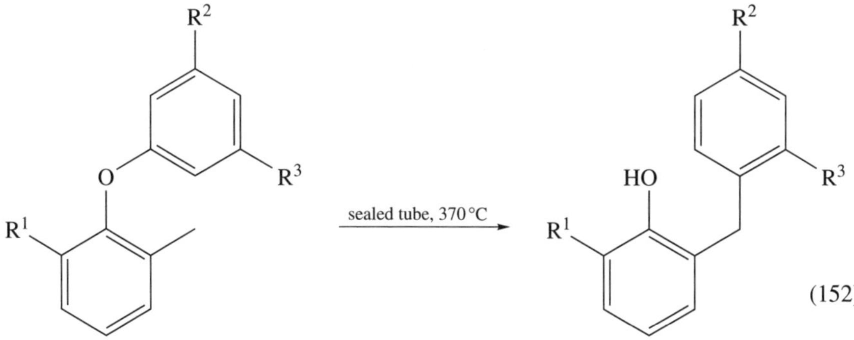

(222) →(sealed tube, 370 °C)→ (223)　　(152)

(a) $R^1 = Me, R^2 = R^3 = H; 70\%$
(b) $R^1 = R^3 = H, R^2 = OMe; 50\%$
(c) $R^1 = Ph, R^2 = R^3 = H; 50\%$
(d) $R^1 = Bn, R^2 = R^3 = H; 50\%$
(e) $R^1 = R^3 = H, R^2 = Me; 50\%$

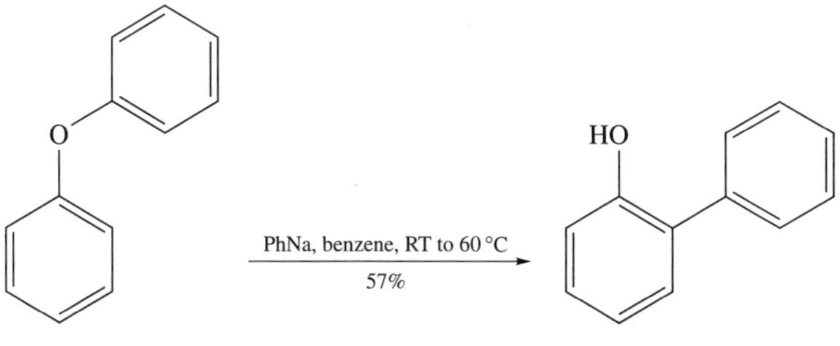

(224) →(PhNa, benzene, RT to 60 °C, 57%)→ (225)　　(153)

The Smiles rearrangement is an intramolecular nucleophilic substitution that follows the pattern given in equation 158[5,295]. The nucleophilic attack normally requires an electron-withdrawing group (e.g. nitro, sulphonyl or halogen) either in the *ortho* or the *para* position on the aromatic ring where the substitution takes place; generally X is a good leaving group (S, SO, SO_2 or O), and Y is a strong nucleophile, usually the conjugate base of OH, NH_2, NHR or SH. The reaction takes place on the carbon directly bonded to the leaving group X[296]. Equations 159–161 show some examples of this rearrangement.

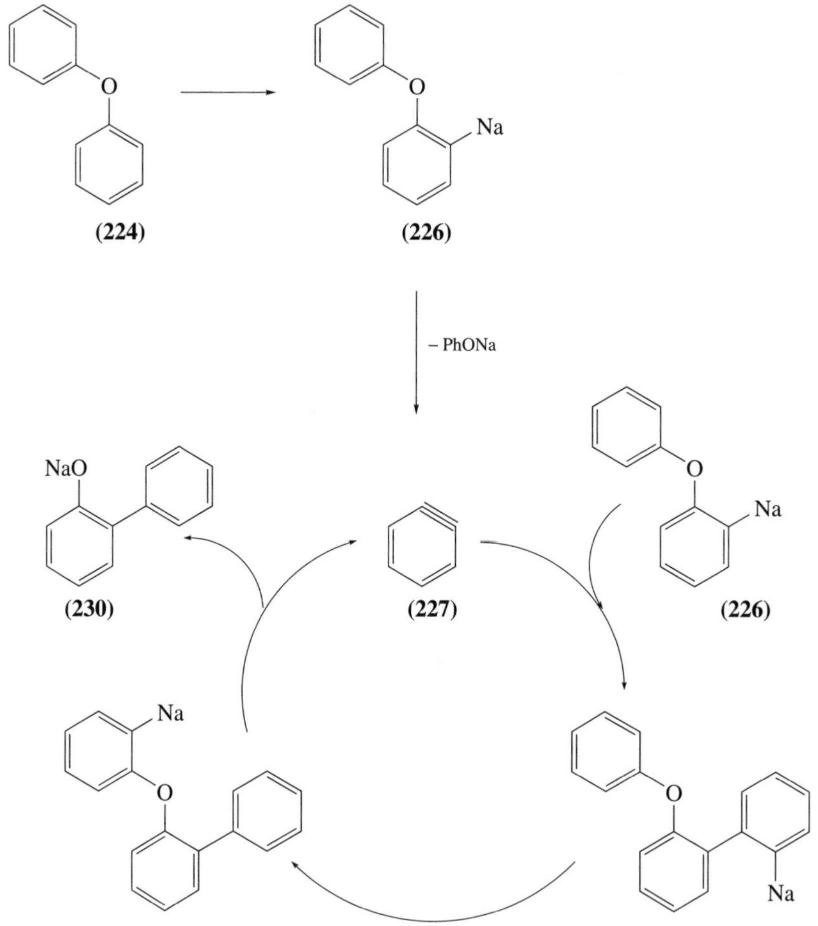

(224) → (226)

− PhONa

(230) ← (227) ← (226)

(229) ← (228)

(154)

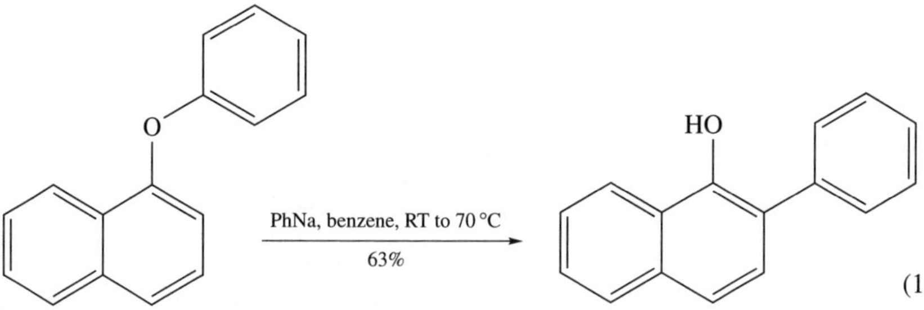

(155)

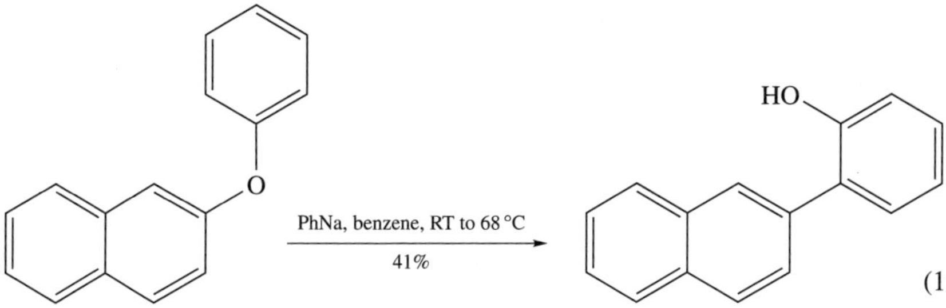

6. Synthesis of phenols

(156) PhNa, benzene, RT to 68 °C, 41%

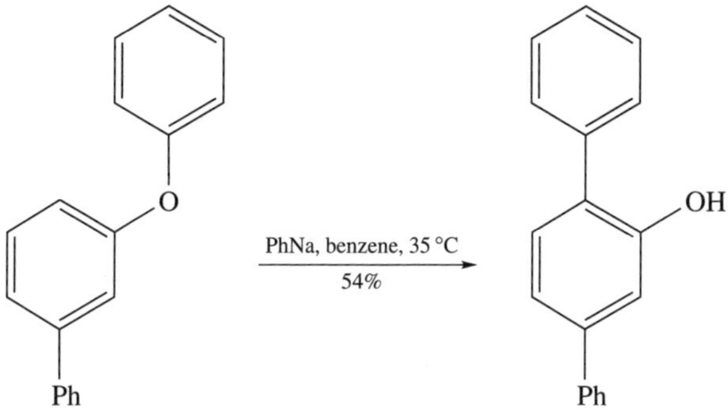

(157) PhNa, benzene, 35 °C, 54%

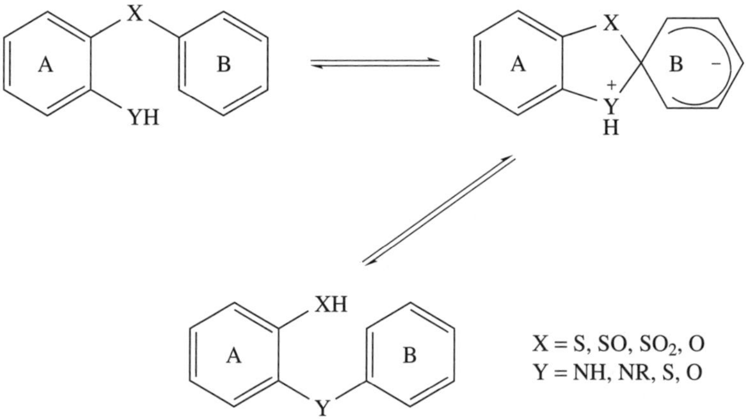

(158) X = S, SO, SO$_2$, O; Y = NH, NR, S, O

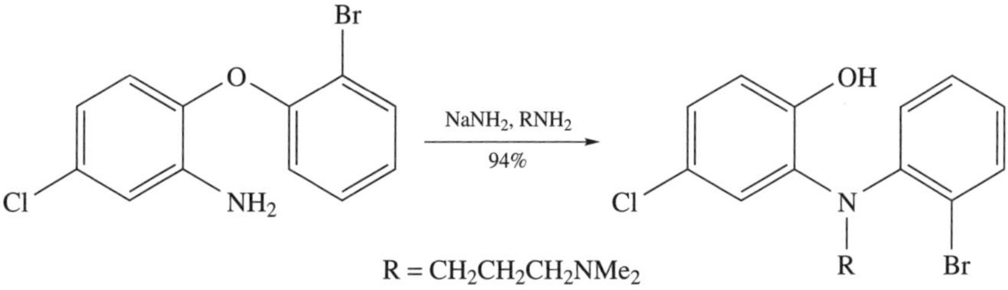

(159)[297] NaNH$_2$, RNH$_2$, 94%; R = CH$_2$CH$_2$CH$_2$NMe$_2$

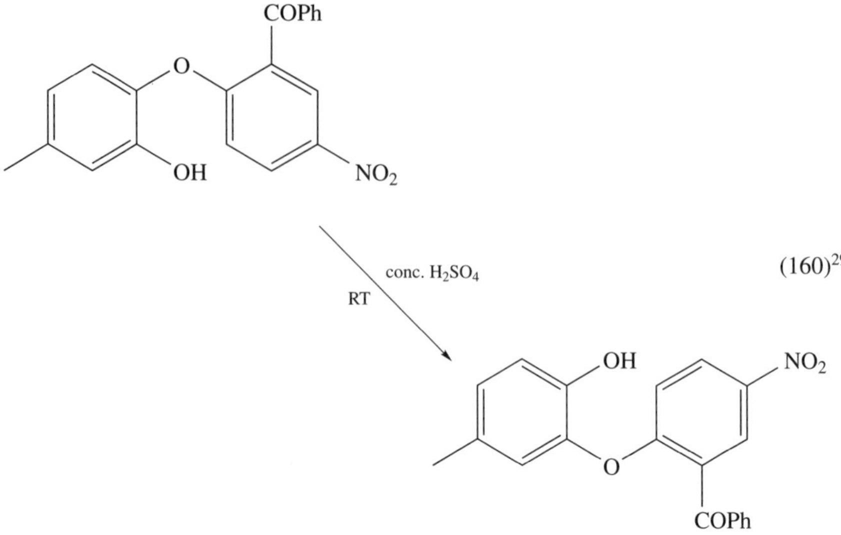

(160)[298]

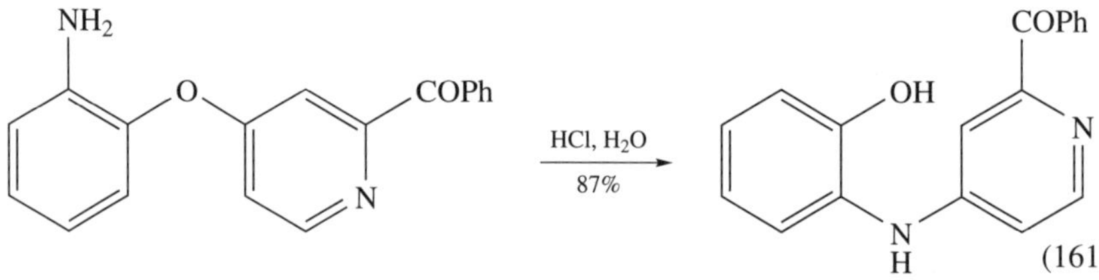

(161)[299]

D. Dienones. Dienone–Phenol Rearrangement

On acid treatment, cyclohexadienones **231** with two alkyl groups in position 4 undergo 1,2 migration of one of these groups to afford phenolic compounds **232** in good yields (equation 162)[300]. This reaction, known as the dienone–phenol rearrangement, is an important method for the preparation of highly substituted phenols that are not readily available by conventional aromatic substitution chemistry. In the overall reaction the driving force is the formation of an aromatic system. Examples of this rearrangement are shown in equations 163–168.

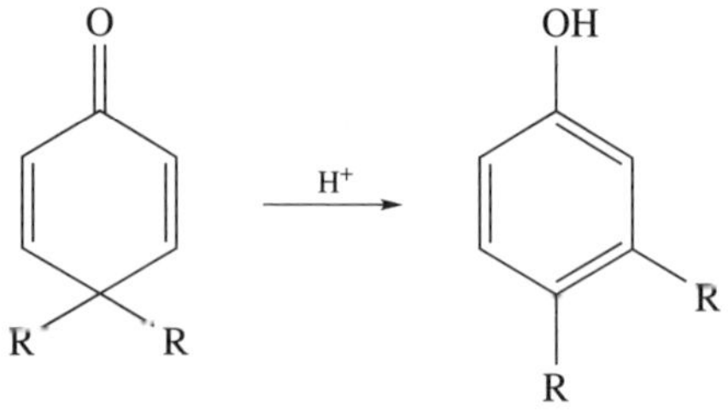

(162)

(231)   **(232)**

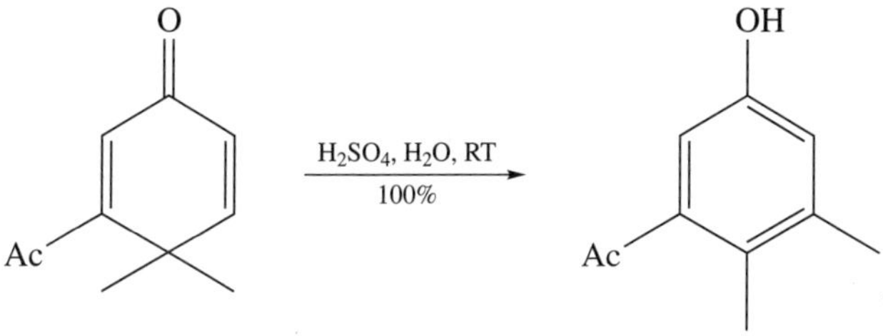

6. Synthesis of phenols

(163)[301]

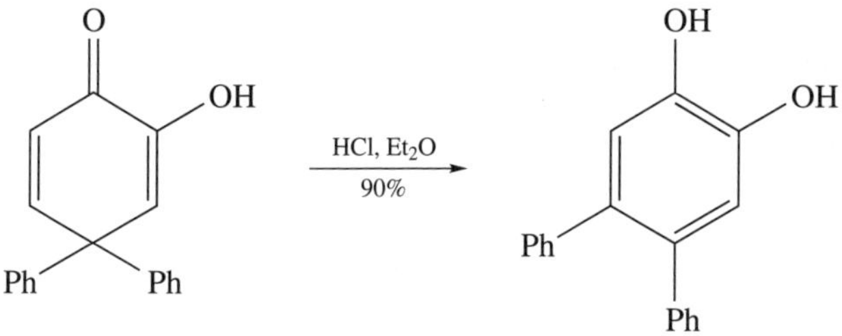

(164)[302]

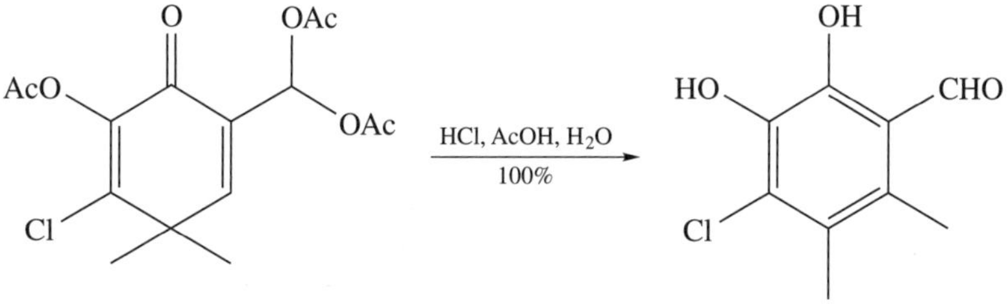

(165)[303]

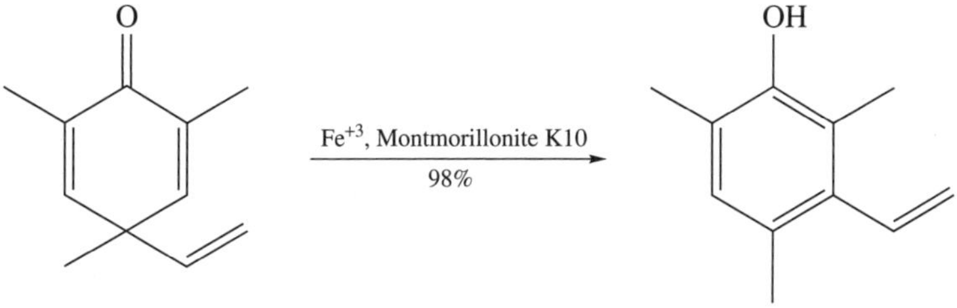

(166)[304]

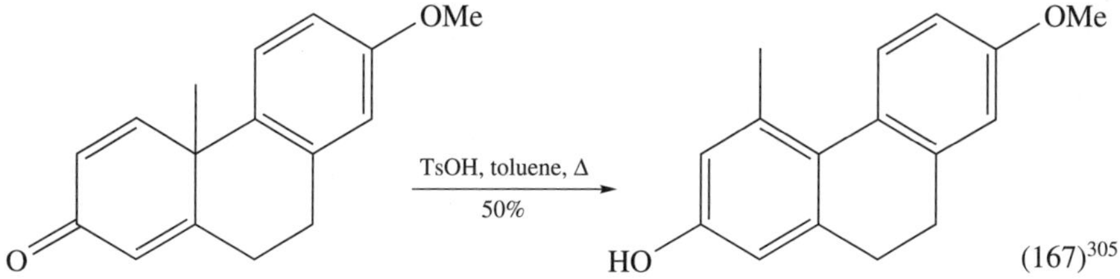

(167)[305]

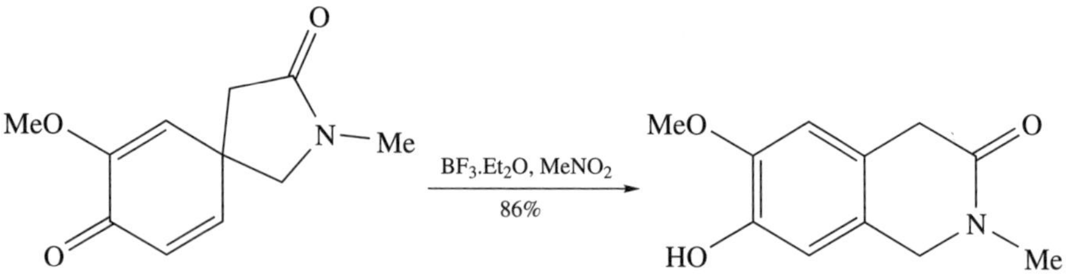

(168)[306]

Dienone–phenol rearrangements can also be achieved photochemically. For instance, cyclohexadienones **233** and **234** rearrange upon irradiation at 366 and 300 nm, respectively, to give phenols **235** and **236**, respectively, in high yields (75–87%) (equation 169)[307].

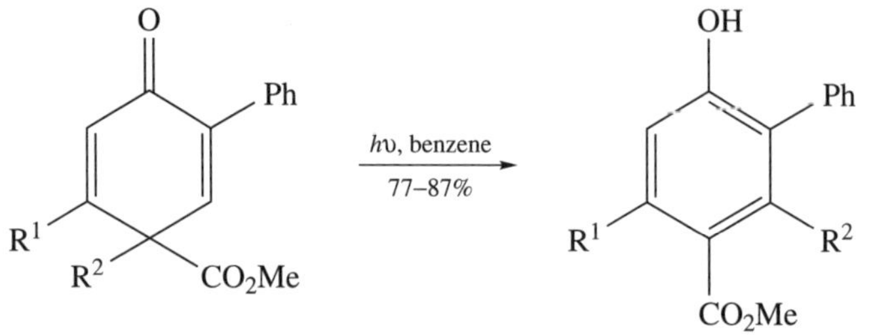

(233) R^1 = H, R^2 = alkyl
(234) R^1 = OMe, R^2 = alkyl

(235) R^1 = H, R^2 = alkyl
(236) R^1 = OMe, R^2 = alkyl

(169)

E. Phenolic Esters. Fries Rearrangement

Phenolic esters **237** can be rearranged under heating with Lewis acids or Brønsted acids in a synthetically useful reaction known as the Fries rearrangement to afford hydroxyaryl ketones **238** and **239** (equation 170)[308]. Among the wide variety of employed acids ($AlCl_3$, $HgCl_2$, $SnCl_4$, $FeCl_3$, BF_3, $AlCl_3$-$ZnCl_2$, $TiCl_4$, TsOH, H_3PO_4, HF, CH_3SO_3H, etc.)[309], $AlCl_3$ has been the most extensively used.

Two mechanistic pathways are proposed in the literature for the Fries rearrangement: (a) intramolecular[310], (b) intermolecular[311]. In the case of aryl benzoates, the Fries rearrangement has been shown to be reversible[312]. Both *ortho*- and *para*-hydroxyaryl ketones can be produced, and conditions can often be selected to enhance the yield of one of the isomers. The *ortho/para* ratio depends on the temperature, the solvent and the amount of catalyst used. Though exceptions are known, lower temperatures generally favour the *para* product and higher temperatures the *ortho* rearrangement product. For instance, benzoate **240** has been transformed at room temperature into a precursor of the coumarin dehydrogeijerin **241** (equation 171)[313]. Similarly, propionate **242** undergoes *para* Fries rearrangement at room temperature to yield the *para*-substituted phenol **243** in 97% yield (equation 172)[314]. Equations 173–177 show four examples of rearrangements at higher temperatures and in all cases *ortho*-phenols have been the main reaction product. One exception is the reaction of acetate **244** with $ZrCl_4$ at room temperature to afford in a highly selective way the corresponding *ortho*-acetylphenol **245** in 97% yield (equation 176)[315].

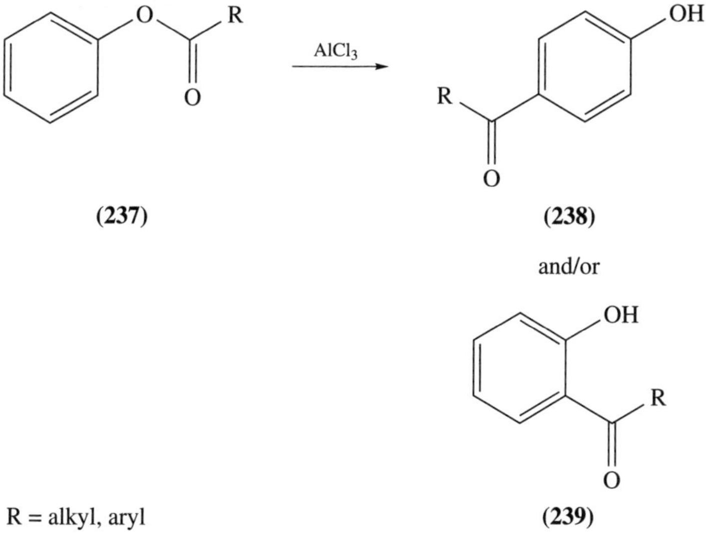

6. Synthesis of phenols

(237) → **(238)** and/or **(239)** (AlCl$_3$)

R = alkyl, aryl

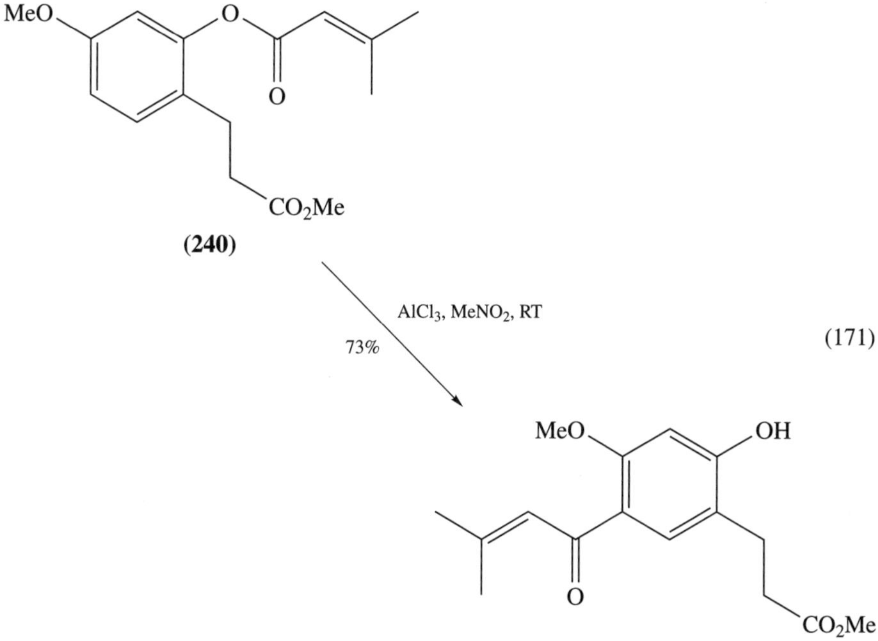

(240) —AlCl$_3$, MeNO$_2$, RT, 73%→ **(241)**

(170)

(171)

Fries rearrangement has generally been carried out using AlCl$_3$ as a promoter in more than a stoichiometric amount because most Lewis acids are deactivated by the free hydroxy groups of the products. Kobayashi and coworkers[309b] reported a catalytic version of this type of reaction using small amount of Sc(OTf)$_3$. Equation 177 shows an example where ketone **247** has been obtained in 85% yield from 1-naphthyl acetate **246** using 5 mol% of $Sc(OTf)_3$ at 100 °C.

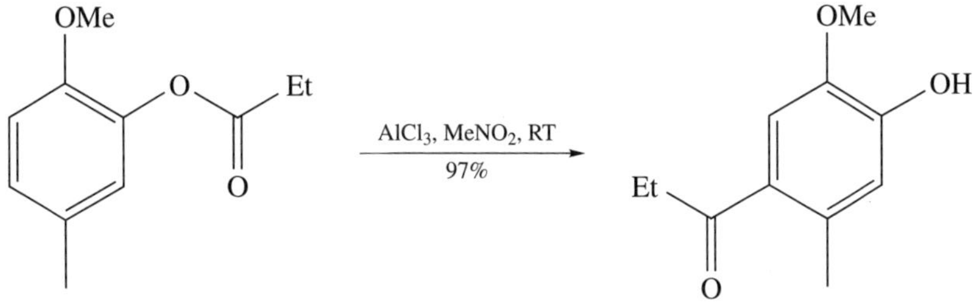
OMe
O
Et
O
AlCl3, MeNO2, RT
97%
OMe
OH
Et
O

(242) **(243)**

(172)

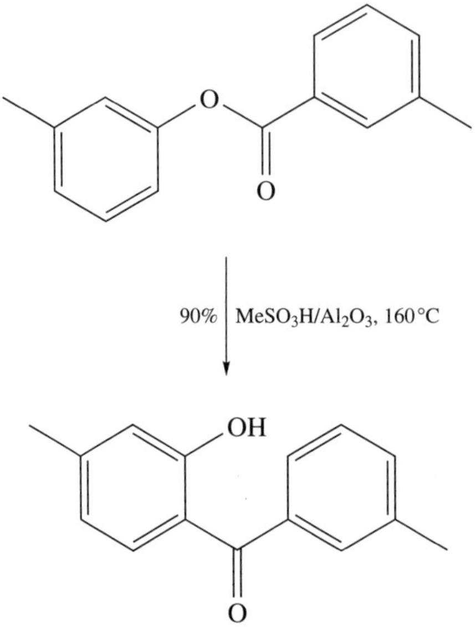
O
O
90%
MeSO3H/Al2O3, 160 °C
OH
O

(173)[316]

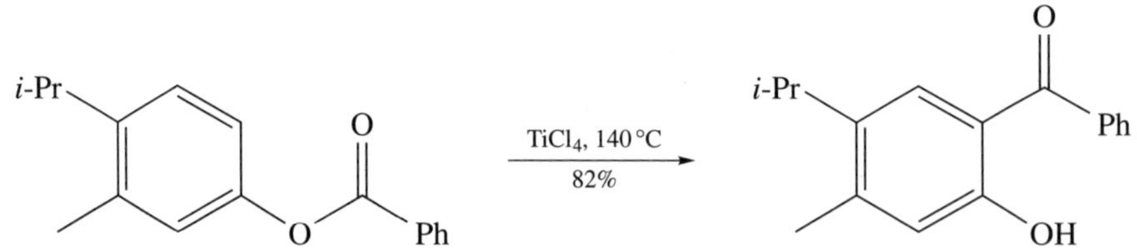
i-Pr
O
O
Ph
TiCl4, 140 °C
82%
i-Pr
O
Ph
OH

(174)[317]

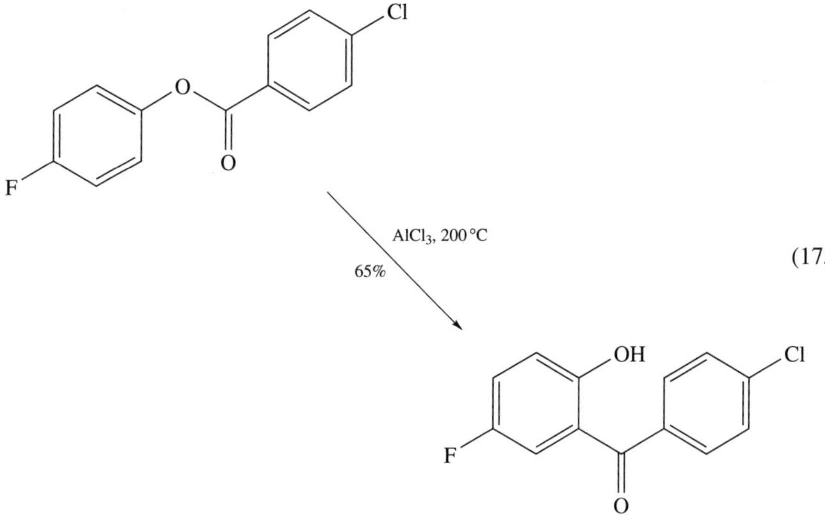

6. Synthesis of phenols

(175)[318]

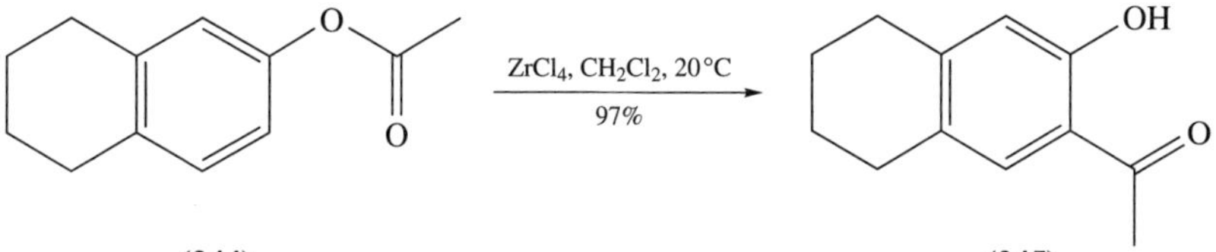

(176)

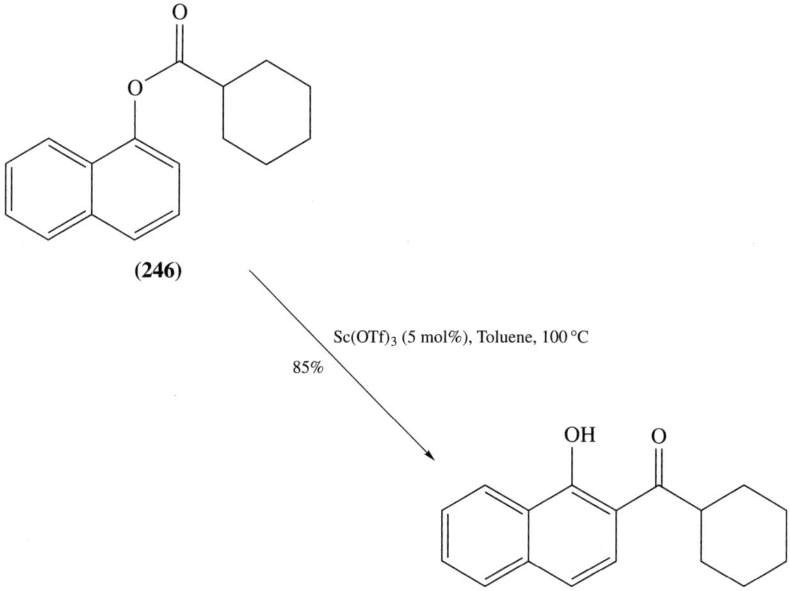

(177)

The Fries reaction can also be metal-promoted to afford, under the proper reaction conditions, good yields of specific *ortho* acyl migration products. For instance, *o*-bromophenyl pivaloate (**248**) has been treated at −95 °C with *s*-butyllithium to afford *o*-hydroxypivalophenone (**249**) in 76% yield (equation 178)[319]. Similarly, benzoate **250** gave *o*-hydroxyketone **251** in 82% yield by treatment with *n*-butyllithium (equation 179)[320].

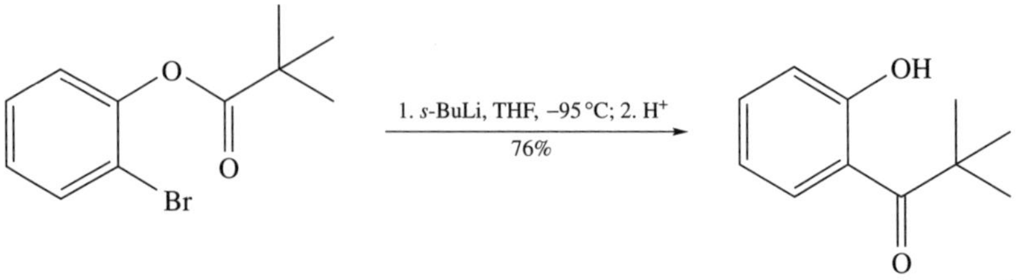

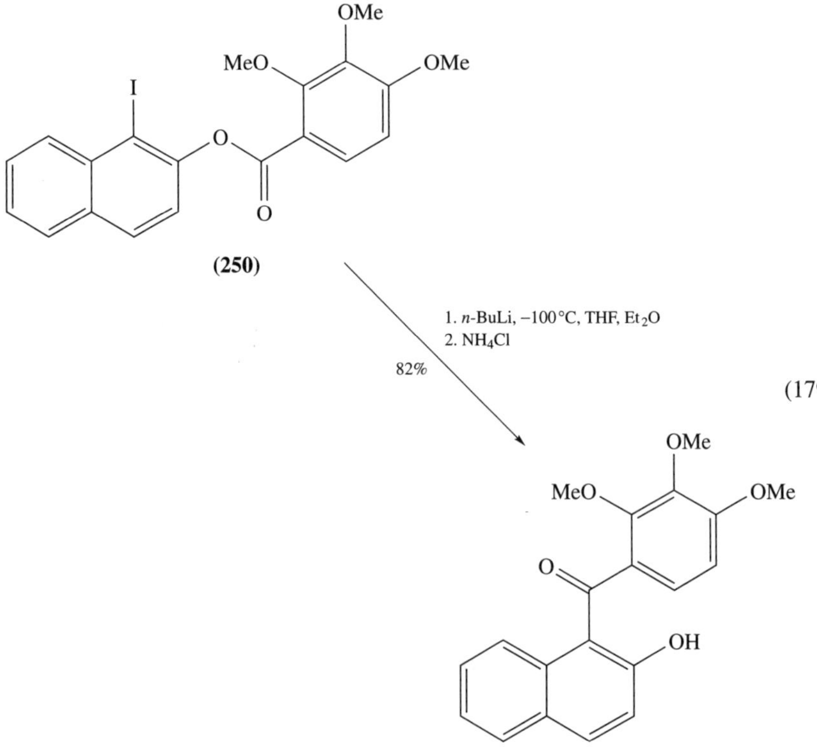

The Fries rearrangement can also be carried out in the absence of a catalyst by photolysis. This reaction, known as the photo-Fries rearrangement[321], is predominantly an intramolecular free-radical process formed by the initial photolysis of the ester[322,323]. Both *ortho* and *para* migrations are observed. The product distribution is strongly dependent on the reaction conditions[324]. Limiting the mobility of the radical pair by increasing the solvent viscosity[325] or modifying mass transfer phenomena (using restricted spaces such as cyclodextrins[324,326], micellar solutions[327] and silica surfaces[328]) allows modification of the *ortho/para* ratio. For instance, irradiation of phenylbenzoate (**252**) in water affords 80% of 2-hydroxybenzophenone (**253**) and 20% of 4-hydroxybenzophenone (**254**) (equation 180)[326c]. Nevertheless, the yield of the *ortho*-phenol **253** can be increased up to 99% using solid β-cyclodextrin (equation 181)[324a]. Similarly, phenol **256** has been obtained as the sole isomer from benzoate **255** (equation 182)[324a]. Equations 183 and 184 are examples of *ortho* and *para* rearrangements, respectively.

(**252**) (**253**), 80% (**254**), 20% (180)

(**252**) (**253**), 99% (**254**), 1% (181)

(**255**) (**256**) (182)

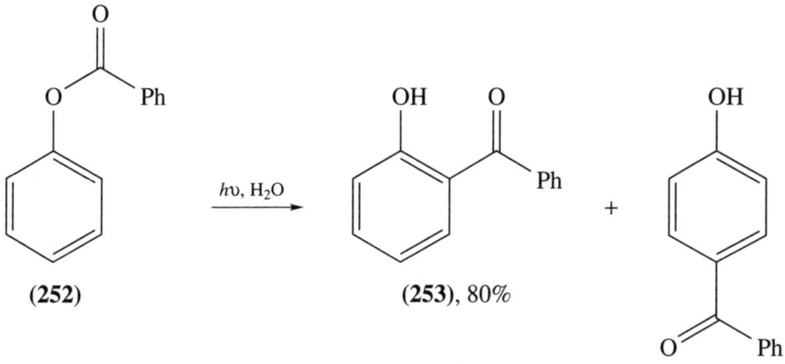

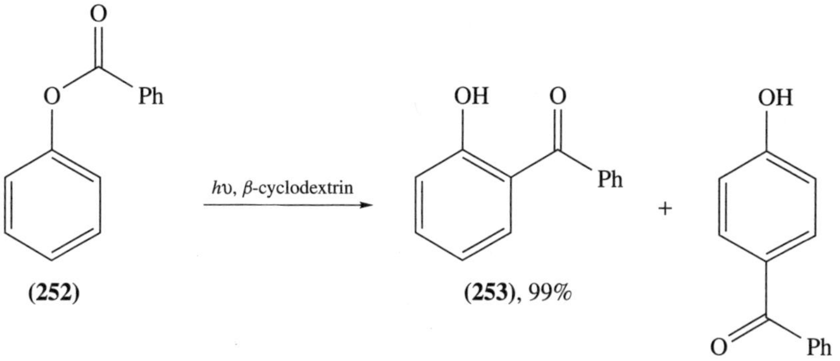

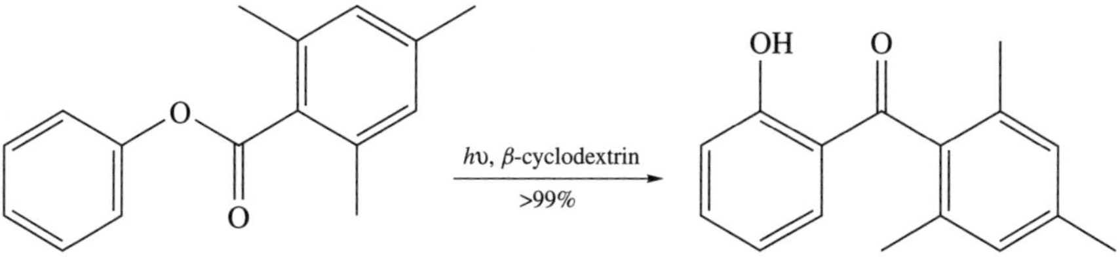

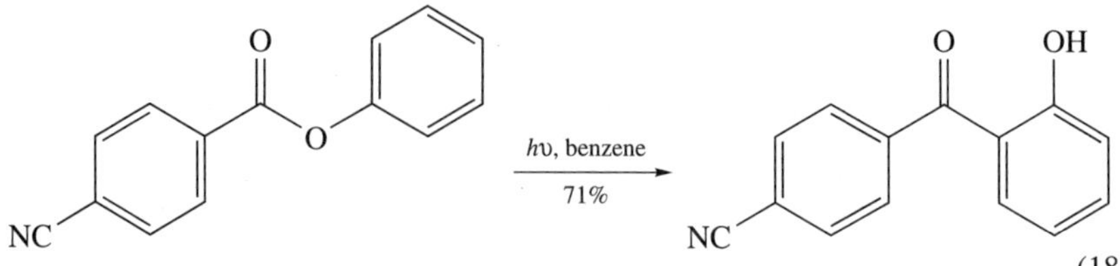

(183)[321]

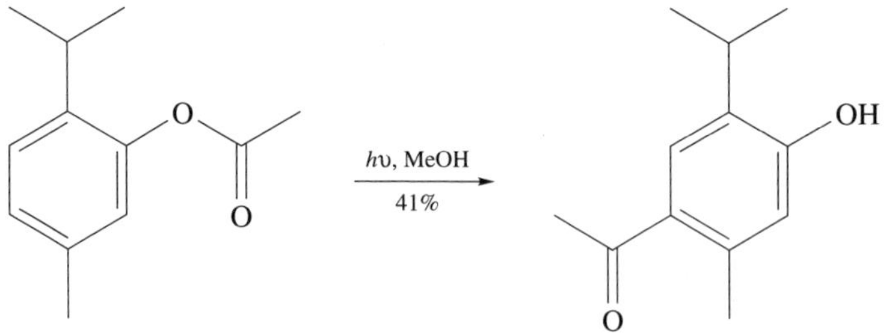

(184)[329]

VI. ACKNOWLEDGEMENT

We are indebted to the Xunta de Galicia (PGIDT99-PX120904B).

VII. REFERENCES

1. D. V. Tischenko and M. A. Churbakov, *J. Appl. Chem.*, **7**, 764 (1934); *Chem. Abstr.*, **29**, 2520 (1935).
2. (a) V. E. Meharg and I. Allen, *J. Am. Chem. Soc.*, **54**, 2920 (1932).
   (b) N. R. Shreve and C. J. Marsel, *Ind. Eng. Chem.*, **38**, 254 (1946).
3. N. Ohta and K. Kayami, *Rep. Gov. Chem. Ind. Res. Inst. Tokyo*, **47**, 327 (1952); *Chem. Abstr.*, **48**, 9941 (1954).
4. Jpn. Patent 06211716 (1994); *Chem. Abstr.*, **122**, 31113 (1994).
5. J. F. Bunnett and R. E. Zahler, *Chem. Rev.*, **49**, 273 (1951).
6. L. Desvergnes, *Chim. Ind.*, **26**, 1271 (1931).
7. J. Zilberman, D. Ioffe and I. Gozlan, *Synthesis*, 659 (1992).
8. A. P. Krapcho and D. Waterhouse, *Synth. Commun.*, **28**, 3415 (1998).
9. (a) S. Oae, N. Furukawa, M. Kise and M. Kawanishi, *Bull. Chem. Soc. Jpn.*, **39**, 1212 (1966).
   (b) L. R. Buzbee, *J. Org. Chem.*, **31**, 3289 (1966).
10. W. W. Hartman, *Org. Synth. Coll. Vol I*, 175 (1976).
11. M. F. Clarke and L. N. Owen, *J. Chem. Soc.*, 2108 (1950).
12. Ger. offen, 19651040 (1998); *Chem. Abstr.*, **129**, 54179 (1999).
13. (a) A. E. Jemmett, S. Horwood and I. Wellings, *J. Chem. Soc.*, 2794 (1958).
    (b) Ger. offen, 615118 (1978); *Chem. Abstr.*, **89**, 215118 (1978).
14. Ger. offen, 2813570 (1979); *Chem. Abstr.*, **90**, 24790 (1979).
15. DE Patent 727109 (1944); *Chem. Abstr.*, **38**, 2667 (1944).
16. D. E. Boswell, J. A. Brennan and P. S. Landis, *Tetrahedron Lett.*, **60**, 5265 (1970).
17. H. Rapoport, T. P. King and J. B. Lavigne, *J. Am. Chem. Soc.*, **73**, 2718 (1951).
18. (a) N. L. Drake, H. C. Harris and C. B. Jaeger, *J. Am. Chem. Soc.*, **70**, 168 (1948).
    (b) M. Baltas, L. Cazaux, A. De Blic, L. Gorrichon and P. Tisnes, *J. Chem. Res. (S)*, **9**, 284 (1988).
19. B. Weinstein, O. P. Crews, M. A. Leaffer, B. R. Baker and L. Goodman, *J. Org. Chem.*, **27**, 1389 (1962).
20. W. W. Hartman, J. R. Byers and J. B. Dickey, *Org. Synth.*, 451 (1943).

6. Synthesis of phenols

21. H. Seeboth, *Angew. Chem., Int. Ed. Engl.*, **6**, 307 (1967).
22. J. W. Cornforth, O. Kauber, J. E. Pike and R. Robinson, *J. Chem. Soc.*, 3348 (1955).
23. H. T. Bucherer, *J. Prakt. Chem.*, **69**, 84 (1904).
24. GB Patent 11300880 (1968); *Chem. Abstr.*, **90**, 575286 (1969).
25. E. S. Lewis and J. M. Insole, *J. Am. Chem. Soc.*, **86**, 32 (1964).
26. (a) I. Szele and H. Zollinger, *J. Am. Chem. Soc.*, **100**, 2811 (1978).
    (b) Y. Hashida, R. G. M. Landells, G. E. Lewis, I. Szele and H. Zollinger, *J. Am. Chem. Soc.*, **100**, 2816 (1978).
27. H. Leader, R. M. Smejkal, C. S. Payne, F. N. Padilla, B. P. Doctor, R. K. Gordon and P. K. Chiang, *J. Med. Chem.*, **32**, 1522 (1989).
28. J. P. Mayer, J. M. Cassady and D. E. Nichols, *Heterocycles*, **31**, 1035 (1990).
29. M. Rey, T. Vergnani and A. S. Dreiding, *Helv. Chim. Acta*, **68**, 1828 (1985).
30. Jpn. Patent 290230 (2000); *Chem. Abstr.*, **133**, 281610 (2000).
31. M. Watanabe, M. Kawada, T. Hirata, Y. Maki and I. Imada, *Chem. Pharm. Bull.*, **32**, 3551 (1984).
32. C. Parkanyi, H. L. Yuan, A. Sappok-Stang, A. R. Gutierrez and S. A. Lee, *Monatsh. Chem.*, **123**, 637 (1992).
33. T. Cohen, A. G. Dietz and J. R. Miser, *J. Org. Chem.*, **42**, 2053 (1977).
34. J. Wellmann and E. Steckhan, *Chem. Ber.*, **110**, 3561 (1977).
35. M. G. Evans and N. Uri, *Trans. Faraday Soc.*, **45**, 224 (1949).
36. D. H. Derbyshire and W. A. Waters, *Nature*, **165**, 401 (1950).
37. (a) J. A. Vesely and L. Schmerling, *J. Org. Chem.*, **35**, 4028 (1970).
    (b) G. A. Olah, T. Keumi and A. P. Fung, *Synthesis*, 536 (1979).
38. G. A. Olah, A. P. Fung and T. Keumi, *J. Org. Chem.*, **46**, 4305 (1981).
39. (a) C. Berrier, J.-C. Jacquesy, M.-P. Jouannetaud and A. Renoux, *Tetrahedron Lett.*, **27**, 4565 (1986).
    (b) J.-C. Jacquesy, M.-P. Jouannetaud, G. Morellet and Y. Vidal, *Tetrahedron Lett.*, **25**, 1479 (1984).
40. M. E. Kurz and G. J. Johnson, *J. Org. Chem.*, **36**, 3184 (1971).
41. (a) A. H. Haines, *Methods for the Oxidation of Organic Compounds*, Academic Press, London, 1985, p. 173.
    (b) J. R. L Smith and R. O. C. Norman, *J. Chem. Soc.*, 2897 (1963).
    (c) C. Walling and R. A. Johnson, *J. Am. Chem. Soc.*, **97**, 363 (1975).
    (d) C. Walling and O. M. Camaioni, *J. Am. Chem. Soc.*, **97**, 1603 (1975).
42. G. A. Hamilton, J. W. Hanifin and J. P. Friedman, *J. Am. Chem. Soc.*, **88**, 5269 (1966).
43. (a) S. Udenfriend, C. T. Clark, J. Axelrod and B. B. Brodie, *J. Biol. Chem.*, **208**, 731 (1954).
    (b) C. T. Clark, R. D. Downing and J. B. Martin, *J. Org. Chem.*, **27**, 4698 (1962).
44. D. Jerina, J. Daly, W. Landis, B. Witkop and S. Undenfriend, *J. Am. Chem. Soc.*, **89**, 3347 (1967).
45. M. Blanchard, C. Bouchoule, G. Djaneye-Boundjou and P. Canesson, *Tetrahedron Lett.*, **29**, 2177 (1988).
46. (a) M. Makosza and J. Winiarski, *Acc. Chem. Res.*, **20**, 282 (1987).
    (b) M. Makosza and K. Wojcierchauski, *Liebigs Ann. Chem.*, 1805 (1997).
    (c) J. Miller, *Aromatic Nucleophilic Substitution*, Elsevier, New York, 1968.
47. M. Makosza and T. Gluika, *J. Org. Chem.*, **48**, 3860 (1983).
48. (a) T. Brose, F. Holzscheiter, G. Mattersteig, W. Pritzkow and V. Voerckel, *J. Prakt. Chem.*, **324**, 497 (1992).
    (b) G. Mattersteig and W. Pritzkow, *J. Prakt. Chem.*, **322**, 569 (1990).
49. (a) M. Makosza and K. Sienkiewicz, *J. Org. Chem.*, **55**, 4979 (1990).
    (b) M. Makosza and K. Sienkiewicz, *J. Org. Chem.*, **63**, 4199 (1998).
    (c) R. G. Pews, Z. Lysenko and P. C. Vosejpka, *J. Org. Chem.*, **62**, 8255 (1997).
50. C. Walling and R. B. Hodgdon, *J. Am. Chem. Soc.*, **80**, 228 (1958).
51. T. Tezuka, N. Narita, W. Ando and S. Oae, *J. Am. Chem. Soc.*, **103**, 3045 (1981).
52. (a) N. A. Milas, *J. Am. Chem. Soc.*, **59**, 2342 (1937).
    (b) N. A. Milas and S. Sussman, *J. Am. Chem. Soc.*, **59**, 2345 (1937).
53. Y. Ogata, Y. Sawaki, K. Tomizawa and T. Ohno, *Tetrahedron*, **37**, 1485 (1981).
54. (a) K. Elbs, *J. Prakt. Chem.*, **48**, 179 (1893).

(b) E. J. Bergmann and B. M. Pitt, *J. Am. Chem. Soc.*, **80**, 3717 (1958).
(c) S. M. Sethna, *Chem. Rev.*, **49**, 91 (1951).
(d) R. Dermer, *Chem. Rev.*, **57**, 103 (1957).
(e) F. Minisci, A. Citterio and C. Giordano, *Acc. Chem. Res.*, **16**, 27 (1983).
(f) E. J. Behrman, *Org. React.*, **35**, 421 (1988).
(g) E. J. Behrman, *J. Am. Chem. Soc.*, **85**, 3478 (1963).
(h) E. J. Behrman and P. P. Walker, *J. Am. Chem. Soc.*, **84**, 3454 (1962).
55. A. H. Haines, *Methods for the Oxidation of Organic Compounds*, Academic Press, London, 1985, p. 180.
56. W. Baker and N. C. Brown, *J. Chem. Soc.*, 2303 (1948).
57. W. Baker and R. Savage, *J. Chem. Soc.*, 1602 (1938).
58. M. Iimura, T. Tanaka and S. Matsuura, *Chem. Pharm. Bull.*, **32**, 3354 (1984).
59. N. Adityachaudhury, C. L. Kirtaniya and B. Mukherjee, *Tetrahedron*, **27**, 2111 (1971).
60. K. E. Schulte and G. Rücker, *Arch. Pharm.*, **297**, 182 (1964).
61. R. B. Desai and S. Sethna, *J. Indian Chem. Soc.*, **28**, 213 (1951).
62. (a) E. Boyland, D. Manson and P. Sims, *J. Chem. Soc.*, 3623 (1953).
(b) E. J. Behrman, *Org. React.*, **35**, 432 (1988).
63. (a) J. T. Edward, *J. Chem. Soc.*, 1464 (1954).
(b) E. Boyland and P. Sims, *J. Chem. Soc.*, 980 (1954).
(c) E. J. Behrman, *J. Org. Chem.*, **57**, 2266 (1992).
64. (a) R. D. Chambers, P. Goggin and W. K. R. Musgrave, *J. Chem. Soc.*, 1804 (1959).
(b) J. D. McClure, *J. Org. Chem.*, **28**, 69 (1963).
65 (a) H. Hart and C. A. Buehler, *J. Org. Chem.*, **29**, 2397 (1964).
(b) H. Hart, *Acc. Chem. Res.*, **4**, 337 (1971).
66. (a) N. L. Weinberg and N. C. Wu, *Tetrahedron Lett.*, 3367 (1975).
(b) US. Patent 4024032, (1978); *Chem. Abstr.*, **103**, 1591332 (1985).
67. Fr. Patent 2422733 (1979); *Chem. Abstr.*, **92**, 146427 (1980).
68. K. Fujimoto, Y. Tokuda, H. Maekawa, Y. Matsubara, T. Mizumo and I. Nishiguchi, *Tetrahedron*, **52**, 3889 (1996).
69. J. B. Powlowski, S. Dagley, V. Massey and D. P. Ballou, *J. Biol. Chem.*, **262**, 69 (1987).
70. (a) A. M. Klibanov, Z. Berman and B. N. Alberti, *J. Am. Chem. Soc.*, **103**, 6263 (1981).
(b) S. G. Burton, A. Boshoff, W. Edwards and P. D. Rose, *J. Mol. Cat. B.: Enzym.*, **5**, 411 (1998).
71. (a) B. J. Auret, S. K. Balani, D. R. Boyd and R. M. E. Greene, *J. Chem. Soc., Perkin Trans. 1*, 2659 (1984).
(b) A. Wiseman and D. J. King, *Topics Enzymol. Ferment. Biotechnol.*, **6**, 151 (1982).
72. D. R. Boyd, R. M. Campbell, H. C. Craig, C. G. Watson, J. W. Daly and D. M. Jerina, *J. Chem. Soc., Perkin Trans. 1*, 2438 (1976).
73. H. A. Schreuder, W. G. J. Hol and J. Drenth, *J. Biol. Chem.*, **263**, 3131 (1988).
74. (a) B. Vigne, A. Archelas and R. Furstoss, *Tetrahedron*, **47**, 1447 (1991).
(b) H. Yoshioka, T. Nagasawa, R. Hasegawa and H. Yamada, *Biotechnol. Lett.*, 679 (1990).
(c) R. J. Theriault and T. M. Longfield, *Appl. Microbiol.*, **25**, 606 (1973).
75. (a) A. L. Hunt, D. E. Hughes and J. M. Lowenstein, *Biochem. J.*, **69**, 170 (1958).
(b) Eur. Patent 559132 (1985); *Chem. Abstr.*, **103**, 1591332 (1985).
(c) T. Nagasawa, B. Hurh and T. Yamane, *Biotechnol. Biochem.*, **58**, 665 (1994).
76. F. M. Pasutto, N. N. Singh, F. Jamali, R. T. Coutts and S. Abuzar, *J. Pharm. Sci.*, **76**, 177 (1987).
77. M. Iwamoto, J.-I. Hirata, K. Matsukami and S. Kagawa, *J. Phys. Chem.*, **87**, 903 (1983).
78. (a) E. Suzuki, K. Makashiro and Y. Ono, *Bull. Chem. Soc. Jpn., Chem. Commun.*, 953 (1988).
(b) USSR Patent 4446646 (1988); *Chem. Abstr.*, **112**, 216427 (1990).
79. A. S. Kharitonov, V. I. Sobolev and G. I. Panov, *Russ. Chem. Rev.*, **61**, 11130 (1992).
80. (a) G. Belussi and C. Perego, *CATTECH*, **4**, 4 (2000).
(b) G. I. Panov, *CATTECH*, **4**, 18 (2000).
(c) P. P. Notté, *Topics Catal.*, **13**, 387 (2000).
81. H. Mimoun, L. Saussine, E. Daire, M. Postel, J. Fischer and R. Weiss, *J. Am. Chem. Soc.*, **105**, 3101 (1983).
82. (a) K. B. Sharpless and T. R. Verhoeven, *Aldrichim. Acta*, **12**, 63 (1979).
(b) F. Di Furia and G. Modena, *Pure Appl. Chem.*, **54**, 1853 (1982).

6. Synthesis of phenols

83. M. Bonchio, V. Conte, F. Di Furia and G. Modena, *J. Org. Chem.*, **54**, 4368 (1989).
84. M. Bianchi, M. Bonchio, V. Conte, F. Coppa, F. Di Furia, G. Modena, S. Moro and S. Standen, *J. Mol. Catal.*, **83**, 107 (1993).
85. M. Bonchio, V. Conte, F. Di Furia, G. Modena and S. Moro, *J. Org. Chem.*, **59**, 6262 (1994).
86. (a) T. Jintoku, H. Taniguchi and Y. Fujiwara, *Chem. Lett.*, 1865 (1987).
    (b) T. Jintoku, K. Takaki, Y. Fujiwara, Y. Fuchita and K. Hiraki, *Bull. Chem. Soc. Jpn.*, **63**, 438 (1990).
87. N. Kitajima, M. Ito, H. Fukui and Y. Moro-oka, *J. Chem. Soc., Chem. Commun.*, 102 (1991).
88. K. Kimura and R. Machida, *J. Chem. Soc., Chem. Commun.*, 499 (1984).
89. (a) T. Kitano, Y. Kuroda, M. Mori, S. Ito and K. Sasaki, *J. Chem. Soc., Perkin Trans. 2*, 981 (1993).
    (b) Y. Kuroda, M. Mori, A. Itoh, F. Kitano, F. Yamaguchi, K. Sasaki and M. Nitta, *J. Mol. Catal.*, **73**, 237 (1992).
    (c) T. Kitano, T. Wani, T. Ohnishi, T. Jiang, Y. Kuroda, A. Kunai and K. Sasaki, *Catal. Lett.*, 11 (1991).
90. (a) M. G. Clerici and P. Ingallina, *Catal. Today*, **41**, 351 (1998).
    (b) Y. Moro-oka and N. Akita, *Catal. Today*, **41**, 327 (1998).
    (c) A. Kunai, K. Ishihata, S. Ito and K. Sasaki, *Chem. Lett.*, 1967 (1998).
    (d) A. Kunai, T. Kitano, Y. Kuroda, L.-F. Jiang and K. Sasaki, *Catal. Lett.*, **4**, 139 (1990).
    (e) K. Otsuka and I. Yumanaka, *Catal. Today*, **57**, 71 (2000).
    (f) T. Tatsumi, K. Yuasa and H. Tominaga, *J. Chem. Soc., Chem. Commun.* 1446 (1992).
    (g) T. Miyake, M. Hamada, Y. Sasaki and M. Oguri, *Appl. Catal. A. Gen.*, **131**, 33 (1995).
    (h) I. Yamanaka, K. Nakagaki, T. Akimoto and K. Otsuka, *J. Chem. Soc., Perkin Trans. 1*, 2511 (1996).
    (i) K. Otsura, I. Yamanaka and K. Hosokawa, *Nature*, **345**, 697 (1990).
    (j) I. Yamanaka, T. Nabeta, S. Takenaka and K. Otsuda, *Stud. Surf. Sci. Catal.*, **130**, 815 (2000).
91. S.-I. Niwa, M. Eswaramoorthy, J. Nair, A. Raj, N. Itoh, H. Shoji, T. Namba and F. Mizukami, *Science*, **295**, 105 (2002).
92. G. Sosnovsky and J. H. Brown, *Chem. Rev.*, **66**, 529 (1966).
93. K. A. Parker and K. A. Kozishi, *J. Org. Chem.*, **52**, 674 (1987).
94. (a) M. Julia, V. P. Saint-Jalmes and J.-N. Verpenaux, *Synlett*, 233 (1993).
    (b) H. W. Gschwend and H. R. Rodríguez, *Org. React.*, **26**, 1 (1979).
    (c) M. Nilsson and T. Norin, *Acta Chim. Scand.*, **17**, 1157 (1963).
95. S. W. Breuer and F. A. Broster, *J. Organomet. Chem.*, **35**, C5 (1972).
96. E. Negishi, in *Comprehensive Organometallic Chemistry* (Eds. G. Wilkinson, F. G. A. Stone and E. W. Abel), Vol. 7, Pergamon Press, Oxford, 1982, pp. 270, 305, 326 and 354.
97. (a) T. L. Yarboro and C. Karr, *J. Org. Chem.*, **24**, 1141 (1959).
    (b) H. G. Kuivilla, *J. Am. Chem. Soc.*, **76**, 870 (1954).
    (c) H. C. Brown, C. Snyder, B. C. S. Rao and G. Zweifel, *Tetrahedron*, **42**, 5505 (1986).
    (d) C. Lane and V. Sniekus, *Synlett*, 1294 (2000).
98. (a) H. C. Brown, M. M. Midland and G. W. Kabalka, *J. Am. Chem. Soc.*, **93**, 1024 (1971).
    (b) H. C. Brown, M. M. Midland and G. W. Kabalka, *Tetrahedron*, **42**, 5523 (1986).
99. G. W. Kabalka, T. M. Shoup and N. M. Goudgaon, *J. Org. Chem.*, **54**, 5930 (1989).
100. G. W. Kabalka, P. P. Wadgaonkar and T. M. Shoup, *Organometallics*, **9**, 1316 (1990).
101. K. S. Webb and D. Levy, *Tetrahedron Lett.*, **36**, 5117 (1995).
102. R. Köster and Y. Morita, *Justus Liebigs Ann. Chem.*, **704**, 70 (1967).
103. (a) G. W. Kabalka and H. C. Hedgecock, *J. Org. Chem.*, **40**, 1776 (1975).
     (b) G. W. Kabalka and S. W. Slayden, *J. Organomet. Chem.*, **125**, 273 (1977).
104. M. F. Hawthorne, *J. Org. Chem.*, **22**, 1001 (1957).
105. (a) R. L. Kidwell and S. D. Darling, *Tetrahedron Lett.*, 531 (1966).
     (b) R. L. Kidwell, M. Murphy and S. D. Darling, *Org. Synth.*, **49**, 90 (1969).
106. (a) M. Iwao, J. N. Reed and V. Sniekus, *J. Am. Chem. Soc.*, **104**, 5531 (1982).
     (b) V. Sniekus, *Chem. Rev.*, **90**, 879 (1990).
107. M. Kawase, A. K. Sinhababu, E. M. McGhee, T. Milby and R. T. Borchardt, *J. Med. Chem.*, **33**, 2204 (1990).
108. E. C. Taylor and A. McKillop, *Acc. Chem. Res.*, **3**, 338 (1970).

109. A. McKillop and E. C. Taylor, in *Comprehensive Organometallic Chemistry* (Eds. G. Wilkinson, F. G. A. Stone and E. W. Abel), Vol. 7, Pergamon, Oxford, 1982, p. 465.
    (b) S. D. Burke and R. L. Danheiser (Eds.), *Oxidizing and Reducing Agents*, in *Handbook of Reagents for Organic Synthesis*, Wiley, Chichester, 1999, p. 450.
110. (a) E. C. Taylor, F. Kienzle, R. L. Robey and A. McKillop, *J. Am. Chem. Soc.*, **92**, 2175 (1970).
    (b) E. C. Taylor, H. W. Atland, R. H. Danforth, G. McGillivray and A. McKillop, *J. Am. Chem. Soc.*, **92**, 3520 (1970).
111. A. McKillop, J. S. Fowler, M. J. Zelesko, J. D. Hunt, E. C. Taylor and G. McGillivray, *Tetrahedron Lett.*, 2423 (1969).
112. S. W. Breuer, G. M. Pickles, J. C. Podesta and F. G. Thorpe, *J. Chem. Soc., Chem. Commun.*, 36 (1975).
113. N. J. Lewis, S. Y. Gabhe and M. R. DelaMater, *J. Org. Chem.*, **42**, 1479 (1977).
114. R. C. Cambie, P. I. Higgs, P. S. Rutledge and P. D. Woodgate, *J. Organomet. Chem.*, **384**, C6 (1990).
115. S. D. Burke and R. L. Danheiser (Eds.), *Oxidizing and Reducing Agents*, in *Handbook of Reagents for Organic Synthesis*, Wiley, Chichester, 1999, p. 262.
116. F. A. Davis, J. Wei, A. C. Sheppard and S. Gubernick, *Tetrahedron Lett.*, **28**, 5115 (1987).
117. F. A. Davis, J. M. Billmers, D. J. Gosciniak, J. C. Towson and R. D. Bach, *J. Org. Chem.*, **51**, 4240 (1986).
118. F. A. Davis, M. S. Haque, T. G. Ulatowski and J. C. Towson, *J. Org. Chem.*, **51**, 2402 (1986).
119. H. Taddei and A. Ricci, *Synthesis*, 633 (1986).
120. J. R. Falck, J.-Y. Lai, D. V. Ramana and S.-G. Lee, *Tetrahedron Lett.*, **40**, 2715 (1999).
121. H. J. Shine, *Aromatic Rearrangements*, Elsevier, Amsterdam, 1967, p. 182.
122. (a) E. Bamberger, *Chem. Ber.*, **27**, 1327 (1894).
    (b) E. Bamberger, *Chem. Ber.*, **33**, 3600 (1900).
123. (a) T. Sone, K. Hamamoto, Y. Seiji, S. Shinkai and O. Manabe, *J. Chem. Soc., Perkin Trans. 2*, 1596 (1981).
    (b) G. Kohnstam, W. A. Petch and D. L. H. Williams, *J. Chem. Soc., Perkin Trans. 2*, 423 (1984).
    (c) L. A. Sternson and R. Chandrasakar, *J. Org. Chem.*, **49**, 4295 (1984).
124. (a) J. C. Fishbein and R. A. McClelland, *J. Am. Chem. Soc.*, **109**, 2824 (1987).
    (b) J. C. Fishbein and R. A. McClelland, *Can. J. Chem.*, **74**, 1321 (1996).
125. J. I. Kroschwitz and M. Howe-Grant (Eds.), *Kirk-Othemer Encyclopedia of Chemical Technology*, 4th Ed., Vol. 2, Wiley, New York, 1996, p. 586.
126. A. Zoran, O. Khodzhaev and Y. Sasson, *J. Chem. Soc., Chem. Commun.*, 2239 (1994).
127. A. Dalenko and E. Buncel, *Can. J. Chem.*, **52**, 623 (1974).
128. (a) H. J. Shine, *Aromatic Rearrangements*, Elsevier, Amsterdam, 1967, pp. 272 and 357.
    (b) E. Buncel, *Can. J. Chem.*, **78**, 1251 (2000).
129. E. Buncel, *Acc. Chem. Res.*, **8**, 132 (1975).
130. J. Singh, P. Singh, J. L. Boivin and P. E. Gagnon, *Can. J. Chem.*, **41**, 499 (1963).
131. I. Shimao and S. Oae, *Bull. Chem. Soc. Jpn.*, **56**, 643 (1983).
132. R. Tanikaga, K. Maruyama, R. Goto and A. Kaji, *Tetrahedron Lett.*, 5925 (1966).
133. A. Albini, E. Fasani, M. Moroni and S. Pietra, *J. Org. Chem.*, **51**, 88 (1986).
134. (a) G. M. Badger and R. G. Buttery, *J. Chem. Soc.*, 2243 (1954).
    (b) D. J. W. Goon, N. G. Murray, J.-P. Schoch and N. J. Bunce, *Can. J. Chem.*, **51**, 3827 (1973).
    (c) R. H. Squire and H. H. Jaffé, *J. Am. Chem. Soc.*, **95**, 8188 (1973).
    (d) H. J. Shine, W. Subotkowski and E. Gruszecka, *Can. J. Chem.*, **64**, 1108 (1986).
135. (a) T. Doppler, H. Schmid and H.-J. Hansen, *Helv. Chim. Acta*, **62**, 271 (1979).
    (b) E. Giovannini, J. Rosales and B. De Sousa, *Helv. Chim. Acta*, **54**, 2111 (1971).
136. (a) H. D. Dakin, *Am. Chem. J.*, **42**, 477 (1909).
    (b) W. Baker, H. F. Bondy, J. Gumb and D. Miles, *J. Chem. Soc.*, 1615 (1953).
137. M. B. Hocking, *Can. J. Chem.*, **51**, 2384 (1973).
138. R. I. Meltzer and J. Doczi, *J. Am. Chem. Soc.*, **72**, 4986 (1950).
139. D. G. Orphanos and A. Taurins, *Can. J. Chem.*, **44**, 1875 (1966).
140. D. G. Crosby, *J. Org. Chem.*, **26**, 1215 (1961).

6. Synthesis of phenols

141. J. Zhu, R. Beugelmans, A. Bigot, G. P. Singh and M. Bois-Choussy, *Tetrahedron Lett.*, **34**, 7401 (1993).
142. (a) A. Roy, K. R. Reddy, P. K. Mohanta, H. Ila and H. Junjappa, *Synth. Commun.*, **29**, 3781 (1999).
     (b) M. Matsumoto, H. Kobayashi and Y. Hotta, *J. Org. Chem.*, **49**, 4740 (1984).
143. R. S. Varma and K. P. Naicker, *Org. Lett.*, **1**, 189 (1999).
144. J. Boeseken, W. D. Cohen and C. J. Kip, *Recl. Trav. Chim. Pays-Bas.*, **55**, 815 (1936).
145. Y. Ogata and Y. Sawaki, *J. Org. Chem.*, **34**, 3985 (1969).
146. F. Camps, J. Coll, A. Messeguer and M. A. Pericàs, *Tetrahedron Lett.*, **22**, 3895 (1981).
147. (a) L. Syper, *Synthesis*, 167 (1989).
     (b) M. E. Jung and T. I. Lazarova, *J. Org. Chem.*, **62**, 1553 (1997).
     (c) V. Bolitt and C. Mioskowski, *J. Am. Chem. Soc.*, **113**, 6320 (1991).
148. G. W. Kabalka, N. K. Reddy and C. Narayana, *Tetrahedron Lett.*, **33**, 865 (1992).
149. A. McKillop and W. R. Sanderson, *Tetrahedron*, **51**, 6145 (1995).
150. C. H. Hassall, *Org. React.*, **9**, 73 (1957).
151. W. H. Saunders, *J. Am. Chem. Soc.*, **77**, 4679 (1955).
152. (a) M. Hawthorne, W. D. Emmons and K. S. McCallum, *J. Am. Chem. Soc.*, **80**, 6393 (1958).
     (b) W. D. Emmons and G. B. Lucas, *J. Am. Chem. Soc.*, **77**, 2287 (1955).
     (c) L. Syper, K. Kloc and J. Mlochowski, *Tetrahedron*, **36**, 123 (1980).
153. (a) K. A. Parker and T. Iqbal, *J. Org. Chem.*, **45**, 1149 (1980).
     (b) W. H. Rastetter, T. J. Richard and M. D. Lewis, *J. Org. Chem.*, **43**, 3163 (1978).
154. (a) I. M. Godfrey, M. V. Sargent and J. A. Elix, *J. Chem. Soc., Perkin Trans. 1*, 1353 (1974).
     (b) R. L. Hannan, R. B. Barber and H. Rapoport, *J. Org. Chem.*, **44**, 2153 (1979).
     (c) S. Bengtssan and T. Högberg, *J. Org. Chem.*, **58**, 3538 (1993).
155. (a) M. Matsumoto, H. Kobayashi and Y. Hotta, *J. Org. Chem.*, **49**, 4740 (1984).
     (b) H. Ishii, K. Harada, T. Ishida, E. Ueda and K. Nakajima, *Tetrahedron Lett.*, 319 (1975).
156. M. F. Hawthorne, W. D. Emmons and K. S. McCallum, *J. Am. Chem. Soc.*, **80**, 6393 (1958).
157. (a) S. L. Friess and A. H. Soloway, *J. Am. Chem. Soc.*, **73**, 3968 (1951).
     (b) E. E. Smissman, J. P. Li and Z. H. Israili, *J. Org. Chem.*, **33**, 4231 (1968).
158. W. E. Doering and L. Speers, *J. Am. Chem. Soc.*, **72**, 5515 (1950).
159. P. Bagchi, F. Bergmann and D. K. Banerjee, *J. Am. Chem. Soc.*, **71**, 989 (1949).
160. D. K. Datta and P. Bagchi, *J. Org. Chem.*, **25**, 932 (1960).
161. M. T. Tetenbaum and C. R. Hauser, *J. Org. Chem.*, **23**, 229 (1958).
162. P. P. Fu and R. G. Harvey, *Org. Prep. Proced. Int.*, **13**, 152 (1981).
163. J. W. Lyga and G. A. Meier, *Synth. Commun.*, **24**, 2491 (1994).
164. Chinese Patent CN 1128747 (1996); *Chem. Abstr.*, **131**, 199517 (1996).
165. J. K. Ray and R. G. Harvey, *J. Org. Chem.*, **48**, 1352 (1983).
166. (a) F. W. Comer, T. Money and A. I. Scott, *J. Chem. Soc., Chem. Commun.*, 231 (1967).
     (b) A. J. Birch and F. W. Donovan, *Aust. J. Chem.*, **6**, 360 (1953).
     (c) T. Money, *Chem. Rev.*, **70**, 553 (1970).
167. T. M. Harris and C. M. Harris, *Tetrahedron*, **33**, 2159 (1977).
168. G. Bringmann and S. Schneider, *Liebigs Ann. Chem.*, 765 (1985).
169. A. B. Smith III and S. N. Kilényi, *Tetrahedron Lett.*, **26**, 4419 (1985).
170. A. G. M. Barrett and R. A. E. Carr, *J. Org. Chem.*, **51**, 4254 (1986).
171. O. Lefebvre, T. Brigaud and C. Portella, *Tetrahedron*, **54**, 5939 (1998).
172. M. Yamaguchi, K. Shibato, H. Nakashima and T. Minami, *Tetrahedron*, **44**, 4767 (1988).
173. M. A. Tius, A. Thurkauf and J. W. Truesdell, *Tetrahedron Lett.*, **23**, 2823 (1982).
174. T. M. Harris and P. J. Wittek, *J. Am. Chem. Soc.*, **97**, 3270 (1975).
175. H. Stetter and S. Vestner, *Chem. Ber.*, **97**, 169 (1964).
176. J. F. Stephen and E. Marcus, *J. Org. Chem.*, **35**, 258 (1970).
177. H. Zak and U. Schmidt, *Chem. Ber.*, **106**, 3652 (1973).
178. U. Schmidt and M. Schwochau, *Monatsh. Chem.*, **98**, 1492 (1967).
179. (a) J. L. Douglas and T. Money, *Tetrahedron*, **23**, 3545 (1967).
     (b) A. I. Scott, H. Guilford, J. J. Ryan and D. Skingle, *Tetrahedron*, **27**, 3025 (1971).
     (c) T. Money, T. H. Qureshi, G. B. Webster and A. I. Scott, *J. Am. Chem. Soc.*, **87**, 3004 (1965).
     (d) T. Money, J. L. Douglas and A. I. Scott, *J. Am. Chem. Soc.*, **88**, 624 (1966).

(e) T. Money, F. W. Comer, G. B. Webster, I. G. Wright and A. I. Scott, *Tetrahedron*, **23**, 1990 (1967).
180. (a) T. M. Harris, M. P. Wachter and G. H. Wiseman, *J. Chem. Soc., Chem. Commun.*, 177 (1969).
     (b) T. M. Harris and M. P. Wachter, *Tetrahedron*, **26**, 5255 (1970).
181. (a) L. Crombie and A. W. G. James, *J. Chem. Soc., Chem. Commun.*, 357 (1966).
     (b) L. Crombie, D. E. Games and M. H. Knight, *J. Chem. Soc., Chem. Commun.*, 355 (1966).
     (c) T. M. Harris and R. L. Carney, *J. Am. Chem. Soc.*, **89**, 6734 (1967).
     (d) H. Guilford, A. I. Scott, D. Skingle and M. Yalpani, *J. Chem. Soc., Chem. Commun.*, 1127 (1968).
     (e) F. W. Comer, T. Money and A. I. Scott, *J. Chem. Soc., Chem. Commun.*, 231 (1967).
182. (a) T. Miyakoshi and H. Togashi, *Synthesis*, **5**, 407 (1990).
     (b) N. Clauson-Kaas, N. Elming and Z. Tyle, *Acta Chem. Scand.*, **9**, 1 (1955).
     (c) J. T. Nielsen, N. Elming and N. Clauson-Kaas, *Acta Chem. Scand.*, **9**, 9 (1955).
     (d) N. Clauson-Kaas, N. Elming and Z. Tyle, *Acta Chem. Scand.*, **9**, 23 (1955).
183. (a) N. Clauson-Kaas and P. Nedenskov, *Acta Chem. Scand.*, **9**, 27 (1955).
     (b) W. R. Boehme and W. G. Scarpf, *J. Org. Chem.*, **26**, 1692 (1961). (c) T. Miyakoshi and H. Togashi, *Synthesis*, **5**, 407 (1990).
184. (a) A. T. Balaban and C. D. Nenitzescu, *J. Chem. Soc.*, 3553 (1961).
     (b) R. Lukes and M. Pergal, *Collect. Czech. Chem. Commun.*, **24**, 36 (1959).
185. H. G. Rajoharison, H. Soltani, M. Arnaud, C. Roussel and J. Metzger, *Synth. Commun.*, **10**, 195 (1980).
186. I. V. Shcherbakova, E. V. Kuznetsov, I. A. Yudilevich, O. E. Kompan, A. T. Balaban, A. H. Abolin, A. V. Polyakov and Y. T. Struchkov, *Tetrahedron*, **44**, 6217 (1988).
187. (a) V. I. Dulenko and S. V. Tolkunov, *Khim. Geterotsikl. Soedin.*, **7**, 889 (1987); *Chem. Abstr.*, **108**, 167343 (1987).
     (b) K.-L. Hoffmann, G. Maas and M. Regitz, *J. Org. Chem.*, **52**, 3851 (1987).
188. (a) G. N. Walker, *J. Org. Chem.*, **23**, 34 (1958).
     (b) D. Leuchs, *Chem. Ber.*, **98**, 1335 (1965).
189. (a) V. Prelog, O. Metzler and O. Jeger, *Helv. Chim. Acta*, **30**, 675 (1947).
     (b) V. Prelog, L. Ruzicka and O. Metzler, *Helv. Chim. Acta*, **30**, 1883 (1947).
     (c) V. Prelog, J. Würsch and K. Königsbacher, *Helv. Chim. Acta*, **34**, 259 (1951).
190. R. Kaushal, *J. Indian Chem. Soc.*, **23**, 16 (1946).
191. L. W. Singh, H. Ila and H. Junjappa, *Synthesis*, 873 (1987).
192. K. Eichinger, P. Nussbaumer, S. Balkan and G. Schulz, *Synthesis*, 1061 (1987).
193. A. R. Katritzky, S. A. Belyakov, Y. Fang and J. S. Kiely, *Tetrahedron Lett.*, **39**, 8051 (1998).
194. A. R. Katritzky, S. A. Belyakov, S. A. Henderson and J. P. Steel, *J. Org. Chem.*, **62**, 8215 (1997).
195. P. N. Gordon, *J. Org. Chem.*, **22**, 1006 (1957).
196. W. Theilacker and W. Schmid, *Ann. Chem.*, **570**, 15 (1950).
197. I. E.-S. El-Kholy, M. M. Mishrikey, F. K. Rafla and G. Soliman, *J. Chem. Soc.*, 5153 (1962).
198. B. B. Sinder and J. J. Patricia, *J. Org. Chem.*, **54**, 38 (1989).
199. J. R. Peterson, R. S. Egler, D. B. Horsley and T. J. Winter, *Tetrahedron Lett.*, **28**, 6109 (1987).
200. T.-H. Chan and P. Brownbridge, *J. Am. Chem. Soc.*, **102**, 3534 (1980).
201. (a) T.-H. Chan and P. Brownbridge, *J. Chem. Soc., Chem. Commun.*, 578 (1979).
     (b) T.-H. Chan and P. Brownbridge, *Tetrahedron*, **37**, Suppl. 1, 387 (1981).
202. T.-H. Chan and P. Brownbridge, *J. Chem. Soc., Chem. Commun.*, 20 (1981).
203. T.-H. Chan and T. Chaly, *Tetrahedron Lett.*, **23**, 2935 (1982).
204. S. Danishefsky, *Acc. Chem. Res.*, **14**, 400 (1981).
205. S. Danishefsky, C.-F. Yan, R. K. Singh, R. B. Gammill, P. M. McCurry, N. Fristch and J. Clardy, *J. Am. Chem. Soc.*, **101**, 7001 (1979).
206. R. Baker, V. B. Rao, P. D. Ravenscroft and C. J. Swain, *Synthesis*, 572 (1983).
207. J. Banville and P. Brassard, *J. Chem. Soc., Perkin Trans. 1*, 1852 (1996).
208. S. Danishefsky, R. K. Singh and R. B. Gammill, *J. Org. Chem.*, **43**, 379 (1978).
209. S. Danishefsky, M. P. Prisbylla and S. Hiner, *J. Am. Chem. Soc.*, **100**, 2918 (1978).
210. S. Danishefsky and S. J. Etheredge, *J. Org. Chem.*, **44**, 4716 (1979).
211. M. Fink, H. Gaier and H. Gerlach, *Helv. Chim. Acta*, **65**, 2563 (1982).

212. P. Langer and B. Kracke, *Tetrahedron Lett.*, **41**, 4545 (2000).
213. A. P. Kozikowski and R. Schmiesing, *Tetrahedron Lett.*, 4241 (1978).
214. D. Obrecht, *Helv. Chim. Acta*, **74**, 27 (1991).
215. (a) J. Savard and P. Brassard, *Tetrahedron*, **40**, 3455 (1984).
    (b) J. Savard and P. Brassard, *Tetrahedron Lett.*, 4911 (1979).
    (c) D. W. Cameron, G. I. Feutrill, G. B. Gamble and J. Stavrakis, *Tetrahedron Lett.*, **27**, 4999 (1986).
216. (a) J.-L. Grandmaison and P. Brassard, *J. Org. Chem.*, **43**, 1435 (1978).
    (b) C. Brisson and P. Brassard, *J. Org. Chem.*, **46**, 1810 (1981).
    (c) J.-P. Gesson, J. C. Jacquesy and B. Renoux, *Tetrahedron*, **40**, 4743 (1984).
    (d) J.-P. Gesson, J. C. Jacquesy and B. Renoux, *Tetrahedron Lett.*, **24**, 2757 (1983).
    (e) J.-P. Gesson, J. C. Jacquesy and B. Renoux, *Tetrahedron Lett.*, **24**, 2761 (1983).
    (f) J.-P. Gesson, J. C. Jacquesy and M. Mondon, *Tetrahedron Lett.*, **22**, 1337 (1981).
    (g) D. W. Cameron, C.-Y. Gan, P. G. Griffiths and J. A. Pattermann, *Aust. J. Chem.*, **51**, 421 (1998).
    (h) D. W. Cameron, P. G. Griffiths and A. G. Riches, *Aust. J. Chem.*, **52**, 1173 (1999).
217. J.-P. Gesson, J. C. Jacquesy and M. Mondon, *Tetrahedron Lett.*, **21**, 3351 (1980).
218. J. G. Bauman, R. B. Barber, R. D. Gless and H. Rapoport, *Tetrahedron Lett.*, **21**, 4777 (1980).
219. J.-P. Gesson and M. Mondon, *J. Chem. Soc., Chem. Commun.*, 421 (1982).
220. J. R. Wiseman, J. J. Pendery, C. A. Otto and K. G. Chiong, *J. Org. Chem.*, **45**, 516 (1980).
221. (a) Y. Kita, R. Okunaka, T. Honda, M. Sindo, M. Taniguchi and M. Saxo, *J. Org. Chem.*, **56**, 119 (1991).
    (b) Y. Kita, R. Okunaka, M. Sasho, M. Taniguchi, T. Honda and Y. Tamura, *Tetrahedron Lett.*, **29**, 5943 (1988).
222. (a) Y. Tamura, F. Fukata, M. Saxo, T. Tsugoshi and Y. Kita, *J. Org. Chem.*, **50**, 2273 (1985).
    (b) Y. Kita, K. Higuchi, Y. Yoshida, K. Iio, S. Kitagaki, K. Ueda, S. Akai and H. Fujioka, *J. Am. Chem. Soc.*, **123**, 3214 (2001).
    (c) Y. Tamura, M. Kirihara, J.-I. Sekihachi, O. Ryuichi, S.-I. Mohri, T. Tsugoshi, S. Akai, M. Saxo and Y. Kita, *Tetrahedron Lett.*, **28**, 3971 (1987).
    (d) Y. Kita, M. Kirihara, Y. Fujii, R. Okunaka, S. Akai, H. Maeda and Y. Tamura, *J. Chem. Soc., Chem. Commun.*, 136 (1990).
    (e) Y. Tamura, M. Kirihara, M. Sasho, S. Akai, J.-I. Sekihachi, R. Okunaka and Y. Kita, *J. Chem. Soc., Chem. Commun.*, 1474 (1987).
    (f) T. Izawa, Z.-G. Wang, Y. Nishimura, S. Kondo and H. Umezawa, *Chem. Lett.*, 1655 (1987).
    (g) H. Fujioka, H. Yamamoto, H. Kondo, H. Annoura and Y. Kita, *J. Chem. Soc., Chem. Commun.*, 1509 (1989).
    (h) Y. Kita, K. Higuchi, Y. Yutaka, K. Iio, S. Kitagaki, S. Akai and H. Fujioka, *Angew. Chem., Int. Ed. Engl.*, **38**, 683 (1999).
    (i) A. P. Marchand, P. Annapurna, W. H. Watson and A. Nagl, *J. Chem. Soc., Chem. Commun.*, 281 (1989).
    (j) N. G. Ramesh, K. Iio, A. Okajima, S. Akai and Y. Kita, *J. Chem. Soc., Chem. Commun.*, 2741 (1999).
223. T. Matsumoto, H. Yamaguchi and K. Suzuki, *Synlett*, 433 (1996).
224. M. D. Shair, T. Y. Yoon, K. K. Mosny, T. C. Chou and S. Danishefsky, *J. Am. Chem. Soc.*, **118**, 9509 (1996).
225. K. Iio, N. G. Ramesh, A. Okajima, K. Higuchi, H. Fujioka, S. Akai and Y. Kita, *J. Org. Chem.*, **65**, 89 (2000).
226. Y. Kita, K. Iio, A. Okajima, Y. Takeda, K.-I. Kawaguchi, B. A. Whelan and S. Akai, *Synlett*, 292 (1998).
227. Y. Tamura, M. Saxo, K. Nakagawa, T. Tsugoshi and Y. Kita, *J. Org. Chem.*, **49**, 473 (1984).
228. E. Caliskan, D. W. Cameron and P. G. Griffiths, *Aust. J. Chem.*, **52**, 1013 (1999).
229. (a) A. Maggiani, A. Tubul and P. Brun, *Synthesis*, 631 (1997).
    (b) A. Maggiani, A. Tubul and P. Brun, *Tetrahedron Lett.*, **39**, 4485 (1998).
    (c) R. D. Chambers, A. J. Roche and M. H. Rock, *J. Chem. Soc., Perkin Trans. 1*, 1095 (1996).
    (d) H. N. C. Wong, Y. D. Xing, Q. Q. Gong and C. Zhang, *Synthesis*, 787 (1984).
230. A. Maggiani, A. Tubul and P. Brun, *J. Chem. Soc., Chem. Commun.*, 2495 (1999).

231. T. Mukaiyama, T. Tsuji and N. Iwasawa, *Chem. Lett.*, 697 (1979).
232. (a) K. Matsumoto and A. Sera, *Synthesis*, 999 (1985).
    (b) L. C. Garver and E. E. van Tamelen, *J. Am. Chem. Soc.*, **104**, 868 (1982).
233. K. T. Potts and E. B. Walsh, *J. Org. Chem.*, **53**, 1199 (1988).
234. (a) P. Brownbridge and T.-H. Chan, *Tetrahedron Lett.*, **21**, 3423 (1980).
    (b) T. M. Balthazor and E. L. Williams, *Synth. Commun.*, **22**, 1023 (1992).
    (c) A. Padwa, M. Dimitroff, A. G. Waterson and T. Wu, *J. Org. Chem.*, **62**, 4088 (1997).
235. G.-D. Zhu, M. A. Staeger and S. A. Boyd, *Org. Lett.*, **2**, 3345 (2000).
236. A. S. K. Hashmi, T. M. Frost and J. W. Bats, *J. Am. Chem. Soc.*, **122**, 11553 (2000).
237. A. S. K. Hashmi, T. M. Frost and J. W. Bats, *Org. Lett.*, **3**, 3769 (2001).
238. H. W. Moore and O. H. W. Decker, *Chem. Rev.*, **86**, 821 (1986).
239. (a) R. L. Danheiser and S. K. Gee, *J. Org. Chem.*, **49**, 1672 (1984).
    (b) M. W. Reed and H. W. Moore, *J. Org. Chem.*, **52**, 3491 (1987).
    (c) R. L. Danheiser, S. K. Gee and J. J. Pérez, *J. Am. Chem. Soc.*, **108**, 806 (1986).
    (d) G. B. Dudley, K. S. Takaki, D. D. Cha and R. L. Danheiser, *Org. Lett.*, **2**, 3407 (2000).
    (e) R. L. Danheiser, A. Nishida, S. Savariar and M. P. Trova, *Tetrahedron Lett.*, **29**, 4917 (1988).
    (f) J. L. Loebach, D. M. Bennett and R. L. Danheiser, *J. Org. Chem.*, **63**, 8380 (1998).
    (g) A. B. Smith III, S. A. Kozmin, C. M. Adams and D. V. Paone, *J. Am. Chem. Soc.*, **122**, 4984 (2000).
    (h) A. B. Smith III, S. A. Kozmin and D. V. Paone, *J. Am. Chem. Soc.*, **121**, 7223 (1999).
    (i) S. L. Xu and H. W. Moore, *J. Org. Chem.*, **54**, 4024 (1989).
    (j) M. Taing and H. W. Moore, *J. Org. Chem.*, **61**, 329 (1996).
240. L. Wolff, *Justus Liebigs Ann. Chem.*, **394**, 23 (1912).
241. (a) R. L. Danheiser, D. S. Casebier and F. Firooznia, *J. Org. Chem.*, **60**, 8341 (1995).
    (b) R. L. Danheiser, D. S. Casebier and A. H. Huboux, *J. Org. Chem.*, **59**, 4844 (1994).
    (c) R. L. Danheiser, R. G. Brisbois, J. J. Kowalczyk and R. F. Miller, *J. Am. Chem. Soc.*, **112**, 3093 (1990).
    (d) R. L. Danheiser and A. L. Helgason, *J. Am. Chem. Soc.*, **116**, 9471 (1994).
    (e) R. L. Danheiser, D. S. Casebier and J. L. Loebach, *Tetrahedron Lett.*, **33**, 1149 (1992).
242. C. J. Kowalski and G. S. Lal, *J. Am. Chem. Soc.*, **110**, 3693 (1988).
243. S. L. Xu and H. W. Moore, *J. Org. Chem.*, **57**, 326 (1992).
244. R. Andriamiadanarivo, B. Pujol, B. Chantegel, C. Deshayes and A. Doutheau, *Tetrahedron Lett.*, **34**, 7923 (1993).
245. F. Zaragoza Dörwald, *Metal Carbenes in Organic Synthesis*, Wiley-VCH, Weinheim, 1999.
246. (a) D. M. Gordon and S. J. Danishefsky, *J. Org. Chem.*, **57**, 7052 (1992).
    (b) M. E. Bos, W. D. Wulf, R. A. Miller, S. Chamberlin and T. A. Brandvold, *J. Am. Chem. Soc.*, **113**, 9293 (1991).
    (c) J. King, P. Quayle and J. F. Malone, *Tetrahedron Lett.*, **31**, 5221 (1990).
    (d) K. A. Parker and C. A. Coburn, *J. Org. Chem.*, **56**, 1666 (1991).
    (e) K. H. Dötz and W. Kuhn, *Angew. Chem., Int. Ed. Engl.*, **22**, 732 (1983).
    (f) D. L. Boger, O. Huter, K. Mbiya and M. Zhang, *J. Am. Chem. Soc.*, **117**, 11839 (1995).
    (g) C. S. Tomooka, H. Liu and H. W. Moore, *J. Org. Chem.*, **61**, 6009 (1996).
    (h) D. L. Danheiser, R. G. Brisbois, J. J. Kowalczyk and R. F. Miller, *J. Am. Chem. Soc.*, **112**, 3093 (1990).
    (i) R. L. Danheiser and M. P. Trova, *Synlett*, 573 (1995).
247. M. A. Hoffman and L. S. Liebeskind, *J. Am. Chem. Soc.*, **112**, 8617 (1990).
248. (a) K. H. Dötz and H. Larbig, *J. Organomet. Chem.*, **405**, C38 (1991).
    (b) D. F. Harvey, E. M. Grenzer and P. K. Gantzel, *J. Am. Chem. Soc.*, **116**, 6719 (1994).
249. I. Merino and L. Hegedus, *Organometallics*, **14**, 2522 (1995).
250. A.-U. Rehman, W. F. K. Schnatter and N. Manolache, *J. Am. Chem. Soc.*, **115**, 9848 (1993).
251. (a) K. H. Dötz, *Angew. Chem., Int. Ed. Engl.*, **14**, 644 (1975).
    (b) K. H. Dötz, *Angew. Chem., Int. Ed. Engl.*, **23**, 587 (1984).
    (c) K. H. Dötz, *Chem. Ber.*, **121**, 655 (1988).
252. S. R. Pulley, S. Sen, A. Vorogushin and E. Swanson, *Org. Lett.*, **1**, 1721 (1999).
253. W. D. Wulff, B. M. Bax, T. A. Brandvold, K. S. Chan, A. M. Gilbert, R. P. Hsung, J. Mitchell and J. Clardy, *Organometallics*, **13**, 102 (1994).
254. A. Yamashita, *Tetrahedron Lett.*, **27**, 5915 (1986).

255. W. D. Wulff, A. M. Gilbert, R. P. Hsung and A. Rahm, *J. Org. Chem.*, **60**, 4566 (1995).
256. J. P. A. Harrity and W. J. Kerr, *Tetrahedron Lett.*, **34**, 2995 (1993).
257. J. P. A. Harrity and W. J. Kerr, *Tetrahedron*, **49**, 5565 (1993).
258. Y. H. Choi, K. S. Rhee, K. S. Kim, G. C. Shin and S. C. Shin, *Tetrahedron Lett.*, **36**, 1871 (1995).
259. K. H. Dötz and T. Leese, *Bull. Soc. Chim. Fr.*, **134**, 503 (1997).
260. G. A. Peterson and W. D. Wulff, *Tetrahedron Lett.*, **38**, 5587 (1997).
261. A.-U. Rehman, W. F. K. Schnatter and N. Manolache, *J. Am. Chem. Soc.*, **115**, 9848 (1993).
262. K. H. Dötz, I. Pruskil and J. Mühlermeier, *Chem. Ber.*, **115**, 1278 (1982).
263. J. Bao, W. D. Wulff, V. Dragisich, S. Wenglowsky and R. G. Ball, *J. Am. Chem. Soc.*, **116**, 7616 (1994).
264. K. H. Dötz and M. Popall, *Tetrahedron*, **41**, 5797 (1985).
265. (a) P. Hoffmann and M. Hämmerle, *Angew. Chem., Int. Ed. Engl.*, **28**, 908 (1989).
     (b) M. M. Gleichmann, K. H. Dötz and B. A. Hess, *J. Am. Chem. Soc.*, **118**, 10551 (1996).
266. M. Torrent, M. Duran and M. Solà, *Organometallics*, **17**, 1492 (1998).
267. M. Torrent, M. Duran and M. Solà, *J. Chem. Soc., Chem. Commun.*, 999 (1998).
268. D. S. Tarbell and J. C. Petropoulos, *J. Am. Chem. Soc.*, **74**, 244 (1952).
269. M. H. Todd and C. Abell, *Tetrahedron Lett.*, **41**, 8183 (2000).
270. C. Venkatachalapathy, K. Pitchumani and S. Sivasubramanian, *Indian J. Chem., Sect. B: Org. Chem. Incl. Med. Chem.*, **37B**, 301 (1998).
271. (a) R. P. Lutz, *Chem. Rev.*, **84**, 215 (1984).
     (b) W. N. White and W. K. Fife, *J. Am. Chem. Soc.*, **83**, 3846 (1961).
272. I. A. Pearl, *J. Am. Chem. Soc.*, **70**, 1746 (1948).
273. (a) T. Capecchi, C. B. De Koning and J. P. Michael, *J. Chem. Soc., Perkin Trans. 1*, 2681 (2000).
     (b) C. F. H. Allen and J. W. Gates, *Org. Synth. Coll. Vol. III*, 418 (1955).
274. F. Lach and C. J. Moody, *Tetrahedron Lett.*, **41**, 6893 (2000).
275. T. R. R. Pettus, M. Inoue, X.-T. Chen and S. J. Danishefsky, *J. Am. Chem. Soc.*, **122**, 6160 (2000).
276. (a) D. E. Fuerst and B. M. Stolz, *Org. Lett.*, **2**, 3521 (2000).
     (b) F. Lach and C. J. Moody, *Tetrahedron Lett.*, **41**, 6893 (2000).
     (c) W. N. White and W. K. Fife, *J. Am. Chem. Soc.*, **83**, 3846 (1961).
     (d) S. Takano, M. Akiyama and K. Ogasawara, *Chem. Lett.*, 505 (1985).
277. G. Fráter, A. Habich, H.-J. Hansen and H. Schmid, *Helv. Chim. Acta*, **52**, 335 (1969).
278. L. Claisen, O. Eisleb and F. Kremers, *Ann.*, **418**, 69 (1919).
279. R. J. Giguere, T. L. Bray, S. M. Duncan and G. Majetich, *Tetrahedron Lett.*, **27**, 4945 (1986).
280. (a) T. Borgulya, R. Madeja, P. Gahrni, H.-J. Hansen, H. Schmid and R. Barner, *Helv. Chim. Acta*, **56**, 14 (1973).
     (b) W. Gerrard, M. F. Lappert and H. B. Silver, *Proc. Chem. Soc.*, 19 (1957).
281. (a) K. Maruyama, N. Nagai and Y. Naruta, *J. Org. Chem.*, **51**, 5083 (1986).
     (b) K. Maruyama, N. Nagai and Y. Naruta, *Tetrahedron Lett.*, **26**, 5149 (1985).
282. F. M. Sonnenberg, *J. Org. Chem.*, **35**, 3166 (1970).
283. (a) K. Narasaka, E. Bald and T. Mukaiyama, *Chem. Lett.*, 1041 (1975).
     (b) K. Maruoka, J. Sato, H. Banno and H. Yamamoto, *Tetrahedron Lett.*, **31**, 377 (1990).
284. U. Koch-Pomeranz, H.-J. Hansen and H. Schmid, *Helv. Chim. Acta*, **56**, 2981 (1973).
285. G. Balavoine, G. Bram and F. Guibe, *Nouv. J. Chim.*, **2**, 207 (1978).
286. R. A. Sheldon, J. A. Elings, S. K. Lee, H. E. B. Lempers and R. S. Downing, *J. Mol. Catal. A: Chem.*, **134**, 129 (1998).
287. B. M. Trost and F. D. Toste, *J. Am. Chem. Soc.*, **120**, 815 (1998).
288. H. Ito, A. Sato and T. Taguchi, *Tetrahedron Lett.*, **38**, 4815 (1997).
289. P. Wipf and S. Ribe, *Org. Lett.*, **3**, 1503 (2001).
290. G. Erker, M. Aulbach, M. Knickmeier, D. Wingbermühle, C. Krüger, M. Nolte and S. Werner, *J. Am. Chem. Soc.*, **115**, 4590 (1993).
291. U. Widmer, H.-J. Hansen and H. Schmid, *Helv. Chim. Acta*, **56**, 2644 (1973).
292. A. Factor, H. Finkbeiner, R. A. Jerussi and D. M. White, *J. Org. Chem.*, **35**, 57 (1970).
293. A. Luttringhaus and G. W. Saaf, *Ann.*, **557**, 25 (1945).
294. A. Luttringhaus and H. Schuster, *Angew. Chem.*, **70**, 438 (1958).
295. (a) W. E. Truce and E. M. Kreider, *Org. React.*, **18**, 99 (1971).

(b) J. Sauer and R. Huisgen, *Angew. Chem.*, **72**, 294 (1960).
296. P. C. H. Eichinger, J. H. Bowie and R. N. Hayes, *J. Am. Chem. Soc.*, **111**, 4224 (1989).
297. (a) G. E. Bonvicino, L. H. Yododzinski and R. A. Hardy, *J. Org. Chem.*, **27**, 4272 (1962).
(b) D. M. Schmidt and G. E. Bonvicino, *J. Org. Chem.*, **49**, 1664 (1984).
298. J. D. Loudon and J. A. Scott, *J. Chem. Soc.*, 265 (1953).
299. R. Radinov, M. Khaimova, S. Simova and E. Simova, *Liebigs Ann. Chem.*, 231 (1988).
300. (a) H. J. Shine, *Aromatic Rearrangements*, Elsevier, Amsterdam, 1967, p. 55.
(b) C. J. Collins and J. F. Eastman, in *The Chemistry of the Carbonyl Group* (Ed. S. Patai), Wiley-Interscience, Chichester, 1966, p. 775.
(c) A. G. Schultz and S. A. Hardinger, *J. Org. Chem.*, **56**, 1105 (1991).
(d) A. G. Schultz, S. A. Hardinger, M. Macielag, P. G. Mehta and A. G. Taveras, *J. Org. Chem.*, **52**, 5482 (1987).
(e) B. Miller, *Acc. Chem. Res.*, **8**, 245 (1975).
301. G. Goodyear and A. J. Waring, *J. Chem. Soc., Perkin Trans. 2*, 103 (1990).
302. A. A. Frimer, V. Marks, M. Sprecher and P. Gilinsky-Sharon, *J. Org. Chem.*, **59**, 1831 (1994).
303. L. De Buyck, H. De Pooter and N. Schamp, *Bull. Soc. Chim. Belg.*, **97**, 55 (1988).
304. S. Chalais, P. Laszlo and A. Mathy, *Tetrahedron Lett.*, **27**, 2627 (1986).
305. A. K. Banerjee, J. A. Castillo-Meléndez, W. Vera, J. A. Azócar and M. S. Laya, *J. Chem. Res. (S)*, **7**, 324 (2000).
306. G. M. Rishton and M. A. Schwartz, *Tetrahedron Lett.*, **29**, 2643 (1988).
307. Z. Guo and A. G. Schultz, *Org. Lett.*, **3**, 1177 (2001).
308. (a) K. Fries and G. Finck, *Chem. Ber.*, **41**, 4271 (1908).
(b) K. Fries and W. Pfaffendorf, *Chem. Ber.*, **43**, 212 (1910).
(c) H. J. Shine, *Aromatic Rearrangements*, Elsevier, Amsterdam, 1967, pp. 72 and 365.
(d) A. M. Blatt, *Org. React.*, **1**, 342 (1942).
(e) G. A. Olah, *Friedel-Crafts Chemistry*, Wiley-Interscience, New York, 1973.
(f) R. Martin, *Org. Prep. Proced. Int.*, **24**, 369 (1992).
309. (a) S. Kobayashi, M. Moriwaki and I. Hachiya, *Tetrahedron Lett.*, **37**, 2053 (1996).
(b) S. Kobayashi, M. Moriwaki and I. Hachiya, *J. Chem. Soc., Chem. Commun.*, 1527 (1995).
(c) D. C. Harrowven and F. R. Dainty, *Tetrahedron Lett.*, **37**, 7659 (1996).
310. (a) Y. Ogata and H. Taguchi, *Tetrahedron*, **20**, 1661 (1964).
(b) A. Furka and T. Széll, *J. Chem. Soc.*, 2312 (1960).
(c) T. Széll and A. Furka, *J. Chem. Soc.*, 2321 (1960).
311. (a) J. L. Gibson and L. S. Hart, *J. Chem. Soc., Perkin Trans. 2*, 1343 (1991).
(b) I. M. Dawson, L. S. Hart and J. S. Littler, *J. Chem. Soc., Perkin Trans. 2*, 1601 (1985).
(c) J. R. Norell, *J. Org. Chem.*, **38**, 1924 (1973).
(d) A. M. El-Abbady, F. G. Baddar and A. Labib, *J. Chem. Soc.*, 1083 (1961).
(e) C. R. Hauser and E. H. Man, *J. Org. Chem.*, **17**, 390 (1952).
(f) A. Schonberg and A. Mustafa, *J. Chem. Soc.*, 642 (1943).
(g) D. S. Tarbell and P. E. Fanta, *J. Am. Chem. Soc.*, **65**, 2169 (1943).
(h) F. Krausz and R. Martin, *Bull. Soc. Chim. Fr.*, 2192 (1965).
(i) R. Martin, *Bull. Soc. Chim. Fr.*, 983 (1974).
(j) A. Warshawsky, R. Kalir and A. Patchornik, *J. Am. Chem. Soc.*, **100**, 4544 (1978).
312. F. Effenberger and R. Gutmann, *Chem. Ber.*, **115**, 1089 (1982).
313. N. Cairns, L. M. Harwood and D. P. Astles, *Tetrahedron Lett.*, **29**, 1311 (1988).
314. R. Martin, J.-P. Gavard, M. Delfly, D. Demerseman and A. Tromelin, *Bull. Soc. Chim. Fr.*, 659 (1986).
315. D. C. Harrowven and R. F. Dainty, *Tetrahedron Lett.*, **37**, 7659 (1996).
316. H. Sharghi and B. Kaboudin, *J. Chem. Res. (S)*, 628 (1998).
317. R. Martin and P. Demerseman, *Synthesis*, 738 (1992).
318. J.-P. Kaplan, B. M. Raizou, M. Desarmenien, P. Feltz, P. M. Headley, P. Worms, K. G. Loyd and G. Bartholini, *J. Med. Chem.*, **23**, 702 (1980).
319. J. A. Miller, *J. Org. Chem.*, **52**, 322 (1987).
320. S. Horne and R. Rodrigo, *J. Org. Chem.*, **55**, 4520 (1990).
321. D. Belluš and P. Hrdlovič, *Chem. Rev.*, **67**, 599 (1967).
322. J. C. Anderson and C. B. Reese, *J. Chem. Soc.*, 1781 (1963).
323. C. E. Kalmus and D. M. Hercules, *Tetrahedron Lett.*, 1575 (1972).
324. (a) M. S. Syamala, B. Nageswer and V. Ramamurthy, *Tetrahedron*, **44**, 7234 (1988).

(b) A. V. Veglia, A. N. Sánchez and R. H. De Rossi, *J. Org. Chem.*, **55**, 4083 (1990).
325. R. Suau, G. Torres and M. Valpuesta, *Tetrahedron Lett.*, **36**, 1311 (1995).
326. (a) M. Ohara and K. Watanabe, *Angew. Chem., Int. Ed. Engl.*, **14**, 820 (1975).
(b) A. V. Veglia and R. H. De Rossi, *J. Org. Chem.*, **58**, 4941 (1993).
(c) R. Chênevert and N. Voyer, *Tetrahedron Lett.*, **25**, 5007 (1984).
327. A. K. Singh and S. M. Sonar, *Synth. Commun.*, **15**, 1113 (1985).
(b) A. K. Singh and T. S. Raghuraman, *Tetrahedron Lett.*, **26**, 4125 (1985).
328. N. J. Turro and W. R. Cherry, *J. Am. Chem. Soc.*, **100**, 7431 (1978).
329. V. P. Pathak and R. N. Khanna, *Synthesis*, 882 (1981).

CHAPTER 7

UV-visible spectra and photoacidity of phenols, naphthols and pyrenols

EHUD PINES

Chemistry Department, Ben-Gurion University of the Negev, P.O.B. 653, Beer-Sheva 84105, Israel
fax: 8-6472-943; e-mail: epines@bgumail.bgu.ac.il

I. INTRODUCTION	491
II. THE THERMODYNAMIC ASPECTS OF PHOTOACIDITY	493
III. ON THE ORIGIN OF PHOTOACIDITY	498
IV. THE ELECTRONIC STRUCTURE OF PHOTOACIDS	507
V. FREE-ENERGY CORRELATIONS BETWEEN PHOTOACIDITY AND REACTIVITY	522
VI. CONCLUDING REMARKS: EVALUATION OF OUR CURRENT UNDERSTANDING OF THE PHOTOACIDITY OF HYDROXYARENES	523
VII. REFERENCES	525

I. INTRODUCTION

Hydroxyarenes become stronger acids upon electronic excitation[1-5]. Such a property of an aromatic molecule is usually described as 'photoacidity', and the molecules undergoing such a transition upon electronic excitation are usually named 'photoacids'. Photoacids are Brønsted acids, and their excited state acidity may be described in terms used for ground state acids as were defined by Brønsted some 80 years ago[6,7]. Following Brønsted, one usually associates acidity with a proton-transfer reaction where a proton is transferred from a proton donor (an acid) to a proton acceptor (a base) (equation 1).

$$\text{AH (acid)} + \text{B (base)} \rightleftharpoons \text{A}^- \text{(base)} + \text{BH}^+ \text{(acid)} \tag{1}$$

The reversible nature of acid–base reactions implies the existence of conjugated acid–base pairs, i.e. A^- is the conjugate base of AH and AH is the conjugate acid of A^-. A more

---

The Chemistry of Phenols. Edited by Z. Rappoport
© 2003 John Wiley & Sons, Ltd ISBN: 0-471-49737-1

modern observation is that proton transfer proceeds most often along a hydrogen bond, formed between the proton donor and the proton acceptor so that the reactive coordinate where proton transfer occurs is usually of the type $A-H^+ \cdots B$. The hydrogen-bonding interaction may be viewed as a relatively weak interaction between the proton donor and the proton acceptor through the sharing of a hydrogen atom.

Proton-transfer reactions and hydrogen-bonding interactions may occur within one molecule or between two molecules (intra- and intermolecular proton transfer, respectively). Photoacids such as the phenols or the naphthols readily undergo intermolecular proton transfer reactions in aqueous solutions[1-5]. When the proton is transferred to a solvent molecule the reaction is sometimes called a 'proton-transfer-to-solvent' reaction (PTTS reaction)[8,9]. In non-aqueous solutions, hydroxyarenes form moderately strong hydrogen bonds of the type: $O-H^+ \cdots O$ which usually do not lead to a full proton-transfer reaction either in the ground or the excited state of the photoacid.

The discovery of photoacidity was made by Förster more then 50 years ago[1-4]. Förster correctly explained the unusual large Stokes shift found in the fluorescence of several classes of aromatic dyes, including 1- and 2-naphthol derivatives as an indication of excited state proton-transfer reaction which results in the formation of the molecular anion still in the excited state. Thus, it become clear that excited-state proton transfer may compete with other radiative and non-radiative decay routes of the photoacid. The main modern-day importance of photoacids lies in their ability to initiate and then to follow acid–base reactions so they may be regarded as optical probes for the study of general proton-transfer reactions.

Over the years the field of photoacids (and photobases) has been reviewed many times[10-18]. The most extensive list of photoacids appeared, so far, in a 1976 review by Ireland and Wyatt[12]. The hydroxyarenes are the most widely used photoacids. In polar solutions they may undergo an excited-state proton-transfer reaction according to the general reaction scheme of equations 2–5.

a. Electronic excitation:

$$ROH \xrightarrow{h\nu} (R^*OH)_{LE} \qquad (2)$$

b. Partial intramolecular charge transfer assisted by the solvent:

$$(R^*OH)_{LE} \longrightarrow (R^{*-}OH^+)_{S_1} \qquad (3)$$

where LE denotes the locally excited singlet state and S_1 denotes the first singlet state of the photoacid in polar solvents. The S_1 state may be directly accessed from the ground state.

c. Formation of a reactive coordinate along a hydrogen bond between the photoacid and a base molecule:

$$(R^{*-}OH^+)_{S_1} + B \longrightarrow (R^{*-}O-H^+\cdots B)_{hb} \qquad (4)$$

The base molecule, B, may either be a solvent molecule or a solute molecule and hb denotes the hydrogen-bonded reaction complex.

This stage may involve some further electronic rearrangement in the photoacid toward the formation of the photobase.

d. Photoacid dissociation and ion-pair recombination:

$$(R^{*-}O-H^+\cdots B)_{hb} \underset{k_r}{\overset{k_d}{\rightleftharpoons}} [R^*O^- \cdots H^+ - B]_{ip} \underset{k_D}{\overset{k_S}{\rightleftharpoons}} R^*O^- + {}^+HB \qquad (5)$$

7. UV-visible spectra and photoacidity of phenols, naphthols and pyrenols   493

where ip denotes the ion-pair state, which may be either solvent separated or a contact pair, k_d and k_r are the 'on-contact' rate constants for the photoacid dissociation and ion-pair recombination, respectively, while k_S and k_D are the diffusion-limited rate constants for ion-pair dissociation to infinite separation and ion-pair formation, respectively[5].

The charge separation stage, hb → ip, may involve considerable electronic rearrangement in the photobase.

Some of the most common hydroxyarenes[5,8–17] used as photoacids are listed in Figures 1–3.

It is the aim of this chapter to describe some of the modern views on the origins of photoacidity of simple hydroxyarenes. Photoacids were extensively studied in the gas phase in clusters of various sizes including small to medium size clusters of ammonia[18–23], water[18,20,24–27] and methanol[28]. A second branch of research was carried out in solution and has been focusing on the various dynamic aspects of the proton-transfer reaction from photoacids observed mainly in aqueous solutions[5,10–17]. Phenol and phenol derivatives (Figure 1) due to their relatively small molecular weight and their relatively high vapor pressure[23,29,30] have been mainly used in gas-phase research. Naphthols and naphthol derivatives (Figure 2), having intermediate molecular weights and strong photoacidities, have been studied both in the gas phase[18–29] and in the liquid phase[30–51]. The pyrenols (Figure 3) have been almost exclusively studied in the liquid phase due to their low vapor pressure and their excellent properties as dye molecules[5,52–66].

These two main branches of the study of photoacids have been carried out mostly in parallel. The effort to converge the two methodologies into one coherent view of photoacidity is not always apparent in the literature and is far from being concluded. Indeed, much of the issues described in this chapter are still in debate or are altogether unresolved.

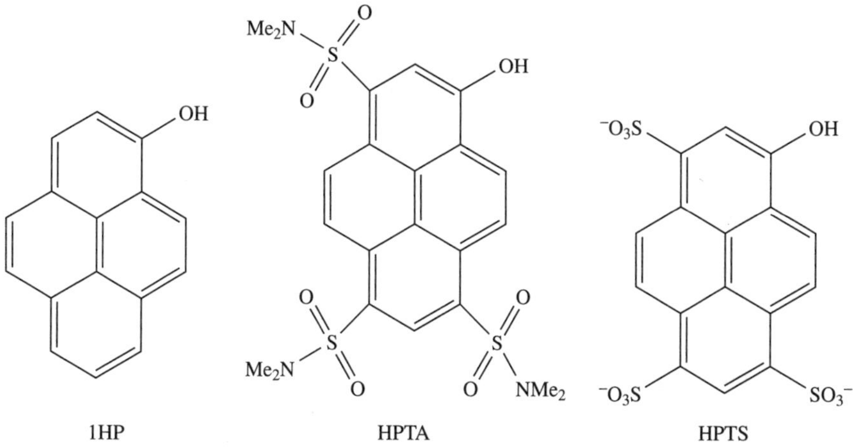

FIGURE 1. Pyrene derivatives used as photoacids: 1-hydroxypyrene (1HP), 8-hydroxy-1,3,6-tris(N,N-dimethylsulfonamido)pyrene (HPTA) and 8-hydroxypyrene 1,3,6-trisulfonate (HPTS)

II. THE THERMODYNAMIC ASPECTS OF PHOTOACIDITY

Photoacidity is most often described by the Förster cycle diagram (Figure 4)[2]. Following Förster, photoacidity is defined in terms of K_a^*, the excited state equilibrium constant for the dissociation reaction of the photoacid.

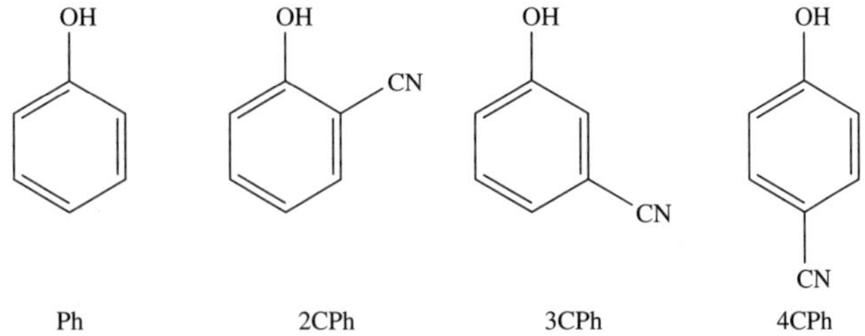

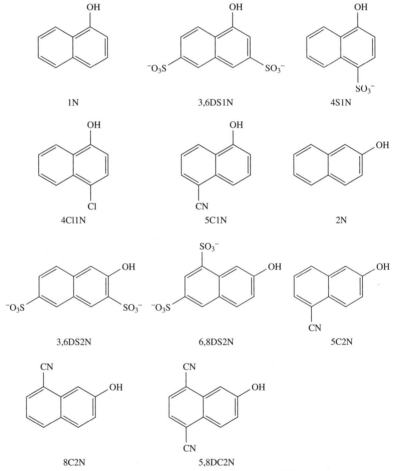

FIGURE 2. Common phenols used as photoacids: phenol (Ph), 2-cyanophenol (2CPh), 3-cyanophenol (3CPh) and 4-cyanophenol (4CPh)

FIGURE 3. Naphthol derivatives used as photoacids: 1-naphthol (1N), 1-naphthol-3,6-disulfonate (3,6S1N), 1-naphthol-4-sulfonate (4S1N), 4-chloro-1-naphthol (4Cl1N), 5-cyano-1-naphthol (5C1N), 2-naphthol (2N), 2-naphthol-3,6-disulfonate (3,6DS2N), 2-naphthol-6,8-disulfonate (6,8DS2N), 5-cyano-2-naphthol (5C2N), 8-cyano-2-naphthol (8C2N) and 5,8-dicyano-2-naphthol (5,8DC2N)

7. UV-visible spectra and photoacidity of phenols, naphthols and pyrenols   495

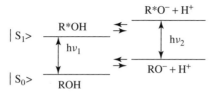

FIGURE 4. Schematic representation of energy levels of a photoacid RO*H and its conjugate base R*O$^-$; $\Delta pK_a = pK_a - pK_a^* = Nh\Delta\nu/[\ln(10)RT]$, where N is the Avogadro constant, h is the Planck constant, $\Delta\nu = \nu_1 - \nu_2$, ν_1 being the 0–0 transition of the acid and ν_2 the 0–0 transition of the anion; $|S_1>$ is the first singlet excited state and $|S_0>$ is the ground state

Photoacidity occurs, per definition, when the excited molecule becomes a stronger acid in the excited state as compared to its ground state acidity, so $pK_a^* < pK_a$, where pK_a is the equilibrium constant for the proton dissociation reaction in the ground state. For phenols, naphthols and pyrenols, the enhancement in the acidity constant K_a is between 5 (1HP) and 12 (3,6DC2N) orders of magnitude, which at room temperature translates into a free-energy increase of 7 to 16 kcal mol^{-1} in favor of the dissociation reaction in the excited state of the photoacid. The Förster cycle is a thermodynamic cycle. It connects between the optical properties of the photoacid and its conjugate photobase and the thermodynamic properties of the excited-state proton-transfer reaction. The main practical use of the Förster cycle is to get a rough estimation (usually, Förster cycle pK_a^* values of hydroxyarenes come within one to two pK_a units of the pK_a^* values found by direct time-resolved measurements) of the excited-state proton acidity of the photoacid but it does not give much clue as to the molecular process(es) which are involved in photoacidity. Nevertheless, the Förster cycle makes an excellent starting point for the discussion of photoacidity, as it allows the estimation of the excited-state acidity of many photoacids from simple, readily conducted optical measurements and establishes the idea that photoacids may be treated from a thermodynamic point of view similarly to ordinary ground-state acids.

Figure 5 shows the absorption spectra of phenol and the phenolate anion in water. The first three electronic transitions are shown for the base form while the same spectral range covers the first two electronic transitions of the acid form. The electronic transitions of the acid are blue-shifted compared to the electronic transition of the base. In both cases the oscillator strength of the S_1 transition is much weaker than that of the S_2 transition. The first two electronic transitions of phenol and phenolate ion are assigned ^1L_b (S_1) and ^1L_a (S_2) transitions according to Platt notations[68]. Fluorescence is from S_1 and obeys the Kasha rule, which states that internal-conversion processes are much faster than the S_n radiative-decay rate back to the ground state. Thus, ordinary Förster-cycle calculations only consider the energies of the S_1 transitions of the photoacid and its conjugate photobase anion. Photoacidity of the first electronic triplet state is not considered in this review.

Figures 6 and 7 show the spectral behavior of HPTA, which is a much stronger photoacid than phenol having $pK_a^* = -0.8$ compared to pK_a^* of about 4 for phenol. The photoacidity of HPTA is sufficiently large for HPTA to dissociate in pure methanol, while proton dissociation of excited phenol is not observed even in water.

Weller[5,32,33] has shown that photoacids may be titrated while in the excited state by monitoring their fluorescence intensity as a function of the pH of the solution. The fluorescence titration curves, after lifetime correction, yield similar information to the information gathered by acid–base titrations in the ground state. Thus, the gradual addition of a strong

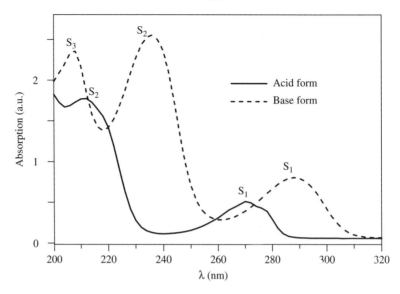

FIGURE 5. Absorption spectra of phenol (———) and phenolate ion (- - -) in water: acid-form pH = 6.0, base-form pH = 12 (from Reference 67)

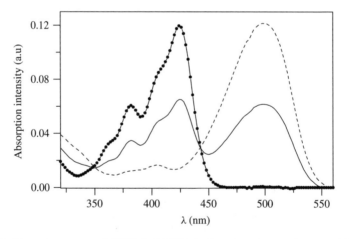

FIGURE 6. Absorption spectra of HPTA in MeOH: the base-form (···) maximum at 499 nm was titrated by trifluoromethanesulfonic acid; the acid-form (- - -) maximum is at 425 nm. Full line: intermediate pH at which both acid and base forms are present in the solution (from Reference 67)

mineral acid such as HCl to a solution of a photoacid gradually shifts the acid–base equilibrium in both the ground and excited state of the photoacid toward the acid form when the titration starts in basic conditions. The shift in the ground-state equilibrium populations of the acid and base forms of HPTA was monitored by absorption spectroscopy (Figure 6), while the corresponding shift in the excited-state population as a result of

7. UV-visible spectra and photoacidity of phenols, naphthols and pyrenols   497

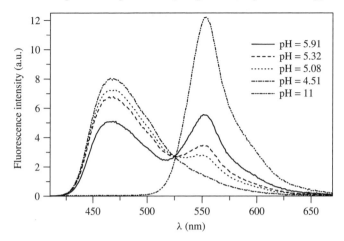

FIGURE 7. Excited-state acid–base equilibrium of HPTA in MeOH followed by the fluorescence titration of the base form. The base was titrated by trifluoromethanesulfonic acid. Acid band maximum is at 466 nm and base band is at 553 nm. Notice the red shift in the fluorescence spectra compared to the absorption spectra shown in Figure 6 (from Reference 67)

the change in the pH of the solution was monitored by fluorescence spectroscopy and is shown in Figure 7. Only S_1 transitions are depicted.

The pK_a^* found in this way may be directly compared with Förster-cycle calculations. However, straightforward utilization of the fluorescence titration method is usually limited to moderately strong photoacids due to partial deactivation processes of the photoacid occurring in very concentrated mineral acid solutions. The most accurate method of finding the pK_a^* of a photoacid is by direct kinetic measurements of the excited-state proton dissociation and recombination rates[58–60]. However, these measurements are not trivial and are limited to a relatively small number of photoacids where accurate measurement of the excited-state reversible dynamics of the proton-transfer reaction is possible.

Förster-cycle calculations thus appear to be the most general way for estimating the pK_a^* values of photoacids. There was some confusion in the past regarding the practical method for estimating the 0–0 transitions of the photoacid in solution. Using either the absorption spectra or the fluorescence spectra alone usually introduces considerable errors into the calculation, each set of data producing a different pK_a^* value. Estimating the 0–0 transitions from the crossing points between the absorption and the fluorescence spectra of the photoacid and the photobase is not always possible. Weller[69] suggested averaging the transition energies of absorption and fluorescence taken at the peak intensities of the transition bands. The averaging procedure is carried out separately for the photoacid transitions and for the base transitions. Förster-cycle calculations with the averaged transition-energy values is usually found to fall within one pK_a^* unit of the true pK_a^* value found by direct measurements.

The absorption maxima of the phenol and phenolate anion in water appear to be at 270 nm and 286 nm, respectively. Introducing these values into the Förster cycle together with the known ground-state pK_a of phenol (9.82)[70] gives a pK_a^* of 5.7, which underestimates the acidity of the excited phenol. Introducing the values of the fluorescence maxima at 229 nm (phenol) and 336 nm (phenolate ion) gives a pK_a^* value of 2.3, which overestimates the photoacidity of phenol. Introducing the averaged transition energies gives a pK_a^* of 4.0, which should be a good estimation for the photoacidity of phenol. The averaging

TABLE 1. pK_a and pK_a^* values of some common hydroxyarene photoacids

Photoacid	pK_a	pK_a^*
Phenol	9.82[70]	4[71]
2-Cyanophenol	6.97[72a]	0.66[72a]
3-Cyanophenol	8.34[72a]	1.89[72a]
4-Cyanophenol	7.74[72a]	3.33[72a]
1-Hydroxypyrene	8.7[12]	4.1[69]
HPTS	8.0[59]	1.4[59]
HPTA	5.6[67]	−0.8[67]
1-Naphthol	9.4[73]	−0.2[45]
1-Naphthol-3,6-disulfonate	8.56[74]	1.1[42]
5-Cyano-1-naphthol	8.5[47]	−2.8[47]
1-Naphthol-4-sulfonate	8.27[75]	−0.1[75]
2-Naphthol	9.6[73]	2.8[9]
5,8-Dicyano-2-naphthol	7.8[9]	−4.5[9]
5-Cyano-2-naphthol	8.75[9]	−0.3[9]
8-Cyano-2-naphthol	8.35[9]	−0.4[9]
2-Naphthol-6,8-disulfonate	8.99[74]	0.7[72b]

procedure seems even to work in cases where the emitting state is thought to be different than the state directly accessed by absorption. An example of this is 1-naphthol, where the averaged Förster-cycle value is about −0.5 compared to the directly measured value of −0.2.

Most of the error in the Förster-cycle calculations appears to be instrumental, i.e. the error introduced by uncertainties in the spectroscopic measurements of the absorption and fluorescence maxima. Misreading or an uncalibrated instrumental reading of 1 nm at 275 nm will result in an error of 0.3 pK_a^* units. The severity of this problem tends to relax by the averaging procedure outlined above, which usually results in 'cancellation of errors'. This is especially important when the reading errors are systematic. Even so, deviations of up to one pK_a^* unit between the spectroscopic data of different laboratories appear to be common. A large data base of pK_a^* values from various sources was summarized elsewhere[12]. The pK_a and pK_a^* in water of the photoacids shown in Figures 1–3 are given in Table 1 and judged to be reliable. The pK_a^* values were found from Förster-cycle calculations unless otherwise stated.

III. ON THE ORIGIN OF PHOTOACIDITY

It makes sense to start the discussion on the molecular-level processes which are responsible for photoacidity by first analyzing Brønsted acidity in general.

The attachment of a proton to a negatively charged molecule or to a neutral molecule is always a very exothermic process in the gas phase, the proton affinity (PA) of most common organic molecules being between 160 and 220 kcal mol^{-1} [76,77], where proton affinity is the energy gained in the gas phase in the process depicted in equation 6

$$M + p^+ \longrightarrow Mp^+ \qquad (6)$$

where M is the isolated molecule in the gas phase and p^+ is the proton. When the proton is attached in the gas phase to a negative ion such as the phenolate anion, one may express the proton affinity of the anion using the sum of the processes given in equation 7.

$$\text{ROH} \longrightarrow \text{RO}\cdot + \text{H}\cdot$$

$$\text{H}\cdot \longrightarrow \text{H}^+ + e^-$$

$$\text{RO}\cdot + e^- \longrightarrow \text{RO}^-$$

$$\overline{\text{ROH} = \text{H}^+ + \text{RO}^-}$$

(7)

It follows that the proton affinity of RO^- in the gas-phase reaction $\text{RO}^- + \text{H}^+ \to \text{ROH}$ may be formally broken down into three separate contributions: the formation of the ROH bond, $D(\text{ROH})$, the attachment of an electron to the proton, $I(\text{H})$, and the ionization of the molecular anion, $E(\text{RO}^-)$, to give the radical. The first two processes are exothermic and the third one is endothermic. The proton affinity is given as their sum in equation 8,

$$\text{PA}(\text{RO}^-) = D(\text{ROH}) + I(\text{H}) - E(\text{RO}^-) \tag{8}$$

where $I(\text{H})$ is equal to the ionization energy of the hydrogen atom and $E(\text{RO}^-)$ is equal to the electron affinity of the RO^- radical. The gas-phase proton affinity of anions is much larger than the proton affinity of their corresponding neutral molecules. Two examples are: $\text{PA}(\text{H}_2\text{O}) = 167$ kcal mol^{-1}, $\text{PA}(\text{OH}^-) = 391$ kcal mol^{-1} and $\text{PA}(\text{HF}) = 117$ kcal mol^{-1}, $\text{PA}(\text{F}^-) = 371$ kcal mol^{-1} [76,77]. The difference between the proton affinities of the neutral molecules and those of their corresponding anions are usually more than 200 kcal mol^{-1} and is attributed mainly to the neutralization of the charge of the proton by the anion. The proton affinity of the phenolate anion is about 350 kcal mol^{-1}, which is significantly less than the typical proton affinities of small anions, the difference between $\text{PA}(\text{OH}^-)$ and $\text{PA}(\text{PhO}^-)$ being 41 kcal mol^{-1}. This is partly due to the stabilization energy of the phenolate anion by resonance in the phenolate ion which shifts some negative charge away from the oxygen atom and delocalizes it on the benzene ring. One may conclude that Brønsted basicity rather than Brønsted acidity is the fundamental property of neutral molecules and negative ions in the gas phase, and that molecular properties and charge distribution affect the inherent gas-phase basicity of molecular anions. In situations where a second base is present in the gas phase, a proton-transfer reaction (equation 9) may occur:

$$\text{ROH} + \text{B} \rightleftharpoons \text{RO}^- + \text{BH}^+ \tag{9}$$

The free-energy change, ΔG, of such a reaction in the gas phase is simply $\text{PA}(\text{RO}^-) - \text{PA}(\text{B})$. When B is OH^- and ROH is phenol, then ΔG_g is $\text{PA}(\text{PhO}^-) - \text{PA}(\text{OH}^-) = -41$ kcal mol^{-1}, so in this reaction the phenol molecule acts as the Brønsted acid and the OH^- anion as the Brønsted base. Clearly, relative proton-affinity values determine the relative acidity scale of molecules in the gas phase.

Brønsted acidity comes into play in condensed phases, where proton dissociation is enhanced by the solvent or by other solute molecules which act as proton acceptors (bases) and stabilize the charge of the bare proton. The acid dissociation of phenol in solution may be written as in equation 10,

$$(\text{PhOH})_s = (\text{PhO}^-)_s + (\text{H}^+)_s \tag{10}$$

where s denotes the fully solvated (equilibrium solvation) species. The overall free-energy change (and hence the proton dissociation constant, see below) following the proton dissociation reaction in a solvent s is given by equation 11,

$$\Delta G_s = \Delta G_g + \Delta G_t(\text{PhO}^-) + \Delta G_t(\text{H}^+) - \Delta G_t(\text{PhOH}) \tag{11}$$

where ΔG_g is the free-energy change upon proton dissociation in the gas phase and $\Delta G_t(X)$ is the free-energy change upon transferring the reactant X from the gas phase to solution.

The conventional thermodynamic description of an acid dissociation in solution is by the equilibrium constant of the dissociation reaction (equation 12),

$$K_a = [\text{RO}^-][\text{H}^+]/[\text{ROH}] \tag{12}$$

where K_a is given by equation 13,

$$K_a = \exp[-\Delta G_s/RT] \tag{13}$$

from which equation 14 follows,

$$pK_a = \Delta G_s/\ln(10)RT = \Delta G_s/2.3RT \tag{14}$$

It is usually extremely difficult to calculate ΔG_s of ground-state acids from first principles with uncertainty of less than several kcal mol^{-1}, which translates into uncertainty of several pK_a units. In the excited state an additional difficulty involves the accurate electronic description of the excited state, which makes the task of calculating the pK_a^* of a photoacid even tougher. A recent attempt[78] to calculate the excited-state pK_a^* of phenol resulted in a value larger by more than 4 pK_a units than the experimental one (a value of $pK_a^*(\text{calc}) = -0.2$, compared to the experimental value of about 4). A very recent calculation of the ground-state dissociation constant of phenol resulted also in overestimation of the dissociation constant, giving 7.2 compared to the experimental value of about 10.0[31].

To have a feeling for the computational difficulties involved in this type of calculation equation 11 may be rewritten as equation 15:

$$\Delta G_s = \text{PA}(\text{PhO}^-) - \Delta G_t(\text{H}^+) + \Delta G_t((\text{PhO}^-) - (\text{phOH})) \tag{15}$$

For phenol, PA (phenolate) = 350 kcal mol^{-1}, ΔG_t of the proton is about 260 kcal mol^{-1} [76] and $\Delta G_t((\text{PhO}^-) - (\text{PhOH}))$ may be roughly estimated assuming that it is mainly given by the Born free-energy of solvation of charged cavities immersed in a dielectric continuum (equation 16),

$$\Delta G_{\text{Born}} = e^2/2(1 - 1/\varepsilon_s)(1/r_B) \tag{16}$$

Here e is the electron charge, ε_s is the static dielectric constant of the solvent and r_B is the radius of the Born cavity around the charge, which may be approximated by the radius of the isolated ion. The solvation (Born) energy is calculated for a transfer from vacuum conditions to the solvent.

Substituting in equation 11 the known experimental parameters for phenol dissociation ($\Delta G_s = 13.8$ kcal mol^{-1} calculated from the ground-state equilibrium constant, $pK_a = 10.0$), $\Delta G_t((\text{PhO}^-) - (\text{PhOH}))$ of the phenolate/phenol system is about -76 kcal mol^{-1}, which is about 10% less than the accepted value for the electrostatic solvation energy of the chloride anion in water, $\Delta G_e(\text{Cl}^-) = -85$ kcal mol^{-1}. These simple considerations imply that the $\Delta G_t((\text{PhO}^-) - (\text{PhOH}))$ contribution to the overall free energy of solvation is largely electrostatic, and that relatively small differences in the gas-phase proton affinity of the base and in specific solvent–solute interactions of the photoacid and the base determine the relatively narrow (in free-energy units) acidity scale in aqueous solution. It

7. UV-visible spectra and photoacidity of phenols, naphthols and pyrenols

is clear that the calculation of the absolute pK_a values in solution from first principles is a formidable task if one insists that the calculated pK_a values should exactly reproduce the experimental ones. A one-pK_a-unit error in the calculated pK_a translates into a mere 1.3 kcal mol^{-1} error in the calculated overall stabilization energies of all species involved in the proton-dissociation reaction, each of these stabilization energies being about two orders of magnitude larger than the desired error bars.

This situation considerably improves if one limits oneself to the calculation of the relative acidity of the excited state compared to the acidity in the ground state. It is clear that photoacidity depends on the difference between ground- and excited-state free energies of solvation. Of all the parameters appearing in equation 12 only $\Delta G_t(H^+)$ does not depend on the electronic state of the photoacid:

$$pK_a^* - pK_a = (\Delta G_s^* - \Delta G_s)/2.3RT = (PA(R^*O^-) - PA(RO^-))$$
$$+ \Delta G_t((R^*O^-) - (RO^-)) + \Delta G_t((ROH) - (R^*OH))/2.3RT \quad (17)$$

Equation 17 may be viewed as an explicit form of the Förster cycle. It depends on both intramolecular and intermolecular factors which determine the extent of the photoacidity. The first factor is the difference between the excited-state and the ground-state proton affinities of the photobase. This difference will be equal to the difference in the intramolecular stabilization of the proton upon the electronic excitation of the acid, and will depend, in general, on the quantum-mechanical properties of the first excited electronic state of the photoacid. The second factor is the difference in the solvation energies of the base and the photoacid upon electronic excitation. The magnitude of the solvation-energy terms will depend in general both on the solvent and the solutes and will depend on the nature of the first electronic state of the photoacid and its conjugate base.

The traditional approach has been to define photoacidity as an intramolecular property of the photoacid[10,12,13,79,80]. In terms of equation 17, this approach places the main reason for photoacidity in the reduced proton affinity of the molecular anion in the electronic excited state. Alternatively, this means that photoacidity is mainly the result of the reduction in the dissociation energy of the photoacid in the gas phase upon electronic excitation. What is the reason behind this reduced proton affinity of the photobase?

There are two views regarding this scenario. The traditional view has been to ascribe photoacidity mainly to the increased reactivity of the photoacid in the excited state brought about by charge migration from the non-bonding electrons of the oxygen atom to the aromatic π system of the photoacid (n–π^* transition), thus weakening the O–H bond and making the photoacid a stronger acid in the excited state. The aromatic residue is viewed in this approach as becoming more electronegative in the excited state, shifting some electron density away from the oxygen atom, thus making it a weaker base[80,81]. This intramolecular charge redistribution following electronic excitation is stabilized by polar solvents. This view of photoacidity is portrayed in Figure 8. The increased acidity of 2-naphthol in the excited state is rationalized by assuming that a partial positive charge develops on the oxygen atom and a partial negative charge develops on the distal aromatic

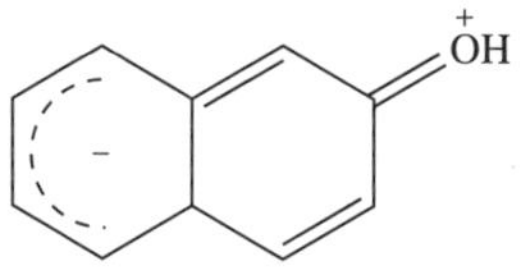

FIGURE 8. A traditional view of the electronic structure of a 'classic' photoacid, 2-naphthol in its first electronic excited state (after Bell[80])

ring of the naphthol. The oxygen atom then becomes partially 'repulsive' toward the proton, which in turn explains the rapid dissociation of the proton observed in the excited state[80,81].

A second, very recent view of photoacidity places the main electronic rearrangements within the product side of the dissociation reaction (the photobase)[66,77,82]. This view is corroborated by *ab initio* and semi-empirical calculations of the electronic distribution of several photoacids and photobases which show the excited state of the photobase to have a much larger charge-transfer character than the corresponding electronic state of the photoacid[66,77]. No significant $n-\pi^*$ transition was observed in the photoacid side of phenol and pyrenol. According to this recent view of photoacidity, photoacids become stronger acids in the excited state because the photobase becomes a much weaker base in the excited state. It is clear from the foregoing discussion that both scenarios fall within the arguments leading to equation 17 and define acidity in general terms. Rather, the two scenarios differ in the details of the molecular mechanism which is responsible for the proton affinity of the photobase being lower in the excited state with respect to the ground state: Is it because the excited base is less reactive toward the proton due to a larger internal stabilization energy of the negative charge and hence the smaller proton affinity (second scenario), or is the proton affinity of the excited anion smaller because the formed photoacid is less stable and more reactive (first scenario)? There is already a debate developing over this second recent scenario[66,82]: Is it or is it not a true revisionist description of photoacidity? Clearly, this question goes back to the basic definition of Brønsted acidity: Is acidity some inherent property of the acid or is it just reflecting the low reactivity of the base toward the proton? In other words, is it possible to define an acidity scale based entirely on the properties of the acid? And, by doing so, is it possible to separate between the actual proton-transfer act, which clearly depends also on the stabilization energy of the base (both internal (gas phase) and external (solvation) energies), and the property we call 'acidity'? Excited-state proton transfer may or may not happen during the lifetime of the excited state, depending on the polarity of the solvent. So should it not be better to concentrate on the intramolecular processes occurring at the acid side regardless whether they lead to an observed proton-transfer reaction? In other words, is there a better way to define photoacidity than by using Brönsted-type terminology and the pK_a^* scale?

Aside from these fundamental questions, some more questions arise from the practical difficulty in exactly calculating the Förster-cycle parameters of a photoacid from first principles (equation 17). From a thermodynamic point of view, in order to justify a product-side-driven reaction it is not sufficient to identify a larger electronic rearrangement in the excited state of the base compared with that found in the excited acid side. One rather has to show that both internal energies and solvation energies of the photobase are larger than in the ground state and are driving the proton dissociation reaction, and so are the main reason behind the enhancement in the acidity of the photoacid. In this stage one cannot conclude with certainty from either theoretical or experimental considerations that this is indeed the general situation which accounts for photoacidity (see the following section). In contrast, it is rewarding to point out several experimental observations which, although they do not prove, point out that both the acid side and the base side are active in determining the extent of photoacidity of hydroxyarenes in solution. The first observation, which probably has led to the traditional view of photoacidity, is that most of the enhanced acidity of excited hydroxyarenes may be traced back to the increase in the dissociation rate of the photoacid and, to a much lesser extent, to the decrease in the rate of the proton recombination to the photobase. Taking HPTS as an example, the dissociation rate of the acid on contact (i.e. excluding the effect of the electrostatic attraction between the proton and the anion) in the excited state increases by about 5 orders of magnitude, from about 10^5 s^{-1} to about 10^{10} s^{-1}. At the same time, the proton recombination rate to the photobase decreases by about 2 orders of magnitudes, from less than 10^{12} s^{-1} to about 3×10^9 s^{-1}. For 1-naphthol the situation is even more extreme. The dissociation rate of the photoacid increases by about 8 orders of magnitude while the recombination rate of the proton with the photobase decreases by less than 2 orders of magnitude. Clearly, the main dynamic effect appears from the photoacid side and not from the photobase side. However, this observation is by no means a general rule of photoacidity. There are good indications that the extreme excited-state acidity of protonated amine photoacids, such as the protonated 1-aminopyrene photoacid[15], comes from a very large reduction in the photobase reactivity, while the dissociation rates of the photoacids do not increase dramatically in the excited state and are typically two orders of magnitudes smaller than the dissociation rates of hydroxyarene photoacids having similar pK_a^* values[83].

The second observation concerns the increase in the hydrogen-bonding interaction of the O−H moiety of the hydroxyarene. Several observations of this effect were reported in the past, for phenol, naphthol and pyrenol derivatives. Perhaps the most direct observation concerns the red shift observed in the IR absorption frequency of the complexed O−H bond. A shift of about 250 cm^{-1} was observed for O−H···O and O−H···N type bonds of 1:1 complexes of 1-naphthol with water and ammonia when 1-naphthol was electronically excited. This shift translates to an about 0.7 kcal mol^{-1} increase in the hydrogen-bonding interaction in the excited state of the photoacid. A similar effect was observed in solution by Weller for the system 1-hydroxypyrene complexed with pyridine in methylcyclohexane[5]. Other observations include phenol and 1- and 2-naphthol complexed with dioxane in isooctane[34], and HPTA complexed with dioxane and DMSO in dichloromethane and dichloroethane[84]. In all cases the hydrogen-bonding interactions of the photoacid were found to increase upon electronic excitation by 0.5−3 kcal mol^{-1}. No proton transfer was observed in these systems.

The increase in the hydrogen-bonding interaction in the electronic excited state of the photoacid is a very convincing indication of stronger hydrogen bonds as compared to the ground-state situation. According to the widely accepted model of Pimentel[85,86] for the effect of the hydrogen-bonding interaction on the electronic transitions from and to the ground electronic state of the chromophore, a situation where both the absorption and the fluorescence spectra are red-shifted, and the fluorescence shift being the larger one, can only arise from the hydrogen bond being stronger in the excited state. This is indeed the situation for 1- and 2-naphthol and HPTA. Finally, the spectral shift of the photoacid due to polar interactions with the solvent may be correlated with empirical solvent parameters in a procedure suggested by Kamlet and Taft and their coworkers (the K−T analysis[87−89], see below). Such correlations usually result in a much larger effect of solvent basicity (β factor) on the fluorescence spectra of the hydroxyarene than the solvent basicity effect on the absorption spectra, indicating again, according to Pimentel's model[85,86], stronger hydrogen bonds in the excited state of the acid. It does appear, then, that the O−H moiety of the hydroxyarenes forms stronger hydrogen bonds in the excited state, implying photoacidity emerging, at least partially, from the photoacid side. A correlation between the aqueous pK_a values of various acids and the strength of the hydrogen-bonding interaction of their acidic proton was demonstrated in the solid state by NMR measurements, giving some direct evidence that stronger acids form stronger hydrogen bonds[90]. The NMR measurements have been mainly carried out in non-polar environments which do not support the ionization process involved in the proton-dissociation reaction of hydroxyarenes.

The relative strength of the hydrogen-bonding interactions may also be estimated indirectly by correlating their effect on the optical transition frequencies of the chromophore. In the Kamlet−Taft (K−T) analysis[87−89], any solvent-influenced property of the solute may be correlated using a multi-parameter fit (equation 18),

$$P_{s\text{-}s} = P^o_{s\text{-}s} + s\pi^* + a\alpha + b\beta \qquad (18)$$

where $P_{s\text{-}s}$ is the measured solvent-influenced property of the solute; $P^o_{s\text{-}s}$ is the numerical value of the chosen solute property in cyclohexane; π^* is the normalized solvent polarity scale; α and β are the solvent-acidity and the solvent-basicity scales, respectively; s, a and b are solute-dependent specific numerical coefficients, which characterize the solute molecule. The π^*, α and β parameters are assumed to be independent of each other (orthogonal) and additive, i.e. an ideal binary mixture of two solvents should correlate according to their combined values of π^*, α and β weighted by their relative composition in the solvent mixture.

Figure 9 shows an example of a correlation of the spectral shift of the peak fluorescence frequency of HPTA photoacid with the K–T parameters of several organic solvents. Most of the investigated solvents did not support proton dissociation within the lifetime of the excited state of HPTA, which is about 3.7 ns.

The K–T analysis, which is corroborated by direct IR measurements of the absorption of the stretching frequency of the O–H bond, shows that HPTA acts as strong hydrogen-bond donor (large b value) through hydrogen-bonding interaction of the type O–H···s[91]. At the same time, there is no evidence (small a value) for the oxygen atom accepting hydrogen bonds of the type O···H–s. This means a large sensitivity of the fluorescence spectra to the basicity of the solvent, and a much smaller sensitivity to the acidity of the solvent. In addition, the photoacid exhibits large sensitivity to the polarity of the solvent (large s values), indicating a relatively large dipole moment of the photoacid in the excited state compared to the ground state.

In an additional set of similar experiments the methoxy derivatives of HPTA, HPTS[91], 1-naphthol[91] and 2-naphthol[49] were examined by the K–T procedure. It was found that replacing the proton by a methyl group almost eliminated the hydrogen-bond interactions of the oxygen atom, so solvent basicity had a much smaller effect on the fluorescence spectra of these methoxy photoacids (Figure 10). At the same time the shape of the spectra, its location and the s values remained almost unchanged, indicating that the intrinsic electronic structure of the methoxy derivative is analogous to that of the photoacid.

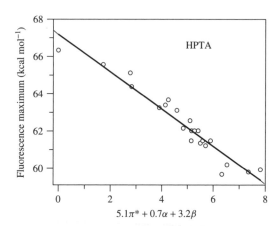

FIGURE 9. Correlation of the fluorescence spectra of HPTA in pure solvents with the Kamlet–Taft solvent-polarity parameters[91]

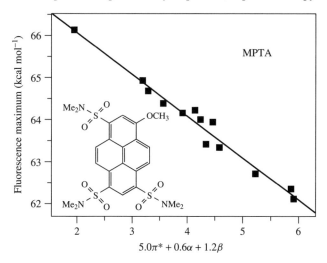

FIGURE 10. Correlation of the fluorescence spectra of 8-methoxy-1,3,6-tris(N,N-dimethylsulfonamido)pyrene (MPTA) measured in pure solvents with Kamlet–Taft parameters. MPTA has similar electronic structure to HPTA but is much less affected by hydrogen-bonding interactions (from Reference 91)

In contrast, it was found by a similar K–T analysis that the conjugate photobase acted as a better hydrogen-bond acceptor in the ground state, accepting a hydrogen bond of the type $RO^-\cdots s^{49}$. This set of observations supports the idea that both the acid side and the base side are generally active in determining the extent of photoacidity of hydroxyarenes, the acid being a stronger acid in the excited state and the base being a stronger base in the ground state.

Figure 11 shows the hydrogen-bond free energy of the interaction (b values) of a series of hydroxyarene photoacids plotted against their photoacidity strength scaled in terms of their free energy of proton dissociation in aqueous solutions. There is a linear correspondence between the two values, indicating that the relative strength of hydroxyarene photoacids in non-polar solvents may be scaled using their relative pK_a^* values in aqueous solutions.

Finally, there appears to be a correlation between the acidity of the photoacid in aqueous solutions and the strength of the hydrogen-bonding interaction (Figure 11)[91]. This observation is in accord with the general observation stated earlier that the stronger the acid, the stronger the hydrogen-bond interactions that it undergoes with a given base.

A general rule may be extracted from these observations. For a given hydrogen-bond donor (the photoacid), the strength of the hydrogen bond that it forms in the excited state with a hydrogen-bond acceptor (a solvent molecule, or an additional base molecule dissolved in the solvent) will increase with increase in the β value of the hydrogen-bond acceptor. A similar observation holds for a given hydrogen-bond acceptor. In this case, the hydrogen-bond strength will increase with the α value of the proton donor.

In conclusion, it is the opinion of this review that photoacidity manifests itself in both the photoacid and the photobase sides, the reactant side becoming a stronger acid and the product side becoming a weaker base in the excited state. It is still a matter of additional experimental and theoretical studies to establish if general rules may be drawn up concerning the relative importance and generality of these processes. Similarities to ground-state

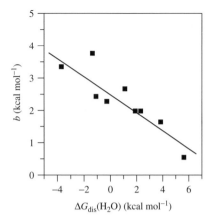

FIGURE 11. Hydrogen-bond interaction of hydroxyarene photoacids (parameter b in equation 18) versus dissociation free energy of dissociation of the photoacids in water (from Reference 91)

acids should also be pointed out. The following remarks concerning the debate about the traditional description of photoacidity may help to clear this issue. Enhanced acidity due to the anion side stabilized by electronic resonance has been a textbook explanation for the marked ground-state acidity of hydroxyarenes. Electronic resonance stabilization of the anionic charge by the aromatic ring is traditionally considered the main effect for the increased acidity of hydroxyarenes compared with the acidity of non-aromatic alcohols and the main reason for strong deviation from Hammett-type structure–acidity correlations. Figure 12 has been used to explain the very large acidity of p-nitrophenol[92]. In this case, the resonance stabilization of the anion is much more important than the resonance stabilization of the acid, leading to a larger increase in the acidity of the substituted phenol as compared with the predicted polar effect of the p-nitro group based on its effect on the ionization of benzoic acid.

Figure 13 shows all the contributing resonance hybrids to the ground states of phenol and phenolate anion. Those of the anion are thought to be more important than those of the phenol molecule by several kcal mol^{-1}. This was used as an argument for the increased acidity of phenol over non-aromatic alcohols[93].

However, having said that, the anion-side scenario was generally overlooked when excited-state photoacidity was considered, even in cases where it has become evident that the photobase undergoes an extensive intramolecular charge-transfer process. As an example, an extensive charge-transfer process has been assumed in the 1-naphtholate anion where a recent *ab initio* calculation showed that roughly 2/3 of a unit charge is transferred from the oxygen atom to the naphthalene ring[46,51]. In contrast, the electronic structure of the photoacid did not show such an extensive charge-transfer process. Those observations have been made without remarking which side contributes more to the overall photoacidity

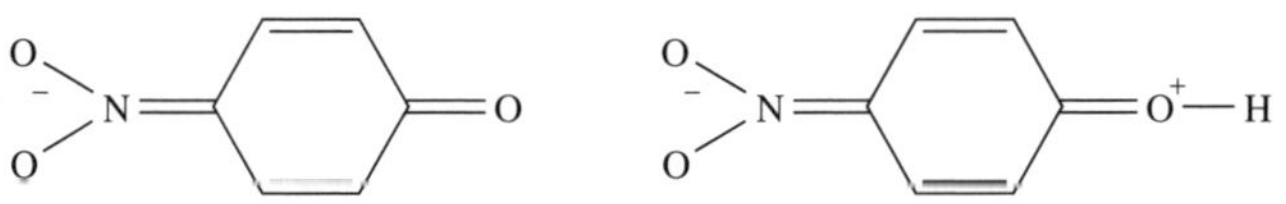

FIGURE 12. The important resonance hybrids of p-nitrophenol and its conjugate anion in the ground state

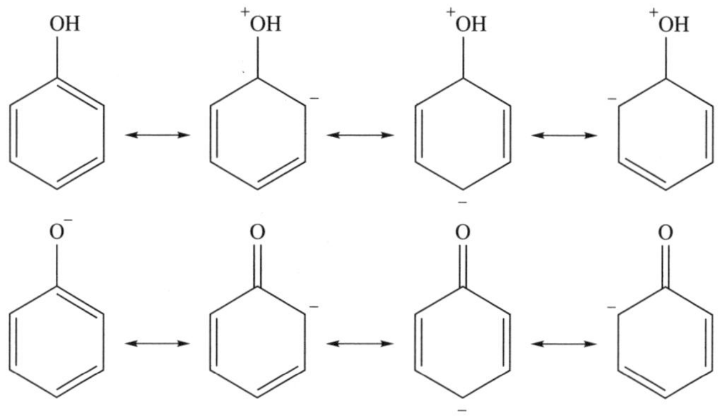

FIGURE 13. Resonance hybrids of phenol and phenolate anion in the ground state

of 1-naphthol. All said, it is perhaps best to refer to Bell's book *The Proton in Chemistry*[80] which, about 30 years after the publication of its 2nd edition, is still arguably the most authoritative contribution written on the physical aspects of acid–base reactions. In Bell's book, almost side by side, ground-state and excited-state acidities of hydroxyarenes are discussed. To account for the considerable ground-state acidity of phenol, Bell invokes the anion-side resonance description of the phenolate anion, which reduces its reactivity as a base by delocalizing part of the negative charge over the aromatic residue. In contrast, the large increase in the excited-state acidity of 2-naphthol is attributed exclusively to a resonance structure of the photoacid similar to that shown in Figure 8. In view of many similar arguments appearing throughout the literature describing photoacidity in terms of the increased acidity of the acid side[12], it is only fair to say that the recent paper by Hynes and coworkers[66] is constructive in stressing the importance of the anion-side charge-transfer reaction in the excited state of photoacids, and by doing so, in a somewhat paradoxical way, making photoacids more like ordinary ground-state acids than, perhaps, what has been traditionally thought previously.

Finally, we believe that a search for a new, more general definition of photoacids is in place, perhaps through their ability to form strong hydrogen bonds in the excited state, regardless of whether or not proton dissociation occurs within the excited-state lifetime of the photoacid. Thus, it may well be rewarding to describe the photoacidity phenomenon in terms not necessarily connected to proton transfer and Brønsted acidity of photoacids. By doing so, it would make the definition of photoacids applicable to a larger group of molecules, extending its application to non-polar environments where no proton transfer occurs within the excited-state lifetime. Clearly, more studies must be carried out before conclusive treatments of these issues may be achieved

IV. THE ELECTRONIC STRUCTURE OF PHOTOACIDS

The origins of the enhanced acidity of hydroxyarenes and other photoacids are clearly due to the differences between the quantum-mechanical properties of the first electronic singlet state (the fluorescence emitting state) and the ground electronic state of the photoacid. Aside from the question whether acid or base is more important in determining the pK_a^* of the excited photoacid, one faces a more fundamental question as to why photoacidity occurs at all. To answer this question one should deal with the electronic structure of the photoacid in the excited state. The electronic structure of both the photoacid and the photobase is important in determining the observed increase in the Brønsted acidity of the photoacid in the excited state. The elucidation of the electronic structure of hydroxyarenes in the excited state has become one of the most intriguing and demanding tasks in photoacid research. Although considerable progress has been achieved, our current understanding of this problem is still far from being conclusive concerning questions of photoacidity. Is there a 'special' electronic state which is responsible for photoacidity? Is this 'special' state accessed directly from the ground state? How long does it take for the electronic state to relax from the locally excited state to this photoacidity state when it is not accessed directly from the ground state? What intra- and intermolecular processes control the rate of this electronic relaxation? Which is the more important electronic rearrangement, the one occurring at the acid side or the one occurring at the anion side? From an experimental and theoretical point of view, these questions should have been approached by first undertaking the task of spectroscopic assignment of the first few electronic transitions covering the relevant absorption and fluorescence spectra of the photoacids in question. Unfortunately, systematic analysis of the electronic spectra of hydroxyarenes has met with great difficulties already in the stage of the spectroscopic assignment. In many cases the electronic spectra of the photoacid is congested and is usually thought to comprise two overlapping transitions, each mixed to a various degree with other, higher-lying electronic states.

Discussion of the theoretical aspects of the electronic structure of optically excited hydroxyarenes has been greatly influenced by the work of Platt and his coworkers at the University of Chicago. A source book of the papers of the Chicago group (1949–1964)[68b] summarizes their considerable contribution to the interpretation of the electronic spectra of simple aromatic systems. Platt's model utilizes the free-electron molecular-orbital method (when applied to conjugate linear chains of alternating single and double bonds as found in some polyenes, this method is sometimes called the 'electron in a box' model). Platt applied this model to aromatic molecules, which may be viewed as having a π electronic system lying on a single closed loop or a 'perimeter'. Platt's 'Perimeter Model' was developed for 'catacondensed' hydrocarbons, whose general formula is $C_{4n}H_{4n+8}$, and their carbon atoms form a single periphery. The general result of the model, which was corroborated by experimental findings, is that there are regularities in the spectra of simple aromatic compounds. These regularities are the energies of their lower electronic levels, the ordering of the levels according to one spectroscopic scheme and the distinctive molecular-orbital characteristics of each level. The energy of these levels changes smoothly, moving from one molecular system to the other.

The lowest four electronic levels common to all catacondensed hydrocarbons are, according to Platt's notation, ^1L_b, ^1L_a, ^1B_b and ^1B_a. The L transitions are generally almost forbidden, (especially the ^1L_b transition) having very small oscillator strength, while the B transitions are strongly allowed, having typically oscillator strengths between one to two orders of magnitude larger than the L transitions. In Platt's notation, subscript 'a' stands for electronic levels having the electron density of the electrons on the atoms and the nodal points (zero electron density points) on the bonds connecting the atoms; subscript 'b' stands for electronic levels having the electron density of the π electrons on the bonds and the nodal points on the atoms. In general, the number of nodal points of the two lowest states, the L states, equals the number of atoms and their dipole moments are expected to be small and similar in magnitude to the ground-state dipoles. The dipoles of the ^1L_b and ^1L_a states are generally orthogonal to each other, the dipole of the ^1L_b state being along the short symmetry axis and the dipole of the ^1L_a state being along the long symmetry axis of the molecule. Also, the ^1L_b state sometimes appears to be more vibronically structured than the ^1L_a state. Clearly, the main idea behind Platt's free-electron model is its simplicity, which allows each electronic level to be described by some characteristic molecular-orbital properties that define its unique physical identity. Experimentally, Platt's approach is strictly valid in a limited number of unsubstituted aromatic systems. The assignment of these levels already becomes less strict in the pyrene system, in which only 14 out of its 16 carbon atoms lie on one peripheral. Substituents and polar interactions with the solvent also affect the simple picture outlined by Platt. However, it is customary to retain Platt's notation in the assignment of the electronic levels of substituted benzene and naphthalene, although the distinctive physical character of these levels become blurred in the substituted molecules. Polar substituents are believed to stabilize the ^1L_a state more than the ^1L_b state, so they lower the transition energy of the ^1L_a state compared to the transition energy of the ^1L_b state. Polar substituents may also enhance the polarity of the ^1L_a and ^1L_b states and mix them. Inversion between the two L states may occur in polar environment, which further stabilizes the ^1L_a state over the ^1L_b state. The two L states may also be coupled to each other by some vibronic modes of the aromatic ring. This may result in the two L states being in a dynamic equilibrium with each other. Moreover, each of the two L states may be mixed to a different degree with the allowed B levels, thus 'borrowing' oscillator strength from these levels and considerably changing their characteristic spectra.

Over the past decade Platt's notations were used extensively to describe the electronic levels of several hydroxyarene photoacids. This was very constructive in bringing to attention, in a qualitative way, the complexity of the electronic structure of some very common photoacids. However, the extent of the quantitative analysis which may be drawn from such considerations is still unclear. Arguably, the most researched and best example for the complexity of the photoacidity phenomenon from the viewpoint of the electronic structure of the photoacid is the 1-naphthol molecule. The ground-state acidity of 1-naphthol is almost identical with the ground-state acidity of the 2-naphthol isomer, yet the excited-state acidity of 1-naphthol is 3 orders of magnitude larger than the corresponding acidity of 2-naphthol. This observation has puzzled researchers for the past 50 years. The spectroscopic scope of this problem is evident when the absorption spectra of 1-naphthol is compared with that of 2-naphthol (Figure 14).

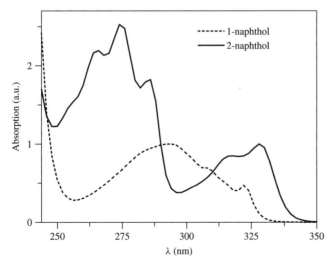

FIGURE 14. Absorption spectra at 20 °C of 1-naphthol (dashed line) and 2-naphthol (solid line) in H_2O[51]

The absorption spectra of 2-naphthol consist of two excitation bands, assigned as ^1L_b (S_1) and ^1L_a (S_2) transitions. In contrast, the absorption spectra of 1-naphthol taken over the same spectral range contains only one absorption band having roughly the same spectral width as the combined spectral widths of the ^1L_b and ^1L_a absorption bands of 2-naphthol. Thus the two L transitions are thought to overlap in the 1-naphthol absorption spectra.

The large difference in the appearance of the electronic absorption spectra of 1- and 2-naphthol seems to indicate that photoacidity may be correlated with some spectral features common to all photoacids of a given family. Such common features, if they indeed exist, may be used as 'fingerprints' for identifying the extent of inherent photoacidity exhibited by the photoacid, regardless of whether or not it can be ionized in the medium. Before addressing the various approaches dealing with this issue in connection with the 1- and 2-naphthol dilemma, it is worthwhile to point out that the general situation is most probably more complicated than what it appears to be from visual inspection of the spectra of 1- and 2-naphthol.

Figure 15 shows the absorption spectra of several 1- and 2-naphthol derivatives. The very broad absorption band of 5-cyano-1-naphthol looks like a red-shifted 1-naphthol spectra where the spectrum of 1,6-dibromo-2-naphthol resembles the two-band absorption spectrum of 2-naphthol shifted to the red by about 20 nm. The broad absorption spectrum of 1-naphthol is retained in the absorption spectra of 1 naphthol-3,6-disulfonate, but the spectrum becomes much more structured. In contrast, the familiar spectral features of 2-naphthol become blurred in the case of 2-naphthol-6,8-disulfonate and 1,6-dibromo-2-naphthol, which are considerably red-shifted and appear wider and almost featureless. In fact, the two spectra resemble each other more than they resemble the spectrum of either the 'parent' 2-naphthol molecule or the 1-naphthol isomer. In addition, no clear correlation exists between the shape of the spectra and the pK_a^* of the photoacid, the three 2-naphthol derivatives and 1-naphthol-3,6-disulfonate all having a pK_a^* that falls within 1 pK_a unit

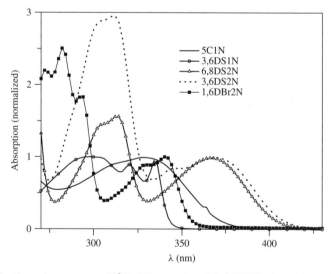

FIGURE 15. Absorption spectra at 20 °C of 5-cyano-1-naphthol (5C1N), 1-naphthol-3,6-disulfonate (3,6DS1N), 2-naphthol-3,6-disulfonate (3,6DS2N), 2-naphthol-6,8-disulfonate (6,8DS2N) and 1,6-dibromo-2-naphthol (1,6DBr2N) measured in water at acidic pH values from 4 to 7[91]

of each other, while 5-cyano-1-naphthol is a much stronger photoacid having a pK_a^* value of about -2.8^{47}. Evidently, substitutions change the spectra of naphthols not in a simple way and the magnitude of the change depends on the number of the substituents, their ring position and their chemical nature.

We thus limit ourselves mainly to a discussion of the electronic spectra of the unsubstituted naphthols and phenol. The very important class of pyrenol photoacids is also largely excluded from our discussion, although the absorption spectra of 1-hydroxypyrene seems to fall within Platt's description exhibiting a typical ^1L_b, ^1L_a, ^1B_b, ^1B_a 4-band structure[103]. This does not mean that, from a pure theoretical background, pyrenols should not be analyzed in terms of Platt's notation, a practice that has been extensively undertaken, very recently, by Hynes and coworkers[65,66]. Our opinion is, rather, that regularities concerning the molecular basis for photoacidity should be drawn only in the face of clear experimental evidence. Considering our current state of knowledge, this does not appear to be the case when most other pyrenols are considered (see also Figure 16).

With the above reservations in mind, we summarize below the different approaches that attempt to elucidate the excited-state acidity of 1- and 2-naphthol by analyzing the structure of their electronic spectra. As already pointed out, there is a considerable difference between the photoacidity of 1- and 2-naphthol (about 3 pK_a units). In contrast, the two naphthol isomers exhibit almost identical ground-state acidities, the difference between the pK_a of the two isomers being less than 0.2 pK_a units ($pK_a = 9.4$ and 9.5 for 1- and 2-naphthol, respectively). This simple observation suggests, although does not prove, that the two isomers differ mainly in their electronic structure in the excited state. Direct comparison between the electronic spectra of the two isomers has provided, arguably, the

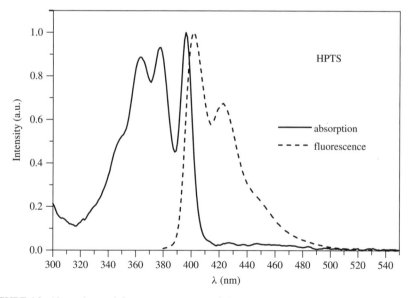

FIGURE 16. Absorption and fluorescence spectra of the HPTA molecule in acetonitrile. The mirror-like symmetry appearing at first sight to exist between the absorption and fluorescence spectra is misleading, the absorption spectra being about 30% wider and more structured. The sharp, vibronic-like spectral features were interpreted as coming from a mixture of ^1L_b and ^1L_a transitions, similar to the 1-naphthol case[65], or alternatively, as originating from strong solvent–solute interactions of a single S_1 state in the case of the methoxy analogue of HPTS, the MPTS molecule[94]

best known case where enhanced photoacidity was tracked to some specific electronic rearrangement in the excited photoacid, namely the ^1L_b to ^1L_a level crossing. At least three different scenarios are attached to this proposed electronic transition. In all scenarios for which the ^1L_b state is assumed, the lower singlet state of the molecules in the gas phase (the S_1 state) and the ^1L_a level is assumed to be higher in energy (the S_2 state) and more polar than the ^1L_b state. Level inversion may occur in polar solvents which stabilize the ^1L_a state more than they stabilize the less polar ^1L_b state. Polar substituents may cause level crossing already in the gas phase. An example for such a substituent effect is found in the 1-naphtholate anion, where the S_1 state in the gas phase is thought to be the ^1L_a state[35,68b] (strictly speaking, Platt's notation describes the unsubstituted naphthalene molecule, so the 1-naphtholate anion should be viewed as a naphthalene molecule with O^- substituent at the 1 position). The enhanced photoacidity of 1-naphthol over 2-naphthol is then explained as the result of level inversion: While the emitting state of 2-naphthol is the directly excited ^1L_b state, level inversion occurs in 1-naphthol where the emitting state is not directly accessed from the ground state and is identified as the more polar ^1L_a state. The three scenarios which make this mechanism their starting point differ by the way they treat the inversion process.

The origins of the first scenario goes back to the classic studies of Shizuka and Tsutsumi[38,39]. In this scenario the ^1L_b and ^1L_a transitions are congested together in the absorption (the absorption spectrum of 1-naphthol in water is shown in Figure 14) and fluorescence spectra of 1-naphthol (Figure 17).

In this scenario, the absorption red edge of 1-naphthol is thought to be mainly the ^1L_b state and the blue edge of the absorption spectrum to be mainly the ^1L_a state[34]. This is the reason suggested for the absorption spectra of 1-naphthol being roughly as wide as the first and second transitions of 2-naphthol combined together. The situation is reversed in the fluorescence spectra of 1-naphthol in polar solutions (Figure 17). Here, as in the absorption spectra, the width of the fluorescence band is roughly twice as large as the

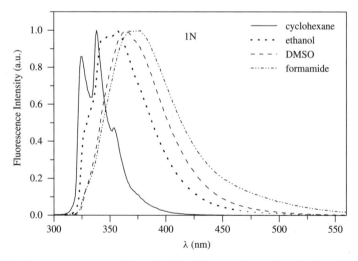

FIGURE 17. Fluorescence spectra of 1-naphthol in different solvents. Moving from formamide to cyclohexane, the fluorescence spectra is considerably shifted to the blue and becomes much narrower and more structured. In formamide and cyclohexane, the emitting state of 1-naphthol is thought to be ^1L_a and ^1L_b, respectively (from Reference 91)

fluorescence band of 2-naphthol. The red edge of the band is assigned to the ^1L_a state and the blue edge is assigned to belong to the ^1L_b state[35,51]. Moving to less polar solvents, the fluorescence spectrum becomes narrower and more structured than the fluorescence spectrum in water. This progression in the various spectra is explained by the two states being strongly coupled and in rapid equilibrium. The relative '^1L_b' or '^1L_a' nature of the fluorescence band is determined by the polarity of the solvent, changing gradually from being mostly ^1L_a type in water and formamide to being mostly ^1L_b type in cyclohexane[51]. Such a gradual change in the structure of the spectrum is not observed in the case of 2-naphthol[51], where much smaller spectral changes are observed as a function of solvent polarity (Figure 18).

A similar conclusion about the emitting state of 1- and 2-naphthol was reached from the K–T analysis of the fluorescence spectra of the two isomers[51] (Figures 19 and 20). The K–T analysis showed much better correlation of the 2-naphthol spectra in various solvents than the corresponding 1-naphthol spectra.

Good correlation ($R = 0.94$) was found when the fluorescence spectrum of 1-naphthol was divided into two emitting states. For the red-edge emitting state (^1L_a) the correlation has yielded a polar state, $2.8\pi^* - 1.3\alpha + 3.1\beta$, and for the blue-edge emitting state (^1L_b) the outcome was a non-polar state, $1.1\pi^* - 0.1\alpha + 0.8\beta$ ($R = 0.95$)[51]. The poor correlation of the position of the fluorescence maximum shown in Figure 20 was attributed to the fluorescence maximum being the combination of two emitting singlet states of different polarity which partially overlap. This indicates non-trivial changes in the 1-naphthol fluorescence spectrum as a function of the polarity of the solvent, the location and the relative weight of each emitting state having different dependence on solvent polarities. The two overlapping fluorescence transitions were assigned ^1L_b and ^1L_a transitions. In this scenario, level dynamics are assumed to be extremely fast and follow the solvation relaxation dynamics of the solvent, so level crossing did not determine the rate of the proton transfer from the photoacids which is assumed to be activated in the solvent.

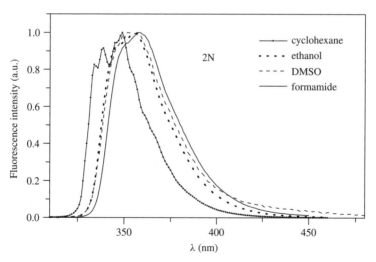

FIGURE 18. Fluorescence spectra of 2-naphthol in several solvents of various polarity. Note the much smaller solvent effect on the fluorescence spectra of 2-naphthol compared to 1-naphthol in the same solvents (Figure 17). The emitting level of 2-naphthol is thought to be the ^1L_b state in all solvents (from Reference 91)

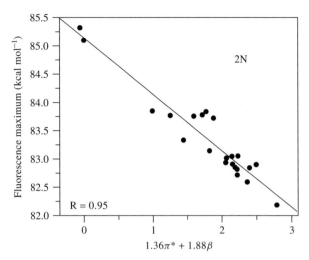

FIGURE 19. The good correlation found between the fluorescence maximum of 2N and solvent polarity using Kamlet–Taft analysis in 22 solvents. In this case no level crossing is evident and the emitting state is assumed to be ^1L_b in all solvents (from Reference 91)

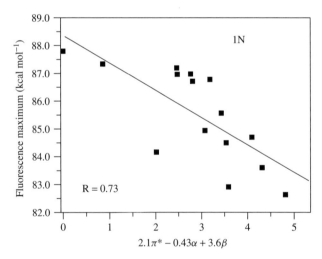

FIGURE 20. The poor correlation found between the fluorescence maximum of 1N and solvent polarity using Kamlet–Taft solvent-polarity parameters of 15 common solvents (from Reference 91)

7. UV-visible spectra and photoacidity of phenols, naphthols and pyrenols 515

The second scenario was developed to describe the situation pertaining to 1-naphthol in the gas phase[18,25,27,28]. In the gas phase, excitation is assumed to be a pure ^1L_b transition. Following excitation, level crossing to the ^1L_a state may occur in 1-naphthol-base gas-phase clusters and is promoted by some vibrational modes of the naphthalene ring which allow the otherwise symmetry-forbidden L_b to L_a transition. Polar interactions in the cluster stabilize the level crossing. Level dynamics was suggested to be the rate-determining step for the onset of the photoacidity of 1-naphthol in gas-phase clusters and aqueous solutions. The characteristic level-crossing time was estimated to be several ps in water clusters.

In the third scenario[66], developed for phenol derivatives on theoretical grounds, enhanced photoacidity was traced to the ^1L_b–^1L_a transition occurring upon proton dissociation. In this intriguing scenario the photoacid is assumed to be in a ^1L_b state the polarity and internal acidity of which resemble that of the ground state. Level crossing to the polar ^1L_a state occurs in the anion which, for that reason, is a much weaker base than the ground-state anion. In this scenario, level crossing does not consist of the rate-limiting step for the proton transfer although such a possibility was not entirely ruled out. An additional activated charge-transfer process was assumed likely to be the rate-limiting step for proton dissociation.

It is unclear if any of these scenarios may be considered a general description of photoacidity. More likely, each of these scenarios describes a possible intramolecular route which may contribute to photoacidity under certain experimental conditions but does not exclusively define photoacidity by itself. One should not rule out situations where the photoacidity state is directly accessed from the ground state and no further level 'switching' or crossing occurs in either the photoacid or the photobase side. This seems to be the case of 2-naphthol and its derivatives (see below). Also, it is unlikely that level dynamics determine the rate of the proton-transfer reaction in solution, the latter being usually a much slower process determined by the overall free-energy change upon proton dissociation. In fact, if we consider the arguments brought up in the first part of this review, even the seemingly clear-cut assignment of the emitting states of 1- and 2-naphthol must raise questions when their overall photoacidity is examined from Förster-cycle considerations. The general rule is that the lowest emitting state of 1-naphthol and 2-naphthol is ^1L_a and ^1L_b, respectively. In order to preserve the logic of the foregoing discussion, one must assume that the 1-naphtholate and 2-naphtholate anions are also ^1L_a and ^1L_b, respectively, since any other situation would not result in 1-naphthol being the strongest photoacid of the two isomers. This indeed appears to be the case[95,96], and suggests that being in the ^1L_a state rather than in the ^1L_b state roughly contributes one-third of the total photoacidity of 1-naphthol. It follows that the increase in the photoacidity due to the photoacid and the base being in the more polar ^1L_a state (1N) rather than being in the relatively non-polar ^1L_b state (2N) causes only one-half of the effect of the photoacid being in the excited state.

If one adopts the idea that regularities are found in the first two electronic levels of unsubstituted hydroxyarenes, then it is clear that the effect of the electronic structure on photoacidity according to the ^1L_b, ^1L_a terminology should increase in the order: ^1L_b to $^1L'_b$, ^1L_a to $^1L'_a$, ^1L_b to $^1L'_a$ and ^1L_a to ^1L_b, where ^1L denotes the electronic level of the photoacid and $^1L'$ denotes the electronic level of the photobase. In the case of 1-naphthol, the acid-side changes from being ^1L_a—like in water—to being ^1L_b—like in non-polar solvents—while the naphtholate anion is probably $^1L'_a$ in all polar and moderately polar solvents. It follows from the above order of photoacidities that the photoacidity of 1-naphthol should increase, moving from water to less polar solvents, where the acid side becomes higher in energy due to electronic rearrangement to form the less polar ^1L_b state. In contrast, 2-naphthol dissociation is either ^1L_b to $^1L'_b$, as usually assumed[95,96], or ^1L_b to

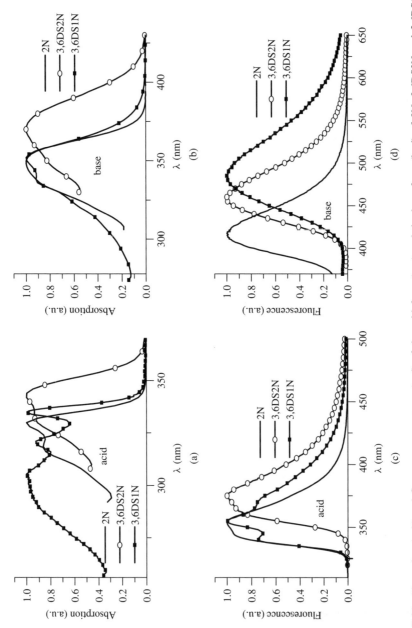

FIGURE 21. Absorption (a, b) and fluorescence spectrum (c, d) of the acid form (a, c) and the base form (b, d) of 2N, 3,6DS2N and 3,6DS1N in methanol[91]

$^1L'_a$ where the $^1L'_a$ is of a greatly reduced charge-transfer nature than the corresponding $^1L'_a$ state of 1-naphthol. It follows that in the case of 2-naphthol, one expects a much smaller solvent effect on the Förster-cycle acidity than the corresponding effect on 1-naphthol acidity. This indeed seems to be the case when the photoacidity of 1-naphthol and 2-naphthol was estimated from Förster-cycle calculations in water and methanol (Tables 1 and 2).

Absorption and fluorescence spectra of the acid and base forms of 2N, 3,6DS2N and 3,6DS1N in methanol used for the Förster-cycle calculations in methanol are shown in Figure 21.

2-Naphthol and its 3,6-disulfonate derivative show consistency in their spectral features in both the photoacid and base sides, while much less consistency is evident in the spectral features of the 1-naphthol derivatives in the acid side, where the fluorescence spectrum appears to be much narrower and more structured than the absorption spectrum (Figure 21). This points to more extensive electronic rearrangements in the acid side of

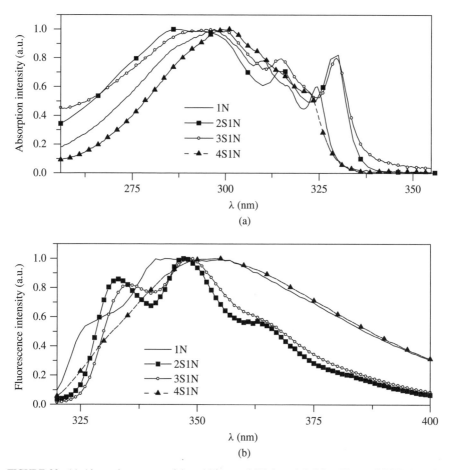

FIGURE 22. (a) Absorption spectra of the acid forms of 1N, 1-naphthol-2-sulfonate (2S1N), 1-naphthol-3-sulfonate (3S1N) and 1-naphthol-4-sulfonate (4S1N) in methanol. (b) Fluorescence spectra of the acid forms of 1N, 2S1N, 3S1N and 4S1N in methanol[91]

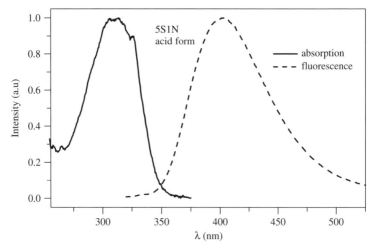

FIGURE 23. Absorption and fluorescence spectra of the acid form of 1-naphthol-5-sulfonate (5S1N) in methanol[91]

1-naphthol as a function of the solvent than in the acid side of 2-naphthol, and that both isomers show relatively small changes in the spectral features at the naphtholate side. This indicates the consistency of the emitting state of the naphtholate anion of both isomers.

Figures 22 and 23 offer a closer look at the absorption and fluorescence spectra of several 1-naphthol derivatives in methanol. In the case of sulfonate-substituted 1-naphthols, the substituent effect on the absorption spectra is relatively small, while a considerable effect is observed in the corresponding fluorescence spectra of the photoacids. This effect resembles the solvent effect on the fluorescence spectrum of the parent 1-naphthol molecule. Using the analogy to the effect of solvent polarity on the spectra, the ring position of the sulfonate group seems to appear more 'polar', moving from the 2- to the 4-position of the naphthol ring system.

The order of the effect of the ring position on the spectra is: $4 > 3 >$ unsubstituted > 2. This means that the further the substituent is from the OH group, the larger is its effect on the polarity of the first emitting state of 1-naphthol, probably through better stabilization of the charge-transfer character of the ^1L_a state. A second mechanism which seems to increase the ^1L_b character of the emitting state is direct hydrogen-bonding interactions between the OH and the sulfonate group at the 2 position and, to a lesser extent, at the 3 position.

Interestingly, Förster-cycle calculations of the pK_a^* in methanol (Table 2) seem to confirm the substituent effect on the polarity of the emitting state of 1-naphthol as discussed above: the less polar the emitting state of the acid compared to the emitting state of its conjugate base, the larger the Förster-cycle acidity of the photoacid. The calculated Förster-cycle difference between the ground-state and excited-state acidities in methanol was 12.3, 11.3, 10.9, 9.3 and 8.8 for the 2-substituted, 3-substituted, unsubstituted 4- and 5-substituted sulfonate photoacids, respectively.

This sort of argument demonstrates the need for defining a photoacidity scale which is independent of whether or not the photoacid is able to dissociate within the excited-state lifetime. The five photoacid derivatives of 1-naphthol discussed above do not dissociate in methanol. The order of their acidity in methanol extracted from Förster-cycle calculations awaits further confirmation. It should be conducted by some other method which would

7. UV-visible spectra and photoacidity of phenols, naphthols and pyrenols 519

TABLE 2. $pK_a^* - pK_a$ values of some common hydroxyarene photoacids from Förster-cycle calculations in methanol[91]

Photoacid	$pK_a^* - pK_a$
1-Naphthol	10.9
1-Naphthol-2-sulfonate	12.3
1-Naphthol-3-sulfonate	11.3
1-Naphthol-4-sulfonate	9.3
1-Naphthol-5-sulfonate	8.8
4-Chloro-1-naphthol	9.8
1-Naphthol-3,6-disulfonate	12.2
2-Naphthol	6.5
2-Naphthol-6,8-disulfonate	7.3
HPTS	7.2
HPTA	6.8

provide a direct measure for their photoacidity as judged by the scaling of some chemical property common to all photoacids in question.

Before such an endeavor is carried out one must rely on circumstantial evidence. Doing so, it appears as if polar substituents affect photoacidity not just by processes identified in ground-state acids, such as the inductive and resonance effects, but, in the case of 1-naphthol, also by systematically affecting the character of its electronic excited state.

It is also encouraging to find that the effect of polar solvents on the electronic spectra of 1-naphthol appears to be qualitatively similar to the effect of polar substituents. This raises hope that the paradigm of the photoacidity of 1-naphthol could be potentially resolved in a general way.

However, as already indicated before, it is very difficult to find regular patterns in the electronic structure within one family of photoacids which directly correlate all their photoacidity related properties. An example of this difficulty is found in the classic paper of Suzuki and Baba[34] on the hydrogen-bonding interactions of phenol and 1- and 2-naphthol.

The two lowest electronic transitions of the three photoacids were assigned in the very non-polar isooctane solvent by analyzing the effect of hydrogen bonding on their respective absorption spectra. In all cases the level ordering was found to be ^1L_b (S_1) and ^1L_a (S_2). The effect of hydrogen bonding was to shift the absorption spectra to the red. For phenol and 1-naphthol the red shift of the absorption of the ^1L_a state was much larger than the red shift of the absorption of the ^1L_b state, an observation which seems to be in harmony with the assignment of ^1L_a as the more polar state of the two. However, the situation was found to be the reverse in 2-naphthol, where the red shift of the ^1L_b state due to hydrogen bonding was found to be three times larger than the red shift of the ^1L_a state. Apparently, the position of the OH group affects the relative polarities of the two lowest electronic states of naphthols. When the absolute magnitude of the red shift was considered, the ordering of the red shift was found to be: ^1L_a(1-naphthol) $\gg$ ^1L_b(2-naphthol) $\gg$ ^1L_b(1-naphthol) = ^1L_a(2-naphthol), in accordance with the order of the photoacidity of the lowest emitting state of the two photoacids in polar solvents: 1-naphthol $\gg$ 2-naphthol. A similar situation is found when the red shift of the single absorption band of 1-naphthol is compared with the relative red shift of the two absorbing bands of 2-naphthol measured in the same solvents. Figure 24 shows the red shift of the absorption spectra of 1-naphthol to be the largest, in agreement with its greater sensitivity to solvent polarity. The S_1 state of 2-naphthol was found to shift, as in the Suzuki and Baba experiment, more than its S_2 state, an observation which seems to oppose the assumption

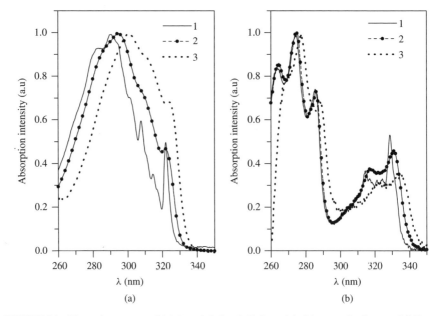

FIGURE 24. Absorption spectra of (a) 1-naphthol and (b) 2-naphthol in several solvents of different polarity: (1) cyclohexane, (2) acetonitrile and (3) DMSO (from Reference 91)

that in this case the S_1 transition is to the less polar ^1L_b state. Evidently, even in this seemingly simple case, the molecular-orbital character of the ^1L_b and ^1L_a states is not directly transferable moving from 1-naphthol to 2-naphthol.

This problem may be tackled by a more systematic analysis of the Stokes shift. Pines and coworkers[51] assumed that the first absorption transition of 1-naphthol and the two absorption transitions of 2-naphthol may be described by Pekarian functions. These functions were analyzed by the Kamlet–Taft analysis (Figures 25 and 26).

The analysis shows the ^1L_a absorption transition of 1-naphthol to be more sensitive to solvent polarity than the ^1L_a or ^1L_b absorption transition of 2-naphthol, indicating that it is the most polar of the three states. Comparison between the ^1L_b and ^1L_a states of 2-naphthol shows the ^1L_b state to be less polar than the ^1L_a state but considerably more sensitive to hydrogen-bonding interactions with the solvent. The greater sensitivity of the ^1L_b state to hydrogen-bonding interactions with bases is in quantitative agreement with the findings of Suzuki and Baba[34] discussed above. In both cases the spectral shift due to hydrogen-bonding interaction with the base was found to be three times larger in the ^1L_b state.

These findings consist an argument against the idea that regularities in the photoacidity behavior of hydroxyarenes may be defined and quantitatively analyzed simply by assuming constancy in the properties of their two lowest electronic singlet states. Indeed, one cannot even rule out situations where the less polar state in terms of its dipole moment and charge-transfer properties is the more acidic one as the 2-naphthol case appears to be, at least when photoacidity is judged by the strength of the hydrogen-bonding interaction of the acidic hydrogen atom of the −OH group.

An additional way to identify level crossing between the two lowest singlet states of hydroxyarenes as opposed to one emitting level gradually changing its properties was

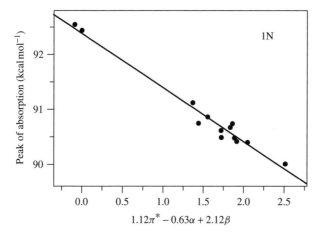

FIGURE 25. Correlation of the peaks of Pekarian functions (energy scale) used to approximate the UV-vis absorption spectra of 1-naphthol (from Reference 91)

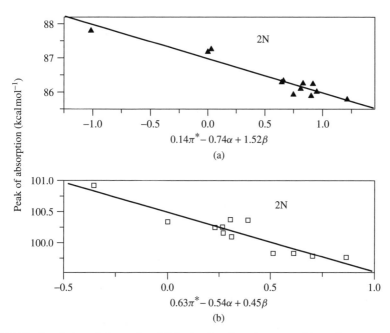

FIGURE 26. Correlation of the peaks of Pekarian functions (energy scale) used to approximate the UV-vis absorption spectra of 2-naphthol: (a) blue band and (b) red band with Kamlet–Taft parameters[91]. See Figure 24 for details of the absorption spectra

suggested by Hynes and coworkers[66]. They argued that ^1L_a to ^1L_b level switching may be demonstrated by comparing the free parameter P_{s-s}^o in the K–T analysis (P_{s-s}^o corresponds to the transition energy of the probe in cyclohexane in equation 18) of the absorption spectra of the photoacid with the P_{s-s}^o found in the K–T analysis of the fluorescence spectra of the photoacid in the same set of solvents. In cases where level switching occurs in the excited state of the photoacid, the absorption transition is assumed to be ^1L_b, while the fluorescence transition is assumed to occur from an ^1L_a state. Assuming that level crossing does not occur in non-polar solvents, one finds the P_{s-s}^o of the fluorescence to be higher in energy than the P_{s-s}^o of the absorption, a situation which cannot happen if absorption and fluorescence are to and from the same electronic level. Hynes and coworkers argued that such a situation occurs in HPTS, although the complexity of this system still resists clear-cut conclusions.

V. FREE-ENERGY CORRELATIONS BETWEEN PHOTOACIDITY AND REACTIVITY

Pines, Fleming and coworkers have utilized a free-energy correlation between the excited-state equilibrium constant of the photoacid and the proton dissociation rate[83,97]. Such correlations are extensions of similar correlations existing between the equilibrium constant and reactivity of ground-state acids (the 'Brønsted relation'[98]).

A 'universal' correlation (equation 19) was suggested to exist between the excited-state proton-transfer rate constant k_p and photoacidity in aqueous solutions:

$$k_p \sim k_o \exp(-(\Delta G_a + w^r)/kT) \tag{19}$$

where w^r is the so-called 'work function' of the work done when separating the two reactants to infinity and k_o is the reaction frequency prefactor, which is assumed to depend on the solvent and to be identical for all photoacids of a given family in a given solvent; ΔG_a is the reaction free-energy given by Marcus' Bond-Energy–Bond-Order (MBEBO) theory (equation 20)[99],

$$\Delta G_a = \Delta G^o/2 + \Delta G_o^\# + \Delta G_o^\# \cosh[\Delta G^o \ln 2/(2\Delta G_o^\#)]/\ln 2 \tag{20}$$

where $\Delta G_o^\#$ is the activation free-energy of the symmetric transfer when the total free-energy change following the proton transfer is equal to zero, i.e. when ΔpK_a between the proton donor and the proton acceptor equals zero.

The semi-empirical model for proton dissociation presented above is supported by recent *ab initio* studies of Kiefer and Hynes[100,101]. Figure 27 shows the good correlation found between the excited-state pK_a^* of hydroxyarene photoacids and their corresponding proton dissociation rate in aqueous solutions. The free-energy correlation seems to indicate that the equilibrium constant of the photoacid gives an excellent measure for its reactivity in the excited state regardless of whether the emitting state is ^1L_b or ^1L_a. This draws a line between the fundamental question as to why a particular photoacid has a particular pK_a^* and the question of how to estimate the reactivity of the photoacid, the latter property of the photoacid being proportional to its pK_a^*. It appears that, as a general rule, one could estimate the relative reactivity of a group of substituted photoacids by using empirical correlations between structure and acidity originally found for the ground-state acids. Such an approach has been successfully utilized by Tolbert and coworkers, who were able to synthesize 'enhanced' photoacids by predicting their pK_a^* values from Hammett's σ value of the introduced substituents[8,9,17].

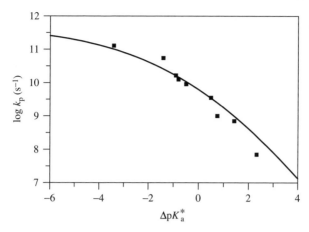

FIGURE 27. The free-energy correlation found for the dissociation reaction of hydroxyarene photoacids in aqueous solutions at room temperature (squares). The parameters of the fits are $\log(k_o) = 11.7$, $\Delta G_o^{\#} = 2.5$ kcal mol^{-1}, $W_r = 0$ (from Reference 91)

VI. CONCLUDING REMARKS: EVALUATION OF OUR CURRENT UNDERSTANDING OF THE PHOTOACIDITY OF HYDROXYARENES

The photoacidity of hydroxyarenes has attracted considerable interest over the past 50 years. Many conventions about photoacidity have their origins in the early studies of photoacidity. These conventions are now being critically examined by a new generation of researchers who have at their disposal new experimental tools and enhanced computational capabilities. A fresh outlook is already emerging from these latest studies, an outlook which appreciates the great complexity of these seemingly simple aromatic molecules. New ingredients have been successfully integrated into the old concepts, which have been used to describe photoacidity. This progress has not yet resulted in a coherent and full understanding of photoacidity, although the field is well prepared and poised for such a development to occur.

Hydroxyarene photoacids may be divided into two groups of molecules, the 1-naphthol-like and the 2-naphthol-like photoacids. The latter resemble ground-state photoacids in that the proton-transfer equilibrium takes place in one electronic level, presumably the ^1L_b state. There are many features common to ground-state acidity and the excited-state acidity of 2-naphthol-like photoacids. Among these are the substituent effect through resonance and inductive interactions whose molecular mechanism does not seem to differ much from their respective mechanism in the ground state, although it is noteworthy that the magnitude of these effects is usually larger in the excited state. Also, ring positions do not necessarily have the same effects on acidity in the ground and the excited state of the photoacid. In addition, solvent polarity seems to affect 2-naphthol-like photoacids in a similar way to how it affects ground-state acids, thus making the effect of the solvent on the reactivity of the photoacid predictable from the corresponding ground-state data. Finally, the photophysics and photochemistry of the first emitting state of 2-naphthol-like photoacids appear to be simple with relatively small deactivation routes other than the radiative decay and adiabatic proton-transfer reaction. One may characterize 2-naphthol-like photoacids as 'well-behaved' photoacids or as 'proper photoacids'. Substituted pyrenols also seem to fall under this category of well-behaved photoacids, although some of their electronic properties are still in debate.

The situation is drastically changed with 1-naphthol-like photoacids of which 1-naphthol is their best representative. 1-Naphthol exhibits enhanced photoacidity, complex absorption and fluorescence spectra which is very sensitive to solvent and ring substituents. The main route for its excited-state deactivation in aqueous solution is proton quenching[102], a very intriguing phenomenon by its own merit which is not discussed in this review. The complexity found in the photophysics and photochemistry of 1-naphthol is attributed to the complex structure of its first two electronic singlet states which is affected by polar interactions with the solvent and intramolecularly by the chemical structure and position of ring substituents. The exact details of these interactions and their effect on the electronic structure of 1-naphthol and its photoacidity await further investigation. However, regularities which are found in the appearance of the 1-naphthol spectra and theoretical considerations from first principles clearly point out the reason for this complexity. It is generally accepted that the lowest emitting state of 1-naphthol is sensitive to polar interactions, changing from being ^1L_b-like in a non-polar environment to being ^1L_a-like in a polar environment. The enhanced acidity of 1-naphthol over its 2-isomer is attributed to the ^1L_a state being more polar and of greater charge-transfer character than the ^1L_b state. Correlation between the appearance of the fluorescence spectra of the photoacid and its excited-state reactivity is expected and indeed observed in the case of 1-naphthol, although it is not clear how general are these observations. Ring substituents seem to introduce a similar effect on the electronic structure of excited 1-naphthol; however, this effect has not yet been studied in detail. An example is shown above in Figures 22 and 23. Förster-cycle calculations and the spectral appearance of the sulfonate-substituted 1-naphthols correlate with the expected inductive effect at each ring position in the excited state; the migration of the electronic charge to the naphthalene ring is expected to be largest in the 5-substituted naphthol (Figure 23) and smallest in the 2-substituted naphthol. This makes excited 5S1N the most ^1L_a-like isomer, with almost featureless absorption and fluorescence spectra, and the excited 2-isomer the most ^1L_b-like isomer, with strong vibrational features in both the absorption and fluorescence spectra. The effect of the substituents on the Förster-cycle acidity of 1-naphthol is resolved in methanol. The order of Förster-cycle photoacidity in methanol is: 5S1N > 4S1N > 1N > 3S1N > 2S1N (Table 2).

The complex electronic structure of 1-naphthol-like photoacids makes them non-conventional photoacids. In this case, photoacidity is influenced by additional factors not present in ground-state acids, namely electronic rearrangements occurring during the lifetime of the excited photoacids. Clearly, electronic rearrangements occurring in the short-lived excited state of the photoacids (typically, the excited-state lifetime in the singlet state is no longer than a few nanoseconds) can affect both the dynamics of the excited-state proton-transfer reaction and the thermodynamics of the photoacids. The expected and observed non-trivial photoacidity of 1-naphthol-like photoacids awaits further investigation.

Our final observation is that, in gross details, photoacids seem to generally resemble ground-state acids in most studied cases where proton-transfer reaction is observed. As in ground-state hydroxyarenes, Brønsted acidity is greatly affected by the stabilization of the conjugate base in polar solvents and by intramolecular charge-transfer processes, shifting some of the anionic charge away from the oxygen atom to the aromatic ring. The charge-transfer process is assisted by inductive and resonance effects at the aromatic ring. Level mixing in the excited state, although very important in 1-naphthol-like acids, is secondary in importance to the acid being in the excited state. Level dynamics, if indeed they exist, do not seem to be the rate-determining step for proton-transfer reaction in polar solvents. Thus, level dynamics do not affect the generality of the above-stated observation. The extension of the photoacidity scale to less-polar environments where no proton transfer is observed within the short lifetime of the excited photoacid is a desirable goal, which may be achieved by scaling the hydrogen-bonding interaction of the photoacids or perhaps

7. UV-visible spectra and photoacidity of phenols, naphthols and pyrenols   525

by calibrating their Förster-cycle acidities using spectral analysis of their lowest optical transitions.

VII. REFERENCES

1. Th. Förster, *Naturwissenschaften*, **36**, 186 (1949).
2. Th. Förster, *Z. Elektrochem., Angew. Phys. Chem.*, **54**, 42 (1950).
3. Th. Förster, *Z. Electrochem., Angew. Phys. Chem.*, **54**, 531 (1950).
4. Th. Förster, *Pure Appl. Chem.*, **24**, 443 (1970).
5. A. Weller, *Prog. React. Kinet.*, **1**, 187 (1961).
6. J. N. Brønsted, *Recl. Trav. Chim. Pays-Bas*, **42**, 718 (1923).
7. J. N. Brønsted, *Chem. Rev.*, **5**, 231 (1928).
8. L. M. Tolbert and J. E. Haubrich, *J. Am. Chem. Soc.*, **112**, 863 (1990).
9. L. M. Tolbert and J. E. Haubrich, *J. Am. Chem. Soc.*, **116**, 10593 (1994).
10. E. Vander Donckt, *Prog. React. Kinet.*, **5**, 273 (1970).
11. S. G. Schulman, in *Modern Fluorescence Spectroscopy* (Ed. E. L Wehry), Vol. 2, Plenum Press, New York, 1976, p. 239.
12. F. Ireland and P. A. H. Wyatt, *Adv. Phys. Org. Chem.*, **12**, 131 (1976).
13. I. Y. Martynov, A. B. Demyashkevich, B. M. Uzhinov and M. G. Kuzmin, *Russ. Chem. Rev. (Engl. Transl.)*, **46**, 1 (1977).
14. H. Shisuka, *Acc. Chem. Res.*, **18**, 41 (1985).
15. L. G. Arnaut and S. J. Formosinho, *J. Photochem. Photobiol., A*, **75**, 21 (1993).
16. P. Van and D. Shukla, *Chem. Rev.*, **93**, 571 (1993).
17. L. M. Tolbert and K. M. Solntsev, *Acc. Chem. Res.*, **35**, 19 (2002).
18. R. Knochenmuss and I. Fischer, *Int. J. Mass Spectrom.*, **220**, 343 (2002).
19. O. Cheshnovsky and S. Leutwyler, *J. Chem. Phys.*, **88**, 4127 (1988).
20. R. Knochenmuss, O. Cheshnovsky and S. Leutwyler, *Chem. Phys. Lett.*, **144**, 317 (1988).
21. J. J. Breen, L. W. Peng, D. M. Willberg, A. Heikal, P. Cong and A. H. Zewail, *J. Chem. Phys.*, **92**, 805 (1990).
22. S. K. Kim, J. K. Wang and A. H. Zewail, *Phys. Lett.*, **228**, 369 (1994).
23. S. K. Kim, J. J. Breen, D. M. Willberg, L. W. Peng, A. Heikal, J. A. Syage and A. H. Zewail, *J. Chem. Phys.*, **99**, 7421 (1995).
24. R. Knochenmuss, G. R. Holtom and D. Ray, *Chem. Phys. Lett.*, **215**, 188 (1993).
25. R. Knochenmuss and D. E. Smith, *J. Chem. Phys.*, **101**, 291 (1994).
26. R. Knochenmuss, P. L. Muiño and C. Wickleder, *J. Phys. Chem.*, **100**, 11218 (1996).
27. R. Knochenmuss, V. Karbach, C. Wickleder, S. Gfaf and S. Leutwyler, *J. Phys. Chem., A*, **102**, 1935 (1998).
28. R. Knochenmuss, I. Fischer, D. Luhus and Q. Lin, *Isr. J. Chem.*, **39**, 221 (1999).
29. R. Knochenmuss, K. M. Solntsev and L. M. Tolbert, *J. Phys. Chem., A*, **105**, 6393 (2001).
30. J. Steadman and J. A. Syage, *J. Chem. Phys.*, **92**, 4630 (1990).
31. G. Pino, G. Gregoire, C. Debonder-Lardeux, C. Jouvet, S. Martrenchard and M. D. Solgadi, *Phys. Chem. Chem. Phys.*, **2**, 893 (2002).
32. W. Urban and A. Weller, *Ber. Bunsenges. Phys. Chem.*, **67**, 787 (1958).
33. A. Weller, *Discuss. Faraday. Soc.*, **27**, 28 (1959).
34. S. Suzuki and H. Baba, *Bull. Chem. Soc. Jpn.*, **40**, 2199 (1967).
35. A. Tramer and M. Zaborovska, *Acta Phys. Pol.*, **5**, 821 (1968).
36. S. Tobita and H. Shizuka, *Chem. Phys. Lett.*, **75**, 140 (1980).
37. H. Shizuka and S. Tobita, *J. Am. Chem. Soc.*, **104**, 6919 (1982).
38. H. Shizuka and K. Tsutsumi, *Bull. Chem. Soc. Jpn.*, **56**, 629 (1983).
39. K. Tsutsumi and H. Shizuka, *Chem. Phys. Lett.*, **52**, 485 (1977).
40. G. W. Robinson, P. J. Thistlethwaite and J. Lee, *J. Phys. Chem.*, **90**, 4224 (1986).
41. G. W. Robinson, *J. Phys. Chem.*, **95**, 10386 (1991).
42. J. Lee, G. W. Robinson, S. P. Webb, L. A. Phillips and J. H. Clark, *J. Am. Chem. Soc.*, **108**, 6538 (1986).
43. J. Lee, G. W. Robinson and M. P. Basses, *J. Am. Chem. Soc.*, **108**, 7477 (1986).
44. A. Masad and D. Huppert, *Chem. Phys. Lett.*, **180**, 409 (1991).
45. E. Pines and G. R. Fleming, *Chem. Phys.*, **183**, 393 (1994).

46. E. Pines, D. Tepper, B.-Z. Magnes, D. Pines and T. Barak, *Ber. Bunsenges. Phys. Chem.*, **102**, 504 (1998).
47. E. Pines, D. Pines, T. Barak, B.-Z. Magnes, L. M. Tolbert and J. E. Haubrich, *Ber. Bunsenges. Phys. Chem.*, **102**, 511 (1998).
48. K. M. Solntsev, D. Huppert, L. M. Tolbert and N. Agmon, *J. Am. Chem. Soc.*, **120**, 9599 (1998).
49. K. M. Solntsev, D. Huppert and N. Agmon, *J. Phys. Chem., A*, **102**, 9599 (1998).
50. K. M. Solntsev, D. Huppert, N. Agmon and L. M. Tolbert, *J. Phys. Chem., A*, **104**, 4658 (2000).
51. B-Z. Magnes, N. Strashnikova and E. Pines, *Isr. J. Chem.*, **39**, 361 (1999).
52. Th. Förster and S. Völker, *Chem. Phys. Lett.*, **34**, 1 (1975).
53. M. Gutman, D. Huppert and E. Pines, *J. Am. Chem. Soc.*, **103**, 3709 (1981).
54. M. J. Politi and H. Fendler, *J. Am. Chem. Soc.*, **106**, 265 (1984).
55. M. J. Politi, O. Brandt and H. Fendler, *J. Chem. Phys.*, **89**, 2345 (1985).
56. E. Pines and D. Huppert, *J. Chem. Phys.*, **84**, 3576 (1986).
57. E. Pines and D. Huppert, *Chem. Phys. Lett.*, **126**, 88 (1986).
58. E. Pines, D. Huppert and N. Agmon, *J. Chem. Phys.*, **88**, 5620 (1988).
59. N. Agmon, E. Pines and D. Huppert, *J. Chem. Phys.*, **88**, 5631 (1988).
60. E. Pines and D. Huppert, *J. Am. Chem. Soc.*, **111**, 4096 (1989).
61. D. Huppert, E. Pines and N. Agmon, *J. Opt. Soc. Am., B: Opt. Phys.*, **7**, 1545 (1990).
62. E. Pines, D. Huppert and N. Agmon, *J. Phys. Chem.*, **95**, 666 (1991).
63. N. Barrash-Shiftan, B. Brauer and E. Pines, *J. Phys. Org. Chem.*, **11**, 743 (1998).
64. T.-H. Tran-Thi, T. Gustavsson, C. Prayer, S. Pommeret and J. T. Hynes, *Chem. Phys. Lett.*, **329**, 421 (2000).
65. T.-H. Tran-Thi, C. Prayer, Ph. Millie, P. Uznanski and J. T. Hynes, *J. Phys. Chem.*, **106**, 2244 (2002).
66. J. T. Hynes, T.-H. Tran-Thi and T. Gustavsson, *J. Photochem. Photobiol., A*, **A154**, 3 (2002).
67. E. Pines, unpublished results.
68. (a) J. R. Platt, *J. Phys. Chem.*, **17**, 484 (1949).
    (b) J. R. Platt, *Systematics of the Electronic Spectra of Conjugated Molecules.*, Wiley, Chicago, 1964.
69. A. Weller, *Z. Electrochem.*, **56**, 662 (1952).
70. N. Mikami, *Bull. Chem. Soc. Jpn.*, **68**, 683 (1995).
71. E. L. Wehry and L. B. Rogers, *J. Am. Chem. Soc.*, **87**, 4234 (1965).
72. (a) S. G. Schulman, W. R. Vincent and W. J. M. Underberg, *J. Phys. Chem.*, **85**, 4068 (1981).
    (b) S. G. Schulman, L. S. Rosenberg and W. R. Vincent, *J. Am. Chem. Soc.*, **101**, 139 (1979).
73. A. Bryson and R. W. Mattews, *Aust. J. Chem.*, **16**, 401 (1963).
74. *Dictionary of Organic Compounds*, (Ed. J. Buckingam) 5th ed., Chapman and Hall, New York, 1982.
75. R. M. C. Henson and A. H. Wyatt, *J. Chem. Soc., Faraday Trans.*, **271**, 669 (1974).
76. *J. Phys. Chem. Ref. Data, Supplement*, 17 (1988).
77. J. A. Jeffrey, *An Introduction to Hydrogen Bonding* (Ed. D. G. Truhlar), Oxford University Press, New York, 1997.
78. G. Granucci, J. T. Hynes, P. Millie and T.-H. Tran-Thi, *J. Am. Chem. Soc.*, **122**, 12243 (2000).
79. N. Mataga and T. Kubota, *Molecular Interactions and Electronic Spectra*, Chap. 7, Dekker, New York, 1970.
80. R. P. Bell, *The Proton in Chemistry*, 2nd ed., Chap. 6, Cornell University Press, 1973.
81. R. Stewart, *The Proton: Application to Organic Chemistry*, Academic Press, Orlando, 1985.
82. N. Agmon, W. Rettig and C. Groth, *J. Am. Chem. Soc.*, **124**, 1089 (2002).
83. E. Pines and G. R. Fleming, *J. Phys. Chem.*, **95**, 10448 (1991).
84. E. Pines, D. Pines, Y. Z. Ma and G. R. Fleming, in preparation.
85. G. C. Pimentel, *J. Am. Chem. Soc.*, **79**, 3323 (1957).
86. G. C. Pimentel and A. L. McClellan *The Hydrogen Bond*, Chap. 7, Freeman, San Francisco, 1960, p. 206–225.
87. R. W. Taft, J.-L. Abboud and M. J. Kamlet, *J. Am. Chem. Soc.*, **103**, 1080 (1981).
88. R. W. Taft, J.-L. Abboud and M. J. Kamlet, *J. Org. Chem.*, **49**, 2001 (1984).

7. UV-visible spectra and photoacidity of phenols, naphthols and pyrenols

89. M. H. Abraham, G., J. Buist, P. L. Grellier, R. A. McGill, D. V. Ptior, S. Oliver, E. Turner, J. J. Morris, P. J. Taylor, P. Nicolet, P.-C. Maria, J-F. Gal, J.-L. M. Abboud, R. M. Doherty, M. J. Kamlet, W. J. Shuely and R. W. Taft, *J. Phys. Org. Chem.*, **2**, 540 (1989).
90. P. Lorente, I. G. Shenderovich, N. S. Golubev, G. S. Denisov, G. Buntowsky and H.-H. Limbach, *Magn. Res. Chem.*, **39**, S18 (2001).
91. B.-Z. Magnes, T. Barak and E. Pines, in preparation.
92. L. P. Hammett, *Physical Organic Chemistry*, 2nd edn., McGraw-Hill, New York, 1970.
93. J. D. Roberts and M. Caserio, *Basic Principles of Organic Chemistry*, Benjamin, Menlo Park, CA, 1965.
94. R. Jimenez, D. A. Case and F. E. Romesberg, *J. Phys. Chem., B*, **106**, 1090 (2002).
95. D. Hercules and L. Rogers, *Spectrochim. Acta*, **15**, 393 (1959).
96. N. K. Zaitsev, A. B. Demyashkevich and M. G. Kuzmin, *High Energy Chem. (Engl. Transl.)*, **13**, 288 (1979).
97. E. Pines, B. Z. Magnes, M. J. Lang and G. R. Fleming, *Chem. Phys. Lett.*, **281**, 413 (1997).
98. J. N. Brønsted and K. J. Pederson, *Z. Phys. Chem.*, **108**, 185 (1924).
99. A. O. Cohen and R. A. Marcus, *J. Phys. Chem.*, **72**, 4249 (1968).
100. P. M. Kiefer and J. T. Hynes, *J. Phys. Chem., A*, **106**, 1834 (2002).
101. P. M. Kiefer and J. T. Hynes, *J. Phys. Chem., A*, **106**, 1850 (2002).
102. C. M. Harris and B. K. Selinger, *J. Phys. Chem.*, **84**, 1366 (1980).
103. M. Vasak, M. R. Whipple, A. Berg and J. Michl, *JACS*, **100**, 872 (1978).

CHAPTER 8

Hydrogen-bonded complexes of phenols

C. LAURENCE, M. BERTHELOT and J. GRATON

Laboratoire de Spectrochimie, University of Nantes, 2, rue de la Houssiniere BP 92208, F-44322 Nantes Cedex 3, France
Fax: (+33) (0)2 51 12 54 12; e-mail: Christian.Laurence@chimie.univ-nantes.fr, Michel.Berthelot@chimie.univ-nantes.fr, Jerome.Graton@chimie.univ-nantes.fr

I. INTRODUCTION	530
II. HYDROGEN-BOND BASICITY OF PHENOLS	531
III. HYDROGEN-BOND ACIDITY OF PHENOLS	535
A. log K_A^H Scale	535
B. log K_α Scale	539
C. Complexation with Pyridine N-oxide	540
D. Solvatochromic Shifts of Reichardt's Betaine Dye	540
E. Hydrogen-bond Acidity from Partition Coefficients	543
IV. SELF-ASSOCIATION OF PHENOLS	546
V. INTRAMOLECULAR HYDROGEN BONDS	551
A. OH···O$_2$N	553
B. OH···O=C	553
C. OH···Halogen	554
D. OH···OH(Me)	556
E. OH···N	556
VI. PROPERTIES OF THE COMPLEXES	557
A. Thermodynamic Properties	557
B. Binding Energy D_e and Dissociation Energy D_o	577
C. Geometry of Phenol Hydrogen Bonds	579
D. Hydrogen-bonding Site(s)	582
VII. CONSTRUCTION OF HYDROGEN-BOND BASICITY SCALES FROM PHENOLS	586
A. Thermodynamic Scales of Hydrogen-bond Basicity	586
B. UV, NMR and IR Spectroscopic Scales	589

The Chemistry of Phenols. Edited by Z. Rappoport
© 2003 John Wiley & Sons, Ltd   ISBN: 0-471-49737-1

VIII. HYDROGEN BONDING AND PROTONATION	593
IX. REFERENCES	596

I. INTRODUCTION

For many years phenols have been well recognized as participants in hydrogen bonding. In the chapter on hydrogen bonding in Pauling's (1939) *The Nature of the Chemical Bond*[1], phenols are said to 'form stronger hydrogen bonds than aliphatic alcohols because of the increase in electronegativity of the oxygen atom' resulting from the n-electron donation of the OH group to the aromatic ring. The three-dimensional structure of crystalline resorcinol (1,3-dihydroxybenzene) is explained by self-association through $\cdots$OH$\cdots$OH$\cdots$ hydrogen bonds. Many *ortho*-substituted phenols, e.g. *o*-nitrophenol or *o*-hydroxyacetophenone, are listed as substances forming strong intramolecular hydrogen bonds. In the proceedings (edited by Hadži[2]) of the first international conference (1957) on hydrogen bonding held in Ljubljana, Yugoslavia, there were studies of hydrogen-bonded complexes of phenols by neutron diffraction, infrared and electronic spectrometry. In the first text devoted entirely to hydrogen bonding, The *Hydrogen Bond* by Pimentel and McClellan (1960)[3], phenols are classed as well-recognized hydrogen-bonding acids and hydrogen-bonding bases. In a table of nearly 300 entries of thermodynamic data (equilibrium constants, enthalpy, entropy) for hydrogen-bond formation, many data concern ArOH$\cdots$Base complexes where phenol, substituted phenols, 1-naphthol and 2-naphthol are hydrogen-bond donors (HBD). A fourth book, *Hydrogen Bonding* by Joesten and Schaad (1974)[4], contains *ab initio* and (mostly) semiempirical calculations of the hydrogen-bond geometry and energy, the thermodynamics of hydrogen bonding and empirical correlations between thermodynamic and spectroscopic properties of hydrogen-bonded complexes. There is also a chapter on intramolecular and homo-intermolecular (self-association) hydrogen bonds, and an appendix of thermodynamic data and A-H stretching frequency shifts with nearly 2000 entries. Hydrogen-bonded phenols are particularly well represented in this book. Other data and discussions on the hydrogen-bonded complexes of phenols are found in the book entitled *Hydrogen Bonding* by Vinogradov and Linell (1971)[5], in the three-volume series entitled *The Hydrogen Bond. Recent Developments in Theory and Experiments*, edited by Schuster, Zundel and Sandorfy (1976)[6], in the review by Rochester (1971)[7] on the *Acidity and inter- and intramolecular H-bonds* of the hydroxyl group and in the multi-author publication (1991) entitled[8] *Intermolecular Forces. An Introduction to Modern Methods and Results*. The theoretical interpretation of hydrogen bonding has been discussed by Scheiner (1997) in *Hydrogen Bonding, A Theoretical Perspective*[9] and *Molecular Interactions, from van der Waals to Strongly Bound Complexes*[10], by Hadži (1997)[11] in *Theoretical Treatments of Hydrogen Bonding* and by Smith (1994)[12] in *Modeling the Hydrogen Bond*. *Hydrogen Bonding in Biological Structures* by Jeffrey and Saenger (1991)[13] and *An Introduction to Hydrogen Bonding* by Jeffrey (1997)[14] focus on general principles and crystal structure studies. Reviews relating to the importance of hydrogen bonding in crystal engineering have been written by Subramanian and Zaworotko[15], Desiraju[16] and Aakeröy[17].

Despite this voluminous literature on hydrogen bonding, there have been very few discussions on the hydrogen-bond basicity of phenols. The ability of phenols to act as hydrogen-bond acceptors is considered in Section II.

The main purpose of Section III is to establish the position of phenols on the scales of hydrogen-bond acidity, either solute ($\log K_A^H$, α_2^H, $\log K_\alpha$) or solvent ($E_T(30)$) scales. Here, the ability of phenols to act as hydrogen-bond donors will be compared to that of other O−H (water, alcohols, carboxylic acids), N−H, S−H and C−H hydrogen-bond donors. It is interesting to note that it was not until 1989 that phenol was found to be a (slightly) better hydrogen-bond donor than acetic acid, in spite of being a worse Brønsted acid by more than 5 pK_a units in water. This illustrates that hydrogen-bonding phenomena have little in common with proton transfer when acids with different functional groups are compared.

Various types of phenol complexes will be examined in Sections IV–VI. Dimers and multimers of self-associated phenols, (ArOH)$_n$, will be considered both in solution and in the solid state (Section IV). The existence and, subsequently, the geometry and energy of intramolecular hydrogen bonds in *ortho*-substituted phenols are discussed in Section V. The most recent thermodynamic, spectroscopic (mainly IR), geometrical and theoretical results on the heterodimers of phenols complexed to Lewis bases, ArOH···B, will be presented in Section VI.

Phenols are among the most useful reference hydrogen-bond donors for building thermodynamic and spectroscopic (NMR, UV and IR) scales of hydrogen-bond basicity. The building of such scales contributes not only to the increasing efforts towards a quantitative description of the hydrogen bond, but also to the difficult and unachieved task of measuring quantitatively the strength of organic Lewis bases. Scales constructed from phenol, 4-fluorophenol and 4-nitrophenol are presented in Section VII.

It is possible to increase the strength of the hydrogen bond by using complexes of Lewis bases with phenol derivatives of increasing hydrogen-bond donor strength, e.g. from polymethylphenols to polynitrophenols. Then the transition from a hydrogen-bonded complex ArOH···B to a proton-transfer complex ArO$^-$···$^+$HB can be observed. Proton transfer in the hydrogen-bonded complexes of phenols is studied in Section VIII.

II. HYDROGEN-BOND BASICITY OF PHENOLS

Laurence and coworkers[18] have measured the equilibrium constant of reaction 1

$$2\ FC_6H_4OH \rightleftharpoons (FC_6H_4OH)_2 \quad (1)$$

in CCl$_4$ at 298 K by following the absorbance variations of the v(OH) infrared band of 4-fluorophenol at 3614 cm^{-1} with increasing concentrations of the phenol. Assuming that the self-association of 4-fluorophenol is limited to the formation of a dimer in the 4 to 50 mmol dm^{-3} range, they find a constant $K = [\text{dimer}]/[\text{monomer}]^2$ value of 0.76 dm^3 mol^{-1}. The measurement, in the same conditions, of the complexation constants of 4-fluorophenol with water and alcohols (reaction 2)[18], ethers (reaction 3)[19] and various organic Lewis bases B (reaction 4)[20]

$$4\text{-}FC_6H_4OH + ROH \rightleftharpoons 4\text{-}FC_6H_4OH\cdots O(R)H \quad (2)$$

$$4\text{-}FC_6H_4OH + ROR' \rightleftharpoons 4\text{-}FC_6H_4OH\cdots O(R)R' \quad (3)$$

$$4\text{-}FC_6H_4OH + B \rightleftharpoons 4\text{-}FC_6H_4OH\cdots B \quad (4)$$

provides a hydrogen-bond basicity scale pK_{HB}[21] (equation 5) (Section VII, A) for 4-fluorophenol, water, alcohols and various Lewis bases.

$$pK_{HB} = \log_{10} K\,(4\text{-}FC_6H_4OH\cdots B,\ CCl_4,\ 298\ K) \quad (5)$$

The scale, illustrated in Figure 1, shows that 4-fluorophenol is a weaker hydrogen-bond acceptor (HBA) than water, alcohols and aliphatic ethers. This is expected since the

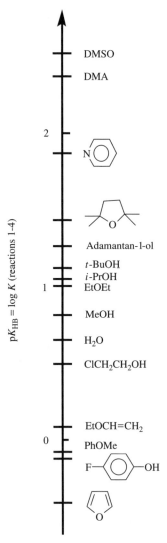

FIGURE 1. Comparison of the HB basicity of 4-fluorophenol to water, alcohols, ethers and miscellaneous Lewis bases

phenyl group withdraws electronic density from the oxygen lone pairs through its field-inductive and resonance effects. Other phenols cannot be studied by this method because of the overlap of their own OH band with the OH band of 4-fluorophenol. Laurence and coworkers[18] then turned to a spectroscopic scale of hydrogen-bond basicity, $\Delta v(OH)$, namely the displacement, on H-bond formation, of the 3618 cm^{-1} OH band of a very strong hydrogen-bond donor, the perfluoroalcohol $(CF_3)_3COH$. Results for 5 phenols and 7 alcohols are reported in column 3 of Table 1. Figure 2 shows that pK_{HB} and $\Delta v(OH)$ are very well correlated. This correlation enables the calculation of the secondary pK_{HB}

TABLE 1. pK_{HB}, $\Delta\nu(OH)$ and β_2^H hydrogen-bond basicity scales for phenols and, for comparison, water and alcohols

ROH	Primary pK_{HB}[a]	$\Delta\nu(OH)$[b]	Secondary pK_{HB}[c]	β_2^H[d]
Adamantan-1-ol	1.27	482	—	0.51
t-BuOH	1.14	468	—	0.49
i-PrOH	1.06	455	—	0.47
EtOH	1.02	438	—	0.44
MeOH	0.82	417	—	0.41
H_2O	0.65	—	—	0.38
$ClCH_2CH_2OH$	0.50	376	—	0.35
$4\text{-}MeC_6H_4OH$	—	304	0.03	0.24
$3\text{-}MeC_6H_4OH$	—	301	0.01	0.24
C_6H_5OH	—	289	−0.07	0.22
$4\text{-}FC_6H_4OH$	−0.12	281	−0.13[e]	0.21
$3\text{-}CF_3C_6H_4OH$	—	ca 248	ca −0.36	ca 0.16
$(CF_3)_2CHOH$	—	ca 161	ca −0.96	ca 0.03

[a]Experimental complexation constants of reactions 1 and 2.
[b]In cm^{-1}. $\Delta\nu(OH) = 3618 - \nu(OH\cdots O)$.
[c]Calculated from the equation $pK_{HB} = 0.692 (\Delta\nu(OH)/100) - 2.07$.
[d]Calculated from equation 6.
[e]The agreement between the primary (experimental) value and the secondary value, calculated from the $\nu(OH\cdots O)$ band, indicates that complexes to the π and F sites can be neglected to a first approximation.

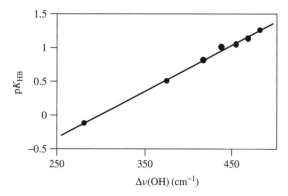

FIGURE 2. Correlation between the thermodynamic pK_{HB} (towards 4-fluorophenol) and the spectroscopic $\Delta\nu(OH)$ (towards perfluoro-t-butyl alcohol) scales of hydrogen-bond basicity ($n = 7$, $r = 0.998$) allowing the calculation of pK_{HB} for very weakly basic alcohols and phenols

values of 4 new phenols, reported in column 4 of Table 1. These pK_{HB} values can be anchored to the empirical β_2^H scale of hydrogen-bond basicity (equation 6)[22] normalized from 0 to 1 ($\beta_2^H = 1$ is for HMPA).

$$\beta_2^H = (pK_{HB} + 1.1)/4.636 \qquad (6)$$

This β_2^H scale constitutes the last column in Table 1. From the correlation of pK_{HB} with the Hammett σ° constant of the ring substituent (equation 7), many other pK_{HB} values can be calculated for *meta*- and *para*-substituted phenols.

$$pK_{HB} = -0.650\sigma^\circ - 0.05 \qquad (7)$$

n(number of points) = 5, r(correlation coefficient) = 0.993,

s(standard deviation) = 0.02

Berthelot and colleagues[23] have estimated the pK_{HB} values of the phenolic OH group in the **intra**molecular hydrogen-bonded systems **1, 2** and **3**. The higher basicity pK_{HB} of **1** and **2** compared to phenol (−0.07) can be explained by cooperative effects[24] involved in hydrogen-bond formation: the oxygen electron pairs are more basic in OH···B than in the free OH group[25]. The push-pull effect shown by the curved arrows in **3** opposes the cooperativity effect and pK_{HB} falls.

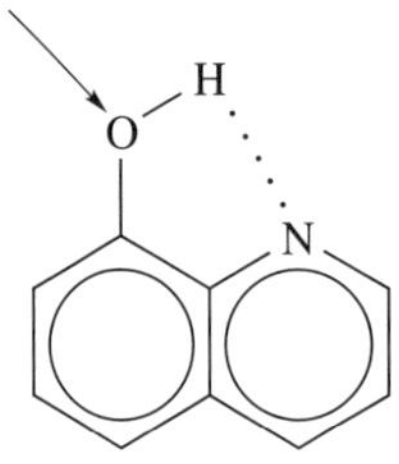

pK_{HB} = 0.48   pK_{HB} = 0.23   pK_{HB} = −0.12

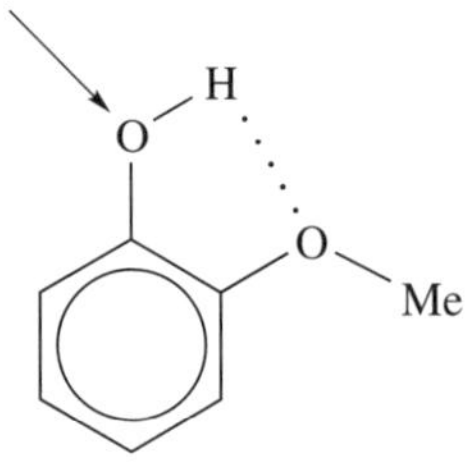

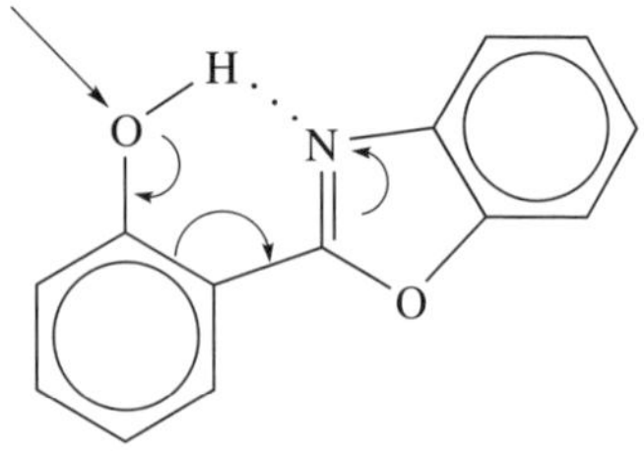

(1)   (2)   (3)

The cooperativity effect can also increase the hydrogen-bond basicity of the phenolic OH group in **inter**molecular hydrogen-bonded systems. For example, the study of phenol-triethylamine systems (equation 8)[26] shows that a large increase in the ratio of initial concentrations [phenol]$_0$/[Et$_3$N]$_0$ leads to a large increase in the apparent complexation constant, which is explained by the formation of complexes of 2 : 1 stoichiometry (reaction 9). The evaluation of the constants of the 1 : 1 equilibrium (K_1) (reaction 8) and the 2 : 1 equilibrium (K_2) (reaction 9) gives K_1 = 62 and K_2 = 40 dm^3 mol^{-1} for 4-fluorophenol, i.e. a pK_{HB} (log K_2) value of 1.60 for a phenolic OH group hydrogen-bonded to NEt$_3$, to be compared to pK_{HB} = −0.12 for a free phenolic OH function. Other complexation constants have been measured for 2 : 1 complexes of phenols hydrogen-bonded to tetramethylurea[27] and tri-n-butylamine[28]. Zeegers-Huyskens[24] has recently reviewed how the displacement of the electronic clouds and of the nuclei, upon the formation of a first hydrogen bond A—H···B, affects the hydrogen-bond basicity and acidity of the other specific sites in the two partners.

$$\text{ArOH} + \text{Et}_3\text{N} \xrightleftharpoons{K_1} \text{ArOH}\cdots\text{NEt}_3 \quad (8)$$

$$\text{ArOH}\cdots\text{NEt}_3 + \text{ArOH} \xrightleftharpoons{K_2} \underset{\text{Ar}}{\text{O}}-\text{H}\cdots\underset{\text{Ar}}{\text{O}}-\text{H}\cdots\text{NEt}_3 \quad (9)$$

When a solute is dissolved in a pure hydrogen-bond donor solvent, such as water or alcohol, all acceptor sites are involved in the solvation phenomenon. For example, the partition of phenols between water and organic phases now depends both on the oxygen and on the π hydrogen-bond basicities, because of the excess of water molecules. From sets of water-solvent partition coefficients, Abraham[29] has constructed a scale of effective or summation hydrogen-bond basicity, $\sum \beta_2^H$, for about 350 solutes, of which 72 are phenols. A few examples are given in Table 2. For the phenols and anisoles,

TABLE 2. Comparison of β_2^H and $\Sigma\beta_2^H$ for phenols, anisoles, alcohols and ethers

Base	$\Sigma\beta_2^H$	β_2^H	Diff.^{d}	Base	$\Sigma\beta_2^H$	β_2^H	Diff.^{e}
Et$_2$O	0.45	0.46^a	−0.01	C$_6$H$_5$OH	0.30	0.22^b	+0.08
i-Pr$_2$O	0.41	0.48^a	−0.07	3-MeC$_6$H$_4$OH	0.34	0.24^b	+0.10
n-Bu$_2$O	0.45	0.43^a	+0.02	4-MeC$_6$H$_4$OH	0.31	0.24^b	+0.07
THF	0.48	0.51^a	−0.03	4-FC$_6$H$_4$OH	0.23	0.21^b	+0.02
THP	0.54	0.50^a	+0.04	C$_6$H$_5$OMe	0.29	0.22^c	+0.07
MeOH	0.44	0.41^b	+0.03	4-ClC$_6$H$_4$OMe	0.24	0.18^c	+0.06
n-C$_8$H$_{17}$OH	0.48	0.46^b	+0.02	2-MeC$_6$H$_4$OMe	0.29	0.21^c	+0.08
		avg	0			avg	+0.07

^{a}Calculated from the pK_{HB} of Reference 19 and equation 6.
^{b}From Table 1.
^{c}From the pK_{HB} values of the oxygen atom (Reference 30) and equation 6.
^{d}Random differences for aliphatic ethers and alcohols.
^{e}Systematic positive difference for phenols and anisoles.

the systematic positive difference between β_2^H, measuring the oxygen basicity alone, and $\Sigma\beta_2^H$, measuring the overall basicity, demonstrates the contribution of the π basicity to $\Sigma\beta_2^H$ values.

Nobeli and colleagues[31] have studied the hydrogen-bond basicity of phenols and anisoles from both the frequency of hydrogen-bond formation in molecular crystal structures and ab initio calculations on their complexes with methanol. The percentage of crystal structures found in the Cambridge Structural Database (CSD)[32] where the oxygen of furan, anisole, tetrahydrofuran or phenol fragments accepts a hydrogen bond from an OH donor is in the order: furan ≪ anisole < THF < phenol. These results do not imply that phenol is a better acceptor than anisole or THF, but are rather explained by the cooperativity effect since, in 25 of the 87 hydrogen bonds accepted by the phenol oxygen, it was simultaneously acting as a donor to an oxygen atom. In vacuo the calculated energy of a MeOH···O hydrogen bond is in the order: furan < phenol ⩽ anisole < THF, in agreement with the pK_{HB} scale (−0.40[19] < −0.07[18] ∼ −0.07[30] < 1.28[19]). The perpendicular conformation of phenols forms hydrogen bonds a few kJ mol^{-1} stronger than the planar one. This can be partly attributed to the change in the oxygen atom charge density caused by delocalization of the lone-pair charge density into the π system of the ring.

III. HYDROGEN-BOND ACIDITY OF PHENOLS

A. log K_A^H Scale

In 1989, Abraham and coworkers[33,34] constructed a scale of solute hydrogen-bond acidity based on the numerous literature results of log K values for the 1 : 1 hydrogen-bond complexation reaction (equations 10 and 11) in which a series of hydrogen-bond acids AH$_i$ complex with a given reference base in dilute solution in CCl$_4$. Such series of log K values were collected against 45 reference bases, e.g. pyridine, triethylamine, tetramethylurea, tetramethylthiourea, N,N-dimethylacetamide, HMPA, acetone, DMSO, THF, acetonitrile, triphenylphosphine oxide, 1-methylimidazole, diethyl sulfide or pyridine N-oxide. By plotting log K values for acids against a given reference base vs. log K values for acids against any other reference base, they obtained a series of straight lines (equation 12) that intersected near a 'magic point' at (−1.1, −1.1). The constants L_B and D_B characterize the 45 reference bases B. The log K_A^H values, computed using a program described in Reference 33, represent the hydrogen-bond acidity of the acids over the 45 equations, and their mean constitutes a scale of solute hydrogen-bond acidity. Experimentally, new K_A^H values might be obtained from complexation constants with a reference base with $L_B = 1$ and $D_B = 0$. Pyridine ($L_B = 1.0151$, $D_B = 0.0139$) and triphenyl phosphate ($L_B = 1.0008$, $D_B = 0.0008$) might be used for such measurements.

$$\text{A-H} + \text{B} \rightleftharpoons \text{A-H} \cdots \text{B} \qquad (10)$$

$$K(\text{dm}^3\,\text{mol}^{-1}) = [\text{AH} \cdots \text{B}]/[\text{AH}][\text{B}] \qquad (11)$$

$$\log K^i(\text{series of acids against base B}) = L_B \log K_A^H + D_B \qquad (12)$$

The $\log K_A^H$ scale is not quite general in the sense that a number of acid–base combinations are excluded from equation 12. However phenols, as well as alcohols and strong NH donors, can be combined with all types of bases.

The $\log K_A^H$ values calculated for 58 phenols are given in Table 3. The comparison of phenols with alcohols, carboxylic acids, NH, CH and SH donors is illustrated in Figure 3. By assuming the magic point to be the origin of the $\log K_A^H$ scale in CCl$_4$, this scale can be moved to the more convenient origin of zero by adding $+1.1$. At the same time, the scale can be compressed somewhat so that values extend from 0 to 1. Equation 13 converts $\log K_A^H$ to an empirical α_2^H scale, which is generally used in linear solvation energy relationships[34,35].

$$\alpha_2^H = (\log K_A^H + 1.1)/4.636 \qquad (13)$$

Within the phenol family there are connections between hydrogen-bond acidities and full proton transfer acidity. Abraham and colleagues[33] found two good correlations between the $\log K_A^H$ scale and a parameter characteristic of proton transfer, the pK_a value in water. Equations 14 and 15 might be valuable in the conversion of pK_a into $\log K_A^H$, or *vice versa*.

$$\log K_A^H \text{ (3-substituted phenols)} = 8.13 - 0.66\,\text{p}K_a \qquad (14)$$

$$n = 11, \quad s = 0.09, \quad r = 0.980$$

$$\log K_A^H \text{ (4-substituted phenols)} = 5.56 - 0.39\,\text{p}K_a \qquad (15)$$

$$n = 14, \quad s = 0.11, \quad r = 0.965$$

However, there is no *general* connection between the $\log K_A^H$ and the pK_a scales, or any other measure of proton transfer. For example, $\log K_A^H$ is larger for phenol than for simple carboxylic acids. This has been attributed[33,36] to resonance stabilization in the carboxylate anion, which will disproportionately favor full proton transfer over hydrogen bonding. An additional stereoelectronic cause has been suggested[33,37]: the lone-pair repulsion between the incoming hydrogen-bond acceptor and the carbonyl group.

The classical front strain steric effect plays a significant role in influencing the hydrogen-bond acidity of *ortho*-substituted phenols. The introduction of one *ortho*-alkyl group into phenol lowers $\log K_A^H$ (Table 3). A 2,6-dialkyl substitution produces a more severe steric inhibition to hydrogen-bond formation (Table 3) and 2,6-di-*i*-propylphenol and 2,6-di-*t*-butylphenol become so weak that they were excluded from the analysis (Figure 4).

Intramolecular hydrogen bonding also leads to a reduction in the $\log K_A^H$ values. If electronic effects cancel out, a rough measure of the effect of intramolecular hydrogen bonding might be the differences between the values for the 2- and 4-substituted

TABLE 3. Hydrogen-bond acidity scale log K_A^H for phenols[33]

	log K_A^H		log K_A^H
2-Naphthol	1.74	4-i-Propyl	1.45
1-Naphthol	1.72	3,4,5-Trimethyl	1.43
Phenol	1.66	3-Ethyl	1.44
		4-Propyl	1.43
Meta- and/or *Para*-substituted phenols		4-Ethyl	1.43
4-Nitro-3-trifluoromethyl	3.33	4-Octyl	1.44
4-Nitro	2.72	3-Dimethylamino	1.31
3,4,5-Trichloro	2.69		
3,5-Di(trifluoromethyl)	2.68	Mono-*ortho*-alkyl-substituted phenols	
4-Cyano	2.55	4-Methyl-2-t-butyl	(1.52)[a]
3-Nitro	2.54	3-Methyl-6-t-butyl	(1.47)[a]
3,5-Dichloro	2.49	2,4-Di-t-Butyl	(1.43)[a]
3-Cyano	2.48	2,5-Dimethyl	1.40
3,4-Dichloro	2.35	2-i-Propyl	1.38
4-Trifluoromethyl	2.25	2,3-Dimethyl	1.37
4-Acetyl	2.25	2,4-Dimethyl	1.37
3-Trifluoromethyl	2.24	2,3,5-Trimethyl	1.31
3-Bromo	2.14	2-t-Butyl	1.22
3-Chloro	2.11		
3-Fluoro	2.04	Di-*ortho*-alkyl-substituted phenols	
4-Iodo	2.05	4-Bromo-2,6-dimethyl	1.05
4-Bromo	2.03	2,6-Dimethyl	0.71
4-Chloro	2.01	2,4,6-Trimethyl	0.63
4-Fluoro	1.82	2-Methyl-6-t-butyl	0.59
4-Phenyl	1.66		
3-Methoxy	1.64	Phenols with intramolecular hydrogen bonds	
4-Methoxy	1.56	2-Cyano	2.32
4-s-Butyl	1.55	2,6-Dichloro-4-nitro	2.17
3-Methyl	1.55	2-Chloro	1.91
4-Methyl	1.54	Pentachloro	1.46
3,5-Dimethyl	1.53	Pentabromo	1.21
3,4-Dimethyl	1.49	2,6-Dichloro	0.39
4-t-Butyl	1.49	2-Methoxy	0.11

[a]Doubtful values, compared to 2-t-butylphenol.

phenol. This difference is small for the C≡N and Cl substituents (0.23 and 0.10 log units, respectively) and very large (1.45 log unit) for the OMe substituent.

For *meta*- and *para*-substituted phenols, log K_A^H values spread over 2 log units from 3-dimethylaminophenol to 4-nitro-3-trifluoromethylphenol. Their order is well explained by classical electronic effects. A dual-substituent parameter analysis gives equations 16 and 17, where σ_F and σ_R are the Taft field-inductive and resonance substituent constants[38], respectively.

$$\log K_A^H \text{ (3-substituted phenols)} = 1.63 + 1.35\sigma_F + 0.63\sigma_R \tag{16}$$

$$n = 11, \quad s = 0.05, \quad r = 0.995$$

$$\log K_A^H \text{ (4-substituted phenols)} = 1.64 + 1.38\sigma_F + 1.01\sigma_R \tag{17}$$

$$n = 14, \quad s = 0.06, \quad r = 0.992$$

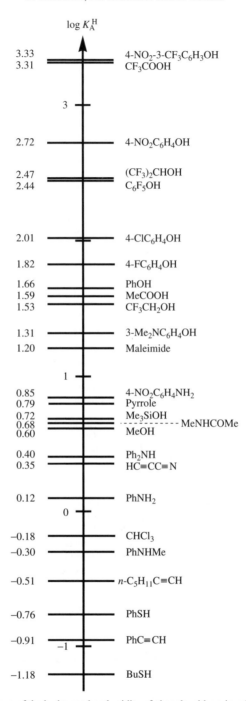

FIGURE 3. Comparison of the hydrogen-bond acidity of phenols with various hydrogen-bond donors

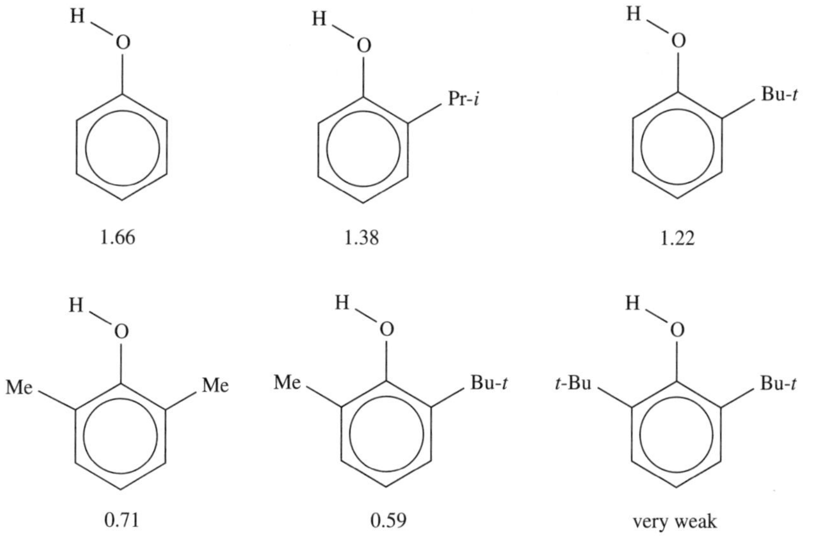

FIGURE 4. log K_A^H values. The order is consistent with the expected relative steric requirements of the phenols

B. log K_α Scale

Hydrogen-bonding equilibrium constants K_α have been measured[37] by titrational calorimetry or IR spectroscopy for sixteen phenols and a large and varied selection of hydrogen-bond donors, against N-methylpyrrolidinone. These have been used to create the log K_α scale (equations 18 and 19) of hydrogen-bond acidity for use in drug design. To this end, they have been measured in 1,1,1-trichloroethane, a solvent whose high polarity is considered a much better model for biological membranes than the previously employed apolar solvent CCl$_4$. Values are reported in Table 4. The extremes of the scale are 4-nitrophenol (3.12) and 2,6-di-t-butylphenol (0). For phenols and alcohols, a reasonable relation (equation 20) is found between log K_A^H and log K_α, although different solvents have been used. There are four main families: carboxylic acids, phenols, alcohols and azoles (pyrroles, indoles, etc.), for the correlation between hydrogen bonding (log K_α) and full proton transfer in water (pK_a). For phenols, equation 21 seems of lower quality than equations 14 and 15. The log K_A^H scale is possibly more reliable than the log K_α scale. In particular, the log K_α values of 4-methoxyphenol (2.18) and 2-t-butylphenol (1.85) appear suspect compared to those of phenol (2.14) and 2-methylphenol (1.75), respectively. They disagree with the well-known electron-donor property of the 4-methoxy substituent[38] and the greater steric effect of the 2-t-butyl group compared to the 2-methyl group.

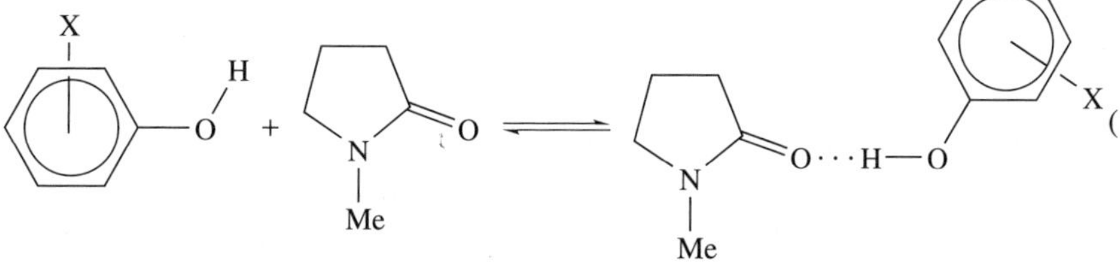

TABLE 4. The log K_α hydrogen-bond acidity scale[37]: comparison of phenols with other hydrogen-bond donors

Meta- and para-substituted phenols	log K_α	Various hydrogen-bond donors	log K_α
4-Nitro	3.12		3.55
4-Trifluoromethyl	2.80		
3-Chloro	2.50	(triazole with Ph)	
4-Methoxy	2.18		
None	2.14		
3-Methyl	1.89		
3-i-Propyl	1.89	Trifluoroacetic acid	ca 3.55
3-N,N-Dimethylamino	1.79	Hexafluoroisopropanol	2.83
		$(CF_3CO)_2NH$	2.63
Ortho-substituted phenols		Acetic acid	2.04
		Trifluoroethanol	2.00
2-Cyano	2.69	Thioacetanilide	1.52
2-Chloro	2.33	Methanol	1.48
2-i-Propyl	1.95	Acetanilide	1.34
2-t-Butyl	1.85	p-Toluene sulfonamide	1.15
2-Methyl	1.75	2-Chloroethanol	1.08
2,6-Dimethyl	1.08	Pyrrole	0.95
2,6-Dichloro	0.98	t-Butyl alcohol	0.78
2,6-Di-i-propyl	ca 0	4-Nitro-N-methylaniline	0.73
2,6-Di-t-butyl	ca 0	Chloroform	ca 0.4

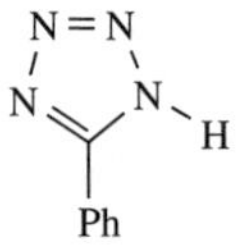

$$K_\alpha(\text{dm}^3\,\text{mol}^{-1}) = [\text{Complex}]/[\text{H-bond donor}][N\text{-methylpyrrolidinone}] \quad (19)$$

$$\log K_\alpha = 0.870 \log K_A^H + 0.70 \quad (20)$$

$$n = 21, \quad s = 0.13, \quad r = 0.986$$

$$\log K_\alpha = 6.25 - 0.40 \text{p}K_a \quad (21)$$

$$n = 9, \quad s = 0.11, \quad r = 0.927$$

C. Complexation with Pyridine N-oxide

A hydrogen-bond acidity scale has been constructed by Frange and coworkers[39] and by Sraïdi[40,41] based on log K values for complexation with pyridine N-oxide in cyclohexane (equations 22 and 23). Values for phenols and, for comparison, thiols, chloroform, pyrrole and alcohols are collected in Table 5. There is a fair measure of agreement between log K and log K_A^H ($n = 16$, $r = 0.992$) and log K_α ($n = 9$, $r = 0.972$).

$$A-H + \text{Py}N\to O \rightleftharpoons \text{Py}N\to O \cdots H-A \quad (22)$$

$$K(\text{dm}^3\,\text{mol}^{-1}) = [\text{Complex}]/[\text{AH}][\text{Pyridine N-oxide}] \quad (23)$$

D. Solvatochromic Shifts of Reichardt's Betaine Dye

Reichardt's dye **4** is the most widely used solvatochromic probe of probe/solvent interactions[42]. The solvatochromic shifts of the longest-wavelength intramolecular

TABLE 5. Hydrogen-bond acidity of hydrogen-bond donors including phenols towards pyridine N-oxide[41]

Hydrogen-bond donors	log K	Substituted phenols	log K
2-Propanethiol	−0.18	4-Methyl	2.86
2-Methyl-2-propanethiol	−0.13	3-Methyl	2.96
Chloroform	0.68	None	2.98
Isopropanol	1.30	4-Fluoro	3.27
Cyclohexanol	1.32	4-Chloro	3.35
1-Octanol	1.41	3-Fluoro	3.34
2-Phenylethanol	1.51	Pentafluoro	3.69
Methanol	1.61		
Pyrrole	1.61		
2-Chloroethanol	1.82		
2-Bromoethanol	1.84		
Propargyl alcohol	1.88		
Trichloroethanol	2.48		
Hexafluoroisopropanol	3.66		

charge-transfer $\pi-\pi^*$ absorption band of this dye provide a quantitative measurement of solvent effects. They are measured in kcal mol^{-1} (1 cal = 4.184 J) using the molar electronic transition energy, $E_T(30)$. For example $E_T(30)$ spreads from 31 kcal mol^{-1} for pentane to 65.3 kcal mol^{-1} for hexafluoroisopropanol[43]. With the phenolate oxygen atom, the dye has a strong hydrogen-bond acceptor center, suitable for interactions with hydrogen-bond donors. Hydrogen bonding to the phenolate oxygen will lead to a stabilization of the π ground state relative to the less basic π^* excited state, and this will be accompanied by an increase in the transition energy (Figure 5).

Coleman and Murray[44] have reported evidence that the dye **4** forms hydrogen bonds with dilute acetonitrile solutions of phenols, alcohols and water. They have measured the equilibrium constants for the complexation of the dye **4** (ArO$^-$) with these hydrogen-bond donors in MeCN (equation 24). For 6 phenols, 2 alcohols and water, the logarithm of these complexation constants is well correlated with log K_A^H ($n = 9$, $r = 0.985$). These results provide quantitative support for suggestions[45,46] that the $E_T(30)$ scale is at least

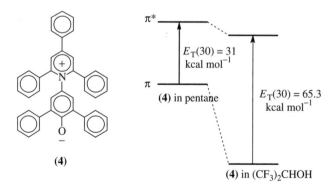

FIGURE 5. Structure of Reichardt's dye **4**, and a schematic diagram showing the influence of a hydrogen-bond donor solvent on the ground and excited states of the intramolecular charge-transfer absorption

as much a measure of solvent hydrogen-bond donor acidity as it is of van der Waals interactions in hydrogen-bond donor solvents.

$$ArO^- \cdots CH_3CN + ROH \cdots NCCH_3 \rightleftharpoons ArO^- \cdots HOR + CH_3CN \cdots CH_3CN \quad (24)$$

Hormadaly and Marcus[47] have measured $E_T(30)$ for 22 liquid and supercooled liquid phenols at room temperature. The $E_T(30)$ values found (Table 6) show the phenols to be better hydrogen-bond donor solvents than alcohols (compare E_T for phenol, 61.4, and methanol, 55.4). Bulky alkyl groups at both *ortho* positions (2,6-di-*t*-butylphenol, 41.1) and intramolecular hydrogen bonds (methyl salicylate, 45.4) decrease $E_T(30)$ drastically. Thus the solvent $E_T(30)$ scale shows the same effects as already found on the solute log K_A^H scale. Figure 6 compares the two scales for phenols, water, alcohols, NH and CH donors. The correlation is statistically significant: for 31 hydrogen-bond donor solvents, the solute hydrogen-bond acidity (log K_A^H) explains 70% of the variance of the solvent $E_T(30)$ scale. Four main reasons might, however, explain the differences between the two scales. First, one is comparing an electronic energy (E_T) to a Gibbs energy (log $K = \Delta G/RT$). Second, $E_T(30)$ measures not only hydrogen bonding but also nonspecific van der Waals interactions. Third, $E_T(30)$ is a solvent scale, taking into account, for example, the self-association of amphiprotic solvents, while log K_A^H is a solute scale for monomeric compounds. Last, the phenolate oxygen in **4** is sterically hindered by two bulky *ortho*-phenyl groups and $E_T(30)$ might be more sensitive than log K_A^H to steric effects.

The $E_T(30)$ values of 55 phenols have been determined[48] by means of a special technique using solutions of the phenols in 1,2-dichloroethane as inert solvent. Surprisingly,

TABLE 6. $E_T(30)$ (kcal mol^{-1}) values of hydrogen-bond donor solvents and, for comparison, log K_A^H

Compound	$E_T(30)$	log K_A^H	Compound	$E_T(30)$	log K_A^H
CH donors			2-*t*-Butylphenol	49.0[b]	1.22
Non-1-yne	33.7[a]	−0.51[c]	1-Butanol	49.7[a]	0.43
Phenylacetylene	37.2[a]	−0.56	1-Propanol	50.5[a]	0.36
Pentafluorobenzene	38.6[a]	−1.06[d]	Benzyl alcohol	50.7[a]	0.72
Chloroform	39.1[a]	−0.18	2,4-Dimethylphenol	50.8[b]	1.37
Dichloromethane	40.7[a]	−0.50	Ethanol	51.8[a]	0.44
Propargyl chloride	41.7[a]	−0.24	2-Methylphenol	52.5[b]	1.30
Ethyl propiolate	45.4[a]	−0.23	Methanol	55.4[a]	0.60
			2,2,2-Trichloroethanol	54.1[a]	1.22
NH donors			2-Chloroethanol	55.1[a]	0.50
Aniline	44.4[a]	0.12	2-Chlorophenol	55.4[b]	1.91
Pyrrole	51.0[a]	0.79	2-Fluoroethanol	56.6[a]	0.73
N-Methylacetamide	52.1[a]	0.68	3-Methylphenol	56.2[a]	1.55
			2,2,2-Trifluoroethanol	59.8[a]	1.53
OH donors			4-Methylphenol	60.8[b]	1.54
t-Butanol	43.3[a]	0.38	Phenol	61.4[b]	1.66
2,6-Dimethylphenol	47.6[b]	0.71	Water	63.1[a]	0.54
2-Propanol	48.4[a]	0.40	Hexafluoroisopropanol	65.3[a]	2.47

[a] Reference 41.
[b] Reference 47.
[c] Value for hept-1-yne.
[d] Value for pentachlorobenzene.

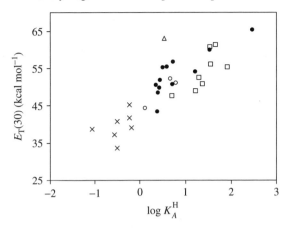

FIGURE 6. Comparison of the solvent $E_T(30)$ scale and the solute $\log K_A^H$ scale for CH (×), NH (○), alcoholic OH (●), phenolic OH (□) hydrogen-bond donors and water (△). Data from Table 6

these values are not correlated with $\log K_A^H$. 4-Methoxyphenol and 4-methylphenol have about the same solute hydrogen-bond acidity (1.56 and 1.54, respectively) but very different $E_T(30)$ values (45.4 and 53.3, respectively). Conversely, 4-cyanophenol ($\log K_A^H = 2.55$) is a much stronger hydrogen-bond donor than 3-ethylphenol ($\log K_A^H = 1.44$), but these phenols have about the same $E_T(30)$ values (52.2 and 51.6 kcal mol^{-1}).

E. Hydrogen-bond Acidity from Partition Coefficients

Partition coefficients can be used to deduce the relative solute–solvent effects, e.g. solute–octanol less solute–water interactions in the case of octanol/water partition coefficients. From experimental octanol/water (o/w) and chloroform/water (Cl/w) partition coefficients Taft and colleagues[49,50] derived an octanol/chloroform (o/Cl) partition coefficient and showed that a number of solute parameters cancel out in the difference. Thus $\log P_{o/Cl}$ depends only on the solute effective hydrogen-bond acidity, $\varepsilon\alpha$, and on the solute volume, V_x (equation 25). In equation 25 solute hydrogen-bond acidity favors octanol, since octanol is a better hydrogen-bond acceptor than chloroform, while solute size favors chloroform, since it is easier for the solute to create a cavity in chloroform, which has less cohesion energy than octanol. The coefficient of V_x is obtained from a set of solutes without hydrogen-bond acidity, while the coefficient of $\varepsilon\alpha$ is calculated from simple mono-hydrogen-bond donors, using known α_2^H values determined from hydrogen-bond complexation constants in CCl$_4$ (see Section IIIA, equation 13). Equation 25 can be rearranged to give equation 26, from which the effective hydrogen-bond acidity $\varepsilon\alpha$ of the solute immersed in pure active solvents can be calculated. Table 7 gives typical $\varepsilon\alpha$ values for some phenols together with values for other hydrogen-bond donors for comparison.

$$\log P_{o/Cl} = \log P_{o/w} - \log P_{Cl/w} = -1.00(0.01 V_x) + 3.20\varepsilon\alpha - 0.03 \quad (25)$$

$$\varepsilon\alpha = [\log P_{o/Cl} + 1.00(0.01 V_x) + 0.03]/3.20 \quad (26)$$

In the same way, Abraham[29] has calculated, from partition coefficients, an effective or summation hydrogen-bonding acidity scale, $\Sigma\alpha_2^H$. However, while $\varepsilon\alpha$ is obtained from one reference partition system (octanol/chloroform), with the hypothesis that a number of solute parameters acting on $\log P$ cancel out, Abraham determines the $\Sigma\alpha_2^H$ values by a back calculation procedure over numerous sets of partition systems, and uses linear solvation energy relationships with a more complete set of solute parameters. Table 7 shows that $\Sigma\alpha_2^H$, $\varepsilon\alpha$ and α_2^H values do not differ much for simple mono hydrogen-bond donors. It is also important to note that the order of acidity:

1-alkynes ≈ thiols ≈ amines < alcohols ⩽ carboxylic acids ⩽ phenols

remains basically the same whether the acidity is calculated from partition coefficients ($\varepsilon\alpha$ or $\Sigma\alpha_2^H$) or from hydrogen-bond complexation constants (α_2^H).

Interestingly, the value of $\varepsilon\alpha$ for bisphenol A (**5**) is nearly twice the value for phenol. This additive effect opens up a large field of investigation for the calculation of hydrogen-bond acidities of complex polyfunctional molecules such as solutes of biological importance which are not accessible by other techniques. Finally, the effective acidity of 2-methoxyphenol (**2**) (guaiacol; $\varepsilon\alpha = 0.20$) is much smaller than that of 3- and 4-methoxyphenol and for 2-nitrophenol (**6**) this acidity is nearly zero ($\varepsilon\alpha = 0.07$).

This reduced acidity is in line with the IR spectra of **2** and **6** in chloroform[51]. 2-Methoxyphenol (Figure 7b) exhibits two absorptions at 3621 and 3544 cm^{-1}, corresponding respectively to a free OH absorption and a weak intramolecular hydrogen-bonded OH band. When 2-nitrophenol (Figure 7c) is dissolved in the same solvent, there is no absorption near 3600 cm^{-1} corresponding to a free phenol such as shown in Figure 7a for *p*-methoxyphenol, and the large shift of the intramolecularly hydrogen-bonded absorption from *ca* 3600 to 3240 cm^{-1} is an indication of a strong chelation leaving no residual acidity to the solute.

More detailed work on the intramolecular hydrogen bond of *ortho*-nitrophenols has recently been carried out by Chopineaux-Courtois and coworkers[52] and by Abraham and

TABLE 7. Comparison between the hydrogen-bond acidities of substituted phenols obtained from partition coefficients ($\varepsilon\alpha$, $\Sigma\alpha_2^H$) and from equilibrium constants (α_2^H) in apolar solvents

Compound	$\varepsilon\alpha^a$	$\Sigma\alpha_2^{H\,b}$	$\alpha_2^{H\,c}$	Compound	$\varepsilon\alpha^a$	$\Sigma\alpha_2^{H\,b}$	$\alpha_2^{H\,c}$
Phenols				2-NO$_2$ phenol	0.07	0.05	—
4-OMe phenol	0.56	0.57	0.57	2-CHO phenol	0.11	0.11	—
3,5-Me$_2$ phenol	0.56	0.57	0.57	Bisphenol A (**5**)	1.26	—	—
3-OMe phenol	0.58	0.59	0.59	**Alcohols**			
Phenol	0.58	0.60	0.60	*t*-Butanol	0.36	0.30	0.32
2-Naphthol	0.66	0.61	0.61	Methanol	0.29	0.43	0.37
4-COOEt phenol	0.66	0.69	0.71	2,2,2-Trifluoroethanol	0.58	0.57	0.57
4-Cl phenol	0.68	0.67	0.67	**Carboxylic acids**			
4-COMe phenol	0.73	—	0.72	Acetic acid	0.60	0.61	0.55
3-NO$_2$ phenol	0.76	0.79	0.78	Trichloracetic acid	—	0.95	0.95
4-NO$_2$ phenol	0.84	0.82	0.82	**NH, SH, CH donors**			
2,5-Me$_2$ phenol	0.57	0.54	0.54	Acetanilide	0.45	0.50	—
2,4-Me$_2$ phenol	0.58	0.53	0.53	Ethylamine	—	0.16	0.00^d
1-Naphthol	0.67	0.61	0.61	Thiophenol	—	0.09	0.07
2-OMe phenol	0.20	0.22	0.26	Phenylethyne	—	0.12	0.12

^{a}Calculated from equation 26.
^{b}Reference 29.
^{c}Calculated from hydrogen-bonding complexation constants in tetrachloromethane (equation 13).
^{d}Estimated value for alkylamines.

8. Hydrogen-bonded complexes of phenols

[Structure (5): HO-C6H4-C(CH3)2-C6H4-OH]

[Structure (6): 2-nitrophenol with intramolecular H-bond]

(5)   (6)

colleagues[53]. In 1,2-dichloroethane/water and cyclohexane/water partition coefficients, the large increase in lipophilicity found for *ortho*-nitrophenol by comparison with the *meta*- and *para*-isomers is due to the loss of hydrogen-bond acidity provoked by the intramolecular hydrogen bond. This increase is not observable in the octanol/water system, which does not depend on the solute hydrogen-bond acidity strength[54]. Abraham and coworkers[53] obtained the overall hydrogen-bond acidities $\Sigma\alpha_2^H$ of several mono-, di- and tri-nitrophenols. In compounds **6, 7** and **8**, the hydrogen-bond acidity strength

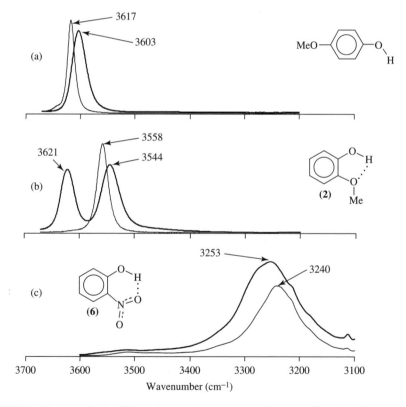

FIGURE 7. IR spectra in the OH stretching region of substituted phenols diluted in CCl$_4$ (———) and CHCl$_3$ (———)[51]

of the phenolic group is ruined by the strong intramolecular hydrogen bond. However, when steric effects between *ortho*-substituents create some distortions from the ideal planar geometry, the intramolecular hydrogen bond is weakened and compounds **9, 10** and **11** partly recover some hydrogen-bond acidity, reducing their lipophilicity in cyclohexane/water and 1,2-dichloroethane/water binary phases.

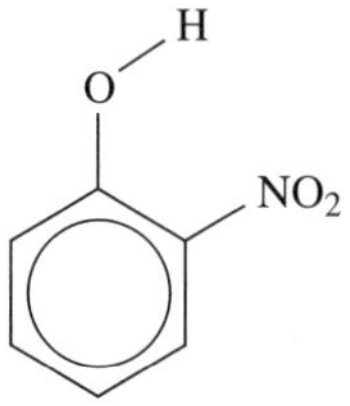

$\Sigma\alpha_2^H = 0.05$
(6)

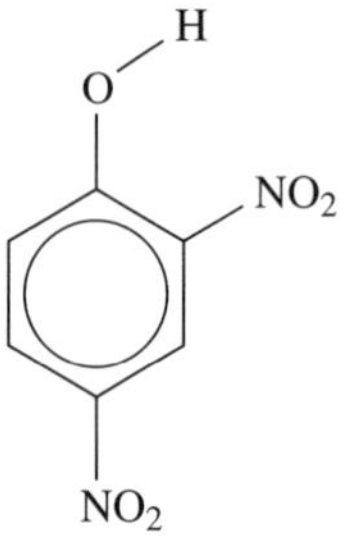

$\Sigma\alpha_2^H = 0.09$
(7)

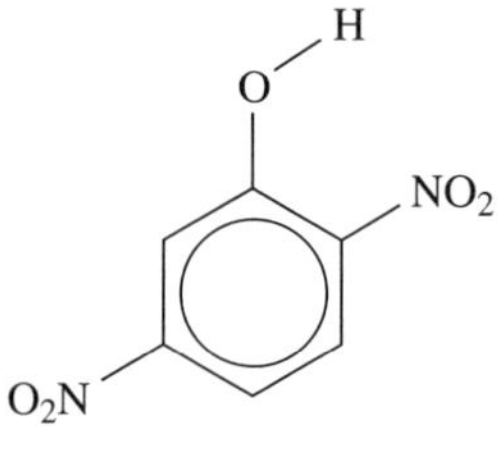

$\Sigma\alpha_2^H = 0.11$
(8)

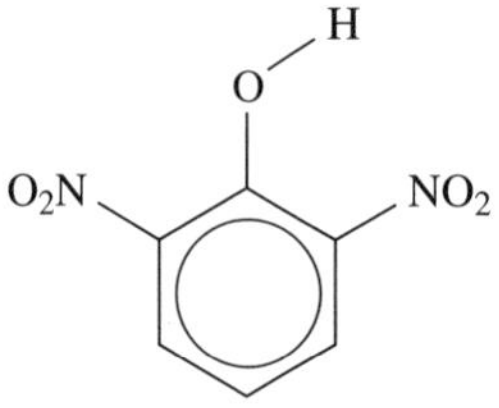

$\Sigma\alpha_2^H = 0.17$
(9)

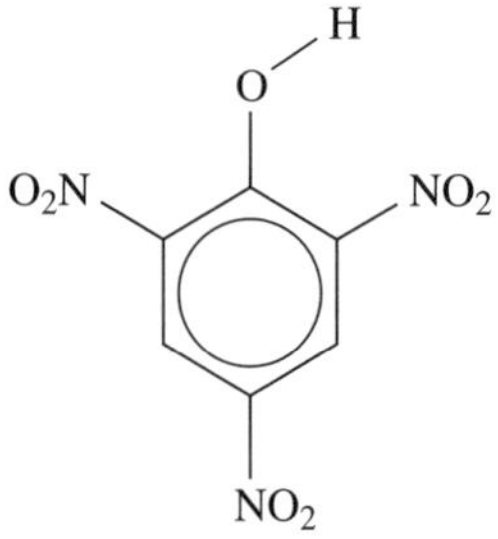

$\Sigma\alpha_2^H = 0.46$
(10)

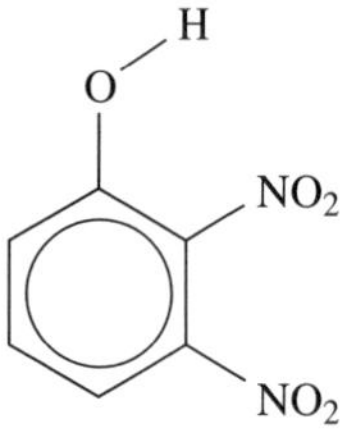

$\Sigma\alpha_2^H = 0.67$
(11)

IV. SELF-ASSOCIATION OF PHENOLS

The self-association of phenols has received little attention in the last few decades so that most of the results in the field have already been gathered in the reviews of Rochester[7] and Joesten and Schaad[4]. The few recent contributions to the analysis of phenol self-association have not greatly clarified the confusion prevailing about the degree of polymerization and the structures of the polymers. By dispersive IR spectroscopy, Frohlich[55] found that the major species in the $5 \times 10^{-3} - 1.6 \times 10^{-2}$ mol dm^{-3} concentration range is the dimer, but proton NMR shifts measured on solutions of phenol of higher concentrations ($10^{-2} - 3$ mol dm^{-3}) were found to be compatible with a trimer formation[56]. Heat capacity measurements carried out on several alcohols and phenols led Pérez-Casas and coworkers[57,58] to the conclusion that tetramers are the most abundant species in apolar solvents in the concentration range $2 \times 10^{-2} - 0.8$ mol dm^{-3}. All these results are not totally incompatible since they refer to different concentration ranges. However, the thermodynamic parameters evaluated by various techniques on the basis of distinct simplifying assumptions need further refinements since unacceptably large differences appear between the results reported by the different authors. Thus, the dimerization equilibrium constant K_{di} of phenol reported by Frohlich (70 dm³ mol⁻¹)[55] is 100 times greater than the (most reasonable) value (0.74) given by Singh and Rao[59] and 5 times larger than the constant found by Huggins and coworkers (13)[60].

Using Fourier Transform IR spectroscopy which allows accurate measurements of small absorbance variations, Laurence and coworkers[18] measured a dimerization constant of 0.76 dm³ mol⁻¹ for 4-fluorophenol in a narrow range of concentration (4 × 10⁻³–5 × 10⁻² mol dm⁻³) where the dimer is the dominant associated species. Their value is in agreement with the substituent effect of a 4-fluorophenyl group on the basicity of a hydroxyl group (see Section II).

In order to show the complexity of the phenol self-association, we have reported in Figure 8 the evolution of its IR spectrum in the domain of the OH stretching vibration as a function of concentration[51]. For the most diluted solutions (4 × 10⁻²–0.2 mol dm⁻³), the predominant band corresponds to the monomer **12** at 3612 cm⁻¹, but an absorption corresponding to the O-dimer (structure **13**) near 3500 cm⁻¹ is already present at 4 × 10⁻² mol dm⁻³. In a 4 mol dm⁻³ solution, the phenol is mainly polymerized as shown by the importance of the broad band at 3350 cm⁻¹. However, monomeric as well as dimeric

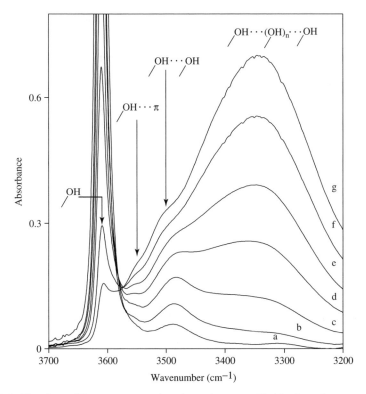

FIGURE 8. Phenol at different concentrations in carbon tetrachloride (a scale expansion has been used for certain spectra for clarity)[51]: (a) 0.04, (b) 0.12, (c) 0.2, (d) 0.4, (e) 0.67, (f) 2, (g) 4 mol dm⁻³

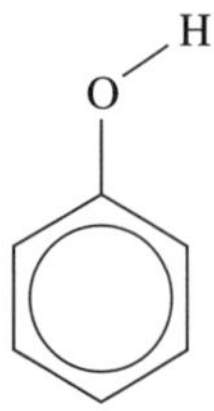

free
ν(OH) = 3612 cm^{-1}
(12)

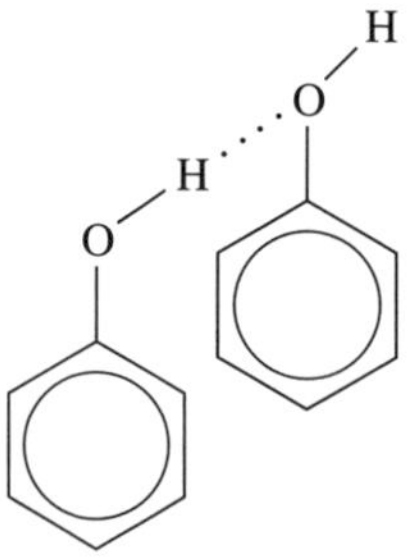

O dimer
ν(OH···O) = 3480 cm^{-1}
(13)

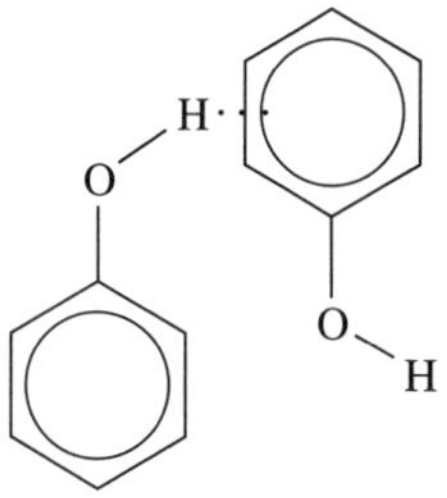

π dimer
ν(OH···π) = 3555-3560 cm^{-1}
(14)

species are still present. While the spectra in Figure 8 give little information on the extent of polymerization or on the linear or cyclic structure of the polymers corresponding to the broad band at 3350 cm^{-1}, they cast some new light on the dimerization equilibria by revealing the presence of an additional dimeric species that has been either neglected or misinterpreted[61]. The weak, but significant, absorption shown at 3550–3560 cm^{-1} must be attributed[51] to the dimeric form **14** where the hydroxyl group of a first phenol molecule is bound to the π cycle of a second phenol molecule.

The importance of this π dimeric form increases when the basicity of the aromatic ring is strengthened by alkyl substitution. This can be seen in Figure 9 where the spectra of 3,4,5-trimethylphenol (Figure 9a) and 4-methylphenol (Figure 9b) clearly present the same characteristics as the spectrum of the heteroassociation of phenol on anisole, where the OH···π and OH···O complexes have already been identified[62] (Figure 9c). Moreover, the OH···π absorption of a phenol has been assigned in the spectrum of the *cis* isomer of 2,2'-dihydroxybiphenyl (**15**), which presents[63] an absorption at 3556 cm^{-1}.

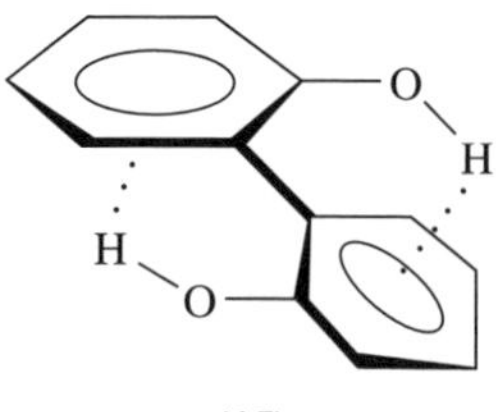

(15)

It is clear that this dimeric structure has never been taken into account in any of the different calculations leading to estimations of dimerization constants. This association is, however, far from being negligible. Its relative importance can be evaluated semi-quantitatively from the 4-fluorophenol-anisole association that has been fully analyzed by Marquis[30]. In dilute carbon tetrachloride solutions, he estimated the two 1 : 1 equilibrium constants on both the π and O sites and found that 34% of the complexation occurs on the π ring of anisole. Misinterpretations of the same kind have appeared for phenols that bear HBA substituents adding more possibilities of association to the phenolic OH acid group. For example, in 4-methoxyphenol a large dimerization constant of 3.00 dm^3 mol^{-1} was found[59] and is due to the simple accumulation of three association constants: (i) on the phenolic oxygen, (ii) on the methoxy oxygen and (iii) on the π cloud.

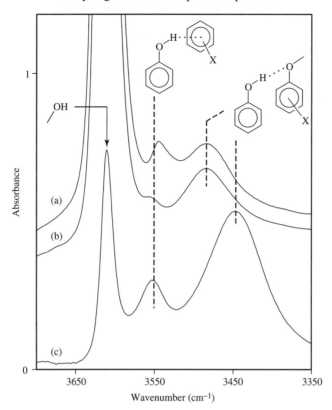

FIGURE 9. Self-association of 3,4,5-trimethylphenol (a) and 4-cresol (b) and heteroassociation of phenol with anisole (c). Taken from Reference 51

Another experimental technique to study the self-association of phenols is to investigate how molecules of phenols pack together in the crystalline state. This type of analysis is made possible by the availability of the computer-based CSD[32]. The CSD contains unit-cell dimensions of more than 230,000 (April 2001 release) three-dimensional crystal-structure determinations that have been studied by X-ray or neutron diffraction. Each crystal structure is identified by a unique six-letter code, called its REFCOD, with an additional two digits for duplicate structures and measurements.

To define the properties relevant to the motif (or synthon) $\cdots$OH$\cdots$OH$\cdots$, we have searched in the CSD for the structure of simple phenols in which the phenolic hydroxy group is the only one capable of forming hydrogen bonds. This condition limits the sample to phenols with only alkyl or hydroxy substituents. The OH group invariably acts as both a hydrogen-bond donor and a hydrogen-bond acceptor and links each molecule to two others. Either hydrogen-bonded chains or hydrogen-bonded rings are formed. Most of the structures consist of infinite chains. The exceptions are hydrogen-bonded cyclic tetramers, e.g. in one form of 4-methylphenol (CRESOL01)[64] and in 2,6-di-i-propylphenol (GAPTOG)[65], and cyclic hexamers, e.g. in 2-i-propyl-5-methylphenol (IPMPEL)[66] and 3,4-dimethylphenol (DPHNOL10)[67]. Of the hydrogen-bonded infinite chains, the most common arrangement is one in which the molecules are related by a two-fold screw axis as in 2,6-dimethylphenol (DMEPOL10)[68]. There are other examples of a helical arrangement based on three- and four-fold screw axes. Catechol (CATCOL12)[69] forms chains of cyclic dimers (with a third intramolecular hydrogen bond) and resorcinol (RESORA13)[70] forms chains of cyclic tetramers (tetrameric helices). The degree of polymerization seems sensitive to the steric crowding of the OH group due to neighboring bulky substituents, varying from infinity for 2,6-dimethylphenol (DMEPOL10), to four for 2,6-di-i-propylphenol (GAPTOG) and to zero for 2,6-di-t-butylphenol (LERFET)[71]. In this structure the phenolic hydrogen is not located but the O···O distance of 3.32 Å shows at most a very weak hydrogen bond, compared to 2.74 Å for 2,6-di-i-propylphenol. In probucol (**16**) (HAXHET)[72], a drug used to control blood-cholesterol levels, intermolecular hydrogen bonding between hydroxyl groups is also prevented in the crystal. Figure 10 illustrates how phenols self-associate in the solid state.

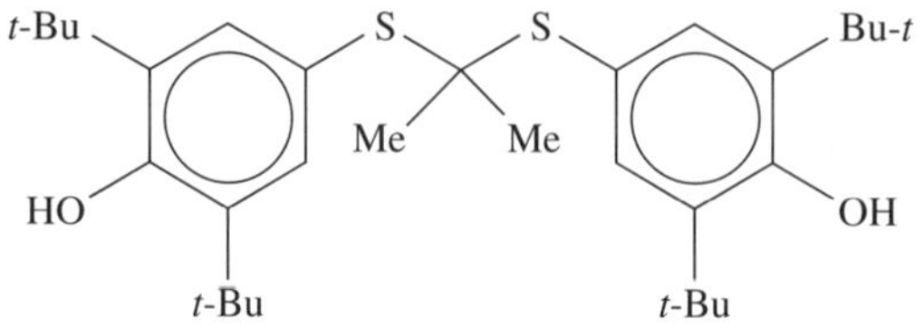

(**16**)

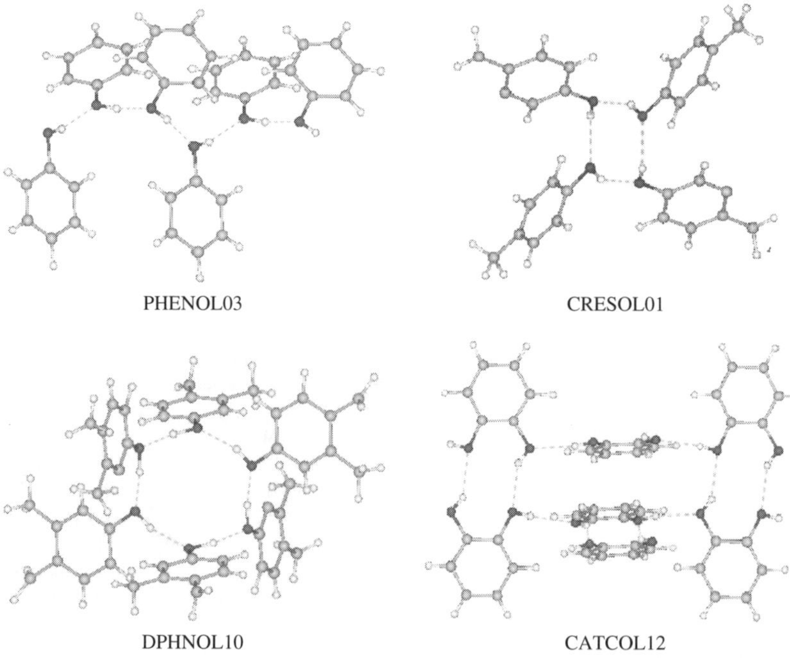

PHENOL03

CRESOL01

DPHNOL10

CATCOL12

FIGURE 10. Self-association of phenols in the solid state. The phenolic OH groups link to form infinite chains in phenol (PHENOL03), cyclic tetramers in 4-methylphenol (CRESOL01), cyclic hexamers in 3,4-dimethylphenol (DPHNOL10) and dimeric helices in catechol (CATCOL12)

V. INTRAMOLECULAR HYDROGEN BONDS

An important effort has recently been devoted to understanding and predicting the internal hydrogen-bond (IHB) structures and strengths of chelated *ortho*-substituted phenols by means of new experimental methods and theoretical calculations. Among the different systems studied, 2-nitrophenols[73-80] and especially 2-benzoylphenols[81-108] (salicylic acid derivatives) have certainly been the most popular models. However, a great wealth of new structural information has also been published for internally hydrogen-bonded phenols with *ortho* substituents such as halogens[109-114], alcohol and ether oxygens and their thio analogues[115-117], and sp^2 and sp^3 nitrogens[118-123]. 2-Thiobenzoyl[86,91,92,124] as well as phosphine or amine oxides[125-129] have also been found to be good acceptor groups for the formation of strong IHBs. In a recent communication, a new type of IHB between the hydroxyl and methyl groups has been detected in the IR spectrum of the 2-cresol cation[130].

The traditional ways of evaluating the IHB characteristics are to assess the vibrational frequencies or intensities of the OH stretching or torsion in the IR spectra and the chemical shifts of the hydroxyl protons in the NMR spectra which are found to be nicely correlated[131-133]. Crystal-structure analysis also provides essential information in this field. Bilton and coworkers[134] carried out a systematic survey of the internally hydrogen-bonded frames in the 200,000 structures of the CSD[32] and gave a general overview of the IHB in the solid state.

These experimental data are now accompanied, or even replaced, by theoretical *ab initio* calculations which can examine the conformers that cannot be observed experimentally. There is now general agreement that the Density Functional Theory (DFT) and, in particular, Becke's three-parameters Lee–Yang–Parr hybrid method (B3LYP) with a 6-31G* or a 6-31G** standard basis set provide cost-effective evaluations of geometries and energies comparable with experimental data[74,79,104,106,107]. However, the selection of a reference conformer for the quantitative evaluation of the hydrogen-bond energy ΔE still raises some questions. In the pioneering works on 2-halophenols[4], it was found that the internally H-bonded *syn* conformer **17** and the *anti* conformer **18** coexist in apolar solvents, allowing the experimental determination of the enthalpy difference between the two isomers (see Section V.A). In spite of the introduction of additional interactions between the two non-bonded *ortho*-substituents, it appears that the reference geometry, reflecting at best the trends in the geometry of 2-benzoylphenols[104], is indeed that of the *anti* isomer where the OH group is rotated by 180° around the CO bond axis. In quantum-chemical calculations of ΔE, Lampert and colleagues[104] performed single-point calculations on the frozen molecule while Palomar and coworkers[106] and Catalán and coworkers[107] optimized the geometry of this isomer. The difference between the two analyses which corresponds to the full geometry relaxation of isomer **18** is in the range 8.5–10 kJ mol^{-1} for planar systems (Table 8).

syn
(17)

anti
(18)

TABLE 8. Calculated strengths of the IHB with (ΔE_O/kJ mol^{-1}) and without (ΔE_{NO}/kJ mol^{-1}) geometry optimization of the *anti* conformer of 2-hydroxybenzoyl compounds 2-HOC$_6$H$_4$COX[a] [107]

X	ΔE_{NO}	ΔE_O	Difference
Cl	48.8	40.2	8.6
CN	54.4	45.9	8.5
OMe	60.5	51.4	9.1
H	61.0	51.6	9.4
Me	69.5	59.5	10.0

[a] At B3LYP/6-31G**.

Table 8 shows that no universal definition of the IHB strengths can be given for all 2-substituted phenols. Nevertheless, comparison of the experimental and theoretical data is sufficient to unravel the different factors affecting the properties of these molecules. These are (i) the hydrogen-bond strength of the OH group, (ii) the HBA ability of the *ortho*-substituent X, (iii) the steric accessibility of the accepting atom and (iv) the cooperative electronic delocalization in the ring formed by the chelation. The influence of the first two factors can be estimated from the analysis of the intermolecular complex formations of phenols (Sections III and VII). The essential role of steric and delocalization effects on the stability of the IHB has been analyzed by Bilton and coworkers[134] in their survey of the CSD. Several thousand structures containing IHB rings of different sizes were examined. They found that the 50 most probable motifs are constituted of 5- and 6-membered rings, and among these motifs the 10 most probable rings are planar and conjugated 6-membered. These findings are in agreement with the concept of Resonance Assisted Hydrogen Bonding (RAHB) introduced by Gilli and colleagues[135]. In this model, the presence of alternate single and double bonds between the phenolic group and the acceptor substituent allows an electron delocalization which strengthens the hydrogen-bond ability of both the OH and the C=X groups in the resonance structures **19 ↔ 20**.

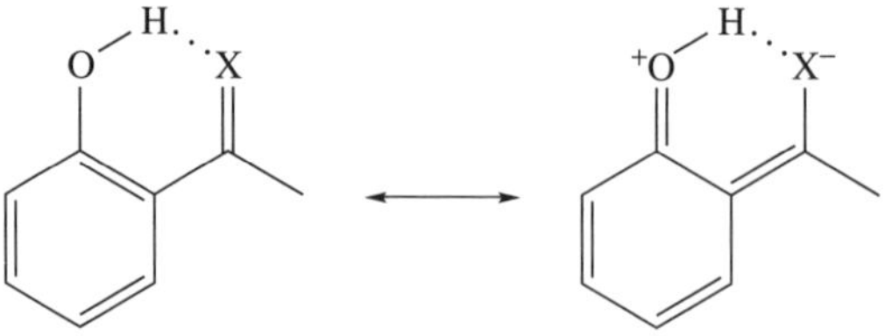

(19)   (20)

Palomar and coworkers[106] calculated that the stabilization energy gained by the chelation on going from an aliphatic compound **21** to an alkene transmitting group (structure **22**) amounts to about 40 kJ mol^{-1}.

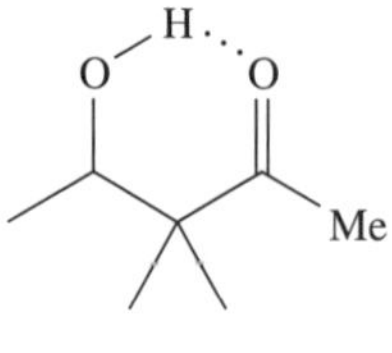

(21) 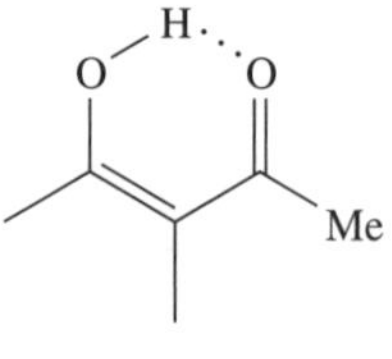   (22)

A. OH···O₂N

Several experimental and theoretical works on 2-nitrophenol (**23**) and 2-nitroresorcinol (**24**) have reaffirmed the planarity of both molecules and the C_{2v} symmetry of **24**. The microwave spectrum of **24** indicates[73] that no proton transfer occurs between the phenolic and nitro oxygen atoms. Kovács and coworkers[74,75] reported the FTIR and FT Raman spectra of **23** and **24** and assigned all the fundamentals by means of a scaled B3LYP/6–31G** density functional force field. Ab initio molecular orbital calculations were also needed to interpret the electron diffraction spectra of these two compounds[76,77]. The IHB lengths found by electron diffraction spectroscopy are 1.72 and 1.76 in **23** and **24**, respectively, corresponding to 66% of the van der Waals radii of the hydrogen and oxygen atoms. This important shortening indicates a strong stabilization by an RAHB mechanism[78] leading to a calculated energy[79] of the hydrogen bond equal to about 42 kJ mol⁻¹ in **24**. Natural abundance ¹⁷O NMR chemical shifts have been measured[80] for a series of 4-substituted-2-nitrophenols (**25**). The ¹⁷O OH signal is more sensitive to the substituent effect than the NO₂ signal and the presence of the NO₂ group in the *ortho* position reduces the substituent effect sensitivity of the OH group by about 25% in comparison with 4-substituted phenols.

(23)   (24)   (25)

B. OH···O=C

The strong intramolecular hydrogen bond that occurs in *ortho*-hydroxybenzoyl compounds (**26**) is still the subject of numerous papers. It is now well established that the large strength of the IHB is due to the synergistic delocalization of electrons between the OH and CO group permitted by the alternation of single and double bonds, the so-called RAHB effect. While the simplest compound **26** (Table 9) remains the most popular model, more complex structures leading to competitive hydrogen bonds such as 2,6-dicarbonylphenols[81] **27** or benzophenone **28** and its tricyclic analogues fluorenones[82] and anthrones[83] give additional information on the IHB strength. Quantum-mechanical calculations on different existing or virtual conformers have led to several papers contributing to a better understanding of the structure of most of the 2-hydroxybenzoyl compounds[84,100–107]. In this series, the influence of the COX substituent on the OH acidity is claimed to be of minor importance[104]. However, Palomar and colleagues[106] found a significant increase in the OH acidity with carbonyl groups such as COCN and CONO₂, whereas the electronic demand of the CN and NO₂ groups appears to be supplied by the phenolic oxygen. The IHB strength is therefore affected by the basicity of the carbonyl oxygen which increases mainly with the resonance donating ability of the substituent X. The great majority of the molecular frames are found to be approximately planar with the important exception of the compound containing the amide group (X = NR₂)[86,87,104]. In Table 9, the IHB strengths of some 2-hydroxybenzoyl compounds are given in increasing order together with the lengths of three bonds directly involved in the IHB. It can be seen

TABLE 9. *Ab initio* calculated parameters for 2-hydroxybenzoyl compounds **26**

Substituent X	d_{OH} (Å)[a]	$d_{H\cdots O}$ (Å)[a]	$d_{O=C}$ (Å)[a]	ΔE (kJ mol^{-1})[b]
Cl	0.982	1.752	1.212	40.3
NO$_2$	0.983	1.709	1.213	40.6
F	0.982	1.793	1.212	42.3
C≡N	0.987	1.724	1.238	46.0
OH	0.987	1.727	1.233	50.9
OMe	0.987	1.721	1.234	51.5
H	0.990	1.729	1.235	51.2
NMe$_2$	0.991	1.686	1.249	(66.5)[c]
Me	0.994	1.654	1.242	59.4
NH$_2$	0.996	1.644	1.246	64.4
NHMe	0.996	1.640	1.249	66.9

[a] References 104 and 106 (B3LYP/6-31G**); d = bond length.
[b] Calculated energy difference between the *syn* isomer with IHB and the *anti* isomer. The energy of the *anti* isomer is optimized[107].
[c] The geometry of the *anti* isomer is not optimized.

from this table that the shortening of the hydrogen bond and the concomitant lengthening of the OH and C=O bond calculated at the B3LYP/6-31G** level are well correlated with the energy of the hydrogen bond.

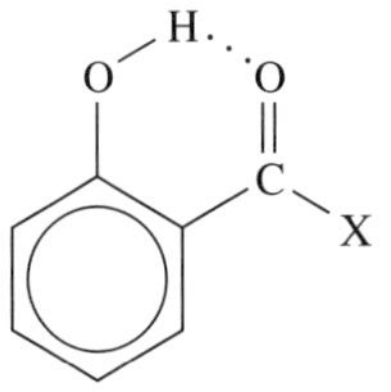

(26) 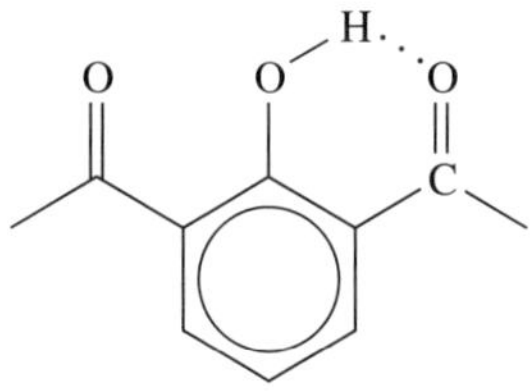   (27) 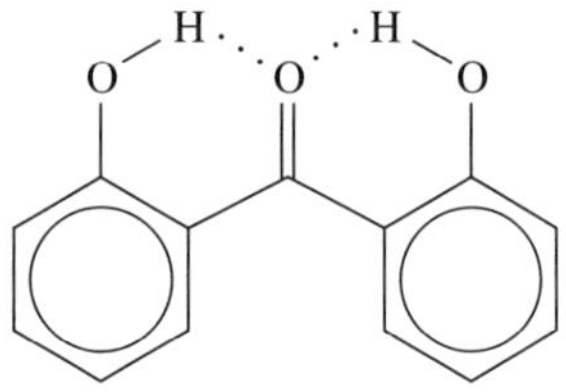   (28)

C. OH···Halogen

2-Halophenols constitute the simplest structural model for the analysis of an intramolecular hydrogen bond since the *syn–anti* isomerization involves only the rotation of the OH proton around the single C–O bond. Furthermore, the two forms coexist in apolar solvents and give two characteristic absorptions in the OH stretching region of the IR spectrum with the notable exception of the fluoro derivative. In a series of papers Okuyama and Ikawa[109,110] have re-examined by FTIR the relative stability of the *syn–anti* isomers by varying the temperature and the pressure. The enthalpies of isomerization are reported in Table 10.

These precise measurements[109] provide a better discrimination between the halogens than the work of Baker and Shulgin[136]. The enthalpies determined for the IHB follow the variation found by Ouvrard and colleagues[138] for the intermolecular association of 4-fluorophenol with halocyclohexanes. The hydrogen-bond acceptor ability of the halogen atom is therefore the main factor affecting the IHB strength in this system. Until recently, the enthalpies calculated by Carlson and coworkers[137] from the torsional frequencies of the OH group in the far IR spectrum were considered[4] as more reliable than the measurements obtained from IR intensities of the fundamental O–H stretching. However, the assignment of the two bands on which the enthalpy determination was made seems to be erroneous[109].

TABLE 10. Enthalpies (kJ mol^{-1}) and IR Δv(OH) frequency shifts (cm^{-1}) for the *syn* (**17**)–*anti* (**18**) isomerization of 2-halophenols and the intermolecular association of 4-fluorophenol with halocyclohexanes

Halogen X	Intramolecular hydrogen bonding				Intermolecular hydrogen bonding	
	$-\Delta H^a$ CCl$_4$	$-\Delta H^b$ CCl$_4$	$-\Delta H^c$ C$_6$H$_{12}$	Δv(OH)^a CCl$_4$	$-\Delta H^d$ CCl$_4$	Δv(OH)^d CCl$_4$
F	—	—	6.0	—	12.6	59
Cl	6.6	6.0	6.8	57	8.8	77
Br	5.7	5.1	6.6	75	7.5	88
I	4.0	4.5	6.1	96	6.2	92

^{a}Reference 109.
^{b}Reference 136.
^{c}Reference 137.
^{d}Reference 138.

It should be noted that the Δv(OH) and ΔH values vary in opposing directions. This apparent contradiction to the Badger–Bauer rule[139], also found in the intermolecular association of phenols with haloalkanes[140], is another example of the family dependence of this rule[138]. For 2-fluorophenol, 2,6-difluorophenol and tetrafluorohydroquinone, gas-phase electron diffraction studies indicate the existence of a weak IHB[111,112]. The lengths of the hydrogen bonds H···F (2.13, 2.05 and 2.02 Å, respectively) are 80% shorter than the sum of the van der Waals radii of the hydrogen and fluorine atoms.

The existence of an IHB in 2-trifluoromethylphenol has long been recognized[141] by the presence of two IR absorptions for this molecule. However, the absorption of the chelated OH group is observed at higher wavenumbers (3624 cm^{-1}) than that of the free OH group (3604 cm^{-1}).

Theoretical calculations[113,114] carried out on different *ortho*-trifluoromethylphenols **29–31** show that the chelation rings are not planar. In **29**, the OH and CF bonds are

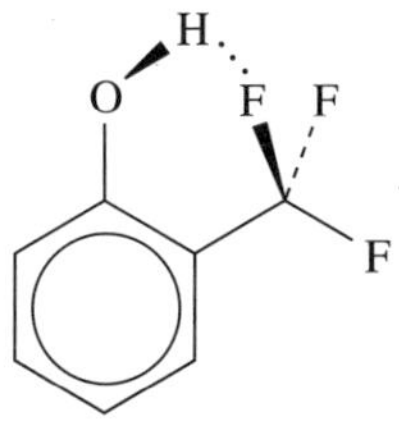

(**29**) 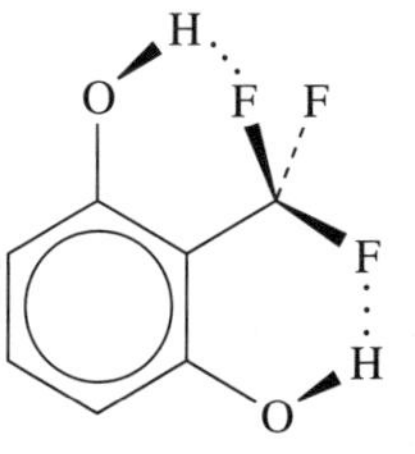   (**30**) 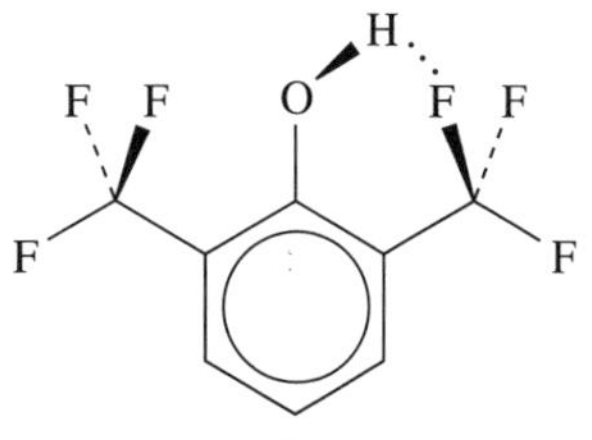   (**31**)

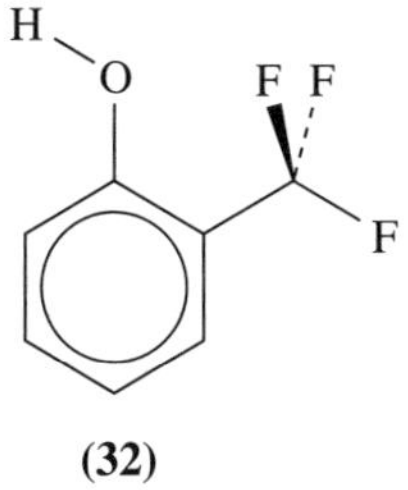

(**32**)

twisted toward the same side of the benzene ring by 14° and 48°, respectively. The calculated energy difference between **29** and **32** is 7.2 kJ mol^{-1} at the MP2/6-31G** level in favor of the chelated form. The hydrogen bond lengths in **29** and **31** are found to be very similar (1.98 Å and 1.97 Å, respectively). In **30**, the global minimum structure presents two slightly different H···F hydrogen bond lengths of 1.88 and 1.84 Å.

D. OH···OH(Me)

Langoor and van der Mass[115] analyzed the IR spectrum in the OH region of several frames containing IHBs with 2 substituents bearing a hydroxy or a methyl group. By using Fourier self-deconvolution and second derivatives spectra, they were able to assign most of the overlapping absorptions and found the following increasing order of IHB strengths for compounds 33–37:

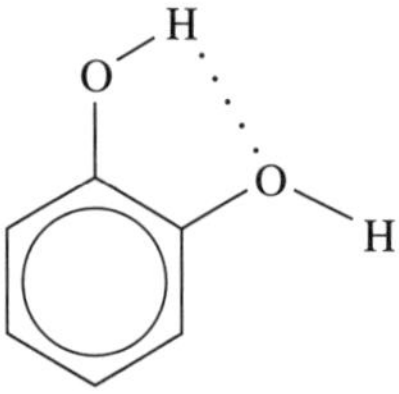

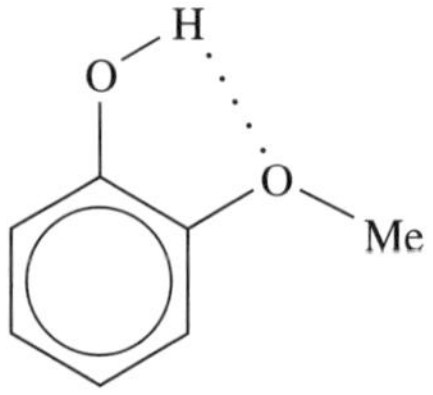

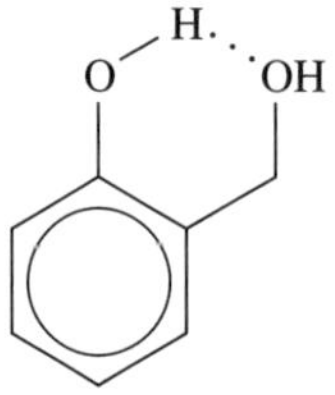

$\nu_{OH\cdots O}$ (cm^{-1}) 3570   3558   3440
                    (33)   (34)   (35)

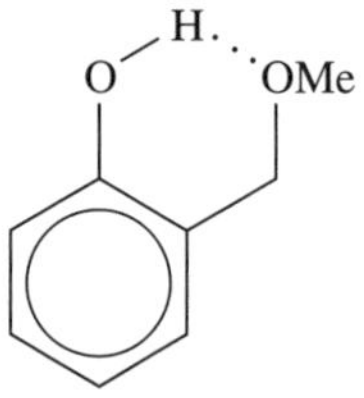

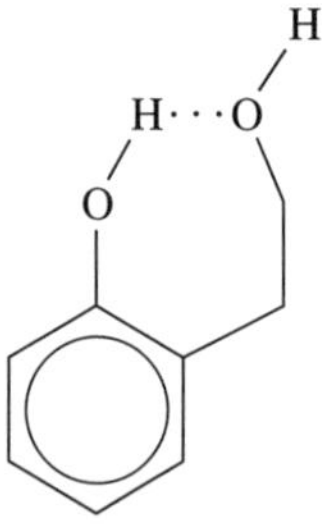

                    3409   3365
                    (36)   (37)

Ab initio quantum-chemical calculations at the DFT/6-31G** level[116] yield a value of 17.4 kJ mol^{-1} for the IHB strength of **33** relative to the optimized *anti* isomer. At the same level of calculation the IHB length is 2.12 Å. The calculated rotation barrier of the methoxy substituent in **34** is 30.5 kJ mol^{-1} [117]. This indicates an important restriction of the rotation due to the chelation by comparison with the barrier in anisole (12.5 kJ mol^{-1}).

E. OH···N

The IHBs of *ortho*-methylamino **38** and *ortho*-iminophenols **39** deserve interest since the amino and imino nitrogens are among the most basic atoms in intermolecular associations of phenols (see Section VII.A). Indeed in *ortho*-Mannich bases (**38**) an intramolecular proton transfer OH···N ⇌ $^-$O···HN$^+$ is observed[118,119] when the difference ΔpK_a between the protonated amino group and the phenol exceeds a value of about 3. Another important feature of *ortho*-Mannich bases is the bent hydrogen bond due to the non-planar IHB rings. When a nitro group is placed in the competing *ortho* position 6, the OH group forms an OH···O$_2$N IHB[120]. In spite of the significant difference in hydrogen-bond basicity between a benzylamine nitrogen and a nitrobenzene oxygen, the oxygen site is preferred since it allows the formation of a planar chelation cycle stabilized by an RAHB effect. However, the IHB with the nitrogen is sufficiently strong to rotate the dimethylamino group of **40** at the expense of its conjugation with the naphthalene ring[121]. UV-visible, IR and ^{1}H NMR spectroscopic data as well as crystal structures and theoretical calculations are available[122,123] for a series of benzalmidines **39**. These compounds form strong planar IHBs and tautomeric equilibria of phenol-imine ⇌ keto-amine may be observed in solution.

(38)   (39)   (40)

VI. PROPERTIES OF THE COMPLEXES

A. Thermodynamic Properties

Most thermodynamic studies of the equilibria between hydrogen-bonded complexes of phenols and their free component molecules have been conducted in a diluting solvent. Binary solutions of phenols (phenol[142,143], *o*-cresol[144]) in the pure base propionitrile have also been studied[142–144] by means of Raman[142] and IR[143,144] spectrometry. Factor analysis of the $\nu(C\equiv N)$ band indicates the formation of a 1 : 1 complex over a large concentration range. However, this procedure is not recommended for the determination of equilibrium constants because these exhibit a strong concentration dependence.

A variety of methods (IR, UV, NMR) have been used in attempts to determine the complexation constants K. The results should provide values for the free energies $\Delta G°$ of complexation, and, from their temperature variation, the corresponding enthalpy and entropy changes, $\Delta H°$ and $\Delta S°$. An alternative method for determining $\Delta H°$ is by direct calorimetry. For a general text on the determination of K, $\Delta H°$ and $\Delta S°$ the reader is referred to the books by Joesten and Schaad[4] and Vinogradov and Linnell[5]. In the first book[4], there is a compilation of results from the literature up to 1974. Results between 1974 and ca 1986–1988 have been treated statistically in the paper on the log K_A^H scale[33] (Section III.A). We shall focus here on the more recent literature but, before describing these results, we want to give a number of comments and caveats.

First, most methods of evaluating K have been based on the assumption that a single 1 : 1 complex is formed. However, the interaction of a phenol with a base may give rise to other complex species such as **41** for monofunctional single lone-pair bases, **42** for monofunctional two lone-pairs bases, **43a**, **43b** and **43c** for polyfuctional bases and **44** for polyphenols. In addition, self-association of phenol (Section IV) and of the base can occur. In an IR study of the complexation of 3,5-dichlorophenol with ketones and ethers[145], the use of a 1-mm optical pathlength obliged the authors to vary the phenol concentration up

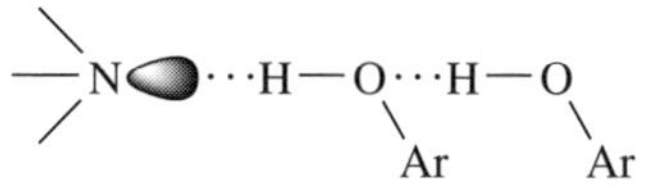

(41) \(\rangle\)N⋯H—O⋯H—O(Ar)(Ar)

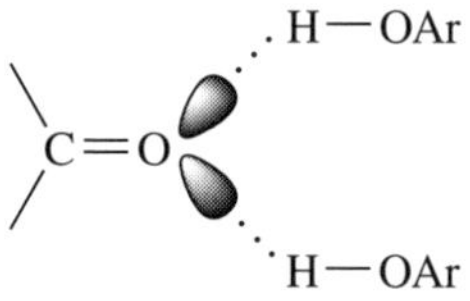

(42) C=O with two H—OAr

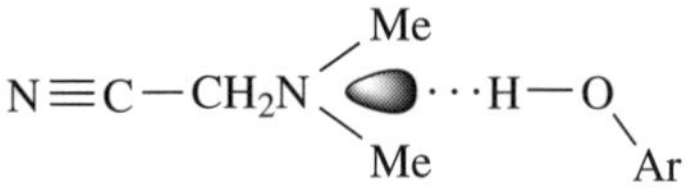

(43a) N≡C—CH₂N(Me)(Me)⋯H—O—Ar

(43b) Ar—O—H⋯N≡C—CH₂N(Me)(Me)

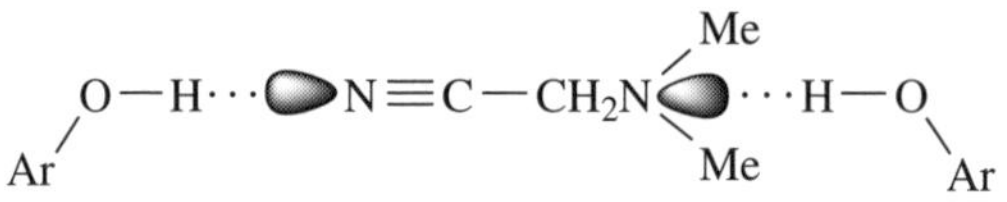

(43c) Ar—O—H⋯N≡C—CH₂N(Me)(Me)⋯H—O—Ar

(44) B⋯HOC₆H₄OH⋯B

to 0.02 dm³ mol⁻¹ and to take into account both phenol self-association and 2 : 1 complex formation in the measurement of the 1 : 1 equilibrium constants. A simpler way would have been to avoid multiple equilibria by adjusting the initial concentrations of phenol and base to the chemistry involved, i.e. to use very dilute solutions of phenol and excess base in order to minimize the phenol self-association and the formation of **41** and/or **42**. This was done for the complexation of 4-fluorophenol with ethers[19] and ketones[146] by choosing a 1-cm-pathlength cell. This cannot be done with NMR signals less sensitive to hydrogen bonding but more sensitive to solvent effects than IR spectrometry. For example, when the chemical shift of the phenolic OH proton is used to evaluate the association of phenol with nitriles and oxygen bases[147], the NMR chemical shift data suggest the presence of 1 : 1 and n : 1 phenol–base complexes, when the ratio of the phenol concentration to that of the base is high. At a low concentration ratio, the 1 : 1 complex is solvated by an enrichment of its solvation shell in base molecules. In the case of polyfunctional bases, e.g. **43**, several 1 : 1 complexes are formed, such as **43a** and **43b**, with individual thermodynamic parameters K_i and ΔH_i°. One must be aware that experimental quantities are only apparent ones (K_{app} and ΔH_{app}), which are related to individual parameters by equation 27.

$$K_{app} = \Sigma_i K_i \quad \Delta H_{app}^\circ = (\Sigma_i K_i \Delta H_i^\circ)/\Sigma_i K_i \quad (27)$$

A second caveat concerns the dependence of the numerical values of the free energies and entropies of complexation on the concentration scale used[148]. ΔH° must be calculated by applying the van't Hoff equation to K_x or K_m values, the complexation constants on the mole fraction or molal concentration scales, respectively. If one uses K_c (molar concentration), enthalpies must be corrected for the thermal expansion of the solvent.

Third, the ΔH° and ΔS° values of many hydrogen-bonded complexes have been obtained from van't Hoff plots where the temperature range ΔT was usually too small. Enthalpies and entropies calculated with $\Delta T = 10°$ for the complexes of 4-nitrophenol with amines[149] are inevitably less reliable than those calculated with $\Delta T = 78°$ for substituted phenols hydrogen-bonded to dimethylacetamide[150] or with $\Delta T = 57°$ for substituted phenols complexed with diphenyl sulfoxide[151], simply because the error in ΔH° is inversely related to ΔT.

8. Hydrogen-bonded complexes of phenols

Last, it is generally believed that a monotonic relationship exists between $\Delta H°$ and $\Delta S°$ for hydrogen-bond formation on the basis that 'a higher value of $-\Delta H$ implies stronger bonding, with a more restricted configuration in the complex, hence greater order, leading to a larger value of $-\Delta S$'[3]. A great number of such correlations have been given for related complexes in the book by Joesten and Schaad[4] on the basis of wrong statistics. Indeed, the apparently simple equation 28

$$\Delta H° = \beta \Delta S° + \text{constant} \qquad (28)$$

hides difficult statistical problems since both $\Delta H°$ and $\Delta S°$ are loaded with correlated errors when they are obtained from van't Hoff plots. Among others, Exner[152] has achieved a statistically correct treatment of equation 28, but we are unaware of its application in the field of hydrogen-bond complexation if we exclude the very recent work of Ouvrard and coworkers[138] on the complexes of 4-fluorophenol with 18 halogenoalkanes. For this system they have established the validity of the extrathermodynamic equation 28. The isoequilibrium temperature β (592 K) is determined with some uncertainty but the confidence interval (529–701 K) does not include the isoentropic relationship ($\beta \to \infty$). In contrast, the hydrogen bonding of 3- and 4-substituted phenols to dimethylacetamide[150] results in an almost isoentropic series; the $\Delta H°$ varies from -23.4 kJ mol^{-1} for 3-dimethylaminophenol to -34.3 kJ mol^{-1} for 4-nitrophenol while the extremes of $T \Delta S°$ values differ only by 0.6 kJ mol^{-1}.

We have assembled in Table 11 the recent determinations of complexation constants K, complexation enthalpies $\Delta H°$ and complexation entropies $\Delta S°$ for hydrogen bonding of phenols with various bases. When many substituted phenols have been complexed to the same base, values are given for the parent compound (unsubstituted phenol) and for the weakest and strongest hydrogen-bond donors. Many hydrogen-bond acceptors have several potential hydrogen-bonding sites; the main interaction site is written in the formula in bold type. The solvent is specified, since thermodynamic constants show a significant dependence on the nature of the solvent[153]. All results were obtained by means of IR spectrometry on the $v(OH)$ phenolic band, except for (i) studies on the $v(C\equiv N)$ band of NBu$_4^+$OCN$^-$ [154] and $v_{as}(N_3^-)$ band of NBu$_4^+$N$_3^-$, (ii) an electron spin resonance study[155], (iii) a ^{13}C NMR determination[156], (iv) the simultaneous use of an FTIR and a calorimetric method[145] and (v) a UV determination on the nitroaromatic chromophore of 3,4-dinitrophenol[157]. The logarithms of the K values are related to the pK_a of the phenols and to the Hammett substituent constants[38] of the phenolic substituent. They are not related to the pK_a of N-heterocyclic bases with two vicinal nitrogen atoms[158]. For these systems log K values are notably higher than predicted from the pK_a of the base in water. Figure 11 shows this peculiar behavior of azaaromatics where the two lone pairs are parallel or are pointing at each other. For the hydrogen-bonded complexes of the phenols with tetraalkylammonium halides[159–161], the complexation entropies are significantly smaller than with neutral bases. This has been tentatively explained[160] by the aggregation of the salts in CCl$_4$. The dimers (NR$_4^+$X$^-$)$_2$ might be separated into ion pairs by the addition of phenol. Consequently the complexation might not effect a significant variation of the number of molecular species in the solution (equation 29).

$$(NR_4^+ X^-)_2 + ArOH \rightleftharpoons NR_4^+ X^- \cdots HOAr + NR_4^+ X^- \qquad (29)$$

Hine and coworkers have measured the complexation constants of 1,8-biphenylenediol in cyclohexane[183] and of 4,5-dinitro-1,8-biphenylenediol in chloroform[184], hydrogen-bonded to various oxygen and nitrogen bases, in order to study their double hydrogen-bonding ability, i.e. the existence of bifurcated hydrogen bonds as in **45**. X-ray crystal structures of the solid complexes of 1,8-biphenylenediol with

TABLE 11. Summary of thermodynamic results for hydrogen-bonded complexes of phenols with various bases at 25 °C in different solvents

HBD	HBA		X	Solvent	$\log K_c$	$-\Delta H°$ (kJ mol^{-1})	$-\Delta S°$ (J mol^{-1} K^{-1})	$\Delta \nu$(OH) (cm^{-1})	Reference
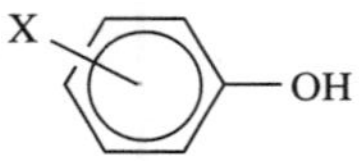 X—⟨⟩—OH	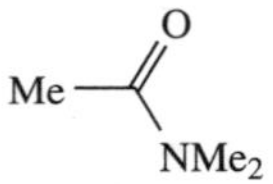 **Carbonyl bases** Me–C(=O)–NMe$_2$		3-NMe$_2$	CCl$_4$	1.86	23.85	24.9	336	150
			H		2.11	25.46	25.6	341	
			4-NO$_2$		3.40	34.87	32.4	431	
Cl—⟨⟩(OH)—Cl	R^1–C(=O)–R^2			c-C$_6$H$_{12}$					145
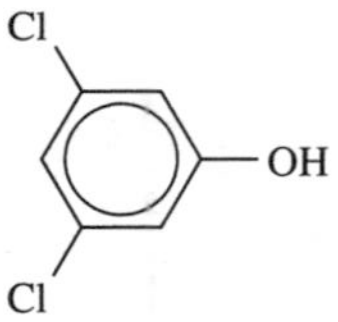	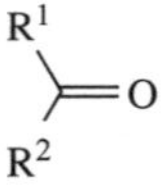 R^1	R^2							
	Me	Me			2.08	29.4	58.6	—	
	Me	Et			2.05	29.3	59.0	—	
	Me	n-Pr			2.07	28.7	56.5	—	
	Me	i-Pr			2.05	29.2	58.6	—	
	Me	n-Bu			2.07	29.0	57.7	—	
	Me	i-Bu			2.04	28.5	56.5	—	
	Me	t-Bu			2.04	29.1	58.6	—	
	Me	n-Hept			2.07	28.5	56.1	—	
	Et	Et			2.00	29.2	59.4	—	
	Et	n-Bu			2.04	28.7	57.3	—	
	n-Pr	n-Pr			2.03	28.7	57.3	—	
	i-Pr	i-Pr			2.00	28.9	58.6	—	
	n-Bu	n-Bu			2.05	28.5	56.5	—	
	t-Bu	t-Bu			1.81	28.3	60.2	—	
	n-Hex	n-Hex			2.09	28.6	55.6	—	
	Cyclopentanone				2.21	29.6	56.9	—	
	Cyclohexanone				2.25	30.2	58.2	—	

Base	X	Solvent					Ref
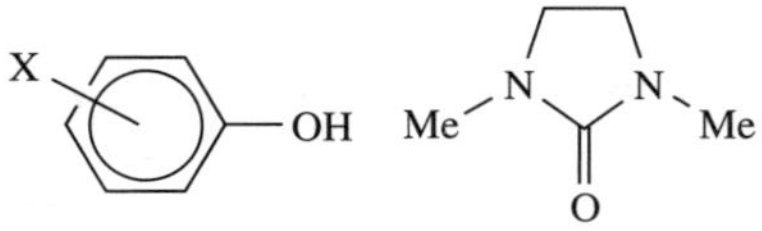 1,3-dimethyl-imidazolidin-2-one	3,4-Me$_2$	CCl$_4$	1.95	25	48	319	162
	H		2.19	27	50	333	
	3,5-Cl$_2$		3.32	36	40	438	
	3,4-Me$_2$	ClCH$_2$CH$_2$Cl	1.34	19	41	334	
	H		1.53	22	43	360	
	3,5-Cl$_2$		2.35	30	55	—	
N-methylsuccinimide	4-Me	CCl$_4$	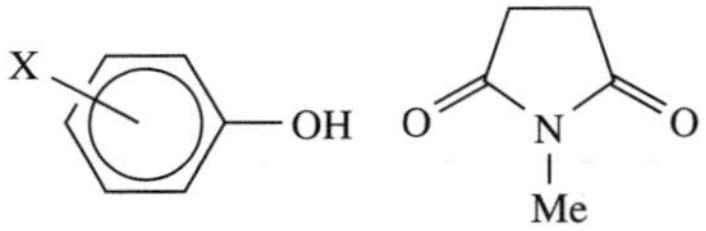 1.20	18	—	188	163
	H		1.27	17	—	193	
	4-Cl		1.46	18.7	—	215	
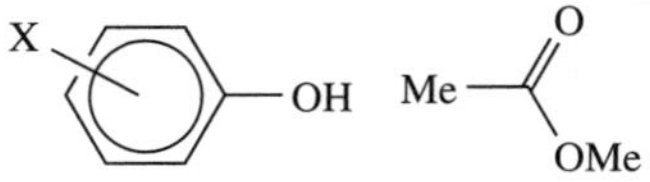 Me–C(O)–OMe	4-Me	CCl$_4$	0.67	16.2	41.6	—	164
	H		0.91	17.3	40.6	—	
	3,5-Cl$_2$		1.38	19.9	40.4	—	
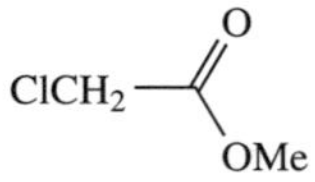 ClCH$_2$–C(O)–OMe	4-Me		0.41	14.2	39.8	—	
	H		0.68	15.4	38.7	—	
	3,5-Cl$_2$		1.13	18.0	38.7	—	
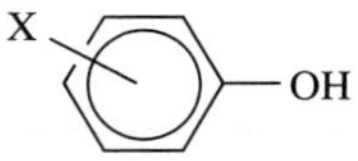 CH$_2$=CH–COOMe	4-Me	CCl$_4$	0.61	10.8	—	—	165
	H		0.73	13.4	—	—	
	3,5-Cl$_2$		1.29	14.2	—	—	

(continued overleaf)

TABLE 11. (*continued*)

HBD	HBA	X	Solvent	log K_c	$-\Delta H°$ (kJ mol^{-1})	$-\Delta S°$ (J mol^{-1} K^{-1})	$\Delta\nu$(OH) (cm^{-1})	Reference
	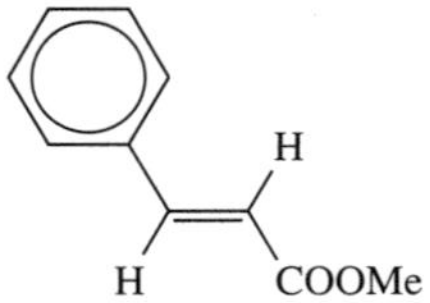 Ph-CH=C(H)(COOMe)	4-Me		0.92	12.8	—	—	
		H		0.97	13.1	—	—	
		3,5-Cl$_2$		1.48	14.6	—	—	
	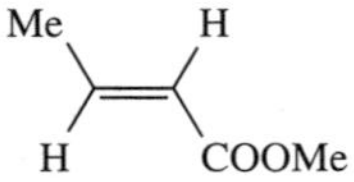 Me-C(H)=C(H)(COOMe)	4-Me		0.87	12.9	—	—	
		H		1.00	13.8	—	—	
		3,5-Cl$_2$		1.53	15.6	—	—	
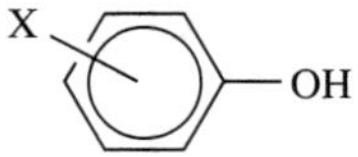 X–C$_6$H$_4$–OH	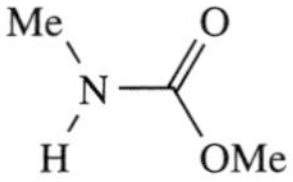 Me-N(H)-C(O)-OMe	4-OMe	CCl$_4$	1.36	22.4	—	222	166
		H		1.51	23.3	—	227	
		3,4,5-Cl$_3$		2.45	29.1	—	306	
	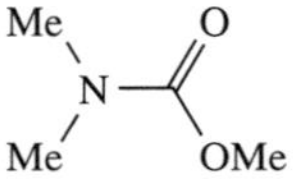 Me-N(Me)-C(O)-OMe	4-OMe		1.46	22.6	—	239	
		H		1.56	23.3	—	247	
		3,4,5-Cl$_3$		2.62	29.3	—	335	
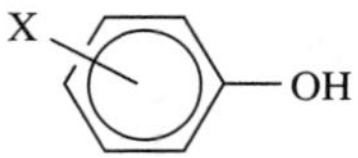 X–C$_6$H$_4$–OH	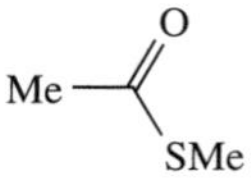 Me-C(O)-SMe	4-OMe	CCl$_4$	0.60	13.9	—	145	167
		H		0.68	14.4	—	145	
		3,4,5-Cl$_3$		1.40	17.2	—	190	

Donor	Acceptor	Solvent					Ref.
X-C6H4-OH (HCONMe2)	4-OMe	CCl4	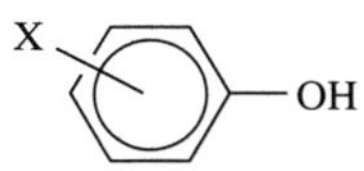 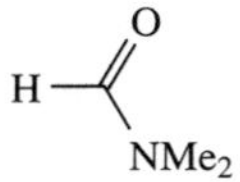 1.67	22.9	—	268	168
	H		1.77	23.9	—	297	
	3,5-Cl2		2.72	26.5	—	381	
PhOH	a (cyclohexanone)	CCl4	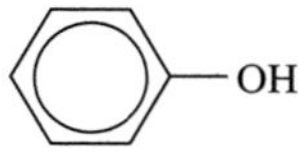 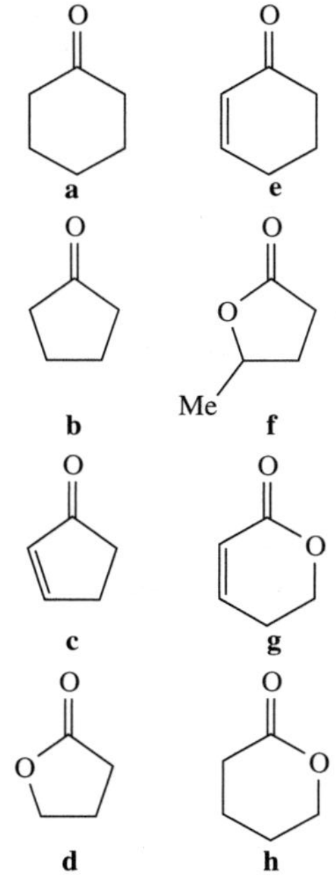 1.26	24.0	56.2	242	169
	b (cyclopentanone)		1.31	25.2	59.6	247	
	c (cyclopentenone)		1.42	26.7	62.5	272	
	d (γ-butyrolactone)		1.43	28.6	68.6	192	
	e (cyclohexenone)		1.43	29.1	70.2	252	
	f (γ-valerolactone)		1.46	29.2	69.9	197	
	g (δ-valerolactone, unsat.)		1.51	29.4	69.5	217	
	h (δ-valerolactone)		1.61	29.7	68.9	221	

(*continued overleaf*)

TABLE 11. (continued)

HBD	HBA		X	Solvent	$\log K_c$	$-\Delta H°$ (kJ mol^{-1})	$-\Delta S°$ (J mol^{-1} K^{-1})	$\Delta \nu$(OH) (cm^{-1})	Reference
3,5-dichlorophenol (Cl, Cl, OH)	**Ethers**			c-C$_6$H$_{12}$					145
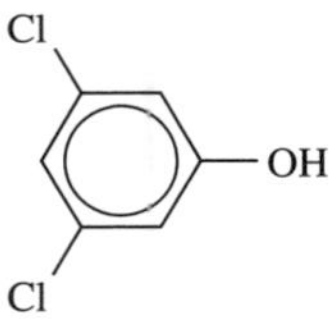	R^1	R^2							
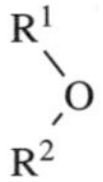	Me	t-Bu			1.96	32.0	69.9	374	
	Et	Et			1.79	29.5	64.9	342	
	n-Pr	n-Pr			1.61	28.3	64.0	351	
	i-Pr	i-Pr			1.95	31.5	68.6	362	
	n-Bu	n-Bu			1.61	28.4	64.4	357	
	i-Bu	i-Bu			1.28	27.8	69.0	353	
	n-Oct	n-Oct			1.69	29.0	65.3	359	
	n-Dec	n-Dec			1.70	29.3	65.7	359	
	Trimethylene oxide				2.32	31.6	61.5	330	
	Tetrahydrofuran				2.17	30.5	60.7	340	
	Tetrahydropyran				2.04	30.0	61.5	347	
	1,4-Dioxane				1.74	25.0	50.6	295	
4-chlorophenol (Cl, OH)				c-C$_6$H$_{12}$					170
	R^1	R^2							
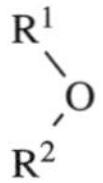	Et	Et			1.43	30.5	—	—	
	Et	n-Pr			1.35	27.2	—	—	
	Me	t-Bu			1.45	34.3	—	—	
	Et	t-Bu			1.46	35.1	—	—	
	Tetrahydrofuran				1.74	30.5	—	—	
	Tetrahydropyran				1.67	28.9	—	—	

Imines, guanidines

Phenol	Base	X	Solvent					Ref
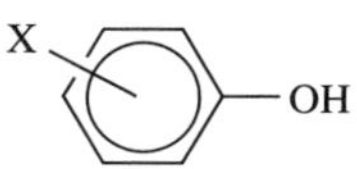 X–C6H4–OH	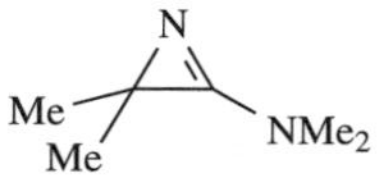 Me₂C=N–NMe₂ (Me, Me)	3,4-Me₂	CCl₄	2.06	29.7	—	433	171
		H		2.32	31.3	—	451	
		3,5-Cl₂		3.40	35.5	—	541	
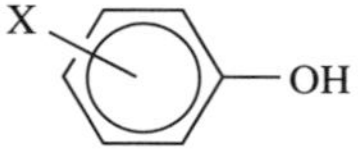 X–C6H4–OH	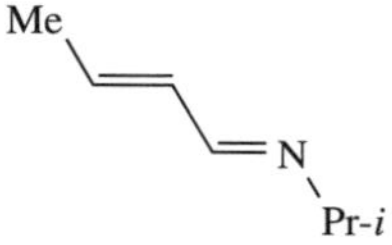 Me–CH=CH–CH=N–Pr-i	4-OMe	CCl₄	1.86	33	—	560	172
		H		1.98	33.3	—	610	
		3,4,5-Cl₃		3.19	37	—	1000	
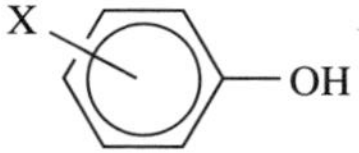 X–C6H4–OH	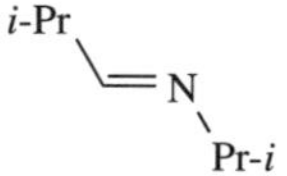 i-Pr–CH=N–Pr-i	4-OMe	CCl₄	1.55	30.4	—	—	173
		H		1.63	31.5	—	—	
		3,4-Cl₂		2.47	39.4	—	—	
X–C6H4–OH	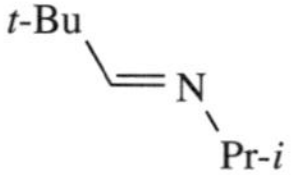 t-Bu–CH=N–Pr-i	4-OMe		1.09	33.5	—	—	
		H		1.21	34.0	—	—	
		3,4-Cl₂		2.11	37.0	—	—	

(continued overleaf)

TABLE 11. (continued)

HBD	HBA	X	Solvent	$\log K_c$	$-\Delta H°$ (kJ mol^{-1})	$-\Delta S°$ (J mol^{-1} K^{-1})	$\Delta \nu$(OH) (cm^{-1})	Reference
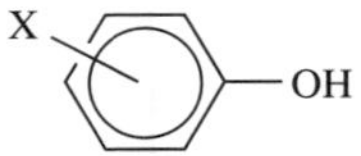 X—C$_6$H$_4$—OH	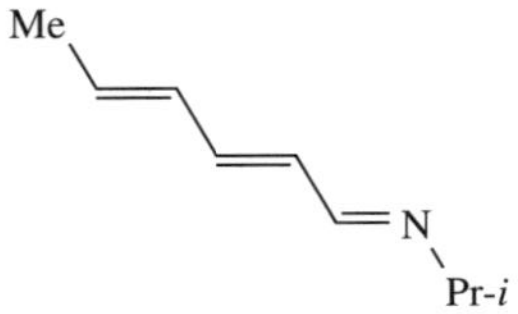 Me—CH=CH—CH=CH—CH=N—Pr-i	3,4-Me$_2$	CCl$_4$	1.74	20	34	—	174
		H		1.95	23	40	—	
		3,4-Cl$_2$		2.72	30	49	—	
	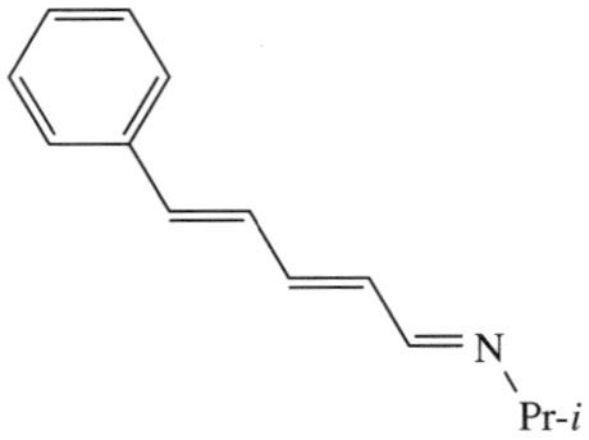 Ph—CH=CH—CH=CH—CH=N—Pr-i	3,4-Me$_2$		1.76	19	30	—	
		H		1.88	20	31	—	
		3,4-Cl$_2$		2.92	33	55	—	
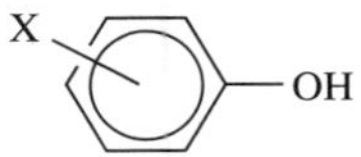 X—C$_6$H$_4$—OH	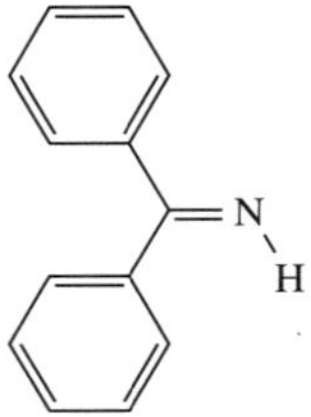 Ph$_2$C=N—H	4-OMe	CCl$_4$	1.41	28.8	66.3	—	175
		H		1.59	29.6	66.3	—	
		3,4,5-Cl$_3$		2.75	33.5	68.7	—	

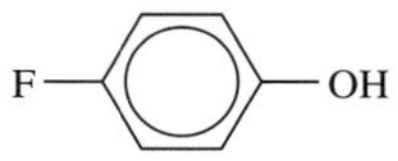

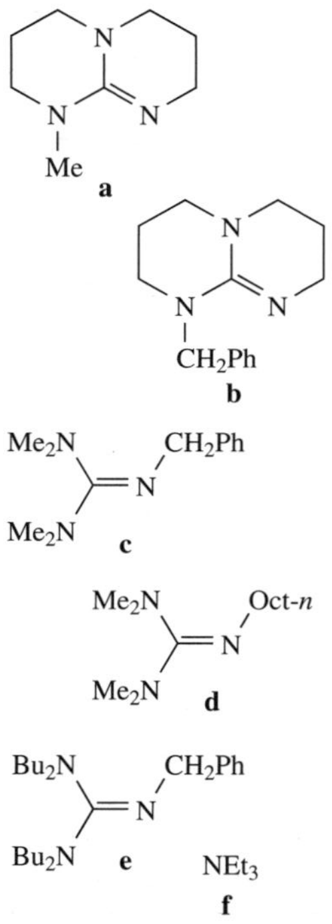

a	CDCl₃	2.03	—	—	—
b		1.89	—	—	—
c		1.81	—	—	—
d		1.68	—	—	—
e		1.64	—	—	156
f		1.43	—	—	

(continued overleaf)

TABLE 11. (continued)

HBD	HBA	X	Solvent	$\log K_c$	$-\Delta H^\circ$ (kJ mol^{-1})	$-\Delta S^\circ$ (J mol^{-1} K^{-1})	$\Delta\nu$(OH) (cm^{-1})	Reference
Anions								
 X—⟨OH⟩	NBu$_4^+$ Br$^-$	3,4-Me$_2$	CCl$_4$	2.87	18.0	—	451	159
		H		3.18	24.7	—	460	
		3-Br		3.91	43.7	—	520	
	NHept$_4^+$ I$^-$	3,4-Me$_2$		2.28	16.0	—	371	160
		H		2.47	16.4	—	385	
		3,5-Cl$_2$		3.25	20.5	—	457	
	NBu$_4^+$ Cl$^-$	3,4-Me$_2$		3.38	29.9	—	~530	
		4-Me		3.43	21.3	—	~560	
		H		3.71	35.8	—	~560	
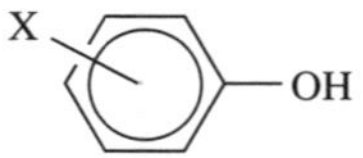 X—⟨OH⟩	NBu$_4^+$ OCN$^-$	2,6-i-Pr$_2$	CCl$_4$	2.48	24.0	—	—	154
		H		4.78	43.0	—	—	—
		4-F		5.08	45.5	—	—	—
	NBu$_4^+$ N$_3^-$	2,6-i-Pr$_2$		1.87	—	—	—	
		H		3.98	—	—	—	
		4-F		4.51	—	—	—	
Aromatic N-heterocycles								
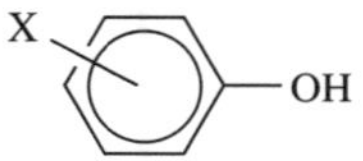 X—⟨OH⟩	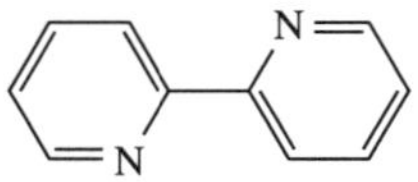 ⟨N=⟩—⟨=N⟩	3,4-Me$_2$	CCl$_4$	1.07	24.1	—	425	176
		H		1.24	25.3	60.5	440	
		3,4,5-Cl$_3$		2.10	30.8	—	550	

	X	Solvent				Ref.
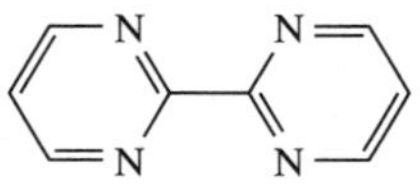 2,2'-bipyrimidine	4-OMe		2.13	28.9	—	285
	H		2.31	29.9	55.6	300
	3,4-Cl$_2$		3.19	35.0	55.8	335
1,10-phenanthroline derivatives (R^1, R^2, R^3, R^4)	H	ClCH$_2$CH$_2$Cl	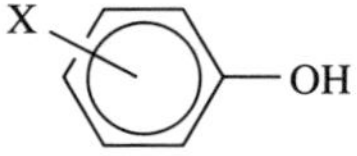 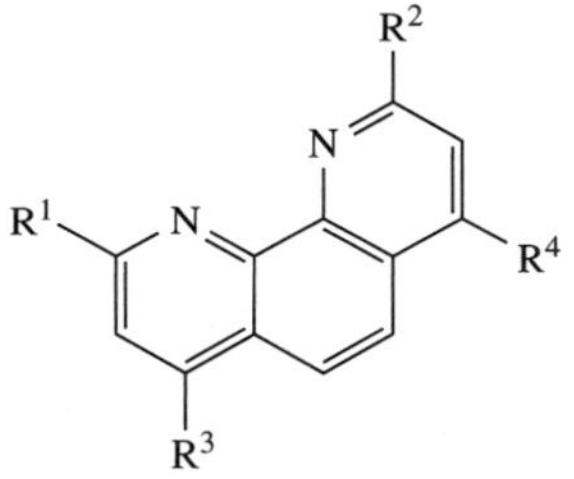 1.56	21.7	42.8	—	177
$R^1 = R^2 = H$, $R^3 = R^4 = H$	H		1.84	24.7	47.6	—	
$R^1 = R^2 = H$, $R^3 = R^4 = Ph$	H		1.67	23.6	47.2	—	
$R^1 = R^2 = Me$, $R^3 = R^4 = H$	4-OMe		1.80	22.4	39.6	—	
$R^1 = R^2 = Me$, $R^3 = R^4 = Ph$	H		2.03	24.0	41.7	—	
	3,4,5-Cl$_3$		3.32	34.8	51.0	—	
4,4'-azopyridine	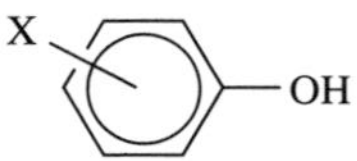 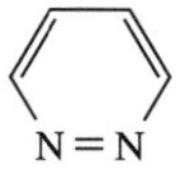 3,4-Me$_2$	CCl$_4$	1.47	24	50	382	178
	H		1.67	25	52	398	
	3-NO$_2$		2.57	32	58	470	

(continued overleaf)

TABLE 11. (continued)

HBD	HBA	X	Solvent	$\log K_c$	$-\Delta H°$ (kJ mol^{-1})	$-\Delta S°$ (J mol^{-1} K^{-1})	$\Delta \nu$(OH) (cm^{-1})	Reference
(phenol, X-substituted)	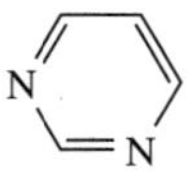 pyridine	3,4-Me$_2$		1.08	21	50	372	
		H		1.23	23	52	385	
		3,5-Cl$_2$		1.91	30	61	462	
	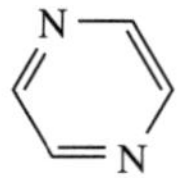 pyrazine	3,4-Me$_2$		0.93	21	52	365	
		H		1.04	22	54	371	
		3,5-Cl$_2$		1.69	27	60	450	
	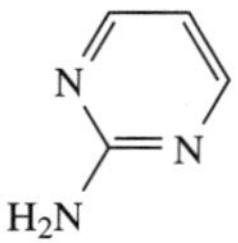 2-aminopyrimidine	3,4-Me$_2$		1.48	23	49	—	
		H		1.62	25	51	—	
		3,5-Cl$_2$		2.34	31	60	—	
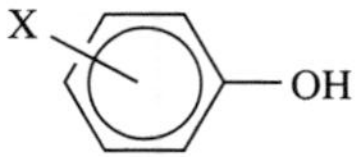	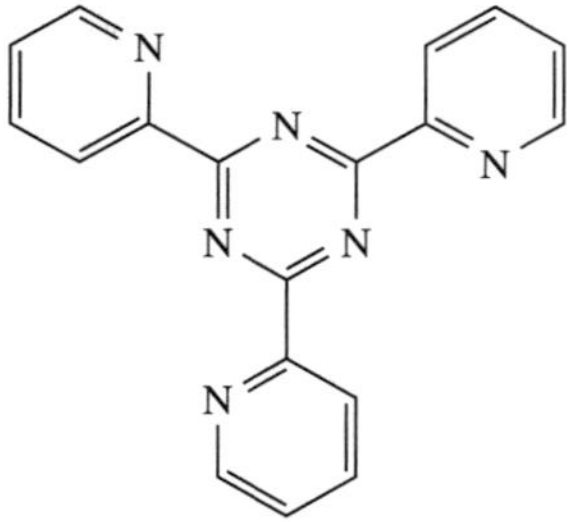 2,4,6-tri(2-pyridyl)-1,3,5-triazine	3,4-Me$_2$	ClCH$_2$CH$_2$Cl	1.46	23	—	440	158
		H		1.60	26	—	470	
		3,5-Cl$_2$		2.50	36	—	540	
	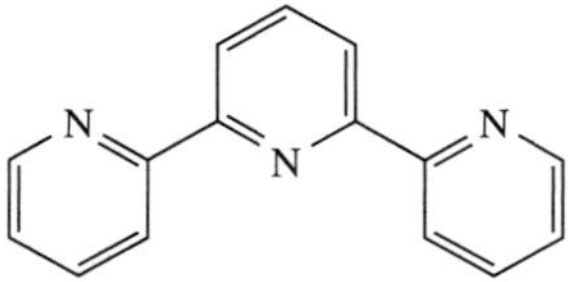 2,2′:6′,2″-terpyridine	3,4-Me$_2$		1.08	17	—	410	
		H		1.20	22	—	425	
		3,5-Cl$_2$		1.90	27	—	495	

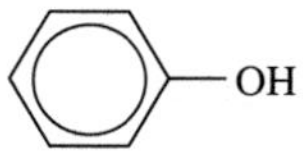

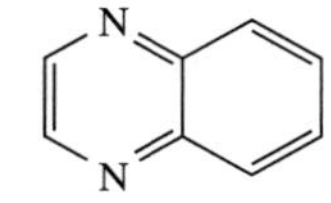

Phenol + quinoxaline (a) / 1,8-naphthyridine (b) / pyrazino[2,3-b]quinoxaline-type (c)	a	ClCH$_2$CH$_2$Cl	0.65	—	—	—	
	b		0.95	—	—	—	
	c		1.04	—	—	—	158
Phenol (X-substituted) + N-methylimidazole	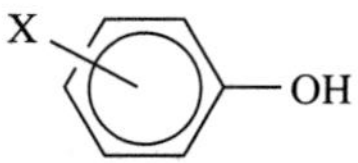 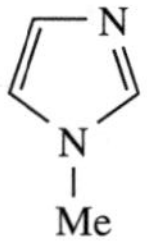 3,4-Me$_2$	CCl$_4$	2.12	—	—	466	
	H		2.37	—	—	487	179
	3,4-Cl$_2$		3.44	—	—	563	

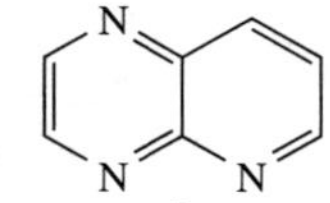

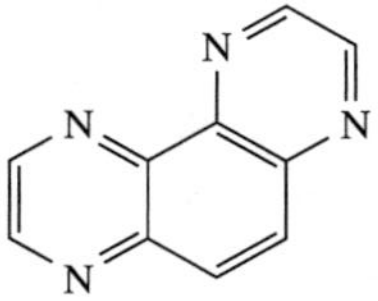

(continued overleaf)

TABLE 11. (continued)

HBD	HBA	X	Solvent	$\log K_c$	$-\Delta H°$ (kJ mol^{-1})	$-\Delta S°$ (J mol^{-1} K^{-1})	$\Delta \nu$(OH) (cm^{-1})	Reference
	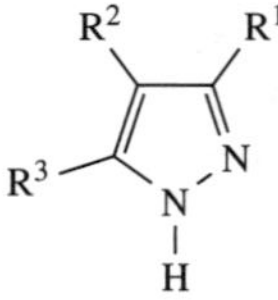 (pyrazole with R^1, R^2, R^3 substituents, NH)							
	$R^1 = R^2 = R^3 = $ Me		c-C$_6$H$_{12}$	4.20	—	—	—	157
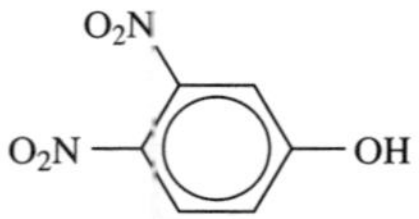	$R^1 = R^3 = $ Me, $R^2 = $ H			4.07	—	—	—	
	$R^1 = $ Me, $R^2 = R^3 = $ H			3.88	—	—	—	
	$R^1 = R^3 = $ H, $R^2 = $ Me			3.73	—	—	—	
	$R^1 = R^2 = R^3 = $ H			3.51	—	—	—	
	$R^1 = $ Me, $R^2 = $ Br, $R^3 = $ H			3.10	—	—	—	
	$R^1 = R^3 = $ H, $R^2 = $ Br			<2.94	—	—	—	
3,4-dinitrophenol (O$_2$N, O$_2$N, OH)								

Base	X	Solvent					Ref
3-aminoquinoline	3,4-Me₂	CCl₄	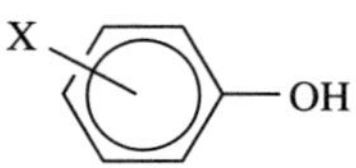 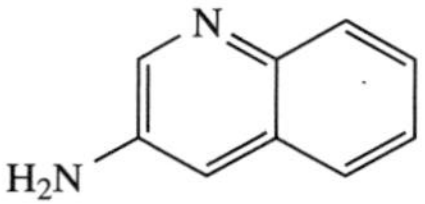 1.56	33	81	550	
	H		1.74	34	81	540	
	3,5-Cl₂		2.58	39	83	650	180
8-aminoquinoline	3,4-Me₂		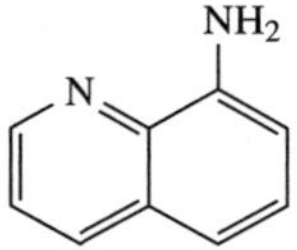 0.78	17	44	375	
	H		0.90	20	50	—	
	3,5-Cl₂		1.49	26	59	480	

Miscellaneous bases

Base	Solvent						Ref
MeCN	CCl₄	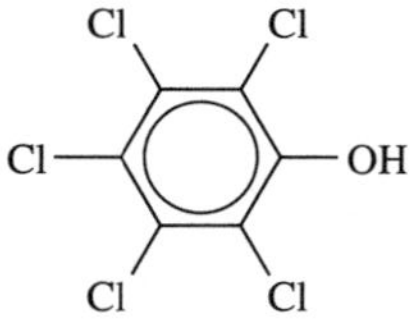 0.46	8.8	—	—		181
cyclohexanone		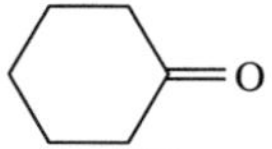 0.89	11.8	—	—		
1,4-dioxane		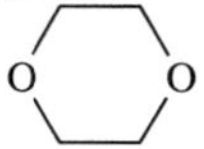 0.74	11.9	—	—		
Bu₂O		0.41	13.6	—	—		
Et₂O		0.54	13.4	—	—		
tetrahydropyran		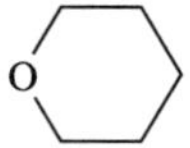 0.85	14.1	—	—		

(continued overleaf)

TABLE 11. (continued)

HBD	HBA	X	Solvent	log K_c	$-\Delta H°$ (kJ mol^{-1})	$-\Delta S°$ (J mol^{-1} K^{-1})	$\Delta\nu$(OH) (cm^{-1})	Reference
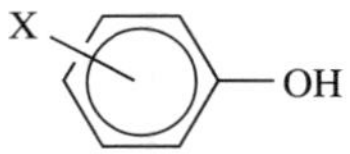 X—⟨OH⟩	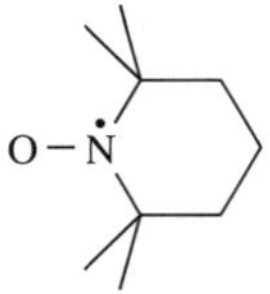 (2,2,6,6-tetramethylpiperidine-N-oxyl)	H		1.29	19.2	—	—	155
		4-F		1.45	23.6	—	—	
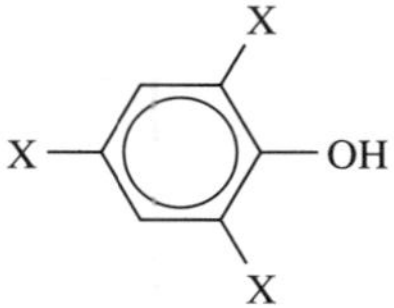 X—⟨OH⟩—X (2,6-di-X phenol)	n-Bu$_3$N	Cl	C$_2$Cl$_4$	1.36	—	—	—	28
		Br		1.58	—	—	—	
		I		1.59	—	—	—	
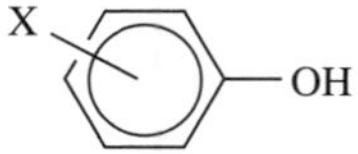 X—⟨OH⟩	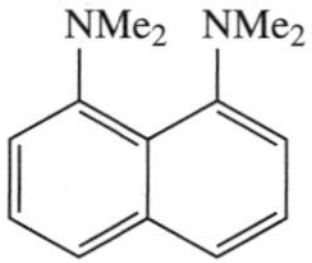 1,8-bis(NMe$_2$)naphthalene	3,4-Cl$_2$	ClCH$_2$CH$_2$Cl	0.32	—	—	—	182
		2,3,4,5,6-Cl$_5$		1.93	—	—	—	

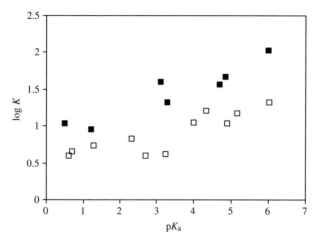

FIGURE 11. The plot of the logarithm of the equilibrium constant of phenol-azaaromatics complexes vs. the pK_a of the base shows: (□) bases with only one N atom (e.g. pyridine or quinoline), two non-vicinal N atoms (e.g. pyrimidine or pyrazine) or pyridazine where the two lone pairs are not pointing to each other; (■) bases with two parallel or pointing to each other nitrogen lone pairs (adapted from Reference 158)

hexamethylphosphoramide, 1,2,6-trimethyl-4-pyridone and 2,6-dimethyl-γ-pyrone show that double hydrogen bonds are formed to an oxygen atom in each of these bases[185]. This does not establish that **45** is always the predominant complex in solution, where the extra internal rotation possible would favor the singly hydrogen-bonded complex **46**. However, the high complexation constants of the 1,8-diols with amides, SO and PO bases, compared to those of phenols of about the same Brønsted acidity, are interpreted by the formation of two hydrogen bonds to the oxygen atom (complex **45**). There appear to be significant amounts of single hydrogen bonding (complex **46**) with the ethers and all the nitrogen bases. Unfortunately, the ΔS values are not precise enough for showing more negative values for the formation of **45** than of **46**.

X = H, 4-NO$_2$

(**45**)   (**46**)

Higher than expected complexation constants are found for the complexes of phenols with amphiprotic thioamides[186,187] and 2-aminopyridine[188]. They have been interpreted by the existence of the cyclic complexes **47** and **48**.

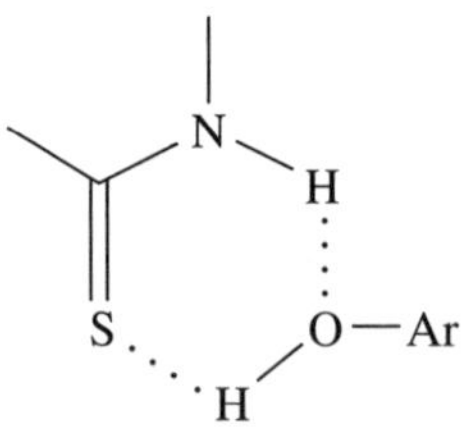

(47) 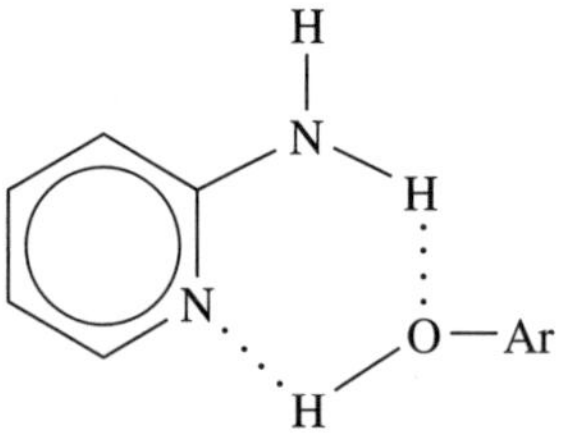   (48)

IR spectrometry shows the existence of two equilibria for the complexation of phenols with carbonyl bases in CCl$_4$ (equations 30 and 31)[146,189]. Two different 1 : 1 stereoisomeric complexes are formed: the planar bent n complex **a** and the planar bidentate linear n complex **b**. The complex **b** has also been given the structure **c** (out-of-plane π complex). Experimentally, an overall complexation constant is determined which is the sum of the individual complexation constants K_n and K_π for each stereoisomeric complex. Massat and coworkers[190] have proposed an IR method for evaluating the constants K_n and K_π of phenol–alkylketone complexes. They have shown that the n vs. π complex competition depends on the alkyl branching, measured by n_α, the number of methyls alpha to the carbonyl, and on the phenol acidity, measured by pK_a (equations 32 and 33).

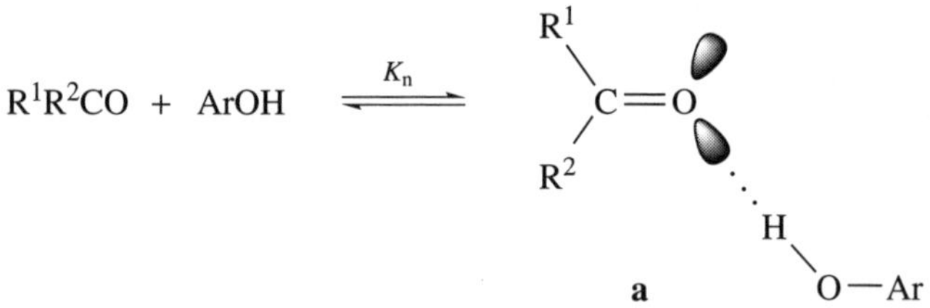

$$R^1R^2CO + ArOH \underset{}{\overset{K_n}{\rightleftharpoons}} \quad \textbf{a} \tag{30}$$

$$R^1R^2CO + ArOH \underset{}{\overset{K_\pi}{\rightleftharpoons}} \quad \textbf{b} \quad \text{or} \quad \textbf{c} \tag{31}$$

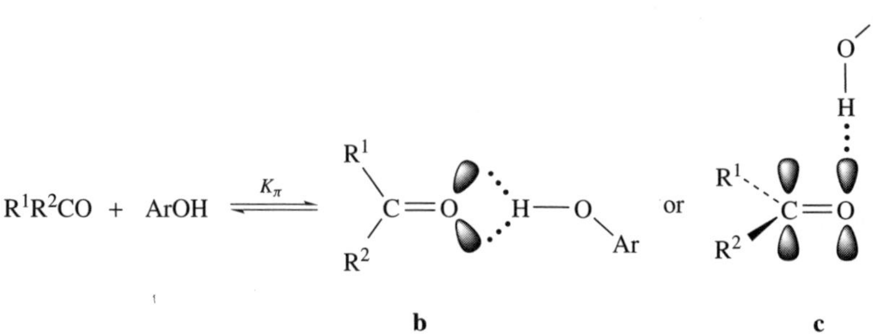

$$\log K_n = 5.3 - 0.44\, \text{p}K_a - 0.08\, n_\alpha \tag{32}$$

$$\log K_\pi = 5.21 - 0.48\, \text{p}K_a \tag{33}$$

Two geometries are also possible in the hydrogen bonding of 4-fluorophenol to epoxides, peroxides and sterically hindered ethers[19]. The most stable complex has geometry **49**, and the least stable one the trigonal geometry **50**.

(49)   (50)

B. Binding Energy D_e and Dissociation Energy D_o

The complexation enthalpies discussed above contain not only the electronic contribution to the interaction energies, but also contributions arising from translational, rotational and vibrational motions of the nuclei. If measured in a solvent or a matrix, they also contain a solvation term. We shall designate D_e the electronic portion of the interaction energy, i.e. the dissociation energy from the equilibrium geometry, or binding energy, or hydrogen-bond energy; D_o would refer to this same quantity, after correction for zero-point vibrational energies. For a stable complex, ΔH (or ΔE after a ΔpV correction) is negative, signifying its formation to be exothermic, while $D_e(D_o)$ is taken as positive since it refers to the energy required to dissociate the complex. A precise knowledge of the binding energies of hydrogen bonds is crucial for the theoretical understanding of this molecular interaction. However, even for small hydrogen-bonded complexes, precise experimental data on hydrogen-bond binding energies are very scarce.

Accurate hydrogen-bond energies were determined in the gas phase for complexes between 1-naphthol or 1-naphthol-d_3 (D at C2, C4, O) and H_2O, CH_3OH, NH_3 and ND_3 using the stimulated emission pumping-resonant two-photon ionization spectroscopy technique in supersonic jets[191]. In these complexes 1-naphthol acts as the hydrogen-bond donor. The dissociation energies (kJ mol^{-1}) obtained for the S_0 electronic ground state are $D_0 = 24.34 \pm 0.83$ for 1-naphthol·H_2O, 31.64 ± 1.63 for 1-naphthol·CH_3OH, 32.07 ± 0.06 for 1-naphthol·NH_3 and 33.51 ± 0.17 for 1-naphthol-d_3·ND_3. Adding the spectral red-shift of the complex relative to the free naphthol yields a dissociation energy in the S_1 first-excited state that is approximately 8% higher. Clearly, 1-naphthol is a stronger hydrogen-bond donor than H_2O, leading to a hydrogen bond with H_2O that is approximately a factor of two stronger than in $H_2O \cdot H_2O$, the water dimer. The larger dissociation energy for the 1-naphthol·CH_3OH than for the 1-naphthol·H_2O complex can be attributed to dispersive interactions between the 1-naphthol moiety and the CH_3 group. Comparing the 1-naphthol·H_2O and 1-naphthol·NH_3 complexes, it can be seen that the hydrogen bond to the stronger hydrogen-bond acceptor NH_3 is 8 kJ mol^{-1} stronger than for H_2O. All these data agree with the pK_{HB} values measuring the hydrogen-bond basicity of H_2O (0.65), CH_3OH (0.82) and NH_3 (1.74) (Section VII.A).

With recent advances in computer technology, it is now possible to carry out *ab initio* calculations on relatively large molecules and to obtain reliable hydrogen-bond energies. However, the large size of phenols has, so far, restricted the calculations to small-size hydrogen-bond acceptor molecules (e.g. H_2O and NH_3). As an illustration of the agreement between the experimental and calculated energies, Table 12 contains a comparison of the D_0 values of 1-naphthol·B (B = H_2O, CH_3OH, NH_3, ND_3). Calculations were performed[191] at the MP2 level on SCF optimized structures with Pople's 6-31G(d, p) standard basis set. These *ab initio* calculations give good values for the dissociation energy, except for methanol. The computed binding energies D_e of the complexes phenol·H_2O and phenol·NH_3[192] (where phenol is the hydrogen-bond donor), at different levels of theory (B3LYP, MP, MCPF) and different Dunning's basis sets (D95*,

TABLE 12. Calculated (MP2/6-31G(d,p) // SCF/6-31G(d,p)) vs. experimental dissociation energies D_0 (kJ mol^{-1})

	1-naphthol·H$_2$O	1-naphthol·CH$_3$OH	1-naphthol·NH$_3$	1-naphthol-d$_3$·ND$_3$
SCF	24.38	26.38	29.55	31.27
Δ(MP2)[a]	8.21	10.60	10.92	10.92
BSSE (SCF)[b]	−4.68	−5.40	−4.47	−4.47
BSSE (MP2)[b]	−4.31	−5.44	−4.44	−4.44
CP (SCF + MP2)[c]	23.60	26.14	31.56	33.28
Difference[d]	−3%	−17%	−2%	−1%

[a] Correlation energy contribution to D_e.
[b] Basis set superposition error.
[c] Counterpoise corrected value.
[d] Difference in percent between calculated and experimental D_0 values.

TABLE 13. Calculated binding energies D_e (kJ mol^{-1}) of C$_6$H$_5$OH···OH$_2$ and C$_6$H$_5$OH···NH$_3$[a]

Level of theory[b,c]	C$_6$H$_5$OH·H$_2$O	C$_6$H$_5$OH·NH$_3$
MP2(D95*) // MP2(D95*)	38.91 (29.71)	50.21 (35.98)
MP4(D95*) // MP2(D95*)	38.07	48.53
B3LYP(D95*) // B3LYP(D95*)	36.40 (31.80)	47.28 (39.75)
MP2(D95++**) // MP2(D95*)	37.66 (25.52)	46.02 (33.89)
B3LYP(D95++**) // B3LYP(D95++**)	31.38 (26.78)	40.58 (35.98)
MCPF(D95++**) // B3LYP(D95++**)	34.73	41.84

[a] In parentheses are counterpoise corrected binding energies.
[b] MPn methods include electron correlation.
[c] B3LYP: Three-parameter hybrid density functional method; the MCPF method is an extension of the singles and doubles configuration interaction approach; D95++** is a Dunning double-zeta plus polarization and diffuse functions quality basis set; D95* is the D95++** basis set in which basis set functions and the polarization functions on the hydrogen atoms have not been included.

D95 + +**), are given in Table 13 in order to show the sensitivity of the hydrogen-bond energies to the method of calculation. The best estimates of D_e at the B3LYP (MCPF) levels are 31.38 (34.73) and 40.58 (41.84) kJ mol^{-1} for the phenol–water and phenol–ammonia complexes, respectively. As a matter of fact, the counterpoise uncorrected B3LYP (MCPF) values can be quite accurate, since the basis set superposition error can partially compensate for the lack of dispersion energy evaluation. Including the B3LYP zero-point correction, the B3LYP (MCPF) dissociation energies D_0 of the phenol–water and phenol–ammonia complexes are 23.43 (26.78) and 32.64 (33.89) kJ mol^{-1}, respectively.

For the hydrogen-bonded phenol·oxirane complex[193], the performance of the SCF and BLYP density functional methods was compared, using the Pople's 6-31G(d, p) and 6-311 + +G(d, p) basis sets. The MP2/6-31G(d, p) hydrogen-bond energy is D_e = 28.9 kJ mol^{-1} and the dissociation energy is D_0 = 23.8 kJ mol^{-1}.

The dissociation energy D_0 of the phenol·methanol complex (phenol as hydrogen-bond donor), calculated with MP2 and B3LYP using the rather small 6-31G(d, p) basis set, is 21.78 and 22.91 kJ mol^{-1}, respectively[194]. The calculated D_0 shows good agreement with the experimental value (25.56 ± 0.75 kJ mol^{-1})[195].

C. Geometry of Phenol Hydrogen Bonds

Whereas energetic data in the gas phase, to which the calculations directly pertain, are hard to obtain, many geometries have been evaluated to high precision, not only in the gas phase but also in the solid adducts.

In the solid state, neutron diffraction studies are the most useful since they allow one to determine the precise location of the hydrogen-bonded hydrogen in the O−H···B moiety. For illustration, we have selected the adduct of 2-methylpyridine with pentachlorophenol[196]. In the crystal, the molecular OH···N, and not the ionic O⁻···H−N⁺, adduct is formed as shown in Figure 12. The length of the hydrogen bond, 1.535(7) Å, is much shorter than the sum of van der Waals radii of H and N (2.6 Å) but still longer than the sum of the covalent radii (1.0 Å). The hydrogen bond is not perfectly linear (OH···N = 167.5(6)° instead of 180°) but is directed almost exactly at the Nsp^2 lone pair (HNC4′ = 172.7(3)°). The elongation of the O−H bond is very large. If bond orders of the O−H and H···N bonds are calculated from the distances, using the Pauling rule and the bond valence model[197], one obtains $S_{OH} = 0.71$ and $S_{H···N} = 0.24$. Although the rule of bond order conservation ($\Sigma S = 1$ around H) is not ideally fulfilled, the quarter of a valence unit of the hydrogen bond means that the incipient proton transfer has already reached an advanced stage in the hydrogen-bonded complex.

Many X-ray diffraction crystal structures of solid phenol adducts have been published and can be found in the CSD database[32]. Several are given in Chapter 2 of the present volume. The reader can search in the CSD for either well-defined hydrogen-bonded complexes or perform a statistical survey of ArOH···B contacts. Examples of the first search are:

(i) 1 : 1 Hydrogen-bonded complexes of pentafluorophenol with Ph$_3$AsO[198], Ph$_3$PO[199], 4,4′-bis(dimethylamino)benzophenone (Michler's ketone)[200] and a 2 : 1 complex of pentafluorophenol with 1,4-dioxane[201]. In the complex involving Michler's ketone, the phenol is hydrogen-bonded to the carbonyl group and not to a nitrogen atom. On the contrary, in the adduct of CF$_3$SO$_2$H with the Michler's ketone, one of the two nitrogen atoms has been protonated by the acid. This illustrates that for polyfunctional hydrogen-bond acceptors the hydrogen-bonding site is not always the protonation site (Section VI.D).

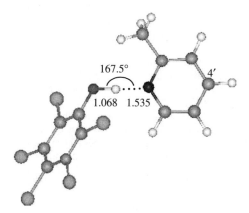

FIGURE 12. Structure, at 30 K, of the solid 1 : 1 adduct 2-methylpyridine-pentachlorophenol viewed on the pyridine plane

(ii) 1 : 1 Hydrogen-bonded complexes of pentachlorophenol with 3-cyanopyridine[202] and 4-acetylpyridine[203]. The comparison of the hydrogen-bond lengths indicates a stronger hydrogen bond to 4-acetylpyridine, in agreement with the pK_{HB} scale[188]. In these adducts, hydrogen bonding occurs at Nsp^2, the protonation site. However, in CCl$_4$ solution, the Nsp of 3-cyanopyridine and the carbonyl of 4-acetylpyridine are also secondary hydrogen-bond acceptor sites[188].

(iii) 1 : 1 Double-hydrogen-bonded adducts of 1,8-biphenylenediol and related compounds with hexamethylphosphoric triamide, 2,6-dimethyl-4-pyrone and 1,2,6-trimethyl-4-pyridone[185,204]. For each of these complexes both OH groups of the diol **45** are hydrogen-bonded to the same basic O atom at the base. In the same vein, in the 1 : 2 complex of 2,6-dimethylpyridine-N-oxide with pentachlorophenol[205] the oxygen atom of the N-oxide group accepts hydrogen bonds from two molecules of pentachlorophenol. This property of oxygen atoms of C=O, P=O and N → O groups to accept simultaneously several hydrogen bonds constitutes a major difference between oxygen and nitrogen atoms as hydrogen-bond acceptors.

(iv) 1 : 1 Complexes of 2,9-dimethyl-1,10-phenanthroline and resorcinol[206] and 1,10-phenanthroline with 1,1'-binaphthyl-2,2'-diol[207] (**51**). These are examples of three-centered hydrogen bonds where an OH group binds in a bifurcated manner to the two N atoms.

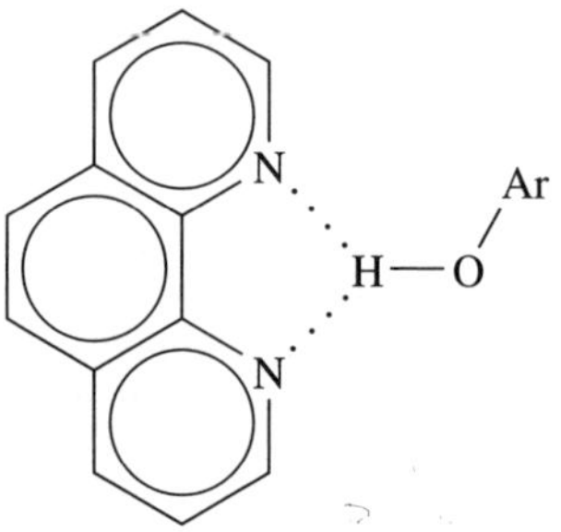

(**51**)

(v) Complexes of phenols, bisphenols and trisphenols with polyamines. These molecules are attractive candidates as building blocks for supramolecular chemistry. Complexes of bisphenols and trisphenols with hexamethylenetetramine generate strings, multiple helices and chains of rings[208]. One-dimensional chains, two-dimensional bilayers and a three-dimensional diamondoid architecture are formed in hydrogen-bonded adducts of 4,4'-biphenol with 1,4-diazabicyclo[2.2.2]octane and 1,2-diaminoethane[209]. Hexamethylene tetramine is a four-fold acceptor of OH···N hydrogen bonds in its 1 : 2 adduct with 2,2'-biphenol[210] (**52**).

The second type of CSD search relies on the fact that a large proportion of crystal structures involve molecules with HBA and/or HBD functional groups. Thus it is possible to perform statistical surveys of hydrogen-bond geometries, directed to specific classes of hydrogen-bond complexes, e.g. the complexes of phenols with nitriles or those with primary amines. Statistical methods lead to averaged radial and angular parameters of the hydrogen bond. These methods are of vital importance because the hydrogen-bond geometry is easily deformed by other interactions in the crystal. If a sufficient number of structures is examined, chemically significant trends may be observed in the averaged data. Table 14 summarizes the results obtained for ArOH···Nsp[211], ArOH···Nsp^2 [212] and

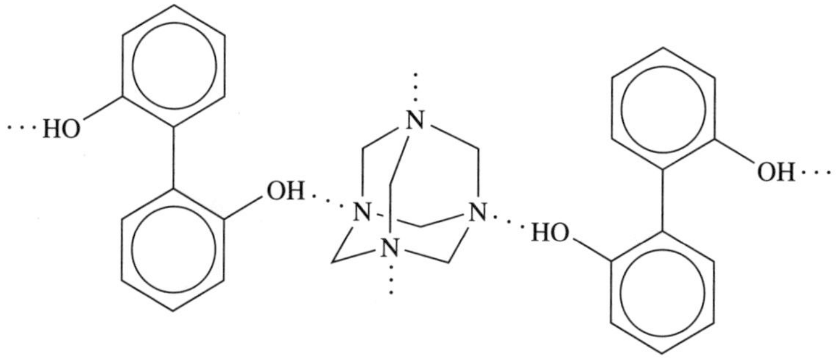

(52)

TABLE 14. Hydrogen-bond lengths d (Å) and angles θ (deg) for phenol complexes ArOH$\cdots$N

HB acceptor	n^a	d(N$\cdots$H)	d(NO)	θ(NHO)	Reference
Nsp (nitriles)	25	1.99	2.87	154	211
Nsp^2	29	1.90	2.76	162	212
Nsp^3 (amines)					
Primary	4	1.78	2.75	170	213
Secondary	16	1.75	2.71	166	213
Tertiary	64	1.83	2.77	161	213

^{a}Number of hydrogen-bonded contacts.

ArOH$\cdots$Nsp^{3} [213] hydrogen bonds. These results indicate that hydrogen bonds are shorter and more linear according to the basicity order: Nsp^3 > Nsp^2 > Nsp. In the family of amines, steric effects are possibly responsible for the longer and less linear hydrogen bonds in tertiary than in primary and secondary amines.

In the gas phase, the structure of the phenol·water complex has been obtained by Gerhards and coworkers[214] and by Berden and coworkers[215] from the fully rotationally resolved spectrum of the $S_0 \to S_1$ origin. Phenol acts as the hydrogen-bond donor, with water oxygen in the plane of the ring and water hydrogens above and below this plane, as shown in Figure 13. In the S_0 ground state, the O—O separation in the hydrogen bond is 2.93 Å and the deviation from linearity is 6.7°. A shorter O—O distance (2.81 Å) but a greater deviation from linearity (14°) is found for the phenol·methanol complex, the structure of which could be determined by rotationally resolved laser-induced fluorescence spectroscopy[216].

The geometries of phenol–NH$_3$[192,217], phenol–(H$_2$O$_2$)$_2$[218], phenol–(H$_2$O)$_3$[219], phenol–(H$_2$O)$_4$[220], phenol–oxirane[193], phenol–HCOOH[221], phenol–(HCOOH)$_2$[221], phenol–CH$_3$COOH[222] and phenol–(CH$_3$COOH)$_2$[222] have also been obtained *in vacuo* by *ab initio* calculations. The structures of phenol–(H$_2$O)$_2$ and phenol–(H$_2$O)$_3$ correspond to cyclic water dimer and tetramer, respectively. The replacement of one of the water molecules by phenol causes no fundamental changes in the geometries. The 'reaction' of phenol with the cyclic formic acid dimer **53** (equation 34) shows that the gain in binding energy by the insertion of a phenol molecule into the cyclic dimer and the formation of an extra hydrogen bond overcompensates for the break of a hydrogen bond in the cyclic dimer **53**.

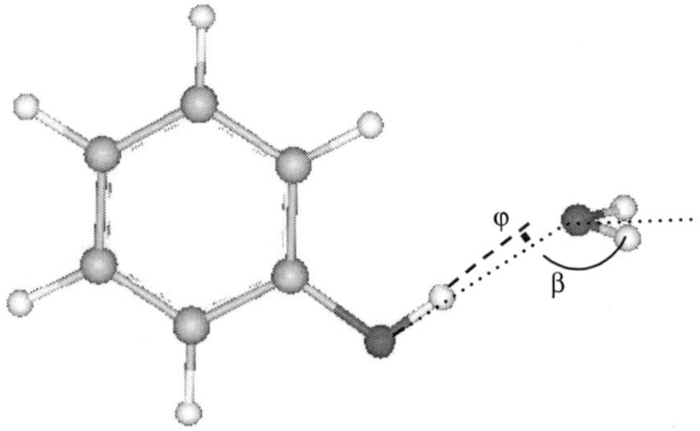

FIGURE 13. Trans-linear structure of phenol–H_2O: $R(OO) = 2.93$ Å. Linearity $\varphi = 6.7°$. Directionality $\beta = 144.5°$

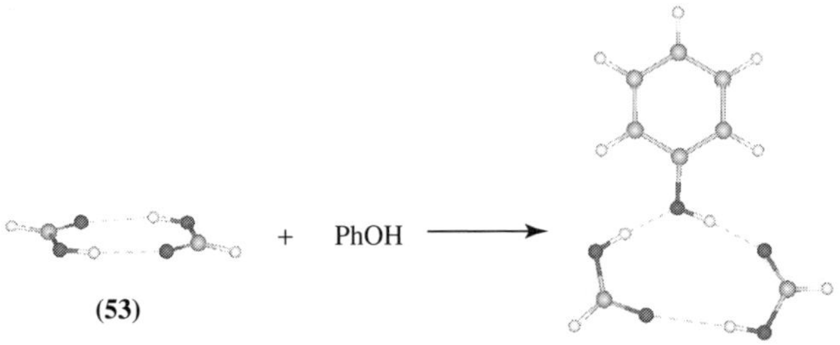

(53) + PhOH ⟶ (34)

This tendency to allow insertion of a phenol molecule is lower for acetic acid since two isomers for phenol–$(CH_3COOH)_2$ are observed, the stabilization energies of phenol inserted in **(54)** and attached to **(55)** $(CH_3COOH)_2$ being comparable.

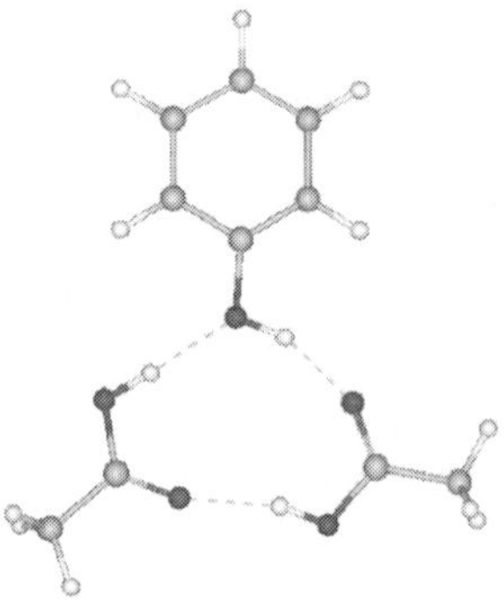

(54) 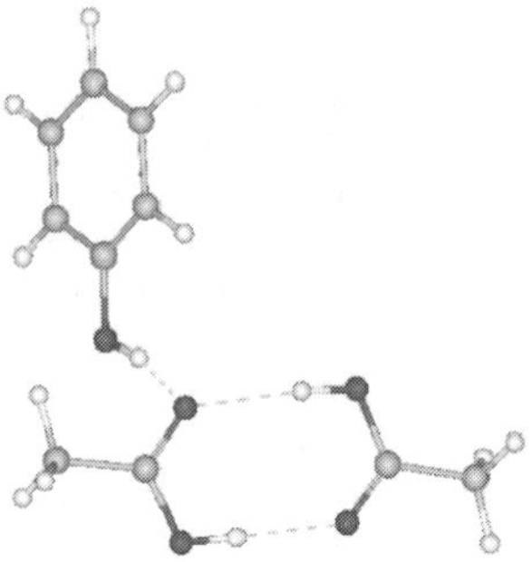 (55)

D. Hydrogen-bonding Site(s)

The majority of organic molecules are characterized by more than one potential HBA site. The site(s) of hydrogen bonding can be determined by various experimental methods

(IR, NMR, X-ray diffraction), theoretical calculations and comparison with one-site models. This can be illustrated on the hydrogen-bond complexes of phenols with progesterone (**56**)[223]. This molecule bears two potential HBA groups corresponding to the oxygens of $C_3=O$ and $C_{20}=O$. In the complex with 4-fluorophenol in CCl_4 solution, the existence of two 1 : 1 hydrogen-bond complexes is shown by the shift to lower wavenumbers of both infrared carbonyl bands. By comparison to the complexes with the models isophorone (**57**) and i-PrCOMe (**58**), it is found that *ca* 80% of the phenol molecules are hydrogen-bonded to $C_3=O$. In the same vein, the complex to O_3 is more stable by 3.6 kJ mol^{-1} on the enthalpic scale, in agreement with theoretical calculations. In the solid state, the X-ray structure of a 1 : 1 progesterone·resorcinol complex[224] also shows that both carbonyl groups accept hydrogen bonds from resorcinol and that the $C_3=O \cdots HO$ hydrogen bond is shorter (stronger) by 0.04 Å than the $C_{20}=O \cdots HO$ bond. Other examples of complexes of phenols with polysite molecules, many of biological interest, are given below.

(**56**)   (**57**)   (**58**)

For phenols of pK_a ranging from 10.3 to 4.5, $OH \cdots O=C$ hydrogen bonds are formed with 3-methyl-4-pyrimidone (**59**). With picric acid ($pK_a = 0.4$) protonation occurs at the N_1 nitrogen atom. For phenols of intermediate pK_a values, there is no preferred site of interaction, both $ArOH \cdots O=C$ and $NH^+ \cdots O^-Ar$ bonds being formed in solution[225].

In a comparative study of complexation enthalpies of phenols with enamino and amino ketones, it is suggested that, unlike the saturated base **60** where the complexation involves the nitrogen atom, hydrogen bonding to the push-pull compound **61** mainly takes place on the carbonyl group[226]. However, when the amino nitrogen and the carbonyl group are separated by only one CH_2 group (**62**), the two sites are hydrogen-bonded to phenols[227].

Phenols ($pK_a = 10.2 - 7.7$) are hydrogen-bonded to the oxygen atom of N,N-diethylnicotinamide (**63**)[228]. Thus the hydrogen-bonding site is not the preferred site of protonation in aqueous solution which is the nitrogen atom of the pyridine ring. This is also the case for the methylated derivative of cytosine **64** and for 1-methyl-2-pyrimidone (**65**), where hydrogen bonding occurs at the oxygen atom while protonation takes place on N_3. In contrast, both protonation and hydrogen bonding occurs on the O_4 oxygen of 1,3-dimethyluracil (**66**)[229].

In the complexes of phenols with the Schiff base (**67**) the hydrogen-bonding site seems governed by the accessibility of the lone pair, which is markedly higher for the Nsp than for the Nsp^2 nitrogen atom[230]. In the same way, in the complexes of phenols with **68–70**, steric factors seems important for the preferred hydrogen-bonding site(s). These are: (i) the N_1 and N_7 atoms for the purine (**68**) complexes[231], (ii) mainly the N_3 atom for the adenine (**69**) complexes[231] and (iii) the oxygen atom for the di-2-pyridyl diketone (**70**) complexes[232].

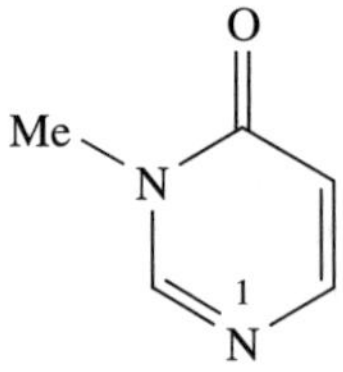

Push-pull and steric effects might explain why phenols are hydrogen-bonded to the C_6=O and C_8=O functions of the methyl derivative **71** of uric acid[233], and to C_6=O and N_7 of N,N-1,9-tetramethylguanine (**72**)[234]. In this field of carbonyl vs. Nsp^2 competition, the hydrogen bonds between metyrapone (**73**) and phenols are predominantly formed on the nitrogen atom of ring **A**[235].

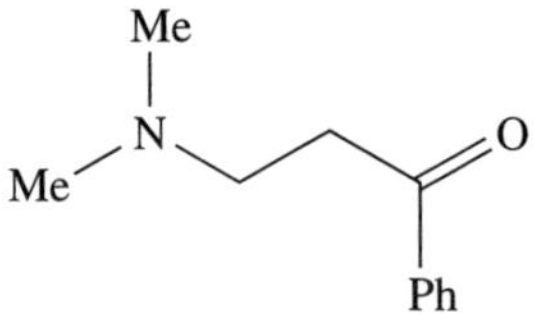

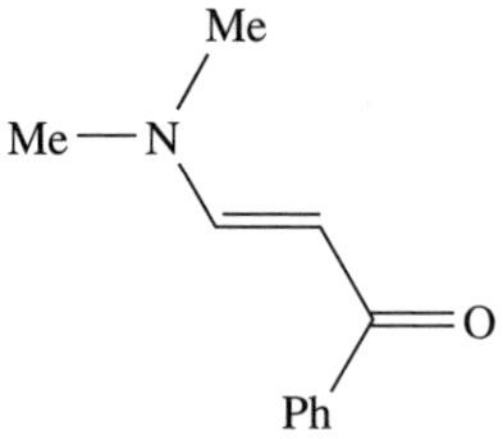

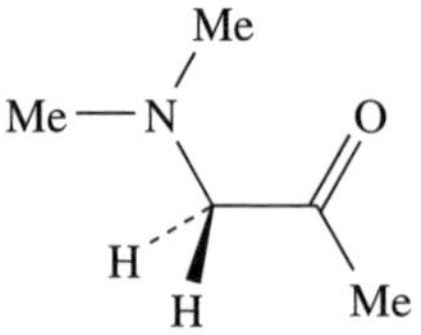

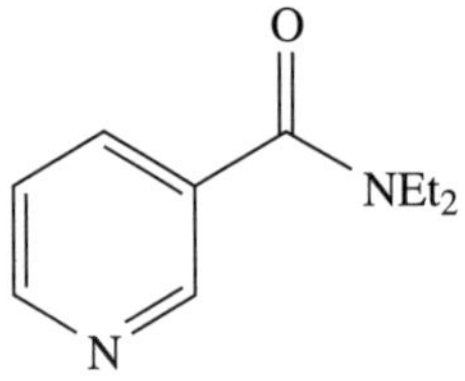

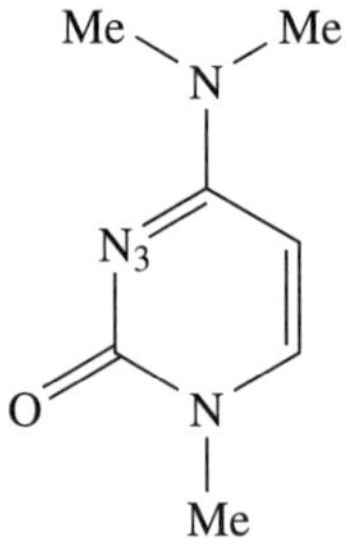

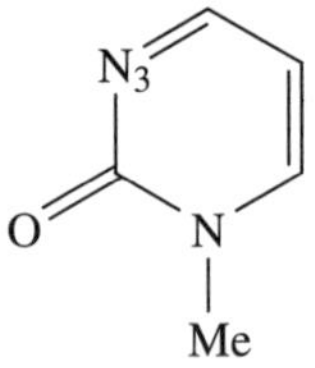

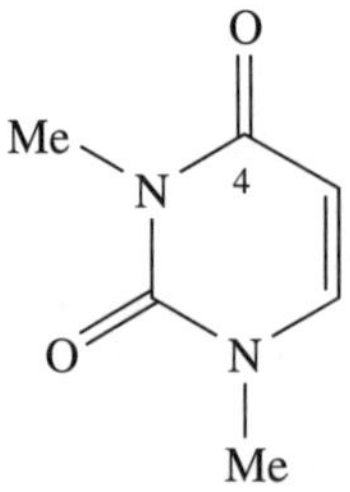

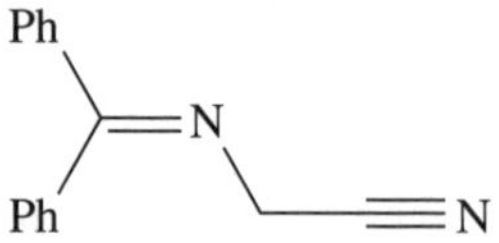

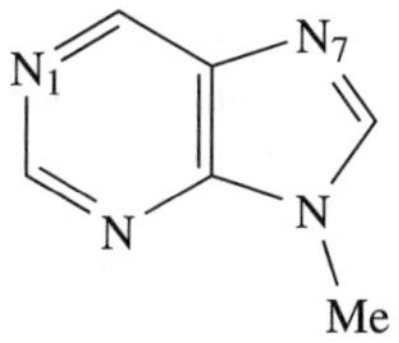

(68) 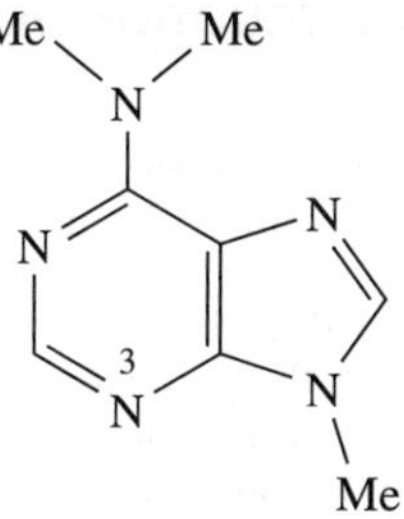 (69) 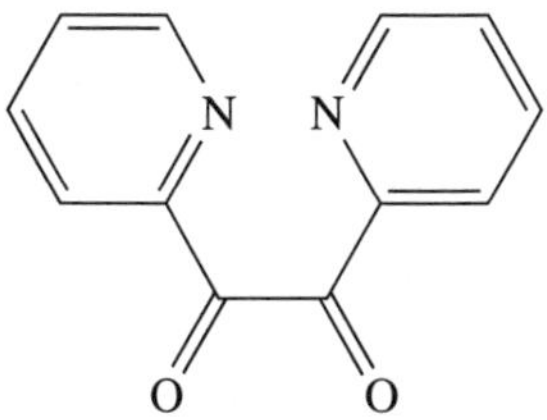 (70) 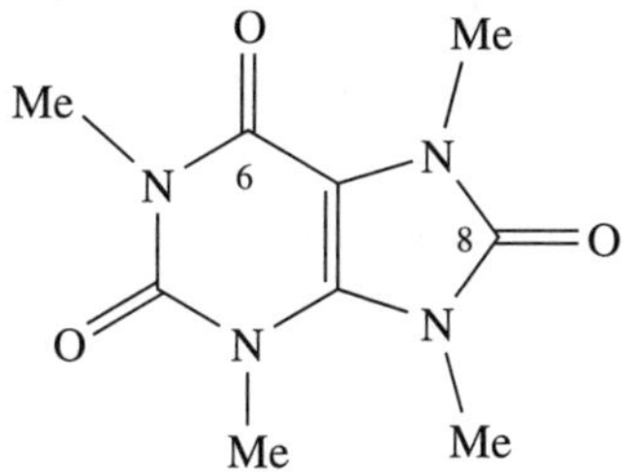 (71) 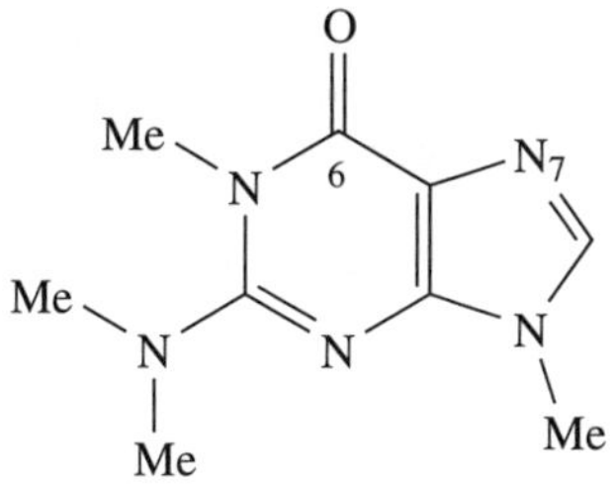 (72) 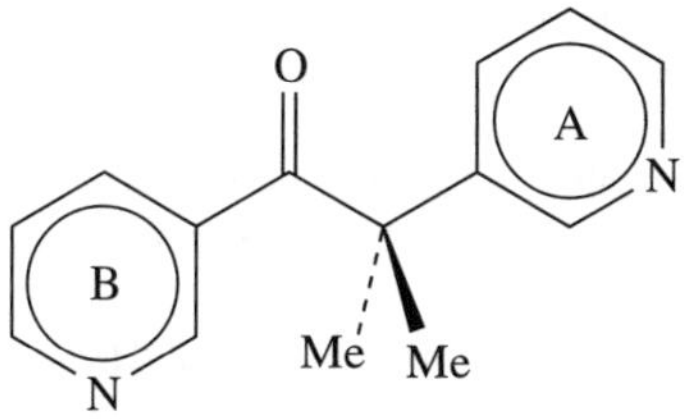 (73) 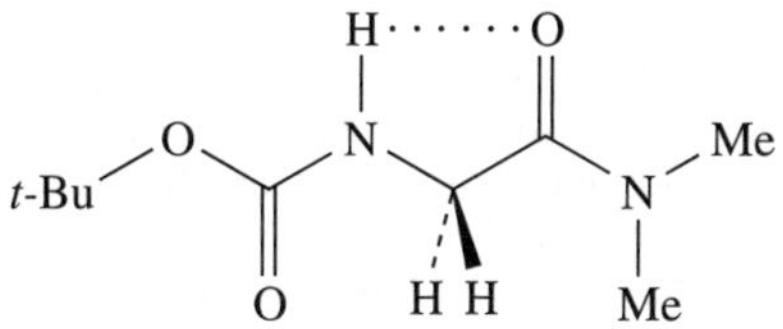 (74)

The hydrogen-bonded complexes of phenols with the model dipeptide **74** have been investigated[236]. Complexation occurs at both the amide and urethane carbonyl groups. About 45% of the complexes are formed on the urethane functions, almost independent of the Brønsted acidity of the phenols. When phenols are attached to the amide group, the intramolecular hydrogen bond seems to be broken.

VII. CONSTRUCTION OF HYDROGEN-BOND BASICITY SCALES FROM PHENOLS

For technical reasons, phenols are convenient reference hydrogen-bond donors for hydrogen-bonding studies. We present below their use for constructing thermodynamic and spectroscopic scales of hydrogen-bond basicity. These scales are either solute scales when the phenol and the base are dissolved in an inert solvent, or solvent scales when the phenol is studied in the pure base. In the latter case, methods such as the solvatochromic comparison method or the calorimetric pure base method have been developed to unravel the hydrogen-bond contribution to the overall solvent effect.

A. Thermodynamic Scales of Hydrogen-bond Basicity

Since the work of Gurka and Taft[20] and Arnett and coworkers[237], 4-fluorophenol has proved to be an excellent reference hydrogen-bond donor for the establishment of a thermodynamic hydrogen-bond basicity scale of organic bases B. This solute scale, denoted by pK_{HB}[21], is defined as the logarithm of the formation constant K of the 1 : 1 hydrogen-bonded complex 4-FC$_6$H$_4$OH$\cdots$B in CCl$_4$ at 25 °C (equations 35–37). The choice of these standard conditions allows the accurate determination of K over a wide basicity range, by measuring equilibrium concentrations from various properties such as the ^{19}F NMR shifts[20], the absorbance of the OH stretching IR band[237] at 3614 cm^{-1} or calorimetric determination of the heat of reaction[237]. The absorbance of the UV band caused by the $\pi \rightarrow \pi^*$ transition at 281 nm can also be used[238]. Fifty-five equilibrium constants were determined by ^{19}F NMR with values ranging from Et$_2$S ($pK_{HB} = 0.11$) to (Me$_2$N)$_3$PO ($pK_{HB} = 3.56$). pK_{HB} values for 20 additional bases were further reported[153]. The study[153] of reaction 35 in several solvents of relative permittivity ranging from 2.02 (c-C$_6$H$_{12}$) to 10.36 (1,2-dichloroethane) shows that linear free-energy relationships (log K in a given solvent vs. pK_{HB} in CCl$_4$) are obeyed by oxygen and Nsp bases. However, Nsp^2 and Nsp^3 bases gain strength relative to oxygen bases as the solvent reaction field rises, probably because of an increase in the extent of proton sharing in hydrogen-bonded complexes permitted by the action of polar solvents.

$$B + 4\text{-FC}_6\text{H}_4\text{OH} \rightleftharpoons 4\text{-FC}_6\text{H}_4\text{OH}\cdots B \tag{35}$$

$$K(\text{dm}^3\,\text{mol}^{-1}) = [4\text{-FC}_6\text{H}_4\text{OH}\cdots B]/[B][4\text{-FC}_6\text{H}_4\text{OH}] \tag{36}$$

$$pK_{HB} = \log_{10} K \tag{37}$$

Few further studies on the pK_{HB} scale were reported between 1972 and 1988, when Laurence, Berthelot and coworkers began to extend systematically the pK_{HB} scale to various families of organic bases. The results were published in a series of papers[18,19,23,146,186,188,239–256] referenced in chronological order in Table 15. These papers give the chemist a database for a range of HBA strengths and a variety of functionalities not previously approached. In Table 16, we have selected a number of pK_{HB} values among the ca 1,000 bases now available. The lowest published K value for reaction 35 is 0.14 dm^3 mol^{-1} ($pK_{HB} = -0.85$)[252] for the very weak π base 2,3-dimethylbut-2-ene. The highest published K values are 4570 dm^3 mol^{-1} ($pK_{HB} = 3.66$) for the neutral base Ph$_3$AsO[238] and 120,000 dm^3 mol^{-1} ($pK_{HB} = 5.08$)[154] for the tetrabutylammonium cyanate ion pair Bu$_4$N$^+$OCN$^-$. Thus, at present, the stability of 4-fluorophenol hydrogen-bonded complexes extends over a range of 6 pK units corresponding to a 35 kJ mol^{-1} Gibbs energy range.

8. Hydrogen-bonded complexes of phenols

TABLE 15. Hydrogen-bonding basicity scale constructed from 4-fluorophenol

Base family	pK_{HB} range	Reference	Base family	pK_{HB} range	Reference
Amidines	1.28 to 3.14	239	Cyanamidate	3.24	250
Water, alcohols and phenols	−0.96 to 1.27	18	Thioamides and thioureas	0.30 to 2.29	186
Acetamidines, benzamidines	0.99 to 2.72	240	Nitramines and nitramidates	0.82 to 1.91	251
Iminologous compounds	1.23 to 2.10	241	π bases (aromatic, ethylenic)	−0.85 to 0.02	252
Formamidines	0.60 to 2.75	242	Chelated compounds	0.09 to 2.48	23
Amides, ureas and lactams	0.75 to 2.79	243	2,6-Di-t-butylpyridine	−0.54	253
Nitriles	−0.26 to 2.24	244	Sulfonyl bases	0.80 to 2.90	254
Super-basic nitriles	1.56 to 2.24	245	Ketones, aldehydes	−0.06 to 2.92	146
Amidines	0.83 to 2.22	246	Pyridines	−0.49 to 2.93	188
Amidates	2.70 to 3.56	247	Ethers, peroxides	−0.53 to 1.98	19
Nitro bases	0.13 to 1.55	248	Primary amines	0.67 to 2.62	255
Esters, lactones and carbonates	0.08 to 2.09	249	Haloalkanes	−0.70 to 0.26	256

TABLE 16. Hydrogen-bonding acceptor strengths of neutral bases

Base	HBA site(s)	pK_{HB}[a]	Base	HBA site(s)	pK_{HB}[a]
Cyclohexene	π	−0.82	N-Methylthioacetamide	CS	1.14
Methyl iodide	I	−0.47	Acetone	CO	1.18
Benzene	π	−0.50	N,N-Dimethylbenzenesulfonamide	SO_2	1.19
p-Xylene	π	−0.30	Tetrahydrofuran	Osp^3	1.28
Butyl bromide	Br	−0.30	γ-Butyrolactone	CO	1.32
Naphthalene	2 π	−0.26	Pyrimidine	2 Nsp^2	1.37
Cyclohexyl chloride	Cl	−0.23	Cyclohexanone	CO	1.39
1-Hexyne	π	−0.22	1-Diethylamino-2-nitroethene	NO_2	1.58
Phenol	π + O	−0.07	Diethylcyanamide	Nsp	1.63
Octyl fluoride	F	0.02	N,N'-Diphenylacetamidine	Nsp^2	1.65
Diphenylamine	2 π + N	0.08	Ammonia	Nsp^3	1.68
Anisole	π + O	0.11	Morpholine	O + Nsp^3	1.86
Pyrrole	Nsp^2	0.15	Pyridine	Nsp^2	1.86
Nitromethane	NO_2	0.27	N-Methylformamide	CO	1.96
Tetrahydrothiophene	Ssp^3	0.30	Triethylamine	Nsp^3	1.99
Methyl salicylate	CO	0.32	Methylamine	Nsp^3	2.15
Aniline	π + N	0.56	N-Methylacetamide	CO	2.30
Water	Osp^3	0.64	Piperidine	Nsp^3	2.35
Ethyl formate	CO	0.66	Tetramethylurea	CO	2.44
2,2,2-Trifluoroethylamine	Nsp^3	0.67	Dimethylacetamide	CO	2.44
Benzaldehyde	CO	0.78	2,6-Dimethyl-γ-pyrone	CO	2.50
1,3,5-Triazine	3 Nsp^2	0.80	1-Methyl-2-pyridone	CO	2.57
Diethyl sulfate	SO_2	0.80	Dimethyl sulfoxide	SO	2.58
Methanol	Osp^3	0.82	Quinuclidine	Nsp^3	2.63
Diethyl carbonate	CO	0.88	Pyridine N-oxide	NO	2.70
Acetonitrile	Nsp	0.91	N-Methylimidazole	Nsp^2	2.72
Ethyl benzoate	CO	0.94	4-N,N-Dimethylaminopyridine	Nsp^2	2.80
Methyl acetate	CO	1.00	Triphenylphosphine oxide	PO	3.16
Diethyl ether	Osp^3	1.01	Tetramethylguanidine	Nsp^2	3.21
1,4-Dioxane	$2Osp^3$	1.03	Hexamethylphosphoramide	PO	3.56
Dimethyltrifluoroacetamide	CO	1.04			

[a] The pK_{HB} values are determined by FTIR spectrometry. Estimated precision: 0.02 pK unit.

In 1989 a log K_β solute hydrogen-bond basicity scale was constructed for 91 bases[37]. It was scaled to 4-nitrophenol as hydrogen-bond donor in 1,1,1-trichloroethane (equations 38 and 39) and was explicitly targeted to the needs of the medicinal chemist. To this end, measurements were made in 1,1,1-trichloroethane, a solvent considered a better model for real biological phases than the non-polar tetrachloromethane. In addition, data are given for molecules of special interest to the medicinal chemist, for example many heterocycles never before investigated. The log K_β and pK_{HB} scales have a similar meaning and it is not unreasonable to find a fair correspondence between 24 common values (equation 40).

$$B + 4\text{-}O_2NC_6H_4OH \rightleftharpoons 4\text{-}O_2NC_6H_4OH \cdots B \tag{38}$$

$$K_\beta (\text{dm}^3 \text{mol}^{-1}) = [4\text{-}O_2NC_6H_4OH \cdots B]/[B][4\text{-}O_2NC_6H_4OH] \tag{39}$$

$$\log K_\beta = 1.27 \text{p} K_{HB} + 0.11 \tag{40}$$

$$n = 24, \quad r = 0.995, \quad s = 0.08$$

In contrast to the good agreement generally found between hydrogen-bonding complexation constants[4,33,257], there is a serious dearth of reliable hydrogen-bond enthalpies. Discrepancies amounting to 5–10 kJ mol^{-1} are often found[4] between the results obtained by different workers studying the same system by the same or different methods. For example, the results collected in Table 17 of sixteen determinations of the phenol·pyridine system vary from -20.9 to -31.8 kJ mol^{-1}. In view of the fact that most hydrogen-bond enthalpies for neutral hydrogen-bond donors and acceptors fall between -10 to -40 kJ mol^{-1}, these discrepancies seriously reduce the usefulness of such measurements.

TABLE 17. Enthalpies (kJ mol^{-1}) for complexation of phenol to pyridine in CCl$_4$

$-\Delta H°$	Method[a]	Reference
20.9	VH	M. Tsuboi, *J. Chem. Soc. Japan, Chem. Sect.*, **72**, 146 (1951).
20.9	VH	N. Fuson, P. Pineau and M. L. Josien, *J. Chim. Phys.*, **55**, 454 (1958).
24.5	VH	V. Sara, J. Moravec and M. Horak, *Collect. Czech. Chem. Comm.*, **44**, 148 (1979).
27.2	VH	J. Rubin and G. S. Panson, *J. Phys. Chem.*, **69**, 3089 (1965).
27.2	VH	H. Dunken and H. Fritzche, *Z. Chem.*, **1**, 249 (1961).
27.2	VH	M. Goethals, K. Platteborze and Th. Zeegers-Huyskens, *Spectrochim. Acta*, **48**, *Part A*, 671 (1992).
27.4	CAL	D. Neerink and L. Lamberts, *Bull. Soc. Chim. Belg.*, **75**, 473 (1966).
28.5	CAL	J. N. Spencer, J. C. Andrefsky, A. Grushow, J. Naghdi, L. M. Patti and J. F. Trader, *J. Phys. Chem.*, **91**, 1673 (1987).
28.6	VH	J. Juffernbruch and H. H. Perkampus, *Spectrochim. Acta, Part A*, **36**, 485 (1980).
29.3	VH	T. Gramstad, *Acta Chem. Scand.*, **16**, 807 (1962).
29.3	VH	R. J. Bishop and L. E. Sutton, *J. Chem. Soc.*, 6100 (1964).
29.3	VH	F. Cruege, G. Girault, S. Constal, J. Lascombe and P. Rumpf, *Bull. Soc. Chim. Fr.*, 3889 (1970).
29.3	CAL	E. M. Arnett, L. Joris, E. J. Mitchell, T. S. S. R. Murty, T. M. Gorie and P. v. R. Schleyer, *J. Am. Chem. Soc.*, **92**, 2365 (1970).
29.7	VH	K. R. Bhaskar and S. Singh, *Spectrochim. Acta, Part A*, **23**, 1155 (1967).
31.4	VH	Ch. Venkat Rama Rao, C. Jacob and A. K. Chaudra, *J. Chem. Soc., Faraday Trans. 1*, **78**, 3025 (1982).
31.8	CAL	J. Mullens, J. Yperman, J. P. François and L. C. van Poucke, *J. Phys. Chem.*, **89**, 2937 (1985).

[a]VH denotes van't Hoff equation and CAL denotes calorimetric method.

In determinations employing a variation of equilibrium constant with temperature, difficulties arise mainly from the use of a too restricted range of temperature variation, while values of $\Delta H°$ determined calorimetrically depend strongly on the reliability of the equilibrium constant. Arnett and coworkers[237] have proposed a pure-base method to avoid the need for accurate equilibrium constants. In this method the base is used as the solvent and the heat produced by van der Waals interactions is corrected by a model compound. Arnett and coworkers[237] used 4-fluorophenol as the hydrogen-bond donor and 4-fluoroanisole as the model compound. A selection of their results[237,258] on the enthalpy of hydrogen bonding of 4-fluorophenol to various bases is collected in Table 18. Enthalpies vary from 5.1 kJ mol^{-1} for the weakest complex with benzene to 39.7 kJ mol^{-1} for the strongest complex with quinuclidine. They constitute a solvent basicity scale that, however, differs little from a solute scale measured in dilute CCl$_4$.

B. UV, NMR and IR Spectroscopic Scales

The sensitivity of the A-H stretching infrared frequency to hydrogen-bond formation is well known[4]. The frequency shift, Δv, is generally represented as the difference between the stretching frequency for the monomeric A-H in an 'inert' solvent and the lowered stretching frequency for A-H$\cdots$B in the same 'inert' solvent. Koppel and Paju[259] have suggested that the phenolic OH shift (equation 41) can be used as a solute hydrogen-bonding basicity scale and have collected literature results for ca 200 bases. Δv(OH) values vary from 14 cm^{-1} for the very weak chloro base CHCl$_3$ to 727 cm^{-1} for the strong nitrogen base N-methylpiperidine. Many of these values must, however, be considered with caution because of (i) their variation with base concentration[260], (ii) overlap with the v(CH) bands and (iii) the great breadth and complicated shape of the v(OH$\cdots$B) band[261]. In fact, phenolic shifts are mainly recommended for measuring the basicity of weak bases as shown for alcohols[18], nitriles[244], nitro bases[248], ethylenic, acetylenic and aromatic π bases[252], sulfonyl bases[254], ethers[19] and haloalkanes[256]. For stronger bases, such as pyridines or amines, methanolic shifts are preferable[262]. IR OH frequency shifts are useful values for predicting hydrogen-bond enthalpies. In 1937, Badger and Bauer[139] proposed that a linear relationship exists between the enthalpy of the hydrogen bond and

TABLE 18. Enthalpies of complexation (kJ mol^{-1}) of 4-fluorophenol with various bases measured by the pure base method[258]

Base	$-\Delta H°$	Base	$-\Delta H°$
Benzene	5.15	Tetrahydrofuran	24.06
1-Iodobutane	6.49	N,N-Dimethylformamide	29.16
1-Bromobutane	7.61	Dimethyl sulfoxide	30.17
1-Chlorobutane	8.08	Pyridine	30.96
Diethyl sulfide	15.19	N,N-Dimethylacetamide	31.13
Tetrahydrothiophene	15.52	4-Picoline	31.76
Acetonitrile	17.57^a	Tetramethylene sulfoxide	31.97
Tetramethylene sulfone	17.78	4-Dimethylaminopyridine	32.64^a
Ethyl acetate	19.83	Hexamethylphosphoramide	36.53
1,4-Dioxane	21.34	Trimethylamine N-oxide	36.82^b
2-Butanone	21.76	Triethylamine	37.32
Cyclopentanone	23.01	Quinuclidine	39.75^c
Cyclohexanone	23.77		

^{a}In CCl$_4$.
^{b}In CH$_2$Cl$_2$.
^{c}In o-C$_6$H$_4$Cl$_2$.

the frequency shift of the A-H stretching vibration. This correlation has been challenged by many research groups[4,145,237] and supported by others[4,263]. Today the consensus seems to be[138,258] that the $\Delta H - \Delta \nu$ correlation is family-dependent. If the domain of validity of the correlation has been clearly established for a given family, reliable ΔH data can be predicted for compounds belonging to this family. For example, the enthalpy of complexation of 4-fluorophenol with any chloroalkane in CCl_4 can be calculated[138] from $\Delta \nu(OH \cdots Cl)$ and equation 42.

$$\Delta \nu(OH)(cm^{-1}) = 3611 - \nu(OH \cdots B) \quad (41)$$

$$-\Delta H°(kJ\,mol^{-1}) = 0.12\,\Delta \nu(OH \cdots Cl)(cm^{-1}) - 0.4 \quad (42)$$

$$n = 5, \quad r = 0.984, \quad s = 0.37\,kJ$$

With hydrogen-bond formation the $S_0 \rightarrow S_1$ transition of a phenol ArOH undergoes a bathochromic shift towards the spectral position of the corresponding transition of the anion. For example, the $\pi \rightarrow \pi^*$ transition of 4-fluorophenol at 281.1 nm in CCl_4 (absorption coefficient ca 3,000 $dm^3\,mol^{-1}\,cm^{-1}$) is shifted to 286.5 nm on hydrogen bonding with Oct_3PO[238], because of the stabilization of the π^* excited state relative to

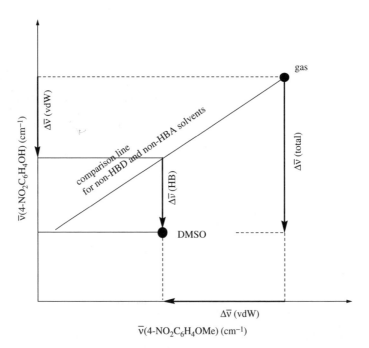

FIGURE 14. The solvatochromic comparison principle. In a plot of the corresponding $\bar{\nu}$ values of a hydrogen-bond (HB) donor probe, $4\text{-}NO_2C_6H_4OH$, vs. a very similar but non-hydrogen-bond donor probe, $4\text{-}NO_2C_6H_4OMe$, non-HBA and non-HBD solvents draw a so-called comparison line with a very high correlation coefficient from the gas phase to polyhalogenated benzenes, because the van der Waals effects of these solvents are similar for the two probes. HBA solvents (e.g. DMSO) are displaced below the comparison line because of an enhanced solvatochromic shift caused by hydrogen bonding. The contribution $\Delta\bar{\nu}(HB)$ to the total solvatochromic shift $\bar{\nu}(gas) - \bar{\nu}(DMSO)$ of $4\text{-}NO_2C_6H_4OH$ is calculated as shown in the figure

8. Hydrogen-bonded complexes of phenols

the less acidic π ground state. Greater shifts are observed when intramolecular charge transfer occurs upon excitation in push-pull compounds such as 4-nitroaniline or 4-nitrophenol[37,264]. Kamlet and Taft[265] have proposed a method for constructing a scale of solvent hydrogen-bond basicity from these shifts of electronic transitions, which they refer to as the solvatochromic comparison method. In this method, 4-nitrophenol is the reference hydrogen-bond donor and the base is used as the solvent, so complete association of 4-$NO_2C_6H_4OH$ can be assumed. The solvatochromic comparison method is outlined in Figure 14. Magnitudes of enhanced solvatochromic shifts in hydrogen-bond acceptor solvents are determined for 4-nitrophenol (**75**) relative to 4-nitroanisole (**76**) for the $\pi \rightarrow \pi^*$ transition of longest wavelength (283.6 nm in heptane) in order that the $\Delta \bar{\nu}$ (HB) contains only the hydrogen-bond contribution to the solvatochromic shift. Nicolet and Laurence[266] have improved the precision and sensitivity of the method through their thermosolvatochromic comparison method. This method takes advantage of variations in solvent properties with temperature (0–105 °C) and of a better-defined comparison line, fixed by the largest possible range of solvents, from the gas phase to the most polar but non-HBA and non-HBD (or very weak HBA and/or HBD) solvents. They have thus

TABLE 19. Solvatochromic shifts $\Delta \bar{\nu}_1$(HB) (cm^{-1}) of 4-nitrophenol attributable to hydrogen bonding[267]

Basic solvents	$\Delta \bar{\nu}$(HB)	Basic solvents	$\Delta \bar{\nu}$(HB)
π bases		Ethyl acetate	993
Benzene	193	Methyl acetate	1033
Toluene	209	Cyclohexanone	1064
p-Xylene	253	2-Butanone	1108
Mesitylene	364	Dimethylformamide	1451
Prehnitene	454	N-Methylpyrrolidinone	1525
Haloalkanes		Tetramethylurea	1558
n-Butyl bromide	223	Dimethylacetamide	1582
n-Butyl chloride	260	**SO and PO bases**	
Thioethers		Sulfolane	657
Trimethylene sulfide	650	Diethyl sulfite	906
Dimethyl sulfide	668	Diethyl chlorophosphate	1131
Tetrahydrothiophene	738	Trimethyl phosphate	1314
Diethyl sulfide	745	Triethyl phosphate	1458
Di-i-propyl sulfide	866	Dimethyl sulfoxide	1466
Di-n-butyl sulfide	875	Tetramethylene sulfoxide	1523
Ethers		Hexamethylenephosphoramide	2000
Anisole	417	**Pyridines**	
Dioxolane	785	2,6-Difluoropyridine	822
Dioxane	919	2-Fluoropyridine	1158
Dibenzyl ether	927	3-Bromopyridine	1457
Tetrahydrofuran	1183	Pyridine	1770
Diethyl ether	1205	4-Methylpyridine	1927
Di-n-butyl ether	1322	2,4,6-Trimethylpyridine	1972
2,2,5,5-Tetramethyltetrahydrofuran	1423	**Amines**	
Nitriles		N,N-Dimethylbenzylamine	2022
Chloroacetonitrile	364	N,N'-Dimethylpiperazine	2029
Benzonitrile	749	Triethylamine	2311
Acetonitrile	771	N,N-Dimethyl-c-hexylamine	2316
Dimethylcyanamide	1092	Tri-n-butylamine	2424
Carbonyl bases			
Diethyl carbonate	908		
Acetone	986		

calculated[267] the solvatochromic hydrogen-bonding shifts of 4-nitrophenol for an extended sample of oxygen, nitrogen, carbon, halogen and sulfur bases. Their results are given in Table 19. Solvatochromic shifts attributable to hydrogen bonding vary from 193 cm^{-1} (2.3 kJ mol^{-1}) for the 4-nitrophenol·benzene complex to 2,424 cm^{-1} (29.0 kJ mol^{-1}) for the 4-nitrophenol·tri-n-butylamine complex. The latter value compares well with the enthalpy of formation of the complex 4-nitrophenol·triethylamine (43.1 kJ mol^{-1}, in c-C$_6$H$_{12}$)[268] insofar as the electronic shifts refer to the difference in hydrogen-bond electronic energies between the ground and the excited states.

(75) (76)

The significance of $\Delta\bar{\nu}$ (HB) as a hydrogen-bonding parameter has been tested by its correlation with complexation constants, NMR shifts, vibrational IR shifts and enthalpies of hydrogen-bond formation[265,267]. Family-dependent correlations are generally found between the above properties[267]. The only significant family-independent correlation, illustrated in Figure 15, is with the enthalpy of hydrogen-bond formation of 4-fluorophenol complexes ($r = 0.992$ for 37 complexes). This correlation follows directly from the similarity principle: not only are 4-nitrophenol and 4-fluorophenol similar OH donors but also both properties (ΔH and $\Delta \nu$) are similar, referring more or less to the energy of the hydrogen bond.

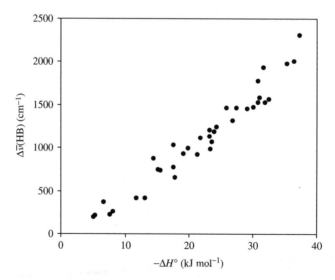

FIGURE 15. Solvatochromic hydrogen-bond shifts for 4-nitrophenol (longest wavelength $\pi \to \pi^*$ transition) in HBA solvents[267] plotted against the enthalpy of hydrogen-bond formation of 4-fluorophenol in pure HBA solvents[258]

In NMR spectroscopy, the hydrogen-bond shift, or the difference in chemical shifts for free and complexed hydrogen-bond donors, can be used as an indication of hydrogen-bond strength. Gurka and Taft[20] have used ^{19}F NMR data of 4-fluorophenol hydrogen-bonded to bases in CCl_4. 4-Fluoroanisole was used as the internal reference to represent intramolecular screening effects similar to 4-fluorophenol so that the chemical shifts observed for 4-fluorophenol would be due entirely to hydrogen-bond formation. Limiting ^{19}F NMR shifts, Δ in ppm, between free 4-fluorophenol and the 1 : 1 complex 4-$FC_6H_4OH\cdots B$ have been obtained[20] for 62 bases of widely different structures in CCl_4 at 25 °C. Additional values are given in Reference 153. A linear correlation between pK_{HB} and Δ was shown to apply to bases without large steric effects. It is particularly significant that the correlation includes bases with substantial variations in the entropies of complexation.

VIII. HYDROGEN BONDING AND PROTONATION

When hydrogen bonds (HBs) of increasing strength are formed in solution (the HB equilibrium is given in equation 43), the attraction of the acceptor B for the proton becomes so great that the latter can leave the phenol molecule to reach the base B, leading to proton transfer (PT) and the formation of an ion pair (the PT equilibrium in equation 43). Depending on the experimental conditions, the ion pair may further dissociate into solvated ions (the D equilibrium in equation 43). However, the new HB formed between the protonated base BH^+ and the phenolate ion ArO^- is generally so strong that no noticeable increase in conductivity can be detected when PT occurs. The two tautomeric forms on each side of the PT equilibrium delimit an important domain where the proton is delocalized between the two accepting species B and ArO^- and/or jumps easily from one to the other well of its potential energy surface corresponding to the covalent O−H and H−B^+ bonds. These intermediary states are characterized by high proton polarizabilities that can be detected by an intense continuum raising the base line of the mid-IR spectrum of the PT adduct. They have been the focus of numerous studies in the last few years and the most recent developments in these HB → PT reactions were reviewed in 1996 by Szafran[269] and in 2000 by Zundel[270].

$$ArOH + B \xrightleftharpoons{HB} ArOH\cdots B \xrightleftharpoons{PT} ArO^-\cdots HB^+ \xrightleftharpoons{D} ArO^- + HB^+ \qquad (43)$$

Among the different HBDs studied in the analysis of PT equilibria, substituted phenols are certainly the most versatile models for several structural reasons:

(i) A large variety of substituents can be added to the five positions of the phenolic ring, enabling minute modifications of the acidity over a wide range of pK_a values. So far, HB → PT reactions have been reported for pK_a values ranging from -0.70 for 3,5-dichloro-2,4,6-trinitrophenol[271] to $pK_a = +10.67$ for 3,4,5-trimethylphenol[272].

(ii) The IR stretching and deformations of the hydroxyl group are highly sensitive to the changes in HB complexation and in PT. Their positions allow a safe identification of the free and H-bonded species[273] and their intensities are good probes for the quantitative estimation of the extent of PT[274]. Moreover, in some substituted phenols, the ring vibrations are good indicators of the PT level[28,275] and specific phenolate C−O vibrations can also be found[274] in the spectrum near 1200–1250 cm^{-1}.

(iii) Due to the presence of benzenic π electrons, molecular (OH$\cdots$B) and ionic ($O^-\cdots HB^+$) HB complexes of phenols may be distinguished from the free molecule by their different $\pi \rightarrow \pi^*$ transition spectra in the 270–400 nm UV region[269,276–279].

(iv) Chlorophenols and especially *ortho*-chlorophenols give good quality crystals that can be grown from non-aqueous solutions[280–282] and used for X-ray diffraction studies.

(v) The rigid frame of phenols permits simple calculations of the hydrogen-bond dipole moments which are vectorial differences $\Delta\vec{\mu}$ between the dipole moments of the complexes and the sum of the moments of the separate free molecules[276,283,284].

There is no doubt that mid-IR spectroscopy is the most appropriate technique for the analysis of HB → PT reactions, since a single spectrum provides precise information not only on the extent of PT from the positions and the intensities of the vibrational peaks but also on the proton polarizability levels that are characterized by the location and the intensities of the broad bands forming the so-called continuum[270,285]. The far-IR domain has also been explored in order to find the HB vibration v_σ in the 150–300 cm^{-1} range[126,286]. In the near-infrared, Rospenk and Zeegers-Huyskens[287] examined the first overtone of the v(OH···N) absorption of the phenol·pyridine system. Whereas no proton transfer occurs in the fundamental and the first excited vibrational states, they found in this first overtone a splitting that they assigned to PT in the second vibrational state. The ^{1}H or ^{13}C NMR shifts of hydrogen-bonded systems[288–290] follow the same trends as the IR frequency shifts. However, the PT reaction is a fast process on the NMR time scale and a lowering of the temperature is always necessary to obtain the decoalescence of the neutral and ionic hydrogen-bonded signals[271,288]. Homoconjugation equilibrium constants corresponding to the HB formation of substituted phenols with their conjugate phenolate ions (equilibrium 44) have been determined[291] by potentiometric titration.

$$\text{ArOH} + {}^-\text{OAr} \rightleftharpoons \text{ArOH}\cdots{}^-\text{OAr} \tag{44}$$

Factors influencing the extent of proton transfer are: (i) Brønsted acidity and basicity of the proton donor and acceptor, (ii) solvent, (iii) temperature and (iv) concentration. In the following, we examine these various factors.

(i) *Acidity and basicity of the proton donor and acceptor.* The degree of PT is clearly related to the differences ΔpK_a between the protonated base and the phenol (equation 45). Table 20 shows that increasing substitution of the phenol moiety by chlorine substituents is sufficient to cover the full range of extent of PT in acetonitrile.

$$\Delta pK_a = pK_a(\text{BH}^+) - pK_a(\text{ArOH}) \tag{45}$$

TABLE 20. Extent of the PT for phenol·n-propylamine systems in CD$_3$CN[292]

Phenol	$K_{PT}{}^a$	% PT^b	$\Delta pK_a{}^c$
H	0	0	0.82
4-Cl	0.010	1	1.53
3-Cl	0.031	3	1.86
2-Cl	0.064	6	2.22
3,5-Cl$_2$	0.176	15	2.79
2,4-Cl$_2$	0.30	23	2.96
2,3-Cl$_2$	5.25	84	3.27
2,4,5-Cl$_3$	99	99	4.71
2,3,4,5,6-Cl$_5$	∞	100	5.45

^{a}Calculated from the intensity of the NH$_3{}^+$ bending vibration. $K_{PT} = [\text{O}^-\cdots\text{HN}^+]/[\text{OH}\cdots\text{N}]$.
^{b}%PT = 100 $K_{PT}/(K_{PT} + 1)$.
^{c}pK_a of n-propylamine in methanol, 10.71.

TABLE 21. Extent of the PT of pentachlorophenol·pyridine complexes in CCl_4[274]

Pyridine	$K_{HB}{}^a$	$K_{PT}{}^b$	% PT^c
H	97	0	0
3-Me	283	0.24	19
3,5-Me$_2$	2097	0.97	49
2,4-Me$_2$	185	1.73	63
2,4,6-Me$_3$	227	6.81	87

$^a K_{HB} = [\text{OH}\cdots\text{N}]/[\text{OH}][\text{N}]$.
^{b}Calculated from the intensity of the δ_{OH} band. $K_{PT} = [\text{O}^-\cdots\text{HN}^+]/[\text{OH}\cdots\text{N}]$.
$^c \%\text{PT} = 100\, K_{PT}/(K_{PT}+1)$.

When the phenol is kept constant, similar variations are observed[274] for a series of methyl-substituted pyridines of increasing basicity (Table 21).

It can be seen in Table 21 that the steric hindrance due to the presence of *ortho*-substituents in the pyridine ring affects strongly the formation of the neutral HB, whereas a regular trend is observed for the PT equilibrium constant.

The literature reveals the use of several partners of substituted phenols. Aliphatic amines have been the most popular[28,271,272,274–278,280,284,285,292]. However, numerous studies refer to other nitrogen bases such as dimethylaniline[293], pyridines[274–276,288], imines, guanidines[277,279,294,295] and 1,8-bis(dimethylamino)naphthalene[182,290,296,297]. Oxygen bases (amines *N*-oxides[273,286,289,298], carboxylate[281,282] and phenolate[291] ions) are also convenient models for the study of HB → PT reactions. As seen in Section V, systematic studies of PT equilibria have also been carried out with *ortho*-substituted phenols presenting intramolecular hydrogen bonds[118,126,299].

(ii) *Solvent effect.* As expected for an equilibrium between neutral and charged forms, the extent of PT is highly dependent of the nature of the HB environment[275,284]. In Table 22, the percentages of PT for the 2,4,6-trichlorophenol·triethylamine complex measured in different solvents are reported. The % PT increases with the increase in solvent polarity measured by its dielectric permittivity ε or by the Onsager function. A further displacement towards the ionic HB is observed when the solvent possesses HBD CH groups as shown in Table 22 for chloroform, dichloromethane and dibromomethane. The shift of the equilibrium towards the ionic tautomer can be explained by the cooperative

TABLE 22. Solvent effect on the PT in 2,4,6-trichlorophenol·triethylamine complexes[275]

Solvent	ε^a	$\dfrac{\varepsilon-1}{2\varepsilon+1}{}^b$	% PT^c
n-C$_7$H$_{14}$	2.1	0.21	10
CCl$_4$	2.2	0.22	12
CH$_3$CCl$_3$	7.2	0.40	25
n-BuCl	7.2	0.40	25
CDCl$_3$	4.6	0.35	45
CH$_2$Br$_2$	7.2	0.40	50
CD$_2$Cl$_2$	8.9	0.42	55

^{a}Dielectric permittivity of the solvent.
^{b}Onsager function.
^{c}Determined from the phenolate band at 1245 cm^{-1}.

TABLE 23. Thermodynamic parameters for HB and PT[279]: phenol and $CH_3(CH=CH)_5CH=NBu$ (B) in methylcyclohexane at 310 K

Equilibrium	$\Delta G°^a$	$\Delta H°^a$	$\Delta S°^b$
$PhOH + B \rightleftharpoons PhOH\cdots B$	−7.5	−36.8	−95
$PhOH\cdots B \rightleftharpoons PhO^-\cdots HB^+$	+15.1	−23.4	−125

^{a}kJ mol^{-1}.
^{b}J mol^{-1} K^{-1}.

HB ($CH\cdots O^-\cdots HN^+$) of the CH donor on the strongly basic negative oxygen which stabilizes the polar form.

(iii) *Temperature effect.* $\Delta H°$ values measured by van't Hoff plots in solution are all negative[270]. However, these enthalpies are the sum of two terms. The first one, intrinsic and positive[270], corresponds to the PT itself, and the larger second one corresponds to a negative solvation enthalpy.

Table 23 provides an example where the two steps, HB and PT, have been treated on the same binary system in an apolar solvent where the solute–solvent interactions are minimized. Owing to the large negative ΔH values, even small decreases of a few tens of a degree shift the HB step to completion and increase notably the extent of PT. The larger negative entropy and the smaller negative enthalpy in the PT compared to the HB are both unfavorable to the ionic form, so that the PT equilibrium constants are smaller than the HB equilibrium constants for identical systems.

(iv) *Concentration effects.* When the phenol and the base are mixed, the HB hetero-complex (equilibria 45) is formed but, depending on the base strength and on the phenol concentration, substantial association is likely to occur on the very basic phenolate oxygen (equilibrium 46)[28,182,290,294−298]. This new HB can be further shifted towards an extended ionic structure (equilibrium 47), which strengthens the polar form by a strong cooperative homoconjugation effect. Similarly, differences in the PT levels arise from the presence or absence of one or more hydrogen atoms on the acceptor B^{272} as a consequence of a homoconjugation of the base in excess (equilibrium 48).

$$OH + O^-\cdots HB^+ \rightleftharpoons OH\cdots O^-\cdots HB^+ \quad (46)$$

$$OH\cdots O^-\cdots HB^+ \rightleftharpoons O^-\cdots HO\cdots HB^+ \quad (47)$$

$$O^-\cdots HBH^+ + B \rightleftharpoons O^-\cdots HBH^+\cdots B \rightleftharpoons O^-\cdots HB\cdots HB^+ \quad (48)$$

In the same way, addition of water to the complex always increases the amount of PT by formation of cooperative polyassociations on the polar structure[292].

IX. REFERENCES

1. L. Pauling, *The Nature of the Chemical Bond*, Cornell University Press, Ithaca, New York, 1939.
2. D. Hadži, *Hydrogen Bonding*, Pergamon Press, London, 1959.
3. G. C. Pimentel and A. L. McClellan, *The Hydrogen Bond*, Freeman, San Francisco, 1960.
4. M. D. Joesten and L. J. Schaad, *Hydrogen Bonding*, M. Dekker, New York, 1974.
5. S. N. Vinogradov and R. H Linell, *Hydrogen Bonding*, Van Nostrand, New York, 1971.
6. P. Schuster, G. Zundel and C. Sandorfy, *The Hydrogen Bond. Recent Developments in Theory and Experiments*, North-Holland, Amsterdam, 1976.

7. C. H. Rochester, in *The Chemistry of the Hydroxyl Group*, Part 1 (Ed. S. Patai), Chap. 7, Wiley, New York, 1971, pp. 327–392.
8. P. L. Huyskens, W. A. P. Luck and Th. Zeegers-Huyskens, *Intermolecular Forces. An Introduction to Modern Methods and Results*, Springer-Verlag, Berlin, 1991.
9. S. Scheiner, *Hydrogen Bonding, A Theoretical Perspective*, Oxford University Press, New York, 1997.
10. S. Scheiner, *Molecular Interactions, from van der Waals to Strongly Bound Complexes*, Wiley, New York, 1997.
11. D. Hadži, *Theoretical Treatments of Hydrogen Bonding*, Wiley, New York, 1997.
12. D. A. Smith, *Modeling the Hydrogen Bond*, ACS Symposium Series, Vol. 569, American Chemical Society, Washington, 1994.
13. G. A. Jeffrey and W. Saenger, *Hydrogen Bonding in Biological Structures*, Springer-Verlag, Berlin, 1991.
14. G. A. Jeffrey, *An Introduction to Hydrogen Bonding*, Oxford University Press, New York, 1997.
15. S. Subramanian and M. J. Zaworotko, *Coord. Chem. Rev.*, **137**, 357 (1994).
16. G. R. Desiraju, *Chem. Commun.*, 1475 (1997).
17. C. B. Aakeröy, *Acta Crystallogr., Sect. B*, **53**, 569 (1997).
18. C. Laurence, M. Berthelot, M. Helbert and K. Sraïdi, *J. Phys. Chem.*, **93**, 3799 (1989).
19. M. Berthelot, F. Besseau and C. Laurence, *Eur. J. Org. Chem.*, 925 (1998).
20. D. Gurka and R. W. Taft, *J. Am. Chem. Soc.*, **91**, 4794 (1969).
21. R. W. Taft, D. Gurka, L. Joris, P. v. R. Schleyer and W. J. Rakshys, *J. Am. Chem. Soc.*, **91**, 4801 (1969).
22. M. H. Abraham, P. L. Grellier, D. V. Prior, J. J. Morris, P. J. Taylor, C. Laurence and M. Berthelot, *Tetrahedron Lett.*, **30**, 2571 (1989).
23. M. Berthelot, C. Laurence, D. Foucher and R. W. Taft, *J. Phys. Org. Chem.*, **9**, 255 (1996).
24. Th. Zeegers-Huyskens, *Recent Research and Developments in Physical Chemistry*, **2**, 1105 (1998).
25. P. L. Huyskens, *J. Am. Chem. Soc.*, **99**, 2578 (1977).
26. D. Clotman, D. Van Lerberghe and Th. Zeegers-Huyskens, *Spectrochim. Acta, Part A*, **26**, 1621 (1970).
27. J. P. Muller, G. Vercruysse and Th. Zeegers-Huyskens, *J. Chim. Phys.*, **69**, 1439 (1972).
28. Z. Pawelka and Th. Zeegers-Huyskens, *Bull. Soc. Chim. Belg.*, **106**, 481 (1997).
29. M. H. Abraham, *J. Phys. Org. Chem.*, **6**, 660 (1993).
30. E. Marquis, Diplome d'Etudes Approfondies, Nantes, 1996.
31. I. Nobeli, S. L. Yeoh, S. L. Price and R. Taylor, *Chem. Phys. Lett.*, **280**, 196 (1997).
32. F. H. Allen and O. Kennard, *Chem. Design Automation News*, **8**, 3137 (1993).
33. M. H. Abraham, P. L. Grellier, D. V. Prior, P. P. Duce, J. J. Morris and P. J. Taylor, *J. Chem. Soc., Perkin Trans. 2*, 699 (1989).
34. M. H. Abraham, *Chem. Soc. Rev.*, 73 (1993).
35. M. H. Abraham, *Pure Appl. Chem.*, **65**, 2503 (1993).
36. M. H. Abraham, P. P. Duce, J. J. Morris and P. J. Taylor, *J. Chem. Soc., Faraday Trans. 1*, **83**, 2867 (1987).
37. M. H. Abraham, P. P. Duce, D. V. Prior, D. G. Barratt, J. J. Morris and P. J. Taylor, *J. Chem. Soc., Perkin Trans. 2*, 1355 (1989).
38. C. Hansch, A. Leo and R. W. Taft, *Chem. Rev.*, **91**, 165 (1991).
39. B. Frange, J. L. M. Abboud, C. Benamou and L. Bellon, *J. Org. Chem.*, **47**, 4553 (1982).
40. K. Sraïdi, Diplôme d'études supérieures, Marrakech (1986).
41. K. Sraïdi, Thèse de Doctorat d'Etat, Marrakech (1986).
42. C. Reichardt, *Chem. Rev.*, **94**, 2319 (1994).
43. C. Laurence, P. Nicolet and C. Reichardt, *Bull. Soc. Chim. Fr.*, 125 (1987).
44. C. A. Coleman and C. J. Murray, *J. Org. Chem.*, **57**, 3578 (1992).
45. R. W. Taft and M. J. Kamlet, *J. Am. Chem. Soc.*, **98**, 2886 (1976).
46. Y. Marcus, *J. Solution Chem.*, **20**, 929 (1991).
47. J. Hormadaly and Y. Marcus, *J. Phys. Chem.*, **83**, 2843 (1979).
48. S. Spange, M. Lauterbach, A. K. Gyra and C. Reichardt, *Justus Liebigs Ann. Chem.*, 323 (1991).

49. R. W. Taft and J. S. Murray, in *Theoretical and Computational Chemistry*, Vol. 1, *Quantitative Treatments of Solute/Solvent Interactions* (Eds. P. Politzer and J. S. Murray), Chap. 3, Elsevier, Amsterdam, 1994, pp. 55–82.
50. R. W. Taft, M. Berthelot, C. Laurence and A. J. Leo, *Chemtech*, **26**, 20 (1996).
51. M. Luçon and M. Berthelot, unpublished work.
52. V. Chopineaux-Courtois, F. Reymond, G. Bouchard, P.-A. Carrupt and H. H. Girault, *J. Am. Chem. Soc.*, **121**, 1743 (1999).
53. M. H. Abraham, C. M. Du and J. A. Platts, *J. Org. Chem.*, **65**, 7114 (2000).
54. M. H. Abraham, H. S. Chadha, J. P. Dixon and A. J. Leo, *J. Phys. Org. Chem.*, **7**, 712 (1994).
55. H. Frohlich, *J. Chem. Educ.*, **70**, A3 (1993).
56. L. Lessinger, *J. Chem. Educ.*, **72**, 85 (1995).
57. S. Pérez-Casas, R. Moreno-Esparza, M. Costas and D. Patterson, *J. Chem. Soc., Faraday Trans.*, **87**, 1745 (1991).
58. S. Pérez-Casas, L. M. Trejo and M. Costas, *J. Chem. Soc., Faraday Trans.*, **87**, 1733 (1991).
59. S. Singh and C. N. R. Rao, *J. Phys. Chem.*, **71**, 1074 (1967).
60. C. M. Huggins, G. C. Pimentel and J. N. Shoolery, *J. Phys. Chem.*, **60**, 1311 (1956).
61. A. Hall and J. L. Wood, *Spectrochim. Acta, Part A*, **23**, 2657 (1967).
62. B. B. Wayland and R. S. Drago, *J. Am. Chem. Soc.*, **86**, 5240 (1964).
63. W. Beckering, *J. Phys. Chem.*, **65**, 206 (1961).
64. C. Bois, *Acta Crystallogr., Sect. B*, **26**, 2086 (1970).
65. K. Prout, J. Fail, R. M. Jones, R. E. Warner and J. C. Emmett, *J. Chem. Soc., Perkin Trans. 2*, 265 (1988).
66. A. Thozet and M. Perrin, *Acta Crystallogr., Sect. B*, **36**, 1444 (1980).
67. M. T. Vandenborre, H. Gillier-Pandraud, D. Antona and P. Becker, *Acta Crystallogr., Sect. B*, **29**, 2488 (1973).
68. D. Antona, F. Longchambon, M. T. Vandenborre and P. Becker, *Acta Crystallogr., Sect. B*, **29**, 1372 (1973).
69. H. Wunderlich and D. Mootz, *Acta Crystallogr., Sect. B*, **27**, 1684 (1971).
70. G. E. Bacon and R. J. Jude, *Z. Kristallog., Kristallg., Kristallp., Kristallch.*, **138**, 19 (1973).
71. G. G. Lazaru, V. L. Kuskov, Ya. S. Lebedev, W. Hiller and M. K. A. Ricker, private communication to Cambridge Structural Database (1999); see text and Reference 32.
72. J. J. Gerber, M. R. Caira and A. P. Lotter, *J. Crystallogr. Spectrosc. Res.*, **23**, 863 (1993).
73. W. Caminati, B. Velino and R. Danieli, *J. Mol. Spectrosc.*, **161**, 208 (1993).
74. A. Kovács, V. Izvekov, G. Keresztury and G. Pongor, *Chem. Phys.*, **238**, 231 (1998).
75. A. Kovács, G. Keresztury and V. Izvekov, *Chem. Phys.*, **253**, 193 (2000).
76. K. B. Borisenko and I. Hargittai, *J. Phys. Chem.*, **97**, 4080 (1993).
77. K. B. Borisenko, C. W. Bock and I. Hargittai, *J. Phys. Chem.*, **98**, 1442 (1994).
78. C. W. Bock and I. Hargittai, *Struct. Chem.*, **5**, 307 (1994).
79. C. Chung, O. Kwon and Y. Kwon, *J. Phys. Chem. A*, **101**, 4628 (1997).
80. W. Boykin, *J. Mol. Struct.*, **295**, 39 (1993).
81. A. Simperler and W. Mikenda, *Monatsh. Chem.*, **130**, 1003 (1999).
82. V. Sharma, B. Bachand, M. Simard and J. D. Wuest, *J. Org. Chem.*, **59**, 7785 (1994).
83. W. Geiger, *Chem. Ber.*, **107**, 2976 (1974).
84. J. Palomar, J. L. G. De Paz and J. Catalán, *Chem. Phys.*, **246**, 167 (1999).
85. A. Mitsuzuka, A. Fujii, T. Ebata and N. Mikami, *J. Phys. Chem. A*, **102**, 9779 (1998).
86. E. Steinwender and W. Mikenda, *Monatsh. Chem.*, **125**, 695 (1994).
87. M. Berthelot, C. Laurence, M. Luçon, C. Rossignol and R. W. Taft, *J. Phys. Org. Chem.*, **9**, 626 (1996).
88. J. Gebicki and A. Krantz, *J. Chem. Soc., Perkin Trans. 2*, 1617 (1984).
89. E. Orton, M. A. Morgan and G. C. Pimentel, *J. Phys. Chem.*, **94**, 7936 (1990).
90. L. A. Peteanu and R. A. Mathies, *J. Phys. Chem.*, **96**, 6910 (1992).
91. W. Mikenda, F. Pertlik and E. Steinwender, *Monatsh. Chem.*, **124**, 867 (1993).
92. W. Mikenda, E. Steinwender and K. Mereiter, *Monatsh. Chem.*, **126**, 495 (1995).
93. J. Valdés-Martínez, M. Rubio, R. Cetina Rosado, J. Salcedo-Loaiza, R. A. Toscano, G. Espinosa-Pérez, S. Hernandez-Ortega and K. Ebert, *J. Chem. Cryst.*, **27**, 627 (1997).
94. D. W. Boykin, A. L. Baumstark and M. Beeson, *J. Org. Chem.*, **56**, 1969 (1991).
95. D. W. Boykin, S. Chandrasekaran and A. L. Baumstark, *Magn. Reson. Chem.*, **31**, 489 (1993).
96. P. E. Hansen, S. N. Ibsen, T. Kristensen and S. Bolvig, *Magn. Reson. Chem.*, **32**, 399 (1994).

97. P. E. Hansen, *J. Mol. Struct.*, **321**, 79 (1994).
98. J. Albigaard, S. Bolvig and P. E. Hansen, *J. Am. Chem. Soc.*, **120**, 9063 (1998).
99. A. U. Acuña, F. Amat-Guerri, J. Catalán and F. Gonzáles-Tablas, *J. Phys. Chem.*, **84**, 629 (1980).
100. J. Rodríguez, *J. Comput. Chem.*, **15**, 183 (1994).
101. M. V. Vener and S. Scheiner, *J. Phys. Chem.*, **99**, 642 (1995).
102. G. Chung, O. Kwon and Y. Kwon, *J. Phys. Chem. A*, **102**, 2381 (1998).
103. G. Alagona and C. Ghio, *J. Mol. Liquids*, **61**, 1 (1994).
104. H. Lampert, W. Mikenda and A. Karpfen, *J. Phys. Chem.*, **100**, 7418 (1996).
105. C. M. Estévez, M. A. Ríos and J. Rodríguez, *Struct. Chem.*, **3**, 381 (1992).
106. J. Palomar, J. L. G. De Paz and J. Catalán, *J. Phys. Chem. A*, **104**, 6453 (2000).
107. J. Catalán, J. Palomar and J. L. G. De Paz, *J. Phys. Chem. A*, **101**, 7914 (1997).
108. W. A. L. K. Al-Rashid and M. F. El-Bermani, *Spectrochim. Acta, Part A*, **47**, 35 (1991).
109. M. Okuyama and S. Ikawa, *J. Chem. Soc., Faraday Trans.*, **90**, 3065 (1994).
110. T. Yasuda, M. Okuyama, N. Tanimoto, S. Sekiguchi and S. Ikawa, *J. Chem. Soc., Faraday Trans.*, **91**, 3379 (1995).
111. E. Vajda and I. Hargittai, *J. Phys. Chem.*, **97**, 70 (1993).
112. E. Vajda and I. Hargittai, *J. Phys. Chem.*, **96**, 5843 (1992).
113. A. Kovács, I. Kolossváry, G. I. Csonka and I. Hargittai, *J. Comput. Chem.*, **17**, 1804 (1996).
114. A. Kovács and I. Hargittai, *J. Mol. Struct. (Theochem)*, **455**, 229 (1998).
115. M. H. Langoor and J. H. van der Maas, *J. Mol. Struct.*, **403**, 213 (1997).
116. G. Chung, O. Kwon and Y. Kwon, *J. Phys. Chem. A*, **101**, 9415 (1997).
117. S. Tsuzuki, H. Houjou, Y. Nagawa and K. Hiratani, *J. Phys. Chem. A*, **104**, 1332 (2000).
118. A. Koll and P. Wolschann, *Monatsch. Chem.*, **130**, 983 (1999).
119. A. Filarovski, A. Koll and T. Glowiak, *J. Chem. Cryst.*, **27**, 707 (1997).
120. K. Rutkowski and A. Koll, *J. Mol. Struct.*, **322**, 195 (1994).
121. E. Grech, J. Nowicka-Scheibe, Z. Olejnik, T. Lis, Z. Pawelka, Z. Malarski and L. Sobczyk, *J. Chem. Soc., Perkin Trans. 2*, 343 (1996).
122. M. Yildiz, Z. Kiliç and T. Hökelek, *J. Mol. Struct.*, **441**, 1 (1998).
123. A. Simperler and W. Mikenda, *Monatsh. Chem.*, **128**, 969 (1997).
124. T. Steiner, *Chem. Commun.*, 411 (1998).
125. J. B. Levy, N. H. Martin, I. Hargittai and M. Hargittai, *J. Phys. Chem. A*, **102**, 274 (1998).
126. B. Brzeziński, A. Rabold and G. Zundel, *J. Chem. Soc., Faraday Trans.*, **90**, 843 (1994).
127. T. Dziembowska, Z. Malarski and B. Szczodrowska, *J. Solution Chem.*, **25**, 179 (1996).
128. T. Dziembowska, Z. Rozwadowski and P. E. Hansen, *J. Mol. Struct.*, **436–437**, 189 (1997).
129. J. B. Levy, *Struct. Chem.*, **9**, 179 (1998).
130. A. Fujii, E. Fujimaki, T. Ebata and N. Mikami, *J. Am. Chem. Soc.*, **120**, 13256 (1998).
131. W. Fateley, G. L. Carlson and F. F. Bentley, *J. Phys. Chem.*, **79**, 199 (1975).
132. T. Schaeffer, *J. Phys. Chem.*, **79**, 1888 (1975).
133. M. Takasuka and Y. Matsui, *J. Chem. Soc., Perkin Trans. 2*, 1743 (1979).
134. C. Bilton, F. H. Allen, G. P. Sheilds and J. A. K. Howard, *Acta Crystallogr., Sect. B*, **56**, 849 (2000).
135. G. Gilli, F. Bellucci, V. Ferretti and V. Bertolasi, *J. Am. Chem. Soc.*, **111**, 1023 (1989).
136. A. W. Baker and A. T. Shulgin, *Can. J. Chem.*, **43**, 650 (1965).
137. G. L. Carlson, W. G. Fateley, A. S. Manocha and F. F. Bentley, *J. Phys. Chem.*, **76**, 1553 (1972).
138. M. Ouvrard, M. Berthelot and C. Laurence, *J. Phys. Org. Chem.*, **14**, 804 (2001).
139. R. M. Badger and S. H. Bauer, *J. Chem. Phys.*, **5**, 839 (1937).
140. R. West, D. L. Powell, L. S. Whatley, M. K. T. Lee and P. v. R. Schleyer, *J. Am. Chem. Soc.*, **84**, 3221 (1962).
141. O. R. Wulf and U. Liddel, *J. Am. Chem. Soc.*, **57**, 1464 (1935).
142. J. C. F. Ng, Y. S. Park and H. F. Shurvell, *J. Raman Spectrosc.*, **23**, 229 (1992).
143. J. C. F. Ng, Y. S. Park and H. F. Shurvell, *Spectrochim. Acta, Part A*, **48**, 1139 (1992).
144. R. B. Girling and H. F. Shurvell, *Vib. Spectrosc.*, **18**, 77 (1998).
145. M. H. Abraham, D. V. Prior, R. A. Schulz, J. J. Morris and P. J. Taylor, *J. Chem. Soc., Faraday Trans.*, **94**, 879 (1998).
146. F. Besseau, M. Luçon, C. Laurence and M. Berthelot, *J. Chem. Soc., Perkin Trans. 2*, 101 (1998).

147. T. Gramstad and L. J. Stangeland, *Acta Chem. Scand.*, **47**, 605 (1993).
148. L. G. Hepler, *Thermochim. Acta*, **50**, 69 (1981).
149. W. Libus, M. Mecik and W. Sulck, *J. Solution Chem.*, **6**, 865 (1977).
150. B. Stymme, H. Stymme and G. Wettermarck, *J. Am. Chem. Soc.*, **95**, 3490 (1973).
151. S. Ghersetti and A. Lusa, *Spectrochim. Acta*, **21**, 1067 (1965).
152. O. Exner, *Prog. Phys. Org. Chem.*, **10**, 411 (1973).
153. L. Joris, J. Mitsky and R. W. Taft, *J. Am. Chem. Soc.*, **94**, 3438 (1972).
154. P. Goralski, M. Berthelot, J. Rannou, D. Legoff and M. Chabanel, *J. Chem. Soc., Perkin Trans. 2*, 2337 (1994).
155. I. J. Brass and A. T. Bullock, *J. Chem. Soc., Faraday Trans. 1*, **74**, 1556 (1978).
156. M. Rappo-Rabusio, M. F. Llauro, Y. Chevalier and P. Le Perchec, *Phys. Chem. Chem. Phys.*, **3**, 99 (2001).
157. G. Guihéneuf, K. Sraïdi, R. Claramunt and J. Elguero, *C. R. Acad. Sci. Paris*, **305**, 567 (1987).
158. M. Goethals, B. Czarnik-Matusewicz and Th. Zeegers-Huyskens, *J. Heterocycl. Chem.*, **36**, 49 (1999).
159. J. B. Rulinda and Th. Zeegers-Huyskens, in *Molecular Spectroscopy of Dense Phases* (Eds. M. Grosmann, S. G. Elkamoss and J. Ringeissen), Elsevier, Amsterdam, 1976, p. 617.
160. J. B. Rulinda and Th. Zeegers-Huyskens, *Advances in Molecular Relaxation and Interaction Processes*, **14**, 203 (1979).
161. J. B. Rulinda and Th. Zeegers-Huyskens, *Bull. Soc. Chim. Belg.*, **84**, 159 (1975).
162. E. Vrolix, M. Goethals and Th. Zeegers-Huyskens, *Spectrosc. Lett.*, **26**, 497 (1993).
163. M. Jarva, Ph. D. Thesis, University of Oulu, Finland, 1978.
164. G. S. F. D'Alva Torres, C. Pouchan, J. J. C. Teixeira-Dias and R. Fausto, *Spectrosc. Lett.*, **26**, 913 (1993).
165. M. D. G. Faria, J. J. C. Teixeira-Dias and R. Fausto, *J. Mol. Struct.*, **263**, 87 (1991).
166. K. Platteborze, J. Parmentier and Th. Zeegers-Huyskens, *Spectrosc. Lett.*, **24**, 635 (1991).
167. A. Smolders, G. Maes and Th. Zeegers-Huyskens, *J. Mol. Struct.*, **172**, 23 (1988).
168. Z. Pawelka and Th. Zeegers-Huyskens, *Vib. Spectrosc.*, **18**, 41 (1998).
169. M.-L. H. Jeng and Y.-S. Li, *Spectrochim. Acta, Part A*, **45**, 525 (1989).
170. L. Bellon, R. W. Taft and J. L. M. Abboud, *J. Org. Chem.*, **45**, 1166 (1980).
171. J. Vaes and Th. Zeegers-Huyskens, *Tetrahedron*, **32**, 2013 (1976).
172. P. Migchels and Th. Zeegers-Huyskens, *J. Mol. Struct.*, **247**, 173 (1991).
173. P. Migchels, Th. Zeegers-Huyskens and D. Peeters, *J. Phys. Chem.*, **95**, 7599 (1991).
174. P. Migchels, N. Leroux and Th. Zeegers-Huyskens, *Vib. Spectrosc.*, **2**, 81 (1991).
175. V. Leunens, P. Migchels, D. Peeters and Th. Zeegers-Huyskens, *Bull. Soc. Chim. Belg.*, **101**, 165 (1992).
176. M. Goethals, K. Platteborze and Th. Zeegers-Huyskens, *Spectrochim. Acta, Part A*, **48**, 671 (1992).
177. G. G. Siegel and Th. Zeegers-Huyskens, *Spectrochim. Acta, Part A*, **45**, 1297 (1989).
178. O. Kasende and Th. Zeegers-Huyskens, *J. Phys. Chem.*, **88**, 2132 (1984).
179. S. Bogaerts, M. C. Haulait-Pirson and Th. Zeegers-Huyskens, *Bull. Soc. Chim. Belg.*, **87**, 927 (1978).
180. N. Leroux, M. Goethals and Th. Zeegers-Huyskens, *Vib. Spectrosc.*, **9**, 235 (1995).
181. B. Czarnik-Matusewicz and Th. Zeegers-Huyskens, *J. Phys. Org. Chem.*, **13**, 237 (2000).
182. Z. Pawelka and Th. Zeegers-Huyskens, *J. Mol. Struct. (Theochem.)*, **200**, 565 (1989).
183. J. Hine, S. Hahn and D. E. Miles, *J. Org. Chem.*, **51**, 577 (1986).
184. J. Hine and K. Ahn, *J. Org. Chem.*, **52**, 2083 (1987).
185. J. Hine, K. Ahn, J. C. Galluci and S. M. Linden, *J. Am. Chem. Soc.*, **106**, 7980 (1984).
186. C. Laurence, M. Berthelot, J.-Y. Le Questel and M. J. El Ghomari, *J. Chem. Soc., Perkin Trans. 2*, 2075 (1995).
187. J. L. M. Abboud, C. Roussel, E. Gentric, K. Sraïdi, J. Lauransan, G. Guihéneuf, M. J. Kamlet and R. W. Taft, *J. Org. Chem.*, **53**, 1545 (1988).
188. M. Berthelot, C. Laurence, M. Safar and F. Besseau, *J. Chem. Soc., Perkin Trans. 2*, 283 (1998).
189. C. Laurence, M. Berthelot and M. Helbert, *Spectrochim. Acta, Part A*, **41**, 883 (1985).
190. A. Massat, A. Cosse-Barbi and J. P. Doucet, *J. Mol. Struct.*, **212**, 13 (1989).
191. T. Bürgi, T. Droz and S. Leutwyler, *Chem. Phys. Lett.*, **246**, 291 (1995).
192. M. Sodupe, A. Oliva and J. Bertran, *J. Phys. Chem.*, **101**, 9142 (1997).

193. A. Inauen, J. Hewel and S. Leutwyler, *J. Chem. Phys.*, **110**, 1463 (1999).
194. J. Küpper, A. Westphal and M. Schmitt, *Chem. Phys.*, **263**, 41 (2001).
195. A. Courty, M. Mons, B. Dimicoli, F. Piuzzi, V. Brenner and P. Millié, *J. Phys. Chem. A*, **102**, 4890 (1998).
196. T. Steiner, C. C. Wilson and I. Majerz, *Chem. Commun.*, 1231 (2000).
197. T. Steiner, *J. Phys. Chem. A*, **102**, 7041 (1998).
198. B. Birknes, *Acta Chem. Scand., Ser. B*, **30**, 450 (1976).
199. T. Gramstad, S. Husebye and K. Maarmann-Moe, *Acta Chem. Scand., Ser. B*, **40**, 26 (1986).
200. T. Gramstad, S. Husebye, K. Maarmann-Moe and J. Saebø, *Acta Chem. Scand., Ser. B*, **41**, 555 (1987).
201. T. Gramstad, S. Husebye and K. Maarmann-Moe, *Acta Chem. Scand., Ser. B*, **39**, 767 (1985).
202. Z. Malarski, I. Majerz and T. Lis, *Acta Crystallogr., Sect. C*, **43**, 1766 (1987).
203. I. Majerz, Z. Malarski and W. Sawka-Dobrowolska, *J. Mol. Struct.*, **249**, 109 (1996).
204. J. Hine, K. Ahn, J. C. Galluci and S. M. Linden, *Acta Crystallogr., Sect. C*, **46**, 2136 (1990).
205. S. Dega-Szafran, Z. Kosturkiewicz, E. Tykarska, M. Szafran, D. Lemanski and B. Nogaj, *J. Mol. Struct.*, **404**, 25 (1997).
206. W. H. Watson, J. Galloy, F. Vögtle and W. M. Müller, *Acta Crystallogr., Sect. C*, **40**, 200 (1984).
207. E. Garcia-Martinez, E. M. Varquez-Lopez and D. G. Tuck, *Acta Crystallogr., Sect. C*, **54**, 840 (1998).
208. P. I. Coupar, C. Glidewell and G. Ferguson, *Acta Crystallogr., Sect. B*, **53**, 521 (1997).
209. G. Ferguson, C. Glidewell, R. M. Gregson, P. R. Mechan and I. L. J. Patterson, *Acta Crystallogr., Sect. B*, **54**, 151 (1998).
210. E. J. McLean, C. Glidewell, G. Ferguson, R. M. Gregson and A. J. Lough, *Acta Crystallogr., Sect. C*, **55**, 1867 (1999).
211. J. Y. Le Questel, M. Berthelot and C. Laurence, *J. Phys. Org. Chem.*, **13**, 347 (2000).
212. A. L. Llamas-Saiz, C. Foces-Foces, O. Mo, M. Yanez and J. Elguero, *Acta Crystallogr., Sect. C*, **48**, 700 (1992).
213. J. Graton, unpublished results.
214. M. Gerhards, M. Schmitt, K. Kleinermanns and W. Stahl, *J. Chem. Phys.*, **104**, 967 (1996).
215. G. Berden, W. L. Meerts, M. Schmitt and K. Kleinermanns, *J. Chem. Phys.*, **104**, 972 (1996).
216. M. Schmitt, J. Kupper, D. Spangenberg and A. Westphal, *Chem. Phys.*, **254**, 349 (2000).
217. A. Schiefke, C. Deusen, C. Jacoby, M. Gerhards, M. Schmitt, K. Kleinermanns and P. Hering, *J. Chem. Phys.*, **102**, 9197 (1995).
218. M. Gerhards and K. Kleinermanns, *J. Chem. Phys.*, **103**, 7392 (1995).
219. T. Burgi, M. Schutz and S. Leutwyler, *J. Chem. Phys.*, **103**, 6350 (1995).
220. Y. Dimitrova, *Recent Res. Phys. Chem.*, **3**, 133 (1999).
221. P. Imhof, W. Roth, C. Janzen, D. Spangenberg and K. Kleinermanns, *Chem. Phys.*, **242**, 141 (1999).
222. P. Imhof, W. Roth, C. Janzen, D. Spangenberg and K. Kleinermanns, *Chem. Phys.*, **242**, 153 (1999).
223. J.-Y. Le Questel, G. Boquet, M. Berthelot and C. Laurence, *J. Phys. Chem. B*, **104**, 11816 (2000).
224. O. Dideberg, L. Dupont and H. Campsteyn, *Acta Crystallogr., Sect. B*, **31**, 637 (1975).
225. O. Kasende and Th. Zeegers-Huyskens, *J. Mol. Struct.*, **75**, 201 (1981).
226. H. van Brabant-Govaerts and P. Huyskens, *Bull. Soc. Chim. Belg.*, **90**, 987 (1981).
227. J. Parmentier and Th. Zeegers-Huyskens, *Bull. Soc. Chim. Belg.*, **101**, 201 (1992).
228. J. De Taeye, G. Maes and Th. Zeegers-Huyskens, *Bull. Soc. Chim. Belg.*, **92**, 917 (1983).
229. O. Kasende and Th. Zeegers-Huyskens, *J. Phys. Chem.*, **88**, 2636 (1984).
230. C. Laureys and Th. Zeegers-Huyskens, *J. Mol. Struct.*, **158**, 301 (1987).
231. S. Toppet, J. De Taeye and Th. Zeegers-Huyskens, *J. Phys. Chem.*, **92**, 6819 (1988).
232. P. Migchels, G. Maes, Th. Zeegers-Huyskens and M. Rospenk, *J. Mol. Struct.*, **193**, 223 (1989).
233. E. Vrolix and Th. Zeegers-Huyskens, *Vib. Spectrosc.*, **5**, 227 (1993).
234. J. De Taeye, J. Parmentier and Th. Zeegers-Huyskens, *J. Phys. Chem.*, **92**, 4556 (1988).
235. P. Migchels and Th. Zeegers-Huyskens, *J. Phys. Org. Chem.*, **8**, 77 (1995).
236. J. Parmentier, C. Samyn, M. van Beylen and Th. Zeegers-Huyskens, *J. Chem. Soc., Perkin Trans. 2*, 387 (1991).

237. E. M. Arnett, L. Joris, E. J. Mitchell, T. S. S. R. Murty, T. M. Gorie and P. v. R. Schleyer, *J. Am. Chem. Soc.*, **92**, 2365 (1970).
238. A. Chardin, Ph. D. Thesis, Nantes (1997).
239. E. D. Raczynska, C. Laurence and P. Nicolet, *J. Chem. Soc., Perkin Trans. 2*, 1491 (1988).
240. E. D. Raczynska and C. Laurence, *J. Chem. Res. (S)*, 148 (1989).
241. C. Laurence, M. Berthelot, E. D. Raczynska, J.-Y. Le Questel, G. Duguay and P. Hudhomme, *J. Chem. Res. (S)*, 250 (1990).
242. E. D. Raczynska, C. Laurence and M. Berthelot, *Can. J. Chem.*, **70**, 2203 (1992).
243. J.-Y. Le Questel, C. Laurence, A. Lachkar, M. Helbert and M. Berthelot, *J. Chem. Soc., Perkin Trans. 2*, 2091 (1992).
244. M. Berthelot, M. Helbert, C. Laurence and J.-Y. Le Questel, *J. Phys. Org. Chem.*, **6**, 302 (1993).
245. M. Berthelot, M. Helbert, C. Laurence, J.-Y. Le Questel, F. Anvia and R.W. Taft, *J. Chem. Soc., Perkin Trans. 2*, 625 (1993).
246. E. D. Raczynska, C. Laurence and M. Berthelot, *Analyst*, **119**, 683 (1994).
247. A. Chardin, M. Berthelot, C. Laurence and D. G. Morris, *J. Phys. Org. Chem.*, **7**, 705 (1994).
248. C. Laurence, M. Berthelot, M. Luçon and D. G. Morris, *J. Chem. Soc., Perkin Trans. 2*, 491 (1994).
249. F. Besseau, C. Laurence and M. Berthelot, *J. Chem. Soc., Perkin Trans. 2*, 485 (1994).
250. A. Chardin, M. Berthelot, C. Laurence and D. G. Morris, *J. Phys. Org. Chem.*, **8**, 626 (1995).
251. A. Chardin, C. Laurence, M. Berthelot and D. G. Morris, *Bull. Soc. Chim. Fr.*, **133**, 389 (1996).
252. F. Besseau, C. Laurence and M. Berthelot, *Bull. Soc. Chim. Fr.*, **133**, 381 (1996).
253. A. Chardin, C. Laurence and M. Berthelot, *J. Chem. Res. (S)*, 332 (1996).
254. A. Chardin, C. Laurence, M. Berthelot and D. G. Morris, *J. Chem. Soc., Perkin Trans. 2*, 1047 (1996).
255. J. Graton, C. Laurence, M. Berthelot, J.-Y. Le Questel, F. Besseau and E. D. Raczynska, *J. Chem. Soc., Perkin Trans. 2*, 997 (1999).
256. C. Ouvrard, M. Berthelot and C. Laurence, *J. Chem. Soc., Perkin Trans. 2*, 1357 (1999).
257. M. H. Abraham, P. L. Grellier, D. V. Prior, J. J. Morris and P. J. Taylor, *J. Chem. Soc., Perkin Trans. 2*, 521 (1990).
258. E. M. Arnett, E. J. Mitchell and T. S. S. R. Murty, *J. Am. Chem. Soc.*, **96**, 3875 (1974).
259. I. A. Koppel and A. I. Paju, *Org. React. (USSR)*, **11**, 121 (1974).
260. A. Allerhand and P. v. R. Schleyer, *J. Am. Chem. Soc.*, **85**, 371 (1963).
261. C. Sandorfy, *Top. Curr. Chem.*, **120**, 42 (1984).
262. L. Joris and P. v. R. Schleyer, *Tetrahedron*, **24**, 5991 (1968).
263. R. S. Drago, *Struct. Bonding*, **15**, 73 (1973).
264. M. J. Kamlet, R. R. Minesinger and W. H. Gilligan, *J. Am. Chem. Soc.*, **94**, 4744 (1972).
265. M. J. Kamlet and R. W. Taft, *J. Am. Chem. Soc.*, **98**, 377 (1976).
266. P. Nicolet and C. Laurence, *J. Chem. Soc., Perkin Trans. 2*, 1071 (1986).
267. C. Laurence, P. Nicolet and M. Helbert, *J. Chem. Soc., Perkin Trans. 2*, 1081 (1986).
268. R. A. Hudson, R. M. Scott and S. N. Vinogradov, *Spectrochim. Acta, Part A*, **26**, 337 (1970).
269. M. Szafran, *J. Mol. Struct.*, **381**, 39 (1996).
270. G. Zundel, *Adv. Chem. Phys.*, **111**, 1 (2000).
271. M. Ilczyszyn and H. Ratajczak, *J. Chem. Soc., Faraday Trans.*, **91**, 3859 (1995).
272. P. L. Huyskens, Th. Zeegers-Huyskens and Z. Pawelka, *J. Solution Chem.*, **28**, 915 (1999).
273. B. Brzeziński, B. Brycki, G. Zundel and T. Keil, *J. Phys. Chem.*, **95**, 8598 (1991).
274. G. Albrecht and G. Zundel, *J. Chem. Soc., Faraday Trans. 1*, **80**, 553 (1984).
275. R. Krämer and G. Zundel, *J. Chem. Soc., Faraday Trans.*, **86**, 301 (1990).
276. Z. Malarski, M. Rospenk, L. Sobczyk and E. Grech, *J. Phys. Chem.*, **86**, 401 (1982).
277. B. Brzeziński, E. Grech, Z. Malarski, M. Rospenk, G. Schroeder and L. Sobczyk, *J. Chem. Res. (S)*, 151 (1997).
278. M. Zhou, H.-W. Zhu, S. Kasham and R. M. Scott, *J. Mol. Struct.*, **270**, 187 (1992).
279. P. E. Blatz and J. A. Tompkins, *J. Am. Chem. Soc.*, **114**, 3951 (1992).
280. I. Majerz, Z. Malarski and L. Sobczyk, *Chem. Phys. Lett.*, **274**, 361 (1997).
281. Z. Dega-Szafran, A. Komasa, M. Grundwald-Wyspiańska, M. Szafran, G. Buczak and A. Katrusiak, *J. Mol. Struct.*, **404**, 13 (1997).

282. M. Szafran, Z. Dega-Szafran, G. Buczak, A. Katrusiak, M. J. Potrzebowski and A. Komasa, *J. Mol. Struct.*, **416**, 145 (1997).
283. P. L. Huyskens, Z. Phys. Chem., *Neue Folge*, **133**, 129 (1982).
284. I. Majerz and L. Sobczyk, *Bull. Pol. Acad. Sci. Chem.*, **39**, 347 (1991).
285. M. Wierzejewska and H. Ratajczak, *J. Mol. Struct.*, **416**, 121 (1997).
286. G. Zundel, *J. Mol. Struct.*, **381**, 23 (1996).
287. M. Rospenk and Th. Zeegers-Huyskens, *J. Phys. Chem. A*, **101**, 8428 (1997).
288. M. Ilczyszyn and H. Ratajczak, *J. Chem. Soc., Faraday Trans.*, **91**, 1611 (1995).
289. B. Brycki and M. Szafran, *J. Mol. Liquids.*, **59**, 83 (1994).
290. S. Toppet, K. Platteborze and Th. Zeegers-Huyskens, *J. Chem. Soc., Perkin Trans. 2*, 831 (1995).
291. J. Magoński, Z. Pawlak and T. Jasiński, *J. Chem. Soc., Faraday Trans.*, **89**, 119 (1993).
292. G. Zundel and A. Nagyrevi, *J. Phys. Chem.*, **82**, 685 (1978).
293. M. Ilczyszyn, *J. Chem. Soc., Faraday Trans.*, **90**, 1411 (1994).
294. B. Brzeziński and G. Zundel, *J. Mol. Struct.*, **380**, 195 (1996).
295. B. Brzeziński, P. Radziejewski and G. Zundel, *J. Chem. Soc., Faraday Trans.*, **91**, 3141 (1995).
296. P. Huyskens, K. Platteborze and Th. Zeegers-Huyskens, *J. Mol. Struct.*, **436–437**, 91 (1997).
297. K. Platteborze-Stienlet and Th. Zeegers-Huyskens, *J. Mol. Struct.*, **378**, 29 (1996).
298. B. Brzeziński, G. Schroeder, G. Zundel and T. Keil, *J. Chem. Soc., Perkin Trans. 2*, 819 (1992).
299. B. Brzeziński, G. Zundel and R. Krämer, *J. Phys. Chem.*, **91**, 3077 (1987).

CHAPTER 9

Electrophilic reactions of phenols

V. PRAKASH REDDY

Department of Chemistry, University of Missouri-Rolla, Rolla, Missouri 65409, USA
Fax: (573)-341-6033; e-mail: preddy@umr.edu

and

G. K. SURYA PRAKASH

Loker Hydrocarbon Research Institute and Department of Chemistry, University of Southern California, Los Angeles, California 90089-1661, USA
Fax: (213)-740-6270; email: gprakash@usc.edu

I. INTRODUCTION	606
II. FRIEDEL–CRAFTS ALKYLATION	606
A. Lewis Acid Catalysis	607
B. Bronsted Acid Catalysis	612
C. Solid Acid Catalysis	612
D. Alkylations under Supercritical Conditions	621
E. Stereoselective Alkylations	621
F. Formylation and Phenol-Formaldehyde Resins	626
III. FRIEDEL–CRAFTS ACYLATIONS	629
IV. NITRATION AND NITROSATION	632
A. Regioselectivity	632
B. Peroxynitrite-induced Nitration and Nitrosation	636
C. Nitrosation by Nitrous Acid	638
D. Nitration by Tetranitromethane	638
E. Nitration by Metal Nitrates	638
V. FRIES AND RELATED REARRANGEMENTS	639
A. Lewis Acid Catalyzed Fries Rearrangements	639
B. Bronsted Acid Catalyzed Fries Rearrangements	642
VI. ELECTROPHILIC HALOGENATION	645
A. Fluorination	646
B. Chlorination	649

The Chemistry of Phenols. Edited by Z. Rappoport
© 2003 John Wiley & Sons, Ltd ISBN: 0-471-49737-1

C. Bromination	649
D. Iodination	651
VII. PHENOL–DIENONE REARRANGEMENTS	651
VIII. REFERENCES	656

I. INTRODUCTION

Phenolic functional groups are often encountered in a variety of pharmaceuticals, agrochemicals and polymer materials. Phenol-formaldehyde resins, the polymers derived from phenols, for example, are the most widely used industrial polymers. Selective functionalization of the aromatic rings of phenols is therefore of great importance[1]. Usually phenols are functionalized through electrophilic aromatic substitution reactions, such as Friedel–Crafts alkylations and acylations, and electrophilic halogenations, nitrations and nitrosations. The Friedel–Crafts alkylation of phenols gives *ortho-* and *para-* alkylphenols, the regioselectivity being dependent on the catalyst used. The alkylations can be initiated by a wide variety of substrates, such as alcohols, alkyl halides and alkenes. Being industrially important chemicals, numerous catalysts have been explored for efficient preparation of the alkylphenols. Both Bronsted and Lewis acids can be used as the catalysts. The homogeneous catalysts are increasingly being replaced by the solid acid catalysts, such as zeolites, Nafion-H and Amberlyst type of catalysts, in order to avoid the environmental problems associated with the product workup. There is some progress toward the use of supercritical water and carbon dioxide as solvents. Stereochemistry of these reactions may be controlled in favorable cases by using chiral catalysts. The Friedel–Crafts acylations are more regioselective than the alkylations and a two-step process involving the acylation followed by reduction of the carbonyl groups may provide a clean route to the alkylphenols.

The nitration and nitrosation of phenols are biologically important phenomena. For example, the oxidative stress induces the formation of peroxynitrite *in vivo*, which effects nitration of the tyrosine residues of the enzymes, causing deleterious effects. Nitration of phenols can be conveniently carried out by Olah's nitronium and nitrosonium salts[2]. Nitrosation followed by oxidation is also a convenient alternative for the preparation of the nitrophenols. The regiochemistry of the electrophilic reactions is dependent on the catalyst used and the reagent. The normal *ortho/para* directing effect of the phenolic hydroxy group in the electrophilic substitution reactions is altered in the presence of superacids, due to the formation of the protonated phenols under these conditions. The electrophilic reactions of phenols including nitrations, nitrosations, alkylations, acylations and halogenations, due to their industrial significance, have received much attention. However, in spite of many reviews detailing these reactions in connection with other topics of interest, the field has not been reviewed in general. The present review focuses on the recent developments in this broad area. An earlier volume of this series reviewed electrophilic halogenations of phenols[3]. We therefore include only recent developments of electrophilic halogenations.

II. FRIEDEL–CRAFTS ALKYLATION

Phenols are highly reactive toward the Friedel–Crafts alkylation reactions involving tertiary alkyl halides. Phenol, 2-methylphenol and 2,6-dimethylphenol react with tertiary alkyl halides such as 1-bromoadamantane in the absence of any external catalyst to give exclusively the *para*-(1-adamantyl)phenols (equation 1). These compounds have found uses in the preparation of certain copolymers[4].

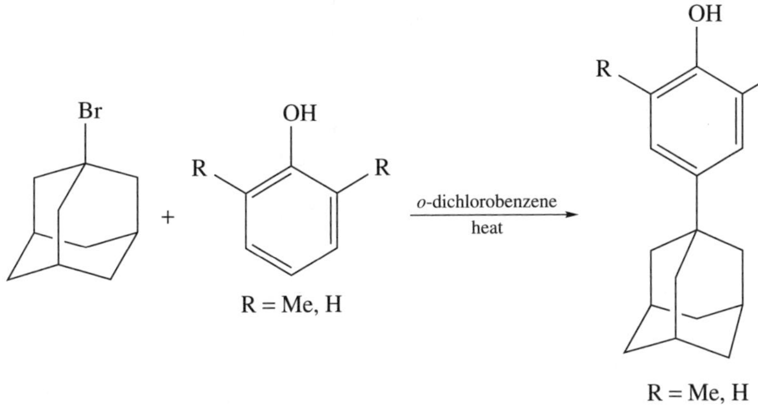

$$\text{1-bromoadamantane} + \text{2,6-R}_2\text{-phenol (R = Me, H)} \xrightarrow[\text{heat}]{o\text{-dichlorobenzene}} \text{4-(1-adamantyl)-2,6-R}_2\text{-phenol (R = Me, H)} \quad (1)$$

The reaction of phenol with secondary alkyl halides, such as 2-bromoadamantane, cyclohexyl bromide and *exo*-2-bromonorbornane, also proceeds noncatalytically to give the corresponding *ortho*- and *para*-alkylated phenols[5]. The Friedel–Crafts alkylations using primary alcohols, however, require catalysts.

The alkylation of phenols is an industrially prominent reaction, as close to one million tons of the alkylated phenols are being produced each year. They find various applications such as antioxidants and polymer stabilizers. The O-alkylated phenols are also used in the manufacture of dyes and agrochemicals.

A variety of catalysts, homogeneous and heterogeneous, have been continually developed for the Friedel–Crafts alkylations. Although it is difficult to classify these catalysts as being exclusively either Bronsted or Lewis acids, for convenience in the organization of the broad material, we have classified the catalysts as: (a) Lewis acidic, (b) Bronsted acidic and (c) solid acid catalysts. It is important, however, to notice that in many cases, such as zeolites, both Lewis and Bronsted acid sites coexist. Catalysis by 100% pure $AlCl_3$ is less effective than in the presence of traces of water, suggesting again that it is impracticable to distinguish a catalyst exclusively as a Lewis or a Bronsted acid. Most of the solid acid catalysts we have considered in this review fit into this category, i.e. they have both Bronsted and Lewis acid sites.

A. Lewis Acid Catalysis

The alkylation of phenols can be achieved using cumyl and *tert*-butyl hydroperoxides using Lewis acids such as $TiCl_4$ or $FeCl_3$ (equation 2). $FeCl_3$ is the preferred catalyst for the alkylation of phenol using *tert*-butyl hydroperoxide. These soft Lewis acids (softer than H^+) preferentially attack the oxygen attached to the tertiary aliphatic carbon, rather than the hydroxyl oxygen, resulting in the formation of the carbocationic intermediate upon its cleavage. The latter readily reacts with the phenols. Reaction of cumene hydroperoxide with phenol in the presence of $FeCl_3$, for example, results in the formation of the 4-cumylphenol. Similarly, *ortho*-cresol gave 2-methyl-4-cumylphenol (51% yield), *para*-cresol gave 2-cumyl-4-methylphenol (27%), 1-naphthol gave 4-cumyl-1-naphthol (81%) and resorcinol gave 4-cumylresorcinol (39%). In case of sterically crowded substrates, radical mechanism may compete, resulting in the formation of dimeric products[6].

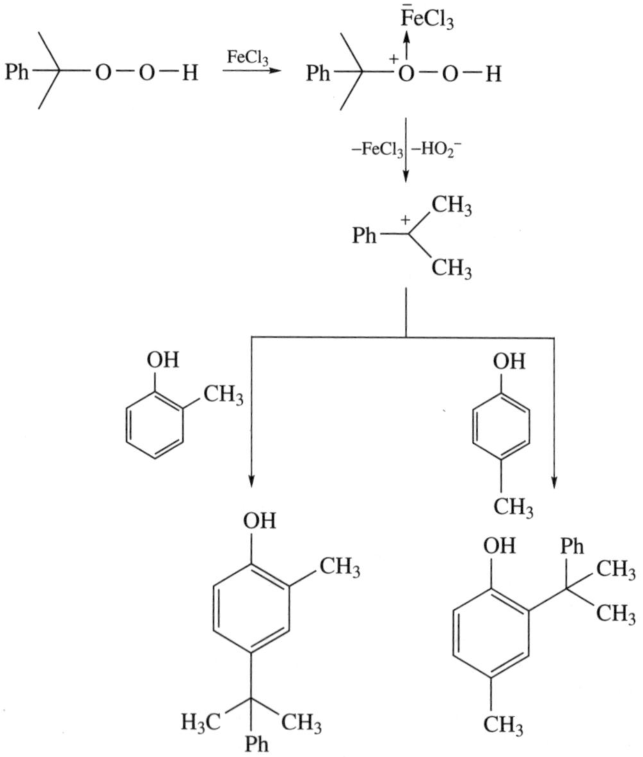

Friedel–Crafts alkylation of dicyclopentadiene with phenol using boron trifluoride-etherate as the catalyst gives 2-[4-(2-hydroxyphenyl)tricyclo[5.2.1.0(2, 6)]dec-8-yl]phenol, which can be used in the preparation of the phenol-formaldehyde resins (equation 3)[7].

Alkylation of phenol with methanol can be effected regioselectively to give *ortho*-cresol in the presence of iron-magnesium oxide catalysts[8]. High selectivity for *ortho*-alkylation can be achieved using $ZnAl_2O_4$, prepared by reacting $Zn(OAc)_2$ with $Al(OPr-i)_3$[9], $AlCl_3$[10] and CeO_2-MgO catalyst[11]. The *ortho*-alkylation is the major pathway for the alkylation of naphthols using methanol or other alcohols in the presence of iron oxide catalyst in the gas-phase reaction[12], whereas predominant O-alkylation occurs using dimethyl carbonate and $AlPO_4$-derived catalysts[13].

In the presence of catalytic amounts of anhydrous potassium carbonate, phenols react with trifluoroacetaldehyde ethyl hemiacetal to give the *para*-alkylated products (C-alkylation of the phenolate anions). Thus, phenol under these conditions gives 4-(2,2,2-trifluoro-1-hydroxyethyl)phenol as the predominant product. The reaction catalyzed by zinc halides predominantly gave the *ortho*-substituted product (equation 4)[14,15].

Aryloxymagnesium bromides react with isatins under extremely mild conditions to provide 3-(2-hydroxyaryl)-3-hydroxyindolones in good yield. The reaction is highly selective for C-alkylation of the ambident phenolate anion[16]. The electron-rich aromatic group undergoes nucleophilic addition to the β-carbonyl group of the isatin to give the intermediate dienone (not isolated), the aromatization of which gives the final product (equation 5). The reaction is highly regioselective and *meta*-substituted phenols undergo alkylation at the less crowded *ortho* position of the phenol. The reaction is applicable to a variety of substituted isatins and phenols.

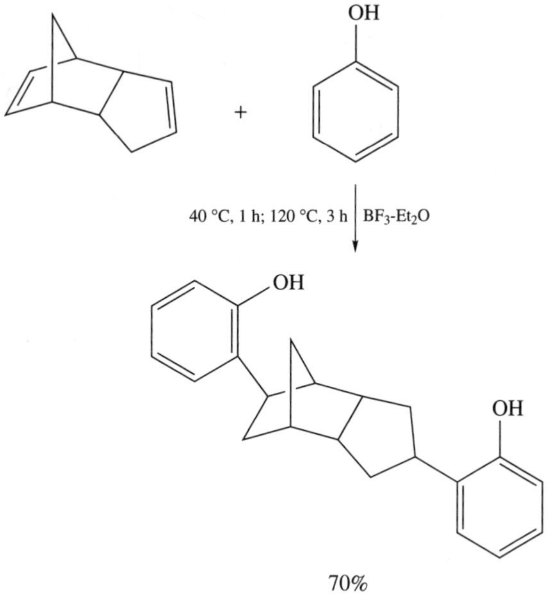

(3)

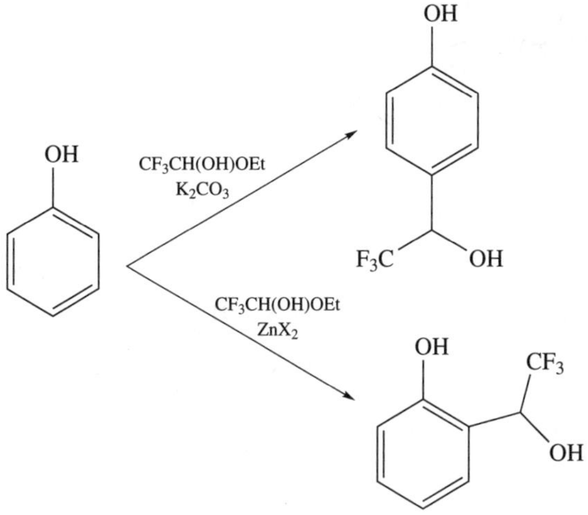

(4)

The reaction of the tetra-O-acetyl-5-thio-α-D-xylopyranosyl-1-O-trichloroacetimidate with phenol in the presence of boron trifluoride-etherate, at low temperatures, gives a mixture of the corresponding O-glycosidation and the electrophilic substitution product, 4-hydroxyphenyl-5-thio-D-xylopyranoside (equation 6)[17].

AlCl$_3$-catalyzed alkylation of calix[8]arenes with isopropyl chloride gives selective upper-rim isopropylation, showing the phenolic nature of the calixarenes (equation 7). The reaction is limited to calix[8]arenes and is not successful with calix[4]arenes and calix[6]arenes[18], in which case mixtures of products are obtained.

(5)

(6)

(7)

B. Bronsted Acid Catalysis

The use of Bronsted acid catalysts such as HF and H_2SO_4 is discouraged in favor of mild solid acid catalysts such as zeolites, montmorillonite and Nafion-H (*vide infra*). Triflic acid and *p*-toluenesulfonic acid can also be used as convenient catalysts for the alkylation of phenols.

para-Toluenesulfonic acid (TsOH) monohydrate is an efficient catalyst for the Friedel–Crafts alkylation of phenols with activated alkyl halides, alkenes or tosylates under mild conditions. In comparison to conventional Friedel–Crafts catalysts such as $AlCl_3$, BF_3, HF and concentrated H_2SO_4, the extent of the formation of undesired products from side reactions such as transalkylation or polymerization was shown to be minimal in the TsOH-catalyzed reactions[19].

Phenol reacts with linear and branched alkenes in the presence of trifluoromethanesulfonic acid (CF_3SO_3H) in chloroform to give the *ortho*- and *para*-alkylphenols, in moderate yields (equation 8)[10]. With branched alkenes, the *para*-alkyl phenols are the major products. The regioselectivity is dramatically altered from entirely *para*-alkylphenol to *ortho*-alkylphenol going from 100% potassium phenolate to 0% potassium phenolate in the presence of the Lewis acid $AlCl_3$.

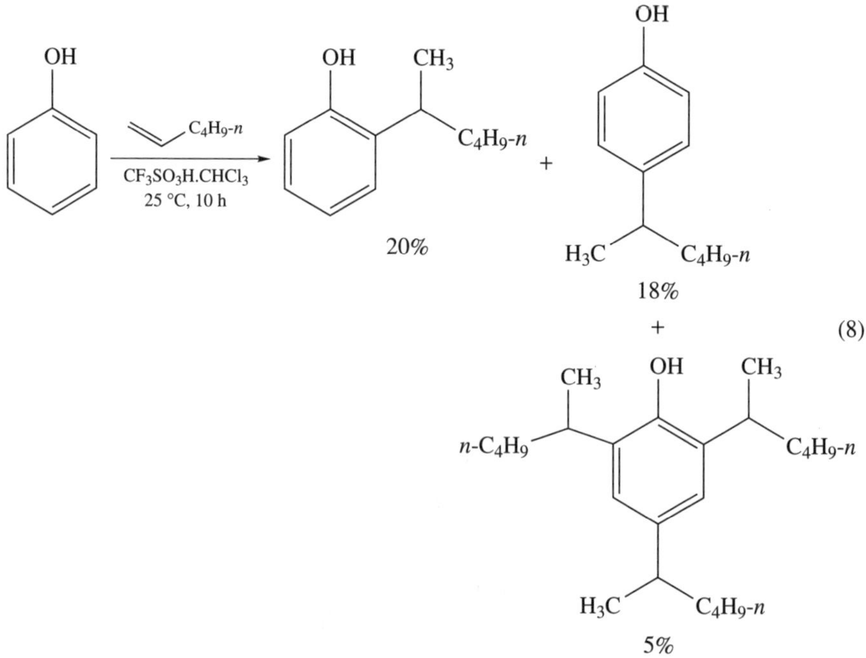

(8)

C. Solid Acid Catalysis

The conventional homogeneous Friedel–Crafts alkylation reactions using HF, H_2SO_4, $AlCl_3$ and BF_3 catalysts are being increasingly replaced by heterogeneous catalysts such as zeolites and Nafion-H in order to minimize the environmental pollution due to the toxic waste water accumulation in the former processes. Alkylation of phenol to give cresols using methanol and zeolites can be achieved at high temperatures[20].

9. Electrophilic reactions of phenols

Nafion-H, a solid perfluorinated resinsulfonic acid, is a strongly acidic reagent and promotes electrophilic alkylation of aromatics under relatively mild conditions. The reactions proceed heterogeneously and workup of the reactions involves a simple filtration of the catalyst at the end of the reaction. Phenols are much more reactive than the alkylbenzenes for the alkylations and high yields of the alkylphenols are obtained. Methylation of phenol using methanol over Nafion-H catalysis gives anisole (37%) and cresols (10%), together with methylanisoles and xylenols. The xylenols are obtained in 15 to 20% yields by methylation of cresols using Nafion-H[21].

The alkylation of phenol with alkyl chloroformates and alkyl oxalates under Nafion-H catalysis proceeds in both liquid and gas-phase conditions in good yields (equation 9). However, these reactions are not regioselective. Importantly, acylation products were not detected under these conditions[22].

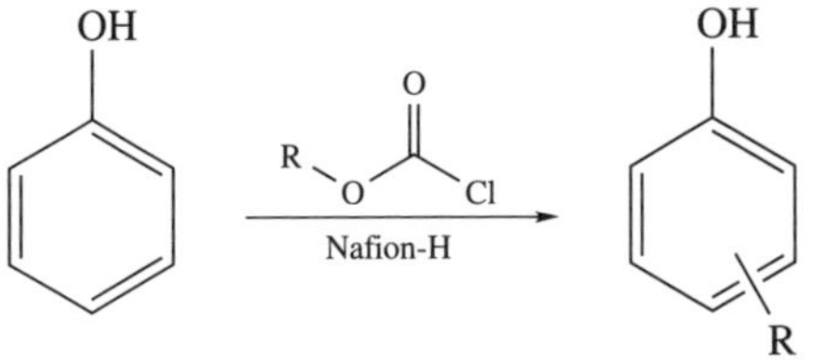

$$\text{(9)}$$

The Claisen rearrangement of allyl phenyl ethers proceeds in the presence of the Nafion-H and silica/Nafion-H nanocomposites[23]. The 2-methyldihydrobenzofuran is formed as a major product (75%) in the presence of the Nafion-H beads; the minor product is the *ortho*-allylphenol (25%) (equation 10). However, the *ortho*-allylphenol is formed as the major product in the presence of the Nafion/H-silica nanocomposites.

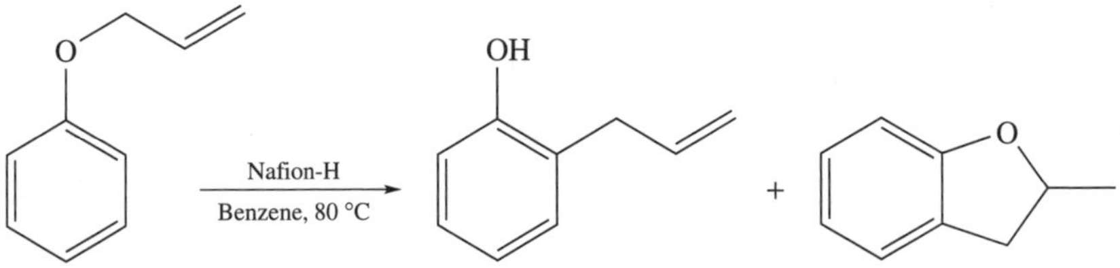

$$\text{(10)}$$

Phenols undergo Friedel–Crafts alkylations with allylic chlorides or allylic alcohols over solid acid catalysts such as acidic K10 clay. For example, 2-buten-1-ol gives 3-aryl-1-butene and 1-aryl-2-butene, albeit in low yields (12%) (equation 11). Allyl carbocations are involved as the reaction intermediates in these reactions[24].

The metal cation-exchanged montmorillonites such as Al^{3+}-montmorillonite can be used for the direct alkylation of phenols using ketones (reductive alkylation). The reaction involves the alkylation, followed by reduction of the intermediate alcohols. Cyclohexanone thus reacts with phenol to give 4-cyclohexylphenol. The reaction of the 4-alkylcyclohexanones with phenols gives almost exclusively the *trans*-(4-alkylcyclohexyl)phenols, useful as the precursors for liquid crystalline materials (equation 12). The deoxygenative reduction of the intermediate tertiary alcohols involves the formation of the carbocation intermediate, which is quenched by a hydride ion apparently derived from the phenol. The use of pentadeuteriophenol in the alkylations results in a significant incorporation of deuterium at the benzylic carbon of the alkylphenols, supporting this hypothesis[25].

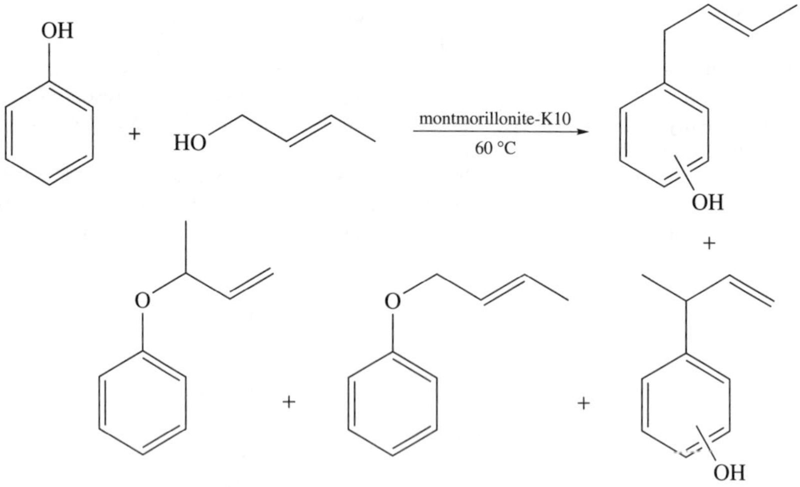

$$(11)$$

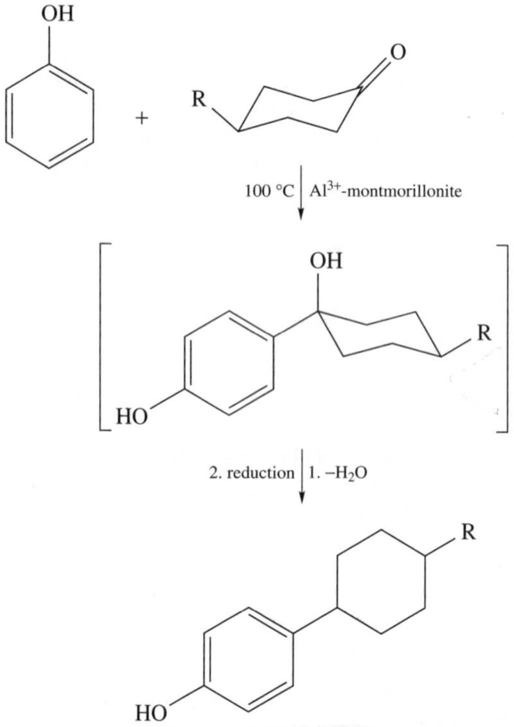

$$(12)$$

1-Naphthol, on the other hand, reacts with 4-alkylcyclohexanones in the presence of Fe^{3+}-montmorillonite to give the tetrahydrobenzonaphthofurans as the major products (equation 13). The Al^{3+}-montmorillonite catalysis also gives the same products, in lower yields. The intramolecular cyclization of the resulting olefin intermediates may account for the observed products.

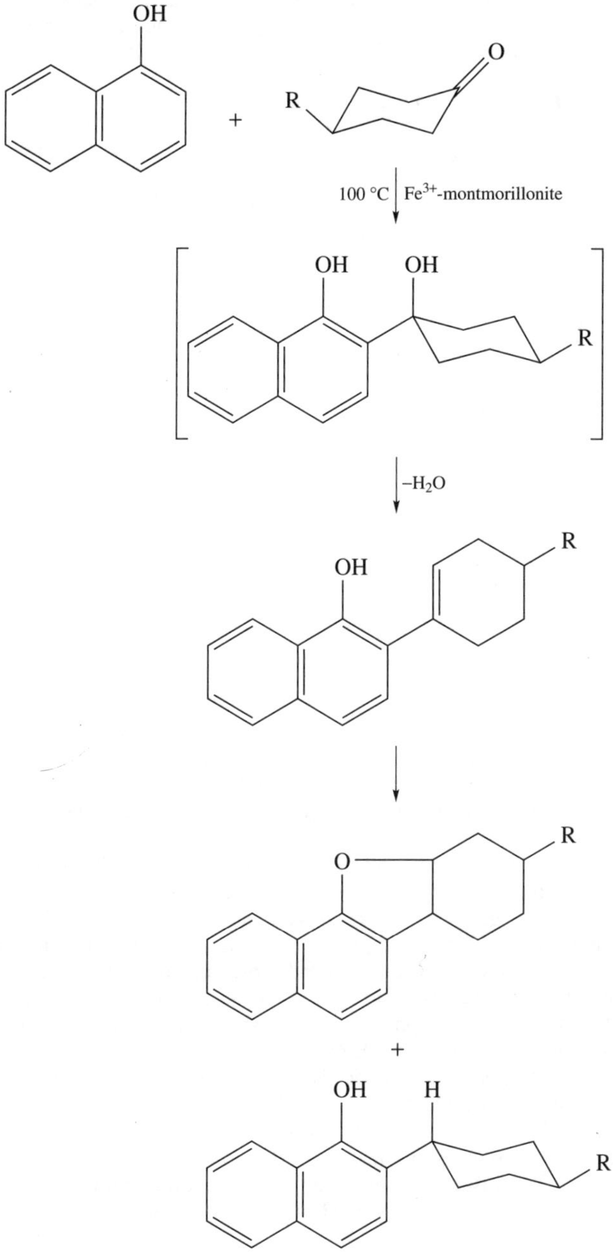

(13)

Major (reduction product)

Montmorillonite-KSF catalyzes the transalkylation of 2,4-di-*tert*-butylphenols in the presence of excess phenol or toluene. The *ortho-tert*-butyl group is preferentially transferred in the process, giving the *para-tert*-butylphenol as the major product (equations 14 and 15). Using xylenes as the solvent at higher temperatures (140 °C) it was possible to transalkylate both of the *tert*-butyl groups. The catalyst can be recycled without loss of reactivity or selectivity[26].

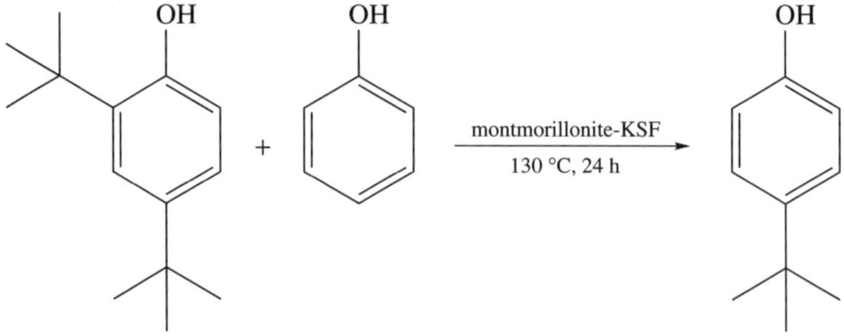

Montmorillonite K10 effects regioselective cyclopentylation (68% *ortho*-selectivity) of phenol using cyclopentanol (equation 16)[27]. The latter serves as starting material for the preparation of optically active (S)-penbutolol, an antihypertensive drug.

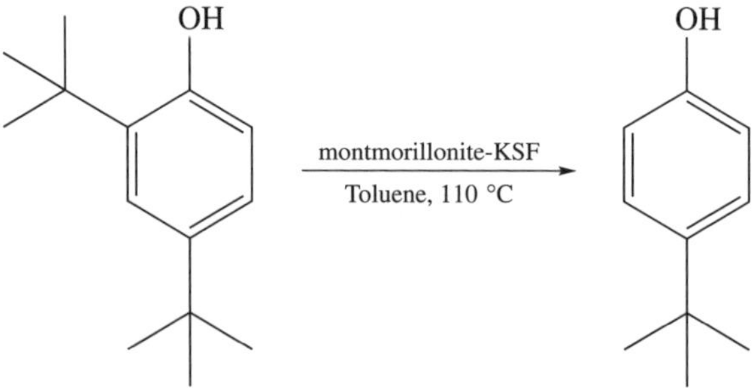

Vapor-phase alkylation of phenol with *tert*-butyl alcohol in the presence of trivalent iron-substituted molecular sieve catalysts (FeMCM-41) gives *para-tert*-butylphenol with high regioselectivity[28,29]. Supported heteropoly acid catalysts have been used in the heterogeneous alkylation reactions of 1-octene or nonene with phenol at 80–100 °C. The catalyst $H_4SiW_{12}O_{40}/SiO_2$ gives 90% *para*-alkylphenol and 10% *ortho*-alkylphenol.

Zeolite catalysis for the alkylation of phenols is an industrially important process[30]. It reduces the cost associated with filtration and disposal of chemical waste generated in the homogeneous catalysis. The alkylation of cresols on zeolites USHY and HZSM-5 in a flow reactor at 380 °C and atmospheric pressure shows the following reactivity order for cresols: *para* > *meta* > *ortho*. The HZSM-5 acid sites are more active than those of USHY, for *para-* and *meta-* but not for *ortho*-cresol, showing that the *ortho*-cresol's access to the active sites in HZSM-5 is the limiting factor. The cresols are transformed through unimolecular isomerization and transalkylation reactions. On HZSM-5, due to its relatively smaller pore size, isomerization is the dominant pathway[31]. Alkylation of phenols with camphene catalyzed by large-porous beta-zeolite yields the corresponding O- and C-alkylated phenols. The C- versus O-alkylations can be partly controlled by the reaction solvent[32].

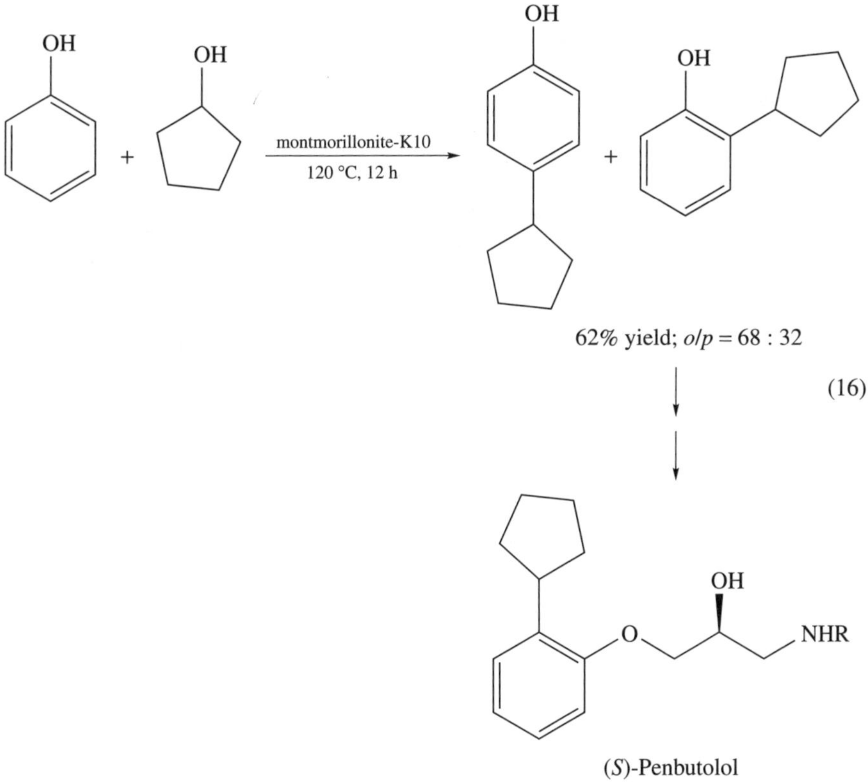

9. Electrophilic reactions of phenols

[Reaction scheme: phenol + cyclopentanol → para-cyclopentylphenol + ortho-cyclopentylphenol, montmorillonite-K10, 120 °C, 12 h; 62% yield; o/p = 68 : 32]

(16)

(S)-Penbutolol

A mixture of C- and O-alkylated phenols was obtained when phenol was treated with cyclohexene over silica-supported boron trifluoride-hydrate catalyst (equation 17)[33]. In these reactions, the O-alkylated compounds are the major products (50–65%). The ring-alkylated products are formed in 20–30% yields, along with 2–5% of the O,C-dialkylated compounds. Fresh catalyst needs to be added during the reaction, due to the possible catalyst poisoning. The reaction may involve the initial formation of the cyclohexyl cation by the protonation of the cyclohexene through the BF_3-coordinated phenol. Whereas the homogeneous boron trifluoride solution causes the rearrangement of the O-alkylated phenols to C-alkylated isomeric compounds, the solid catalyst, due to its relatively milder Lewis acidity, does not promote such rearrangement. Thus alkyl phenyl ethers could be ring-alkylated with the latter reagent without involving the cleavage of the ether moiety.

Alkylation of phenol with methanol has been carried out over Lewis acid ion-exchanged Y-zeolites, FeY, ZnY, CdY and LaY at temperatures of 523, 573, 623, 673 and 698 K to give *ortho*-cresol, 2,6-xylenol and anisole. Selectivity to *ortho*-cresol decreases with increase of temperature, as it further reacts to give 2,6-xylenol[34].

Phenols react with deactivated carbonyl compounds such as chloral (2,2,2-trichloroethanal) in the presence of different dealuminated protonic zeolites (Y-FAU, MOR, MFI and BEA) to give the corresponding carbinols. A high *para*-selectivity was achieved using HBEA zeolite (Si : Al = 12.5) (equation 18)[35].

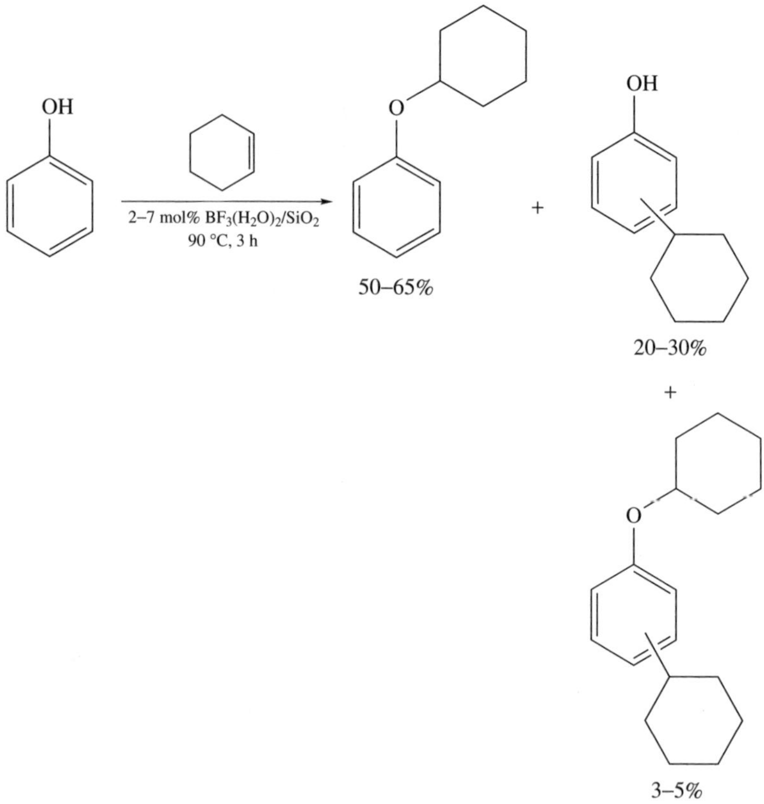

[Structural reaction scheme with phenol + cyclohexene, 2–7 mol% BF$_3$(H$_2$O)$_2$/SiO$_2$, 90 °C, 3 h, giving cyclohexyl phenyl ether (50–65%), cyclohexylphenol (20–30%), and cyclohexyl(cyclohexylphenyl) ether (3–5%)] (17)

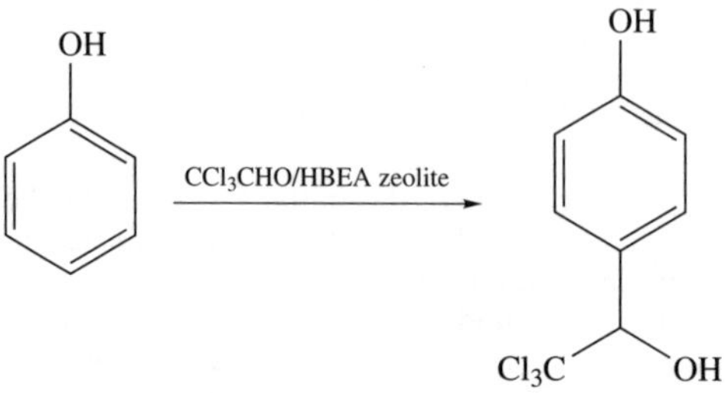

[Phenol + CCl$_3$CHO/HBEA zeolite → 4-(1-hydroxy-2,2,2-trichloroethyl)phenol] (18)

The vapor-phase catalytic alkylation of phenol with methanol and dimethyl carbonate on CrPO$_4$ and CrPO$_4$-AlPO$_4$ catalysts gives a mixture of O- and C-alkylation products, the latter being predominantly *ortho*-isomers (equation 19)[36].

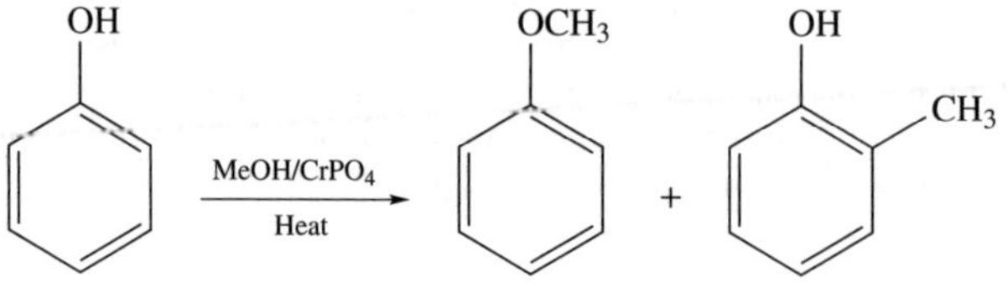

[Phenol + MeOH/CrPO$_4$, Heat → anisole + o-cresol] (19)

9. Electrophilic reactions of phenols

1,2-Tungstophosphoric acid (HPW) and its Cs and ammonium salts encapsulated into the channels of MCM-41 molecular sieves were useful for the conversion of phenol and acetone to Bisphenol-A[37]. The Cs-HPW/MCM system was more selective to the *p,p'*-isomer than that of zeolites ZSM-5 and H-Y. The Bisphenol-A is useful industrially in the production of polymeric resins. Various other catalysts such as Amberlyst resins were used for this purpose (equation 20)[38]. The latter catalyst gave a 90% selectivity for the *p,p'*-isomer[39]. The MCM-41 encapsulated catalyst was shown to have superior thermal characteristics compared to that of the Amberlyst catalyst.

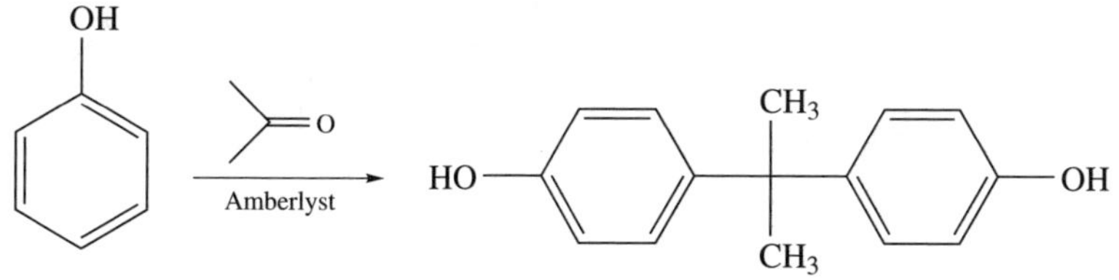

(20)

Highly acidic Al-MCM-41, U-MCM-41 and Th-MCM-41 catalysts have been used for the Friedel–Crafts alkylation of 2,4-di-*tert*-butylphenol with cinnamyl alcohol to give the corresponding substituted benzopyran (equation 21)[40]. The reaction involves an initial *ortho*-alkylation, followed by an acid-catalyzed intramolecular cyclization. Loss of the 2-*tert*-butyl group results in minor byproducts.

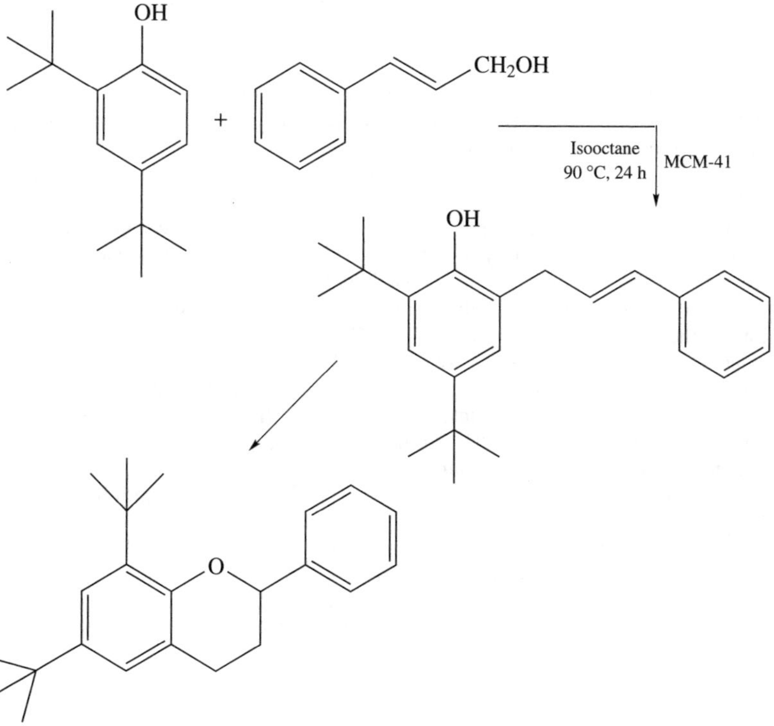

(21)

The effects of various parameters on the *tert*-butylation of phenol on the Zeolite-H-beta have been studied[41]. Alkylation of phenol in the vapor phase using Zeolite SAP-11 and *tert*-butyl alcohol gives the *ortho*- and *para-tert*-butylphenols, together with the 2, 6-di-*tert*-butylphenol (equation 22)[42]. Vapor-phase alkylation of phenol with *tert*-butyl alcohol over solid superacid catalysts, such as sulfated zirconia[43] and mesoporous H-AlMCM-41[44], gives *para-tert*-butylphenol as a major product in high regioselectivity.

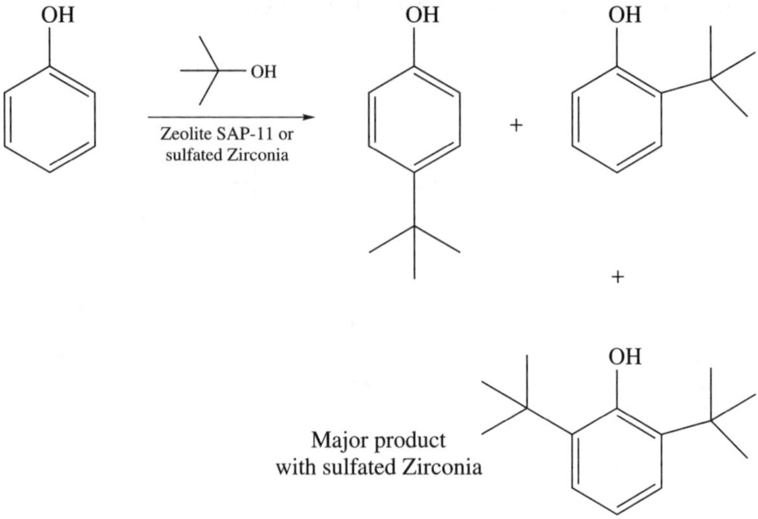

(22)

Modified HY zeolites, with increased pore size distribution, are shown to be efficient catalysts for the alkylation of phenol with long-chain olefins[45]. The enhanced activity results by improved accessibility of active acid sites on the zeolite for the long alkyl chains. The modified zeolites are potentially valuable catalysts in the petroleum industry[46].

Solid acid catalysts, consisting of polysiloxane bearing alkylsulfonic acid groups (MCM-41), are comparable in their catalytic activity to those of the polystyrene-based cation exchange resins. These catalysts can be used in the preparation of *para*-Bisphenol-A by the alkylation of phenol with acetone. Other application of these catalysts lie in the alkylation of phenol with isobutene at 90–130 °C[47].

Lewis acids immobilized on ionic liquids have been used as the acid catalysts for the alkylation of phenols. The catalytic activities of the immobilized ionic liquids were found to be higher than those for the zeolites. Typically, ionic liquids such as butylmethylimidazolium halides are treated with $AlCl_3$ to give the ionic liquids with halogenoaluminates as the counter anions. They show enhanced Lewis acid character and promote predominantly C-alkylation of phenols over O-alkylation. The alkylation of phenol with dodecene, for example, in the presence of these immobilized ionic liquids results in up to 70% of C-alkylated products (*ortho* and *para* products) and 30% of O-alkylated product, comparable to zeolite catalysis (equation 23). The rates of alkylation of phenols are slower than those of arenes due to the complexation of the phenolic group with the Lewis acidic ionic liquids. At higher temperatures conversions of up to 99% could be achieved.

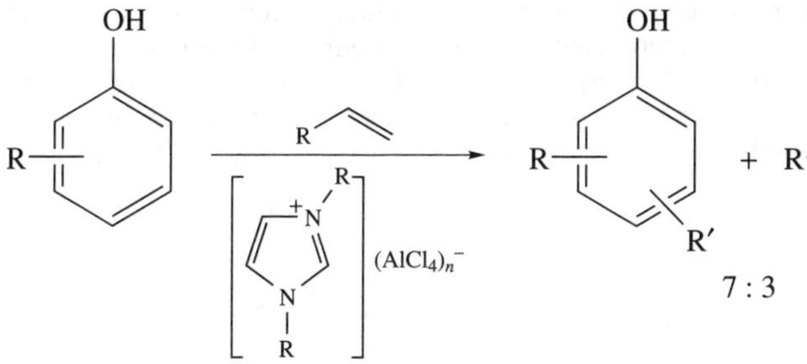

9. Electrophilic reactions of phenols

(23)

D. Alkylations under Supercritical Conditions

Alkylation of phenols using primary, secondary and tertiary alcohols was achieved using supercritical water (at the near-critical region, 250–350 °C). This process eliminates the need for environmentally hazardous organic solvents and acid catalysts[48]. Both *ortho*- and *para*-alkylphenols were formed in these reactions, their ratio being dependent on the temperature of the reaction mixture[49].

Supercritical carbon dioxide can be used as a solvent in the $BF_3 - Et_2O$-catalyzed alkylation of phenols. Under these conditions phenol reacts with 2-chloro-2,4, 4-trimethylpentane and poly(isobutylene)-Cl (PIB-Cl) to give the corresponding *para*-alkylated phenols (equations 24 and 25)[50].

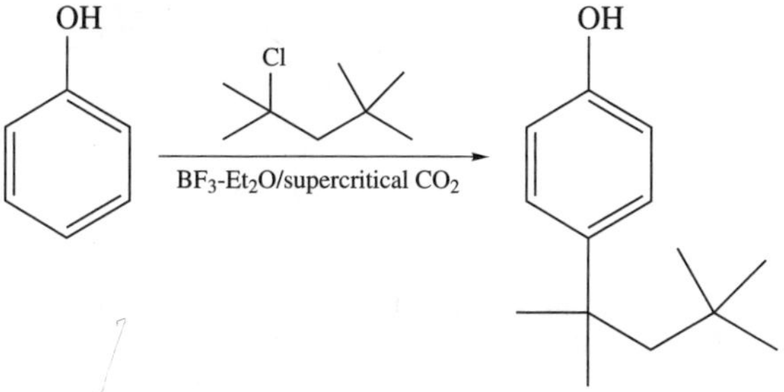

(24)

(25)

E. Stereoselective Alkylations

The hydroxyalkylation of phenolates with N-protected α-amino aldehydes gives β-amino-*ortho*-hydroxybenzyl alcohols with good to excellent diastereoselection. For example, the reaction of 4-methoxyphenol with ethylmagnesium bromide followed by reaction with N-protected α-amino aldehydes gives ephedrine-like compounds with high diastereoselectivity in good yields (equation 26)[51]. The stereochemistry of the reaction can be controlled by modulating the nature of the reactive complex, e.g. by varying the Grignard reagents and other reaction conditions.

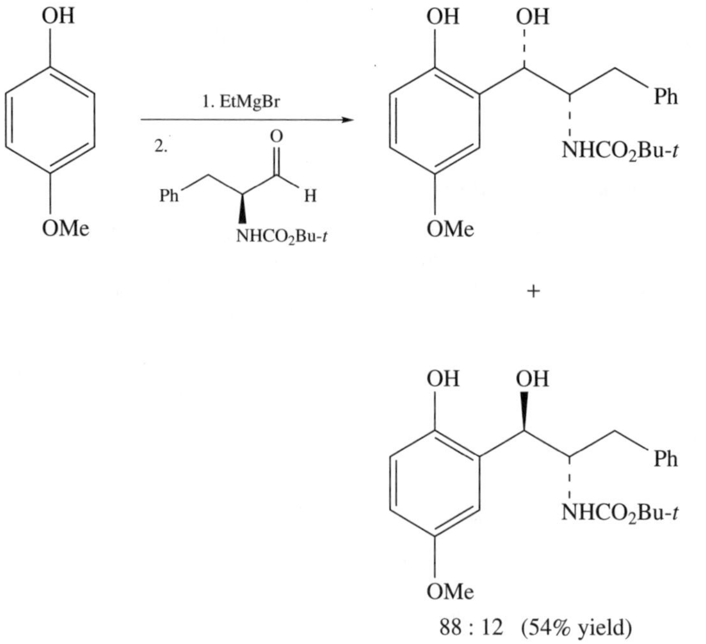

(26)

88 : 12 (54% yield)

The alkylation of phenoxymagnesium halides with N-(*tert*-butoxycarbonyl)-α-amino aldehydes also gives excellent diastereoselection. 2-*tert*-Butylphenoxymagnesium bromide, for example, has been found to react with N-(*tert*-butoxycarbonyl)-L-prolinal to give exclusively the *syn* diastereomer regio- and stereoselectively (equation 27)[52].

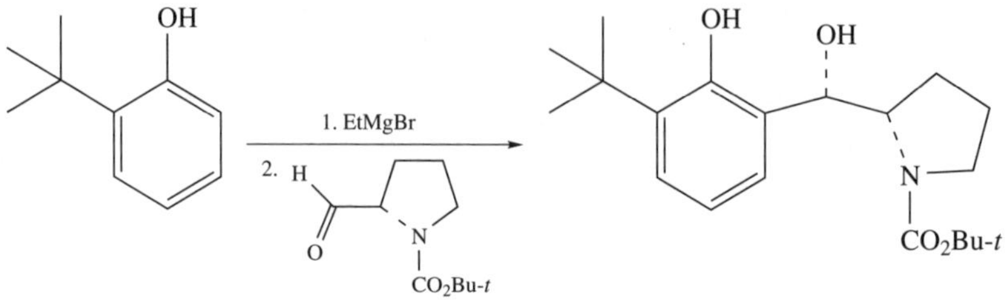

(27)

The crystal structures of the bromomagnesium phenolate and its complex with *para*-isopropylbenzaldehyde further demonstrate the chelation control as the factor for the regioselective alkylations. In this process the metal coordination sphere would be expanded from 4 to 5[53].

The hydroxyalkylation of phenols with chiral glyoxylates, followed by hydrolysis, gives regioselectively 2-hydroxymandelic acids with high enantioselectivity (equation 28). The crystal-structure determination of the titanium phenoxide complex shows evidence for chelation-controlled reaction giving the observed high enantioselectivities[54].

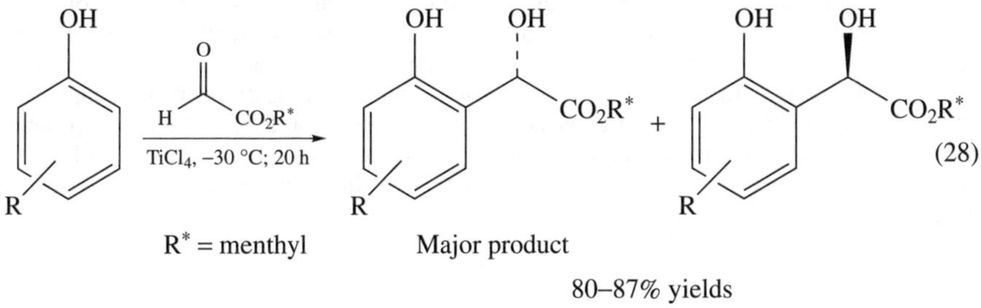

$$\text{(28)}$$

R* = menthyl   Major product

80–87% yields

The synthesis of analogues of the spiroketal-containing pyranonaphthoquinone antibiotic griseusin A can be achieved by the regio- and stereoselective hydroxyalkylation of 4,8-dimethoxy-1-naphthol (equation 29)[55].

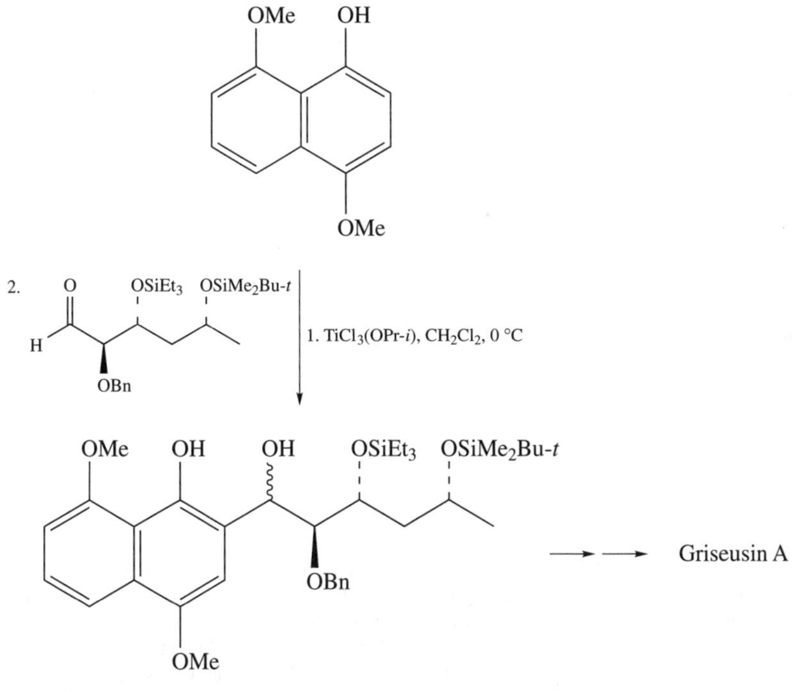

$$\text{(29)}$$

→ Griseusin A

The reaction of 5 equivalents of substituted phenols with 1 eq of (S)- or (R)-methyl 7-(4-fluorophenyl)-7-hydroxyheptanoate afforded *ortho*-alkylated phenol derivatives enantioselectively in 33 to 42% chemical yield and 90 to 93% enantiomeric excess. These derivatives are used as the non-prostanoic thromboxane A(2) receptor antagonists[56].

Trost and Toste have developed asymmetric O- and C-alkylations of phenols[57], with enantioselectivities ranging from 80 to 97%. The reaction of various substituted phenols with five- to seven-membered cyclic allyl carbonates in the presence of chiral salen catalysts leads to the formation of chiral O-allylphenols with up to 94% enantioselectivity. The latter undergo Claisen rearrangement upon heating to 50 °C in the presence of Lewis acid, Eu(fod)$_3$, with excellent chirality transfer from the substrate to give *ortho*-allylphenols. Other Lewis acids such as BCl$_3$ or Et$_2$AlCl lead to products with significant racemization. The reaction is also applicable to acyclic allylic carbonates, in which case a mixture of *cis*- and *trans*-isomers of the *ortho*-allylphenols are formed (equation 30).

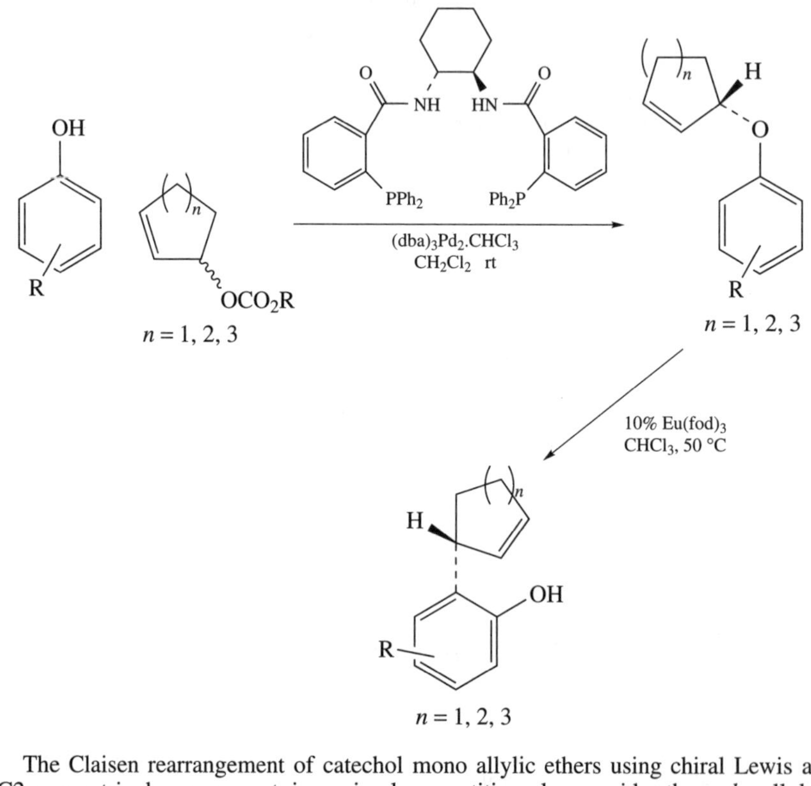

(30)

The Claisen rearrangement of catechol mono allylic ethers using chiral Lewis acidic, C2-symmetric, boron reagent, in equimolar quantities, also provides the *ortho*-allyl products with high enantioselectivity (equations 31 and 32)[58]. Under these reaction conditions, 2-hexenylphenyl ether and O-methyl protected catechol monoallyl ethers, both of which do not have a free *ortho*-hydroxy group, did not undergo the Claisen rearrangement. Thus, the complexation of the boron with the free hydroxyl group is essential for the chiral boron reagent to act as the catalyst for the Claisen rearrangement. O-allyl ethers of salicylic acid undergoes the Claisen rearrangement, but the enantioselectivity and the yield of the *ortho*-allyl product were low (equation 33). The *para*-isomer is also formed as a byproduct.

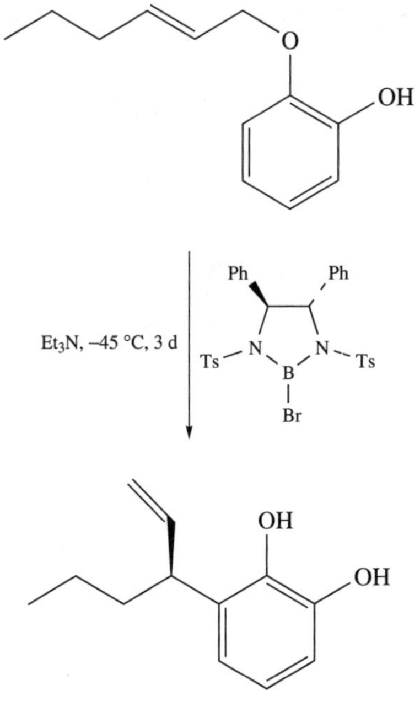

9. Electrophilic reactions of phenols

(31)

89% yield; ee: 93%

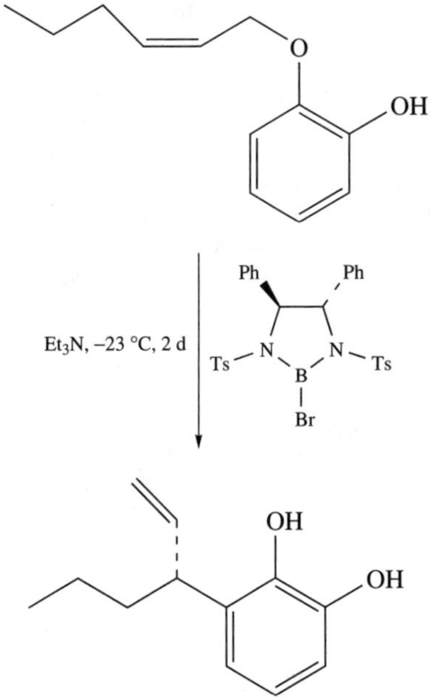

(32)

97% yield; ee: 95%

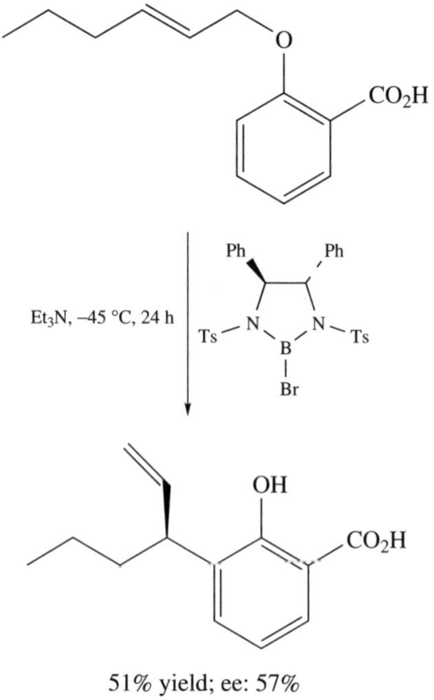

(33)

51% yield; ee: 57%

F. Formylation and Phenol-Formaldehyde Resins

Phenols can be electrophilically formylated by a variety of reagents. Formaldehyde/$SnCl_4$/Bu_3N gives salicylaldehydes from phenols with high yields and selectivity (equation 34)[59]. Selective *ortho*-formylation of phenols has also been achieved using paraformaldehyde and magnesium chloride/triethylamine (equation 35). Alkyl-substituted phenols give excellent yields of the corresponding salicylaldehydes. Similar results have been obtained with chlorophenols and 3- and 4-methoxyphenols. 2-Methoxyphenol is unreactive under these conditions[60]. Other reagents for the formylations of phenols include HCN/$AlCl_3$, DMF/$POCl_3$, $MeOCHCl_2$/$TiCl_4$ and $CHCl_3$/NaOH[61].

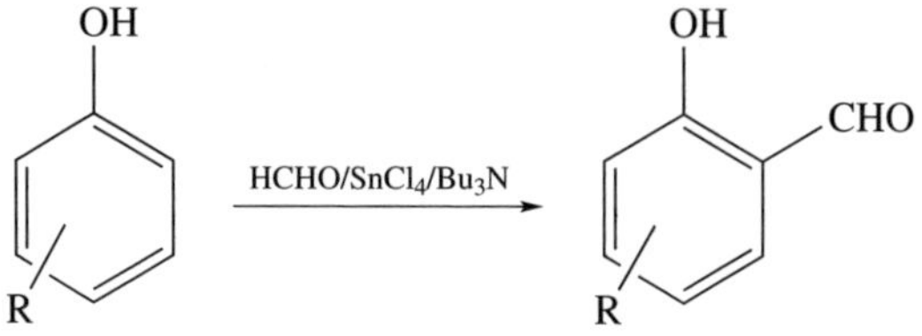

(34)

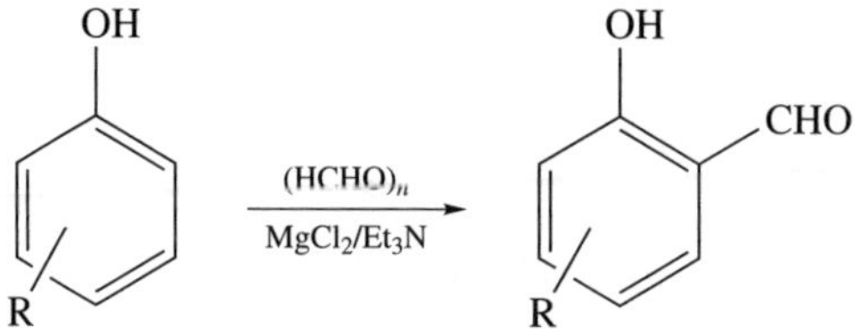

(35)

9. Electrophilic reactions of phenols

The reactions of phenols with formaldehyde in the presence of montmorillonite KSF-Et$_3$N as a heterogeneous catalyst give the substituted salicylaldehydes in high yields (equation 36)[62].

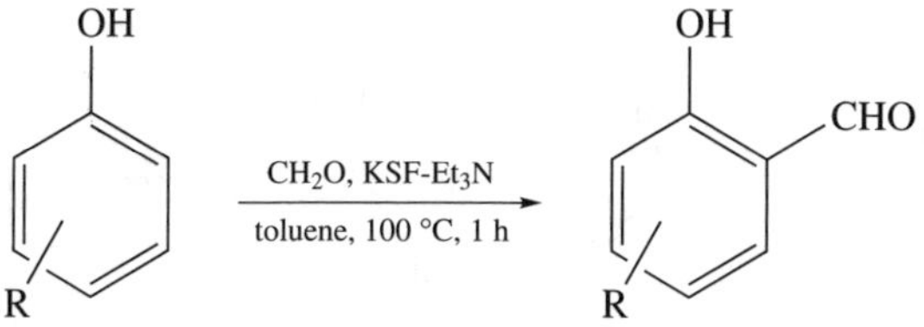

$$\underset{R}{\text{phenol}} \xrightarrow[\text{toluene, 100 °C, 1 h}]{\text{CH}_2\text{O, KSF-Et}_3\text{N}} \underset{R}{\text{2-hydroxybenzaldehyde}} \qquad (36)$$

The hexamethylenetetramine-trifluoroacetic acid system was shown to introduce three aldehyde groups into phenol. Thus, 2-hydroxy-1,3,5-benzenetricarbaldehyde was synthesized from phenol conveniently in one step by this method (equation 37)[63].

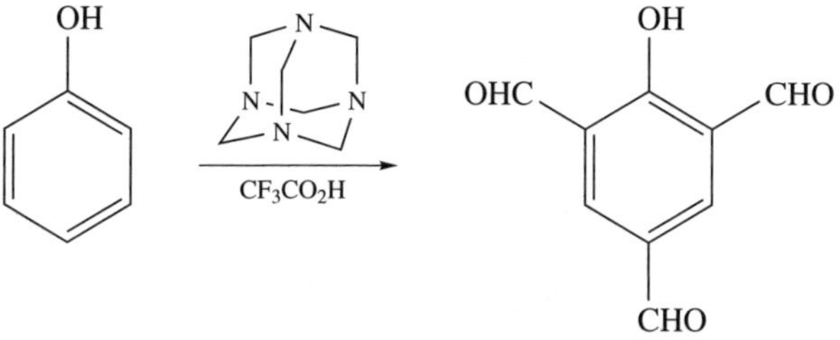

$$\text{PhOH} \xrightarrow[\text{CF}_3\text{CO}_2\text{H}]{\text{(CH}_2)_6\text{N}_4} \text{2-hydroxy-1,3,5-benzenetricarbaldehyde} \qquad (37)$$

Phenol reacts readily with formaldehyde to give trimethylolphenol (2,4,6-tris(hydroxymethyl)phenol), which undergoes further alkylative polymerization in the presence of acid catalysts (equation 38). Thus-formed phenol-formaldehyde resins (prepolymers) can be used to crosslink a variety of polymers. This is a broad area of industrial significance.

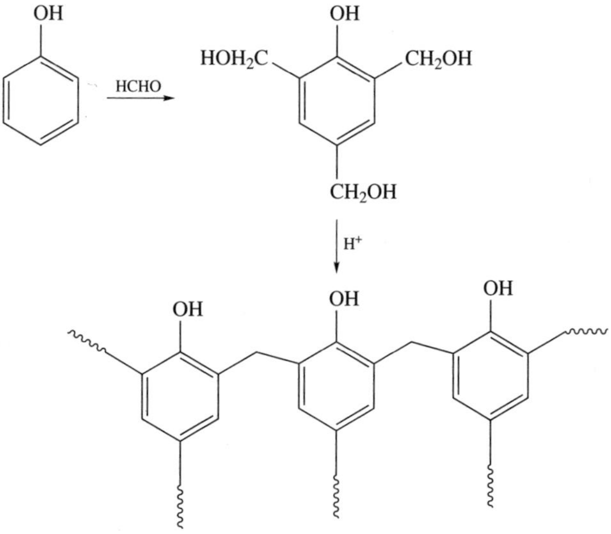

$$\text{PhOH} \xrightarrow{\text{HCHO}} \text{2,4,6-tris(hydroxymethyl)phenol} \xrightarrow{\text{H}^+} \text{phenol-formaldehyde resin} \qquad (38)$$

phenol-formaldehyde resin

The phenol-formaldehyde prepolymers were polymerized with 4-(1-phenylethyl)phenol (*para*-styrenated phenol) (equation 39). The sulfonation of the resulting polymer gave a cation exchange resin, which is useful as an acid catalyst[64].

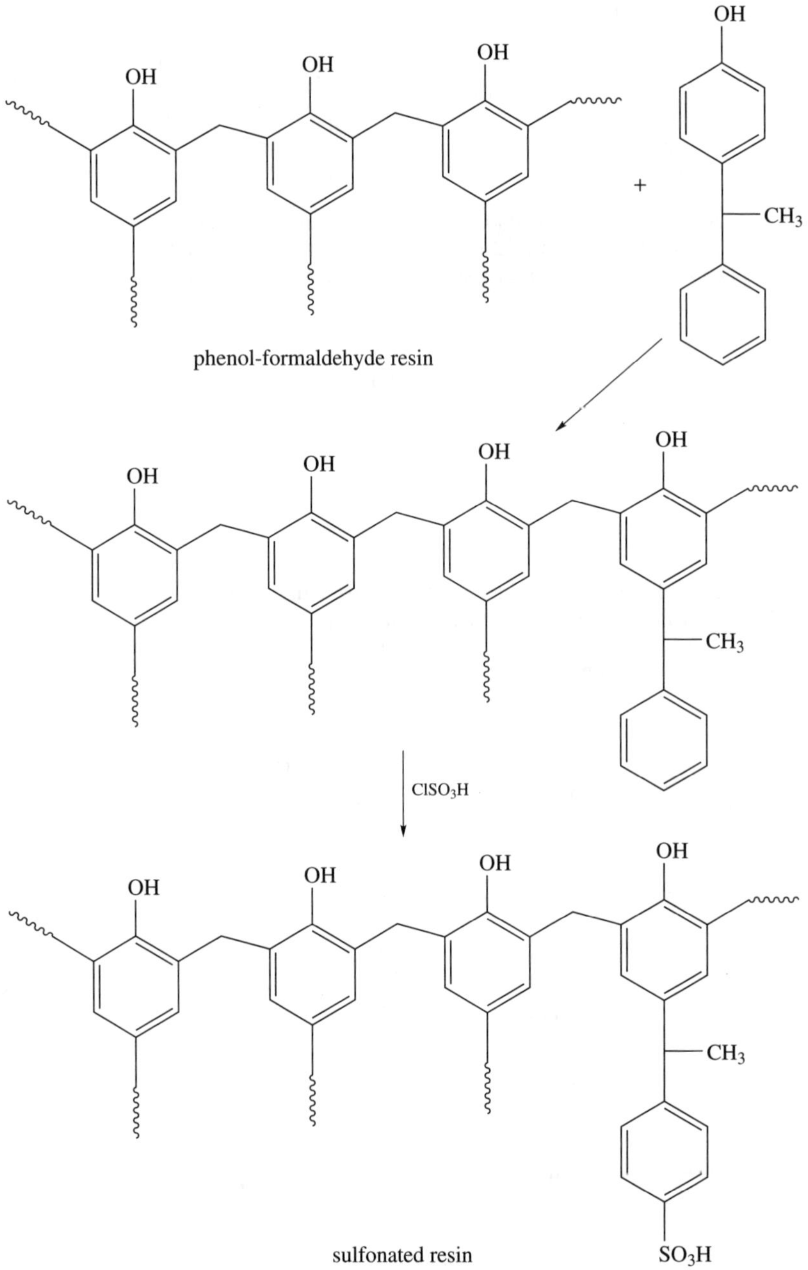

III. FRIEDEL–CRAFTS ACYLATIONS

Regioselective direct acylation of phenol and naphthol derivatives with acid chlorides was achieved by using hafnium triflate, $Hf(OTf)_4$ (5 to 20 mol%), as a catalyst (equations 40 and 41)[65].

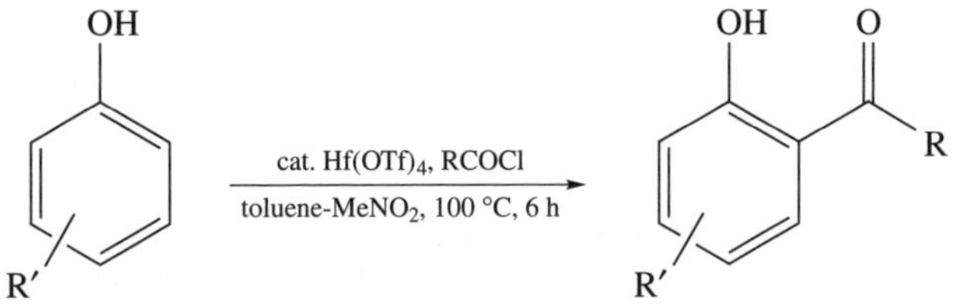

$$\text{phenol derivative} \xrightarrow[\text{toluene-MeNO}_2, 100\,°C, 6\,h]{\text{cat. Hf(OTf)}_4, \text{RCOCl}} \text{ortho-acyl phenol} \quad (40)$$

$R' = OMe, H, Me; R = Me$ 53–84%

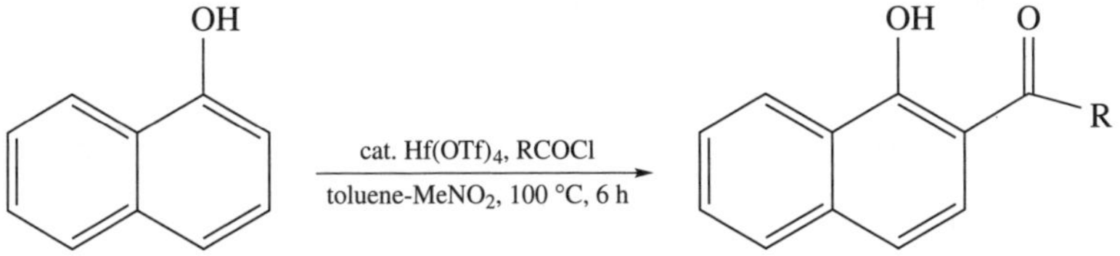

$$\text{1-naphthol} \xrightarrow[\text{toluene-MeNO}_2, 100\,°C, 6\,h]{\text{cat. Hf(OTf)}_4, \text{RCOCl}} \text{2-acyl-1-naphthol} \quad (41)$$

$R = Me, c\text{-}C_6H_{11}$ ca 90%

The $Hf(OTf)_4$, is also effective in the *ortho*-acylation of phenols using carboxylic acids instead of the acid chlorides, although somewhat lower yields are obtained and larger amounts of the catalyst are required (equations 42 and 43)[66].

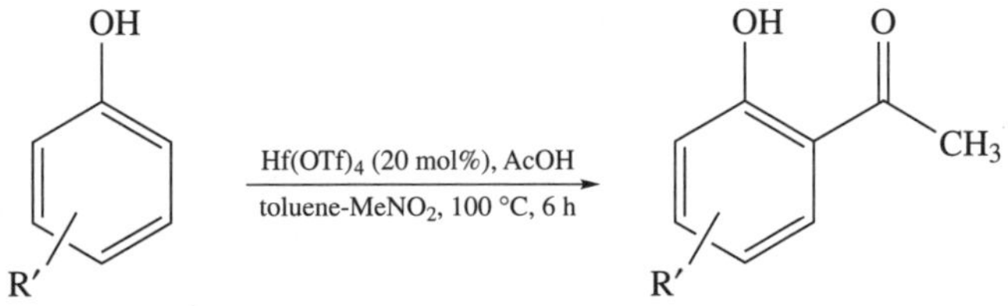

$$\text{phenol derivative} \xrightarrow[\text{toluene-MeNO}_2, 100\,°C, 6\,h]{Hf(OTf)_4\,(20\,\text{mol\%}),\,AcOH} \text{2'-hydroxyacetophenone} \quad (42)$$

$R' = OMe, H, Me, \text{etc.}$ 55–72%

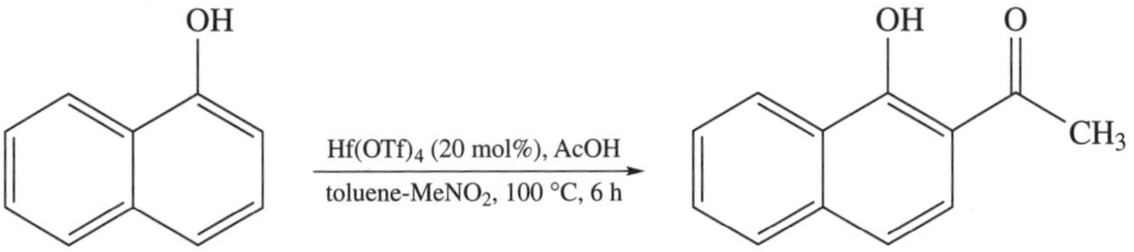

$$\text{1-naphthol} \xrightarrow[\text{toluene-MeNO}_2, 100\,°C, 6\,h]{Hf(OTf)_4\,(20\,\text{mol\%}),\,AcOH} \text{2-acetyl-1-naphthol} \quad (43)$$

81%

Similarly, scandium triflate ($Sc(OTf)_3$), zirconium triflate ($Zr(OTf)_4$) and titanium chloro(tris)triflate ($TiCl(OTf)_3$) were also used for the *ortho*-acylation of phenols and 1-naphthols using acid chlorides[67,68].

By using suitable protecting groups, *meta*-acylation of phenols and ansioles was made possible using acetyl chloride and aluminium chloride[69].

Phenol reacts with acetic anhydride to give 4-methylcoumarin in a process involving O-/C-diacylation and cyclization over CeNaY zeolite in high yields (equation 44)[70].

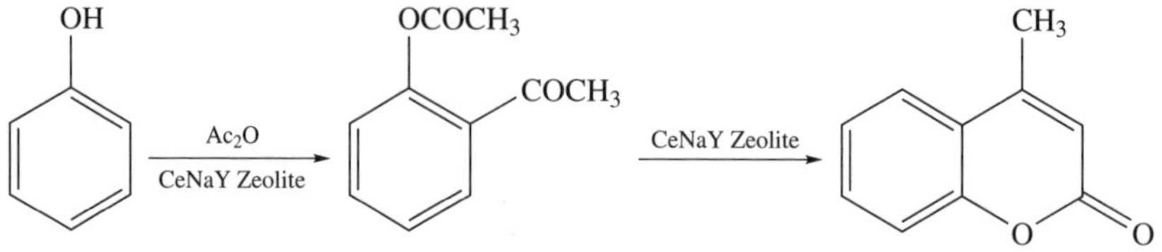

Acetylation of 2-methoxynaphthalene by acetic anhydride over HBEA zeolite gives 1-acetyl-2-methoxynaphthalene, 2-acetyl-6-methoxynaphthalene and a small amount of 1-acetyl-7-methoxynaphthalene (equation 45)[71]. The 1-acetyl-2-methoxynaphthalene rearranges to the other isomers under longer contact times, probably involving both intermolecular transacylation and intramolecular rearrangements (equation 46).

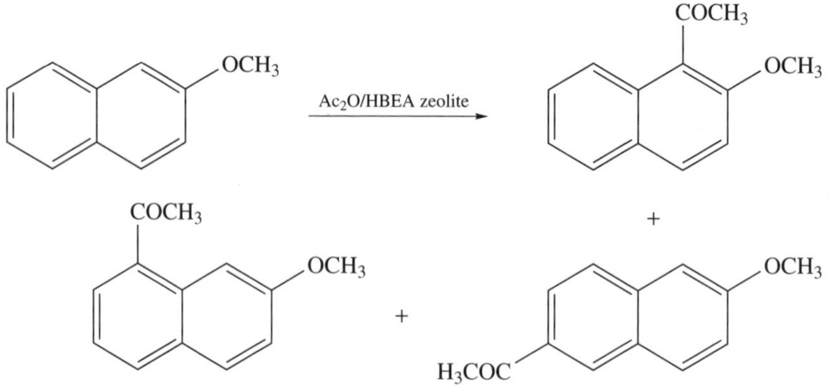

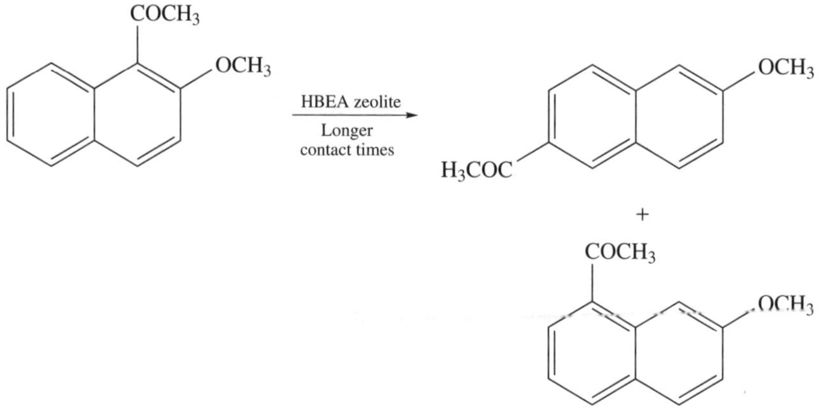

9. Electrophilic reactions of phenols 631

Gas-phase acetylation of phenol using β-zeolites gives phenyl acetate rapidly, which rearranges (see Fries rearrangement, *vide infra*) to *ortho*-hydroxyacetophenone and *para*-hydroxyacetophenone. The *o/p* ratio is high under these conditions[72].

Zeolites such as HZSM-5 were used for the acylation of phenol using acetic anhydride or acyl halides[73-77]. Cobalt, copper and cerium ions show a promoting effect on the zeolite-catalyzed acylation reactions[73]. The Friedel–Crafts acetylation of phenol over acidic zeolites involves initial formation of the phenyl ester, followed by the Fries rearrangement, both being catalyzed by the zeolites[74]. Usually, high *para*-selectivity for the acylation is observed for the zeolite catalysis. However, modification of the zeolites involving dealumination of the outer surface of the crystallites gives high *ortho*-selectivity[78].

Pyridine-catalyzed acylation of phenols using benzoyl chloride and benzoyl bromide was reported[79]. Acylation of phenols using acetyl chloride or benzoyl chloride can be achieved using triflic acid as the catalyst[80] in nonpolar solvents such as methylene chloride. The role of pyridine in these reactions seems to be the intermittent formation of the benzoylpyrimidinium ions as the reactive species. The activated phenolic compounds such as resorcinol, on the other hand, could be acylated in near-supercritical water (250–300 °C) without using any external Lewis acid catalysts (equation 47)[81]. The equilibrium conversions in water, however, are to the extent of about 4%. Running the same reactions in neat acetic acid causes a tenfold increase in yield.

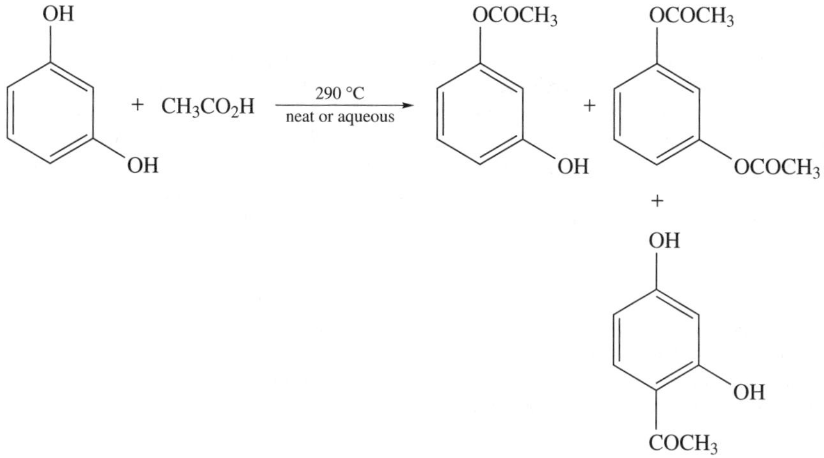

$$\text{resorcinol} + CH_3CO_2H \xrightarrow[\text{neat or aqueous}]{290\,°C} \text{products} \quad (47)$$

Whereas the Friedel–Crafts alkylations require only catalytic quantities of the Lewis acidic $AlCl_3$ catalyst, Friedel–Crafts acylations of phenols require excess Lewis acids, due to the complex formation of the Lewis acids with the hydroxyl group[82]. Boron trifluoride-phosphoryl chloride, in stoichiometric amounts, is used for the Fridel–Crafts reaction of phenol with β,β-dimethylacrylic acid to give the acrylophenone[83].

It was shown that acetic anhydride/zinc chloride is an efficient C-acylating reagent for phenol and polyphenols (such as resorcinol, phloroglucinol, catechol and pyrogallol) resulting in improved yields of the corresponding hydroxyacetophenones[84]. With resorcinol, the isomeric diacetyl derivatives are formed in excellent yields in a single step, while catechol and hydroquinone give only monoacetyl derivatives. Pyrogallol gives a monoacetyl derivative, while phloroglucinol gives both mono- and diacetyl derivatives but not triacetyl derivatives.

Montmorillonite K10 and KSF are highly efficient for the O-acetylation of phenols and naphthols (equation 48)[85]. The reaction can be achieved in solvents such as CH_2Cl_2 or under solvent-free conditions.

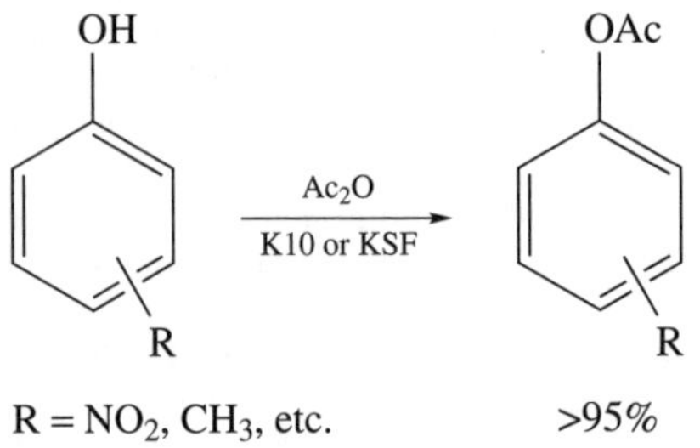

$$\underset{R = NO_2, CH_3, \text{etc.}}{\underset{R}{\text{OH}}} \xrightarrow[\text{K10 or KSF}]{Ac_2O} \underset{>95\%}{\underset{R}{\text{OAc}}} \qquad (48)$$

Phenols undergo Friedel–Crafts type reaction with RSCN in the presence of BCl_3 to give the *ortho*-imino products which, upon hydrolysis, give thiocarboxylic esters[86].

IV. NITRATION AND NITROSATION

A. Regioselectivity

Phenol reacts with $NaNO_2$ on wet SiO_2 at room temperature to give mono- or dinitrosation products, which are *in situ* oxidized by oxone to give the *ortho*- and *para*-nitrophenols in high yields, depending on the reaction conditions[87].

Phenols and alkylaromatics can be nitrated with 100% nitric acid on MoO_3/SiO_2, WO_3/SiO_2, TiO_2/SiO_2 and TiO_2-WO_3/SiO_2 systems in over 90% yields[88]. In these reactions, the most active catalysts showed *para*-selectivity for nitration.

Due to the industrial importance of the 2-nitrophenol, extensive research has been focused on enhancing the regioselectivity of the nitration of phenol[89]. The regiochemistry of the nitration is dramatically increased, giving an *ortho/para* ratio of 13.3 with acetyl nitrate as the reagent, when dry silica gel was used for the catalysis. In chloroform solvent, in the absence of silica gel, a normal *ortho/para* ratio of 1.8 was obtained[90]. 2-Naphthol gives 1-nitro-2-naphthol exclusively, under these conditions. 4-Hydroxy-3-methoxybenzaldehyde (vanillin) gives the expected product, 4-hydroxy-3-methoxy-5-nitrobenzaldehyde in high yields (equation 49).

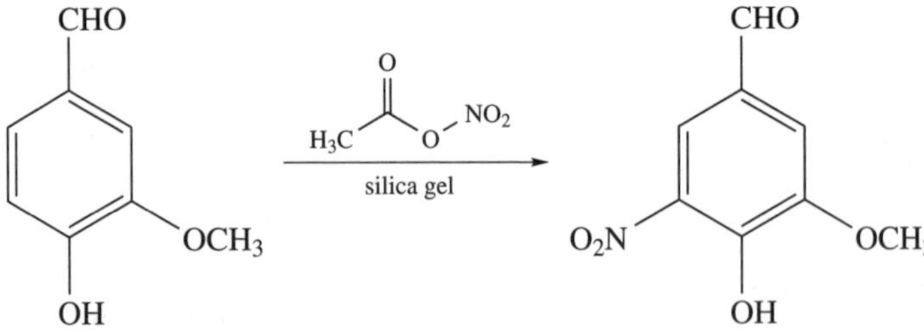

$$\underset{\text{OH}}{\underset{\text{OCH}_3}{\text{CHO}}} \xrightarrow[\text{silica gel}]{H_3C-C(O)-O-NO_2} \underset{\text{OH}}{\underset{O_2N \quad \text{OCH}_3}{\text{CHO}}} \qquad (49)$$

The acidic hydrogen of phenol may participate in the formation of a phenolacetyl nitrate–silica complex, in which the nitro group is well positioned in a six-membered transition state, for the *ortho*-attack. In other words, the initially formed oxonium ion is stabilized through the H-bonding interactions with the silica gel (equation 50).

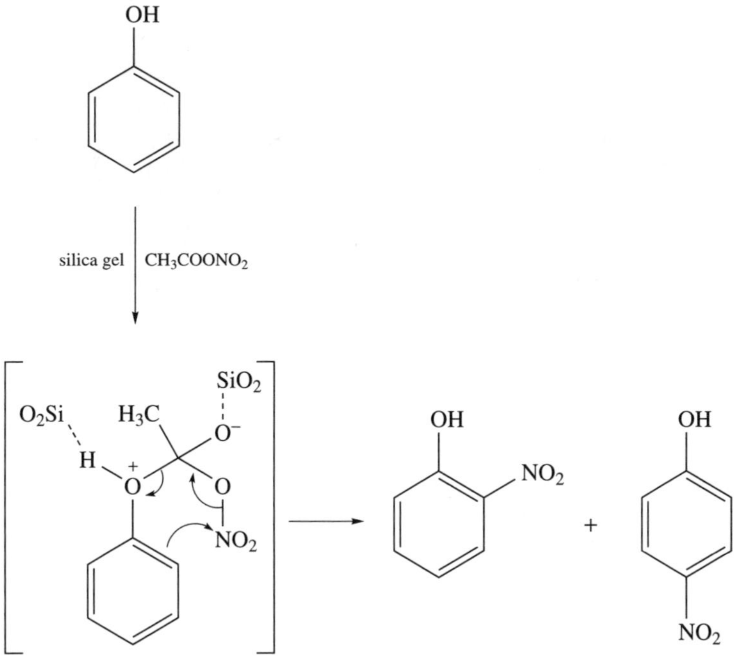

$$\text{(50)}$$

The use of supported catalysts, such as zeolites, usually provides the *para*-isomer as the predominant product. Nitration of phenols using 'claycop'[91], a reagent consisting of an acidic montmorillonite impregnated with anhydrous cupric nitrate and montmorillonite impregnated with bismuth nitrate[92] proceeds highly regioselectively, giving predominantly *ortho*-nitration in high yields. Even higher regioselectivity for *ortho* nitrations was observed using nitronium tetrafluoroborate under micellar catalysis[93].

Pyridinium salts bearing carboxylate side chains and pyridones react with $NOBF_4$ to give the corresponding O-nitrates, which are effective nitrating agents for phenols. These nitration reactions proceed with high regioselectivity to give the predominant *ortho*-products, in quantitative yields[93]. The nitration of some substituted phenols leads to mixtures of mononitrated products. Dinitro products are obtained for the activated phenols such as *para*-methoxyphenol and naphthols. Spectroscopic evidence shows that intermolecular association between pyridinium salts and the phenols leads to the observed regioselectivity.

The direct nitration of calix[6]arene was not successful. However, sulfonation followed by nitration of the calix[6]arene gave *para*-nitrocalix[6]arene (equation 51)[94]. The *para*-calix[*n*]arene ($n = 1, 3, 5$) sulfonic acids, prepared by treatment of the corresponding calixarenes with H_2SO_4, are reacted with $HNO_3-H_2SO_4$ to give *para*-nitro calixarenes in 15–25% yields. The electron-withdrawing sulfonic acid groups in these compounds prevent *ortho*-nitration and favor *ipso*-nitration[95].

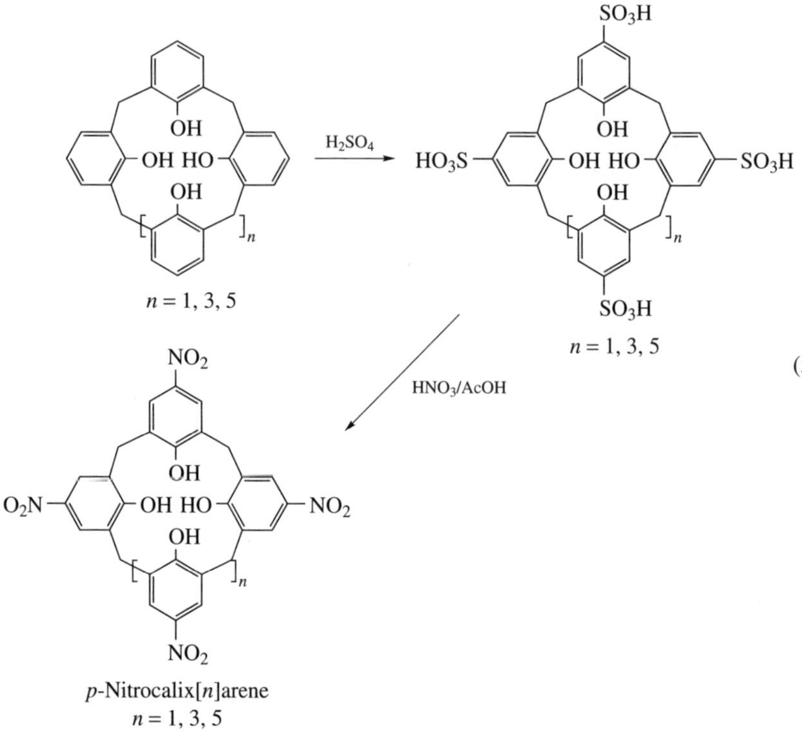

(51)

p-Nitrocalix[*n*]arene
$n = 1, 3, 5$

The direct nitration of calix[4]arene, obtained by Lewis acid catalyzed dealkylation of *tert*-butylcalix[4]arene, with $HNO_3/AcOH$ in benzene, however, has been reported to give 88% yield of *para*-nitrocalix[4]arene (equation 52)[96]. The method has been extended to other calix[*n*]arenes ($n = 4, 6, 8$), providing a convenient one-step method for the preparation of *para*-nitrocalix[*n*]arenes[97,98]. Calix[*n*]arenes have also been directly nitrated with $KNO_3/AlCl_3$ to give *para*-nitrocalix[*n*]arenes in good yields[99].

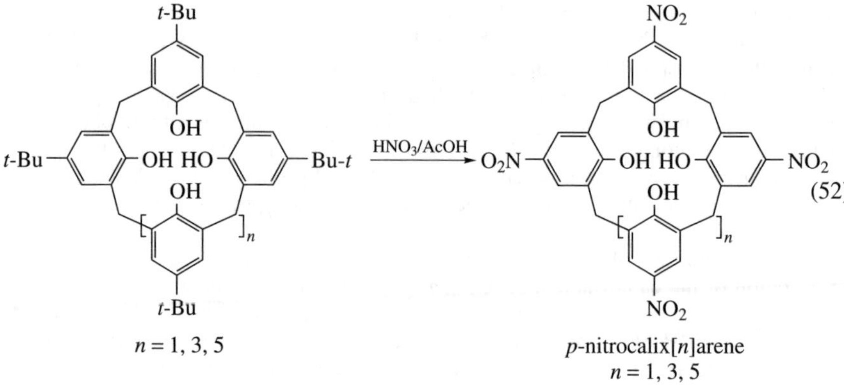

(52)

p-nitrocalix[*n*]arene
$n = 1, 3, 5$

The 1,3-diether derivatives of *tert*-butylcalix[4]arene can be selectively nitrated at the *para*-position of the phenolic units to give calix[4]arenes bearing *tert*-butyl and nitro groups at the upper rim in alternating sequence, in yields up to 75% (equation 53). The structures of the products were established by single-crystal X-ray analysis[100]. Partly O-alkylated *para-tert*-butylcalix[4]arenes are converted into mono-, di-, tri-, and tetranitrocalix[4]arenes via *ipso*-nitration using $HNO_3/AcOH$ in CH_2Cl_2 (equation 54)[101].

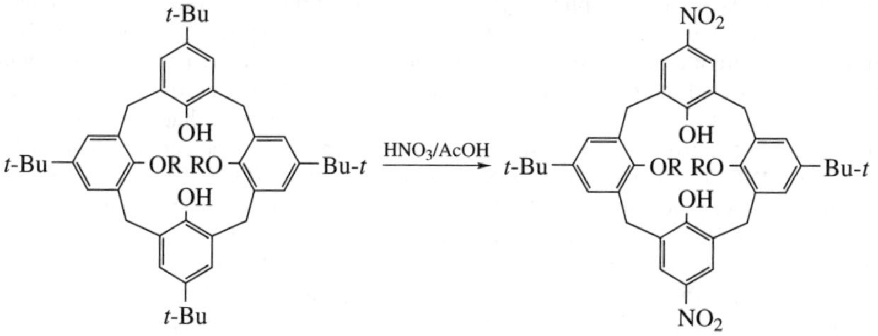

$R = CH_2CH_2OEt$ $R = CH_2CH_2OEt$ (53)

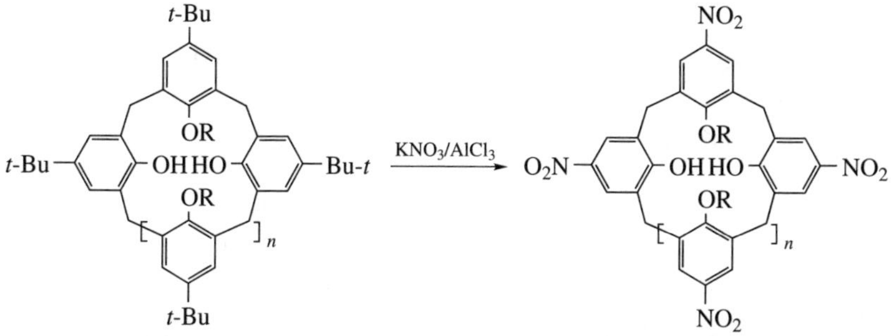

e.g., $R = (CH_2)_4Me$, CH_2Ph; $n = 1, 3, 5$ e.g., $R = (CH_2)_4Me$, CH_2Ph; $n = 1, 3, 5$ (54)

Selectively mono- and 1,3-dinitrated calix[4]arenes have been prepared by the nitration of tribenzoyl and 1,3-dibenzoylcalix[4]arenes using $HNO_3/AcOH$, followed by the deprotection of the benzoyl group with NaOH/EtOH (equation 55)[102].

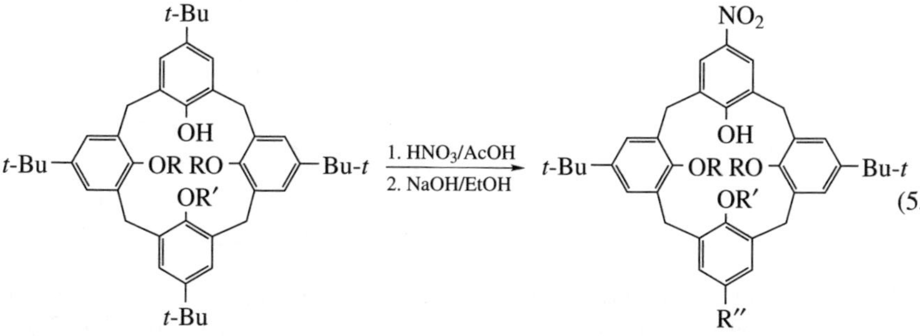

(55)

(1) $R' = R' = COPh$
(2) $R = COPh$, $R' = H$

(3) $R' = R' = COPh$; $R'' = t$-Bu
(4) $R = COPh$, $R' = H$, $R'' = NO_2$

B. Peroxynitrite-induced Nitration and Nitrosation

Nitration is the major reaction for phenols using peroxynitrite, whereas aqueous solutions of nitric oxide give mixtures of nitro and nitroso derivatives depending upon the nature of the phenol[103], the acidity of the medium and the presence of CO_2/carbonate salts. Nitrosation occurs on phenol substrates bearing a free *para*-position with respect to the OH group, with the exception of 1-naphthol, affording a 1 : 1 mixture of the 2- and the 4-nitroso derivatives. 4-Methoxyphenol gives 80% yield of the *ortho*-nitro derivative using NO as the reagent, whereas under similar conditions 2,6-dimethylphenol gives 65% of *para*-nitroso derivative and 30% of *para*-nitro derivative. Chroman derivatives (analogues of tocopherols) showed the highest reactivity with nitric oxide and peroxynitrite, suggesting that they can act as efficient scavengers of these toxic intermediates; in both cases the corresponding 5-nitro derivative was the only reaction product detected (equation 56). Peroxynitrite ($ONOO^-$), the product of $NO^{\bullet}$ with superoxide ($O_2^{\bullet -}$), being more stable than the nitric oxide, gives only about 10% yields of the nitrated chromans, as compared to over 60% with NO. The nitric oxide overproduction causes various pathologies such as neuronal degeneration, diabetes and atherosclerosis and thus the chromans can be useful substrates to capture these species *in vivo*. Another metabolic end product of NO, the nitrite ion (NO_2^-), was also shown to be involved in the nitration of the tyrosine residues of the enzymes[104].

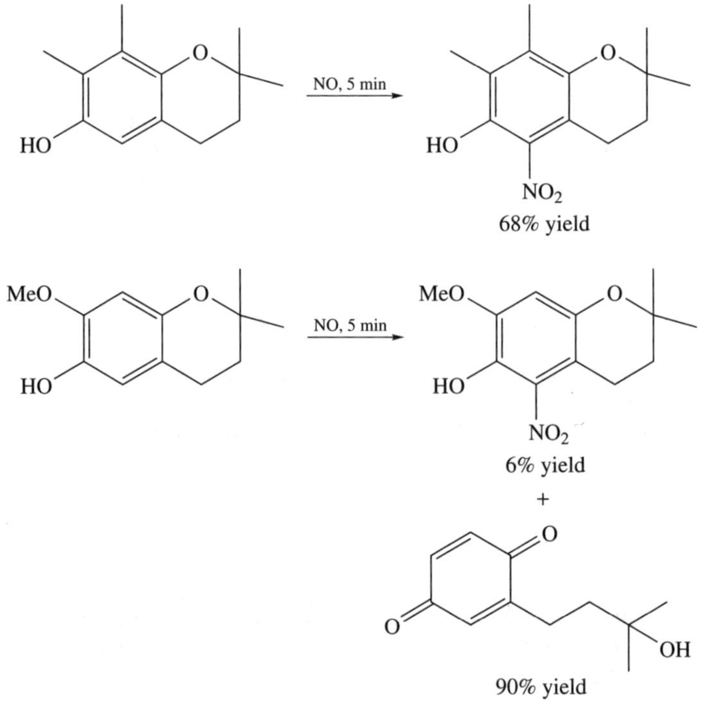

(56)

The reaction of peroxynitrite ($ONOO^-$) was also investigated with a series of *para*-substituted phenols in phosphate buffer solutions[105]. The corresponding 2-nitro derivative and the 4-substituted catechol were the major products. The reaction exhibits good correlation with Hammett σ_p^+ and half-wave reduction potentials, suggesting a possible one-electron transfer process involving the nitrosoniun ion (NO^+) as initial electrophile generated from peroxynitrous acid. ^{15}N CIDNP studies also resulted in similar conclusions[106].

Pryor and coworkers have shown that peroxynitrite-mediated nitrosations and nitrations of phenols are modulated by CO_2. The reaction was found to be first order with respect to peroxynitrite and zero order with respect to phenol, showing that an activated intermediate of peroxynitrite, perhaps the peroxynitrite anion-CO_2 adduct ($O=N-OO-CO_2^-$), is involved as the intermediate (equation 57)[107,108]. At pH higher than 8.0, 4-nitrosophenol is the major product, whereas in acidic media significant amounts of the 2- and 4-nitrophenols were formed. Peroxynitrite also induces biological nitration of tyrosine residues of the proteins. The detection of 3-nitrotyrosine is routinely used as an *in vivo* marker for the production of the cytotoxic species peroxynitrite ($ONOO^-$). It was shown that nitrite anion (NO_2^-) formed *in situ* by the reaction of nitric oxide and hypochlorous acid (HOCl) is similarly able to nitrate phenolic substrates such as tyrosine and 4-hydroxyphenylacetic acid[109].

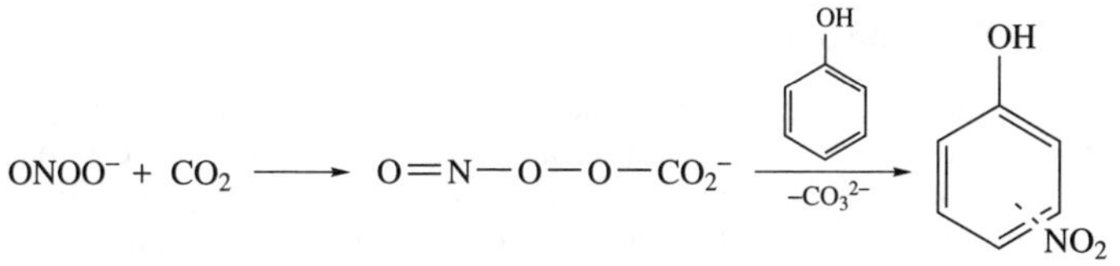

$$ONOO^- + CO_2 \longrightarrow O=N-O-O-CO_2^- \xrightarrow{-CO_3^{2-}} \text{PhOH} + \text{4-NO}_2\text{-PhOH} \tag{57}$$

It was shown that tryptophan is also nitrated by peroxynitrite in the absence of transition metals to one predominant isomer of nitrotryptophan, as determined from spectral characteristics and liquid chromatography-mass spectrometry analysis. Typical hydroxyl radical scavengers partially inhibited the nitration[110]. The yields of the nitration of tyrosine and salicylate by peroxynitrite are significantly improved by the Fe(III)–EDTA complex[111,112].

Sterically hindered phenols react with nitric oxide under basic conditions to give either cyclohexadienone diazenium diolates or oximates. Phenols with 2,6-di-*tert*-butyl and 4-methyl (butylated hydroxytoluene, BHT), 4-ethyl or 4-methoxymethyl substituents yield the corresponding 2,6-di-*tert*-butyl-2,5-cyclohexadienone-4-alkyl-4-diazenium diolate salts (equation 58)[113].

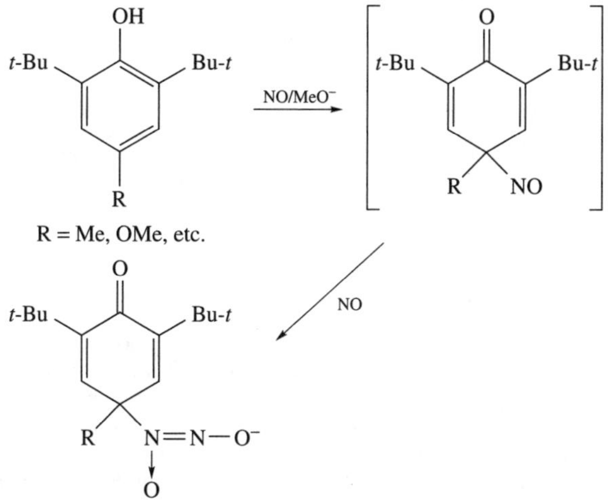

(58)

C. Nitrosation by Nitrous Acid

The reaction of phenols with nitrous acid gives the *ortho-* and *para-*nitroso products, which are formed through a neutral dienone intermediate, the proton loss from the latter being the rate-limiting step[114,115]. It has been shown that the nitrous acid can act as a catalyst for the formation of the nitro derivatives. Thus the conventional preparation of nitro compounds by the oxidation of nitroso compounds may be replaced by methods using an electron-transfer pathway in certain cases. In the latter method, the phenoxide reacts with nitrosonium ion to give the phenoxy radical and nitric oxide radical. The nitric oxide radical is in equilibrium with the nitronium radical by reaction with nitronium ion. The reaction of the phenoxy radical with the nitronium radical results in the formation of the *ortho-* and *para-*nitro products[116]. Leis and coworkers carried out kinetic studies on the reaction of phenolate ions with alkyl nitrites and found that the initially formed product is the O-nitrite ester, which evolves by a complex mechanism to give the *ortho-* and the *para-*nitro products[117].

D. Nitration by Tetranitromethane

Tetranitromethane ($C(NO_2)_4$) reacts under mild reaction conditions with phenols. The first reactions of tetranitromethane with unsaturated hydrocarbons were initially carried out by Ostromyslenkskii[118] and Werner[119]. Titov suggested that the reactions follow an ionic mechanism with the initial formation of a π-complex[120]. A radical mechanism or an electron-transfer mechanism may also operate in these reactions, depending on the reaction conditions. It is a convenient reagent for the nitration of phenols in biological systems. For example, nitration of the tyrosyl residues in the lipase from Pseudomonas cepacia (CPL) was achieved using tetranitromethane (equation 59). The modified enzyme showed better enantioselectivity in the hydrolysis reactions of esters, due to increased acidity of the phenolic group[121]. Further studies are needed to understand the scope of this and related reactions using hexanitroethane[122].

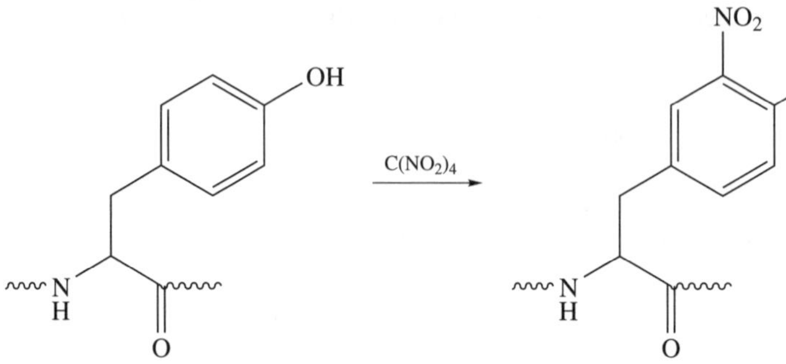

E. Nitration by Metal Nitrates

Nitration of phenol and its derivatives with $Cu(NO_3)_2$, $Fe(NO_3)_3$ and $Cr(NO_3)_3$ salts in different anhydrous organic solvents was examined. It was found that solvents have a major effect on the regioselectivity as well as on the competitive formation of the 2,4-dinitro derivatives. Salt effects ($LiClO_4$) on the rates of reaction were also observed[123]; i.e. the rates of nitrations increased in the presence of inorganic salts such as lithium perchlorate. Several derivatives of phenol were nitrated by lanthanide(III) nitrates in ethyl acetate. The regioselectivity of the nitration was based only on the phenolic OH group and was independent of the substituents. Thus, 3-substituted 5-nitrophenols were the only products observed under the conditions employed[124].

Nitrosation of phenols using metal nitrites in acidic media results in the formation of the *ortho*- and the *para*-nitroso phenols, which are subsequently spontaneously oxidized to the corresponding nitro compounds. High yields of the *ortho*- and *para*-nitrophenols (85%–90%) have been obtained under the appropriate reaction conditions. Thus 2-chlorophenol gave a quantitative *para* substitution at pH 3.5 in the absence of oxygen using 3 equivalents of nitrite salt. In acetate-buffered solutions, 4-*tert*-butylphenol gave 90% of 2-nitro-4-*tert*-butylphenol, whereas resorcinol gave 85% of 2,4-dinitrosoresorcinol (nonoxidized product). The initially formed nitroso compounds in these reactions may be oxidized to nitro derivatives by oxygen, or by reactions involving the reduction of the nitrous acid to nitric oxide, as shown below[125]. The possibility of reaction of the phenoxy radical directly with NO_2 radical has also been proposed (equations 60 and 61)[126].

$$2 HNO_2 \rightleftharpoons NO_2^{\bullet} + NO^{\bullet} + H_2O \qquad (60)$$

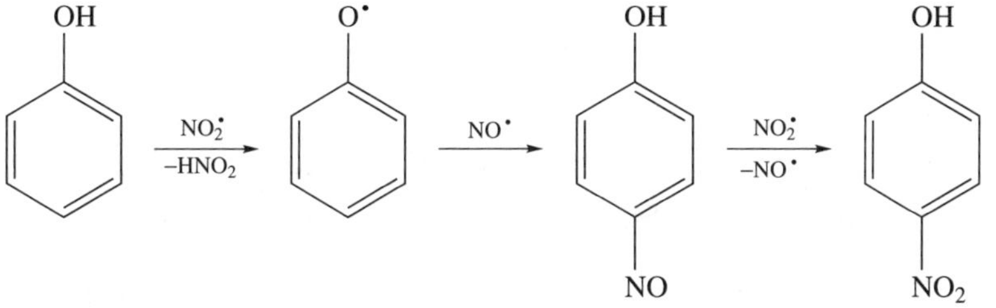

(61)

V. FRIES AND RELATED REARRANGEMENTS

A. Lewis Acid Catalyzed Fries Rearrangements

The O-acyl derivatives of phenols (phenyl esters), in the presence of Lewis acids, undergo rearrangement to give *ortho*- and *para*-acyl phenols, which is generally known as the Fries rearrangement[127]. Fries rearrangement of phenyl esters followed by a Wolff–Kishner reduction provides a convenient procedure for the preparation of alkylated phenols[128]. The rearrangement involves the reversible formation of the Wheland intermediates from the Lewis acid-complexed substrate and is useful for the isomerization of the acylated phenols under appropriate reaction conditions (equation 62). The Lewis acid may complex both oxygens of the ester group when used in excess. The reaction proceeds by both intermolecular as well as intramolecular rearrangement pathways, depending on the substrate, reaction temperature and solvent[129,130].

A variety of catalysts, such as $TiCl_4$, $AlCl_3$, BF_3 and CF_3SO_3H, may be used for the Fries rearrangement. Hafnium trifluoromethanesulfonate, $Hf(OTf)_4$ (5 to 20 mol%), was recently used as an efficient catalyst in the Fries rearrangement of acyloxy benzene or naphthalene derivatives[65]. Scandium triflate ($Sc(OTf)_3$), zirconium triflate ($Zr(OTf)_4$) and titanium chlorotriflate ($TiCl(OTf)_3$) were also used for the Fries rearrangements of phenyl and naphthyl acetates[67]. A silica-supported heteropoly acid has been used as the catalyst for the conversion of phenol to phenyl acetate and its subsequent Fries rearrangement to 4-hydroxyacetophenone. The esterification proceeds at 140 °C and the Fries rearrangement of the ester proceeds at 200 °C on the same catalyst, with 90% regioselectivity to give the *para*-isomer (10% yields)[131]. *ortho*-Acetyl- and benzoylhydroxy[2.2]paracyclophanes have been prepared from 4-hydroxy[2.2]paracyclophane using $TiCl_4$-catalyzed Fries rearrangement with high yields[132]. Alumina/methanesulfonic acid has been used to prepare *ortho*-hydroxyaryl ketones, by acylation of phenol and naphthol derivatives with carboxylic acids, followed by Fries rearrangement of the resulting phenolic esters[133]. The Fries rearrangement of phenyl acetate has been studied over various zeolites, among which Zeolite H-Beta was found to be the superior catalyst[74]. In the same studies, MCM-41 zeolitic material was also developed as an efficient catalyst for esters with sterically hindered groups. Alkylphenols were O-acylated using γ-chlorobutyroyl chloride, which undergoes Fries rearrangement with $AlCl_3$ to give hydroxyaryl ketones[134].

The regiochemistry of the Fries rearrangement is dependent on the reaction conditions. For example, the reaction of *meta*-cresyl acetate with $AlCl_3$ gives the *para*-acetyl-*meta*-cresol as the major product at low temperatures, while the *ortho*-acetyl-*meta*-cresol is formed as the major product at high temperatures (equation 63)[135].

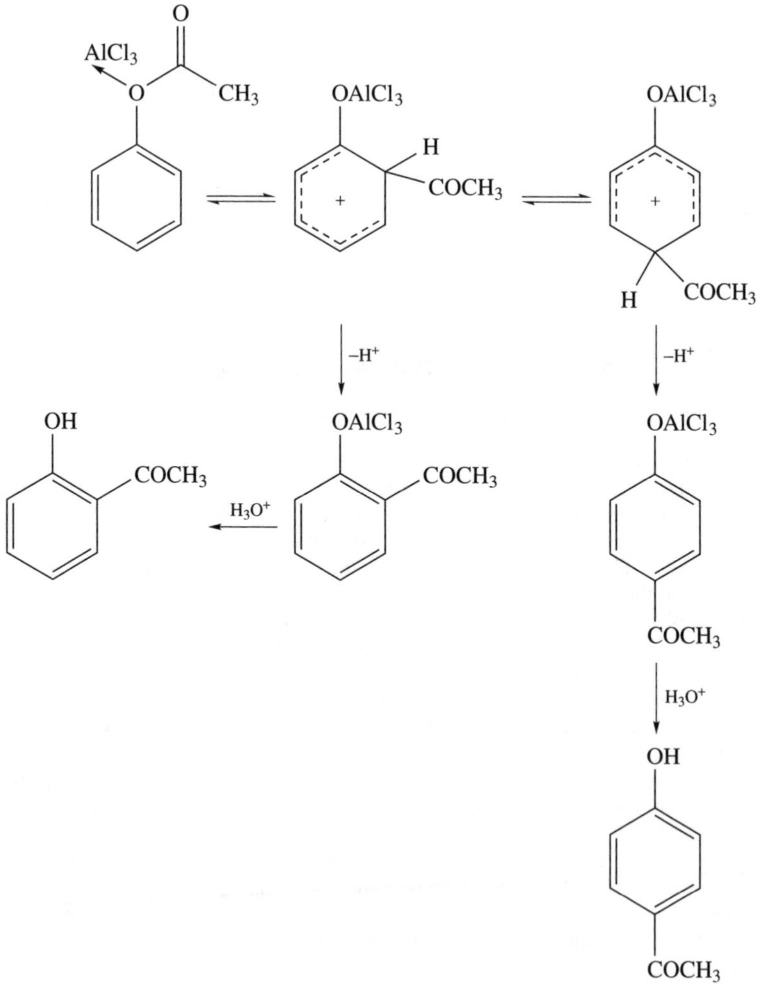

(62)

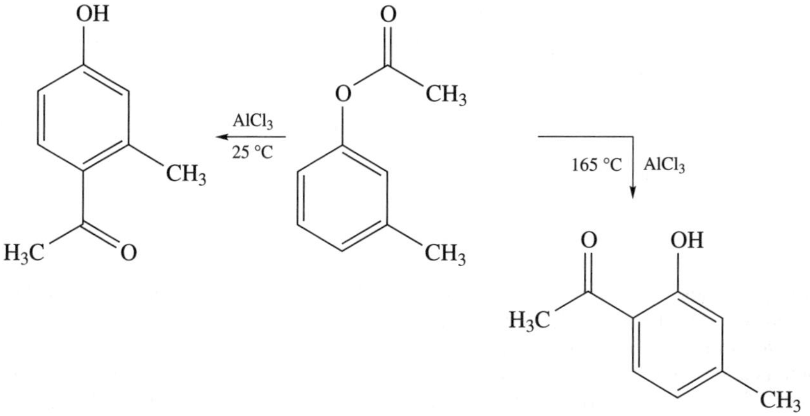

9. Electrophilic reactions of phenols 641

(63)

O-glycopyranosides of 1- and 2-naphthols undergo a Fries type of rearrangement using Lewis acids such as BF_3-Et_2O to give 2- and 1-C-glycopyranosides, respectively (equation 64). O-2-tetrahydropyranyl phenols and naphthols also rearrange under the BF_3-Et_2O catalysis to give the corresponding *ortho*-alkylated phenols (equation 65)[136]. Such aryl C-glycosides are anti-tumour agents. Several versions of these compounds have also been prepared by Friedel–Crafts reaction using Zr-complexes/silver perchlorate[137].

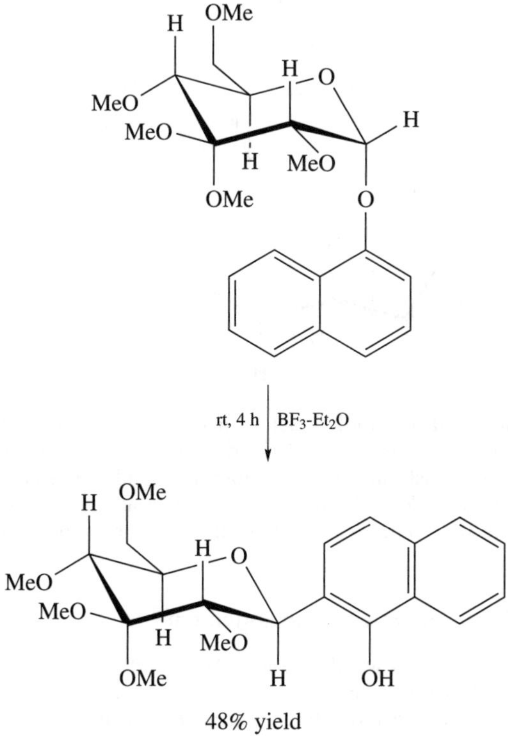

(64)

48% yield

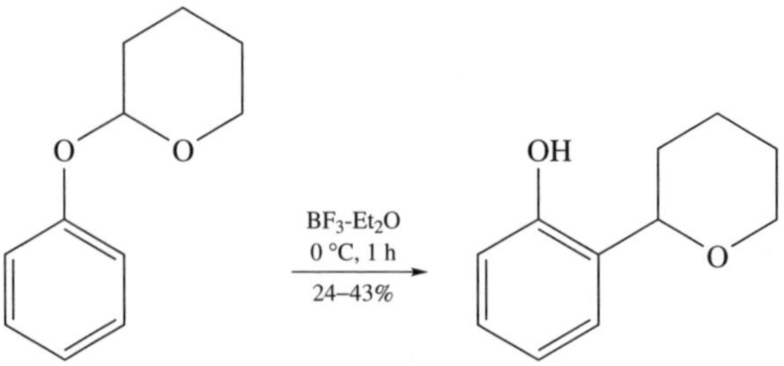

Pivalophenones were prepared by the Fries reaction of Ph, cresyl and xylyl pivalates in the presence of HCl−SnCl$_4$ and by the Friedel−Crafts acylation of the phenols by Me$_3$CCOCl in the presence of SnCl$_4$[138].

Phenols and naphthols also react with unprotected α- and β-glycosides, directly, in the presence of trimethylsilyl triflate catalyst under mild conditions (equation 66)[139].

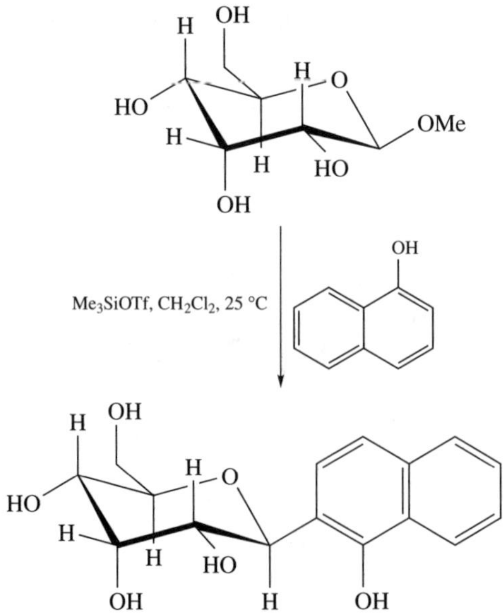

A variety of phenols and naphthols react with mannose and glucosyl phosphates to give the α-O-glucosyl or α-O-mannosyl derivatives in the presence of the trimethylsilyl triflate, which spontaneously undergo a regiospecific and stereospecific Fries type of rearrangement to give the *ortho*-β-C-glucosyl and β-C-mannosyl phenols[140], which are useful intermediates for the synthesis of biologically active compounds (equation 67).

B. Bronsted Acid Catalyzed Fries Rearrangements

Olah and coworkers have shown that Nafion-H, a perfluorinated resinsulfonic acid, acts as an efficient catalyst for the Fries rearrangement of aryl benzoates. For example, *meta*-chlorophenyl benzoate undergoes Fries rearrangement in the presence of Nafion-H

9. Electrophilic reactions of phenols

in 75% yield with an *ortho/para* ratio of 1 : 2.6 (equation 68)[141].

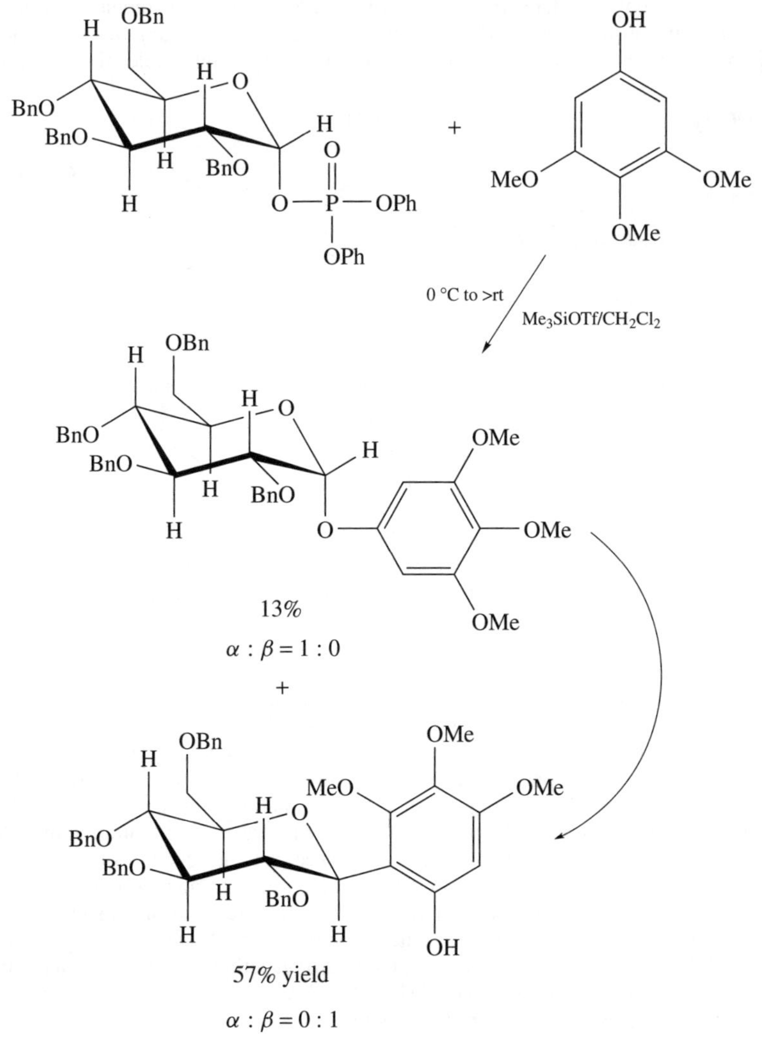

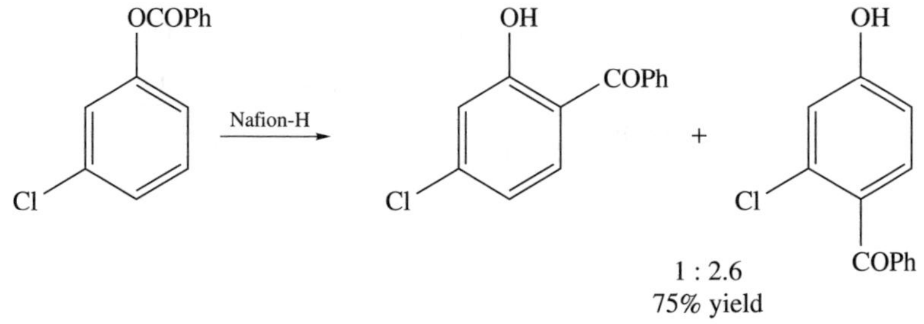

Nafion-H-silica nanocomposite (13% Nafion-H) catalyzed Fries rearrangement of phenyl acetate at high temperatures gives phenol, *ortho*-acetylphenol, *para*-acetylphenol and *para*-acetylphenyl acetate in a ratio of 45 : 5 : 24.5 : 25.5 (equation 69). The rearrangement in the presence of added phenol gives exclusively the *para*-acetylphenol, showing that the Fries rearrangement under these conditions is intermolecular in nature[23].

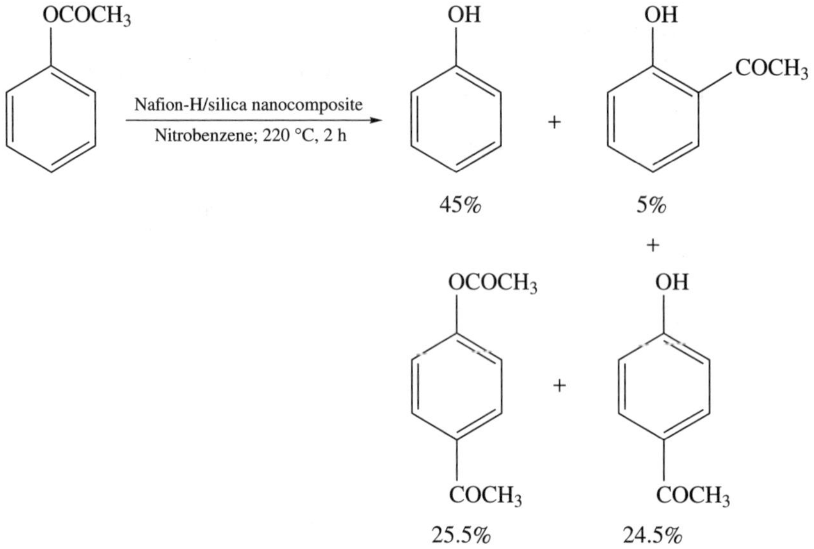

$$\text{PhOCOCH}_3 \xrightarrow[\text{Nitrobenzene; 220 °C, 2 h}]{\text{Nafion-H/silica nanocomposite}} \text{PhOH} + o\text{-HOC}_6\text{H}_4\text{COCH}_3 + p\text{-CH}_3\text{COOC}_6\text{H}_4\text{COCH}_3 + p\text{-HOC}_6\text{H}_4\text{COCH}_3 \quad (69)$$

45% : 5% : 25.5% : 24.5%

Hoelderich and coworkers systematically compared the catalytic activities of zeolites, Nafion-H and Nafion-H-silica nanocomposite catalysts for the Fries rearrangements[142]. They have found that the acidic zeolite H-BEA is the most selective catalyst and the products *para*-hydroxyacetophenone and *ortho*-hydroxyacetophenone are obtained in a ratio of 4.7 : 1 (equation 70). Although Nafion-H-silica nanocomposite is a more efficient catalyst than the Nafion-H beads, its performance decreases as the concentration of Nafion-H in the resin is decreased. They have also observed that the change of solvent from cumene to phenol in the Fries rearrangement of phenyl acetate increases the conversion significantly. More recently it has been shown that the Fries rearrangement of phenyl acetate catalyzed by the pentasil-type zeolite T-4480 affords 2-hydroxyacetophenone in good yield (73.6%), with a minor product of 4-hydroxyacetophenone (o/p selectivity = 26.6 : 2.9). Similar reactions using H-ZSM yields these products with much less o/p selectivity in low yields ($o : p = 2.9$; 7.3% yield)[143].

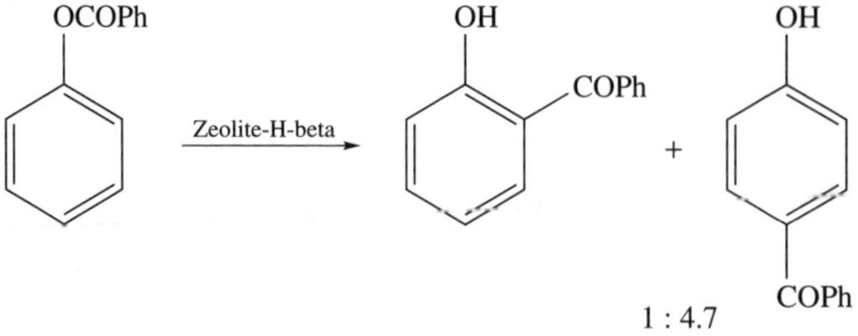

$$\text{PhOCOPh} \xrightarrow{\text{Zeolite-H-beta}} o\text{-HOC}_6\text{H}_4\text{COPh} + p\text{-HOC}_6\text{H}_4\text{COPh} \quad (70)$$

1 : 4.7

The industrially significant 2,4-dihydroxybenzophenone can be prepared in 88% yield by the Fries rearrangement of the resorcinol benzoate formed *in situ* by the reaction of benzoic acid and resorcinol using zeolite-H-beta catalyst (equation 71). A variety of solvents such as butylbenzene and *n*-decane are used successfully for these reactions[144].

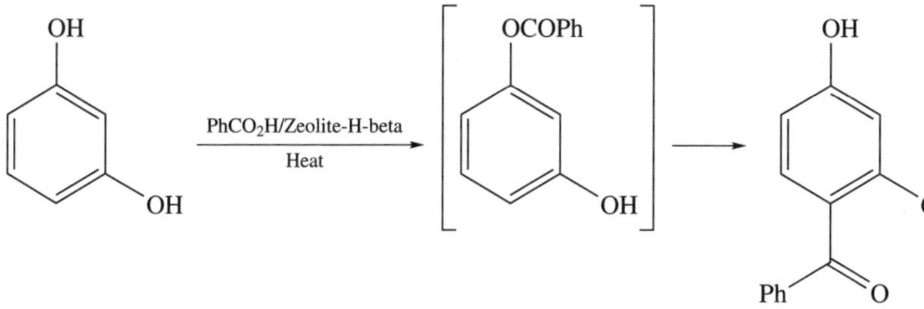

$$\text{(71)}$$

The propionylation of phenol with propionyl chloride can be carried out over zeolite-H-beta, Re-Y, H-Y, mordenite, H-ZSM-5 and $AlCl_3$ at 140 °C to give *para*-hydroxypropiophenone and *ortho*-hydroxypropiophenone as the major products. Among these catalysts, the zeolite-H-beta is the most efficient. The product distribution depends upon the reaction conditions and acidity of the zeolite catalysts[145]. The reaction involves the initial O-propionylation of the phenol followed by its rapid Fries rearrangement.

VI. ELECTROPHILIC HALOGENATION

The electrophilic halogenation of phenols give rise to mixtures of *ortho*- and *para*-substituted phenols. Phenols are more reactive than alkylaromatics in these reactions due to the enhanced resonance stabilization of the carbocationic intermediates (equation 72). However, in superacidic solutions, the oxygen protonation of the phenols leads to the deactivated substrate for halogenation and *meta*-halo products are obtained (equation 73)[21].

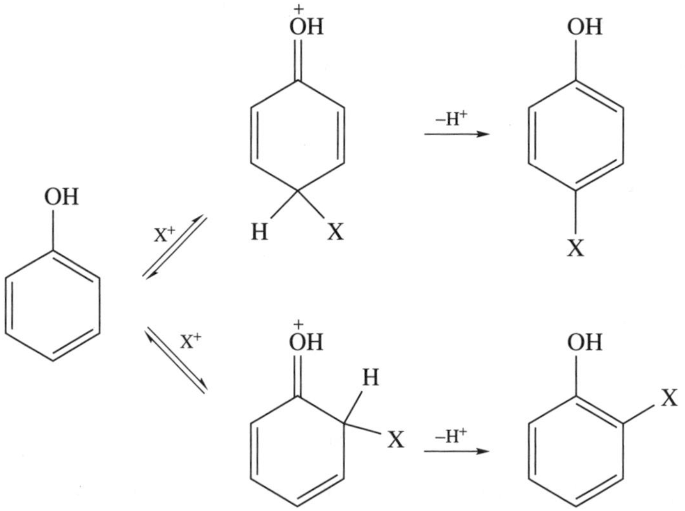

$$\text{(72)}$$

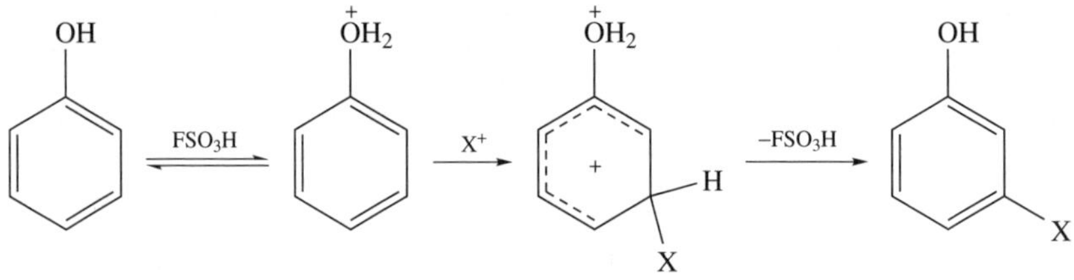

$$\underset{\text{OH}}{\text{C}_6\text{H}_5} \xrightleftharpoons{\text{FSO}_3\text{H}} \underset{\overset{+}{\text{OH}_2}}{\text{C}_6\text{H}_5} \xrightarrow{X^+} \left[\underset{\overset{+}{\text{OH}_2}}{\underset{X}{\text{C}_6\text{H}_4\text{-H}}}\right] \xrightarrow{-\text{FSO}_3\text{H}} \underset{\text{OH}}{\underset{X}{\text{C}_6\text{H}_4}} \quad (73)$$

X = F, Cl, Br

A. Fluorination

Phenols can be fluorinated using F_2/N_2 solutions in solvents such as chloroform or trifluoroacetic acid at low temperatures to give high conversions to *ortho*- and *para*-fluorinated phenols, with minimal regioselectivity. The *ortho*-isomer predominated by about 1.5 : 1 (equation 74)[146].

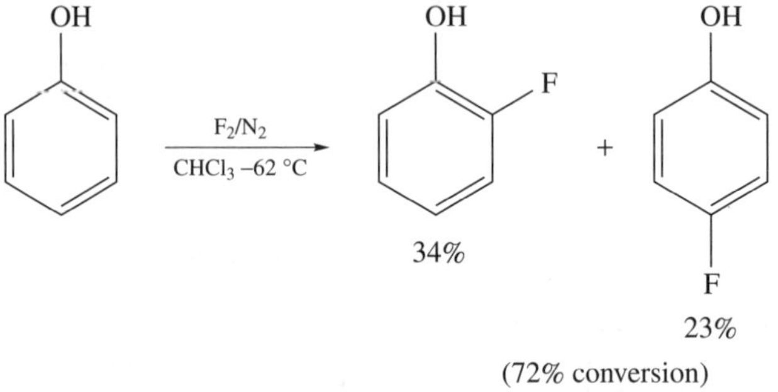

$$\text{PhOH} \xrightarrow[\text{CHCl}_3 \; -62\,°\text{C}]{F_2/N_2} \text{o-F-C}_6\text{H}_4\text{OH} \; (34\%) + \text{p-F-C}_6\text{H}_4\text{OH} \; (23\%) \quad (74)$$

(72% conversion)

It was found that increasing polarity of the solvent increased the yields of the reaction, in the following order: $CF_3CO_2H > CF_3CH_2OH > CH_3OH > CHCl_3 > CFCl_3$, which indicates the electrophilic substitution mechanism. Fluorine solutions in hydroxyl group containing solvents give ROF species, which is a source of electrophilic F^+ species. The fluorination of benzoic acid in a variety of hydroxylic solvents, such as trifluoroacetic acid, 2,2,2-trifluoroethanol and methanol, gave *meta*-fluorobenzoic acid as the major product, further confirming the electrophilic nature of these reactions. Highest regioselectivity is observed in 2,2,2-trifluoroethanol: 74 (*m*-fluorobenzoic acid) : 19 (*o*-fluorobenzoic acid) (equation 75).

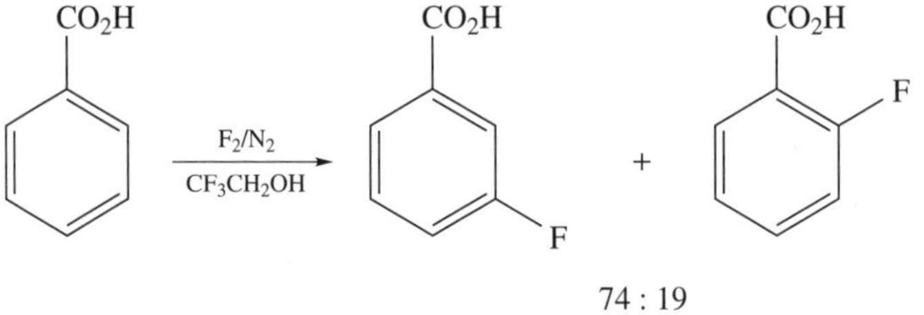

$$\text{PhCO}_2\text{H} \xrightarrow[\text{CF}_3\text{CH}_2\text{OH}]{F_2/N_2} \text{m-F-C}_6\text{H}_4\text{CO}_2\text{H} + \text{o-F-C}_6\text{H}_4\text{CO}_2\text{H} \quad (75)$$

74 : 19

Electrophilic fluorinating reagents such as Selectfluor and related compounds can be used for the ring fluorination of phenols. The reaction of phenol with 1,3-bis(4-fluoro-1,4-diazoniabicyclo[2.2.2]oct-1-yl-)propane tetratriflate in methanol gives moderate yields

(59%) of 2-fluoro- and 4-fluorophenols in a ratio of 1.5 : 1 (equation 76). 2-Naphthol similarly gave 1-fluoro-2-naphthol with this reagent at a reaction temperature of 80 °C in acetonitrile solvent (equation 77). These reactions were dramatically improved using the more reactive reagent, Selectfluor[147].

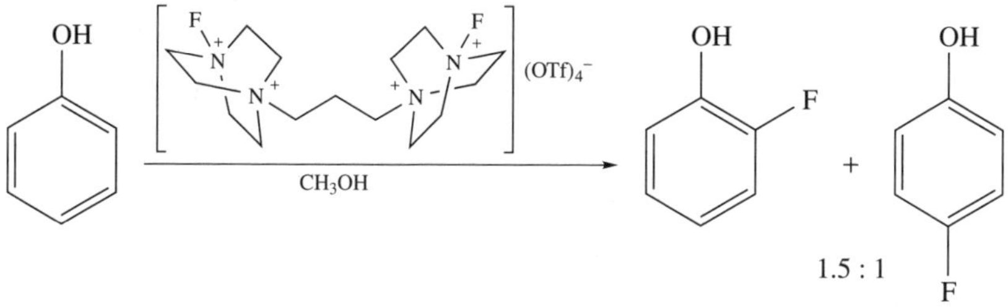

Other reagents such as perfluoro-[N-fluoro-N-(4-pyridyl)acetamide][148] and N-(R)-N-fluoro-1,4-diazoniabicyclo[2.2.2]octane salts (R = CH$_3$, CH$_2$Cl, C$_2$H$_5$, CF$_3$CH$_2$, C$_8$H$_{17}$)[149] also readily fluorinate phenol to give 2- and 4-fluorophenols under mild conditions. The DesMarteau sulfonimide ((CF$_3$SO$_2$)$_2$NF) and N-fluorocarboxamides are powerful electrophilic fluorinating agents, potentially suitable for the electrophilic fluorination of phenols[150,151]. Banks and coworkers have prepared analogous N-fluoro compounds, perfluoro-N-fluoro-N-(4-pyridyl)methanesulfonamide and perfluoro-(N-fluoro-N-(4-pyridyl)acetamide as electrophilic fluorinating agents[148]. Using the latter reagent it was shown that phenol gives 2-fluorophenol and 4-fluorophenol (1 : 1) in 91% yield.

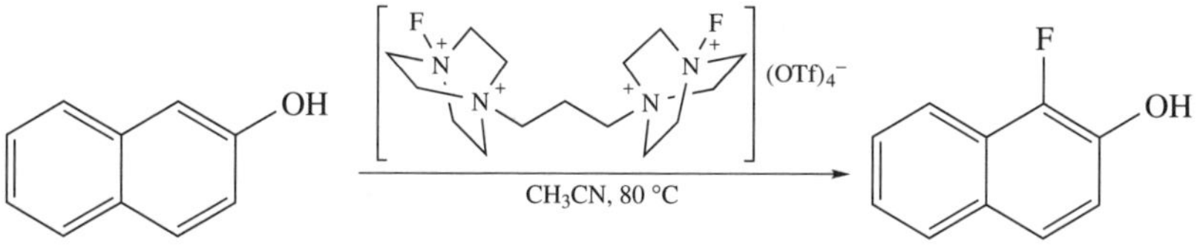

A series of alkyl- or (trifluoromethyl)-substituted N-fluoropyridinium-2-sulfonates were found to be suitable for the electrophilic fluorination of phenol, naphthol and the trimethylsilyl ether of phenol, highly regioselectively. Exclusive or predominant *ortho*-fluorination could be achieved by these reagents (equations 78 and 79)[152]. The observed regioselectivity was explained as due to the H-bonding interaction of the 2-sulfonate anion with the hydroxy groups of the phenol derivatives, in which the 'F$^+$' of the reagent is in closer proximity to the *ortho*-position of the phenolic OH group. The *ortho*-fluoro cyclohexadienone is formed as an intermediate in agreement with this mechanism. These reactions proceed highly regioselectively in nonpolar solvents such as dichloromethane or 1,2-dichloroethane. Phenol under these conditions gives 80% of *ortho*-fluorophenol (equation 79) and only 2% of *para*-fluorophenol. 1-Naphthol similarly gives predominantly the *ortho*-fluoronaphthol. Polar solvents such as hexafluoroisopropanol, (CF$_3$)$_2$CHOH, diminish the H-bonding interaction of the reagent, making the reaction less regioselective. In the latter solvent, phenol gives 57% of *ortho*-fluorophenol and 13% of *para*-fluorophenol and 6% of 2,4-difluorophenol. ^{18}F-labeled fluorophenols may be readily available by these methods. Conventionally, the ^{18}F-labeled fluorophenols are obtained by a Baeyer–Villiger oxidation of fluorobenzaldehydes and fluoroacetophenones[153].

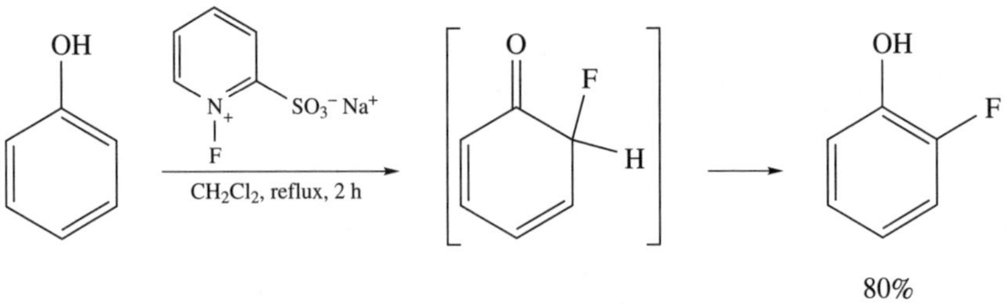

[Scheme (78): Fluorination of phenol with N-fluoropyridinium-2-sulfonate sodium salt in CH$_2$Cl$_2$, reflux, 2 h, via a 6-fluoro-6H-cyclohexa-2,4-dienone intermediate, gives 2-fluorophenol in 80% yield.]

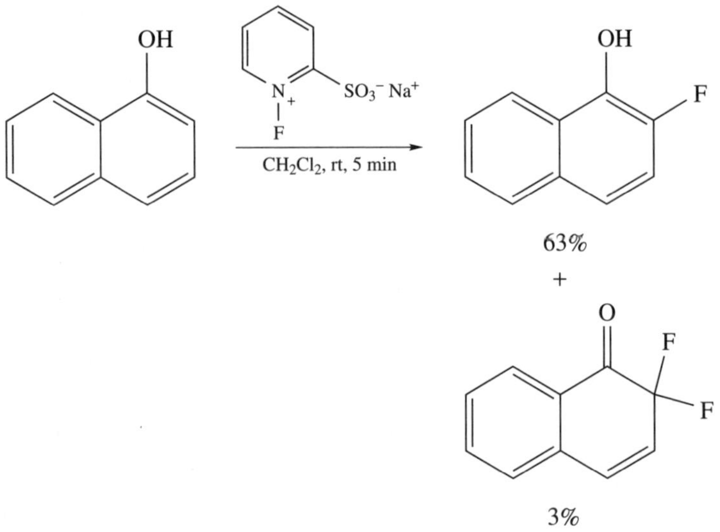

[Scheme (79): Fluorination of 1-naphthol with N-fluoropyridinium-2-sulfonate sodium salt in CH$_2$Cl$_2$, rt, 5 min, gives 2-fluoro-1-naphthol (63%) and 2,2-difluoro-2H-naphthalen-1(2H)-one (3%).]

Anodic fluorination of phenols in the presence of Et$_3$N/5HF readily afforded 4,4-difluorocyclohexa-2,5-dien-1-ones, which could be converted to *para*-fluorophenols in good yields by a subsequent reduction with Zn in aqueous acidic solutions (equation 80)[154].

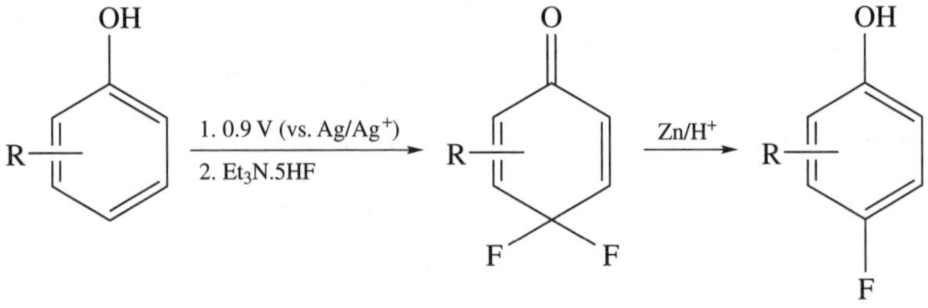

[Scheme (80): R-C$_6$H$_4$-OH → (1. 0.9 V vs. Ag/Ag$^+$; 2. Et$_3$N·5HF) → 4-R-4,4-difluorocyclohexa-2,5-dien-1-one → (Zn/H$^+$) → 4-R-4-fluorophenol.]

The oxidative fluorination of 4-alkylphenols to give the 4-fluoro-4-alkylcyclohexa-2,5-dien-1-ones can be achieved by using hypervalent iodine reagents, such as phenyliodo bis(trifluoroacetate) or phenyliodine diacetate in the presence of pyridinium polyhydrogen fluoride (equation 81) (*vide infra*)[155].

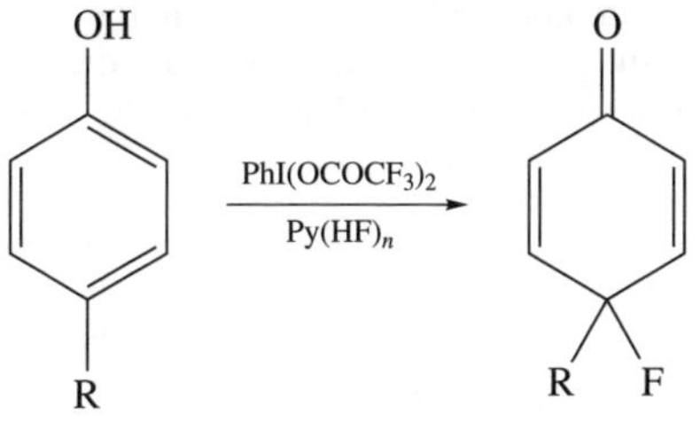

9. Electrophilic reactions of phenols

$$\text{[Phenol with R] } \xrightarrow{\text{PhI(OCOCF}_3)_2} \text{[cyclohexadienone with R, F]} \qquad (81)$$
$$\text{Py(HF)}_n$$

B. Chlorination

Phenols are monochlorinated regioselectively using sulfuryl chloride and amines (such as di-*sec*-butylamine) as the catalysts in nonpolar solvents (equation 82). In a typical experiment an *ortho/para* ratio of 22 was obtained with yields of about 90%[156].

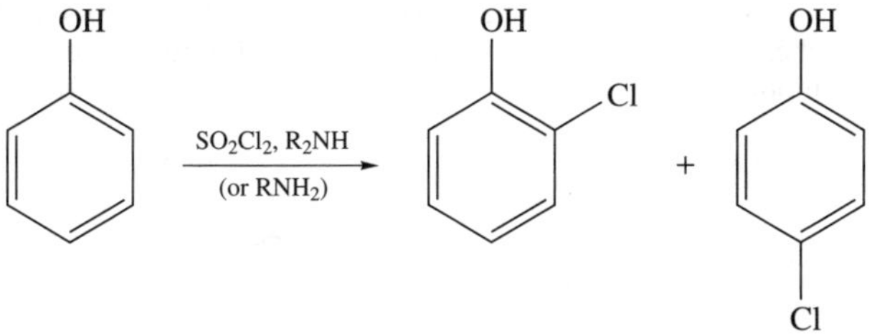

$$\text{PhOH} \xrightarrow[\text{(or RNH}_2)]{\text{SO}_2\text{Cl}_2,\ \text{R}_2\text{NH}} \text{2-chlorophenol} + \text{4-chlorophenol} \qquad (82)$$

In the absence of the amines, the yields of the chlorinated products are very low. Thus the addition of 8 mol% of the primary or secondary amines increased the conversion of phenol from 7.2 to 97.5% and the reaction was complete in less than one hour. The reaction is highly regioselective, giving almost exclusively the *ortho*-chlorinated products. The highest *o/p* ratio of 65.9 was observed when di-isobutylamine was used as the catalyst. The reaction is completely nonregioselective in the absence of the amine catalysts. The use of two equivalents of sulfuryl chloride resulted in the formation of 2,6-dichlorophenol as the predominant product (equation 83). Tertiary amines such as triethylamine, on the other hand, gave low *o/p* ratios, typically ranging from 0.5 to 1.3.

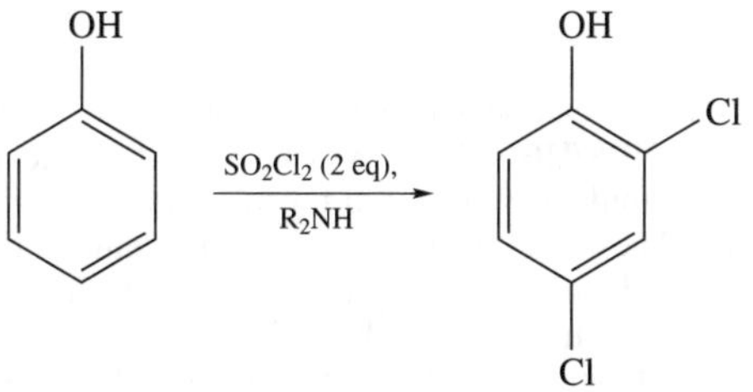

$$\text{PhOH} \xrightarrow[\text{R}_2\text{NH}]{\text{SO}_2\text{Cl}_2\ (2\ \text{eq}),} \text{2,4-dichlorophenol} \qquad (83)$$

C. Bromination

There are numerous procedures for the bromination of phenolic compounds and the regioselectivity in these reactions has been frequently achieved by varying the nature of the solvent system[3]. Controlled monobromination of phenols can be achieved using N-bromosuccinimide (NBS) on silica gel[157].

The involvement of the bromocyclohexadienones as the reaction intermediates in the electrophilic bromination of phenols is confirmed by the isolation of the 4-alkoxy-cyclohexa-2,5-dienones, from the reaction of phenols with Br_2/ROH in the presence of $AgClO_4$ and Na_2CO_3 (equation 84)[158].

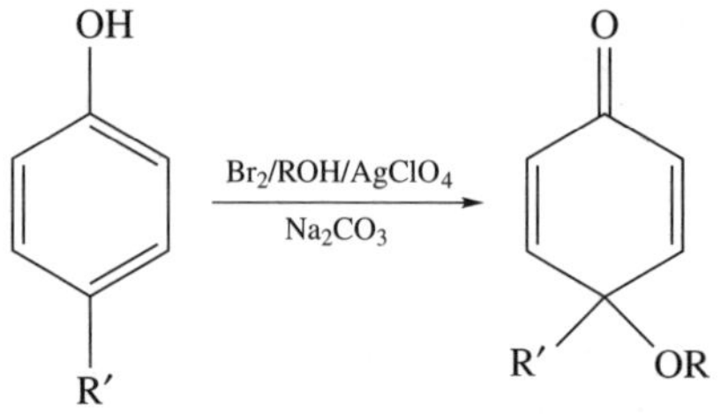

$$\underset{R'}{\underset{|}{C_6H_4}}\text{-OH} \xrightarrow[Na_2CO_3]{Br_2/ROH/AgClO_4} \text{4-alkoxycyclohexa-2,5-dienone} \quad (84)$$

The regioselectivity of the bromination of the phenols is enhanced in the presence of adjacent O-glycosylated groups. The bromination of O-(2,3,4,6-tetra-O-acetyl-β-D-glucopyranosyl) phenols gives the *para*-bromo isomer highly regioselectively[159], perhaps due to the steric hindrance for the *ortho* substitution (equations 85 and 86).

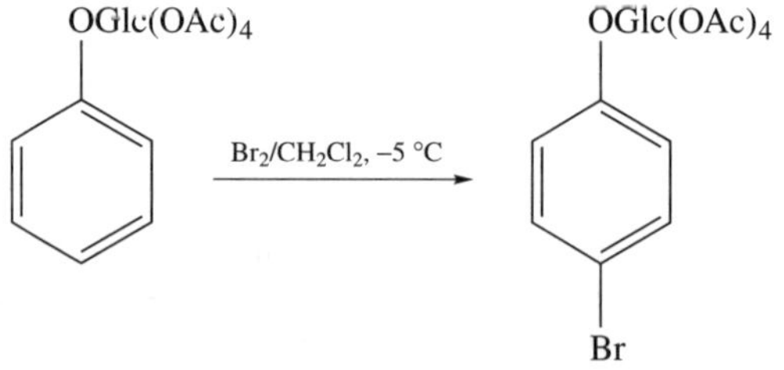

$$\text{PhOGlc(OAc)}_4 \xrightarrow{Br_2/CH_2Cl_2, -5\,°C} \text{4-Br-C}_6\text{H}_4\text{-OGlc(OAc)}_4 \quad (85)$$

$$\text{2-OGlc(OAc)}_4\text{-phenol} \xrightarrow{Br_2/CH_2Cl_2, -5\,°C} \text{4-Br-2-OGlc(OAc)}_4\text{-phenol} \quad (86)$$

The bromination of phenols can be achieved in high yields using N-bromosuccinimide (NBS)/HCl in acetone[160]. NBS/HBF_4-Et_2O was also used as the brominating agent for phenols[161]. N-bromosuccinimide was also used for the regioselective bromination of naphthols as well as phenols. The regioselectivity was dependent on the solvent used in the reaction; acetonitrile as a solvent gave the *para*-isomers, whereas carbon disulfide gave *ortho* isomers[162]. Thus bromination of 1-naphthol in acetonitrile solvent gave 4-bromo-1-naphthol, whereas in CS_2 solvent it results in the formation of 2-bromonaphthol. Similarly, regioselective *ortho* brominations were observed using NBS or Br_2 in the presence of primary or secondary amines[163,164]. Importantly, bromination at the benzylic position was not observed in the case of methylphenols. Solid state electrophilic bromination has also been achieved by NBS[165]. The bromination of phenol to *ortho*-bromophenol was achieved on a large scale using trans-alkylation strategy. Thus *para-tert*-butyl phenol was brominated exclusively at the *ortho* position and the *tert*-butyl group is transferred to toluene in the presence of $AlCl_3$ catalyst. The *para-* and *ortho-tert*-butyltoluenes are then reconverted to *para-tert*-butylphenol using excess phenol and Engelhard F-24 catalyst[166].

9. Electrophilic reactions of phenols

The solid-phase bromination of hindered phenols using NBS was reported to give the corresponding brominated cyclohexadienones[167].

The *para*-selective bromination of phenol can be achieved by using a variety of reagents such as DBU hydrobromide perbromide[168], tetrabromocyclohexadienone[169], tetraalkylammonium tribromides[170], hexamethylenetetramine tribromide[171] and NBS/HBF$_4$.Et$_2$O (equation 87)[161]. In the latter reagent it was suggested that bromonium tetrafluoroborate (BrBF$_4$) is the actual brominating agent.

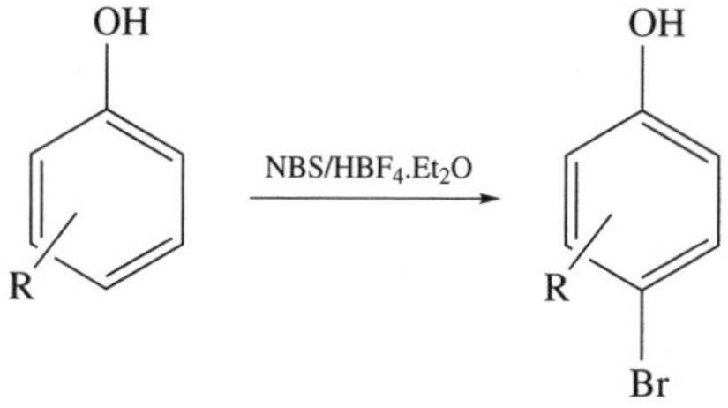

For example, R = H, CN, Cl

D. Iodination

Selective solid-phase iodination of phenolic groups could be achieved using bis(pyridinium) iodotetrafluoroborate[172], which does not react with O-protected phenols under these mild conditions. Using this reagent, it was shown that in peptides containing multiple tyrosine residues (e.g. the analgesic peptide dermorphin) selective O-protection of the tyrosine residues could be used to chemoselectively iodinate the unprotected tyrosine residues (equation 88).

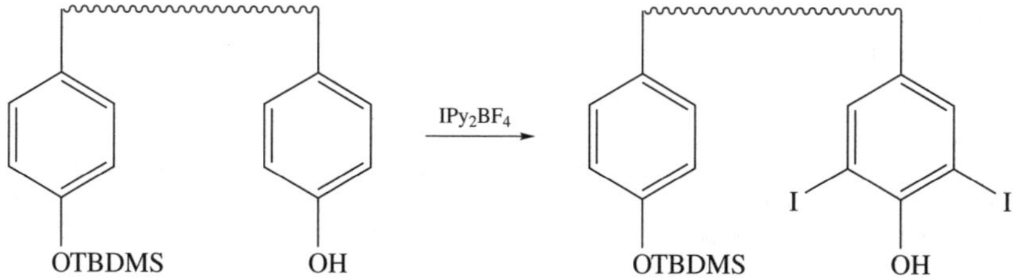

VII. PHENOL–DIENONE REARRANGEMENTS

The reversible conversion of phenols to dienone intermediates is an important transformation in the synthesis of natural products. This rearrangement occurs efficiently in superacid solutions[173–181]. The corresponding version for the halophenols to give halodienones has been reviewed in an earlier volume of this series[3]. 4-Bromo-2,4,6-trialkylcyclohexa-2,5-dienones have recently been synthesized by electrophilic bromination[182] of the corresponding phenols.

The reaction of mono- and polycyclic 4-alkylphenyl ethers using the hypervalent iodine compound, PhI(OCOCF$_3$)$_2$, in the presence of chloride and fluoride ions gives the corresponding 4-chloro- and 4-fluoro-cyclohexa-2,5-dienones[155,183]. The corresponding oxidative reactions in the presence of the alcohols give 4-alkoxycyclohexadienones. These reactions may be used in the preparation of fluoro- and alkoxy-substituted hydroindolenones and hydroquinolenones, which are the precursors of various biologically active compounds (equations 89 and 90).

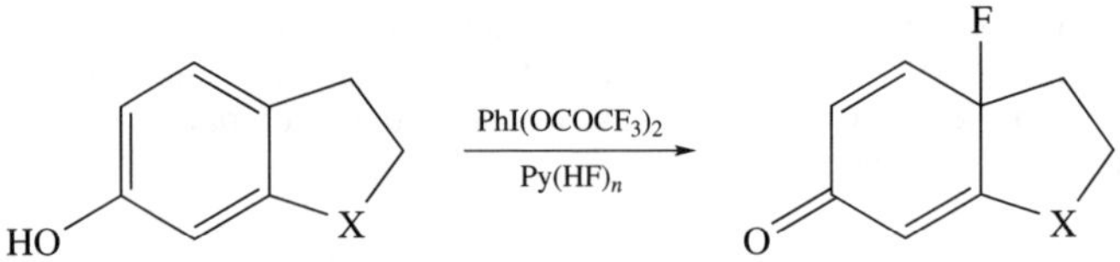

$$X = CH_2, NCO_2Me \xrightarrow[Py(HF)_n]{PhI(OCOCF_3)_2} \text{30-70\% yields} \quad (89)$$

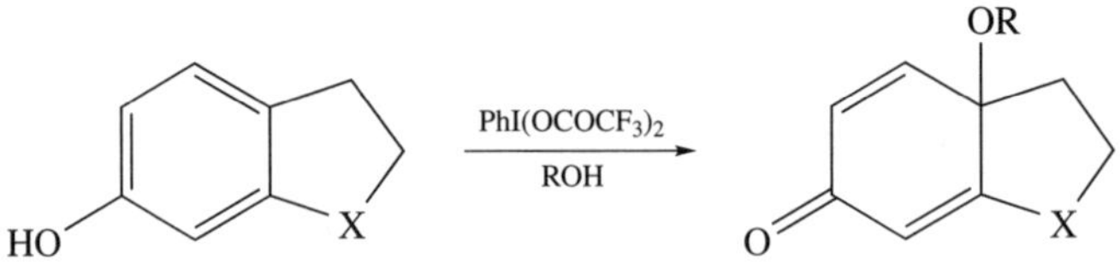

$$X = CH_2, NCO_2Me \xrightarrow[ROH]{PhI(OCOCF_3)_2} \text{30-70\% yields} \quad (90)$$

The reaction of estrone derivatives with $HF-SbF_5$ or FSO_3H-SbF_5 gives estra-4,9-dien-3,7-dione (equation 91). The intermediate tricationic species and their isomers have been characterized by ^1H NMR spectroscopy[184].

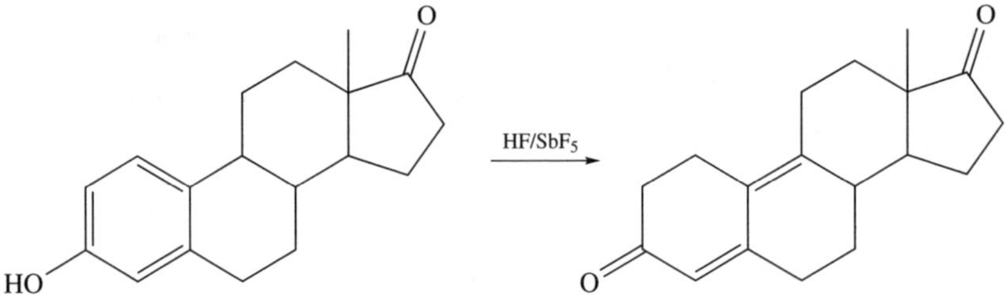

$$\xrightarrow{HF/SbF_5} \quad (91)$$

In the presence of suitable hydride donors the dienone intermediates formed in these reactions can be further reduced to the corresponding ketones. 3-Hydroxytetralin in the presence of $HF-SbF_5$ and methylcyclopentane gives 3-oxodecaline (equation 92)[185].

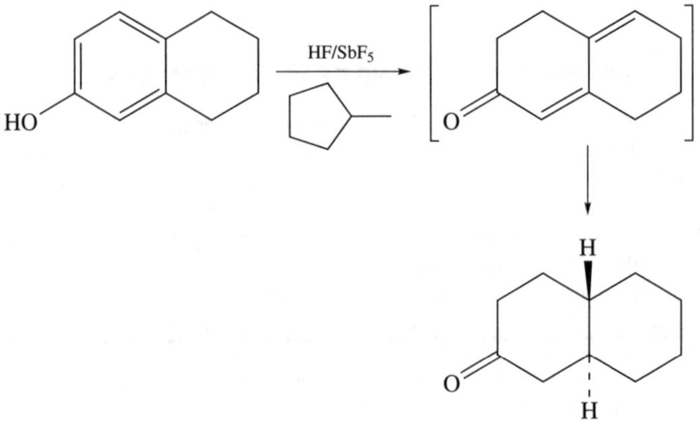

$$\xrightarrow{HF/SbF_5} \quad (92)$$

9. Electrophilic reactions of phenols

In superacidic media, phenols and anisoles are diprotonated to give the superelectrophilic O,C-diprotonated gitonic dications. The latter react readily with aromatics to give regioselectively arylated 4-aryl-2-cyclohexenones, which slowly isomerize to the 3-aryl-2-cyclohexenones under the reaction conditions. At longer reaction times, the latter are the predominant products. 4-Methylphenol, for example, in the presence of benzene and HF/SbF$_5$ gives initially a mixture of 4-methyl-4-phenyl-2-cyclohexenone (29%) and 4-methyl-3-phenyl-2-cyclohexenone (33%) after 1.5 min and after 15 min; the latter rearranged product can be isolated in 90% yield (equation 93). The possible superacid catalyzed route is shown in equation 94. A variety of aromatics such as benzene, naphthalene and tetrahydroquinoline can be used as the arylating agents in these reactions[186].

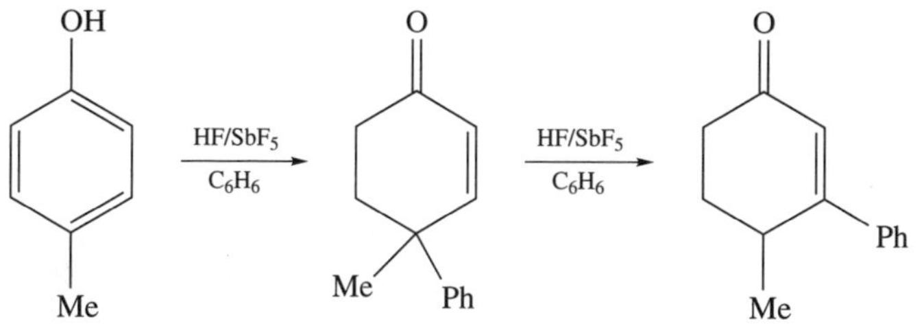

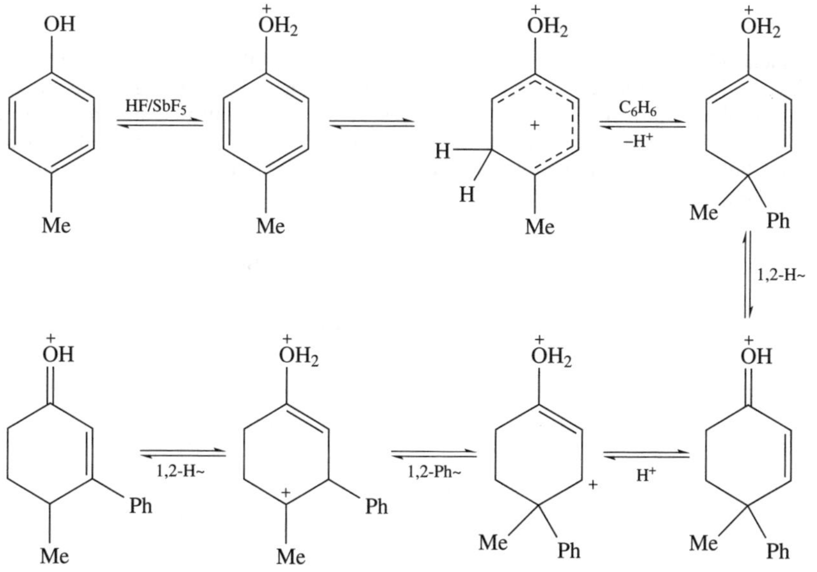

The synthesis of the spirocyclic cyclohexadienone ring system of the schiarisanrin family of natural products were based on the Lewis acid-promoted C-alkylation of the corresponding phenols or their derivatives[187]. The dibenzodioxepen, for example, when reacted with Lewis acids such as AlCl$_3$, or Me$_3$SiOTf, give the intermediate oxomethylene ylides, which undergo cyclization to give the spirocyclic cyclohexadienone (equation 95). The latter serves as a convenient intermediate for the schiarisanrin family of natural products, which exhibit cytotoxicities at $\mu g\, ml^{-1}$ levels against several standard cell lines.

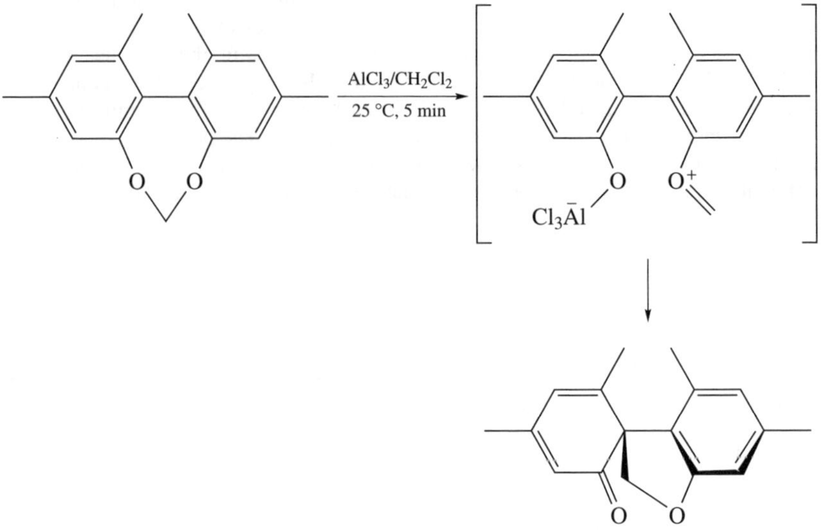

(95)

The phenol-dienones could be conveniently prepared directly from phenols by reaction with Br_2 in the presence of $AgClO_4$ and Na_2CO_3. These dienones are transformed efficiently to the 4-alkoxycyclohexa-2,5-dienones by the silver ion mediated reaction in the presence of the corresponding alcohols (equation 96)[158,188]. The solid-phase bromination of *tert*-butyl-substituted phenols with N-bromosuccinimide also affords halogenated cyclohexadienones[167].

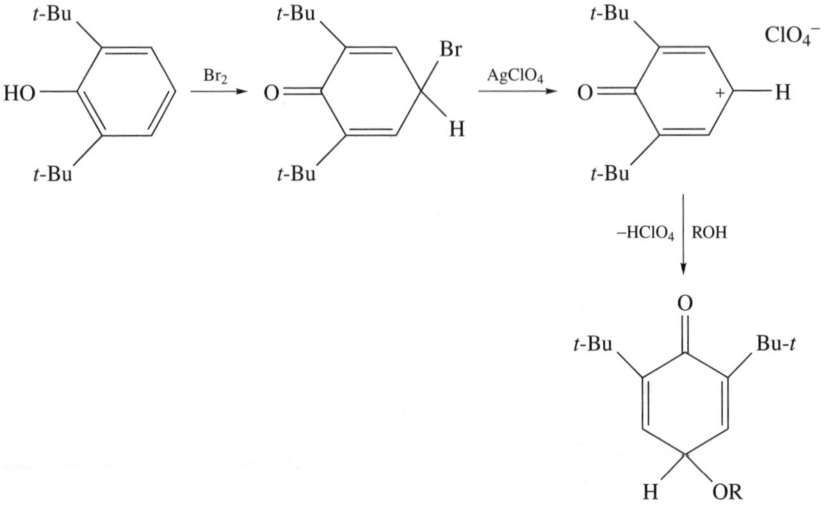

(96)

For example, R = Me, Et, *i*-Pr, *t*-Bu, *t*-Am

The halodienones can also be reacted with other phenols to give the biphenol derivatives in the presence of silver perchlorate (equation 97)[189].

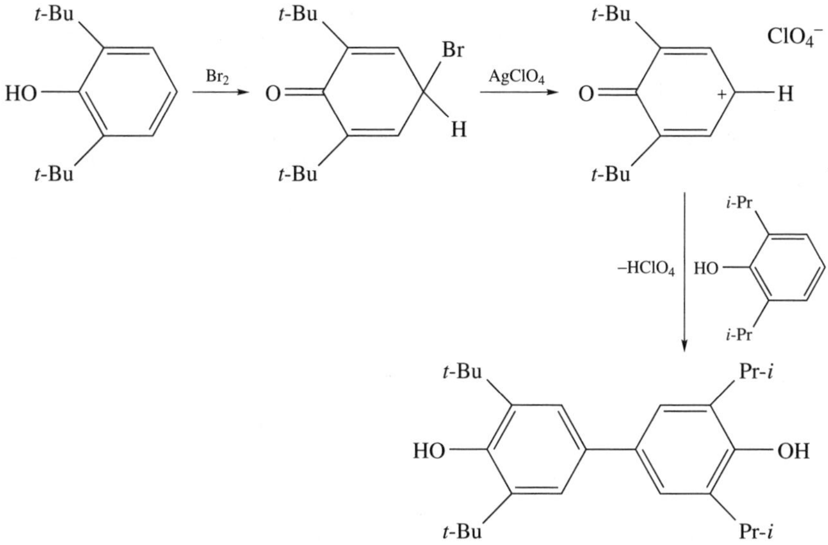

4-Fluorocyclohexa-2,5-dienone derivatives were obtained in high yield by reaction of para-substituted phenols with 1-fluoro-4-chloromethyl-1,4-diazoniabicyclo[2.2.2]octane bis(tetrafluoroborate) (Selectfluor™; F-TEDA-BF$_4$) or its 4-hydroxy analogue (Accufluor™; NFTh) in acetonitrile (equation 98). Estrogen steroids were readily converted to β-fluoro-1,4-estradien-3-one derivatives in high yields using this method (equation 99)[190].

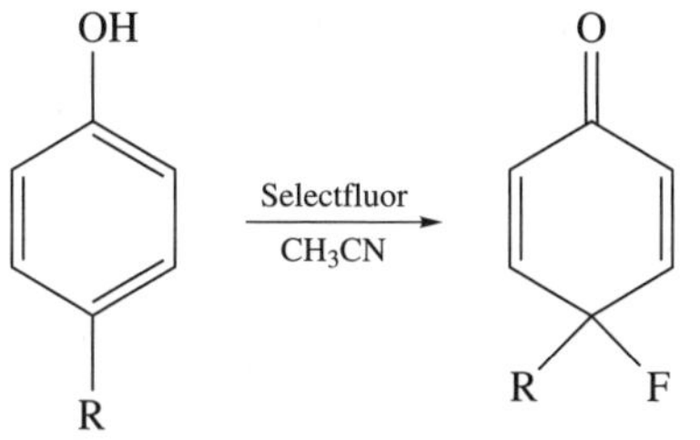

For example, R = Me, i-Pr

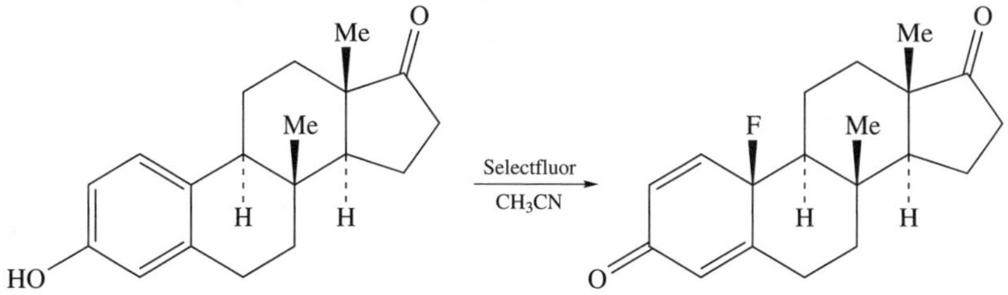

VIII. REFERENCES

1. J. H. P. Tyman, *Synthetic and Natural Phenols*, Elsevier, New York, 1996.
2. G. A. Olah, R. Malhotra and S. C. Narang, in *Nitration, Methods and Mechanisms (Organic Nitro Chemistry Series)* (Ed. H. Feuer), VCH, New York, 1989.
3. J. M. Brittain and P. B. D. De La Mare, in Supplement D. *The Chemistry of Halides, Pseudo-Halides and Azides* (Eds. S. Patai and Z. Rappoport), Wiley, New York, 1983.
4. L. J. Mathias, J. Jensen, K. Thigpen, J. McGowen, D. McCormick and L. Somlai, *Polymer*, **42**, 6527 (2001).
5. Y. Arredondo, M. Moreno-Manas and R. Pleixats, *Synth. Commun.*, **26**, 3885 (1996).
6. L. Liguori, H. R. Bjorsvik, F. Fontana, D. Bosco, L. Galimberti and F. Minisci, *J. Org. Chem.*, **64**, 8812 (1999).
7. G. H. Hsiue, S. J. Shiao, H. F. Wei, W. J. Kuo and Y. A. Sha, *J. Appl. Polym. Sci.*, **79**, 342 (2001).
8. H. Grabowska, J. Jablonski, W. Mista and J. Wrzyszcz, *Res. Chem. Intermed.*, **22**, 53 (1996).
9. H. Grabowska, W. Mista, J. Trawczynski, J. Wrzyszcz and M. Zawadzki, *Res. Chem. Intermed.*, **27**, 305 (2001).
10. G. Sartori, F. Bigi, R. Maggi and A. Arienti, *J. Chem. Soc., Perkin Trans. 1*, 257 (1997).
11. S. Sato, M. Koizumi and F. Nozaki, *J. Catal.*, **178**, 264 (1998).
12. J. Wrzyszcz, H. Grabowska, W. Mista, L. Syper and M. Zawadzki, *Appl. Catal. A, Gen.*, **166**, L249 (1998).
13. F. M. Bautista, J. M. Campelo, A. Garcia, D. Luna, J. M. Marinas and A. A. Romero, *React. Kinet. Catal. Lett.*, **63**, 261 (1998).
14. Y. Gong, K. Kato and H. Kimoto, *Bull. Chem. Soc. Jpn.*, **74**, 377 (2001).
15. Y. Gong, K. Kato and H. Kimoto, *Synlett*, 1058 (2000).
16. P. Hewawasam and M. Erway, *Tetrahedron Lett.*, **39**, 3981 (1998).
17. M. Baudry, V. Barberousse, Y. Collette, G. Descotes, J. Pires, J.-P. Praly and S. Samreth, *Tetrahedron*, **54**, 13783 (1998).
18. B. Yao, J. Bassus and R. Lamartine, *Bull. Soc. Chim. Fr.*, **134**, 555 (1997).
19. M. P. D. Mahindaratne and K. Wimalasena, *J. Org. Chem.*, **63**, 2858 (1998).
20. A. A. Agaev, D. B. Tagiev and Z. M. Pashaev, *Russ. J. Appl. Chem.*, **67**, 924 (1994).
21. G. A. Olah, G. K. S. Prakash and J. Sommer, *Superacids*, Wiley, New York, 1985.
22. G. A. Olah, D. Meidar, R. Malhotra, J. A. Olah and S. C. Narang, *J. Catal.*, **61**, 96 (1980).
23. B. Torok, I. Kiricsi, A. Molnar and G. A. Olah, *J. Catal.*, **193**, 132 (2000).
24. K. Smith, G. M. Pollaud and I. Matthews, *Green Chem.*, **1**, 75 (1999).
25. T. Nishimura, S. Ohtaka, A. Kimura, E. Hayama, Y. Haseba, H. Takeuchi and S. Uemura, *Appl. Catal. A, Gen.*, **194**, 415 (2000).
26. F. Bigi, M. Lina, R. Maggi, A. Mazzacani and G. Sartori, *Tetrahedron Lett.*, **42**, 6543 (2001).
27. P. Phukan and A. Sudalai, *J. Chem. Soc., Perkin Trans. 1*, 3015 (1999).
28. S. K. Badamali, A. Sakthivel and P. Selvam, *Catal. Lett.*, **65**, 153 (2000).
29. S. K. Badamali, A. Sakthivel and P. Selvam, *Catal. Today*, **63**, 291 (2000).
30. P. S. E. Dai, *Catal. Today*, **26**, 3 (1995).
31. F. E. Imbert, M. Guisnet and S. Gnep, *J. Catal.*, **195**, 279 (2000).
32. V. V. Fomenko, D. V. Korchagina, N. F. Salakhutdinov, I. Y. Bagryanskaya, Y. V. Gatilov, K. G. Ione and V. A. Barkhash, *Russ. J. Org. Chem.*, **36**, 539 (2000).
33. K. Wilson, D. J. Adams, G. Rothenberg and J. H. Clark, *J. Mol. Catal. A, Chem.*, **159**, 309 (2000).
34. V. V. Balasubramanian, V. Umamaheshwari, I. S. Kumar, M. Palanichamy and V. Murugesan, *Proc. Indian Acad. Sci., Chem. Sci.*, **110**, 453 (1998).
35. N. Barthel, A. Finiels, C. Moreau, R. Jacquot and M. Spagnol, *Top. Catal.*, **13**, 269 (2000).
36. F. M. Bautista, J. M. Campelo, A. Garcia, D. Luna, J. M. Marinas, A. A. Romero and M. R. Urbano, *React. Kinet. Catal. Lett.*, **62**, 47 (1997).
37. K. Nowinska and W. Kaleta, *Appl. Catal. A, Gen.*, **203**, 91 (2000).
38. A. P. Singh, *Catal. Lett.*, **16**, 431 (1992).
39. G. D. Yadav and N. Kirthivasan, *Appl. Catal. A, Gen.*, **154**, 29 (1997).
40. R. Tismaneanu, B. Ray, R. Khalfin, R. Semiat and M. S. Eisen, *J. Mol. Catal. A, Chem.*, **171**, 229 (2001).

41. K. Zhang, C. H. Huang, H. B. Zhang, S. H. Xiang, S. Y. Liu, D. Xu and H. X. Li, *Appl. Catal. A, Gen.*, **166**, 89 (1998).
42. S. Subramanian, A. Mitra, C. V. V. Satyanarayana and D. K. Chakrabarty, *Appl. Catal. A, Gen.*, **159**, 229 (1997).
43. A. Sakthivel, N. Saritha and P. Selvam, *Catal. Lett.*, **72**, 225 (2001).
44. A. Sakthivel, S. K. Badamali and P. Selvam, *Microporous Mesoporous Mat.*, **39**, 457 (2000).
45. X. W. Li, M. Han, X. Y. Liu, Z. F. Pei and L. Q. She, *Progress in Zeolite and Microporous Materials, Pts A–C, Stud. Surf. Sci. Catal.*, **105**, 1157 (1997).
46. A. N. Vasilyev and P. N. Galich, *Pet. Chem.*, **37**, 1 (1997).
47. S. Wieland and P. Panster, *Heterogeneous Catalysis and Fine Chemicals IV, Stud. Surf. Sci. Catal.*, **108**, 67 (1997).
48. K. Chandler, F. H. Deng, A. K. Dillow, C. L. Liotta and C. A. Eckert, *Ind. Eng. Chem. Res.*, **36**, 5175 (1997).
49. K. Chandler, C. L. Liotta, C. A. Eckert and D. Schiraldi, *AIChE J.*, **44**, 2080 (1998).
50. G. Deak, T. Pernecker and J. P. Kennedy, *J. Macromol. Sci., Pure Appl. Chem.*, **A32**, 979 (1995).
51. F. Bigi, G. Casnati, G. Sartori and G. Araldi, *Gazz. Chim. Ital.*, **120**, 413 (1990).
52. F. Bigi, G. Casnati, G. Sartori, G. Araldi and G. Bocelli, *Tetrahedron Lett.*, **30**, 1121 (1989).
53. G. Bocelli, A. Cantoni, G. Sartori, R. Maggi and F. Bigi, *Chem., Eur. J.*, **3**, 1269 (1997).
54. F. Bigi, G. Bocelli, R. Maggi and G. Sartori, *J. Org. Chem.*, **64**, 5004 (1999).
55. M. A. Brimble, M. R. Nairn and J. S. O. Park, *J. Chem. Soc., Perkin Trans. 1*, 697 (2000).
56. S. Fukumoto, Z. Terashita, Y. Ashida, S. Terao and M. Shiraishi, *Chem. Pharm. Bull.*, **44**, 749 (1996).
57. B. M. Trost and F. D. Toste, *J. Am. Chem. Soc.*, **120**, 815 (1998).
58. H. Ito, A. Sato and T. Taguchi, *Tetrahedron Lett.*, **38**, 4815 (1997).
59. G. Casiraghi, G. Casnati, G. Puglia, G. Sartori and M. G. Terenghi, *J. Chem. Soc., Perkin Trans. 1*, 1862 (1980).
60. N. U. Hofslokken and L. Skattebol, *Acta Chem. Scand.*, **53**, 258 (1999).
61. B. S. Furniss, A. J. Hannaford, P. W. G. Smith and A. R. Tatchell. *Vogel's Textbook of Practical Organic Chemistry*, Wiley, New York, 1989.
62. F. Bigi, M. L. Conforti, R. Maggi and G. Sartori, *Tetrahedron*, **56**, 2709 (2000).
63. A. A. Anderson, T. Goetzen, S. A. Shackelford and S. Tsank, *Synth. Commun.*, **30**, 3227 (2000).
64. V. C. Malshe and E. S. Sujatha, *React. Funct. Polym.*, **43**, 183 (2000).
65. S. Kobayashi, M. Moriwaki and I. Hachiya, *Tetrahedron Lett.*, **37**, 2053 (1996).
66. S. Kobayashi, M. Moriwaki and I. Hachiya, *Tetrahedron Lett.*, **37**, 4183 (1996).
67. S. Kobayashi, M. Moriwaki and I. Hachiya, *Bull. Chem. Soc. Jpn.*, **70**, 267 (1997).
68. S. Kobayashi, M. Moriwaki and I. Hachiya, *Synlett*, 1153 (1995).
69. B. Bennetau, F. Rajarison and J. Dunogues, *Tetrahedron*, **50**, 1179 (1994).
70. Y. V. S. Rao, S. J. Kulkarni, M. Subrahmanyam and A. V. R. Rao, *J. Chem. Soc., Chem. Commun.*, 1456 (1993).
71. E. Fromentin, J. M. Coustard and M. Guisnet, *J. Catal.*, **190**, 433 (2000).
72. E. V. Sobrinho, E. Falabella, S. Aguiar, D. Cardoso, F. Jayat and M. Guisnet, *Proc. Int. Zeolite Conf., 12th*, **2**, 1463 (1999).
73. Y. V. S. Rao, S. J. Kulkarni, M. Subrahmanyam and A. V. R. Rao, *Appl. Catal. A, Gen.*, **133**, L1 (1995).
74. H. Van Bekkum, A. J. Hoefnagel, M. A. Vankoten, E. A. Gunnewegh, A. H. G. Vogt and H. W. Kouwenhoven, *Zeolites and Microporous Crystals*, **83**, 379 (1994).
75. M. E. Davis, P. Andy, J. Garcia-Martinez, G. Lee, H. Gonzalez and C. W. Jones, *PCT Int. Appl.*, WO 0132593 (2001); *Chem. Abstr.*, **134**, 353177 (2001).
76. A. K. Pandey and A. P. Singh, *Catal. Lett.*, **44**, 129 (1997).
77. I. Neves, F. R. Ribeiro, J. P. Bodibo, Y. Pouilloux, M. Gubelmann, P. Magnoux, M. Guisnet and G. Perot, *Proc. Int. Zeolite Conf., 9th*, **2**, 543 (1993).
78. M. Guisnet, D. B. Lukyanov, F. Jayat, P. Magnoux and I. Neves, *Ind. Eng. Chem. Res.*, **34**, 1624 (1995).
79. L. M. Litvinenko, I. N. Dotsenko, A. I. Kirichenko and L. I. Bondarenko, *Tezisy Dokl.—Ukr. Resp. Konf. Fiz. Khim., 12th*, 139 (1977); *Chem. Abstr.*, **93**, 7226 (1980).
80. F. Effenberger and H. Klenk, *Chem. Ber.*, **107**, 175 (1974).

81. J. S. Brown, R. Glaser, C. L. Liotta and C. A. Eckert, *Chem. Commun.*, 1295 (2000).
82. A. V. Golounin, *Russ. J. Org. Chem.*, **36**, 134 (2000).
83. N. Jain and H. G. Krishnamurty, *Indian J. Chem.*, Sect B, **38**, 1237 (1999).
84. A. S. R. Anjaneyulu, U. V. Mallavadhani, Y. Venkateswarlu and A. V. R. Prasad, *Indian J. Chem., Sect. B*, **26**, 823 (1987).
85. T. S. Li and A. X. Li, *J. Chem. Soc., Perkin Trans. 1*, 1913 (1998).
86. M. Adachi, H. Matsumura and T. Sugasawa, *Eur. Pat. Appl.*, 266849 (1988).
87. M. A. Zolfigol, E. Ghaemi and E. Madrakian, *Synth. Commun.*, **30**, 1689 (2000).
88. T. Milczak, J. Jacniacki, J. Zawadzki, M. Malesa and W. Skupinski, *Synth. Commun.*, **31**, 173 (2001).
89. D. A. Conlon, J. E. Lynch, F. W. Hartner, R. A. Reamer and R. P. Volante, *J. Org. Chem.*, **61**, 6425 (1996).
90. J. A. R. Rodrigues, A. P. de Oliveira, P. J. S. Moran and R. Custodio, *Tetrahedron*, **55**, 6733 (1999).
91. B. Gigante, A. O. Prazeres and M. J. Marcelocurto, *J. Org. Chem.*, **60**, 3445 (1995).
92. S. Samajdar, F. F. Becker and B. K. Banik, *Tetrahedron Lett.*, **41**, 8017 (2000).
93. H. Pervez, S. O. Onyiriuka, L. Rees, J. R. Rooney and C. J. Suckling, *Tetrahedron*, **44**, 4555 (1988).
94. S. Shinkai, T. Tsubaki, T. Sone and O. Manabe, *Tetrahedron Lett.*, **26**, 3343 (1985).
95. S. Shinkai, K. Araki, T. Tsubaki, T. Arimura and O. Manabe, *J. Chem. Soc., Perkin Trans. 1*, 2297 (1987).
96. K. No and Y. Noh, *Bull. Korean Chem. Soc.*, **7**, 314 (1986).
97. S. Kumar, N. D. Kurur, H. M. Chawla and R. Varadarajan, *Synth. Commun.*, **31**, 775 (2001).
98. P. S. Wang, R. S. Lin and H. X. Zong, *Synth. Commun.*, **29**, 2225 (1999).
99. W. C. Zhang, Y. S. Zheng and Z. T. Huang, *Synth. Commun.*, **27**, 3763 (1997).
100. O. Mogck, V. Bohmer, G. Ferguson and W. Vogt, *J. Chem. Soc., Perkin Trans. 1*, 1711 (1996).
101. W. Verboom, A. Durie, R. J. M. Egberink, Z. Asfari and D. N. Reinhoudt, *J. Org. Chem.*, **57**, 1313 (1992).
102. K. C. Nam and D. S. Kim, *Bull. Korean Chem. Soc.*, **15**, 284 (1994).
103. S. Yenes and A. Messeguer, *Tetrahedron*, **55**, 14111 (1999).
104. J. P. Eiserich, M. Hristova, C. E. Cross, A. D. Jones, B. A. Freeman, B. Halliwell and A. van der Vliet, *Nature*, **391**, 393 (1998).
105. N. Nonoyama, K. Chiba, K. Hisatome, H. Suzuki and F. Shintani, *Tetrahedron Lett.*, **40**, 6933 (1999).
106. M. Lehnig, *Arch. Biochem. Biophys.*, **368**, 303 (1999).
107. R. M. Uppu, J. N. Lemercier, G. L. Squadrito, H. W. Zhang, R. M. Bolzan and W. A. Pryor, *Arch. Biochem. Biophys.*, **358**, 1 (1998).
108. J. N. Lemercier, S. Padmaja, R. Cueto, G. L. Squadrito, R. M. Uppu and W. A. Pryor, *Arch. Biochem. Biophys.*, **345**, 160 (1997).
109. J. P. Eiserich, C. E. Cross, A. D. Jones, B. Halliwell and A. vander Vliet, *J. Biol. Chem.*, **271**, 19199 (1996).
110. B. Alvarez, H. Rubbo, M. Kirk, S. Barnes, B. A. Freeman and R. Radi, *Chem. Res. Toxicol.*, **9**, 390 (1996).
111. M. S. Ramezanian, S. Padmaja and W. H. Koppenol, *Chem. Res. Toxicol.*, **9**, 232 (1996).
112. M. S. Ramezanian, S. Padmaja and W. H. Koppenol, *Methods Enzymol.*, 269 (Nitric Oxide pt B), 195, (1996).
113. D. S. Bohle and J. A. Imonigie, *J. Org. Chem.*, **65**, 5685 (2000).
114. B. C. Challis and R. J. Higgins, *J. Chem. Soc., Perkin Trans. 2*, 2365 (1972).
115. B. C. Challis, R. J. Higgins and A. J. Lawson, *J. Chem. Soc., Perkin Trans. 2*, 1831 (1972).
116. J. H. Ridd, *Chem. Soc. Rev.*, **20**, 149 (1991).
117. J. R. Leis, A. Rios and L. Rodriguez-Sanchez, *J. Chem. Soc., Perkin Trans. 2*, 2729 (1998).
118. I. I. Ostromyslenskii, *Abstr. Pap. Am. Chem. Soc.*, **43**, 97 (1910).
119. A. Werner, *Chem. Ber.*, **42**, 4324 (1908).
120. A. I. Titov, *Usp. Khim.*, **27**, 877 (1958); *Chem. Abstr.*, **53**, 6313 (1959).
121. W. V. Tuomi and R. J. Kazlauskas, *J. Org. Chem.*, **64**, 2638 (1999).
122. A. G. Mayants, K. G. Pyreseva and S. S. Gordeichuk, *Zh. Org. Khim.*, **24**, 884 (1988); *Chem. Abstr.*, **111**, 134045 (1989).

123. H. Firouzabadi, N. Iranpoor and M. A. Zolfigol, *Iran J. Chem. Chem. Eng., Int. Engl. Ed.*, **16**, 48 (1997).
124. S. X. Gu, H. W. Jing, J. G. Wu and Y. M. Liang, *Synth. Commun.*, **27**, 2793 (1997).
125. M. L. delaBreteche, M. A. Billion and C. Ducrocq, *Bull. Soc. Chim. Fr.*, **133**, 973 (1996).
126. L. Eberson and F. Radner, *Acc. Chem. Res.*, **20**, 53 (1987).
127. A. H. Blatt, *Org. React.*, **1**, 342 (1942).
128. J. Becht and W. Gerhardt, *Abh. Akad. Wiss. DDR*, **1**, 143 (1977).
129. M. J. S. Dewar and L. S. Hart, *Tetrahedron*, **26**, 973 (1970).
130. F. Jayat, M. J. S. Picot and M. Guisnet, *Catal. Lett.*, **41**, 181 (1996).
131. R. Rajan, D. P. Sawant, N. K. K. Raj, I. R. Unny, S. Gopinathan and C. Gopinathan, *Indian J. Chem. Technol.*, **7**, 273 (2000).
132. V. Rozenberg, T. Danilova, E. Sergeeva, E. Vorontsov, Z. Starikova, K. Lysenko and Y. Belokon, *Eur. J. Org. Chem.*, 3295 (2000).
133. H. Sharghi and B. Kaboudin, *J. Chem. Res. (S)*, 628 (1998).
134. N. N. Yusubov, *Russ. Chem. Bull.*, **45**, 978 (1996).
135. T. Laue and A. Plagens. *Named Organic Reactions*, Wiley, New York, 1998.
136. T. Kometani, H. Kondo and Y. Fujimori, *Synthesis*, 1005 (1988).
137. T. Matsumoto, H. Katsuki, H. Jona and K. Suzuki, *Tetrahedron Lett.*, **30**, 6185 (1989).
138. R. Martin, *Bull. Soc. Chim. Fr.*, 373 (1979).
139. K. Toshima, G. Matsuo and M. Nakata, *J. Chem. Soc., Chem. Commun.*, 997 (1994).
140. E. R. Palmacci and P. H. Seeberger, *Org. Lett.*, **3**, 1547 (2001).
141. G. A. Olah, M. Arvanaghi and V. V. Krishnamurthy, *J. Org. Chem.*, **48**, 3359 (1983).
142. A. Heidekum, M. A. Harmer and W. F. Hoelderich, *J. Catal.*, **176**, 260 (1998).
143. V. Borzatta, E. Poluzzi and A. Vaccari, *PCT Int. Appl.*, WO 012339 (2001); *Chem. Abstr.*, **134**, 266099 (2001).
144. A. J. Hoefnagel and H. van Bekkum, *Appl. Catal., A*, **97**, 87 (1993).
145. V. D. Chaube, P. Moreau, A. Finiels, A. V. Ramaswamy and A. P. Singh, *J. Mol. Catal. A, Chem.*, **174**, 255 (2001).
146. L. Conte, G. P. Gambaretto, M. Napoli, C. Fraccaro and E. Legnaro, *J. Fluorine Chem.*, **70**, 175 (1995).
147. R. E. Banks, M. K. Besheesh, S. N. Mohialdin-Khaffaf and I. Sharif, *J. Fluorine Chem.*, **81**, 157 (1997).
148. R. E. Banks, M. K. Besheesh and E. Tsiliopoulos, *J. Fluorine Chem.*, **78**, 39 (1996).
149. R. E. Banks, M. K. Besheesh, S. N. Mohialdin-Khaffaf and I. Sharif, *J. Chem. Soc., Perkin Trans. 1*, 2069 (1996).
150. D. D. DesMarteau and M. Witz, *J. Fluorine Chem.*, **7**, 52 (1991).
151. S. T. Purrington and W. A. Jones, *J. Org. Chem.*, **48**, 761 (1983).
152. T. Umemoto and G. Tomizawa, *J. Org. Chem.*, **60**, 6563 (1995).
153. I. Ekaeva, L. Barre, M. C. Lasne and F. Gourand, *Appl. Radiat. Isot.*, **46**, 777 (1995).
154. T. Fukuhara, M. Sawaguchi and N. Yoneda, *Electrochem. Commun.*, **2**, 259 (2000).
155. O. Karam, A. Martin, M. P. Jouannetaud and J. C. Jacquesy, *Tetrahedron Lett.*, **40**, 4183 (1999).
156. J. M. Gnaim and R. A. Sheldon, *Tetrahedron Lett.*, **36**, 3893 (1995).
157. A. Ghaffar, A. Jabbar and M. Siddiq, *J. Chem. Soc. Pak.*, **16**, 272 (1994).
158. K. Omura, *J. Org. Chem.*, **61**, 7156 (1996).
159. S. Mabic and J. P. Lepoittevin, *Tetrahedron Lett.*, **36**, 1705 (1995).
160. B. Andersh, D. L. Murphy and R. J. Olson, *Synth. Commun.*, **30**, 2091 (2000).
161. T. Oberhauser, *J. Org. Chem.*, **62**, 4504 (1997).
162. M. C. Carreno, J. L. G. Ruano, G. Sanz, M. A. Toledo and A. Urbano, *Synlett*, 1241 (1997).
163. D. E. Pearson, R. D. Wysong and C. V. Breder, *J. Org. Chem.*, **32**, 2358 (1967).
164. S. Fujisaki, H. Eguchi, A. Omura, A. Okamoto and A. Nishida, *Bull. Chem. Soc. Jpn.*, **66**, 1576 (1993).
165. J. Sarma and A. Nagaraju, *J. Chem. Soc., Perkin Trans. 2*, **6**, 1113 (2000).
166. N. R. Trivedi and S. B. Chandalia, *Org. Process Res. Dev.*, **3**, 5 (1999).
167. V. B. Voleva, I. S. Belostotskaya, N. L. Komissarova and V. V. Ershov, *Izv. Akad. Nauk, Ser. Khim.*, 1310 (1996); *Engl. Trans. Russ. Chem. Bull.*, **45**, 1249 (1996).
168. H. A. Muathen, *J. Org. Chem.*, **57**, 2740 (1992).

169. D. J. Cram, I. B. Dicker, M. Lauer, C. B. Knobler and K. N. Trueblood, *J. Am. Chem. Soc.*, **106**, 7150 (1984).
170. S. Kajigaeshi, T. Kakinami, T. Okamoto, H. Nakamura and M. Fujikawa, *Bull. Chem. Soc. Jpn.*, **60**, 4187 (1987).
171. S. C. Bisarya and R. Rao, *Synth. Commun.*, **23**, 779 (1993).
172. G. Arsequell, G. Espuna, G. Valencia, J. Barluenga, R. P. Carlon and J. M. Gonzalez, *Tetrahedron Lett.*, **40**, 7279 (1999).
173. J. C. Gesson, J. C. Jacquesy and R. Jacquesy, *Tetrahedron Lett.*, 4733 (1971).
174. J. M. Coustard, J. P. Gesson and J. C. Jacquesy, *Tetrahedron Lett.*, 4929 (1972).
175. J. M. Coustard, J. P. Gesson and J. C. Jacquesy, *Tetrahedron Lett.*, 4929 (1972).
176. J. M. Coustard and J. C. Jacquesy, *Tetrahedron Lett.*, 1341 (1972).
177. J. M. Coustard and J. C. Jacquesy, *Bull. Soc. Chim. Fr.*, 2098 (1973).
178. J. P. Gesson and J. C. Jacquesy, *Tetrahedron*, **29**, 3631 (1973).
179. J. P. Gesson, J. C. Jacquesy and R. Jacquesy, *Bull. Soc. Chim. Fr.*, 1433 (1973).
180. J. P. Gesson, J. C. Jacquesy and R. Jacquesy, *Tetrahedron Lett.*, 4119 (1974).
181. J. P. Gesson, J. C. Jacquesy, R. Jacquesy and G. Joly, *Bull. Soc. Chim. Fr.*, 1179 (1975).
182. A. F. Hegarty and J. P. Keogh, *J. Chem. Soc., Perkin Trans. 2*, 758 (2001).
183. O. Karam, M. P. Jouannetaud and J. C. Jacquesy, *New J. Chem.*, **18**, 1151 (1994).
184. J. C. Jaquesy, R. Jacquesy and J. F. Patoiseau, *J. Chem. Soc., Chem. Commun.*, 785 (1973).
185. J. M. Coustard, M. Douteau, J. C. Jacquesy and R. Jacquesy, *Tetrahedron Lett.*, 2029 (1975).
186. J. C. Jacquesy, J. M. Coustard and M. P. Jouannetaud, *Top. Catal.*, **6**, 1 (1998).
187. R. S. Coleman, J. M. Guernon and J. T. Roland, *Org. Lett.*, **2**, 277 (2000).
188. K. Omura, *J. Org. Chem.*, **61**, 2006 (1996).
189. K. Omura, *J. Org. Chem.*, **63**, 10031 (1998).
190. S. Stavber, M. Jereb and M. Zupan, *Synlett*, 1375 (1999).

CHAPTER **10**

Synthetic uses of phenols

MASAHIKO YAMAGUCHI

Department of Organic Chemistry, Graduate School of Pharmaceutical Sciences, Tohoku University, Aoba, Sendai, 980-8578 Japan
Fax: 81-22-217-6811; e-mail: yama@mail.pharm.tohoku.ac.jp

I. INTRODUCTION	661
II. DERIVATIZATION OF PHENOLS	662
A. C—O Bond Formation.	662
1. O-alkylation	662
2. O-arylation	670
3. O-glycosidation	673
B. C—C Bond Formation.	676
1. C-alkylation: Bond formation with sp^3 carbon	676
2. C-alkenylation and C-phenylation: Bond formation with sp^2 carbon atoms.	680
3. C-ethynylation: Bond formation with sp carbon atoms	682
4. C-hydroxyalkylation and related reactions	682
5. C-formylation, C-acylation and C-carboxylation	685
C. C-fluorination	687
II. METAL PHENOXIDE AS REAGENT IN ORGANIC SYNTHESIS	688
A. Organic Synthesis Using Metal Complexes of Monophenol	688
B. Organic Synthesis Using Metal Complexes of Biphenol: BINOL and Derivatives.	691
C. Organic Synthesis Using Metal Complexes of Salicylaldehyde Imines	697
IV. REFERENCES	704

I. INTRODUCTION

The purpose of this review is to provide an overview on the recent advances in synthetic chemistry of phenols since 1980. In organic synthesis, phenols are important both as substrates and as reagents. Phenols can be derivatized either at the hydroxy group or the aromatic moiety, for which many classical methods have been employed both in industry

---

The Chemistry of Phenols. Edited by Z. Rappoport
© 2003 John Wiley & Sons, Ltd   ISBN: 0-471-49737-1

and in the laboratory. The first part of this review describes recent work to enhance the efficiency of these processes, particularly on the C−O bond formation and the C−C bond and C−F bond formation. Metalated phenols have become very important reagents in organic synthesis, and notable is the use of chiral phenols as ligand in asymmetric synthesis. The second part of this review treats synthetic reactions using metal phenoxide reagents. The notation 'cat' is provided in equations in order to discriminate catalytic reactions from stoichiometric reactions.

II. DERIVATIZATION OF PHENOLS

This section treats the synthetic reactions of phenols leading to C−O bond formation at the hydroxy group and C−C bond formation at the aromatic nuclei. Reactions similar to those of aliphatic alcohols, aromatic hydrocarbons or anisole are in general excluded. Oxidation reactions and replacement reactions of the hydroxy group are treated in other chapters of this book.

A. C−O Bond Formation

1. O-alkylation

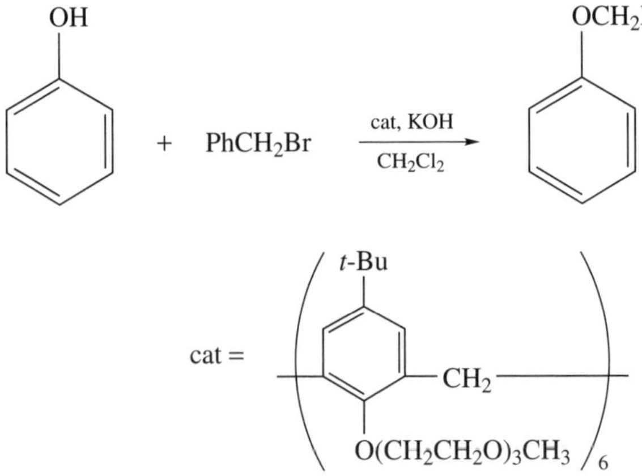

The classical Williamson synthesis treats alkali metal phenoxides and alkyl halides or alkyl sulfates in organic solvents to give O-alkylphenols[1,2]. The problem of insolubility of the phenoxides can be overcome by phase transfer catalysis using tetraalkylammonium salts, crown ethers or poly(ethyleneglycol)s in the presence of alkali metal hydroxides or fluorides[3−6]. Both solid−liquid and liquid−liquid biphasic systems are employed. The phase transfer reaction is dramatically accelerated by microwave irradiation using a domestic oven, which can be conducted without organic solvents on the solid support such as sodium hydroxide, alumina, zeolite or sodium carbonate[7−10]. Often, the reactions are completed within 1 min. A calix[6]arene equipped with poly(oxyethylene) group at the oxygen atoms catalyzes the phenol O-alkylation in a solid−liquid system (equation 1)[11,12]. The catalyst is more effective than benzyltrimethylammonium chloride, polyethyleneglycol diethyl ether and 18-crown-6 in terms of the reaction rate and catalyst loading. Micelles formed from cetyltrimethylammonium bromide in water were used for the O-alkylation of 2,6-disubstituted phenols[13]. Cs_2CO_3 is effective for the Williamson synthesis in organic solvents because of its higher solubility than K_2CO_3 or Na_2CO_3[14]. Notably, $Ni(acac)_2$ promotes t-alkylation of phenol in the presence of $NaHCO_3$ (equation 2)[15]. In order to avoid the formation of metal halides as a byproduct of the Williamson synthesis, use of dimethyl carbonate for the methylating reagent was examined[16–19]. The reagent is non-toxic, and produces only methanol and carbon dioxide as byproduct (equation 3). Sennyey and coworkers[16] and Lee and Shimizu[17] recommended the use of a catalytic amount of pentaalkylguanidine or Cs_2CO_3 as the base rather than K_2CO_3 or Na_2CO_3. The phase transfer method employing solid K_2CO_3 and tetrabutylammonium bromide was also reported for the carbonate O-alkylation[20].

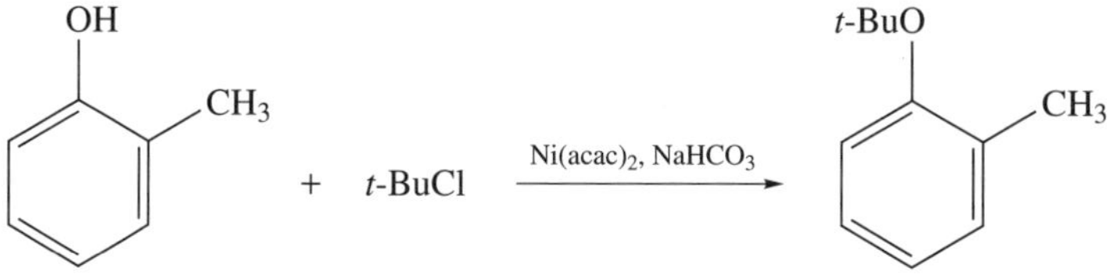

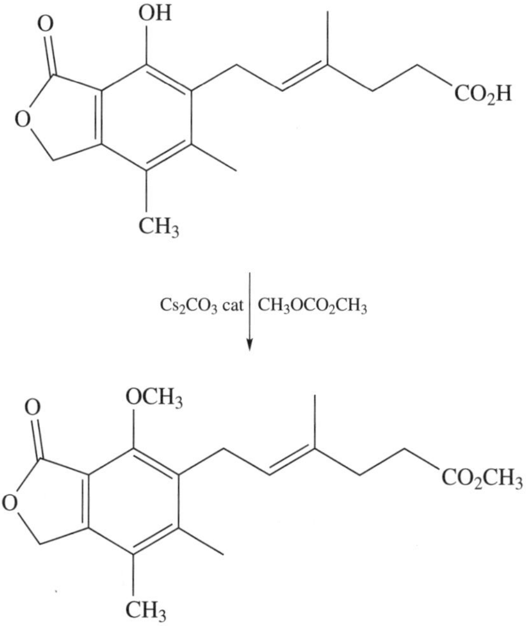

The Mitsunobu reaction proved to be useful for the synthesis of aryl alkyl ethers from alcohols and phenols[21]. The method proceeds under mild conditions and tolerates many functional groups with inversion of configuration, as exemplified by the reactions of lactate and $endo$-5-norbornen-2-ol (equations 4 and 5)[22,23]. Neighboring group participation, however, was observed in the reactions of exo-5-norbornen-2-ol (equation 6) and $trans$-1-hydroxy-2-aminoindane with phenol[23–25]. The Mitsunobu reaction of a tertiary propargylic alcohol takes place at the hindered carbon via S_N2[26].

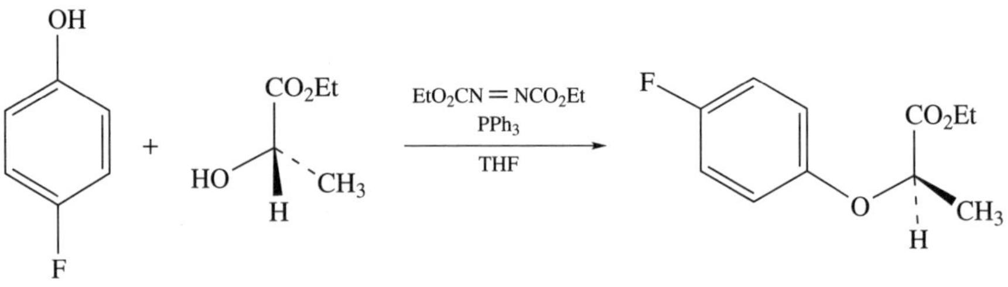

(4)

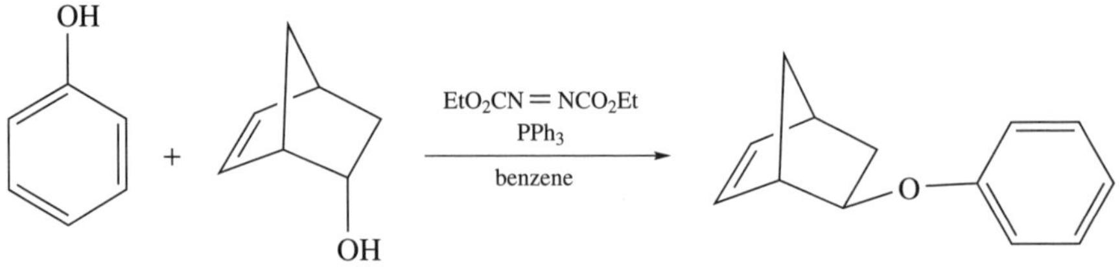

(5)

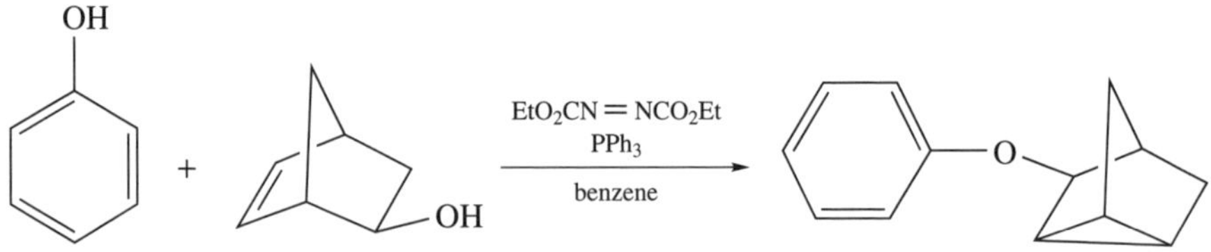

(6)

Phenol serves as an excellent oxygen nucleophile in transition metal catalyzed reactions[27]. O-allylation catalyzed by palladium, rhodium or ruthenium proceeds via π-allyl metal complexes. Phenol itself as well as sodium phenoxide, stannyl phenoxide, silyl phenoxide and phenyl carbonate is employed in the presence or absence of bases such as triethylamine, potassium fluoride or alumina[28–30]. Allyl carbonates are generally employed as the precursor of π-allyl metals, and allyl acetate or vinyl epoxide is used in some cases[31–33]. Miura and coworkers used allylic alcohol in the presence of $Ti(OPr-i)_4$[34]. π-Allylpalladium species generated by C–C bond cleavage of methylenecyclopropanes or by the C–C bond formation such as the Heck reaction also undergo phenoxylation[35,36]. Sinou and coworkers examined the regio- and stereochemistry of palladium catalyzed phenol O-allylation with allyl carbonates; acyclic primary allyl carbonates give primary phenyl ethers as the thermodynamic products; under kinetic control the selectivity was influenced by the steric and electronic nature of the allyl carbonates[37]. The phenoxylation of a 2-cyclohexenol carbonate proceeds with retention of configuration, which is consistent with the known π-allylpalladium chemistry (equation 7). Evans and Leahy attained preferential formation of secondary phenyl ethers from secondary allylic carbonates using a rhodium catalyst with net retention of configuration (equation 8)[38]. Palladium complexes derived from propargylic carbonates undergo phenol addition at the central carbon atom[39],

10. Synthetic uses of phenols

and Ihara and coworkers utilized the addition reaction followed by fragmentation of the cyclobutane ring for the stereoselective synthesis of cyclopentanones (equation 9)[40]. A catalytic amount of copper salt effectively promotes the O-alkylation by 2-methyl-3-butyn-2-ol trifluoroacetate in S_N2 regioselectivity[41].

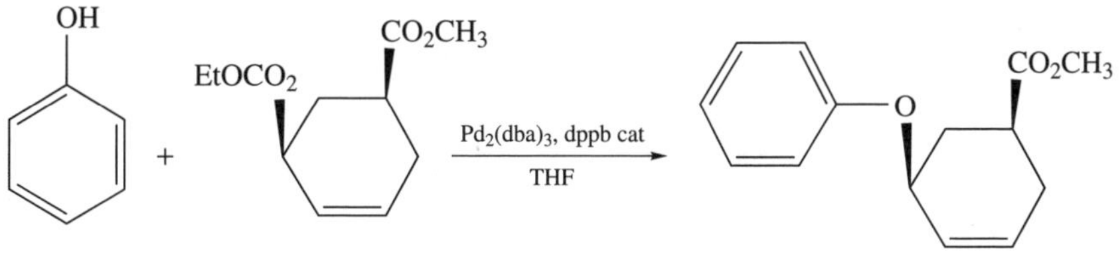

(7)

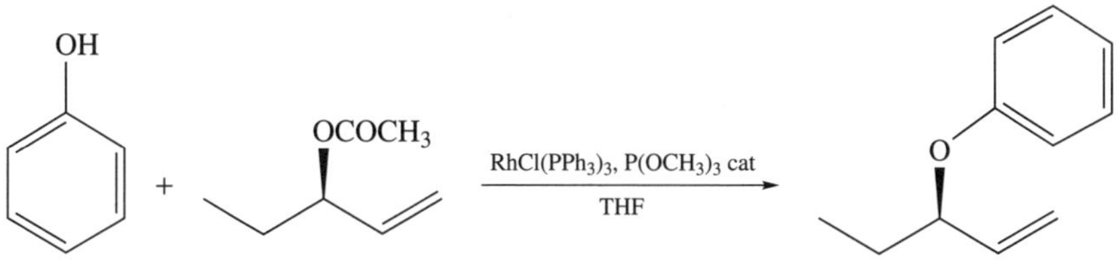

(8)

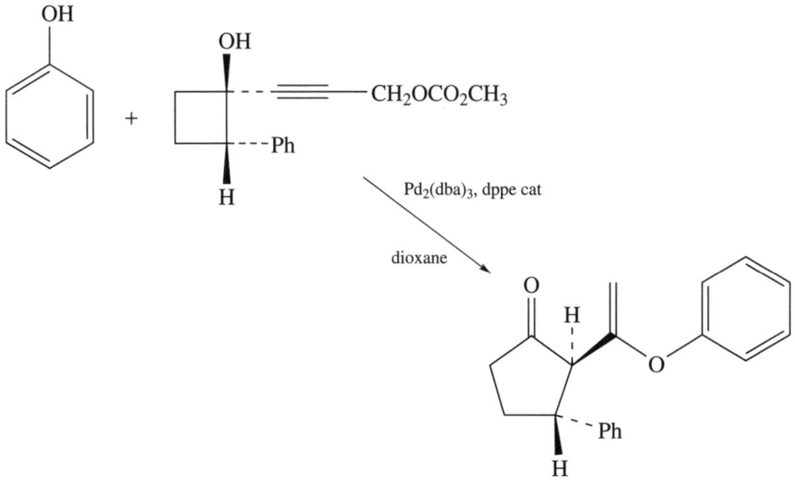

(9)

Phenols are used as the nucleophile in the asymmetric allylation of π-allylpalladium complexes. Trost and Toste attained asymmetric phenyl ether formation in high enantiomeric excess (ee) using diphosphine ligand derived from chiral 1,2-cyclohexanediamine (equation 10)[42]. Dynamic kinetic resolution of the racemic secondary allylic carbonate is conducted in the presence of tetrabutylammonium chloride, which increases the rate of $\pi-\sigma-\pi$ isomerization of the π-allyl palladium intermediate (equation 11)[43]. Lautens and coworkers cleaved *meso*-oxabicyclic alkenes with phenol in the presence of a catalytic amount of a chiral ferrocenyldiphosphine and a rhodium complex (equation 12)[44].

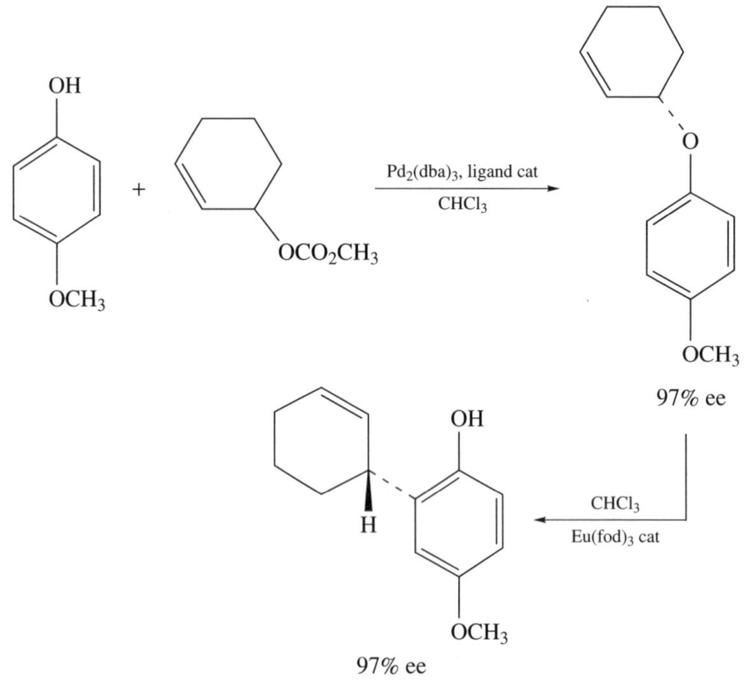

(10)

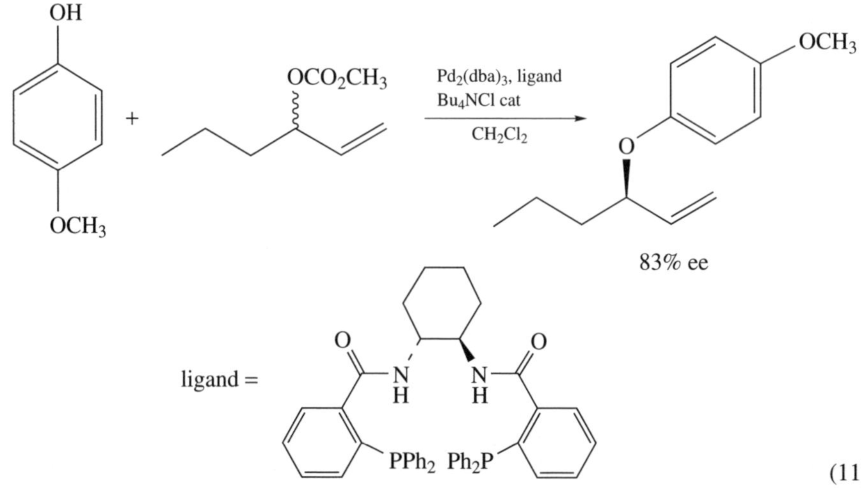

(11)

Excess diazomethane has been used to convert phenols to methyl ethers in the presence or absence of acids[1]. Employment of transition metal derivatives, typically $Rh_2(OAc)_4$[45–48] and recently CH_3ReO_3[49], allows one to react functionalized diazo compounds in an intramolecular or intermolecular O-alkylation (equation 13). The stability of diazo compounds derived from active methylene compounds toward OH insertion was compared using 2-propanol and the order will probably apply also to phenols: $PO(OEt)_2 >$ $Ph_2PO > EtO_2C \sim PhSO_2 > Me_2NCO > CN > PhCH_2 > Ph \sim H$. The reactivity should be reversed.

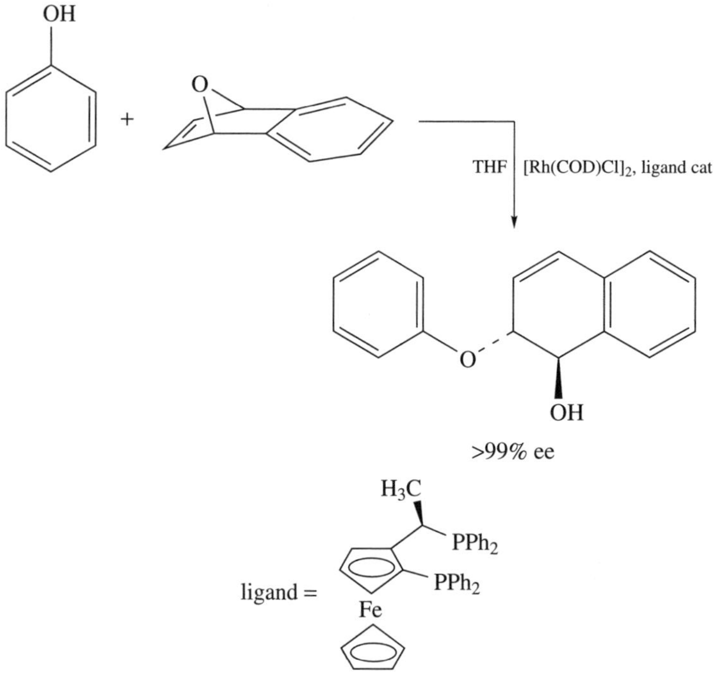

(12)

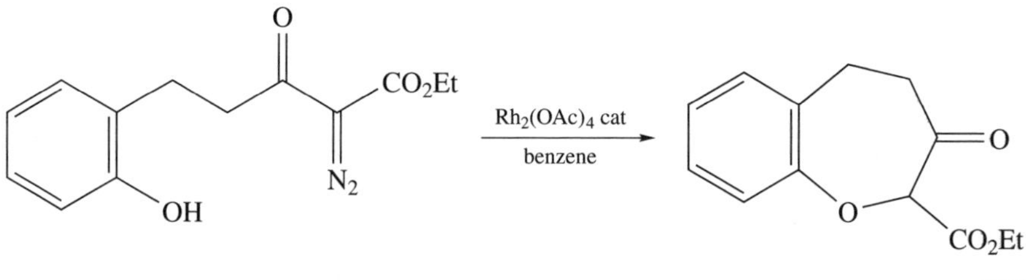

>99% ee

ligand = [ferrocene with H₃C, PPh₂, PPh₂ substituents]

(13)

Shibasaki and coworkers developed gallium lithium bis(naphthoxide) (GaLB) for the asymmetric cleavage of *meso*-epoxides with *p*-methoxyphenol giving optically active hydroxy ethers (equation 14)[50]. The 6,6'-bis(triethylsilylethynyl) derivative of GaLB improved the stability of the catalyst, resulting in higher chemical yields. The ring opening of cyclohexene oxide with phenol did not take place using conventional bases (BuLi, NaOBu-*t*, KOBu-*t*, K_2CO_3 or Cs_2CO_3) or Lewis acids (BF_3, $ZnCl_2$), which indicates the efficiency of the bimetallic catalyst with Brønsted basicity and Lewis acidity. An *N,N'*-ethylenebis(salicylideneamine) (salen) cobalt complex developed by Ready and Jacobsen catalyzes the kinetic resolution of a racemic epoxide (equation 15)[51]. Since epibromohydrin epimerizes in the presence of bromide anion, kinetic dynamic resolution gives the optically active phenoxy alcohol in 74% chemical yield (equation 16). Employing the cooperative nature of the Jacobsen catalyst, i.e. two molecules of the complex are involved in the transition state, acceleration of the rate and decrease in the catalyst loading was attained using oligomeric salen cobalt complexes[52,53]. Jung and Starkey developed a reaction of epoxyketones with phenols under phase transfer conditions to give α-phenoxyenones, which were converted to biaryl ethers after dehydrogenation (equation 17)[54].

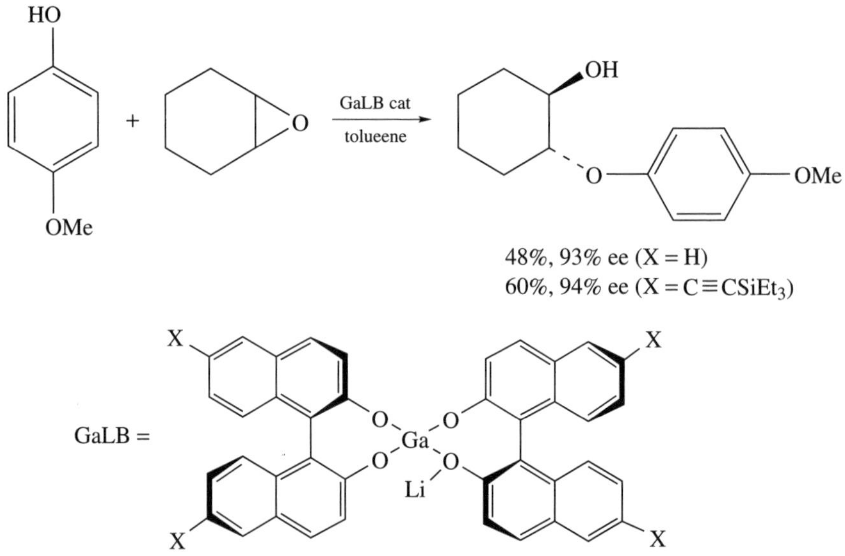

48%, 93% ee (X = H)
60%, 94% ee (X = C≡CSiEt$_3$)

(14)

99% ee

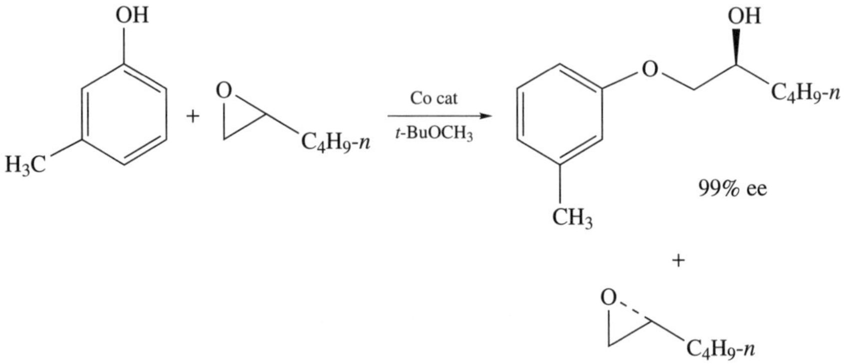

(15)

Addition of phenols to activated C−C multiple bonds is another method for O-alkylation. Conjugated carbonyl compounds with β-leaving groups react with metal phenoxides, giving the substituted products via addition−elimination, and the resulted β-aryloxylated carbonyls are versatile intermediates for synthesis of heterocyclic compounds[55−58]. Addition

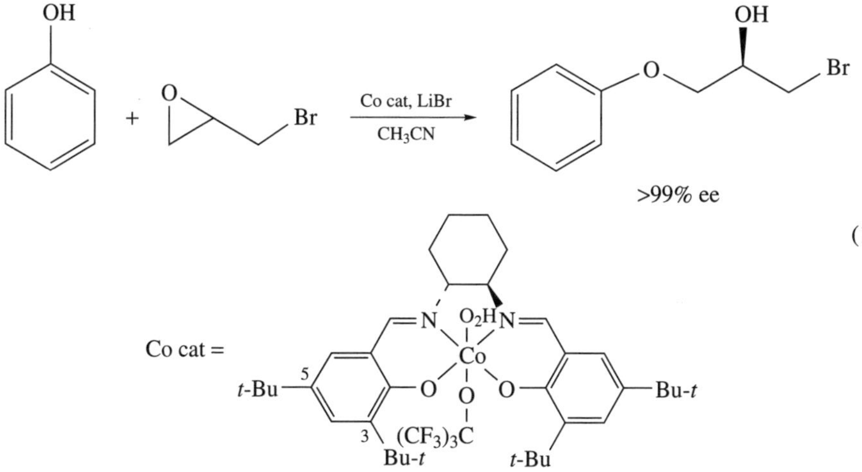

10. Synthetic uses of phenols

[Structure diagram for equation (16): phenol + epibromohydrin with Co cat, LiBr in CH₃CN gives aryl glycidyl bromide product, >99% ee]

(16)

Co cat = [Co-salen type catalyst with t-Bu groups and (CF₃)₃C substituent]

to acetylenic compound is another O-alkylation method, where no β-leaving group is necessary. Even (1-alkynyl)carbene tungsten complex can be used as the acceptor of phenol in the presence of triethylamine (equation 18)[59]. Addition of sodium phenoxides to tetrafluoroethylene generates carbanions stable to β-elimination, which can be trapped with carbon dioxide (equation 19)[60]. Vinyl ethers $CF_2=CFOR$ undergo the addition giving $PhOCF_2CFHOR$[61]. Phenol adds to $PhC\equiv CCF_3$ in an *anti*-stereochemistry under both kinetic and thermodynamic control[62]. Addition to unactivated olefin occurs in the presence of strong electrophilic reagents. The non-nucleophilic selenium reagent m-$O_2NC_6H_4SO_3SePh$ derived from PhSeSePh and (m-$O_2NC_6H_4SO_3)_2$ was used for the phenoxyselenation of simple alkenes[63]. The asymmetric version of palladium catalyzed intramolecular phenol addition to alkene (the Wacker-type reaction) was initially studied by Hosokawa and Murahashi[64,65], and Uozumi and Hayashi later attained high ee (equation 20)[66,67].

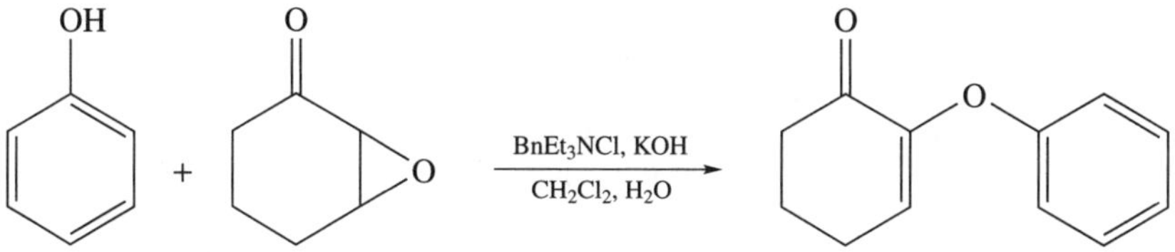

[Scheme for equation (17): phenol + cyclohexenone epoxide with BnEt₃NCl, KOH in CH₂Cl₂, H₂O gives 2-phenoxycyclohexenone]

(17)

[Scheme for equation (18): phenol + (CO)₅W=C(OEt)C≡CPh with Et₃N in CH₂Cl₂ gives (CO)₅W=C(OEt)CH=C(OPh)Ph]

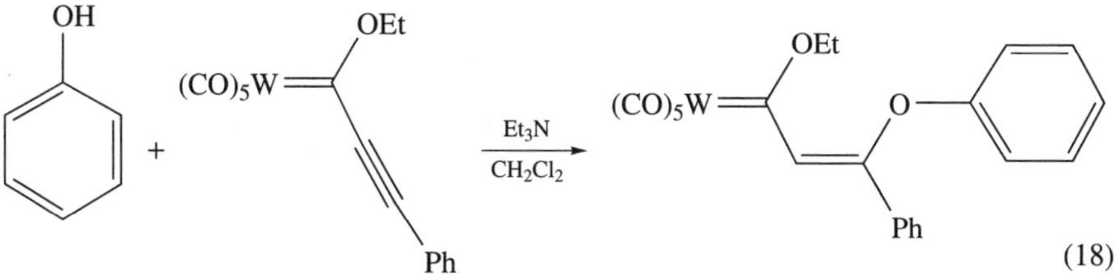

(18)

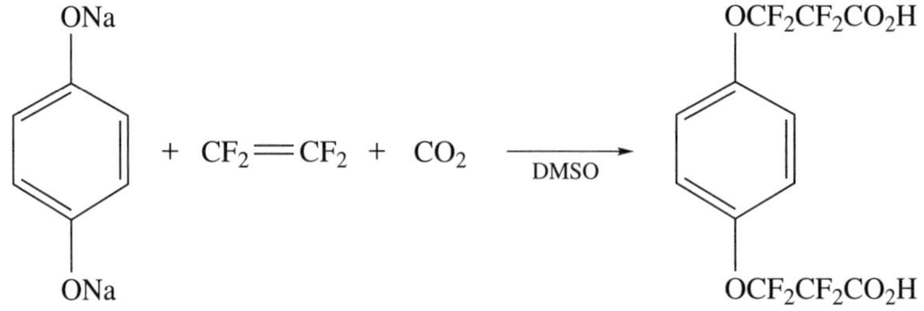

$$\text{(ArONa)}_2 + CF_2{=}CF_2 + CO_2 \xrightarrow{\text{DMSO}} \text{Ar(OCF}_2CF_2CO_2H)_2 \quad (19)$$

(Reaction 19: disodium hydroquinone + $CF_2{=}CF_2$ + CO_2 in DMSO → 1,4-bis(OCF$_2$CF$_2$CO$_2$H)benzene)

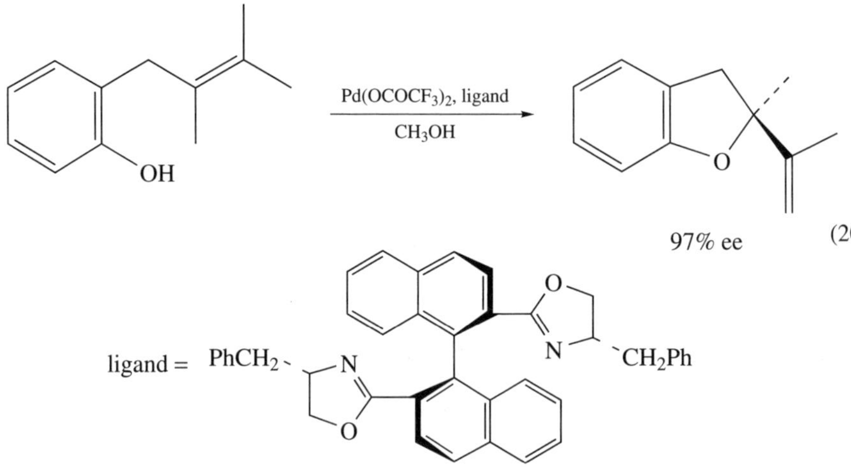

(20) 2-(2-methylallyl-2-methyl)phenol → benzofuran derivative with $Pd(OCOCF_3)_2$, ligand, CH_3OH, 97% ee

ligand = (S)-BINAP-bis(benzyl oxazoline) [PhCH$_2$-substituted bis-oxazoline on binaphthyl]

2. O-arylation

Total synthesis of vancomycin and related antibacterial substances active against MRSA required effective diaryl ether formation. Diaryl ethers are also important in polymer synthesis. In the classical Ullmann reaction aryl halides are heated with alkali metal phenoxides at high temperature, typically at about 200 °C, in the presence of copper powder or copper salts. New methods which will conduct the coupling at lower reaction temperature and possess broader applicability must therefore be developed[68–70]. Nicolaou and coworkers[71] and Snieckus and coworkers[72] used activated aryl halides with o-triazene and o-carbamyl groups as the substrate. Boger and Yohannes conducted the intramolecular coupling in a non-polar solvent, and suppressed the racemization of a phenylalanine derivative (equation 21)[73]. Buchwald and coworkers, using CuOTf and Cs$_2$CO$_3$, eliminated the prior preparation of metal phenoxides and coupled aryl iodides at 110 °C in toluene (equation 22)[74]. Addition of a catalytic amount of 1-naphthoic acid accelerated the reactions of less reactive aryl bromides. A library screening for the amine ligand in the copper catalyzed reaction revealed 8-hydroxyquinoline and 2-(N,N-dimethylamino)methyl-3-hydroxypyridine to be effective (equation 23)[75–78]. Palladium complexes also catalyze the diaryl ether formation as indicated by Mann and Hartwig employing 1,1′-diphenylphosphinoferrocene (DPPF) ligand. The method can couple aryl bromides with electron-withdrawing groups and sodium phenoxides[79]. Biphenylphosphine and binaphthylphosphine were used by Buchwald and coworkers who coupled less reactive aryl bromide, chloride and triflate possessing electron-donating groups (equation 24)[80]. A modified Ullmann reaction was reported by Barton and coworkers using arylbismuth in the presence of a catalytic amount of a copper complex (equation 25)[81,82]. Evans and coworkers and others found that the reaction of phenol and arylboronic acid in the presence of stoichiometric amounts of Cu(OAc)$_2$ gave aryl ethers at room temperature (equation 26)[83–85]. Jung and coworkers employed intramolecular Pummerer-type rearrangement for the diaryl ether synthesis[86].

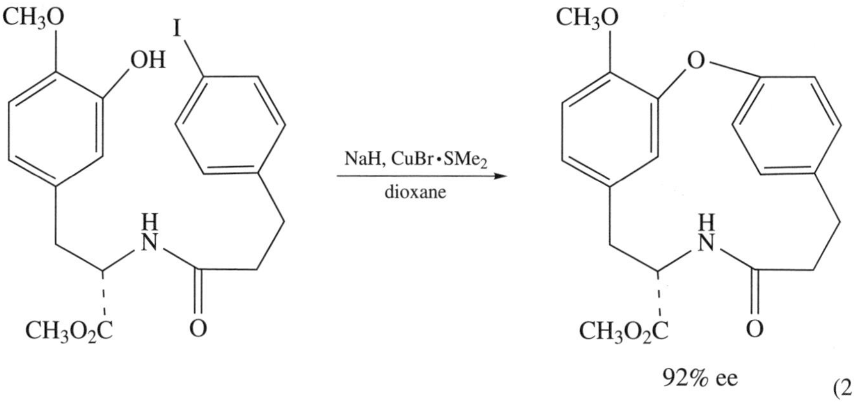

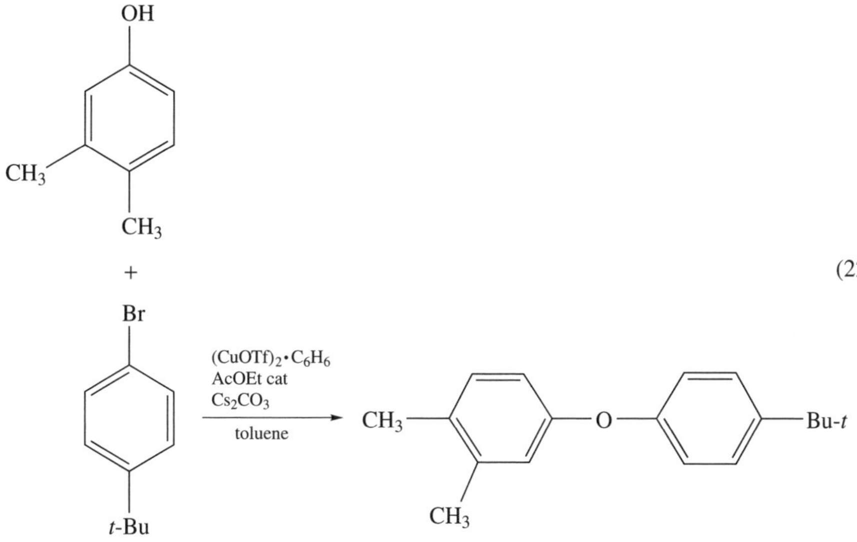

(23)

(24)

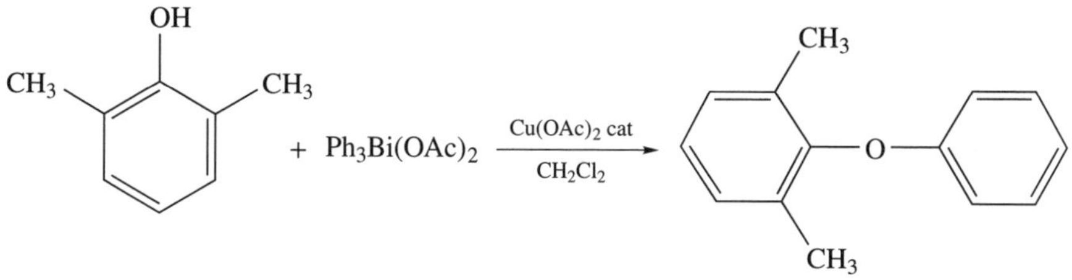

(25)

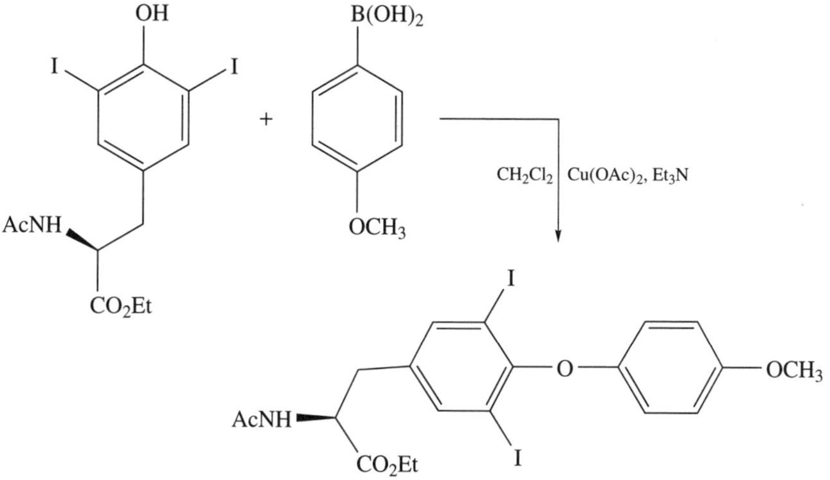

(26)

The S_NAr reaction is another attractive method for diaryl ether synthesis, and reactions of o-nitro- and o-cyanofluorobenzenes with phenols were reported[87,88]. π-Complexation of aryl halides with transition metals activates the aromatic nuclei toward S_NAr. Segal employed a ruthenium chlorobenzene complex in the poly(aryl ether) synthesis[89], and the methodology was extensively studied by Pearson, Rich and their coworkers using manganese complex and later iron and ruthenium complexes in natural product synthesis[90–94]. The intramolecular substitution of an aromatic chloride with a phenylalanine derivative takes place at room temperature without racemization (equation 27).

3. O-glycosidation

In relation to the synthesis of natural products and biologically active unnatural compounds, O-glycosidation of phenol has been studied; it is an acetal formation reaction at the sugar anomeric position. The classical König–Knorr method treats phenol with glycosyl bromide or chloride in the presence of metal promoters such as mercury or cadmium halides[95]. Yields, however, were not satisfactory, and several effective methods were developed (equation 28). Phase transfer methods using alkali metal bases are effective for the O-glycosidation of phenols giving thermodynamically stable β-anomers either from perbenzyl or peracetyl glycosyl bromides and even from N-acetylglucosamine[96–98]. Unreactive o-hydroxyacetophenones are O-glycosidated using K_2CO_3 and benzyltrimethylammonium chloride. Glycosyl fluorides are excellent substrates, since fluorophilic activation differentiates many other oxygen functionalities in a sugar molecule. Suzuki and

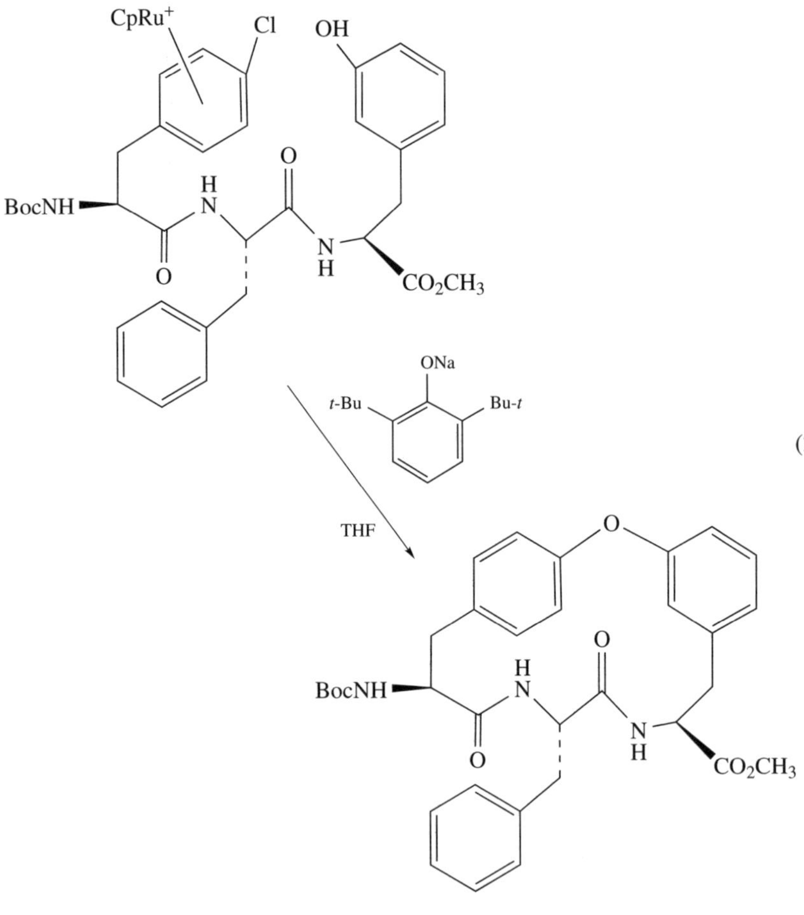

(27)

coworkers found that benzyl protected glucopyranosyl fluorides reacted with phenol in the presence of Cp_2HfCl_2–$AgClO_4$ promoter giving the α-anomers selectively[99]. Reaction of peracetylated glucopyranosyl fluorides was examined by Yamaguchi and coworkers: $BF_3 \cdot OEt_2$ gave the α-anomers, while addition of a guanidine base provided the β-anomers[100]. 1-Acetyl sugars, being stable and readily available, are used for the glycosidation with silylated or stannylated phenols in the presence of Lewis acids, giving the β-anomers[101–103]. 1-Trifluoroacetyl sugars react with phenol itself in the presence of $BF_3 \cdot OEt_2$[104]. Inazu and coworkers used 1-dimethylphosphonothioate in the presence of $AgClO_4$[105]. Kahn and coworkers employed sulfoxide in the presence of trifluoromethanesulfonic anhydride via Pummerer rearrangement, where the stereochemistry is controlled by changing the solvent; the α-anomers are formed in toluene and the β-anomers in dichloromethane[106]. 1-Trimethylsilyl ethers were used by Tieze and coworkers[107]. Free 1-hydroxy sugars were glycosylated by *in situ* formed *p*-nitrobenzenesulfonate giving the α-anomers[108], and the Mitsunobu reaction was employed by Roush giving the β-anomers[109].

Danishefsky and coworkers used a 1,2-α-epoxyglucose as an glycosyl donor, which is derived from a glucal by dioxirane oxidation[110,111]. Under basic conditions, the configuration at the anomeric center inverts giving the β-anomers (equation 29), while Lewis

activator = R$_4$NOH, Cp$_2$HfCl$_2$-AgClO$_4$, BF$_3$·OEt$_2$, BF$_3$·OEt$_2$-tetramethylguanidine, SnCl$_4$, AgClO$_4$, Tf$_2$O, PPh$_3$-EtO$_2$CN=NCO$_2$Et, etc.

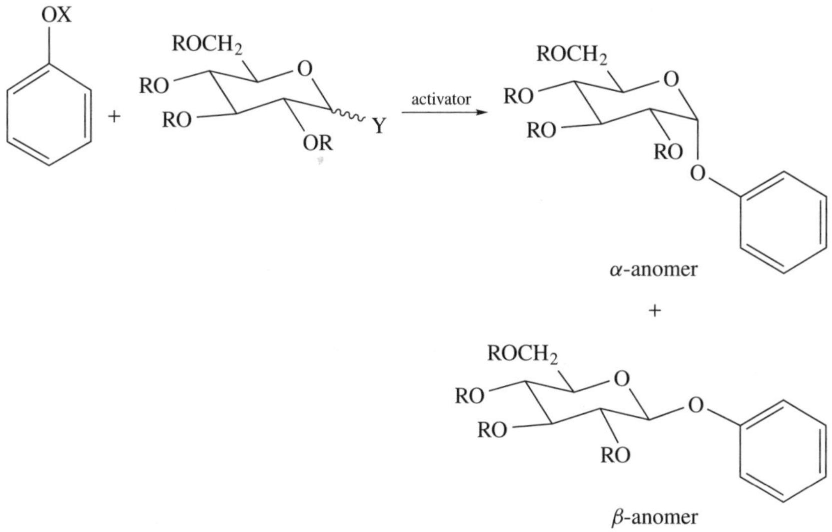

X = H, SiMe$_3$, SnBu$_3$, etc.

Y = F, Cl, Br, OAc, OPSMe$_2$, SOPh, OH, etc.

X = Ac, PhCO, PhCH$_2$, etc. (28)

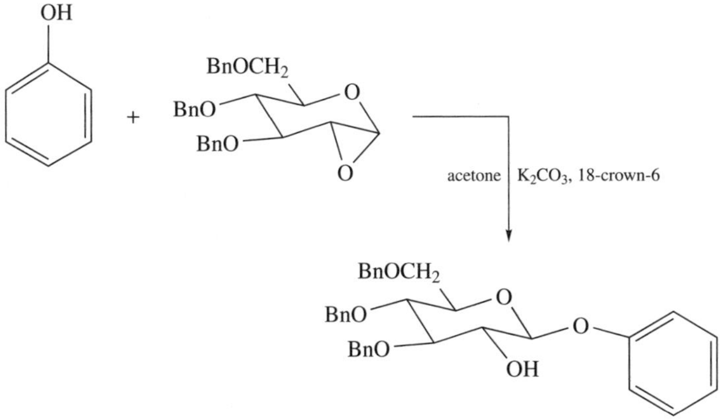

(29)

acid promotion gives the α-anomers predominantly[112]. Lewis acids, BF$_3$·OEt$_2$, InCl$_3$ and Yb(OTf)$_3$, promote the Ferrier reaction, which is the S$_N$2′ reaction of glycal and phenols giving phenyl 2-unsaturated glucosides[113,114]. Microwave irradiation without solvent was reported to promote the Ferrier reaction[115]. 1,2-Cyclopropanated glucose was used for the donor in the presence of [Pt(C$_2$H$_4$)Cl$_2$]$_2$ giving 2-methylated α-glucosides with concomitant cyclopropane cleavage[116].

B. C−C Bond Formation

Among many C−C bond forming reactions of phenols, the classical Friedel–Crafts method is still important, since it possesses an advantage of converting aromatic C−H bonds to C−C bonds without any synthetic intermediate. However, there are drawbacks of (i) employing strong Lewis acids often in stoichiometric amounts, and (ii) being effective only for C-alkylation or C-acylation and not, for example, C-ethenylation or C-ethynylation. Alkylation of halogenated phenols has become useful based on the development of various organometallic reactions such as the Suzuki coupling, the Heck reaction, the Stille coupling and the Sonogashira coupling. This methodology, however, requires extra steps for the preparation of halogenated phenols in a regioselective manner, and the hydroxy group generally must be protected prior to the organometallic reaction. A demand therefore exists for new synthetic methodologies, which directly convert aromatic C−H bonds of phenol to C−C bonds, employing catalytic amounts of reagents. The hydroxy group of phenol serves as a directing group in such aromatic C−H bond activation and C−C bond formation.

1. C-alkylation: Bond formation with sp^3 carbon

The C−C bond formation between the aromatic sp^2 carbons of phenol and sp^3 carbons can be conducted either under basic or acidic conditions, and can compete with O-alkylation. The o/p-selectivity is another matter of interest. Although the classical reaction of alkali metal phenoxides with alkyl halides generally takes place at the oxygen atom, C-alkylation occurs in some cases depending on solvent, counter cation and heterogeneity of the reaction system[117]. The regio- and stereoselectivity of the phenol alkylation was examined using a chiral *ortho*-ester under acidic conditions (equation 30)[118]. The C- and O-alkylation is controlled by the electronic nature of the substituent on the *ortho*-ester; an electron-rich aryl group induces the C-alkylation. The stereochemistry at the benzyl carbon is retention of configuration for the *trans*-isomer and inversion for the *cis*-isomer, which is explained by the involvement of free carbocation. Heating a mixture of

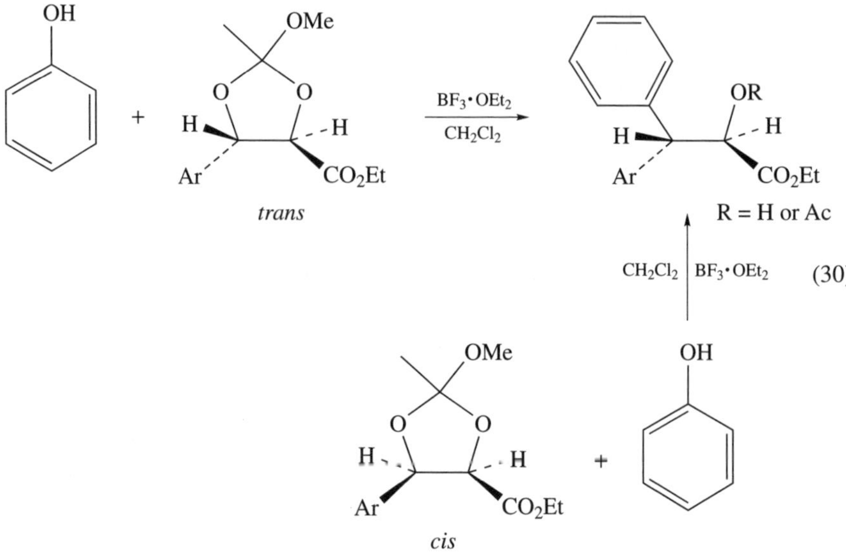

10. Synthetic uses of phenols

phenol and 1-adamantyl halides at 100–200 °C for several hours to days gives C-alkylated phenols, in which a small amount of the acid formed during the reaction may be catalyzing the generation of the tertiary carbocations (equation 31)[119]. Adamantyl bromide gives predominantly the *p*-isomer while the chloride gives the *o*-isomer. Secondary alkyl halides such as 2-bromoadamantane, bromocyclohexane or 2-bromonorbornane can also be used.

(31)

Several methods were reported for C-allylation of phenol, which is an alternative to the Claisen rearrangement[120,121]. The reaction sometimes competes with chroman formation by the addition of the phenol hydroxy group to the olefin. Potassium phenoxides in the presence of $ZnCl_2$ react with allyl halides giving *o*-allylphenols (equation 32)[122]. In the absence of the zinc salt, a modest yield of O-allylated phenol is obtained. Stoichiometric amounts of copper metal and copper(II) perchlorate also promote the *o*-allylation[123]. Molybdenum complexes $[Mo(CO)_4Br_2]_2$ or $Mo(CO)_3(CH_3CN)_2(SnCl_3)Cl$ catalyze allylation of phenol with allyl acetate, and $(acac)_2Mo(SbF_6)_2$ allyl alcohol, in which the formation of a π-allyl molybdenum complex is proposed[124,125]. Treatment of allyl alcohols with Brønsted acids in general provides chromans, for which two mechanisms are suggested: (i) initial O-allylation followed by cyclization, and (ii) initial C-allylation[126–128]. Conjugated 1,3-dienes are used for the C-allylation in the presence of Lewis acid, zeolite or a transition metal complex[129–131]. Rhodium catalysis found by Bienaymé and coworkers couples β-springene and a phenol giving the C-allylated phenol (equation 33)[132]. Cleavage of a vinylcyclopropanecarboxylate with tin phenoxide was reported to give an *o*-allylated product[133].

(32)

Inoue, Sato and coworkers studied a [2.3]sigmatropic rearrangement for the synthesis of *o*-alkylphenol. Reaction of a sulfoxide and phenol in the presence of dehydrating reagents such as thionyl chloride or benzenesulfonyl chloride provides phenoxysulfonium salts, which on treatment with triethylamine are converted to *o*-(α-alkylthioalkyl)phenols[134,135]. Since the benzylic thio group can be readily removed, the overall transformation provides the *o*-alkylated phenols. Later, a method to treat sulfides with sulfuryl chloride was developed for the same transformation (equation 34)[136], which is more effective than the original Gassmann's method employing *N*-chlorosuccinimide.

Posner and Canella used the directed metalation technology for phenol C-alkylation (equation 35); phenol was dimetalated at both the hydroxy group and the *o*-position with *t*-butyllithium, and treatment with methyl iodide gave *o*-cresol[137]. Brandsma and coworkers employed a complex reagent of butyllithium, N,N,N',N'-tetramethylethylenediamine, and potassium *t*-butoxide for the metalation[138]. Bates and Siahaan metalated cresols with butyllithium and potassium *t*-butoxide, and the *o*- and *m*-isomers gave the organometallic intermediate in good yield, while the yield was fair for the *p*-isomer[139]. The Simmons–Smith reagent is effective for the *o*-methylation of phenol, which is considered to involve iodomethylzinc phenoxide (equation 36)[140].

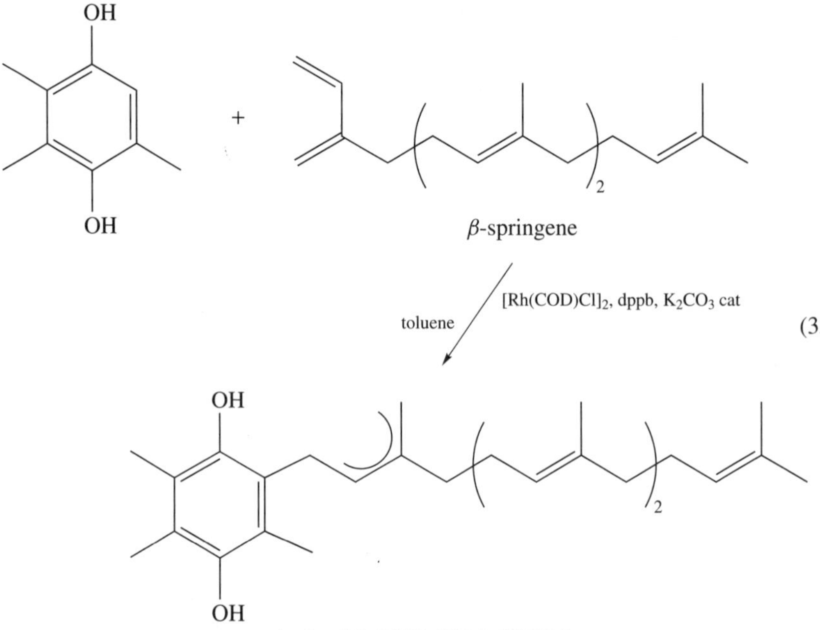

(33)

dppb = Ph$_2$PCH$_2$(CH$_2$)$_2$CH$_2$PPh$_2$

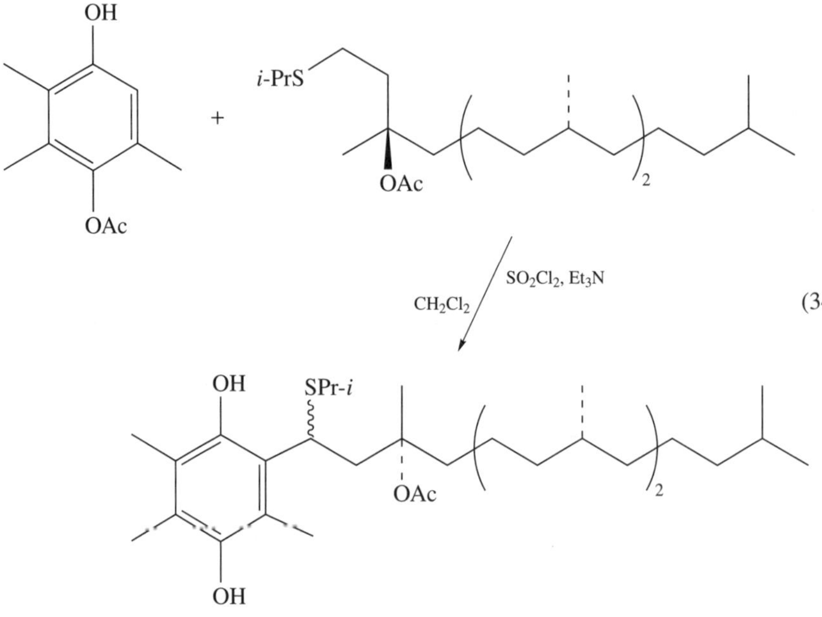

(34)

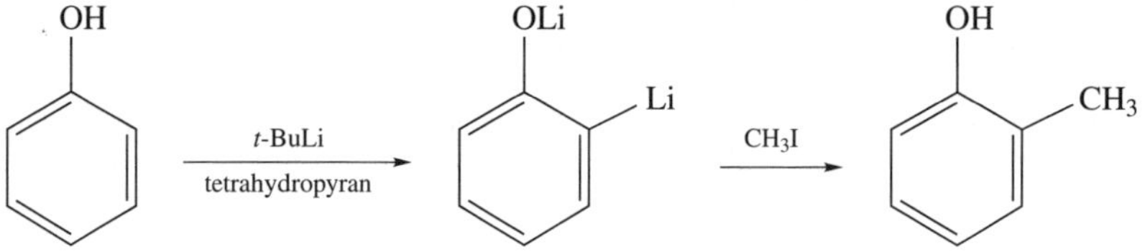

10. Synthetic uses of phenols

(equations 35, 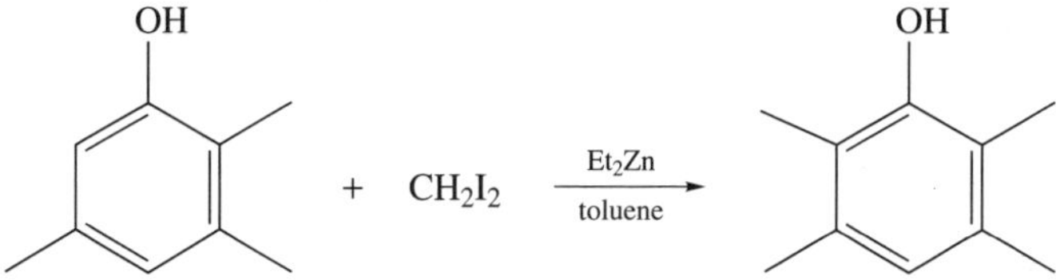 36, 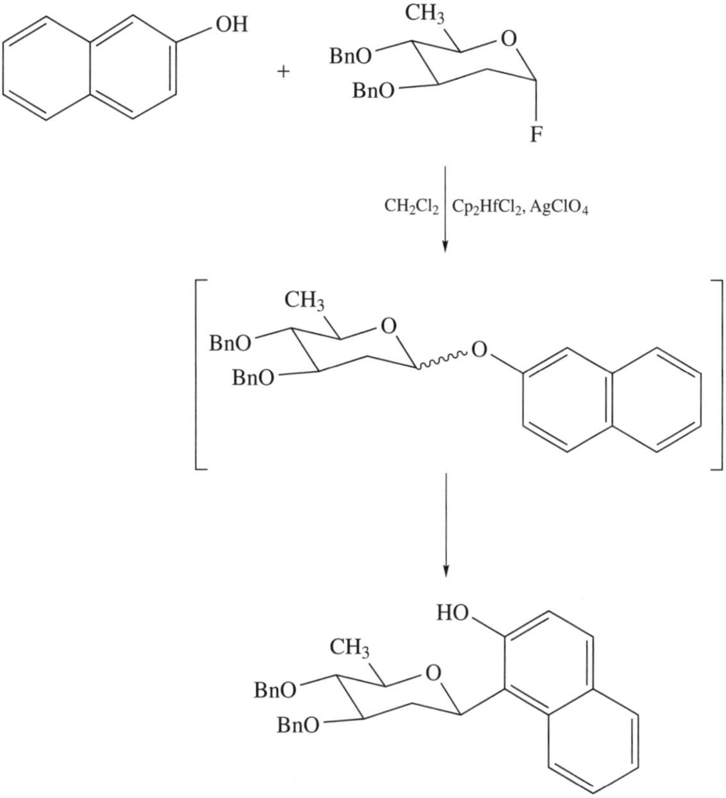 37 shown as schemes)

Total synthesis of a group of antibiotics containing aryl C-glycoside linkage, many of which possess C—C bonds between phenol o/p-positions and sugar anomeric centers, have attracted much interest during the last two decades. Suzuki and coworkers showed that the initial aryl O-glycosidation followed by a rearrangement to the C-glycoside (O- to C-glycosyl rearrangement) provides convenient access to this C—C bond formation (equation 37)[141–143]. Glycosyl fluoride and phenol are reacted in the presence of the

Cp_2HfCl_2–$AgClO_4$ reagent giving the C-glycosidated phenol at the *o*-position. Kometani and coworkers reported the use of $BF_3 \cdot OEt_2$ for this transformation[144]. The stereochemistry of the glycosyl center is dependent on the Lewis acid, and the stronger Lewis acid Cp_2HfCl_2–$AgClO_4$ gives the thermodynamically stable β-anomer from glucopyranosides. The rearrangement of the kinetically favorable α-anomer to the thermodynamically stable β-isomer is observed with a weaker Lewis acid. Suzuki and coworkers[145] and Toshima and coworkers[146] indicated that glycosyl esters and ethers can also be used as the glycosyl donor. 2-Unsaturated glucose also undergoes such C–O rearrangement with phenol in the presence of Lewis acids[147,148].

2. C-alkenylation and C-phenylation: Bond formation with sp^2 carbon atoms

Nulceophilic attack of phenol on a carbonyl followed by dehydration has been generally used to attach alkenyl sp^2 carbon atoms to the phenol nuclei. The methodology works well when the dehydration reaction can be controlled as in the classical Pechmann reaction, which is the condensation of β-ketoesters and phenols to give coumarins[149]. The reaction is accelerated by applying microwave irradiation or using an ionic liquid as the solvent[150,151]. A zeolite catalyst allows the synthesis of coumarins from acetic anhydride and phenols with concomitant Claisen condensation[152]. A modified Pechmann reaction employs propiolic acid in place of a keto ester under microwave irradiation[153]. Zeolite HSZ-360 catalyzes the reaction of phenol and a propargyl alcohol to give chromen (equation 38) in which an enyne compound generated by dehydration is considered to be the intermediate[154].

(38)

Addition reactions of phenols to acetylenes which provide a direct access to C-alkenylated phenols have recently been developed. The method giving such non-cyclized alkenylphenols requires in some cases devices to avoid the decomposition of the products. Sartori and coworkers reported alkenylation of phenol with phenylacetylene in the presence of HSZ-360 catalyst[155]. Yamaguchi and coworkers found that ethenylation (C_2-olefination) of phenol can be conducted using acetylene in the presence of stoichiometric amounts of $SnCl_4$ and tributylamine (equation 39)[156,157]. The ethenylation takes place exclusively at the *o*-position. The reaction tolerates electron donating and withdrawing groups on phenol and is relatively insensitive to steric hindrance; *m*-substituted phenols give mixtures of regioisomers in comparable amounts even in the case of *m*-(*t*-butyl)phenol. Modifications of the reaction conditions give 2,6-divinylphenols[158]. The mechanism involves a carbometalation of tin phenoxide and ethynyltin (carbostannylation) followed by protodestannylation under aqueous base conditions[159]. The stannylated alkene structure is considered to protect the ethenylphenols from decomposition. Use of butyllithium as the base in place of tributylamine allows one to conduct the ethenylation with trimethylsilylacetylene in a catalytic mode in regard to the metal reagents (equation 40)[160]. Gallium phenoxides also react with the silylacetylene giving o-(β-silylethenyl)phenols[161]; the organogallium compound undergoes similar carbometalation with the organotin compound. This is an interesting example of organometallic reagents of elements arranged diagonally in the periodic table that exhibit similar reactivities. Trost and Toste developed a palladium catalyzed reaction of phenol and alkyl propiolates in the presence of carboxylic acid (equation 41)[162]. The electron-withdrawing group is essential to activate the alkyne, and the phenol needs electron-donating groups.

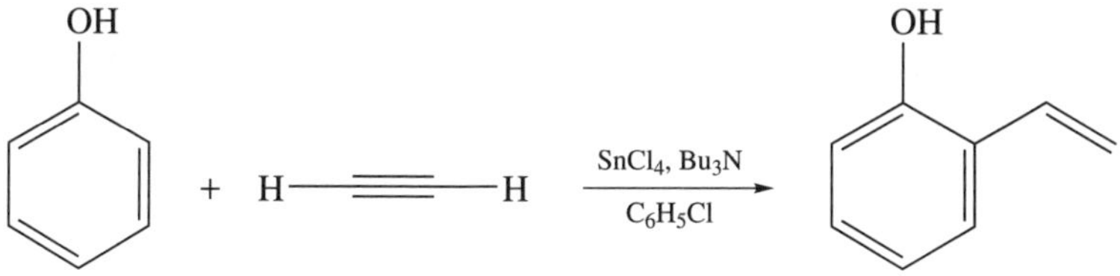

$$\text{PhOH} + \text{H}\mathord{-}\!\!\equiv\!\!\mathord{-}\text{H} \xrightarrow[\text{C}_6\text{H}_5\text{Cl}]{\text{SnCl}_4, \text{Bu}_3\text{N}} \text{o-vinylphenol} \quad (39)$$

$$\text{PhOH} + \text{H}\mathord{-}\!\!\equiv\!\!\mathord{-}\text{SiMe}_3 \xrightarrow[\text{C}_6\text{H}_5\text{Cl}]{\text{SnCl}_4, \text{BuLi cat}} \text{o-vinylphenol} \quad (40)$$

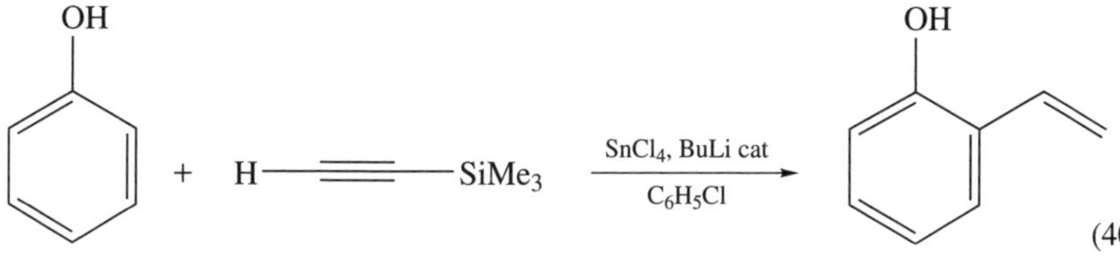

$$\text{3,5-(CH}_3\text{O)}_2\text{C}_6\text{H}_3\text{OH} + \text{H}\mathord{-}\!\!\equiv\!\!\mathord{-}\text{CO}_2\text{CH}_3 \xrightarrow[\text{Pd}_2(\text{dba})_3, \text{NaOAc cat}]{\text{HCO}_2\text{H}} \text{product} \quad (41)$$

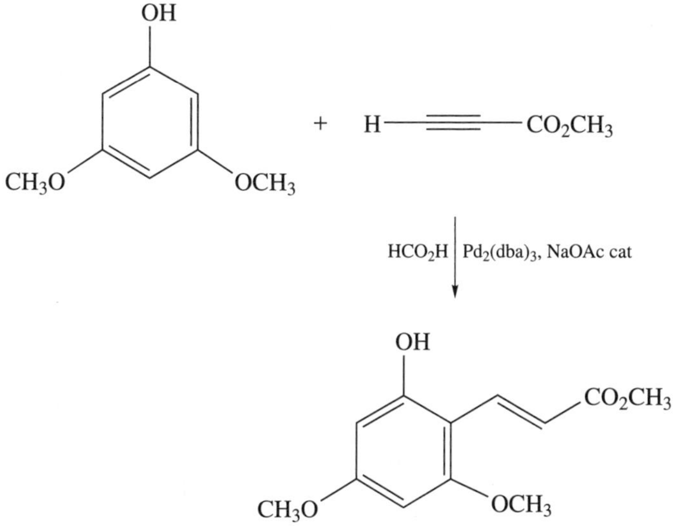

Barton and coworkers indicated that phenols are directly C-phenylated with Ph_4BiX or Ph_3BiX_2 (X = $OCOCF_3$ etc.) reagents under basic conditions (equation 42)[163]. Yamamoto and coworkers later developed an asymmetric version of the reaction in the presence of optically active amines[164]. Jung and coworkers developed a Pummerer-type rearrangement of 2-sulfinylphenol giving α-ketosulfonium salt, which was attacked by the phenol giving biphenols (equation 43)[165].

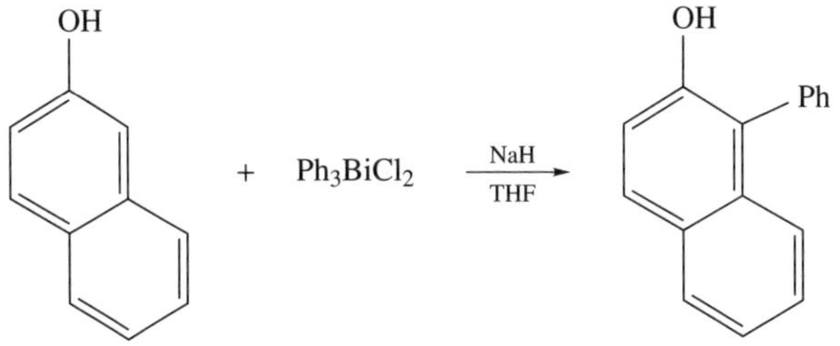

3. C-ethynylation: Bond formation with sp carbon atoms

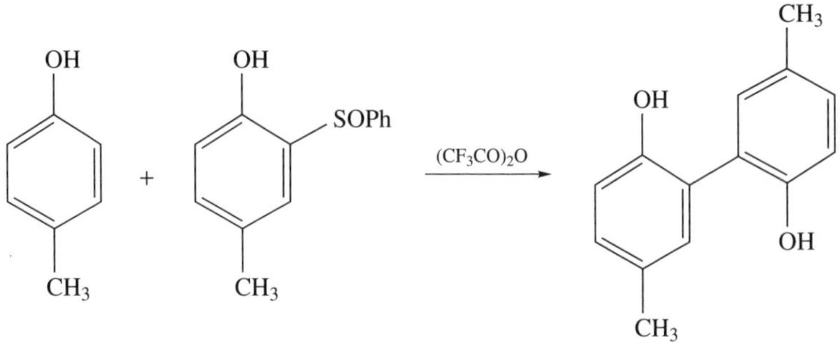

Ethynylation of phenol has been conducted using the Sonogashira coupling reaction of a halogenated phenol and terminal alkyne. It was recently found by Yamaguchi and coworkers that phenol itself can be ethynylated at the *o*-position using triethylsilylethynyl chloride; the reaction is catalyzed by $GaCl_3$, butyllithium and 2,6-di(*t*-butyl)-4-methylpyridine (equation 44)[166]. The reaction takes place via carbogallation of the phenoxygallium and the silylacetylene, followed by β-elimination regenerating $GaCl_3$. O-alkylation of phenol with $Cl_2C=CF_2$ under phase transfer conditions followed by treatment with excess butyllithium also gives *o*-ethynylphenols[167].

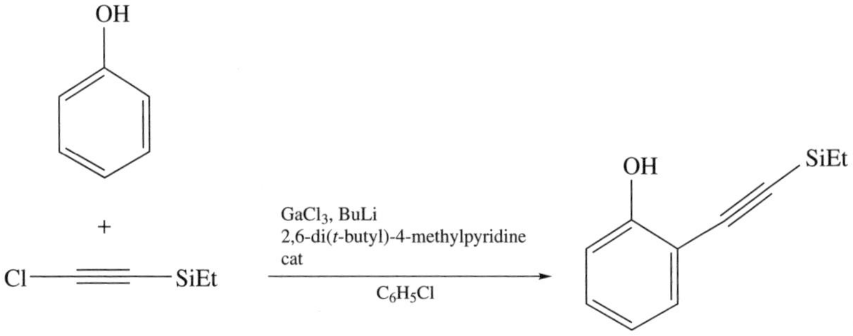

4. C-hydroxyalkylation and related reactions

Metal phenoxides are structurally related to metal enolates, and undergo aldol reaction to give C-hydroxyalkylated phenols. Reaction of formaldehyde and phenol to give phenol resins is of industrial importance, and occurs under either basic or acidic conditions. Casiraghi and coworkers observed an uncatalyzed reaction of phenols and paraformaldehyde giving salicyl alcohols in the presence of 1 equivalent of 1,2-dimethoxyethane (equation 45)[168]. The reactions of magnesium, titanium and aluminum phenoxides which take place at the *o*-position of the phenol hydroxy group were extensively studied by Casnati and coworkers[169]. Applications to heterocyclic carbonyl compounds have appeared[170,171]. Reaction of trifluoroacetaldehyde hemiacetal and phenols gives the *o*- and *p*-isomers depending on the promoters; K_2CO_3 gives the *p*-isomers and ZnI_2 the *o*-isomers[172,173]. Phenylboronic acid or dichlorophenylborate in the presence of triethylamine reacts with phenols and aldehydes giving 1,3,2-dioxaborins, which are hydrolyzed oxidatively with hydrogen peroxide (equation 46)[174,175].

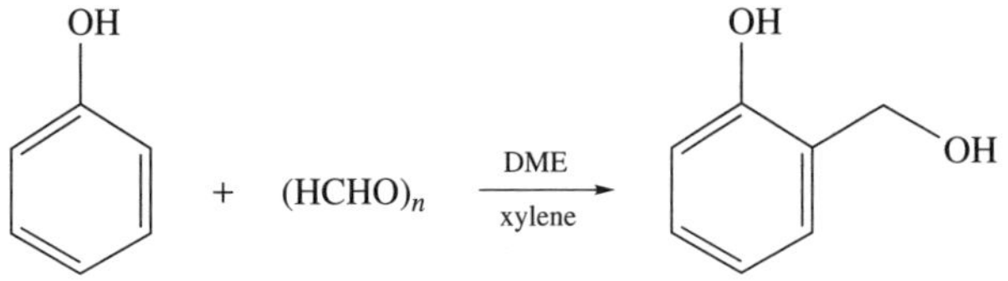

$$\text{PhOH} + (\text{HCHO})_n \xrightarrow[\text{xylene}]{\text{DME}} \text{2-(hydroxymethyl)phenol} \quad (45)$$

$$\text{3-methoxyphenol} + C_2H_5\text{CHO} \xrightarrow[\text{CH}_2\text{Cl}_2]{\text{PhBCl}_2, \text{Et}_3\text{N}} \text{dioxaborin} \quad (46)$$

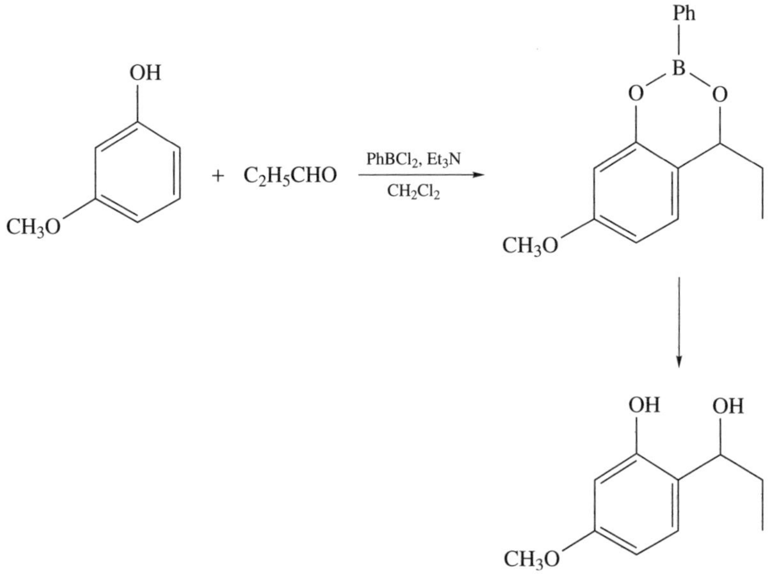

The stereochemistry of the aldol reaction between phenols and aldehydes was studied in detail by Italian chemists. Addition of magnesium phenoxides to chiral aldehydes with α-heteroatoms such as glyceraldehyde, sugar aldehyde and aminoaldehyde gives uniformly the *syn*-isomers, while titanium phenoxides give the *anti*-isomers (equation 47)[176–179]. The magnesium phenoxides are considered to form chelation intermediate and the titanium Cram-model intermediate. 8-Phenylmenthyl ester is an excellent chiral auxiliary for the diastereoselective addition of titanium phenoxides to glyoxylate and pyruvate[180–182]. Enantioselective addition of phenol to chloral using a stoichiometric amount of chiral menthyloxyaluminum promoter gives the adducts in 80% ee[183]. Double asymmetric induction in the addition of phenol to menthyl pyruvate employing a menthyloxyaluminum promoter indicated that the use of the same configuration of menthyl derivative provided higher stereoselectivity (matched pair)[184]. Erker developed a catalytic asymmetric reaction of methyl pyruvate and 1-naphthol using 1 mol% of a chiral zirconium cyclopentadienyl complex giving the adduct in 84% ee (equation 48)[185].

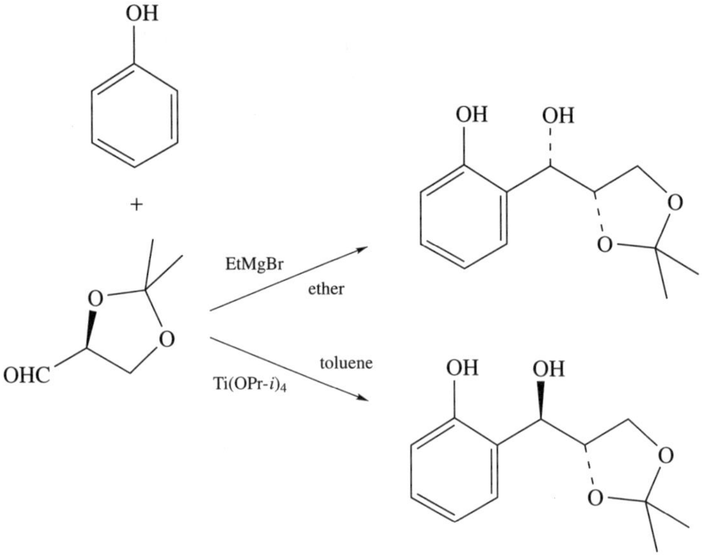

The classical Mannich reaction converts phenols to aminomethylated phenols. The reaction involves the addition of phenols to C=N bonds of imines or iminium salts formed from formaldehyde and primary or secondary amines, respectively[186,187]. Recent modifications employ the reaction of an aminal in the presence of SO_3, which gives a sulfonate ester, followed by o-aminomethylation (equation 49)[188,189]; Sc(OTf)$_3$ catalyzed three-component reactions of phenol, glyoxylates and amine[190]. Addition of a titanium phenoxide generated from TiCl$_4$ and the phenol to activated C=N bonds of a chiral glyoxylate imine exhibits high diastereoselectivity (equation 50)[191,192]. Fukuyama utilized Lewis acid promotion for the stereoselective Mannich reaction of phenols and cyclic acylimines[193].

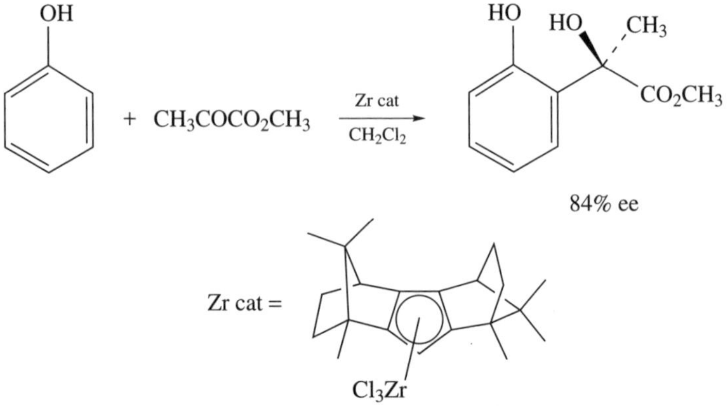

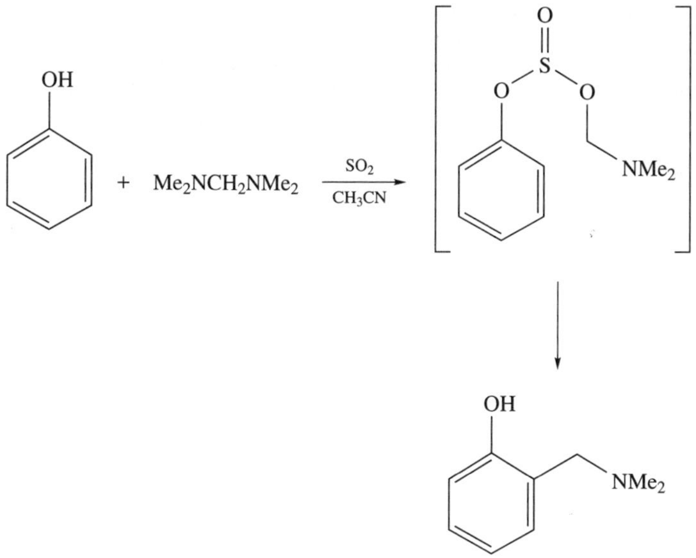

[Reaction scheme showing phenol + Me$_2$NCH$_2$NMe$_2$ with SO$_2$/CH$_3$CN giving an intermediate phenyl sulfite with NMe$_2$, then leading to 2-(dimethylaminomethyl)phenol] (49)

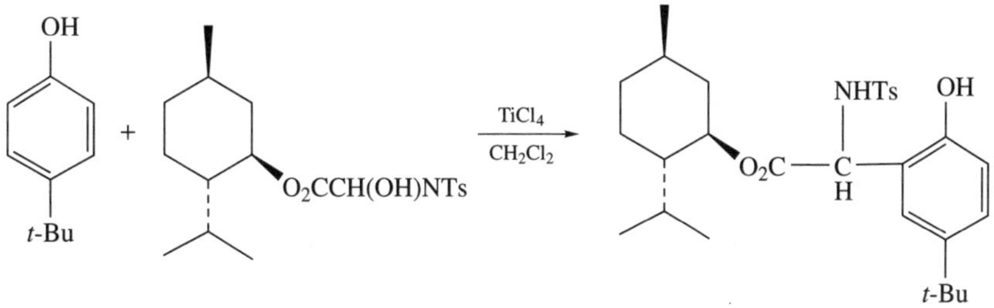

[Reaction scheme showing 4-t-butylphenol + menthyl O$_2$CCH(OH)NTs with TiCl$_4$/CH$_2$Cl$_2$ giving menthyl ester of 2-hydroxy-5-t-butylphenyl-CH(NHTs)-CO$_2$- product, 99% de] (50)

5. C-formylation, C-acylation and C-carboxylation

Attaching C=O groups to phenol nuclei has been conducted using classical methods such as the Reimer–Tieman reaction, the Duff reaction (formylation), the Friedel–Crafts acylation, the Fries rearrangement (acylation) and the Kolbe–Schmidt reaction (carboxylation)[194–196]. New methods employing various metal derivatives were developed to improve the efficiency of the processes. The Reimer–Tieman reaction conducted with chloroform under basic conditions can be accelerated by ultrasound irradiation[197,198]. Jacobsen and coworkers employed the modified Duff reaction, treating hexamethylenetetramine in trifluoroacetic acid for large-scale preparation of substituted salicylaldehydes[199]. Phenols are conveniently formylated at the *o*-position by treating paraformaldehyde with tin or magnesium phenoxides (generated from the phenols with either SnCl$_4$–tributylamine or a Grignard reagent) which involves the Canizzaro oxidation of initially formed salicyl alcohols (equation 51)[200,201].

Phenols are C-acylated either by electrophilic substitution under acidic conditions or by nucleophilic acylation under basic conditions. Advances in the chemistry of strong acids and Lewis acids provided novel aspects to catalytic Fries rearrangement and Friedel–Crafts acylation. Effenberger and Gutermann used a catalytic amount of trifluoromethanesulfonic acid for the Fries rearrangement and obtained the *o*-isomer as the thermodynamic product[202]. Kobayashi and coworkers reported the catalytic Fries rearrangement using $Hf(OTf)_4$ or $Sc(OTf)_3$, which are Lewis acids relatively insensitive to oxygen functionalities, including water[203]. While phenol is 4-acylated by this method, *m*-substituted phenols and 1-naphthol are 2-acylated. The Friedel–Crafts acylation is conducted using carboxylic acid in the presence of $Hf(OTf)_4$ (equation 52)[204] or zeolite HZSM[205]. The latter exhibits very high *o*-selectivity.

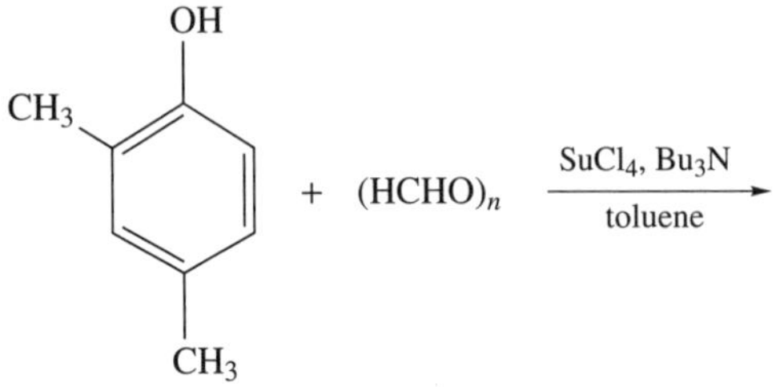

$$\text{2,4-dimethylphenol} + (HCHO)_n \xrightarrow[\text{toluene}]{SuCl_4, Bu_3N} \text{3-CHO-2,4-dimethylphenol} \quad (51)$$

$$\text{1-naphthol} + CH_3CO_2H \xrightarrow[\text{toluene-methanol}]{Hf(OTf)_4 \text{ cat}} \text{2-acetyl-1-naphthol} \quad (52)$$

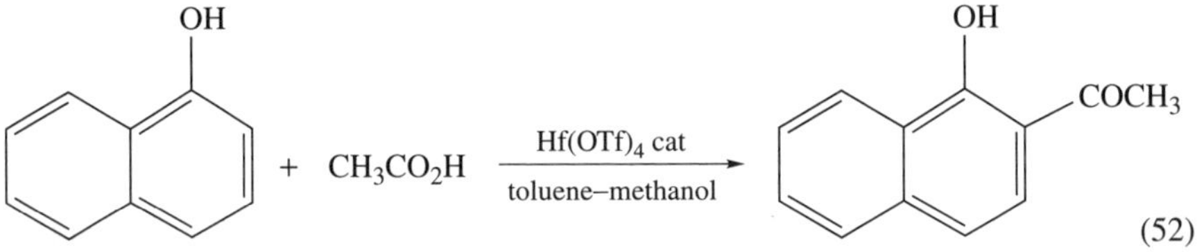

Sartori and coworkers indicated that magnesium phenoxides can be C-acylated with unsaturated acid chloride and oxalyl chloride[206,207]. The effect of the metal on the acylation of *o*-(*t*-butyl)phenoxide with chloroacetyl chloride was also examined in regard to the O/C-selectivity and *o/p*-selectivity. Alkali metal phenoxides give O-acylated product exclusively; aluminum and titanium phenoxides, and to some extent magnesium phenoxide, exhibit a tendency to C-acylation[208]. As for the reaction site, the exclusive *o*-acylation was observed for $(ArO)_3Al$, $(ArO)_4Ti$ and ArOMgBr, while $ArOAlCl_2$ and $ArOTiCl_3$ were relatively *p*-selective. The results were ascribed to the higher coordinating ability of magnesium metal. Sugasawa and Piccolo and their coworkers showed that BCl_3 is effective for the *o*-acylation of phenols with acid chlorides (equation 53)[209,210].

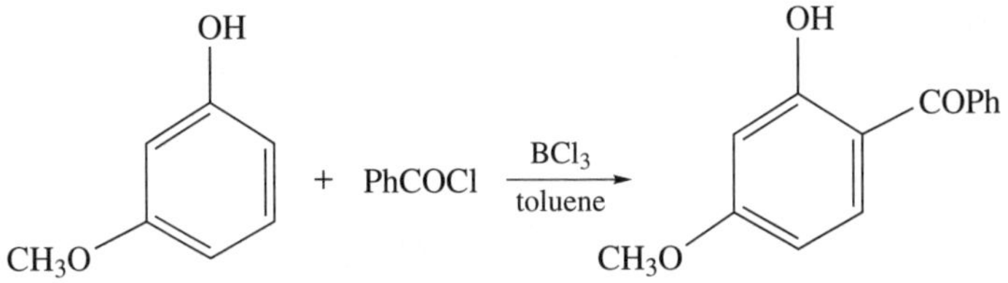

$$\text{3-methoxyphenol} + PhCOCl \xrightarrow[\text{toluene}]{BCl_3} \text{2-benzoyl-5-methoxyphenol} \quad (53)$$

The classical Kolbe–Schmidt reaction treats alkali metal phenoxides and carbon dioxide at higher than atmospheric pressure, giving salicylic acid. Hirao and Kato developed several modifications for industrial production[211]. Recently, phenol phosphate was enzymatically carboxylated, giving *p*-hydroxybenzoic acid[212]. As for related reactions, Sartori and coworkers conducted *o*-carbamoylation of aluminum or boron phenoxides with alkyl isocyanate[213], and Adachi and Sugasawa *o*-cyanated phenols using methyl thioisocyanate in the presence of BCl_3 (equation 54)[214].

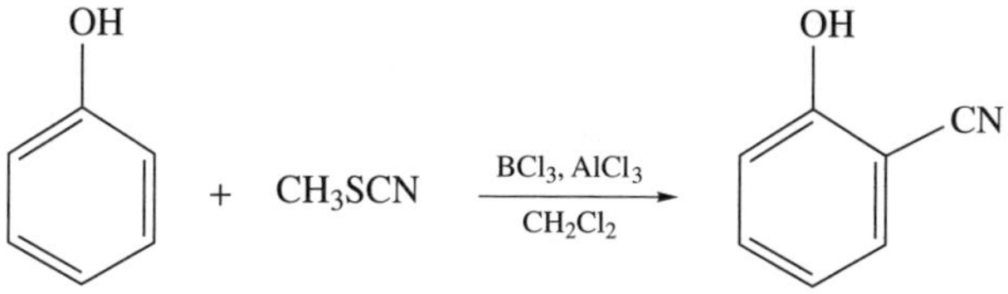

[Reaction scheme: phenol + CH₃SCN, BCl₃, AlCl₃, CH₂Cl₂ → 2-cyanophenol]   (54)

C. C-fluorination

Organofluorine compounds have become very important in relation to the development of novel biologically active substances[215,216]. Since the direct treatment of fluorine and organic molecules results in an explosive reaction, modified methods has been developed for effective aromatic fluorination. Use of 11% molecular fluorine diluted with nitrogen was examined by Misaki for fluorination of phenols[217]; phenol gave predominantly *o*-fluorophenol (equation 55), *p*-cresol gave a considerable amount of 4-fluoro-2,5-cyclohexadienone and salicylic acid was fluorinated at the 4-position. The presence of a Lewis acid such as BCl₃ or AlCl₃ increases the yield and the percentage of the *p*-isomer[218]. In order to control the reactivity and attain selectivity of fluorination, reagents containing O−F bonds such as CsSO₄F were developed[219]. Later, N−F compounds were also studied and were shown to have the advantage of controlling the reactivity by changing the nitrogen substituents. Barnette used *N*-fluorosulfoneamide CF₃SO₂N(*t*-Bu)F, which reacted with potassium salt of 1-naphthol to give the 2-fluoro derivative[220]. Des-Marteau and coworkers developed more reactive *N*-fluorosulfoneimides (CF₃SO₂)₂NF, which directly fluorinated phenol[221]. *N*-Fluoropyridinium salts were studied extensively by Umemoto and coworkers and, for example, the reaction can be promoted by introducing electron-withdrawing groups on the pyridine[222–225]. Very high *o*-selectivity was attained when a betaine was employed. Treatment of 4-hydroxyphenyl acetate with 2,6-di(methoxycarbonyl)pyridinium salt gave a considerable amount of the 4-fluoro derivative along with the 2-derivative (equation 56).

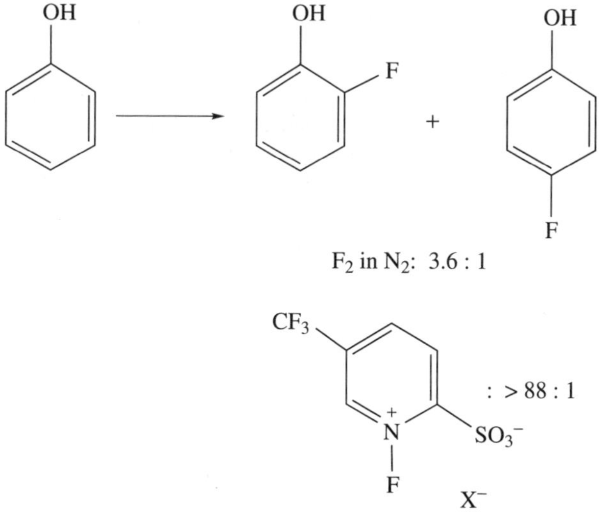

[Reaction scheme: phenol → 2-fluorophenol + 4-fluorophenol]

F₂ in N₂: 3.6 : 1   (55)

[Structure: 3-CF₃-pyridinium-N-F with SO₃⁻ substituent, X⁻]

: > 88 : 1

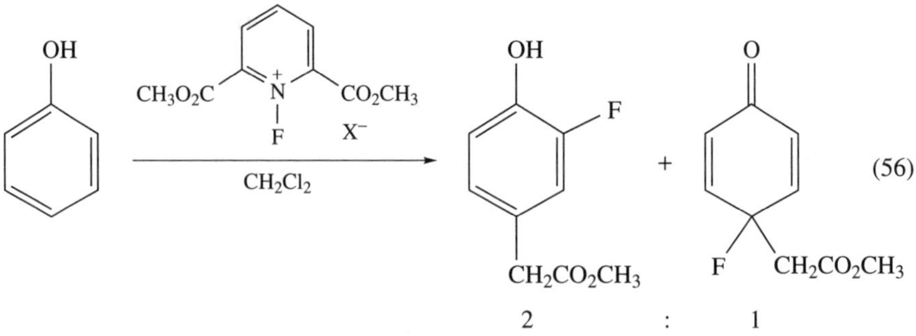

[Equation 56 scheme]

(56)

2 : 1

II. METAL PHENOXIDE AS REAGENT IN ORGANIC SYNTHESIS

Metal phenoxides are utilized extensively in organic synthesis as reagents, since they can readily be prepared from phenols and appropriate metal reagents, and the phenol moiety can easily be modified either sterically or electronically. Particularly, 2,2'-dihydroxy-1,1'-binaphthyl (BINOL), salicylideneamine and N,N'-ethylenebis(salicylideneamine) (salen) proved to be excellent phenol ligands for asymmetric synthesis. Since some of their reactions have recently been reviewed[226], it may not be appropriate to reproduce all of them. Instead, this section concentrates on the effect of the phenol moiety on the chemical reactivity and selectivity, and tries to provide structure–activity relationships for the metal phenoxide reagents. Metalated derivatives of monophenols, biphenols and salicylaldehyde imines are discussed separately.

A. Organic Synthesis Using Metal Complexes of Monophenol

Maruoka and Yamamoto introduced aluminum phenoxide reagents in organic synthesis[227]. Aluminum phenoxides are sufficiently Lewis acidic to interact with oxygen functionalities such as carbonyl or ether, and to change the reaction site or the stereochemistry. High selectivity in the axial attack of methyllithium addition to 4-(t-butyl)cyclohexanone was attained using bulky aluminum reagents such as methylaluminum bis(2,4,6-tri(t-butyl)phenoxide) (MAT) (equation 57)[228]. In the absence of the reagent, modest selectivity for the axial attack was observed. Analogously, the presence of MAT directs the addition to α-methyl substituted aldehydes in a high anti-Cram manner (equation 58), and the addition to conjugate enones at the γ-position. Aluminum tris(2,6-diphenylphenoxide) (ATPH) gives mostly 1,4-adducts even from unsaturated aldehydes (equation 59)[229]. These aluminum reagents were used in several selective syntheses which otherwise could not be conducted, such as 1,6-addition to

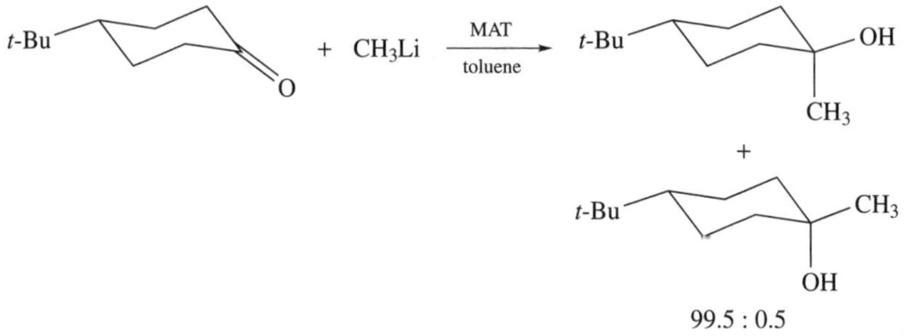

99.5 : 0.5

(57)

10. Synthetic uses of phenols

acetophenone[230,231], enolate formation from unsaturated aldehydes and aldol reaction at the remotest nucleophilic center[232] and selective alkylation of hindered aldehydes in the presence of less hindered aldehydes and ketones[233]. Use of appropriate phenoxides controls the double bond stereochemistry in the Claisen rearrangement of allyl vinyl ethers (equation 60)[234], which was extended to asymmetric synthesis using a binaphthyl derivative ATBN-F (equation 61)[235].

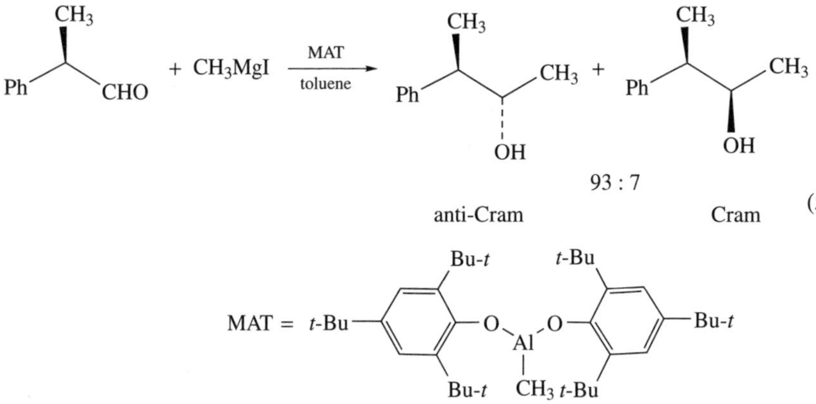

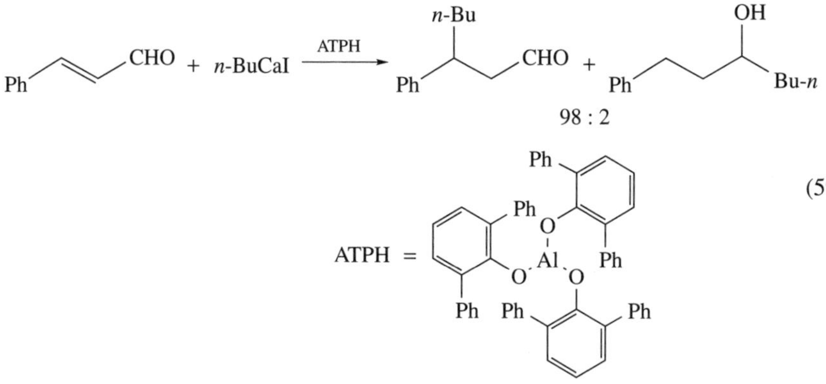

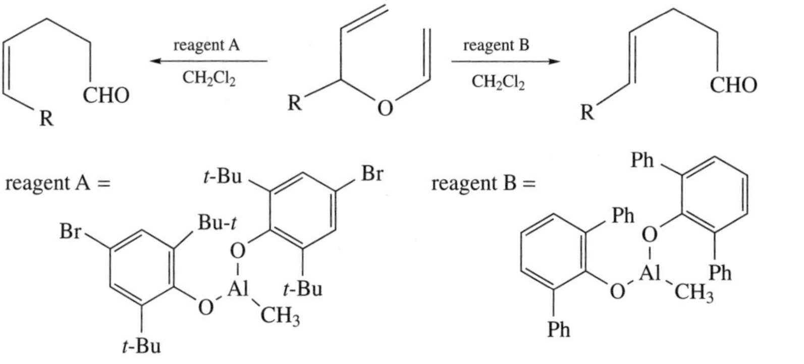

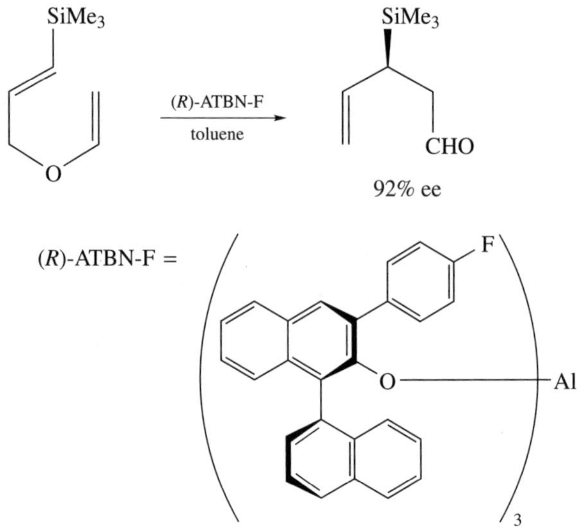

$$\text{allyl-O-CH}_2\text{-CH=CH-SiMe}_3 \xrightarrow[\text{toluene}]{(R)\text{-ATBN-F}} \text{CH}_2\text{=CH-CH(SiMe}_3\text{)-CH}_2\text{-CHO}$$

92% ee

(R)-ATBN-F = (binaphthyl-O)$_3$Al with 4-F-phenyl substituent

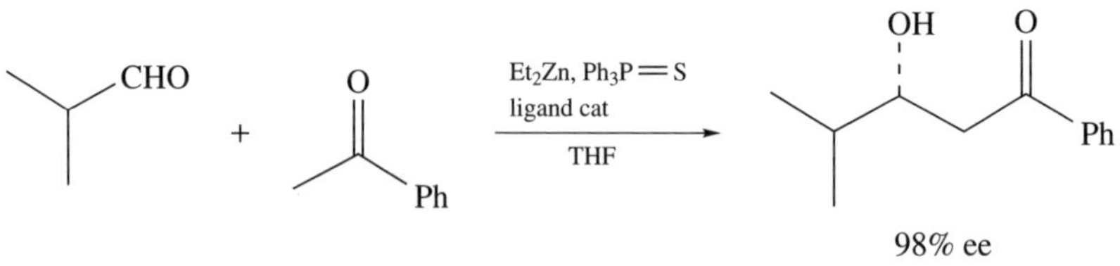

(61)

$$\text{iPr-CHO} + \text{CH}_3\text{C(O)Ph} \xrightarrow[\text{THF}]{\text{Et}_2\text{Zn, Ph}_3\text{P=S, ligand cat}} \text{iPr-CH(OH)-CH}_2\text{-C(O)Ph}$$

98% ee

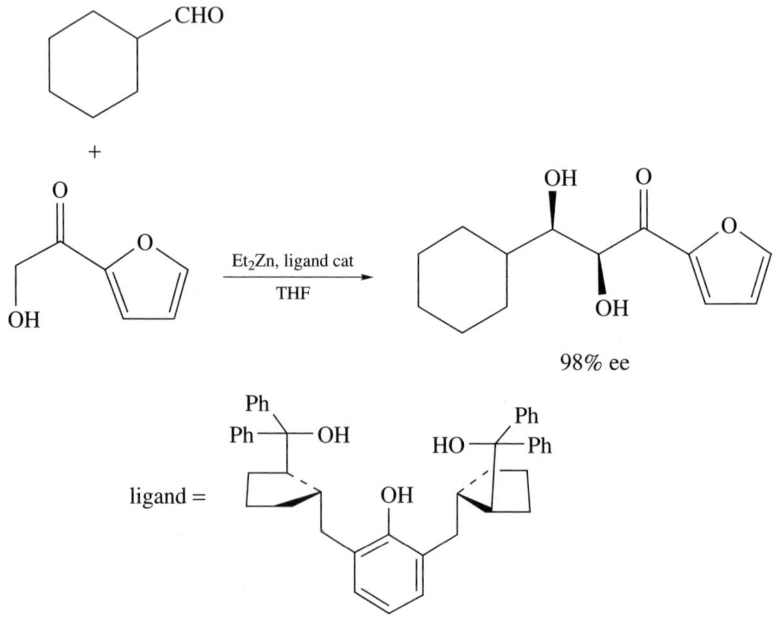

(62)

Cy-CHO + HOCH$_2$-C(O)-(2-furyl) $\xrightarrow[\text{THF}]{\text{Et}_2\text{Zn, ligand cat}}$ Cy-CH(OH)-CH(OH)-C(O)-(2-furyl)

98% ee

(63)

ligand = bis(α,α-diphenyl-cyclohexylmethanol) phenol

Trost and coworkers developed a chiral zinc phenoxide for the asymmetric aldol reaction of acetophenone or hydroxyacetophenone with aldehydes (equations 62 and 63)[236,237]. This method does not involve the prior activation of the carbonyls to silyl enol ethers as in the Mukaiyama aldol reactions. Shibasaki and coworkers employed titanium phenoxide derived from a phenoxy sugar for the asymmetric cyanosilylation of ketones (equation 64)[238]. 2-Hydroxy-2′-amino-1,1′-binaphthyl was employed in the asymmetric carbonyl addition of diethylzinc[239], and a 2′-mercapto derivative in the asymmetric reduction of ketones and carbonyl allylation using allyltin[240–242].

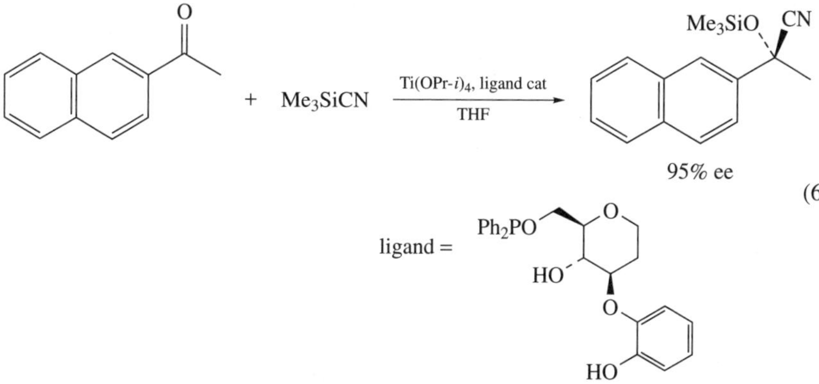

(64)

Yamamoto and coworkers protonated silyl enol ethers with a stoichiometric amount of a complex derived from BINOL and $SnCl_4$ giving optically active α-alkyl ketones[243]. A catalytic reaction was developed employing another tin complex derived from BINOL monomethyl ether (LBA), in which 2,6-dimethylphenol was used as the proton source (equation 65)[244].

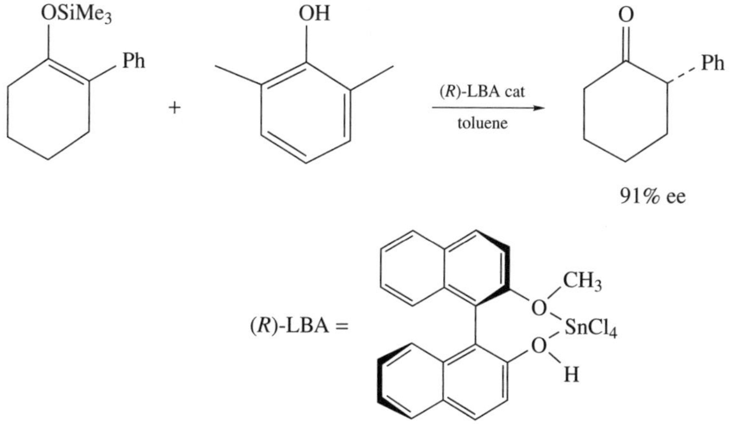

(65)

B. Organic Synthesis Using Metal Complexes of Biphenol: BINOL and Derivatives

The most common metal biphenoxide used in organic synthesis is that derived from chiral BINOL[245,246], the aluminum hydride complex of which was employed in asymmetric carbonyl reduction by Noyori and coworkers[247]. Since then, its potential has been demonstrated in a variety of stoichiometric and catalytic asymmetric reactions: the Diels–Alder reaction, ene-reaction, carbonyl addition reaction, conjugate addition reaction, epoxide cleavage reaction or enolate protonation. The effect of the substituents into the BINOL moiety is discussed here.

The earliest work of a modified BINOL in asymmetric synthesis was conducted by Yamamoto and coworkers, who employed a stoichiometric amount of 10,10'-dihydroxy-9,9'-biphenanthrene aluminum hydride complex in the reduction of phenyl ketones (equation 66)[248]. Higher enantiomeric excess (ee) was attained compared with the original BINOL. Introduction of the 3,3'-substituents into the BINOL generally results in higher ee in the Diels–Alder reaction, provided that the group does not interfere with the reaction. Kelly and coworkers reported the reaction of juglone and 1-methoxy-1,3-cyclohexadiene in the presence of a stoichiometric amount of 3,3'-diphenyl-1,1'-binaphthylborane derivative in >98% ee (equation 67)[249]. The higher selectivity compared

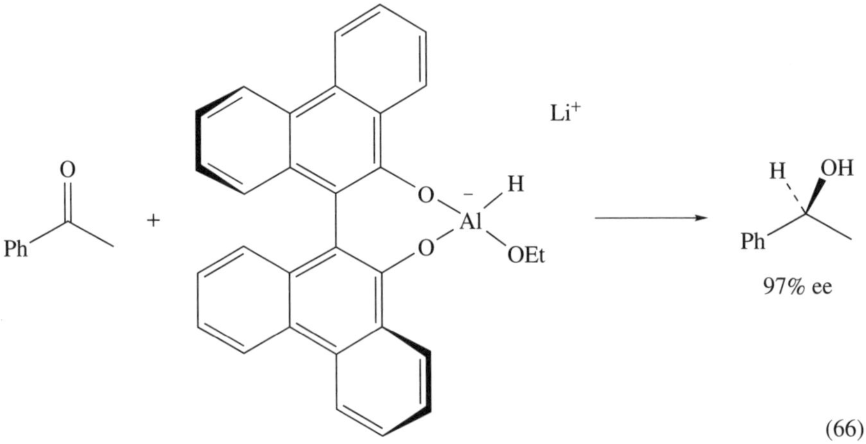

97% ee

(66)

R = Ph: >98% ee
R = CH$_3$: 70% ee

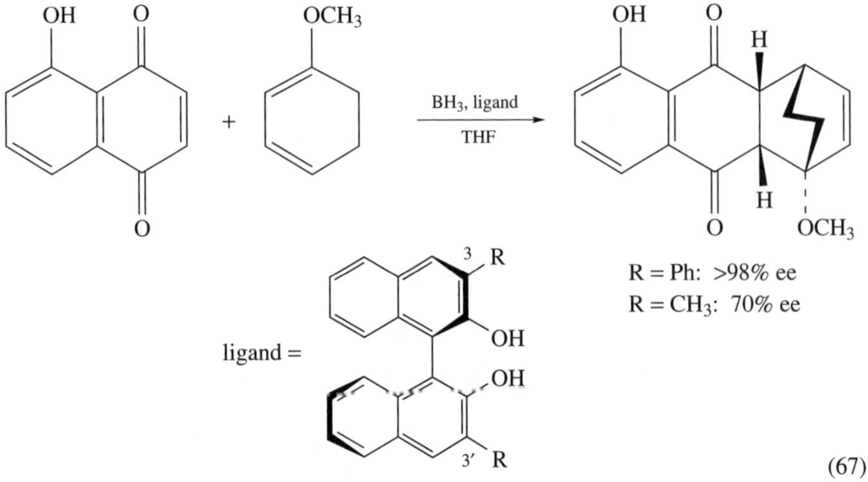

(67)

with the 3,3′-dimethyl derivative (70% ee) was attributed to the effective shielding of an enantioface by the phenyl group. Yamamoto and coworkers employed 2,2′-dihydroxy-3,3′-bis(triarylsilyl)-1,1′-binaphthyl aluminum complex in the asymmetric hetero-Diels–Alder reaction and found the tris(3,5-xylyl)silyl derivative to exhibit higher ee than triphenylsilyl (equation 68)[250–252]. Wulff and coworkers employed 2,2-diphenyl-4,4′-dihydroxy-3,3′-diphenanthryl (VAPOL) aluminum complex possessing a deeper pocket, and attained 97.8% ee with a turnover number of 200[253,254]. The asymmetric Claisen rearrangement of 1-trimethylsilylvinyl cinnamyl ether was promoted by 3,3′-bis(t-butyldiphenylsilylated) BINOL aluminum complex (equation 69)[255], and the asymmetric ene-reaction of 2-phenylthiopropene and pentafluorobenzaldehyde by the triphenylsilyl derivative[256].

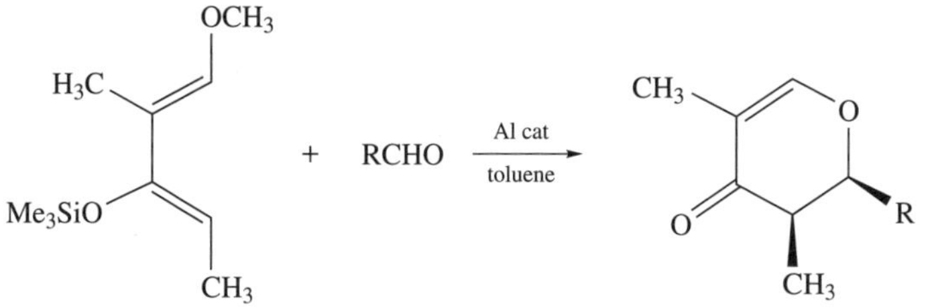

$Ar_3 = Ph_3$: 92% ee
$Ar_3 = (3,5\text{-xylyl})_3$: 97% ee

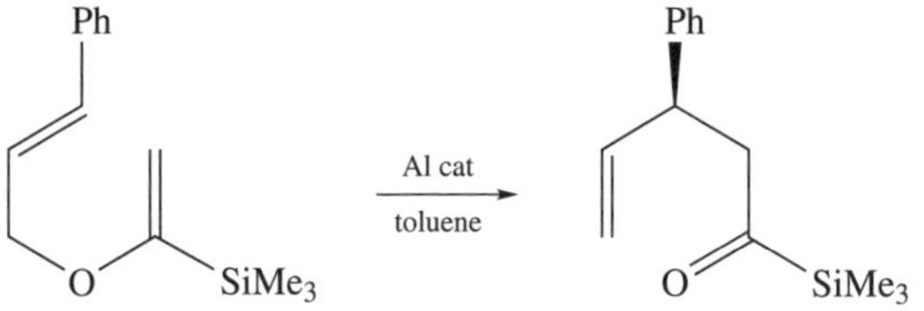

(68)

$Ar_3 = Ph_3$: 80% ee
$Ar_3 = t\text{-BuPh}_2$: 88% ee

(69)

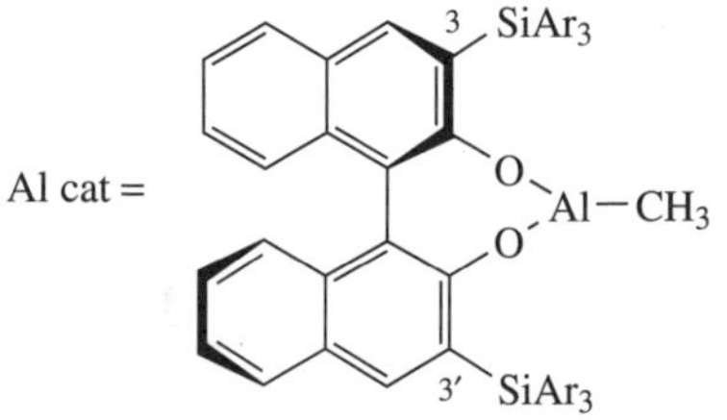

Their 3,3′-substituents are utilized not only for their steric bulk, but also for the coordination to metals. Yamamoto and coworkers employed a boron complex of 3,3′-bis(2-hydroxyphenyl) BINOL in the asymmetric Diels–Alder reaction of cyclopentadiene and acrylaldehyde (equation 70)[257–261]. The ligand possesses two additional hydroxy groups and forms a helical structure on coordination. The catalyst is considered to function as a chiral Brønsted acid and a Lewis acid. The complex was also used in the Diels–Alder reactions and aldol reactions of imines. Although addition of diethylzinc to aldehydes gives low ee using BINOL itself or its 3,3′-diphenyl derivative, the selectivity can be increased when coordinating groups are introduced at the 3,3′-positions. Katsuki and coworkers developed 3,3'-bis(dialkylcarbamoyl) BINOL for highly selective addition to aromatic and unsaturated aldehydes, in which the amide group is considered to form a rigid chelated structure to the zinc metal (equation 71)[262]. The same catalyst is effective for the asymmetric Simmons–Smith cyclopropanation of allylic alcohols[263]. Pu and coworkers introduced 2,5-dialkoxyphenyl group at the 3,3'-positions, and attained very high ee even for aliphatic acyclic aldehydes, in which the oxygen functionality is likely to play an important role[264,265]. A polymeric catalysts containing the functionalized BINOL were also developed[266].

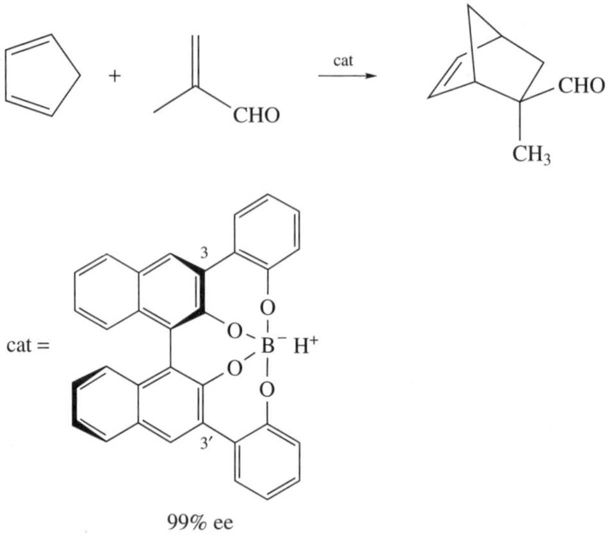

(70)

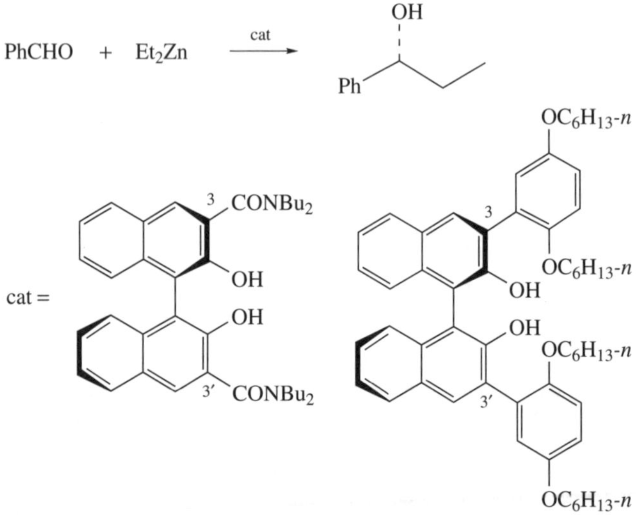

(71)

Shibasaki and coworkers employed 3,3'-bis(diarylphosphonoylmethyl) BINOL aluminum complex for the asymmetric silylcyanation of aldehydes (equation 72)[267]. The

10. Synthetic uses of phenols

phosphonate group is designed for the nucleophilic activation of the silyl cyanide without affecting the Lewis acidic aluminum center. Accordingly, the phosphinoylethyl derivative with the C_2-tether between BINOL and phosphinoyl moiety exhibits very low activity. Tuning of the substituents led to the development of a successful asymmetric Reisert reaction, in which a 2-methylphenyl derivative exhibited higher reactivity and stereoselectivity than a phenyl derivative (equation 73)[268,269]. A 3-hydroxymethyl BINOL lanthanum complex catalyzes the asymmetric epoxidation of conjugated ketones with cumene hydroperoxide[270,271]. Reetz and coworkers showed the reversal of the absolute configuration in the asymmetric oxidation of tolyl methyl sulfide with t−butyl hydroperoxide, when BINOL titanium complex and 6,6′-dinitro-1,1′,2,2′,3,3′,4,4′-octahydro BINOL complex were used. A very low asymmetric induction was observed in the absence of the nitro group[272]. Shibasaki and coworkers linked two BINOL moieties at the 3-position and used zinc complex of the product in the direct aldol reaction of hydroxyacetophenone and aldehydes[273,274].

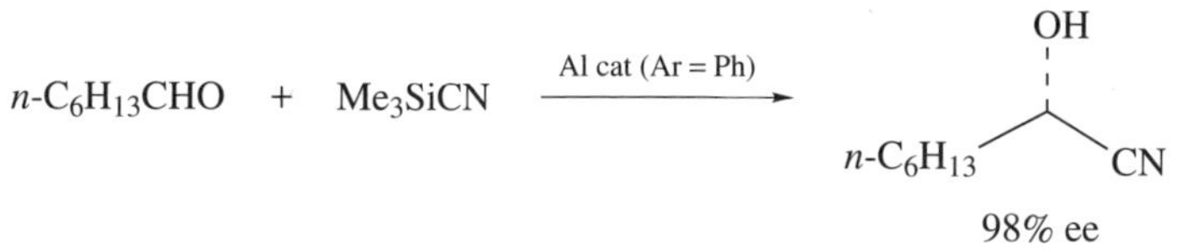

$$n\text{-}C_6H_{13}CHO + Me_3SiCN \xrightarrow{\text{Al cat (Ar = Ph)}} \underset{98\% \text{ ee}}{n\text{-}C_6H_{13}\overset{OH}{\underset{}{\text{–}}}CN} \qquad (72)$$

[structure of 4-chloroquinoline with N(allyl)₂] + Me₃SiCN $\xrightarrow[\text{CH}_2\text{Cl}_2]{\text{RCOCl} \\ \text{Al cat (Ar} = o\text{-CH}_3\text{C}_6\text{H}_4)}$ [product with CN, COR, 96% ee]

R = 2-furyl

Al cat = [BINOL structure with POAr₂ at 3,3′ positions and O–AlCl bridge]

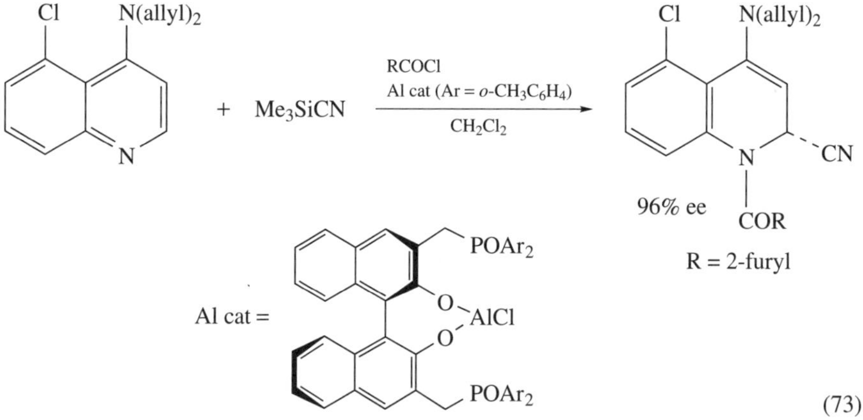

(73)

6,6′-Substituents on BINOL also affect the reaction course. Mikami and coworkers, employing 6,6′-dibromo BINOL titanium complex, enhanced the stereoselectivity in the ene-reactions of trisubstituted olefins (equation 74), which was attributed to the compression of the internal bond angle Cl-Ti-Cl[275]. Kobayashi and coworkers conducted the asymmetric addition of silyl enol ethers to imines catalyzed by BINOL zirconium complex, in which the introduction of the 6,6′-dibromo group increased the ee from 70% to 90% (equation 75)[276–278]. Shibasaki and coworkers employed the same dibromo BINOL lanthanum complex in the Diels–Alder reaction of cyclopentadiene and acryloyloxazolidone[279], and the higher ee in the reaction compared to the original BINOL was ascribed to the increased Lewis acidity. In the nitroaldol reaction, the use of 6,6′-diethynyl BINOL lanthanum complex attained higher diastereoselectivity and enantioselectivity (equation 76, also see equation 14)[280].

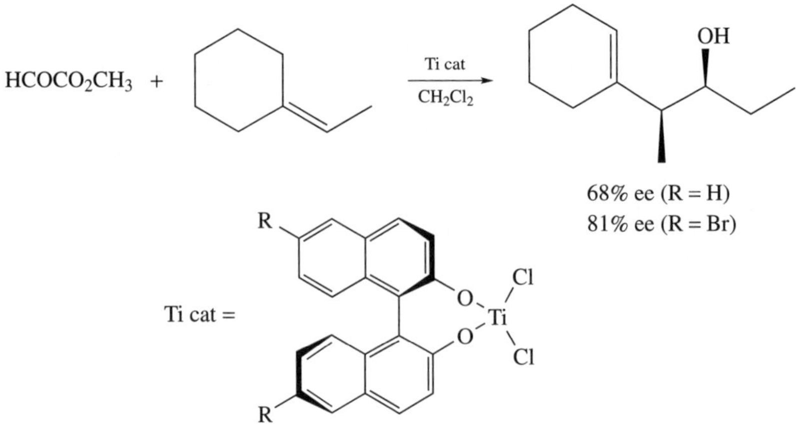

HCOCO$_2$CH$_3$ + [cyclohexylidene-ethane] $\xrightarrow[\text{CH}_2\text{Cl}_2]{\text{Ti cat}}$ [product]

68% ee (R = H)
81% ee (R = Br)   (74)

Ti cat = [(R-substituted BINOL)TiCl$_2$ complex]

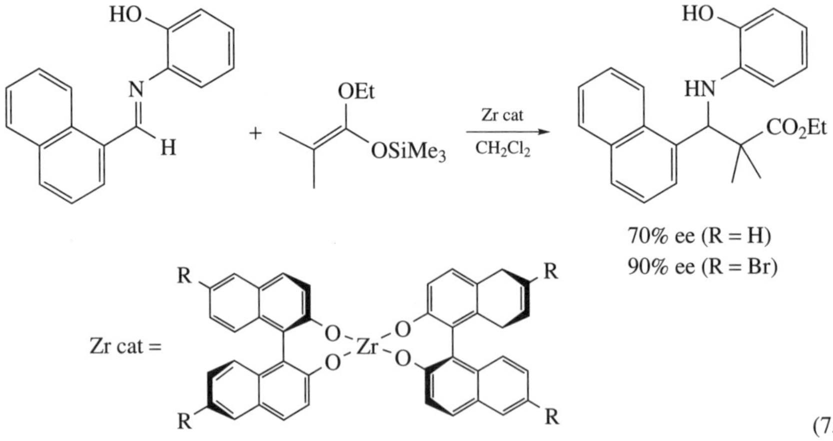

[1-naphthyl-CH=N-(2-hydroxyphenyl)] + [CH$_2$=C(Me)(OEt)(OSiMe$_3$) — actually Me$_2$C=C(OEt)(OSiMe$_3$)] $\xrightarrow[\text{CH}_2\text{Cl}_2]{\text{Zr cat}}$ [naphthyl-CH(NH-o-C$_6$H$_4$OH)-C(Me)$_2$-CO$_2$Et]

70% ee (R = H)
90% ee (R = Br)

Zr cat = [bis(BINOLate)Zr complex]

(75)

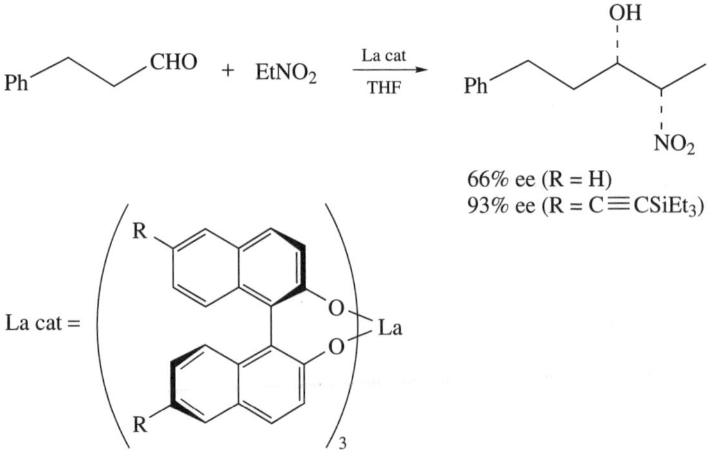

Ph$\sim$CHO + EtNO$_2$ $\xrightarrow[\text{THF}]{\text{La cat}}$ Ph$\sim$CH(OH)-CH(NO$_2$)-CH$_3$

66% ee (R = H)
93% ee (R = C≡CSiEt$_3$)   (76)

La cat = ([BINOLate-La])$_3$

10. Synthetic uses of phenols 697

Other use of the functionalized chiral BINOL includes the 5,5',6,6',7,7',8,8'-octahydro derivative developed by Chan and coworkers, the titanium complex of which is more effective than BINOL in the enantioselective addition of triethylaluminum and diethylzinc[281,282]; a 4,4',6,6'-tetrakis(perfluorooctyl) BINOL ligand developed for easy separation of the product and catalyst using fluorous solvents for the same zinc reaction[283]; an aluminum complex of 6,6'-disubstituted-2,2'-biphenyldiols used by Harada and coworkers in the asymmetric Diels–Alder reaction[284]; a titanium complex of (S)-5,5',6,6',7,7',8,8'-octafluoro BINOL employed by Yudin and coworkers in the diethylzinc addition, in the presence of which the reaction of the enantiomeric (R)-BINOL is promoted[285].

C. Organic Synthesis Using Metal Complexes of Salicylaldehyde Imines

Like BINOL, salicylaldehyde imines have become very important in asymmetric catalysis and a variety of polydentate ligands prepared from chiral monoamines and diamines are employed in oxidation reactions, carbenoid reactions and Lewis acid catalyzed reactions. As in the previous section, this section emphasizes the effect of the phenol moiety on the asymmetric catalysis. An imine derived from a chiral 1-phenethylamine and salicylaldehyde was employed in the copper catalyzed asymmetric cyclopropanation by Nozaki, Noyori and coworkers in 1966, which is the first example of the asymmetric catalysis in a homogeneous system[286]. Salicylaldehyde imines with ethylenediamine (salen) have been studied extensively by Jacobsen and Katsuki and their coworkers since 1990 in asymmetric catalysis. Jacobsen and coworkers employed the ligands prepared from chiral 1,2-diamines and Katsuki and coworkers sophisticated ligands possess chirality not only at the diamine moiety but also at the 3,3'-positions.

Asymmetric cyclopropanation of styrene developed by Noyori was extended by Aratani to the industrial production of chrysanthemic acid (equation 77)[287,288]. Fukuda and Katsuki using salen cobalt complex prepared from chiral 1,2-diphenylamine attained high ee in the cyclopropanation of styrene with diazoacetate esters (equation 78)[289]. Unlike epoxidation (*vide infra*), introduction of *t*-butyl groups to 3,3'-positions results in low catalytic activity. However, a complex possessing 5,5'-dimethoxy groups exhibits high ee as well as high *trans*-selectivity. The same complex is used in the [2.3]sigmatropic rearrangement of S-ylide derived from allyl aryl sulfide and *t*-butyl diazoacetate[290].

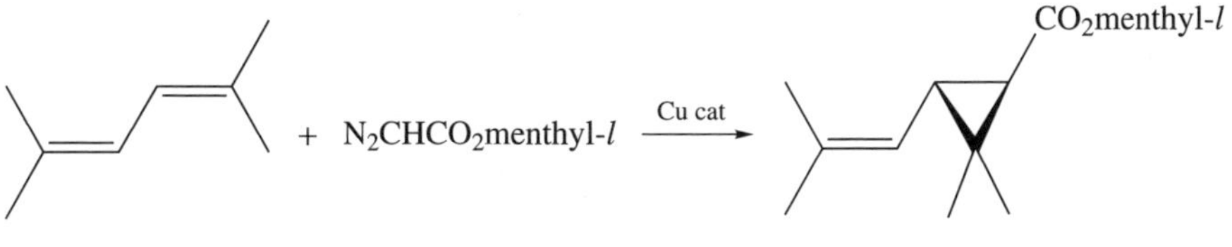

$$+ \text{N}_2\text{CHCO}_2\text{menthyl-}l \xrightarrow{\text{Cu cat}} \quad \text{CO}_2\text{menthyl-}l$$

94% ee (77)

Cu cat =

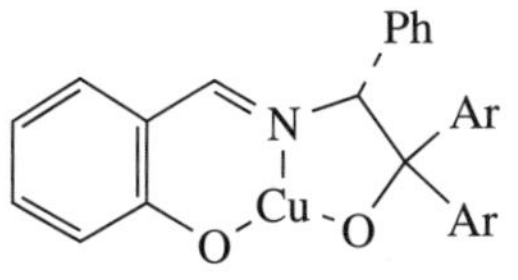

Ar = 2-octyloxy-4-(*t*-butyl)phenyl

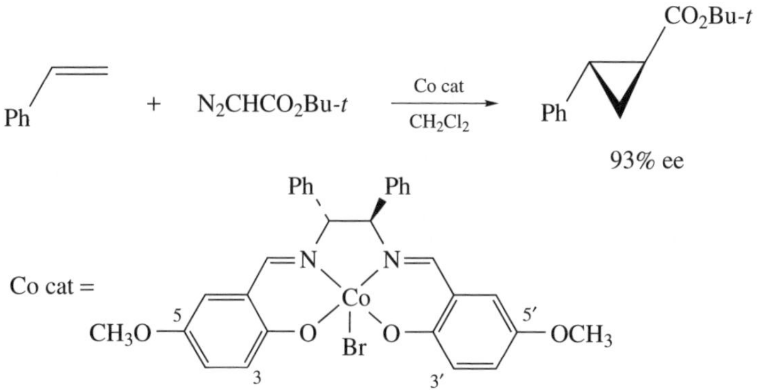

(78)

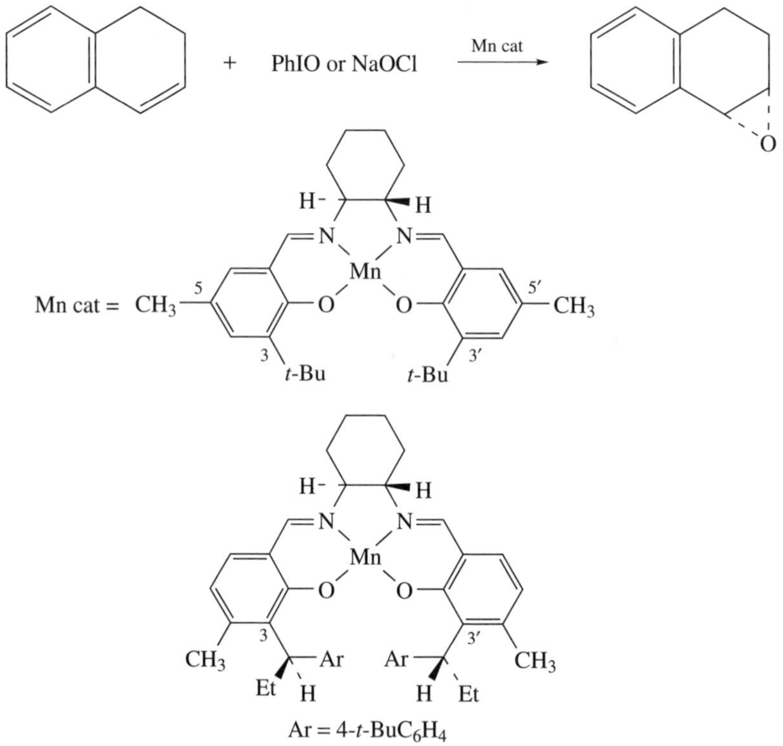

(79)

Ar = 4-t-BuC$_6$H$_4$

The stability of the salicylaldehyde imine ligand under oxidative conditions lead to the application in asymmetric oxidation reactions[291,292]. Jacobsen and coworkers employed a manganese salen complex with t-butyl groups at the 3,3'-positions and attained especially high ee for the epoxidation of disubstituted *cis*-alkenes (equation 79)[293–297]. The role of bulky groups was ascribed to blocking the side-on attack to the manganese oxo-intermediate. Electron-donating groups at the 5,5'-positions also enhance the ee. Katsuki and coworkers examined salen manganese complexes derived from chiral diamines and chiral aldehydes, which possess 1-phenylpropyl or 1-naphthyl group at the 3,3'-positions.

The diastereomeric complexes containing stereogenic centers at the diamine moiety and at the 3,3'-substituent exhibit different behaviors in asymmetric catalysis[298-301]. For example, asymmetric epoxidation of dihydronaphthalene using PhIO gave the product in high ee, when a ligand derived from (S,S)-2,3-diphenyl-2,3-butanediamine and (R)-aldehyde was employed (equation 80). It was also observed that the stereochemistry in the asymmetric epoxidation of *cis*-alkenes is mainly governed by the configuration at the diamine moiety rather than by the 1-phenylpropyl moiety, and that the stereochemistry of the *trans*-alkene epoxidation is governed by the configuration at the 1-phenylpropyl moiety.

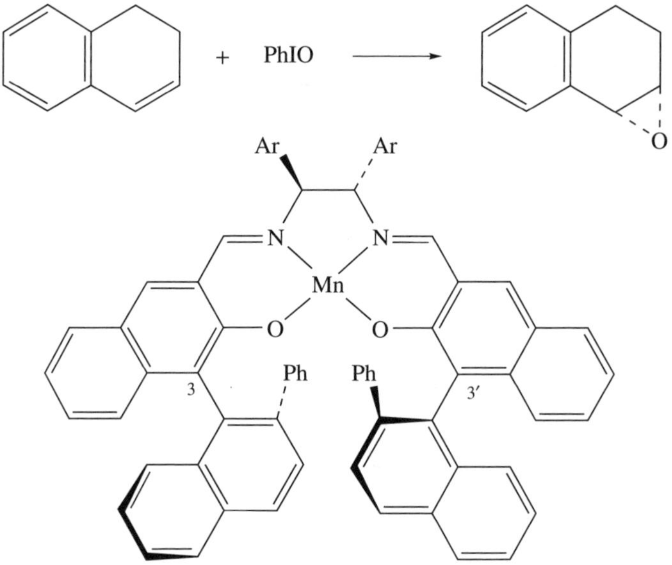

$$\text{(80)}$$

Ar = 3,5-xylyl

Asymmetric sulfide oxidation giving optically active sulfoxide has also been studied using metal complexes of salicylaldehyde imines (equation 81)[302]. Fujita and coworkers examined a vanadium salen complex I derived from (R,R)-1,2-diaminocyclohexane and obtained (S)-sulfoxide in 40% ee from phenyl methyl sulfide using *t*-butyl hydroperoxide as oxidant[303]. The selectivity is higher for a ligand equipped with 3,3'-dimethoxy groups than that without 3,3'-substituents or that with 3,3'-di(*t*-butyl) groups. Bolm and Bienwald improved the ee up to 85% by employing salicylaldehyde *t*-leucinol imine vanadium complex II with aqueous hydrogen peroxide, where the introduction of a 6-(*t*-butyl) group and *t*-butyl or nitro group at the 4-position enhances the enantioselectivity[304]. A disulfide or dithioketal can also be oxidized asymmetrically[305]. Although titanium salen complexes were not quite effective[306-308], Katsuki and coworkers improved the effectivity by using a complex III possessing (R,R)-diamine moiety and (S)-axis chiral moiety at the 3,3'-positions[309,310]. As for the manganese salen complex, Jacobsen and coworkers found that the selectivity of the sulfide oxidation can be markedly increased by employing complex IV with bulky substituents at the 3,3'-positions and electron-donating groups at the 5,5'-positions[311]. Katsuki and coworkers further improved the ee using manganese complex V with axis chiral groups, in which the matched pair was (R,R)-1,2-diaminocyclohexane and (S)-axis configuration[312,313]. Notably, the opposite combination is the matched pair for the epoxidation (*vide supra*).

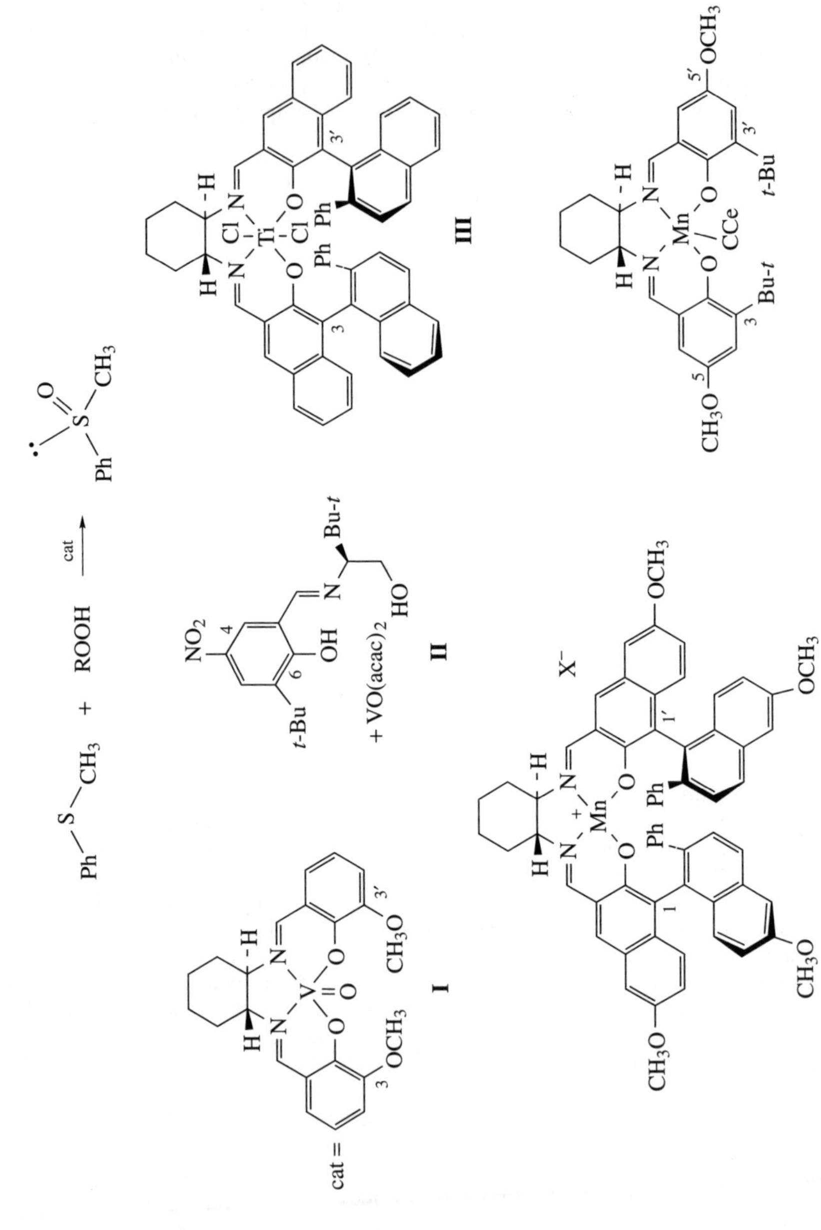

$$\text{Ph}\!-\!\text{S}\!-\!\text{CH}_3 + \text{ROOH} \xrightarrow{\text{cat}} \underset{\text{Ph}}{\overset{\overset{\displaystyle O}{\|}}{S}}\!-\!\text{CH}_3$$

(81)

cat =

I: V(=O) complex with salen-type ligand from (1R,2R)-cyclohexanediamine and 3-OCH₃ salicylaldehyde (3′-OCH₃, 3-OCH₃).

II: 5-NO₂-3-t-Bu salicylaldehyde Schiff base with (S)-t-Bu-glycinol + VO(acac)₂.

III: Ti(Cl)₂ complex with binaphthyl-salen ligand (3,3′-positions bearing 2-Ph-naphthyl substituents).

IV: Mn⁺ salen complex with binaphthyl backbone bearing OCH₃ groups (1, 1′ positions; multiple CH₃O and Ph substituents), X⁻ counterion.

V: Mn complex with salen ligand from (1R,2R)-cyclohexanediamine and 3-t-Bu-5-OCH₃ salicylaldehyde, axial CCe ligand.

10. Synthetic uses of phenols

Bolm and coworkers developed a chiral copper complex from an oxazoline and salicylic acid for the Baeyer–Villiger oxidation employing oxygen and an aldehyde for the oxidant, and high ee was obtained with a 4-nitro-6-(t-butyl)salicylic acid derivative (equation 82)[314–317]. Salicylaldehyde itself can be used as a catalyst ligand for the Baeyer–Villiger oxidation as indicated by Strukul and coworkers. Reaction of K_2PtCl_4 and 6-methoxysalicylaldehyde in the presence of 2,2′-bis(diphenylphosphino)-1,1′-binaphthyl (BINAP) gives acylplatiumn complexes, which in the presence of perchloric acid catalyzes the asymmetric Baeyer–Villiger oxidation with hydrogen peroxide (equation 83)[318].

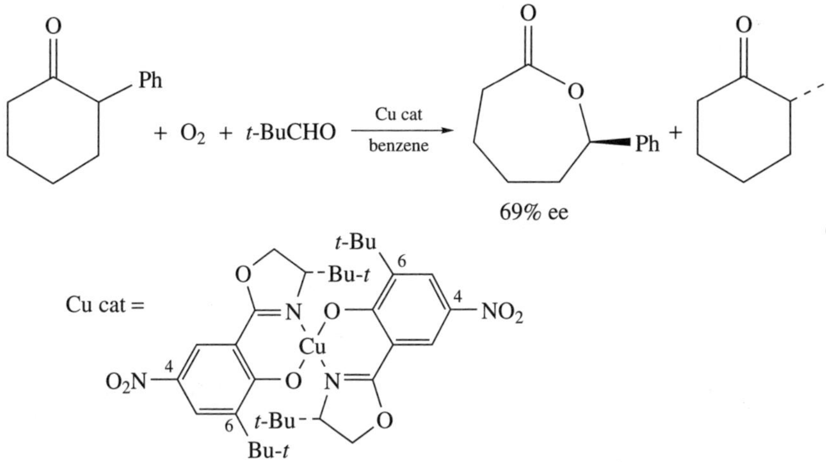

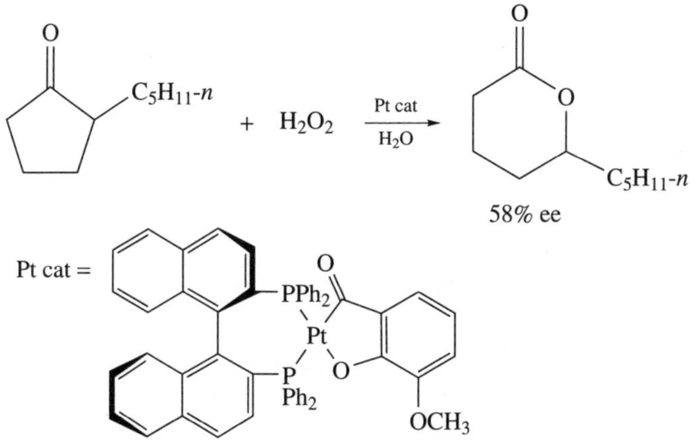

Manganese salen complex catalyzes C–H oxidation of organic molecules with NaOCl or PhIO, giving alcohols[319]. Larrow and Jacobsen observed kinetic resolution in the benzylic hydroxylation[320]. Katsuki and coworkers used the axis chiral salen manganese complexes for the benzyl hydroxylation and ether hydroxylation, and attained higher ee with the ligand possessing (R,R)-diamine and (R)-axis chirality (equation 84)[321–323].

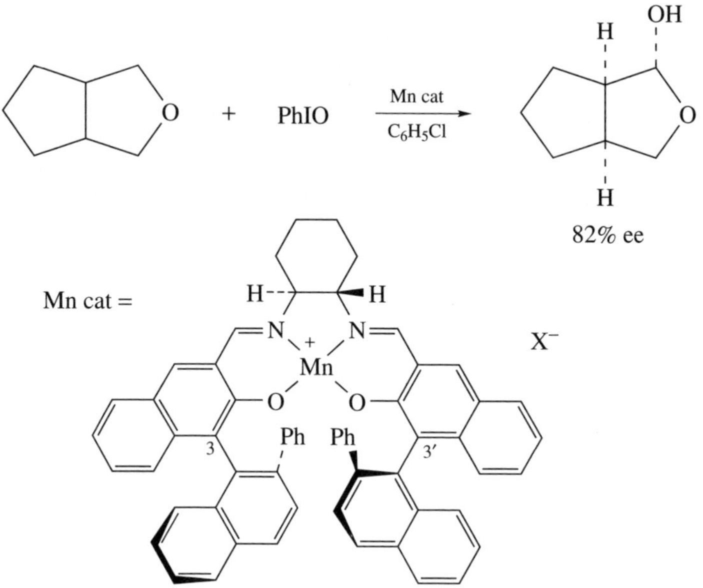

(84)

Although asymmetric aziridination of styrenes was attempted by Burrow and Katsuki and their coworkers using manganese salen complexes in the presence of PhI=NTs, low asymmetric induction was observed[324-326]. Nishikori and Katsuki later employed a salen complex synthesized from (R,R)-2,3-diaminobutane and (S)-biphenol, and found that the chirality at the 3,3′-positions is more important for the asymmetric induction (equation 85)[327]. Carreira conducted the stoichiometric amination of enol ethers and alkenes using a manganese nitride salen complex[333]. Komatsu extended the methodology to the catalytic process and attained 94% ee for aziridination of β-isopropylstyrene[332].

Titanium complexes of chiral imines derived from salicylaldehydes are employed not only for oxidation reactions, but also for carbonyl addition reactions. Asymmetric silylcyanation of aldehydes can be catalyzed by a titanium complex (equation 86)[333-336]. Introduction of the 6-(t-butyl) group at the salicylaldehyde moiety enhanced the selectivity and at the same time reverses the absolute configuration; the bulky group may be inhibiting the approach of cyanide from the re-face[333]. Bolm and Müller employed a sulfoximine in the presence of Ti(OPr-i)$_4$ for the stoichiometric cyanation of aldehydes[337]. Titanium imine complex was also used for the Mukaiyama asymmetric aldol reaction by Oguni and coworkers[338] and Carreira and coworkers[339-341]. Carreira employed salicylaldehyde imine derived from 2-amino-2′-hydroxy-1,1′-binaphthyl (equation 87). Asymmetric organometal alkylation of epoxide and aziridine was examined using the related titanium complex[342,343]. Inoue and Mori treated aldehydes with hydrogen cyanide in the presence of Ti(OPr-i)$_4$ and imines, which were derived from either (S)-valyl-(S)-tryptophan/2-hydroxy-1-naphthaldehyde imine or (S)-valine/3,5-dibromosalicylaldehyde imine (equation 88)[344]; the complexes provide enantiomeric cyanohydrins. Snapper and Hoveyda screened similar dipeptides by a combinatorial method for finding an effective ligand for enantioselective cleavage of $meso$-epoxides (equation 89)[345,346].

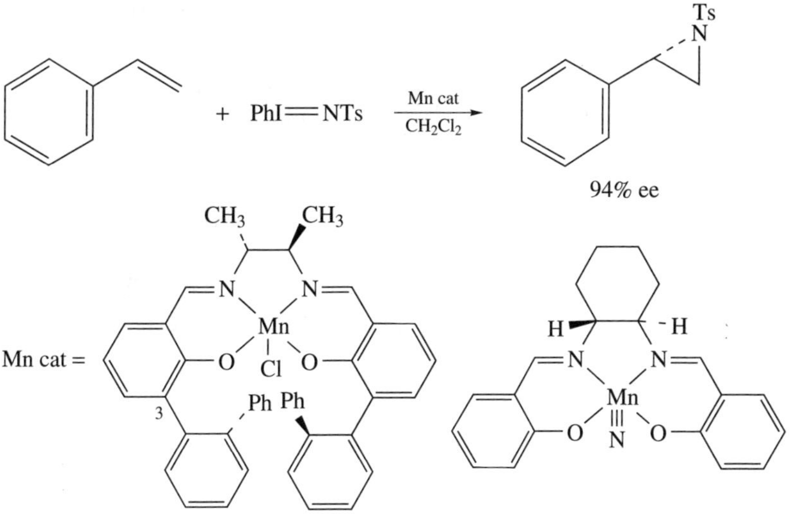

10. Synthetic uses of phenols

$$\text{PhCH=CH}_2 + \text{PhI=NTs} \xrightarrow[\text{CH}_2\text{Cl}_2]{\text{Mn cat}} \text{Ph-aziridine-NTs} \quad 94\% \text{ ee} \tag{85}$$

Mn cat = [chiral Mn salen complexes shown]

$$\text{(CH}_3\text{)C=CH-CHO (tiglaldehyde)} + \text{Me}_3\text{SiCN} \xrightarrow[\text{CH}_2\text{Cl}_2]{\text{Ti cat}} \text{allylic cyanohydrin} \quad 96\% \text{ ee} \tag{86}$$

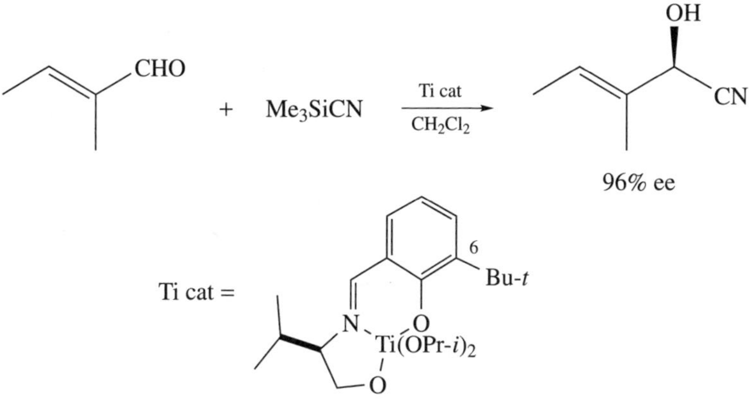

Ti cat = [chiral Schiff base Ti(OPr-i)$_2$ complex with t-Bu substituent]

$$\text{CH}_3\text{CH=CH-CHO} + \text{CH}_2\text{=C(OSiMe}_3\text{)(OCH}_3\text{)} \xrightarrow[\text{ether}]{\text{Ti cat}} \text{CH}_3\text{CH=CH-CH(OH)-CH}_2\text{-CO}_2\text{CH}_3 \quad 97\% \text{ ee} \tag{87}$$

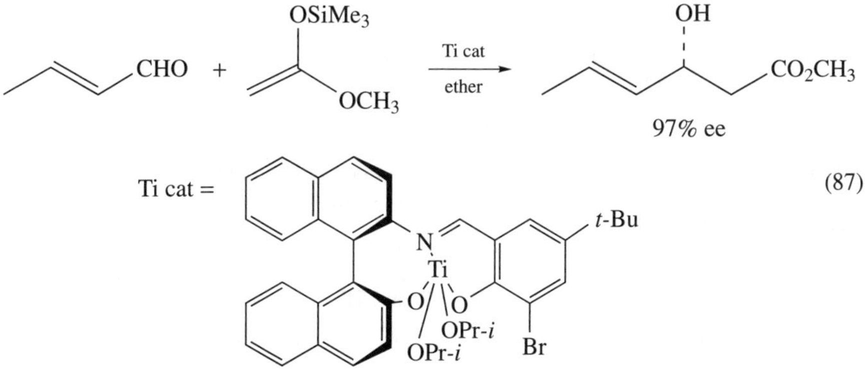

Ti cat = [chiral binaphthyl Schiff base Ti(OPr-i)$_2$ complex with t-Bu and Br substituents]

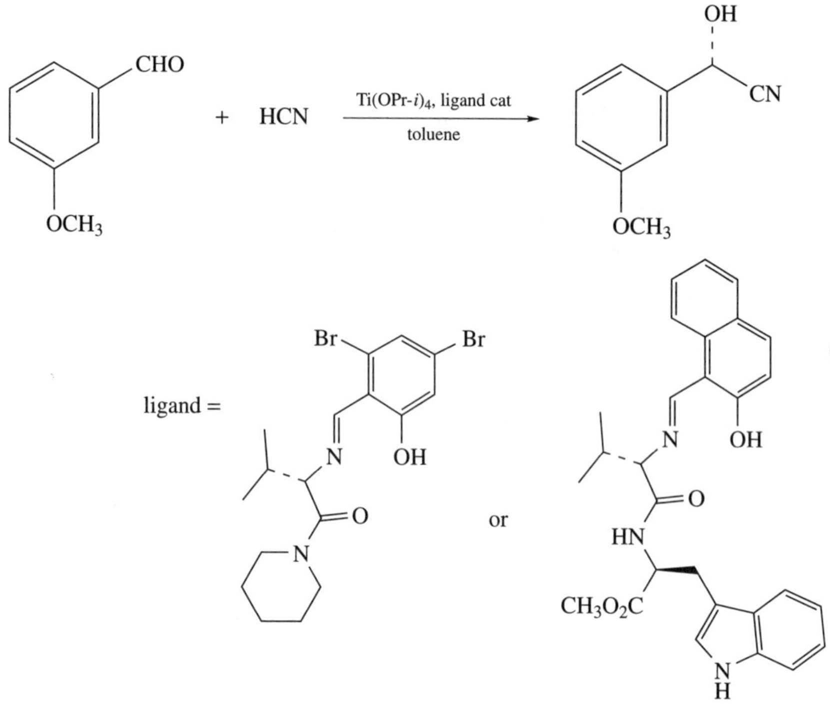

[Scheme (88)]

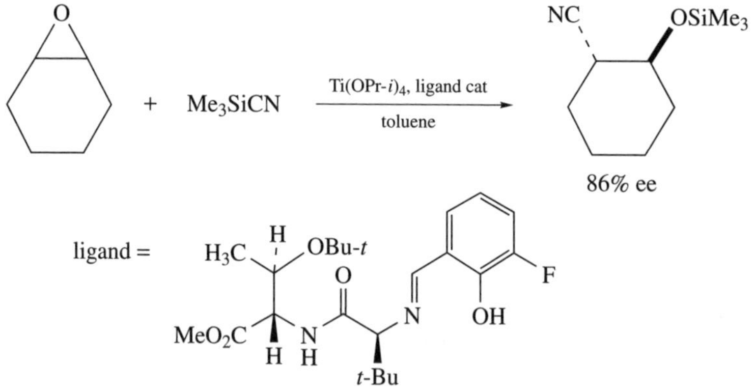

[Scheme (89), 86% ee]

Derivatives of phenols are becoming more important in industrial use containing drugs, materials, catalysts etc. Consequently, the development of more efficient methods is very necessary from a synthetic point of view.

IV. REFERENCES

1. T. W. Greene and P. G. M. Wuts, *Protective Groups in Organic Synthesis*, Wiley, New York, 1999.
2. W. E. Keller, *Phase-Transfer Reactions*, Georg Thieme Verlag, Stuttgart, 1986.

10. Synthetic uses of phenols

3. A. Loupy, J. Sansoulet and F. Vaziri-Zand, *Bull. Soc. Chim. Fr.*, 1027 (1987).
4. A. Loupy, J. Sansoulet, E. Dîez-Barra and J. R. Carrillo, *Synth. Commun.*, **21**, 1465 (1991).
5. E. Reinholz, A. Becker, B. Hagenbruch, S. Schäfer and A. Schmitt, *Synthesis*, 1069 (1990).
6. T. Ando, J. Yamawaki, T. Kawate, S. Sumi and T. Hanafusa, *Bull. Chem. Soc., Jpn.*, **55**, 2504 (1982).
7. J.-X. Wang, M. Zhang, Z. Xing and Y. Hu, *Synth. Commun.*, **26**, 301 (1996).
8. D. Bogdal, J. Pielichowski and A. Boron, *Synth. Commun.*, **28**, 3029 (1998).
9. G. Nagy, S. V. Filip, E. Surducan and V. Surducan, *Synth. Commun.*, **27**, 3729 (1997).
10. B. M. Khadilkar and P. M. Bendale, *Synth. Commun.*, **27**, 2051 (1997).
11. H. Taniguchi and E. Nomura, *Chem. Lett.*, 1773 (1988); H. Taniguchi, Y. Otsuji and E. Nomura, *Bull. Chem. Soc. Jpn.*, **68**, 3563 (1995).
12. Y. Okada, Y. Sugitani, Y. Kasai and J. Nishimura, *Bull. Chem. Soc. Jpn.*, **67**, 586 (1994).
13. B. Jursic, *Tetrahedron*, **44**, 6677 (1988).
14. J. C. Lee, J. Y. Yuk and S. H. Cho, *Synth. Commun.*, **25**, 1367 (1995).
15. M. Lissel, S. Schmidt and B. Neuman, *Synthesis*, 382 (1986).
16. G. Barcelo, D. Grenouillat, J.-P. Senet and G. Sennyey, *Tetrahedron*, **46**, 1839 (1990).
17. Y. Lee and I. Shimizu, *Synlett*, 1063 (1998).
18. A. Perosa, M. Selva, P. Tundo and F. Zordan, *Synlett*, 272 (2000).
19. S. Ouk, S. Thiebaud, E. Borredon, P. Legars and L. Lecomte, *Tetrahedron Lett.*, **43**, 2661 (2002).
20. F. Camps, J. Coll and M. Moretó, *Synthesis*, 186 (1982).
21. D. L. Hughes, *Org. React.*, **42**, 335 (1992).
22. N. L. Dirlam, B. S. Moore and F. J. Urban, *J. Org. Chem.*, **52**, 3587 (1987).
23. R. S. Subramanian and K. K. Balasubramanian, *Tetrahedron Lett.*, **31**, 2201 (1990).
24. J. Freedman, M. J. Vaal and E. W. Huber, *J. Org. Chem.*, **56**, 670 (1991).
25. K. C. Santhosh and K. K. Balasubramanian, *Synth. Commun.*, **24**, 1049 (1994).
26. R. S. Subramanian and K. K. Balasubramanian, *Synth. Commun.*, **19**, 1255 (1989).
27. J. Tsuji, *Palladium Reagents and Catalysts*, Wiley, Chichester, 1995.
28. J. Muzart, J.-P. Genêt and A. Denis, *J. Organomet. Chem.*, **326**, C23 (1987).
29. D. R. Deardorff, D. C. Myles and K. D. MacFerrin, *Tetrahedron Lett.*, **26**, 5615 (1985).
30. D. R. Deardorff, S. Shambayati, R. G. Linde II and M. M. Dunn, *J. Org. Chem.*, **53**, 189 (1988).
31. R. C. Larock and N. H. Lee, *Tetrahedron Lett.*, **32**, 6315 (1991).
32. S.-K. Kang, D.-C. Park, J.-H. Jeon, H.-S. Rho and C.-M. Yu, *Tetrahedron Lett.*, **35**, 2357 (1994).
33. S.-K. Kang, D.-Y. Kim, R.-K. Hong and P.-S. Ho, *Synth. Commun.*, **26**, 3225 (1996).
34. T. Satoh, M. Ikeda, M. Miura and M. Nomura, *J. Org. Chem.*, **62**, 4877 (1997).
35. D. H. Camacho, I. Nakamura, S. Saito and Y. Yamamoto, *Angew. Chem., Int. Ed. Engl.*, **38**, 3365 (1999); *J. Org. Chem.*, **66**, 270 (2001).
36. R. C. Larock and X. Han, *J. Org. Chem.*, **64**, 1875 (1999).
37. C. Goux, M. Massaxcret, P. Lhoste and D. Sinou, *Organometallics*, **14**, 4585 (1995).
38. P. A. Evans and D. K. Leahy, *J. Am. Chem. Soc.*, **122**, 5012 (2000).
39. J.-R. Labrosse, P. Lhoste and D. Sinou, *J. Org. Chem.*, **66**, 6634 (2001).
40. M. Yoshida, H. Nemoto and M. Ihara, *Tetrahedron Lett.*, **40**, 8583 (1999).
41. J. D. Godfrey, Jr., R. H. Mueller, T. C. Sedergran, N. Soundararajan and V. J. Colandrea, *Tetrahedron Lett.*, **35**, 6405 (1994).
42. B. M. Trost and F. D. Toste, *J. Am. Chem. Soc.*, **120**, 815, 9074 (1998).
43. B. M. Trost and F. D. Toste, *J. Am. Chem. Soc.*, **121**, 4545 (1999).
44. M. Lautens, K. Fagnou and M. Taylor, *Org. Lett.*, **2**, 1677 (2000).
45. C. J. Moody and R. J. Taylor, *J. Chem. Soc., Perkin Trans. 1*, 721 (1989).
46. D. Haigh, *Tetrahedron*, **50**, 3177 (1994).
47. G. G. Cox, D. J. Miller, C. J. Moody, E.-R. H. B. Sie and J. J. Kulagowski, *Tetrahedron*, **50**, 3195 (1994).
48. G. Shi, Z. Cao and W. Cai, *Tetrahedron*, **51**, 5011 (1995).
49. Z. Zhu and J. H. Espenson, *J. Am. Chem. Soc.*, **118**, 9901 (1996).
50. T. Iida, N. Yamamoto, S. Matunaga, H.-G. Woo and M. Shibasaki, *Angew. Chem., Int. Ed. Engl.*, **37**, 2223 (1998).
51. J. M. Ready and E. N. Jacobsen, *J. Am. Chem. Soc.*, **121**, 6086 (1999).

52. J. M. Ready and E. N. Jacobsen, *J. Am. Chem. Soc.*, **123**, 2687 (2001).
53. D. A. Annis and E. N. Jacobsen, *J. Am. Chem. Soc.*, **121**, 4147 (1999).
54. M. E. Jung and L. S. Starkey, *Tetrahedron Lett.*, **36**, 7363 (1995); *Tetrahedron*, **53**, 8815 (1997).
55. O. E. O. Hormi, M. R. Moisio and B. C. Sund, *J. Org. Chem.*, **52**, 5272 (1987).
56. X. Huang, P. He and G. Shi, *J. Org. Chem.*, **65**, 627 (2000).
57. M. L. Purkayastha, M. Chandrasekharam, J. N. Vishwakarma, H. Ila and H. Junjappa, *Synthesis*, 245 (1993).
58. A. Padwa, W. H. Bullock, A. D. Dyszlewski, S. W. McCombie, B. B. Shankar and A. K. Ganguly, *J. Org. Chem.*, **56**, 3556 (1991).
59. R. Aumann, R. Fröhlich and S. Kotila, *Organometallics*, **15**, 4842 (1996).
60. R. Arnold-Stanton and D. M. Lemal, *J. Org. Chem.*, **56**, 151 (1991).
61. C. L. Bumgardner, J. E. Bunch and M.-H. Whangbo, *Tetrahedron Lett.*, **27**, 1883 (1986).
62. A. E. Feiring and E. R. Wonchoba, *J. Org. Chem.*, **57**, 7014 (1992).
63. M. Yoshida, S. Sasage, K. Kawamura, T. Suzuki and N. Kamigata, *Bull. Chem. Soc. Jpn.*, **64**, 416 (1991).
64. T. Hosokawa, T. Uno, S. Inui and S.-I. Murahashi, *J. Am. Chem. Soc.*, **103**, 2318 (1981).
65. T. Hosokawa, C. Okuda and S.-I. Murahashi, *J. Org. Chem.*, **50**, 1282 (1985).
66. Y. Uozumi, K. Kato and T. Hayashi, *J. Am. Chem. Soc.*, **119**, 5063 (1997); *J. Org. Chem.*, **63**, 5071 (1998).
67. Y. Uozumi, H. Kyota, K. Kato, M. Ogasawara and T. Hayashi, *J. Org. Chem.*, **64**, 1620 (1999).
68. J. Lindley, *Tetrahedron*, **40**, 1433 (1984).
69. J. Zhu, *Synlett*, 133 (1997).
70. J. S. Sawyer, *Tetrahedron*, **56**, 5045 (2000).
71. K. C. Nicolaou, C. N. C. Boddy, S. Natarajan, T.-Y. Yue, H. Li, S. Bräse and J. M. Ramanjulu, *J. Am. Chem. Soc.*, **119**, 3421 (1997).
72. A. V. Kalinin, J. F. Bower, P. Riebel and V. Snieckus, *J. Org. Chem.*, **64**, 2986 (1999).
73. D. L. Boger and D. Yohannes, *J. Org. Chem.*, **56**, 1763 (1991).
74. J.-F. Marcoux, S. Doye and S. L. Buchwald, *J. Am. Chem. Soc.*, **119**, 10539 (1997).
75. P. J. Fagan, E. Hauptman, R. Shapiro and A. Casalnuovo, *J. Am. Chem. Soc.*, **122**, 5043 (2000).
76. R. F. Pellón, R. Carrasco, V. Milián and L. Rodés, *Synth. Commun.*, **25**, 1077 (1995).
77. R. K. Gujadhur, C. G. Bates and D. Venkataraman, *Org. Lett.*, **3**, 4315 (2001).
78. E. Buck, Z. J. Song, D. Tschaen, P. G. Dormer, R. P. Volante and P. J. Reider, *Org. Lett.*, **4**, 1623 (2002).
79. G. Mann and J. F. Hartwig, *Tetrahedron Lett.*, **38**, 8005 (1997).
80. A. Aranyos, D. W. Old, A. Kiyomori, J. P. Wolfe, J. P. Sadighi and S. L. Buchwald, *J. Am. Chem. Soc.*, **121**, 4369 (1999).
81. D. H. R. Barton, J.-C. Blazejewski, B. Charpiot and W. B. Motherwell, *J. Chem. Soc., Chem. Commun.*, 503 (1981).
82. D. H. R. Barton, J.-P. Finet, J. Khamsl and C. Pichon, *Tetrahedron Lett.*, **27**, 3619 (1986).
83. D. M. T. Chan, K. L. Monaco, R.-P. Wang and M. P. Winters, *Tetrahedron Lett.*, **39**, 2933 (1998).
84. D. A. Evans, J. L. Katz and T. R. West, *Tetrahedron Lett.*, **39**, 2937 (1998).
85. J. Simon, S. Salzbrunn, G. K. S. Prakash, N. A. Petasis and G. A. Olah, *J. Org. Chem.*, **66**, 633 (2001).
86. M. E. Jung, D. Jachiet, S. I. Khan and C. Kim, *Tetrahedron Lett.*, **36**, 361 (1995).
87. E. A. Schmittling and J. S. Sawyer, *J. Org. Chem.*, **58**, 3229 (1993).
88. J. S. Sawyer, E. A. Schmittling, J. A. Palkowitz and W. J. Smith, III, *J. Org. Chem.*, **63**, 6338 (1998).
89. J. A. Segal, *J. Chem. Soc., Chem. Commun.*, 1338 (1983).
90. A. J. Pearson, P. R. Bruhn and S.-Y. Hau, *J. Org. Chem.*, **51**, 2137 (1986).
91. A. J. Pearson and H. Shin, *Tetrahedron*, **48**, 7527 (1992).
92. A. J. Pearson, J. G. Park and P. Y. Zhu, *J. Org. Chem.*, **57**, 3583 (1992).
93. J. W. Janetka and D. H. Rich, *J. Am. Chem. Soc.*, **119**, 6488 (1997).
94. A. J. Pearson and G. Bignan, *Tetrahedron Lett.*, **37**, 735 (1996).

10. Synthetic uses of phenols 707

95. R. R. Schmidt, 'Synthesis of Glycosides', in *Comprehensive Organic Synthesis* (Ed. B. M. Trost), Vol. 6, Chap. 1.2, Pergamon Press, Oxford, 1991, p. 33.
96. D. Dess, H. P. Kleine, D. V. Weinberg, R. J. Kaufman and R. S. Sidhu, *Synthesis*, 883 (1981).
97. R. Roy and F. Tropper, *Synth. Commun.*, **20**, 2097 (1990).
98. M. Hongu, K. Saito and K. Tsujihara, *Synth. Commun.*, **29**, 2775 (1999).
99. T. Matsumoto, M. Katsuki and K. Suzuki, *Chem. Lett.*, 437 (1989).
100. M. Yamaguchi, A. Horiguchi, A. Fukuda and T. Minami, *J. Chem. Soc., Perkin Trans. 1*, 1079 (1990).
101. K. Oyama and T. Kondo, *Synlett*, 1627 (1999).
102. F. Clerici, M. L. Gelmi and S. Mottadelli, *J. Chem. Soc., Perkin Trans. 1*, 985 (1994).
103. H. Müller and C. Tschierske, *J. Chem. Soc., Chem. Commun.*, 645 (1995).
104. Z.-J. Li, L.-N. Cai and M.-S. Cai, *Synth. Commun.*, **22**, 2121 (1992).
105. T. Yamanoi, A. Fujioka and T. Inazu, *Bull. Chem. Soc. Jpn.*, **67**, 1488 (1994).
106. D. Kahne, S. Walker, Y. Cheng and D. V. Engen, *J. Am. Chem. Soc.*, **111**, 6881 (1989).
107. L.-F. Tieze, R. Fischer and H.-J. Guder, *Tetrahedron Lett.*, **23**, 4661 (1982).
108. S. Koto, N. Morishima, M. Araki, T. Tsuchiya and S. Zen, *Bull. Chem. Soc. Jpn.*, **54**, 1895 (1981).
109. W. R. Roush and X.-F. Lin, *J. Org. Chem.*, **56**, 5740 (1991); *J. Am. Chem. Soc.*, **117**, 2236 (1995).
110. J. Gervay and S. Danishefsky, *J. Org. Chem.*, **56**, 5448 (1991).
111. R. G. Dushin and S. J. Danishefsky, *J. Am. Chem. Soc.*, **114**, 3471 (1992).
112. C. Chiappe, G. L. Moro and P. Munforte, *Tetrahedron*, **53**, 10471 (1997).
113. M. Takhi, A. A.-H. Abdel-Rahman and R. R. Schmidt, *Synlett*, 427 (2001).
114. B. S. Babu and K. K. Balasubramanian, *Tetrahedron Lett.*, **41**, 1271 (2000).
115. S. Sowmya and K. K. Balasubramanian, *Synth. Commun.*, **24**, 2097 (1994).
116. J. Beyer, P. R. Skaanderup and R. Madsen, *J. Am. Chem. Soc.*, **122**, 9575 (2000).
117. H. O. House, *Modern Synthetic Reactions*, Chap. 9, W. A. Benjamin, Menlo Park, 1972, p. 520.
118. J. J. Bozell, D. Miller, B. R. Hames and C. Loveless, *J. Org. Chem.*, **66**, 3084 (2001).
119. Y. Arredondo, M. Moreno-Manas and R. Pleixats, *Synth. Commun.*, **26**, 3885 (1996) and references cited therein.
120. P. Wipf, 'Claisen Rearrangement', in *Comprehensive Organic Synthesis* (Ed. B. M. Trost), Vol. 5, Chap. 7.2, Pergamon Press, Oxford, 1991, p. 827.
121. S. J. Rhoads and N. R. Raulins, *Org. React.*, **22**, 1 (1975).
122. F. Bigi, G. Casiraghi, G. Casnati and G. Sartori, *Synthesis*, 310 (1981).
123. J. B. Baruah, *Tetrahedron Lett.*, **36**, 8509 (1995).
124. A. V. Malkov, S. L. Davis, I. R. Baxendale, W. L. Mitchell and P. Kocovsky, *J. Org. Chem.*, **64**, 2751 (1999).
125. A. V. Malkov, P. Spoor, V. Vinader and P. Kocovsky, *J. Org. Chem.*, **64**, 5308 (1999).
126. V. K. Ahluwalia, K. K. Arora and R. S. Jolly, *J. Chem. Soc., Perkin Trans. 1*, 335 (1982).
127. F. M. D. Ismail, M. J. Hilton and M. Stefinovic, *Tetrahedron Lett.*, **33**, 3795 (1992).
128. M. Matsui, N. Karibe, K. Hayashi and H. Yamamoto, *Bull. Chem. Soc., Jpn.*, **68**, 3569 (1995).
129. K. Ishihara, M. Kubota and H. Yamamoto, *Synlett*, 1045 (1996).
130. M. Matsui and H. Yamamoto, *Bull. Chem. Soc. Jpn.*, **68**, 2663 (1995); **69**, 137 (1996).
131. F. Bigi, S. Carloni, R. Maggi, C. Muchetti, M. Rastelli and G. Sartori, *Synthesis*, 301 (1998).
132. H. Bienaymé, J.-E. Ancel, P. Meilland and J.-P. Simonato, *Tetrahedron Lett.*, **41**, 3339 (2000).
133. G. Sartori, F. Bigi, G. Casiraghi and G. Castani, *Tetrahedron*, **39**, 1716 (1983).
134. K. Sato, S. Inoue, K. Ozawa and M. Tazaki, *J. Chem. Soc., Perkin Trans. 1*, 2715 (1984).
135. K. Sato, S. Inoue, K. Ozawa, T. Kobayashi, T. Ota and M. Tazaki, *J. Chem. Soc., Perkin Trans. 1*, 1753 (1987).
136. S. Inoue, H. Ikeda, S. Sato, K. Hori, T. Ota, O. Miyamoto and K. Sato, *J. Org. Chem.*, **52**, 5495 (1987).
137. G. H. Posner and K. A. Canella, *J. Am. Chem. Soc.*, **107**, 2571 (1985).
138. H. Andringa, H. D. Verkruijsse, L. Brandsma and L. Lochmann, *J. Organomet. Chem.*, **393**, 307 (1990).
139. R. B. Bates and T. J. Siahaan, *J. Org. Chem.*, **51**, 1432 (1986).
140. E. K. Lehnert, J. S. Sawyer and T. L. Macdonald, *Tetrahedron Lett.*, **30**, 5215 (1989).

141. T. Matumoto, M. Katuki and K. Suzuki, *Tetrahedron Lett.*, **29**, 6935 (1988).
142. T. Matumoto, M. Katuki, H. Jona and K. Suzuki, *Tetrahedron Lett.*, **30**, 6185 (1989).
143. T. Matumoto, M. Katuki, H. Jona and K. Suzuki, *J. Am. Chem. Soc.*, **113**, 6982 (1991).
144. T. Kometani, H. Kondo and Y. Fujimoto, *Synthesis*, 1005 (1988).
145. T. Matumoto, T. Hosoya and K. Suzuki, *Tetrahedron Lett.*, **31**, 4629 (1990).
146. K. Toshima, G. Matsuo and K. Tatsuta, *Tetrahedron Lett.*, **33**, 2175 (1992).
147. N. G. Ramesh and K. K. Balasubramanian, *Tetrahedron Lett.*, **33**, 3061 (1992).
148. G. Casiraghi, M. Cornia, G. Rassu, L. Zetta, G. G. Fava and M. F. Belicchi, *Tetrahedron Lett.*, **29**, 3323 (1988).
149. S. Sethna and R. Phadke, *Org. React.*, **7**, 1 (1953).
150. S. Frére, V. Thiéry and T. Besson, *Tetrahedron Lett.*, **42**, 2791 (2001).
151. A. C. Khandekar and B. M. Khadilkar, *Synlett*, 152 (2002).
152. Y. V. Subba Rao, S. J. Kulkarni, M. Subramanyam and A. V. Rama Rao, *J. Chem. Soc., Chem. Commun.*, 1456 (1993).
153. F. Bigi, S. Carloni, R. Maggi, C. Muchetti and G. Sartori, *J. Org. Chem.*, **62**, 6024 (1997).
154. A. de la Hoz, A. Moreno and E. Vázquez, *Synlett*, 608 (1999).
155. G. Sartori, F. Bigi, A. Pastorio, C. Porta, A. Arienti, R. Maggi, N. Moretti and G. Gnappi, *Tetrahedron Lett.*, **36**, 9177 (1995).
156. M. Yamaguchi, A. Hayashi and M. Hirama, *J. Am. Chem. Soc.*, **117**, 1151 (1995).
157. M. Yamaguchi, M. Arisawa, K. Omata, K. Kabuto, M. Hirama and T. Uchimaru, *J. Org. Chem.*, **63**, 7298 (1998).
158. M. Yamaguchi, M. Arisawa, Y. Kido and M. Hirama, *Chem. Commun.*, 1663 (1997).
159. M. Yamaguchi, K. Kobayashi and M. Arisawa, *Synlett*, 1317 (1998).
160. K. Kobayashi and M. Yamaguchi, *Org. Lett.*, **3**, 241 (2001).
161. M. Yamaguchi, K. Kobayashi and M. Arisawa, *Inorg. Chim. Acta*, **296**, 67 (1999).
162. B. M. Trost and F. D. Toste, *J. Am. Chem. Soc.*, **118**, 6305 (1996).
163. D. H. R. Barton, N. Y. Bhatnagar, J.-C. Blazejewski, B. Charpiot, J.-P. Finet, D. J. Lester, W. B. Motherwell. M. T. B. Papoula and S. P. Stanforth, *J. Chem. Soc., Perkin Trans. 1*, 2657 (1985).
164. S. Saito, T. Kano, H. Muto, M. Nakadai and H. Yamamoto, *J. Am. Chem. Soc.*, **121**, 8943 (1999); T. Kano, Y. Oyabu, S. Saito and H. Yamamoto, *J. Am. Chem. Soc.*, **124**, 5365 (2002).
165. M. E. Jung, C. Kim and L. von der Bussche, *J. Org. Chem.*, **59**, 3248 (1994).
166. K. Kobayashi, M. Arisawa and M. Yamaguchi, *J. Am. Chem. Soc.*, **124**, 8528 (2002).
167. R. Subramanian and F. Johnson, *J. Org. Chem.*, **50**, 5430 (1985).
168. G. Casiraghi, G. Casnati, G. Puglia and G. Sartori, *Synthesis*, 124 (1980).
169. G. Casnati, G. Casiraghi, A. Pochini, G. Sartori and R. Ungaro, *Pure Appl. Chem.*, **55**, 1677 (1983).
170. G. Sartori, R. Maggi, F. Bagi, A. Arienti, C. Porta and G. Predieri, *Tetrahedron*, **50**, 10587 (1994).
171. P. Hewawasam and M. Erway, *Tetrahedron Lett.*, **39**, 3981 (1998).
172. Y. Gong, K. Kato and H. Kimoto, *Synlett*, 1403 (1999).
173. Y. Gong, K. Kato and H. Kimoto, *Bull. Chem. Soc. Jpn.*, **74**, 377 (2001).
174. W. Nagata, K. Okada and T. Aoki, *Synthesis*, 365 (1979).
175. C. K. Lau, M. Mintz, M. A. Berstein and C. Dufresne, *Tetrahedron Lett.*, **34**, 5527 (1993).
176. G. Casiraghi, M. Cornia, G. Castani, G. G. Fava, M. F. Belicchi and L. Zetta, *J. Chem. Soc., Chem. Commun.*, 794 (1987).
177. G. Casiraghi, M. Cornia and G. Rassu, *J. Org. Chem.*, **53**, 4919 (1988).
178. M. Cornia and G. Casiraghi, *Tetrahedron*, **45**, 2869 (1989).
179. F. Bigi, G. Castani, G. Sartori, G. Araldi and G. Bocelli, *Tetrahedron Lett.*, **30**, 1121 (1989).
180. O. Piccolo, L. Filippini, L. Tinucci, E. Valoti and A. Citterio, *Helv. Chim. Acta*, **67**, 739 (1984).
181. F. Bigi, G. Castani, G. Sartori, C. Dalprato and R. Bortolini, *Tetrahedron: Asymmetry*, **1**, 861 (1990).
182. F. Bigi, G. Bocelli, R. Maggi and G. Sartori, *J. Org. Chem.*, **64**, 5004 (1999).
183. F. Bigi, G. Casiraghi, G. Castani and G. Sartori, *J. Org. Chem.*, **50**, 5018 (1985).
184. G. Sartori, F. Bigi, G. Castani, G. Sartori, P. Soncini, G. G. Fava and M. F. Belicchi, *J. Org. Chem.*, **53**, 1779 (1988).
185. G. Erker and A. A. H. van der Zeijden, *Angew. Chem., Int. Ed. Engl.*, **29**, 512 (1990).

186. M. Tramontini and L. Angiolini, *Tetrahedron*, **46**, 1791 (1990).
187. H. Heaney, in *Comprehensive Organic Synthesis* (Ed. B. M. Trost), Vol. 2, Pergamon Press, Oxford, 1991, p. 953.
188. R. A. Fairhurst, H. Heaney, G. Papageorgiou and R. F. Wilkins, *Tetrahedron Lett.*, **29**, 5801 (1988).
189. H. Heaney, G. Papageorgiou and R. F. Wilkins, *Tetrahedron*, **53**, 13361 (1997).
190. T. Huang and C.-J. Li, *Tetrahedron Lett.*, **41**, 6715 (2000).
191. Y.-J. Chen, C.-S. Ge and D. Wang, *Synlett*, 1429 (2000).
192. C.-S. Ge, Y.-J. Chen and D. Wang, *Synlett*, 37 (2002).
193. S. Tohma, A. Endo, T. Kan and T. Fukuyama, *Synlett*, 1179 (2000).
194. W. E. Truce, *Organic React.*, **9**, 37 (1957).
195. H. Wynberg and E. W. Meijer, *Org. React.*, **28**, 1 (1982).
196. A. Behr, *Chem. Eng. Technol.*, **10**, 16 (1987).
197. A. Thoer, G. Denis, M. Delmas and A. Gaset, *Synth. Commun.*, **18**, 2095 (1988).
198. J. C. Cochran and M. G. Melville, *Synth. Commun.*, **20**, 609 (1990).
199. J. F. Larrow, E. N. Jacobsen, Y. Gao, Y. Hong, X. Nie and C. M. Zepp, *J. Org. Chem.*, **59**, 1939 (1994).
200. G. Casiraghi, G. Casnati, G. Puglia, G. Sartori and G. Terenghi, *J. Chem. Soc., Perkin Trans. 1*, 1862 (1980).
201. R. X. Wang, X. Z. You, Q. J. Meng, E. A. Mintz and X. R. Bu, *Synth. Commun.*, **24**, 1757 (1994).
202. F. Effenberger and R. Gutmann, *Chem. Ber.*, **115**, 1089 (1982).
203. S. Kobayashi, M. Moriwaki and I. Hachiya, *Tetrahedron Lett.*, **37**, 2053 (1996); *Bull. Chem. Soc. Jpn.*, **70**, 267 (1997).
204. S. Kobayashi, M. Moriwaki and I. Hachiya, *Tetrahedron Lett.*, **37**, 4183 (1996).
205. I. Neves, F. Jayat, P. Magnoux, G. Pérot, F. R. Ribeiro, M. Gubelmann and M. Guisnet, *J. Chem. Soc., Chem. Commun.*, 717 (1994).
206. F. Bigi, G. Casiraghi, G. Castani, S. Marchesi and G. Sartori, *Tetrahedron*, **40**, 4081 (1984).
207. F. Bigi, G. Casiraghi, G. Castani and G. Sartori, *J. Chem. Soc., Perkin Trans. 1*, 2655 (1984).
208. G. Sartori, G. Castani, F. Bigi and G. Predieri, *J. Org. Chem.*, **55**, 4371 (1990).
209. T. Toyoda, K. Sasakura and T. Sugasawa, *J. Org. Chem.*, **46**, 189 (1981).
210. O. Piccolo, L. Filippini, L. Tinucci, E. Valoti and A. Citterio, *Tetrahedron*, **42**, 885 (1986).
211. I. Hirao and T. Kato, *Bull. Chem. Soc. Jpn.*, **46**, 3470 (1974).
212. M. Aresta, E. Quaranta, R. Liberio, C. Dileo and I. Tommasi, *Tetrahedron*, **54**, 8841 (1998).
213. G. Balduzzi, F. Bigi, G. Casiraghi, G. Castani and G. Sartori, *Synthesis*, 879 (1982).
214. M. Adachi and T. Sugasawa, *Synth. Commun.*, **20**, 71 (1990).
215. S. T. Purrington, B. S. Kagen and T. B. Patrick, *Chem. Rev.*, **86**, 997 (1986).
216. G. S. Lal, G. O. Pez and R. G. Syvret, *Chem. Rev.*, **96**, 1737 (1996).
217. S. Misaki, *J. Fluorine Chem.*, **17**, 159 (1981); **21**, 191 (1982).
218. S. T. Purrington and D. L. Woodward, *J. Org. Chem.*, **56**, 142 (1991).
219. S. Stavber and M. Zupan, *Chem. Lett.*, 1077 (1996).
220. W. E. Barnette, *J. Am. Chem. Soc.*, **106**, 452 (1984).
221. S. Singh, D. D. DesMarteau, S. S. Zuberi, M. Witz and H.-N. Huang, *J. Am. Chem. Soc.*, **109**, 7194 (1987).
222. T. Umemoto, K. Kawada and K. Tomita, *Tetrahedron Lett.*, **27**, 4465 (1986).
223. T. Umemoto, S. Fukami, G. Tomizawa, K. Harasawa, K. Kawada and K. Tomita, *J. Am. Chem. Soc.*, **112**, 8563 (1990).
224. T. Umemoto and G. Tomizawa, *J. Org. Chem.*, **60**, 6563 (1995).
225. T. Umemoto, M. Nagayoshi, K. Adachi and G. Tomizawa, *J. Org. Chem.*, **63**, 3379 (1998).
226. For many examples see: E. N. Jacobsen, A. Pfartz and H. Yamamoto (Eds.), *Comprehensive Asymmetric Catalysis*, Springer-Verlag, Berlin, 1999.
227. K. Maruoka and H. Yamamoto, *Tetrahedron*, **44**, 5001 (1988).
228. K. Maruoka, T. Itoh and H. Yamamoto, *J. Am. Chem. Soc.*, **107**, 4573 (1985); K. Maruoka, T. Itoh, M. Sakurai, K. Nonoshita and H. Yamamoto, *J. Am. Chem. Soc.*, **110**, 3588 (1988).
229. K. Maruoka, H. Imoto, S. Saito and H. Yamamoto, *J. Am. Chem. Soc.*, **116**, 4131 (1994).
230. K. Maruoka, M. Ito and H. Yamamoto, *J. Am. Chem. Soc.*, **117**, 9091 (1995).
231. S. Saito, K. Shimada, H. Yamamoto, E. M. de Marigorta and I. Fleming, *Chem. Commun.*, 1299 (1997).

232. K. Maruoka, S. Saito, A. B. Concepcion and H. Yamamoto, *J. Am. Chem. Soc.*, **115**, 1183 (1993).
233. S. Saito, M. Shiozawa, M. Ito and H. Yamamoto, *J. Am. Chem. Soc.*, **120**, 813 (1998).
234. K. Maruoka, K. Nonoshita, H. Banno and H. Yamamoto, *J. Am. Chem. Soc.*, **110**, 7922 (1988); K. Maruoka, H. Banno, K. Nonoshita and H. Yamamoto, *Tetrahedron Lett.*, **30**, 1265 (1989); K. Nonoshita, H. Banno, K. Maruoka and H. Yamamoto, *J. Am. Chem. Soc.*, **112**, 316 (1990).
235. K. Maruoka, H. Banno and H. Yamamoto, *J. Am. Chem. Soc.*, **112**, 7791 (1990); K. Maruoka, S. Saito and H. Yamamoto, *J. Am. Chem. Soc.*, **117**, 1165 (1995).
236. B. M. Trost and H. Ito, *J. Am. Chem. Soc.*, **122**, 12003 (2000).
237. B. M. Trost, H. Ito and E. R. Silcoff, *J. Am. Chem. Soc.*, **123**, 3367 (2001).
238. Y. Hamashima, M. Kanai and M. Shibasaki, *J. Am. Chem. Soc.*, **122**, 7412 (2000).
239. S. Vyskocil, S. Jaracz, M. Smrcina, M. Sticha, V. Hanus, M. Polasek and P. Kocovsky, *J. Org. Chem.*, **63**, 7727 (1998).
240. S. M. Azad, S. M. W. Bennett, S. M. Brown, J. Green, E. Sinn, C. M. Topping and S. Woodward, *J. Chem. Soc., Perkin Trans. 1*, 687 (1997).
241. A. J. Blake, A. Cunningham, A. Ford, S. J. Teat and S. Woodward, *Chem. Eur. J.*, **6**, 3586 (2000).
242. A. Cunningham and S. Woodward, *Synlett*, 43 (2002).
243. K. Ishihara, M. Kaneeda and H. Yamamoto, *J. Am. Chem. Soc.*, **116**, 11179 (1994).
244. K. Ishihara, S. Nakamura, M. Kaneeda and H. Yamamoto, *J. Am. Chem. Soc.*, **118**, 12854 (1996).
245. L. Pu, *Chem. Rev.*, **98**, 2405 (1998).
246. G. J. Rowlands, *Tetrahedron*, **57**, 1865 (2001).
247. R. Noyori, I. Tomino and Y. Tanimoto, *J. Am. Chem. Soc.*, **101**, 3129 (1979).
248. K. Yamamoto, H. Fukushima and M. Nakazaki, *J. Chem. Soc., Chem. Commun.*, 1490 (1984).
249. T. R. Kelly, A. Whiting and N. S. Chandrakumar, *J. Am. Chem. Soc.*, **108**, 3510 (1986).
250. K. Maruoka, T. Itoh, T. Shirasaka and H. Yamamoto, *J. Am. Chem. Soc.*, **110**, 310 (1988).
251. K. Maruoka and H. Yamamoto, *J. Am. Chem. Soc.*, **111**, 789 (1989).
252. K. Maruoka, A. B. Concepcion and H. Yamamoto, *Bull. Chem. Soc. Jpn.*, **65**, 3501 (1992).
253. J. Bao, W. D. Wulff and A. L. Rheingold, *J. Am. Chem. Soc.*, **115**, 3814 (1993).
254. D. P. Heller, D. R. Goldberg and W. D. Wulff, *J. Am. Chem. Soc.*, **119**, 10551 (1997).
255. K. Maruoka, H. Banno and H. Yamamoto, *Tetrahedron: Asymmetry*, **2**, 647 (1991).
256. K. Maruoka, Y. Hoshino, T. Shirasaka and H. Yamamoto, *Tetrahedron. Lett.*, **29**, 3967 (1988).
257. K Ishihara and H. Yamamoto, *J. Am. Chem. Soc.*, **116**, 1561 (1994).
258. K. Maruoka, N. Murase and H. Yamamoto, *J. Org. Chem.*, **58**, 2938 (1993).
259. K. Ishihara, S. Kondo, H. Kurihara and H. Yamamoto, *J. Org. Chem.*, **62**, 3026 (1997).
260. K. Ishihara, H. Kurihara and H. Yamamoto, *J. Am. Chem. Soc.*, **118**, 3049 (1996).
261. K. Ishihara, H. Kurihara, M. Matsumoto and H. Yamamoto, *J. Am. Chem. Soc.*, **120**, 6920 (1998).
262. H. Kitajima, K. Ito and T. Katsuki, *Chem. Lett.*, 343 (1996).
263. H. Kitajima, Y. Aoki, K. Ito and T. Katsuki, *Chem. Lett.*, 1113 (1995); *Bull. Chem. Soc. Jpn.*, **70**, 207 (1997).
264. W.-S. Huang, Q.-S. Hu and L. Pu, *J. Org. Chem.*, **63**, 1364 (1998); **64**, 7940 (1999).
265. W.-S. Huang and L. Pu, *J. Org. Chem.*, **64**, 4222 (1999).
266. Q.-S. Hu, W.-S. Huang, D. Vitharana, X.-F. Zheng and L. Pu, *J. Am. Chem. Soc.*, **119**, 12454 (1997).
267. Y. Hamashima, D. Sawada, M. Kanai and M. Shibasaki, *J. Am. Chem. Soc.*, **121**, 2641 (1999).
268. G. Manickam, H. Nogami, M. M. Kanai, H. Gröger and M. Shibasaki, *Synlett*, 617 (2001).
269. M. Takamura, K. Funabashi, M. Kanai and M. Shibasaki, *J. Am. Chem. Soc.*, **122**, 6327 (2000); **123**, 6801 (2001).
270. M. Bougauchi, S. Watanabe, T. Arai, H. Sasai and M. Shibasaki, *J. Am. Chem. Soc.*, **119**, 2329 (1997).
271. C. Qian, C. Zhu and T. Huang, *J. Chem. Soc., Perkin Trans. 1*, 2131 (1998).
272. M. T. Reetz, C. Merk, G. Naberfeld, J. Rudolph, N. Griebenow and R. Goddard, *Tetrahedron Lett.*, **38**, 5273 (1997).
273. N. Yoshikawa, N. Kumagai, S. Matsunaga, G. Moll, T. Ohshima, T. Suzuki and M. Shibasaki, *J. Am. Chem. Soc.*, **123**, 2466 (2001).

274. N. Kumagai, S. Matsunaga, N. Yoshikawa, T. Ohshima and M. Shibasaki, *Org. Lett.*, **3**, 1539 (2001).
275. M. Terada, Y. Motoyama and K. Mikami, *Tetrahedron Lett.*, **35**, 6693 (1994).
276. H. Ishitani, M. Ueno and S. Kobayashi, *J. Am. Chem. Soc.*, **119**, 7153 (1997).
277. S. Kobayashi, H. Ishitani and M. Ueno, *J. Am. Chem. Soc.*, **120**, 431 (1998).
278. S. Kobayashi, K. Kusakabe, S. Komiyama and H. Ishitani, *J. Org. Chem.*, **64**, 4220 (1999).
279. H. Sasai, T. Tokunaga, S. Watanabe, T. Suzuki, N. Itoh and M. Shibasaki, *J. Org. Chem.*, **60**, 7388 (1995).
280. T. Morita, T. Arai, H. Sasai and M. Shibasaki, *Tetrahedron: Asymmetry*, **9**, 1445 (1998).
281. A. S. Chan, F.-Y. Zhang and C.-W. Yip, *J. Am. Chem. Soc.*, **119**, 4080 (1997).
282. F.-Y. Zhang and A. S. C. Chan, *Tetrahedron: Asymmetry*, **8**, 3651 (1997).
283. Y. Tian and K. S. Chan, *Tetrahedron Lett.*, **41**, 8813 (2000).
284. T. Harada, M. Takeuchi, M. Hatsuda, S. Ueda and A. Oku, *Tetrahedron: Asymmetry*, **7**, 2479 (1996).
285. S. Pandiaraju, G. Chen, A. Lough and A. K. Yudin, *J. Am. Chem. Soc.*, **123**, 3850 (2001).
286. H. Nozaki, S. Moriuti, H. Takaya and R. Noyori, *Tetrahedron Lett.*, 5239 (1966); A. Pfaltz, in *Comprehensive Asymmetric Catalysis* (Eds. E. N. Jacobsen, A. Pfartz and H. Yamamoto), Chap. 16.1, Springer-Verlag, Berlin, 1999 p. 513.
287. T. Aratani, *Pure Appl. Chem.*, **57**, 1839 (1996).
288. T. Aratani, in *Comprehensive Asymmetric Catalysis* (Eds. E. N. Jacobsen, A. Pfartz and H. Yamamoto), Chap. 41.3, Springer-Verlag, Berlin, 1999, p. 1451.
289. T. Fukuda and T. Katsuki, *Synlett*, 825 (1995); *Tetrahedron*, **53**, 7201 (1997).
290. T. Fukuda and T. Katsuki, *Tetrahedron Lett.*, **38**, 3435 (1997).
291. E. N. Jacobsen and M. H. Wu, in *Comprehensive Asymmetric Catalysis* (Eds. E. N. Jacobsen, A. Pfartz and H. Yamamoto), Chap. 18.2, Springer-Verlag, Berlin, 1999, p. 649.
292. T. Katsuki, *J. Synth. Org. Chem., Jpn.*, **53**, 940 (1995).
293. W. Zhang, J. L. Loebach, S. R. Wilson and E. N. Jacobsen, *J. Am. Chem. Soc.*, **112**, 2801 (1990).
294. E. N. Jacobsen, W. Zhang and M. L. Güler, *J. Am. Chem. Soc.*, **113**, 6703 (1991).
295. E. N. Jacobsen, W. Zhang, A. R. Muci, J. R. Ecker and L. Deng, *J. Am. Chem. Soc.*, **113**, 7063 (1991).
296. P. J. Pospisil, D. H. Carsten and E. N. Jacobsen, *Chem. Eur. J.*, **2**, 974 (1996).
297. B. D. Brandes and E. N. Jacobsen, *Tetrahedron Lett.*, **36**, 5123 (1995).
298. R. Irie, K. Noda, Y. Ito, N. Matsumoto and T. Katsuki, *Tetrahedron Lett.*, **31**, 7345 (1990); *Tetrahedron: Asymmetry*, **2**, 481 (1991).
299. N. Hosoya, R. Irie and T. Katsuki, *Synlett*, 261 (1993).
300. N. Hosoya, A. Hatayama, R. Irie, H. Sasaki and T. Katsuki, *Tetrahedron*, **50**, 4311 (1994).
301. N. Hosoya, R. Irie, T. Hamada, K. Suzuki and T. Katsuki, *Tetrahedron*, **50**, 11827 (1994).
302. C. Bolm, K. Muniz and J. P. Hildebrand, in *Comprehensive Asymmetric Catalysis* (Eds. E. N. Jacobsen, A. Pfartz and H. Yamamoto), Chap. 19, Springer-Verlag, Berlin, 1999, p. 697.
303. K. Nakayama, M. Kojima and J. Fujita, *Chem. Lett.*, 1483 (1986).
304. C. Bolm and F. Bienwald, *Angew. Chem., Int. Ed. Engl.*, **34**, 2640 (1995); *Synlett*, 1327 (1998).
305. G. Liu, D. A. Cogan and J. A. Ellman, *J. Am. Chem. Soc.*, **119**, 9913 (1997).
306. K. Nakajima, C. Sasaki, M. Kojima, T. Aoyama, S. Ohba, Y. Saito and J. Fujita, *Chem. Lett.*, 2189 (1987).
307. S. Colonna, A. Manfredi, M. Spadoni, L. Casell and M. Gullotti, *J. Chem. Soc., Perkin Trans. 1*, 71 (1987).
308. C. Sasaki, K. Nakajima, M. Kojima and J. Fujita, *Bull. Chem. Soc. Jpn.*, **64**, 1318 (1991).
309. C. Ohta, H. Shimizu, A. Kondo and T. Katsuki, *Synlett*, 161 (2002).
310. B. Saito and T. Katsuki, *Tetrahedron Lett.*, **42**, 3873 (2001).
311. M. Palucki, P. Hanson and E. N. Jacobsen, *Tetrahedron Lett.*, **33**, 7111 (1992).
312. K. Noda, N. Hosoya, R. Irie, Y. Yamashita and T. Katsuki, *Tetrahedron*, **50**, 9609 (1994).
313. C. Kokubo and T. Katsuki, *Tetrahedron*, **52**, 13895 (1996).
314. C. Bolm and O. Beckmann, in *Comprehensive Asymmetric Catalysis* (Eds. E. N. Jacobsen, A. Pfartz and H. Yamamoto), Chap. 22, Springer-Verlag, Berlin, 1999, p. 803.
315. C. Bolm, G. Schlingloff and K. Weickhardt, *Tetrahedron Lett.*, **34**, 3405 (1993).
316. C. Bolm, G. Schlingloff and K. Weickhardt, *Angew. Chem., Int. Ed. Engl.*, **33**, 1848 (1994).

317. C. Bolm and G. Schlingloff, *J. Chem. Soc., Chem. Commun.*, 1247 (1995).
318. A. Gusso, C. Baccin, F. Pinna and G. Strukul, *Organometallics*, **13**, 3442 (1994).
319. T. Katsuki, in *Comprehensive Asymmetric Catalysis* (Eds. E. N. Jacobsen, A. Pfartz and H. Yamamoto), Chap. 21, Springer-Verlag, Berlin, 1999, p. 791.
320. J. F. Larrow and E. N. Jacobsen, *J. Am. Chem. Soc.*, **116**, 12129 (1994).
321. K. Hamachi, R. Irie and T. Katsuki, *Tetrahedron Lett.*, **37**, 4979 (1996).
322. T. Hamada, R. Irie, J. Mihara, K. Hamachi and T. Katsuki, *Tetrahedron*, **54**, 10017 (1998).
323. A. Miyafuji and T. Katsuki, *Tetrahedron*, **54**, 10339 (1998).
324. E. N. Jacobsen, in *Comprehensive Asymmetric Catalysis* (Eds. E. N. Jacobsen, A. Pfartz and H. Yamamoto), Chap. 17, Springer-Verlag, Berlin, 1999, p. 607.
325. K. J. O'Connor, S.-J. Wey and C. J. Burrows, *Tetrahedron Lett.*, **33**, 1001 (1992).
326. K. Noda, N. Hosoya, R. Irie, Y. Irto and T. Katsuki, *Synlett*, 469 (1993).
327. H. Nishikori and T. Katsuki, *Tetrahedron Lett.*, **37**, 9245 (1996).
328. J. D. Bois, C. S. Tomooka, J. Hong, E. M. Carreira and M. W. Day, *Angew. Chem., Int. Ed. Engl.*, **36**, 1645 (1997).
329. J. D. Bois, C. S. Tomooka, J. Hong and E. M. Carreira, *Acc. Chem. Res.*, **30**, 364 (1997).
330. J. D. Bois, J. Hong, E. M. Carreira and M. W Day, *J. Am. Chem. Soc.*, **118**, 915 (1996).
331. J. D. Bois, C. S. Tomooka, J. Hong and E. M. Carreira, *J. Am. Chem. Soc.*, **119**, 3179 (1997).
332. S. Minakata, T. Ando, M. Nishimura, I. Ryo and M. Komatsu, *Angew. Chem., Int. Ed. Engl.*, **37**, 3392 (1998).
333. A. Mori and S. Inoue, in *Comprehensive Asymmetric Catalysis* (Eds. E. N. Jacobsen, A. Pfartz and H. Yamamoto), Chap. 28, Springer-Verlag, Berlin, 1999, p. 983.
334. M. Hayashi, Y. Miyamoto, T. Inoue and N. Oguni, *J. Chem. Soc., Chem. Commun.*, 1752 (1991); *J. Org. Chem.*, **58**, 1515 (1993).
335. J. Yaozhong, Z. Xiangge, H. Wanhao, W. Lanjun and M. Aqiao, *Tetrahedron: Asymmetry*, **6**, 405 (1995).
336. J. Yaozhong, Z. Xiangge, H. Wanhao, L. Zhi and M. Aqiao, *Tetrahedron: Asymmetry*, **6**, 2915 (1995).
337. C. Bolm and P. Müller, *Tetrahedron Lett.*, **36**, 1625 (1995).
338. M. Hayashi, T. Inoue, Y. Miyamoto and N. Oguni, *Tetrahedron*, **50**, 4385 (1994).
339. E. M. Carreira, R. A. Singer and W. Lee, *J. Am. Chem. Soc.*, **116**, 8837 (1994).
340. E. M. Carreira, W. Lee and R. A. Singer, *J. Am. Chem. Soc.*, **117**, 3649 (1995).
341. R. A. Singer and E. M. Carreira, *J. Am. Chem. Soc.*, **117**, 12360 (1995).
342. N. Oguni, Y. Miyagi and K. Itoh, *Tetrahedron Lett.*, **39**, 9023 (1998).
343. P. Müller and P. Nury, *Helv. Chim. Acta*, **84**, 662 (2001).
344. A. Mori, H. Nitta, M. Kudo and S. Inoue, *Tetrahedron Lett.*, **32**, 4333 (1991); H. Nitta, D. Yu, M. Kudo, A. Mori and S. Inoue, *J. Am. Chem. Soc.*, **114**, 7969 (1992).
345. B. M. Cole, K. D. Shimizu, C. A. Krueger, J. P. A. Harrity, M. L. Snapper and A. H. Hoveyda, *Angew. Chem., Int. Ed. Engl.*, **35**, 1668 (1996).
346. K. D. Shimizu, B. M. Cole, C. A. Krueger, K. W. Kuntz, M. L. Snapper and A. H. Hoveyda, *Angew. Chem., Int. Ed. Engl.*, **36**, 1704 (1997).

CHAPTER 11

Tautomeric equilibria and rearrangements involving phenols

SERGEI M. LUKYANOV and ALLA V. KOBLIK

ChemBridge Corporation, Malaya Pirogovskaya str., 1, 119435 Moscow, Russia
e-mail: semiluk@chembridge.ru

I. INTRODUCTION	714
II. TAUTOMERISM IN PHENOLS	714
A. Keto–Enol Tautomerism in Phenols	714
1. Monocyclic phenols	715
2. Polycyclic phenols	719
3. Phenols bearing nitrogen-containing substituents	721
a. Nitrosophenol–quinone oxime tautomerism	721
b. Arylazophenol–quinone arylhydrazone tautomerism	724
c. Tautomerism in Schiff bases	726
4. Metal-coordinated phenols	731
5. Phenols inserted into conjugated systems	735
B. Ring-chain Tautomerism	737
C. Tautomer Transformations in Side Chains	740
III. REARRANGEMENTS OF PHENOLS	743
A. *Cis–trans*-Isomerizations and Conformational Transformations	743
B. Phenol–Dienone Conversions	745
C. Hydroxy Group Migrations	748
D. Isomerizations of Alkylphenols	749
E. Isomerizations of Phenols Containing Unsaturated Side Chains	750
1. Double bond migration	750
2. Allylphenol–coumaran rearrangement	753
IV. REARRANGEMENTS OF O-SUBSTITUTED PHENOL DERIVATIVES	759
A. Rearrangements of Alkyl and Aryl Phenolic Ethers	759
B. Rearrangements of Allyl Aryl Ethers	761

The Chemistry of Phenols. Edited by Z. Rappoport
© 2003 John Wiley & Sons, Ltd ISBN: 0-471-49737-1

1. Claisen rearrangement................................ 761
2. Other isomerizations of allyloxy arenes 769
C. Rearrangements of Propargyl Aryl Ethers.................... 770
D. Rearrangements of Phenolic Esters 773
   1. Fries rearrangement 773
   2. Baker–Venkataraman rearrangement 778
   3. Anionic *ortho*-Fries rearrangement 778
V. REARRANGEMENTS OF FUNCTIONALIZED ARENES 799
A. Transformations of Peroxides 799
B. Isomerizations of *N*-Arylhydroxylamines 801
VI. REARRANGEMENTS OF NON-AROMATIC CARBOCYCLES 806
A. Dienone–Phenol Rearrangements 806
B. Rearrangements Involving Ring Expansion 816
C. Rearrangements Involving Ring Contraction 818
VII. REARRANGEMENTS OF HETEROCYCLIC COMPOUNDS 823
A. Five-membered Heterocycles 823
B. Six-membered Heterocycles 823
C. Transformations of Oxepines 824
VIII. REFERENCES 830

I. INTRODUCTION

Rearrangements involving phenols are no less various than the phenolic systems themselves. Indeed, any compound can be regarded as 'phenol' if an aromatic ring in its structure is connected directly to one or more hydroxy groups. The aromaticity of phenols is responsible for the fact that phenols turn out to be the end products of most rearrangements discussed here whereas the rearrangements of phenols themselves are comparatively rare.

The rearrangements to form phenolic systems have been known for a long time and are in essence the methods for synthesis of phenols. The literature concerning these methods is too voluminous to review in detail. Therefore, this chapter contains only a concise survey of these reactions which were described previously in many reviews. Most attention is devoted to the recently discovered or modified rearrangements, in which the phenols serve as reactants or are formed as isolable products, or are believed to participate as intermediates.

II. TAUTOMERISM IN PHENOLS

The tautomeric transformations of phenols can be subdivided into two groups: (i) keto–enol tautomerism which is accompanied by loss of the aromatic character of the ring, and (ii) tautomeric equilibrium involving participation of substituents where the aromatic phenol nucleus is conserved.

A. Keto–Enol Tautomerism in Phenols

The tautomerism of hydroxyarenes occupies a particular position among keto–enol tautomer transformations of various organic compounds because of the aforementioned loss of aromaticity. In contrast to carbonyl compounds (e.g. the keto form **1** is more energetically favored than the enol form **2** by 42 kJ mol$^{-1, 1a}$, equation 1), the phenols **3** are much more stable than their keto tautomers (**4** or **5**) because the energy gained by the **3 → 4** or **3 → 5** conversions is offset in plenty by the simultaneous large decrease in resonance energy (*ca* 151 kJ mol^{-1}) (equation 2)[1b,2].

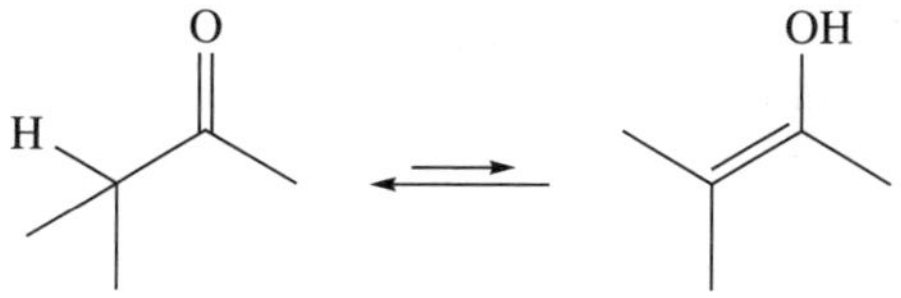

$$\underset{(1)}{\text{pivaldehyde-like keto}} \rightleftharpoons \underset{(2)}{\text{enol}} \quad (1)$$

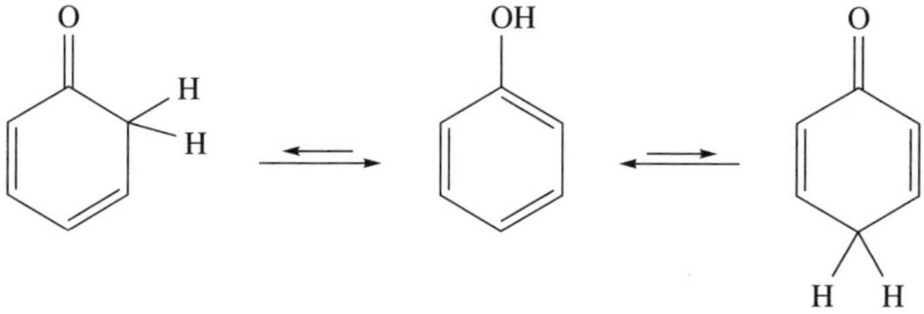

$$\underset{(4)}{\text{2,4-cyclohexadienone}} \rightleftharpoons \underset{(3)}{\text{phenol}} \rightleftharpoons \underset{(5)}{\text{2,5-cyclohexadienone}} \quad (2)$$

Therefore, the keto–enol tautomerism of phenols becomes significant only if there are additional factors which in one way or another reduce the difference between enolization and aromatic π-conjugation energies. These factors are: (i) an increase in the number of hydroxy groups leads to equalization of the aromatic conjugation energy to the total enolization energy of several carbonyl groups; (ii) one or more aryl rings annulated with the phenolic cycle decrease the total aromatic conjugation energy; (iii) electron-withdrawing substituents in the *ortho-* and *para-*positions of phenol give rise to a redistribution of the electron density in the system and result in lowering the aromatization energy; besides, the nature of the keto–enol tautomerism in this situation can change since the proton migrates to the electronegative substituent but not to the aromatic ring; (iv) bulky groups in the *ortho-*positions of phenol create a steric hindrance which stabilizes a quinoid structure[2]; (v) formation of phenolate anions as well as metal coordination facilitate the fixation of the keto form owing to delocalization of the negative charge into the aromatic ring.

The influence of these factors, either separately or together, was described in detail in several reviews[1-3].

1. Monocyclic phenols

More than 100 years ago Thiele[4] and Lapworth[5] put forward the hypothesis that the exclusive substitution of phenol at the *ortho-* and *para-*positions might be attributed to rapid equilibration of phenol **3** with the transient keto forms **4** and **5**. Since that time, the keto–enol equilibrium ratio in phenol itself has been estimated repeatedly and by application of various research methods. Thus, *ab initio* 6-31G* basis set calculations were recently carried out on the structures of phenol **3**, and its keto tautomers 2,4-cyclohexadienone **4** and 2,5-cyclohexadienone **5**[6]. Energy calculations were carried out by using the all-electron *ab initio* Hartree–Fock formalism (RHF) as well as 2nd-order Moller–Plesset formalism (MP2) on the RHF-optimized geometries. It was shown that phenol **3** is significantly more stable than dienones **4** and **5** by 47.4 and 42.5 kJ mol^{-1} (RHF) as well as 72.5 and 70.6 kJ mol^{-1} (MP2), respectively. An equilibrium constant '**3** ⇌ **4**' was estimated as 1.98×10^{-13}, i.e. in excellent agreement with experimental results as shown below.

The two keto tautomers of phenol **3**, i.e. **4** and **5**, were generated by flash photolysis of polycyclic precursors **6–8** in aqueous solution, and the pH–rate profiles of their **4** → **3** and **5** → **3** enolization reactions were measured[7]. The rates of the reverse reactions, **3** → **4** and **3** → **5**, were determined from the rates of acid-catalyzed hydrogen exchange at the *ortho*- and *para*-positions of phenol **3** (equation 3).

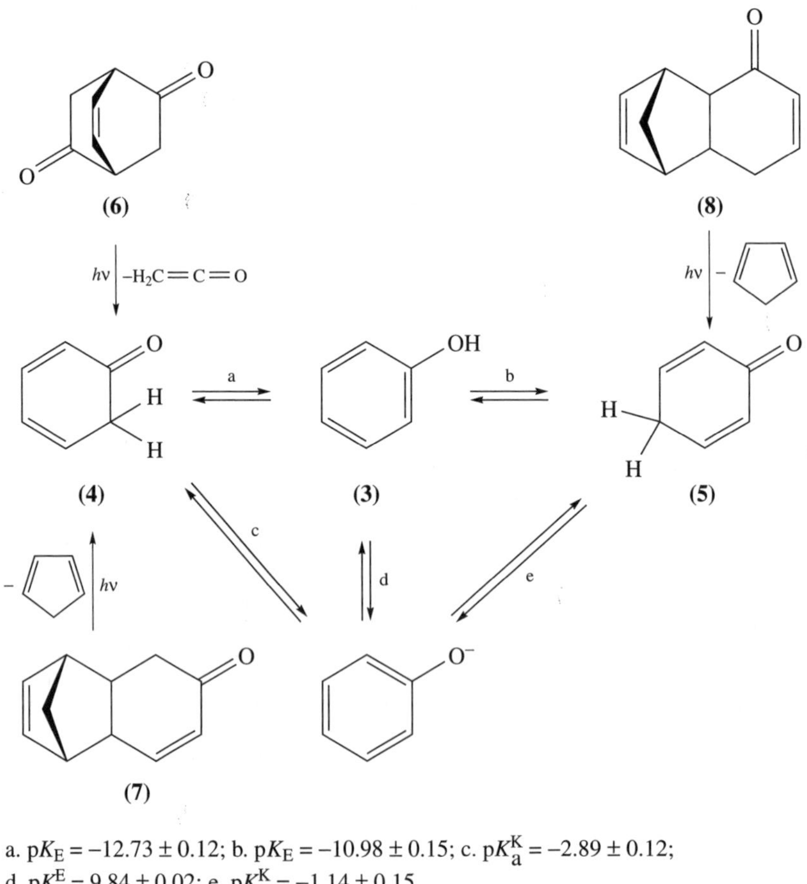

a. $pK_E = -12.73 \pm 0.12$; b. $pK_E = -10.98 \pm 0.15$; c. $pK_a^K = -2.89 \pm 0.12$; d. $pK_a^E = 9.84 \pm 0.02$; e. $pK_a^K = -1.14 \pm 0.15$

From enolization constants of the dienones **4** and **5** at 25 °C and the acidity constant of phenol **3** ($pK_a^E = 9.84 \pm 0.02$ at 298 K), the C–H-acidity constants of ketones **4** and **5** can be calculated ($pK_a^K = pK_E + pK_a^E$). It turns out that ketones **4** and **5** can be ranked among the strongest carbon acids[7] as shown in equation 3. They disappear by proton transfer to the solvent with lifetimes $\tau(\mathbf{4}) = 260$ μs and $\tau(\mathbf{5}) = 13$ ms, and they are insensitive to pH in the range from 3 to 10. The magnitude of the kinetic isotope effect was also assessed (see also Reference 8).

The keto tautomers of monohydric phenols are frequently invoked as reactive intermediates in many reactions, such as the Reimer–Tiemann[9] and Kolbe–Schmitt reactions[10], electrophilic substitution (e.g. bromination[5,11]) as well as the photo-Fries rearrangement (see Section IV.D and also Refs. 16–18 cited in Reference 7). In certain cases the keto

11. Tautomeric equilibria and rearrangements involving phenols

forms turn out to be the products of such reactions. For example, a strategy of a 'blocked tautomer' has been used as a method for the introduction of the angular methyl group[12] (equation 4).

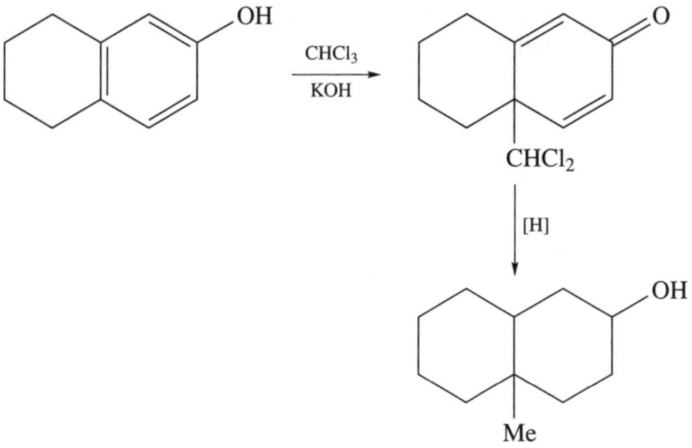

$$\text{(4)}$$

Such a 'phenol keto-tautomer equivalent strategy' was used for conjugate reduction of cyclic enones[13] (equation 5). The quinone monoketals **9** and *para*-quinol ethers **10** were used as precursors to keto-tautomer equivalents of substituted phenols, namely enones **11**, which were prepared by action of bis(2,6-di-*tert*-butyl-4-methylphenoxy)methylaluminium (MAD), followed by addition of lithiumtri-*sec*-butyl borohydride (L-Selectride). The enones **11** obtained are reasonably stable at a freezer temperature without aromatization[13].

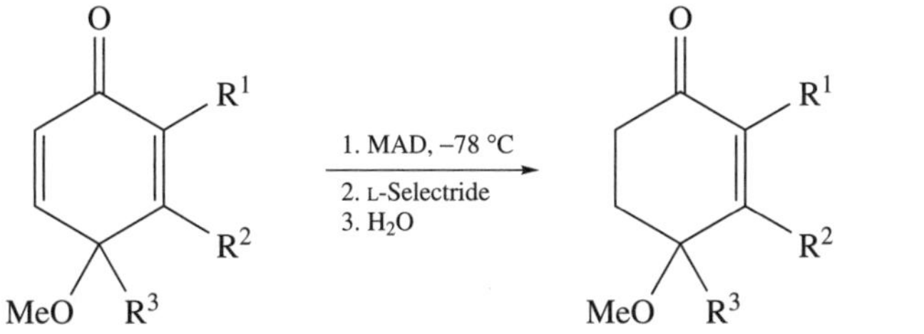

$$\text{(5)}$$

(9) R³ = OMe
(10) R³ = Bu, Ph, 4-MeC₆H₄

(11)
R¹ = H, Me; R² = H, Me, OMe

The influence of bulky *ortho*-substituents on the tautomerism of phenols can be illustrated by the recently reported generation and isolation of 4-alkoxy-2,6-di-*tert*-butylcyclohexa-2,5-dienones **13**. They were generated efficiently by the Ag ion mediated reaction of 4-bromocyclohexa-2,5-dienone **12** with simple alcohols (equation 6). All the dienones **13** were proved to be very susceptible to a prototropic rearrangement to form the phenols **14** under catalysis with bases, acids or SiO₂[14].

The introduction of additional hydroxy groups into the phenolic ring assists the development of a ketonic character because the energy released by formation of multiple keto groups compensates for the loss of resonance stabilization. There are many reports concerning the ability of polyhydric phenols to react as tautomeric keto forms[2]. For instance, the conversion of phenol into aniline proceeds under very drastic conditions (350–450 °C, 50–60 bar) and the substitution of one hydroxy group in resorcinol by an amino group occurs quite readily at 200 °C, whereas phloroglucinol gives 3,5-dihydroxyaniline and 3,5-diaminophenol in almost quantitative yield under very mild conditions (long storage at room temperature with ethanolic solution of ammonia)[1b,2]. Phloroglucinol **15** is the most typical example of tautomerism in polyhydric phenols. Thus, **15** reacts with hydroxylamine to produce trioxime **16**[15] and gives hexamethylcyclohexane-1,3,5-trione **17** in reaction with excess of methyl iodide[16] (equation 7).

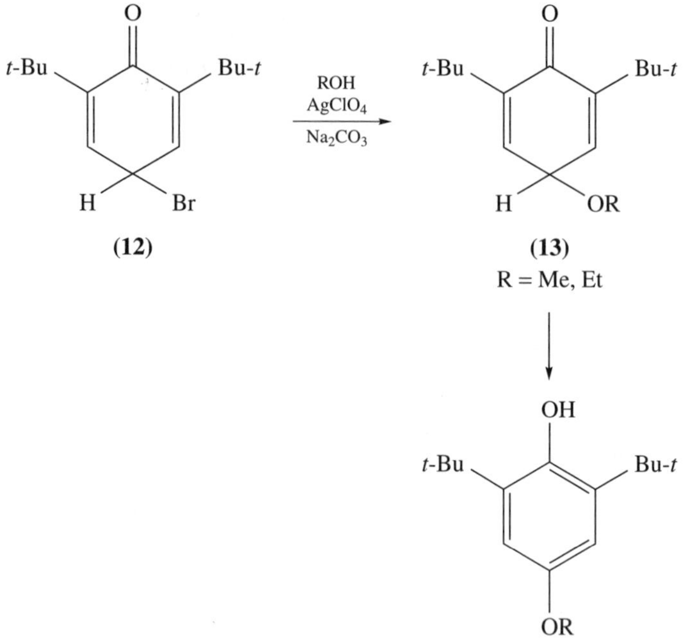

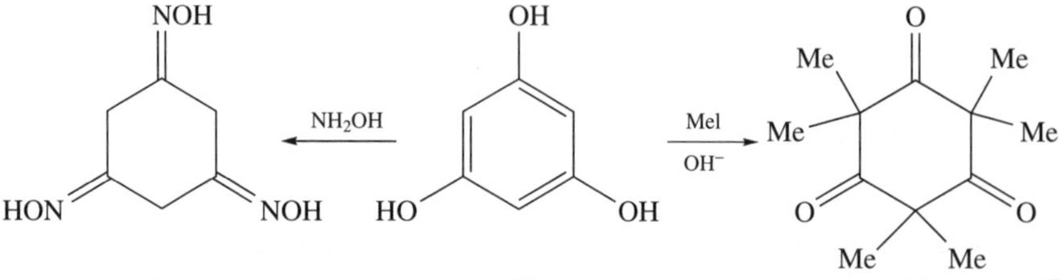

It was shown by all-electron *ab initio* Hartree–Fock (RHF) calculations that the enolic form, i.e. 1,3,5-benzenetriol **15**, is by far more stable than the keto form, i.e. 1,3,5-cyclohexanetrione[17]. On the other hand, the latter is more abundant in the phloroglucinol system than is the keto form of phenol (i.e. 2,4-cyclohexadien-1-one **4**) in the phenol system. Nevertheless, the keto form of phloroglucinol cannot be observed by spectral methods both in solutions and in the solid state. It is now thought that phloroglucinol behaves like a polyketone due to tautomeric transformations of its anions **18** and **19** (equation 8).

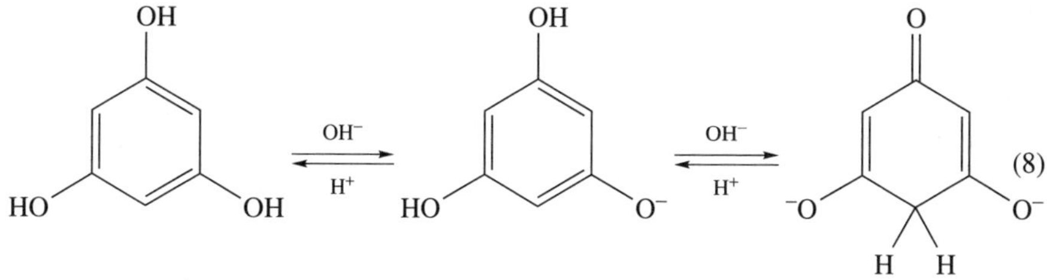

$$(15) \xrightleftharpoons[H^+]{OH^-} (18) \xrightleftharpoons[H^+]{OH^-} (19) \quad (8)$$

The existence of the dianion **19** was proved by means of NMR spectroscopy[18]. Thus, the ¹H NMR spectrum of phloroglucinol in aqueous solution contains a single resonance of aromatic protons (δ 6.05) which shows a small shift to high field (δ 6.02) after addition of one mole of an alkali. However, the addition of a second mole of alkali results in the disappearance of the aromatic proton signal. Instead of this, olefinic proton signals (δ 5.03) as well as signals of the methylene protons (δ 3.0) appear.

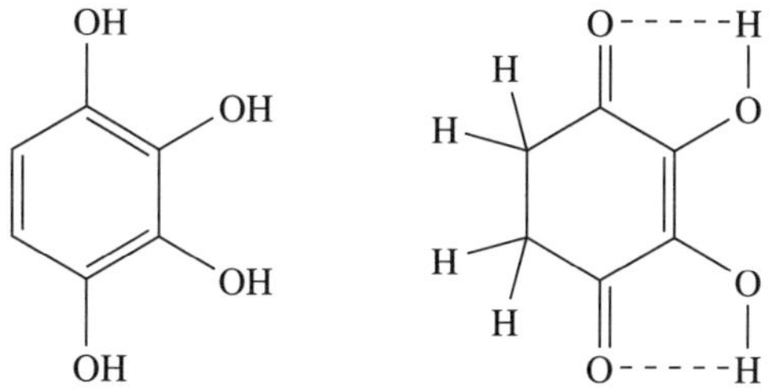

(20) (21)

A polyphenol such as 1,2,3,4-tetrahydroxybenzene can exist in the two isomeric forms **20** and **21**[19]. An addition of acid to an alkaline solution of phenol **20** results in the formation of the solid diketo form **21** that is stable at room temperature owing to intramolecular hydrogen bonds. The aromatization **21** → **20** occurs only by heating of 2,3-dihydroxycyclohex-2-en-1,4-dione **21** in acidic solution.

Consequently, it can be concluded that the tautomerism does not involve a mobile equilibrium in the series of monocyclic phenols containing one to four hydroxy groups.

2. Polycyclic phenols

The tautomeric properties of hydroxynaphthalenes show in the most unambiguous manner that the naphthalene system is less aromatic than that of benzene. The benzoannelation appreciably destabilizes the aromatic tautomers not only among phenols but also in the arene series[20]. Therefore, even the monohydroxy naphthalenes display in their chemical reactions properties typical for the tautomeric keto form.

However, a real tautomerism in monohydric phenols appears only if the phenolic ring is fused with at least two aromatic rings. Thus, 9-hydroxyanthracene (anthrol) **22** undergoes a reversible conversion into ketone **23** (anthrone) in which two separate aromatic rings are conjugated with a carbonyl group. This conjugation stabilizes very much the keto form (equation 9)[1b,2]. The keto tautomer becomes increasingly stable in the higher polycyclic phenols. For example, the keto form of hydroxynaphthacene **24** shows very little tendency for enolization, whereas in the pentacene series **25** the phenolic forms are unknown. On the other hand, another isomer of hydroxynaphthacene exists as the two separable forms **26** and **27** (equation 10)[1b,2].

(22) **(23)** (9)

(24)

(25)

(10)

(26) **(27)**

(28) **(29)** **(30)** (11)

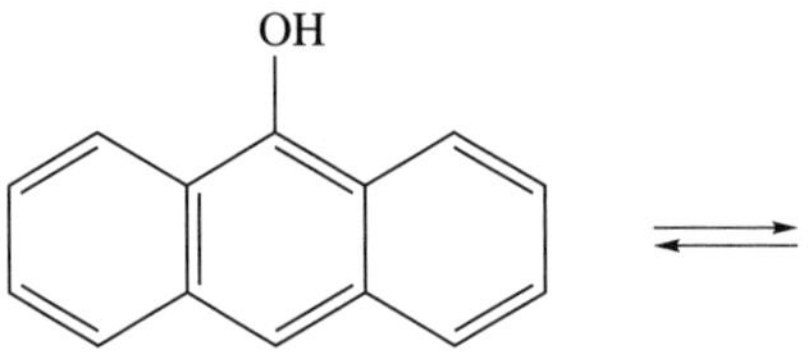

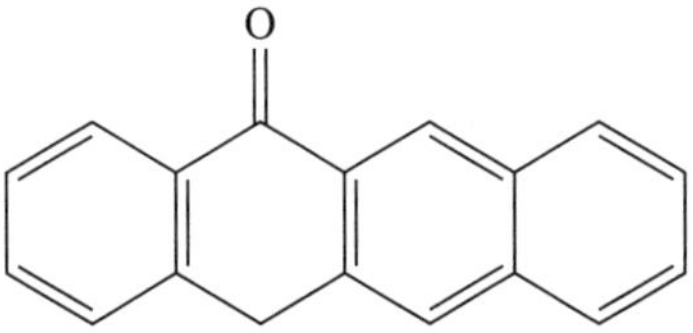

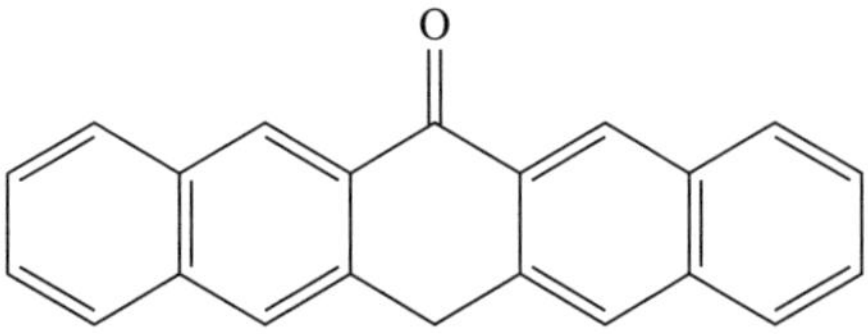

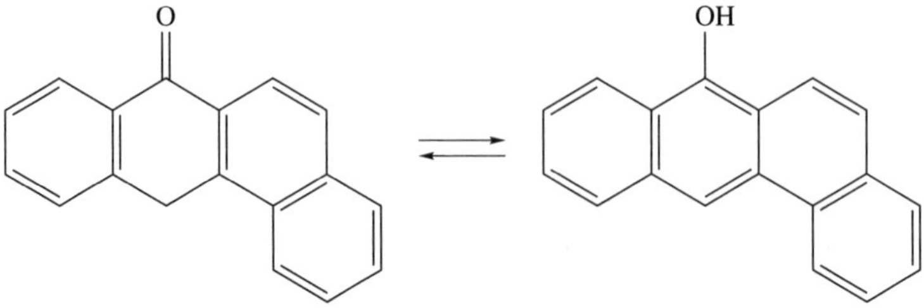

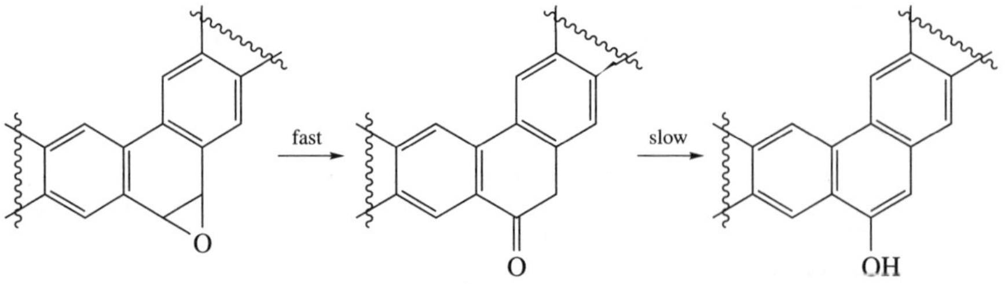

11. Tautomeric equilibria and rearrangements involving phenols

It was shown recently that K-region* arene oxides can rearrange to phenols in two steps: (i) rapid rearrangement of the arene oxide **28** forming the positionally isomeric keto tautomers **29** of the K-region phenols **30**, followed by slow enolization to **30**[22] (equation 11).

The kinetic characteristics were measured for the rearrangements of arene oxides of benzo[a]anthracene, its methyl-substituted derivatives as well as for other polycyclic arene oxides (for transformations of arene oxides into phenols, see Section VII.C). The mechanism of these acid-catalyzed rearrangements and the isotope effects in these reactions were discussed[22].

Very interesting tautomeric properties are inherent in polycyclic systems that contain annulated phenol and quinone rings. The simplest model for these compounds is naphthazarin **32** which can exist, both in solution and in the solid state, as a fast equilibrium mixture of several tautomers (**32a–32c**) where forms **32a** and **32b** (i.e. a degenerate tautomeric pair of identical 1,4-diones) predominate (equation 12).

In contrast to keto–enol tautomerism, such enol–enol tautomerism is characterized by extremely rapid hydrogen transfers. It was shown by *ab initio* calculations[23,24] that structures **32a** and **32b** are more stable than the degenerate tautomeric forms **32c** and **32e** by 104.7 kJ mol^{-1} as well as by 117 kJ mol^{-1} than symmetric structure **32d**. According to these calculations, a synchronous tunneling of two protons must occur in the naphthazarine molecule **32** between the identical structures **32a** and **32b** with a frequency of 20 to 40 MHz, i.e. approximately 10^{11} to 10^{13} migrations of hydrogen from one oxygen atom to another per second take place.

Related systems to the naphthazarines are perylenequinones **33**, which are biologically active pigments obtainable from natural sources. These compounds are of interest not only because of their peculiar structure features, but also owing to their photodynamic activity.

The keto–enol tautomerism of the dihydroxy perylenequinones **33a–d** was studied by ^{1}H, ^{2}H and ^{13}C NMR spectroscopy[25,26] (equation 13). The most important factors determining the tautomeric equilibrium in these helix-shaped systems are the substituent effects, the strength of intramolecular phenol–quinone hydrogen bonds, the distortion from planarity of the perylenequinone structure and solvation as well as aggregation effects.

3. Phenols bearing nitrogen-containing substituents

a. Nitrosophenol–quinone oxime tautomerism. The introduction of electron-withdrawing substituents into the *ortho*- and *para*-positions of phenol results in reducing the

---

* The terminology '*K-region*', '*non-K-region*' and '*bay-region* arene oxide' can be illustrated by reference to the phenanthrene ring **31**[21]. The addition of an oxygen atom to the C=C double bonds gives: (a) K-region, (b) non-K-region and (c) bay-region arene oxides.

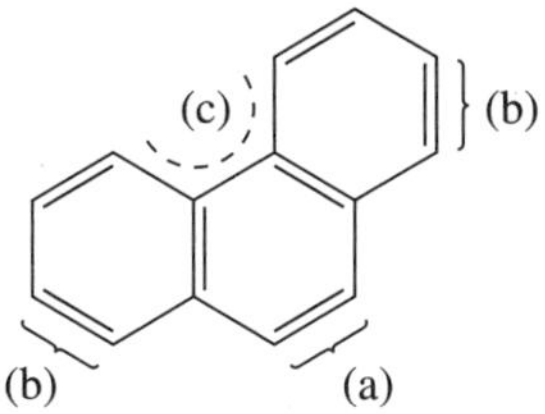

(31)

energy of aromatic conjugation and in a strong polarization of the oxygen–hydrogen bond in the hydroxy group due to redistribution of the electron density within the molecule. These changes facilitate the dissociation of the O–H bond and promote the appearance of a keto–enol tautomerism. The above-named effects are most typical for nitrosophenol **34** since the nitroso group possesses the greatest negative conjugative effect[2]. Besides, it is able to add a proton by rearrangement to form an oxime moiety (equation 14).

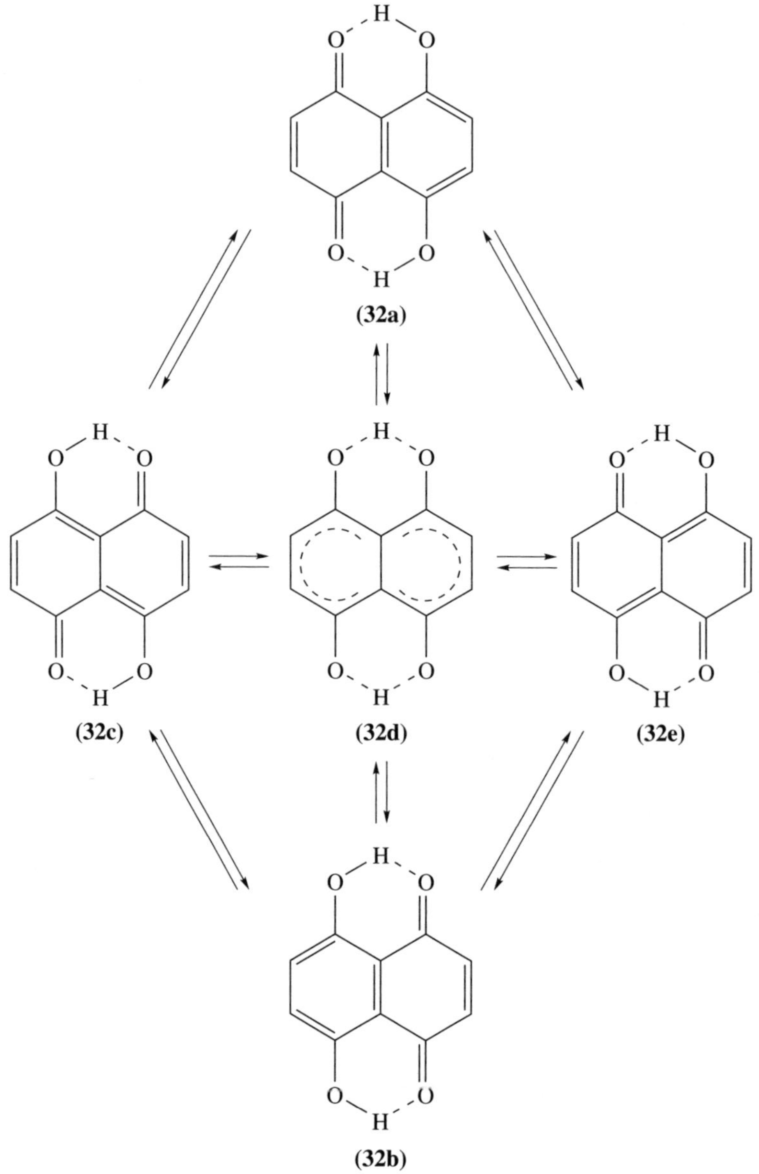

(12)

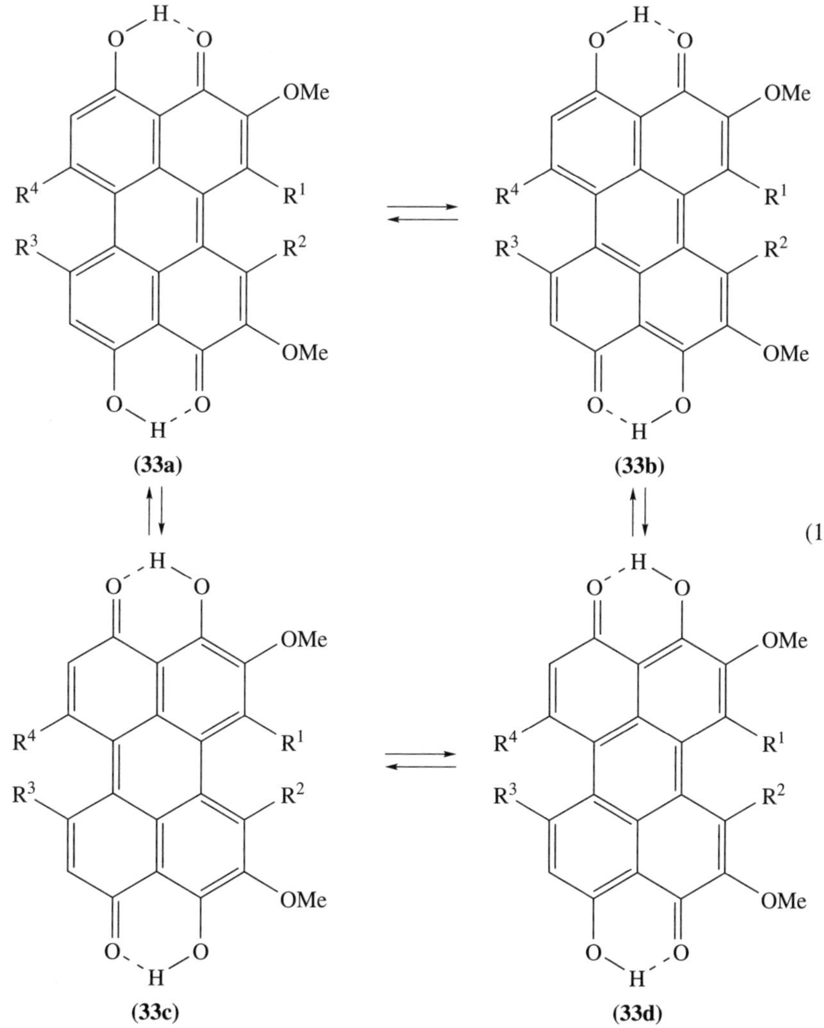

11. Tautomeric equilibria and rearrangements involving phenols

(33a) ⇌ (33b)

⇅ ⇅   (13)

(33c) ⇌ (33d)

$R^1 = R^2 = MeCH(OH)CH_2,$ $R^3 = R^4 = MeO; R^3R^4 = OCH_2O$

$R^1R^2 =$ (COMe, COMe) ; (COMe, MeCH(OH)) ; (MeCH(OH), MeCH(OH)) ; (CMe(OH)COMe, Me)

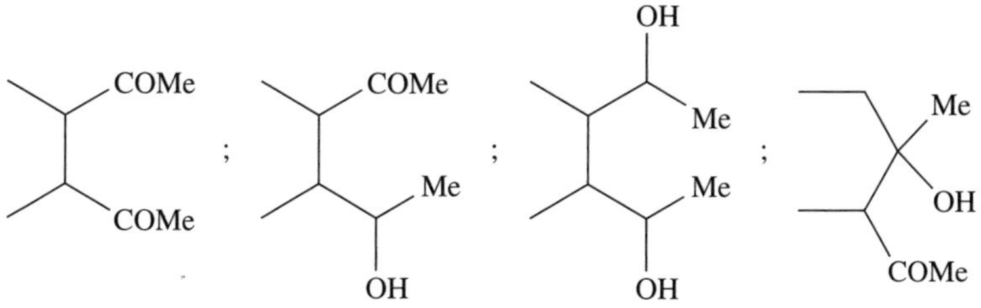

The tautomerism in nitrosophenols has been reviewed in detail[2,3].

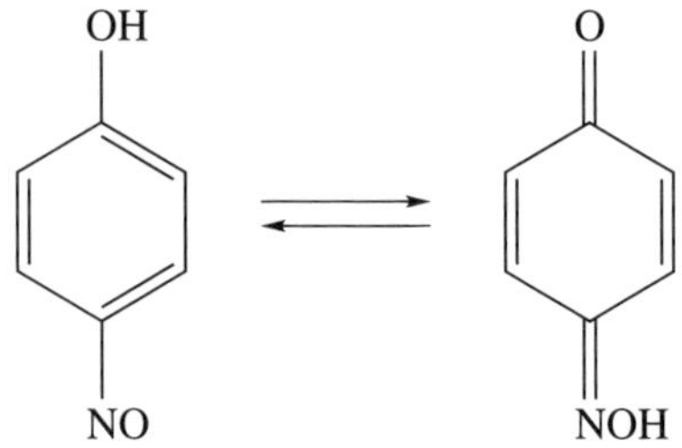

$$\text{(34)} \quad \rightleftharpoons \quad \tag{14}$$

b. Arylazophenol–quinone arylhydrazone tautomerism. The tautomeric equilibrium between *para*-arylazophenols **35** and *para*-quinone arylhydrazones **36** has been investigated extensively using phenol, anthranol and naphthol derivatives (equation 15). The results obtained were summarized in several reviews[27].

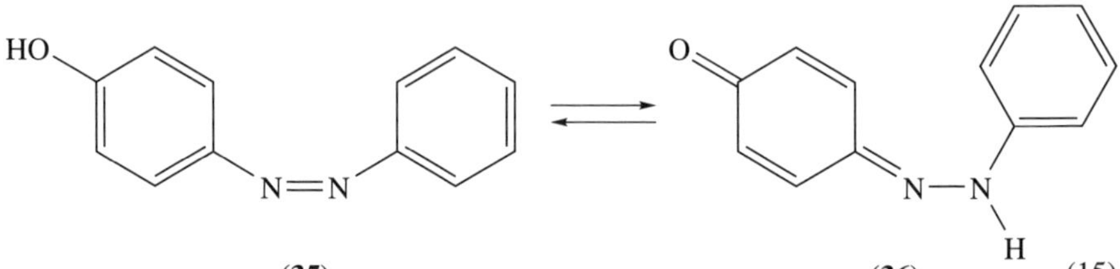

$$\text{(35)} \quad \rightleftharpoons \quad \text{(36)} \tag{15}$$

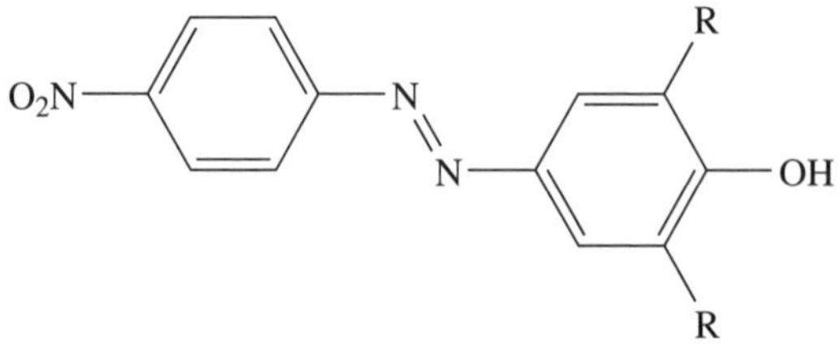

(37) R = OMe, CH$_2$OCH$_3$,

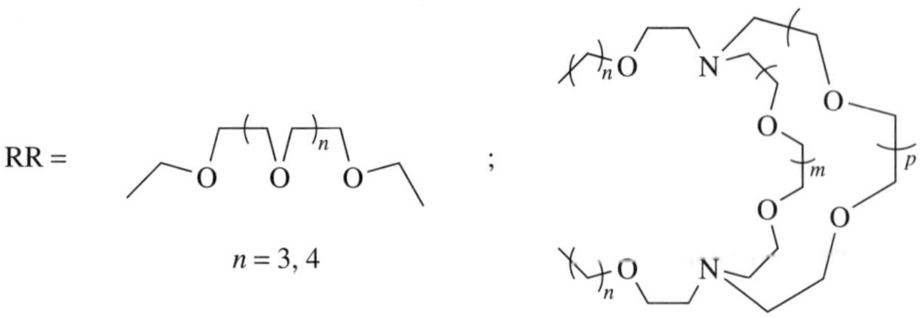

RR = ... ; n = 3, 4

n = 0, 1; m = 1, 2; p = 0, 1, 2

11. Tautomeric equilibria and rearrangements involving phenols 725

The phenylazophenol—quinone phenylhydrazone tautomerism in a series of azobenzene derivatives **37** was investigated by UV-visible spectroscopy[27]. The results revealed a high sensitivity of this tautomerism to substituent variation. It was found that system **37** exists in the azo form only for R = CH_2OCH_3, but in the hydrazone form only for R = CH_3O. The change between these compounds may be attributed to intramolecular hydrogen bonding of the phenolic group (**37**, R = CH_2OCH_3) with the more basic oxygens of the ether groups.

The UV, IR and NMR spectroscopy methods were used to investigate the tautomeric equilibrium of the benzoylhydrazones **38** in solvents having different polarity[28]. Compounds **38** exhibited 1,3-, 1,7- and 1,9-prototropic shifts. The fraction of azophenols **38a** increased on increasing the solvent polarity and the redox potential of the quinoid form. These compounds can exist as the three tautomeric structures **38a–c**. It was shown that the azo form (**38a**) is absent when the substituents R^1 and R^2 are isopropyl or *tert*-butyl, i.e. the keto forms are relatively stabilized by the presence of bulky *ortho*-substituents[28].

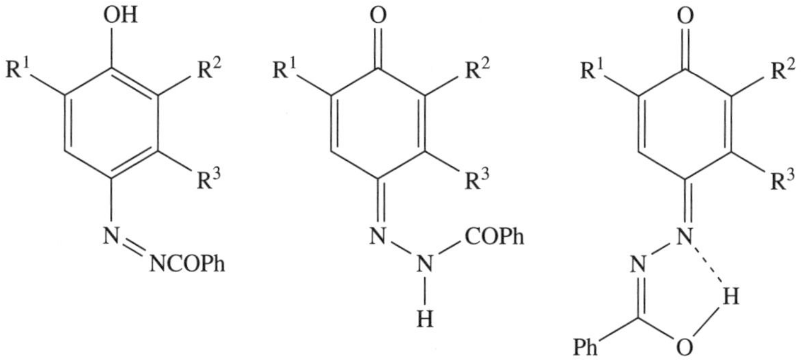

(**38a**) (**38b**) (**38c**)

R^1, R^2 = H, Me, *i*-Pr, *t*-Bu, Br; R^3 = H, Me

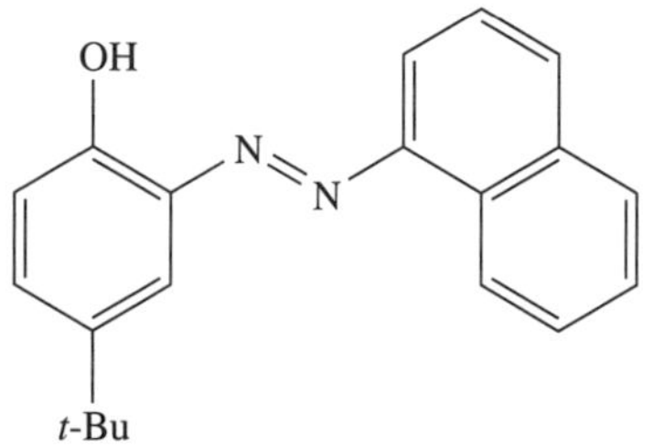

(**39**) 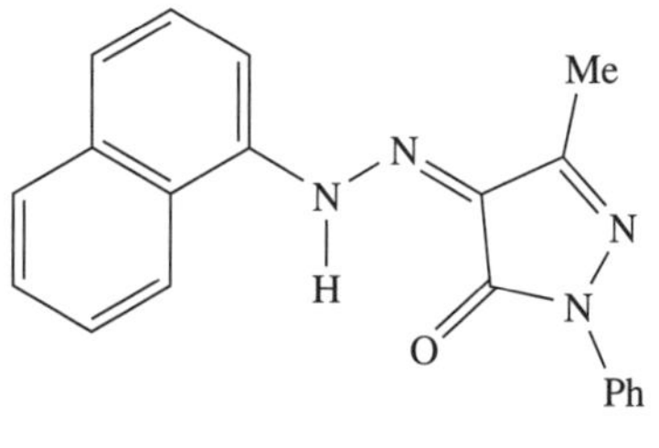 (**40**)

An ^1H, ^{13}C and ^{15}N spectral investigation of four 1-naphthylazo compounds **39–42**, which were prepared by coupling 1-naphthalenediazonium chloride with the appropriate passive components, was reported[29].

It was found that compounds **39** and **42** exist almost completely as the tautomeric azo forms whereas compound **40** is completely in the hydrazone structure, and compound **41** exists predominantly in the hydrazone form. In that way, the annelation of the benzene ring in the active component has, contrary to the annelation of the benzene ring in the passive component, practically negligible influence on the azo–hydrazone equilibrium[29].

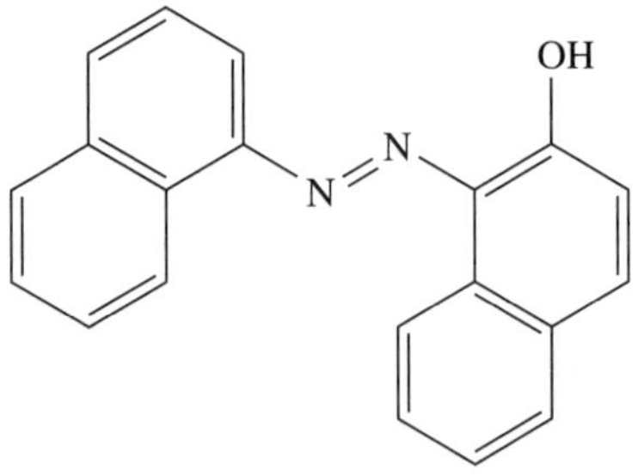

(41) 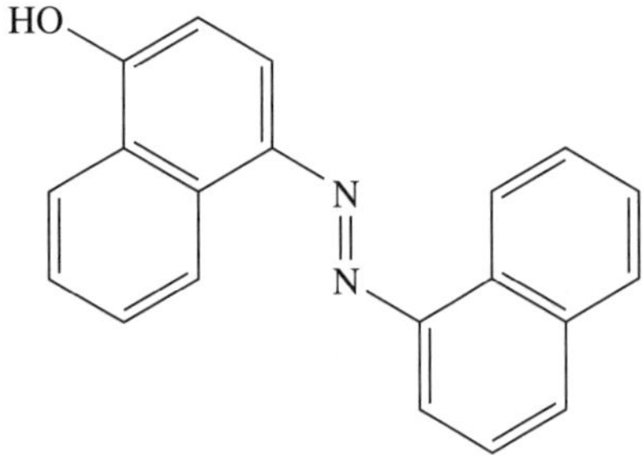 **(42)**

c. Tautomerism in Schiff bases. The Schiff bases formed by condensation of aromatic amines and aromatic aldehydes containing an *ortho*-hydroxy group can exist in two tautomeric forms, namely the phenol-imine **43** and the keto-enamine **44** (equation 16). For adducts formed from anilines and salicylaldehyde, the keto tautomer is found to be highly disfavored owing to the loss of aromaticity. However, in the case of Schiff bases formed from 2-hydroxynaphthaldehyde, the keto–enamine tautomer **44** is present to a significant extent in a rapid exchange equilibrium with the phenol-imine structure **43**. These conclusions were drawn from results of ^{1}H, ^{13}C and ^{15}N NMR spectroscopy[30–33].

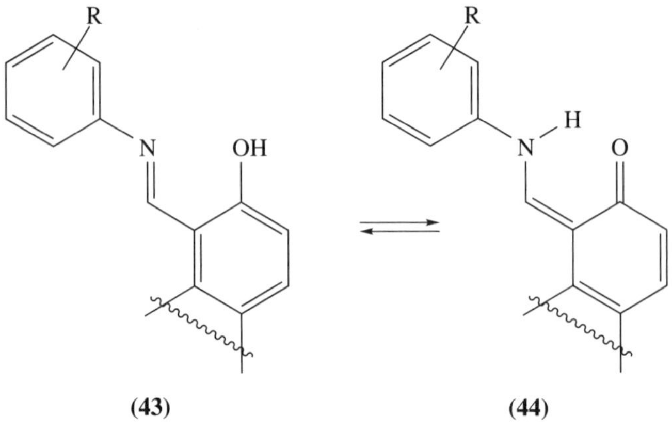

(43) **(44)** (16)

The equilibrium in the case of Schiff bases prepared from salicylaldehyde and 2-amino-, 2,3-diamino-, 2,6-diamino- and 3-aminomethylpyridine was studied by means of NMR, UV and IR spectroscopy and X-ray crystallography[34,35]. It was shown that the enol-imines were the predominant form in non-polar solvents, whereas in polar solvents a rapid tautomeric interconversion between the enol-imines and the keto-enamines as well as a slow hydrolysis were observed. The tendency to tautomeric interconversion was significant for the 2-(3-pyridylmethyliminomethyl)phenol **45** while in the case of other Schiff bases it was very low.

Because of the contradictory literature reports, the physical and spectral properties of *N*-salicylidene-1,2-diaminobenzene **46** were reinvestigated[36]. In the solid state compound **46** exists as a phenol–imine tautomer, wherein the phenolic hydrogen atom is hydrogen-bonded to the imine nitrogen atom.

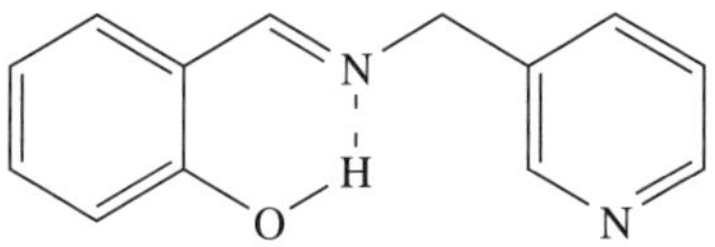

(45) 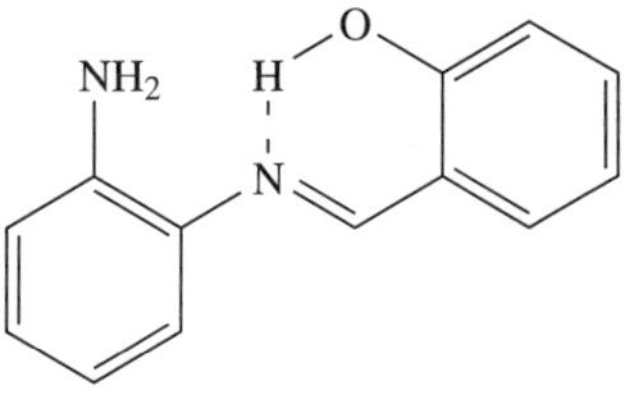 (46)

The introduction of a second hydroxy group into an *ortho*-position of the phenolic fragment in the Schiff base influences significantly the tautomeric equilibrium. Thus, in the series of *N*-(2,3-dihydroxybenzylidene)amine derivatives **47a–e** all the compounds are characterized by the presence of a strong intramolecular O−H···N bond which determines the formation of a six-membered pseudocycle. Except for compound **47b**, all the molecules are associated as dimers with two intermolecular O−H···OH bonds which are included into a ten-membered pseudocycle[37,38]. In contrast to the *N*-(2-hydroxybenzylidene)amines for which the phenolic tautomer prevails considerably, in compounds **47** the quinonic form is present in significant amounts and is dominant even for compound **47d**.

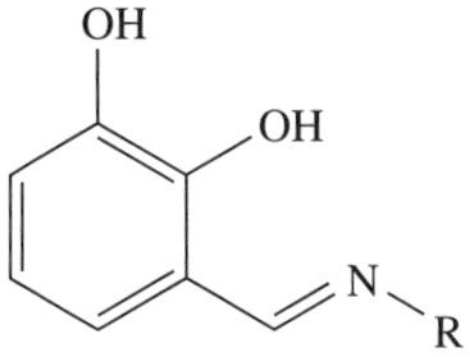

(47)

(a) R = Ph; (b) R = 4-MeC$_6$H$_4$; (c) R = 2-ClC$_6$H$_4$;
(d) R = *i*-Pr; (e) R = cyclopropyl

The tautomeric equilibrium between the phenol imine structure (OH···N form) and the keto-enamine structure (O···HN form) was determined by UV-Vis spectroscopy in polar solvents for the bis(crown ether) ligands **48** which contain recognition sites for Na and Ni guest cations[39].

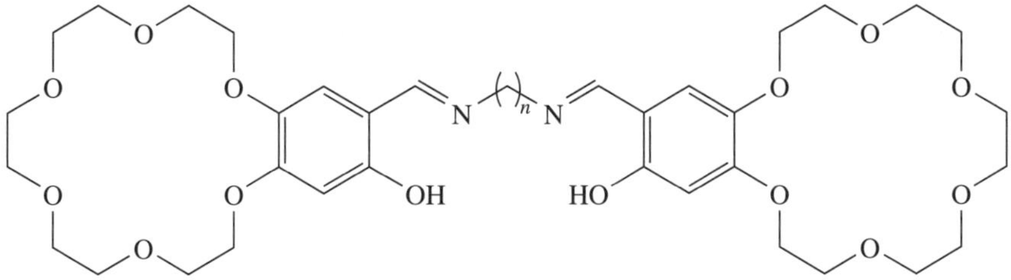

(48) *n* = 2–4

The azomethines considered above have a hydroxy group which is attached to an arylcarbaldehyde fragment. At the same time the keto–enol tautomerism was reported also for systems containing the hydroxy group in the arylamine fragment.

Thus, 4-[(4-dimethylamino)phenyl)imino]-2,5-cyclohexadien-1-one (DIA), also known as Phenol Blue, is a merocyanine dye that exists in two extreme resonance hybrids of a keto and a phenolate form (**49, 50**). Hybrid **49** is expected to contribute more in the solid state, whereas hybrid **50**, owing to its larger dipole moment, is believed to contribute more in polar solvents[40].

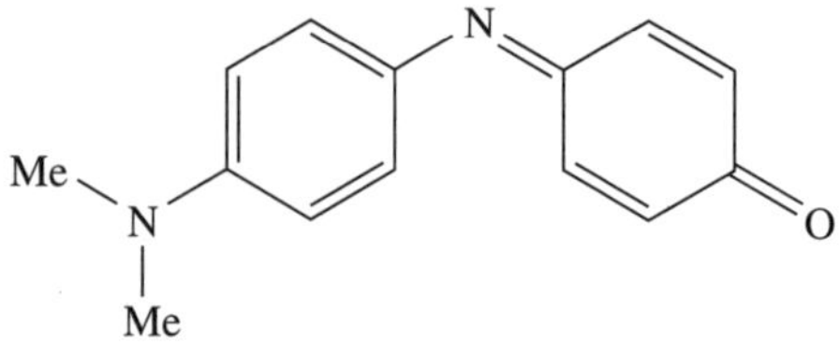

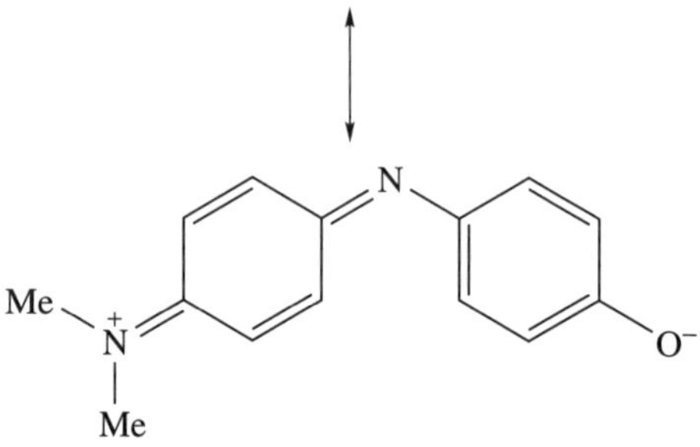

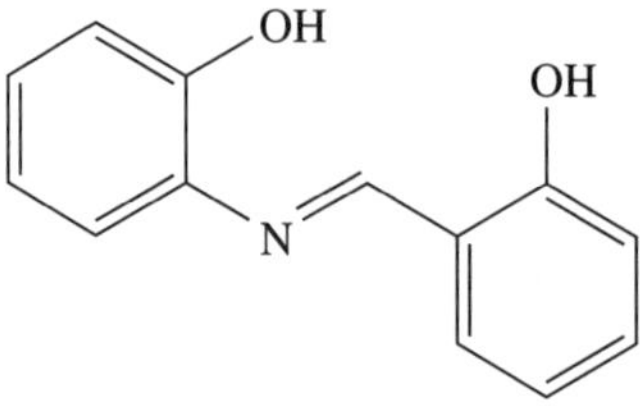

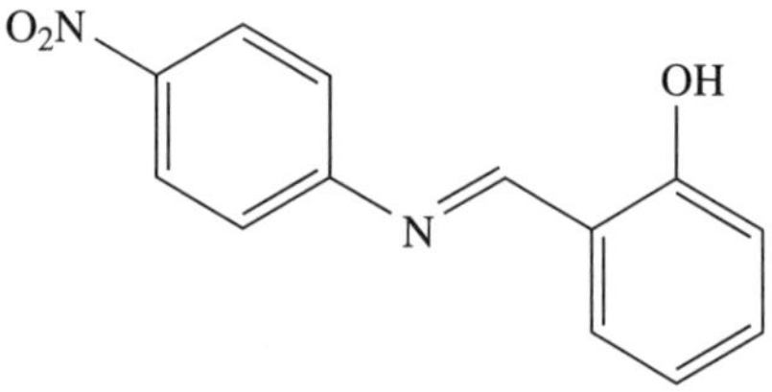

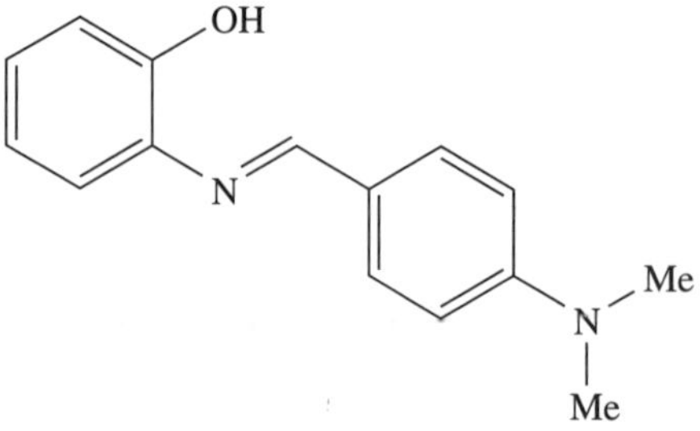

11. Tautomeric equilibria and rearrangements involving phenols

However, many reports in this field describe the intramolecular hydrogen bonding and tautomerism in Schiff bases bearing the hydroxy groups in both fragments.

The ^{13}C cross-polarization magic-angle-spinning NMR spectra of three structures (**51–53**) have shown the keto–hydroxy tautomerism in compound **51** but not in **52** and **53**. This was confirmed by a single-crystal X-ray diffraction study of compound **51**. The results revealed that the distinct molecules in the unit cell are linked by intermolecular hydrogen bonds[41].

A series of substituted salicylaldimines **54** was prepared by the condensation of various hydroxy and methoxy salicylaldehydes and 2,6-di-*tert*-butyl-4-aminophenol. It was shown by UV-Vis and ^{1}H NMR spectroscopy investigations that compounds **54** exist in solutions both in the phenol-imine and keto-enamine tautomeric forms[42].

(**54**)

R = 3-OH, 3-OMe, 4-OH, 4-OMe, 5-OH, 5-OMe, 4,6-(OH)$_2$

It should be noted that the keto-enamine tautomers of Schiff bases are observed always when the latter are derived from 2-hydroxynaphthaldehyde and aniline. However, in Schiff bases derived from salicylaldehyde and aniline, the new band at >400 nm in UV-Vis spectra was not observed in both polar and non-polar solvents, but it appeared in acidic media[43]. In this work[43] which contains a quite good survey of the investigations of tautomerism in Schiff bases, the effects of the solvent polarity and acidic media on the phenol-imine ⇌ keto-amine tautomeric equilibrium in systems **55** and **56** were reported. It was shown by ^{1}H NMR and UV-Vis spectra that compound **55** is in tautomeric equilibrium of structures **55a** and **55b** in both polar and non-polar solvents (equation 17), whereas the tautomer **56b** was not observed for compound **56** (equation 18)[43].

The tautomerism and photochromism of 2-[(2-hydroxyphenyl)aminomethylene]-2*H*-benzo[*b*]thiophenone **57** and its acetyl derivatives were studied by UV-Vis spectroscopy[44]. The acylation of compound **57** with Ac$_2$O affords under different conditions the mono-acetyl (**58**) and diacetyl (**59**) derivatives (equation 19). It was found that a mobile equilibrium of three forms takes place in solutions of compound **58** (equation 20). The equilibrium is shifted to the left (i.e. to form **58**) in solvents of low polarity (hydrocarbons, ethers, acetone, acetonitrile) while polar solvents such as DMF, DMSO or HMPA stabilize the more acidic form **60**. The latter undergo a rearrangement upon irradiation by sunlight (equation 21)[44].

Compounds showing excited-state intramolecular proton transfer (ESIPT) were proposed as efficient materials to protect against UV radiation damage and to store information at the molecular level. The ESIPT involves the intramolecular transfer of proton from a hydroxy or amino group to an accepting site on the molecule such as carbonyl oxygen or another nitrogen while the molecule is in the excited state. Among these compounds are 2-(2′-hydroxyphenyl)imidazole (**61**) and (2′-hydroxyphenyl)benzimidazole

(**62**) derivatives[45] (equation 22). Quantum-chemical calculations of the phototautomerization in these and related systems were carried out recently[46].

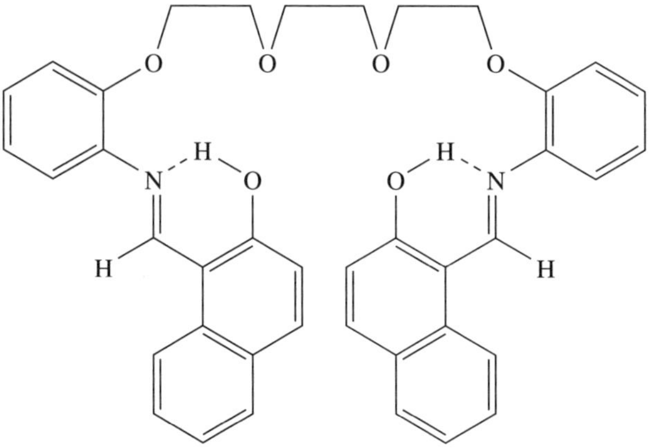

(**55a**)

⇅ (17)

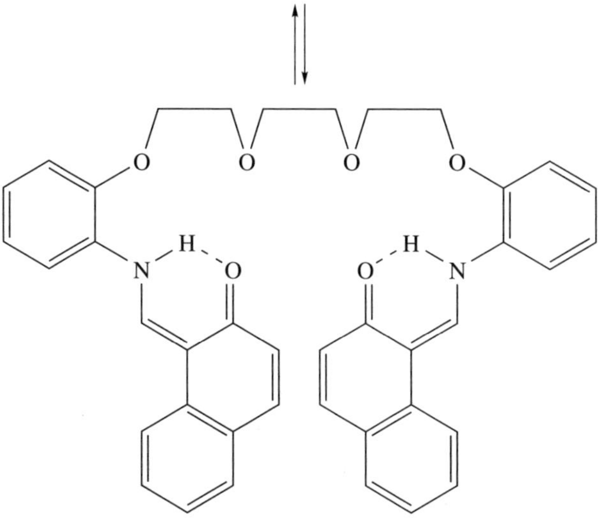

(**55b**)

Analogous ESIPT properties in competition with ESICT (excited-state intramolecular charge transfer) were observed in the pyrazole series (**63**), in which the spectral characteristics can be fine-tuned by substituent variations as well as by solvent effects (equation 23)[47].

The 4-aminopyrimidinoanthrones **64** were shown to have the amino-ketone structure in the crystal state and in neutral organic solvents, whereas in acidic or basic media the tautomeric equilibrium was shifted toward the ionic forms of the imino-phenol structure[48] (equation 24).

11. Tautomeric equilibria and rearrangements involving phenols

The tautomerism in phenols containing other substituents (CHO, COR, CH=CHCOOH etc.) were described in detail in another review[2].

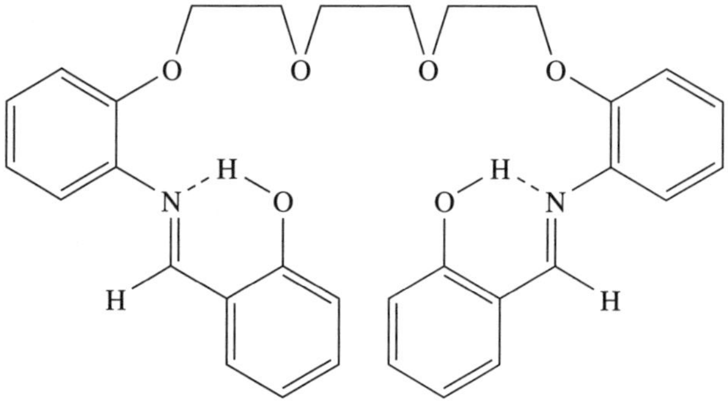

(56a)

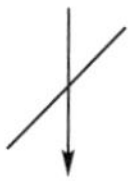

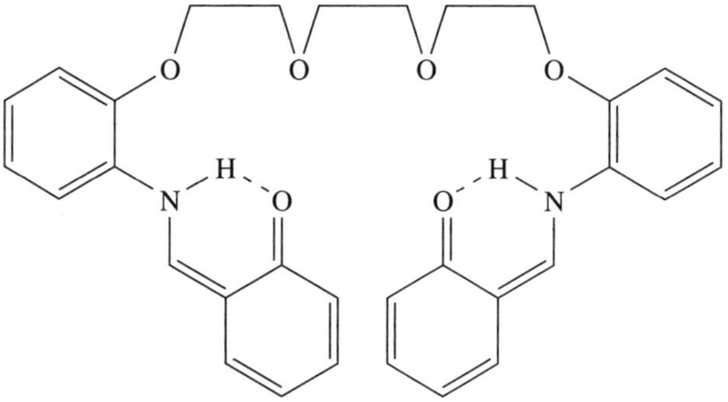

(56b)

(18)

4. Metal-coordinated phenols

The coordination of transition metals is known to influence the keto–enol tautomerism in the condensed phase[49]. The effect of coordination of bare Fe^+ ions on the keto–enol equilibrium of phenol was investigated by means of generation of various cyclic $[Fe, C_6, H_6, O]^+$-isomers. These isomers were characterized by collisional activation (CA) and Fourier transform ion cyclotron resonance (FTICR) mass spectrometry[49]. It was shown that the energy difference between the phenol–iron complex **65** and the keto isomer **66** is not perturbed by the presence of the iron cation in comparison with the uncomplexed isomers **3** and **4** (equation 25). Thus, the energy difference for both the neutral and the Fe^+-coordinated systems amounts to ca 30 kJ mol^{-1} in favor of the phenolic tautomer.

Furthermore, it was also found that the dissociation of the Fe^+-complex of the valence tautomers benzene oxide $\rightleftarrows$ oxepin proceeds via a [phenol-Fe^+] complex **65** rather than via the [2,4-cyclohexadien-1-one-Fe^+] species **66** (see also Section VII.C).

The effect of η^2 coordination on the arenes was studied in the context of the phenol–ketodiene equilibrium[50]. It was shown that this equilibrium for the free ligands favors heavily the phenol tautomer (*vide supra*) whereas for the complexes [Os(NH$_3$)$_5$-2,3-η^2-arene)]$^{2+}$ (arene = phenol; 2-, 3-, 4-methylphenol; 3,4-dimethylphenol) the corresponding equilibrium constants approach unity (20 °C). The conversion of phenol **67** into the 2,4-cyclohexadien-1-one **68** was kinetically favored over the formation of the 2,5-isomer **69**, although the latter is the thermodynamically favored product (equation 26). It was assumed that osmium rehybridizes the C(5) and C(6) atoms to form a metallocyclopropane. This removes much of the resonance energy and therefore destabilizes the enolic form of the free ligand. The free energies of ketonization (25 °C) for the η^2-phenol complex in comparison with free phenol are shown in equations 27 and 28[50].

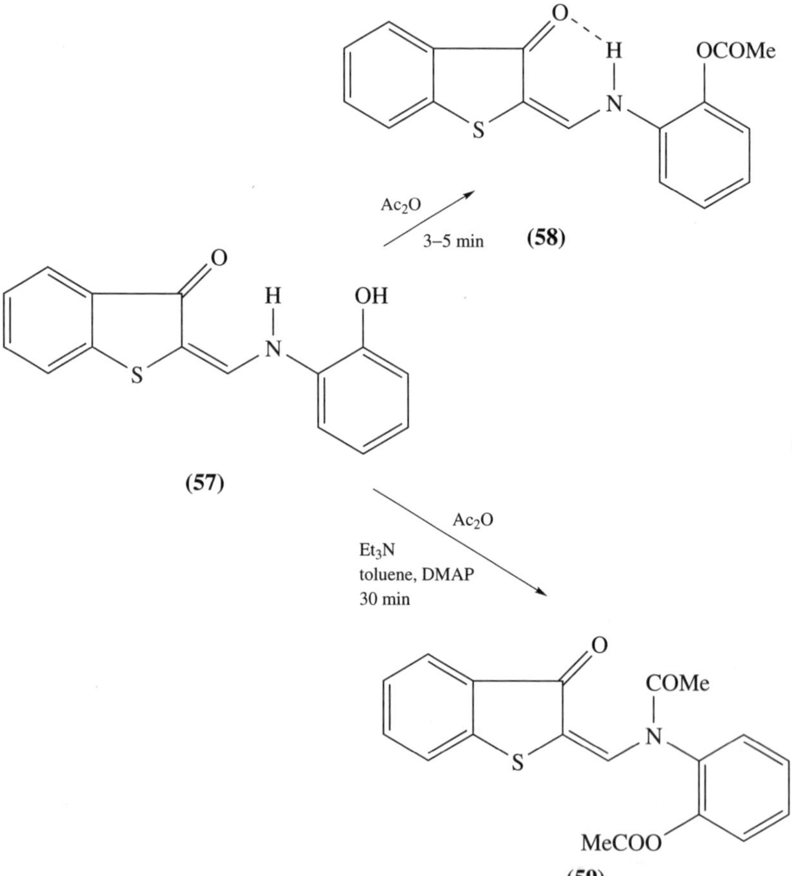

(19)

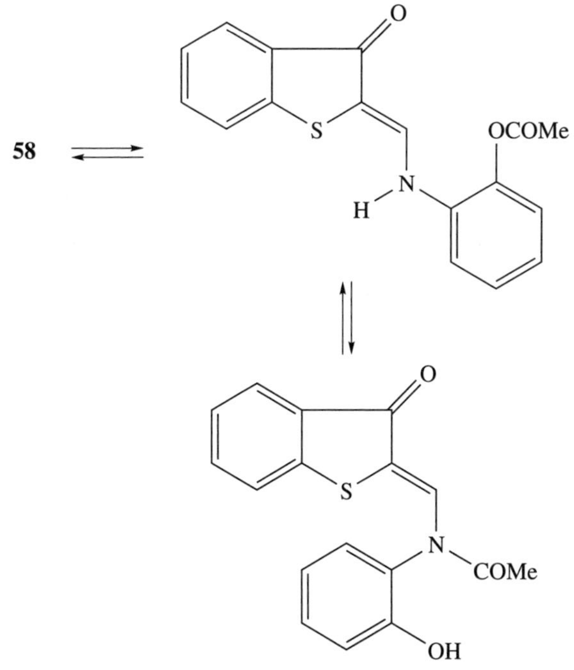

11. Tautomeric equilibria and rearrangements involving phenols

(20)

(60)

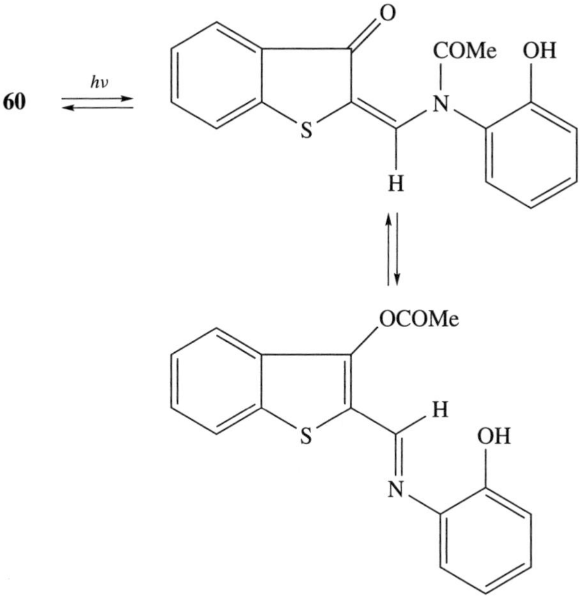

(21)

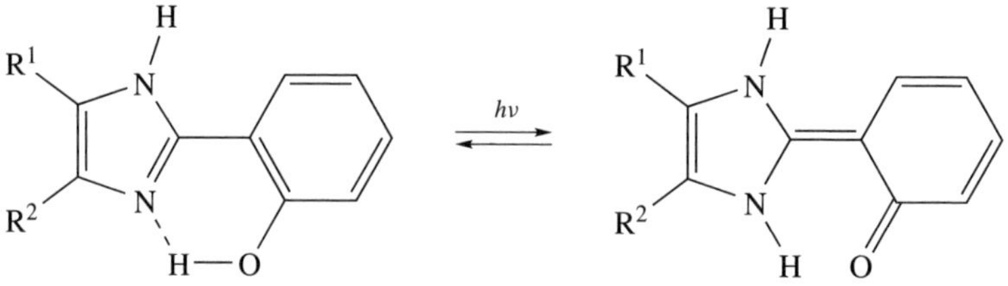

(61) R^1, R^2 = Me, Ph
(62) R^1R^2 = [benzo]

$h\nu$-XeCl laser (308 nm)

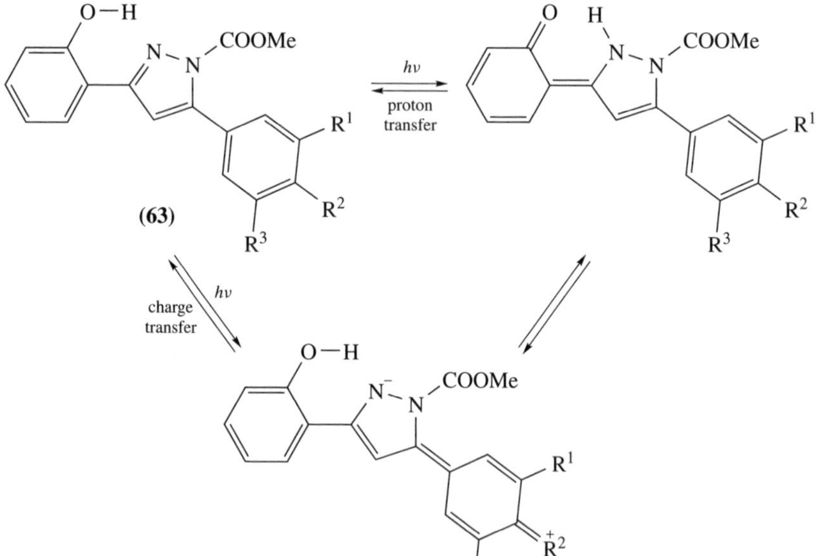

(22)

(63)

R^1, R^3 = H, OMe
R^2 = H, Me, Cl, Br, OMe, NH_2, NMe_2

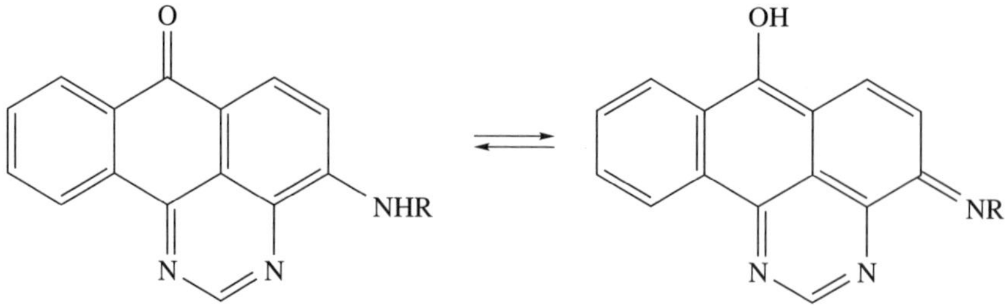

(23)

(64)   R = H, Bu, Ph

(24)

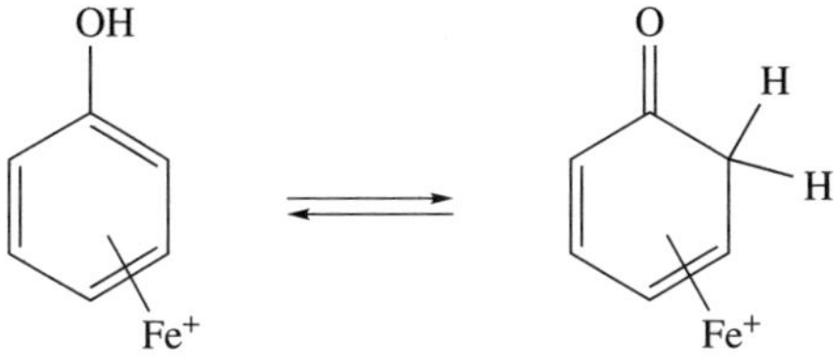

11. Tautomeric equilibria and rearrangements involving phenols

(25)

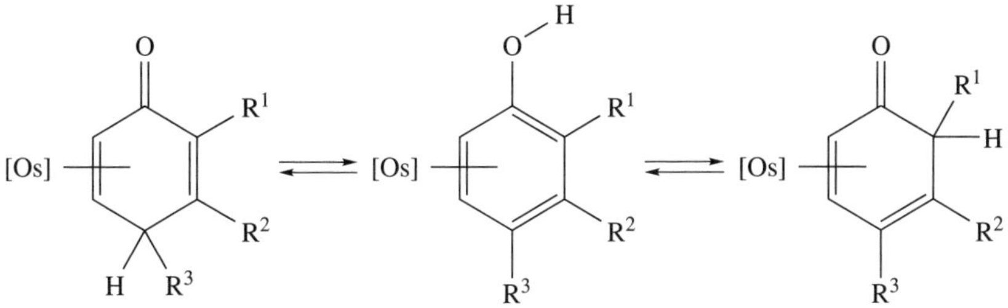

(26)

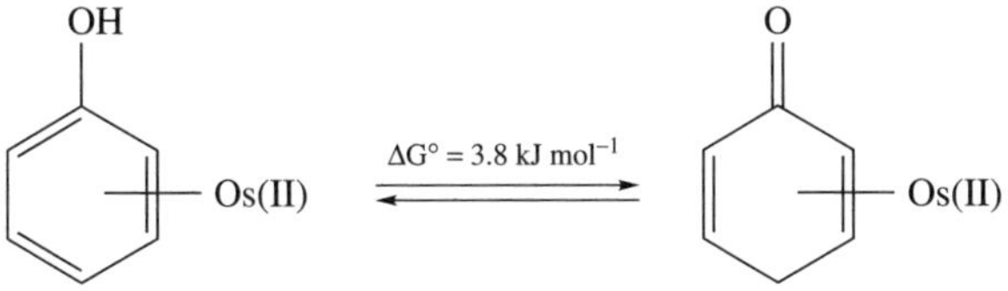

(27)

(28)

Such dearomatization of the arene ligand activates it toward an electrophilic addition. Thus, osmium(II) was used as a dearomatization agent for the direct 10β-alkylation of β-estradiol **70**[51] (equation 29). When the tautomeric mixture **71** ⇌ **72** was placed in acidic methanol and reprecipitated, a 3:1 equilibrium ratio of the phenolic **71** and dienone **72** tautomers was observed[51]. This intermolecular Michael addition to the C(10) position of the aromatic steroid was unprecedented.

An application of molybdenum and ruthenium complexes for synthesis of substituted phenols was also reported recently[52–55].

5. Phenols inserted into conjugated systems

An interesting situation arises when a tendency of ketodiene tautomer to transform into phenol results in a disturbance of the conjugation system in the whole structure in which this tautomer is a fragment. Thus, in the series of porphyrinoids **74** containing a semiquinone moiety, the macrocycle achieves the aromatization by undergoing a keto–enol tautomerization, whereby the phenolic subunit in structure **73** is transformed in such a way that the inner three carbon atom moiety becomes part of the 18 π-electron

(29)

11. Tautomeric equilibria and rearrangements involving phenols

aromatic core, whereas the outer carbon atoms generate an enone unit[56] (equation 30). This 'keto–enol' tautomerization would still result in the loss of the arene subunit, but the formation of a thermodynamically favorable aromatic aza[18]annulene would compensate for this loss.

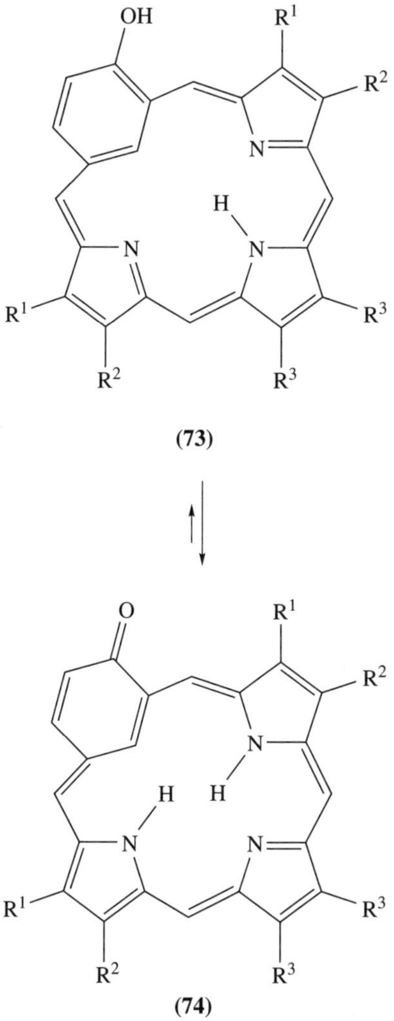

(30)

R^1 = Me, R^2 = Et, R^3 = Et; R^3R^3 = $(CH_2)_4$

B. Ring-chain Tautomerism

A classical example for a tautomeric equilibrium between the cyclic (lactone-phenolic) and open-chain (quinoid) forms is the behavior of phenolphthalein **75** as a function of the pH[57,58] (equation 31).

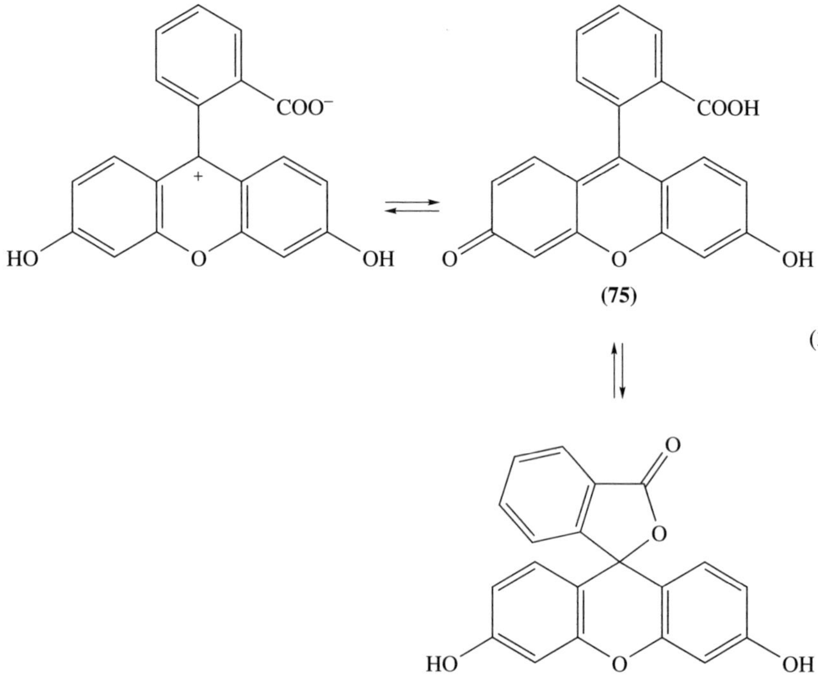

(75)

(31)

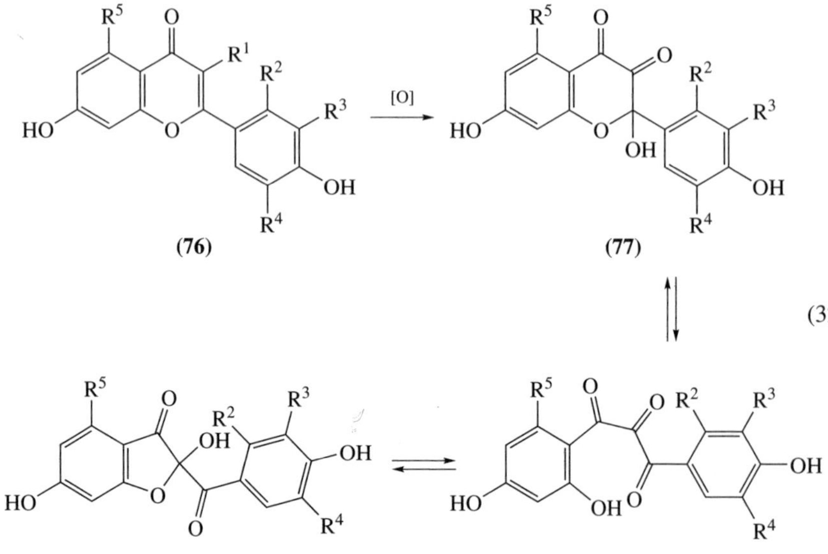

(76) (77) (78)

(32)

R^1 = OH, OMe; R^2, R^3, R^4, R^5 = H, OH

11. Tautomeric equilibria and rearrangements involving phenols   739

It should be noted, however, that most of the ring-chain tautomeric transformations of phenols proceed without loss of aromaticity of the arene cycle. The metal [Cu(II), Fe(II), Fe(III)] catalyzed oxidation of flavonols **76** gives the 2-(hydroxybenzoyl)-2-hydroxybenzofuran-3(2H)-ones **78** which are in an equilibrium with the initially formed 2-(hydroxyphenyl)-2-hydroxybenzopyran-3,4-diones **77**[59] (equation 32).

The hydroformylation of *ortho*-propenylphenols **79** gives the cyclic hemiacetals **80** in yields varying from 70 to 100%[60] (equation 33).

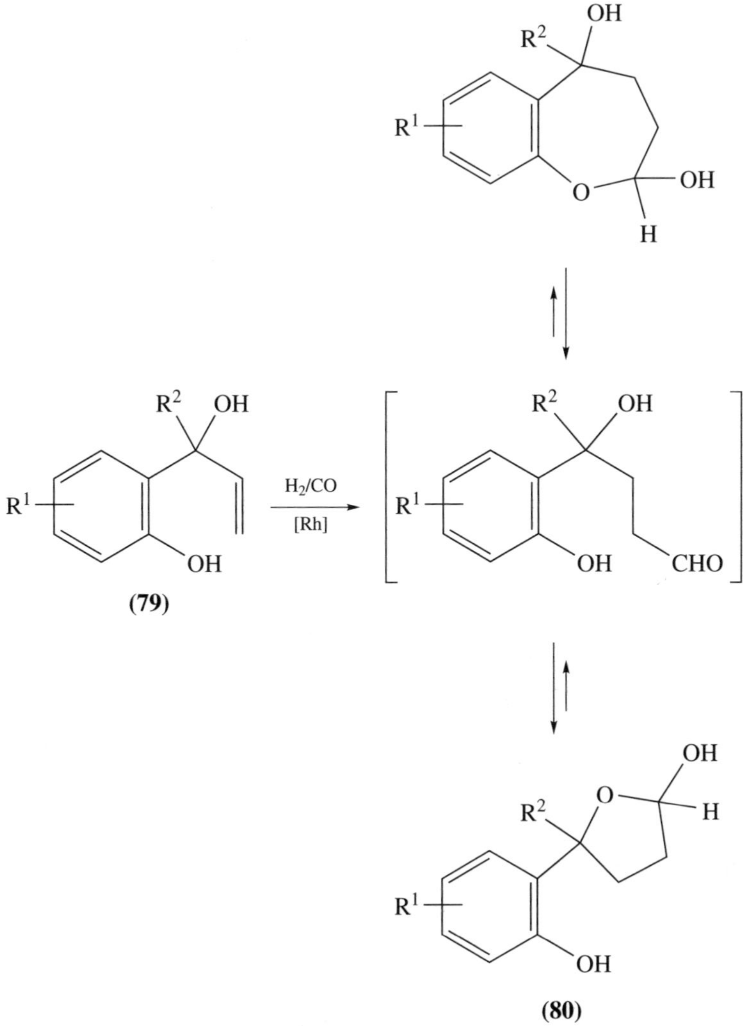

$R^1 = H, 4\text{-Me}, 5\text{-Cl}$
$R^2 = H, Me, Ph, 4\text{-MeC}_6H_4$

The hydroformylation of the *ortho*-prop-2-enylphenol **81** which contains no benzylic hydroxy group gives a mixture of the open-chain aldehyde **82** and the seven-membered cyclic hemiacetal **83** in a **82:83** ratio of approximately 40:60 (equation 34)[60]. The benzofuran epoxide **85** and its valence-isomeric quinone methide **86**, both readily obtainable from benzofuran **84**, rearrange thermally above $-20\,°C$ to form the allylic alcohol **87** and the tautomeric phenol **88** (equation 35)[61].

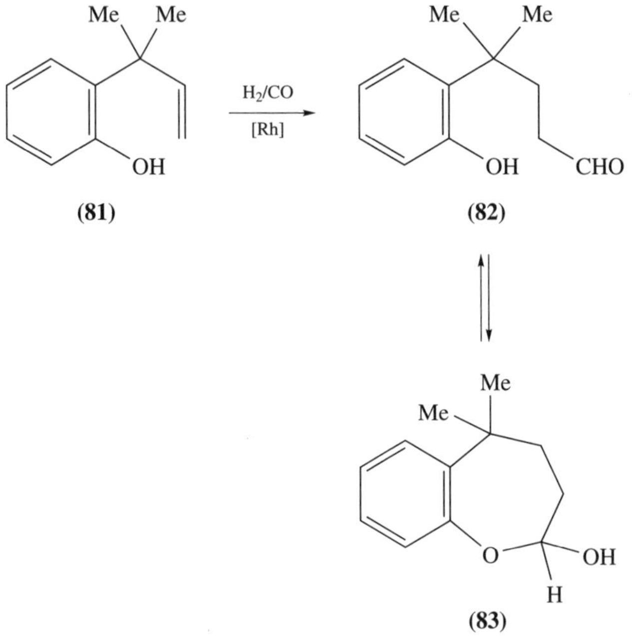

(34)

The addition of trichlorotitanium 4-*tert*-butylphenolate **89** to phthalaldehyde gives the intermediate **90**, which undergoes a ring–chain tautomerism to afford the cyclic isomer **91**. The latter reacts with the second molecule of **89** to yield the final product **92** via replacement of the acetalic OH group by the *p-t*-butylphenol moiety[62] (equation 36).

Many examples of ring–chain tautomerism in phenols are described in a recent review[63]. It should be mentioned in conclusion that tautomerism can also take place in the substituents at the phenolic ring.

C. Tautomer Transformations in Side Chains

In the presence of a second ionogenic group having a basic character, a prototropic tautomeric equilibrium is observed between the neutral **93** and the zwitterionic **94** forms[64,65] (equation 37).

Acylation of 2-hydroxyacetophenone **95** with RCOCl gives the esters **96**, which undergo a Baker–Venkataraman rearrangement (see Section IV.D.2) in the presence of *t*-BuOK to afford the phenolic β-diketones **97**. The enol tautomers **97a** and **97b** were observed by means of ^{1}H NMR spectroscopy[66,67] (equation 38).

(35)

A cooperative proton motion was observed within the hydrogen-bonded structure of 4-substituted phenolic N-oxides **98**[68] (equation 39).

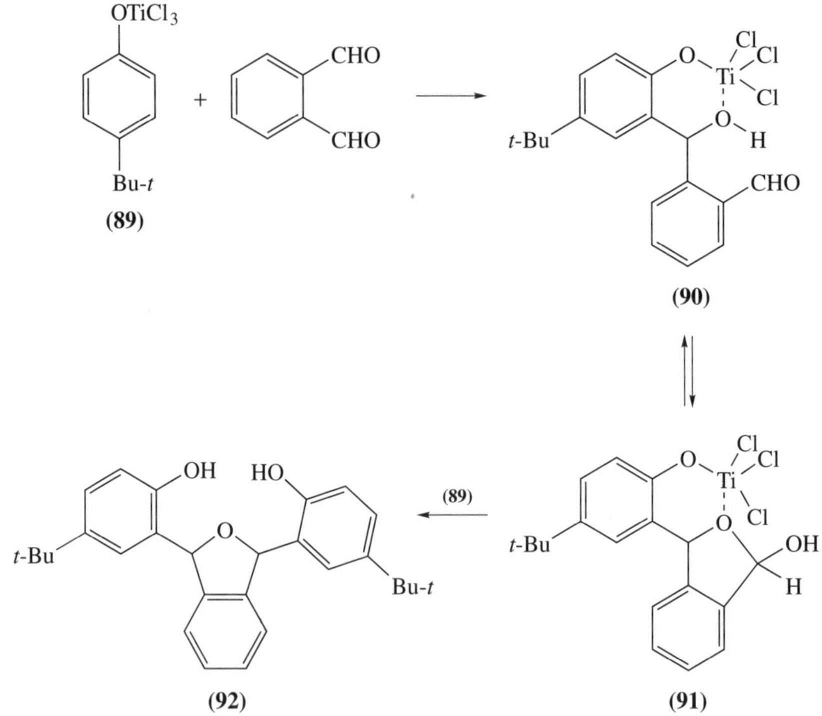

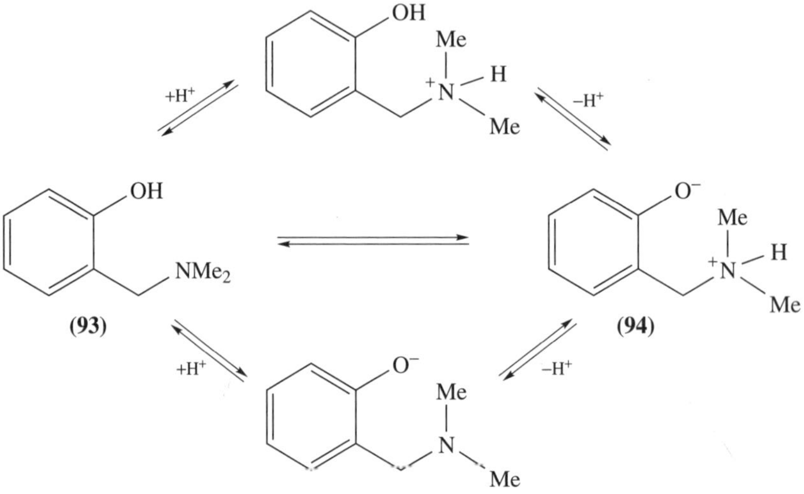

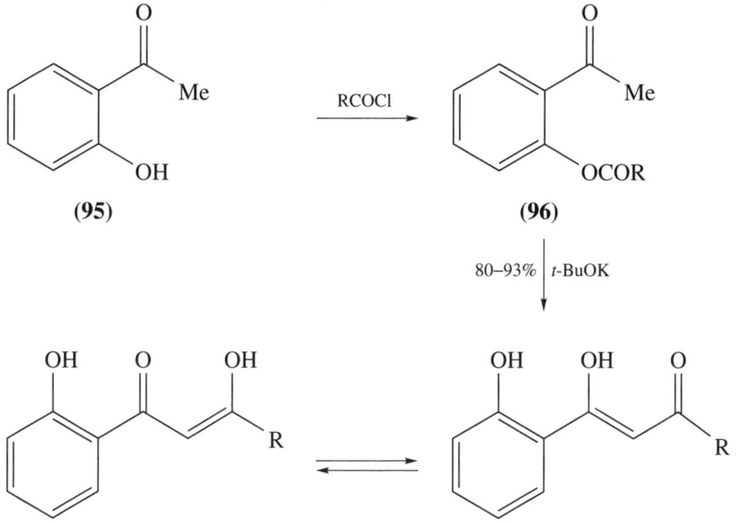

11. Tautomeric equilibria and rearrangements involving phenols   743

(38)

R = Ph, 4-MeOC$_6$H$_4$, 4-ClC$_6$H$_4$, 4-FC$_6$H$_4$, t-Bu

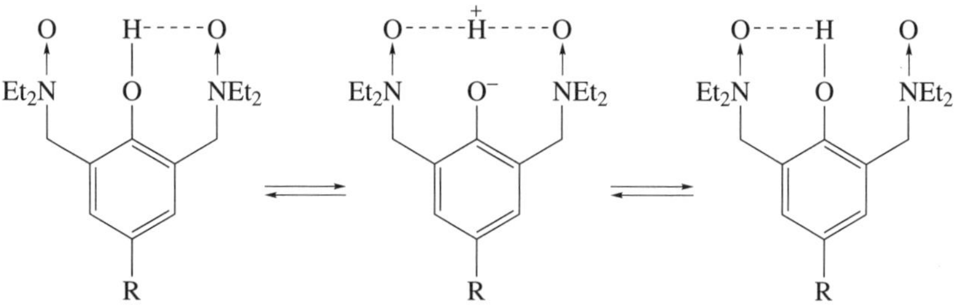

(39)

R = Me, Ph, t-Bu, F, Cl, COOEt, COOMe, CN, NO$_2$, 3,4-(NO$_2$)$_2$

Many examples of tautomeric transformations as well as rearrangements in the phenol series were considered in detail in a book[69].

III. REARRANGEMENTS OF PHENOLS

A. Cis–trans-Isomerizations and Conformational Transformations

The titled structural changes in phenols deserve attention as much as the hydroxy group affects the geometry of the molecule. Thus, it was shown by ^{1}H and ^{19}F NMR spectroscopies as well as by X-ray diffraction that 2-(2,2,2-trifluoro-1-iminoethyl)phenols **99** exist exclusively as the E-isomers with intermolecular hydrogen bonding in the solid state whereas these compounds isomerize to give a mixture of 66% Z- and 34% E-isomers in chloroform solutions[70].

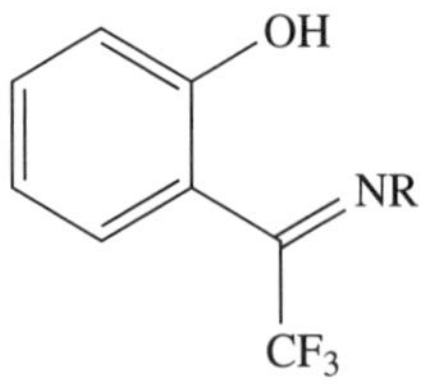

(99) R = Me, Pr

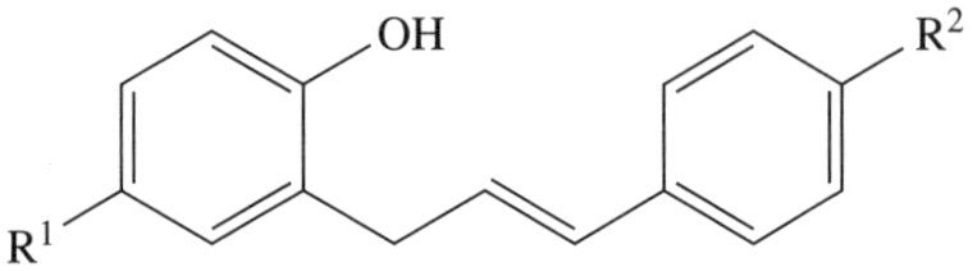

(100) R^1, R^2 = H, Cl, Br, Ph

Cis–trans-Isomerization together with dehalogenation reactions and cyclizations were observed upon irradiation (125-W medium-pressure Hg lamp, argon, 1 h) of trans-2-cinnamylphenols **100**, which are bichromophoric systems[71,72].

The rotational and conformational isomerism in dimeric proanthocyanidines **101** was studied by NMR spectroscopy. It was found that the geometry of these important polyflavanoids depends on the nature of the solvent (in organic solvents and water)[73]. The effect of the Y atom and the substituents X on the planarity and the barrier to internal rotation about the aryl–Y bond were estimated by semiempirical quantum-chemical calculations of the 4-XC$_6$H$_4$YH (X = H, NO$_2$, NMe$_2$; Y = O, S, Se) systems[74].

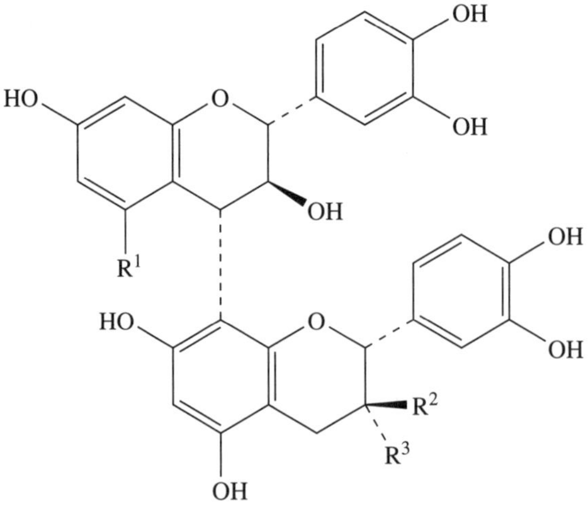

(101)

R^1, R^2, R^3 = OH, H, OH; OH, OH, H; H, OH, H

Z,E-isomerism was shown by ^{13}C NMR spectroscopy in the series of 4-X-2-methoxynaphthalenonium ions **102**. It was found that electron-donating substituents X stabilize the Z-isomer (equation 40). A Z,E-isomerism around the C–O bond in the corresponding 2-hydroxy-(**102**, R = H) and 2,4-dimethoxynaphthalenonium ions (**102**, R = Me, X = OMe) was not observed[75,76].

It should be noted that a large variety of conformations is typical for the cyclic polyphenols-*calixarenes*[77], which are considered in Chapter 19.

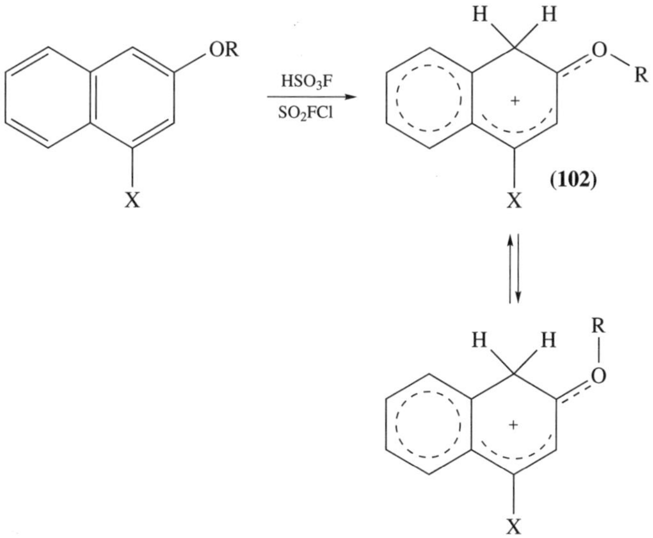

11. Tautomeric equilibria and rearrangements involving phenols

(40)

B. Phenol–Dienone Conversions

It is generally known that the processes of reversible oxidation of phenols, i.e. the conversions of phenolic systems into quinone structures and vice versa, are of great importance in biochemical reactions. The reaction partners mentioned above can serve as donors and acceptors of electrons and protons, i.e. as antioxidant systems. The conversions of phenols into cyclohexadienones are accompanied by the loss of aromaticity and in essence are not rearrangements, although the term 'phenol–dienone rearrangement' is found in the literature[78]. A review which summarizes in detail the oxidation reactions of phenols under conditions of halogenation, nitration and alkylation as well as radical reactions appeared[78]. The various transformations of phenols upon oxidation with nickel peroxide were also reviewed[79]. Therefore, only recent reports concerning the phenols-to-quinones conversions are described in this section.

The quinone monoketals **9** and *para*-quinol ethers **10** mentioned above[13] (Section II.A.1) can be obtained by anodic oxidation of the corresponding O-protected phenols **103**[80] (equation 41) or upon oxidation of substituted phenols **104** with one equivalent of phenyliodonium diacetate (PIDA) at an ambient temperature[81] (equation 42).

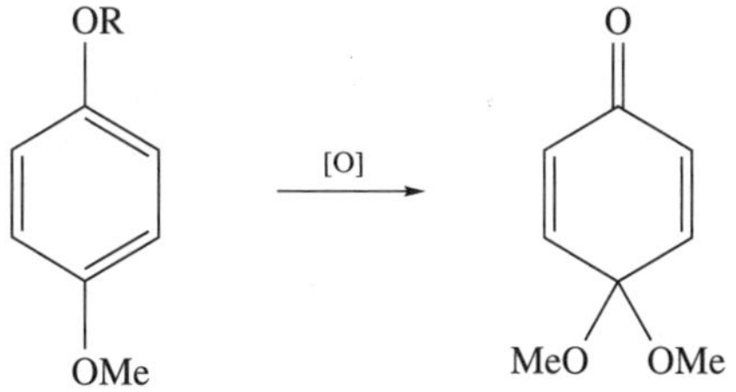

(41)

$R = Me_3Si, t\text{-}BuMe_2Si, MeOCH_2, MeO(CH_2)_2OCH_2$

(42)

(104) **(10)**

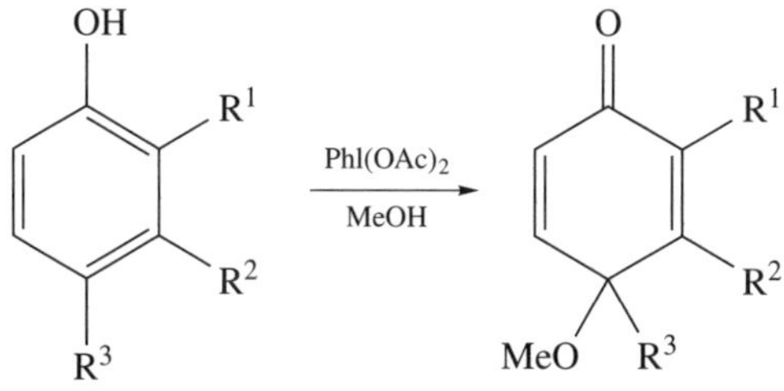

$R^1, R^2 = H, Me; R^3 = Me, t\text{-Bu}, Ph$

The annulation of these oxidation products **9, 10** with the anion of 3-cyanophthalide **105** affords access to a range of anthraquinones **106**[80,81] (equation 43).

(105)

(9) $R^3 = OMe$

(10) $R^3 = Me, t\text{-Bu}, Ph$

(43)

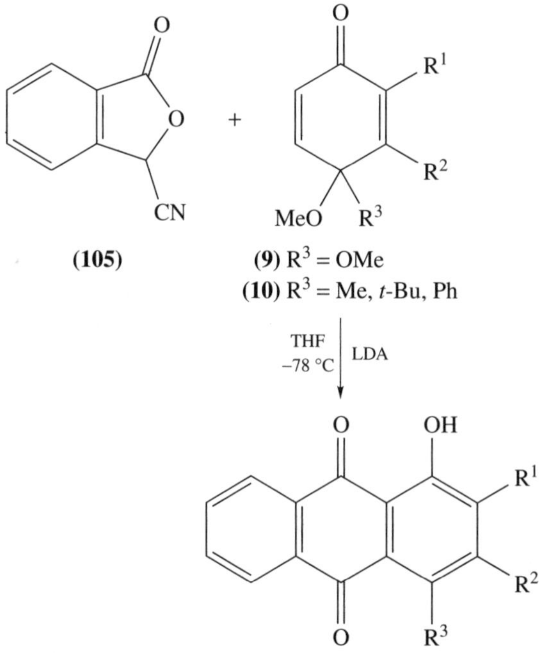

(106) $R^1, R^2 = H, Me;$
$R^3 = OMe, Me, t\text{-Bu}, Ph$

Hydroxylation is one of the most widespread conversions of phenols in redox reactions. This conversion occurs under a wide range of conditions, namely, at various pH, in organic and aqueous solutions as well as in the solid phase, due to the participation of quinoid intermediates that are prone to both ionic and radical transformations. Thus, the oxidation of 3,6-di-*tert*-butylpyrocatechol **107** in protic media is accompanied by the formation of 3,6-di-*tert*-butyl-2-hydroxy-*para*-benzoquinone **108** (equation 44). Hydroxylation of the 3,5-isomer **109** results in dealkylation (by an ionic or a radical route) and

11. Tautomeric equilibria and rearrangements involving phenols 747

isomerization with formation of 6-*tert*-butyl-2-hydroxy-*para*-benzoquinone **110** as well as compound **108** (equation 45). It was found that hydroxylation is of great importance for heterophase redox reactions and is closely connected with the formation of nitrogen-containing organic compounds where the nitrogen comes from nitrogen compounds in the air (equation 46)[82,83].

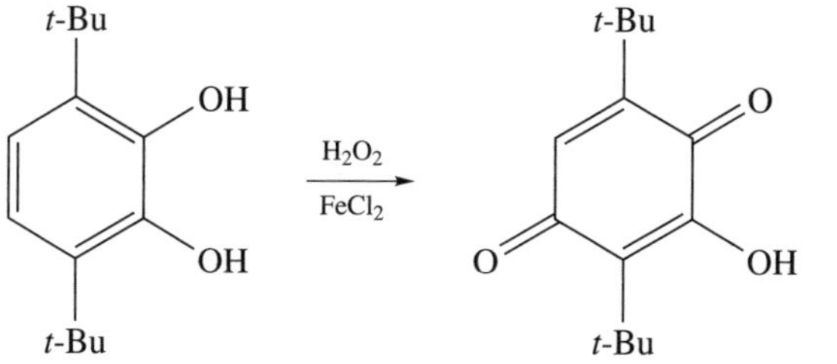

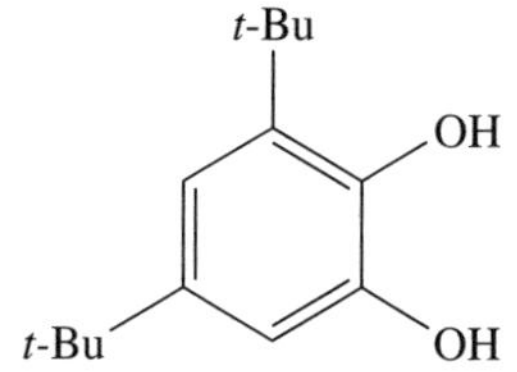

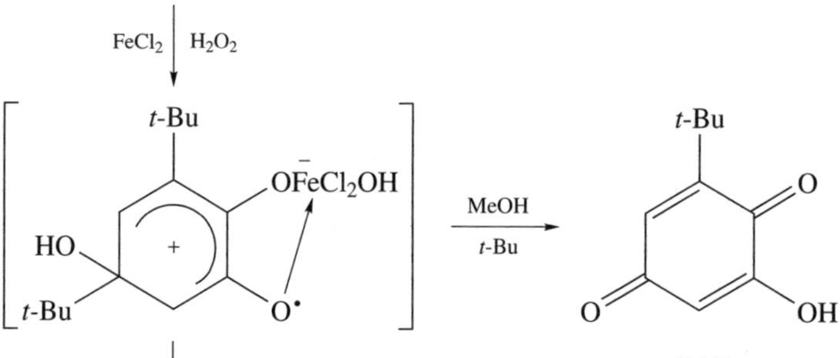

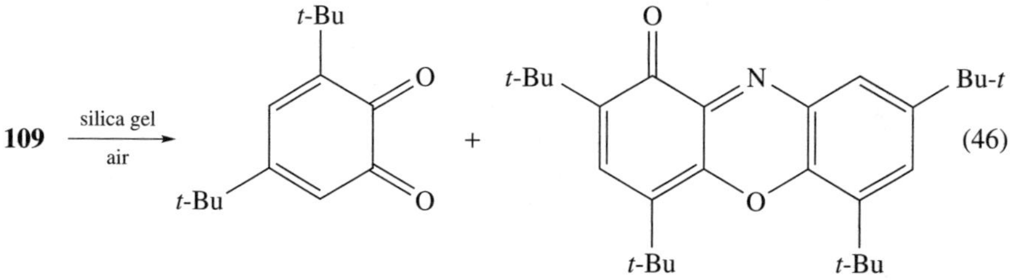

An efficient regio- and stereoselective organometallic method to *nucleophilic* phenol *ortho*-functionalization promoted by a cyclopentadienyl iridium cation ([Cp*Ir]$^{2+}$, where Cp* is C$_5$Me$_5$) was reported by Amouri and coworkers[84–86] (equation 47) (the *electrophilic* phenol functionalization by means of electron-rich moiety [Os(NH$_3$)$_5$]$^{2+}$ was mentioned above[51], see Section II.A.4).

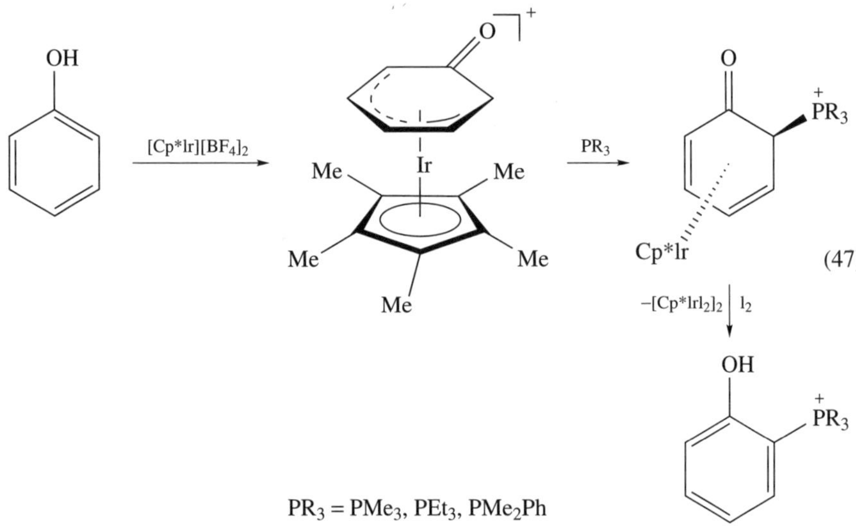

$$PR_3 = PMe_3, PEt_3, PMe_2Ph$$

The mushroom tyrosinase-catalyzed oxidative decarboxylation of 3,4-dihydroxyphenyl mandelic acid (**111**, R = H) and α-(3,4-dihydroxyphenyl) lactic acid (**111**, R = Me) proceeds via the quinone methide intermediate **112**. The coupled dienone–phenol rearrangement and keto–enol tautomerism transforms the quinone methide **112** into 1-acyl-3,4-dihydroxyphenyl compounds **113** (equation 48)[87,88].

The structures and properties of quinone methides were recently reviewed[89]. Inter alia, the microbial tyrosine phenol lyase (TPL) catalyzes the α, β-elimination of L-tyrosine to phenol and ammonium pyruvate. It is assumed that the process includes three steps, the second of which is tautomerization of the aromatic moiety which converts it into a good leaving group (equation 49)[90].

Various isomerizations were reported, including the tautomeric transformations of 2,6-disubstituted phenols which involve participation of phenoxy radicals and cation radicals[91,92].

C. Hydroxy Group Migrations

The rearrangements of phenols which are accompanied by hydroxy group transpositions are called the *Wessely–Moser reaction*[93,94] (equations 50 and 51). In essence, these rearrangements are recyclizations of flavonoides **114** via the ring-opened form **115** to give the novel structures **116**. Compounds that can participate in these rearrangements are flavones (**114**, R^2 = H, R^3 = Aryl), flavonoles (**114**, R^2 = OH, R^3 = Aryl), isoflavones (**114**, R^2 = Aryl, R^3 = H), chromones (**114**, R^2 = H, R^3 = Alkyl), chromonoles (**114**, R^2 = OH, R^3 = Alkyl), xanthones (**114**, R^{2}R^3 = benzo) as well as benzopyrylium salts (e.g. see Reference 95).

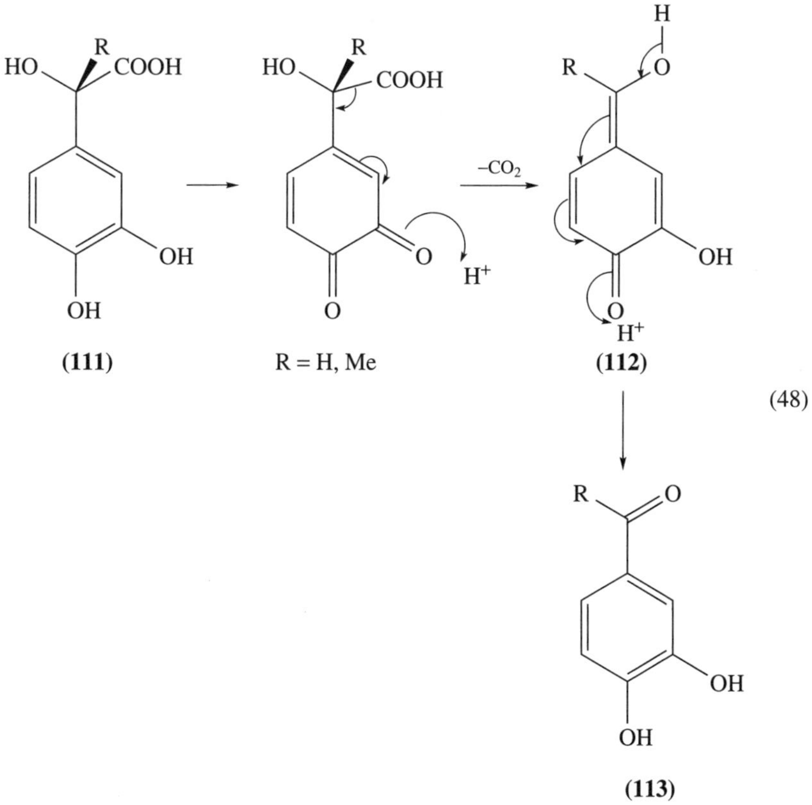

D. Isomerizations of Alkylphenols

Information about the transformations of alkylphenols upon heating and action of acid catalysts is too voluminous and is concentrated mainly in the patent literature (for a review see Reference 96). Thus, the higher n-alkylphenols undergo alkyl group elimination, transalkylation and transposition of side chains under acid catalysis conditions. The isopropyl and *tert*-butyl groups have the greatest migration ability. For example, 2-methyl-6-isopropylphenol rearranges readily to afford 2-methyl-4-isopropylphenol by action of catalytic amounts of H_2SO_4 at ca 60 °C[97]. 2-*tert*-Butylphenol rearranges almost quantitatively into 4-*tert*-butylphenol already at −40 °C in liquid HF solution[98].

In spite of such an abundant literature concerning the alkylphenol conversions, investigations in this field are still progressing. The Amberlyst 15-catalyzed alkylation of phenol or catechol with olefins, capable of forming the stable *tert*-alkylcarbenium ions, results in the corresponding *tert*-alkylphenols at 25–130 °C, with the *para*-isomer being the favored product. However, the alkylation at 140–150°C leads to *sec*-alkylphenols, with both *ortho*- and *para*-isomers in almost equal amounts[99]. It was found that 2-*tert*-butylphenol isomerizes easily to 4-*tert*-butylphenol during the alkylation of phenol with *tert*-butanol in the vapor phase on an SAPO-11 catalyst (silicoaluminophosphate molecular sieves)[100].

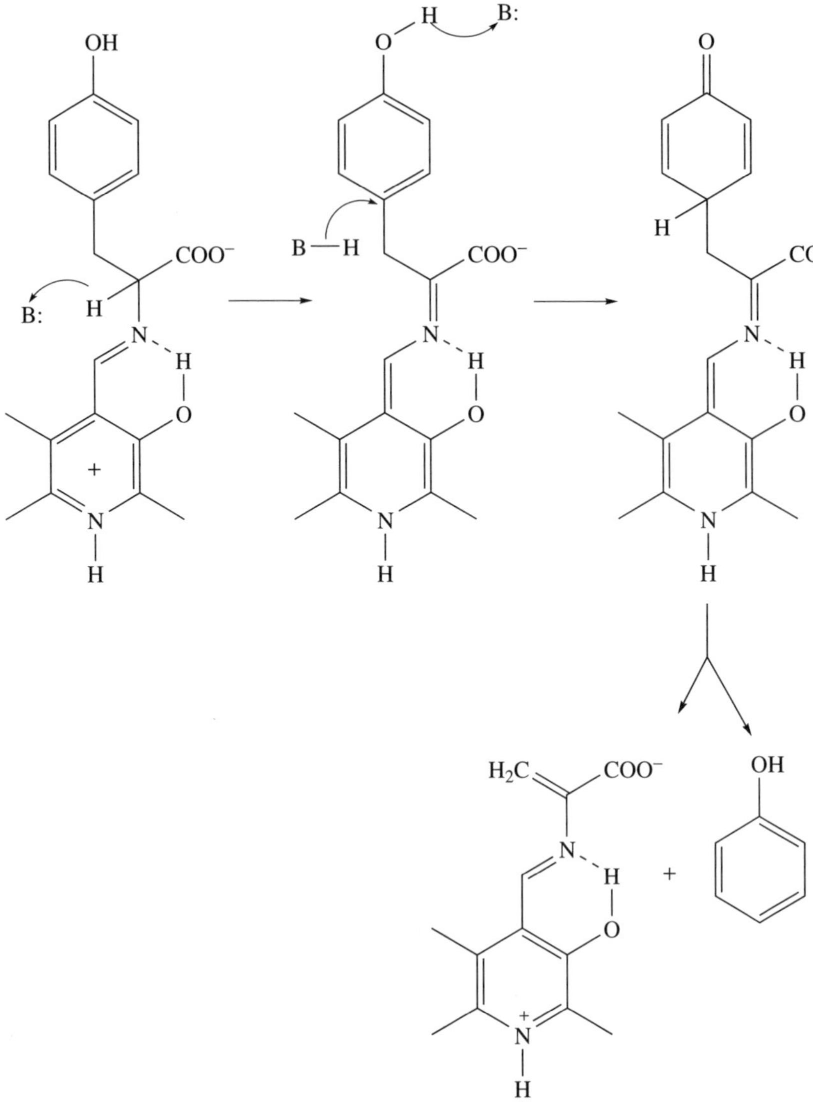

(49)

E. Isomerizations of Phenols Containing Unsaturated Side Chains

1. Double bond migration

The classic example for conversion of allylphenols to propenylphenols is the base-catalyzed rearrangement of eugenol **117** to isoeugenol **118** (equation 52)[101]. Silyl protected phenolic tertiary cinnamyl alcohols **119** undergo a lithium-ammonia induced hydrogenolysis with concomitant double bond migration. This reaction serves as a unique approach to prenyl-substituted aromatic compounds **120** (equation 53)[102,103].

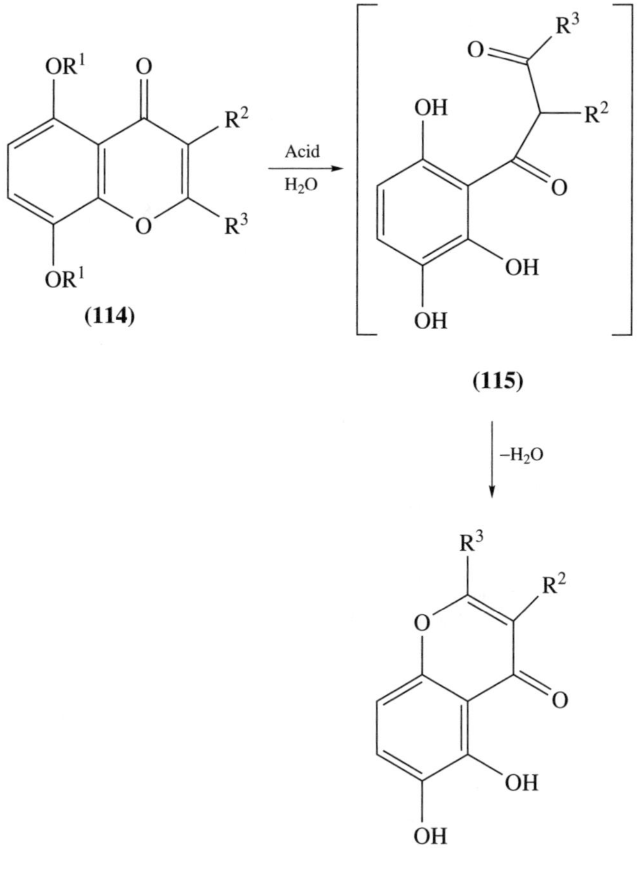

11. Tautomeric equilibria and rearrangements involving phenols 751

R^1 = H, Alkyl; R^2 = H, OH, Aryl; R^3 = H, Alkyl, Aryl
Acid = HI, H_2SO_4, Py · HCl

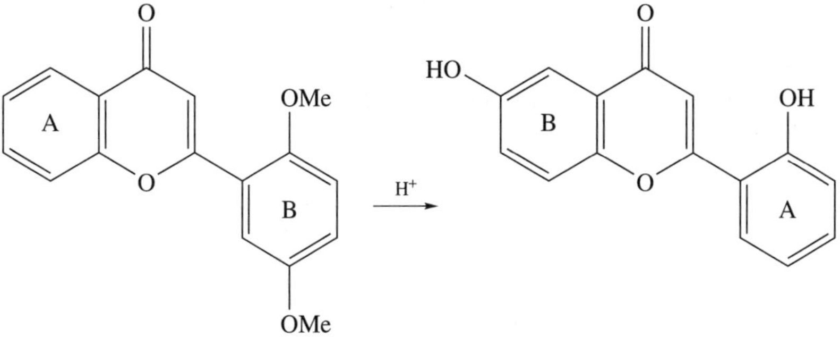

(50)

(51)

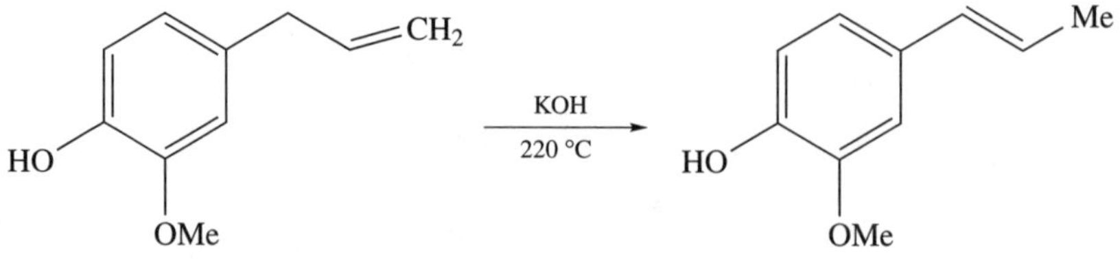

<img of structures>

(117) → (118) KOH, 220 °C (52)

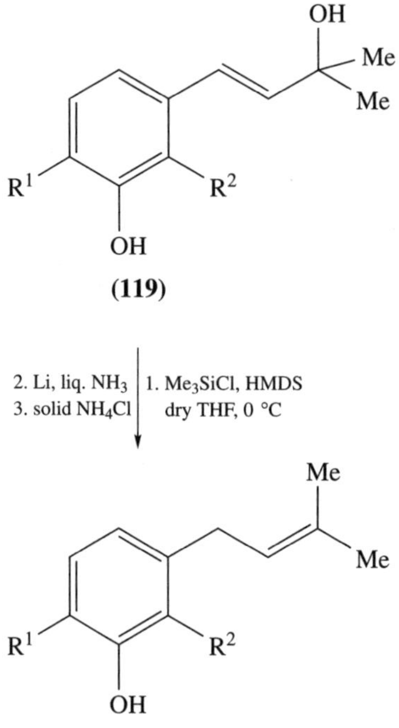

(119)

1. Me$_3$SiCl, HMDS, dry THF, 0 °C
2. Li, liq. NH$_3$
3. solid NH$_4$Cl

(120) (53)

R^1 = H, Me, OMe; R^2 = H, OMe
HMDS = hexamethyldisilazane

The isomerization of 2-allylphenol **121** to 2-propenylphenol **122** catalyzed by the *ortho*-metallated complex Rh[P(OPh)$_3$]$_3$[P(OPh)$_2$(OC$_6$H$_4$)] produces only one isomer (equation 54)[104].

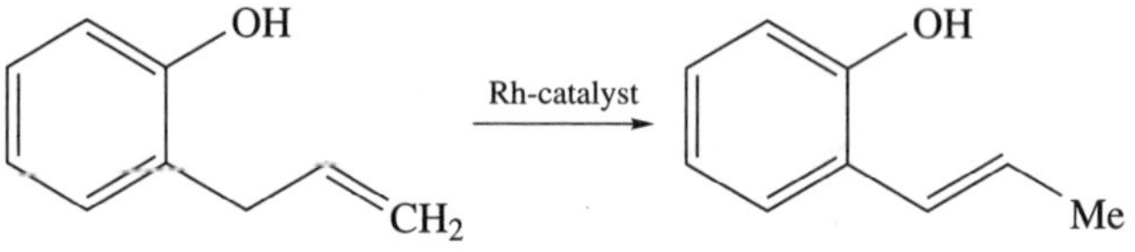

(121) → (122) Rh-catalyst (54)

2. Allylphenol–coumaran rearrangement

The abnormal Claisen rearrangement (see also Section IV.B.1) of 2-allylphenols **123** leads to spirodienones **124** and **125** (equation 55). This reaction is a [1,5s]-homosigmatropic process that is accompanied by transfer of hydrogen atom from the hydroxy group to the γ-C-atom of the C=C bond. Compounds **124** and **125** can undergo further transformations, namely, a reverse conversion into phenols **123**, *trans–cis*-isomerization, isomerization of the side chain ($R^2 =$ Me) (with the exception of isomers **125**) and a [1,3]-sigmatropic rearrangement into coumaranes **126** and **127** (equations 55 and 56). The formation of isomer **126** occurs especially readily if the substituent R^2 in the intermediates **124** and **125** is a vinyl or an aryl group[105]. The 2-(1'-arylallyl)phenols **128** were transformed on heating in N,N-diethylaniline at 225 °C to the *trans*-2-aryl-3-methylcoumaranes **129** in excellent yields[105] (equation 57).

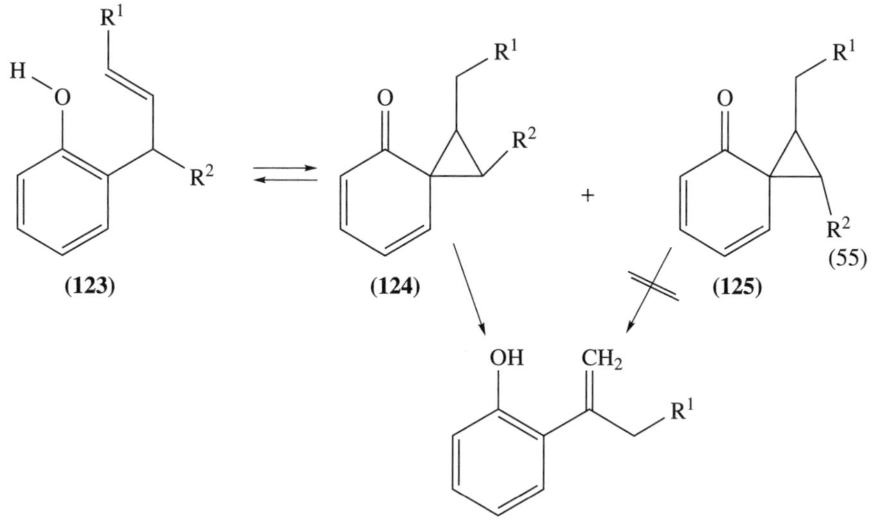

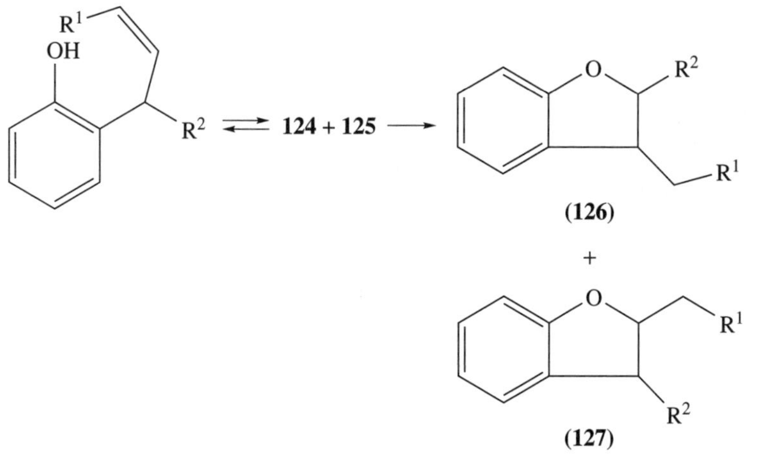

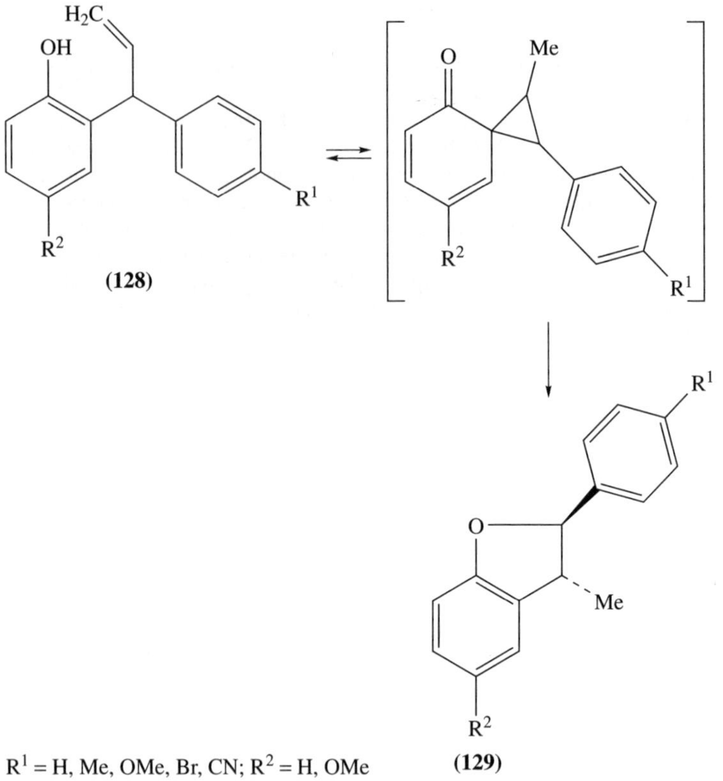

R^1 = H, Me, OMe, Br, CN; R^2 = H, OMe   (129)

(57)

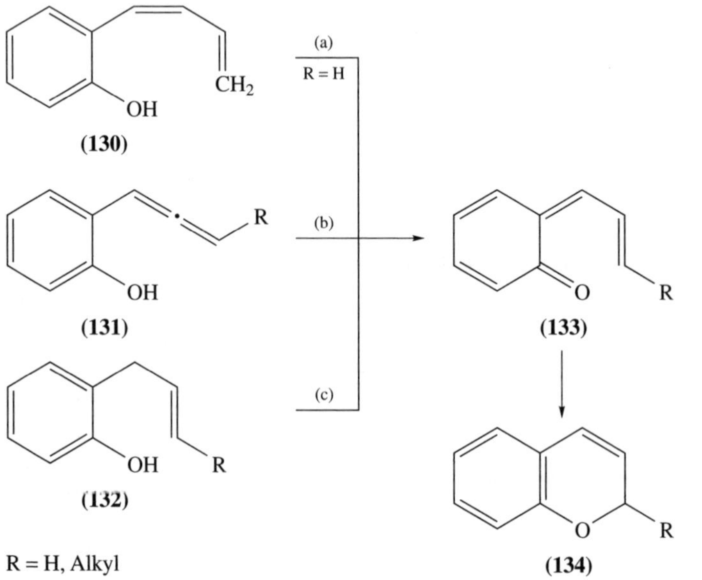

R = H, Alkyl

(58)

11. Tautomeric equilibria and rearrangements involving phenols 755

2*H*-Chromenes **134** were obtained via cyclization of the unstable intermediates—vinyl-*o*-quinone methides **133**—which can be formed by various paths: (a) from *ortho*-(*cis*-buta-1,3-dienyl)phenols **130** by thermal [1,7*a*]-hydrogen shift; (b) from *ortho*-allenylphenols **131** (which are intermediates in the Claisen rearrangement of propargyl phenyl ethers, see Section IV.C) by [1,5s]-hydrogen shift; and (c) by dehydration of *ortho*-allylphenols **132** with dichlorodicyanobenzoquinone (DDQ) (equation 58)[106]. The *ortho*-quinomethanes **136** were prepared by thermolysis of *ortho*-hydroxyphenyl carbinols **135**[106] (equation 59).

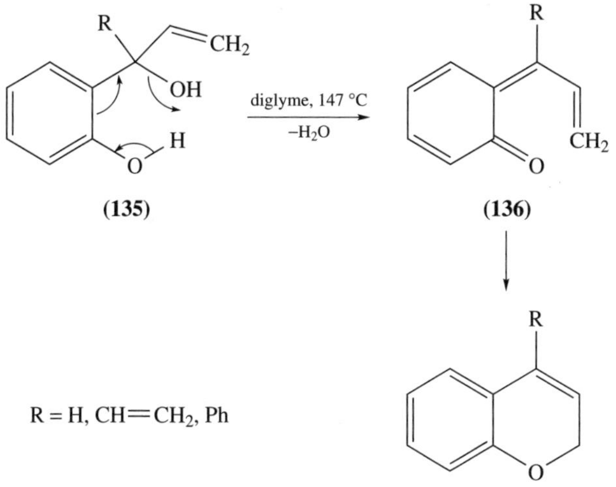

R = H, CH=CH$_2$, Ph

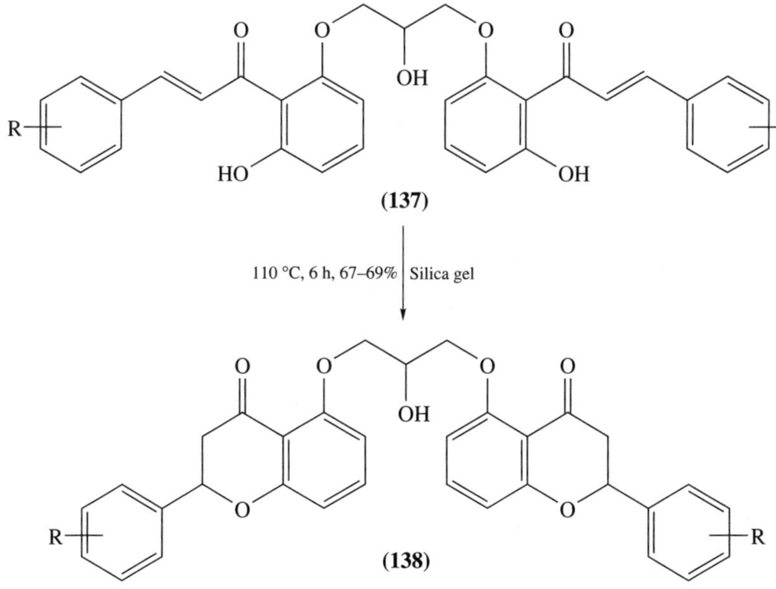

(59)

(60)

R = H, 3-OMe, 4-OMe

(61)

11. Tautomeric equilibria and rearrangements involving phenols 757

Isomerization of phenols **137** over silica gel in the solid phase furnishes the corresponding 2,3-dihydro-4-oxo-4*H*-1-benzopyrane derivatives **138** (equation 60)[107]. The cascades of the charge-accelerated rearrangements of the *ortho*-(1,1-dimethylpropenyl)phenol **139** catalyzed by Brönsted acid (e.g. trifluoroacetic acid, equation 61) as well as by Lewis acids (anhydrous AlCl$_3$ or TiCl$_4$, equations 62 and 63) proceed via the common intermediate **140**[108].

The analogous isomerization of *ortho*-hydroxyaryl phenylethynyl ketone **141** leads to 6-methoxyflavone **142** and 5-methoxyaurone **143** (equation 64)[109].

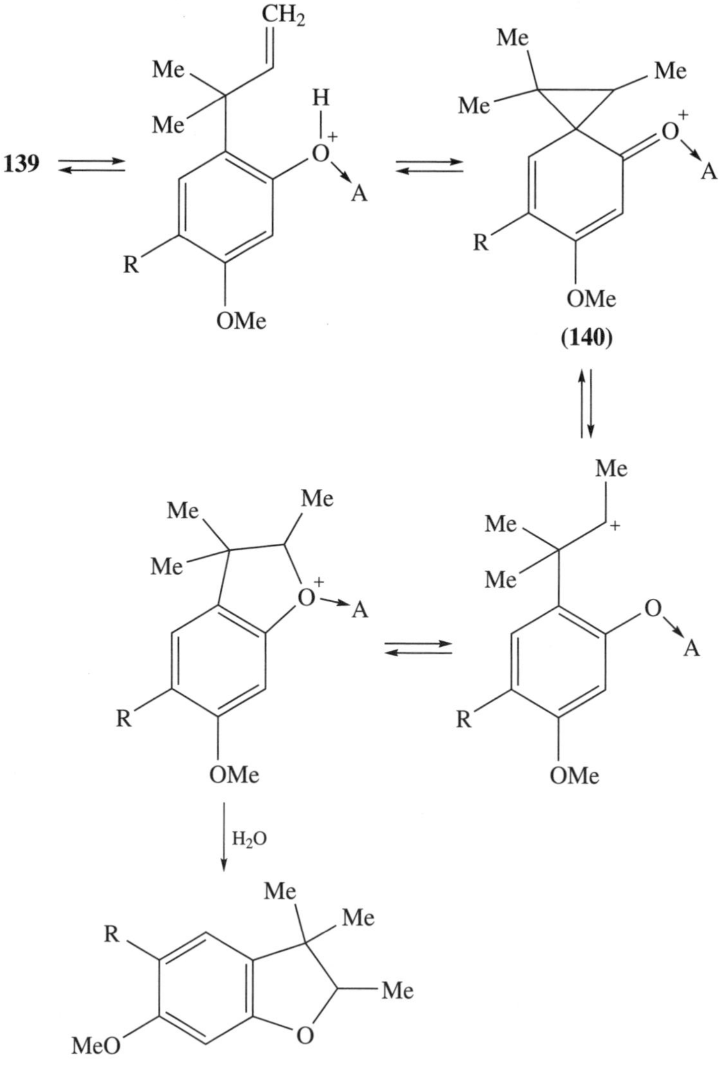

(62)

A = Lewis acid (AlCl$_3$, TiCl$_4$)

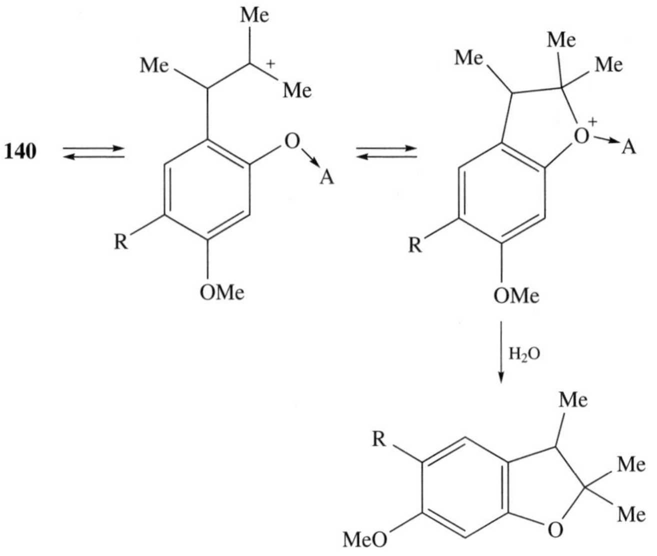

$$R = \text{(structures shown)} \quad \text{and} \quad \text{(structure shown)}$$

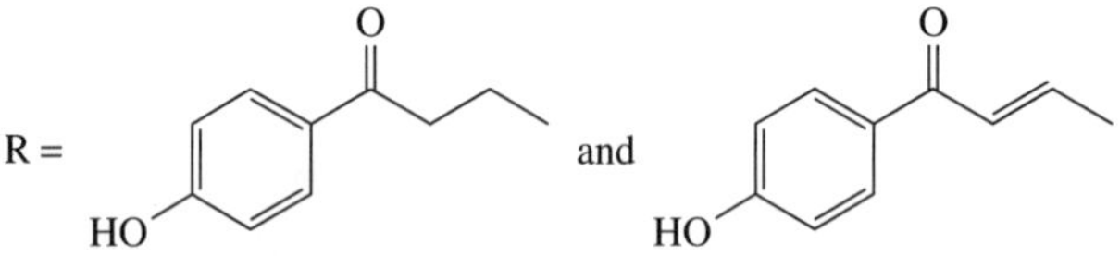

A = Lewis acid (AlCl$_3$, TiCl$_4$)

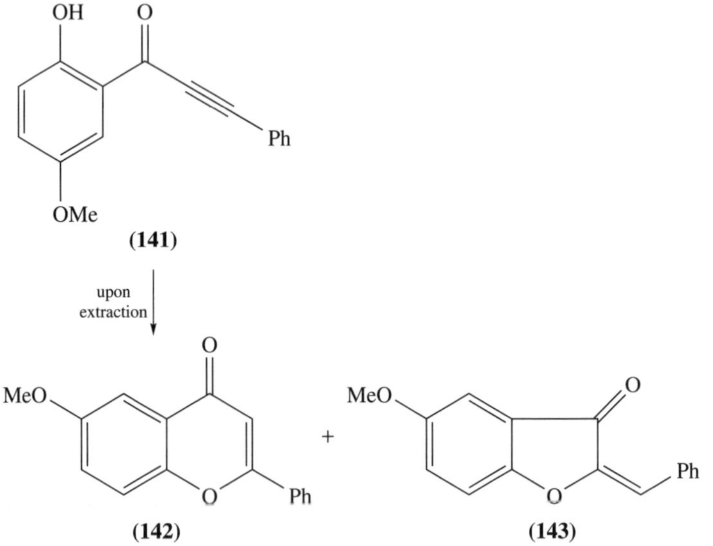

(63)

(64)

11. Tautomeric equilibria and rearrangements involving phenols 759

IV. REARRANGEMENTS OF O-SUBSTITUTED PHENOL DERIVATIVES

A. Rearrangements of Alkyl and Aryl Phenolic Ethers

Alkyl aryl ethers are quite stable on heating. Phenyl benzyl ether isomerizes slowly at 250 °C to afford 4-benzylphenol and its *ortho*-isomer as a minor product[110]. The conditions of isomerizations of O-alkylated and O-aralkylated phenols were reviewed[111].

The rearrangements of diaryl ethers are more useful for organic synthesis. The most known reaction in this field is the *Smiles rearrangement* (equation 65)[112,113]. Electron-donating substituents R^2 facilitate this rearrangement which often turns out to be reversible.

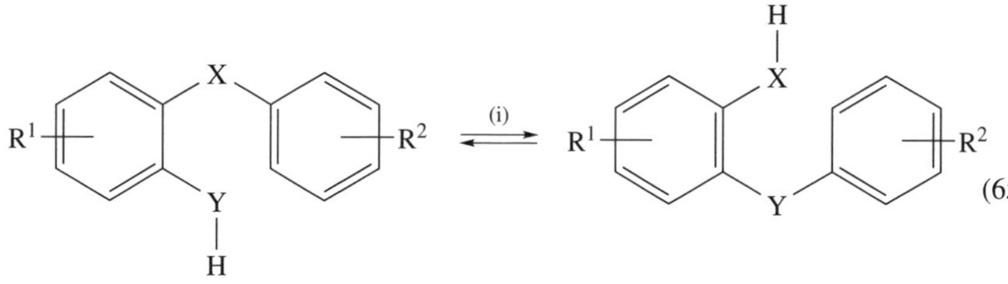

(i) NaOH, KOH, $NaNH_2$; H_2O, MeOH, EtOH, C_6H_6, DMF; 50–100 °C
R^1 = H, Me, Hal; R^2 = H, NO_2, Hal; X = O, S, SO_2, COO; Y = O, S, NH, SO_2

The isomerization of diaryl ethers to *ortho*-arylphenols in the presence of phenylsodium is known as the *Lüttringhaus rearrangement*[114] (equation 66).

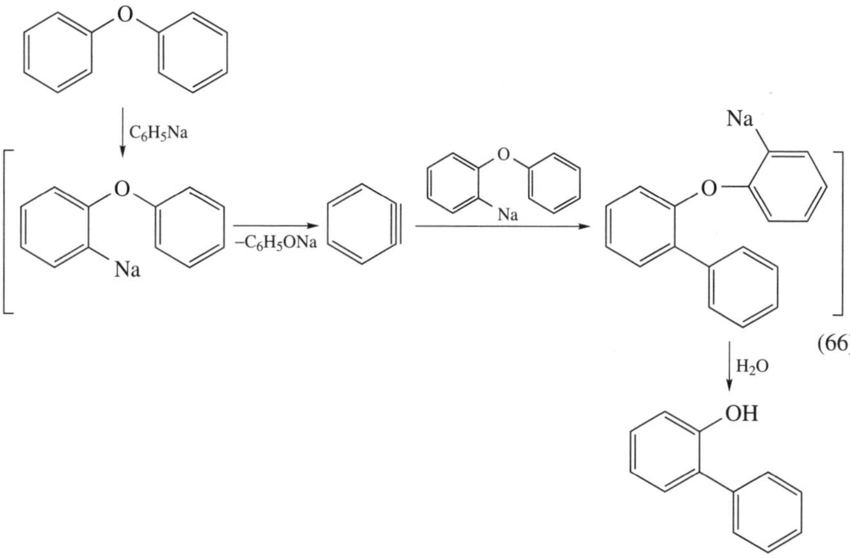

An unusual ring-contraction reaction occurs on the acid-catalyzed interaction of trimethylhydroquinone **144** with cycloalkane-1,2-diols (e.g. **145**) to form the spiro compounds **146** (equation 67)[115]. Besides two isomers of cyclohexane-1,2-diols **145**, this rearrangement was also described for cyclopentane-, cycloheptane- and cyclooctane-1,2-diols[115].

(67)

11. Tautomeric equilibria and rearrangements involving phenols

It should be mentioned here that very interesting constitutional and translational isomerism is observed in the series of catenanes and rotaxanes which contain phenol derivatives such as macrocyclic phenylene-crown components as well as phenolic polyether chains[116-118] (see also Lehn's recently published book[119]).

B. Rearrangements of Allyl Aryl Ethers

Among the isomerizations of phenolic ethers, the rearrangements of allyloxyarenes occupy a special position because of the wide variety of pathways and the great synthetic significance.

1. Claisen rearrangement

The overwhelming majority of literature devoted to isomerizations of allyl aryl ethers is connected with the aromatic Claisen rearrangement and is summarized in detail in many reviews[120-124]. Although the [3,3]-sigmatropic isomerization of phenol ethers to the corresponding C-alkylated derivatives has enjoyed widespread application in organic synthesis for over seventy years, it continues to be a very important reaction for the construction of a carbon–carbon bond. This section presents only recent reports.

In general, the aromatic Claisen rearrangement can be illustrated by equation 68. The initial step in the thermal Claisen rearrangement of an allyl aryl ether leads to an *ortho*-dienone which usually enolizes rapidly to form the stable product, an *ortho*-allylphenol (so-called *ortho*-Claisen rearrangement, 147 → 148 → 149). However, if the rearrangement proceeds to an *ortho*-position bearing a substituent, a second [3,3]-rearrangement step, followed by enolization, occurs to afford the *para*-allylphenol (*para*-Claisen rearrangement, 147 → 150 → 151 → 152). The temperature range for typical reactions is 150 °C to 225 °C[121].

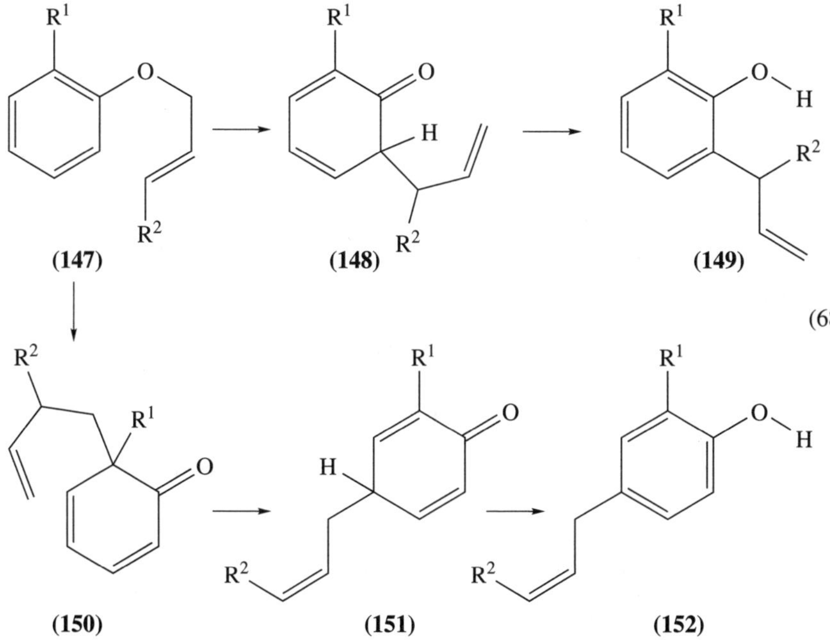

(68)

The intramolecular nature of the rearrangement was established by means of ^{14}C-labeled allyl phenyl ether as well as by a crossover experiment. *Ab initio* calculations were performed to determine the transition-state structures and the energetics of aromatic Claisen rearrangement as well as in related isomerizations[125]. It was shown during an investigation of the solvent effects on the thermal Claisen rearrangement that isomerization of cinnamyloxybenzene **153** in diethylene glycol gives, in addition to 'normal products' **154** and **155**, also 2-cinnamylphenol **157** and diethylene glycol monocinnamyl ether **158**. The formation of the ether **158** was ascribed to the acidic and high dielectric properties of the glycol solvent that allows generation and capture of the cinnamyl cationic intermediate **156** (equation 69)[126].

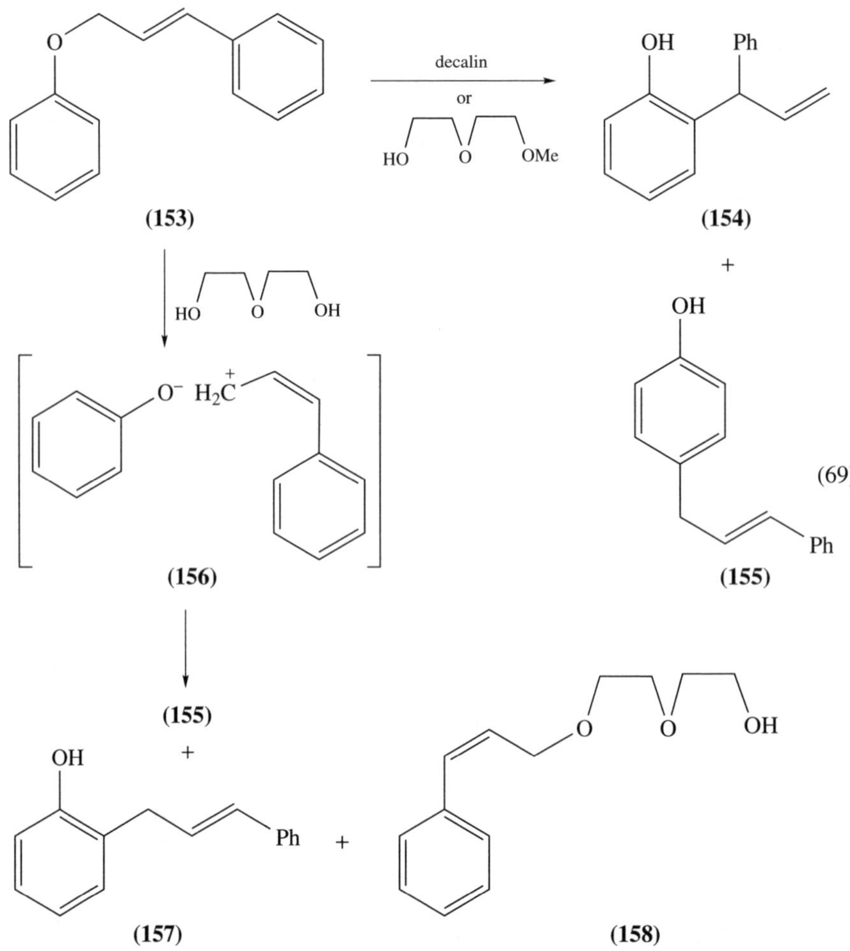

The preparative Claisen rearrangement was studied in aqueous media at temperatures up to 300 °C. The experiments were conducted in the recently created pressurized microwave batch reactor and in conventional heated autoclaves. It was found that allyl phenyl ether isomerizes in water during 10 min at 240 °C to give the *ortho*-Claisen rearrangement product in 84% conversion[127].

11. Tautomeric equilibria and rearrangements involving phenols 763

The Claisen rearrangement can be effectively catalyzed by Lewis acids, Brönsted acids, bases, Rh(I) and Pt(0) complexes as well as by silica[121]. Several reviews were published recently in which the application of zeolites and acid-treated clays as catalysts for the Claisen rearrangement was described[128–130]. Thus, it was shown that the rearrangement conditions for phenolic allyl ethers can be dramatically milder if this reaction is carried out by thermolysis of a substrate immobilized on the surface of previously annealed silica gel for chromatography. For example, the thermolysis of ether **159** on silica gel (in a **159**: SiO_2 ratio of 1:10 w/w) at 70 °C gives the phenol **160** in 95% yield after 3.5 hours[131] (equation 70). An additional example is shown in equation 71[131].

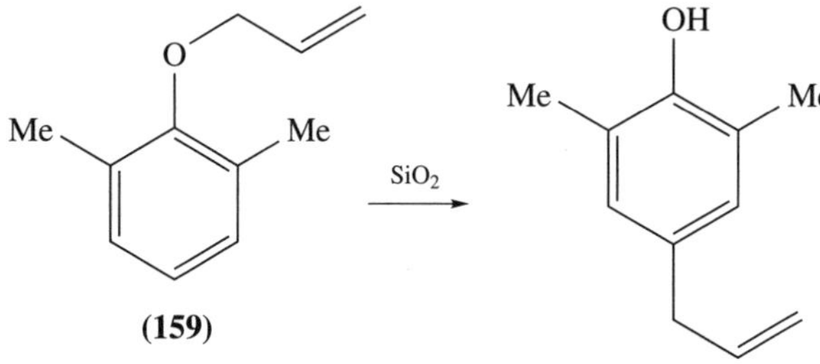

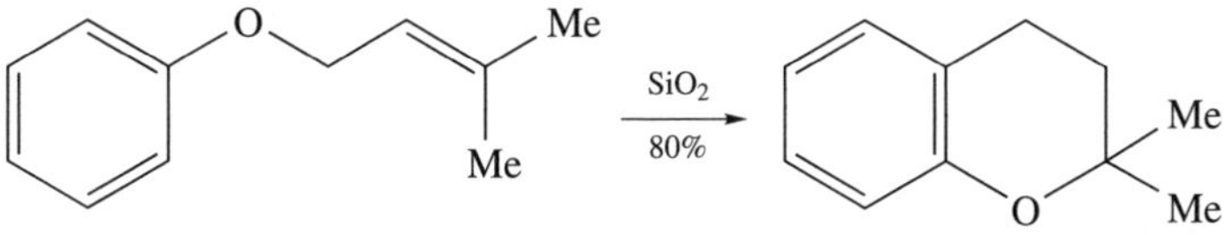

An unusual [1,3]-rearrangement of aryl 2-halocyclohexenylmethyl ethers **161** was promoted by trifluoroacetic acid[132] (since the thermal rearrangement failed because the ethers **161** are stable up to 240 °C). When the ethers **161** were exposed to TFA at room temperature, an extremely facile reaction afforded the products **162** in good yields (65–80%). However, no products of Claisen rearrangement were formed (equation 72)[132].

On the contrary, the acid-catalyzed rearrangement of the allyl ether **163** failed owing to acidolysis. The ether **163** was rearranged on heating in N,N-diethylaniline (equation 73)[133]. It is interesting that the reaction of phenol with methylenecyclopropane **164** proceeds smoothly to give the phenol **165** by an addition/ring-opening reaction followed by Claisen rearrangement in 56% yield (equation 74)[134].

A novel class of purely thermally activated dyes which became colored only upon heating (i.e. without any other components) was created by using the fact that the neutral and colorless allyl aryl ethers **166** generate an acidic group upon heating due to Claisen rearrangement. The phenol groups thus formed undergo an intramolecular acid–base reaction, which in turn causes the opening of the lactone ring and the coloration (equation 75)[135].

In the studies of syringin, an active component in traditional Chinese medicine, it was shown that 4-hydroxy-3,5-dimethoxybenzoic acid **167** reacted with allyl bromide under basic conditions to produce a mixture of O- and C-allylated compounds **168, 169**. After the mixture was subjected to heating at about 200 °C, a *para*-Claisen rearrangement took place to form the main product **169** in 71% yield (equation 76)[136].

Claisen rearrangement is widely used in organic synthesis. Thus, to obtain *ortho*-methoxylated phenethylamino derivatives as potent serotonin agonist **170**, the strategy employed was based on the Claisen rearrangement and isomerization of the allyl fragment. The bromine atom was attached at an *ortho*-position to the hydroxy group in order to force a regiospecificity on the Claisen rearrangement (equation 77)[137].

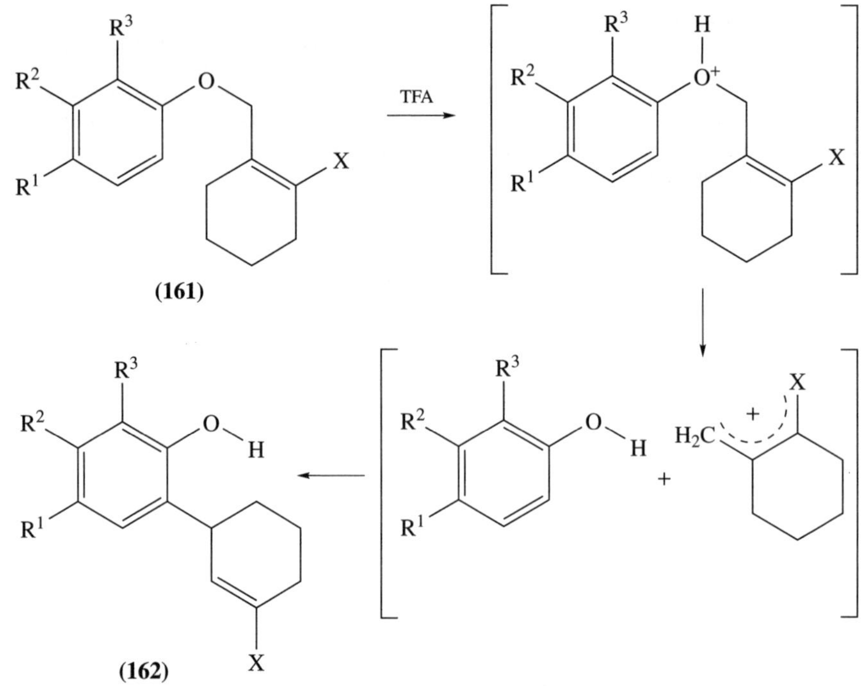

$$X = Cl, Br; R^1 = H, Me, OMe, Cl$$
$$R^1 = H, R^2R^3 = benzo$$
$$R^3 = H, R^1R^2 = benzo$$

(72)

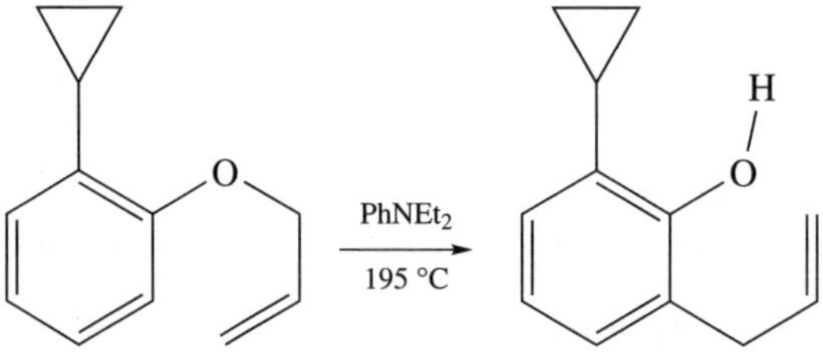

(73)

The *ortho*-allylphenols 171 and 172 which were used for the synthesis of coumaranes (Section III.E.2) were obtained by means of a thermal Claisen rearrangement (equations 78 and 79)[105].

It was found that molybdenum hexacarbonyl effectively catalyzes a tandem Claisen rearrangement—cyclization reaction of allyl aryl ethers 173 to produce the dihydrobenzofurans 174 in good yields (equation 80)[138]. However, the methallyl aryl ethers 175 under the same conditions (40 mol% of catalyst Mo(CO)$_6$ in refluxing toluene for 55 hours) gave good yields of the corresponding 2,2-dimethylchromans 176 (equation 81)[139].

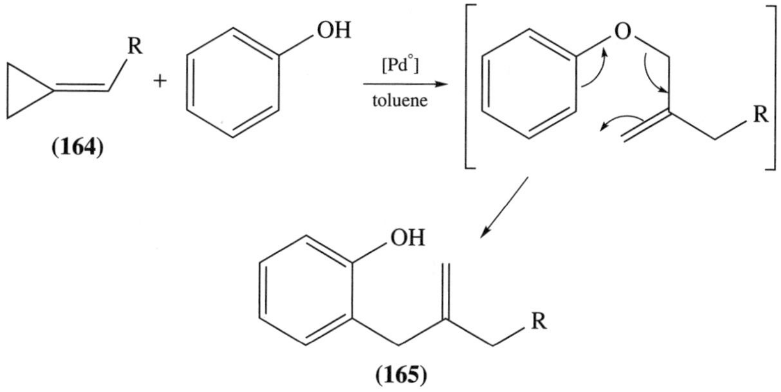

11. Tautomeric equilibria and rearrangements involving phenols

(74)

R = n-C$_7$H$_{15}$;
[Pd^0] = [Pd(PPh$_3$)$_4$](5 mol%) + P(o-tolyl)$_3$(10 mol%)

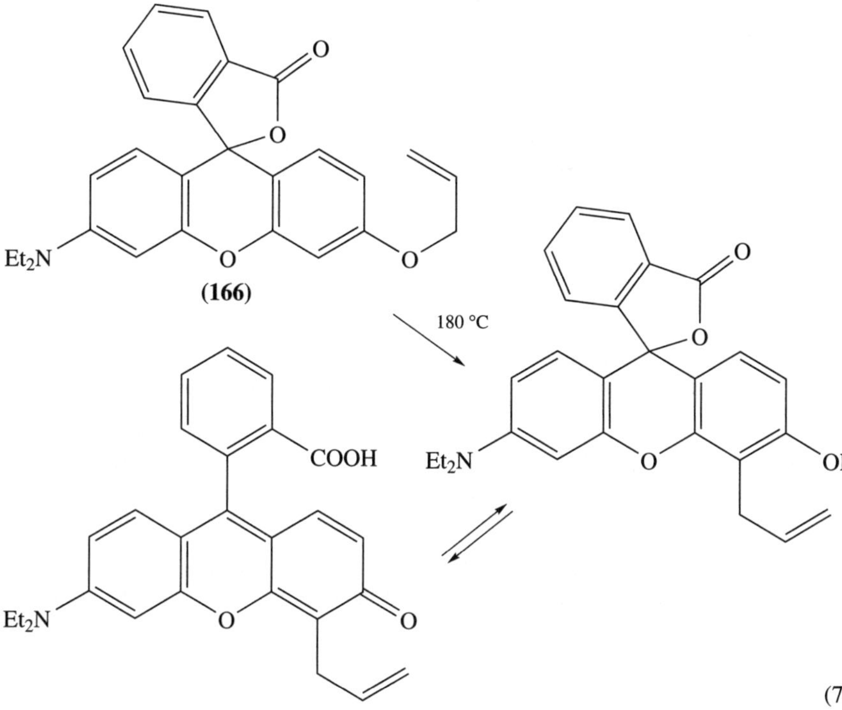

(75)

The *ortho*-Claisen rearrangement was employed in the synthesis of dihydrobenzopyrans **179** using aqueous trifluoroacetic acid as the catalyst for both the condensation of the phenols **177** with allyl alcohols **178** and the rearrangement which was followed by cyclization (equation 82)[140].

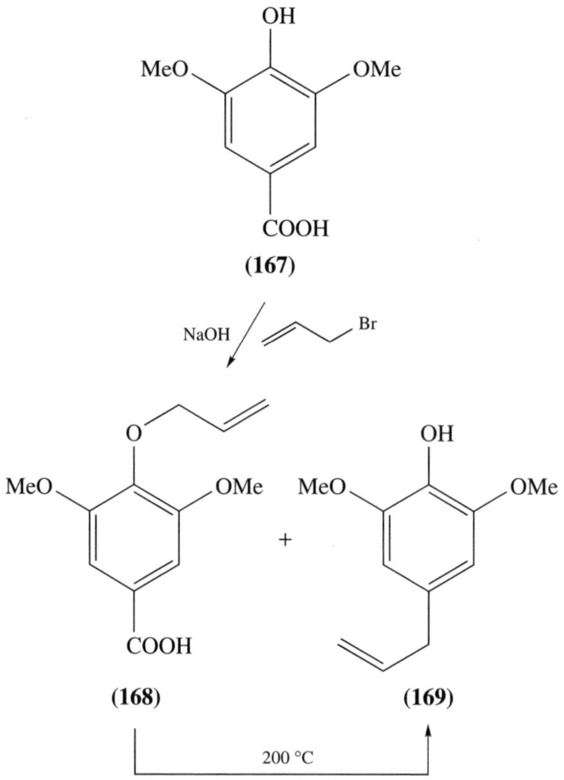

[Structure **(167)**: phenol with OH, two OMe (ortho), and COOH para]

NaOH, allyl bromide (CH₂=CHCH₂Br)

(76)

[Structure **(168)**: O-allyl ether] + [Structure **(169)**: rearranged allyl phenol]

200 °C

    The Claisen rearrangement was also used for the preparation of coumarins and their derivatives. Thus, alkyl 3-acetoxy-2-methylenebutanoate **180** reacts with phenol to afford the ether **181**, which rearranges into methylenecoumarin **182** (equation 83)[141].

    The original 'tandem Claisen rearrangement' promoted by Et_2AlCl and 2-methyl-2-butene was utilized for synthesis of a new type of macrocyclic derivatives **186** from the corresponding macrocyclic polyethers **185** which were formed via **183** and **184** (equations 84 and 85)[142]. This very rapid reaction results in good yields of potential host molecules and supramolecular building blocks under mild conditions, instead of the thermal treatment.

    The aromatization of intermediates under thermal Claisen rearrangement conditions can affect also the alicyclic fragments annelated with the phenolic ring. Thus, the rearrangement of the naphthalene derivative **187** is accompanied by a retro-Diels–Alder reaction involving de-ethylenation (equation 86)[143]. This strategy was used for a high yield synthesis of racemic hongconin **190** (equation 87). The key intermediate **189** was prepared starting from the Diels–Alder adduct **188** in three steps including a Fries rearrangement (see Section IV.D.1)[144].

    While the aliphatic Claisen rearrangement[123] has proven to be a major synthetic tool for controlling the stereochemistry in a C–C bond formation, the aromatic Claisen rearrangement has not been exploited as an asymmetric aryl alkylation protocol[145]. A facile

11. Tautomeric equilibria and rearrangements involving phenols 767

asymmetric O-alkylation of phenols is required in order to carry out the catalytic Claisen rearrangement that proceeds with excellent chirality transfer. These two aims were achieved recently by application of the chiral catalyst **191** for asymmetric O-alkylation[145]. In addition, a new catalytic version of aromatic Claisen rearrangement was proposed where the selectivity for *ortho*-migration and the high chirality transfer are provided by a lanthanide catalyst (equation 88)[145].

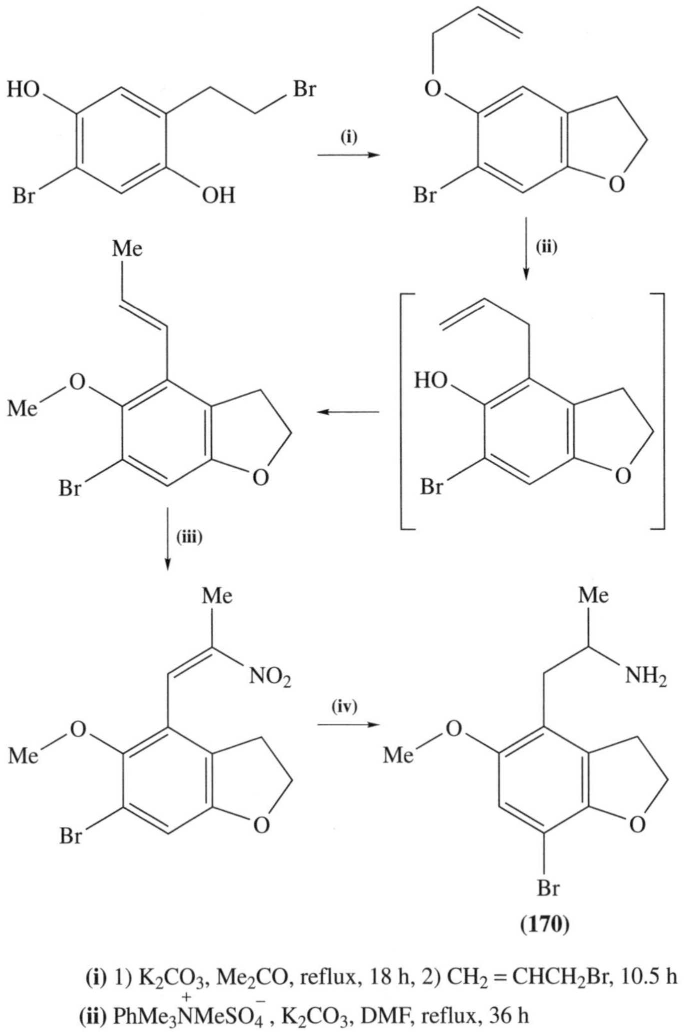

(77)

(i) 1) K_2CO_3, Me_2CO, reflux, 18 h, 2) $CH_2=CHCH_2Br$, 10.5 h
(ii) $PhMe_3\overset{+}{N}MeSO_4^-$, K_2CO_3, DMF, reflux, 36 h
(iii) $AgNO_2$, I_2, pyridine
(iv) 1) $LiAlH_4$, THF, H_2O, 2) $Br_2/AcOH$

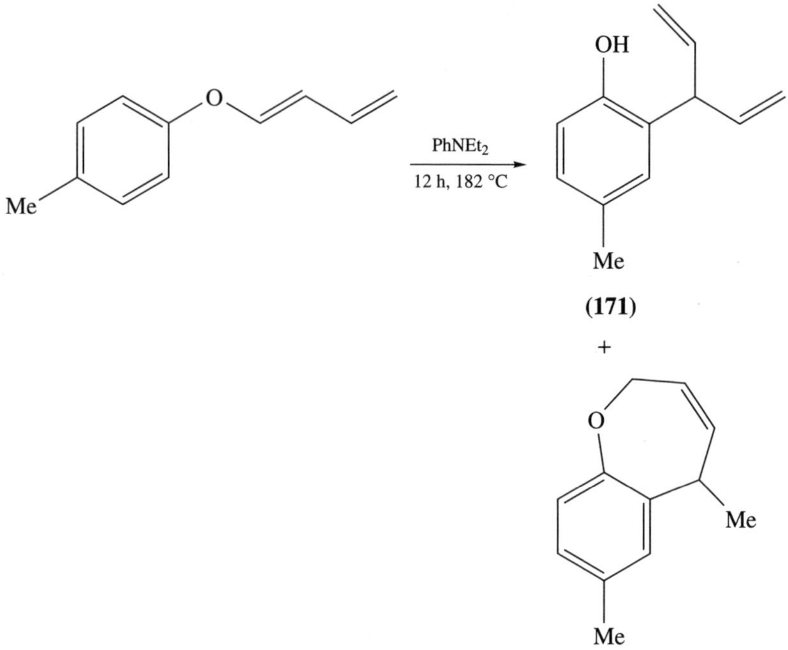

(171) + (78)

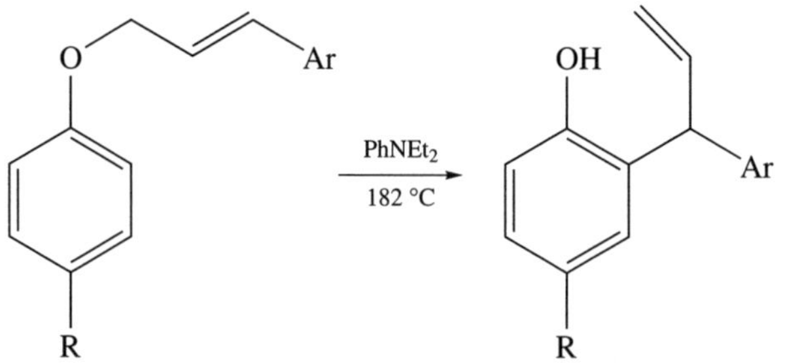

(172) (79)

R = H, Me, OMe; Ar = 4-XC_6H_4, X = H, Me, MeO, Br, CN

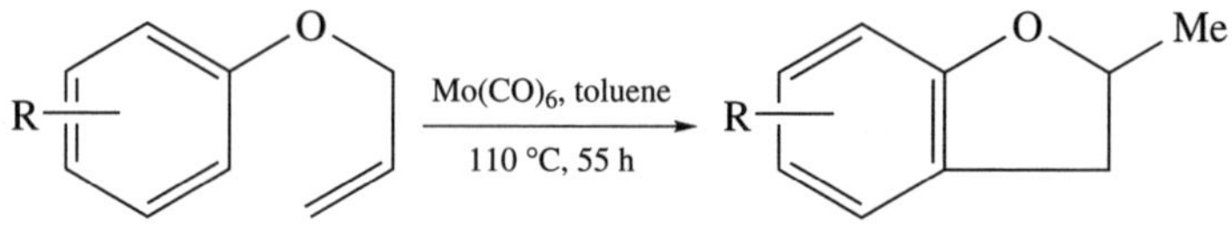

(173) (174) (80)

R = H, 4-Me, 4-MeO, 4-Cl

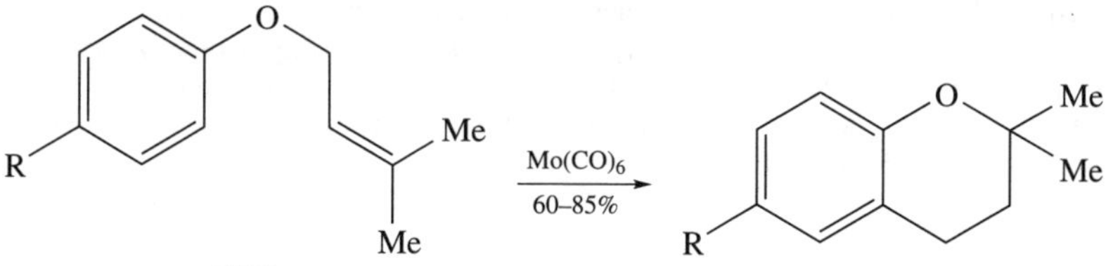

11. Tautomeric equilibria and rearrangements involving phenols

(175) → (176) R = H, Me, MeO, Et, Br, CHO Mo(CO)₆, 60–85% (81)

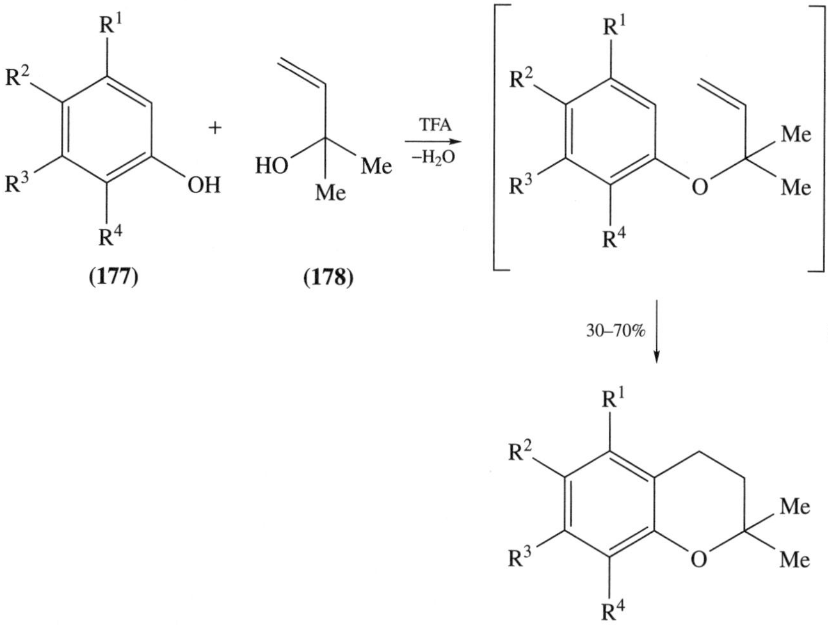

(177) + (178) → [intermediate] → (179) TFA, −H₂O, 30–70% (82)

R¹, R³, R⁴ = H, Me; R² = H, OH

A highly enantioselective and regioselective aromatic Claisen rearrangement was carried out using the reaction of catechol monoallyl ethers **192** with the chiral boron reagent **193**. This reaction occurs without the formation of either the *para*-rearrangement or the abnormal Claisen rearrangement products (equation 89)[146].

The aromatic Claisen rearrangement was employed in the synthesis of building blocks for various macrocyclic compounds, such as pendant-capped porphyrins[147], multidentate macrocycles containing 1,3,4-oxadiazole, imine and phenol subunits[148], as well as to prepare longithorone B, a sixteen-membered farnesylated *para*-benzoquinone[149].

2. Other isomerizations of allyloxy arenes

A new convenient synthesis of alkyl and aryl 1-propenyl ethers in good to excellent yields was developed. The aryl allyl ethers obtained can be smoothly isomerized to the desired 1-propenyl ethers by refluxing in a basic ethanolic solution containing pentacarbonyliron as a catalyst[150,151]. The interesting isomerization of 2-(allyloxy)phenyllithium **194** in the presence of tetramethylethylene diamine (TMEDA) occurs with a new domino cyclization–elimination sequence to afford the 2-(cyclopropyl)phenol **195** (equation 90)[152].

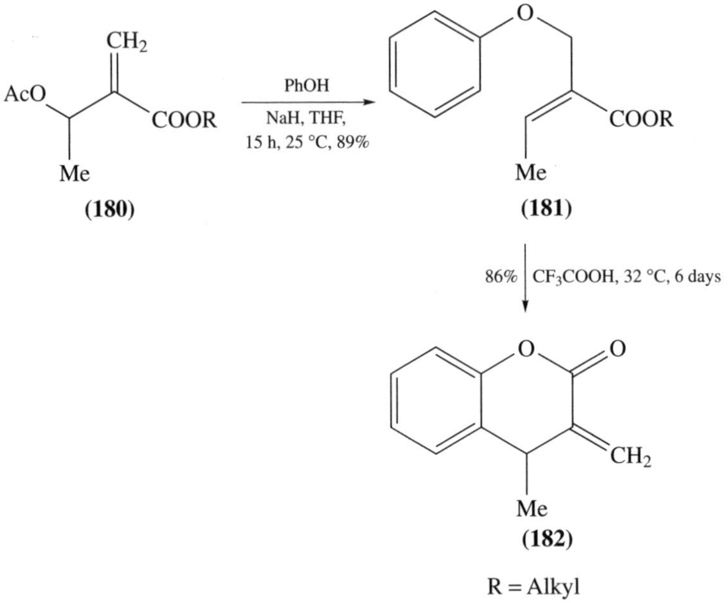

$$
\begin{array}{c}
\text{(180)} \xrightarrow[\text{15 h, 25 °C, 89%}]{\text{PhOH, NaH, THF}} \text{(181)} \xrightarrow[\text{86%}]{\text{CF}_3\text{COOH, 32 °C, 6 days}} \text{(182)}
\end{array}
\tag{83}
$$

R = Alkyl

A unique rearrangement of 2-bromophenyl allyl ethers **196** proceeds as a completely regio- and stereospecific process without any migration to the *para*-position and with conservation of the regiochemistry in the allyl substituents of the phenolic products **198**. It was assumed that the reaction occurs via the π-allyl complexes **197** (equation 91)[153].

The reversible migrations of aryloxy groups along the perimeter of the pentaphenyl-substituted cyclopentadiene system **199** also deserve attention here (equation 92)[154].

A more detailed description of such circumambulatory rearrangements was published recently[155].

C. Rearrangements of Propargyl Aryl Ethers

The titled reactions are employed for synthesis of benzopyrane derivatives. Thus, the racemic cordiachromene **202** (from the cannabinoid class) was prepared starting from 6-methylhept-5-en-2-one **200** using the Claisen rearrangement of the intermediate propargyl ether **201** in an overall yield of 50% (equation 93)[156].

New photochromic chromenes **204** and **205** annulated with a furan ring were obtained using the Claisen rearrangement of propargyl ethers **203** (equation 94)[157]. It is interesting that the Claisen rearrangement of aryl propargyl ether **206**, which was carried out by heating in *N,N*-diethylaniline at 215 °C, gave naphthopyran **207** whereas naphthofuran **208** was obtained as a sole product under the same conditions but in the presence of cesium fluoride (equation 95)[158]. The addition of components other than CsF (e.g. CsCl, KF, RbF, CaF_2, BaF_2) lead to chromene **207** in yields of 84–97% whereas the

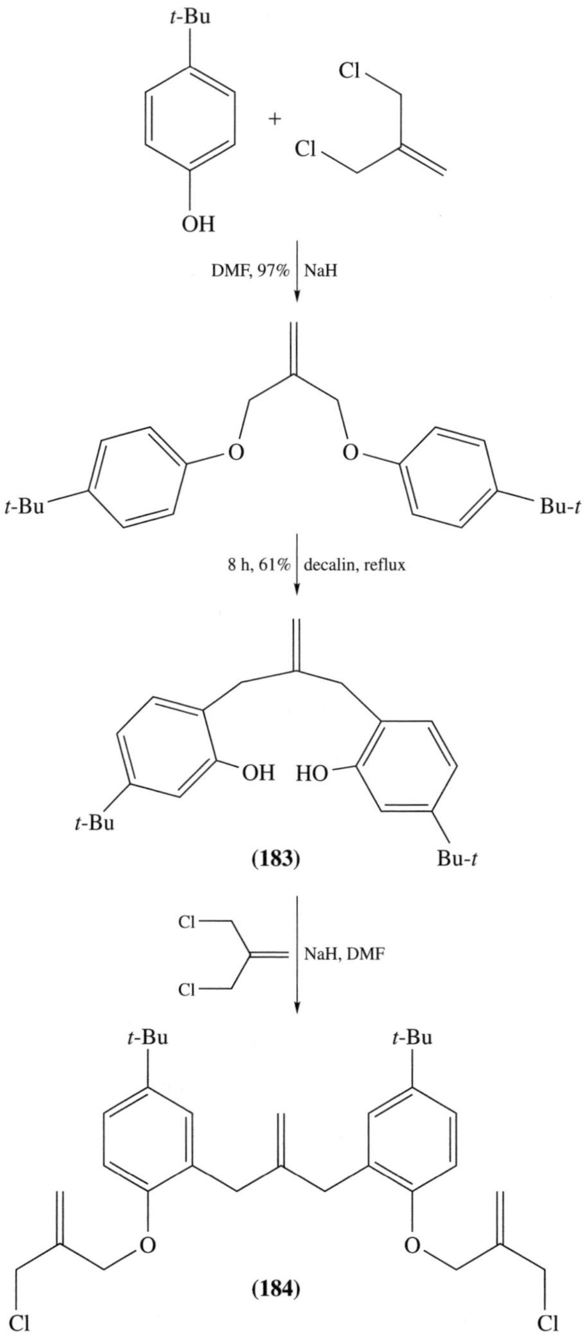

11. Tautomeric equilibria and rearrangements involving phenols

(84)

(183)

(184)

183 + 184

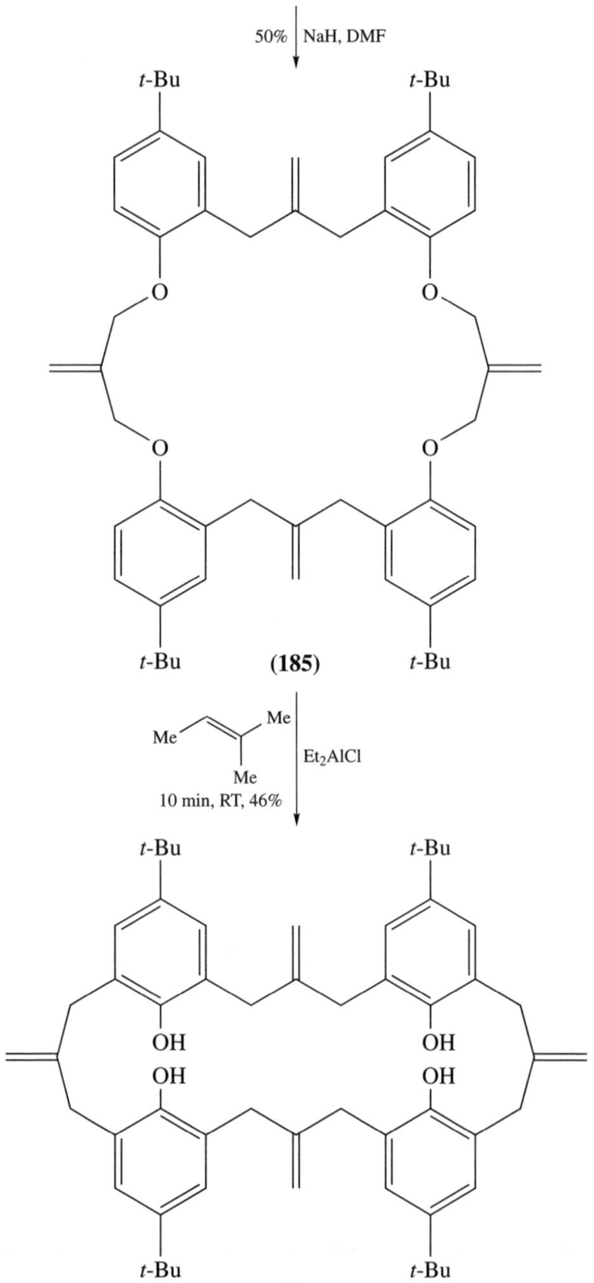

$\downarrow$ 50% NaH, DMF

(185)

$\downarrow$ Et$_2$AlCl, 10 min, RT, 46%

(186)

(85)

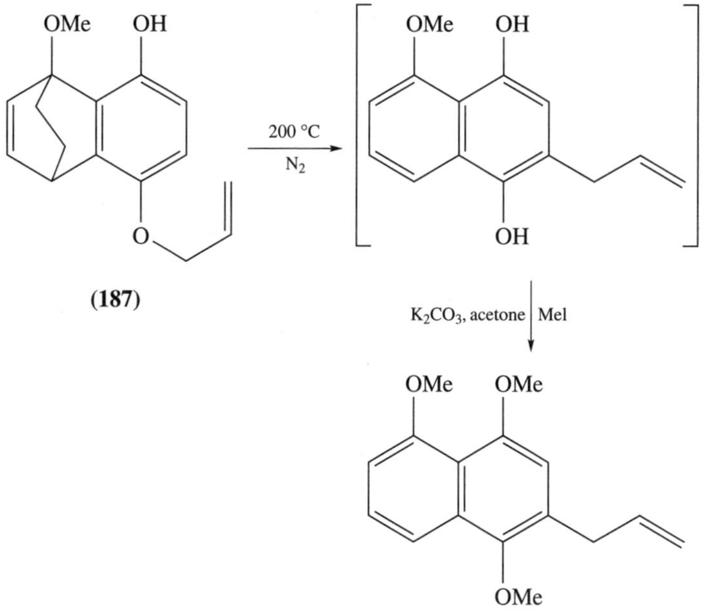

reaction in the presence of CsF (0.1–10 mol equiv) results in the furan **208** (86–87%) and chromene **207** as a byproduct (2.2–6.8%). Such pathway change can be explained by the fact that the formation of benzopyrans **213** occurs via enolization step **209** → **210** while cesium fluoride acts as a soft base providing the abstraction of α-hydrogen atom from the α-allenylketone **209** to give the enolate anion **211** that cyclizes to the benzofuran **212** (equation 96)[158–160].

The benzofuran derivatives **215** and **217** were obtained also by Claisen rearrangement of 2-phenylsulfinyl-2-propenyl phenyl ethers **214** (refluxing in mesitylene in the presence of SiO_2, 180 °C, 22 h) (equation 97)[161] as well as of aryl β-chloroallyl ethers **216**[162] (equation 98). These aryl ethers act here as the synthetic equivalents of aryl propargyl ethers.

D. Rearrangements of Phenolic Esters

1. Fries rearrangement

The Fries rearrangement used for the preparation of aryl ketones from phenolic esters is now one of the most significant reactions in the synthetic chemistry of aromatic compounds, both in the classical version (equation 99) and in the newest modifications (see Section IV.D.3).

In general, high reaction temperatures favor the *ortho*-rearrangement whereas low temperatures favor the *para*-rearrangement, although many exceptions are known. The mechanistic aspects, scope, procedures and synthetic applications of the *ortho*- and *para*-Fries rearrangement are generalized in detail in many reviews[163–167]. The use of rare-earth element (Sc, Hf, Zr) complexes as water-compatible catalyst (Lewis acids) in the Fries rearrangement was described in a recent survey[168] as well as in a series of papers[169–173]. Novel efficient catalysts, such as a mixture of methanesulfonic acid and phosphorus oxychloride (MAPO)[174], various zeolites[175–179] as well as silica composite catalysts[180–183], were proposed for the Fries rearrangement. Studies of Fries rearrangement under microwave irradiation conditions were also reported[184–188]. The Fries rearrangement was efficiently carried out in liquid hydrogen fluoride[98,189], which was also employed as a medium for the cleavage of ω-amino acids from a Merrifield resin in peptide synthesis[190]. The kinetics and mechanisms of a Fries rearrangement catalyzed by $AlCl_3$ in different solvents were discussed in a series of papers by Japanese chemists[191–194].

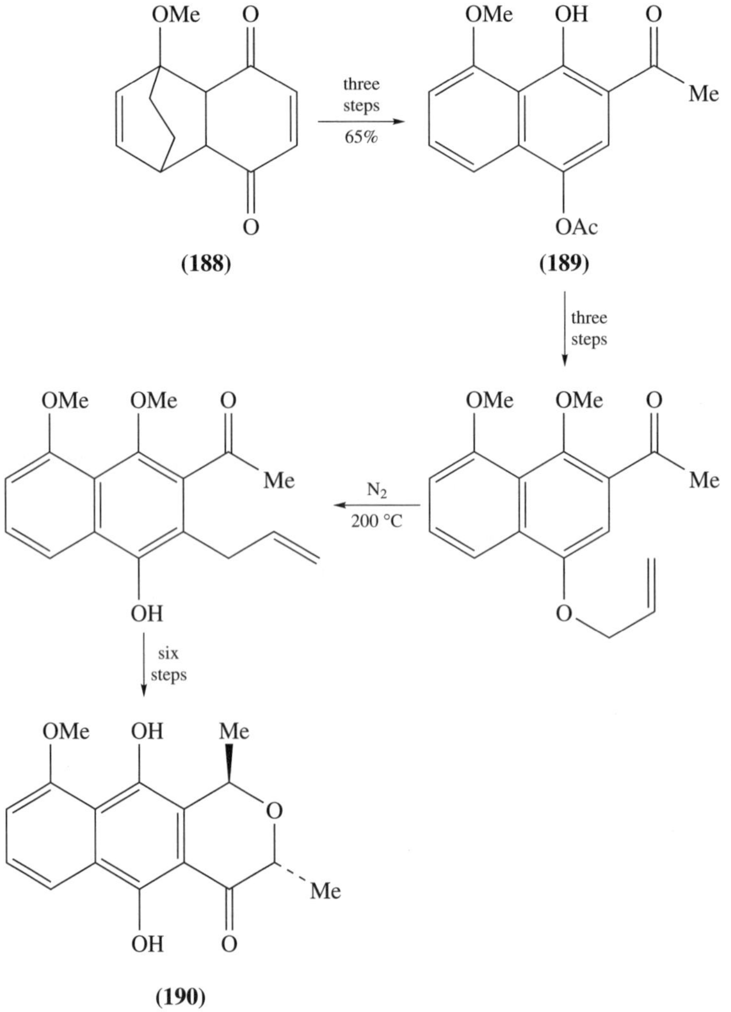

(87)

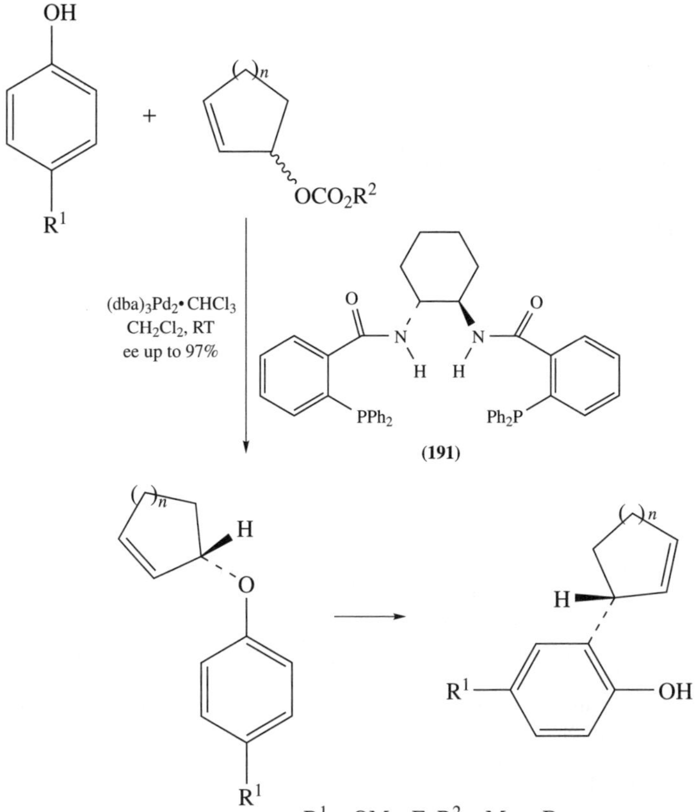

11. Tautomeric equilibria and rearrangements involving phenols

$R^1 = OMe, F; R^2 = Me, t\text{-Bu};$
$n = 1–3$

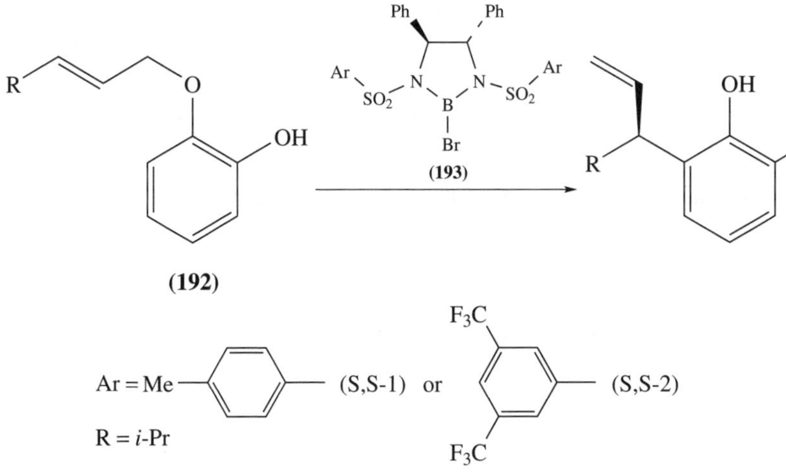

(88)

(89)

Ar = Me—⟨⟩— (S,S-1) or 3,5-(F$_3$C)$_2$C$_6$H$_3$ (S,S-2)

R = i-Pr

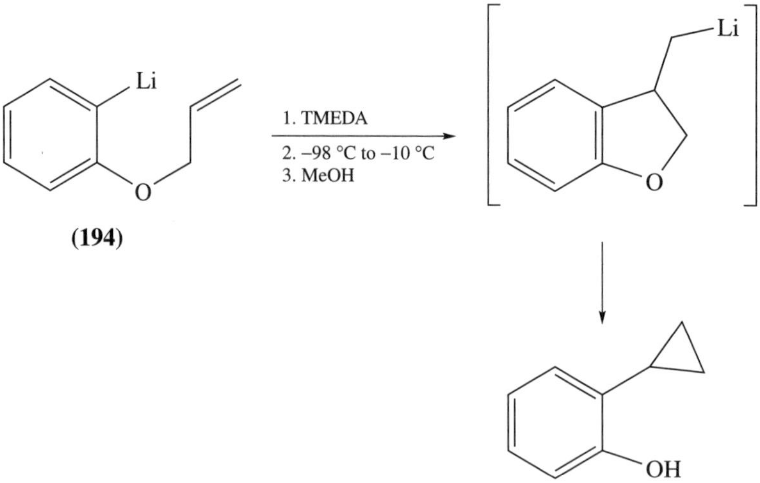

(194) [structures] (90) → (195)

A new approach to the synthesis of 3-acetyl-5-methoxynaphthoquinone **221** involves the pyrolysis of the polycycle **218** derived from the Diels–Alder adduct **188** (Section IV.B.1). The regiospecific Fries rearrangement of diacetoxynaphthalene **219** leads to the naphthol **220** whose oxidation gives the desired product **221** as a key intermediate for the synthesis of naturally occurring antibiotic pyranoquinones (equation 100)[195].

A specific Fries rearrangement takes place when 1-aroyloxy-5-methoxynaphthalenes **222** undergo an intramolecular acyl transfer to form *peri*-hydroxynaphthoyl aryl ketones **223** under mild conditions in the presence of trifluoroacetic anhydride and boron trifluoride etherate (equation 101)[196].

The first synthesis of a dimeric pyranonaphthoquinone **225** which is related to naturally occurring biologically active compounds such as actinorhodin and crisamicin includes the double Fries rearrangement of the bis-ether **224** as one of the stages[197,198] (equation 102).

The bicoumarin **229** was obtained using a double Fries rearrangement of the diacetate **226** promoted by TiCl$_4$ as a Lewis acid, and a subsequent cyclization of the dicarbonate **228** derived from the diketone **227** (equation 103)[199]. The Fries rearrangement of hydroxycoumarin chloroacetates **230** provides a new short pathway to furocoumarins **231** (equation 104)[200].

The Fries rearrangement was efficiently used for the synthesis of O- and C-glycosides. Thus, the 'O → C-glycoside rearrangement' as an access to C-glycosides is a two-stage reaction which proceeds in a one pot in the presence of a Lewis acid. The first step is the low-temperature O-glycosidation of the 1-fluoro sugar **232**, X = F to form the O-glycoside **233**, which is further converted *in situ* to *ortho*-C-glycoside **234** simply by raising the temperature[201,202] (equation 105). An analogous approach to aryl C-glycosides was proposed by Schmidt and coworkers[203–205] (equation 106) (see also Reference 206).

The so-called 'thia-Fries rearrangement' occurs upon treatment of aryl phenylsulfinates **235** (obtained by reaction of phenols with phenylsulfinyl chloride) with AlCl$_3$ at 25 °C to afford the (phenylsulfinyl)phenols **236** in good yields (equation 107)[207].

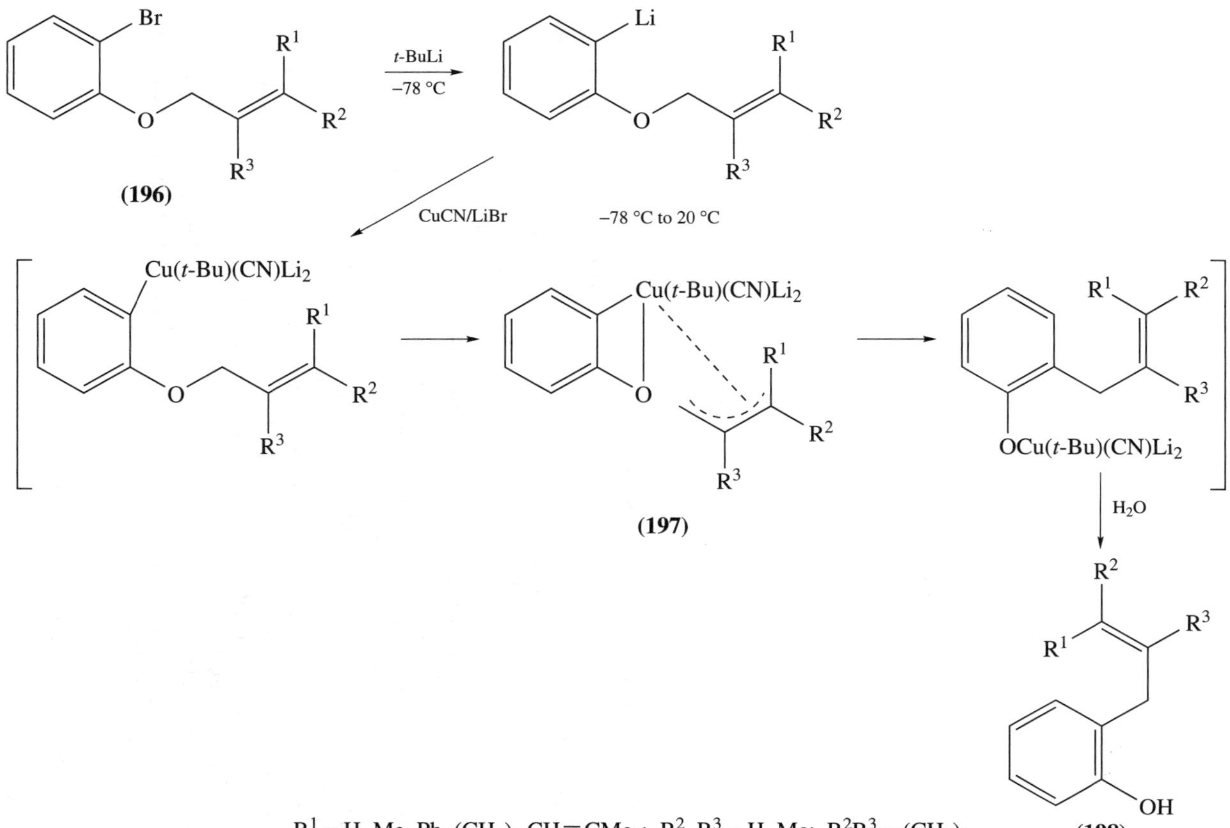

(91)

R^1 = H, Me, Ph, $(CH_2)_2$CH=CMe_2; R^2, R^3 = H, Me; R^2R^3 = $(CH_2)_4$

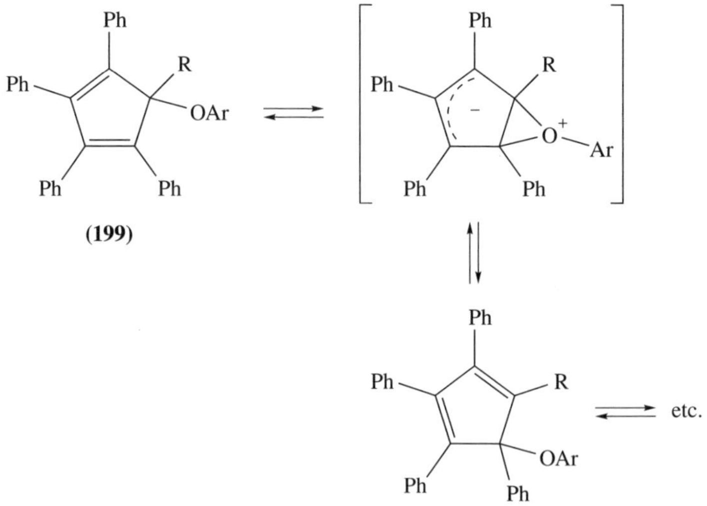

$$\text{(199)} \rightleftharpoons [\ldots] \rightleftharpoons \ldots \rightleftharpoons \text{etc.} \qquad (92)$$

Ar = 4-XC$_6$H$_4$; X = H, Me, MeO, Cl, NO$_2$, NH$_2$,
R = 4-MeC$_6$H$_4$

2. Baker–Venkataraman rearrangement

The rearrangement of *ortho*-aroyloxyacetophenones **237** to *ortho*-hydroxybenzoylmethanes **238** in the presence of basic reagents is known as the *Baker–Venkataraman rearrangement* (for a review see Reference 208) (equation 108).

There are scanty reports about the Baker–Venkataraman rearrangement which is used in synthesis very seldom. Thus, in the approach mentioned in equation 106 the C-glycoside **239** undergo O-benzoylation to afford the ester **240**, which rearranges into the 1,3-dicarbonyl compound **241** formed as a keto–enol mixture in 48% yield (equation 109)[209].

In another approach the same starting C-glycoside **239** was acylated with *para*-anisoyl chloride to form the ester **242**, which was treated with lithium diisopropylamide (LDA) to give the enol of a dibenzoylmethane **243** (equation 110)[209].

A brief survey (5 papers from 1933 to 1950) was given and the conditions of Baker–Venkataraman rearrangement were investigated elsewhere[210] (equation 111). It was found that sodium ethoxide in benzene was the best catalyst for this reaction. It was also shown that this rearrangement failed in the case of the ester **244**.

The Baker–Venkataraman rearrangement was used as a key step in syntheses of trihydroxyflavanones **245** (equation 112)[211] as well as isoflavones **246** (equation 113)[212].

An interesting example of Baker–Venkataraman rearrangement was reported for *peri*-acyloxyketones **247** (equation 114)[213].

3. Anionic ortho-Fries rearrangement

Side by side with the wide application of the classical Fries rearrangement in organic synthesis, a new approach is developing lately. This method represents an anionic

(93)

(94)

$R_1 = (CH)_4, (CH_2)_4, (CH_2)_5; R^2 = Me, Ph$

equivalent of the *ortho*-Fries rearrangement which is based on the so-called *directed ortho metalation reaction* (DoM) (equation 115). The DoM reaction comprises the deprotonation of molecule **248** in an *ortho*-position to the heteroatom-containing directed metalation group (DMG) by a strong base such as alkyllithium to form the *ortho*-lithiated intermediate **249**. The latter upon treatment with electrophilic reagents gives 1,2-disubstituted products **250**. 40 DMGs are known, over half of which, including the $CONR_2$ and $OCONR_2$ groups, have been introduced into synthetic practice during the last twenty years. A comprehensive review of DoM reactions, of which only a small part is represented by anionic rearrangements, was published a decade ago[214].

(95)

(206) **(207)** **(208)**

(209) **(210)** **(211)** **(212)** **(213)**

(96)

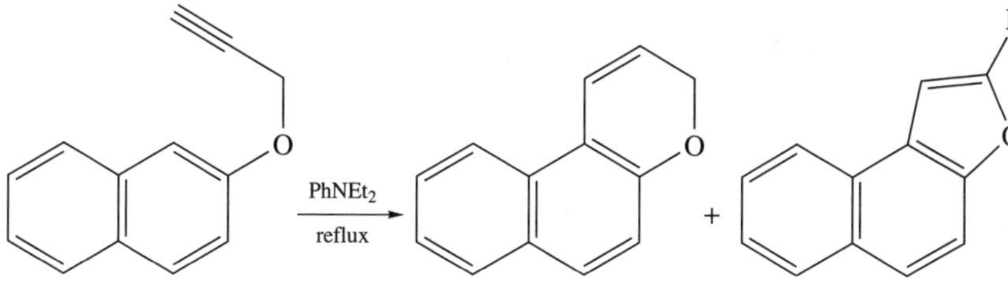

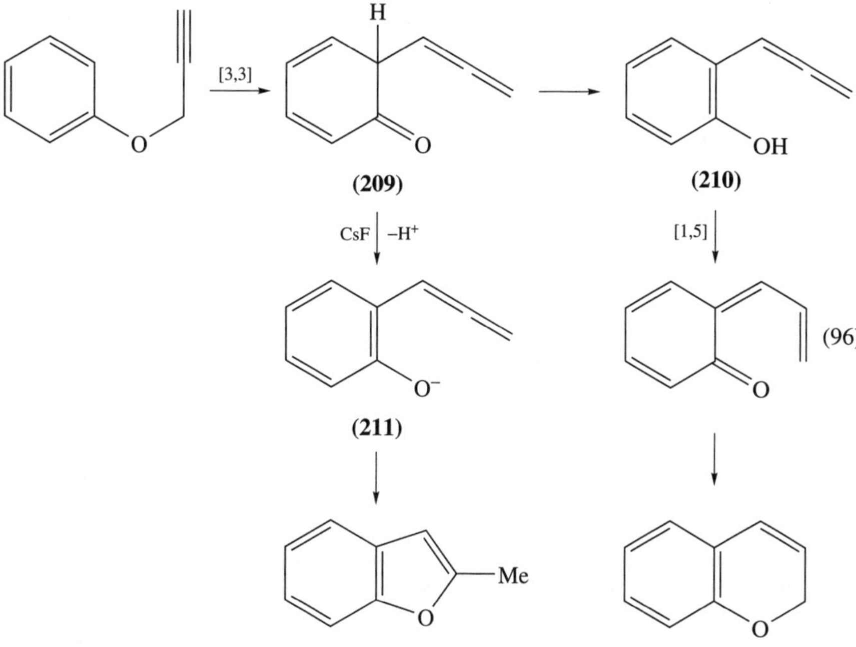

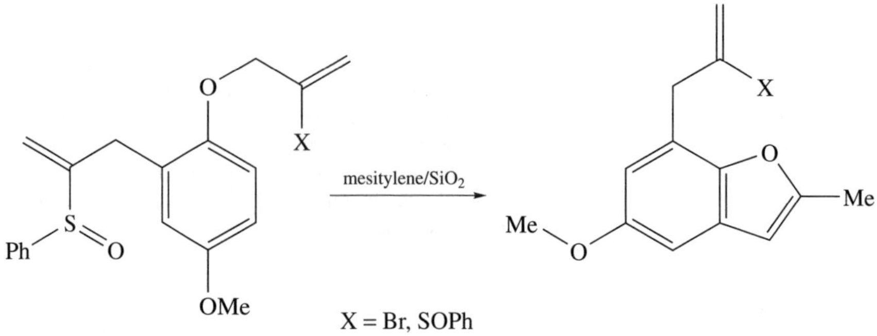

(214) → (215) X = Br, SOPh (97)

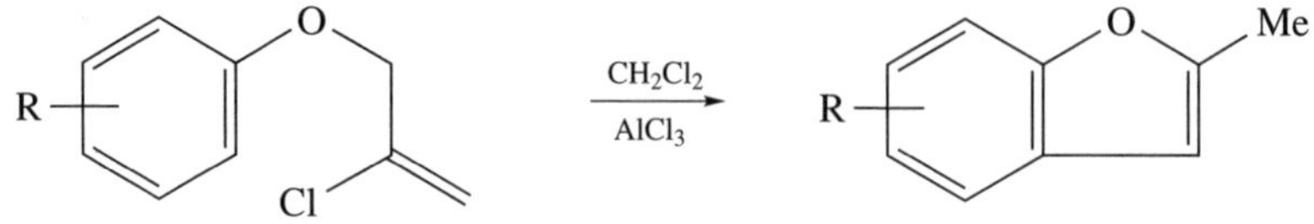

(216) R = H, 4-Cl, 4-NO₂, 2,4-Cl₂, 2,3,5-Me₃ (217) (98)

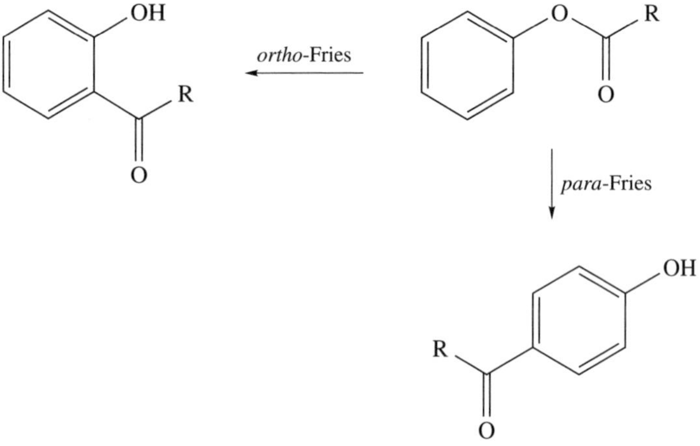

(99)

An unprecedented O → C 1,3-carbamoyl migration of the *ortho*-lithiated species **252** in the course of a directed metalation reaction of carbamates **251** to give the salicylamides **253** was first reported by Sibi and Snieckus (equation 116)[215]. This approach was afterwards developed in a series of investigations[216–218].

The dicarbamates **254** were smoothly lithiated by using *t*-BuLi—TMEDA at −80 °C and then allowed to warm to room temperature over 16 hours. Under these conditions a smooth anionic *ortho*-Fries rearrangement gave the diamido derivatives **255** in fair yields (25–80%) (equation 117)[219] (see also Reference 220). A similar rearrangement was also described for [2,2]-paracyclophanes[221].

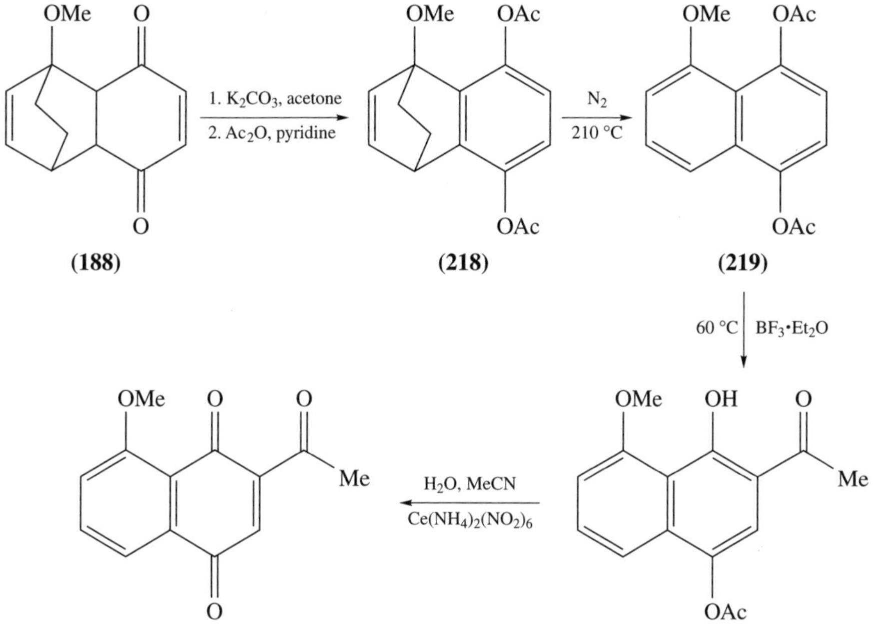

11. Tautomeric equilibria and rearrangements involving phenols 783

(188) (218) (219)

(221) (220) (100)

Anionic *ortho*-Fries rearrangement which involves a 1,3-transposition of a carbamoyl group occurred also in the chromium complex **256** on warming the lithium intermediate **257** to $-20\,^\circ$C (equation 118)[222]. The lithio benzo[*b*]thiophene **258** obtained at $-78\,^\circ$C was allowed to attain room temperature, and when it was left stirring for 12 hours it gave the salicylamide **259** (equation 119)[223].

A transformation called 'metallo-Fries rearrangement' was described for lithiation of O-substituted *ortho*-nitrophenols **260** (equation 120)[224]. Analogous migrations of SiR$_3$ groups were reported for reactions of the bromine-substituted O-silylated phenols with *t*-BuLi[225].

An anionic *ortho*-Fries rearrangement has also been observed in the naphthyl-, phenanthryl-, pyridyl- and quinolinylcarbamate series. It was found that the rate of anionic *ortho*-Fries rearrangement is highly sensitive to N-substitution and temperature, and was shown by crossover experiments to proceed by an intramolecular mechanism.

However, the real Fries rearrangement, i.e. a transformation of aryl esters into *ortho*-hydroxyketones accompanied by migrations of acyl groups, can also be a metal-promoted reaction to produce, under the proper reaction conditions, good yields of *ortho*-specific acyl migration products. Thus, *ortho*-bromophenyl pivaloate (**261**, R^1 = *t*-Bu, R^2 = R^3 = H) affords *ortho*-hydroxypivalophenone (**262**, R^1 = *t*-Bu, R^2 = R^3 = H) in 76% yield (equation 121), whereas the phenyl pivaloate reacts with AlCl$_3$ (refluxed in dichloroethane for 18 h) to form *para*–*tert*-butylphenol (25%) and phenol (65%)[226]. The same work[226] described also the so-called anionic *homo*-Fries rearrangement, namely, a series of pivaloates **263**–**265** having the ester functionality separated from the aromatic nucleus by a carbon chain gave, under the same conditions, different products depending on the length of this chain (equations 122–124).

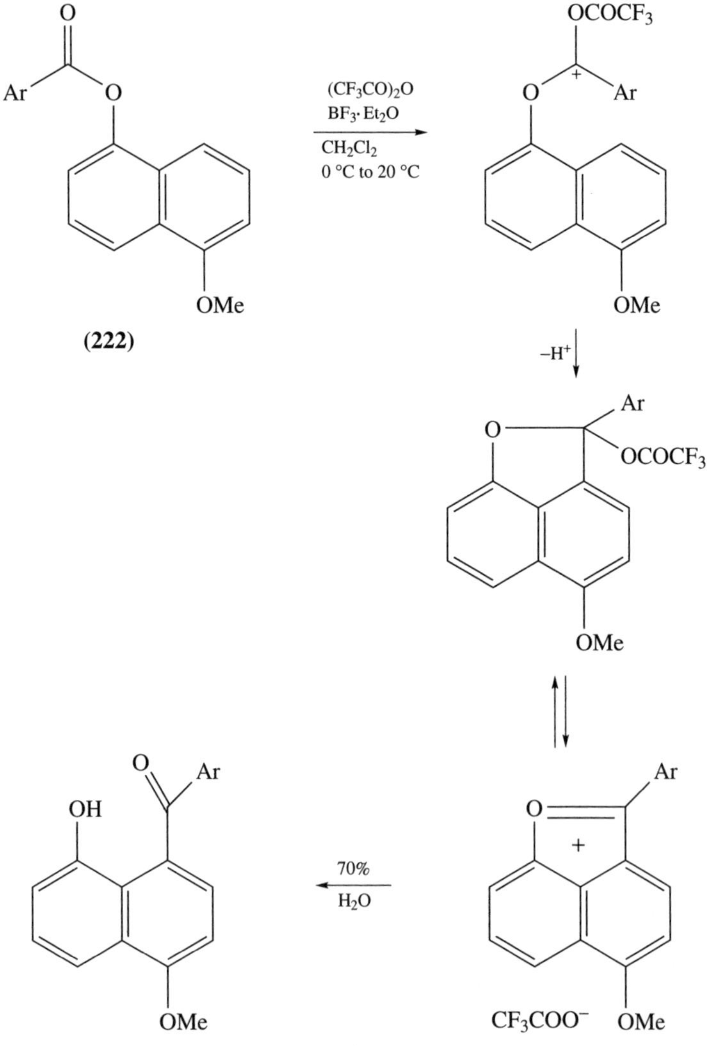

(222)

(101)

(223) Ar = Ph, 4-MeOC$_6$H$_4$, 4-O$_2$NC$_6$H$_4$

ortho-Hydroxymethylated benzophenones **267**, key intermediates in the synthesis of the phenolic alkaloids (±)-cherylline and (±)-latifine, were obtained by anionic Fries rearrangement of the ester precursors **266** (equation 125)[227].

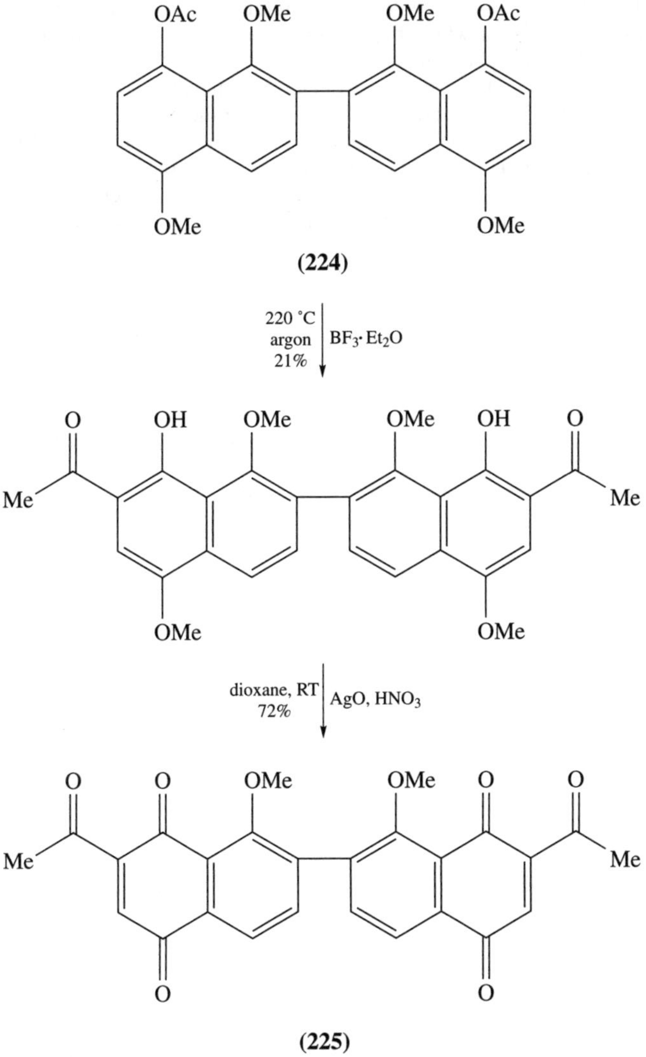

11. Tautomeric equilibria and rearrangements involving phenols 785

(224)

220 °C, argon, 21% | BF$_3$·Et$_2$O

(102)

dioxane, RT, 72% | AgO, HNO$_3$

(225)

The above-named anionic *homo*-Fries rearrangement was employed for developing a general approach to substituted hydroxyphthalans **269** as precursors to isobenzofurans[228]. In this approach the treatment of benzyl esters **268** with BuLi in a 4:1:1 THF–Et$_2$O–hexane mixture at −100 °C was followed by immediate quenching with NH$_4$Cl, and the crude material **269** was treated with dimethyl acetylenedicarboxylate (cat. AcOH, 100 °C, 30 min) to give the intermolecular Diels–Alder adducts **270** in yields of 40–80% (equation 126)[228].

(−)-Balanol, a fungal metabolite with potent protein kinase C inhibitory activity, was prepared in a total synthesis in which the anionic *homo*-Fries rearrangement was used as a key step to form the benzophenone subunit **271** (equation 127)[229,230].

One more variant of the anionic Fries rearrangement, namely a *lateral* Fries rearrangement, constitutes an O → C carbamoyl transposition and thereby provides a regiospecific and general route to 2-hydroxyphenyl acetamides **273**, which are precursors to the benzo- and naphthofuranones **274**. This reaction proceeds via migration of a carbamoyl group in the starting carbamate **272** to a side chain but not to the aromatic nucleus (equation 128)[231]. The analogous 2-hydroxyphenyl acetamides were also described elsewhere[232].

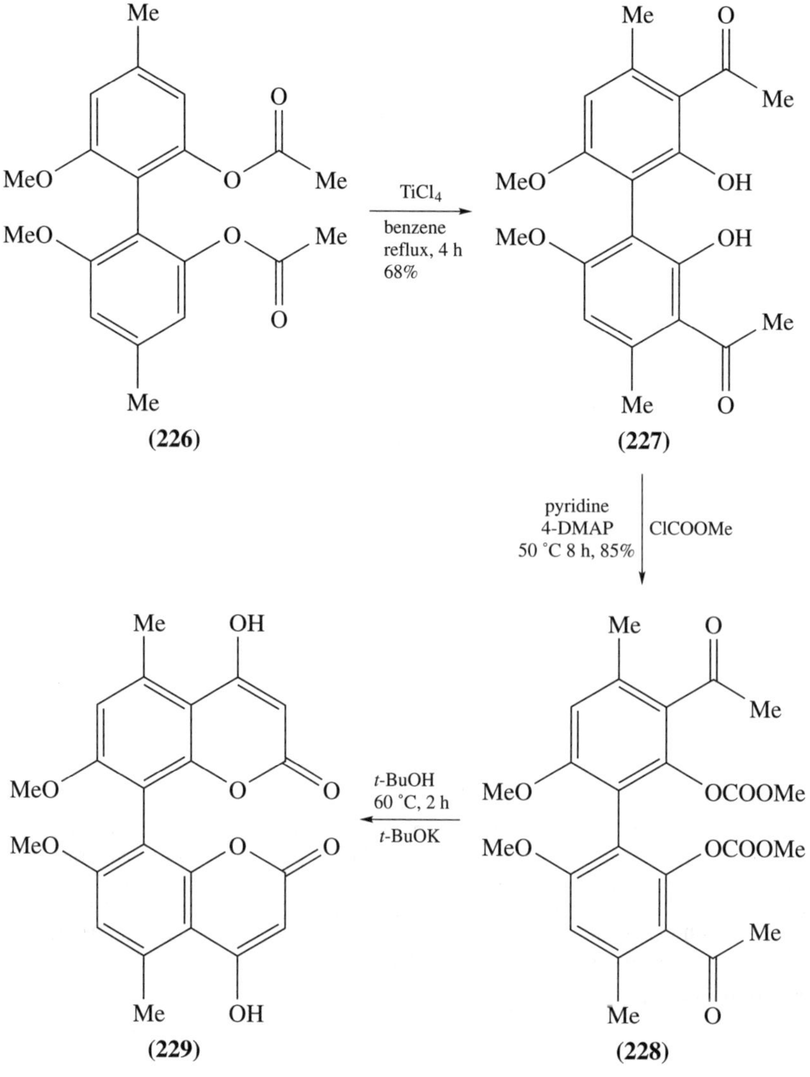

(103)

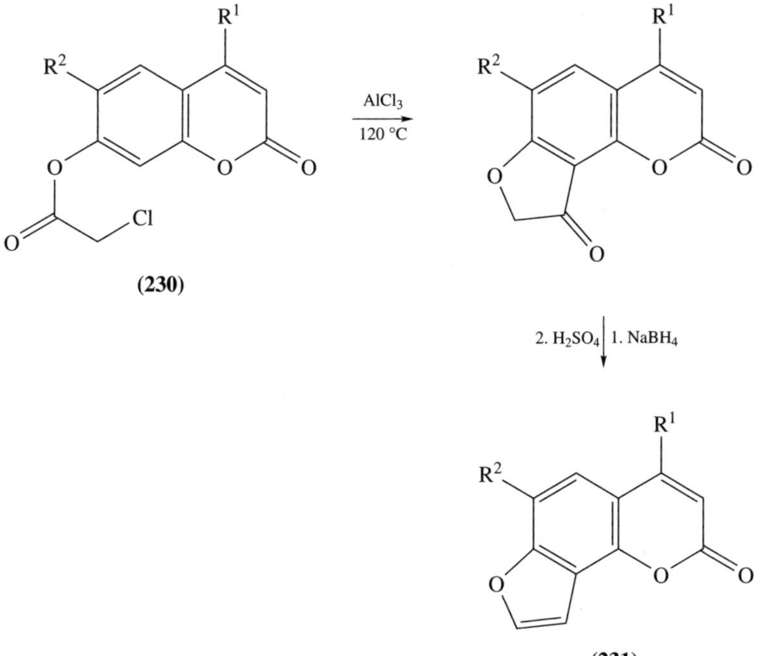

11. Tautomeric equilibria and rearrangements involving phenols

(230) → (231)

AlCl₃, 120 °C

2. H₂SO₄ | 1. NaBH₄   (104)

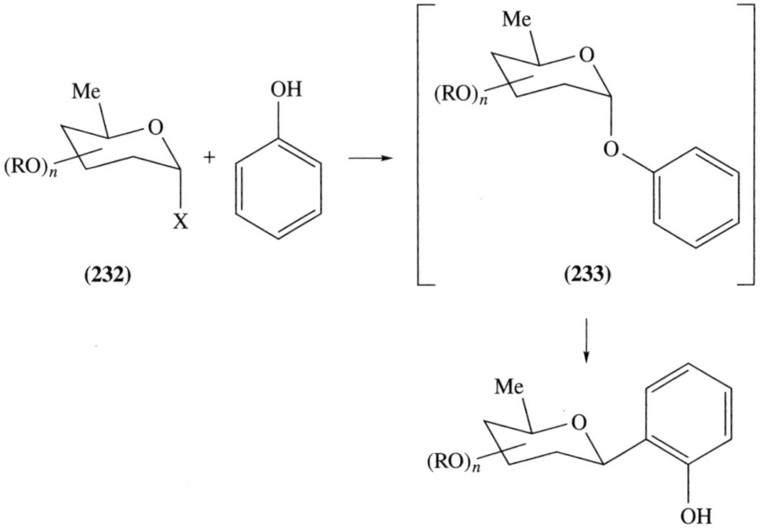

(232) + PhOH → [(233)] → (234)   (105)

R = Me, PhCH₂
X = F, OAc

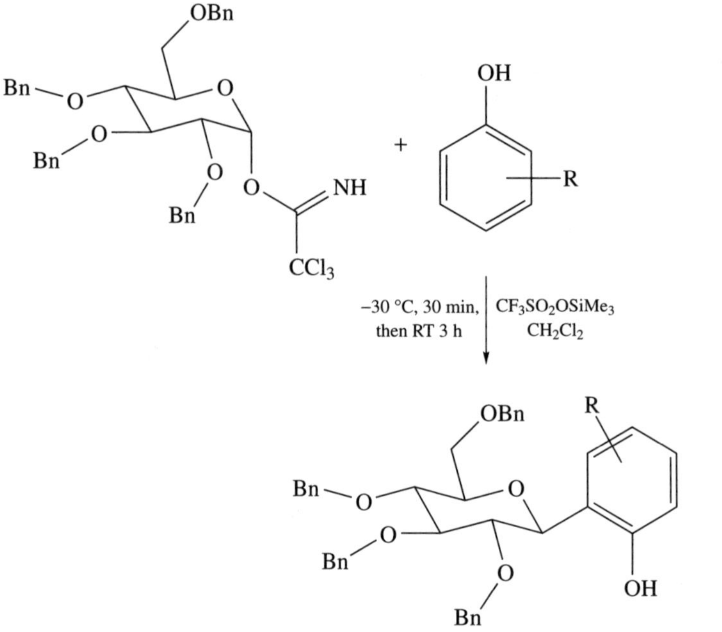

$$\text{(106)}$$

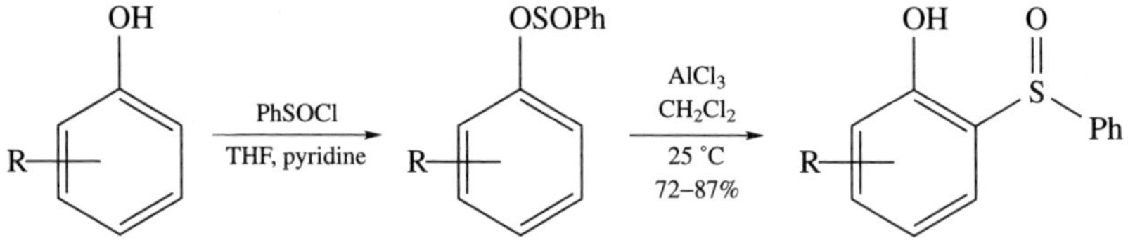

$$\text{(107)}$$

R = 4-Me, 4-MeO, 3-Cl, 3-Me, 2-Me

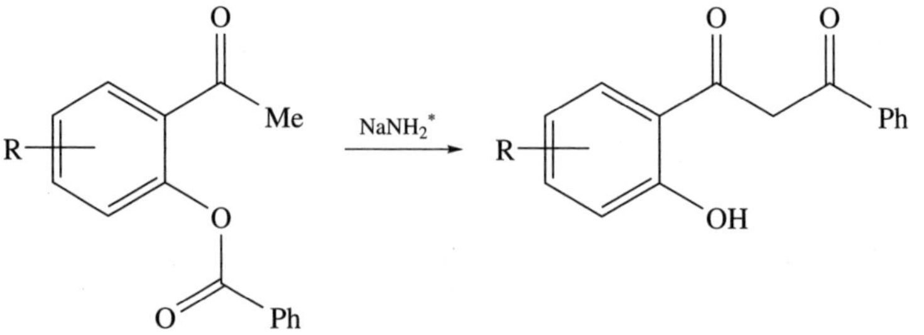

$$\text{(108)}$$

R = Alk, OH, OAlk
* also EtONa, Na, K$_2$CO$_3$, NaOH

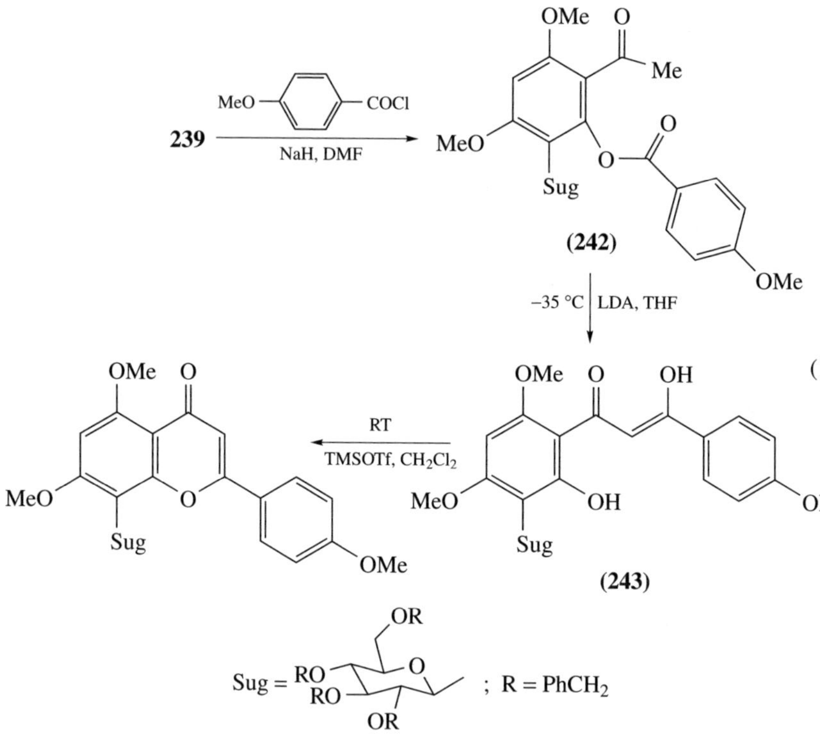

(110)

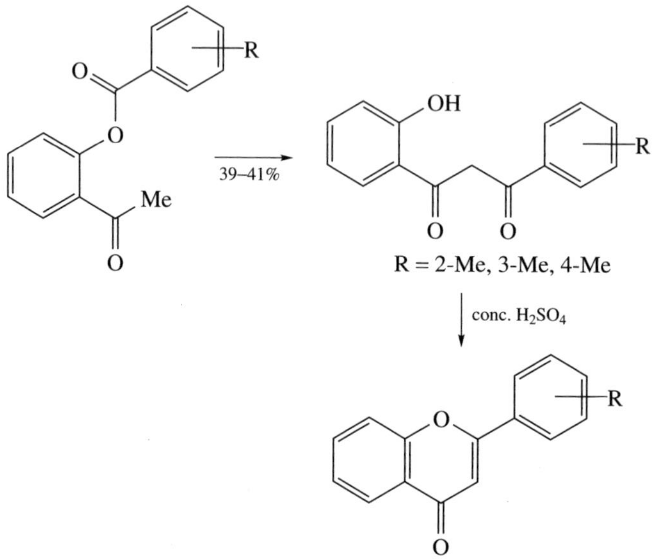

Sug = tetra-O-benzyl sugar ; R = PhCH$_2$

(111)

R = 2-Me, 3-Me, 4-Me

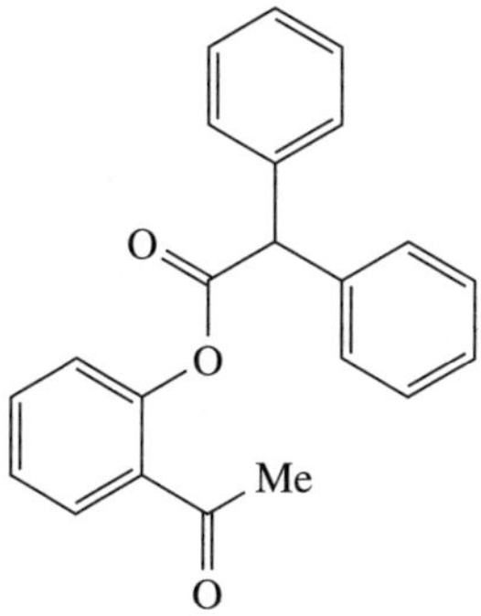

11. Tautomeric equilibria and rearrangements involving phenols

(244)

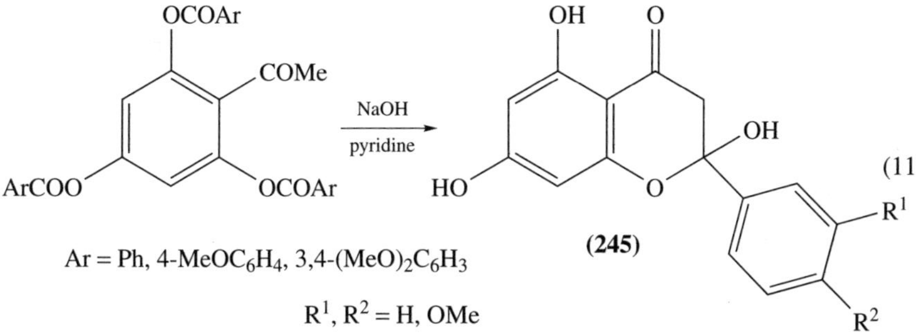

(112)

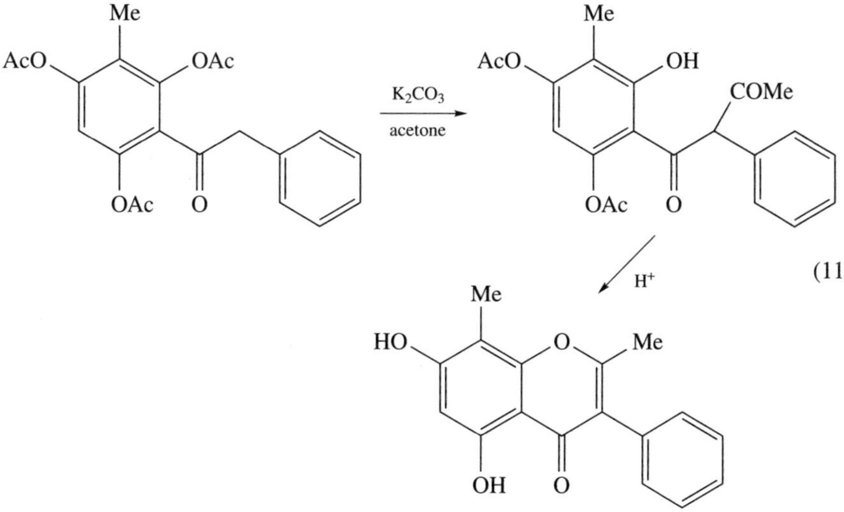

Ar = Ph, 4-MeOC$_6$H$_4$, 3,4-(MeO)$_2$C$_6$H$_3$

R^1, R^2 = H, OMe

(245)

(113)

(246)

(247)

(114)

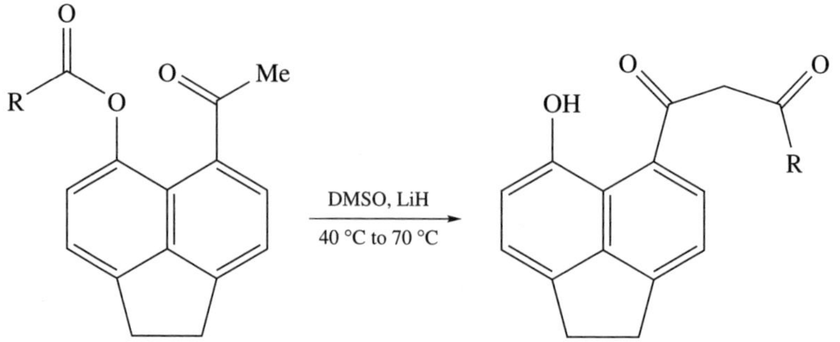

R = Me, Ph, 4-MeC_6H_4, 4-$MeOC_6H_4$, 4-$O_2NC_6H_4$

(248) (249) (250)

(115)

DMG = directed metalation group

(251) (252)

R = H, Me, Cl, OMe (18–75%)

(116)

(253)

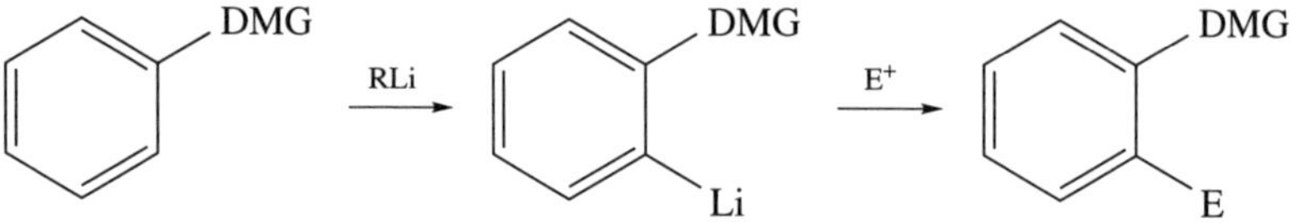

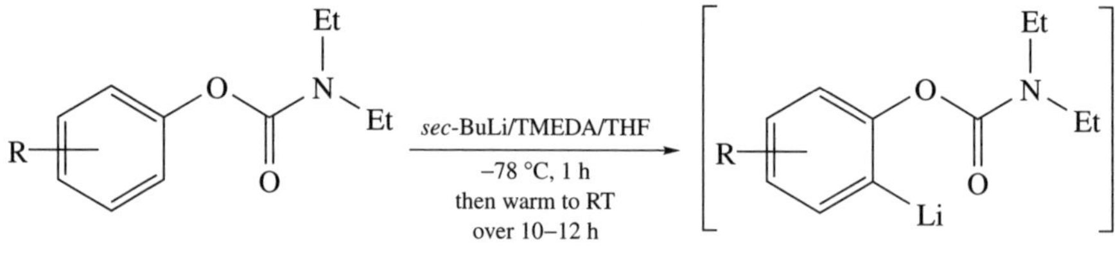

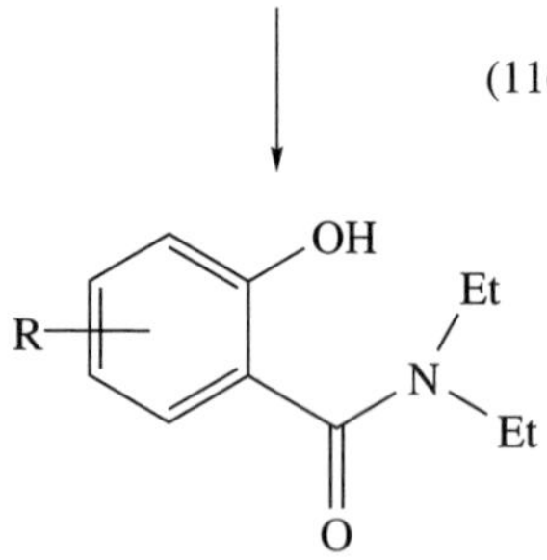

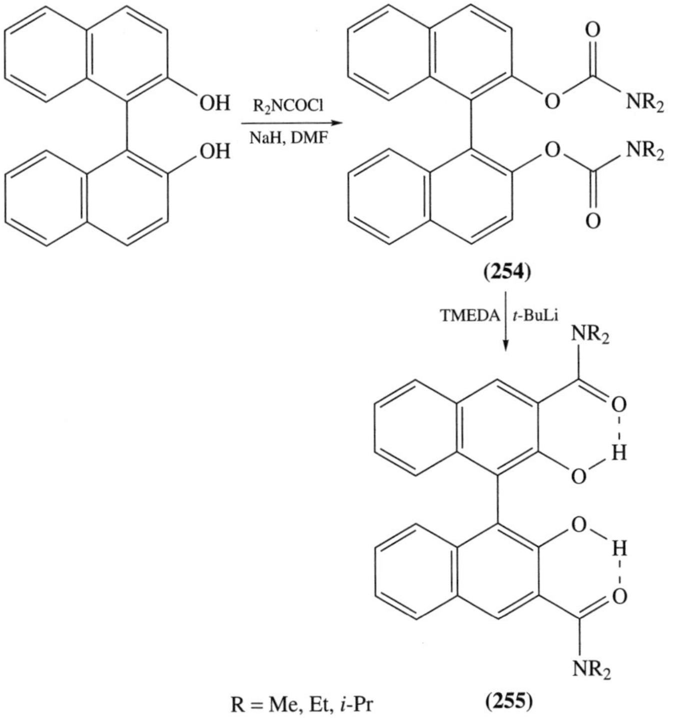

11. Tautomeric equilibria and rearrangements involving phenols

(254)

R = Me, Et, i-Pr   (255)

(117)

(256)

(257)

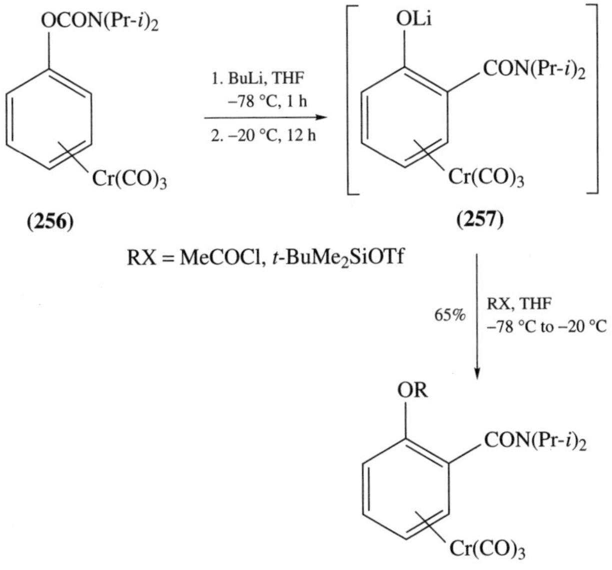

RX = MeCOCl, t-BuMe$_2$SiOTf

R = MeCO, t-BuMe$_2$Si

(118)

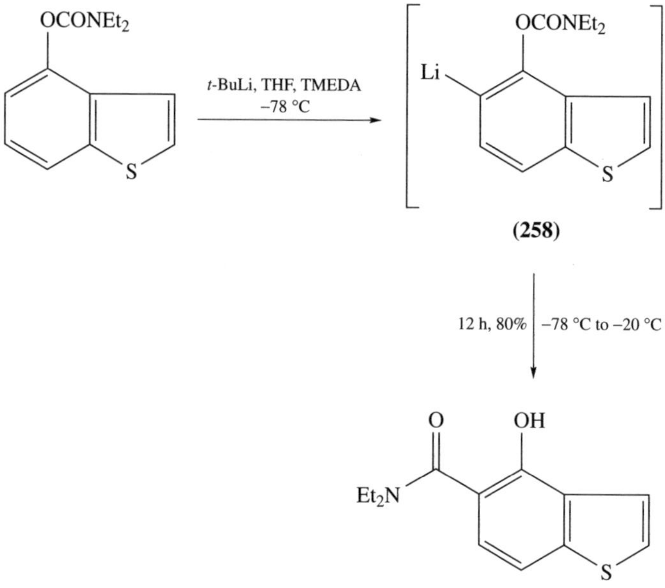

$$\text{(119)}$$

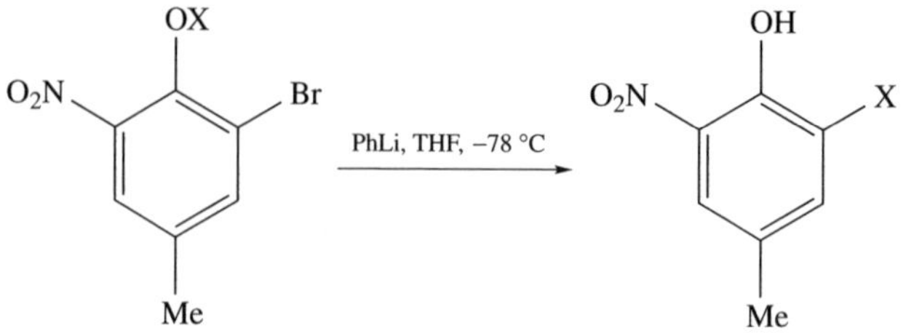

Reaction scheme: starting benzothiophene with OCONEt$_2$ → *t*-BuLi, THF, TMEDA, −78 °C → lithiated intermediate **(258)** → 12 h, 80%, −78 °C to −20 °C → product **(259)** bearing Et$_2$NCO and OH groups.

$$\text{(120)}$$

Compound **(260)**: 2-OX, 3-Br, 6-O$_2$N, 4-Me phenyl → PhLi, THF, −78 °C → product with 2-OH, 3-X, 6-O$_2$N, 4-Me.

X = CONEt$_2$, COCHMe$_2$, COOCH$_2$Ph (49–65%);
SiMe$_3$, SiMe$_2$Bu-*t*, Si(Pr-*i*)$_3$, SiPh$_2$Bu-*t* (53–72%)

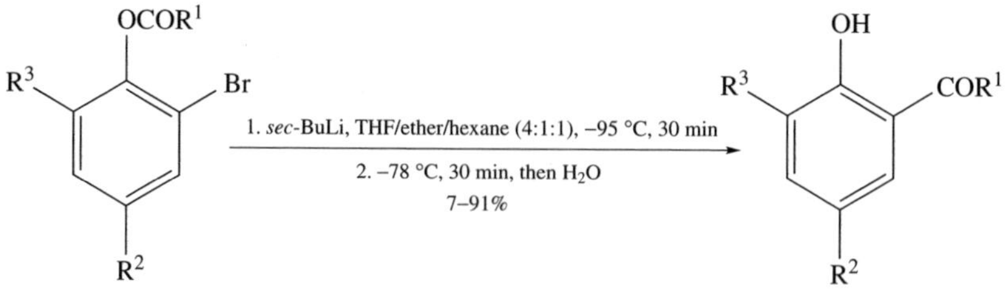

$$\text{(121)}$$

Compound **(261)** (OCOR^1, Br, R^2, R^3 substituted phenyl):
1. *sec*-BuLi, THF/ether/hexane (4:1:1), −95 °C, 30 min
2. −78 °C, 30 min, then H$_2$O
7–91%
→ **(262)** (OH, COR^1, R^2, R^3)

R^1 = Me, Et, *i*-Pr, *t*-Bu, Ph, 1-adamantyl, *t*-pentyl, CMe$_2$CH$_2$Cl
R^2, R^3 = H, Me, *t*-Bu

(263)

(122)

(264)

(123)

1 : 1

(265)

(124)

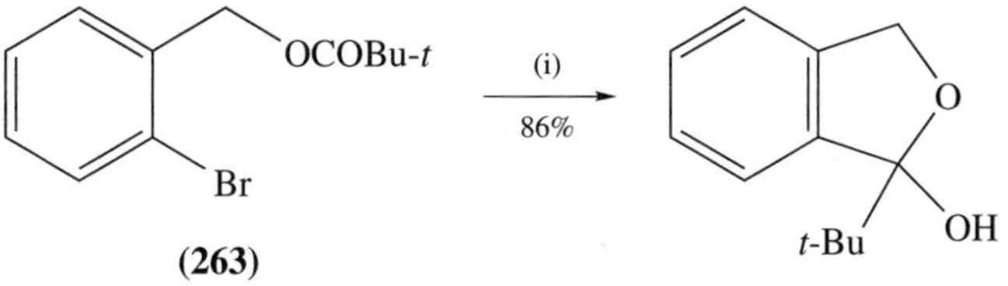

(i) – for reaction conditions see equation 121

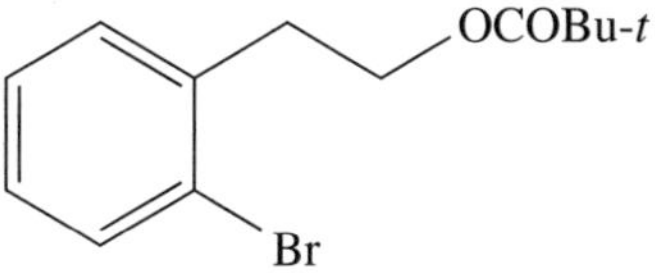

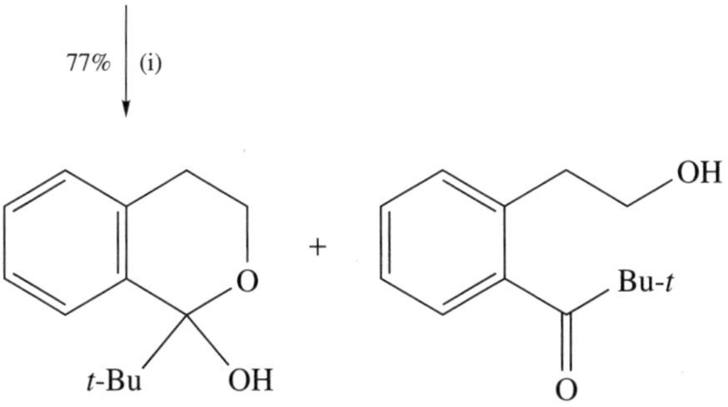

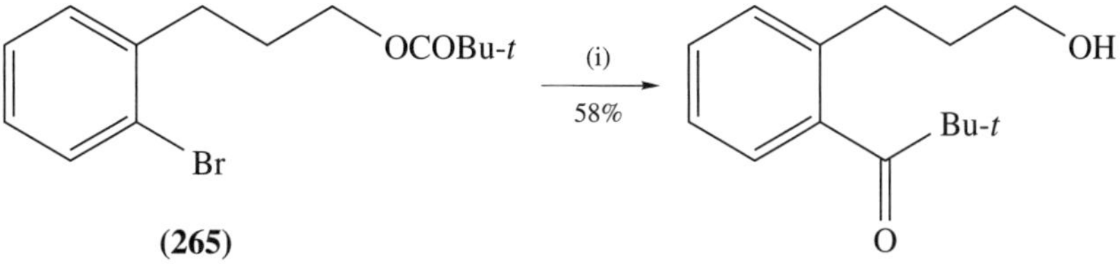

A new carbanion-induced ring-to-ring carbamoyl transfer reaction **275** → **276**, formally a *remote* anionic Fries rearrangement, proceeds upon the directed metalation of biaryl *ortho*-carbamates **275** containing a protecting group (PG) at the *ortho*-position[233] (equation 129). Tandem remote anionic Fries rearrangement and anionic Friedel–Crafts reactions were observed on *ortho*-carbamoyl- as well as carbamoyloxytriarylphosphane oxides **277**, which were converted into P-phenyl functionalized phosphininones **278**[234] (equation 130).

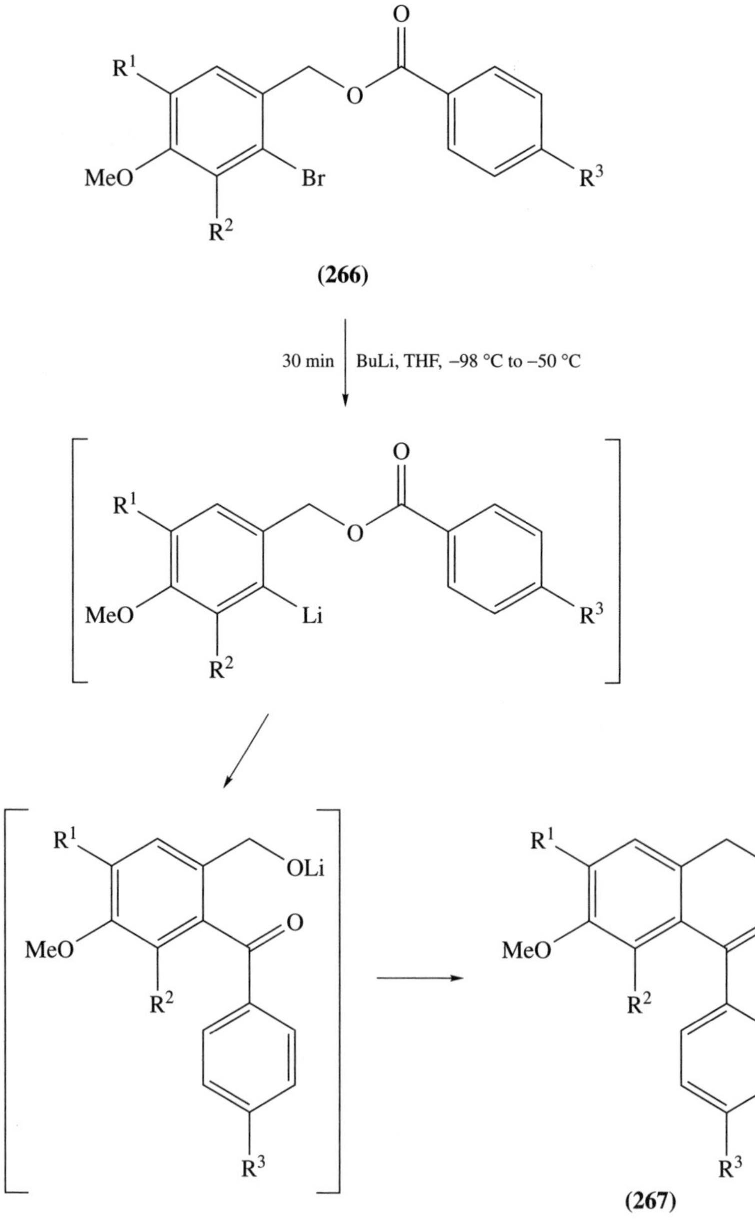

(125)

R^1, R^2 = H, OCH_2Ph; R^3 = OCH_2Ph

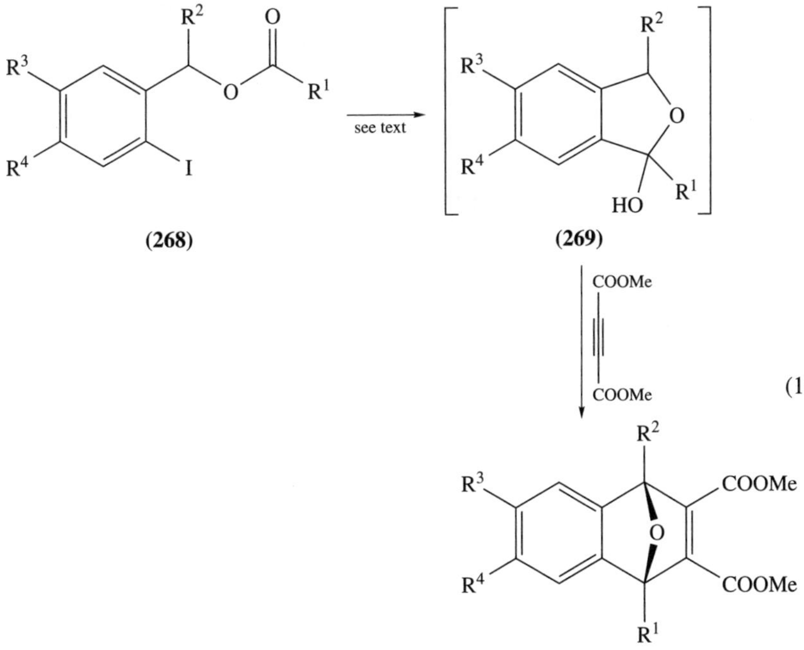

11. Tautomeric equilibria and rearrangements involving phenols

(268) → (269) → (270) (126)

R^1 = (CH$_2$)$_2$CH=CH$_2$, 3,4-(methylenedioxy)phenyl, (CH$_2$)$_3$C≡CH, (CH$_2$)$_3$C≡CTMS, Ph, Et, Me,
R^2 = H, (CH$_2$)$_3$CH=CH$_2$, R^3, R^4 = H, OMe

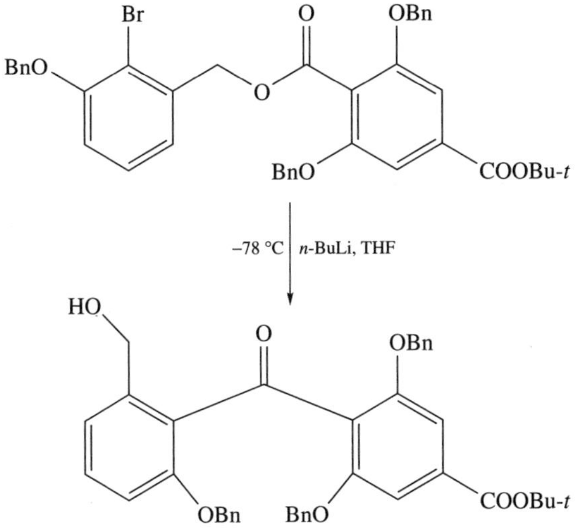

(271) (127)

$-78\,°C$ | n-BuLi, THF

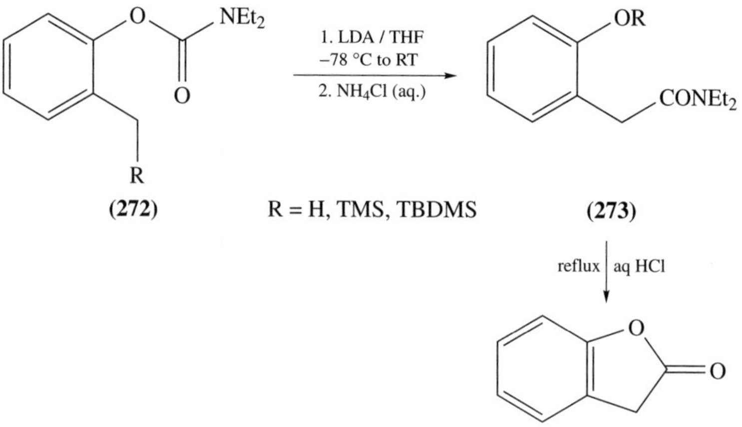

(272) → (273) (128)

R = H, TMS, TBDMS

reflux | aq HCl

(274)

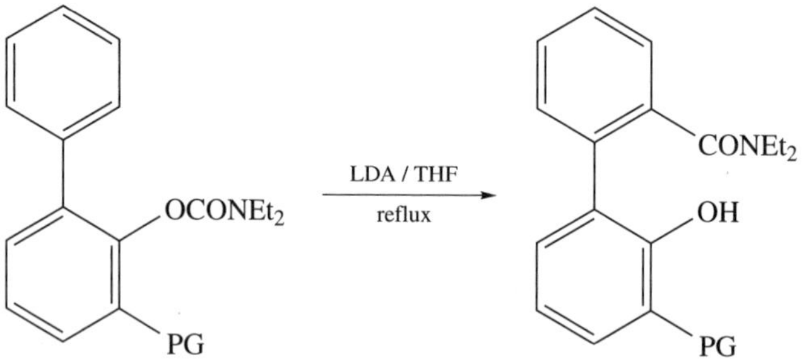

(275) → (276) (129)

PG = OMe, SiEt$_3$ (protecting groups for 'normal' metalation)

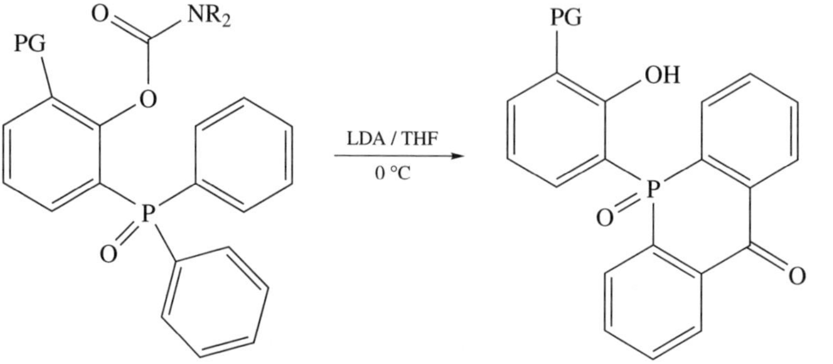

(277) → (278) (130)

PG = OMe, TMS; R = Et, i-Pr

11. Tautomeric equilibria and rearrangements involving phenols

V. REARRANGEMENTS OF FUNCTIONALIZED ARENES

This section covers the isomerizations of aromatic derivatives bearing oxygen- and nitrogen-containing functional groups which lead to phenols. Because these reactions are widely known, only a brief survey will be presented here concerning the most typical examples of these transformations. More detailed information can be found elsewhere[3].

A. Transformations of Peroxides

α-Aryl alkyl hydroxyperoxides **279** derived from aromatic hydrocarbons bearing branched side chains (isopropylbenzene, diarylmethanes, etc.) rearrange in the presence of strong acids to give phenols and carbonyl compounds (*Hock–Sergeev reaction*)[235] (equation 131). In general, a similar process is the *Baeyer–Villiger oxidation*[236] that occurs as oxidative rearrangement of aromatic aldehydes and aryl alkyl ketones **280**. These compounds form the esters **281** under the influence of hydrogen peroxide or peracids (equation 132).

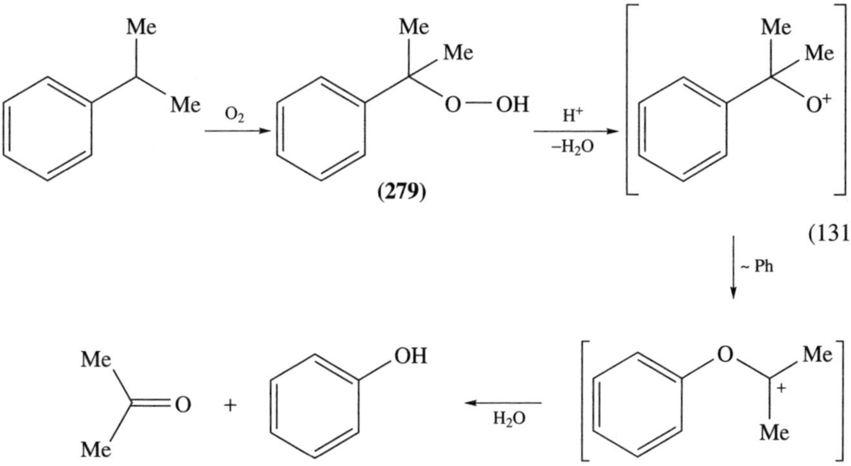

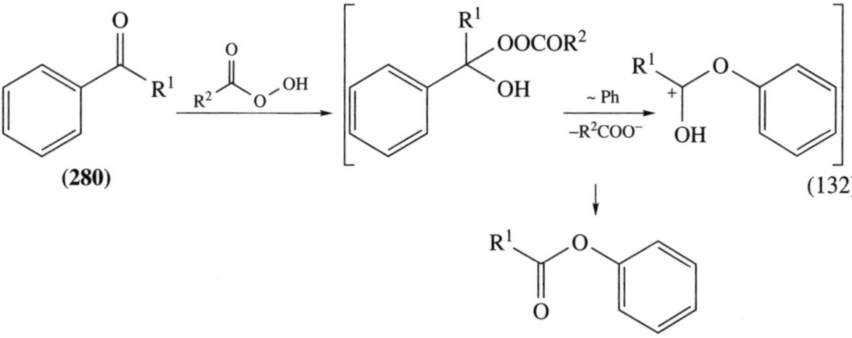

This list has to be continued by the *Dakin rearrangement*, which is the oxidation of aromatic *ortho*- or *para*-hydroxyaldehydes with H_2O_2 in the presence of alkali to afford polyhydric phenols[237] (equation 133).

(133)

(134)

11. Tautomeric equilibria and rearrangements involving phenols 801

Diacyl peroxides, which are known as radical sources, can decompose by an ionic mechanism in the presence of strong acids. Thus, benzoyl peroxide **282** can be converted into phenyl benzoate in a process whose first step involves a Lewis acid catalyzed carboxy inversion reaction to the mixed carbonate **283** (equation 134)[238].

B. Isomerizations of *N*-Arylhydroxylamines

The action of mineral acids brings about the rearrangements of *N*-arylhydroxylamines **284** into *para*-aminophenols **285**[3,113,239] (equation 135). This intermolecular transformation is known as the *Bamberger rearrangement*. If the *para*-position is occupied by an alkyl group, the imine intermediate **286** cannot be aromatized by deprotonation but it undergoes hydrolysis to form the quinole **287** in which an alkyl migration occurs (equation 136)[240] (see also Section VI.A). A very interesting rearrangement takes place upon treatment of 2-naphthyl hydroxylamine **288** with pyridine/SO_3 in acetone[241,242] (equation 137). These reactions are related to the *Boyland–Sims rearrangement*[243,244].

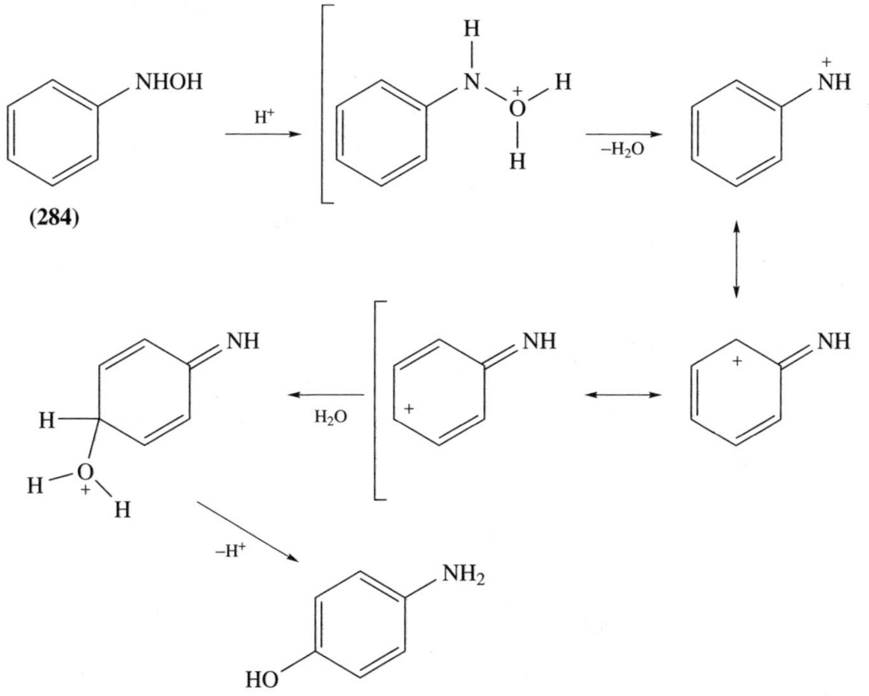

(135)

The *N*-aryl-*N*-acylhydroxylamines **289** and **290** rearrange to aminophenol derivatives in the presence of sulfonyl chlorides[245] (equation 138) as well as of iodonium salts in a reaction similar to the benzidine rearrangement[246] (equation 139). The *N*-aryl-N,O-diacylhydroxylamine **291** undergoes isomerization on heating to produce dibenzoylated aminophenol **292** (equation 140)[245,247]. The *Wallach rearrangement* consists of isomerization of aromatic azoxy compounds **293** to form the hydroxyazobenzenes **294** on heating in the presence of strong acids[248] (equation 141). A similar rearrangement proceeds upon the sulfonation of nitrones **295**[245] (equation 142) as well as acylation of dialkylaryl-N-oxides **296**[249] (equation 143).

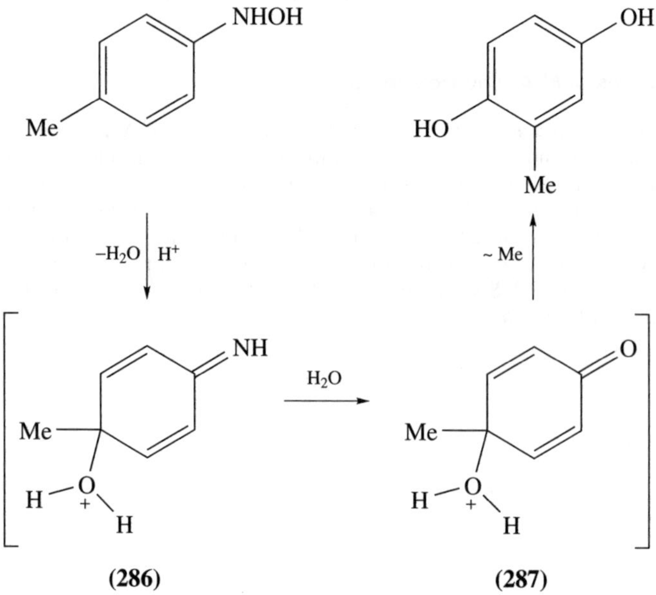

(136)

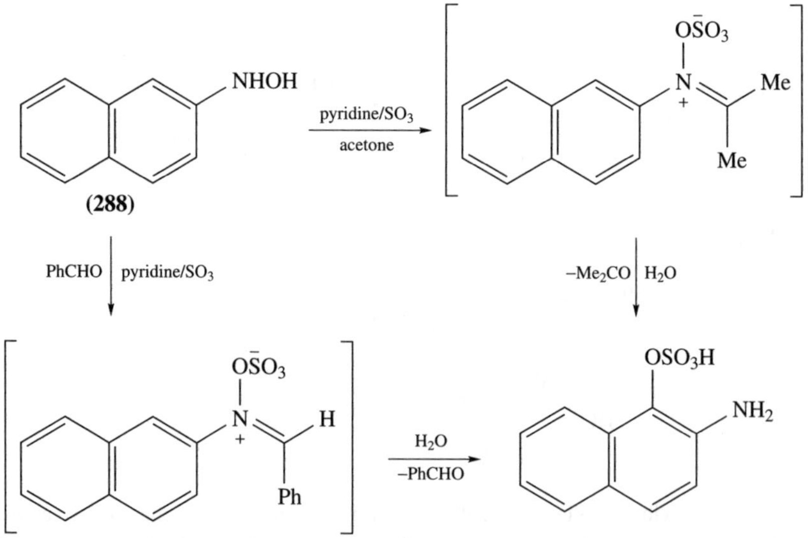

(137)

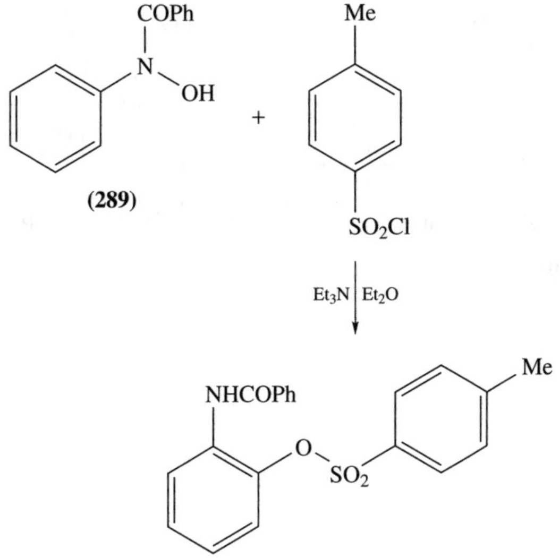

11. Tautomeric equilibria and rearrangements involving phenols

(138)

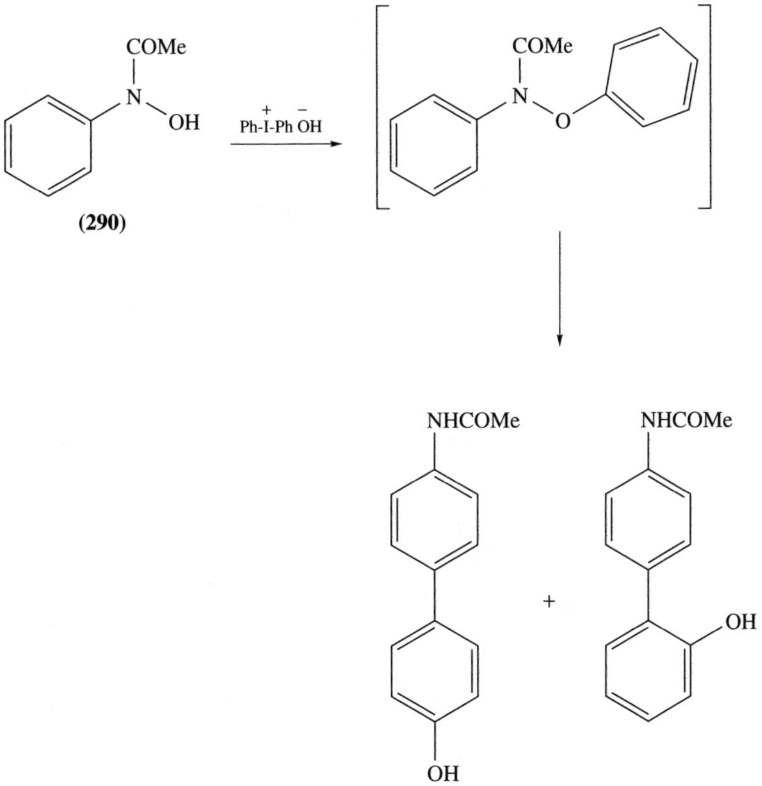

(289) + p-tolyl-SO₂Cl →(Et₃N, Et₂O) 2-(NHCOPh)-C₆H₄-O-SO₂-C₆H₄-Me

(290) + Ph-I⁺-Ph OH⁻ → [PhN(COMe)-O-Ph] → 4-NHCOMe-C₆H₄-C₆H₄-4-OH + 4-NHCOMe-C₆H₄-C₆H₄-2-OH

(139)

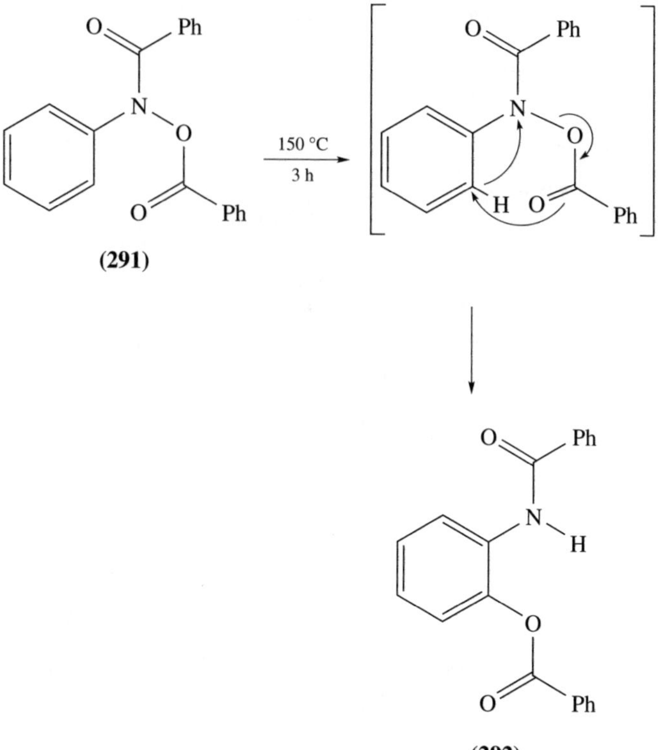

(291)

(292)

(140)

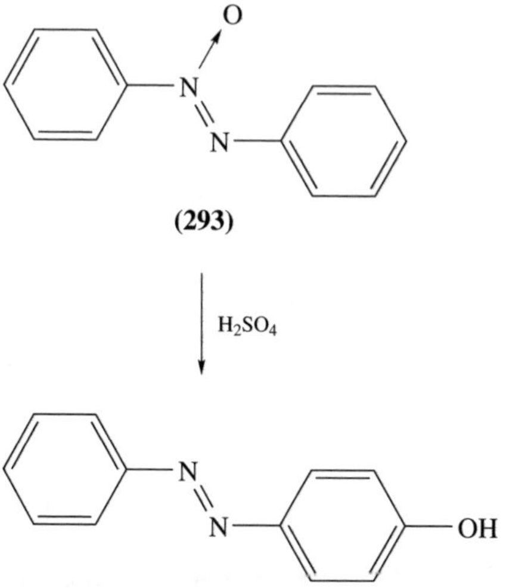

(293)

(294)

(141)

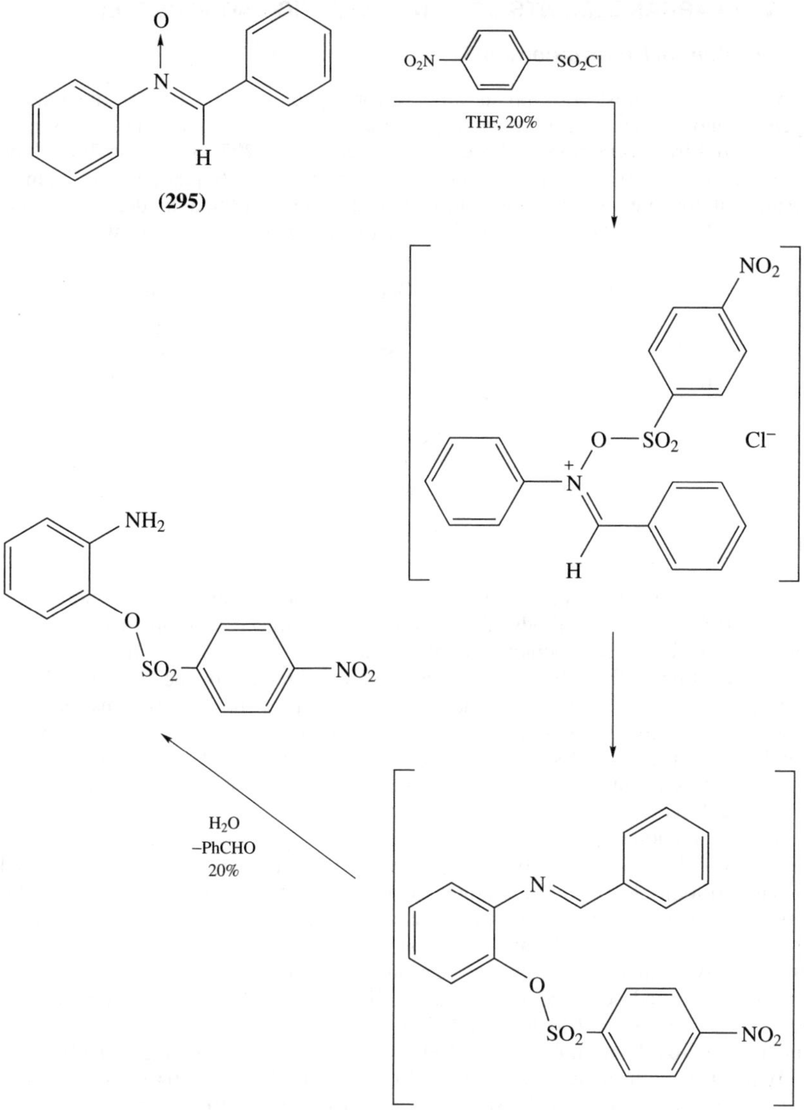

11. Tautomeric equilibria and rearrangements involving phenols

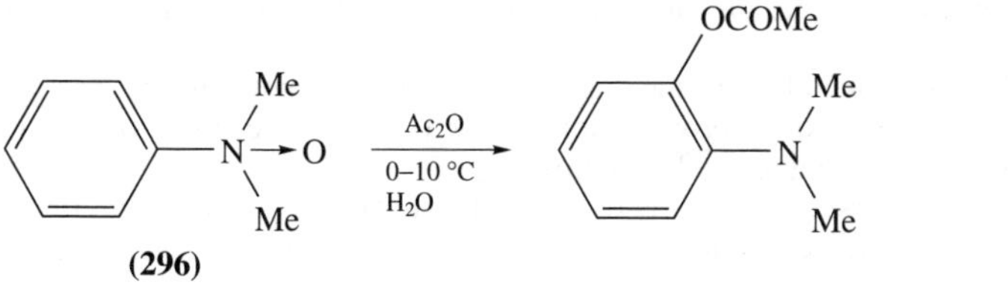

(142)

(143)

VI. REARRANGEMENTS OF NON-AROMATIC CARBOCYCLES

A. Dienone–Phenol Rearrangements

Perhaps one of the most widespread ways to form phenols is by the rearrangements of alicyclic dienones. The simplest variant of these isomerizations can be represented by acid-catalyzed transformation of 2,5-cyclohexadien-1-ones **297** to phenols **298** which proceeds with migration of a group R and aromatization of the ring (equation 144). Numerous versions of the dienone–phenol rearrangement were described in detail in many reviews[250–253]. A list of other surveys and original papers can be found elsewhere[254].

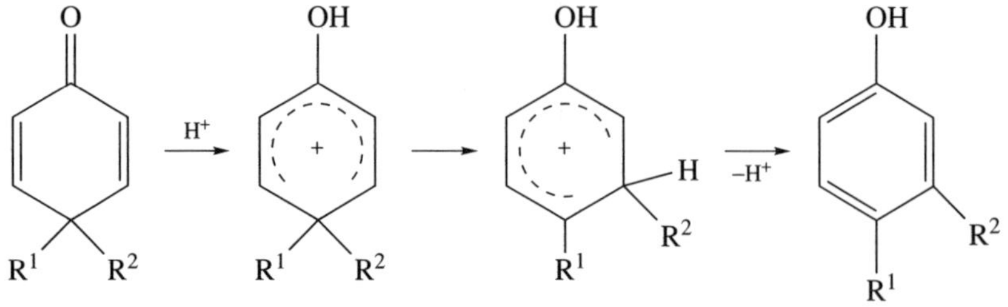

(144)

(**297**)   (**298**)

The mechanism of the dienone–phenol rearrangement was investigated very thoroughly by many authors[255–258]. The methods of deuterium isotope effects[259], competitive [1,2] and [1,5] migrations of benzylic groups[260] and others, were used to study this mechanism. A theoretical evaluation of the substituent influence on the direction of the dienone–phenol rearrangement was carried out[261]. In general, the first step of these transformations is a protonation (or coordination with Lewis acid) of the carbonyl oxygen to form a cyclohexadienyl cation. The second step includes a migration of an alkyl or aryl group to the adjacent electron-deficient carbon atom. Subsequent elimination of proton leads to the stable phenol **298** (equation 144).

As a rule, the dienone–phenol rearrangements are catalyzed by strong acids (H_2SO_4, HCl, CF_3COOH)[251], but other catalytic systems were also reported. Thus, the Fe^{3+}-doped acidic montmorillonite K10 clay accelerates greatly (by factors of 10^5 to 10^6) the cyclohexadienone–phenol rearrangement which occurs in a few minutes at room temperature according to [1,2] and [3,3] pathways (equation 145)[262].

Unusual catalysis in dienone–phenol rearrangements were also described, e.g. the first example of antibody-catalyzed 1,2-isomerization of C—C bonds[263] as well as base-catalyzed rearrangements of 2-hydroxyanilinium salts **299**[264]. The latter reaction includes the formation of 2-oxidoanilinium ylides **300**, which rearrange on heating (40 °C) to the ethers **301** together with the dienones **302** and the phenols **303** and **304** (equation 146). It should be noted that a Claisen [3,3] rearrangement of ethers **301** to form the phenols **303** can be excluded because the ethers **301** prepared beforehand fail to rearrange even at 80 °C.

The dienone–phenol rearrangement can be induced not only by protonation of the oxygen atom, but also by bromination of the C=C double bond via the generation of carbocation intermediates[265,266] (equation 147).

A series of papers devoted to rearrangements of cyclohexadienone intermediates formed upon bromination and chlorination of phenols was published[267–270]. The migration tendency

11. Tautomeric equilibria and rearrangements involving phenols

of the different atoms (e.g. bromine) and groups was investigated in detail using various cyclohexadienone systems[271–273]. It was shown that the migration aptitude can be as follows: Me < Et < vinyl as 1:50:12,000. The cyano group migrates extremely slowly or does not migrate at all[273]. Rearrangements with migration of acetoxy groups were also reported[274].

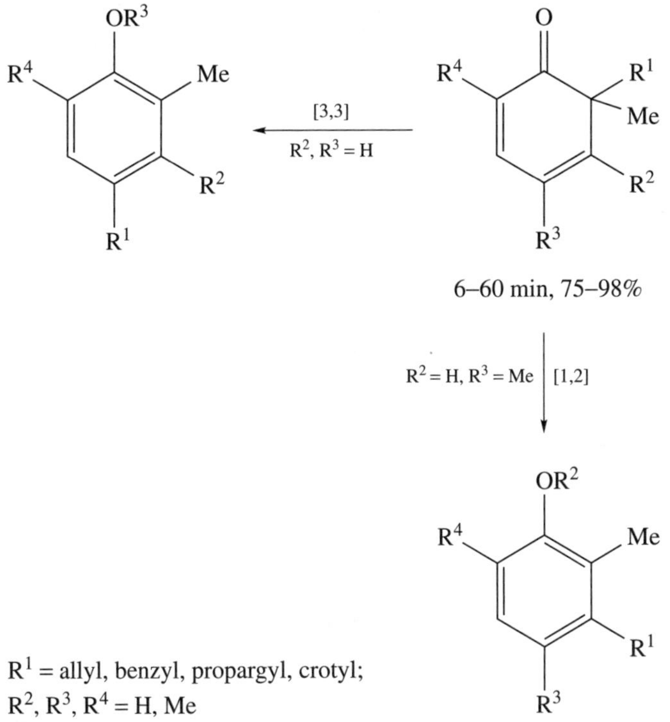

R^1 = allyl, benzyl, propargyl, crotyl;
R^2, R^3, R^4 = H, Me

(145)

6–60 min, 75–98%

R^2 = H, R^3 = Me   [1,2]

The dienone–phenol rearrangement is widely employed for the synthesis of many polycyclic structures, the formation of which demands an expansion of one of the cycles. Thus, the aporphine-type plant alkaloids **307–310** can be obtained via the dienone–phenol rearrangement of orientalinone **305** and dienol–benzene rearrangement of orientalinol **306**[275] (equation 148).

Analogous rearrangements were reported for similar systems such as proaporphines[276], the alkaloids of *Croton sparsiflorus Morong*[277] and cannabinoids[278]. The dienone–phenol and dienol–benzene rearrangements were studied in the eupodienone-1 series (a constituent of *Eupomatia laurina R. Br.*). These compounds **311** were transformed under a variety of acidic conditions into dibenzocyclooctene derivatives **312**[279–281] (equation 149). It is remarkable that the rearrangement proceeds with migration of the C—C bond which connects two six-membered rings, i.e. in essence a migration of the aryl group but not of the alkyl group occurs.

Various dienone–phenol rearrangements were carried out in spirocyclic[282] and bicyclic systems[283–285]. It was shown during investigations of cyclohexa-2,5-dienones bearing acyl

(146)

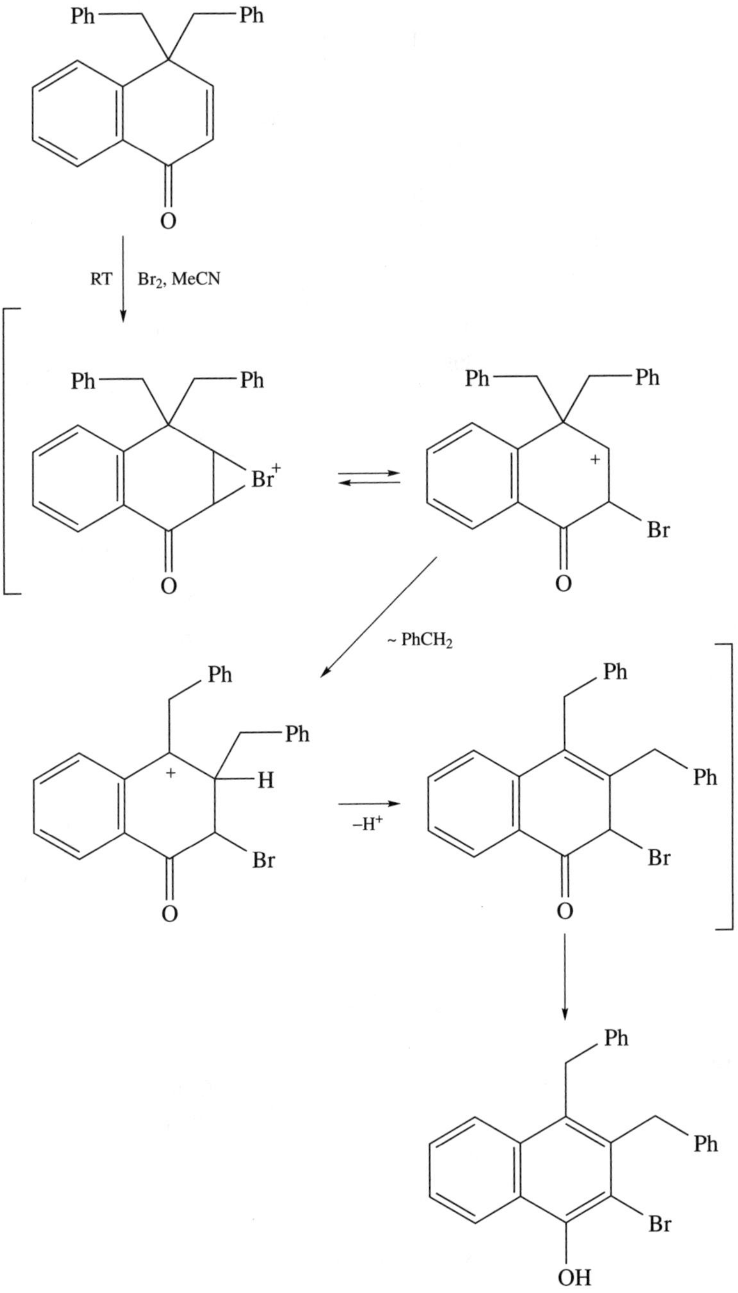

11. Tautomeric equilibria and rearrangements involving phenols

(147)

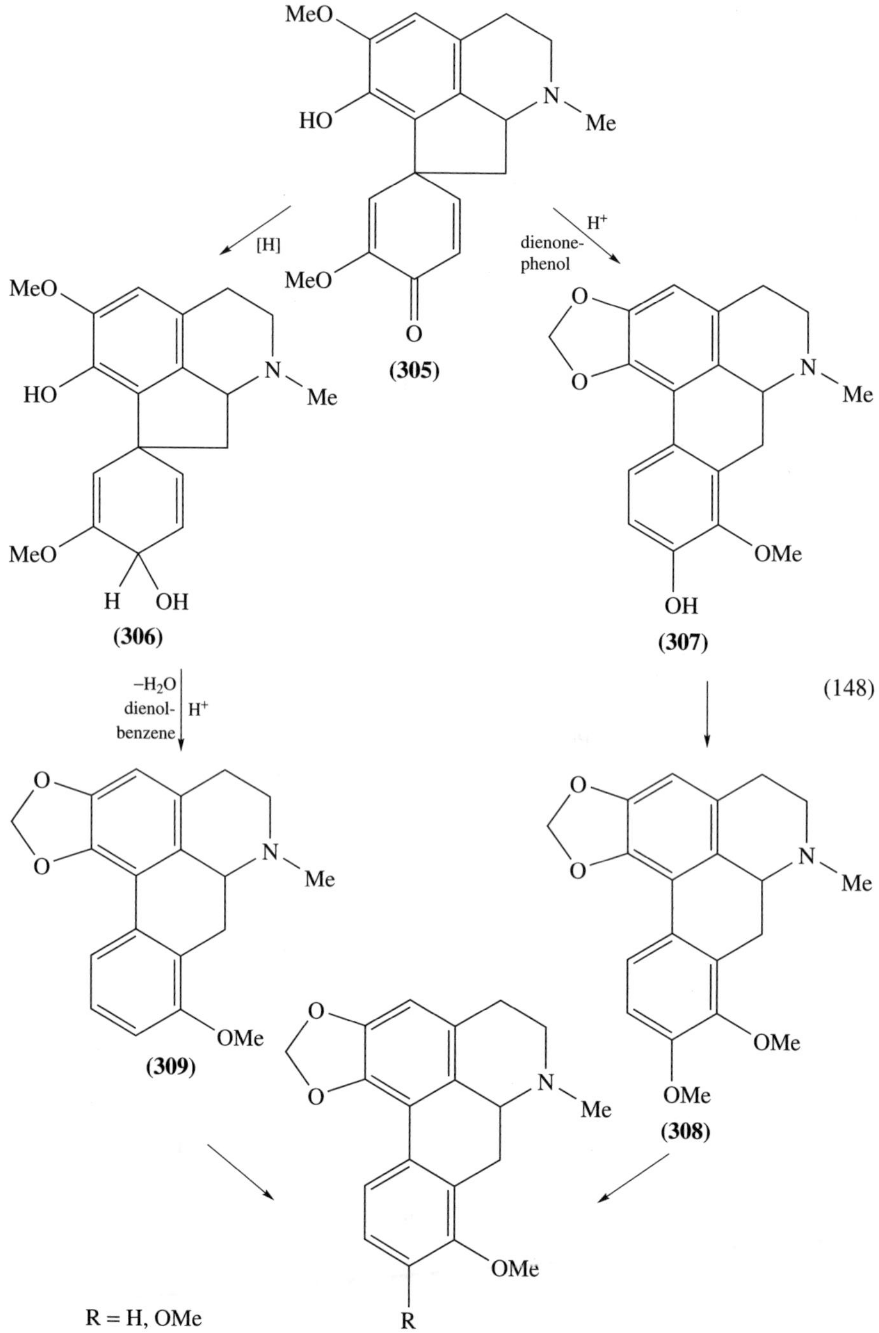

[H]

dienone-
phenol

(305)

(306)

(307)

−H₂O
dienol-
benzene H⁺

(148)

(309)

(308)

R = H, OMe

(310)

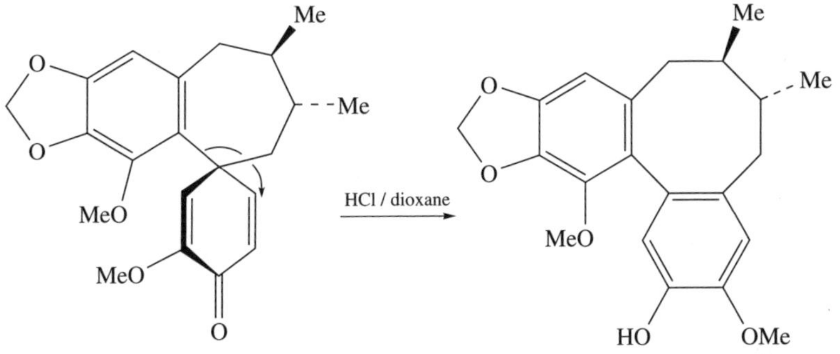

11. Tautomeric equilibria and rearrangements involving phenols 811

(311) → HCl / dioxane → (312)　　(149)

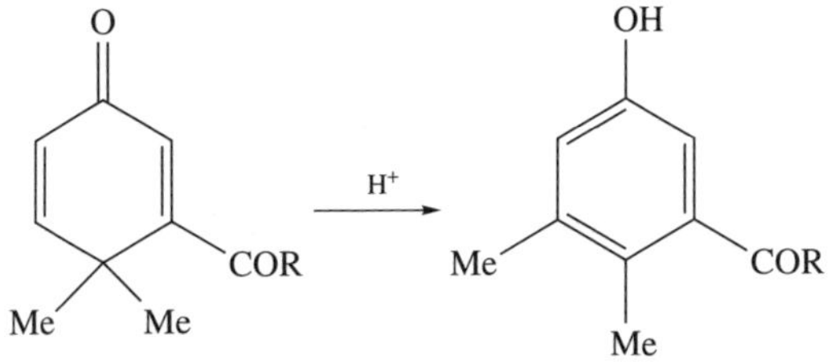

(313) R = Me, EtO → H⁺ → (314)　　(150)

groups that under acidic conditions 3-acetyl- (**313**, R = Me) and 3-ethoxycarbonyl-4,4-dimethylcyclohexa-2,5-dienones (**313**, R = EtO) rearrange to the 3-acyl- (**314**, R = Me) and 3-ethoxycarbonyl-4,5-dimethylphenols (**314**, R = EtO) via a methyl migration from position 4 to position 5[286] (equation 150). However, the treatment of 4-benzoylcyclohexa-2,5-dienone **315** with acids failed to realize the desired dienone–phenol rearrangement with a [1,2] acyl migration from C(4) to C(3) but instead gave 4-methylphenyl benzoate **316** by a retro-Fries rearrangement[287–289]. Crossover experiments suggest strongly that this rearrangement is at least partly intermolecular (via path b) (equation 151).

Interesting transformations occur in systems where the cyclohexa-2,5-dienone fragment is in a spiro connection with heterocycles. Thus, treatment of griseofulvin derivative **317** with magnesium iodide results in the xanthone derivative **318** via a dienone–phenol rearrangement[290] (equation 152).

Rearrangement of spirodienones **319** gave substituted 6H-dibenzo[b,d]pyran-6-ones **320** and **321**[291] (equation 153). The rearrangement in aqueous sulfuric acid [path (i)] consistently affords high yield of the O-migration product **321** whereas rearrangement using aqueous sodium hydroxide can involve lactone hydrolysis followed by a rearrangement regarded as formal C-migration.

It was found that the treatment of spirodienone **322** with a H_2SO_4/AcOH mixture (1:50 v/v) results in an isomerization to form the cinnamic acid derivative **323** instead of the classical dienone–phenol rearrangement product[292] (equation 154).

The quinoline **325** was obtained in a dienone–phenol rearrangement of azaspirodienone **324** under vigorous conditions in the presence of an oxidizing agent[293] (equation 155). The treatment of the spirodienone **326** with $BF_3 \cdot Et_2O$ gives the 1,3-diazepine derivative **327** (73%) via a dienone–phenol rearrangement[294] (equation 156).

(151)

(152)

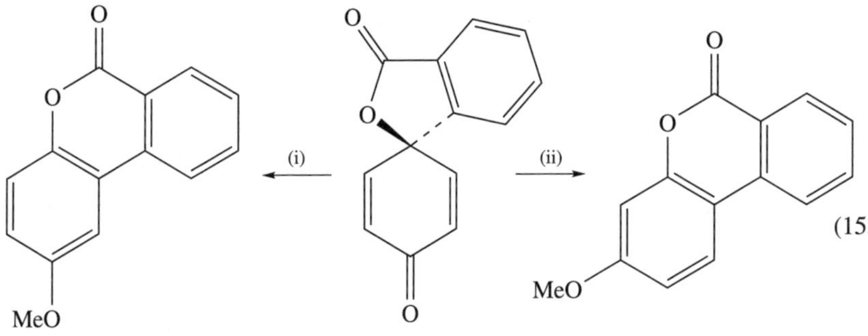

11. Tautomeric equilibria and rearrangements involving phenols

(i) 1) 50% aqueous H$_2$SO$_4$, 2) K$_2$CO$_3$, Me$_2$SO$_4$, 88%
(ii) 1) 10% aqueous NaOH, 2) K$_2$CO$_3$, Me$_2$SO$_4$, 42% [**321**:**320** = 95:5]

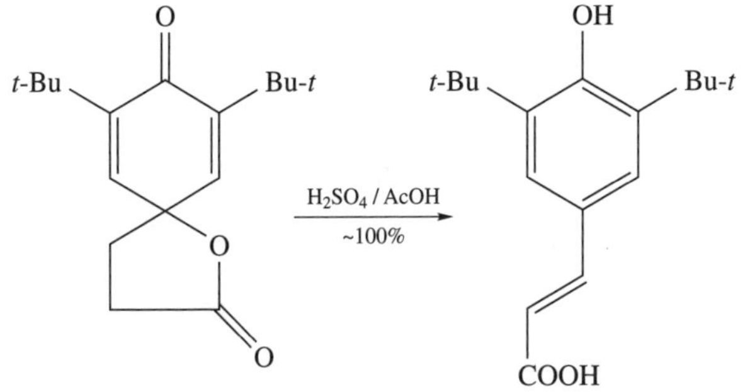

(153)

(154)

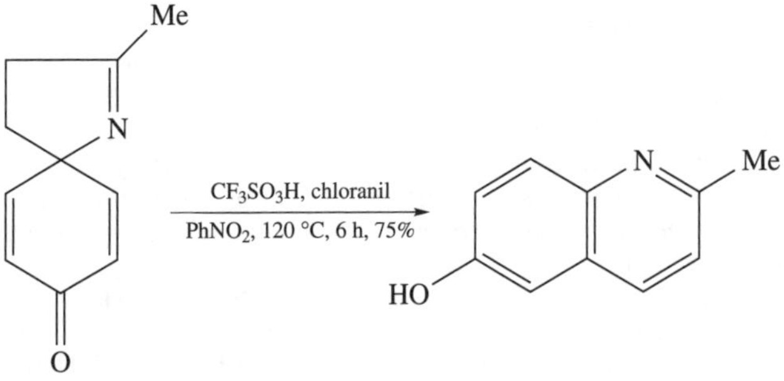

(155)

Application of the dienone–phenol rearrangement in steroid chemistry has been reported in many publications[295–300].

Other polycyclic systems such as naphthoquinones also undergo the dienone–phenol rearrangement. Thus, acetylation of naphthalene-1,4,5(8H)trione **328** with Ac$_2$O containing an acid (H$_2$SO$_4$, HClO$_4$) resulted in a rearrangement yielding naphthoquinone **329**[301–303] (equation 157).

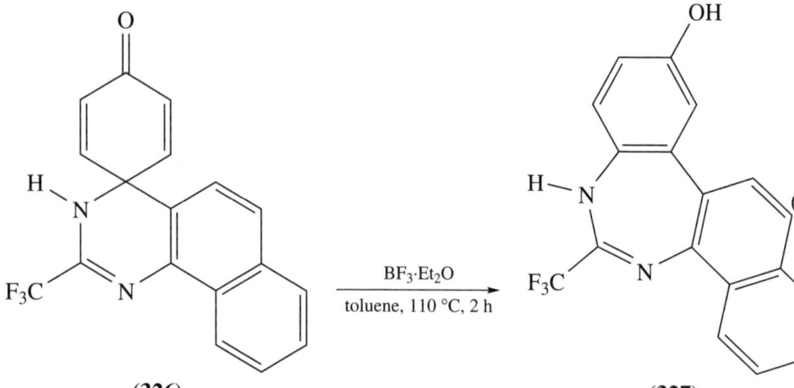

(326) → **(327)** (156)

BF$_3$·Et$_2$O, toluene, 110 °C, 2 h

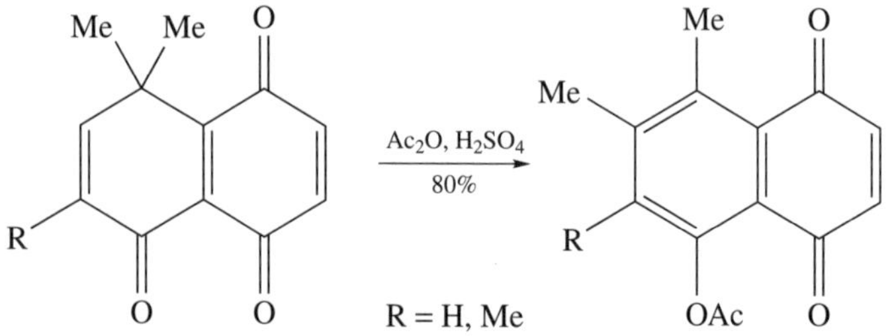

(328) → **(329)** (157)

Ac$_2$O, H$_2$SO$_4$, 80%

R = H, Me

The acid-catalyzed dienone–phenol rearrangement of 2-hydroxy- and 2-alkoxycyclohexa-2,5-dien-1-ones **330** proceeds with regioselective migration of the C(4) substituent to the C(5) position only to form the corresponding phenols **331**[254] (equation 158). Such regioselectivity can be simply explained by considering the relative electron density at C(3) versus C(5) positions in the protonated form of the dienone **330**.

The key step in the regiospecific synthesis of phenolic bis-glycosides is a regiocontrolled dienone–phenol-type rearrangement of cyclohexadienediols **332** to disubstituted phenols **333** in which a glycal fragment migrates in a 1,2-shift[304] (equation 159). Competitive dienone–phenol-type rearrangements were observed in the synthesis of the 2,4-disubstituted naphthols **334** and **335**[305] (equation 160). In principle, this regioselectivity is determined by the fact that phenyl, sec-butyl and n-butyl substituents migrate preferentially compared to methyl.

Protonation of cyclohexadienediol **336** produced the cation **337** which can follow a 'normal' dienone–phenol rearrangement pathway when the substituents R^1 are Me and Ph, and the t-Bu substituent can be eliminated in the last step **337** → **338**. However, when R^1 was a substituted phenyl, the cationoid intermediate **337** cyclized to the oxonium cation **339**, which then underwent deprotonation to give the oxepine **340**[306] (equation 161).

A dienone–phenol rearrangement occurs also as a migration of hydrogen atoms in systems containing exocyclic C=C double bonds in the six-membered rings of **341** and **342**[3,307] (equations 162 and 163).

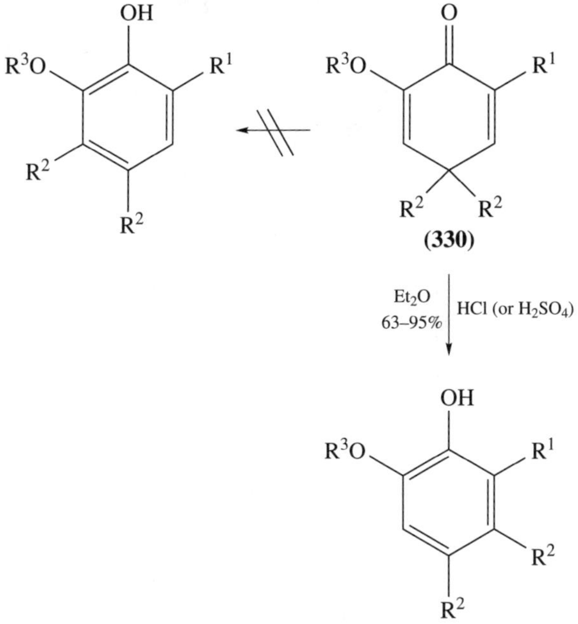

11. Tautomeric equilibria and rearrangements involving phenols 815

(330)

$$\text{Et}_2\text{O}, \text{HCl (or H}_2\text{SO}_4\text{)}, 63\text{–}95\%$$

(331)

R^1 = H, Me, *t*-Bu, Ph; R^2 = Me, Ph; R^3 = H, Me

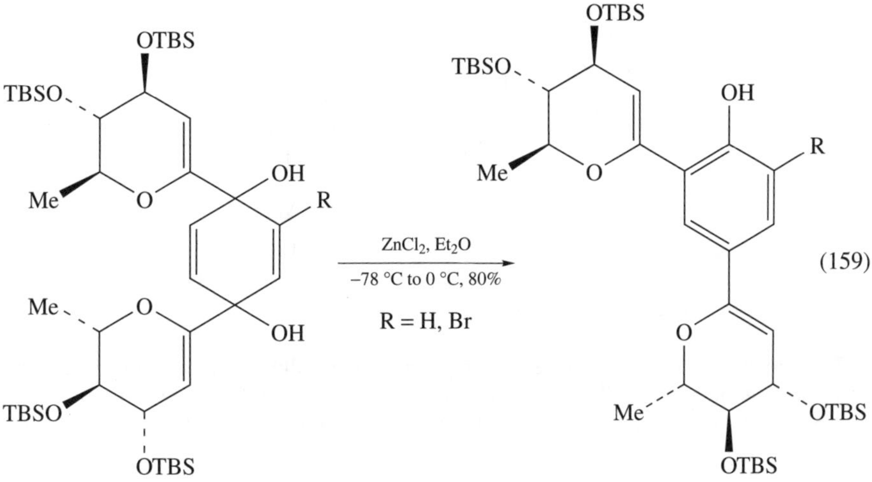

(158)

(332) → (333)

ZnCl$_2$, Et$_2$O
−78 °C to 0 °C, 80%

R = H, Br

(159)

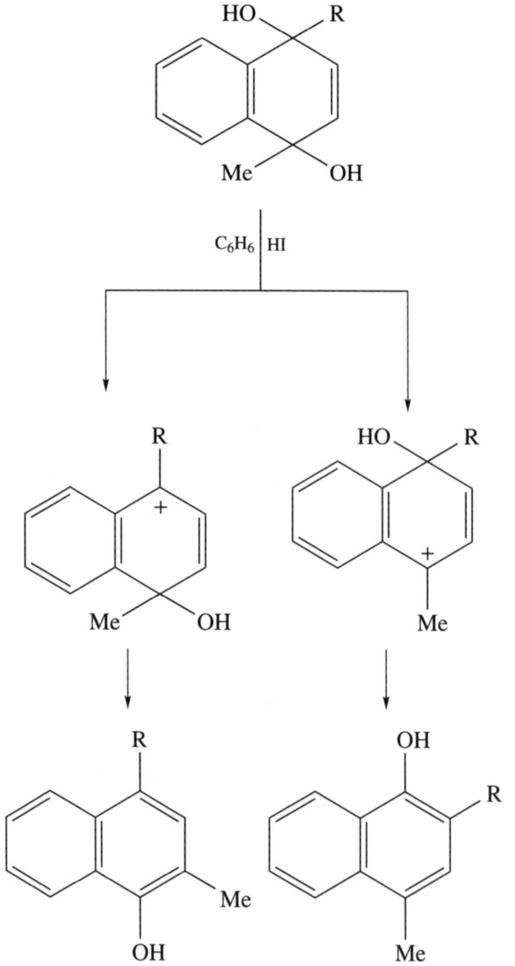

(160)

R = *n*-Bu, *s*-Bu, Ph

The quinone methides **343** undergo a rapid and practically quantitative rearrangement on neutral alumina at 70–80 °C to afford the alkenylphenols **344**[308] (equation 164).
A series of dienone–phenol photorearrangements was also reported[309–311].

B. Rearrangements Involving Ring Expansion

Phenols can be formed also in rearrangements of small carbocyclic rings, starting from cyclopropane derivatives. For instance, the reaction of benzoylcyclopropene **345** with acetylenes **346** in the presence of 10 mol% of [ClRh(CO)$_2$]$_2$ results in the oxepines **347** and phenols **348**.

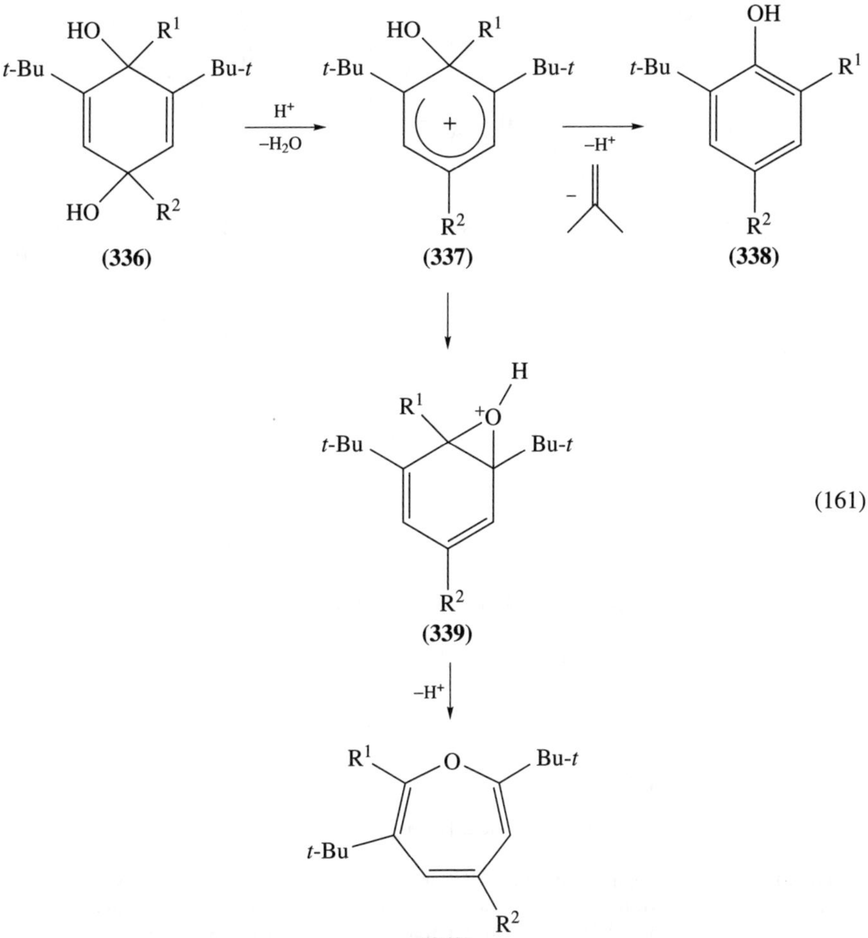

11. Tautomeric equilibria and rearrangements involving phenols

$R^1 = Me, Ph, Ar; R^2 = t\text{-}Bu, Ph, Ar; Ar = p\text{-tolyl}, m\text{-tolyl}, p\text{-Me}_2NC_6H_4$

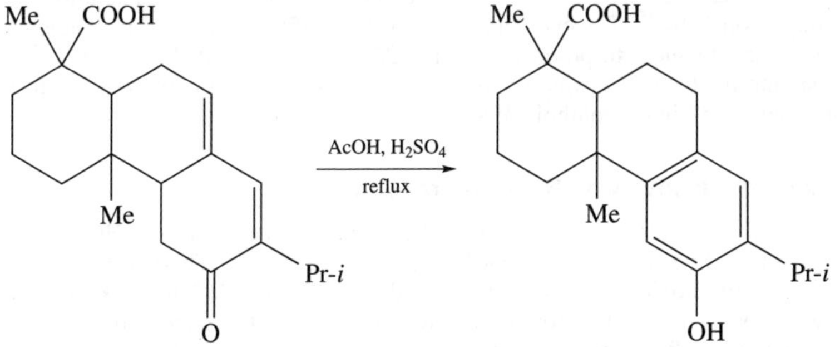

(161)

(162)

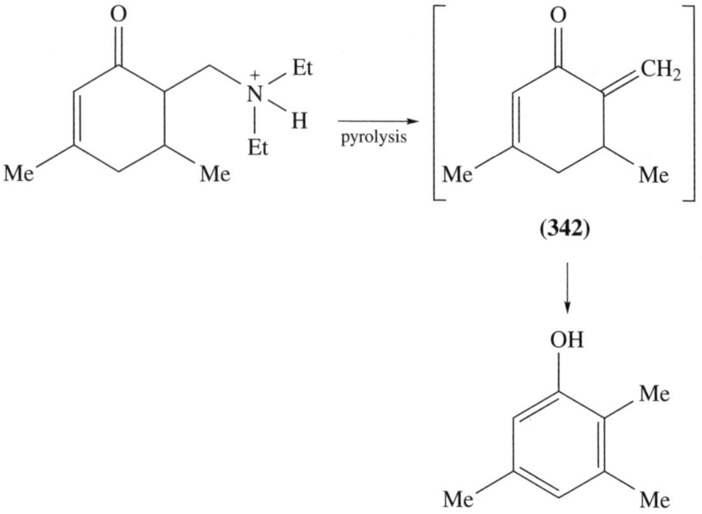

Treatment of the oxepines **347** with HCl at 40 °C brings about a practically quantitative rearrangement to afford the isomeric phenols **349** (equation 165)[312,313] (for rearrangements of oxepines, see Section VII.C). A similar reaction occurs if an alkyne fragment is connected to a cyclopropene ring in one molecule[313] (equations 166 and 167). Liquid-phase thermolysis of the cyclopropane **350** as well as oxirane **352** leads to the same phenol **351** (equation 168)[314].

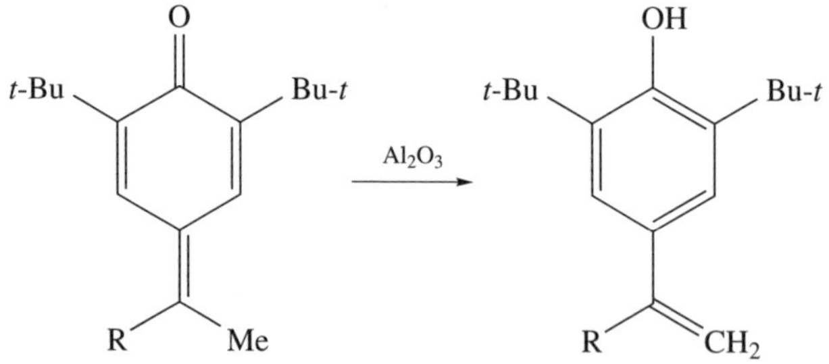

The thermal rearrangement of diarylcyclobutenones **353** gives the naphthol derivatives **354** (equation 169)[315]. The cyclobutane derivatives **355** undergo a retro-Diels–Alder reaction and rearrangement to produce naphthol **356** (equation 170)[316]. The formation of cyclobutene intermediate **358** was assumed for the transformation of trans-α-diazo-β-ketophosphonates **357** into naphthols **359** (equation 171)[317].

C. Rearrangements Involving Ring Contraction

The hydrolysis of the tropolone methyl ether **360** with concentrated HCl in boiling EtOH results in the hydroxyfluorenone **361** (equation 172)[318–320]. Thermal rearrangement with loss of sulfur dioxide occurs on heating the γ-sultones **362** in dioxane, DMSO, dioxane–water or THF at 90 °C for 6–10 h to give 90% of the styrene derivatives **363** in a highly stereospecific manner (equation 173)[321].

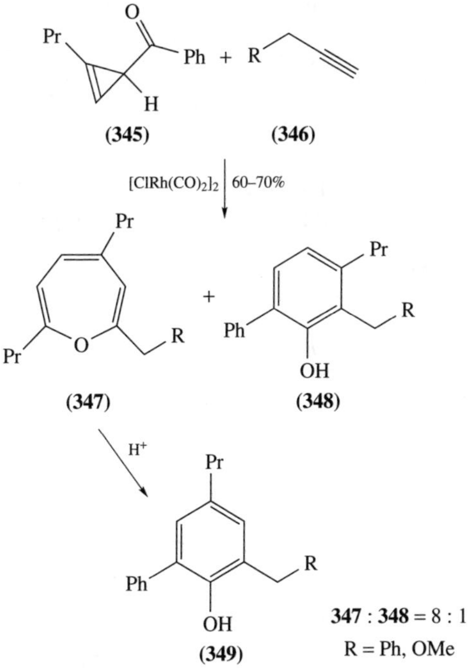

11. Tautomeric equilibria and rearrangements involving phenols

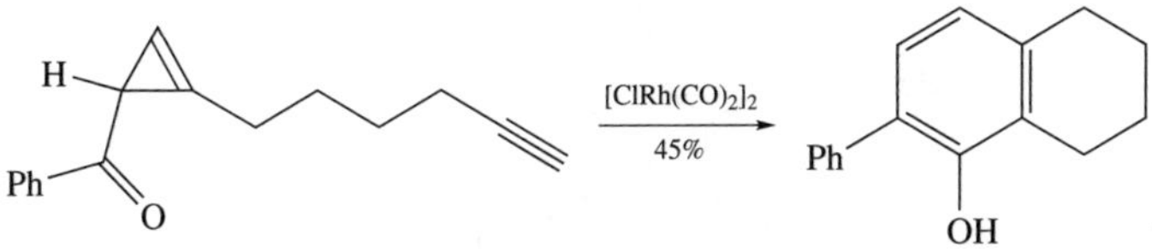

347 : 348 = 8 : 1
R = Ph, OMe

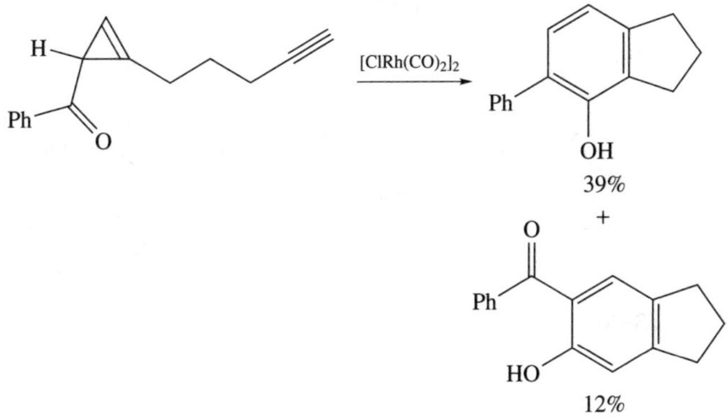

(165)

(166)

(167)

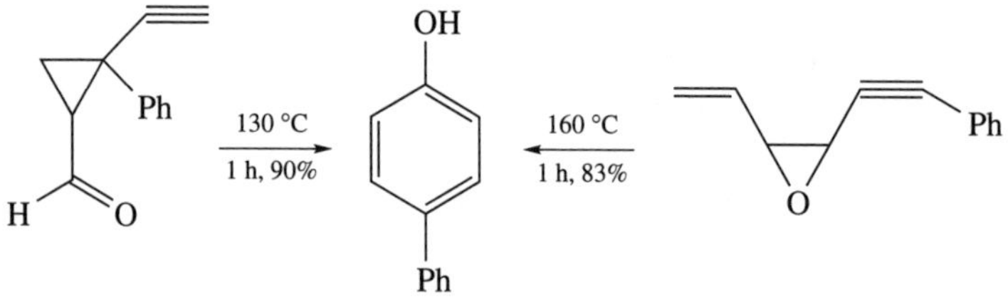

(350) →[130 °C, 1 h, 90%] (351) ←[160 °C, 1 h, 83%] (352) (168)

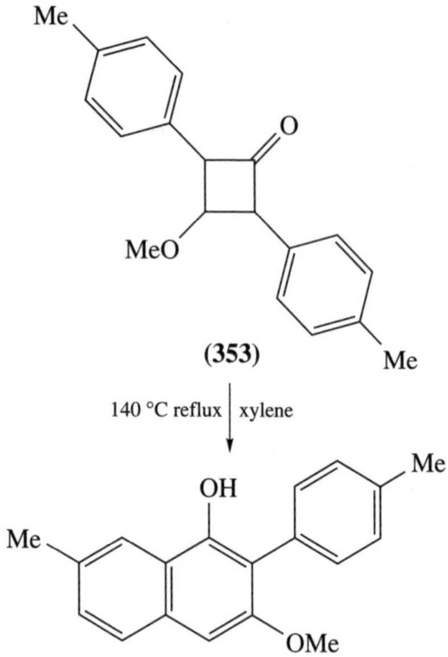

(353) →[140 °C reflux, xylene] (354) (169)

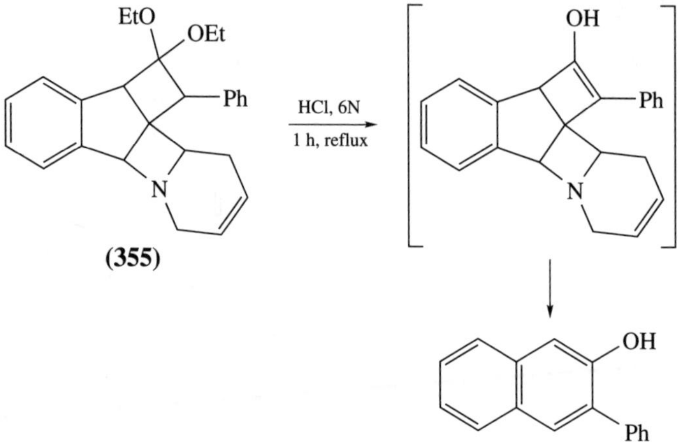

(355) →[HCl, 6N, 1 h, reflux] [intermediate] → (356) (170)

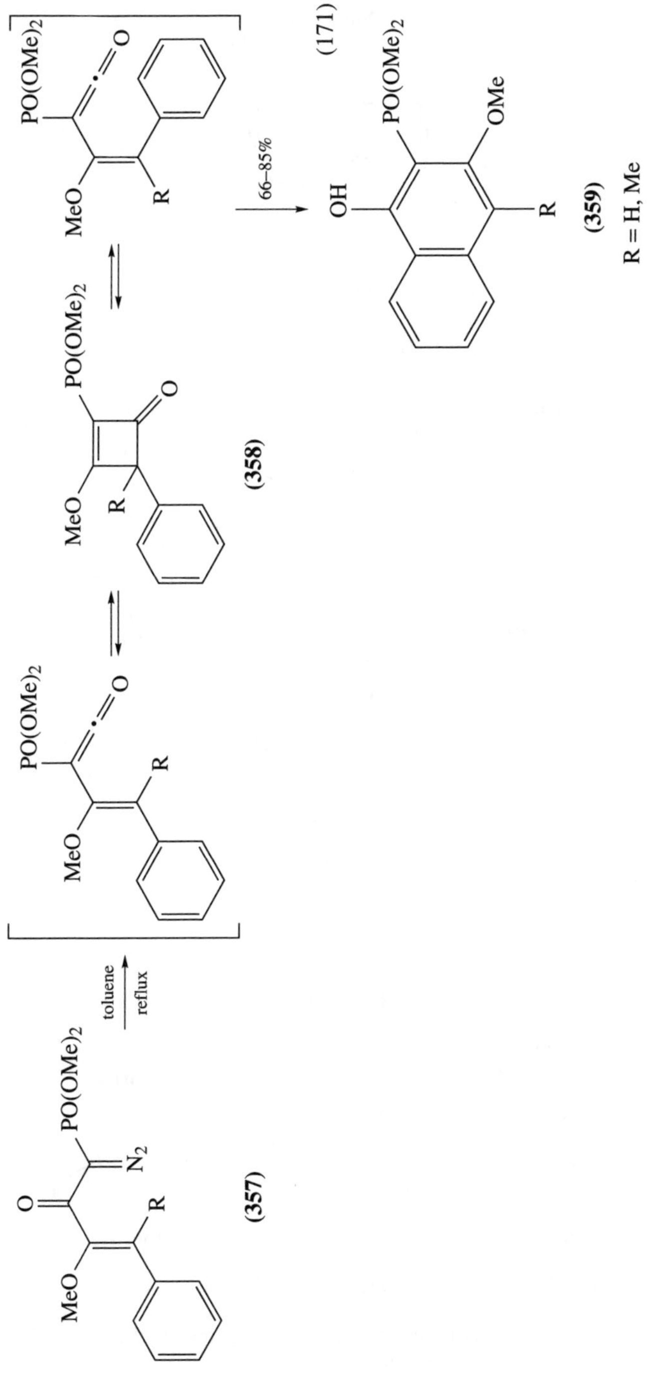

(357) ⇌ (toluene reflux) [intermediate] ⇌ (358) ⇌ [intermediate] → (66–85%) (359)

R = H, Me

(171)

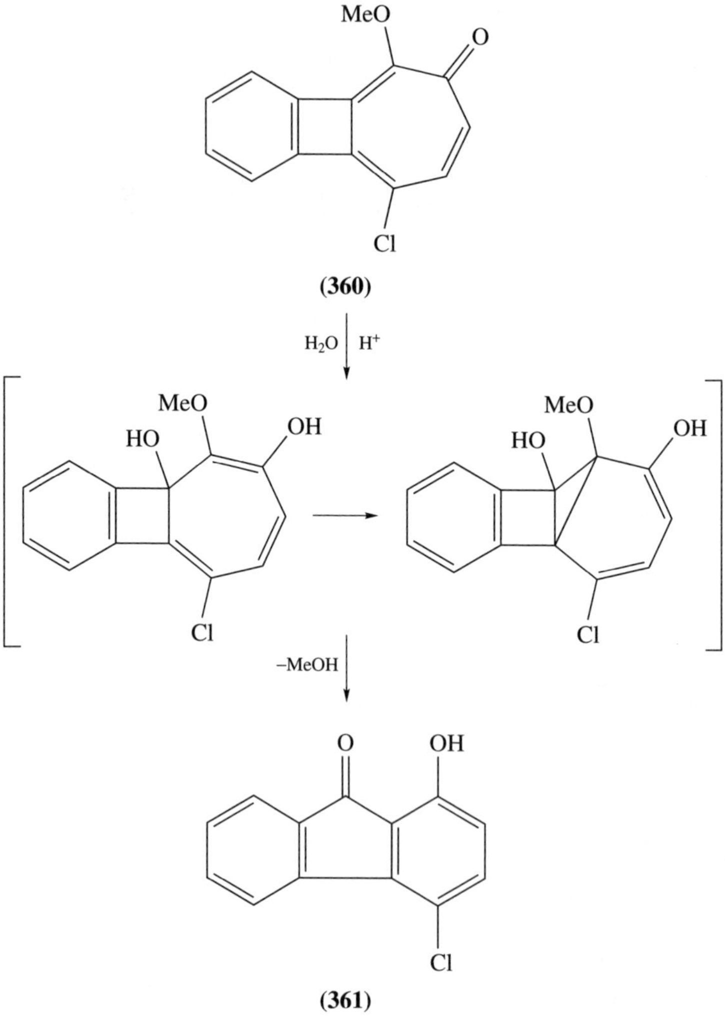

$$\text{(360)} \xrightarrow{\text{H}_2\text{O} \mid \text{H}^+} [\text{intermediates}] \xrightarrow{-\text{MeOH}} \text{(361)} \qquad (172)$$

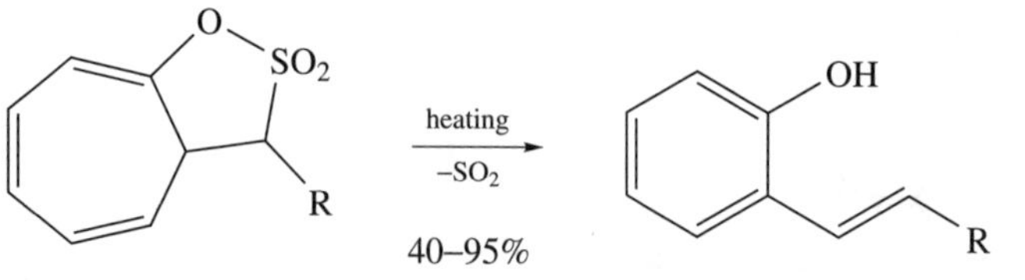

$$\text{(362)} \xrightarrow[\substack{-\text{SO}_2 \\ 40-95\%}]{\text{heating}} \text{(363)} \qquad (173)$$

R = Ph, 4-ClC$_6$H$_4$, 4-O$_2$NC$_6$H$_4$, PhCO, CN, SO$_2$Me, CH=CH$_2$

Many examples of acid- and base-catalyzed rearrangements of tropone derivatives into phenols have been described in several reviews[3,322,323].

VII. REARRANGEMENTS OF HETEROCYCLIC COMPOUNDS

Phenols can be formed also by rearrangements of oxygen- and nitrogen-containing heterocycles. As a rule, these transformations involve recyclizations to produce a benzene ring bearing the hydroxy group.

A. Five-membered Heterocycles

2-Acylfurans **364** react with secondary amines (piperidine, pyrrolidine, morpholine, dibutylamine) in the presence of catalytic amounts of acids (AcOH, HCl) to give enamines **365**, which rearrange during distillation into 2-aminophenols **366**[324,325] (equation 174).

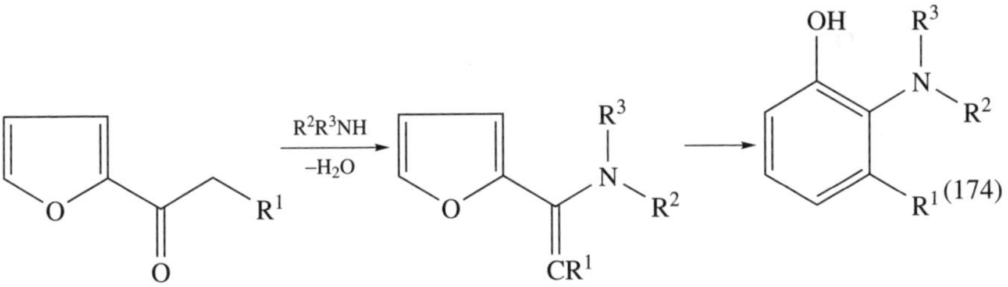

$R^1 = H, Me, Ph; R^2 = R^3 = Bu, R^2R^3 = (CH_2)_4, (CH_2)_5, (CH_2)_2O(CH_2)_2$

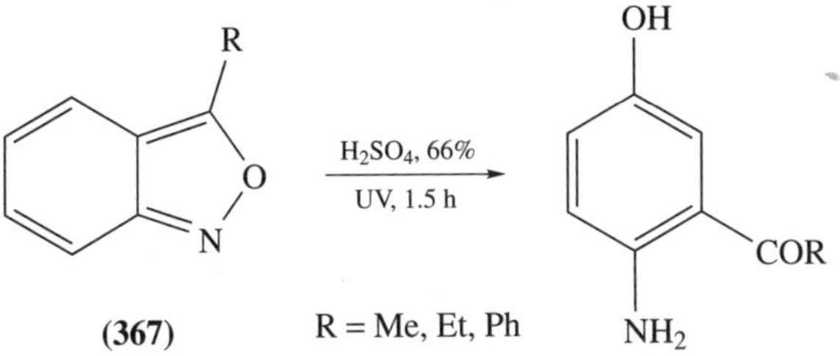

Benzo[c]-1,2-oxazoles **367** transform in the presence of strong acids and, upon UV-irradiation, into 3-acylaminophenols **368**[326] (equation 175).

B. Six-membered Heterocycles

$2H$-Pyrans **370** derived from pyrylium salts **369** undergo during phase transfer catalysis conditions a recyclization on refluxing in Ac_2O to give the acetylsalicylic acid derivatives **371**[327] (equation 176). An interesting rearrangement occurs on prolonged refluxing of 6,6-dimethyl-4-phenyl-$6H$-dibenzo[b,d]pyran **372** in trifluoroacetic acid to afford 4-hydroxy-9,9-dimethyl-3-phenylfluorene **373**[328] (equation 177).

Some examples of transformations of pyran systems into phenol derivatives have been reviewed[329].

It has long been known that reactions of the 2-methylpyrylium salts **374** with oxygen nucleophiles are accompanied by recyclizations into phenols **375**[330] (equation 178).

It was shown that benzo[*b*]pyrylium salts **376** are capable also of recyclizations to produce the phenol derivatives **377**[331,332] (equation 179). The 1,3-benzodioxanes **378** undergo an acid-induced fragmentation to xanthylium salts **380** via phenolic intermediate **379**[333] (equation 180).

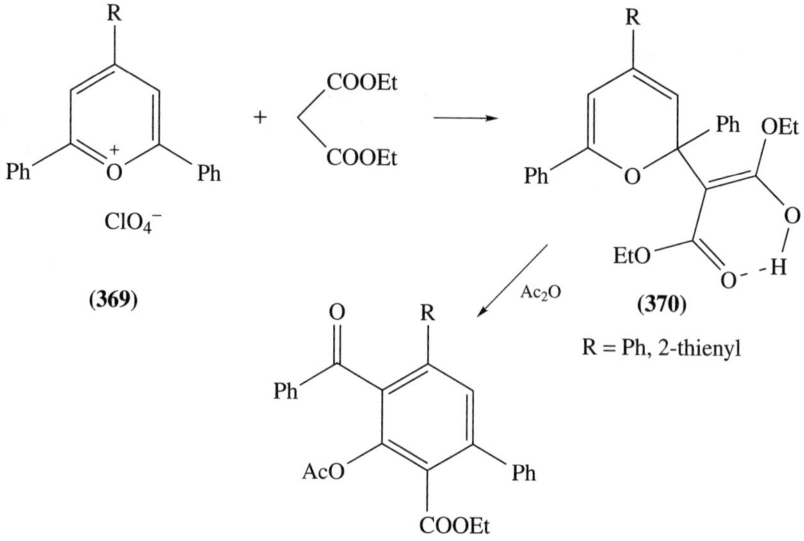

(176)

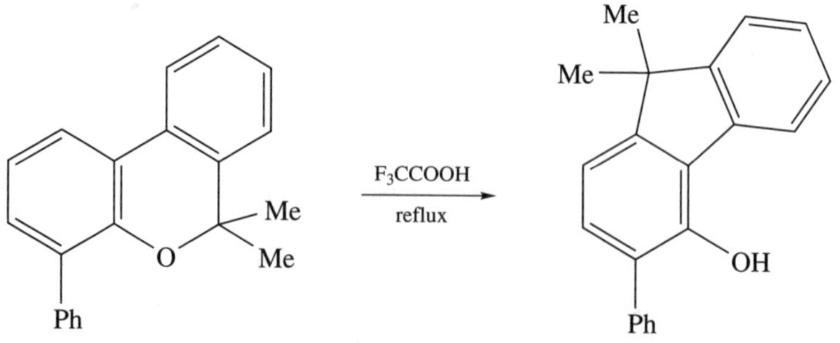

R = Ph, 2-thienyl

(177)

C. Transformations of Oxepines

Among the transformations of heterocycles, the rearrangement 'oxepines (**381**)→ benzene oxides (**382**)→ phenols (**383**)' is best known (equation 181). This rearrangement was described in detail in several surveys[21,334–336]. The most studied aspect of the arene oxide chemistry is the ring expansion to oxepines, on the one hand, and aromatization reaction to phenols, on the other.

Semiempirical and *ab initio* calculations were carried out to investigate the relative stabilities of O-protonated benzene oxide and its related carbenium ions and to obtain further insight into the mechanism of the acid-catalyzed isomerization of the benzene

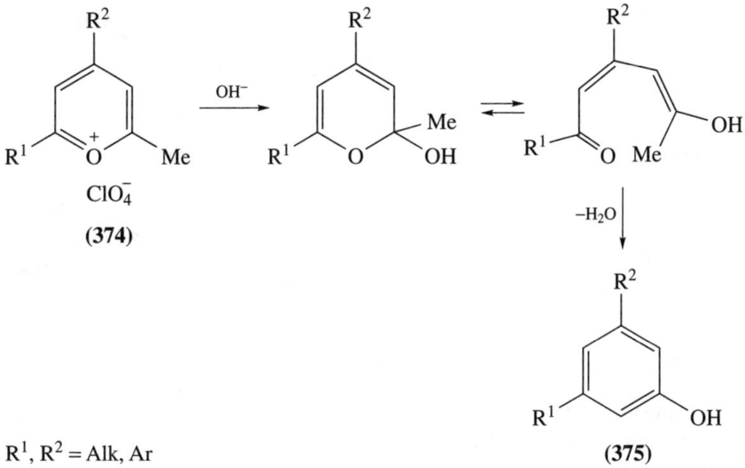

11. Tautomeric equilibria and rearrangements involving phenols

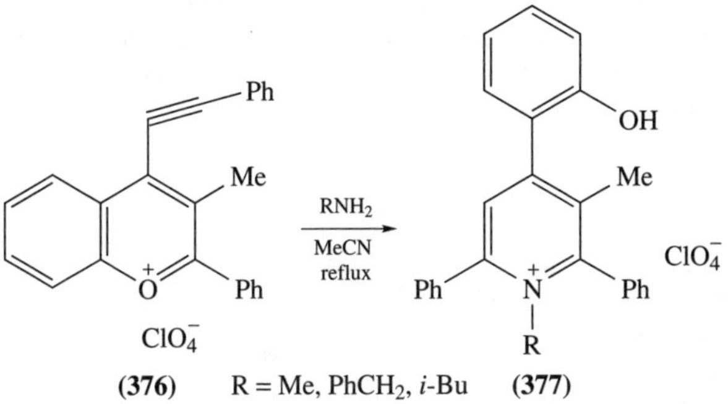

$R^1, R^2 = Alk, Ar$

(374) → (375)　(178)

(376) → (377)　R = Me, PhCH$_2$, i-Bu　(179)

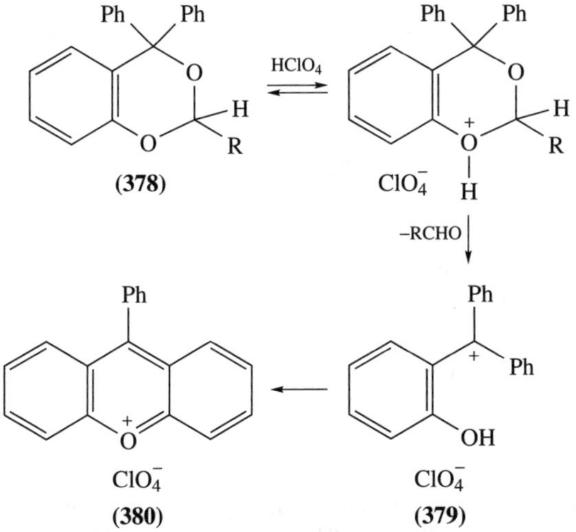

(378) ⇌ → (379) → (380)　(180)

(381) (382) (383)

(181)

(382) (384) (385)

(182)

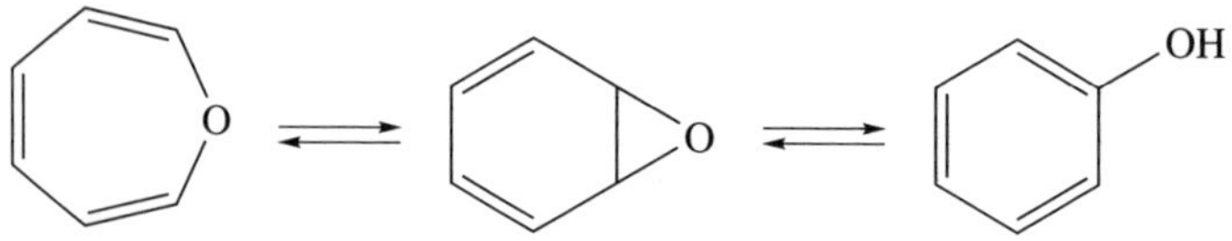

oxide **382** to phenol[337]. The results suggest that the O-protonated oxide **384** is not on the main reaction pathway in the **382** → **383** process but that *para*-quinonoid ions **385** are formed directly upon protonation (equation 182).

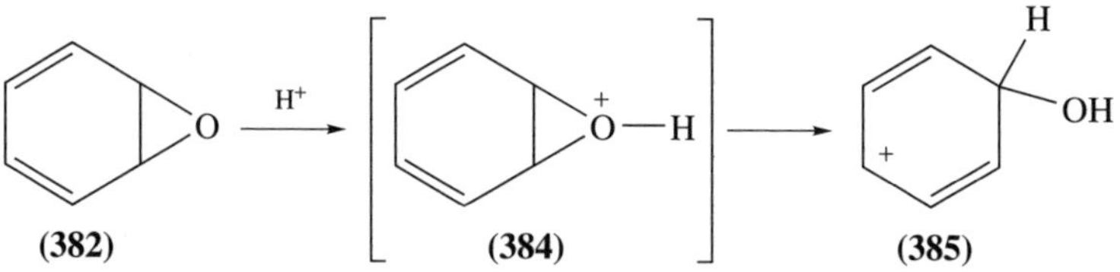

The hexafluorobenzene oxide **386** having no hydrogen atoms rearranges spontaneously to hexafluorocyclohexa-2,4-dienone **387** in polar solvents (acetonitrile, acetone) at room temperature as well as in non-polar solvents at elevated temperatures. Benzene oxide **386** is reduced under very mild conditions (sodium iodide in acetone at RT) to pentafluorophenol **388**[338] (equation 183).

(387) (386) (388)

(183)

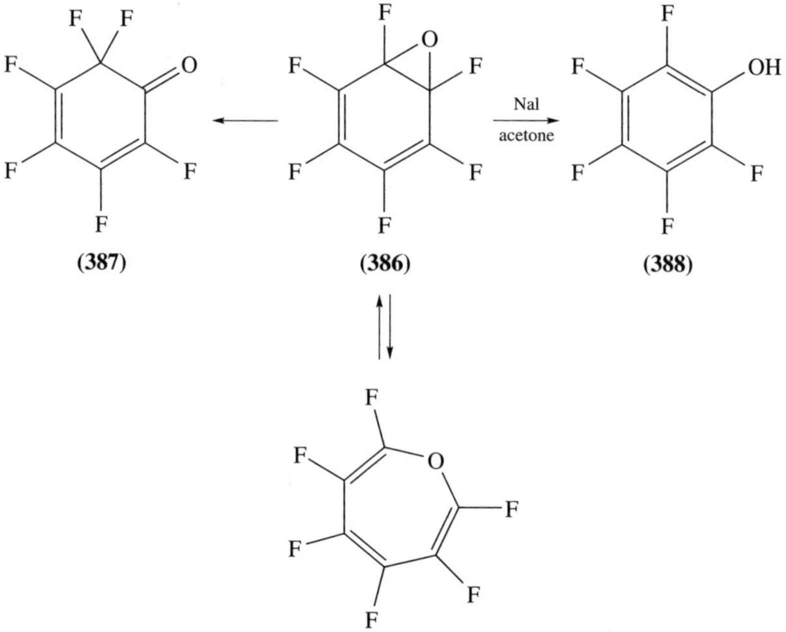

The mixture of valence isomers **389** and **390** undergo aromatization on a silica gel chromatography column to afford the phenols **391**, which are in equilibrium with dihydrofurans **392**[339] (equation 184). The vinylbenzene 1,2-oxides **394** and **397** are in equilibrium with

11. Tautomeric equilibria and rearrangements involving phenols 827

their valence isomers **393** and **396** in aprotic solvents (*n*-hexane, CCl$_4$) but they undergo rapid conversion in the presence of water or methanol to the vinylphenol rearrangement products **395** and **398**[340] (equations 185 and 186). Three isomeric amino-substituted arene oxides **399–401** serving as models for the postulated involvement of amino-acid-derived arene oxides during the biosynthesis of various fungal metabolites rearrange to the corresponding phenols **402** and **403** rather than give the amine/epoxide cyclization products[341] (equations 187–189).

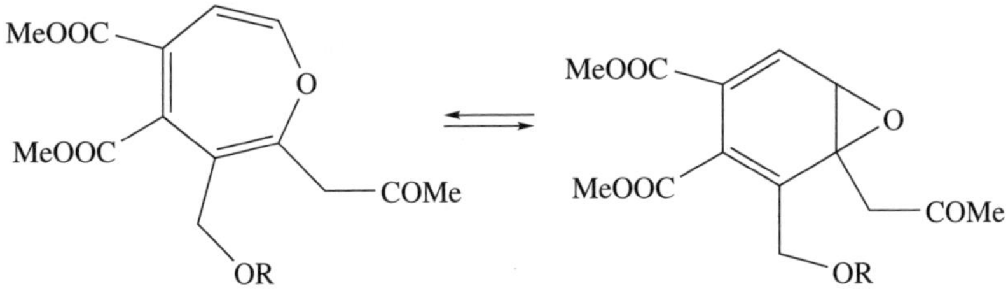

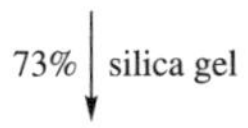

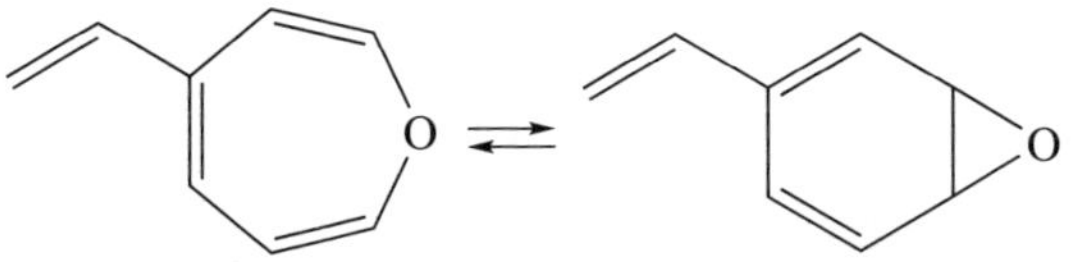

(184)

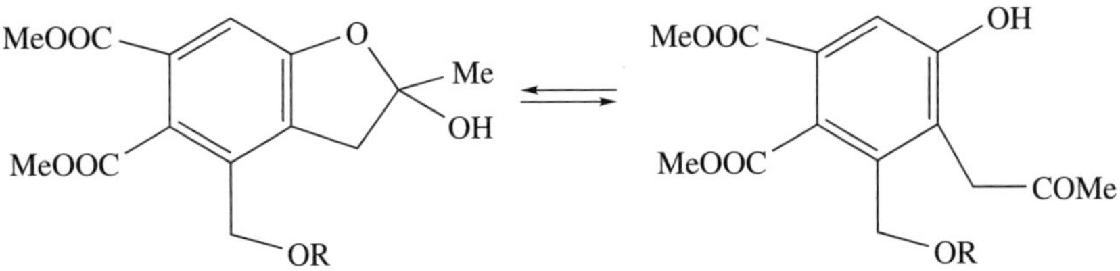

(185)

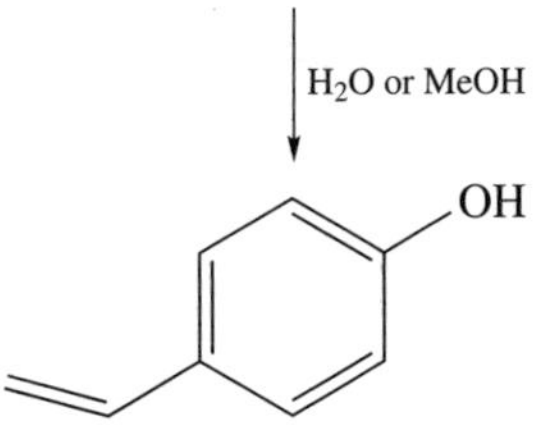

R = Me, Et

(186)

(187)

(188)

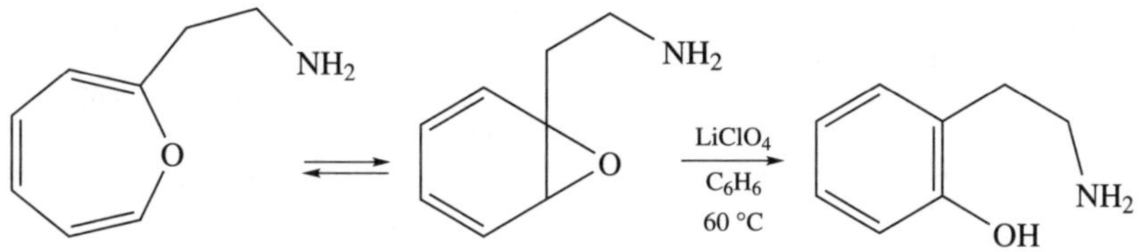

11. Tautomeric equilibria and rearrangements involving phenols

(401) ⇌ → (LiClO₄, C₆H₆, 60 °C) → (402) (189)

The percentage of arene oxide component in the mixture decreases along the series **399 > 400 > 401**, and compound **401** exists largely in the oxepin form.

The mechanism of aromatization of arene 1,2-oxides was studied using a series of model compounds such as 1-carboxy-, 1-carbomethoxy-, 1-formyl- and 1-(hydroxymethyl)benzene oxides[342]. The results obtained support the literature suggestions that arene 1,2-oxides may be intermediates in hydroxylation reactions of biological systems[343,344].

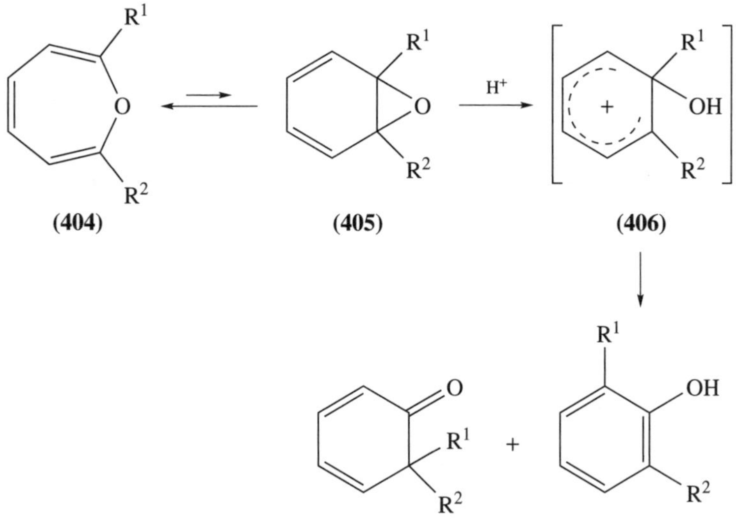

(404) ⇌ (405) →(H⁺) [(406)] → (408) + (407) (190)

$R^1 = R^2 = Me$, **407** (21%) + **408** (27%)
$R^1 = COOMe, R^2 = Me$, **407** (35%) + **408** (65%)
$R^1 = R^2 = COOMe$, **407** (100%)
$R^1 = R^2 = Ph$, **407** (100%)

Acid-catalyzed isomerization of 2,7-disubstituted oxepins **404** leads to products **407** and **408**, depending on the nature of the substituent[345] (equation 190). It was found that the oxepin valence tautomer **404** is more stable than the oxide valence tautomer **405** in 1,2-disubstituted arene 1,2-oxides. The isomerization proceeds via the so-called NIH shift (NIH = National Institute of Health, Bethesda, MD, USA) which involves the migration of the R^1 substituent in the intermediate cation **406** to either of the adjacent carbon atoms to form the products **407** and **408**.

Treatment of ketone **409** with lithium diisopropylamide (LDA) results in the ethyl 1,2-dihydroxybenzoate **410** in a 74% yield (equation 191)[346]. The acid-catalyzed isomerization of diarene oxides derived from benz[*a*]anthracene, chrysene and benzo[*c*]phenanthrene gives mixtures of isomeric polycyclic phenols[347]. Finally, it should be mentioned that dibenzo[*b*,*e*]oxepin **411** undergoes an interesting rearrangement to 2-hydroxyphenylindene **412**[348] (equation 192).

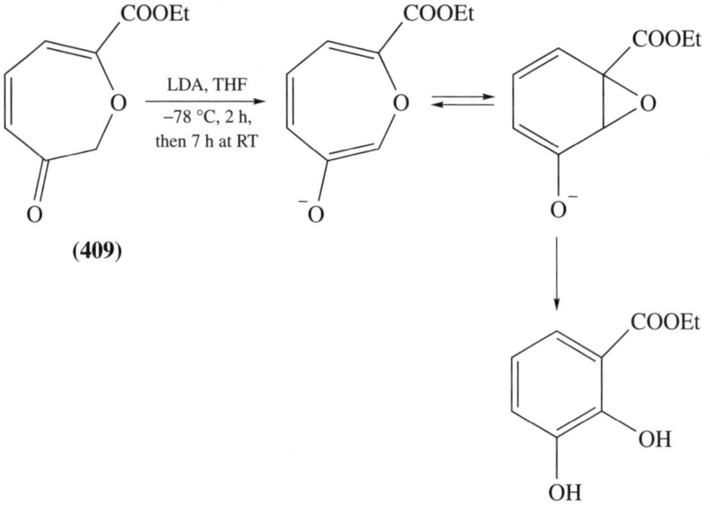

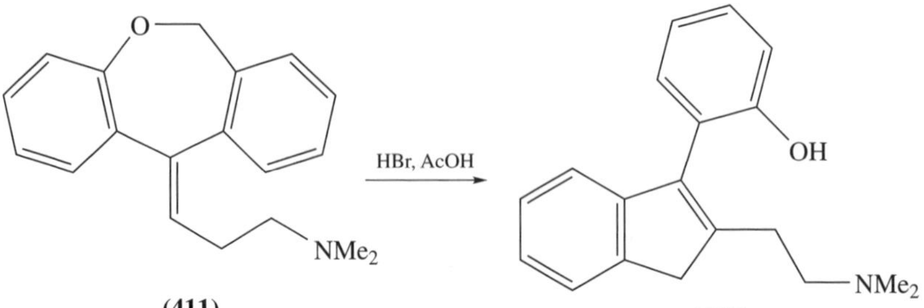

VIII. REFERENCES

1. (a) J. Toullec, in *The Chemistry of Enols* (Ed. Z. Rappoport), Chap. 6, Wiley, Chichester, 1990.
   (b) R. H. Thomson, *Quart. Rev.*, **10**, 27 (1956).
2. V. V. Ershov and G. A. Nikiforov, *Usp. Khim.*, **35**, 1953 (1966); *Chem. Abstr.*, **66**, 115367e (1967).
3. K.-F. Wedemeyer, in *Methoden der Organischen Chemie (Houben-Weyl)*, Phenole, Band VI/1c, Parts 1–2, Georg Thieme Verlag, Stuttgart, 1976.
4. J. Thiele, *Justus Liebigs Ann. Chem.*, **306**, 87 (1899).
5. A. Lapworth, *J. Chem. Soc.*, 1265 (1901).
6. T. A. Gadosy and R. A. McClelland, *THEOCHEM*, **369**, 1 (1996).

7. M. Capponi, I. G. Gut, B. Hellrung, G. Persy and J. Wirz, *Can. J. Chem.*, **77**, 605 (1999).
8. V. M. Sol, R. Louw and P. Mulder, *Recl. Trav. Chim. Pays-Bas*, **109**, 346 (1990).
9. H. Winberg, *Chem. Rev.*, **60**, 169 (1960).
10. A. S. Lindsey and H. Jeskey, *Chem. Rev.*, **57**, 583 (1957).
11. O. S. Tee, N. R. Iyengar and J. M. Bennett, *J. Org. Chem.*, **51**, 2585 (1986).
12. R. B. Woodward, *J. Am. Chem. Soc.*, **62**, 1208 (1940).
13. B. J. Doty and G. W. Morrow, *Tetrahedron Lett.*, **31**, 6125 (1990).
14. K. Omura, *J. Org. Chem.*, **61**, 7156 (1996).
15. A. Baeyer, *Ber.*, **19**, 159 (1886).
16. J. Herzig and B. Erthal, *Monatsh. Chem.*, **31**, 827 (1910).
17. K. Mandix, A. Colding, K. Elming, L. Sunesen and I. Shim, *Int. J. Quantum Chem.*, **46**, 159 (1993).
18. R. Highet and T. Batterham, *J. Org. Chem.*, **29**, 475 (1964).
19. W. Mayer and R. Weiss, *Angew. Chem.*, **68**, 680 (1956).
20. J. E. Bartmess and S. S. Griffith, *J. Am. Chem. Soc.*, **112**, 2931 (1990).
21. D. R. Boyd and D. M. Jerina, in *The Chemistry of Heterocyclic Compounds* (Eds. A. Weissenberger and E. C. Taylor), Vol. 42, Part 3, '*Small Ring Heterocycles*' (Ed. A. Hassner), Chap. II, Wiley, New York, Chichester, 1985, pp. 197–282.
22. N. T. Nashed, J. M. Sayer and D. M. Jerina, *J. Am. Chem. Soc.*, **115**, 1723 (1993).
23. J. R. de la Vega, J. H. Busch, J. H. Schaube, K. L. Kunze and B. E. Haggert, *J. Am. Chem. Soc.*, **104**, 3295 (1982).
24. K. B. Andersen, *Acta Chem. Scand.*, **53**, 222 (1999).
25. A. Arnone, L. Merlini, R. Mondelli, G. Nasini, E. Ragg, L. Scaglioni and U. Weiss, *J. Chem. Soc., Perkin Trans. 2*, 1447 (1993).
26. S. Mazzini, L. Merlini, R. Mondelli, G. Nasini, E. Ragg and L. Scaglioni, *J. Chem. Soc., Perkin Trans. 2*, 2013 (1997).
27. E. Chapoteau, B. P. Czech, C. R. Gebauer, A. Kumar, K. Leong, D. T. Mytych, W. Zazulak, D. H. Desai, E. Luboch, J. Krzykawski and R. A. Bartsch, *J. Org. Chem.*, **56**, 2575 (1991) and references cited therein.
28. K. S. Burmistrov, N. V. Toropin, S. I. Burmistrov, T. V. Gosteminskaya, V. I. Savich, S. S. Artemchenko and N. V. Baranova, *Zh. Org. Khim.*, **29**, 735 (1993); *Chem. Abstr.*, **120**, 269528d (1994).
29. A. Lycka, *Dyes Pigm.*, **43**, 27 (1999).
30. S. R. Salman, R. D. Farrant and J. C. Lindon, *Spectrosc. Lett.*, **24**, 1071 (1991).
31. S. R. Salman, J. C. Lindon, R. D. Farrant and T. A. Carpenter, *Magn. Reson. Chem.*, **31**, 991 (1993).
32. S. R. Salman and N. A. I. Saleh, *Spectrosc. Lett.*, **31**, 1179 (1998).
33. S. H. Alarcon, A. C. Olivieri and M. Gonzales-Sierra, *J. Chem. Soc., Perkin Trans. 2*, 1067 (1994).
34. Z. Cimerman, R. Kiralj and N. Galic, *J. Mol. Struct.*, **323**, 7 (1994).
35. N. Galic, Z. Cimerman and V. Tomisic, *Anal. Chim. Acta*, **343**, 135 (1997).
36. M. Kwiatkowski, E. Kwiatkowski, A. Olechnowicz, B. Kosciuszko-Panek and D. M. Ho, *Pol. J. Chem.*, **68**, 85 (1994).
37. M. Carles, F. Mansilla-Koblavi, J. A. Tenon, T. Y. N'Guessan and H. Bodot, *J. Phys. Chem.*, **97**, 3716 (1993).
38. F. Mansilla-Koblavi, J. A. Tenon, S. T. N. Ebby, J. Lapasset and M. Carles, *Acta Crystallogr., Sect. C: Cryst. Struct. Commun.*, **C15**, 1595 (1995).
39. Z. Hayvali, N. Gunduz, Z. Kilic and E. Weber, *J. Prakt. Chem.*, **341**, 568 (1999).
40. A. Serrano and S. Canuto, *Int. J. Quantum Chem.*, **70**, 745 (1998).
41. D. Maciejewska, D. Pawlak and V. Koleva, *J. Phys. Org. Chem.*, **12**, 875 (1999).
42. V. T. Kasumov, F. Ucun, I. Kartal and F. Koksal, *Spectrosc. Lett.*, **32**, 485 (1999).
43. M. Yildiz, Z. Kilic and T. Hökelek, *J. Mol. Struct.*, **441**, 1 (1998).
44. V. P. Rybalkin, V. A. Bren, V. I. Minkin and G. D. Palui, *Zh. Org. Khim.*, **28**, 2310 (1992); *Chem. Abstr.*, **120**, 8115h (1994).
45. A. Douhal, F. Amat-Guerri, M. P. Lillo and A. U. Acuna, *J. Photochem. Photobiol., A: Chem.*, **78**, 127 (1994).
46. M. Fores, M. Duran, M. Sola and L. Adamowicz, *J. Phys. Chem., A*, **103**, 4413 (1999).

47. M. E. Rampey, C. E. Halkyard, A. R. Williams, A. J. Angel, D. R. Hurst, J. D. Townsend, A. E. Finefrock, C. F. Beam and S. L. Studer-Martinez, *Photochem. Photobiol.*, **70**, 176 (1999).
48. B. E. Zaitsev, Yu. N. Zaitseva, M. A. Ryabov, G. V. Sheban, O. A. Zotova and M. V. Kazankov, *Khim. Geterocycl. Soedin.*, 1109 (1994); *Chem. Abstr.*, **122**, 132470a (1995).
49. H. Becker, D. Schröder, W. Zummack and H. Schwarz, *J. Am. Chem. Soc.*, **116**, 1096 (1994) and references cited therein.
50. M. E. Kopach, W. G. Hipple and W. D. Harman, *J. Am. Chem. Soc.*, **114**, 1736 (1992).
51. M. E. Kopach, L. P. Kelsh, K. C. Stork and W. D. Harman, *J. Am. Chem. Soc.*, **115**, 5322 (1993).
52. W. D. Wulff, B. M. Bax, T. A. Brandvold, K. S. Chan, A. M. Gilbert, R. P. Hsung, J. Mitchell and J. Clardy, *Organometallics*, **13**, 102 (1994).
53. M. E. Bos, W. D. Wulff and K. J. Wilson, *Chem. Commun.*, 1863 (1996).
54. P. Ghosh, N. Bag and A. Chakravorty, *Organometallics*, **15**, 3042 (1996).
55. P. Ghosh, A. Pramanic and A. Chakravorty, *Organometallics*, **15**, 4147 (1996).
56. T. D. Lash, S. T. Chaney and D. T. Richter, *J. Org. Chem.*, **63**, 9076 (1998).
57. W. Hückel, *Organische Chemie*, Akademische Verlaggesellschaft Geest & Portig K.-G., Leipzig, 1955, S. 547–551.
58. N. O. Mchedlov-Petrosyan and V. N. Kleshchevnikova, *Zh. Obshch. Khim.*, **60**, 900 (1990); *Chem. Abstr.*, **113**, 134163r (1990).
59. G. Jungbluth, I. Rühling and W. Ternes, *J. Chem. Soc., Perkin Trans. 2*, 1946 (2000).
60. D. Anastasiou and W. R. Jackson, *Aust. J. Chem.*, **45**, 21 (1992).
61. W. Adam, G. Käb and M. Sauter, *Chem. Ber.*, **127**, 433 (1994).
62. G. Sartori, F. Bigi, R. Maggi, A. Pastorio, C. Porta and G. Bonfanti, *J. Chem. Soc., Perkin Trans. 1*, 1879 (1994).
63. R. E. Valters, F. Fülöp and D. Korbonits, *Adv. Heterocycl. Chem.*, **64**, 251 (1995).
64. I. S. Ryzhkina, L. A. Kudryavtseva, V. E. Bel'skii, I. E. Ismaev, V. I. Morozov, A. V. Il'yasov and B. E. Ivanov, *Zh. Obshch. Khim.*, **69**, 820 (1990); *Chem. Abstr.*, **113**, 132338c (1990).
65. R. A. Shagidullina, I. S. Ryzhkina, A. B. Mirgorodskaya, L. A. Kudryavtseva, V. E. Bel'skii and B. E. Ivanov, *Izv. Akad. Nauk, Ser. Khim.*, 1215 (1994); *Chem. Abstr.*, **122**, 80636n (1995).
66. R. Ahmed, M. A. Malik and M. Zia-Ul-Haq, *J. Chem. Soc. Pak.*, **12**, 352 (1990); *Chem. Abstr.*, **115**, 71041t (1991).
67. S. W. Ng, A. H. Othman and S. N. Abdul Malek, *J. Chem. Crystallogr.*, **24**, 331 (1994).
68. B. Brzezinski, H. Maciejewska-Urjasz and G. Zundel, *J. Mol. Struct.*, **319**, 177 (1994).
69. V. I. Minkin, L. P. Olekhnovich and Yu. A. Zhdanov, *Understanding Chemical Reactivity: Molecular Design of Tautomeric Compounds*, Kluwer Academic Publ., Dordrecht, 1987.
70. R.-M. Schoth, E. Lork and G.-V. Roeschenthaler, *J. Fluorine Chem.*, **78**, 187 (1996).
71. M. C. Jimenez, M. A. Miranda and R. Tormos, *J. Org. Chem.*, **63**, 1323 (1998).
72. M. C. Jimenez, F. Marquez, M. A. Miranda and R. Tormos, *J. Org. Chem.*, **59**, 197 (1994).
73. T. Hatano and R. W. Hemingway, *J. Chem. Soc., Perkin Trans. 2*, 1035 (1997).
74. G. A. Chmutova and H. Ahlbrecht, *Russ. J. Gen. Chem.*, **68**, 1775 (1998).
75. I. B. Repinskaya, M. M. Shakirov, K. Yu. Koltunov and V. A. Koptyug, *Zh. Org. Khim.*, **28**, 1005 (1992); *Chem. Abstr.*, **118**, 191052f (1993).
76. I. B. Repinskaya, K. Yu. Koltunov, M. M. Shakirov and V. A. Koptyug, *Zh. Org. Khim.*, **28**, 1013 (1992); *Chem. Abstr.*, **118**, 168608h (1993).
77. S. Shinkai, *Tetrahedron*, **49**, 8933 (1993).
78. V. V. Ershov, A. A. Volod'kin and G. N. Bogdanov, *Usp. Khim.*, **32**, 154 (1963); *Chem. Abstr.*, **59**, 3800h (1963).
79. M. V. George and K. S. Balachandran, *Chem. Rev.*, **75**, 491 (1975).
80. R. A. Russell, A. I. Day, B. A. Pilley, P. J. Leavy and R. N. Warrener, *J. Chem. Soc., Chem. Commun.*, 1631 (1987).
81. A. S. Mitchell and R. A. Russell, *Tetrahedron Lett.*, **34**, 545 (1993).
82. V. B. Vol'eva, T. I. Prokof'eva, I. S. Belostotskaya, N. L. Komissarova and V. V. Ershov, *Russ. Chem. Bull.*, **47**, 1952 (1998).
83. V. B. Vol'eva, A. I. Prokof'ev, A. Yu. Karmilov, N. D. Komissarova, I. S. Belostotskaya, T. I. Prokof'eva and V. V. Ershov, *Izv. Akad. Nauk, Ser. Khim.*, 1975 (1998); *Chem. Abstr.*, **130**, 204205h (1999).
84. J. Le Bras, H. Amouri and J. Vaissermann, *J. Organomet. Chem.*, **567**, 57 (1998).
85. J. Le Bras, H. Amouri and J. Vaissermann, *Inorg. Chem.*, **37**, 5056 (1998).

86. H. Amouri, J. Le Bras and J. Vaissermann, *Organometallics*, **17**, 5850 (1998).
87. M. Sugumaran, H. Dali and V. Semensi, *Biochem. J.*, **277**, 849 (1991).
88. M. Sugumaran, H. Dali and V. Semensi, *Biochem. J.*, **281**, 353 (1992).
89. J. Pospisil, S. Nespurek and H. Zweifel, *Polym. Degrad. Stab.*, **54**, 7 (1996).
90. N. G. Faleev, S. N. Spirina, T. V. Demidkina and R. S. Phillips, *J. Chem. Soc., Perkin Trans. 2*, 2001 (1996).
91. K. Omura, *J. Org. Chem.*, **56**, 921 (1991).
92. S. Domagala, V. Steglinska and J. Dziegiec, *Monatsh. Chem.*, **129**, 761 (1998).
93. P. De Mayo, *Molecular Rearrangements*, Interscience, New York–London, 1963, p. 652.
94. S. Raychaudhuri, T. R. Seshadri and S. K. Mukerjee, *Indian J. Chem.*, **11**, 1228 (1973).
95. J. A. Joule and G. F. Smith, *Heterocyclic Chemistry*, Van Nostrand Reinhold, London, 1972.
96. K.-F. Wedemeyer, in *Methoden der Organischen Chemie (Houben-Weyl)*, Phenole, Band VI/1c, Parts 1–2, Georg Thieme Verlag, Stuttgart, 1976, p. 1076.
97. H. Hart and E. A. Haglund, *J. Org. Chem.*, **15**, 396 (1950).
98. J. R. Norell, *J. Org. Chem.*, **38**, 1929 (1973).
99. C. B. Campbell, A. Onopchenko and D. C. Young, *Ind. Eng. Chem. Res.*, **29**, 642 (1990).
100. S. Subramanian, A. Mitra, C. V. V. Satyanarayama and D. K. Chakrabarthy, *Appl. Catal., A*, **159**, 229 (1997).
101. K.-F. Wedemeyer, in *Methoden der Organischen Chemie (Houben-Weyl)*, Phenole, Band VI/1c, Parts 1–2, Georg Thieme Verlag, Stuttgart, 1976, p. 1081.
102. P. Ballester, M. Capo and J. M. Saa, *Tetrahedron Lett.*, **31**, 1339 (1990).
103. P. Ballester, M. Capo, X. Garcias and J. M. Saa, *J. Org. Chem.*, **58**, 328 (1993).
104. A. M. Trzeciak and J. J. Ziolkowski, *Gazz. Chim. Ital.*, **124**, 403 (1994).
105. E. Schmid, G. Frater, H.-J. Hansen and H. Schmid, *Helv. Chim. Acta*, **55**, 1625 (1972).
106. R. Hug, H.-J. Hansen and H. Schmid, *Helv. Chim. Acta*, **55**, 1675 (1972).
107. N. K. Sangwan, B. S. Verma and K. S. Dhindsa, *Indian J. Chem.*, **31B**, 590 (1992).
108. S. F. Nielsen, C. E. Olsen and S. B. Christensen, *Tetrahedron*, **53**, 5573 (1997).
109. A. Selva, *Gazz. Chim. Ital.*, **122**, 291 (1992).
110. F. M. Elkobaisi and W. J. Hickinbottom, *J. Chem. Soc.*, 1873 (1959).
111. K.-F. Wedemeyer, in *Methoden der Organischen Chemie (Houben-Weyl)*, Phenole, Band VI/1c, Parts 1–2, Georg Thieme Verlag, Stuttgart, 1976, p. 492.
112. K.-F. Wedemeyer, in *Methoden der Organischen Chemie (Houben-Weyl)*, Phenole, Band VI/1c, Parts 1–2, Georg Thieme Verlag, Stuttgart, 1976, p. 553.
113. J. March, *Advanced Organic Chemistry. Reactions, Mechanisms and Structure*, Chap. 13, Wiley, New York, 1985.
114. M. Schlosser, *Angew. Chem.*, **76**, 132 (1964).
115. L. Novak, P. Kovacs, P. Kolonits, O. Opovecz, J. Fekete and C. Szantay, *Synthesis*, 809 (2000).
116. D. B. Amabilino, P. R. Ashton, S. E. Boyd, M. Gomez-Lopez, W. Hayes and J. F. Stoddart, *J. Org. Chem.*, **62**, 3062 (1997).
117. D. B. Amabilino, P.-L. Anelli, P. R. Ashton, G. R. Brown, E. Cordova, L. A. Godinez, W. Hayes, A. E. Kaifer, D. Philp, A. M. Z. Slawin, N. Spenser, J. F. Stoddart, M. S. Tolley and D. J. Williams, *J. Am. Chem. Soc.*, **117**, 11142 (1995).
118. P. R. Ashton, R. Ballardini, V. Balzani, A. Credi, M. T. Gandolfi, M. Venturi, L. Perez-Garcia, L. Prodi, J. F. Stoddart, M. Venturi, A. J. P. White and D. J. Williams, *J. Am. Chem. Soc.*, **117**, 11171 (1995).
119. J.-M. Lehn, *Supramolecular Chemistry. Concepts and Perspectives*, VCH Verlagsgesellschaft mbH, Weinheim, 1995.
120. S. J. Rhoads and N. R. Raulins, 'The Claisen and Cope Rearrangements', in *Organic Reactions*, Vol. 22, Wiley, New York, 1975, p. 1.
121. R. P. Lutz, *Chem. Rev.*, **84**, 205 (1984).
122. C. J. Moody, *Adv. Heterocycl. Chem.*, **42**, 203 (1987).
123. S. M. Lukyanov and A. V. Koblik, 'Rearrangements of Dienes and Polyenes', in *The Chemistry of Dienes and Polyenes* (Ed. Z. Rappoport), Vol. 2, Chap. 10, Wiley, Chichester, 2000, p. 861.
124. K.-F. Wedemeyer, in *Methoden der Organischen Chemie (Houben-Weyl)*, Phenole, Band VI/1c, Parts 1–2, Georg Thieme Verlag, Stuttgart, 1976, p. 502.
125. S. Yamabe, S. Okumoto and T. Hayashi, *J. Org. Chem.*, **61**, 6218 (1996).

126. T. Hayashi, Y. Okada, K. Arite and H. Kuromizu, *Nippon Kagaku Kaishi*, 255 (1997); *Chem. Abstr.*, **126**, 250958w (1997).
127. J. An, L. Bagnell, T. Cablewski, C. R. Strauss and R. W. Trainor, *J. Org. Chem.*, **62**, 2505 (1997).
128. J. A. Elings, R. S. Downing and R. A. Sheldon, *Stud. Surf. Sci. Catal.*, **94** (*Catalysis by Microporous Materials*), 487 (1995).
129. P.-S. E. Dai, *Catal. Today*, **26**, 3 (1995).
130. S. R. Chitnis and M. M. Sharma, *React. Funct. Polym.*, **32**, 93 (1997).
131. W. A. Smit, S. I. Pogrebnoi, Yu. B. Kalyan and M. Z. Krimer, *Izv. Akad. Nauk SSSR, Ser. Khim.*, 1934 (1990); *Chem. Abstr.*, **114**, 23481u (1991).
132. N. Geetha and K. K. Balasubramanian, *Tetrahedron Lett.*, **39**, 1417 (1998).
133. S. S. Mochalov, S. S. Pogodin, Yu. S. Shabarov and N. S. Zefirov, *Zh. Org. Khim.*, **32**, 138 (1996); *Chem. Abstr.*, **130**, 38218t (1999).
134. D. H. Camacho, I. Nakamura, S. Saito and Y. Yamamoto, *Angew. Chem., Int. Ed. Engl.*, **38**, 3365 (1999).
135. M. Inouye, K. Tsuchiya and T. Kitao, *Angew. Chem., Int. Ed. Engl.*, **31**, 204 (1992).
136. G. Li, Z. Li and X. Fang, *Synth. Commun.*, **26**, 2569 (1996).
137. S. R. Waldman, A. P. Monte, A. Bracey and D. E. Nichols, *Tetrahedron Lett.*, **37**, 7889 (1996).
138. A. M. Bernard, M. T. Cocco, V. Onnis and P. P. Piras, *Synthesis*, 41 (1997).
139. A. M. Bernard, M. T. Cocco, V. Onnis and P. P. Piras, *Synthesis*, 256 (1998).
140. F. M. D. Ismail, M. J. Hilton and M. Stefinovic, *Tetrahedron Lett.*, **33**, 3795 (1992).
141. S. E. Drewes, N. D. Emslie, N. Karodia and G. Loizou, *Synth. Commun.*, **20**, 1437 (1990).
142. H. Uzawa, K. Hiratani, N. Minoura and T. Takahashi, *Chem. Lett.*, 307 (1998).
143. I. R. Green, S. Nefdt, V. I. Hugo and P. W. Snijman, *Synth. Commun.*, **24**, 3189 (1994).
144. I. R. Green, V. I. Hugo, F. Oosthuizen and R. G. F. Giles, *Synth. Commun.*, **26**, 867 (1996).
145. B. M. Trost and F. D. Toste, *J. Am. Chem. Soc.*, **120**, 815 (1998).
146. H. Ito, A. Sato and T. Taguchi, *Tetrahedron Lett.*, **38**, 4815 (1997).
147. B. Garcia, C.-H. Lee, A. Blasko and T. C. Bruice, *J. Am. Chem. Soc.*, **113**, 8118 (1991).
148. M. A. Perez and J. M. Bermejo, *J. Org. Chem.*, **58**, 2628 (1993).
149. T. Kato, K. Nagae and M. Hoshikawa, *Tetrahedron Lett.*, **40**, 1941 (1999).
150. J. V. Crivello and S. Kong, *J. Polym. Sci., Part A: Polym. Chem.*, **37**, 3017 (1999).
151. J. V. Crivello, S. Kong and L. Harvilchuck, *J. Macromol. Sci., Pure Appl. Chem.*, **A36**, 1123 (1999).
152. W. F. Bailey and E. R. Punzalan, *Tetrahedron Lett.*, **37**, 5435 (1996).
153. J. Barluenga, R. Sanz and F. J. Fananas, *Tetrahedron Lett.*, **38**, 6103 (1997).
154. I. E. Mikhailov, V. I. Minkin, G. A. Dushenko, I. A. Kamenetskaya and L. P. Olekhnovich, *Dokl. Akad. Nauk SSSR*, **299**, 1399 (1988); *Chem. Abstr.*, **110**, 23044w (1989).
155. S. M. Lukyanov and A. V. Koblik, 'Rearrangements of Dienes and Polyenes', in *The Chemistry of Dienes and Polyenes* (Ed. Z. Rappoport), Vol. 2, Chap. 10, Wiley, Chichester, 2000, p. 781.
156. P. H. Kahn and J. Cossy, *Tetrahedron Lett.*, **40**, 8113 (1999).
157. J.-L. Pozzo, A. Samat, R. Guglliemetti, V. Lokshin and V. I. Minkin, *Can. J. Chem.*, **74**, 1649 (1996).
158. H. Ishii, T. Ishikawa, S. Takeda, S. Ueki, M. Suzuki and T. Harayama, *Chem. Pharm. Bull.*, **38**, 1775 (1990).
159. H. Ishii, T. Ishikawa, S. Takeda, S. Ueki and M. Suzuki, *Chem. Pharm. Bull.*, **40**, 1148 (1992).
160. T. Ishikawa, A. Mizutani, C. Miwa, Y. Oku, N. Komano, A. Takami and T. Watanabe, *Heterocycles*, **45**, 2261 (1997).
161. M. A. Khan, *Bull. Chem. Soc. Japan*, **64**, 3682 (1991).
162. J. N. Kim, S. Y. Shin and E. K. Ryu, *Bull. Korean Chem. Soc.*, **13**, 361 (1992).
163. A. Gerecs, 'The Fries Reaction', in *Friedel-Crafts and Related Reactions* (Ed. G. A. Olah), Vol. III, Part 1, Interscience Publishers, New York, 1964, p. 499.
164. H. Kwart and K. King, in *The Chemistry of Functional Groups, The Chemistry of Carboxylic Acids and Esters* (Ed. S. Patai), Chap. 8, Interscience Publishers, New York, 1969, p. 341.
165. D. Häbich and F. Effenberger, *Synthesis*, 841 (1979).
166. H. Heaney, 'The Bimolecular Aromatic Friedel-Crafts Reactions', in *Comprehensive Organic Synthesis* (Eds. B. M. Trost, I. Fleming and L. A. Paquette), Vol. 2, Chap. 3.2, Pergamon Press, Oxford, 1991, p. 733.

167. R. Martin, *Org. Prep. Proc. Int.*, **24**, 373 (1992).
168. S. Kobayashi, I. Hachiya, H. Ishitani, M. Moriwaki and S. Nagayama, *Zh. Org. Khim.*, **32**, 214 (1996); *Chem. Abstr.*, **126**, 143950r (1997).
169. S. Kobayashi, M. Morivaki and I. Hachiya, *Synlett*, 1153 (1995).
170. S. Kobayashi, M. Morivaki and I. Hachiya, *J. Chem. Soc., Chem. Commun.*, 1527 (1995).
171. S. Kobayashi, M. Morivaki and I. Hachiya, *Tetrahedron Lett.*, **37**, 2053 (1996).
172. S. Kobayashi, M. Morivaki and I. Hachiya, *Bull. Chem. Soc. Japan*, **70**, 267 (1997).
173. D. C. Harrowven and R. F. Dainty, *Tetrahedron Lett.*, **37**, 7659 (1996).
174. B. Kaboudin, *Tetrahedron*, **55**, 12865 (1999).
175. A. Vogt, H. W. Kouwenhoven and R. Prins, *Appl. Catal., A*, **123**, 37 (1995).
176. F. Jayat, M. J. S. Picot and M. Guisnet, *Catal. Lett.*, **41**, 181 (1996).
177. M. Sasidharan and R. Kumar, *Stud. Surf. Sci. Catal.*, **105B** (*Progress in Zeolite and Microporous Materials, Pt. B*), 1197 (1997).
178. U. Freese, F. Heinrich and F. Roessner, *Catal. Today*, **49**, 237 (1999).
179. H. van Bekkum, A. J. Hoefnagel, M. A. van Koten, E. A. Gunnewegh, A. H. G. Vogt and H. W. Kouwenhoven, *Stud. Surf. Sci. Catal.*, **83** (*Zeolites and Microporous Crystals*), 379 (1994).
180. A. Heidekum, M. A. Harmer and W. F. Hoelderich, *Prepr. Am. Chem. Soc., Div. Pet. Chem.*, **42**, 763 (1997).
181. A. Heidekum, M. A. Harmer and W. F. Hoelderich, *J. Catal.*, **176**, 260 (1998).
182. J. M. Campelo, R. Chakraborty, J. M. Marinas and A. A. Romero, *React. Kinet. Catal. Lett.*, **65**, 107 (1998).
183. C. Venkatachalapathy and K. Pitchumani, *Tetrahedron*, **53**, 17171 (1997).
184. V. Sridar and V. S. S. Rao, *Indian J. Chem., Sect. B*, **33B**, 184 (1994).
185. G. L. Kad, I. R. Trehan, J. Kaur, S. Nayyar, A. Arora and J. S. Brar, *Indian J. Chem., Sect. B*, **35B**, 734 (1996).
186. I. R. Trehan, J. S. Brar, A. K. Arora and G. L. Kad, *J. Chem. Educ.*, **74**, 324 (1997).
187. B. M. Khadilkar and V. R. Madyar, *Synth. Commun.*, **29**, 1195 (1999).
188. F. M. Moghaddam, M. Ghaffarzadeh and S. H. Abdi-Oskoui, *J. Chem. Res. Synop.*, 574 (1999).
189. J. R. Norell, *J. Org. Chem.*, **38**, 1924 (1973).
190. L. P. Miranda, A. Jones, W. D. F. Meutermans and P. F. Alewood, *J. Am. Chem. Soc.*, **120**, 1410 (1998).
191. J. Yamamoto, H. Tamura, K. Takahara, H. Yamana and K. Maeta, *Nippon Kagaku Kaishi*, 707 (1994); *Chem. Abstr.*, **121**, 255379q (1994).
192. J. Yamamoto, H. Yamana, Y. Haraguchi and H. Sasaki, *Nippon Kagaku Kaishi*, 747 (1996); *Chem. Abstr.*, **125**, 195120u (1996).
193. J. Yamamoto, Y. Haraguchi, T. Iwaki, H. Yamana and H. Sasaki, *Nippon Kagaku Kaishi*, 909 (1996); *Chem. Abstr.*, **125**, 328254q (1996).
194. Y. Isota, N. Ohkubo, M. Takaoka and J. Yamamoto, *Nippon Kagaku Kaishi*, 421 (1999); *Chem. Abstr.*, **131**, 116055z (1999).
195. V. I. Hugo, J. L. Nicholson, P. W. Snijman and I. R. Green, *Synth. Commun.*, **24**, 23 (1994).
196. V. V. Mezheritskii and O. M. Golyanskaya, *Russ. J. Org. Chem.*, **35**, 910 (1999).
197. M. A. Brimble, D. Neville and L. J. Duncalf, *Tetrahedron Lett.*, **39**, 5647 (1998).
198. M. A. Brimble, L. J. Duncalf and D. Neville, *J. Chem. Soc., Perkin Trans. 1*, 4165 (1998).
199. G.-Q. Lin and M. Zhong, *Tetrahedron: Asymmetry*, **8**, 1369 (1997).
200. V. F. Traven, D. V. Kravtchenko, T. A. Chibisova, S. V. Shorshnev, R. Eliason and D. H. Wakefield, *Heterocycl. Commun.*, **2**, 345 (1996).
201. T. Matsumoto, T. Hosoya and K. Suzuki, *Tetrahedron Lett.*, **31**, 4629 (1990).
202. T. Matsumoto, M. Katsuki and K. Suzuki, *Tetrahedron Lett.*, **29**, 6935 (1988).
203. J.-A. Mahling and R. R. Schmidt, *Synthesis*, 325 (1993).
204. J.-A. Mahling and R. R. Schmidt, *Justus Liebigs Ann. Chem.*, 467 (1995).
205. E. El Telbani, S. El Desoky, M. A. Hammad, A. R. H. A. Rahman and R. R. Schmidt, *Eur. J. Org. Chem.*, 2317 (1998).
206. K. Mori, Z.-H. Qian and S. Watanabe, *Justus Liebigs Ann. Chem.*, 485 (1992).
207. M. E. Jung and T. I. Lazarova, *Tetrahedron Lett.*, **37**, 7 (1996).
208. R. Levine and W. C. Fernelius, *Chem. Rev.*, **54**, 493 (1954).
209. J.-A. Mahling, K.-H. Jung and R. R. Schmidt, *Justus Liebigs Ann. Chem.*, 461 (1995).

210. F. Cramer and G. H. Elschnig, *Chem. Ber.*, **89**, 1 (1956).
211. M. Hauteville, M. Chadenson and M. J. Chopin, *C.R. Acad. Sci. Paris, Ser. C*, **278**, 471 (1974).
212. W. Rahman and K. T. Nasim, *J. Org. Chem.*, **27**, 4215 (1962).
213. V. V. Mezheritskii, A. L. Pikus and N. G. Tregub, *Zh. Org. Khim.*, **27**, 2198 (1991); *Chem. Abstr.*, **116**, 173990q (1992).
214. V. Snieckus, *Chem. Rev.*, **90**, 879 (1990).
215. M. P. Sibi and V. Snieckus, *J. Org. Chem.*, **48**, 1935 (1983).
216. A. S. Parsons, J. M. Garcia and V. A. Snieckus, *Tetrahedron Lett.*, **35**, 7537 (1994).
217. F. Beaulieu and V. Snieckus, *J. Org. Chem.*, **59**, 6508 (1994).
218. M. Skowronska-Ptasinska, W. Verboom and D. N. Reinhoudt, *J. Org. Chem.*, **50**, 2690 (1985).
219. M. R. Dennis and S. Woodward, *J. Chem. Soc., Perkin Trans. 1*, 1081 (1998).
220. S. M. W. Bennett, S. M. Brown, G. Conole, M. R. Dennis, P. K. Fraser, S. Radojevic, M. McPartlin, C. M. Topping and S. Woodward, *J. Chem. Soc., Perkin Trans. 1*, 3127 (1999).
221. H. Hopf and D. G. Barrett, *Justus Liebigs Ann. Chem.*, 449 (1995).
222. A. Quattropani, G. Bernardinelli and E. P. Kündig, *Helv. Chim. Acta*, **82**, 90 (1999).
223. S. S. Mandal, S. S. Samanta, C. Deb and A. De, *J. Chem. Soc., Perkin Trans. 1*, 2559 (1998).
224. I. R. Hardcastle and P. Quayle, *Tetrahedron Lett.*, **35**, 1749 (1994).
225. G. Simchen and J. Pfletschinger, *Angew. Chem.*, **88**, 444 (1976).
226. J. A. Miller, *J. Org. Chem.*, **52**, 322 (1987).
227. A. Couture, E. Deniau, S. Lebrun and P. Grandclaudon, *J. Chem. Soc., Perkin Trans. 1*, 789 (1999).
228. S. Horne and R. Rodrigo, *J. Chem. Soc., Chem. Commun.*, 164 (1992).
229. J. W. Lampe, P. F. Hughes, C. K. Biggers, S. H. Smith and H. Hu, *J. Org. Chem.*, **59**, 5147 (1994).
230. J. W. Lampe, P. F. Hughes, C. K. Biggers, S. H. Smith and H. Hu, *J. Org. Chem.*, **61**, 4572 (1996).
231. A. V. Kalinin, M. A. J. Miah, S. Chattopadhyay, M. Tsukazaki, M. Wicki, T. Nguen, A. L. Coelho, M. Kerr and V. Snieckus, *Synlett*, 839 (1997).
232. J. Clayden, J. H. Pink, N. Westlund and F. X. Wilson, *Tetrahedron Lett.*, **39**, 8377 (1998).
233. W. Wang and V. Snieckus, *J. Org. Chem.*, **57**, 424 (1992).
234. M. Gray, B. J. Chapell, N. J. Taylor and V. Snieckus, *Angew. Chem., Int. Ed. Engl.*, **35**, 1558 (1996).
235. H. Hock and H. Kropf, *Angew. Chem.*, **69**, 319 (1957).
236. C. H. Hassell, *Org. React.*, **9**, 73 (1957).
237. J. B. Lee and B. C. Uff, *Quart. Rev.*, **21**, 453 (1967).
238. D. B. Denney and D. Z. Denney, *J. Am. Chem. Soc.*, **84**, 2455 (1962).
239. E. D. Hughes and C. K. Ingold, *Quart. Rev.*, **6**, 45 (1952).
240. E. Bamberger, *Ber.*, **33**, 3600 (1900).
241. E. Boyland and D. Manson, *Biochem. J.*, **101**, 84 (1966).
242. D. Manson, *J. Chem. Soc., C*, 1508 (1971).
243. E. J. Behrman, *J. Am. Chem. Soc.*, **89**, 2424 (1967).
244. E. Boyland and P. Sims, *J. Chem. Soc.*, 4198 (1958).
245. G. T. Tisue, M. Grassmann and W. Lwowski, *Tetrahedron*, **24**, 999 (1968).
246. J. R. Cox and M. F. Dunn, *Tetrahedron Lett.*, 985 (1963).
247. L. Horner and H. Steppan, *Justus Liebigs Ann. Chem.*, **606**, 24 (1957).
248. E. Buncel, *Acc. Chem. Res.*, **8**, 132 (1975).
249. R. Huisgen, F. Bayerlein and W. Heydkamp, *Chem. Ber.*, **92**, 3223 (1959).
250. C. J. Collins and J. F. Eastman, in *The Chemistry of the Carbonyl Group* (Ed. S. Patai), Chap. 15, Wiley-Interscience, Chichester, 1966, p. 761.
251. B. Miller, *Acc. Chem. Res.*, **8**, 245 (1975).
252. D. A. Whiting, 'Dienone–Phenol Rearrangements and Related Reactions', in *Comprehensive Organic Synthesis* (Eds. B. M. Trost, I. Fleming and G. Pattenden), Vol. 3, Chap. 3.5, Pergamon Press, Oxford, 1991, p. 803.
253. S. M. Lukyanov and A. V. Koblik, 'Rearrangements of Dienes and Polyenes', in *The Chemistry of Dienes and Polyenes* (Ed. Z. Rappoport), Vol. 2, Chap. 10, Wiley, Chichester, 2000, p. 793.
254. A. A. Frimer, V. Marks, M. Sprecher and P. Gilinsky-Sharon, *J. Org. Chem.*, **59**, 1831 (1994).

255. V. P. Vitullo, *J. Org. Chem.*, **34**, 224 (1969).
256. V. P. Vitullo and E. A. Logue, *J. Org. Chem.*, **37**, 3339 (1972).
257. V. P. Vitullo and E. A. Logue, *J. Chem. Soc., Chem. Commun.*, 228 (1974).
258. B. Hagenbruch and S. Hünig, *Chem. Ber.*, **116**, 3884 (1983).
259. V. P. Vitullo and N. Grossman, *Tetrahedron Lett.*, 1559 (1970).
260. B. Miller, *J. Am. Chem. Soc.*, **92**, 432 (1970).
261. V. A. Koptyug and V. I. Buraev, *Zh. Org. Khim.*, **16**, 1882 (1980); *Chem. Abstr.*, **94**, 14891u (1980).
262. S. Chalais, P. Laszlo and A. Mathy, *Tetrahedron Lett.*, **27**, 2627 (1986).
263. Y. Chen, J.-L. Reymond and R. A. Lerner, *Angew. Chem.*, **106**, 1694 (1994).
264. W. D. Ollis, R. Somanathan and I. O. Sutherland, *J. Chem. Soc., Perkin Trans. 1*, 2930 (1981).
265. X. Shi, R. Day and B. Miller, *J. Chem. Soc., Perkin Trans. 1*, 1166 (1989).
266. B. Miller and X. Shi, *J. Org. Chem.*, **57**, 1677 (1992).
267. J. M. Brittain, P. B. D. de la Mare and P. A. Newman, *J. Chem. Soc., Perkin Trans. 2*, 32 (1981).
268. A. Fischer and G. N. Henderson, *Can. J. Chem.*, **61**, 1045 (1983).
269. S. Husain and M. Kifayatullah, *Indian J. Chem., Sect. B*, **24B**, 711 (1985).
270. R. G. Coombes, *J. Chem. Soc., Perkin Trans. 2*, 1007 (1992).
271. L. H. Klemm and D. R. Taylor, *J. Org. Chem.*, **45**, 4320 (1980).
272. J. N. Marx and Y.-S. P. Hahn, *J. Org. Chem.*, **53**, 2866 (1988).
273. J. N. Marx, J. Zuerker and Y.-S. Hahn, *Tetrahedron Lett.*, **32**, 1921 (1991).
274. S. Quideau, M. A. Looney, L. Pouysegu, S. Ham and D. M. Birney, *Tetrahedron Lett.*, **40**, 615 (1999).
275. J. Kunitomo, M. Oshikata and M. Akasu, *Yakugaku Zasshi*, **101**, 951 (1981); *Chem. Abstr.*, **96**, 31640r (1982).
276. H. Guinaudeau, V. Elango, M. Shamma and V. Fajardo, *J. Chem. Soc., Chem. Commun.*, 1122 (1982).
277. D. S. Bhakuni and S. Jain, *Tetrahedron*, **37**, 3175 (1981).
278. L. Crombie and M. L. Crombie, *J. Chem. Soc., Perkin Trans. 1*, 1455 (1982).
279. B. F. Bowden, R. W. Read and W. C. Taylor, *Aust. J. Chem.*, **34**, 799 (1981).
280. R. W. Read and W. C. Taylor, *Aust. J. Chem.*, **34**, 1125 (1981).
281. A. R. Carroll and W. C. Taylor, *Aust. J. Chem.*, **43**, 1871 (1990).
282. A. S. Kende, K. Koch and C. A. Smith, *J. Am. Chem. Soc.*, **110**, 2210 (1988).
283. A. J. Waring, J. H. Zaidi and J. W. Pilkington, *J. Chem. Soc., Perkin Trans. 1*, 1454 (1981).
284. A. J. Waring, J. H. Zaidi and J. W. Pilkington, *J. Chem. Soc., Perkin Trans. 2*, 935 (1981).
285. A. J. Waring and J. H. Zaidi, *J. Chem. Soc., Perkin Trans. 1*, 631 (1985).
286. G. Goodyear and A. J. Waring, *J. Chem. Soc., Perkin Trans. 2*, 103 (1990).
287. L. B. Jackson and A. J. Waring, *J. Chem. Soc., Perkin Trans. 1*, 1791 (1988).
288. L. B. Jackson and A. J. Waring, *J. Chem. Soc., Perkin Trans. 2*, 907 (1990).
289. L. B. Jackson and A. J. Waring, *J. Chem. Soc., Perkin Trans. 2*, 1893 (1990).
290. T. Oda, Y. Yamaguchi and Y. Sato, *Chem. Pharm. Bull.*, **34**, 858 (1986).
291. D. J. Hart, A. Kim, R. Krishnamurthy, G. H. Merriman and A. M. Waltos, *Tetrahedron*, **48**, 8179 (1992).
292. A. A. Volod'kin, R. D. Malysheva and V. V. Ershov, *Izv. Akad. Nauk SSSR, Ser. Khim.*, 715 (1982); *Chem. Abstr.*, **97**, 38613t (1982).
293. H. Kusama, K. Uchiyama, Y. Yamashita and K. Narasaka, *Chem. Lett.*, 715 (1995).
294. M. Kobayashi, K. Uneyama, N. Hamada and S. Kashino, *Tetrahedron Lett.*, **35**, 5235 (1994).
295. J. R. Hanson, D. Raines and S. G. Knights, *J. Chem. Soc., Perkin Trans. 1*, 1311 (1980).
296. A. G. Avent, J. R. Hanson and L. Yang-zhi, *J. Chem. Soc., Perkin Trans. 1*, 2129 (1985).
297. T. M. Zydowsky, C. E. Totten, D. M. Piatak, M. J. Gasic and J. Stankovic, *J. Chem. Soc., Perkin Trans. 1*, 1679 (1980).
298. A. Planas, J. Tomas and J.-J. Bonet, *Tetrahedron Lett.*, **28**, 471 (1987).
299. J. Y. Satoh, A. M. Haruta, T. Satoh, K. Satoh and T. T. Takahashi. *J. Chem. Soc., Chem. Commun.*, 1765 (1986).
300. F. Mukawa, *J. Chem. Soc., Perkin Trans. 1*, 457 (1988).
301. R. Cassis, M. Scholz, R. Tapia and J. A. Valderrama, *J. Chem. Soc., Perkin Trans. 1*, 2855 (1987).
302. C. Goycoolea, J. G. Santos and J. A. Valderrama, *J. Chem. Soc., Perkin Trans. 2*, 1135 (1988).

303. R. Cassis, M. Scholz, R. Tapia and J. A. Valderrama, *Tetrahedron Lett.*, **26**, 6281 (1985).
304. K. A. Parker and Y. Koh, *J. Am. Chem. Soc.*, **116**, 11149 (1994).
305. J. A. Dodge and A. R. Chamberlin, *Tetrahedron Lett.*, **29**, 4827 (1988).
306. A. Rieker, *Angew. Chem.*, **83**, 449 (1971).
307. R. N. Moore and R. V. Lawrence, *J. Am. Chem. Soc.*, **81**, 458 (1959).
308. D. Braun and B. Meier, *Angew. Chem.*, **83**, 617 (1971).
309. H. E. Zimmerman and D. C. Lynch, *J. Am. Chem. Soc.*, **107**, 7745 (1985).
310. A. G. Schultz and S. A. Hardinger, *J. Org. Chem.*, **56**, 1105 (1991).
311. A. G. Schultz and N. J. Green, *J. Am. Chem. Soc.*, **114**, 1824 (1992).
312. A. Padwa and S. L. Xu, *J. Am. Chem. Soc.*, **114**, 5881 (1992).
313. A. Padwa, J. M. Kassir and S. L. Xu, *J. Org. Chem.*, **62**, 1642 (1997).
314. F. Bourelle-Wargnier, M. Vincent and J. Chuche, *J. Org. Chem.*, **45**, 428 (1980).
315. Z. Zubovics and H. Wittmann, *Justus Liebigs Ann. Chem.*, **765**, 15 (1972).
316. D. L. Fields, *J. Org. Chem.*, **36**, 3002 (1971).
317. D. Collomb, C. Deshayes and A. Doutheau, *Tetrahedron*, **52**, 6665 (1996).
318. M. Sato, S. Ebine and J. Tsunetsugu, *J. Chem. Soc., Chem. Commun.*, 847 (1974).
319. M. Sato, A. Uchida, J. Tsunetsugu and S. Ebine, *Tetrahedron Lett.*, 2151 (1977).
320. M. Sato, K. Inaba, S. Ebine and J. Tsunetsugu, *Bull. Chem. Soc. Japan*, **53**, 2334 (1980).
321. W. E. Truce and C.-I. M. Lin, *J. Am. Chem. Soc.*, **95**, 4426 (1973).
322. F. Pietra, *Chem. Rev.*, **73**, 293 (1973).
323. F. Pietra, *Acc. Chem. Res.*, **12**, 132 (1979).
324. L. Birkofer and G. Daum, *Angew. Chem.*, **72**, 707 (1960).
325. L. Birkofer and G. Daum, *Chem. Ber.*, **95**, 183 (1962).
326. E. Doppler, H.-J. Hansen and H. Schmid, *Helv. Chim. Acta*, **55**, 1730 (1972).
327. G. N. Dorofeenko, A. V. Koblik and K. F. Suzdalev, *Zh. Org. Khim.*, **17**, 1050 (1981); *Chem. Abstr.*, **95**, 150338s (1981).
328. J. L. Webb and W. L. Hall, *J. Org. Chem.*, **38**, 1621 (1973).
329. J. Kuthan, *Adv. Heterocycl. Chem.*, **34**, 145 (1983).
330. A. T. Balaban, A. Dinculescu, G. N. Dorofeenko, G. W. Fischer, A. V. Koblik, V. V. Mezheritskii and W. Schroth, *Adv. Heterocycl. Chem.*, Suppl. 2, 102 (1982).
331. L. A. Muradyan, A. V. Koblik, D. S. Yufit, Yu. T. Struchkov, V. G. Arsenyev and S. M. Lukyanov, *Zh. Org. Khim.*, **32**, 925 (1996); *Chem. Abstr.*, **126**, 171457d (1997).
332. S. M. Lukyanov, A. V. Koblik and L. A. Muradyan, *Russ. Chem. Rev.*, **67**, 817 (1998).
333. S. M. Lukyanov, L. N. Etmetchenko and G. N. Dorofeenko, *Zh. Org. Khim.*, **14**, 399 (1978); *Chem. Abstr.*, **88**, 190536s (1978).
334. D. M. Harrison, in *Comprehensive Organic Chemistry* (Eds. Sir Derek Barton and W. D. Ollis), Vol. 4, Chap. 18.6, Pergamon Press, Oxford, 1979.
335. D. R. Boyd, in *Comprehensive Heterocyclic Chemistry* (Eds. A. R. Katritzky and C. W. Rees), Vol. 7, Chap. 5.17, Pergamon Press, Oxford, 1984, p. 547.
336. E. Vogel and H. Günther, *Angew. Chem.*, **79**, 429 (1967).
337. P. George, C. W. Bock and J. P. Glusker, *J. Phys. Chem.*, **94**, 8161 (1990).
338. N. E. Takenaka, R. Hamlin and D. M. Lemal, *J. Am. Chem. Soc.*, **112**, 6715 (1990).
339. H. Glombik, C. Wolff and W. Tochtermann, *Chem. Ber.*, **120**, 775 (1987).
340. T. Watabe, A. Hiratsuka, T. Aizawa and T. Sawahata, *Tetrahedron Lett.*, **23**, 1185 (1982).
341. W. H. Rastetter and L. J. Nummy, *J. Org. Chem.*, **45**, 3149 (1980).
342. D. R. Boyd and G. A. Berchtold, *J. Am. Chem. Soc.*, **101**, 2470 (1979).
343. H. S.-I. Chao and G. A. Berchtold, *J. Am. Chem. Soc.*, **103**, 898 (1981).
344. H. S.-I. Chao, G. A. Berchtold, D. R. Boyd, J. N. Dynak, J. E. Tomaszewski, H. Yagi and D. M. Jerina, *J. Org. Chem.*, **46**, 1948 (1981).
345. M. J. McManus, G. A. Berchtold, D. R. Boyd, D. A. Kennedy and J. F. Malone, *J. Org. Chem.*, **51**, 2784 (1986).
346. E. Wenkert, R. S. Greenberg and H. S. Kim, *Helv. Chim. Acta*, **70**, 2159 (1987).
347. S. K. Agarwal, D. R. Boyd, M. R. McGuckin, W. B. Jennings and O. W. Howarth, *J. Chem. Soc., Perkin Trans. 1*, 3073 (1990).
348. K.-F. Wedemeyer, in *Methoden der Organischen Chemie (Houben-Weyl)*, Phenole, Band VI/lc, Parts 1–2, Georg Thieme Verlag, Stuttgart, 1976, p. 326.